13 GROUP IIIA	14 GROUP IVA	15 GROUP VA	16 GROUP VIA	17 GROUP VIIA	18 GROUP VIIIA
								2 **He** 4.00
			5 **B** 10.81	6 **C** 12.01	7 **N** 14.01	8 **O** 16.00	9 **F** 19.00	10 **Ne** 20.18
10 GROUP →	11 GROUP IB	12 GROUP IIB	13 **Al** 26.98	14 **Si** 28.09	15 **P** 30.97	16 **S** 32.07	17 **Cl** 35.45	18 **Ar** 39.95
28 **Ni** 58.69	29 **Cu** 63.55	30 **Zn** 65.38	31 **Ga** 69.72	32 **Ge** 72.59	33 **As** 74.92	34 **Se** 78.96	35 **Br** 79.90	36 **Kr** 83.80
46 **Pd** 106.42	47 **Ag** 107.87	48 **Cd** 112.41	49 **In** 114.82	50 **Sn** 118.71	51 **Sb** 121.75	52 **Te** 127.60	53 **I** 126.90	54 **Xe** 131.29
78 **Pt** 195.08	79 **Au** 196.97	80 **Hg** 200.59	81 **Tl** 204.38	82 **Pb** 207.2	83 **Bi** 208.98	84 **Po** (209)	85 **At** (210)	86 **Rn** (222)

Metals —— —— Nonmetals

63 **Eu** 151.96	64 **Gd** 157.25	65 **Tb** 158.93	66 **Dy** 162.50	67 **Ho** 164.93	68 **Er** 167.26	69 **Tm** 168.93	70 **Yb** 173.04	71 **Lu** 174.97
95 **Am** (243)	96 **Cm** (247)	97 **Bk** (247)	98 **Cf** (251)	99 **Es** (252)	100 **Fm** (257)	101 **Md** (258)	102 **No** (259)	103 **Lr** (260)

FUNDAMENTALS OF CHEMISTRY

General, Organic, and Biological

H. Stephen Stoker
Edward B. Walker
Weber State College

Allyn and Bacon, Inc.

Boston London Sydney Toronto

Cover Administrator: Linda Dickinson
Cover Designer: Susan Slovinsky
Composition Buyer: Linda Cox
Manufacturing Buyer: William Alberti
Production Editor: Kathy Smith
Art Editor: Mary Hill

Copyright © 1988 by Allyn and Bacon, Inc.
A Division of Simon & Schuster
160 Gould Street
Needham, Massachusetts 02194

All rights reserved. No part of the material protected by this copyright notice may be reproduced or utilized in any form or by any means, electronic or mechanical, including photocopying, recording, or by any information storage and retrieval system, without the written permission of the copyright owner.

Library of Congress Cataloging-in-Publication Data

Stoker, H. Stephen (Howard Stephen), 1939–
    Fundamentals of chemistry: general, organic, and biological/H. Stephen Stoker, Edward B. Walker.
    p.   cm.
    Includes index.
    ISBN 0-205-10293-X
    1. Chemistry.   I. Walker, Edward B.   II. Title.
QD33.S855 1988
540--dc 19
                                                                87-35162
                                                                   CIP

Printed in the United States of America

10 9 8 7 6 5 4 3 2   93 92 91 90 89

Credits appear on page I-15, which constitutes a continuation of the copyright page.

Brief Contents

1 Some Fundamental Concepts 1
2 The Language and Symbolism of Chemistry 34
3 Structure of the Atom and the Periodic Table 50
4 Chemical Calculations: Formula Weights, Moles, and Chemical Equations 68
5 The Electronic Structure of Atoms 90
6 Chemical Bonding 112
7 Gases, Liquids, and Solids 146
8 Water, Solutions, and Colloids 174
9 Reaction Rates and Chemical Equilibrium 212
10 Acids, Bases, and Salts 226
11 Oxidation–Reduction Processes 256
12 Nuclear Chemistry and Radioactivity 268
13 Chemistry of Several Common Elements and Compounds 290
14 Saturated Hydrocarbons 308
15 Unsaturated Hydrocarbons 340
16 Aromatic Hydrocarbons 370
17 Alcohols, Phenols, and Ethers 390
18 Aldehydes and Ketones 422
19 Carboxylic Acids and Esters 440
20 Amines and Amides 468
21 Stereoisomerism 506
22 Carbohydrates 528
23 Lipids 568
24 Proteins 602
25 Enzymes: Biological Catalysts 644
26 Nucleic Acids 674
27 Vitamins, Minerals, and Hormones 720
28 Metabolism and Nutrition: An Overview 754
29 Carbohydrate Metabolism 778
30 Mitochondrial Oxidation and Phosphorylation Pathways 804
31 Lipid Metabolism 830

32 Amino Acid Metabolism   850
33 Extracellular Fluids   864
Appendix   A1
Answers to Selected Problems   A2
Index   I1

Contents

Preface xi

1 Some Fundamental Concepts 1

1.1 The Scope of Chemistry 1
1.2 The Composition and Structure of Matter 1
1.3 Properties of Matter 2
1.4 Changes in Matter 3
1.5 Observations and Measurements 5
1.6 Precision and Accuracy 5
1.7 Significant Figures 7
1.8 Exponential Notation 8
1.9 Exponential Notation and Mathematical Operations 11
1.10 Systems of Measurement 13
1.11 Conversion Factors and Dimensional Analysis 17
1.12 Density and Specific Gravity 23
1.13 Temperature Scales 25
1.14 Types and Forms of Energy 27
1.15 Heat Energy and Specific Heat 28
Exercises and Problems 30

2 The Language and Symbolism of Chemistry 34

2.1 Pure Substances and Mixtures 35
2.2 Heterogeneous and Homogeneous Mixtures 36
2.3 The Characterization of Pure Substances 39
2.4 The Subdivision of Pure Substances 39
2.5 Elements and Compounds 41
2.6 Discovery and Abundance of the Elements 42
2.7 Names and Symbols of the Elements 43
2.8 Chemical Formulas 45
Exercises and Problems 47

3 Structure of the Atom and the Periodic Table 50

3.1 How Chemists Discover Things—The Scientific Method 51
3.2 Atomic Theory of Matter 53
3.3 Chemical Evidence Supporting the Atomic Theory 54
3.4 Subatomic Particles: Electrons, Protons, and Neutrons 56
3.5 Atomic Number and Mass Number 58
3.6 Isotopes 59
3.7 Atomic Weights 61
3.8 The Periodic Law 62
3.9 The Periodic Table 63
Exercises and Problems 65

4 Chemical Calculations: Formula Weights, Moles, and Chemical Equations 68

4.1 Formula Weights 69
4.2 The Mole: The Chemist's Counting Unit 70
4.3 The Mass of a Mole 73
4.4 The Mole and Chemical Formulas 75
4.5 The Mole and Chemical Calculations 77
4.6 Writing and Balancing Equations 80
4.7 Chemical Equations and the Mole Concept 83
4.8 Chemical Calculations Using Chemical Equations 84
Exercises and Problems 87

5 The Electronic Structure of Atoms 90

5.1 The Energy of an Electron 91
5.2 Electron Shells 92
5.3 Electron Subshells 93
5.4 Electron Orbitals 95

5.5 Writing Electron Configurations 97
5.6 Electron Configurations and the Periodic Law 100
5.7 Electron Configurations and the Periodic Table 101
5.8 Classification Systems for the Elements 106
5.9 Metals, Nonmetals, and Living Organisms 108
Exercises and Problems 110

6 Chemical Bonding 112

6.1 Types of Chemical Bonds 113
6.2 Valence Electrons and Electron-Dot Structures 113
6.3 The Octet Rule 115
6.4 The Ionic-Bond Model 116
6.5 Formulas for Ionic Compounds 118
6.6 Structure of Ionic Compounds 121
6.7 Nomenclature for Ionic Compounds 122
6.8 The Covalent-Bond Model 124
6.9 Multiple-Covalent Bonds 127
6.10 Coordinate-Covalent Bonds 129
6.11 Electronegativity and Bond Polarity 130
6.12 Molecular Polarity 133
6.13 Nomenclature for Covalent Compounds 136
6.14 Polyatomic Ions 138
6.15 Formulas and Names for Ionic Compounds Containing Polyatomic Ions 139
Exercises and Problems 141

7 Gases, Liquids, and Solids 146

7.1 Property Differences between Physical States 147
7.2 The Kinetic Molecular Theory of Matter 147
7.3 The Solid State 149
7.4 The Liquid State 149
7.5 The Gaseous State 150
7.6 A Comparison of Solids, Liquids, and Gases 150
7.7 Gas Laws and Gas Law Variables 150
7.8 Boyle's Law: A Pressure–Volume Relationship 152
7.9 Charles' Law: A Temperature–Volume Relationship 156
7.10 The Combined-Gas Law 158
7.11 Dalton's Law of Partial Pressures 159
7.12 Changes of State 163
7.13 Evaporation of Liquids 163
7.14 Vapor Pressure of Liquids 165
7.15 Boiling and the Boiling Point 166
7.16 Intermolecular Forces in Liquids 168
7.17 Sublimation and Melting 170
7.18 Decomposition 171
Exercises and Problems 171

8 Water, Solutions, and Colloids 174

8.1 Abundance and Distribution of Water 175
8.2 Properties of Water 177
8.3 Boiling and Freezing Points of Water 177
8.4 Thermal Properties of Water 178
8.5 Water's Temperature–Density Relationship 179
8.6 Components of a Solution 181
8.7 Types of Solutions 182
8.8 Terminology Used in Describing Solutions 183
8.9 Solution Formation 184
8.10 Solubility Rules 186
8.11 Solution Concentrations 188
8.12 Concentration: Percentage of Solute 188
8.13 Concentration: Molarity 192
8.14 Concentration: Milliequivalents 195
8.15 Dilution 198
8.16 Colligative Properties of Solutions 200
8.17 Osmosis and Osmotic Pressure 201
8.18 Osmosis and the Human Body 205
8.19 Colloidal Dispersions 207
8.20 Dialysis 208
Exercises and Problems 209

9 Reaction Rates and Chemical Equilibrium 212

9.1 Conditions Necessary for a Chemical Reaction 213
9.2 Exothermic and Endothermic Chemical Reactions 214
9.3 Factors that Influence Reaction Rates 216
9.4 Chemical Equilibrium 219
9.5 Position of Equilibrium 221
9.6 Le Châtelier's Principle 221
Exercises and Problems 224

10 Acids, Bases, and Salts 226

10.1 Acid–Base Definitions 227
10.2 Strengths of Acids and Bases 229
10.3 Polyprotic Acids 230
10.4 Salts 232
10.5 Electrolytes 233

10.6	Neutralization 235		13.8	Soft Water 305
10.7	Ion Formation in Pure Water 235		13.9	Floridated Water 306
10.8	The pH Scale 237			Exercises and Problems 307
10.9	Acid–Base Titrations 239			
10.10	Acid–Base Titration Calculations 241		**14**	**Saturated Hydrocarbons 308**
10.11	Hydrolysis of Salts 245		14.1	Organic Chemistry—An Historical Perspective 309
10.12	Buffer Solutions 249		14.2	Hydrocarbons 310
10.13	Buffers in the Human Body 250		14.3	Saturated Hydrocarbons 311
	Exercises and Problems 252		14.4	Alkanes 311
			14.5	Structural Isomerism 312
11	**Oxidation–Reduction Processes 256**		14.6	Structural Formulas 315
			14.7	IUPAC Nomenclature for Alkanes 318
11.1	Oxidation–Reduction Terminology 257		14.8	Branched-Alkyl Groups 323
11.2	Oxidation Numbers 258		14.9	Cycloalkanes 324
11.3	Classes of Chemical Reactions 262		14.10	Natural Sources of Alkanes 327
11.4	Some Important Oxidation–Reduction Processes 265		14.11	Physical Properties of Alkanes 330
	Exercises and Problems 266		14.12	Chemical Properties of Alkanes 330
			14.13	Alkanes in Living Organisms 334
12	**Nuclear Chemistry and Radioactivity 268**			Exercises and Problems 335
12.1	Atomic Nuclei 269		**15**	**Unsaturated Hydrocarbons 340**
12.2	The Nature of Emissions from Radioactive Materials 270		15.1	Unsaturated Hydrocarbons 341
12.3	Equations for Nuclear Reactions 271		15.2	Alkenes 341
12.4	Bombardment Reactions and Artificial Radioactivity 274		15.3	IUPAC Nomenclature for Alkenes 342
			15.4	The Nature of Carbon–Carbon Multiple Bonds 346
12.5	Rate of Radioactive Decay 275		15.5	Geometric Isomers 348
12.6	Synthetic Elements 277		15.6	Physical Properties and Natural Occurrence of Alkanes 353
12.7	The Ionizing Effects of Radiation 278		15.7	Chemical Reactions of Alkenes 354
12.8	The Biological Effects of Various Types of Radiation 279		15.8	Oxidation and Reduction in Organic Reactions 359
12.9	Detection of Radiation 281		15.9	Polymerization of Alkenes 359
12.10	Sources of Radiation Exposure 282		15.10	Alkynes 362
12.11	Nuclear Medicine 282		15.11	Nomenclature of Alkynes 363
12.12	Nuclear Fission and Nuclear Fusion 285		15.12	Physical Properties and Chemical Reactions of Alkynes 364
12.13	A Comparison of Nuclear and Chemical Reactions 287			Exercises and Problems 365
	Exercises and Problems 288			
			16	**Aromatic Hydrocarbons 370**
13	**Chemistry of Several Common Elements and Compounds 290**		16.1	Aromatic Hydrocarbons 371
			16.2	Benzene: Structure and Bonding 372
13.1	Abundances of the Elements 291		16.3	Nomenclature of Benzene Derivatives 373
13.2	The Atmosphere 295		16.4	Chemical Reactions of Benzene 377
13.3	Carbon Monoxide 297		16.5	Uses of Benzene and Benzene Derivatives 382
13.4	Sulfur Dioxide 298		16.6	Fused-Ring Aromatic Compounds 384
13.5	Nitrogen Monoxide and Nitrogen Dioxide 300		16.7	Sources of Aromatic Hydrocarbons 385
13.6	The Hydrosphere 302		16.8	Heterocyclic Compounds 385
13.7	Hard Water 304			

16.9 Hydrocarbons: A Summary   386
Exercises and Problems   387

17 Alcohols, Phenols, and Ethers   390

17.1 Structural Features of Alcohols, Phenols, and Ethers   391
17.2 Nomenclature of Alcohols and Phenols   392
17.3 Classifications of Alcohols   396
17.4 Physical Properties of Alcohols   397
17.5 Preparation of Alcohols   399
17.6 Reactions of Alcohols   400
17.7 Important Commonly Used Alcohols   404
17.8 Properties and Uses of Phenols   409
17.9 Phenols in Nature   410
17.10 Classification and Nomenclature of Ethers   411
17.11 Properties and Uses of Ethers   412
17.12 Epoxides   413
17.13 Thiols and Disulfides   414
Exercises and Problems   415

18 Aldehydes and Ketones   422

18.1 The Carbonyl Group   423
18.2 Structures of Aldehydes and Ketones   424
18.3 Nomenclature of Aldehydes and Ketones   425
18.4 Properties and Uses of Aldehydes and Ketones   427
18.5 Preparation of Aldehydes and Ketones   430
18.6 Reduction of Carbonyl Compounds   431
18.7 Hemiacetal and Acetal Formation   432
18.8 Oxidation of Aldehydes   434
Exercises and Problems   436

19 Carboxylic Acids and Esters   440

19.1 The Structure of Carboxylic Acids   441
19.2 Nomenclature of Carboxylic Acids   443
19.3 Physical Properties of Carboxylic Acids   446
19.4 Preparation and Reactions of Carboxylic Acids   447
19.5 Acidity of Carboxylic Acids   448
19.6 Salts of Carboxylic Acids   449
19.7 The Ester Functional Group   451
19.8 Preparation of Esters   451
19.9 Nomenclature of Esters   453
19.10 Properties of Esters   454
19.11 Reactions of Esters   456
19.12 Polyesters   457
19.13 Esters of Inorganic Acids   459
19.14 Acid Anhydrides   460
Exercises and Problems   463

20 Amines and Amides   468

20.1 Structure and Classification of Amines   469
20.2 Nomenclature of Amines   470
20.3 Properties, Uses and Occurrences of Amines   472
20.4 Preparation of Amines   476
20.5 Basicity of Amines   477
20.6 Chemical Reactions of Amines   480
20.7 Heterocyclic Amines   483
20.8 Structure and Classification of Amides   490
20.9 Nomenclature of Amides   490
20.10 Properties of Amides   491
20.11 Preparation of Amides   492
20.12 Hydrolysis of Amides   492
20.13 Selected Amides and Their Uses   493
20.14 Alkaloids   495
20.15 Polyamides and Polyurethanes   496
Exercises and Problems   499

21 Stereoisomerism   506

21.1 Types of Isomerism   507
21.2 Optical Isomerism   508
21.3 Chiral Carbon Atoms   510
21.4 Projection Formulas for Enantiomers   513
21.5 Nomenclature for Enantiomers   515
21.6 Optical Activity   516
21.7 Specific Rotation   518
21.8 Racemic Mixtures   519
21.9 Diastereomers and Meso Compounds   520
21.10 Biological Implications of Chirality   522
Exercises and Problems   524

22 Carbohydrates   528

22.1 Types of Biochemical Substances   529
22.2 Occurrence of Carbohydrates   530
22.3 Definition and Classification of Carbohydrates   530
22.4 Classifications and Structures for Monosaccharides   532
22.5 Important Monosaccharides   535
22.6 Cyclic Forms of Monosaccharides   539
22.7 Reactions of Monosaccharides   543
22.8 Disaccharides   548
22.9 Artificial Sweeteners   555
22.10 Polysaccharides   556
Exercises and Problems   562

23 Lipids   568

23.1 Classification of Lipids   569
23.2 Fatty Acids   569

23.3 Fats and Oils 573
23.4 Chemical Reactions of Fats and Oils (Triglycerides) 577
23.5 Phosphoglycerides 581
23.6 Waxes 584
23.7 Sphingolipids 585
23.8 Steroids 588
23.9 Prostaglandins 590
23.10 Cell Membranes 593
Exercises and Problems 598

24 Proteins 602

24.1 Functions and Characteristics of Proteins 603
24.2 Amino Acids—Building Blocks for Proteins 605
24.3 Essential Amino Acids 609
24.4 Stereoisomerism in Alpha-Amino Acids 610
24.5 Acid–Base Properties of Amino Acids 610
24.6 Peptide Formation 614
24.7 Proteins 618
24.8 Primary Structure of Proteins 618
24.9 Secondary Structure of Proteins 620
24.10 Tertiary Structure of Proteins 625
24.11 Quarternary Structure of Proteins 629
24.12 Protein Hydrolysis 631
24.13 Protein Denaturation 631
24.14 Immunoglobulins: An Example of Protein Structure and Function 636
Exercises and Problems 639

25 Enzymes: Biological Catalysts 644

25.1 Importance of Enzymes 645
25.2 Enzyme Nomenclature and Associated Terminology 646
25.3 Enzyme Cofactors 647
25.4 Zymogens or Proenzymes 649
25.5 General Properties of Enzymes 649
25.6 Models of Enzyme Action 651
25.7 Factors Affecting Enzyme Activity 654
25.8 Control of Enzyme Activity 658
25.9 Protein-Digesting Enzymes 663
25.10 Antibiotics that Inhibit Enzyme Activity 664
25.11 Isoenzymes 668
25.12 Medical Applications and Uses of Enzymes 668
Exercises and Problems 670

26 Nucleic Acids 674

26.1 Nucleic Acids 675
26.2 Nucleotides 676
26.3 Primary Nucleic Acid Structure 680
26.4 Double Helix Structure of DNA 683
26.5 DNA Replication 687
26.6 Protein Synthesis 691
26.7 Ribonucleic Acids 692
26.8 Transcription: RNA Synthesis 694
26.9 The Genetic Code 697
26.10 Translation: Protein Synthesis 700
26.11 Mutations 706
26.12 Viruses 708
26.13 Recombinant DNA and Genetic Engineering 711
Exercises and Problems 716

27 Vitamins, Minerals, and Hormones 720

27.1 Characteristics of Vitamins 721
27.2 Water-Soluble Vitamins 722
27.3 Fat-Soluble Vitamins 729
27.4 Minerals 733
27.5 The Endocrine System and Hormone Function 735
27.6 Hormone Structure and Classification 738
27.7 Molecular Mechanisms of Hormone Action 742
27.8 Sex Hormones and the Menstrual Cycle 746
Exercises and Problems 750

28 Metabolism and Nutrition: An Overview 754

28.1 Chemical Energy 755
28.2 Oxidation and Reduction Reactions in Cells 758
28.3 ATP: Universal Energy Currency of the Cell 760
28.4 Electron Carriers 762
28.5 Caloric Content of Foods and Principles of Nutrition 765
28.6 Major Pathways of Oxidative Metabolism 767
28.7 Mechanisms that Regulate Metabolism 769
28.8 Transmembrane Potentials 771
28.9 Photosynthesis 772
Exercises and Problems 775

29 Carbohydrate Metabolism 778

29.1 Glucose Absorption 779
29.2 Glycolysis 780
29.3 Metabolic Fates of Pyruvate 785
29.4 Gluconeogenesis 788
29.5 Pentose Phosphate Pathway 791

29.6 Glycogen Metabolism 794
29.7 Regulation of Carbohydrate Metabolism 799
29.8 Free Energy Changes in Carbohydrate Metabolism 801
Exercises and Problems 802

30 Mitochondrial Oxidation and Phosphorylation Pathways 804

30.1 The Mitochondrion 805
30.2 The Tricarboxylic Acid Cycle 806
30.3 The Electron Transport Chain 813
30.4 Transmembrane Proton Potentials: The Chemiosmotic Hypothesis 818
30.5 ATP Synthesis: Oxidative Phosphorylation 820
30.6 ATP Accounting 822
30.7 Regulation of the TCA Cycle and Oxidative Phosphorylation 825
Exercises and Problems 827

31 Lipid Metabolism 830

31.1 Metabolically Important Sources of Triglycerol Lipids 831
31.2 Lipid Absorption and Transport 832
31.3 Fatty Acid Mobilization and Oxidation 833
31.4 Fatty Acid Biosynthesis 837
31.5 ATP Accounting in Lipid Metabolism 842
31.6 Regulation of Fatty Acid Metabolism 843
31.7 The Relationship between Carbohydrate and Lipid Metabolism 845
Exercises and Problems 847

32 Amino Acid Metabolism 850

32.1 Overview of Amino Acid Metabolism 851
32.2 Carbon Skeletons of Amino Acids 852
32.3 Deamination and Transamination 855
32.4 The Urea Cycle 856
32.5 Amino Acid Biosynthesis 859
32.6 Control of Amino Acid Metabolism 860
Exercises and Problems 861

33 Extracellular Fluids 864

33.1 Extracellular and Intracellular Body Fluids 865
33.2 Blood 866
33.3 The Lymphatic System 869
33.4 The Urinary System 870
33.5 Fluid and Electrolyte Balance 873
33.6 Regulation of pH 874
33.7 Drug Absorption and Excretion 876
Exercises and Problems 879

Appendix A1

Answers to Selected Problems A2

Index I1

Preface

Fundamentals of Chemistry provides an introduction to a broad array of topics in general chemistry, organic chemistry, and biochemistry. It is intended for students in two-semester or three-quarter courses in applied science areas of study, including nursing, allied health disciplines such as medical and inhalation technology, biological sciences, agricultural sciences, and public health. Thus, chemical principles are presented along with numerous applications specifically designed to interest students preparing for careers in professional health sciences.

Vocabulary building is an important function of this course. Hence, the important terms in each chapter are highlighted in color in the margin and are defined within the text using bold and italic type. This makes them easy to locate and remember, and is an aid to the student when reviewing for tests and exams. Color is also used extensively in chemical reactions and metabolic pathways to emphasize the ions, atoms, molecules, or compounds directly involved.

Another feature of this text are the many learning tools. Each chapter opens with a list of learning objectives, keyed to each section of the chapter, and a chapter introduction, which provides a brief overview of the material covered in the chapter. Because we believe students need help in applying theories and concepts as quickly as possible, we have included over 120 worked-out examples. Each chapter ends with a large selection of exercises and problems, also keyed to each section of the chapter. Answers to many of these exercises and problems are in the Answer Section at the end of the text. Exercises whose answers are included in the answer section are highlighted in color.

Visualization is an important part of any science course, and this text includes an unusually large number of photographs and line illustrations carefully selected and designed to enhance pedagogical effectiveness. Particularly in the biochemistry section the artwork has been drawn in such a way as to show, as clearly and simply as possible, where these molecular processes are taking place.

To help both teacher and student, a full set of ancillary materials is available, including a *Study Guide*, an *Instructor's Manual*, a *Test Bank*, and a *Laboratory Manual*. The *Instructor's Manual* contains complete solutions to all text exercises and transparency masters of selected text figures. The *Test Bank* contains 1300 multiple-choice questions, each categorized with a subject heading and keyed to the textbook by section number. The *Study Guide* provides chapter summaries in outline form, worked-out solutions and explanations for those solutions, and student practice tests. The *Laboratory Manual* contains approximately 50 experiments, each with detailed procedures and data-and-report sheets.

We wish to thank our many colleagues who read and commented on the text manuscript at various stages:

Graeme Baker
University of Central Florida

Michael Carlo
Angelo State University

James Chickos
University of Missouri at St. Louis

Peter D. Gardner
University of Utah

Leland Harris
University of Arizona

Robert Lindquist
San Francisco State University

Marjorie Melville
San Antonio College

Stephen Metzner
University of South Dakota

Robert O'Malley
Boston College

M. Larry Peck
Texas A&M University

Thomas I. Pynadath
Kent State University

William Schultz
Eastern Kentucky University

William M. Scovell
Bowling Green State University

Will Sprain
San Jose State University

Donald Titus
Temple University

Vaughn Vandergrift
Murray State University

Hans Zimmer
University of Cincinnati

Their comments were very helpful.

 We hope that all who use these materials will find them understandable, useful and scientifically correct. Because we are sincerely interested in improving them and our courses, we welcome suggestions, criticisms, or other comments.

FUNDAMENTALS OF CHEMISTRY

1 Some Fundamental Concepts

Objectives

After completing Chapter 1, you will be able to:

1.1 Define the term *matter* and list the aspects of matter that are of particular concern to chemists.
1.2 Define the terms *composition* and *structure*.
1.3 Classify a property of a substance as a physical or a chemical property.
1.4 Classify the changes that occur in matter as physical or chemical changes.
1.5 Recognize the difference between qualitative and quantitative observations and define the term *measurement*.
1.6 Explain the difference between the precision and the accuracy of a set of measurements.
1.7 Understand the rule governing how many digits are recorded for a given measurement and know what significant figures are.
1.8 Convert numbers from decimal notation to exponential notation, and vice versa.
1.9 Carry out the mathematical operations of multiplication and division with all numbers expressed in exponential notation.
1.10 Recognize units of the metric system by name and abbreviation and know the numerical meanings associated with various metric system prefixes.
1.11 Set up and work simple metric system–metric system, metric system–English system, and English system–metric system unit conversion problems using dimensional analysis.
1.12 Distinguish between density and specific gravity, including the units of each. You will also be able to use density as a conversion factor between mass and volume, and vice versa.
1.13 List and use the interrelationships among the Fahrenheit, Celsius, and Kelvin temperature scales.
1.14 Distinguish among various forms of energy and various types of energy.
1.15 Understand the relationships among the units calories, kilocalories, and Calories and use specific heat as a conversion factor in problem solving.

INTRODUCTION

One of the first questions that is often asked by individuals considering the study of chemistry is "What exactly is chemistry about?" We will consider that question in this chapter. In addition, we will review some fundamental principles concerning mathematical notation and units of measurement. Much of this material may be familiar to you; nevertheless, a brief review of these topics will provide a foundation for material covered in later chapters.

1.1 THE SCOPE OF CHEMISTRY

An operational definition for chemistry, given in terms of the areas of major concern to chemists, will serve as our starting point. **Chemistry** *is the science concerned with: (1) the composition and structure of matter, (2) the properties of matter, and (3) the changes matter undergoes and the conditions necessary to cause or prevent these changes.*

chemistry

This definition has meaning only when the key scientific words used within it are understood. The term *matter* occurs three times in the definition. What is matter? Intuitively, most people consider matter to be the materials of the physical universe; it is the "stuff" from which the universe is made. This interpretation is a correct one.

A more scientific definition of matter than the "stuff of the universe" can be given. **Matter** *is anything that occupies space and possesses mass.* The property of occupying space is usually easily perceivable by our senses of sight and touch. However, a substance need not be visible to the naked eye in order to occupy space. Mass, a property to be discussed in detail in Section 1.10, is a measure of the total quantity of matter present in an object. Examples of matter include wood, paper, stone, the food we eat, the air we breathe, the fluids of our bodies, our bodies themselves, our clothing, and our shelter. Matter does not include various forms of energy such as heat, light, and electricity; in addition, wisdom, courage, ideas, thoughts, anger, and love are not included.

matter

Our definition of chemistry mentions three aspects of matter of particular concern to chemists. These areas comprise the subject matter for the next three sections of this chapter.

1.2 THE COMPOSITION AND STRUCTURE OF MATTER

The first of the three major concern areas for chemists is the composition and structure of matter. The **composition** *of a substance is known when the identity and amount of each of its components have been determined.* It is not enough to know what is present; the amount of each component present must be specified before the composition can be known. **Structure** *is the manner in which the constituent parts of a substance are put together; that is, the order in which the constituent parts are arranged relative to each other.*

composition

structure

Knowledge of composition and structure can often be put to practical use. For example, the chemical composition of certain body fluids can often be used by a physician to pinpoint the cause of illness in a patient. In particular, the absence or excessive amount of certain substances can provide vital information to the physician.

Structural information about a substance often provides insights into its chemical behavior. In some cases, these insights allow this behavior to be built into other substances. Detergents, substances with cleaning properties similar to those of soap, have structures patterned after that of soap. Numerous prescription drugs used in the field of medicine today (Figure 1–1) have structures similar to those of chemicals that the human body naturally produces. A knowledge of the chemical structures of these naturally occurring substances was the starting point for the development of these drugs.

1.3 PROPERTIES OF MATTER

properties

The properties of matter comprise the second aspect of matter that is of concern to chemists (Section 1.1). **Properties** *are the distinguishing characteristics of a substance that are used in its identification and description.* Just as we recognize a friend by characteristics such as hair color, walk, tone of voice, or shape of nose, we recognize various chemical substances by their properties. Each chemical substance has a unique set of properties that distinguishes it from all other substances. If two samples of pure materials have identical properties, they are necessarily the same substance.

The properties of substances can be used in a number of practical ways, such as

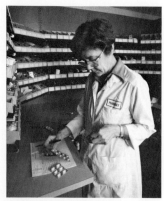

FIGURE 1–1 A knowledge of chemical structures was instrumental in the development of many of the medicines and drugs that are used today.

1. *Identifying an unknown substance.* Identifying a confiscated drug sample such as marijuana involves comparing the properties of the drug with those of known marijuana samples.
2. *Distinguishing between different substances.* A dentist can quickly tell the difference between a real tooth and a false tooth because of property differences.
3. *Characterizing a newly discovered substance.* Any new substance must have a unique set of properties; they must be different from those of any previously characterized substance.
4. *Predicting the usefulness of a substance for specific applications.* A substance that causes hair to fall out obviously should not be used in an anti-dandruff formulation.

physical properties

Two general categories of properties of matter exist: physical properties and chemical properties. **Physical properties** *are properties that are observable without changing a substance into another substance.* Color, odor, taste, size, physical state (solid, liquid, or gas), boiling point, melting point, and density are all examples of physical properties.

During the process of determining a physical property, the physical appearance of a substance may change, but the substance's identity will not. For example, measuring the melting point of a solid cannot be accomplished without changing the solid to a liquid. Although the liquid's appearance is much different from that of the solid, the substance is still the same; its chemical identity has not changed. Hence, melting point is a physical property.

chemical properties

Chemical properties *are properties that matter exhibits as it undergoes changes in chemical composition.* Often, these composition changes result from the interaction (reaction) of the matter with other substances. When copper objects are exposed to moist air for long periods of time, they turn green; this is a chemical property of copper. The green coating formed on the copper is a new substance; it results from the reaction of copper metal with the oxygen, carbon dioxide, and water in air. The properties of this new substance are very different from those of metallic copper.

TABLE 1–1 Selected Physical and Chemical Properties of Oxygen

Physical Properties	Chemical Properties
1. odorless	1. reacts with hydrogen at elevated temperatures to produce water
2. colorless in the gaseous state	2. reacts with fats in foods, causing a rancid odor and taste
3. pale blue color in the liquid state	3. combines with hemoglobin in blood to form oxyhemoglobin
4. melting point of $-219°C$	4. does not react with gold at room temperature
5. boiling point of $-183°C$	

Sometimes, under proper conditions, a single substance will undergo chemical change in the absence of any other substance in a process called *decomposition*. For example, hydrogen peroxide, in the presence of either heat or light, breaks down into the substances water and oxygen.

The *failure* of a substance to undergo change in the presence of another substance is also considered a chemical property. Flammability and nonflammability are both chemical properties.

Table 1–1 contrasts selected physical and chemical properties of the substance oxygen. Note how chemical properties cannot be described without reference to other substances. It does not make sense to say that a substance reacts spontaneously. The substance that it combines with must be specified because it might interact with many different substances.

1.4 CHANGES IN MATTER

The third aspect of matter that is of particular interest to chemists (Section 1.1) is changes in matter. Changes in matter are common and familiar occurrences. Changes take place, for example, when food is cooked, paper is burned, and a pencil is sharpened.

Like properties (Section 1.3), changes in matter are classified as physical or chemical. A **physical change** is *a process that does not alter the basic nature (chemical composition) of the substance under consideration*. No new substances are ever formed as a result of physical change. A **chemical change** is *a process that involves a change in the basic nature (chemical composition) of the substance*. These changes always involve conversion of the material or materials under consideration into one or more new substances having distinctly different properties and composition from those of the original materials.

A change in physical state is the most common type of physical change. Melting, freezing, evaporation, and condensation all represent changes of state. In any of these processes, the composition of the substance undergoing change remains the same even though its physical state and appearance change. The melting of ice does not produce a new substance; the substance is water before and after the change. Similarly, the steam produced from boiling water is still water.

Changes in size, shape, and state of subdivision are examples of physical changes that are not changes of state. Pulverizing an aspirin tablet into a fine powder and cutting a piece of adhesive tape into small pieces are examples of physical changes that involve only the solid state.

The appearance of one or more new substances is always a characteristic of a chemical change. Consider, for example, the rusting of iron objects left exposed to moist air. The reddish-brown substance formed (the rust) is a new substance

FIGURE 1-2 As the result of chemical change, a once bright and shiny shovel acquires a dull rusty appearance when left exposed to moist air.

with obviously different chemical properties from those of the original iron (see Figure 1-2).

Some chemical changes that take place in matter are beneficial because the resulting products are more useful than the starting materials. Other changes, such as the rusting of iron, are undesirable because the resulting products are not useful. By studying the nature of changes in matter, chemists learn how to bring about favorable changes and prevent undesirable ones. The control of chemical change has been a major factor in reaching our modern standard of living. The many plastics, synthetic fibers, and prescription drugs now in common use are the result of controlled chemical changes. Table 1-2 classifies a number of changes of matter as being either physical or chemical.

Most changes in matter can easily be classified as either physical or chemical. However, not all changes are black or white; there are some gray areas. For example, the formation of certain solutions falls in the gray area. Common salt

TABLE 1-2 Classification of Changes as Physical or Chemical

Change	Classification
crushing a dry leaf	physical
rotting of a tree stump	chemical
melting of snow	physical
digesting food	chemical
taking a bite of food	physical
burning a chemistry book	chemical
slicing of an onion	physical
tarnishing of silverware	chemical
souring of milk	chemical
scabbing over of a skin cut	chemical

dissolves easily in water to form a solution of salt water. The salt can easily be recovered by the physical process of evaporating the water. When gaseous hydrogen chloride is dissolved in water, again a solution results; however, in this case, the starting materials cannot be recovered by evaporation. The formation of salt water is considered a physical change because the original components can be recovered in an unchanged form using physical methods. The second solution presents classification problems because of the possibility that a chemical reaction took place.

The changes involved in cooking an egg also present classification problems. The cooked egg contains the same structural units as the uncooked egg. However, some changes in structural arrangement have taken place, so is the change physical or chemical? Despite the existence of gray areas, we will continue to use the concepts of physical and chemical change because their usefulness far outweighs the problems created by a few exceptions.

1.5 OBSERVATIONS AND MEASUREMENTS

Chemists, as well as other scientists, rely upon observations to explain the nature of the substances they study. Two general types of observations exist: qualitative observations and quantitative observations. A **qualitative observation** *is an observation made with the senses and is usually expressed using words instead of numbers.* Possible qualitative observations of a person who is sick in the hospital might be that the person is breathing rapidly, has a high temperature, and is very thin.

A **quantitative observation** *is an observation that requires a numerical measurement and describes something in terms of "how much"*. The quantitative observation that a person has a body temperature of 103.6 °F is much more useful information than just knowing that the person has a high fever. Scientists prefer quantitative observations.

One or more measurements is always a part of any quantitative observation. A **measurement** *determines the dimensions, capacity, quantity, or extent of something.* The most common types of measurements made in chemical laboratories are those of mass, volume, length, temperature, pressure, and concentration.

The next few sections of this chapter deal with the language that scientists use to describe measurements. Measurements need to be both *accurate* and *precise* (Section 1.6). Measurements always consist of two parts: a *number* that tells the amount of the quantity measured and a *unit* that tells the nature of the quantity being measured. Sections 1.7 through 1.9 deal with the numerical part of measurement and Section 1.10 discusses measurement units. An understanding of both the numerical notation and unit systems used by scientists in recording measurements is essential to an understanding of the significance of scientific measurements.

1.6 PRECISION AND ACCURACY

Scientific measurements need to be precise and accurate. Although the terms *precise* and *accurate* are used somewhat interchangeably in nonscientific discussion, they have distinctly different meanings in science. **Precision** *refers to how close multiple measurements of the same quantity are to each other.* **Accuracy** *refers to how close a measurement (or the average of multiple measurements) comes to the true or accepted value.* A simple analogy that does not directly involve measurement—shooting at a target—illustrates the difference between these two terms (see Figure 1–3, p. 6). Accuracy depends on how close the shots are to the center, or bull's eye, of the target. Precision depends on how close the shots are to each other.

FIGURE 1-3 The difference between precision and accuracy.

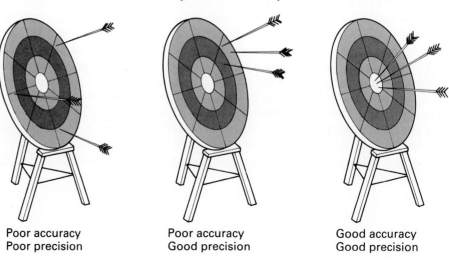

Poor accuracy
Poor precision

Poor accuracy
Good precision

Good accuracy
Good precision

The preciseness of a measurement is directly related to the actual physical measuring device used; that is, precision is an inherent part of any measuring device. Let us contrast the precision of time measurement using three different timepieces: a large clock on a tower, a wristwatch with a second hand, and a digital sports watch (see Figure 1–4). The tower clock is the least precise of the timepieces. With this instrument, time is estimated to the nearest minute. The analog wristwatch enables one to estimate time to the nearest second. Finally, the digital watch is even more precise because here time can be measured in tenths of a second.

In contrast to precision, accuracy depends not only on the measuring device used but also on the technical skill of the person making the measurement. How well can that person read the numerical scale of the instrument? How well can the person calibrate the instrument prior to its use?

Normally, high precision also results in high accuracy. However, high precision and low accuracy are also possible. Results obtained using a high-precision, poorly

FIGURE 1-4 The precision with which time may be measured depends on what type of timepiece is used to measure the time.

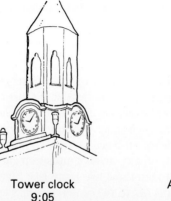

Tower clock
9:05

Analog wrist watch
9:05:22

Digital watch
9:05:23.4

calibrated instrument would give precision, but not accuracy. All measurements would be off by a constant amount as a result of the improper calibration.

1.7 SIGNIFICANT FIGURES

The form in which the numerical part of a measurement is written down must indicate the precision with which the measurement was made. Preciseness is indicated by reporting only the digits that are significant figures. **Significant figures** *in a measurement are those digits that are certain, plus a last digit that has been estimated.*

To illustrate the concept of significant figures, let us consider how two different rulers, shown in Figure 1–5, give measurements of differing precision when used to measure the length of a metal rod. Using Ruler A, we can say with certainty that the length of the rod is between 3 and 4 centimeters. We can also say that the actual length is closer to 4 centimeters and estimate the length of the rod to be 3.7 centimeters. Other people using Ruler A to measure the rod would also have to estimate the length beyond 3 centimeters and might record a value of 3.8 centimeters. The answer of 3.8 centimeters is just as acceptable as 3.7 centimeters. It is understood by scientists that the last digit in any measurement is always estimated and that some variance in this estimated digit will occur. The measurements 3.7 and 3.8 both contain two significant figures. One digit, the 3, we know for certain and one digit, the 7 or 8, is estimated.

Ruler B has more subdivisions on its scale than Ruler A. It is marked off in tenths of a centimeter instead of in centimeters. Using Ruler B, we can definitely say that the length of the rod is between 3.7 and 3.8 centimeters and can estimate it to be 3.74 centimeters. This measurement, 3.74 centimeters, contains three significant figures; the 3 and the 7 are certain digits, and the 4 is estimated digit. Note that you should never record more than one estimated digit in any measurement.

The measurement of the length of the rod obtained by using Ruler B is considered to be more precise than the measurement obtained by using Ruler A. The value of 3.74 centimeters is measured to the nearest 0.01 centimeter, but the value of 3.7 or 3.8 is known only to the nearest 0.1 centimeter. Preciseness for a given measurement is determined by the number of significant figures in the measurement. Thus, the term *precision* not only refers to the degree of reproducibility of repeated measurements (Section 1.6), but also to the number of significant figures in a given measurement.

It is very important to keep track of the number of significant figures in a measurement when recording measurements and using these measurements in calcula-

significant figures

FIGURE 1–5 The scale on a measuring device determines how many significant figures can be recorded for that measurement. Using Ruler A, the length of the object is recorded to two significant figures; using Ruler B, the length of the object is recorded to three significant figures.

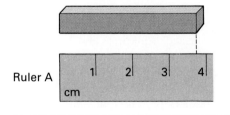
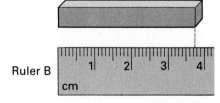

tions. Appendix 1 explains how to keep track of significant figures in mathematical calculations.

1.8 EXPONENTIAL NOTATION

In scientific work, we frequently encounter very large and very small numbers. As an illustration of this situation, consider the following information about human blood, a substance whose major component is water (92% by mass). In one drop of blood (1/20 mL), there are approximately

$$1,600,000,000,000,000,000,000$$

molecules of water. A molecule of water is the smallest possible unit of water. (See Section 2.4.) Obviously, molecules must be very small entities if there are that many of them in a single drop of blood. Also, a single water molecule has a mass of

$$0.00000000000000000000030 \text{ gram}$$

These large and small numbers are difficult to use. Recording them is not only a time-consuming task, but also one that is very prone to error; often too many or too few zeros are recorded. Also, these numbers are awkward to work with in calculations. Consider the problem of multiplying the preceding two numbers together; handling all of the zeros is a mind boggling endeavor.

A method exists for expressing cumbersome, multidigit numbers in compact form. This method is called *exponential notation*, and it eliminates the need to write all the zeros. **Exponential notation** *is a system in which an ordinary decimal number is expressed as the product of a number between 1 and 10 times 10 raised to a power*. The two previously cited numbers that deal with molecules of water are expressed in exponential notation as

$$1.6 \times 10^{21} \text{ molecules}$$

and

$$3.0 \times 10^{-22} \text{ gram}$$

respectively. Note that in exponential form, the numbers now involve six digits, as compared to 22 and 23 previously.

Exponents

As the name implies, exponential notation involves the use of exponents. An **exponent** *is a number written as a superscript that follows another number and indicates how many times the first number is to be multiplied by itself.* The following examples illlustrate the use of exponents.

$$4^3 = 4 \times 4 \times 4 = 64$$
$$2^4 = 2 \times 2 \times 2 \times 2 = 16$$
$$10^5 = 10 \times 10 \times 10 \times 10 \times 10 = 100,000$$

Exponents are also frequently referred to as *powers* of numbers. Thus, 4^3 may be verbally read as "four to the third power", and 2^4 as "two to the fourth power". Raising a number to the second power is often called "squaring" and to the third power "cubing".

Negative as well as positive exponents exist. A negative sign in front of an exponent means that the number and the power to which it is raised is in the denominator of a fraction in which 1 is the numerator. The following examples illustrate this interpretation.

$$10^{-2} = \frac{1}{10^2} = \frac{1}{10 \times 10} = \frac{1}{100} = 0.01$$

$$10^{-4} = \frac{1}{10^4} = \frac{1}{10 \times 10 \times 10 \times 10} = \frac{1}{10,000} = 0.0001$$

Writing Numbers in Exponential Notation

A number written in exponential notation has two parts: (1) a *coefficient*, which is written first, and is a number between 1 and 10, and (2) an *exponential term*, which is 10 raised to a power. The coefficient part is always multiplied by the exponential term. Using the exponential notation form of the number 276 as an example, we have

$$\underbrace{2.76}_{\text{coefficient}} \times \underbrace{10^{\overbrace{2}^{\text{exponent}}}}_{\text{exponential term}}$$

(multiplication sign)

The rules for converting numbers from decimal to exponential notation are very simple.

Rule 1: The value of the exponent is determined by counting the number of places the original decimal point must be moved to give the coefficient, which must be a number between 1 and 10.

Rule 2: If the decimal point is moved to the *left* to get the coefficient, the exponent is a *positive number*. If the decimal point is moved to the *right* to get the coefficient, the exponent is a *negative number*.

Example 1.1. illustrates the use of these rules.

Example 1.1

Express the following numbers in exponential notation.
a. 731   b. 2230.7   c. 0.444   d. 0.000733

Solution
a. The coefficient, which must be a number between 1 and 10, is obtained by moving the decimal point in 731 two places to the left.

This gives us 7.31 as the coefficient. The value of the exponent in the exponential term is +2 because the decimal point was shifted two places to the left. A shift of the decimal point to the left will always result in a positive exponent. Multiplying the coefficient by the exponential term, 10^2, gives us the exponential notation form of the number.

$$7.31 \times 10^2$$

b. The coefficient is obtained by moving the decimal point in 2230.7 three places to the left.

$$2230.7$$

The power of ten is +3, indicating that the decimal point was moved three places to the left to obtain the coefficient. Multiplying the coefficient and exponential terms together gives

$$2.2307 \times 10^3$$

c. Because the number 0.444 has a value of less than 1, the decimal point must be moved to the right in order to obtain a coefficient with a value between 1 and 10.

$$0.444$$

Moving the decimal point to the right means that the exponent will have a negative sign; in this case it is -1. Thus, in exponential notation we have

$$4.44 \times 10^{-1}$$

d. For the number 0.000733, a decimal point shift of four places to the right is needed to give a coefficient with a value between 1 and 10.

$$0.000733$$

Combining the power of ten, -4, and the coefficient together gives us an answer of

$$7.33 \times 10^{-4}$$

Converting from Exponential Notation to Decimal Notation

In order to convert a number in exponential notation such as 6.02×10^{23}, into a regular decimal number, we start by examining the exponent. The value of the exponent tells how many places the decimal point must be moved. If the exponent is positive, the decimal point is moved to the right to give a number greater than one; if it is negative, the decimal point is moved to the left to give a number less than one. Zeros may have to be added to the number as the decimal point is moved.

Example 1.2

Express the following exponential numbers in ordinary decimal notation.
   a. 3.76×10^{-3} b. 6.2×10^5 c. 5.3×10^2 d. 1.11×10^{-7}

Solution

a. The exponent -3 tells us the decimal point will be located three places to the left of where it is in 3.76. We add two zeros to accommodate for the decimal point change.

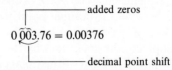

b. The exponent $+5$ tells us the decimal point will be located five places to the right of where it is in 6.2. We add four zeros to mark the new decimal place correctly.

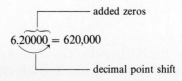

c. The decimal point will be moved two places to the right of where it is in 5.3. This gives the number 530 as the answer.

$$5.30 = 530$$

d. The decimal point will be moved seven places to the left. We add zeros to mark the new decimal place correctly.

$$0.0000001.11 = 0.000000111$$

1.9 EXPONENTIAL NOTATION AND MATHEMATICAL OPERATIONS

A major advantage of writing numbers in exponential notation is that it greatly simplifies the mathematical operations of multiplication and division.

Multiplication in Exponential Notation

The multiplication of two or more numbers expressed in exponential notation involves two separate steps.

Step 1: The coefficients (the decimal numbers between 1 and 10) are multiplied together in the usual manner.

Step 2: The exponents of the powers of ten are *added* algebraically.

Carry out the following multiplications in exponential notation.

a. $(2.33 \times 10^3) \times (1.55 \times 10^4)$
b. $(1.13 \times 10^3) \times (5.81 \times 10^{-6})$
c. $(1.25 \times 10^3) \times (1.85 \times 10^1) \times (2.05 \times 10^6)$

Example 1.3

Solution
a. Multiplying the two coefficients together gives

$$2.33 \times 1.55 = 3.61$$

Multiplication of the two powers of ten to give the exponential part of the answer requires that we add the exponents to give a new exponent.

$$10^3 \times 10^4 = 10^{(3)+(4)} = 10^7$$

Combining the new coefficient with the new exponential term gives the answer

$$3.61 \times 10^7$$

b. Multiplying the two coefficients together gives

$$1.13 \times 5.81 = 6.57$$

When multiplying the exponential terms, we will have to add exponents with different signs. To do this, we first determine the larger number (6 is larger than 3), and then subtract the smaller number from it (6 − 3 = 3). The sign is always the sign of the larger number, which is minus in this case. Thus

$$(3) + (-6) = -3$$
$$10^3 \times 10^{-6} = 10^{(3)+(-6)} = 10^{-3}$$

Combining the coefficient and the exponential term gives

$$6.57 \times 10^{-3}$$

c. Multiplying the three coefficients together gives

$$1.25 \times 1.85 \times 2.05 = 4.74$$

All of the exponential terms have the same sign—positive. When adding numbers of the same sign—positive or negative—just add the numbers and place the common sign in front of the sum.

$$10^3 \times 10^1 \times 10^6 = 10^{(3)+(1)+(6)} = 10^{10}$$

Combining the coefficient and the exponential term gives

$$4.74 \times 10^{10}$$

Division in Exponential Notation

The division of numbers in exponential notation, like multiplication, involves two steps.

Step 1: The coefficients are divided in the usual manner.
Step 2: The exponents of the powers of ten are *subtracted* algebraically. The exponent in the denominator (bottom) is always subtracted from the exponent in the numerator (top).

Note that in multiplication the exponents are *added* and in division the exponents are *subtracted*.

Example 1.4

Carry out the following divisions in exponential notation.

a. $\dfrac{8.42 \times 10^6}{3.02 \times 10^4}$ b. $\dfrac{4.20 \times 10^{-3}}{1.19 \times 10^{-7}}$ c. $\dfrac{9.44 \times 10^{-10}}{8.23 \times 10^6}$

Solution
a. Performing the indicated division of the coefficients gives

$$\frac{8.42}{3.02} = 2.79$$

The division of exponential terms involves the algebraic subtraction of exponents.

$$\frac{10^6}{10^4} = 10^{(+6)-(+4)} = 10^2$$

Algebraic subtraction involves changing the sign of the number to be subtracted and then following the rules for addition. In this problem, the number to be subtracted, $(+4)$, becomes (-4) upon changing the sign. Then we add $(+6)$ and (-4). The answer is $(+2)$, as shown. Combining the coefficient and the exponential term gives

$$2.79 \times 10^2$$

b. The new coefficient is obtained by dividing 4.20 by 1.19.

$$\frac{4.20}{1.19} = 3.53$$

The exponential part of the answer is obtained by subtracting (-7) from (-3). Changing the sign of the number to be subtracted gives $(+7)$. Adding $(+7)$ and (-3) gives $(+4)$. Therefore,

$$\frac{10^{-3}}{10^{-7}} = 10^{(-3)-(-7)} = 10^4$$

Combining the coefficient and the exponential terms gives

$$3.53 \times 10^4$$

c. Dividing the number 9.44 by 8.23 will give the new coefficient.

$$\frac{9.44}{8.23} = 1.15$$

The exponential term division involves subtracting $(+6)$ from (-10). Changing the sign of the number to be subtracted gives (-6). Adding (-6) and (-10) gives (-16).

$$\frac{10^{-10}}{10^6} = 10^{(-10)-(+6)} = 10^{-16}$$

Combining the two parts of the problem yields an answer of

$$1.15 \times 10^{-16}$$

1.10 SYSTEMS OF MEASUREMENT

A unit is a label that describes something that is being measured or counted. It can be almost anything: 4 quarts, 4 dimes, 4 dozen frogs, 4 bushels, 4 inches, or 4 pages. Two formal systems of units of measure are in use in the United States today. Common measurements of commerce, such as those used in a grocery store, are made in the *English system*. The units of this system include the familiar inch, foot, pound, quart, and gallon. A second system, the *metric system*, is used in scientific work. Units in this system include the gram, meter, and liter. The United States is one of a few countries that use different unit systems for commerce and scientific work. The metric system is used in most countries for both commercial and scientific work.

The United States is in the process of a voluntary conversion to the metric system. Many metric system units are now appearing on consumer products. Soft drinks can now be bought in 2-liter containers. Road signs in some states display distances in both miles and kilometers. Canned and packaged goods such as cereals and mixes on grocery store shelves now have their content masses listed in grams as well as in pounds and ounces.

Why should the United States convert to the metric system? The answer is simple. The metric system is superior to the English system. Its superiority lies in the area of interrelationships between units of the same type such as volume or length. Metric unit interrelationships are less complicated than English unit interrelationships because the metric system is a decimal unit system. In the metric system, conversion from one unit size to another can be accomplished simply by moving the decimal point to the right or left an appropriate number of places. The metric system is no more precise than the English system; it is simply more convenient.

The metric system was updated in 1960. By international agreement, a revision of the traditional metric system called the *International System of Units* (abbre-

TABLE 1-3 Metric System Prefixes and Their Mathematical Meanings

Prefix	Symbol	Exponential Number	Common Number	
giga-	G	10^9	1,000,000,000	prefixes for multiple units
mega-	M	10^6	1,000,000	
kilo-	k	10^3	1,000	
hecto-	h	10^2	100	
deca-	da	10^1	10	
deci-	d	10^{-1}	0.1	prefixes for subunits
centi-	c	10^{-2}	0.01	
milli-	m	10^{-3}	0.001	
micro-	μ	10^{-6}	0.000001	
nano-	n	10^{-9}	0.000000001	

(Mathematical Meaning spans Exponential Number and Common Number columns.)

viated *SI* after the French name *Le Systeme International d'Unites*) was adopted by scientists. Acceptance of the SI units has varied within the scientific community. In many applied sciences, including the health sciences, movement toward SI unit use has been very slow.

Because the differences between the two unit systems are not of major importance for the subjects to be covered in this text, we will continue to use the more familiar traditional metric system units. Switching to SI units sometime in the future when their use is more extensive will present no major problem to the student who properly understands the traditional metric system because the two systems are so similar.

Metric System Prefixes

In the metric system, there is one base unit for each type of measurement—length, volume, mass, etc. These base units are multiplied by appropriate powers of ten to form smaller or larger units. The names of the larger and smaller units are constructed from the base unit name by attaching prefixes that tell which power of ten is involved. These prefixes are given in Table 1-3, along with their symbols or abbreviations and mathematical meanings. The prefixes in color are the most frequently used.

The use of numerical prefixes should not be new to you. Consider the use of the prefix *tri-* in the words *tri*angle, *tri*cycle, *tri*o, *tri*nity, and *tri*ple. Each of these words conveys the idea of three of something. The metric system prefixes are used in the same way.

The meaning of a metric system prefix always remains constant; it is independent of the base it modifies. For example, a kilosecond is one thousand seconds; a kilowatt is one thousand watts; and a kilocalorie is one thousand calories. The prefix *kilo-* will always mean one thousand.

Metric Units of Length

meter

The **meter** *is the basic unit of length in the metric system.* Other units of length in the metric system are derived from the meter by using the prefixes listed in Table 1-3. The kilometer (km) is 1000 times larger than the meter (m), whereas the centimeter (cm) and millimeter (mm) are, respectively, 100 and 1000 times smaller than the meter. Figure 1-6 relates the metric units of meter, centimeter, and millimeter to objects and situations we encounter in everyday life.

FIGURE 1-6 Metric units of length in the everyday realm.

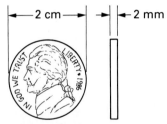

A nickel coin is 2 cm in diameter and 2 mm thick.

A millimeter is the diameter of the wire used in paper clips.

A piece of chalk is about 1 cm in diameter.

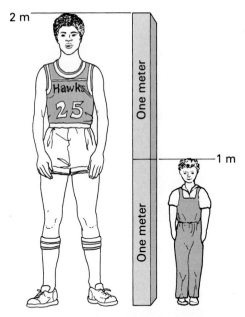

A "normal-sized" basketball player (6'7") is 2 m tall. A 3–4 year old is 1 m tall.

A comparison of metric lengths with the commonly used English system lengths of mile, yard, and inch reveals that a meter is slightly larger than a yard, a kilometer is approximately five-eighths of a mile, and a centimeter is slightly less than one-half of an inch.

Metric Units of Mass

The **gram** *is the basic unit of mass in the metric system.* It is a very small unit compared to the commonly used English mass units of ounce and pound. It takes approximately 28 grams to equal 1 ounce and nearly 454 grams to equal 1 pound. Because of the small size of the gram (g), the kilogram (kg) is a very commonly used unit. A kilogram is equivalent to an English mass of slightly more than 2 pounds. Figure 1–7 (p. 16) relates the mass units of milligram (mg), gram, and kilogram to everday objects.

The terms mass and weight are frequently used interchangeably in discussions. Although in most cases this practice does no harm, technically it is incorrect to interchange the terms and we will not do it in this text. Mass and weight refer to different properties of matter and their difference in meaning should be understood.

Mass *is a measure of the total quantity of matter in an object.* **Weight** *is a measure of the force exerted on an object by the pull of gravity.* The mass of a substance is a constant; the weight of an object is a variable dependent upon the object's geographical location.

FIGURE 1-7 Metric units of mass in the everyday realm

A single staple has a mass of about 3 mg.

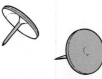
2 thumbtacks have a mass of about 1 g.

A nickel has a mass of about 5 g.

1 unit of donated blood has a mass of 450 g.

1 quart of milk in a cardboard container has a mass of 1 kg.

A 220-lb football player has a mass of 100 kg.

Matter at the equator weighs less than it would at the North Pole because the pull of gravity is less at the equator. The Earth is not a perfect sphere, but bulges at the equator; as a result, the magnitude of gravitational attraction is less at the equator. An object would weigh less on the moon than on Earth because of the smaller size of the moon and the correspondingly lower gravitational attraction. Quantitatively, a 22.0-pound mass weighing 22.0 pounds at the Earth's North Pole would weigh 21.9 pounds at the Earth's equator and only 3.7 pounds on the moon. In outer space an astronaut may be weightless but never massless. In fact, he or she has the same mass in space as on Earth.

Metric Units of Volume

It is frequently faster and more convenient to measure the volume rather than the mass of a substance. This is particularly true for liquids and gases. For example, volume is used instead of mass to determine the amount of gasoline put into an automobile fuel tank. The **liter** *is the basic unit of volume in the metric system.* A liter is abbreviated by using a capital L. If a lowercase l were used for the abbreviation, it might be confused with the number 1.

As with the units of length and mass, the basic unit of volume is modified with prefixes to represent smaller or larger units. The most commonly used prefixed volume unit is the milliliter (mL), which is 1/1000 of a liter. A milliliter is much smaller than a fluid ounce; it takes approximately 30 milliliters to equal 1 fluid

FIGURE 1–8 The metric volume unit of milliliter in the everyday realm.

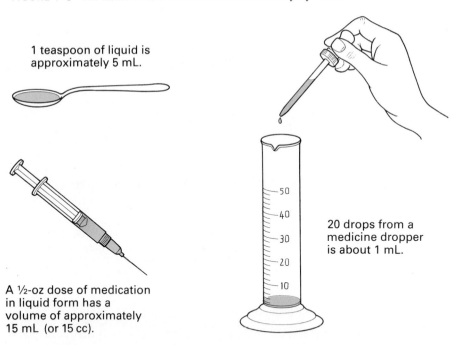

ounce. A quart and a liter are almost the same volume, but a quart is slightly smaller. Figure 1–8 relates the unit of milliliters to everyday situations. In clinical laboratory work, the volume unit cubic centimeters, which is abbreviated cm^3 or simply cc, is frequently used. What is its relationship to the units of liter and milliliter? Cubic centimeter is simply an alternate name for a milliliter, that is,

$$1 \text{ milliliter} = 1 \text{ cubic centimeter}$$

Consequently, the units milliliter and cubic centimeter can be used interchangeably.

1.11 CONVERSION FACTORS AND DIMENSIONAL ANALYSIS

Many times a need arises to change the units of a quantity or measurement to different units. The new units needed can either be in the same measurement system as the old ones or in a different one. With two unit systems in common use in the United States, the need to change measurements from one system to their equivalent in the other system frequently occurs.

The mathematical tool we will use to accomplish this task is a general method of problem solving called *dimensional analysis*. Central to the use of the dimensional analysis problem-solving method is the concept of conversion factors. A **conversion factor** *is a ratio with a value of unity that specifies how one unit of measurement is related to another.*

conversion factor

Let us construct some conversion factors to see how they originate and why they always have values of unity. The quantities "1 hour" and "60 minutes" both describe the same amount of time. We can thus write the following equation.

$$1 \text{ hour} = 60 \text{ minutes}$$

This equation, which describes a fixed relationship, can be used to construct a pair of conversion factors that relate hours and minutes. (Conversion factors always occur in pairs.) Dividing both sides of our hour–minute equation by the quantity 1 hour gives

$$\frac{1 \text{ hour}}{1 \text{ hour}} = \frac{60 \text{ minutes}}{1 \text{ hour}}$$

Because the numerator and denominator of the fraction on the left are identical, this fraction has a value of unity.

$$1 = \frac{60 \text{ minutes}}{1 \text{ hour}}$$

The fraction on the right side of the equation is our conversion factor, and its value is one. Note that the numerator and denominator of the conversion factor describe the same amount of time.

Two conversion factors are always obtainable from any given equality. For the equality we are considering (1 hour = 60 minutes) the second conversion factor is

$$\frac{1 \text{ hour}}{60 \text{ minutes}}$$

This is obtained by dividing both sides of the equality by 60 minutes instead of by 1 hour. The two conversion factors are reciprocals.

$$\frac{60 \text{ minutes}}{1 \text{ hour}} \quad \text{and} \quad \frac{1 \text{ hour}}{60 \text{ minutes}}$$

In general, we will always be able to construct a set of two conversion factors, each with a value of unity, from any two terms that describe the same amount of whatever we are considering.

Metric–Metric Conversion Factors

Metric–metric conversion factors are used to change one metric unit into another metric unit. Both the numerator and the denominator of these conversion factors involve metric system units. Metric system prefix meanings are used to derive these conversion factors. For example, the set of conversion factors involving kilometer and meter is derived from the meaning of the prefix *kilo-*, which is 10^3. The two conversion factors are

$$\frac{10^3 \text{ m}}{1 \text{ km}} \quad \text{and} \quad \frac{1 \text{ km}}{10^3 \text{ m}}$$

Note the reciprocal relationship between the two conversion factors of the set.

The conversion factors relating microgram and gram involve the number 10^{-6}, the mathematical equivalent of micro-, and are

$$\frac{1 \text{ μg}}{10^{-6} \text{ g}} \quad \text{and} \quad \frac{10^{-6} \text{ g}}{1 \text{ μg}}$$

Note the placement of the number 10^{-6} within the conversion factors. The numerical equivalent of the prefix always goes with the base (unprefixed) unit.

Metric–English and English–Metric Conversion Factors

Conversion factors that relate metric units to English units and vice versa are not exact defined quantities because they involve two different systems of measure-

TABLE 1-4 Equalities and Conversion Factors that Relate the English and Metric Systems of Measurement

	Metric to English	English to metric
Length		
1 inch = 2.54 centimeters	$\dfrac{1.00 \text{ in.}}{2.54 \text{ cm}}$	$\dfrac{2.54 \text{ cm}}{1.00 \text{ in.}}$
1 meter = 39.4 inches	$\dfrac{39.4 \text{ in.}}{1.00 \text{ m}}$	$\dfrac{1.00 \text{ m}}{39.4 \text{ in.}}$
1 kilometer = 0.621 mile	$\dfrac{0.621 \text{ mi}}{1.00 \text{ km}}$	$\dfrac{1.00 \text{ km}}{0.621 \text{ mi}}$
Mass		
1 pound = 454 grams	$\dfrac{1.00 \text{ lb}}{454 \text{ g}}$	$\dfrac{454 \text{ g}}{1.00 \text{ lb}}$
1 kilogram = 2.20 pounds	$\dfrac{2.20 \text{ lb}}{1.00 \text{ kg}}$	$\dfrac{1.00 \text{ kg}}{2.20 \text{ lb}}$
1 ounce = 28.3 grams	$\dfrac{1.00 \text{ oz}}{28.3 \text{ g}}$	$\dfrac{28.3 \text{ g}}{1.00 \text{ oz}}$
Volume		
1 quart = 946 milliliters	$\dfrac{1.00 \text{ qt}}{946 \text{ mL}}$	$\dfrac{946 \text{ mL}}{1.00 \text{ qt}}$
1 liter = 1.06 quarts	$\dfrac{1.06 \text{ qt}}{1.00 \text{ L}}$	$\dfrac{1.00 \text{ L}}{1.06 \text{ qt}}$
1 milliliter = 0.034 fluid ounce	$\dfrac{0.034 \text{ fl. oz.}}{1.00 \text{ mL}}$	$\dfrac{1.00 \text{ mL}}{0.034 \text{ fl. oz.}}$

ment. The numbers associated with these conversion factors must be determined experimentally. Table 1-4 lists commonly encountered relationships between metric and English system units. These few factors are sufficient to solve most of the problems that we will encounter.

Dimensional Analysis

Dimensional analysis *is a general problem-solving method that uses the units associated with numbers as a guide in setting up calculations. In this method, units are treated in the same way as numbers; that is, they can be multiplied, divided, or canceled.* For example, just as

$$5 \times 5 = 5^2 \quad (5 \text{ squared})$$

we have

$$\text{mL} \times \text{mL} = \text{mL}^2 \quad (\text{mL squared})$$

Also, just as the 3's cancel in the expression

$$\frac{\cancel{3} \times 5 \times 7}{\cancel{3} \times 2}$$

the centimeters cancel in the expression

$$\frac{\cancel{(\text{cm})} \times (\text{inch})}{\cancel{(\text{cm})}}$$

Like units found in the numerator and denominator of a fraction will always cancel, just as like numbers do.

The following steps show how to set up a problem using dimensional analysis.

Step 1: Identify the known or given quantity (both numerical value and units) and the units of the new quantity to be determined.

This necessary information, which serves as the starting point for setting up the problem, will always be found in the statement of the problem. Write an equation with the given quantity on the left and the units of the desired quantity on the right.

Step 2: Multiply the given quantity by one or more conversion factors in such a manner that the unwanted (original) units are canceled out, leaving only the desired units.

The general format for the multiplication is

$$(\text{information given}) \times (\text{conversion factors}) = (\text{information sought})$$

The number of conversion factors used depends on the individual problem. Except in the simplest problems, it is a good idea to formally predetermine the sequence of unit changes that will be used. This sequence will be called the unit pathway.

Step 3: Perform the mathematical operations indicated by the conversion factor setup.

When performing the calculation, double-check to make sure that all units except the desired set have canceled out.

Now let us work a number of sample problems using dimensional analysis and the steps just outlined. Our first two examples involve only metric–metric conversion factors. The first problem (Example 1.5) is very simple. Even though you can do this problem in your head, let us formally set it up using the steps of dimensional analysis. Much can be learned about dimensional analysis from this sample problem.

Example 1.5

A standard aspirin tablet contains 324 mg of aspirin. How many grams of aspirin are in a standard aspirin tablet?

Solution
Step 1: The given quantity is 324 mg, the mass of aspirin in the tablet. The unit of the desired quantity is grams.

$$324 \text{ mg} = ? \text{ g}$$

Step 2: Only one conversion factor will be needed to convert from milligrams to grams, one that relates milligrams to grams. Two forms of this factor exist.

$$\frac{1 \text{ mg}}{10^{-3} \text{ g}} \quad \text{and} \quad \frac{10^{-3} \text{ g}}{1 \text{ mg}}$$

The second factor is used because it allows for the cancellation of the milligram units, leaving us with grams as the new units.

$$324 \text{ mg} \times \left(\frac{10^{-3} \text{ g}}{1 \text{ mg}}\right) = ? \text{ g}$$

For cancellation, a unit must appear in both the numerator and the denominator. Because the given quantity (324 mg) has milligrams in the numerator, the conversion factor used must be the one with milligrams in the denominator.

If the other conversion factor had been used, we would have

$$324 \text{ mg} \times \left(\frac{1 \text{ mg}}{10^{-3} \text{ g}}\right)$$

No unit cancellation is possible in this setup. Multiplication gives mg²/g as the final units, which is certainly not what we want. In all cases, only one of the two conversion factors of a reciprocal pair will correctly fit into a dimensional-analysis setup.

Step 3: Step 2 takes care of the units. All that is left is to combine numerical terms to get a final answer; we still have to do the arithmetic. Collecting the numerical terms gives

$$\left(324 \times \frac{10^{-3}}{1}\right) \text{g} = 0.324 \text{ g}$$

number from first factor — numbers from second factor

Example 1.6

Healthy men average 71.4 mL of blood per kilogram of body mass. This translates to 7286 mL of blood in a 225-lb man. How many kiloliters of blood does an average 225-lb man possess?

Solution
Step 1: From the problem statement, we identify 7286 mL of blood as the given quantity and kiloliters of blood as the units of the desired quantity.

$$7286 \text{ mL} = ? \text{ kL}$$

Step 2: When dealing with metric–metric unit changes where both the original and desired units carry prefixes, it is recommended that you always channel units through the basic unit (the unprefixed unit). If this is done, you do not need to deal with any conversion factors other than those resulting from prefix definitions. Following this recommendation, the unit pathway for this problem is

$$\text{mL} \longrightarrow \text{L} \longrightarrow \text{kL}$$

In the setup for this problem, we will need two conversion factors, one for the milliliters to liters change and one for the liters to kiloliters change.

$$7286 \text{ mL} \times \left(\frac{10^{-3} \text{ L}}{1 \text{ mL}}\right) \times \left(\frac{1 \text{ kL}}{10^3 \text{ L}}\right) = ? \text{ kL}$$

this conversion factor converts mL to L — this conversion factor converts L to kL

Note how all the units cancel except for kiloliters.

Step 3: Carrying out the indicated numerical calculation gives

$$\left(7286 \times \frac{10^{-3}}{1} \times \frac{1}{10^3}\right) \text{kL} = 0.007286 \text{ kL}$$

number from first factor — numbers from second factor — numbers from third factor

Example 1.7

Our next two sample problems involve the use of both the English and metric systems of units. As mentioned previously, conversion factors between these two systems do not arise from definitions, but rather are determined experimentally. Some of these experimentally determined factors were given in Table 1–4.

In Example 1.6 we learned that an average 225-lb man possesses 7286 mL of blood. What is the volume of this blood in quarts?

Solution

Step 1: The given quantity is 7286 mL of blood and the units of the desired quantity are quarts of blood.

$$7286 \text{ mL} = ? \text{ qt}$$

Step 2: The given units are metric units and the desired units are English units. Going to the information in Table 1–4, we note that one of the three conversion factor sets in the volume area involves the units milliliters and quarts.

$$\frac{946 \text{ mL}}{1 \text{ qt}} \quad \text{and} \quad \frac{1 \text{ qt}}{946 \text{ mL}}$$

The second of these conversion factors is the one we will need to convert from milliliters to quarts.

$$7286 \text{ mL} \times \left(\frac{1 \text{ qt}}{946 \text{ mL}}\right) = ? \text{ qt}$$

Step 3: Performing the indicated arithmetic gives

$$\left(\frac{7286 \times 1}{946}\right) \text{qt} = 7.70 \text{ qt}$$

Example 1.8

Capillaries, the microscopic vessels that carry blood from small arteries to small veins, are on the average only 1 mm long. What is the average length of a capillary in inches?

Solution

Step 1: The given quantity is 1 mm and the units of the desired quantity are inches.

$$1 \text{ mm} = ? \text{ in.}$$

Step 2: None of the length conversion factors given in Table 1–4 will directly affect the conversion we need. Consequently, we will need to use a two-step pathway. First, we will convert the millimeters to meters. Then one of the conversion factors of Table 1–4 (meters to inches) can be used.

$$\text{mm} \longrightarrow \text{m} \longrightarrow \text{in.}$$

The correct conversion factor setup is

$$1 \text{ mm} \times \left(\frac{10^{-3} \text{ m}}{1 \text{ mm}}\right) \times \left(\frac{39.4 \text{ in.}}{1 \text{ m}}\right) = ? \text{ in.}$$

All of the units except for inches cancel, which is what is needed. The information for the middle conversion factor was obtained from the meaning for the prefix *milli-*.

This setup illustrates the fact that sometimes the given units must be changed to intermediate units before common conversion factors (Table 1–4) are applicable.

Step 3: Collecting the numerical factors and performing the indicated math gives

$$\left(\frac{1 \times 10^{-3} \times 39.4}{1 \times 1}\right) \text{in.} = 0.0394 \text{ in.}$$

1.12 DENSITY AND SPECIFIC GRAVITY

Density *is the ratio of the mass of an object to the volume occupied by that object.*

density

$$\text{density} = \frac{\text{mass}}{\text{volume}}$$

People often speak of a substance as being heavier or lighter than another substance. What they actually mean is that the two substances have different densities; a specific volume of one substance is heavier or lighter than the same volume of the second substance.

A correct density expression includes a number, a mass unit, and a volume unit. Although any mass and volume units can be used, densities are usually expressed in grams per cubic centimeter (g/cm^3) for solids, grams per milliliter (g/mL) for liquids, and grams per liter (g/L) for gases. The use of these units avoids the problem of having density values that are extremely small or large numbers. Table 1–5 gives density values for a number of substances. Note that temperature must be specified with density values because substances expand and contract with changes in temperature. Similarly, the pressure of gases is given with their density values.

In a mathematical sense, density can be thought of as a conversion factor that relates the volume of a substance to its mass. This use of density enables the volume of a substance to be calculated if its mass is known. Conversely, the mass can be calculated if the volume is known.

TABLE 1–5 Densities of Selected Substances

Solids (25°C)			
gold	19.3 g/cm³	table salt	2.16 g/cm³
lead	11.3 g/cm³	bone	1.7–2.0 g/cm³
copper	8.93 g/cm³	table sugar	1.59 g/cm³
aluminum	2.70 g/cm³	wood, pine	0.30–0.50 g/cm³
Liquids (25°C)			
mercury	13.55 g/mL	water	0.997 g/mL
milk	1.028–1.035 g/mL	olive oil	0.92 g/mL
blood plasma	1.027 g/mL	ethyl alcohol	0.79 g/mL
urine	1.003–1.030 g/mL	gasoline	0.56 g/mL
Gases (25°C and 1 atmosphere pressure)			
chlorine	3.17 g/L	nitrogen	1.25 g/L
carbon dioxide	1.96 g/L	methane	0.66 g/L
oxygen	1.42 g/L	hydrogen	0.08 g/L
air (dry)	1.29 g/L		

Example 1.9

Blood plasma has a density of 1.027 g/mL at 25 °C. What volume, in milliliters, would 125 g of plasma occupy?

Solution

Step 1: The given quantity is 125 g of blood plasma. The units of the desired quantity are milliliters. Thus, our starting point is

$$125 \text{ g} = ? \text{ mL}$$

Step 2: The conversion from grams to milliliters can be accomplished in one step because density, used as a conversion factor, directly relates grams to milliliters.

$$125 \text{ g} \times \left(\frac{1 \text{ mL}}{1.027 \text{ g}}\right) = ? \text{ mL}$$

Note that the actual conversion factor used is the reciprocal of the density. This was necessary so that the grams units would cancel.

Step 3: Doing the necessary arithmetic gives us our answer.

$$\left(\frac{125 \times 1}{1.027}\right) \text{mL} = 122 \text{ mL}$$

Specific Gravity

specific gravity

Specific gravity is a quantity that is closely related to density. The **specific gravity** *of a solid or liquid is the ratio of the density of that substance to the density of water at 4°C.*

$$\text{specific gravity of solid or liquid} = \frac{\text{density of substance}}{\text{density of water at } 4°C}$$

At 4°C, the density of water is 1.000 g/mL. For gases, specific gravity involves a density comparison with air rather than water. At 25°C, the density of dry air is 1.29 g/L.

When calculating the specific gravity of a substance, both densities must be expressed in the same units. Specific gravity is a *unitless* quantity because the identical sets of density units will always cancel.

Example 1.10

The density of a urine sample is 1.013 g/mL. What is the specific gravity of this urine?

Solution

The specific gravity of the urine sample is equal to the density of the urine sample divided by the density of water at 4°C.

$$\text{specific gravity of urine sample} = \frac{1.013 \text{ g/mL}}{1.000 \text{ g/mL}} = 1.013$$

Note how all the units cancel to make specific gravity a unitless quantity. The urine is 1.013 times as dense as water.

Although the determination of the specific gravity of a solid or gas usually requires both a mass and a volume measurement, it is possible to measure the specific gravity of a liquid directly by using a hydrometer. A hydrometer is a glass float that has a weighted bottom and calibrations on its stem (see Figure 1–9). The higher the density of the liquid, the higher the hydrometer tube will float.

A type of hydrometer called a urinometer is used extensively in clinical laboratories to determine urine sample specific gravities. The specific gravity of urine increases with the amount of solid waste present in the urine, and normal readings fall in the range between 1.018 and 1.025. Readings outside this range indicate certain abnormal body conditions to physicians. For example, low urine specific gravity often is the result of improperly functioning kidneys.

1.13 TEMPERATURE SCALES

The instrument that is most commonly used to measure temperature is the mercury thermometer, which consists of a glass bulb containing mercury sealed to a slender glass capillary tube. A small change in the volume of the mercury in the glass bulb caused by expansion or contraction of the liquid leads to a large change in the height of the mercury in the capillary tube. Graduations on the capillary tube indicate the extent of mercury expansion in terms of defined units called *degrees*.

Three different degree scales (temperature scales) are in common use—Celsius, Kelvin, and Fahrenheit (see Figure 1–10). Both the Celsius and Kelvin scales are part of the metric measurement system and the Fahrenheit scale belongs to the English measurement system. Different size degrees and different reference points are what produce the various temperature scales.

The *Celsius scale* is the scale most commonly encountered in scientific work. The normal boiling and freezing points of water serve as reference points on this scale, with the former having a value of 100° and the latter, 0°. Thus, there are 100 degree intervals between the two reference points.

The *Kelvin scale* is a close relative of the Celsius scale. Both have the same size of degree, and the number of degrees between the freezing and boiling points of water is the same. The two scales differ only in the numerical degree values assigned to the reference points. On the Kelvin scale, the boiling point of water is 373° and the freezing point of water is 273°. On the Kelvin scale, the zero point falls at 273° below that on the Celsius scale. The choice of these reference points, a shift

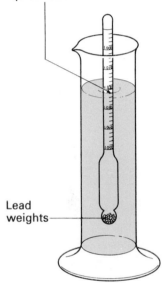

The scale reading at the intersection of the liquid surface and the stem is the specific gravity of the liquid: 1.022.

Lead weights

FIGURE 1–9 Specific gravity measurement using a hydrometer.

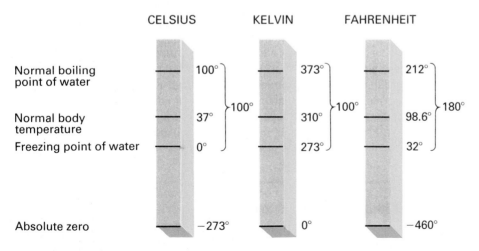

FIGURE 1–10 The relationships among the Celsius, Kelvin, and Fahrenheit temperature scales.

upward by 273 degrees compared to the Celsius scale, makes all temperature readings on the Kelvin scale positive values. A temperature of zero on the Kelvin scale corresponds to the lowest temperature known to scientists. In calculations involving the properties of gases (Section 7.7) temperature must always be specified on the Kelvin scale.

The *Fahrenheit scale* has a smaller degree size than the other two temperature scales. On this scale, there are 180 degrees between the freezing and boiling points of water as contrasted to 100 degrees on the other two scales. Thus, the Celsius (and Kelvin) degree size is 1.8 times larger than the Fahrenheit degree. Reference points on the Fahrenheit scale are 32° for the freezing point of water and 212° for the boiling point of water.

Because the size of the degree is the same, the relationship between the Kelvin and Celsius scales is very simple. No conversion factors are needed; all that is required is an adjustment for the differing numerical scale values. The adjustment factor is 273, the number of degrees by which the two scales are offset from each other.

$$K = °C + 273$$
$$°C = K - 273$$

Note that the symbol for degrees Kelvin is K, not °K.

The relationship between the Fahrenheit and Celsius scales can also be stated in an equation format.

$$°F = \frac{9}{5}(°C) + 32$$

$$°C = \frac{5}{9}(°F - 32)$$

Example 1.11 illustrates the use of these equations.

Example 1.11

Normal body temperature on the Fahrenheit scale is 98.6°F. What temperature is this equivalent to on

a. the Celsius scale
b. the Kelvin scale

Solution
a. Plugging the value 98.6° for the Fahrenheit temperature into the equation

$$°C = \frac{5}{9}(°F - 32)$$

and then solving for °C gives

$$°C = \frac{5}{9}(98.6 - 32)$$

$$= \frac{5}{9}(66.6)$$

$$= 37.0$$

b. Using the answer from part a and the equation

$$K = °C + 273$$

we get, by substitution

$$K = 37.0 + 273$$
$$= 310$$

1.14 TYPES AND FORMS OF ENERGY

On some days you wake up feeling very energetic. On these days you usually accomplish a great deal. By the end of the day you are tired, you have no energy left, and you do not feel like doing any more work. The scientific definition for energy closely parallels the ideas presented in the previous three sentences. **Energy** *is the capacity to do work.*

Energy can exist in any one of several forms. Common forms include radiant (light) energy, chemical energy, thermal (heat) energy, electrical energy, and mechanical energy. These forms of energy are interconvertible. Heating a home is a process that illustrates energy interconversion. As the result of burning natural gas or some other fuel, chemical energy is converted into heat energy. In large conventional power plants that are used to produce electricity, the heat energy obtained from burning coal is used to change water into steam, which can then turn a turbine (mechanical energy) to produce electricity (electrical energy).

During energy interconversions, the law of conservation of energy is always obeyed. The **law of conservation of energy** *states that in any chemical or physical change, energy can be converted from one form to another, but it is neither created nor destroyed.* Studies of energy changes in numerous systems have shown that no system acquires energy except at the expense of energy possessed by another system.

Almost all of our energy on Earth originates from the sun in the form of radiant or light energy. Green plants convert this radiant energy into chemical energy by means of a process called *photosynthesis* (Section 28.9); the chemical energy is stored within the living plant. **Chemical energy** *is energy stored in substances that can be released during a chemical change.* Energy for the human body is obtained, either directly or indirectly, from plants when they are consumed as food.

Chemical change can be used to produce other forms of energy. Chemical changes in an automobile battery produce the electrical energy needed to start a car. Chemical changes that occur in a magnesium flash bulb generate the light energy needed for photographic purposes. The burning of fuels releases both heat and light energy. The energy that runs our life processes, for example, breathing, muscle contractions, and blood circulation, is produced by chemical changes that occur within the cells of the body. The energy required for or generated by chemical changes will be an important point of focus in many discussions in later chapters of this text.

In addition to the various *forms* of energy, there are two *types* of energy: potential energy and kinetic energy. The basis for determining energy types depends on whether the energy is available but not being used or is actually in use. **Potential energy** *is stored energy that results from an object's position, condition, and/or composition.* Water that is backed up behind a dam represents potential energy because of its position. When the water is released, it can be used to produce electrical energy at a hydroelectric plant. A compressed spring can spontaneously expand and do work as the result of potential energy associated with its condition. Chemical energy, such as that stored in gasoline, is potential energy arising from its composition. This stored energy is released when the gasoline is burned.

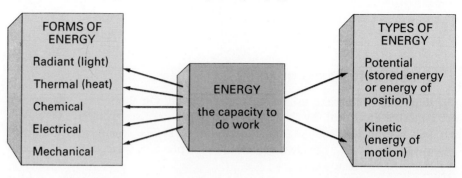

FIGURE 1–11 Energy can be classified by form and by type.

kinetic energy

Kinetic energy *is energy that matter possesses because of its motion.* An object that is in motion has the capacity to do work; if it collides with another object, it will do work on that object. A hammer held in the air above an object possesses potential energy of position. As the hammer moves downward toward an object, this potential energy becomes kinetic energy that can be used to drive a nail into a board. When water behind a dam is released and allowed to flow, its potential energy of position becomes kinetic energy. During the operation of a hydroelectric plant, some of this kinetic energy becomes mechanical energy and electrical energy. Figure 1–11 summarizes the forms and types of energy discussed in this section.

The concepts of potential and kinetic energy will play a part in many discussions in future chapters. For example, in Chapter 7, we will explain the differences between the solid, liquid, and gaseous states of matter in terms of relative amounts of potential and kinetic energy. The pressure that a gas exerts (Section 7.8) is related to kinetic energy.

1.15 HEAT ENERGY AND SPECIFIC HEAT

The form of energy that is most often required for or released by the chemical and physical changes considered in this text is *heat energy*. For this reason, we will consider further particulars about this form of energy.

Units for Heat Energy

calorie

The most commonly used unit of measurement for heat energy is the calorie, which is abbreviated cal. A **calorie** *is the amount of heat energy needed to raise the temperature of one gram of water by one degree Celsius.* For large amounts of heat energy, the measurement is usually expressed in kilocalories.

$$1 \text{ kilocalorie} = 1000 \text{ calories}$$

The SI unit (Section 1.10) of heat energy is the *joule*. The relationship between the joule and the calorie is

$$1 \text{ calorie} = 4.184 \text{ joules}$$

Heat energy values in calories can be converted to joules by multiplying the calorie value by 4.184.

In discussions involving nutrition, the energy content of foods, or dietary tables, the term *Calories* is also used. The dietetic Calorie (spelled with a capital C) is actually 1 kilocalorie (1000 calories). The statement that an oatmeal raisin cookie contains 60 Calories means that 60 kilocalories (60,000 calories) of energy are released when the cookie is metabolized (undergoes chemical change) within the body. Dieters who are aware that a banana split supplies 1500 Calories of energy might be less apt to succumb to temptation if they noted that the banana split contains 1,500,000 calories (calories spelled with a lowercase c).

$$1500 \text{ Calories} = 1500 \text{ kilocalories} = 1,500,000 \text{ calories}$$

Specific Heat

The **specific heat** *of a substance is the quantity of heat energy, in calories, that is necessary to raise the temperature of one gram of the substance by one degree Celsius.* Specific heats for a number of substances in various states are given in Table 1.6.

The higher the specific heat of a substance, the less its temperature will change when it absorbs a given amount of heat. For liquids, water has a relatively high specific heat; it is thus a very effective coolant. The moderate climates of geographical areas where large amounts of water are present, for example, the Hawaiian Islands, are related to water's ability to absorb large amounts of heat without undergoing drastic temperature changes. Desert areas, areas that lack water, are the areas where the extremes of high temperature are encountered on the Earth. The temperature of a living organism remains relatively constant because of the large amounts of water present in it.

Specific heat is an important quantity because it can be used to calculate the number of calories required to heat a known mass of a substance from one temperature to another. It can also be used to calculate how much the temperature of a substance increases when it absorbs a known number of calories of heat. The equation used for such calculations is

$$\text{heat absorbed} = \text{specific heat} \times \text{mass} \times \text{temperature change}$$

If any three of the four quantities in this equation are known the fourth quantity can be calculated. If the units for specific heat are cal/g°C, the units for mass are grams, and the units for temperature change are °C, then the heat absorbed will have the units of calories.

TABLE 1.6 Specific Heat of Selected Common Substances

Substance	Specific Heat (cal/g°C)
water, liquid	1.00
ethyl alcohol	0.58
olive oil	0.47
wood	0.42
aluminum	0.21
glass	0.12
silver	0.057
gold	0.031

Example 1.12

If a hot-water bottle contains 1200 g of water at 65°C, how much heat will it have supplied to a person's "aching back" by the time it has cooled to 37°C (assuming all of the heat energy goes into the person's back)?

Solution

We will substitute known quantities into the equation

$$\text{heat absorbed} = \text{specific heat} \times \text{mass} \times \text{temperature change}$$

Table 1.6 shows that the specific heat of liquid water is 1.00 cal/g °C. The mass of the water is given as 1200 grams. The temperature change in going from 65°C to 37°C is 28°C. Substituting these values into the preceding equation gives

$$\text{heat absorbed} = \left(\frac{1.00 \text{ cal}}{g \, °C}\right) \times (1200 \text{ g}) \times (28 \, °C)$$
$$= 33,600 \text{ cal}$$

Example 1.13

A serving of zucchini has a caloric value of approximately 2500 calories (25 Calories). What would be the temperature change in a cup of water at 20°C (237 g of water) if the same amount of energy was added to it?

Solution
The equation

$$\text{heat absorbed} = \text{specific heat} \times \text{mass} \times \text{temperature change}$$

is rearranged to isolate temperature change on the left side of the equation

$$\text{change in temperature (°C)} = \frac{\text{heat absorbed (cal)}}{\text{mass (g)} \times \text{specific heat (cal/g°C)}}$$

Substituting the known values into the equation gives

$$\text{change in temperature (°C)} = \frac{2500 \text{ cal}}{237 \text{ g} \times 1.00 \text{ cal/g°C}}$$

$$= 11°C$$

The temperature of the water would increase from 20°C to 31°C.

EXERCISES AND PROBLEMS

Chemistry: The Study of Matter

1.1 Define the term *matter*.

1.2 Classify each of the following as matter or nonmatter.
a. air
b. pizza
c. sound
d. light
e. virus
f. gold

1.3 List the three aspects of matter that are of particular concern to chemists.

Composition and Structure of Matter

1.4 Define the terms *composition* and *structure*.

1.5 What are the two types of information necessary to specify the composition of a substance?

1.6 What type of information is central to specifying the structure of a substance?

Properties of Matter

1.7 How does a physical property differ from a chemical property?

1.8 The following are properties of the substance magnesium. Classify them as physical or chemical properties.
a. Solid at room temperature.
b. Ignites upon heating in air.
c. Silvery-white in color.
d. Does not react with cold water.
e. Melts at 651°C.
f. Finely divided form burns in oxygen with a dazzling white flame.
g. Hydrogen gas is produced when it is dissolved in acids.
h. Has a density of 1.738 g/cm³ at 20°C.

1.9 Indicate whether each of the following statements describes a physical or a chemical property.
a. Silver salts discolor the skin by reacting with skin protein.
b. Hemoglobin molecules have a red color.
c. Beryllium metal vapor is extremely toxic to humans.
d. Aspirin tablets can be pulverized with a hammer.
e. Diamonds are very hard substances.
f. Gold metal will not dissolve in either hydrochloric acid or nitric acid.
g. Lithium metal is light enough to float on water.
h. Mercury is a liquid at room temperature.

Changes in Matter

1.10 What is the difference between a physical change and a chemical change?

1.11 Classify each of the following changes as physical or chemical.
a. Rusting of iron.
b. Sharpening a pencil.
c. Burning gasoline.
d. Breaking of glass.
e. Water evaporating from a lake.
f. Cutting grass in your front yard.
g. Forming a bar of copper into wire.
h. Fashioning a table leg from a piece of wood.

1.12 Does each of the following statements describe a physical change or a chemical change?
a. An Alka-Seltzer tablet is dropped into water.
b. A piece of dry ice
c. A board is sawed in two.

d. A rubber band is stretched.
e. Frozen orange juice is reconstituted by added water to it.
f. A container of grape juice that was left standing for an extended period of time is found to be fermented.
g. Ski goggles become fogged.
h. Light is emitted by a flashbulb.

Observations and Measurements

1.13 Classify each of the following observations about a patient in the hospital as qualitative or quantitative.
a. Blood pressure is high.
b. Temperature is 100.2°F.
c. Pulse is normal.
d. Urine specific gravity is 1.020.
e. Face is very pale.
f. Blood sugar level is 150 mg/dL.

1.14 With a very precise measuring device, the length of an object is determined to be 6.32141 cm. Three students, I, II, and III, are asked to determine the length of the same object using a less precise measuring device. Each student measures the object's length in centimeters six times, with the following results.

I: 6.54, 6.53, 6.55, 6.55, 6.56, and 6.52 (average = 6.54)
II: 7.31, 5.55, 5.82, 6.32, 6.67, and 7.92 (average = 6.60)
III: 6.31, 6.31, 6.33, 6.32, 6.34, and 6.31 (average = 6.32)

Characterize each student's performance in terms of accuracy (good or poor) and precision (good or poor).

1.15 Explain how a measurement can be precise but not accurate.

1.16 The actual mass of a certain object is 3.5011 g. Which of the following five measurements is the most precise and which is the most accurate?
a. 3.34 g
b. 3.5553 g
c. 3.50 g
d. 3.51 g
e. 3.483 g

Exponential Notation

1.17 Express the following numbers in exponential notation.
a. 634
b. 54,713
c. 34.13
d. 21.001
e. 34,000
f. 93,000,000
g. 3,513,000,000
h. 1,000,001

1.18 Express the following numbers in exponential notation.
a. 0.012
b. 0.1111
c. 0.00001
d. 0.2304
e. 0.0310
f. 0.000051
g. 0.00000000018
h. 0.000000000000009

1.19 Change the following numbers from exponential notation to decimal (ordinary) notation.
a. 2.3×10^4
b. 9.76×10^1
c. 2.3×10^{-4}
d. 3.3333334×10^5
e. 3.01×10^{-8}
f. 8×10^{27}
g. 6.32×10^{10}
h. 9.3×10^{-12}

1.20 Change the following numbers from exponential notation to decimal (ordinary) notation.
a. 6×10^3
b. 6×10^{-3}
c. 4.3×10^{-4}
d. 4.3×10^4
e. 7.3×10^2
f. 7.3×10^3
g. 3.0003×10^2
h. 3.0003×10^{10}

Mathematical Operations and Exponential Notation

1.21 Carry out the following multiplications of exponential terms.
a. $10^2 \times 10^3$
b. $10^2 \times 10^7$
c. $10^{-2} \times 10^{-3}$
d. $10^{-2} \times 10^{-7}$
e. $10^2 \times 10^{-3}$
f. $10^2 \times 10^{-7}$
g. $10^{-2} \times 10^3$
h. $10^2 \times 10^{-7} \times 10^4$

1.22 Carry out the following multiplications in exponential notation.
a. $(4 \times 10^2) \times (2 \times 10^{-4})$
b. $(3 \times 10^3) \times (3 \times 10^3)$
c. $(4.5 \times 10^{10}) \times (2.0 \times 10^1)$
d. $(2.6 \times 10^{-6}) \times (2.0 \times 10^{-12})$
e. $(3.0 \times 10^{-23}) \times (2.0 \times 10^{20})$
f. $(3.0 \times 10^3) \times (3.0 \times 10^{-3})$
g. $(2.0 \times 10^2) \times (2.0 \times 10^4) \times (2.0 \times 10^6)$
h. $(1.3 \times 10^{-3}) \times (2.0 \times 10^{-7}) \times (1.0 \times 10^{12})$

1.23 Carry out the following divisions of exponential terms.
a. $\dfrac{10^4}{10^2}$
b. $\dfrac{10^4}{10^{-2}}$
c. $\dfrac{10^{-4}}{10^2}$
d. $\dfrac{10^{-4}}{10^{-2}}$
e. $\dfrac{10^{-9}}{10^3}$
f. $\dfrac{10^8}{10^6}$
g. $\dfrac{10^2}{10^{-6}}$
h. $\dfrac{10^{-5}}{10^{-7}}$

1.24 Carry out the following divisions in exponential notation.
a. $\dfrac{4.0 \times 10^{-4}}{2.0 \times 10^6}$
b. $\dfrac{2.0 \times 10^{10}}{1.0 \times 10^{-6}}$
c. $\dfrac{9.0 \times 10^1}{3.0 \times 10^{-4}}$
d. $\dfrac{8.0 \times 10^{-3}}{2.0 \times 10^8}$
e. $\dfrac{3.3 \times 10^6}{1.1 \times 10^9}$
f. $\dfrac{5.0 \times 10^{-10}}{2.5 \times 10^{-12}}$
g. $\dfrac{2.0 \times 10^6}{4.0 \times 10^5}$
h. $\dfrac{4.0 \times 10^5}{3.3 \times 10^{-10}}$

1.25 Perform the following mathematical operations involving exponential terms.
a. $\dfrac{10^3 \times 10^4}{10^5}$
b. $\dfrac{10^{-3} \times 10^{-4}}{10^5}$
c. $\dfrac{10^5 \times 10^{-3}}{10^2 \times 10^4}$
d. $\dfrac{10^6 \times 10^{-8}}{10^5 \times 10^{10}}$

1.26 Perform the following mathematical operations in exponential notation.

a. $\dfrac{(4.0 \times 10^6) \times (1.0 \times 10^{-2})}{(2.0 \times 10^{-5})}$

b. $\dfrac{(9.0 \times 10^7)}{(3.0 \times 10^6) \times (2.0 \times 10^3)}$

c. $\dfrac{(4.5 \times 10^6) \times (4.0 \times 10^{-4})}{(2.0 \times 10^{-6}) \times (2.0 \times 10^{-2})}$

d. $\dfrac{(3.0 \times 10^6) \times (3.0 \times 10^{-7}) \times (2.0 \times 10^{-2})}{(8.0 \times 10^8)}$

Metric System Units

1.27 Write the name of the metric system prefix associated with each of the following mathematical meanings.

a. 10^3 c. 10^{-6} e. $1/1000$ g. 10^{-9}
b. 10^{-3} d. 10^{-2} f. $1/100$ h. 10^6

1.28 Write out the names of the metric system units that have the following abbreviations.

a. cm c. kL e. μL g. ng
b. mg d. Gg f. Mm h. mL

1.29 Arrange each of the following in an increasing sequence, from smallest to largest.

a. milligram, centigram, nanogram
b. gigameter, megameter, kilometer
c. microliter, deciliter, hectoliter
d. milligram, kilogram, microgram
e. kiloliter, nanoliter, centiliter
f. nanometer, centimeter, kilometer

Unit Conversions within the Metric System

1.30 Give both forms of the conversion factor that you would use to relate the following sets of metric units to each other.

a. gram and kilogram
b. liter and centiliter
c. meter and nanometer
d. gram and decagram
e. liter and gigaliter
f. meter and millimeter
g. liter and milliliter
h. gram and milligram

1.31 An aspirin tablet weighs 475 mg. Express its mass in the following units.

a. grams d. centigrams
b. kilograms e. gigagrams
c. micrograms f. nanograms

1.32 Convert each of the following measurements to centimeters.

a. 1.6×10^3 m d. 3×10^8 mm
b. 0.003 km e. 0.123 m
c. 24 nm f. 24,000 Mm

1.33 The human stomach produces approximately 2500 mL of gastric juice per day. What is the volume, in liters, of gastric juice produced?

1.34 Bacteria that cause pneumonia have an average diameter of 0.0000009 m. Express this diameter in centimeters.

1.35 A patient received intravenous fluids in the amounts of 250 mL, 500 mL, 25 cL, and 0.25 L. How many mLs of fluids did the patient receive?

1.36 A patient needs a 1.5 mL injection of a certain drug. If the syringe to be used for the injection is graduated in cubic centimeters, to what mark on the syringe would you draw in the drug?

1.37 A patient needs 0.030 g of a drug. The stock on hand of the drug is in the form of 10-mg tablets. How many tablets should be given?

1.38 A drop of blood has a volume of 0.05 mL. How many drops of blood are there in an adult body that has 5.3 L of blood?

Metric-English and English-Metric Unit Conversions

1.39 For each of the pairs of units listed, indicate which quantity is larger.

a. 1 centimeter, 1 inch e. 1 kilogram, 1 pound
b. 1 meter, 1 yard f. 1 liter, 1 quart
c. 1 kilometer, 1 mile g. 1 microliter, 1 ounce
d. 1 gram, 1 pound h. 1 milliliter, 1 pint

1.40 The mass of premature babies is customarily determined in grams. If a premature baby weighs 1550 g, what is its mass in pounds?

1.41 Typical normal values for daily output of water for the human body are as follows:

Kidneys (urine)	1400 mL
Lungs (water in expired air)	350 mL
Skin (by diffusion and sweat)	450 mL
Intestines (in feces)	200 mL

What daily output of water, in gallons, is this equivalent to?

1.42 The smallest bone in the human body, which is in the ear, has a mass of approximately 3 mg. What is the approximate mass of this bone in pounds?

1.43 Most human cells are bounded by a plasma membrane of approximately 9×10^{-7} cm thickness. What is this thickness in inches?

1.44 An individual weighs 83 kg and is 1.92 m tall. What are his equivalent measurements in pounds and feet?

1.45 An astronaut who drinks 2.0 L of water per day discharges 2.4 L of liquid water per day. (The additional water is produced by metabolism of food.) How many quarts of water does the astronaut discharge per day?

1.46 Assume that your heart beats 70 times per minute and that each time it beats, it pushes 60 mL of blood into the aorta for circulation throughout the body. What volume of blood, in quarts, circulates through the body in an hour?

1.47 The daily dosage of ampicillin to treat an ear infection is 100 mg per kilogram of body mass. What would be the daily dosage, in ounces, for a 63-lb child?

1.48 The current speed limit on United States highways is 55 miles per hour. What is this speed limit in kilometers per hour?

1.49 Some airlines limit a person's baggage to 20 kg on international flights. What is this limit in pounds?

1.50 Diamonds are weighed in terms of carats. One carat is equal to 200 mg. How much does a 5.0-carat diamond weigh in pounds?

Density and Specific Gravity

1.51 A sample of olive oil weighs 46 g. It occupies a volume of 50 mL. What is the density of the olive oil?

1.52 A patient's urine sample has a density of 1.021 g/mL. If this person eliminates 1350 mL of urine on a particular day, how many grams of urine were eliminated?

1.53 Acetone, the solvent in nail polish remover, has a density of 0.791 g/mL. What is the volume of 12.0 g of acetone?

1.54 The density of homogenized milk is 1.03 g/mL. How much does 1 cup (236 mL) of homogenized milk weigh, in pounds?

1.55 A copper penny weighs 3.21 g. If 8 pennies of this mass occupy 2.83 cm^3, what is the density of copper?

1.56 The density of silver is 10.40 g/cm^3. What is the specific gravity of silver?

1.57 A certain sample of urine has a specific gravity of 1.050. What is the density, in g/mL, of this urine?

1.58 What is the specific gravity of lithium, the least dense of all metals, if 5.34 g of lithium occupies a volume of 10.0 cm^3?

Temperature-Scale Conversions

1.59 A doctor has indicated that you should call him if your child's temperature goes above 40°C. You find that the child's temperature is 102°F. Should you call the doctor?

1.60 Mercury freezes at −38.9°C. What is the coldest temperature, in degrees Fahrenheit, that can be measured using a mercury thermometer?

1.61 An oven for baking pizza operates at approximately 525°F. What is this temperature in degrees Celsius?

1.62 Helium has the lowest boiling point of any liquid. It boils at 4 K. What is its boiling point in °C and °F?

1.63 A comfortable temperature for bathtub water is 95°F. What temperature is this in degrees Kelvin?

1.64 Birds have higher body temperatures than humans. A chickadee, for example, maintains a body temperature of 41.0°C. How many Fahrenheit degrees higher than normal human body temperature (98.6°F) is the body temperature of the chickadee?

Types and Forms of Energy

1.65 What is the scientific definition for *energy*?

1.66 State the Law of Conservation of Energy.

1.67 List the predominant *forms* of energy produced when each of the following processes occur.
a. an electric light bulb is turned on
b. a log is burned in a fireplace
c. a green plant grows
d. a bicycle is pedaled
e. a flashlight is turned on
f. a photographer's flash bulb goes off

1.68 What is the difference between potential energy and kinetic energy?

1.69 Identify the principal *type* of energy (kinetic or potential) that is exhibited by each of the following.
a. a car parked on a hill
b. a car traveling at 60 miles per hour
c. chemical energy
d. water behind a dam
e. a falling rock
f. a coiled spring

Heat Energy and Specific Heat

1.70 What is the relationship between a calorie and a
a. kilocalorie
b. Calorie
c. joule

1.71 Cheddar cheese has an energy value of approximately 4.8 Cal per gram. How many kilocalories of energy will you gain by eating one-half pound of cheddar cheese?

1.72 Assume that your mass remains constant on an intake of 2500 Cal/day. How long would it take to lose a pound of fat (7770 Calories of energy) if you started a diet in which you reduced your daily intake of food to 1600 Cal/day?

1.73 If it takes 18.6 cal of heat to raise the temperature of 12.0 g of a substance by 10.0°C, what is the specific heat of the substance?

1.74 How many calories of heat energy are required to raise the temperature of 50.0 g of each of the following substances from 20°C to 40°C?
a. liquid water
b. olive oil
c. aluminum
d. gold

1.75 How many calories are evolved if 40.0 g of silver are cooled from 50°C to 25°C?

1.76 The body contains approximately 5.7 L of blood. Assuming that the density of blood is 1.06 g/mL and that the specific heats of blood and water are the same, how many kilocalories of energy are required to raise the temperature of this amount of blood by 1.0°C?

2 The Language and Symbolism of Chemistry

Objectives

After completing Chapter 2, you will be able to:

2.1 Explain the major differences between pure substances and mixtures.
2.2 Distinguish between the characteristics of heterogeneous and homogeneous mixtures.
2.3 List the criteria for characterizing pure substances.
2.4 Discuss the relationship between the limits of physical and chemical subdivision and define the terms *atoms* and *molecules*.
2.5 Distinguish between the characteristics of elements and compounds.
2.6 State general trends concerning the discovery and abundance of the elements.
2.7 Write the names, when given the symbols, or the symbols when given the names, of the more common elements.
2.8 Interpret a chemical formula in terms of the number of elements and the number of atoms present.

INTRODUCTION

Like all other sciences, chemistry has its own language that must be clearly understood. Restricted meanings for specific words are necessary so that all chemists can understand a given description of a chemical phenomenon in the same way. In this chapter, we will look at some of the special terminology and symbolism fundamental to the study of chemistry. Much of this terminology will be introduced in the context of considering classifications of matter.

2.1 PURE SUBSTANCES AND MIXTURES

All specimens of matter can be divided into the categories of pure substances and mixtures. What are the distinctions between these two classifications of matter?

A **pure substance** *is a form of matter that always has a definite and constant composition.* This constancy of composition is reflected in the properties of pure substances. All samples of a pure substance, regardless of source, must have the same properties under the same conditions. Collectively, these definite and constant physical and chemical properties of a pure substance form a set that is not duplicated by any other pure substance. This unique set of properties provides the identification for the pure substance (Section 1.3).

A pure substance is exactly what its name implies—a single, uncontaminated type of matter. All samples of a pure substance contain only that pure substance; thus, pure water is water and nothing else.

Note that there is a significant difference between the terms *substance* and *pure substance*. *Substance* is a general term used to denote any variety of matter. *Pure substance* is a specific term that applies only to matter with the characteristics just noted.

A **mixture** *is a physical combination of two or more pure substances in which the pure substances retain their identity.* Because a mixture is a physical rather than a chemical combination of components, each component substance retains its own identity and properties. Consider, for example, a mixture of salt and pepper. Close examination will show distinct particles of salt and pepper, with no obvious interaction between them. The salt particles in the mixture are identical in properties and composition to the salt particles in the salt container, and the pepper particles in the mixture are no different from those in the pepper bottle.

Once a mixture is made up, its composition is constant. However, mixtures of the same components with different compositions can also be made up; thus, mixtures are characterized as having variable compositions. Consider the large number of salt–pepper mixtures that could be produced by varying the amounts of the two substances present.

An additional characteristic of any mixture is that its components can be retrieved intact from the mixture by physical means, without a chemical change. In many cases, the differences in properties of the various components make the separation relatively easy. For example, in our salt–pepper mixture, the two components could be separated manually by picking out all of the pepper grains. Alternatively, the separation could be carried out by dissolving the salt particles in water, removing the insoluble pepper particles, and then evaporating the water

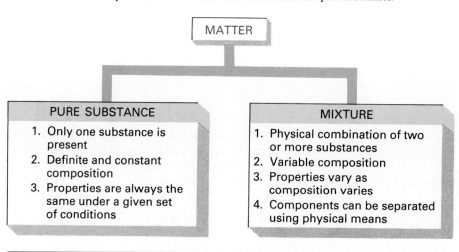

FIGURE 2–1 A comparison of the characteristics of mixtures and pure substances.

to recover the salt. In theory, physical separation is possible for the components of all mixtures. In practice, however, separation is sometimes very difficult or nearly impossible. Consider the logistics involved in trying to separate out the components of a ready-made, uncooked pizza.

Figure 2–1 summarizes the differences between mixtures and pure substances. Further discussion about these two categories of matter will be given in Sections 2.2 through 2.5.

2.2 HETEROGENEOUS AND HOMOGENEOUS MIXTURES

Sometimes the fact that a sample of matter is a mixture can be determined just by looking at it; this is the case with a salt–pepper mixture. The individual components can be visually distinguished as separate entities. However, visual identification of the components of a mixture is not always possible. Sometimes the separate ingredients of a mixture cannot be seen, even with the aid of a microscope. A mixture prepared by dissolving some sugar in water falls in this category. Outwardly, this mixture has the same appearance as pure water.

heterogeneous mixture

Mixtures can be classified as being either heterogeneous or homogeneous, based on the visual recognizability of the components present. A **heterogeneous mixture** *contains visibly different parts or phases, each of which has different properties.* The phases present in a heterogeneous mixture have distinct boundaries and are usually easily observed. A pepperoni and cheese pizza is obviously a heterogeneous mixture that has numerous identifiable components. Even a piece of pepperoni contains a number of phases. Common materials such as rocks and wood are also heterogeneous mixtures; different parts of these materials clearly have different properties, such as hardness and color (see Figure 2–2).

The phases in a heterogeneous mixture may or may not be in the same physical state. Set concrete contains a number of phases, all of which are in the solid state. A mixture of sand and water contains two phases, and each is in a different state (solid and liquid). It is possible to have heterogeneous mixtures in which all components are liquids. In order for these mixtures to occur, the mixed liquids must have limited or no solubility in each other. When this is the case, the mixed liquids form separate layers, with the least dense liquid on top. An oil-and-vinegar salad

FIGURE 2–2 Common materials such as rocks and wood are heterogeneous mixtures.

dressing is an example of such a liquid–liquid mixture (see Figure 2–3a). Oil-and-vinegar dressing *always* consists of two phases (oil and vinegar), regardless of whether the mixture consists of two separate layers or oil droplets dispersed throughout the vinegar after vigorous shaking (see Figure 2–3b). All of the oil droplets together are considered to be a single phase.

A **homogeneous mixture** *contains only one phase, which has uniform properties throughout it*. This type of mixture can have only one set of properties, which are associated with the single phase present. A spoonful of sugar–water taken from the surface of a homogeneous sugar–water mixture is just as sweet to the taste as one taken from the bottom of the container. If this were not the case, the mixture would not be truly homogeneous.

homogeneous mixture

All components present in a homogeneous mixture must be in the same state; otherwise heterogeneity would result. Homogeneous mixtures for all three states are common. Air is a homogeneous mixture of gases; motor oil and gasoline are multicomponent homogeneous mixtures of liquids; and metal alloys such as 14-karat gold (a mixture of copper and gold) are examples of solid homogeneous mixtures.

FIGURE 2–3 Oil-and-vinegar salad dressing—a two-phase heterogeneous mixture. Shaking the mixture does not change the number of phases.

 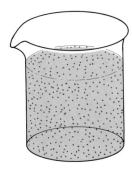

(a) Before shaking   (b) After shaking

A thorough intermingling of the components in a homogeneous mixture is required in order for only a single phase to exist. Sometimes this occurs almost instantaneously during the preparation of the mixture, as in the addition of alcohol to water. At other times, an extended period of mixing or stirring is required. For example, when a hard sugar cube is added to a container of water, it does not instantaneously dissolve to give a homogeneous solution. Only after much stirring does the sugar completely dissolve. Prior to that point, the mixture was heterogeneous because a solid phase (the sugar cube) was present.

A useful summary of the major concepts developed in both this section and Section 2.1 is presented in Figure 2–4. This summary is based on the interplay between the terms *heterogeneous* and *homogeneous* and the terms *chemical* and *physical*. From this interplay come the new expressions *chemically homogeneous*, *chemically heterogeneous*, *physically homogeneous*, and *physically heterogeneous*.

All pure substances are chemically homogeneous. Only one substance can be present in a chemically homogeneous material. Mixtures, which by definition must contain two or more substances, will always be chemically heterogeneous. The term *physically homogeneous* describes materials consisting of only one phase. If two or more phases are present, then the term *physically heterogeneous* applies.

FIGURE 2–4 Samples of matter described using the terms *chemically homogeneous*, *chemically heterogeneous*, *physically homogeneous*, and *physically heterogeneous*.

(a) Pure water	
One substance and One phase	Chemically homogeneous and Physically homogeneous
(b) Sugar water	
Two substances and One phase	Chemically heterogeneous and Physically homogeneous
(c) Oil and water	
Two substances and Two phases (Two liquid phases)	Chemically heterogeneous and Physically heterogeneous
(d) Ice and water	
One substance and Two phases (Solid phase, Liquid phase)	Chemically homogeneous and Physically heterogeneous

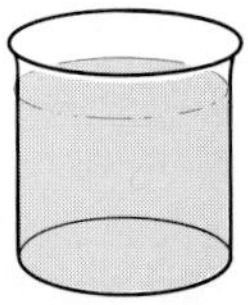

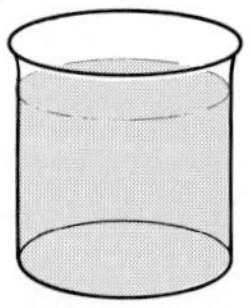

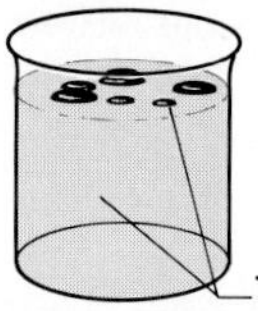

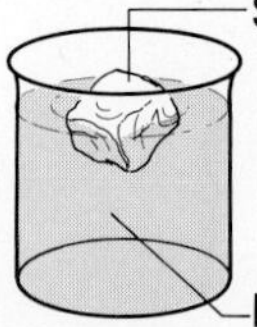

Pure water (Figure 2–4a) is both chemically and physically homogeneous because only one substance and one phase are present. Sugar water (Figure 2–4b) is physically homogeneous, with only one phase present, but is chemically heterogeneous because two substances, sugar and water, are present. A mixture of oil and water (Figure 2–4c) is both chemically and physically heterogeneous because it contains two substances and two phases. Ice cubes in liquid water (Figure 2–4d) represent the somewhat unusual situation of chemical homogeneity and physical heterogeneity. This type of combination occurs only when two or more phases of the same substance are present, as in the case of ice and water.

2.3 THE CHARACTERIZATION OF PURE SUBSTANCES

Today, many sophisticated physical techniques are available to separate mixtures into their component pure substances. After these techniques have been used, the question still arises as to whether the isolated substances are really pure; that is, how complete the separation into components really is. A substance is called *pure* when its properties, under a given set of conditions, cannot be changed by further attempts at purification. As many properties as possible must be examined and numerous separation techniques must be used before concluding that a substance is pure.

Suppose, for example, that powdered iron and sulfur are mixed together. The iron is gray and the sulfur is yellow. We can physically separate the components of the mixture by stirring it with a magnet; the iron clings to the magnet, leaving the sulfur behind. After one stirring, the remaining sulfur is still not pure; this is indicated by a gray–yellow color. Repeated stirrings with more powerful magnets eventually produce bright yellow sulfur, whose color does not change with further treatments. The fact that the color no longer changes is significant, but not absolute, evidence that the sulfur is pure. The property of color should not be used as the sole basis for determining purity; other properties also have to be checked.

A mixture of table salt and fine white sand appears to be a pure substance if only the color is used as a criterion for purity. However, a check of other properties quickly indicates a mixture. One part (salt) melts at a lower temperature than the other part (sand). One part (salt) dissolves in water while the other part (sand) does not dissolve.

2.4 THE SUBDIVISION OF PURE SUBSTANCES

In this section, we consider the processes of both physically and chemically subdividing a sample of a pure substance into smaller and smaller amounts. Throughout this discussion, it is important to remember the distinction between the meanings of the terms *physical* and *chemical* when they are used as modifiers of other words. The term *physical* is related to the idea that the composition of the substances involved does not change, and the term *chemical* is associated with composition change. Thus, physical subdivision is a subdivision process in which no composition change occurs; chemical subdivision, on the other hand, is a subdivision process during which composition change does occur.

Pure Substances and Physical Subdivision

Suppose a one-pound sample of table sugar is divided into one-half-pound portions. Then, the one-half-pound portions are cut in half again to give one-quarter-pound portions, and the subdivision process continues. Is there a limit to

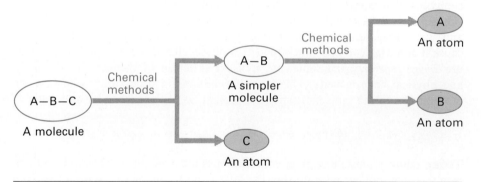

FIGURE 2–5 The breakdown of a molecule containing three atoms (A, B, and C) to yield the constituent atoms.

the number of times this physical subdivision process can be carried out on the sugar sample? The answer is yes. There is a limit beyond which you cannot go and still have sugar. You have to stop when only one unit of the substance called *sugar* remains.

A limit of physical subdivision exists for all pure substances. The smallest unit of a pure substance that maintains all of its properties is called a *molecule*. A **molecule** *is the smallest unit of a pure substance obtainable through physical subdivision of the pure substance.*

Every pure substance has a *unique* molecule as its smallest characteristic unit.* If two pure substances had the same molecule as a basic unit, they would both have the same properties; thus, they would have to be classified as the same pure substance.

Pure Substances and Chemical Subdivision

A molecule is the limit of physical subdivision for a pure substance. Molecules can be broken down into simpler units if chemical, rather than physical, subdivision methods are used. These techniques destroy the original identity of the pure substance and chemical changes occur. The ultimate limit of chemical subdivision for a pure substance is the unit of matter called an *atom*. An **atom** *is the smallest unit of a pure substance obtainable through chemical subdivision of the pure substance.*

All molecules, except the few that contain only one atom, can be broken down into their smallest units if appropriate chemical methods are used. In practice, the breakdown of molecules often occurs in steps, with smaller molecules resulting from the intermediate steps. Ultimately, however, the limit of atoms is reached (see Figure 2–5). It is important to remember that, by definition, chemical subdivision of a molecule destroys the identity of the molecule. A sugar molecule can be broken up into atoms, but the resulting atoms do not show the properties characteristic of sugar.

The atoms present in a molecule can all be the same kind, or two or more kinds can be present. On the basis of this observation, molecules are classified into the categories of *homoatomic* and *heteroatomic*. **Homoatomic molecules** *are*

---
* The concept of a molecule as described here will be modified when certain types of solids are considered (Section 6.6).

molecules in which all atoms present are the same kind. Only one kind of atom can be produced from the chemical breakup of a homoatomic molecule. **Heteroatomic molecules** *are molecules in which two or more different kinds of atoms are present.* Chemical breakup of heteroatomic molecules always produces two or more kinds of atoms. Classification of molecules as either homoatomic or heteroatomic provides the basis for a further classification of pure substances into the categories of elements and compounds; this is the topic of Section 2.5.

heteroatomic molecules

2.5 ELEMENTS AND COMPOUNDS

There are two kinds of pure substances: elements and compounds. An **element** *is a pure substance whose basic unit is a homoatomic molecule.* Elements cannot be broken down into simpler substances by chemical means because they contain only one kind of atom. A **compound** *is a pure substance whose basic unit is a heteroatomic molecule.* Compounds can be broken down into two or more simpler substances by chemical means because at least two different kinds of atoms are present.

element

compound

Presently, 109 pure substances are classified as elements. These elements, which are the simplest known substances, are considered the building blocks of all other types of matter. Every object, regardless of its complexity, is a collection of substances that are made up of these 109 elements. Compared to the total number of compounds characterized by chemists, the number of known elements is extremely small. Over 6 million different compounds are known; each is a definite combination of two or more of the known elements. Figure 2–6 summarizes the differences between elements and compounds.

The difference between compounds and mixtures is very important. Compounds are not mixtures, even though two or more simpler substances can be obtained from them. A compound is the result of the *chemical* combination of two or more elements; a mixture is formed by the *physical* mixing of two or more substances

FIGURE 2–6 A comparison of the characteristics of elements and compounds.

```
                    PURE
                  SUBSTANCE
              ┌───────┴───────┐
           ELEMENT          COMPOUND
```

ELEMENT	COMPOUND
1. Basic unit is a homoatomic molecule	1. Basic unit is a heteroatomic molecule
2. Cannot be broken down into simpler substances by chemical or physical means	2. A chemical combination of two or more elements
3. The building blocks for all other types of matter	3. Can be broken down into constituent elements using chemical, but not physical, means
4. 109 elements are known	4. Has a definite, constant elemental composition

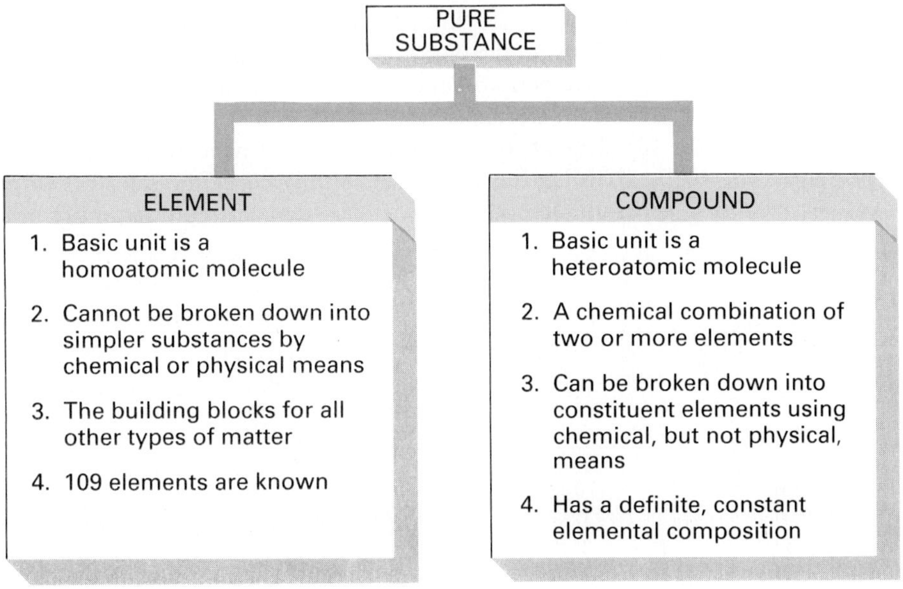

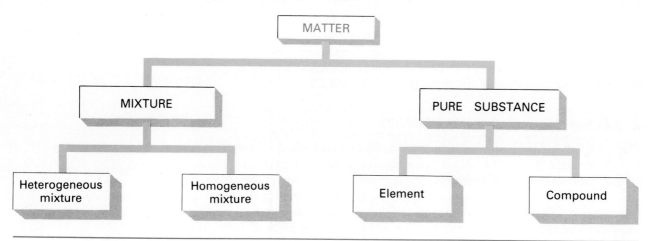

FIGURE 2–7 Categories of classification for matter.

(elements or compounds). Compounds always have properties that are distinctly different from those of the elements used in producing them. Mixtures retain the properties of their individual components.

There are other differences between compounds and mixtures. Compounds have a definite composition, which is a property of all pure substances, and mixtures have variable compositions. Also, because mixtures are only physical combinations of substances, they can be separated by physical means. The separation of a compound into its constituent elements always requires chemical means. Compounds will never yield their constituent elements through only physical separation techniques. Figure 2–7 summarizes the overall scheme developed in Sections 2.1 through 2.5 for classifying matter.

2.6 DISCOVERY AND ABUNDANCE OF THE ELEMENTS

The discovery and isolation of the 109 known elements have taken place over a period of several centuries. Most of the discoveries have occurred since 1700, with the 1800s being the most active period. Table 2–1 shows the relationship between time and the number of known elements.

Eighty-eight of the 109 elements occur naturally, and 21 have been synthesized in the laboratory by bombarding the atoms of naturally occurring elements with small particles. It is generally accepted by scientists that no more naturally oc-

TABLE 2–1 Number of Elements Discovered During Various Fifty-Year Time Periods

Time Period	Number of Elements Discovered During Time Period	Total Number of Elements Known at End of Time Period
ancient–1700	13	13
1701–1750	3	16
1751–1800	18	34
1801–1850	25	59
1851–1900	23	82
1901–1950	16	98
1951–present	11	109

TABLE 2-2 Abundance of the Elements from Various Viewpoints

Element	Abundance (atom %)	Element	Abundance (atom %)
Universe		*Atmosphere*	
hydrogen	91	nitrogen	78.3
helium	9	oxygen	21.0
Earth (including core)		*Hydrosphere*	
oxygen	49.3	hydrogen	66.4
iron	16.5	oxygen	33
silicon	14.5	*Human body*	
magnesium	14.2	hydrogen	63
Earth's crust		oxygen	25.5
oxygen	55.1	carbon	9.5
silicon	16.3	nitrogen	1.4
hydrogen	16.0	*Vegetation*	
aluminum	4.9	hydrogen	49.8
sodium	2.0	oxygen	24.9
iron	1.5	carbon	24.9
calcium	1.5		
magnesium	1.4		
potassium	1.1		

curring elements will be found, although it is possible that additional elements may be prepared synthetically. The last of the naturally occurring elements was discovered in 1925. The elements synthesized by scientists are all unstable (radioactive) and usually change rapidly into stable elements as the result of radioactive emissions (see Section 12.3).

The naturally occurring elements are not evenly distributed on Earth and in the universe. What is startling is the nonuniformity of the distribution. A small number of elements account for the majority of atoms. One must define the area to be considered when answering the question of which elements are the most common. The abundances of elements in the Earth's crust are considerably different from those of the entire Earth; they are even more different from the abundances of elements for the universe. When living entities such as vegetation or the human body are considered, an altogether different perspective emerges. Table 2-2 gives information on element abundances from a number of different viewpoints. The numbers in the table are the percentages of total atoms of the given element. Only element abundances greater than 1% are listed.

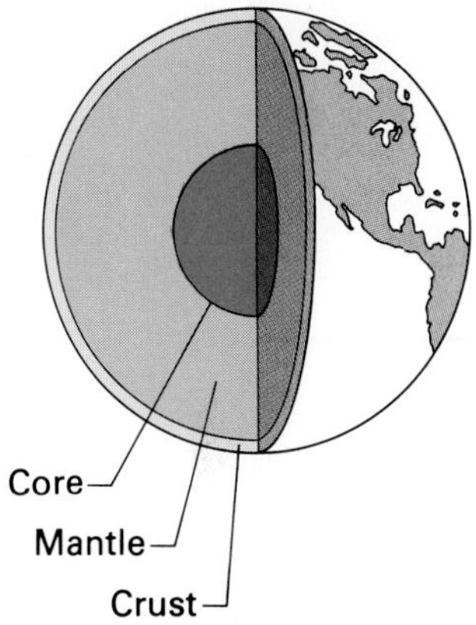

Note from Table 2-2 how the most abundant elements in the universe are not the same as the most abundant ones on Earth. The cosmic figures reflect the composition of stars that are almost entirely made up of hydrogen and helium. Significant differences exist between the composition of the Earth as a whole and its crust. These differences reflect the fact that the figures for the Earth's crust—its waters, atmosphere, and outer covering—do not take into account the composition of the Earth's core or mantle (see Figure 2-8). Note that in both the atmosphere and hydrosphere only two elements occur in large amounts (greater than 1% of atoms). Also note how the relative numbers of carbon atoms increase and hydrogen atoms decrease in vegetation as compared to human body composition.

2.7 NAMES AND SYMBOLS OF THE ELEMENTS

Each element has a unique name that, in most cases, was selected by its discoverer. A wide variety of rationales for choosing a name can be found when studying name origins. Some elements bear geographical names; germanium was

FIGURE 2-8 The structure of the Earth. The Earth's core, with a 3500-km (2200 mi) radius, is believed to be composed mainly of iron. Surrounding the core is fluid material called the mantle, which is 2900 km (1800 mi) thick and is thought to be composed primarily of silicon, oxygen, iron, and magnesium. The very thin, solid, outer layer is the crust, which is 10–50 km (8–26 mi) thick.

named after the native country of its German discoverer and the elements francium and polonium acquired their names in a similar manner. The elements mercury, uranium, neptunium, and plutonium are all named for planets. Helium gets its name from the Greek word *helios*, for sun, because it was first observed spectroscopically in the sun's corona during an eclipse. Some elements carry names that relate to specific properties of the element or the compounds that contain it. Chlorine's name is derived from the Greek *chloros*, denoting greenish-yellow, the color of chlorine gas. Iridium gets its name from the Greek *iris*, meaning rainbow; this alludes to the varying colors of the compounds from which it was isolated.

In the early 1800s, chemists adopted the practice of assigning chemical symbols to the elements. **Chemical symbols** *are a shorthand notation for the names of the elements and are often simple abbreviations of their names.* Chemical symbols are used more frequently when referring to the elements than the names are. A complete list of all the known elements and their symbols is given in Table 2–3. The symbols and names of the more frequently encountered elements are shown in color in this table. You should learn the symbols of these more common elements.

TABLE 2–3 The Elements and Their Chemical Symbols

Symbol	Name	Symbol	Name	Symbol	Name
Ac	actinium	He	helium	Ra	radium
Ag	silver*	Hf	hafnium	Rb	rubidium
Al	aluminum	Hg	mercury*	Re	rhenium
Am	americium	Ho	holmium	Rh	rhodium
Ar	argon	I	iodine	Rn	radon
As	arsenic	In	indium	Ru	ruthenium
At	astatine	Ir	iridium	S	sulfur
Au	gold*	K	potassium*	Sb	antimony*
B	boron	Kr	krypton	Sc	scandium
Ba	barium	La	lanthanum	Se	selenium
Be	beryllium	Li	lithium	Si	silicon
Bi	bismuth	Lu	lutetium	Sm	samarium
Bk	berkelium	Lr	lawrencium	Sn	tin*
Br	bromine	Md	mendelevium	Sr	strontium
C	carbon	Mg	magnesium	Ta	tantalum
Ca	calcium	Mn	manganese	Tb	terbium
Cd	cadmium	Mo	molybdenum	Tc	technetium
Ce	cerium	N	nitrogen	Te	tellurium
Cf	californium	Na	sodium*	Th	thorium
Cl	chlorine	Nb	niobium	Ti	titanium
Cm	curium	Nd	neodymium	Tl	thallium
Co	cobalt	Ne	neon	Tm	thulium
Cr	chromium	Ni	nickel	U	uranium
Cs	cesium	No	nobelium	Une	unnilennium
Cu	copper*	Np	neptunium	Unh	unnilhexium
Dy	dysprosium	O	oxygen	Uno	unniloctium
Er	erbium	Os	osmium	Unp	unnilpentium
Es	einsteinium	P	phosphorus	Unq	unnilquadium
Eu	europium	Pa	protactinium	Uns	unnilseptium
F	fluorine	Pb	lead*	V	vanadium
Fe	iron*	Pd	palladium	W	tungsten*
Fm	fermium	Pm	promethium	Xe	xenon
Fr	francium	Po	polonium	Y	yttrium
Ga	gallium	Pr	praseodymium	Yb	ytterbium
Gd	gadolinium	Pt	platinum	Zn	zinc
Ge	germanium	Pu	plutonium	Zr	zirconium
H	hydrogen				

* These elements have symbols that were derived from non-English sources.

Learning them is a key to having a successful experience in studying chemistry. Fourteen elements have one-letter symbols, 89 have two-letter symbols and 6 have three-letter symbols. If a symbol consists of a single letter, it is capitalized. In all double-letter symbols, the first letter is capitalized, but the second letter is not. Double-letter symbols usually involve the first letter of the element's name. The second letter of the symbol is frequently, but not always, the second letter of the name. Consider the elements terbium, technetium, and tellurium, whose symbols are respectively Tb, Tc, and Te. Obviously, a variety of choices of second letters is necessary because the first two letters of each element's name are the same. In triple-letter symbols, the first letter of the symbol is capitalized and the other two letters are not. The six elements with triple-letter symbols are all synthetic elements and were the last six elements to be produced. These symbols come from a recently introduced systematic method for naming elements.

Eleven elements have strange symbols; that is, their symbols bear no relationship to the element's English language name. In ten of these cases, the symbol is derived from the Latin name of the element; in the case of the element tungsten, a German name is the symbol source. Most of the elements with strange symbols have been known for hundreds of years and date back to the time when Latin was the language of scientists. Elements whose symbols are derived from non-English names are marked with an asterisk in Table 2–3.

2.8 CHEMICAL FORMULAS

The most important piece of information about a compound is its composition. Chemical formulas represent a concise manner for specifying compound composition. A **chemical formula** *is a notation made up of the symbols of the elements present in a compound and numerical subscripts (located to the right of each symbol) that indicate the number of atoms of each element present in a unit of the compound.*

chemical formula

The chemical formula for the compound we call *aspirin* is $C_9H_8O_4$. From this formula, we know the following information about an aspirin molecule: three different elements are present—carbon (C), hydrogen (H), and oxygen (O); and 21 atoms are present—9 carbon atoms, 8 hydrogen atoms, and 4 oxygen atoms.

When only one atom of a particular element is present in a molecule of a compound, that element's symbol is written without a numerical subscript in the formula for the compound. The formula for rubbing alcohol, C_3H_6O, reflects this practice, for the element oxygen, of not writing the subscript 1.

In order to write formulas correctly, it is necessary to follow the capitalization rules for elemental symbols (Section 2.7). Making the error of capitalizing the second letter of an element's symbol can dramatically alter the meaning of a chemical formula. The formulas $CoCl_2$ and $COCl_2$ illustrate this point; the symbol Co stands for the element cobalt, whereas CO stands for one atom of carbon and one atom of oxygen.

Sometimes chemical formulas contain parentheses, as in the formula $Ca(NO_3)_2$. The interpretation of this formula is straightforward—the grouping of atoms within the parentheses, NO_3, is repeated twice. The subscript following the parentheses always indicates the number of units of the multi-atom entity inside the parentheses. As another example of parentheses use, consider the compound $Pb(C_2H_5)_4$; in this compound, four units of C_2H_5 are present. In terms of atoms present, the formula $Pb(C_2H_5)_4$ represents 29 atoms—1 lead (Pb) atom, 8 (4 × 2) carbon (C) atoms, and 20 (4 × 5) hydrogen (H) atoms. The formula $Pb(C_2H_5)_4$ could be (but is not) written as PbC_8H_{20}. Both versions of the formula convey the same information in terms of atoms present; however, the former gives some information about structure (C_2H_5 units are present) that the latter does not. For this reason, the former is the preferred way of writing the formula. Further information concerning the use of parentheses will be presented in Section 6.15. The important concern now is to be able to interpret formulas that contain parentheses in terms of total atoms present. Example 2.1 refines this skill in greater detail.

Example 2.1

Interpret each of the following formulas in terms of how many atoms of each element are present in one structural unit of the substance.

a. $C_{18}H_{21}NO_3$—codeine, a pain-killing drug
b. $Al(OH)_3$—aluminum hydroxide, an ingredient in numerous antacid formulations
c. $Ca_3(PO_4)_2$—calcium phosphate, an abrasive found in some toothpaste formulations

Solution

a. Simply look at the subscripts following the symbols for the elements, and remember that the subscript 1 is implied when no subscript is written. This formula indicates that 18 carbon atoms, 21 hydrogen atoms, 1 nitrogen atom, and 3 oxygen atoms are present in one molecule of compound.
b. The subscript following the parentheses, 3, indicates that 3 OH units are present. We therefore have 3 oxygen atoms and 3 hydrogen atoms present in addition to the 1 aluminum atom.
c. The amounts of phosphorus and oxygen are affected by the subscript 2 outside the parentheses. In one unit of compound we have 3 calcium atoms, 2 (2 × 1) phosphorus atoms and 8 (2 × 4) oxygen atoms.

In addition to formulas, compounds have names. Naming compounds is not as simple as naming elements. While the nomenclature of elements (Section 2.7) has been largely left up to the imagination of their discoverers, extensive sets of systematic rules exist for naming compounds. Rules must be used because of the large number of compounds that exist. Parts of Chapters 6 and 14 are devoted to rules for naming compounds, and we will not worry about naming compounds until that time. For the time being, our focus will be on the meaning and significance of chemical formulas; knowing how to name the compounds that the formulas represent is not a prerequisite for understanding the meanings of formulas.

EXERCISES AND PROBLEMS

Pure Substances and Mixtures

2.1 Consider the following classes of matter: *heterogeneous mixtures*, *homogeneous mixtures*, and *pure substances*.

a. In which of these classes must two or more substances be present?
b. Which of these classes could not possibly have a variable composition?
c. For which of these classes is separation into simpler substances by physical means impossible?

2.2 Assign each of the following descriptions of matter to one of the following categories: *heterogeneous mixture*, *homogeneous mixture*, or *pure substance*.

a. Two substances present, two phases present
b. Two substances present, one phase present
c. One substance present, one phase present
d. One substance present, two phases present
e. Three substances present, four phases present

2.3 Classify each of the following as a *heterogeneous mixture*, a *homogeneous mixture*, or a *pure substance*. Also indicate how many phases are present. (All substances are assumed to be present in the same container.)

a. water
b. water and dissolved table salt
c. water and white sand
d. water and oil
e. water and ice
f. water, ice, and oil
g. carbonated water (soda water) and ice
h. oil, ice, salt water solution, sugar water solution, and pieces of copper metal.

2.4 The phrases *chemically homogeneous*, *chemically heterogeneous*, *physically homogeneous* and *physically heterogeneous* are often used in describing samples of matter. In each of the following situations, two of these phrases apply. Select the correct two phrases for each case.

a. pure water
b. tap water
c. river water
d. pure water and ice
e. oil–and–vinegar salad dressing with spices
f. oil–and–vinegar salad dressing without spices
g. carbonated beverage
h. blood

2.5 Explain how the operational definition of a pure substance allows for the possibility that it is not actually pure.

2.6 Every sample taken from a homogeneous mixture has the same composition as every other sample. The same can be said about samples taken from a pure substance. Why then, are homogeneous mixtures not considered to be pure substances?

2.7 How do the terms *substance* and *pure substance* differ in meaning?

Elements and Compounds

2.8 Explain the difference between an element and a compound.

2.9 Based on the information given, classify each of the pure substances A through K as elements or compounds, or indicate that no such classification is possible because of insufficient information.

a. Analysis with an elaborate instrument indicates that *Substance A* contains two elements.
b. *Substance B* and *Substance C* react to give a new *Substance D*. (Give three answers—one for B, one for C, and one for D.)
c. *Substance E* decomposes upon heating to give *Substance F* and *Substance G*. (Give three answers.)
d. Heating *Substance H* to 1000°C causes no change in it.
e. *Substance I* cannot be broken down into simpler substances by chemical means.
f. *Substance J* cannot be broken down into simpler substances by physical means.
g. Heating *Substance K* to 500°C causes it to change from a solid to a liquid.

2.10 Indicate whether each of the following statements is true or false.

a. Both elements and compounds are pure substances.
b. A compound results from the physical combination of two or more elements.
c. In order for matter to be heterogeneous, at least two compounds must be present.
d. Pure substances cannot have a variable composition.
e. Compounds, but not elements, can have a variable composition.
f. Compounds can be separated into their constituent elements using chemical means.
g. A compound must contain at least two elements.

2.11 Is it possible to have a mixture of two elements and also to have a compound of the same two elements? Explain. Can you think of an example?

2.12 Compounds are classified as pure substances even though two or more substances (elements) are present. Explain.

2.13 What is the difference between heteroatomic molecules and homoatomic molecules?

2.14 Assign each of the following descriptions of matter to one of the following categories: *mixture*, *element*, or *compound*.

a. one substance present, heteroatomic molecules present.
b. one phase present, both heteroatomic and homoatomic molecules present.
c. one substance present, two phases present, homoatomic molecules present.
d. two substances present, one phase present, heteroatomic molecules present.

2.15 Indicate whether each of the following statements is true or false.

a. Molecules must contain three or more atoms if they are heteroatomic.
b. The smallest characteristic unit of a pure substance is an atom.
c. A molecule of a compound must be heteroatomic.
d. A molecule of an element can be homoatomic or heteroatomic, depending on which element is involved.
e. The limit of physical subdivision for a pure substance is a molecule.
f. There is only one kind of molecule for any given compound.
g. Heteroatomic molecules do not maintain the properties of their constituent elements.
h. The main difference between the molecules of elements and the molecules of compounds is the number of atoms they contain.

Discovery and Abundance of the Elements

2.16 How many elements were discovered in each of the following time periods?
a. ancient–1700
b. 1701–1800
c. 1801–1900
d. 1901–1950
e. 1951–date

2.17 What are the two most abundant elements in each of the following realms of matter?
a. universe
b. Earth (including core)
c. Earth's crust
d. atmosphere
e. hydrosphere
f. human body
g. vegetation

Names and Symbols of the Elements

2.18 The elements found in the human body include C, H, O, N, P, S, I, Na, Cl, K, Ca, Fe, Br, Cu, Zn, Co, and Mg. What are the names of these elements?

2.19 There are 12 elements that have an abundance of 1000 parts per million or greater (by mass) in the Earth's crust: hydrogen, oxygen, sodium, magnesium, aluminum, silicon, phosphorus, potassium, calcium, titanium, manganese, and iron. What are the symbols for these elements?

2.20 What chemical elements are represented by the following symbols?
a. Ne
b. Hg
c. Pb
d. Ar
e. Au
f. Ba
g. Sn
h. Cd

2.21 What are the chemical symbols of the following elements?
a. lithium
b. helium
c. silver
d. boron
e. fluorine
f. beryllium
g. nickel
h. arsenic

2.22 Make a list of the elements that have symbols that
a. begin with a letter other than the first letter of the English name for the element.
b. are the first two letters of the element's English name.
c. contain only one letter.

2.23 For which letters of the alphabet are there at least five elements whose chemical symbol begins with that letter?

Chemical Formulas

2.24 The formula HF stands for a compound (hydrogen fluoride), not an element. How can we tell this from the formula itself?

2.25 On the basis of its formula, classify each of the following substances as element or compound.
a. $LiClO_3$
b. CO
c. Co
d. $CoCl_2$
e. $COCl_2$
f. BN
g. S_8
h. Sn

2.26 From the information given about one molecule of each of these compounds, write the chemical formula for the compound.
a. Table sugar contains 12 atoms of carbon, 22 atoms of hydrogen, and 11 atoms of oxygen.
b. Vitamin C contains 6 atoms of carbon, 8 atoms of hydrogen, and 6 atoms of oxygen.
c. Caffeine contains 8 atoms of carbon, 10 atoms of hydrogen, 4 atoms of nitrogen, and 2 atoms of oxygen.

2.27 How many atoms of hydrogen are represented in each of the following formulas?
a. H_2SO_4
b. $HClO_3$
c. $KHCO_3$
d. CH_4
e. $(NH_4)_2SO_4$
f. C_4H_{10}
g. $Al(OH)_3$
h. $Ca(C_2H_3O_2)_2$

2.28 How many atoms of each kind are represented by the following formulas?
a. CO_2
b. HCN
c. NH_4ClO
d. $MgCO_3$
e. Na_3PO_4
f. $Al_2(SO_4)_3$
g. $(NH_4)_3BO_3$
h. $Be(CN)_2$

3 Structure of the Atom and the Periodic Table

Objectives

After completing Chapter 3, you will be able to:

3.1 Understand how chemical knowledge is developed through repeated use of the procedures of the scientific method and distinguish among the terms *experiment*, *fact*, *law*, *hypothesis*, and *theory*.
3.2 Explain current scientific thought concerning atoms as stated in the statements of the atomic theory of matter.
3.3 State the laws of conservation of matter and constant composition, show how data is treated to verify these laws, and explain how atomic theory accounts for these laws.
3.4 Name the three major subatomic particles that make up an atom, tell where each is located within the atom, and indicate the electrical charge and relative mass associated with each particle.
3.5 Define *atomic number* and *mass number* and, given these two numbers, know how to determine the number of protons, neutrons, and electrons present in an atom.
3.6 Define the term *isotope* and write the symbol for an isotope.
3.7 Calculate the atomic weight of an element from the isotopic weights and percentage abundances of its isotopes.
3.8 Understand and state the periodic law.
3.9 Understand the rationale behind the organization of the periodic table, relate the terms *group* and *period* to the periodic table, and list the general information about each element given in the periodic table.

INTRODUCTION

In Chapter 2 we defined a number of terms that dealt with the classification of matter. One term was *atom*, which is the limit of chemical subdivision for a pure substance. We noted that 109 different kinds of atoms are known, and all matter is made from them. In this chapter, we will consider the topic of atoms in greater detail.

Modern day scientists have been able to uncover a great deal of information about atoms through research on various types of matter. All of this research has one thing in common—experimentation. Chemistry is an experimental science; that is, chemical discoveries are made as the result of experimentation. This feature distinguishes chemistry and other sciences from other types of intellectual activity.

3.1 HOW CHEMISTS DISCOVER THINGS— THE SCIENTIFIC METHOD

A majority of the scientific and technological advances of the twentieth century are the result of systematic experimentation using a method of problem solving known as the *scientific method*. The **scientific method** *is a set of specific procedures used for acquiring knowledge and explaining phenomena*. Procedural steps in the scientific method are:

    scientific method

1. identifying a problem and carefully planning procedures to obtain information about all aspects of this problem;
2. collecting data concerning the problem through observation and experimentation;
3. analyzing and organizing the data in terms of general statements (generalizations) that summarize the experimental observations;
4. suggesting probable explanations for the generalizations; and
5. experimenting further to prove or disprove the proposed explanations.

There are special vocabulary terms associated with the scientific method and its use. This vocabulary includes the terms *experiment*, *fact*, *law*, *hypothesis*, and *theory*. Understanding the relationships among these terms is the key to a real understanding of how to obtain chemical knowledge.

The beginning step in the search for chemical knowledge is the identification of a problem concerning some chemical system that needs study. After determining what other chemists have already learned about the selected problem, a chemist sets up procedures for obtaining more information by designing experiments. An **experiment** *is a well-defined, controlled procedure for obtaining information about a system under study*. The exact conditions under which an experiment is carried out must always be noted because conditions such as temperature and pressure affect results.

    experiment

A chemist obtains new, first-hand information about the chosen chemical system by actually carrying out the experimental procedures; that is, new facts about the system are obtained. A **fact** *is a valid observation about some natural phenomenon*. Facts are reproducible pieces of information. If a given experiment is repeated under exactly the same conditions, the same facts should be obtained. In order to be acceptable, all facts must be verifiable by anyone who has the time, means,

    fact

and knowledge needed to repeat the experiments that led to their discovery. It is important that scientific data be published so that other scientists have the opportunity to critique and double-check both the data and experimental design.

As a next step, the chemist makes an effort to determine ways in which the facts about a given chemical system relate both to each other and to facts known about similar chemical systems. Repeating patterns often emerge among the collected facts. These patterns lead to generalizations that are called *laws*, about how chemical systems behave under specific conditions. A **law** *is a generalization that summarizes facts about natural phenomena.*

Chemists and other scientists are not content with merely knowing about natural laws; they want to know *why* a certain type of observation is always made. Thus, after a law is discovered, scientists work out plausible, tentative explanations of the behavior encompassed by the law. These explanations are called *hypotheses*. A **hypothesis** *is a tentative model or picture that offers an explanation for a law.*

Once a hypothesis has been proposed, experimentation begins once again to try to validate the hypothesis. If, after extensive testing, the reliability of a hypothesis is still very high, confidence in it increases to the extent of its acceptance by the scientific community at large. After further lapses of time and the additional accumulation of positive support, the hypothesis assumes the status of a theory. A **theory** *is a hypothesis that has been tested and validated over a long period of time.* The dividing line between a hypothesis and a theory is arbitrary and cannot be precisely defined. There is not a set number of supportive experiments that must be performed in order to give theory status to a hypothesis.

Theories serve two important purposes: (1) they allow scientists to predict what will happen in experiments that have not yet been run, and (2) they simplify the very real problem of being able to remember all of the scientific facts that have already been discovered.

Figure 3–1 summarizes the sequence of steps scientists use when applying the scientific method to a given research problem. It also shows the central role that experimentation plays in the scientific method.

FIGURE 3–1 Chemical knowledge is developed through repeated use of the procedures of the scientific method.

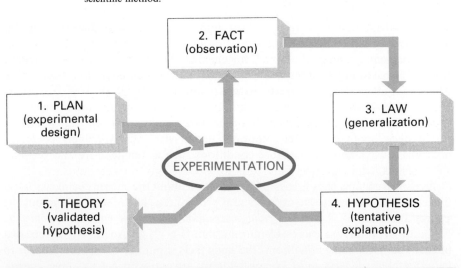

3.2 ATOMIC THEORY OF MATTER

The concept of an atom is an old one, dating back to ancient Greece. Records indicate that around 460 B.C., the Greek philosopher Democritus suggested that continued subdivision of matter would ultimately yield small, indivisible particles that he called *atoms* (from the Greek word *atomos* meaning "uncut or indivisible"). However, Democritus's ideas about matter were forgotten during the Middle Ages.

It was not until the beginning of the nineteenth century that the concept of the atom was rediscovered. English school teacher John Dalton (1776–1844) proposed in a series of papers published in the period 1803–7 that the fundamental building block for all kinds of matter was the atom. Dalton's proposal had as its basis scientific studies carried out by a number of researchers. This was in marked contrast to the early Greek concept of atoms, which was based solely on philosophical speculation. Because of its experimental basis, Dalton's idea received wide attention and stimulated new research and thought about the ultimate building blocks of matter.

Additional research carried out by many scientists has now validated Dalton's basic conclusion that the building blocks for all types of matter are atomic in nature. Some of the details of Dalton's original proposals have had to be modified in the light of more sophisticated experiments, but the basic concept of atoms remains.

Among today's scientists, the concept that atoms are the building blocks for matter is a foregone conclusion. The large amount of supporting evidence for atoms is impressive. Key concepts about atoms, in terms of current knowledge, are found in what is known as the atomic theory of matter. The **atomic theory of matter** *is a set of five statements that summarizes modern day scientific thought about atoms.* These five statements are:

atomic theory of matter

1. All matter is made up of small particles called atoms, of which 109 different "types" are known, and each "type" corresponds to atoms of a different element.
2. All atoms of a given type are similar to each other and significantly different from all other types.
3. The relative number and arrangement of the different types of atoms found within a pure substance determine its identity.
4. Chemical change is a union, separation, or rearrangement of atoms to give new substances.
5. Only whole atoms can participate in or result from any chemical change because atoms are considered to be indestructible during these changes.

FIGURE 3–2 An electron microscope photograph of a piece of graphite. Bright bands are layers of carbon atoms that are only 341 pm apart. This corresponds to a magnification of about 15 million times.

Atoms are incredibly small particles. No one has seen or ever will see an atom with the naked eye. The question may thus be asked: "How can you be absolutely sure that something as minute as an atom really exists?" The achievements of twentieth-century scientific instrumentation have gone a long way toward removing any doubt about the existence of atoms. Electron microscopes, which are capable of producing magnification factors in the millions, have made it possible to photograph images of individual atoms. An example of such an "atom picture" is shown in Figure 3–2.

Just how small is an atom? Although atomic dimensions and masses are not directly measurable, they are known quantities that are obtained by calculation. The data used for the calculations come from measurements made on macroscopic amounts of pure substances.

The diameter of an atom is approximately 10^{-8} cm. If you arranged atoms of diameter 1×10^{-8} cm in a straight line, it would take 10 million of them to extend a length of 1 mm and 254 million of them to reach 1 in. (see Figure 3–3).

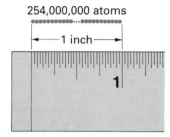

FIGURE 3–3 Comparison of atomic diameters and the common measuring unit of 1 in.

The mass of an atom is obviously also very small. For example, the mass of a uranium atom, which is one of the heaviest of the known kinds of atoms, is 4×10^{-22} g or 9×10^{-25} lb. It would require 1×10^{24} atoms of uranium to give a mass of 1 lb.

3.3 CHEMICAL EVIDENCE SUPPORTING THE ATOMIC THEORY

The law of conservation of matter and the law of constant composition are typical of the evidence that supports atomic theory. These are two of a number of known laws that summarize the results of many thousands of different investigations into the chemical behavior of matter (Section 3.1).

Both of these laws were formulated during the late 1700s. It was during that century that scientists developed experimental methods for accurately measuring the volume and mass of chemical substances. These methods enabled chemists to *quantitatively* study chemical change for the first time.

The Law of Conservation of Matter

Countless experimental studies of chemical reactions have shown that the substances produced in a chemical reaction have the same total mass as the starting materials. The fact that the total amount of matter present does not change during a chemical reaction is known as the *law of conservation of matter*. The **law of conservation of matter** *states that there is no experimentally detectable gain or loss in total mass for the substances involved in a chemical reaction.*

The validity of the law of conservation of matter can be checked by carefully determining the masses of all reactants (the starting materials) and all products (the substances formed) in a chemical reaction. As an illustration, let us consider the reaction of known masses of the elements copper (Cu) and sulfur (S) to form the compound copper sulfide (CuS). As illustrated in Figure 3–4, 66.5 g of copper reacts *exactly* with 33.5 g of sulfur. After that reaction, no copper or sulfur remains in elemental form; the only substance present is the product copper sulfide, the chemically combined copper and sulfur. The mass of this product is 100.0 g,

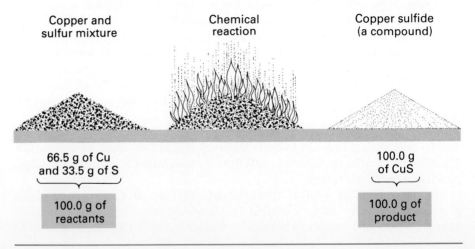

FIGURE 3–4 Experimental verification of the law of conservation of matter. The combined mass of the reactants (copper and sulfur) is the same as the mass of the product (copper sulfide).

which is the same as the combined mass of the reactants. Thus, the law of conservation of matter has been obeyed.

In addition, when the 100.0 g of the product copper sulfide is heated to a high temperature in the absence of air, it decomposes into copper and sulfur, producing 66.5 g of copper and 33.5 g of sulfur. Once again, no detectable change in mass is observed; the mass of the reactant is equal to the masses of the products.

$$\underbrace{100.0 \text{ g CuS}}_{100.0 \text{ g of reactant}} \longrightarrow \underbrace{66.5 \text{ g Cu} + 33.5 \text{ g S}}_{100.0 \text{ g of products}}$$

The law of conservation of matter is consistent with the statements of atomic theory (Section 3.2). Because all reacting chemical substances are made up of atoms (Statement 1), each with its unique identity (Statement 2), and these atoms can neither be created nor destroyed in a chemical reaction, but merely rearranged (Statement 4), it follows that the total mass after the reaction must equal the total mass before the reaction. We have the same number of each kind of atom after the reaction as we had at the start of the reaction.

The Law of Constant Composition

We can determine the composition of a compound by decomposing a weighed amount of the compound into its elements and then determining the mass of each element present. Alternatively, we can also determine composition by weighing a compound that is formed by the combination of known masses of elements.

Studies of composition data for many compounds have led to the conclusion that the percentage by mass of each element present in a given compound does not vary. The law of constant composition is a formal statement of this conclusion.

FIGURE 3–5   Samples of pure water from different sources will have the same composition by mass in accordance with the law of constant composition.

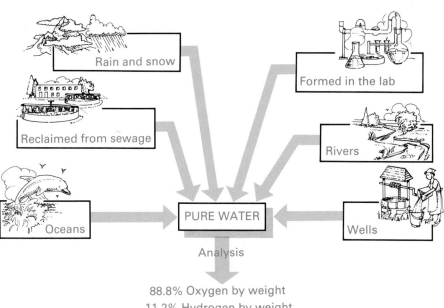

TABLE 3–1 Data Illustrating the Law of Constant Composition

Mass of Ca Used (g)	Mass of S Used (g)	Mass of CaS Formed (g)	Mass of Excess Unreacted Sulfur (g)	Ratio in Which Substances React
55.6	44.4	100.0	none	1.25
55.6	50.0	100.0	5.6	1.25
55.6	100.0	100.0	55.6	1.25
55.6	200.0	100.0	155.6	1.25
111.2	88.8	200.0	none	1.25

law of constant composition

The **law of constant composition** *states that in a pure compound the proportion by mass of the elements present is always the same.* The law of constant composition is also sometimes referred to as the *law of definite proportions.*

In accordance with this law, once we determine the composition of a pure compound, we know that the composition of all other pure samples of the compound will be the same. For example, water has a composition by mass of 88.8% oxygen and 11.2% hydrogen. All samples of pure water, regardless of their sources, will have this composition (see Figure 3–5, p. 55).

An alternate way of examining the constancy of composition of compounds involves looking at the mass ratios in which elements combine to form compounds. From this viewpoint, constancy in composition can easily be demonstrated experimentally.

Suppose we make an attempt to combine various masses of sulfur with a fixed mass of calcium to make calcium sulfide (CaS). Possible experimental data for this attempt is given in the first four lines of Table 3–1. Notice that, regardless of the mass of sulfur present, only a certain amount, 44.4 g, reacts with 55.6 g of calcium. The excess sulfur is left over in an unreacted form. The data illustrate that calcium and sulfur will react in only one fixed-mass ratio (55.6/44.4 = 1.25) to form calcium sulfide. Note also that if the amount of calcium used is doubled (as in line five of Table 3–1), the amount of sulfur with which it reacts also doubles (compare lines one and five of Table 3–1). Nevertheless, the ratio in which the substances react (111.2/88.8) still remains at 1.25.

The data in Table 3–1 is consistent with the statements of atomic theory according to the following line of reasoning:

1. A 55.6 g sample of calcium contains a certain number of calcium atoms.
2. A 44.0 g sample of sulfur contains a number of sulfur atoms that is sufficient to react with the number of calcium atoms present.
3. Because the atoms of calcium and sulfur always react in the same ratio according to the law of constant composition, any amount of sulfur greater than 44.4 g will contain more atoms than are needed and some of the atoms will go unreacted.

3.4 SUBATOMIC PARTICLES: ELECTRONS, PROTONS, AND NEUTRONS

Until the closing decades of the nineteenth century, scientists believed that atoms were solid, indivisible spheres that did not have substructure. Today, this concept is known to be incorrect. Evidence from a variety of sources indicates that atoms are made up of smaller, more fundamental particles called *subatomic particles.* **Subatomic particles** *are very small particles that are the building blocks from which atoms are made.*

subatomic particles

Three major types of subatomic particles exist: electrons, protons, and neutrons.

TABLE 3-2 Charges and Masses of the Various Kinds of Subatomic Particles

	Electron	Proton	Neutron
charge	−1	+1	0
actual mass (g)	9.11×10^{-28}	1.672×10^{-24}	1.675×10^{-24}
relative mass (based on the electron being one unit)	1	1836	1839
relative mass (based on the neutron being one unit)	0 (1/1839)	1	1

These subatomic particles can be distinguished from each other on the basis of their charges. **Electrons** *are subatomic particles that possess a negative (−) charge.* Electrons are the smallest of the subatomic particles in terms of mass. **Protons** *are subatomic particles that possess a positive (+) charge.* The amount of charge associated with a proton is the same as that for an electron; however, the character of the charge is opposite (positive versus negative). The fact that electrons and protons bear opposite charges is extremely important because of the way in which charged particles interact. It is a natural scientific law that particles of opposite (or unlike) charge attract each other and particles of the same (or like) charge repel each other. The third type of subatomic particle, the neutron, lacks charge. **Neutrons** *are subatomic particles that have no charge associated with them; that is, they are neutral.*

All of the subatomic particles are extremely small. The lightest of the three types, the electron, has a mass of 9.11×10^{-28} g. Both protons and neutrons are massive particles compared to the electron. Protons are 1836 times heavier (1.672×10^{-24} g) and neutrons are 1839 times heavier (1.675×10^{-24} g). For most purposes, the masses of protons and neutrons can be considered equal. The mass of an electron is almost negligible compared to the masses of the other, heavier subatomic particles. The properties of the three types of subatomic particles are summarized in Table 3-2.

The arrangement of subatomic particles within an atom is not haphazard. As is shown in Figure 3-6, an atom is composed of two regions: a nuclear region and an extranuclear region, the space outside the nucleus.

The nuclear region of an atom is called the *nucleus* of the atom. The nucleus is an extremely small region located at the center of the atom. The **nucleus** *is the very small, dense, positively charged core of an atom.* All protons and neutrons present in the atom are found within the nucleus. A nucleus always carries a positive charge because of the presence of the positively charged protons within it. Almost all (over 99.9%) of the mass of an atom is concentrated in its nucleus; all of the heavy subatomic particles (protons and neutrons) are there. The smallness of the nucleus, coupled with the large amount of mass there, causes nuclear material to be extremely dense material. Protons and neutrons are often called *nucleons*; they can be found nowhere other than in the nucleus.

The extranuclear region of an atom contains all of the electrons. It is an extremely large region comprising mostly empty space, in which the electrons move rapidly about the nucleus. The motion of the electrons in this extranuclear region determines the volume (size) of the atom in the same way that the blade of a fan determines a volume by its circular motion. The volume occupied by the electrons is sometimes referred to as the *electron cloud*. Because electrons are negatively charged, the electron cloud is also negatively charged.

An atom as a whole is neutral; that is, it has no net charge. How can it be that an entity possessing positive charge (protons) and negative charge (electrons) can

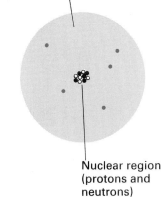

FIGURE 3-6 Arrangement of subatomic particles in an atom.

end up neutral overall? In order for this to occur, the same amount of positive and negative charge must be present in the atom because equal amounts of positive and negative charge cancel each other. Thus, it is always the case that an equal number of protons and electrons are present in an atom. This relationship between protons and electrons is a key relationship.

$$\text{number of protons} = \text{number of electrons}$$

It is important that the size relationships between the parts of an atom be correctly visualized. Again, the nucleus is extremely small compared to the total atom size. A nonmathematical, conceptual model will help you visualize this relationship. Visualize a large, spherical or nearly spherical, fluffy cloud in the sky. Let the cloud represent the negatively charged extranuclear region of the atom. Buried deep within the cloud is the positively charged nucleus, which is the size of a small pebble. A more quantitative example involves the enlargement (magnification) of the nucleus until it would be the size of a baseball (7.4 cm in diameter). If the nucleus were this large, the whole atom would have a diameter of approximately 2.5 mi. Within the large extranuclear region (2.5 mi-diameter), the electrons (smaller than the periods used to end sentences in this text) move about at random. The volume of the nucleus is only 1/100,000 of the atom's total volume; that is, almost all (over 99.99%) of the volume of an atom involves the electron cloud.

The fact that almost all of the mass of an atom is concentrated in the nucleus can best be illustrated by the following example. If a copper penny contained copper nuclei (copper atoms stripped of their electrons) rather than copper atoms (which are mostly empty space), the penny would weigh 190,000,000 tons. Nuclei are indeed very dense matter.

Our discussion of the makeup of atoms in terms of subatomic particles is based on the existence of three types of subatomic particles: protons, neutrons, and electrons. Despite the existence of subatomic particles, we will continue to use the concept of atoms as the fundamental building blocks for all types of matter (Section 2.4). Subatomic particles do not lead an independent existence for any appreciable length of time; they gain stability by joining together to form atoms.

3.5 ATOMIC NUMBER AND MASS NUMBER

The number of protons, neutrons, and electrons in a given atom is specified by two numbers: the atomic number and the mass number.

atomic number

The **atomic number** *is equal to the number of protons in the nucleus of an atom.* Atomic numbers are always integers (whole numbers). All atoms of a given element must contain the same number of protons. Thus, all atoms of a given element must have the same atomic number. Because an atom has the same number of electrons as protons, the atomic number also specifies the number of electrons present.

$$\text{atomic number} = \text{number of protons} = \text{number of electrons}$$

mass number

A second necessary quantity in specifying atomic identities is the mass number. The **mass number** *is equal to the number of protons plus the number of neutrons in the nucleus of the atom; that is, the total number of nucleons present.* The mass of an atom is almost totally accounted for by the protons and neutrons (Section 3.4); hence the term *mass number*. Like the atomic number, the mass number is always an integer.

$$\text{mass number} = \text{number of protons} + \text{number of neutrons}$$

The atomic number and mass number of an atom uniquely specify the atom's makeup in terms of subatomic particles. The following equations show the relationship between subatomic particle makeup and the two numbers.

number of protons = atomic number

number of electrons = atomic number

number of neutrons = mass number − atomic number

Example 3.1

An atom has an atomic number of 9 and a mass number of 19. Determine

a. the number of protons present.
b. the number of neutrons present.
c. the number of electrons present.

Solution

a. There are nine protons because the atomic number is always equal to the number of protons present.
b. There are ten neutrons because the number of neutrons is always obtained by subtracting the atomic number from the mass number.

$$\underbrace{(\text{protons} + \text{neutrons})}_{\text{mass number}} - \underbrace{\text{protons}}_{\text{atomic number}} = \text{neutrons}$$

c. There are nine electrons because the number of protons and the number of electrons is always the same in a neutral atom.

An alphabetical listing of the 109 known elements, along with selected information about each element, is printed on the inside back cover of this text. One of the pieces of information given for each element is its atomic number. If you carefully checked the atomic number data, you would find that an element exists for each atomic number in the numerical sequence 1 through 109. There are no missing numbers in the sequence 1–109. The existence of an element that corresponds to each of these numbers is an indication of the order existing in nature. Thus, elements exist with any number of protons (or electrons) up to and including 109.

Mass numbers are not tabulated in a manner similar to atomic numbers because, as we shall see in the next section, most elements lack a unique mass number.

3.6 ISOTOPES

The identity of an atom is determined by its atomic number, which is the number of protons that it contains. All atoms of an element must contain the same number of protons.

The chemical properties of an atom, which are the basis for its identification, are determined by the number and arrangement of electrons about the nucleus. All atoms that have the same atomic number will have the same number of electrons and the same chemical properties.

All of the atoms of an element do not have to be identical. They can differ from each other in the number of neutrons they have. The presence of one or more extra neutrons in the tiny nucleus of an atom has essentially no effect on the way the atom behaves chemically. For example, all silicon atoms have fourteen pro-

isotope

tons and fourteen electrons. Most silicon atoms also contain fourteen neutrons. However, some silicon atoms contain fifteen neutrons and others contain sixteen neutrons. Thus, three different kinds of silicon atoms exist, each with the same chemical properties; three silicon *isotopes* are said to exist. **Isotopes** *are atoms that have the same number of protons and electrons, but have different numbers of neutrons.* The isotopes of an element will always have the same atomic number and differing mass numbers.

When it is necessary to distinguish between isotopes, the following notation is used.

$$^{\text{mass number}}_{\text{atomic number}}\text{SYMBOL}$$

The atomic number, whose general symbol is Z, is written as a subscript to the left of the elemental symbol for the atom. The mass number, whose general symbol is A, is also written to the left of the elemental symbol, but as a superscript.

TABLE 3–3   Naturally Occurring Isotopic Abundances for Some Common Elements*

Element	Isotope	Percent Natural Abundance	Isotopic Mass (amu)
hydrogen	$^{1}_{1}\text{H}$	99.98	1.01
	$^{2}_{1}\text{H}$	0.02	2.01
carbon	$^{12}_{6}\text{C}$	98.89	12.00
	$^{13}_{6}\text{C}$	1.11	13.00
nitrogen	$^{14}_{7}\text{N}$	99.63	14.00
	$^{15}_{7}\text{N}$	0.37	15.00
oxygen	$^{16}_{8}\text{O}$	99.76	15.99
	$^{17}_{8}\text{O}$	0.04	17.00
	$^{18}_{8}\text{O}$	0.20	18.00
sulfur	$^{32}_{16}\text{S}$	95.00	31.97
	$^{33}_{16}\text{S}$	0.76	32.97
	$^{34}_{16}\text{S}$	4.22	33.97
	$^{36}_{16}\text{S}$	0.01	35.97
chlorine	$^{35}_{17}\text{Cl}$	75.53	34.97
	$^{37}_{17}\text{Cl}$	24.47	36.97
copper	$^{63}_{29}\text{Cu}$	69.09	62.93
	$^{65}_{29}\text{Cu}$	30.91	64.93
titanium	$^{46}_{22}\text{Ti}$	7.93	45.95
	$^{47}_{22}\text{Ti}$	7.28	46.95
	$^{48}_{22}\text{Ti}$	73.94	47.95
	$^{49}_{22}\text{Ti}$	5.51	48.95
	$^{50}_{22}\text{Ti}$	5.34	49.94
uranium	$^{234}_{92}\text{U}$	0.01	234.04
	$^{235}_{92}\text{U}$	0.72	235.04
	$^{238}_{92}\text{U}$	99.27	238.05

* Twenty elements have only one naturally occurring form: $^{9}_{4}\text{Be}$, $^{19}_{9}\text{F}$, $^{23}_{11}\text{Na}$, $^{27}_{13}\text{Al}$, $^{31}_{15}\text{P}$, $^{45}_{21}\text{Sc}$, $^{55}_{25}\text{Mn}$, $^{59}_{27}\text{Co}$, $^{75}_{33}\text{As}$, $^{89}_{39}\text{Y}$, $^{93}_{41}\text{Nb}$, $^{103}_{45}\text{Rh}$, $^{127}_{53}\text{I}$, $^{133}_{55}\text{Cs}$, $^{141}_{59}\text{Pr}$, $^{159}_{65}\text{Tb}$, $^{165}_{67}\text{Ho}$, $^{169}_{69}\text{Tm}$, $^{197}_{79}\text{Au}$, $^{209}_{83}\text{Bi}$.

By this symbolism, the three previously mentioned silicon isotopes would be designated, respectively, as

$$^{28}_{14}\text{Si}, \quad ^{29}_{14}\text{Si}, \quad \text{and} \quad ^{30}_{14}\text{Si}$$

Most naturally occurring elements are mixtures of isotopes. The various isotopes of a given element are of varying abundance; usually one isotope is predominant. Typical of this situation is the element silicon, which exists in nature in three isotopic forms: $^{28}_{14}\text{Si}$, $^{29}_{14}\text{Si}$, and $^{30}_{14}\text{Si}$. The percentage abundances for these three isotopes are 92.21%, 4.70%, and 3.09% respectively. Percentage abundances are number percentages (number of atoms) rather than mass percentages. A sample of 10,000 silicon atoms contains 9221 $^{28}_{14}\text{Si}$ atoms, 470 $^{29}_{14}\text{Si}$ atoms, and 309 $^{30}_{14}\text{Si}$ atoms. Table 3-3 gives natural isotopic abundances and isotopic masses for selected elements. The units used for specifying the mass of the various isotopes in the last column of Table 3-3 will be discussed in Section 3.7. Only 20 elements exist in nature in only one form, and they are listed at the bottom of Table 3-3.

The four most abundant elements in the human body are hydrogen, oxygen, carbon, and nitrogen (Section 2.6). Note in Table 3-3 that isotopic forms exists for all four of these elements.

3.7 ATOMIC WEIGHTS

An atom of a specific element can have several different masses if the element exists in isotopic forms. For example, silicon atoms can have any one of three masses because there are three isotopes of silicon. It might seem necessary to specify isotopic identity every time masses of atoms are encountered; however, this is not the case. In practice, isotopes are seldom mentioned in discussions about the masses of atoms because atoms of an element are treated as if they all had a single mass. The mass used is an average mass that takes into account the existence of isotopes. The use of this average mass reduces the number of masses needed for calculations from many hundreds (one for each isotope) to 109 (one for each element).

The term *atomic weight* is used to describe the average mass of the isotopes of an element*. An **atomic weight** *is an average mass for the isotopes of an element that is expressed on a scale using atoms of* $^{12}_{6}\text{C}$ *as the reference.* Three pieces of information are needed to calculate an atomic weight:

atomic weight

1. The number of isotopes that exist for the element.
2. The masses of the isotopes on the $^{12}_{6}\text{C}$ scale.
3. The percentage abundance of each isotope.

Prior to actually calculating an atomic weight, let us first consider the $^{12}_{6}\text{C}$ scale that is mentioned in the definition of atomic weight. It is impossible to weigh just one atom because atoms are so small. However, it is possible to experimentally determine how many times heavier one type of atom is than another. Using this information, scientists have set up a *relative scale* for masses of atoms. The arbitrary reference point on this scale is the mass of atoms of a particular isotope of carbon—$^{12}_{6}\text{C}$—which has a defined mass of exactly 12.00 *atomic mass units* (*amu*). The masses of all other atoms are then determined relative to that of $^{12}_{6}\text{C}$. For example, if an atom is twice as heavy as $^{12}_{6}\text{C}$, its mass is 24.00 amu on the scale, and if an atom weighs half as much as an atom of $^{12}_{6}\text{C}$, its scale mass is 6.00 amu.

---

* Because of convention we continue to use the term "atomic weight" although the term "atomic mass" is technically more correct.

Values for the masses of isotopes for selected elements on the atomic mass scale are given as the last column of data in Table 3–3.

Example 3–2 shows how information from the amu ($^{12}_{6}C$) scale, percentage abundances, and number of isotopes is used to obtain an atomic weight.

Example 3.2

Naturally occurring chlorine exists in two isotopic forms, $^{35}_{17}Cl$ and $^{37}_{17}Cl$. The relative mass of $^{35}_{17}Cl$ is 34.97 amu, and its abundance is 75.53%; the relative mass of $^{37}_{17}Cl$ is 36.97 amu and it has a 24.47% abundance. What is the atomic weight of chlorine?

Solution

The atomic weight is calculated by multiplying the relative mass of each isotope by its fractional abundance and then totaling the products.

The fractional abundance is the percentage abundance converted to decimal form (divided by 100).

$$^{35}_{17}Cl: \left(\frac{75.53}{100}\right) \times 34.97 \text{ amu} = (0.7553) \times 34.97 \text{ amu} = 26.41 \text{ amu}$$

$$^{37}_{17}Cl: \left(\frac{24.47}{100}\right) \times 36.97 \text{ amu} = (0.2447) \times 36.97 \text{ amu} = 9.04 \text{ amu}$$

$$\text{atomic weight of chlorine} = 35.45 \text{ amu}$$

This calculation involved an element containing just two isotopes. A similar calculation for an element having three isotopes would be carried out the same way, but it would have three terms in the final sum; an element possessing four isotopes would have four terms in the final sum.

The alphabetical listing of the known elements printed inside the back cover of this text gives the calculated atomic weight for each of the elements in the last column of numbers. Atomic weights are also given on the periodic table (Section 3.9) printed inside the front cover. There, the atomic weight is the number underneath each element's symbol.

3.8 THE PERIODIC LAW

During the early part of the nineteenth century, an abundance of chemical facts became available from detailed studies of the known elements. With the hope of providing a systematic approach to the study of chemistry, scientists began to look for order in the increasing amount of chemical information. They were encouraged in their search by the unexplained, but well-known fact that certain elements had properties that were very similar to those of other elements. Numerous attempts were made to find reasons for these similarities and to use them as a way of providing a method for arranging or classifying the elements.

In 1869, these efforts culminated in the discovery of what is now called the *periodic law*. Proposed independently by both Russian chemist Dmitri Mendeleev and German chemist Julius Lothar Meyer, this law is one of the most important chemical laws. Given in its modern form, the **periodic law** states that *when elements are arranged in order of increasing atomic numbers, elements with similar properties occur at periodic (regularly recurring) intervals.*

An illustration of this regularly repeating pattern is given in Figure 3–7 for the sequence of elements with atomic numbers 3 through 20. The elements within

FIGURE 3–7 Periodicity in properties when elements are arranged in order of increasing atomic number.

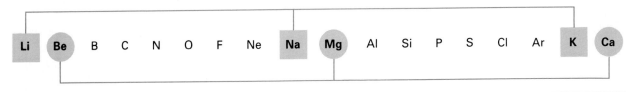

similar geometric symbols (circles and squares) have similar chemical properties. For the sake of simplicity, only two of the periodic relationships are shown; for the elements listed, similar properties are found in every eighth element.

3.9 THE PERIODIC TABLE

A **periodic table** *is a graphical representation of the behavior described by the periodic law.* In this table, the elements are arranged according to increasing atomic number in such a way that the similarities predicted by the periodic law become readily apparent. The most commonly used form of the periodic table is shown in Figure 3–8 (p. 64 and also inside the front cover of the text). In this periodic table, the vertical columns contain elements with similar chemical properties. Each element is represented by a rectangular box, which contains the symbol, atomic number, and atomic weight of the element, as shown in Figure 3–9 (p. 64).

Special chemical terminology exists for specifying the position (location) of an element within the periodic table. This terminology involves the use of the words *group* and *period*. A **group** *in the periodic table is a vertical column of elements.* There are two notations in use for designating individual periodic table groups. In the first notation, which has been in use for many years, groups are designated using Roman numerals and the letters A and B. In the second notation, which has been recently recommended for use by an international scientific commission, the Arabic numbers 1 through 18 are used. Notice, by referring to the periodic table (Figure 3–8), that both group notations are given at the top of each group. The elements with atomic numbers 8, 16, 34, 52 and 84 (O, S, Se, Te, and Po) constitute Group VIA (old notation), or Group 16 (new notation). Because it may be some time before the new group numbering system becomes widely accepted, we will not use it in succeeding chapters of the text.

A **period** *in the periodic table is a horizontal row of elements.* For identification purposes, the periods are numbered sequentially with Arabic numbers, starting at the top of the periodic table. (These period numbers are not explicitly shown on the periodic table.) The elements Na, Mg, Al, Si, P, S, Cl, and Ar are all members of Period 3. Period 3 is the third row of elements, Period 4 the fourth row of elements, and so forth. There are only two elements in Period 1: H and He.

The location of any element in the periodic table is specified by giving its group number and its period number. The element gold, with an atomic number of 79, belongs to Group IB (or 11) and is in Period 6. The element nitrogen, with an atomic number of 7, belongs to Group VA (or 15) and is in Period 2.

A careful study of the periodic table shows that there is one area in the table where the practice of arranging the elements according to increasing atomic numbers seems to be violated. It involves the area where elements 57 and 89, both located in Group IIIB (or 3) of the table, are found. Element 72 follows element 59 and element 104 follows element 89. The missing elements, 58–71 and 90–103,

FIGURE 3–8 The modern periodic table, based on $^{12}_{6}C$. Numbers in parentheses are the mass numbers of the most stable isotopes of radioactive elements.

1 GROUP IA																	18 GROUP VIIIA
1 H 1.01	2 GROUP IIA											13 GROUP IIIA	14 GROUP IVA	15 GROUP VA	16 GROUP VIA	17 GROUP VIIA	2 He 4.00
3 Li 6.94	4 Be 9.01											5 B 10.81	6 C 12.01	7 N 14.01	8 O 16.00	9 F 19.00	10 Ne 20.18
11 Na 22.99	12 Mg 24.30	3 GROUP IIIB	4 GROUP IVB	5 GROUP VB	6 GROUP VIB	7 GROUP VIIB	8 GROUP	9 GROUP VIIIB	10 GROUP	11 GROUP IB	12 GROUP IIB	13 Al 26.98	14 Si 28.09	15 P 30.97	16 S 32.07	17 Cl 35.45	18 Ar 39.95
19 K 39.10	20 Ca 40.08	21 Sc 44.96	22 Ti 47.88	23 V 50.94	24 Cr 52.00	25 Mn 54.94	26 Fe 55.85	27 Co 58.93	28 Ni 58.69	29 Cu 63.55	30 Zn 65.38	31 Ga 69.72	32 Ge 72.59	33 As 74.92	34 Se 78.96	35 Br 79.90	36 Kr 83.80
37 Rb 85.47	38 Sr 87.62	39 Y 88.91	40 Zr 91.22	41 Nb 92.91	42 Mo 95.94	43 Tc (98)	44 Ru 101.07	45 Rh 102.91	46 Pd 106.42	47 Ag 107.87	48 Cd 112.41	49 In 114.82	50 Sn 118.71	51 Sb 121.75	52 Te 127.60	53 I 126.90	54 Xe 131.29
55 Cs 132.91	56 Ba 137.33	57 La 138.91	72 Hf 178.49	73 Ta 180.95	74 W 183.85	75 Re 186.21	76 Os 190.2	77 Ir 192.22	78 Pt 195.08	79 Au 196.97	80 Hg 200.59	81 Tl 204.38	82 Pb 207.2	83 Bi 208.98	84 Po (209)	85 At (210)	86 Rn (222)
87 Fr (223)	88 Ra (226)	89 Ac (227)	104 Unq (261)	105 Unp (262)	106 Unh (263)	107 Uns (262)	108 Uno (265)	109 Une (266)									

Metals ⎯⎯⎯ Nonmetals

58 Ce 140.12	59 Pr 140.91	60 Nd 144.24	61 Pm (145)	62 Sm 150.36	63 Eu 151.96	64 Gd 157.25	65 Tb 158.93	66 Dy 162.50	67 Ho 164.93	68 Er 167.26	69 Tm 168.93	70 Yb 173.04	71 Lu 174.97
90 Th (232)	91 Pa (231)	92 U (238)	93 Np (237)	94 Pu (244)	95 Am (243)	96 Cm (247)	97 Bk (247)	98 Cf (251)	99 Es (252)	100 Fm (257)	101 Md (258)	102 No (259)	103 Lr (260)

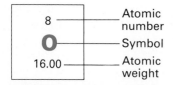

FIGURE 3–9 Arrangement of information within the periodic table.

are located in two rows at the bottom of the periodic table. Technically, the elements at the bottom of the table should be included in the body of the table as shown in Figure 3–10. However, in order to have a more compact table, they are placed at the bottom of the table. This arrangement should present no problems to the user of the periodic table as long as it is recognized for what it is—a space-saving device.

The mass number, which is one of the fundamental identifying characteristics of an atom (Section 3.5), is not part of the information given on a periodic table.

FIGURE 3–10 The periodic table with elements 58–71 and 90–103 (in color) in their proper positions. To conserve space, the periodic table is usually drawn with elements 58–71 and 90–103 placed at the bottom as in Figure 3–8.

1																	2
3	4											5	6	7	8	9	10
11	12											13	14	15	16	17	18
19	20	21	22	23	24	25	26	27	28	29	30	31	32	33	34	35	36
37	38	39	40	41	42	43	44	45	46	47	48	49	50	51	52	53	54
55	56	57	58	59	60	61	62	63	64	65	66	67	68	69	70	71	72	73	74	75	76	77	78	79	80	81	82	83	84	85	86
87	88	89	90	91	92	93	94	95	96	97	98	99	100	101	102	103	104	105	106	107	108	109

The reason for this is that mass number is not a unique quantity for most elements; different isotopes of an element have different mass numbers (Section 3.6). By using the information on a periodic table, you can quickly determine the number of protons and electrons for atoms of an element. However, you cannot derive the actual number of neutrons present in a given isotope from the table.

For many years, the periodic law and periodic table were considered to be empirical, there was no explanation available either for the law or for why the periodic table had the shape that it did. We now know that the theoretical basis for both the periodic law and table is found in electronic theory. The properties of the elements repeat themselves in a periodic manner because the arrangement of electrons about the nucleus of an atom follows a periodic pattern. Electron arrangements and an explanation of the periodic law and periodic table in terms of electronic theory are the subject matter of Chapter 5.

EXERCISES AND PROBLEMS

Scientific Method

3.1 Describe the five steps of the scientific method.

3.2 Define the following terms.
a. hypothesis   d. fact
b. experiment   e. law
c. theory

3.3 What is the difference between a hypothesis and a theory?

3.4 What are two important purposes that theories serve?

3.5 Indicate whether each of the following statements is true or false.
a. The word *hypothesis* describes untested ideas that are proposed to explain a law.
b. A theory is a hypothesis that has not yet been subjected to experimental testing.
c. A law is a general statement that summarizes a number of experimental facts.
d. A hypothesis is a tentative law.
e. A theory is a summary of experimental observations.
f. A law is an explanation of why a particular natural phenomenon occurs.

Atomic Theory of Matter

3.6 List the five statements of modern day atomic theory.

3.7 Which of the following statements are not consistent with the statements of modern day atomic theory?
a. Atoms are the basic building blocks for all kinds of matter.
b. Different "types" of atoms exist.
c. All atoms of a given "type" are identical.
d. Only whole atoms can participate in chemical reactions.
e. Atoms change identity during chemical change processes.
f. 113 different "types" of atoms are known.

3.8 What is the approximate size of an atom (in centimeters)? What is the approximate mass of an atom (in grams)?

Laws of Chemical Change

3.9 Consider a hypothetical reaction in which A and B are reactants and C and D are products. If 10 g of A completely reacts with 7 g of B to produce 14 g of C, how many grams of D will be produced?

3.10 A 4.2 g sample of sodium hydrogen carbonate is added to a solution of acetic acid weighing 10.0 g. The two substances react, releasing carbon dioxide gas to the atmosphere. After reaction, the contents of the reaction vessel weigh 12.0 g. What is the mass of carbon dioxide given off during the reaction?

3.11 When a log burns, the ashes weigh less than the log. When sodium reacts with sulfur, the sodium–sulfur compound formed weighs the same as the reacted sodium and sulfur. When phosphorus burns in air, however, the compound formed weighs more than the original phosphorus. Explain these observations in terms of the law of conservation of matter.

3.12 Two different samples of a pure compound containing elements A and B were analyzed with the following results:
Sample I: 15.8 g of compound yielded 9.8 g of A and 6.0 g of B.
Sample II: 25.0 g of compound yielded 15.5 g of A and 9.5 g of B.
Show that this data is consistent with the law of constant composition.

Subatomic Particles: Protons, Neutrons, and Electrons

3.13 List the two kinds of subatomic particles found in the nucleus. Compare their properties with each other and with those of an electron.

3.14 What is meant by the statement: "Atoms are electrically neutral"?

3.15 Indicate which subatomic particle (proton, neutron, or electron) correctly matches each of the following statements. More than one particle can be used as an answer.

a. possesses a negative charge
b. has no charge
c. has a mass slightly less than that of a neutron
d. has a charge equal to but opposite in sign to an electron
e. is not found in the nucleus
f. has a positive charge
g. can be called a nucleon
h. is the heaviest of the three particles
i. has a relative mass of 1836 if the mass of an electron is 1

3.16 Indicate whether each of the following statements about the nucleus of an atom is true or false.

a. The nucleus of an atom is neutral.
b. The nucleus of an atom contains only neutrons.
c. The number of nucleons present in the nucleus is equal to the number of electrons present outside the nucleus.
d. The nucleus accounts for almost all of the mass of an atom.
e. The nucleus accounts for almost all of the volume of an atom.
f. The nucleus can be positively or negatively charged, depending on the identity of the atom.

3.17 Explain why atoms are considered to be the fundamental building blocks of matter even though smaller particles (subatomic particles) exist.

Atomic Number and Mass Number

3.18 What information about the subatomic makeup of an atom is given by:

a. atomic number
b. mass number
c. mass number − atomic number
d. mass number + atomic number

3.19 Determine the number of protons, neutrons, and electrons present in each of the following atoms.

a. $^{53}_{24}Cr$ c. $^{256}_{101}Md$ e. $^{67}_{30}Zn$ g. $^{40}_{20}Ca$
b. $^{103}_{44}Ru$ d. $^{34}_{16}S$ f. $^{9}_{4}Be$ h. $^{3}_{1}H$

3.20 Write complete symbols, with the help of the information listed inside the back cover, for atoms with the following characteristics.

a. contains 28 protons and 30 neutrons
b. contains 22 protons and 21 neutrons
c. contains 15 electrons and 19 neutrons
d. oxygen atom with 10 neutrons
e. chromium atom with a mass number of 54
f. has an atomic number of 11 and a mass number of 25
g. gold atom that contains 276 subatomic particles
h. beryllium atom that contains 10 nucleons

3.21 The notation $^{7}_{3}Li$ and ^{7}Li can be used interchangeably. Why is this allowable? Can the notations $^{7}_{3}Li$ and $_{3}Li$ be used interchangeably without loss of meaning? Why or why not?

3.22 Can the atomic number of an element be larger than its mass number?

Isotopes

3.23 Identify the atoms that are isotopes in each of the following sets of four atoms.

a. $^{14}_{6}X$, $^{14}_{7}X$, $^{15}_{7}X$, $^{15}_{8}X$
b. $^{20}_{10}X$, $^{21}_{10}X$, $^{22}_{10}X$, $^{23}_{10}X$
c. $^{41}_{19}X$, $^{41}_{20}X$, $^{41}_{21}X$, $^{41}_{22}X$

3.24 What restrictions on the numbers of protons, neutrons, and electrons apply to each of the following?

a. atoms of different isotopes of the same element
b. atoms of a particular isotope of an element
c. atoms of different elements with the same mass number
d. atoms of the same element with different mass numbers

3.25 How do you write the symbol for an atom of an element if you wish to specify a particular isotope of that element?

3.26 The following are selected properties for the most abundant isotope of a particular element. Which of these properties would also be the same for the second most abundant isotope of the element?

a. mass number is 70
b. atomic number is 31
c. melting point is 29.8°C
d. isotopic mass is 69.92 amu
e. reacts with chlorine to give a trichloride

3.27 Write complete symbols, with the help of the information listed inside the back cover, for atoms with the following characteristics.

a. an isotope of boron that contains two fewer neutrons than $^{10}_{5}B$
b. an isotope of boron that contains three more subatomic particles than $^{9}_{5}B$
c. an isotope of boron that contains the same number of neutrons as $^{14}_{7}N$
d. an isotope of boron that contains an equal number of protons, neutrons, and electrons

Atomic Weights

3.28 A certain isotope of silver is 8.91 times heavier than $^{12}_{6}C$. What is the mass of this silver isotope on the amu scale?

3.29 What are the three types of information needed to calculate an atomic weight?

3.30 A sample of lithium of natural origin is found to consist of two isotopes: $^{6}_{3}Li$ (abundance 7.42%) and $^{7}_{3}Li$ (abundance 92.58%). The masses of these isotopes on the amu scale are 6.01 and 7.02, respectively. Calculate the atomic weight of lithium.

3.31 A sample of silicon of natural origin is found to consist of three isotopes: $^{28}_{14}Si$ (abundance 92.21%), $^{29}_{14}Si$ (abundance 4.70%), and $^{30}_{14}Si$ (abundance 3.09%). Their relative masses are 27.98, 28.98, and 29.97 amu, respectively. Calculate the atomic weight of silicon.

3.32 How many times heavier, on the average, is an atom of gold than an atom of beryllium?

3.33 The atomic weight of fluorine is 18.998 amu and the atomic weight of iron is 55.847 amu. All fluorine atoms have a mass of 18.998 amu and not a single iron atom has a mass of 55.847 amu. Explain.

Periodic Law and Periodic Table

3.34 Give a modern day statement of the periodic law.

3.35 How do the terms *group* and *period* relate to the periodic table?

3.36 Give the symbol of the element that occupies each of the following positions in the periodic table.
a. Period 4, Group IIIA
b. Period 5, Group IVB
c. Group IA, Period 2
d. Group VIIA, Period 3
e. Period 1, Group 18
f. Period 6, Group 10
g. Group 3, Period 4
h. Group 14, Period 5

3.37 For each of the following sets of elements, choose the two that would be expected to have similar chemical properties.
a. $_{11}$Na, $_{14}$Si, $_{23}$V, $_{55}$Cs
b. $_{13}$Al, $_{19}$K, $_{32}$Ge, $_{50}$Sn
c. $_{37}$Rb, $_{38}$Sr, $_{54}$Xe, $_{56}$Ba
d. $_{2}$He, $_{6}$C, $_{8}$O, $_{10}$Ne

3.38 What are the three basic pieces of information about each element that are found on a modern day periodic table?

3.39 The following statements either define or are closely related to the terms *periodic law*, *period*, and *group*. Match the terms to the appropriate statements.
a. This is a vertical arrangement of elements in the periodic table.
b. This is a horizontal arrangement of elements in the periodic table.
c. The properties of the elements repeat in a regular way as the atomic numbers increase.
d. Element 19 begins this arrangement in the periodic table.
e. The chemical properties of elements 12, 20, and 38 demonstrate this principle.
f. The element carbon is the first member of this arrangement.
g. Elements 24 and 33 belong to this arrangement.
h. Elements 10, 18, and 36 belong to this arrangement.

4

Chemical Calculations: Formula Weights, Moles, and Chemical Equations

Objectives

After completing Chapter 4, you will be able to:

4.1 Calculate the formula weight of a substance, given its formula and a table of atomic weights.
4.2 Interpret the mole as a counting unit and calculate the number of particles in a given number of moles of a substance.
4.3 Interpret the mole as a variable mass unit and calculate the mass of a given number of moles of a substance.
4.4 Interpret the subscripts in a chemical formula in terms of the number of moles of the various elements present in one mole of the substance.
4.5 Calculate any of the following items: (1) mass of the substance, (2) moles of the substance, (3) number of particles of the substance, (4) atoms of any element in the substance, (5) mass of any element in the substance, or (6) moles of any element in the substance, given the chemical formula and one of the preceding pieces of information about a pure substance.
4.6 Understand the conventions used in writing chemical equations, recognize whether or not an equation is balanced, and balance a chemical equation, given the formulas of all the reactants and products.
4.7 Interpret coefficients in balanced chemical equations in terms of moles.
4.8 Use a balanced chemical equation and other appropriate information to calculate the quantities of reactants consumed and products produced in a chemical reaction.

INTRODUCTION

In this chapter, we will discuss chemical arithmetic, the quantitative relationships between elements and compounds. Anyone dealing with chemical processes needs an understanding of the simpler aspects of this topic. All chemical processes, regardless of where they occur—in the human body, at a steel mill, on top of the kitchen stove, or in a clinical laboratory setting—are governed by the same mathematical rules.

We have already presented some information about chemical formulas (Section 2.8). In this chapter, we will discuss formulas again; this time we will look beyond describing the composition of molecules in terms of constituent atoms. A new unit, the mole, will be introduced and its usefulness will be discussed. Chemical equations will be considered for the first time. We will learn how to represent chemical reactions using chemical equations and how to derive quantitative relationships from these equations.

4.1 FORMULA WEIGHTS

Our entry point into the realm of chemical arithmetic will be a discussion of the quantity called *formula weight*. Once the formula of a substance has been established, it is a simple matter to calculate its formula weight. The **formula weight** of a substance is the sum of the atomic weights of the atoms in its formula. Formula weights, like the atomic weights from which they are calculated, are relative masses based on the $^{12}_{6}C$ relative mass scale (Section 3.7). Example 4.1 illustrates how formula weights are calculated.

formula weight

Calculate the formula weight for each of the following substances.

a. SnF_2—stannous fluoride (a toothpaste additive)
b. $C_{10}H_{14}N_2$—nicotine (a common stimulant)
c. $Pb(C_2H_5)_4$—tetraethyl lead (a gasoline additive)

Example 4.1

Solution

Formula weights are obtained simply by adding the atomic weights of the constituent elements, counting each atomic weight as many times as the symbol for the element occurs in the formula.

a. A formula unit of SnF_2 contains three atoms: one atom of Sn and two atoms of F. The formula weight, the collective mass of these three atoms, is calculated as follows.

$$1 \text{ atom Sn} \times \left(\frac{118.7 \text{ amu}}{1 \text{ atom Sn}}\right) = 118.7 \text{ amu}$$

$$2 \text{ atoms F} \times \left(\frac{19.0 \text{ amu}}{1 \text{ atom F}}\right) = 38.0 \text{ amu}$$

$$\text{formula weight} = \overline{156.7 \text{ amu}}$$

We derive the conversion factors in the calculation from the atomic weights listed on the inside back cover of the text. Our rules for the use of conversion factors are the same of those discussed in Section 1.11.

Conversion factors are not usually explicitly shown in a formula-weight calculation, as they are in our example; the calculation is simplified as follows.

$$\text{Sn:} \quad 1 \times 118.7 \text{ amu} = 118.7 \text{ amu}$$
$$\text{F:} \quad 2 \times 19.0 \text{ amu} = \underline{38.0 \text{ amu}}$$
$$\text{formula weight} = 156.7 \text{ amu}$$

b. Similarly, for nicotine ($C_{10}H_{14}N_2$) we calculate the formula weight as

$$\text{C:} \quad 10 \times 12.0 \text{ amu} = 120.0 \text{ amu}$$
$$\text{H:} \quad 14 \times 1.0 \text{ amu} = 14.0 \text{ amu}$$
$$\text{N:} \quad 2 \times 14.0 \text{ amu} = \underline{28.0 \text{ amu}}$$
$$\text{formula weight} = 162.0 \text{ amu}$$

c. The formula for this compound contains parentheses. Improper interpretation of parentheses (see Section 2.8) is a common error made by students doing formula-weight calculations. In the formula $Pb(C_2H_5)_4$, the subscript 4 outside the parentheses affects all of the symbols inside the parentheses. Thus, we have

$$\text{Pb:} \quad 1 \times 207.2 \text{ amu} = 207.2 \text{ amu}$$
$$\text{C:} \quad 8 \times 12.0 \text{ amu} = 96.0 \text{ amu}$$
$$\text{H:} \quad 20 \times 1.0 \text{ amu} = \underline{20.0 \text{ amu}}$$
$$\text{formula weight} = 323.2 \text{ amu}$$

4.2 THE MOLE: THE CHEMIST'S COUNTING UNIT

The quantity of material in a sample of a substance can be specified either in terms of units of mass or in terms of units of *amount*. Mass is specified in terms of units such as grams, kilograms, or pounds. The amount of a substance is specified by indicating the number of objects present—3, 17, or 437.

We all use both units of mass and units of amount on a daily basis. We work well with this dual system. Sometimes it doesn't matter which type of unit is used; other times one system is preferred over the other. When buying potatoes at the grocery store we can decide on quantity in either mass units (10-lb bag or 20-lb bag) or amount units (7 potatoes or 13 potatoes). When buying eggs, amount units are used almost exclusively—12 eggs (1 dozen) or 24 eggs (2 dozen). We do not ordinarily buy 2 pounds of eggs. On the other hand, grapes are almost always purchased in weighed quantities. It is impractical to count the number of grapes in a bunch. Very few people go to the store with the idea that they will buy 189 grapes.

In chemistry, as in everyday life, both the mass and amount methods of specifying quantity are used. Again, the specific situation dictates the method used. In laboratory work, practicality dictates working with quantities of known mass. Counting out a given number of atoms for a laboratory experiment is somewhat impractical because we cannot see individual atoms.

When performing chemical calculations after the laboratory work has been done, it is often useful and even necessary to think of the quantities of substances present in terms of amounts such as numbers of atoms or molecules instead of mass. The problem that exists when this is done is that very large numbers are always encountered. Any macroscopic-sized sample of a chemical substance contains many trillions of atoms or molecules.

In order to cope with this large number problem, chemists have found it convenient to use a special unit when counting atoms and molecules. The employment of such a unit should not surprise you, as specialized counting units are used in many areas. The two most common counting units are *dozen* and *pair*. Other more specialized counting units exist; for example, at an office supply store, paper is sold by the *ream* (500 sheets), pencils by the *gross* (144), and stencils by the *quire* (24). (See Figure 4–1a–d).

The chemist's counting unit is called a *mole*. What is unusual about the mole is its magnitude. **A mole** is 6.02×10^{23} *objects*. The extremely large size of the mole unit is necessitated by the extremely small size of atoms and molecules (see Figure 4–1e).

mole

To the chemist, *one mole* always means 6.02×10^{23} objects, just as *one dozen* always means 12 objects. Two moles of objects would be two times 6.02×10^{23} objects and five moles of objects would be five times 6.02×10^{23} objects.

The number of objects in a mole, 6.02×10^{23}, has a special name. **Avogadro's number** *is the name given to the numerical value* 6.02×10^{23}. This designation honors Amedeo Avogadro (1776–1856), an Italian physicist whose pioneering work on gases later proved to be valuable in determining the number of particles present in given volumes of substances. When solving problems dealing with the number of objects (atoms or molecules) present in a given number of moles of a substance, Avogadro's number becomes part of the conversion factor used to relate the number of objects present to the number of moles present.

Avogadro's number

From the definition

$$1 \text{ mole} = 6.02 \times 10^{23} \text{ objects}$$

two conversion factors can be derived.

FIGURE 4–1 Examples of counting units used to denote quantities in terms of groups of various sizes.

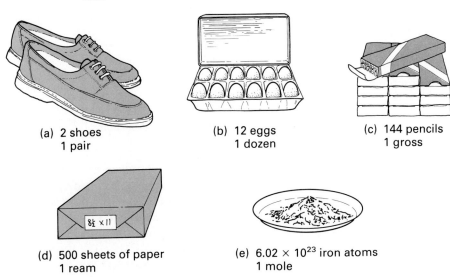

(a) 2 shoes
1 pair

(b) 12 eggs
1 dozen

(c) 144 pencils
1 gross

(d) 500 sheets of paper
1 ream

(e) 6.02×10^{23} iron atoms
1 mole

$$\frac{6.02 \times 10^{23} \text{ objects}}{1 \text{ mole}} \quad \text{and} \quad \frac{1 \text{ mole}}{6.02 \times 10^{23} \text{ objects}}$$

Example 4.2 illustrates the use of these conversion factors in solving problems.

Example 4.2

How many objects are there in each of the following quantities?
a. 0.23 mole of aspirin molecules
b. 1.6 moles of oxygen atoms
c. 0.75 mole of watermelon seeds

Solution

Dimensional analysis (Section 1.11) will be used to solve each of these problems. All of the problems are similar in that we are given a certain number of moles of substance and want to find the number of objects present in the given number of moles. We will need Avogadro's number to solve each of these "moles-to-particles" problems.

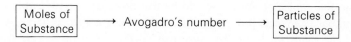

a. The objects of concern are molecules of aspirin. The given quantity is 0.23 mole of aspirin molecules and the desired quantity is the number of aspirin molecules.

$$0.23 \text{ mole aspirin molecules} = ? \text{ aspirin molecules}$$

The setup, using dimensional analysis, involves the use of a single conversion factor, one that relates moles and molecules.

$$0.23 \text{ mole aspirin molecules} \times \left(\frac{6.02 \times 10^{23} \text{ aspirin molecules}}{1 \text{ mole aspirin molecules}} \right)$$
$$= 1.4 \times 10^{23} \text{ aspirin molecules}$$

b. This time we are dealing with atoms instead of molecules. This switch does not change the way we work the problem. We will need the same conversion factor.

The given quantity is 1.6 moles of oxygen atoms and the desired quantity is the actual number of oxygen atoms present.

$$1.6 \text{ moles oxygen atoms} = ? \text{ oxygen atoms}$$

The setup is

$$1.6 \text{ moles of oxygen atoms} \times \left(\frac{6.02 \times 10^{23} \text{ oxygen atoms}}{1 \text{ mole oxygen atoms}} \right)$$
$$= 9.6 \times 10^{23} \text{ oxygen atoms}$$

c. Use of the mole as a counting unit is usually found only in a chemical context. Technically, however, any type of object can be counted in units of moles, even watermelon seeds. One mole denotes 6.02×10^{23} objects; it does not matter what the objects are. Just as we can talk about 3 dozen watermelon seeds, we can talk about 3 moles, or any other number of moles, of watermelon seeds.

The given quantity is 0.75 mole of watermelon seeds and the desired quantity is the number of watermelon seeds. The setup for the problem, using the same conversion factor as in the previous parts, is

$$0.75 \; \cancel{\text{mole watermelon seeds}} \times \left(\frac{6.02 \times 10^{23} \text{ watermelon seeds}}{1 \; \cancel{\text{mole watermelon seeds}}} \right)$$

$$= 4.5 \times 10^{23} \text{ watermelon seeds}$$

This number of seeds would certainly be sufficient to plant an above-average size patch of watermelons.

4.3 THE MASS OF A MOLE

How much does a mole weigh? Are you uncertain about the answer to that question? Let us consider a similar but more familiar question first: "How much does a dozen weigh?" Your response is now immediate. "A dozen what?" you reply. The mass of a dozen identical objects obviously depends on the identity of the object. For example, the mass of a dozen elephants will be somewhat greater than the mass of a dozen peanuts. The mole, like the dozen, is a counting unit. Similarly, the mass of a mole, like the mass of a dozen, will depend on the identity of the object. Thus, the mass of a mole, or *molar mass*, is not a set number; it varies, and is different for each chemical substance. This is in direct contrast to the *molar number*, Avogadro's number, which is the same for all chemical substances.

The **molar mass** *of a substance is the mass in grams that is numerically equal to the substance's formula weight.* For example, the formula weight (atomic weight) of the element sodium is 23.0 amu; therefore, 1 mole of sodium would weigh 23.0 g. In Example 4.1, we calculated that the formula weight of nicotine is 162.0 amu; therefore, 1 mole of nicotine molecules would weigh 162.0 g. We can obtain the actual mass in grams of one mole of any substance by computing its formula weight (atomic weight for elements) and writing "grams" after it. Thus, when we add atomic weights to get the formula weight (in amu's) of a compound, we are simultaneously finding the mass of one mole of that compound (in grams).

Figure 4–2 shows molar quantities of a number of common substances. Note how the numerical value of the mass of a mole varies from substance to substance.

molar mass

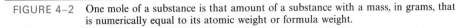

FIGURE 4–2 One mole of a substance is that amount of a substance with a mass, in grams, that is numerically equal to its atomic weight or formula weight.

It is not a coincidence that the molar mass of a substance and its formula weight or atomic weight match numerically. Avogadro's number has the value that it has in order to cause this relationship to exist. The numerical match between molar mass and atomic or formula weight makes the calculation of the mass of any given number of moles of a substance a very simple procedure. When solving problems of this type, the numerical value of the molar mass becomes part of the conversion factor used to convert from moles to grams.

For example, for the compound CO_2, which has a formula weight of 44.0 amu, we can write the equality

$$44.0 \text{ g } CO_2 = 1 \text{ mole } CO_2$$

From this statement (equality), two conversion factors can be written

$$\frac{44.0 \text{ g } CO_2}{1 \text{ mole } CO_2} \quad \text{and} \quad \frac{1 \text{ mole } CO_2}{44.0 \text{ g } CO_2}$$

Example 4.3 illustrates the use of gram–mole conversion factors like these in solving problems.

Example 4.3

What is the mass, in grams, of each of the following molar quantities of chemical substances?

a. 0.34 mole of glucose ($C_6H_{12}O_6$)
b. 2.07 moles of arsenic (As)

Solution
We will use dimensional analysis to solve each of these problems. The relationship between molar mass and atomic or formula weight will serve as a conversion factor in the setup of each problem.

$$\boxed{\text{Moles of Substance}} \xrightarrow{\text{molar mass}} \boxed{\text{Grams of Substance}}$$

a. The given quantity is 0.34 mole of glucose ($C_6H_{12}O_6$) and the desired quantity is grams of glucose.

$$0.34 \text{ mole } C_6H_{12}O_6 = ? \text{ grams } C_6H_{12}O_6$$

The calculated formula weight of $C_6H_{12}O_6$ is 180 amu. Thus,

$$180 \text{ grams } C_6H_{12}O_6 = 1 \text{ mole } C_6H_{12}O_6$$

With this relationship as a conversion factor, the setup for the problem becomes

$$0.34 \text{ mole } C_6H_{12}O_6 \times \left(\frac{180 \text{ g } C_6H_{12}O_6}{1 \text{ mole } C_6H_{12}O_6}\right) = 61 \text{ g } C_6H_{12}O_6$$

b. The given quantity is 2.07 moles of arsenic and the desired quantity is grams of arsenic.

$$2.07 \text{ moles As} = ? \text{ grams As}$$

Here, we are dealing with an element rather than a compound. Thus, 74.9 amu, the atomic weight of arsenic, is used in the mole–gram equality statement.

$$74.9 \text{ grams As} = 1 \text{ mole As}$$

With this relationship as a conversion factor, the setup becomes

$$2.07 \text{ moles As} \times \left(\frac{74.9 \text{ g As}}{1 \text{ mole As}}\right) = 155 \text{ g As}$$

4.4 THE MOLE AND CHEMICAL FORMULAS

A chemical formula has two meanings or interpretations: a microscopic level interpretation and a macroscopic level interpretation. The first interpretation was discussed in Section 2.8. At a microscopic level, a chemical formula indicates the number of atoms of each element present in one molecule or formula unit of a substance. *The numerical subscripts in the formula give the number of atoms of the various elements present in one formula unit of the substance.* The formula N_2O_4, interpreted at the microscopic level, conveys the information that two atoms of nitrogen and four atoms of oxygen are present in one molecule of N_2O_4.

Now that the mole concept has been introduced, a macroscopic interpretation of chemical formulas is possible. At a macroscopic level, a chemical formula indicates the number of moles of atoms of each element present in one mole of a substance. *The numerical subscripts in the formula give the number of moles of atoms of the various elements present in one mole of the substance.* The designation *macroscopic* is given to this molar interpretation because moles are laboratory-sized quantities of atoms. The formula N_2O_4, interpreted at the macroscopic level, conveys the information that two moles of nitrogen atoms and four moles of oxygen atoms are present in one mole of N_2O_4 molecules. Thus, the subscripts in a formula always carry a dual meaning: atoms at the microscopic level and moles of atoms at the macroscopic level.

The validity of the molar interpretation for subscripts in a formula derives from the following line of reasoning. In x molecules of N_2O_4, where x is any number, there are $2x$ atoms of nitrogen and $4x$ atoms of oxygen. Regardless of the value of x, there must always be twice as many nitrogen atoms as molecules and four times as many oxygen atoms as molecules; that is,

$$\text{number of molecules} = x$$
$$\text{number of N atoms} = 2x$$
$$\text{number of O atoms} = 4x$$

Now let x equal 6.02×10^{23}, the value of Avogadro's number. With this x value, the following statements are true.

$$\text{number of molecules} = 6.02 \times 10^{23}$$
$$\text{number of N atoms} = 2 \times (6.02 \times 10^{23})$$
$$\text{number of O atoms} = 4 \times (6.02 \times 10^{23})$$

Because 6.02×10^{23} is equal to 1 mole, the preceding statements can be changed by substitution to read

$$\text{number of molecules} = 1 \text{ mole}$$
$$\text{number of N atoms} = 2 \text{ moles}$$
$$\text{number of O atoms} = 4 \text{ moles}$$

Thus, the mole ratio is the same as the subscript ratio: two to four. When it is necessary to know the number of moles of a particular element *within* a compound, the subscript of that element in the chemical formula becomes part of the conversion factor used to convert from moles of compound to moles of element *within* the compound. Using N_2O_4 as our chemical formula, we can write the following as conversion factors.

$$\text{For N:} \quad \frac{2 \text{ moles N atoms}}{1 \text{ mole } N_2O_4 \text{ molecules}} \quad \text{or} \quad \frac{1 \text{ mole } N_2O_4 \text{ molecules}}{2 \text{ moles N atoms}}$$

$$\text{For O:} \quad \frac{4 \text{ moles O atoms}}{1 \text{ mole } N_2O_4 \text{ molecules}} \quad \text{or} \quad \frac{1 \text{ mole } N_2O_4 \text{ molecules}}{4 \text{ moles O atoms}}$$

Example 4.4 illustrates the use of this type of conversion factor in a problem-solving context.

Example 4.4

How many moles of each type of atom are present in each of the following molar quantities?

a. 0.65 mole of H_2O molecules   b. 1.64 moles of B_4H_{10} molecules

Solution

a. One mole of H_2O contains two moles of hydrogen atoms and one mole of oxygen atoms. We obtain the following conversion factors from this statement.

$$\left(\frac{2 \text{ moles H atoms}}{1 \text{ mole } H_2O \text{ molecules}}\right) \quad \text{and} \quad \left(\frac{1 \text{ mole O atoms}}{1 \text{ mole } H_2O \text{ molecules}}\right)$$

Using the first conversion factor, the moles of hydrogen atoms present are calculated as follows.

$$0.65 \text{ mole } H_2O \text{ molecules} \times \left(\frac{2 \text{ moles H atoms}}{1 \text{ mole } H_2O \text{ molecules}}\right)$$

$$= 1.3 \text{ moles H atoms}$$

Similarly, using the second conversion factor, the moles of oxygen atoms present are calculated as follows.

$$0.65 \text{ mole } H_2O \text{ molecules} \times \left(\frac{1 \text{ mole O atoms}}{1 \text{ mole } H_2O \text{ molecules}}\right)$$

$$= 0.65 \text{ mole O atoms}$$

b. Interpreting the formula B_4H_{10} in terms of moles, we obtain the following conversion factors.

$$\frac{4 \text{ moles B atoms}}{1 \text{ mole } B_4H_{10} \text{ molecules}} \quad \text{and} \quad \frac{10 \text{ moles H atoms}}{1 \text{ mole } B_4H_{10} \text{ molecules}}$$

The setup for calculating the moles of boron atoms present is

$$1.64 \text{ moles } B_4H_{10} \text{ molecules} \times \left(\frac{4 \text{ moles B atoms}}{1 \text{ mole } B_4H_{10} \text{ molecules}}\right)$$

$$= 6.56 \text{ moles B atoms}$$

The setup for calculating the moles of hydrogen atoms present is

$$1.64 \text{ moles } B_4H_{10} \text{ molecules} \times \left(\frac{10 \text{ moles H atoms}}{1 \text{ mole } B_4H_{10} \text{ molecules}}\right)$$

$$= 16.4 \text{ moles H atoms}$$

4.5 THE MOLE AND CHEMICAL CALCULATIONS

In this section, we will combine the major points we have learned about moles to produce a general approach to problem solving that is applicable to a variety of types of chemical situations. In Section 4.2, we learned that *Avogadro's number* provides a relationship between the number of particles of a substance and the number of moles of that same substance.

In Section 4.3, we learned that *molar mass* provides a relationship between the number of grams of a substance and the number of moles of that substance.

In Section 4.4, we learned that *molar interpretation of chemical formula subscripts* provides a relationship between the number of moles of a substance and the number of moles of its component parts.

Moles of Compound ⟶ Formula subscripts ⟶ Moles of Element within Compound

The preceding three concepts can be combined into a single diagram that is very useful in problem solving. This diagram, Figure 4–3, can be viewed as a "road map" from which conversion factor sequences (pathways) may be obtained. It gives all of the needed relationships for solving two general types of problems:

1. calculations where information (moles, particles, or grams) is given about a particular substance and additional information (moles, particles, or grams) is needed concerning the *same* substance; and
2. calculations where information (moles, particles, or grams) is given about a particular substance and information is needed concerning a *component* of that same substance.

For the first type of problem, only the left side of Figure 4–3 (the "A" boxes) is needed. For problems of the second type, both sides of the diagram (both "A" and "B" boxes) are used.

FIGURE 4–3 Useful relationships in chemical formula-based problem solving situations.

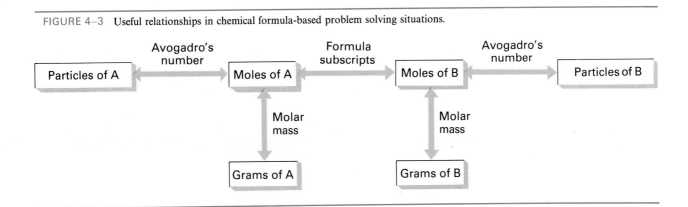

The thinking pattern needed to use Figure 4–3 is very simple.

1. Determine which box in the diagram represents the *given* quantity in the problem.
2. Next, locate the box that represents the *desired* quantity.
3. Finally, follow the indicated pathway that takes you from the given quantity to the desired quantity. This involves simply following the arrows. There will always be only one pathway possible for the needed transition.

Examples 4.5 through 4.7 indicate some of the types of problems that can be solved using the relationships shown in Figure 4–3.

Example 4.5

Vitamin C has the formula $C_6H_8O_6$. Calculate the number of vitamin C molecules present in a 0.250 g (250 mg) tablet of vitamin C.

Solution

We will solve this problem by using the three steps of dimensional analysis (Section 1.11) and Figure 4–3.

Step 1: The given quantity is 0.250 g of $C_6H_8O_6$ and the desired quantity is molecules of $C_6H_8O_6$.

$$0.250 \text{ g } C_6H_8O_6 = ? \text{ molecules } C_6H_8O_6$$

In terms of Figure 4–3, this is a "grams of A" to "particles of A" problem. We are given grams of a substance, A, and desire to find molecules (particles) of that same substance.

Step 2: Figure 4–3 gives us the pathway needed to solve this problem. Starting with "grams of A" we convert to "moles of A" and finally reach "particles of A". The arrows between the boxes along our path give the type of conversion factor needed for each step.

$$\text{grams A} \xrightarrow{\text{molar mass}} \text{moles A} \xrightarrow{\text{Avogadro's number}} \text{particles A}$$

Using dimensional analysis, the setup for this sequence of conversion factors is

$$0.250 \text{ g } C_6H_8O_6 \times \left(\frac{1 \text{ mole } C_6H_8O_6}{176 \text{ g } C_6H_8O_6}\right) \times \left(\frac{6.02 \times 10^{23} \text{ molecules } C_6H_8O_6}{1 \text{ mole } C_6H_8O_6}\right)$$

$$\text{g } C_6H_8O_6 \longrightarrow \text{moles } C_6H_8O_6 \longrightarrow \text{molecules } C_6H_8O_6$$

The number 176 that is used in the first conversion factor is the formula weight of $C_6H_8O_6$. It was not given in the problem, but had to be calculated using atomic weights and the method for calculating formula weights shown in Example 4.1.

Step 3: The solution to the problem, obtained by doing the arithmetic is

$$\frac{0.250 \times 1 \times 6.02 \times 10^{23}}{176 \times 1} \text{ molecules } C_6H_8O_6 = 8.55 \times 10^{20} \text{ molecules } C_6H_8O_6$$

Example 4.6

What is the mass, in grams, of a single aspirin molecule? The formula for aspirin is $C_9H_8O_4$.

Solution

Step 1: The given quantity is a single aspirin molecule, that is, one aspirin molecule. The desired quantity is grams of aspirin.

$$1 \text{ C}_9\text{H}_8\text{O}_4 \text{ molecule} = ? \text{ g C}_9\text{H}_8\text{O}_4$$

In terms of Figure 4–3, this is a "particles of A" to "grams of A" problem.

Step 2: The pathway for this problem is the exact reverse of the one used in the previous example. We are given particles and asked to find grams of the same substance.

$$\text{particles A} \xrightarrow{\text{Avogadro's number}} \text{moles A} \xrightarrow{\text{molar mass}} \text{grams A}$$

Using dimensional analysis, the setup is

$$1 \text{ molecule C}_9\text{H}_8\text{O}_4 \times \left(\frac{1 \text{ mole C}_9\text{H}_8\text{O}_4}{6.02 \times 10^{23} \text{ molecules C}_9\text{H}_8\text{O}_4}\right) \times \left(\frac{180 \text{ g C}_9\text{H}_8\text{O}_4}{1 \text{ mole C}_9\text{H}_8\text{O}_4}\right)$$

The formula weight for aspirin that was used in the second conversion factor was calculated using the atomic weights of carbon, hydrogen, and oxygen and the formula for aspirin.

Step 3: The final answer is obtained by doing the arithmetic.

$$\left(\frac{1 \times 1 \times 180}{6.02 \times 10^{23} \times 1}\right) \text{g C}_9\text{H}_8\text{O}_4 = 2.99 \times 10^{-22} \text{ g C}_9\text{H}_8\text{O}_4$$

Example 4.7

The hallucinogenic drug LSD has the formula $C_{23}H_{30}N_3O$. How many grams of nitrogen are present in a 0.10-g sample of LSD?

Solution

Step 1: There is an important difference between this problem and the preceding two; here we are dealing with not one but two substances, LSD and nitrogen. The given quantity is grams of LSD (substance A) and we are asked to find the grams of nitrogen (substance B). This is a "grams of A" to "grams of B" problem.

$$0.10 \text{ g C}_{23}\text{H}_{30}\text{N}_3\text{O} = ? \text{ g N}$$

Step 2: The appropriate set of conversions for a "grams of A" to "grams of B" problem, from Figure 4–3, is

$$\text{grams A} \xrightarrow{\text{molar mass}} \text{moles A} \xrightarrow{\text{formula subscripts}} \text{moles B} \xrightarrow{\text{molar mass}} \text{grams B}$$

The mathematical setup is

$$0.10 \text{ g LSD} \times \left(\frac{1 \text{ mole LSD}}{376 \text{ g LSD}}\right) \times \left(\frac{3 \text{ moles N}}{1 \text{ mole LSD}}\right) \times \left(\frac{14.0 \text{ g N}}{1 \text{ mole N}}\right)$$

The number 376 that is used in the first conversion factor is the formula weight for LSD. The conversion from "moles of A" to "moles of B" (the second conversion factor) is made using the information contained in the formula $C_{23}H_{30}N_3O$. One mole of LSD contains three moles of nitrogen. The number 14.0 in the final conversion factor is the atomic weight of nitrogen.

Step 3: Collecting the numbers from the various conversion factors together and doing the arithmetic gives us our answer.

$$\left(\frac{0.10 \times 1 \times 3 \times 14.0}{376 \times 1 \times 1}\right) \text{g N} = 0.011 \text{ g N}$$

4.6 WRITING AND BALANCING EQUATIONS

chemical equation

A **chemical equation** *is a written statement that uses symbols and formulas instead of words to describe the changes that occur in a chemical reaction.* The following example shows the contrast between a word description of a chemical reaction and a chemical equation for the same reaction.

Word description: calcium sulfide reacts with water to produce calcium oxide and hydrogen sulfide.

Chemical equation:

$$CaS + H_2O \longrightarrow CaO + H_2S$$

In the same way that chemical symbols are considered the *letters* of chemical language and formulas the *words* of the language, chemical equations can be considered the *sentences* of chemical language.

The conventions used in writing chemical equations are:

1. The correct formulas of the *reactants* (starting materials) are always written on the *left* side of the equation.

$$\boxed{CaS} + \boxed{H_2O} \longrightarrow CaO + H_2S$$

2. The correct formulas of the *products* (substances produced) are always written on the *right* side of the equation.

$$CaS + H_2O \longrightarrow \boxed{CaO} + \boxed{H_2S}$$

3. The reactants and products are separated by an arrow pointing toward the products.

$$CaS + H_2O \boxed{\longrightarrow} CaO + H_2S$$

4. Plus signs are used to separate different reactants or different products from each other.

$$CaS \boxed{+} H_2O \longrightarrow CaO \boxed{+} H_2S$$

When reading chemical equations, plus signs on the reactant side of the equation mean "reacts with", the arrow means "to produce"; and plus signs on the product side mean "and".

A *valid* chemical equation must satisfy two conditions.

1. *It must be consistent with experimental facts.* Only the reactants and products that are actually involved in a reaction are shown in an equation. An accurate formula must be used for each of these substances. Elements in solid and liquid states are represented in equations by the chemical symbol for the element. Elements that are gases at room temperature are represented by the molecular formula denoting the form in which they actually occur in nature. The following monatomic, diatomic, and tetraatomic elemental gases are known.
   Monatomic: He, Ne, Ar, Kr, Xe
   Diatomic: H_2, O_2, N_2, F_2, Cl_2, Br_2 (vapor), I_2 (vapor)
   Tetraatomic: P_4 (vapor), As_4 (vapor)*

2. *It must be consistent with the law of conservation of matter* (Section 3.3). There must be the same number of product atoms of each kind as there are reactant atoms of each kind because atoms are neither created nor destroyed in an ordinary chemical reaction. Equations that satisfy the conditions of this law are said to be *balanced*. Using the conventions previously listed for writing equations does not guarantee a balanced equation.

---

*The four elements listed as vapors are not gases at room temperature but vaporize at slightly higher temperatures. The resultant vapors contain molecules with the formulas indicated.

We will now consider how an unbalanced equation is brought into balance. A **balanced chemical equation** has the same number of atoms of each element involved in the reaction on each side of the equation. An unbalanced equation is brought into balance by adding *coefficients* to the equation, with the coefficients adjusting the number of reactant or product molecules present. A **coefficient** *is a number placed to the left of the formula of a substance that changes the amount, but not the identity, of the substance.* In the notation 2 H_2O, the "2" on the left is a coefficient; 2 H_2O means two molecules of H_2O and 3 H_2O means three molecules of H_2O. Thus, coefficients tell how many formula units of a given substance are present.

balanced chemical equation

coefficient

The following is a balanced chemical equation with the coefficients shown in color.

$$4\,NH_3 + 3\,O_2 \longrightarrow 2\,N_2 + 6\,H_2O$$

The message of this balanced equation is: Four NH_3 molecules react with three O_2 molecules to produce two N_2 molecules and six H_2O molecules. A coefficient of "1" in a balanced equation is not explicitly written; it is considered to be understood. Both Na_2SO_4 and Na_2S have understood coefficients of 1 in the following balanced equation.

$$Na_2SO_4 + 2\,C \longrightarrow Na_2S + 2\,CO_2$$

A coefficient placed in front of a formula applies to the whole formula. In contrast, subscripts, which are also present in formulas, affect only parts of a formula.

$$2\,H_2O$$

— coefficient (affects both H and O)
— subscript (affects only H)

The preceding notation denotes two molecules of H_2O; it also denotes a total of four H atoms and two O atoms.

We will now proceed to the mechanics involved in determining the proper coefficients needed to balance an equation. They will be introduced in Example 4.8. This example should be studied carefully because detailed commentary is given concerning the ins and outs of balancing equations.

Balance the equation

$$FeI_2 + Cl_2 \longrightarrow FeCl_3 + I_2$$

Example 4.8

Solution
Step 1: Examine the equation and pick one element to balance first. It is often convenient to start with the compound that contains the greatest number of atoms, whether a reactant or product, and key in on the element in that compound that has the greatest number of atoms. Using this guideline, we select $FeCl_3$ and the element chlorine within it.

We note that there are three chlorine atoms on the right side of the equation and two atoms of chlorine on the left (in Cl_2). In order for the chlorine atoms to balance, we will need six on each side; six is the lowest number that both three and two will divide into evenly. In order to obtain six atoms on each side of the equation, we place the coefficient 3 in front of Cl_2 and the coefficient 2 in front of $FeCl_3$.

$$FeI_2 + \boxed{3}\,Cl_2 \longrightarrow \boxed{2}\,FeCl_3 + I_2$$

We now have six chlorine atoms on each side of the equation.

$$3\,Cl_2: \quad 3 \times 2 = 6$$
$$2\,FeCl_3: \quad 2 \times 3 = 6$$

Step 2: Now pick a second element to balance. We will balance the iron next. The number of iron atoms on the right side has already been set at two by the coefficient previously placed in front of $FeCl_3$. We will need two iron atoms on the reactant side of the equation instead of the one iron atom now present. This is accomplished by placing the coefficient 2 in front of FeI_2.

$$\boxed{2}\,FeI_2 + 3\,Cl_2 \longrightarrow 2\,FeCl_3 + I_2$$

It is always wise to pick as the second element to balance one whose amount is already set on one side of the equation by a previously determined coefficient. If we had chosen iodine as the second element to balance instead of iron, we would have run into problems. Because neither the coefficient for FeI_2 nor I_2 had been determined, we would have had no guidelines for deciding upon the amount of iodine needed.

Step 3: Now pick a third element to balance. Only one element is left to balance—iodine. The number of iodine atoms on the left side of the equation is already set at four ($2\,FeI_2$). In order to obtain four iodine atoms on the right side of the equation, we place the coefficient 2 in front of I_2

$$2\,FeI_2 + 3\,Cl_2 \longrightarrow 2\,FeCl_3 + \boxed{2}\,I_2$$

The addition of the coefficient 2 in front of I_2 completes the balancing process; all of the coefficients have been determined.

Step 4: As a final check on the correctness of the balancing procedure, count atoms on each side of the equation. The following table can be constructed from our balanced equation.

$$2\,FeI_2 + 3\,Cl_2 \longrightarrow 2\,FeCl_3 + 2\,I_2$$

Atom	Left Side	Right Side
Fe	$2 \times 1 = 2$	$2 \times 1 = 2$
I	$2 \times 2 = 4$	$2 \times 2 = 4$
Cl	$3 \times 2 = 6$	$2 \times 3 = 6$

All elements are in balance: two iron atoms on each side, four iodine atoms on each side, and six chlorine atoms on each side.

Notice that the elements chlorine and iodine in the equation are written in the form of diatomic molecules (Cl_2 and I_2). This is in accordance with the guideline given at the start of this section on the use of molecular formulas for elements that are gases at room temperature.

Some additional comments and guidelines concerning equations in general and the process of balancing in particular are:

1. The coefficients in a balanced equation are always the *smallest set of whole numbers* that will balance the equation. We mention this because more than one set of coefficients will balance an equation. Consider the following three equations.

$$2\,H_2 + O_2 \longrightarrow 2\,H_2O$$
$$4\,H_2 + 2\,O_2 \longrightarrow 4\,H_2O$$
$$8\,H_2 + 4\,O_2 \longrightarrow 8\,H_2O$$

All three of these equations are mathematically correct; there are equal numbers of hydrogen and oxygen atoms on each side of the equation. However, the first equation is considered the conventional form because the coefficients used there are the smallest set of whole numbers that will balance the equation. The coefficients in the second equation are two times those in the first equation, and the third equation has coefficients that are four times those of the first equation.

2. At this point, you are not expected to be able to write down the products for a chemical reaction when given the reactants. After learning how to balance equations, students sometimes get the mistaken idea that they ought to be able to write down equations from scratch. This is not so. You will need more chemical knowledge before attempting this task. At this stage, you should be able to balance simple equations, given *all* of the reactants and *all* of the products.

4.7 CHEMICAL EQUATIONS AND THE MOLE CONCEPT

The coefficients in a balanced chemical equation, like the subscripts in a chemical formula (Section 4.4), have two levels of interpretation—a microscopic level of meaning and a macroscopic level of meaning.

The microscopic level of interpretation was used in the previous section. At this level, a balanced chemical equation gives the relative number of formula units of the various reactants and products involved in a chemical reaction. *The coefficients in the equation give the numerical relationships among formula units consumed (used up) or produced in the chemical reaction.* Interpreted at the microscopic level, the equation

$$N_2 + 3\,H_2 \longrightarrow 2\,NH_3$$

conveys the information that one molecule of N_2 reacts with three molecules of H_2 to produce two molecules of NH_3.

At the macroscopic level of interpretation, chemical equations are used to relate mole-sized quantities of reactants and products to each other. At this level, *the coefficients in the equation give the fixed molar ratios between substances consumed or produced in the chemical reaction.* Interpreted at the macroscopic level, the equation

$$N_2 + 3\,H_2 \longrightarrow 2\,NH_3$$

conveys the information that one mole of N_2 reacts with three moles of H_2 to produce two moles of NH_3.

The validity of the molar interpretation of coefficients in an equation can be derived in a straightforward manner, with the microscopic level of interpretation as a starting point. A balanced chemical equation remains valid (mathematically correct) when each of its coefficients is multiplied by the same number. (If molecules react in a three-to-one ratio, they will also react in a six-to-two or a nine-to-three ratio.) Thus, using the previous equation and multiplying by y, where y is any number, we have

$$(y)\,N_2 + (3y)\,H_2 \longrightarrow (2y)\,NH_3$$

The situation where y is equal to 6.02×10^{23} is of particular interest because 6.02×10^{23} is equal to one mole. Using y is equal to one mole (6.02×10^{23}), we have by substitution

$$\boxed{1 \text{ mole}} \text{ N}_2 + \boxed{3 \text{ moles}} \text{ H}_2 \longrightarrow \boxed{2 \text{ moles}} \text{ NH}_3$$

As with the subscripts in formulas, the coefficients in equations carry a dual meaning; at the microscopic level, coefficients mean number of formula units and at the macroscopic level, they mean moles of formula units.

The coefficients in an equation can be used to generate conversion factors to be used in solving problems. Numerous conversion factors are obtainable from a single balanced equation. Consider the following balanced equation.

$$4 \text{ Fe} + 3 \text{ O}_2 \longrightarrow 2 \text{ Fe}_2\text{O}_3$$

Three mole-to-mole relationships are obtainable from this equation.

$$4 \text{ moles of Fe produces 2 moles of Fe}_2\text{O}_3$$
$$3 \text{ moles of O}_2 \text{ produces 2 moles of Fe}_2\text{O}_3$$
$$4 \text{ moles of Fe reacts with 3 moles of O}_2$$

From these three macroscopic level relationships, six conversion factors can be written.

From the first relationship:

$$\left(\frac{4 \text{ moles Fe}}{2 \text{ moles Fe}_2\text{O}_3}\right) \quad \text{and} \quad \left(\frac{2 \text{ moles Fe}_2\text{O}_3}{4 \text{ moles Fe}}\right)$$

From the second relationship:

$$\left(\frac{3 \text{ moles O}_2}{2 \text{ moles Fe}_2\text{O}_3}\right) \quad \text{and} \quad \left(\frac{2 \text{ moles Fe}_2\text{O}_3}{3 \text{ moles O}_2}\right)$$

From the third relationship:

$$\left(\frac{4 \text{ moles Fe}}{3 \text{ moles O}_2}\right) \quad \text{and} \quad \left(\frac{3 \text{ moles O}_2}{4 \text{ moles Fe}}\right)$$

All balanced chemical equations are the source of numerous conversion factors. The more reactants and products there are in the equation, the greater the number of derivable conversion factors. The next section details how conversion factors like those in the preceding illustration are used in solving problems.

4.8 CHEMICAL CALCULATIONS USING CHEMICAL EQUATIONS

If the information contained in a chemical equation is combined with the concepts of molar mass (Section 4.3) and Avogadro's number (Section 4.2), many useful types of chemical calculations can be carried out. A typical chemical equation-based calculation gives information about one reactant or product of a reaction (number of grams, moles, or particles) and requests information about another reactant or product of the same reaction. The substances involved in such a calculation may both be reactants or products, or may be a reactant and a product.

The conversion factor relationships needed to solve problems of this general type are given in Figure 4–4. This diagram should seem very familiar to you; it is almost identical to Figure 4–3, which you worked with to solve problems based on chemical formulas. There is only one difference between the two diagrams. In Figure 4–3, the subscripts in a chemical formula are listed as the basis for relating "moles of A" to "moles of B". In Figure 4–4, the same two quantities are related using the coefficients of a balanced chemical equation.

FIGURE 4–4 Useful relationships in chemical equation-based problem solving situations.

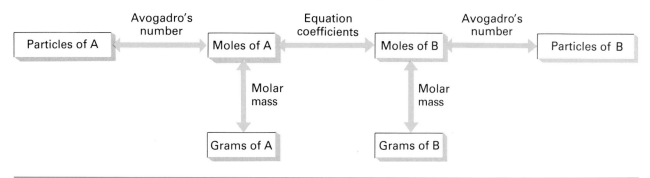

The most common type of chemical equation-based calculation is a "grams of A" to "grams of B" problem. In this type of problem, the mass of one substance involved in a chemical reaction (either reactant or product) is given and information is requested about the mass of another substance involved in the reaction (either reactant or product). This type of problem is frequently encountered in laboratory settings. For example, a chemist may have a certain number of grams of a chemical available and wants to know how many grams of another substance can be produced from it or how many grams of a third substance are needed to react with it. Examples 4.9 and 4.10 illustrate this type of problem.

Example 4.9

The human body converts the glucose, $C_6H_{12}O_6$, contained in foods to carbon dioxide, CO_2, and water, H_2O. The equation for the chemical reaction is

$$C_6H_{12}O_6 \longrightarrow 6\,CO_2 + 6\,H_2O$$

If a candy bar containing 14.2 g (1/2 oz) of glucose is eaten by a person, how many grams of water will the body produce from the ingested glucose, assuming all of the glucose undergoes reaction?

Solution
Step 1: The given quantity is 14.2 g of glucose. The desired quantity is grams of water.

$$14.2\text{ g }C_6H_{12}O_6 = ?\text{ g }H_2O$$

In terms of Figure 4–4, this is a "grams of A" to "grams of B" problem.
Step 2: Using Figure 4–4 as a road map, we determine that the pathway for this problem is

$$\text{grams A} \xrightarrow{\text{molar mass}} \text{moles A} \xrightarrow{\text{equation coefficients}} \text{moles B} \xrightarrow{\text{molar mass}} \text{grams B}$$

The mathematical setup for this problem is

$$14.2\text{ g }C_6H_{12}O_6 \times \left(\frac{1\text{ mole }C_6H_{12}O_6}{180\text{ g }C_6H_{12}O_6}\right) \times \left(\frac{6\text{ moles }H_2O}{1\text{ mole }C_6H_{12}O_6}\right) \times \left(\frac{18.0\text{ g }H_2O}{1\text{ mole }H_2O}\right)$$

$$\text{g }C_6H_{12}O_6 \longrightarrow \text{moles }C_6H_{12}O_6 \longrightarrow \text{moles }H_2O \longrightarrow \text{g }H_2O$$

The 180 g in the first conversion factor is the molar mass of glucose, the 6 and 1 in the second conversion factor are the coefficients, respectively, of H_2O and

$C_6H_{12}O_6$ in the balanced chemical equation, and the 18.0 g in the third conversion factor is the molar mass of H_2O.

Step 3: The solution to the problem obtained by doing the arithmetic after all of the numerical factors have been collected is

$$\left(\frac{14.2 \times 1 \times 6 \times 18.0}{180 \times 1 \times 1}\right) \text{g } H_2O = 8.52 \text{ g } H_2O$$

Example 4.10

The active ingredient in many commercial antacids is magnesium hydroxide, $Mg(OH)_2$, which reacts with stomach acid (HCl) to produce magnesium chloride ($MgCl_2$) and water. The equation for the reaction is

$$Mg(OH)_2 + 2\ HCl \longrightarrow MgCl_2 + 2\ H_2O$$

How many grams of $Mg(OH)_2$ are needed to react with 0.30 g of HCl?

Solution

Step 1: This problem, like Example 4.9, is also a "grams of A" to "grams of B" problem. It differs from the previous problem in that both the given and desired quantities involve reactants.

$$0.30 \text{ g HCl} \longrightarrow ? \text{ g } Mg(OH)_2$$

Step 2: The pathway used to solve it will be the same as in Example 4-9.

$$\text{grams A} \xrightarrow{\text{molar mass}} \text{moles A} \xrightarrow{\text{equation coefficients}} \text{moles B} \xrightarrow{\text{molar mass}} \text{grams B}$$

The dimensional-analysis setup is

$$0.30 \text{ g HCl} \times \left(\frac{1 \text{ mole HCl}}{36.5 \text{ g HCl}}\right) \times \left(\frac{1 \text{ mole } Mg(OH)_2}{2 \text{ moles HCl}}\right) \times \left(\frac{58.3 \text{ g } Mg(OH)_2}{1 \text{ mole } Mg(OH)_2}\right)$$

$$\text{g HCl} \longrightarrow \text{moles HCl} \longrightarrow \text{moles } Mg(OH)_2 \longrightarrow \text{g } Mg(OH)_2$$

The balanced chemical equation for the reaction is used as the bridge that enables us to go from HCl to $Mg(OH)_2$. The numbers in the second conversion factor are coefficients from this equation.

Step 3: The solution obtained from combining all of the numbers in the manner indicated in the setup is

$$\left(\frac{0.30 \times 1 \times 1 \times 58.3}{36.5 \times 2 \times 1}\right) \text{g } Mg(OH)_2 = 0.24 \text{ g } Mg(OH)_2$$

To put our answer in perspective, we note that a common brand of antacid tablets has tablets containing 0.10 g of $Mg(OH)_2$.

EXERCISES AND PROBLEMS

Formula Weights

4.1 Calculate the formula weight of each of the following substances. Obtain the atomic weights from the inside back cover of the text.
a. $NaHCO_3$ (baking soda)
b. $C_{12}H_{22}O_{11}$ (table sugar)
c. $(NH_4)_2SO_4$ (a lawn fertilizer)
d. $C_7H_5NO_3S$ (saccharin, an artificial sweetener)
e. $Na_2S_2O_3$ (a photographic chemical)
f. $C_9H_8O_4$ (naphthalene, an ingredient in mothballs)
g. H_2SO_4 (an industrial acid)
h. $(NH_4)_2B_4O_7$ (a fire-retardant for textiles)

4.2 Calculate the formula weight of each of the following substances. Obtain the atomic weights from the inside back cover of the text.
a. $C_{14}H_9Cl_5$ (DDT, an insecticide)
b. $C_{20}H_{30}O$ (vitamin A)
c. $C_8H_{10}N_4O_2$ (caffeine)
d. $C_2H_4Cl_2$ (a dry-cleaning solvent)
e. $KC_{16}H_{31}O_2$ (a component of liquid soaps)
f. $C_6H_8N_2$ (a black dye)
g. P_4S_3 (tip of some matches)
h. Ag_2S (silverware tarnish)

The Mole as a Counting Unit

4.3 What is the relationship between Avogadro's number and the mole counting unit?

4.4 How many molecules are present in each of the following amounts of substance?
a. 2.00 moles of H_2O molecules
b. 3.25 moles of CO_2 molecules
c. 0.356 mole of CO molecules
d. Avogadro's number of CO molecules

4.5 You are given a sample containing 0.542 mole of a substance.
a. How many atoms are present if the sample is copper metal?
b. How many atoms are present if the sample is iron metal?
c. How many molecules are present if the sample is nitric acid, HNO_3?
d. How many molecules are present if the sample is aspirin, $C_9H_8O_4$?

Molar Mass

4.6 State the definition for the term *molar mass*.

4.7 How much does 1.00 mole of each of the following substances weigh in grams?
a. CO (carbon monoxide)
b. CO_2 (carbon dioxide)
c. H_2O (water)
d. H_2O_2 (hydrogen peroxide)
e. NaCl (table salt)
f. $C_{12}H_{22}O_{11}$ (table sugar)
g. NaCN (sodium cyanide)
h. KCN (potassium cyanide)

4.8 What is the mass, in grams, of each of the following quantities of matter?
a. 2.00 moles of gold atoms
b. 3.75 moles of gold atoms
c. 0.0034 mole of gold atoms
d. 0.0034 mole of silver atoms
e. 3.00 moles of oxygen atoms
f. 3.00 moles of oxygen molecules (O_2)
g. 2.25 moles of NH_3 molecules
h. 2.25 moles of H_2O molecules

4.9 How many moles are present in a sample of each of the following substances if each sample weighs 5.00 g?
a. CO molecules
b. CO_2 molecules
c. B_4H_{10} molecules
d. U atoms
e. Cl atoms
f. Cl_2 molecules
g. O_2 molecules
h. O_3 molecules

4.10 Explain why an easier number to work with, such as the numbers 1.0×10^{23} or 1.0×10^{24}, was not selected as Avogadro's number.

4.11 The mole is the heart of chemical calculations. Why hasn't a balance been invented that directly determines the number of moles in a sample, rather than the number of grams?

Molar Interpretation of Chemical Formulas

4.12 Construct conversion factors that relate the numbers of moles of atoms of hydrogen, sulfur, and oxygen to one mole of H_2SO_4 molecules.

4.13 List the number of moles of each type of atom that are present in each of the following molar quantities.
a. 2.00 moles of SO_2 molecules
b. 2.00 moles of SO_3 molecules
c. 3.00 moles of NH_3 molecules
d. 3.00 moles of N_2H_4 molecules
e. 4.00 moles of B_4H_{10} molecules
f. 3.25 moles of $Al(OH)_3$ formula units
g. 6.00 moles of O_3 molecules
h. 1.02 moles of $C_{12}H_{22}O_{11}$ molecules

4.14 How many *total* moles of *atoms* are present in the following amounts of substances?
a. 2.00 moles of H_2O
b. 2.00 moles of H_2O_2
c. 2.00 moles of $Ca(NO_3)_2$
d. 2.00 moles of $Al(NO_3)_3$
e. 2.00 moles of $Fe_2(SO_4)_3$
f. 2.00 moles of P_4
g. 2.00 moles of NH_3
h. 2.00 moles of HN_3

Chemical Formula-Based Calculations

4.15 Determine the number of atoms in each of the following quantities of an element.
a. 10.8 g of B
b. 40.1 g of Ca
c. 2.0 g of Ne
d. 7.0 g of N
e. 10.0 g of S
f. 3.2 g of U

4.16 Determine the mass in grams of each of the following quantities of substances.
a. 6.02×10^{23} copper atoms
b. 3.01×10^{23} copper atoms
c. 4.7×10^{15} copper atoms
d. 27 copper atoms
e. 6.7×10^{23} molecules of H_2O
f. 1 molecule of H_2O

4.17 Determine the number of moles of substance present in each of the following quantities.
a. 10.0 g of He
b. 10.0 g of N_2O
c. 10.0 g of Al_2S_3
d. 4.0×10^{10} atoms of P
e. 4.0×10^{20} atoms of P
f. 4.0×10^{30} atoms of P

4.18 Determine the number of atoms of sulfur present in each of the following quantities.
a. 10.0 g H_2SO_4
b. 20.0 g SO_3
c. 30.0 g S_4N_4
d. 40.0 g $Al_2(SO_4)_3$
e. 2.00 moles S_8
f. 0.35 mole S_2O

4.19 Determine the number of grams of oxygen present in each of the following quantities.
a. 3.01×10^{23} SO_2 molecules
b. 3.01×10^{23} S_2O molecules
c. 3.01×10^{23} SO_3 molecules
d. 3.01×10^{23} SO_2Cl_2 molecules
e. 2.00 moles S_2O molecules
f. 2.00 moles SO_2 molecules

4.20 The active ingredient in the illegal drug marijuana is the compound tetrahydrocannabinol, which is commonly called THC. The formula for THC is $C_{21}H_{36}O_2$. Determine the number of grams of THC that you need to obtain
a. 1 million (1.0×10^6) molecules
b. 1 million (1.0×10^6) C atoms
c. 1 million (1.0×10^6) H atoms
d. 1 million (1.0×10^6) O atoms
e. 1 million (1.0×10^6) total atoms

4.21 Enflurane, which has the formula $C_3H_2ClF_5O$, is widely used as a surgical anesthetic. A 1.00-g sample of this compound contains how many
a. moles of the compound?
b. individual molecules of the compound?
c. moles of fluorine atoms?
d. carbon atoms?
e. grams of hydrogen?
f. total atoms?

4.22 Two of the most widely used antianxiety drugs are the tranquilizers Librium and Valium. Structurally, they are very similar, with formulas of $C_{16}H_{14}N_3OCl$ and $C_{16}H_{13}N_2OCl$ respectively. Comparing 10.00-g samples of these two compounds, which sample contains the
a. greater number of moles of molecules?
b. greater number of carbon atoms?
c. greater mass, in grams, of nitrogen?
d. smaller mass, in grams, of hydrogen?

Writing and Balancing Chemical Equations

4.23 The formulas used in equations for gaseous elements should reflect their molecular makeup. Which gases (or vapors) should be written as
a. monatomic molecules?
b. diatomic molecules?
c. tetraatomic molecules?

4.24 What symbol is used in a chemical equation to represent the phrase "to produce"?

4.25 Show that the law of conservation of matter is observed for the equation

$$4\,NH_3 + 5\,O_2 \longrightarrow 4\,NO + 6\,H_2O$$

4.26 Balance the following equations.
a. $H_2 + O_2 \longrightarrow H_2O$
b. $NO + O_2 \longrightarrow NO_2$
c. $Al + O_2 \longrightarrow Al_2O_3$
d. $Fe_2O_3 \longrightarrow Fe + O_2$
e. $CaCO_3 \longrightarrow CaO + CO_2$
f. $KClO_3 \longrightarrow KCl + O_2$

4.27 Balance the following equations.
a. $CH_4 + O_2 \longrightarrow CO_2 + H_2O$
b. $C_2H_4 + O_2 \longrightarrow CO_2 + H_2O$
c. $C_3H_8 + O_2 \longrightarrow CO_2 + H_2O$
d. $C_6H_{12} + O_2 \longrightarrow CO_2 + H_2O$

4.28 Balance the following equations.
a. $PbO + NH_3 \longrightarrow Pb + N_2 + H_2O$
b. $SO_2Cl_2 + HI \longrightarrow H_2S + H_2O + HCl + I_2$
c. $Fe(OH)_3 + H_2SO_4 \longrightarrow Fe_2(SO_4)_3 + H_2O$
d. $Na_2CO_3 + Mg(NO_3)_2 \longrightarrow MgCO_3 + NaNO_3$

Chemical Equations and the Mole Concept

4.29 Write the twelve mole–to–mole conversion factors that can be derived from each of the following balanced equations.
a. $2\,Ag_2CO_3 \longrightarrow 4\,Ag + 2\,CO_2 + O_2$
b. $N_2H_4 + 2\,H_2O_2 \longrightarrow N_2 + 4\,H_2O$

4.30 How many moles of the first listed product in each of the following equations could be obtained by reacting 0.750 mole of the first listed reactant with an excess of the other reactant?
a. $FeO + CO \longrightarrow Fe + CO_2$
b. $3\,O_2 + CS_2 \longrightarrow CO_2 + 2\,SO_2$

c. $6\ HCl + 2\ Al \longrightarrow 3\ H_2 + 2\ AlCl_3$
d. $4\ Fe_3O_4 + O_2 \longrightarrow 6\ Fe_2O_3$

4.31 Using each of the following equations, calculate the number of moles of CO_2 that can be obtained from 2.00 moles of the first listed reactant and an excess of all other reactants.

a. $C_7H_{16} + 11\ O_2 \longrightarrow 7\ CO_2 + 8\ H_2O$
b. $2\ HCl + CaCO_3 \longrightarrow CaCl_2 + CO_2 + H_2O$
c. $Na_2SO_4 + 2\ C \longrightarrow Na_2S + 2\ CO_2$
d. $Fe_3O_4 + CO \longrightarrow 3\ FeO + CO_2$

Chemical Equation-Based Calculations

4.32 How many grams of the first reactant in each of the following equations would be needed to produce 20.0 g of N_2 gas?

a. $4\ NH_3 + 3\ O_2 \longrightarrow 2\ N_2 + 6\ H_2O$
b. $(NH_4)_2Cr_2O_7 \longrightarrow Cr_2O_3 + N_2 + 4\ H_2O$
c. $N_2H_4 + 2\ H_2O_2 \longrightarrow N_2 + 4\ H_2$
d. $2NH_3 \longrightarrow N_2 + 3\ H_2$

4.33 The principal constituent of natural gas is methane, which burns in air according to the reaction

$$CH_4 + 2\ O_2 \longrightarrow CO_2 + 2\ H_2O$$

How many grams of CO_2 are produced when 2.0 g of CH_4 is burned in air?

4.34 The catalytic converter now required on American automobiles converts carbon monoxide (CO) to carbon dioxide (CO_2) by the reaction

$$2\ CO + O_2 \longrightarrow 2\ CO_2$$

a. What mass of O_2 in grams is needed to completely react with 100 g of CO?
b. What mass of CO_2 in grams is produced when 20.0 g of CO undergo reaction?

4.35 Both water and sulfur dioxide are products from the reaction of sulfuric acid (H_2SO_4) with copper as shown by the equation

$$2\ H_2SO_4 + Cu \longrightarrow SO_2 + 2\ H_2O + CuSO_4$$

How many grams of H_2O will be produced at the same time that 10.0 g of SO_2 is produced?

4.36 A common laboratory method for preparing oxygen gas (O_2) involves decomposing potassium chlorate ($KClO_3$) as shown by the equation

$$2\ KClO_3 \longrightarrow 2\ KCl + 3\ O_2$$

Based on this equation, how many grams of $KClO_3$ must be decomposed to produce

a. 3.00 moles O_2?
b. 3.00 g O_2?
c. 3.0×10^{23} molecules O_2?
d. 6.25 moles KCl?
e. 0.350 g KCl?
f. 4.2×10^{20} formula units KCl?

5 The Electronic Structure of Atoms

Objectives

After Completing Chapter 5, you will be able to:

5.1 Explain what is meant by the phrase "the energy of an electron is quantized".
5.2 Understand how the term *shell* is used in describing electron arrangements.
5.3 Understand how the term *subshell* is used in describing electron arrangements.
5.4 Understand how the term *orbital* is used in describing electron arrangements.
5.5 Write the electron configuration for any element using the Aufbau principle.
5.6 Explain the relationship between the periodic law and electron configurations.
5.7 Understand the relationship between the shape of the periodic table and electron configurations, and also be able to use the periodic table as a guide in writing electron configurations.
5.8 Classify a given element as a noble gas, representative element, transition element, or inner-transition element, and also be able to classify a given element as a metal or nonmetal.
5.9 Know which metals and nonmetals are represented among the twenty-six elements that are necessary to sustain life.

INTRODUCTION

In this chapter we will consider how an atom's electrons are arranged about its nucleus. This is an important subject because it is the arrangement of electrons about the nucleus that determines an element's chemical properties, that is, its behavior toward other substances.

In Section 3.4 we learned the following facts about electrons:

1. They are one of the three fundamental subatomic particles.
2. They have very little mass in comparison to protons and neutrons.
3. They are found outside the nucleus of an atom.
4. They move rapidly about the nucleus in a volume that defines the size of the atom.

Much more must be known about electrons in order to understand the chemical behavior of the various elements.

5.1 THE ENERGY OF AN ELECTRON

Information about the behavior of electrons as they move about the nucleus comes from a very complex mathematical theory called *quantum mechanics*. The theory of quantum mechanics, which deals with the laws that govern the motion of very small particles like electrons, is beyond the scope of an introductory course. However, some of the concepts obtained from this theory are simple enough for us to easily understand. These concepts will enable us to understand the basis of the periodic law and periodic table, as well as the basic rules governing compound formation (Chapter 6).

Quantum-mechanical theory describes the arrangement of an atom's electrons in terms of their energies. Indeed, when considering an electron's behavior about the nucleus, its energy is its most important property. A significant characteristic of an electron's energy is that it is a quantized property. A **quantized property** *is a property that can have only certain values*; that is, not all values are allowed. Because an electron's energy is quantized, it can have only certain specific energies.

quantized property

Quantization is a phenomenon that is not commonly encountered in the macroscopic world. A somewhat analogous condition to quantization is encountered during the process of a person climbing a flight of stars. In Figure 5–1a (p. 92) you see six steps between ground level and the level of the entrance door. Note that as a person climbs these stairs, there are only six permanent positions he or she can occupy (with both feet together). Thus, the person's position (height above ground level) is quantized; only certain heights are allowed. The opposite of quantization is a *continuum*. A person climbing a ramp up to the entrance (Figure 5–1b) would be able to assume a continuous set of heights above ground level; in this case, all values are allowed.

The energy of an electron determines its behavior about the nucleus. Because electron energies are quantized, only certain behavior patterns are allowed. Descriptions of these behavior patterns involve the use of the terms *shell*, *subshell*, and *orbital*. Sections 5.2–5.4 consider the meaning of these terms and the interrelationships among them.

FIGURE 5–1 A stairway with quantized position levels versus a ramp with continuous position levels.

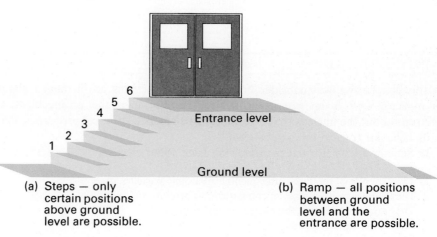

(a) Steps — only certain positions above ground level are possible.

(b) Ramp — all positions between ground level and the entrance are possible.

5.2 ELECTRON SHELLS

shell

Electrons can be grouped into *shells* or *main energy levels*, based on (energy)–(distance-from-the-nucleus) considerations. A **shell** *contains electrons that have approximately the same energy and spend most of their time approximately the same distance from the nucleus.*

Electron shells are identified by a number *n* that can have the values 1, 2, 3, ... The lowest energy shell is assigned an *n* value of 1, the next high a 2, then 3, and so on. Only values of $n = 1$–7 are used at present in designating electron shells. No known atom has electrons that are farther from the nucleus than the seventh main energy level ($n = 7$).

The maximum number of electrons that can be found in an electron shell varies; the higher the energy of the shell, the more electrons the shell can accommodate. The farther away electrons are from the nucleus (a high energy shell), the greater is the volume of space available for electrons; hence, the larger number of electrons in the shell.

The lowest energy shell ($n = 1$) accommodates a maximum of two electrons. In the second, third, and fourth shells, 8, 18, and 32 electrons are allowed, respectively.

TABLE 5–1 Important Characteristics of Electron Shells

Shell	Number Designation (n)	Electron Capacity ($2n^2$)
1st	1	$2 \times 1^2 = 2$
2nd	2	$2 \times 2^2 = 8$
3rd	3	$2 \times 3^2 = 18$
4th	4	$2 \times 4^2 = 32$
5th	5	$2 \times 5^2 = 50$[a]
6th	6	$2 \times 6^2 = 72$[a]
7th	7	$2 \times 7^2 = 98$[a]

[a] The maximum number of electrons in this shell has never been attained in any element now known.

A very simple mathematical equation can be used to calculate the maximum number of electrons allowed in any given shell.

$$\text{shell electron capacity} = 2n^2 \quad (\text{where } n = \text{shell number})$$

For example, when $n = 4$, the value is

$$2n^2 = 2(4)^2 = 32$$

This is the number that was previously given for the number of electrons allowed in the fourth shell. Although a maximum electron-occupancy level exists for each shell (group of electrons), these main energy levels can hold less than the allowable number of electrons in a given situation.

Table 5–1 summarizes the information about electron shells presented in this section.

5.3 ELECTRON SUBSHELLS

Each electron shell (main energy level) is divided into energy sublevels called *subshells*. A **subshell** *contains electrons that all have the same energy*. The number of subshells within a shell depends on the n value for the shell; in each shell there are n subshells.

$$\text{subshells in a shell} = n \quad (\text{where } n = \text{shell number})$$

Each successive shell has one more subshell than the previous one; shell 1 contains one subshell, shell 2 contains two subshells, shell 3 three subshells, and shell 4 four subshells.

Subshells are identified with both a number and a lowercase letter. The notations $3s$, $2p$, $4d$, and $5f$ all represent subshell notation. The number indicates the shell to which the subshell belongs. Electrons are found in four types of subshells, which are denoted by the lowercase letters s, p, d, and f. These four letters, in the order listed, denote subshells of increasing energy within a shell. The lowest energy subshell within a shell is always the s subshell, the next highest is the p subshell, then the d, and finally the f.

FIGURE 5–2 The energy of electron shells increases with increasing n, and the energy of subshells within a shell increases in the order s, p, d, and f.

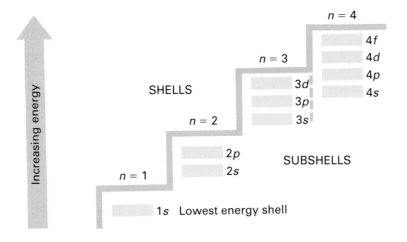

TABLE 5–2 Subshell Arrangements Within Shells

Shell number (n)	Subshells					
1	1s					
2	2s	2p				
3	3s	3p	3d			
4	4s	4p	4d	4f		
5	5s	5p	5d	5f	—	
6	6s	6p	6d	—	—	—
7	7s	—	—	—	—	—

Not all types of subshells are found in all shells. The lowest energy shell ($n = 1$) has only one subshell, the 1s. The second shell ($n = 2$) has two subshells: 2s and 2p. The 2p subshell is of higher energy than the 2s. The third subshell ($n = 3$) has three subshells: 3s, 3p, and 3d. The 3d subshell is of higher energy than the 3p, which in turn is of higher energy than the 3s. It is not until the fourth shell ($n = 4$) that we encounter all four types of subshells. In order of increasing energy, the 4s, 4p, 4d, and 4f subshells are in the fourth shell. Figure 5–2 (p. 93) summarizes the shell-subshell relationships for the first four shells.

A close look at the energy relationships depicted in Figure 5–2 shows that there is an energy overlap between shells 3 and 4. Although the *average* energy of the electrons in the third shell is lower than that of electrons in shell 4, the energies of the different subshells are such that there is one subshell in shell 4, the 4s, that is lower in energy than the highest energy subshell of shell 3, the 3d. The ramifications of this energy overlap will be discussed in Section 5.5.

In Section 5.2 we noted that the seventh shell is the highest numbered shell we will need to describe the electron arrangements of the 109 known elements. Table 5–2 gives information about the subshells found in the first seven shells. Note that number–letter designations are not given for all of the subshells in shells 5, 6, and 7. This is because some subshells in these shells (the dashes) will not be needed, as will be shown in Section 5.5.

The maximum number of electrons that a subshell can hold varies from 2 to 14, depending on whether the subshell is an s, p, d, or f. An s subshell can accommodate only two electrons. The shell in which the s subshell is located does not affect the maximum electron-occupancy figure of two. Subshells of the p, d, and f types can accommodate a maximum of 6, 10, and 14 electrons, respectively. Again, the maximum number of electrons in these subshell types depends only on the type of subshell and is independent of shell number. Table 5–3 summarizes information about distribution of electrons in subshells for shells 1 through 4. Notice the consistency between the numbers in columns 3 and 4 of Table 5–3. Within a shell, the sum of the subshell electron occupancies is the same as the shell electron occupancy ($2n^2$). For example, in shell 4, an s subshell containing 2

TABLE 5–3 Distribution of Electrons Within Subshells

Shell	Number of Subshells Within Shell	Maximum Number of Electrons Within Each Subshell				Maximum Number of Electrons Within Shell ($2n^2$)
		s	p	d	f	
$n = 1$	1	2				2
$n = 2$	2	2	6			8
$n = 3$	3	2	6	10		18
$n = 4$	4	2	6	10	14	32

electrons, a *p* subshell containing 6 electrons, a *d* subshell containing 10 electrons, and an *f* subshell containing 14 electrons add up to a total of 32 electrons, which is the maximum occupancy of shell 4 as calculated by the $2n^2$ formula.

5.4 ELECTRON ORBITALS

The term *orbital* is the last and most basic of the three terms used to describe electron arrangements about nuclei. An **orbital** *is the region of space about a nucleus where an electron with a specific energy is most likely to be found.*

An analogy for the relationship between shells, subshells, and orbitals can be found in the physical layout of a highrise condominium complex. In our analogy, a shell is the counterpart of a floor of the condominium. Just as each floor will contain apartments of varying sizes, a shell contains subshells of varying sizes. In addition, just as apartments contain rooms, subshells contain orbitals. An apartment is a collection of rooms; a subshell is a collection of orbitals.

The following are characteristics of orbitals, the rooms in our electron apartment house.

1. The number of orbitals in a subshell varies; there is one for an *s* subshell, three for a *p* subshell, five for a *d* subshell, and seven for an *f* subshell.
2. The maximum number of electrons found in an orbital does not vary; it is always two.
3. The notation used to designate orbitals is the same as that used for subshells. Thus, the seven orbitals in the 4*f* subshell are called 4*f* orbitals.

We have already noted (Section 5.3) that all electrons in a subshell have the same energy. Thus, all electrons in orbitals of the same subshell also have the same energy. This means that *shell and subshell locations are sufficient to specify the energy of an electron.* This statement will be of great importance in the discussions of Section 5.5.

Orbitals have a definite size and shape that is related to the type of subshell in which they are found. (Remember, an orbital is a region of space. We are not talking about the size and shape of an electron, but rather the size and shape of the region of space in which an electron is found.) An electron can be at only one point in an orbital at any given time, but, because of its rapid movement throughout the orbital, it occupies the entire orbital. An analogy would be the definite volume that is occupied by a rotating fan blade. Typical *s*, *p*, *d*, and *f* orbital shapes are given in Figure 5–3. Notice that the shapes increase in complexity in

FIGURE 5–3 The shapes of atomic orbitals. In order to improve the perspective, the *f* orbital is shown within a cube to illustrate that the lobes of this type of orbital are directed toward the corners of a cube.

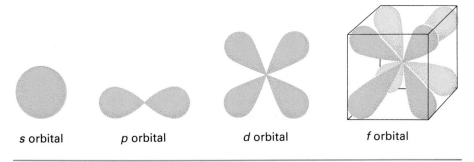

FIGURE 5–4 The orientation of the three orbitals in a *p* subshell.

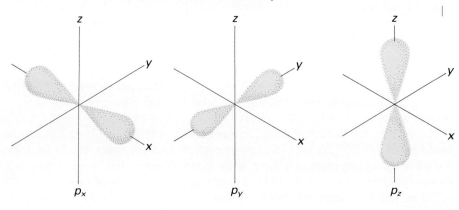

FIGURE 5–5 Size relationships among *s* orbitals located in different subshells.

the order *s*, *p*, *d*, and *f*. Some of the more complex *d* and *f* orbitals have shapes related to, but not identical to, those shown in Figure 5–3.

Orbitals that are in the same subshell differ mainly in their orientation. For example, in Figure 5–4, the three 2*p* orbitals look the same, but they are aligned in different directions—along the *x*, *y*, and *z* axes in a Cartesian coordinate system.

When orbitals of the same type are found in different subshells, such as 1*s*, 2*s*, and 3*s*, they have the same general shape but differ in size (volume), as shown in Figure 5–5.

Example 5.1

Determine the following for the third electron shell (third main energy level) of an atom.

a. the number of subshells it contains
b. the designation used to describe each subshell
c. the number of orbitals in each subshell
d. the maximum number of electrons that could be contained in each subshell
e. the maximum number of electrons that could be contained in the shell

Solution

a. The number of subshells in a shell is the same as the value of *n* for the subshell. Therefore, the third shell ($n = 3$) contains three subshells.
b. The lowest of the three subshells in terms of energy is designated 3*s*, the next high is 3*p*, and the final subshell, with the highest energy, 3*d*.
c. The number of orbitals in a given type (*s*, *p*, *d*, or *f*) of subshell is independent of the shell number. All *s* subshells (1*s*, 2*s*, or 3*s*) contain one orbital; all *p* subshells contain three orbitals; and all *d* subshells contain five orbitals.
d. Regardless of type, an orbital can contain a maximum of only two electrons. Therefore, the 3*s* subshell (one orbital) contains two electrons; the 3*p* subshell (three orbitals) contains six electrons; and the 3*d* subshell (five orbitals) contains ten electrons.
e. The maximum number of electrons in a shell is given by the formula $2n^2$, where *n* is the shell number. Because $n = 3$ in this problem, $2n^2 = 2(3)^2 = 18$. Alternatively, from part (c), we note that shell 3 contains nine orbitals (1 + 3 + 5). Because each orbital can hold two electrons, the maximum number of electrons will be 18 (2 × 9).

5.5 WRITING ELECTRON CONFIGURATIONS

An **electron configuration** *is a statement of how many electrons an atom has in each of its subshells.* Because subshells group electrons according to energy (Section 5.3), electron configurations indicate how many electrons of various energies an atom has.

Electron configurations are not written out in words; a shorthand system with symbols is used. Subshells containing electrons, listed in order of increasing energy, are designated using number–letter combinations (1s, 2s, and 2p). A superscript following each subshell designation indicates the number of electrons in that subshell. The electron configuration for nitrogen using this shorthand notation is

$$1s^2 2s^2 2p^3 \text{ (read "one-}s\text{-two, two-}s\text{-two, two-}p\text{-three")}$$

Thus, a nitrogen atom has an electron arrangement of two electrons in the 1s subshell, two electrons in the 2s subshell, and three electrons in the 2p subshell.

In order to find the electron configuration for an atom, a procedure called the *Aufbau principle* (German *aufbauen*, to build) is used. The **Aufbau principle** states *that electrons normally occupy the lowest energy subshell available.* This guideline brings order to what could be a very disorganized situation. There are many orbitals about the nucleus of any given atom. Electrons do not occupy these orbitals in a random, haphazard fashion; a very predictable pattern, governed by the Aufbau principle, exists for electron orbital occupancy. Orbitals are filled in order of increasing energy.

Use of the Aufbau principle requires knowledge about the electron capacities of orbitals and subshells (Section 5.4) and their relative energies.

Figure 5–6 gives the order in which electron subshells about a nucleus acquire electrons. Note that the sequence of subshell filling is not as simple a pattern as you probably would have predicted. All subshells within a given shell do not necessarily have lower energies than all subshells of higher numbered shells. Because of energy overlaps, beginning with shell 4, one or more lower-energy subshells of a specific shell have energies that are lower than the upper subshells of a preceding shell; thus, they acquire electrons first. For example, the 4s subshell acquires

FIGURE 5–6 Relative energies and filling order for the electron subshells.

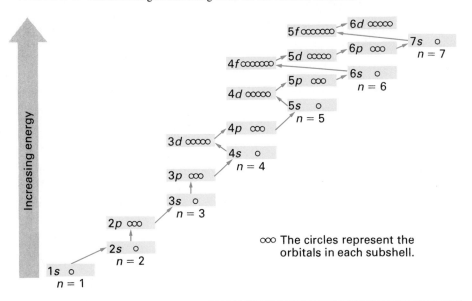

Aufbau diagram

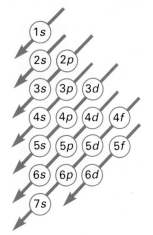

FIGURE 5-7 The Aufbau diagram, which illustrates subshell filling order.

electrons before the 3d subshell (see Figure 5-6). As another example, the s subshell of the sixth energy level fills before the d subshell of the fifth energy level or the f subshell of the fourth energy level (Figure 5-6).

The sequence in which subshells acquire electrons must be learned before electron configurations can be written. A useful mnemonic (memory) device called an *Aufbau diagram* helps considerably with this learning process. An **Aufbau diagram** *is a listing of subshells in the order in which electrons occupy them*. This diagram is illustrated in Figure 5-7. In this diagram, all s subshells are located in column 1; all p subshells are in column 2, and so on. Subshells belonging to the same shell are found in the same row. The order of subshell filling is given by following the diagonal arrows, starting with the top one. The 1s subshell fills first. The second arrow points to (goes through) the 2s subshell, which fills next. The third arrow points to both the 2p and 3s subshells. The 2p fills first; the 3s fills next. Any time a single arrow points to more than one subshell, you should start at the tail of the arrow and work toward its head to determine the proper filling sequence. An Aufbau diagram is an easy way to catalog the information given in Figure 5-6.

We are now ready to write electron configurations. Let us systematically consider electron configurations for the first few elements in the periodic table.

Hydrogen (Z = 1) has only one electron, which goes into the 1s subshell; this subshell has the lowest energy of all subshells. Hydrogen's electron configuration is written as

$$1s^1$$

Helium (Z = 2) has two electrons, both of which occupy the 1s subshell. (Remember, an s subshell contains one orbital and an orbital can accommodate two electrons.) Helium's electron configuration is

$$1s^2$$

Lithium (Z = 3) has three electrons, and the third electron cannot enter the 1s subshell because its maximum capacity is two electrons. (All s subshells are completely filled with two electrons—Section 5.3). The third electron is placed in the next highest energy subshell, the 2s. The electron configuration for lithium is,

$$1s^2 2s^1$$

For beryllium (Z = 4), the additional electron is placed in the 2s subshell, which is now completely filled, giving beryllium the electron configuration

$$1s^2 2s^2$$

For boron (Z = 5), the 2p subshell, which is the subshell of next highest energy (Figures 5-6 or 5-7), becomes occupied for the first time. Boron's electron configuration is

$$1s^2 2s^2 2p^1$$

A p subshell can accommodate six electrons because there are three orbitals within it (Section 5.3). The 2p subshell can thus accommodate the additional electrons found in C, N, O, F, and Ne. The electron configurations of these elements are

C (Z = 6): $1s^2 2s^2 2p^2$
N (Z = 7): $1s^2 2s^2 2p^3$
O (Z = 8): $1s^2 2s^2 2p^4$

F (Z = 9): $1s^2 2s^2 2p^5$

Ne (Z = 10): $1s^2 2s^2 2p^6$

With sodium (Z = 11), the 3s subshell acquires an electron for the first time.

Na (Z = 11): $1s^2 2s^2 2p^6 3s^1$

Note the pattern that is developing in the electron configurations we have written so far. Each element has an electron configuration that is the same as the one just before it except for the addition of one electron.

Electron configurations for other elements are obtained by simply extending the principles we have just illustrated. A subshell of lower energy is always filled before electrons are added to the next highest subshell; this continues until the correct number of electrons have been accommodated.

Example 5.2

Write out the electron configuration for

a. strontium (Z = 38)   b. lead (Z = 82)

Solution

a. The number of electrons in a strontium atom is 38. Remember, the atomic number (Z) gives the number of electrons (Section 3.5). We will need to fill subshells, in order of increasing energy, until 38 electrons have been accommodated.

The 1s, 2s, and 2p subshells fill first, accommodating a total of ten electrons among them.

$$1s^2 2s^2 2p^6 \ldots$$

Next, according to Figures 5–6 and 5–7, the 3s subshell fills and then the 3p subshell.

$$1s^2 2s^2 2p^6 \widehat{3s^2 3p^6} \ldots$$

We have accommodated 18 electrons at this point. We still need to add 20 more electrons to get our desired number of 38.

The 4s subshell fills next, followed by the 3d subshell, giving us 30 electrons at this point.

$$1s^2 2s^2 2p^6 3s^2 3p^6 \widehat{4s^2 3d^{10}} \ldots$$

Note that the maximum electron population for d subshells is ten electrons.

Eight more electrons are needed, which are added to the next two higher subshells, the 4p and the 5s. The 4p subshell can accommodate six electrons and the 5s, two electrons.

$$1s^2 2s^2 2p^6 3s^2 3p^6 4s^2 3d^{10} \widehat{4p^6 5s^2}$$

To double-check that we have the correct number of electrons, 38, we add the superscripts in our final electron configuration.

$$2 + 2 + 6 + 2 + 6 + 2 + 10 + 6 + 2 = 38$$

The sum of the superscripts in any electron configuration should add up to the atomic number if the configuration is for a neutral atom.

b. To write this configuration, we continue along the same lines as in part (a), remembering that the maximum electron subshell populations are $s = 2$, $p = 6$, $d = 10$, and $f = 14$.

Lead, with an atomic number of 82, contains 82 electrons that are distributed among the various subshells. Adding 82 electrons to subshells takes place in the following order. The line of numbers underneath the electron configuration is a running total of added electrons and is obtained by adding the superscripts up to that point. We stop when we hit 82 electrons.

running total of electrons added $\quad 1s^2 2s^2 2p^6 3s^2 3p^6 4s^2 3d^{10} 4p^6 5s^2 4d^{10} 5p^6 6s^2 4f^{14} 5d^{10} 6p^2$
$2 4 10 12 18 20 30 \phantom{d^{10}}36 38 48 \phantom{d^{10}}54 56 70 \phantom{f^{14}}80 \phantom{d^{10}}82$

Notice, in the electron configuration, that the 6p subshell contains only two electrons, even though it can hold a maximum of six. We only put two electrons in this subshell because that is sufficient to give 82 total electrons. If we had completely filled this subshell, we would have had 86 total electrons, which is too many.

Note that for a few elements in the middle of the periodic table, the actual distribution of electrons within subshells differs slightly from that obtained by using the Aufbau principle and Aufbau diagram. These exceptions are caused by very small energy differences between some subshells and are not important in the uses we shall make of electronic configurations.

5.6 ELECTRON CONFIGURATIONS AND THE PERIODIC LAW

A knowledge of the electron configurations of the elements provides an explanation for the periodic law. You will recall, from Section 3.8, that the periodic law points out that the properties of the elements repeat themselves in a regular manner when the elements are ordered in sequence of increasing atomic number. The elements that have similar chemical properties are placed under one another in vertical columns (groups) in the periodic table.

Groups of elements have similar chemical properties because of similarities that exist in their electron configurations. *Chemical properties repeat themselves in a regular manner among the elements because electron configurations repeat themselves in a regular manner among the elements.*

To illustrate this similar-chemical-property–similar-electron-configuration correlation, let us look at the electron configurations of two groups of elements known to have similar chemical properties.

We begin with the elements lithium, sodium, potassium, and rubidium, all members of Group IA of the periodic table. The electron configurations for these elements are

$_3$Li: $\quad 1s^2 \, \boxed{2s^1}$
$_{11}$Na: $\quad 1s^2 2s^2 2p^6 \, \boxed{3s^1}$
$_{19}$K: $\quad 1s^2 2s^2 2p^6 3s^2 3p^6 \, \boxed{4s^1}$
$_{37}$Rb: $\quad 1s^2 2s^2 2p^6 3s^2 3p^6 4s^2 3d^{10} 4p^6 \, \boxed{5s^1}$

Note that each of these elements has one outer electron (shown in color) in an s subshell, the last one added by the Aufbau principle. It is this similarity in outer-shell electron arrangements that causes these elements to have similar chemical properties. In general, elements with similar outer shell electron configurations have similar chemical properties.

Let us consider another group of elements known to have similar chemical properties: fluorine, chlorine, bromine, and iodine of Group VIIA of the periodic table. The electron configurations for these four elements are

$_9$F: $1s^2\,(2s^2 2p^5)$
$_{17}$Cl: $1s^2 2s^2 2p^6\,(3s^2 3p^5)$
$_{35}$Br: $1s^2 2s^2 2p^6 3s^2 3p^6\,(4s^2)\,3d^{10}\,(4p^5)$
$_{53}$I: $1s^2 2s^2 2p^6 3s^2 3p^6 4s^2 3d^{10} 4p^6\,(5s^2)\,4d^{10}\,(5p^5)$

Once again similarities in electron configurations are readily apparent. This time, the repeating pattern involves an outermost s and p subshell containing seven electrons (shown in color).

Section 6.2 will consider in depth the fact that the electrons that are the most important in controlling chemical properties are found in the outermost shell of an atom.

5.7 ELECTRON CONFIGURATIONS AND THE PERIODIC TABLE

One of the strongest pieces of supporting evidence for the assignment of electrons to shells, subshells, and orbitals is the periodic table itself. The basic shape and structure of this table, which was determined many years before electrons were even discovered, is consistent with and can be explained by electron configurations. Indeed, the specific location of an element in the periodic table can be used to obtain information about its electron configuration.

The concept of *distinguishing electrons* is the key to obtaining electron configuration information from the periodic table. The **distinguishing electron** *for an element is the last electron that is added to its electron configuration when the configuration is written using the Aufbau principle.* This last electron is the one that causes an element's electron configuration to differ from that of the element immediately preceding it in the periodic table.

distinguishing electron

As the first step in linking electron configurations to the periodic table, let us analyze the general shape of the periodic table in terms of columns of elements. As shown in Figure 5–8, on the extreme left of the table there are two columns of elements; in the center there is an area containing ten columns of elements; to

FIGURE 5–8 Structure of the periodic table in terms of columns of elements.

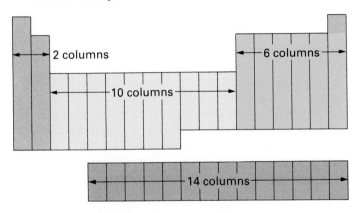

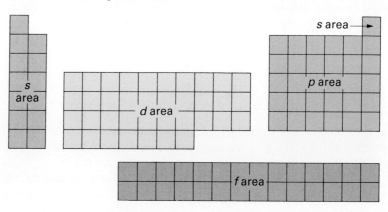

FIGURE 5–9 Areas of the periodic table.

the right there is a block of six columns of elements; and, in two rows at the bottom of the table, there are fourteen columns of elements.

The number of columns of elements in the various regions of the periodic table (Figure 5–8)—2, 6, 10, and 14—is the same as the maximum number of electrons that the various types of subshells can accommodate. We will see shortly that this is a very significant observation; the number matchup is no coincidence. The various columnar regions of the periodic table are called the *s* area (two columns), the *p* area (six columns), the *d* area (ten columns), and the *f* area (fourteen columns), as shown in Figure 5–9.

For all elements located in the *s* area of the periodic table, the distinguishing electron is always found in an *s* subshell. All *p*-area elements have distinguishing electrons in *p* subshells. Similarly, elements in the *d* and *f* areas of the periodic table have distinguishing electrons located in *d* and *f* subshells, respectively. Thus, the area location of an element in the periodic table can be used to determine the type of subshell that contains the distinguishing electron. Note that the element helium belongs to the *s* rather than the *p* area of the periodic table even though its table position is on the right side. (The reason for the placement of helium will be explained in Section 6.3.)

The extent that the subshell containing an element's distinguishing electron is filled can also be determined from the element's position in the periodic table. All elements in the first column of a specific area contain only one electron in the subshell; all elements in the second column contain two electrons in the subshell, and so on. Thus, all elements in the first column of the *p* area (Group IIIA) have an electron configuration ending in p^1. Elements in the second column of the *p* area (Group IVA) have electron configurations ending in p^2, and so on. Similar relationships hold in other areas of the table, as shown in Figure 5–10.

We can also use the periodic table to determine the shell in which the distinguishing electron is located. The relationship that is used involves the number of the period in which the element is found. In the *s* and *p* areas, the period number directly gives the shell number. In the *d* area, the period number minus one is equal to the shell number. (Remember that the 3*d* subshell is filled during the fourth period.) For similar reasons, in the *f* area, the period number minus two equals the shell number. Thus, the subshell that contains the distinguishing electron for elements of period 6 could be the 6*s*, 6*p*, 5*d* (period number minus one), or the 4*f* subshell (period number minus two), depending on the location of the element in period 6. Remember that even though the *f* area is located at the bot-

FIGURE 5-10 The extent of subshell filling is a function of periodic table position.

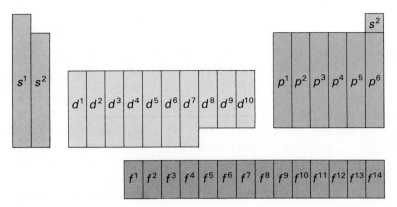

FIGURE 5-11 Relationships of period numbers to shell numbers for distinguishing electrons.

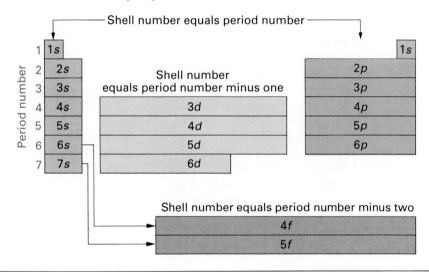

tom of the table, it correctly belongs in periods 6 and 7. The complete matchup between period number and shell number for the distinguishing electron is given in the periodic table of Figure 5–11.

Using the periodic table and Figures 5–9 through 5–11, determine the following for the elements potassium (Z = 19), cobalt (Z = 27), and tin (Z = 50).

a. the type of subshell in which the distinguishing electron is found
b. the extent that the subshell containing the distinguishing electron is filled
c. the shell in which the subshell containing the distinguishing electron is found

Solution

a. Knowing the area of the periodic table in which an element is found (Figure 5–9) is sufficient to determine the type of subshell in which the distinguishing electron is found.

Example 5.3

Potassium: Because this element is found in the s area of the periodic table, the distinguishing electron will be in an s subshell.

Cobalt: Because this element is found in the d area of the periodic table, the distinguishing electron will be in a d subshell.

Tin: Because this element is found in the p area of the periodic table, the distinguishing electron will be in a p subshell.

b. The extent that the subshell containing the distinguishing electron is filled is determined by noting the column in the given area that the element occupies (see Figure 5–10).

Potassium: Because this element is in the first column of the s area, the s subshell involved contains one electron (s^1).

Cobalt: Because this element is in the seventh column of the d area, the d subshell involved contains seven electrons (d^7).

Tin: Because this element is in the second column of the p area, the p subshell involved contains two electrons (p^2).

c. The shell number of the subshell containing the distinguishing electron is obtained from the period number—sometimes directly and sometimes with modifications (see Figure 5–11).

Potassium: Because this element is a period 4 s-area element, the s subshell involved is the $4s$; period number and shell number always match for s area elements. Thus, the electron configuration for potassium ends in $4s^1$.

Cobalt: Because this element is a period 4 d-area element, the d subshell involved is the $3d$; the shell number is always one less than the period number for d area elements. Thus, the electron configuration for cobalt ends in $3d^7$.

Tin: Because this element is a period 5 p-area element, the p subshell involved is the $5p$; period number and shell number always match for p area elements. Thus, the electron configuration for tin ends in $5p^2$.

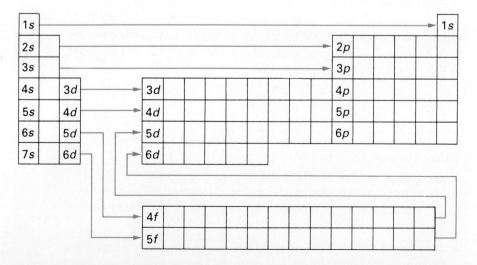

FIGURE 5–12 Using the periodic table as a guide to obtain the order in which subshells are occupied by electrons.

TABLE 5–4 Atomic Number Ranges and Subshell Filling Order.

Atomic Number Range	Subshell Involved
1–2	1s
3–4	2s
5–10	2p
11–12	3s
13–18	3p
19–20	4s
21–30	3d
31–36	4p
37–38	5s
39–48	4d
49–54	5p
55–56	6s
57	5d
58–71	4f
72–80	5d
81–86	6p
87–88	7s
89	6d
90–103	5f
104–109	6d

In order to write complete electron configurations, you must know the order in which the various electron subshells are filled. Up until now, we have obtained this filling order by using an Aufbau diagram (Figure 5–7). We can also obtain this information directly from a periodic table. To obtain it, we merely follow a path of increasing atomic number through the table, noting the various subshells as they are encountered.

Figure 5–12 illustrates this way of using the periodic table. Note particularly how the f area of the periodic table is worked into the scheme. The results of working our way through the periodic table in terms of atomic numbers are summarized in Table 5–4.

Write the complete electron configuration for the element mercury (Z = 80) using the periodic table as your sole guide for subshell filling order.

Example 5.4

Solution
We will obtain subshell filling order by working our way through the periodic table. We will start at hydrogen (Z = 1) and go from box to box, in order of increasing atomic number, until we arrive at position 80, mercury. Every time we traverse the s area we will fill an s subshell, the p area a p subshell, and so on. We will remember that when filling s and p subshells, the shell number is given by the period number; that one must be subtracted from the period number for d subshells; and that two must be subtracted from the period number for f subshells.

Let us begin our journey through the periodic table. As we cross period 1 we encounter H and He, which are both $1s$ elements. Thus, we add the $1s$ electrons

Hg: $1s^2$ . . .

Traversing period 2, we pass through the s area (elements 3 and 4) and the p area (elements 5–10), in that order. We add $2s$ and $2p$ electrons

Hg: $1s^2 2s^2 2p^6 \ldots$

Our trip through period 3 is very similar to that of period 2. We encounter only s-area elements (11 and 12) and p-area elements (13–18). We add s and p electrons, $3s$ and $3p$ because we are in period 3.

Hg: $1s^2 2s^2 2p^6 3s^2 3p^6 \ldots$

Passing through period 4, we go through the s area (elements 19 and 20), the d area (elements 21–30), and the p area (elements 31–36). We add electrons in the $4s$, $3d$ (period number minus one), and $4p$ subshells, in that order.

Hg: $1s^2 2s^2 2p^6 3s^2 3p^6 4s^2 3d^{10} 4p^6 \ldots$

In period 5, we encounter the s area (elements 37 and 38), the d area (elements 39–48), and the p area (elements 49–54). Hence, the $5s$, $4d$ (period number minus one), and $5p$ subshells are filled in order.

Hg: $1s^2 2s^2 2p^6 3s^2 3p^6 4s^2 3d^{10} 4p^6 5s^2 4d^{10} 5p^6 \ldots$

Our journey ends in period 6. We go through the $6s$ area (elements 55 and 56), the $4f$ area (elements 58–71), and the $5d$ area (elements 57 and 72–80). We will completely fill the $6s$, $4f$ (period number minus two), and $5d$ (period number minus one) subshells.

Hg: $1s^2 2s^2 2p^6 3s^2 3p^6 4s^2 3d^{10} 4p^6 5s^2 4d^{10} 5p^6 6s^2 4f^{14} 5d^{10}$

There are a few, slightly irregular electron configurations in the d and f areas of the periodic table. The generalizations in this section do not address this problem. We will be working mostly with s and p area elements in future chapters, and there are no irregularities in electron configurations for these elements.

5.8 CLASSIFICATION SYSTEMS FOR THE ELEMENTS

The elements can be classified in several ways. The two most common classification systems are

1. A system based on the electron configurations of the elements, in which elements are described are *noble-gas*, *representative*, *transition*, or *inner-transition elements*.
2. A system based on selected physical properties of the elements, in which elements are described as *metals* or *nonmetals*.

The classification scheme based on electron configurations of the elements is depicted in Figure 5–13. The groupings of elements resulting from this classification scheme will be used in discussions in subsequent chapters.

noble-gas elements

The **noble-gas elements** *are found in the far right column of the periodic table.* They are all gases at room temperature, and they have little tendency to form chemical compounds. With one exception, the distinguishing electron for a noble gas completes the p subshell; therefore, they have electron configurations ending in p^6. The exception is helium, in which the distinguishing electron completes the first shell—a shell that has only two electrons. Helium's electron configuration is $1s^2$.

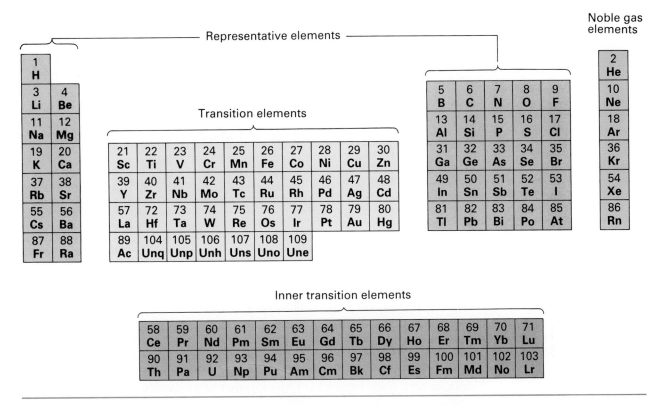

FIGURE 5-13 Elemental classification scheme based on the electron configurations of the elements.

The **representative elements** *are all of the elements of the s and p areas of the periodic table, with the exception of the noble gases.* The distinguishing electron in these elements partially or completely fills an *s* subshell or partially fills a *p* subshell. The representative elements include most of the more common elements.

The **transition elements** *are all of the elements of the d area of the periodic table.* The common feature in the electron configurations of the transition elements is the presence of the distinguishing electron in a *d* subshell.

The **inner transition elements** *are all of the elements of the f area of the periodic table.* The characteristic feature of their electron configurations is the presence of the distinguishing electron in an *f* subshell. There is very little variance in the properties of either the 4*f* or 5*f* series of inner transition elements.

On the basis of selected physical properties of the elements, the second of the two classification schemes divides the elements into the categories of metals and nonmetals. A **metal** *is an element that has the characteristic properties of luster, thermal conductivity, electrical conductivity, and malleability.* With the exception of mercury, all metals are solids at room temperature (25°C). Metals are good conductors of heat and electricity. Most metals are ductile (can be drawn into wires) and malleable (can be rolled into sheets). Most metals have high luster (shine), high density, and high melting points. Among the more familiar metals are the elements iron, aluminum, copper, silver, and gold.

A **nonmetal** *is an element characterized by the absence of the properties of luster, thermal conductivity, electrical conductivity, and malleability.* Many of the non-metals such as hydrogen, oxygen, nitrogen, and the noble gases are gases. The

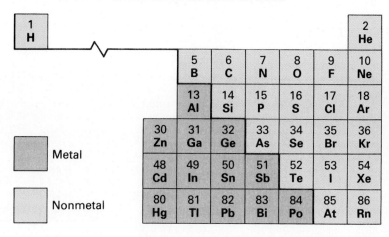

FIGURE 5-14 A portion of the periodic table showing the dividing line between metals and nonmetals. All elements that are not shown are metals.

only nonmetal that is a liquid at room temperature is bromine. Solid nonmetals include carbon, sulfur, and phosphorus. In general, the nonmetals have lower densities and melting points than metals.

The majority of the elements are metals. Only 22 elements are nonmetals; the rest (87) are metals. It is not necessary to memorize which elements are nonmetals and which are metals, as this information is obtainable from a periodic table. As can be seen from Figure 5-14, the location of an element in the periodic table correlates directly with its classification as a metal or nonmetal. Note the steplike heavy line that runs through the right third of the periodic table. This line separates the metals from the nonmetals; the metals are on the left and the nonmetals are on the right. Note also that the element hydrogen is a nonmetal.

The fact that the vast majority of elements are metals in no way indicates that metals are more important than nonmetals. Most nonmetals are rather common and are found in many important compounds. On the other hand, the majority of metals are quite rare and few are found in important compounds.

An analysis of the previously given abundance of the elements in the Earth's crust (Table 2-2) in terms of metals and nonmetals shows that the three most abundant elements, which account for 87% of all atoms, are nonmetals—oxygen, silicon, and hydrogen. The four most abundant elements in the human body (Table 2-2), which comprise over 99% of all atoms in the body, are nonmetals—hydrogen, oxygen, carbon, and nitrogen.

5.9 METALS, NONMETALS, AND LIVING ORGANISMS

Twenty-six elements are known to be essential for the proper functioning of living organisms. Twelve of these elements are nonmetals and fourteen are metals. These essential elements can be divided into three categories based on the amounts needed. The *macronutrients* are needed in large amounts, the *micronutrients* are needed in small amounts, and the *trace elements* are needed in minute amounts.

We have previously noted that the macronutrients are carbon, hydrogen, oxygen, and nitrogen. These nonmetals are found in most of the compounds present within the human body. Chapters 14-20 deal almost exclusively with compounds involving these four elements.

TABLE 5-5 The Most Abundant Elements Present in the Human Body (Nonmetals are given in bold type.)

Element	Percent of Total Number of Atoms in the Human Body	Number of Grams in a 70-kg Man
Macronutrients		
hydrogen	63	6,580
oxygen	25.5	43,550
carbon	9.5	12,590
nitrogen	1.4	1,815
Micronutrients		
calcium	0.31	1,700
phosphorus	0.22	680
potassium	0.06	250
sulfur	0.05	100
chlorine	0.03	115
sodium	0.03	70
magnesium	0.01	42

The micronutrients are a group of four metals and three nonmetals. As can be seen from the numbers in Table 5-5, the amounts of these elements in the body vary from 0.31 atom percentage for calcium to 0.01 atom percentage for magnesium. Note also from Table 5-5 how much smaller the amounts are compared to those for the macronutrients. Collectively, the micronutrients constitute less than one percent of the total atoms in the human body. Despite these small numbers, life could not go without these elements. Body fluid chemistry (Chapter

TABLE 5-6 Trace Elements Known to be Essential in Humans and Animals (Nonmetals are given in bold type.)

Trace Element	Year Need Was Established	Source of Trace Element
Need in Humans Established and Quantified		
iron	17th century	meat, liver, fish, poultry, beans, peas, raisins, prunes
iodine	1850	iodized table salt, shellfish, kelp
zinc	1934	meat, liver, eggs, shellfish
Need in Humans Established But Not Yet Quantified		
copper	1928	nuts, liver, shellfish
manganese	1931	nuts, fruits, vegetables, whole-grain cereals
cobalt	1935	meat, dairy products
molybdenum	1953	organ meats, green leafy vegetables, legumes
selenium	1957	meat, seafood
chromium	1959	meat, beer, unrefined wheat flour
Need Established in Animals But Not Yet in Humans		
tin	1970	*
vanadium	1971	*
fluorine	1972	fluoridated water
silicon	1972	*
nickel	1974	*
arsenic	1975	*

* Present in trace quantities in many foods and in the environment.

33) is an area in which the micronutrients are particularly important. Sulfur and phosphorus are also important in the makeup of the molecular building blocks of living organisms. These two elements, together with the four macroelements, are the basis for amino acids (Chapter 24), sugars (Chapter 22), fats and oils (Chapter 23), and nucleotides (Chapter 26).

Ten of the 15 trace elements are metals. Although they are needed in only minute amounts (much less than 0.01 atom %) by living organisms, they still are of absolute importance for the proper functioning of an organism. Table 5–6 (p. 109) gives information about the sources and functions of these elements. Chapter 27 contains more information about the biochemical functions of selected trace elements in the human body.

EXERCISES AND PROBLEMS

Terminology Associated with Electron Arrangements

5.1 What does the term *quantization* mean, and what property of an electron is quantized?

5.2 The following statements define or are closely related to the terms *shell*, *subshell*, and *orbital*. Match the terms to the appropriate statements.

a. In terms of electron capacity, this unit is the smallest of the three.
b. This unit can contain a maximum of two electrons.
c. This unit can contain as many as or more electrons than either of the other two.
d. The term *subenergy level* is closely associated with this unit.
e. This unit is designated by just a number.
f. The formula $2n^2$ gives the maximum number of electrons that can occupy this unit.
g. This unit is designated in the same way as the orbitals that comprise it.

5.3 Determine the following for the fifth electron shell (the fifth main energy level) of an atom.

a. The number of subshells it contains.
b. The designation used to describe each of the first four subshells.
c. The number of orbitals in each of the first four subshells.
d. The maximum number of electrons that can occupy this fifth electron shell.
e. The maximum number of electrons that can occupy each of the first four subshells.

5.4 Fill in the numerical value(s) that correctly complete(s) each of the following statements.

a. The maximum number of electrons in the second electron shell is _____.
b. The maximum number of electrons in the sixth electron main energy level is _____.
c. A 4f subshell holds a maximum of _____ electrons.
d. A 3d orbital holds a maximum of _____ electrons.
e. A 2s orbital holds a maximum of _____ electrons.
f. The fourth shell contains _____ subshells, _____ orbitals, and a maximum of _____ electrons.

5.5 Describe the general shape of each of the following orbitals.

a. 2s orbital
b. 3s orbital
c. 4p orbital
d. 3d orbital

5.6 Characterize the similarities and differences between

a. a 3s orbital and 3d orbital
b. a 2p orbital and 3p orbital
c. the third shell and fourth shell

5.7 Give the maximum number of electrons that can occupy each of the following units.

a. 3d subshell
b. 2s orbital
c. second shell
d. shell with $n = 4$
e. 4p subshell
f. 4p orbital

Writing Electron Configurations

5.8 Explain the difference between the Aufbau principle and an Aufbau diagram.

5.9 Within any given shell, how do the energies of the *s*, *p*, *d*, and *f* subshells compare?

5.10 For each of the following sets of subshells, determine the one of lowest energy and the one of highest energy.

a. 2s, 3s, 4s
b. 4p, 4d, 4f
c. 3d, 4d, 4f
d. 6s, 3d, 5f

5.11 Explain what each number and letter means in the notations $6s^2$ and $4f^{14}$.

5.12 With the help of an Aufbau diagram, write the complete electron configuration for each of the following atoms.

a. $_9F$ c. $_{50}Sn$ e. $_{44}Ru$ g. $_{86}Rn$
b. $_{17}Cl$ d. $_{20}Ca$ f. $_{24}Cr$ h. $_{36}Kr$

5.13 What is wrong with each of the following attempts to write electron configurations?

a. $1s^2 1p^6 2s^2 2p^6$
b. $1s^2 2s^2 3s^2$
c. $1s^2 2s^2 2p^6 3s^2 3p^6 3d^{10}$
d. $1s^2 2s^2 2p^6 2d^5$

The Periodic Law and Electron Configurations

5.14 Group the following six electron configurations of elements into pairs that you would expect to show similar chemical properties.

a. $1s^2 2s^2 2p^6 3s^1$
b. $1s^2 2s^2 2p^6 3s^2 3p^5$
c. $1s^2 2s^2 2p^6 3s^2 3p^6 4s^2 3d^{10} 4p^5$
d. $1s^2 2s^2 2p^6 3s^2 3p^6 4s^2 3d^{10} 4p^3$
e. $1s^2 2s^2 2p^6 3s^2 3p^6 4s^2 3d^{10} 4p^6 5s^1$
f. $1s^2 2s^2 2p^6 3s^2 3p^3$

The Periodic Table and Electron Configurations

5.15 Indicate the position in the periodic table where each of the following occurs. (Give the symbol of the element.)

a. The $3p$ subshell begins filling.
b. The $2s$ subshell begins filling.
c. The $4p$ subshell begins filling.
d. The $4d$ subshell begins filling.
e. The $5f$ subshell begins filling.
f. The $3d$ subshell begins filling.

5.16 Indicate the position in the periodic table where each of the following occurs. (Give the symbol of the element.)

a. The $4p$ subshell becomes completely filled.
b. The $2s$ subshell becomes half-filled.
c. The $3d$ subshell becomes half-filled.
d. The fourth shell begins filling.
e. The fourth shell becomes completely filled.
f. The third shell becomes half-filled.

5.17 For each of the following elements, tell:

(1) Whether the distinguishing electron is in the s, p, d or f subshell.
(2) The extent that the subshell containing the distinguishing electron is filled.
(3) The shell in which the distinguishing electron is found.

a. $_{26}$Fe   b. $_{54}$Xe   c. $_{68}$Er   d. $_{56}$Ba

5.18 Each of the following is the outer subshell electron configuration of an element after the distinguishing electron has been added. Using the periodic table, identify each of the elements.

a. $5s^1$ c. $6d^2$ e. $4f^{10}$ g. $5d^9$
b. $4p^4$ d. $5p^6$ f. $7s^2$ h. $2p^1$

5.19 Using only the periodic table, determine the electron configurations of the following elements.

a. $_{30}$Zn   b. $_{79}$Au   c. $_{53}$I   d. $_{105}$Unp

Classification Systems for the Elements

5.20 Identify each of the following as a noble gas, representative element, transition element, or inner transition element.

a. $_{54}$Xe   c. $_2$He   e. $_{16}$S   g. $_{19}$K
b. $_{45}$Rh   d. $_{106}$Unh   f. $_3$Li   h. $_{95}$Am

5.21 Classify each of the following elements as metals or nonmetals.

a. $_{15}$P   c. $_{78}$Pt   e. $_1$H   g. $_{33}$As
b. $_4$Be   d. $_{10}$Ne   f. $_{47}$Ag   h. $_{13}$Al

5.22 Twenty-six elements are known to be essential for living organisms. List, by name, the essential elements that are

a. metals
b. nonmetals
c. representative elements
d. representative metals
e. representative nonmetals
f. transition elements
g. transition metals
h. inner-transition elements

6 Chemical Bonding

Objectives

After completing Chapter 6, you will be able to:

6.1 Define the term *chemical bond* and describe the two major types of chemical bonds.
6.2 Determine the number of valence electrons a representative element has, given its location in the periodic table or its electron configuration, and write an electron-dot structure for the element.
6.3 State the octet rule and understand the basis for the rule.
6.4 Define the term *ion* and understand the notation used to denote ions.
6.5 Use electron-dot structures to describe the bonding in simple ionic compounds and write the formula for an ionic compound, given the charges on the ions that are present.
6.6 Understand the structural characteristics of ionic compounds.
6.7 Write the name of a binary ionic compound, given its formula, or vice versa.
6.8 Use electron-dot structures to describe the bonding in simple covalent molecules.
6.9 Define the terms *single-covalent bond*, *double-covalent bond*, and *triple-covalent bond*.
6.10 Define the term *coordinate-covalent bond*.
6.11 Understand the relationship between the magnitude of an element's electronegativity and its position in the periodic table, and classify, with the help of a table of electronegativities, a given bond as nonpolar covalent, polar covalent, or ionic.
6.12 Predict whether a molecule is polar or nonpolar, given its geometry.
6.13 Write the name of a binary covalent compound, given its formula.
6.14 Recognize the names and formulas for common polyatomic ions.
6.15 Write formulas for compounds that contain polyatomic ions and name them.

INTRODUCTION

As scientists study living organisms and the world in which we live, they rarely encounter free isolated atoms. Instead, under normal conditions of temperature and pressure, they almost always find atoms associated in aggregates or clusters ranging in size from two atoms to numbers too large to count. In this chapter, we will discuss reasons why atoms tend to collect together into larger units and provide information about the binding forces (chemical bonds) involved in these units.

As we examine the nature of attractive forces between atoms, we will discover that both the tendency and capacity of an atom to be attracted to other atoms is dictated by its electron configuration. Thus, the concepts introduced in Chapter 5 will be used extensively in treating the subject of chemical bonding.

6.1 TYPES OF CHEMICAL BONDS

Chemical bonds *are the attractive forces that hold atoms together in more complex units*. Current chemical theory describes chemical bonds in terms of two major types: ionic bonds and covalent bonds. The major difference between these types of bonds is the perceived mechanism through which bond formation occurs.

An **ionic bond** *results from the* transfer *of one or more electrons from one atom or group of atoms to another*. This bond model of electron transfer is particularly useful when describing the bonding in compounds that contain both metallic and nonmetallic elements. A **covalent bond** *results from the* sharing *of one or more pairs of electrons between atoms*. This bond model of electron sharing is used in bonding situations where all of the atoms are nonmetals.

A consideration of the properties of numerous compounds suggests the existence of two types of chemical bonds. Some compounds such as sodium chloride (table salt), ammonium nitrate, and calcium sulfate, have rigid crystalline structures, are brittle, and have relatively high melting points (above 300°C). Additionally, such compounds conduct an electric current in the liquid state or in aqueous solution. Bonding in compounds with the preceding general properties is now known to involve ionic bonds.

In contrast, many other compounds have general properties quite different from those previously mentioned. Among these latter compounds are water, carbon dioxide, ammonia, ethyl alcohol, sucrose (table sugar), and aspirin. Such compounds have much lower melting points and tend to be gases, liquids, or low melting solids. They do not conduct an electric current in the liquid state or in aqueous solution as do compounds containing ionic bonds. Bonding in this second type of compound is now known to involve covalent bonding.

In Sections 6.4 and 6.8, we will describe the characteristics of ionic bonds and covalent bonds. Prior to this, however, we will discuss two fundamental concepts that are common to and necessary for understanding both bonding models. These two concepts are:

1. Certain electrons, called valence electrons, are the electrons involved in bonding.
2. Certain arrangements of electrons are more stable than other arrangements of electrons. This concept is known as the *octet rule*.

chemical bonds

ionic bond

covalent bond

6.2 VALENCE ELECTRONS AND ELECTRON-DOT STRUCTURES

Certain electrons, which are called *valence electrons*, are particularly important in determining the bonding characteristics of a given atom. For representative and noble-gas elements (Section 5.8), **valence electrons** *are the electrons in the outermost electron shell, which is the shell with the highest shell number* (n). Valence electrons will always be found in either *s* or *p* subshells. Note the restriction on the use of this definition; it applies only to representative and noble-gas elements. Most of the common elements are representative elements; thus, this definition is quite useful. We will not consider the more complicated valence electron definitions for transition or inner transition elements (Section 5.8) in this text; the presence of incompletely filled *inner d* or *f* subshells is the complicating factor in the definitions involving these elements.

Example 6.1

Determine the number of valence electrons present in atoms of each of the following elements.

a. $_{12}$Mg   b. $_{14}$Si   c. $_{33}$As

Solution

a. The element magnesium has two valence electrons, as can be seen by examining its electron configuration.

$$1s^2 2s^2 2p^6 \underset{\text{highest value of the electron shell number}}{3}s\overset{\text{number of valence electrons}}{2}$$

The highest value of the electron shell number is $n = 3$. Only two electrons are found in shell 3: the two electrons in the 3s subshell.

b. The element silicon has four valence electrons.

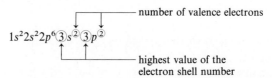

Electrons in two different subshells can simultaneously be valence electrons. The highest shell number is 3 and both the 3s and 3p subshells belong to shell number 3. Hence, all of the electrons in both of these subshells are valence electrons.

c. The element arsenic has five valence electrons.

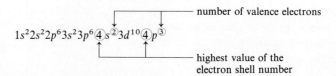

The 3d electrons are not counted as valence electrons because the 3d subshell is in shell 3 and this shell does not have maximum *n* value. Shell 4 is the outermost shell and has maximum *n* value.

FIGURE 6–1 Electron-dot structures for selected elements.

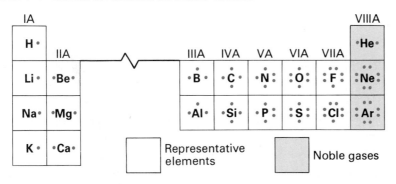

Scientists have developed a shorthand system for designating the number of valence electrons. This system involves the use of electron dot structures. An **electron-dot structure** *consists of an element's symbol with one dot for each valence electron placed around the elemental symbol.* Electron-dot structures for the first 20 elements (all representative or noble-gas elements), arranged as in the periodic table, are given in Figure 6–1.

Note that the location of the dots around the elemental symbols is not critical. The following notations all have the same meaning

$$\overset{\cdot}{\text{Ca}}\cdot \quad \underset{\cdot}{\text{Ca}}\cdot \quad \cdot\overset{\cdot}{\text{Ca}} \quad \cdot\text{Ca}\cdot \quad \underset{\cdot}{\overset{\cdot}{\text{Ca}}}$$

Three important generalizations about valence electrons can be drawn from a study of the structures shown in Figure 6–1.

1. *Representative elements in the same group of the periodic table have the same number of valence electrons.* This should not be surprising to you. Elements in the same group in the periodic table have similar chemical properties as a result of their similar outershell electron configurations (Section 5.6). The electrons in the outermost shell are the valence electrons.
2. *The number of valence electrons for representative elements in a group is the same as the Roman numeral periodic table group number.* For example, the electron-dot structures for oxygen and sulfur, which are both members of Group VIA, show six dots. Similarly, the electron-dot structures of hydrogen, lithium, sodium, and potassium, which are all members of Group IA, show one dot.
3. *The maximum number of valence electrons for any element is eight.* Only the noble gases (Section 5.8), beginning with neon, have the maximum number of eight electrons. Helium, which has only two valence electrons, is the exception in the noble gas family. Obviously, an element with a grand total of two electrons cannot have eight valence electrons. Although shells with n greater than two are capable of holding more than eight electrons, they do so only when they are no longer the outermost shell and are thus not the valence shell. For example, arsenic (Example 6–1c) has 18 electrons in its third shell; however, shell 4 is the valence shell for arsenic.

6.3 THE OCTET RULE

A key concept in modern elementary bonding theory is that certain arrangements of valence electrons are more stable than others. The term *stable* as used here refers to the idea that a system, which in this case is an arrangement of electrons, does not easily undergo spontaneous change.

The valence-electron configurations of the noble gases (helium, neon, argon, krypton, xenon, and radon—Section 5.8) are considered to be the *most stable of all valence-electron configurations*. All of the noble gases except helium possess eight valence electrons, which is the maximum number possible. Helium's valence-electron configuration is $1s^2$. All of the other noble gases possess ns^2np^6 valence-electron configurations, where n has the maximum value found in the atom.

He: $(1s^2)$
Ne: $1s^2\ (2s^22p^6)$
Ar: $1s^2\ 2s^22p^6\ (3s^23p^6)$
Kr: $1s^22s^22p^63s^23p^6\ (4s^2)\ 3d^{10}\ (4p^6)$
Xe: $1s^22s^22p^63s^23p^64s^23d^{10}4p^6\ (5s^2)\ 4d^{10}\ (5p^6)$
Rn: $1s^22s^22p^63s^23p^64s^23d^{10}4p^65s^24d^{10}5p^6\ (6s^2)\ 4f^{14}5d^{10}\ (6p^6)$

Except for helium, each of the noble-gas valence-electron configurations has the common characteristic of having the outermost s and p subshells *completely filled*.

The conclusion that an ns^2np^6 ($1s^2$ for helium) configuration is the most stable of all valence-electron configurations is based on the chemical properties of the noble gases. The noble gases are the *most unreactive* of all the elements. They are the only elemental gases found in nature in the form of individual uncombined atoms. There are no known compounds of helium, neon, and argon and only a very few compounds of krypton, xenon, and radon. The noble gases appear to be "happy" the way they are; they have little or no desire to form bonds to other atoms.

Atoms of many elements that lack this very stable, noble-gas valence-electron configuration tend to attain this configuration through chemical reactions that result in compound formation. This observation is known as the *octet rule* because of the eight valence electrons that are possessed by all of the noble gases except helium. A formal statement of the **octet rule** is: *In compound formation, atoms of elements lose, gain, or share electrons in such a way as to produce a noble-gas electron configuration for each of the atoms involved.*

octet rule

Applications of the octet rule to many different systems have shown that it has value in correctly predicting the observed combining ratios of atoms. For example, it explains why two hydrogen atoms, instead of some other number, are bonded to one oxygen atom in the compound water. It explains why the formula of the ionic compound sodium chloride is NaCl rather than $NaCl_2$, $NaCl_3$, or Na_2Cl.

There are exceptions to the octet rule, but it is still used because of the large amount of information that it is able to correlate. It is particularly effective in explaining compound formation involving only representative elements.

6.4 THE IONIC-BOND MODEL

The concept of transferring one or more electrons between two or more atoms is central to the ionic-bond model. This electron-transfer process produces charged particles called ions. An **ion** *is an atom (or group of atoms) that is electrically charged as the result of the loss or gain of electrons.* Neutrality of atoms is the result of the number of protons (positive charges) being equal to the number of electrons (negative charges). Loss or gain of electrons destroys this proton–electron balance and leaves a net charge on the atom.

ion

If one or more electrons is gained by an atom, a negatively charged ion is produced; excess negative charge is present because electrons now outnumber protons. The loss of one or more electrons by an atom results in the formation of a positively charged ion; more protons are now present than electrons, resulting in excess positive charge. Note that the excess positive charge associated with a positive ion is never caused by proton gain but always by electron loss. If the number of protons remains constant, and the number of electrons decreases, the result is net positive charge. The number of protons, which determines the identity of an element, never changes during ion formation.

The charge on an ion is directly correlated with the number of electrons lost or gained. Loss of one, two, or three electrons gives ions with $+1$, $+2$, and $+3$ charges, respectively. Similarly, a gain of one, two, or three electrons gives ions with -1, -2, and -3 charges, respectively. (Ions that have lost or gained more than three electrons are very seldom encountered.)

The notation for charges on ions is a superscript placed to the right of the elemental symbol. Some examples of ion symbols are

Positive ions: Na^+, K^+, Ca^{2+}, Mg^{2+}, Al^{3+}
Negative ions: Cl^-, Br^-, O^{2-}, S^{2-}, N^{3-}

Note that a single plus or minus sign is used to denote a charge of one, instead of using the notation $^{1+}$ or $^{1-}$. Also note that in multicharged ions, the number precedes the charge sign; that is, the notation for a charge of plus two is $^{2+}$ rather than $^{+2}$.

Example 6.2

Give the symbol for each of the following ions.
a. The ion formed when a barium atom loses two electrons.
b. The ion formed when a phosphorus atom gains three electrons.

Solution
a. A neutral barium atom contains 56 protons and 56 electrons. The barium ion formed by the loss of two electrons would still contain 56 protons, but would have only 54 electrons because two electrons were lost.

$$\begin{array}{r} 56 \text{ protons} = 56 + \text{charges} \\ \underline{54 \text{ electrons} = 54 - \text{charges}} \\ \text{Net charge} = \;\; 2 + \end{array}$$

The symbol of the barium ion is thus Ba^{2+}

b. The atomic number of phosphorus is 15. Thus, 15 protons and 15 electrons are present in a neutral phosphorus atom. A gain of three electrons raises the electron count to 18.

$$\begin{array}{r} 15 \text{ protons} = 15 + \text{charges} \\ \underline{18 \text{ electrons} = 18 - \text{charges}} \\ \text{Net charge} = \;\; 3 - \end{array}$$

The symbol for the ion is P^{3-}.

So far our discussion about electron transfer and ion formation has focused on the loss or gain of electrons by isolated individual atoms. In reality, loss and gain of electrons are partner processes; that is, one does not occur without the other also occurring. Ion formation occurs only when atoms of two elements are present— an element to donate electrons (electron loss) and an element to accept electrons (electron gain). The electrons lost by the one element are the same ones gained by the other element. Thus, positive and negative ion formation must always occur together.

The mutual attraction between positive and negative ions that results from electron transfer constitutes the force that holds the ions together as an ionic compound. This force is referred to as an *ionic bond*. An **ionic bond** *is the attractive force between positive and negative ions that causes them to remain together as a group.*

6.5 FORMULAS FOR IONIC COMPOUNDS

A simple example of ionic bonding occurs between the elements sodium and chlorine in the compound NaCl. Sodium atoms lose (transfer) one electron to chlorine atoms, producing Na^+ and Cl^- ions. The ions combine in a one-to-one ratio to give the compound NaCl.

Why do sodium atoms form Na^+ ions and not Na^{2+} or Na^- ions? Why do chlorine atoms form Cl^- ions rather than Cl^{2-} or Cl^+ ions? In general, what determines the specific number of electrons lost or gained in electron transfer processes? The octet rule (Section 6.3) provides very simple and straightforward answers to these questions. *Atoms tend to gain or lose electrons until they have obtained an electron configuration that is the same as that of a noble gas.*

Consider the element sodium, which has the electron configuration

$$1s^2 2s^2 2p^6 3s^1$$

It can attain a noble-gas configuration by losing one electron (to give it the electron configuration of neon) or by gaining seven electrons (to give it the electron configuration of argon).

$$Na\ (1s^2 2s^2 2p^6 3s^1) \xrightarrow{\text{loss of }1e^-} Na^+\ (1s^2 2s^2 2p^6) \quad \text{electron configuration of neon}$$

$$Na\ (1s^2 2s^2 2p^6 3s^1) \xrightarrow{\text{gain of }7e^-} Na^{7-}\ (1s^2 2s^2 2p^6 3s^2 3p^6) \quad \text{electron configuration of argon}$$

The first process, the loss of one electron, is more energetically favorable than the gain of seven electrons and is the process that occurs. The process that involves the least number of electrons will always be the more energetically favorable process and will be the process that occurs.

Consider the element chlorine, which has the electron configuration

$$1s^2 2s^2 2p^6 3s^2 3p^5$$

It can attain a noble-gas configuration by losing seven electrons, which gives it a neon electron configuration, or by gaining one electron, which gives it an argon electron configuration. The latter occurs for the reason cited previously.

$$Cl\ (1s^2 2s^2 2p^6 3s^2 3p^5) \xrightarrow{\text{loss of }7e^-} Cl^{7+}(1s^2 2s^2 2p^6) \quad \text{electron configuration of neon}$$

$$Cl\ (1s^2 2s^2 2p^6 3s^2 3p^5) \xrightarrow{\text{gain of }1e^-} Cl^-(1s^2 2s^2 2p^6 3s^2 3p^6) \quad \text{electron configuration of argon}$$

The type of considerations we have just used for the elements sodium and chlorine lead to the following generalizations.

1. Metal atoms containing one, two, or three valence electrons (the metals in Groups IA, IIA, and IIIA of the periodic table) tend to lose electrons to acquire a noble-gas electron configuration. The noble gas involved is the one preceding the metal in the periodic table.
    Group IA metals form +1 ions
    Group IIA metals form +2 ions
    Group IIIA metals form +3 ions
2. Nonmetal atoms containing five, six, or seven valence electrons (the nonmetals in Groups VA, VIA, and VIIA of the periodic table) tend to gain electrons to acquire a noble-gas configuration.
    Group VIIA nonmetals form −1 ions
    Group VIA nonmetals form −2 ions
    Group VA nonmetals form −3 ions

The use of electron-dot structures will help you to visualize the formation of ionic compounds through electron-transfer processes. In these compounds, the ions produced achieve noble-gas electron configurations. Let us consider, again, the reaction between sodium, which has one valence electron, and chlorine, which has seven valence electrons, to give NaCl. This reaction can be represented as follows with electron-dot structures.

$$Na\cdot + \cdot\ddot{\underset{\cdot\cdot}{Cl}}: \longrightarrow Na^+[:\ddot{\underset{\cdot\cdot}{Cl}}:]^- \longrightarrow NaCl$$

The loss of an electron by sodium empties its valence shell. The next inner shell, which contains eight electrons (a noble-gas configuration), then becomes the valence shell. After the valence shell of chlorine gains one electron, it then has the desired eight valence electrons.

When sodium, which has one valence electron, combines with oxygen, which has six valence electrons, the oxygen atom requires two sodium atoms to meet its need of two additional electrons.

Note how the need of oxygen for two additional electrons dictates that two sodium atoms are required per oxygen atom; hence the formula Na_2O.

An opposite situation occurs in the reaction between calcium, which has two valence electrons, and chlorine, which has seven valence electrons. Here, two chlorine atoms are required to accommodate electrons transferred from one calcium atom because a chlorine atom can accept only one electron. (It has 7 valence electrons and needs only 8.)

Example 6.3

Show the formation of the following ionic compounds using electron-dot structures.
a. Na_3N b. MgO

Solution
a. Sodium (a Group IA element) has one valence electron, which it would like to lose. Nitrogen, (a Group VA element) has five valence electrons and would

thus like to acquire three more. Three sodium atoms will be required to supply enough electrons for one nitrogen atom.

$$\begin{array}{c} \text{Na} \cdot \\ \text{Na} \cdot + \cdot \ddot{\text{N}} : \\ \text{Na} \cdot \end{array} \longrightarrow \begin{array}{c} \text{Na}^+ \\ \text{Na}^+ \; [:\ddot{\text{N}}:]^{3-} \\ \text{Na}^+ \end{array} \longrightarrow \text{Na}_3\text{N}$$

b. Magnesium (a Group IIA element) has two valence electrons and oxygen (a Group VIA element) has six valence electrons. The transfer of the two magnesium valence electrons to an oxygen atom will result in each atom having a noble-gas electron configuration. Thus, these two elements combine in a one-to-one ratio.

$$\text{Mg} \cdot + \cdot \ddot{\text{O}} : \longrightarrow \text{Mg}^{2+}[:\ddot{\text{O}}:]^{2-} \longrightarrow \text{MgO}$$

It is not always necessary or convenient to draw out electron-dot structures when determining the formula for an ionic compound. Formulas for ionic compounds can be written directly by using the charges associated with the ions being combined and the fact that the total amount of positive and negative charge must add up to zero. Electron loss always equals electron gain in an electron-transfer process. Consequently, ionic compounds are always neutral; no net charge is present. The total positive charge present on the ions that have lost electrons always is exactly counterbalanced by the total negative charge on the ions that have gained electrons. Thus, *the ratio in which positive and negative ions combine is the ratio that achieves charge neutrality for the resulting compound.*

The correct combining ratio when K^+ ions and S^{2-} ions combine is two-to-one. Two K^+ ions (each of plus-one charge) will be required to balance the charge on a single S^{2-} ion.

$$\begin{array}{rl} 2(K^+): & (2 \text{ ions}) \times (\text{charge of } +1) = +2 \\ S^{2-}: & (1 \text{ ion}) \times (\text{charge of } -2) = -2 \\ \hline & \text{net charge} = 0 \end{array}$$

Example 6.4 gives further illustrations of the procedures needed to determine correct combining ratios between ions and write correct ionic formulas from the combining ratios. Three items to note about all ionic formulas are:

1. The symbol for the positive ion is always written first.
2. The charges on the ions that are present are *not* shown in the formula. Knowledge of charges is necessary to determine the formula, but once it is determined, the charges are not explicitly written.
3. The numbers in the formula (the subscripts) give the combining ratio for the ions.

Example 6.4

Determine the formula for the compound that is formed when each of the following pairs of ions interact.

a. Na^+ and P^{3-} b. Be^{2+} and P^{3-}

Solution

a. The Na^+ and P^{3-} ions will combine in a three-to-one ratio because this combination will cause the total charge to add up to zero. Three Na^+ ions give a total positive charge of 3. One P^{3-} ion results in a total negative charge of 3. Thus, the formula for the compound will be Na_3P.

b. The numbers in the charges for these ions are 2 and 3. The lowest common multiple of 2 and 3 is 6 ($3 \times 2 = 6$). Thus, we will need 6 units of positive charge and 6 units of negative charge. Three Be^{2+} ions are needed to give the 6 units of positive charge and two P^{3-} ions are needed to give the 6 units of negative charge. The combining ratio of ions is three-to-two, and the formula is Be_3P_2.

The strategy of finding the lowest common multiple of the numbers in the charges of the ions will always work.

Before leaving the subject of ions, ionic bonds, and formulas for ionic compounds, let us quickly review some of the key principles about ionic bonding.

1. Ionic compounds usually contain both a metallic and a nonmetallic element.
2. The metallic element atoms lose electrons to produce positive ions and the nonmetallic element atoms gain electrons to produce negative ions.
3. The electrons lost by the metal atoms are the same ones that are gained by the nonmetal atoms. Electron loss must always equal electron gain.
4. The ratio in which positive metal ions and negative nonmetal ions combine is the ratio that achieves charge neutrality for the resulting compound.
5. Metals from Groups IA, IIA, and IIIA of the periodic table form ions with charges of $+1$, $+2$, and $+3$, respectively. Nonmetals of Groups VIIA, VIA, and VA of the periodic table form ions with charges of -1, -2, and -3, respectively.

6.6 STRUCTURE OF IONIC COMPOUNDS

The term *molecule* is not appropriate for describing the smallest unit of an ionic compound. In the solid state, ionic compounds consist of an extended array of alternating positive and negative ions. **Ionic solids** *consist of positive and negative ions arranged in such a way that each ion is surrounded by nearest neighbors of the opposite charge.* Any given ion is bonded by electrostatic (positive–negative) attractions to all of the other ions of opposite charge immediately surrounding it. Figure 6–2 shows a two-dimensional cross-section and a three-dimensional view of the arrangement of ions for the ionic compound NaCl (table salt).

ionic solids

We can see in Figure 6–2 that discrete molecules do not exist in an ionic solid. Therefore, the formulas for these solids (Section 6.5) cannot represent the composition of a molecule of the substance. Such formulas represent the simplest ratio in which the atoms combine. For example, in NaCl, there is no single partner for a sodium ion; there are six immediate neighbors (chloride ions) that are equidistant

FIGURE 6–2 Two-dimensional cross-section and three-dimensional view of the ionic solid NaCl.

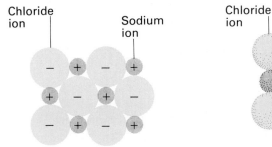
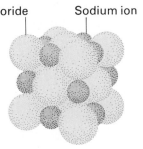

from it. A chloride ion in turn has six immediate sodium neighbors. The formula NaCl represents the fact that in this solid, sodium and chloride ions are present in a one-to-one ratio.

Although the formulas for ionic solids only represent ratios, they are used in equations and in chemical calculations such as molecular weights in the same way as the formulas for molecular species. Remember that they cannot be interpreted as indicating that molecules exist for these substances. Thus, the molecule is not the smallest unit capable of a stable existence for *all* pure substances (Section 2.4)—ionic solids are exceptions.

6.7 NOMENCLATURE FOR IONIC COMPOUNDS

All of the examples of ionic compounds in Sections 6.4 and 6.5 were examples of binary ionic compounds; that is, ionic compounds in which only two elements are present. Names for this type of compound are assigned by using the following rule: *The full name of the metallic element is given first, followed by a separate word containing the stem of the nonmetallic element name and the suffix -ide.* Thus, in order to name the compound NaF, we start with the name of the metal (sodium), follow it with the stem of the name of the nonmetal (fluor-), and then add the suffix -ide. The name becomes *sodium fluoride*.

The stem of the name of the nonmetal is always the first few letters of the nonmetal's name, that is, the name of the nonmetal with its ending chopped off. Table 6–1 gives the stem part of the name for the most common nonmetallic elements. The name of the metal ion present is always exactly the same as the name of the metal itself; the metal's name is never shortened. Example 6.5 illustrates the use of this rule for naming binary ionic compounds.

Example 6.5

Name the following binary ionic compounds.

a. MgO   b. Al_2S_3 c. K_3N d. $CaCl_2$

Solution

The general pattern for naming binary ionic compounds is

$$\text{name of metal} + \text{stem of name of nonmetal} + \text{-ide}$$

a. The metal is magnesium and the nonmetal is oxygen. Thus, the compound name is magnesium *ox*ide.
b. The metal is aluminum and the nonmetal is sulfur; the compound name is aluminum *sulf*ide. Note that no mention is made of the subscripts present in the formula—the 2 and the 3. The name of an ionic compound never contains any reference to formula subscript numbers. There is only one ratio in which aluminum and sulfur atoms combine. Thus, just telling the names of the elements present in the compound is adequate nomenclature.
c. Potassium (K) and nitrogen (N) are present in the compound, and its name is potassium *nitr*ide.
d. The compound name is calcium *chlor*ide.

So far in our discussion of ionic compounds, it has been assumed that the only behavior allowable for an element is predicted by the octet rule. This is a good assumption for nonmetals and most representative element metals. However, there are many other metals that exhibit a less predictable behavior because they are

TABLE 6-1 Names of Some Common Nonmetal Ions

Element	Stem	Name of Ion	Formula of Ion
bromine	brom-	bromide	Br^-
carbon	carb-	carbide	C^{4-}
chlorine	chlor-	chloride	Cl^-
fluorine	fluor-	fluoride	F^-
hydrogen	hydr-	hydride	H^-
iodine	iod-	iodide	I^-
nitrogen	nitr-	nitride	N^{3-}
oxygen	ox-	oxide	O^{2-}
phosphorus	phosph-	phosphide	P^{3-}
sulfur	sulf-	sulfide	S^{2-}

able to form more than one type of ion. For example, iron forms both Fe^{2+} ions and Fe^{3+} ions, depending on chemical circumstances. All of the inner-transition elements, most of the transition elements, and a few representative elements (some metals in the *p* area of the periodic table) exhibit this variable ionic charge behavior.

When naming compounds that contain metals with variable ionic charges, the charge on the metal ion must be incorporated in the name. This is done by using Roman numerals. For example, the chlorides of Fe^{2+} and Fe^{3+} ($FeCl_2$ and $FeCl_3$, respectively) are named iron(II) chloride and iron(III) chloride. Likewise, CuO is named copper(II) oxide. If you are uncertain about the charge on the metal ion in an ionic compound, use the charge on the nonmetal ion (which does not vary) to calculate it. For example, in order to determine the charge on the copper ion in CuO, you can note that the oxide ion carries a -2 charge because oxygen is in Group VIA. This means that the copper ion must have a $+2$ charge to counterbalance the -2 charge.

Example 6.6

Name the following binary ionic compounds, each of which contains a metal whose ionic charge can vary.

a. AuCl   b. Fe_2O_3

Solution

We will need to indicate the magnitude of the charge on the metal ion in the name of each of these compounds by means of a Roman numeral.

a. To calculate the metal ion charge, use the fact that total ionic charge (both positive and negative) must add up to zero.

$$(\text{gold charge}) + (\text{chlorine charge}) = 0$$

The chloride ion has a -1 charge (Section 6.5). Therefore,

$$(\text{gold charge}) + (-1) = 0$$

Thus,

$$\text{gold charge} = +1$$

Therefore, the gold ion present is Au^+, and the name of the compound is gold(I) chloride.

b. For charge balance in this compound we have the equation

$$2 \, (\text{iron charge}) + 3 \, (\text{oxygen charge}) = 0$$

Note that we have to take into account the number of each kind of ion present (2 and 3 in this case). Oxide ions carry a -2 charge (Section 6.5). Therefore,

$$2 \text{ (iron charge)} + 3(-2) = 0$$
$$2 \text{ (iron charge)} = +6$$
$$\text{iron charge} = +3$$

Here, we are interested in the charge on a single iron ion ($+3$) and not in the total positive charge present ($+6$). The compound is named iron(III) oxide because Fe^{3+} ions are present. As is the case for all ionic compounds, the name does not contain any reference to the numerical subscripts in the compound's formula.

In order to correctly name binary ionic compounds, we must know which metals form only one type of ion and which ones form more than one type of ion. In the former case, we do not use a Roman numeral in the compound's name, but in the latter case, we do. To know when to use Roman numerals in ionic compound names, learn the metals that form only one type of ion; their number is much smaller. Figure 6–3 shows the metals that always form a single type of ion in ionic compound formation. Ionic compounds that contain these metals are the only ones without Roman numerals in their names.

6.8 THE COVALENT-BOND MODEL

In binary ionic compounds, the two atoms involved in a given ionic bond are a metal and a nonmetal; these atoms are quite *dissimilar*. These dissimilar atoms are complementary to each other; one atom (the metal) likes to lose electrons and the other atom (the nonmetal) likes to gain electrons. The net result is the transfer of one or more electrons.

Covalent bonds are formed between *similar* or even *identical* atoms. Most often,

FIGURE 6–3 Periodic table showing the metals (in color) that form only one type of ion. The names for ionic compounds containing these metals do not need a Roman numeral.

H																	He
Li	Be											B	C	N	O	F	Ne
Na	Mg											Al	Si	P	S	Cl	Ar
K	Ca	Sc	Ti	V	Cr	Mn	Fe	Co	Ni	Cu	Zn	Ga	Ge	As	Se	Br	Kr
Rb	Sr	Y	Zr	Nb	Mo	Tc	Ru	Rh	Pd	Ag	Cd	In	Sn	Sb	Te	I	Xe
Cs	Ba	La	Hf	Ta	W	Re	Os	Ir	Pt	Au	Hg	Tl	Pb	Bi	Po	At	Rn
Fr	Ra	Ac	Unq	Unp	Unh	Uns	Uno	Une									

Ce	Pr	Nd	Pm	Sm	Eu	Gd	Tb	Dy	Ho	Er	Tm	Yb	Lu
Th	Pa	U	Np	Pu	Am	Cm	Bk	Cf	Es	Fm	Md	No	Lr

two nonmetal atoms are involved. It is not reasonable to suppose that one atom would give up electrons to another atom when the atoms are identical or very similar. The concept of *electron sharing* explains bonding between similar or identical atoms. In electron sharing, two nuclei attract the same electrons, and the resulting attractive forces hold the two nuclei together. The formation of a covalent bond always involves this process of electron sharing.

The hydrogen molecule (H_2) is the simplest covalent bonding situation that exists. Hydrogen, with just one $1s$ electron, needs one more electron to obtain the noble-gas configuration of helium ($1s^2$). A hydrogen atom accomplishes this by sharing its lone electron with another hydrogen atom, which in turn reciprocates, sharing its electron with the first hydrogen atom. The net result is the formation of an H_2 molecule. The two shared electrons in an H_2 molecule do double duty, helping each of the hydrogen atoms achieve a noble-gas configuration.

Covalent bonds are represented by electron-dot structures in much the same way that we used them for ionic bonds. A pair of dots placed between the symbols of the bonded atoms indicates the shared pair of electrons. The electron-dot notation for H_2 is

$$\text{H} \cdot \quad \cdot \text{H} \longrightarrow \text{H:H}$$

shared electron pair

An alternate way of representing the shared electron pair in a covalent bond is to draw a dash between the symbols of the bonded atoms.

$$\text{H—H}$$

Both of the atoms in H_2 have access to the two electrons of the shared electron pair. The concept of overlap of orbitals will help you to visualize this. As shown in Figure 6–4a, suppose two hydrogen atoms are moving toward each other to eventually form H_2. As long as the atoms are well separated, the $1s$ electrons on each of the two atoms are independent of each other. As the atoms get closer together, the orbitals containing the electrons eventually overlap and create an orbital common to both atoms (Figure 6–4b). When this happens, the two electrons move throughout the overlap region between the nuclei and are *shared* by both nuclei.

A situation where two hydrogen atoms in an H_2 molecule are sharing two electrons between them is more stable than one where two separate hydrogen atoms exist. Thus, hydrogen atoms are always found in pairs (as H_2 molecules) in samples of elemental hydrogen.

Using the octet rule and electron-dot structures, let us consider some other simple molecules where covalent bonding is present. The element chlorine, which

FIGURE 6–4 Overlap of orbitals creates a situation where electron sharing can take place.

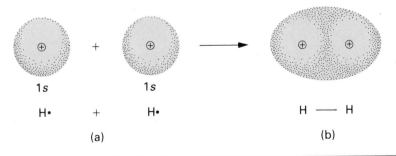

is located in Group VIIA of the periodic table, has seven valence electrons. Its electron-dot structure is

$$\cdot \ddot{\underset{..}{\text{Cl}}}:$$

Chlorine needs only one electron to achieve the octet of electrons that enable it to have a noble-gas electron configuration. When chlorine bonds to other nonmetals, the octet of electrons is completed by means of electron sharing. The molecules HCl, Cl$_2$ and BrCl, whose electron-dot structures follow, are representative of this situation.

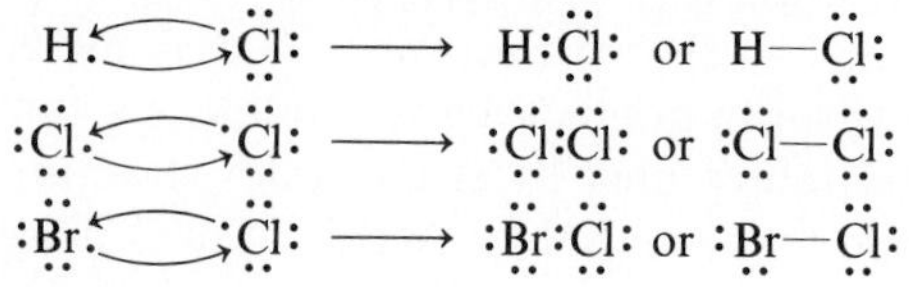

H:⌒:Ċl: ⟶ H:Ċl: or H—Ċl:
:Ċl:⌒:Ċl: ⟶ :Ċl:Ċl: or :Ċl—Ċl:
:Br:⌒:Ċl: ⟶ :Br:Ċl: or :Br—Ċl:

Note that in each of these molecules, chlorine atoms have four pairs of electrons around them (an octet) and only one of the four pairs is involved in electron sharing. The three pairs of electrons on each chlorine atom that are not taking part in the bonding are called *nonbonding electron pairs* or *unshared electron pairs*.

The number of covalent bonds that an atom forms is equal to the number of electrons it needs to achieve a noble-gas configuration. Note that the chlorine atoms in HCl, Cl$_2$, and BrCl each formed only one covalent bond. For chlorine, seven valence electrons plus one electron acquired by electron sharing (one bond) gives the eight valence electrons needed for a noble-gas electronic configuration. The elements oxygen, nitrogen, and carbon have six, five, and four valence electrons, respectively. Therefore, these elements form two, three, and four covalent bonds, respectively. The number of covalent bonds that these elements form is reflected in the formulas of their simplest hydrogen compounds—H$_2$O, NH$_3$, and CH$_4$. Electron-dot diagrams for these molecules follow.

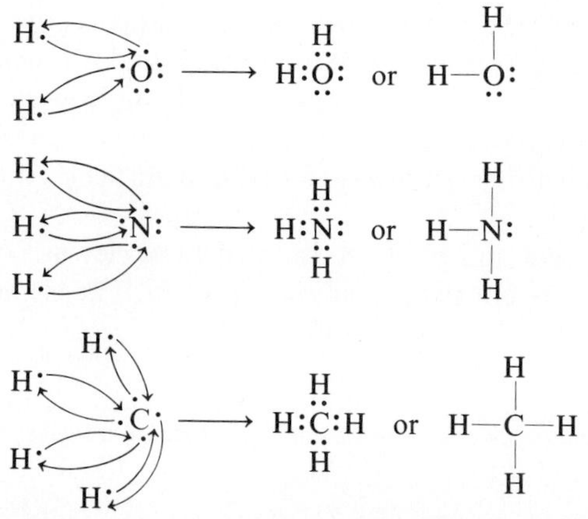

We see here that, just as the octet rule was useful in determining the ratio of ions in ionic compounds, it can also be used to predict formulas in covalent compounds. Example 6.7 gives additional illustrations of the use of the octet rule to determine formulas for molecular compounds.

Example 6.7

Write electron-dot structures for the simplest binary compound that can be formed from the following pairs of elements.

a. nitrogen and iodine   b. sulfur and hydrogen

Solution

a. Nitrogen is in Group VA of the periodic table and has five valence electrons. It will want to form three covalent bonds. Iodine, in Group VIIA of the periodic table, has seven valence electrons and will want to form only one covalent bond. Therefore, three iodine atoms will be needed to meet the needs of one nitrogen atom. The electron-dot structure for this molecule is

Each atom in NI_3 has an octet of electrons, which is circled in color in the following diagram.

b. Sulfur has six valence electrons and hydrogen has one valence electron. Thus, sulfur will form two covalent bonds (6 + 2 = 8) and hydrogen will form one covalent bond (1 + 1 = 2). Remember that for hydrogen an "octet" is two electrons; the noble gas that hydrogen mimics is helium, which has only two valence electrons.

6.9 MULTIPLE-COVALENT BONDS

In our discussion of covalent bonding in Section 6.8, all of the molecules that were chosen as examples to illustrate various aspects of bonding contained only *single-covalent bonds*. A **single-covalent bond** is *a bond where a single pair of electrons is shared between two atoms.*

Many molecules exist where two atoms share two or three pairs of electrons in order to provide a complete octet of electrons for each atom involved in the bonding. These bonds are called double-covalent bonds and triple-covalent bonds. A **double-covalent bond** is *a bond where two atoms share two pairs of electrons.* A double-covalent bond is stronger than a single-covalent bond, but not twice as strong, because the two electron pairs repel each other and cannot become fully concentrated between the two atoms. (Bond strength is a measure of how much energy it takes to break a bond.) A **triple-covalent bond** is *a bond where two atoms share three pairs of electrons.* A triple-covalent bond is stronger than a single-covalent bond or a double-covalent bond, but not three times as strong as a single bond for the reason previously mentioned. Now let us consider some molecules where double or triple bonds are present.

A diatomic N_2 molecule, the form in which nitrogen occurs in the atmosphere, contains a triple-covalent bond. It is the simplest known triple-covalent bond. A nitrogen atom has five valence electrons and needs three additional electrons to complete its octet.

In an N_2 molecule, the only sharing that can take place is between two nitrogen atoms because they are the only atoms present. Thus, in order to acquire a noble-gas electron configuration, each nitrogen atom must share three of its electrons with the other nitrogen atom.

$$:\!\ddot{N}\!\cdot \rightleftarrows \cdot\!\ddot{N}\!: \longrightarrow \;:\!N\!:\!:\!:\!N\!: \;\;\text{or}\;\; :\!N\!\equiv\!N\!:$$

Notice how all three shared electron pairs are placed in the area between the two nitrogen atoms in this bonding diagram. Note also that three lines are used to denote a triple-covalent bond, paralleling the use of one line to denote a single-covalent bond.

When "bookkeeping" electrons in an electron-dot structure to make sure that all atoms in the molecule have achieved their octet of electrons, *all* electrons in a double or triple bond are considered to "belong" to *both* of the atoms involved in that bond. The bookkeeping for the N_2 molecule would be

Each of the circles around a nitrogen atom contains eight valence electrons. Again, all of the electrons in a double or triple bond are considered to belong to each of the atoms in the bond. Circles are never drawn to include just some of the electrons in a double or triple bond.

A slightly more complicated molecule that contains a triple-covalent bond is the molecule C_2H_2. A carbon–carbon triple bond is present, as well as two carbon–hydrogen single bonds. The arrangement of valence electrons in C_2H_2 is

$$H\!:\!\dot{C}\!\cdot \rightleftarrows \cdot\!\dot{C}\!:\!H \longrightarrow \;H\!:\!C\!:\!:\!:\!C\!:\!H \;\;\text{or}\;\; H\!-\!C\!\equiv\!C\!-\!H$$

The two atoms in a triple-covalent bond are usually the same element. They do not, however, have to be the same element. The molecule HCN contains a heteroatomic triple bond.

$$H\!:\!C\!:\!:\!:\!N\!: \;\;\text{or}\;\; H\!-\!C\!\equiv\!N\!:$$

Double-covalent bonds are found in numerous molecules. A common molecule that contains bonding of this type is carbon dioxide (CO_2). In fact, there are two carbon–oxygen double-covalent bonds present in CO_2.

$$:\!\ddot{O}\!\cdot \rightleftarrows \cdot\!\dot{C}\!\cdot \rightleftarrows \cdot\!\ddot{O}\!: \longrightarrow \;:\!\ddot{O}\!:\!:\!C\!:\!:\!\ddot{O}\!: \;\;\text{or}\;\; :\!\ddot{O}\!=\!C\!=\!\ddot{O}\!:$$

Note in the following diagram how the circles are drawn for the octet of electrons around each of the atoms in CO_2.

Not all elements can form double- or triple-covalent bonds. There must be at least two vacancies in an atom's valence electron shell prior to bond formation if it is to participate in a double bond and at least three vacancies are necessary for triple bond formation. This requirement eliminates Group VIIA elements (fluorine, chlorine, bromine, iodine) and hydrogen from participating in such bonds. The Group VIIA elements have seven valence electrons and one vacancy, and hydrogen has one valence electron and one vacancy. All covalent bonds formed by these elements are single-covalent bonds.

Example 6.8

Write an electron-dot structure to describe the bonding in the covalent compound C_2H_4 given the following arrangement of atoms in the compound.

$$\begin{array}{cc} H & H \\ C\ C \\ H & H \end{array}$$

Solution

We will follow the following four steps to determine the electron-dot structure.

Step 1: Determine the number of valence electrons each atom in the molecule possesses.

Step 2: Determine the needs of each atom; that is, the number of electrons each atom must obtain from other atoms through sharing in order to obtain a noble-gas electron configuration.

Step 3: Determine how the needs of each atom will be met. When doing this, we must take into account the arrangement of the atoms within the molecule. We need to know which atoms are bonded to each other.

Step 4: Write an electron-dot structure that is consistent with our needs analysis.

Application of these steps gives the following results for C_2H_4.

Step 1: Each carbon atom (Group IVA) has four valence electrons. Each hydrogen atom (Group IA) has one valence electron.

Step 2: Each hydrogen atom has one valence electron and needs two. Each carbon atom has four valence electrons and needs eight.

Step 3: The one electron each hydrogen atom needs must be obtained from the carbon atom to which it is bonded. Each carbon atom is bonded to three other atoms, each of which shares electrons with the carbon. Two hydrogen atoms each share one electron with the carbon and the other carbon atom shares two electrons with it.

Step 4: The electron-dot structure consistent with the information in Step 3 is

$$\begin{array}{cc} H & H \\ \overset{..}{C}::\overset{..}{C} \\ H & H \end{array}$$

Each carbon atom has eight electrons and each hydrogen atom has two electrons.

Using lines to denote bonds, we can also write the bonding structure for C_2H_4 as

$$\begin{array}{cc} H & H \\ \diagdown\diagup \\ C=C \\ \diagup\diagdown \\ H & H \end{array}$$

6.10 COORDINATE-COVALENT BONDS

So far, in our examples of single-, double-, and triple-covalent bonding, each of the bonded atoms has contributed an equal number of electrons to the bond— one each for a single-covalent bond, two each for a double-covalent bond, and three each for a triple-covalent bond. A few molecules exist in which all of the covalent bonding within the molecule cannot be explained in this manner; instead, the concept of coordinate covalency must be invoked.

A **coordinate-covalent bond** *is a bond in which both electrons of a shared pair come from one of the two atoms involved in the bond.* Coordinate-covalent bonding allows an atom that has two or more vacancies in its valence shell to share a pair of nonbonding electrons that are located on another atom.

Once a coordinate-covalent bond is formed, there is no way to distinguish it from any of the other covalent bonds in a molecule; all electrons are identical regardless of their source. The main use of the concept of coordinate covalency is to help rationalize the existence of certain molecules and ions whose electron-bonding arrangement would otherwise present problems. Again, once a coordinate-covalent bond is formed, it is no different from any other covalent bond.

The molecule N_2O contains a single coordinate-covalent bond.

$$:N:::N:\overset{\times\times}{\underset{\times\times}{O}}: \quad \text{coordinate-covalent bond}$$

The nitrogen–nitrogen triple bond in N_2O is a normal covalent bond; the nitrogen–oxygen bond is a coordinate-covalent bond, where both electrons are supplied by the nitrogen atom.

The triple-covalent bond joining carbon and oxygen in carbon monoxide (CO) is a coordinate-covalent bond.

$$:C:::O:$$

Four of the six electrons in the triple bond can be considered to have come from the oxygen atom. Because carbon has only four valence electrons before bonding, it must share four electrons (from oxygen) in order to achieve an octet of electrons. On the other hand, oxygen has six valence electrons before bonding and thus needs to share only two electrons (from carbon).

6.11 ELECTRONEGATIVITY AND BOND POLARITY

The ionic and covalent models for bonding seem to represent two very distinct forms of bonding. Actually, the two models are closely related to each other; they are the extremes of a broad continuum of bonding patterns. The close relationship between these models is apparent when the concept of electronegativity is considered.

The electronegativity concept has its origins in the fact that the atoms of various elements have differing abilities to attract shared electrons (in a bond) to themselves. Some elements are better electron-attractors than other elements. **Electronegativity** *is a measure of the relative attraction that an atom has for the shared electrons in a bond.*

The actual numerical values of electronegativity for the more common representative elements are given in Figure 6–5. Note that electronegativity values are unitless numbers on a relative scale that runs between 0 and 4. Fluorine, the most electronegative of all the elements, has a value of 4.0 on the scale (the maximum value possible). Other very good electron-attractors are the elements whose periodic table positions are very close to that of fluorine (oxygen, nitrogen, chlorine, sulfur, and so on). *The higher the electronegativity value for an element is, the greater the electron-attracting ability of atoms of that element for shared electrons in bonds is.* Note the trends in electronegativity values that are evident in the data of

FIGURE 6–5 Relative electronegativity values for selected representative elements.

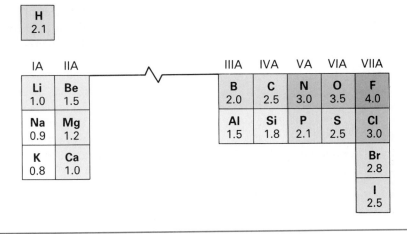

Figure 6–5. Electronegativity increases from left to right across a period of the periodic table and decreases from top to bottom in a group of the periodic table. These two trends result in nonmetals generally having higher electronegativities than metals. This fact is consistent with our previous generalization (Section 6.5) that metals tend to lose electrons and nonmetals tend to gain electrons when an ionic bond is formed. Metals (low electronegativities, poor electron attractors) will give up electrons to nonmetals (high electronegativities, good electron attractors).

It is important to note for future consideration how the electronegativity of the element hydrogen (located to the far left in the periodic table) compares with that of the period 2 elements. Hydrogen's value of electronegativity is between that of boron and carbon.

Li	Be	B	H	C	N	O	F
1.0	1.5	2.0	2.1	2.5	3.0	3.5	4.0

increasing electronegativity →

Electronegativity for an element is not a directly measureable quantity. Rather, electronegativity values are calculated from bond energy information and other related experimental data. Values differ from element to element because of differences in atom size, nuclear charge, and number of inner-shell (non-valence) electrons.

When two identical atoms (atoms of equal electronegativity) share one or more pairs of electrons, each atom exerts the same attraction for the electrons, which results in the electrons being *equally* shared. This type of bond is called a *nonpolar-covalent bond*. A **nonpolar-covalent bond** is one in which there is equal sharing of electrons.

When the two atoms involved in the bond are not identical, and thus have different electronegativities, the electron-sharing situation is quite different. The atom that has highest electronegativity will attract the electrons more strongly than the other atom; this results in an *unequal* sharing of electrons. This type of covalent bond is called a polar-covalent bond. A **polar-covalent bond** is one in which there is unequal sharing of bonding electrons.

nonpolar-covalent bond

polar-covalent bond

(a) Equal sharing

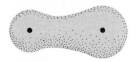

(b) Slightly unequal sharing

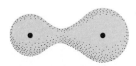

(c) Very unequal sharing

Positive ion   Negative ion

(d) Electron transfer

FIGURE 6–6 The continuum of bonding types.

The significance of a polar-covalent bond is that it creates partial positive and negative charges on atoms. Although both atoms involved in the bond were initially uncharged, the uneven sharing of electrons produces a partial negative charge, a charge of less than one unit, on one bonded atom (the more electronegative one) and an equivalent partial positive charge on the other bonded atom. This means that one end of the bond is negative with respect to the other end.

The partial positive and negative charges associated with the atoms involved in a polar-covalent bond are often indicated by using a notation that involves the lowercase Greek letter δ (delta). A δ^- symbol, meaning a "partial negative charge" is placed above the relatively negative atom of the bond and a δ^+ symbol, meaning a "partial positive charge," is placed above the relatively positive atom.

Using delta notation, the bond in hydrogen fluoride (HF) would be depicted as

$$\overset{\delta^+}{\text{H}} - \overset{\delta^-}{\text{F}}$$

Fluorine is the more electronegative of the two elements; it dominates the electron-sharing process and draws the electrons closer to itself. Hence, the fluorine end of the bond has the δ^- designation; the more electronegative element will always have the δ^- designation.

We now see that most chemical bonds are not 100% covalent (equal sharing) or 100% ionic (no sharing). Instead, most bonds are somewhere in between (unequal sharing). A pictorial representation of the continuum of bonding types that are possible because of the occurrence of unequal sharing is shown in Figure 6–6. Figure 6–6a shows the equal sharing situation that results when both atoms are identical. The sharing must be equal because identical nuclei must affect the bonding electrons in the same way.

Whenever two nuclei differ in their abilities to attract a pair of bonding electrons (electronegativity), unequal sharing results. Figure 6–6b shows the situation where the electronegativity difference is small. Electron sharing will be close to being equal, but not exactly equal. Note that the electron distribution is no longer symmetrical. The bonding electrons spend more time associated with the nucleus that has the greater electronegativity. This is the nucleus on the right side in Figure 6–6b.

Figure 6–6c depicts electron density distribution when a relatively large difference in electron-attracting ability exists between nuclei. The sharing of electrons here can be described as being very unequal.

Finally, when the electron-attracting ability difference is very large, one atom wins the battle. Electron transfer is said to occur, and the situation depicted in Figure 6–6d results. This situation corresponds to what we have previously called an ionic bond.

We can obtain an estimate of the "degree of inequality" of electron sharing in a chemical bond from the electronegativity values of the bonded atoms. The greater the difference in electronegativity is, the greater the inequality in the electron-sharing process is.

It is still convenient to use the terms *ionic* and *covalent* when describing chemical bonds, based on the following guidelines, which take into account electronegativity differences and the resulting unequal sharing of electrons.

Rule 1: When there is zero difference in electronegativity between bonded atoms, the bond is called a *nonpolar-covalent bond*.

Rule 2: When the electronegativity difference between bonded atoms is greater than zero but less than 1.7, the bond is called a *polar-covalent bond*.

Rule 3: When the difference in electronegativity between bonded atoms is 1.7 or greater, the bond is called an *ionic bond*.

The preceding guidelines are the basis for deciding whether to formulate the bonding description for a compound in terms of an ionic electron-dot structure or a covalent electron-dot structure.

Example 6.9

Indicate whether each of the following bonds should be designated as ionic, polar covalent, or nonpolar covalent. Also designate the direction of polarity using delta notation for any polar covalent bonds.

a. N—O   b. Na—F

Solution

a. The electronegativities of nitrogen and oxygen are 3.0 and 3.5, respectively. The electronegativity difference, which is obtained by subtracting the smaller electronegativity value from the larger value, is

$$\text{electronegativity difference} = 3.5 - 3.0 = 0.5$$

The bond is *polar covalent* because the electronegativity difference is greater than zero, but less than 1.7.

The oxygen end of the bond will be negative relative to the nitrogen end because oxygen is the more electronegative of the two elements. Thus, the direction of polarity is

$$\overset{\delta^+ \delta^-}{\text{N}-\text{O}}$$

b. The electronegativity of fluorine is 4.0 and that of sodium is 0.9. Thus, the electronegativity difference is 3.1. The term *ionic* is used to describe bonds where the electronegativity difference is 1.7 or greater. The polarity of the Na—F bond can best be described in terms of ions—complete electron transfer.

6.12 MOLECULAR POLARITY

Molecules, as well as bonds, can have polarity. As we shall see shortly, just because a molecule contains polar bonds does not mean that the molecule as a whole is polar. Molecular polarity depends on the polarity of the bonds within a molecule and the geometry of the molecule (when three or more atoms are present).

Molecular geometry *describes the way in which atoms in a molecule are arranged in space relative to each other.* All molecules containing three or more atoms have characteristic three-dimensional shapes. For example, the triatomic CO_2 molecule is linear; that is, its three atoms lie in a straight line (Figure 6–7a). On the

molecular geometry

FIGURE 6–7 Molecular geometries of selected molecules.

(a) CO_2 – a linear molecule

(b) H_2O – a nonlinear or angular molecule

(c) NH_3 – a trigonal pyramidal molecule

other hand, a H_2O molecule, which is also a triatomic molecule, has a nonlinear or bent geometry (Figure 6–7b). An NH_3 molecule has a trigonal pyramidal geometry, with the nitrogen atom at the apex and the hydrogen atoms at the base of the pyramid (Figure 6–7c).

Determining the molecular polarity of a *diatomic* molecule is simple because only one bond is present. If that bond is nonpolar, the molecule is nonpolar; if the bond is polar, the molecule is polar.

The collective effect of individual bond polarities must be considered to determine molecular polarity for molecules containing more than one bond (triatomic molecules, tetratomic molecules, and so on). Molecular geometry plays an important role in determining this collective effect. In some instances, because of the symmetrical nature of the geometry of the molecule, the effects of the polar bonds are canceled and a nonpolar molecule results. Let us consider the polarities of three specific triatomic molecules—CO_2, H_2O, and HCN.

In the linear CO_2 molecule (Figure 6–7a), both bonds are polar (oxygen is more electronegative than carbon). Despite the presence of these polar bonds, CO_2 molecules are *nonpolar*. The effects of the two polar bonds are canceled out as a result of the oxygen atoms being arranged symmetrically around the carbon atom. The shift of electronic charge toward one oxygen atom is exactly compensated by the shift of electronic charge toward the other oxygen atom. Thus, one end of the molecule is not negatively charged relative to the other end (a requirement for polarity) and the molecule is nonpolar. This cancellation of individual bond polarities, with arrows used to denote the polarities, is diagrammed as follows.

$$\overset{\leftarrow \quad \rightarrow}{O=C=O}$$

The two individual bond polarities are of equal magnitude (each oxygen affects the carbon atom in the same way) but are opposite in direction, so they cancel each other.

The nonlinear (bent) triatomic H_2O molecule (Figure 6–7b) is polar. The bond polarities associated with the two hydrogen–oxygen bonds do not cancel each other because of the nonlinearity of the molecule.

$$\underset{H \quad \quad H}{\overset{O}{\diagup \diagdown}}$$

As a result of their orientation, both bonds contribute to an accumulation of negative charge on the oxygen atom. The two bond polarities are equal in magnitude but are not opposite in their direction.

The generalization that linear triatomic molecules are nonpolar and nonlinear triatomic molecules are polar, which you might be tempted to make on the basis of our discussion of CO_2 and H_2O molecular polarities, is not valid. The linear molecule HCN, which is polar, invalidates this statement. Both bond polarities contribute to nitrogen acquiring a partial negative charge relative to hydrogen in HCN.

$$\overset{\rightarrow \quad \rightarrow}{H-C\equiv N}$$

(Note that the two polarity arrows were drawn in the same direction because nitrogen is more electronegative than carbon and carbon is more electronegative than hydrogen.)

Molecules that contain four and five atoms commonly have trigonal planar and tetrahedral geometries, respectively. The arrangement of atoms associated

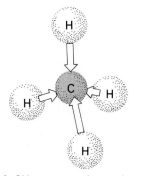

(a) CH_4 — nonpolar molecule

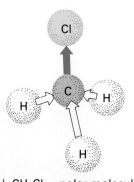

(b) CH_3Cl — polar molecule

FIGURE 6–8 Nonpolar and polar tetrahedral molecules.

with these two geometries is

Trigonal planar   Tetrahedral

Trigonal planar and tetrahedral molecules in which all of the atoms attached to the central atom are identical, such as BF_3 (trigonal planar) and CH_4 (tetrahedral), are *nonpolar*. The individual bond polarities cancel as the result of the highly symmetrical arrangement of atoms around the central atom. (Proof that cancellation does occur involves some trigonometric considerations; it is not an obvious situation.)

If two or more kinds of atoms are attached to the central atom in a trigonal planar or tetrahedral molecule, the molecule will be polar. The high degree of symmetry required for cancellation of the individual bond polarities is no longer present. For example, if one of the hydrogen atoms in CH_4 (a nonpolar molecule) is replaced by a chlorine atom, a polar molecule results, even though the resulting CH_3Cl is still a tetrahedral molecule. A carbon–chlorine bond has a greater polarity than a carbon–hydrogen bond; chlorine has an electronegativity of 3.0 and hydrogen has an electronegativity of only 2.1. Figure 6–8 contrasts the polar CH_3Cl and nonpolar CH_4 molecules. Note that the direction of polarity of the carbon–chlorine bond is opposite to that of the carbon–hydrogen bonds.

Example 6.10

Predict the polarity of each of the following molecules

a. NH_3: trigonal pyramidal   c. N_2O: linear

     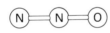

b. H_2S: bent   d. SO_3: trigonal planar

Solution

Knowledge of molecular geometry, which is given for each molecule in this example, is a prerequisite for predicting molecular polarity.

a. Noncancellation of the individual bond polarities in the trigonal pyramidal NH_3 molecule results in it being a *polar* molecule.

The bond polarity arrows all point toward the nitrogen atom because nitrogen is more electronegative than hydrogen.

b. For the bent H_2S molecule, the shift in electron density in the polar sulfur–hydrogen bonds will be toward the sulfur atom because sulfur is more electronegative than hydrogen. The H_2S molecule as a whole is *polar* because of the noncancellation of the individual sulfur–hydrogen bond polarities.

c. The structure of the linear N_2O molecule is unsymmetrical; a nitrogen atom rather than the oxygen atom is the central atom.

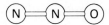

The nitrogen–nitrogen bond is nonpolar; the nitrogen–oxygen bond is polar. The molecule as a whole is *polar* because of the polarity of the nitrogen–oxygen bond.

d. The SO_3 molecule is a trigonal planar molecule. All of the sulfur–oxygen bonds are polar because sulfur and oxygen differ in electronegativity. However, the molecule as a whole is *nonpolar* because of the symmetrical arrangement of the oxygen atoms around the central sulfur atom; this causes the individual bond polarities to cancel.

6.13 NOMENCLATURE FOR COVALENT COMPOUNDS

The simplest and most common type of covalent compound is one that contains two nonmetals (Section 6.8). The names of these compounds are derived by using a rule that is very similar to the one used for naming ionic compounds (Section 6.6). However, one major difference exists. Names for covalent compounds always contain Greek numerical prefixes that give the number of each type of atom present, in addition to the names of the elements present. This is in direct contrast to ionic nomenclature, where formula subscripts are never mentioned when naming compounds.

The basic rule used when constructing the name of a covalent compound is: *The full name of the nonmetal with the lower electronegativity is given first, followed by a separate word containing the stem of the name of the more electronegative nonmetal and the suffix "-ide". Greek numerical prefixes precede the names of both nonmetals.*

Prefix use is necessary because many different compounds exist for most pairs

TABLE 6-2 Greek Numerical Prefixes for the Numbers 1 through 10

Greek Prefix	Number
mono-	1
di-	2
tri-	3
tetra-	4
penta-	5
hexa-	6
hepta-	7
octa-	8
ennea-	9
deca-	10

of nonmetals. For example, all of the following nitrogen–oxygen compounds exist: NO, NO_2, N_2O, N_2O_3, N_2O_4, and N_2O_5. Such diverse behavior between two elements relates to the fact that single-, double-, and triple-covalent bonds exist. The prefixes used are the standard Greek numerical prefixes, which are given for the numbers 1 through 10 in Table 6–2. Example 6.11 shows how these prefixes are used in binary covalent compound nomenclature.

Name the following binary covalent compounds.

a. N_2O_5 b. S_2Cl_2 c. CO_2 d. P_4S_{10} e. CBr_4

Example 6.11

Solution

The names of each of these compounds will consist of two words. These words will have the following general formats.

First word: (prefix) + (full name of least electronegative nonmetal)

Second word: (prefix) + (stem of name of more electronegative nonmetal) + (ide)

a. The elements present are nitrogen and oxygen. The two portions of the name, before adding Greek numerical prefixes, are *nitrogen* and *oxide*. Adding the prefixes gives *dinitrogen* (two nitrogen atoms are present) and *pentoxide* (five oxygen atoms are present). (When an element name begins with a vowel, an *a* or *o* at the end of the Greek prefix is dropped for phonetic reasons, as in pentoxide instead of pentaoxide). Thus, the name of this compound is *dinitrogen pentoxide*.

b. The elements present are sulfur and chlorine. This time, the two portions of the name (including Greek prefixes) are *disulfur* and *dichloride*, which are combined to give the name *disulfur dichloride*.

c. When there is only one atom of the first nonmetal present, it is common to omit the initial prefix of *mono-*. Following this convention, the name of this compound is *carbon dioxide*.

d. The prefix for four atoms is *tetra-* and for ten atoms *deca-*. This compound has the name *tetraphosphorus decasulfide*.

e. Omitting the initial *mono-* (see part c), we name this compound *carbon tetrabromide*.

TABLE 6–3 Selected Covalent Compounds That Have Common Names

Compound Formula	Accepted Common Name
H_2O	water
H_2O_2	hydrogen peroxide
NH_3	ammonia
N_2H_4	hydrazine
CH_4	methane
C_2H_4	ethane
PH_3	phosphine
AsH_3	arsine

There is one standard exception to the use of Greek numerical prefixes when naming covalent compounds. Covalent compounds with hydrogen listed as the first element in the formula are named without prefix use. Thus, the compounds H_2S and HCl are named hydrogen sulfide and hydrogen chloride, respectively.

A few covalent compounds have names that are completely unrelated to the rules we have been discussing. They have common names that were coined prior to the development of systematic rules. At one time, in the early history of chemistry, all compounds had common names. With the advent of systematic nomenclature, most common names were discontinued. A few, however, have persisted and are now officially accepted. The most famous example of this is the compound H_2O, which has the systematic name dihydrogen monoxide (or hydrogen oxide if prefixes are not used). Neither of these names is ever used; H_2O is *water*, a name that is not going to change. Table 6–3 lists additional examples of compounds for which common names are used in preference to systematic names.

6.14 POLYATOMIC IONS

polyatomic ions

A **polyatomic ion** *is a group of covalently bonded atoms that have acquired a charge through the loss or gain of electrons.* Numerous ionic compounds exist in which the positive or negative ion (sometimes both) is polyatomic. Compounds that contain polyatomic ions offer an interesting combination of both ionic and covalent bonding; covalent bonding occurs *within* the polyatomic ion and ionic bonding occurs *between* it and the other ion.

Polyatomic ions are very stable species that generally maintain their identity during chemical reactions. However, they are not molecules; they never occur alone like molecules do. Instead, they always associate with other ions that have opposite charges. Polyatomic ions are *pieces* of compounds, not compounds. Ionic compounds require the presence of both positive and negative ions and neutral overall. Polyatomic ions are always charged species.

Table 6–4 lists the names and formulas of some of the more common polyatomic ions. The table is organized around key elements (other than oxygen) that are present in the polyatomic ions. Note from the table that almost all the polyatomic ions listed contain oxygen atoms. The names, but not necessarily the formulas, of some of these common polyatomic ions should be familiar to you. The polyatomic ions sulfate, hydrogen phosphate, and hydrogen carbonate are all present in the three major types of body fluids—blood plasma, interstitial fluid, and intracellular fluid. Many of the polyatomic ions are found in commercial products such as fertilizers (phosphates, sulfates, nitrates), baking powder (hydrogen carbonate), and building materials (carbonates, sulfates).

There is no easy way to learn the common polyatomic ions; memorization is

TABLE 6-4 Formulas and Names of Some Common Polyatomic Ions

Key Element Present	Formula	Name of Ion
nitrogen	NO_3^-	nitrate
	NO_2^-	nitrite
	NH_4^+	ammonium
sulfur	SO_4^{2-}	sulfate
	HSO_4^-	bisulfate or hydrogen sulfate
	SO_3^{2-}	sulfite
	HSO_3^-	bisulfite or hydrogen sulfite
phosphorus	PO_4^{3-}	phosphate
	HPO_4^{2-}	hydrogen phosphate
	$H_2PO_4^-$	dihydrogen phosphate
	PO_3^{3-}	phosphite
carbon	CO_3^{2-}	carbonate
	HCO_3^-	bicarbonate or hydrogen carbonate
	$C_2O_4^{2-}$	oxalate
	$C_2H_3O_2^-$	acetate
	CN^-	cyanide
chlorine	ClO_4^-	perchlorate
	ClO_3^-	chlorate
	ClO_2^-	chlorite
	ClO^-	hypochlorite
hydrogen	H_3O^+	hydronium
	OH^-	hydroxide
metals	MnO_4^-	permanganate
	CrO_4^{2-}	chromate
	$Cr_2O_7^{2-}$	dichromate

required. The charges and formulas for the various polyatomic ions cannot be easily related to the periodic table, as was the case for many of the monoatomic ions (Section 6.5). The inability to recognize the presence of polyatomic ions in a compound is a major stumbling block for many chemistry students. You must put forth the effort to learn the names and formulas of these polyatomic ions in order to avoid this obstacle.

Note from Table 6-4 the following relationships concerning polyatomic ions.

1. Most of the ions have a negative charge, which can vary from −1 to −3. Only two positive ions are listed in the table: NH_4^+ (ammonium) and H_3O^+ (hydronium).
2. Two of the polyatomic ions, OH^- (hydroxide) and CN^- (cyanide), have names ending in -ide. These names represent exceptions to the rule that the suffix -ide is reserved for use in naming binary compounds (Section 6.6).
3. A number of -ate, -ite pairs of ions exist, as in SO_4^{2-} (sulfate) and SO_3^{2-} (sulfite). The -ate ion will always have one more oxygen atom than the -ite ion. Both the -ate and -ite ions carry the same charge.
4. A number of pairs of ions exist where one member of the pair differs from the other by having a hydrogen atom present, as in CO_3^{2-} (carbonate) and HCO_3^- (hydrogen carbonate or bicarbonate). In such pairs, the charge on the ion containing hydrogen is always one less than that on the other ion.

6.15 FORMULAS AND NAMES FOR IONIC COMPOUNDS CONTAINING POLYATOMIC IONS

Formulas for ionic compounds containing polyatomic ions are determined in the same way that they are for ionic compounds containing monoatomic ions (Section 6.5). The total positive and negative charge present must add up to zero.

Two conventions not encountered previously in formula writing often arise when writing formulas containing polyatomic ions.

1. When more than one polyatomic ion of a given kind is required in a formula, the polyatomic ion is enclosed in parentheses and a subscript placed outside the parentheses is used to indicate the number of polyatomic ions needed.
2. To preserve the identity of polyatomic ions, the same elemental symbol may be used more than once in a formula.

Example 6.12 contains examples illustrating the use of both of these new conventions.

Example 6.12

Determine the formulas for the ionic compounds containing the following pairs of ions.

a. Na^+ and SO_4^{2-} b. Mg^{2+} and NO_3^- c. NH_4^+ and CN^-

Solution

a. In order to equalize the total positive and negative charge, we need two sodium ions ($+1$ charge) for each sulfate ion (-2 charge). We indicate the presence of two Na^+ ions with the subscript 2 following the symbol of this ion. The formula of the compound is Na_2SO_4. The convention that the positive ion is always written first in the formula still holds when polyatomic ions are present.
b. Two nitrate ions (-1 charge) are required to balance the charge on one magnesium ion ($+2$ charge). Because more than one polyatomic ion is needed, the formula will contain parentheses, $Mg(NO_3)_2$. The subscript 2 outside the parentheses indicates two of what is inside the parentheses. If parentheses were not used, the formula would appear to be $MgNO_{32}$, which is not intended and conveys false information. The formula $Mg(NO_3)_2$ indicates a formula unit containing one magnesium atom, two nitrogen atoms, and six oxygen atoms; the formula $MgNO_{32}$ indicates a formula unit containing one magnesium atom, one nitrogen atom, and thirty-two oxygen atoms.
c. In this compound, both ions are polyatomic, which is a perfectly legal situation. Because the ions have equal but opposite charges, they will combine in a one-to-one ratio. Thus, the formula is NH_4CN. No parentheses are necessary because we only need one polyatomic ion of each type in a formula unit. The appearance of the symbol for the element nitrogen (N) at two locations in the formula could be prevented by combining the two nitrogens, resulting in the formula N_2H_4C. The formula N_2H_4C does not convey the message that NH_4^+ and CN^- ions are present. Thus, when writing formulas that contain polyatomic ions, we always maintain the identities of these ions even if it means having the same elemental symbol at more than one location in the formula.

The names of ionic compounds containing polyatomic ions are derived in a manner that is similar to those for binary ionic compounds (Section 6.7). The rule for naming binary ionic compounds is: Give the name of the metallic element first (including, when needed, a Roman numeral indicating ion charge), and then give a separate word containing the stem of the nonmetallic name and the suffix *-ide*.

For our present situation, *if the polyatomic ion is positive, its name is substituted for that of the metal. If the polyatomic ion is negative, its name is substituted for the nonmetal stem plus* -ide. In the case where both positive and negative ions are polyatomic, dual substitution occurs and the resulting name includes just the names of the polyatomic ions. Example 6.13 illustrates the use of these rules.

Example 6.13

Name the following compounds, which contain one or more polyatomic ions.
a. $Ca_3(PO_4)_2$ b. $Fe_2(SO_4)_3$ c. $(NH_4)_2CO_3$

Solution
a. The positive ion present is the calcium ion (Ca^{2+}). We will not need a Roman numeral to specify the charge on a Ca^{2+} ion because it is always a +2. The negative ion is the polyatomic phosphate ion (PO_4^{3-}). The name of the compound is *calcium phosphate*. As in naming binary ionic compounds, subscripts in the formula are not incorporated into the name.
b. The positive ion present is iron(III). The negative ion is the polyatomic sulfate ion (SO_4^{2-}). The name of the compound is *iron(III) sulfate*. The determination that iron is present as iron(III) involves the following calculation dealing with charge balance.

$$2 \text{ (iron charge)} + 3 \text{ (sulfate charge)} = 0$$

The sulfate charge is -2. (You had to memorize that.) Therefore,

$$2 \text{ (iron charge)} + 3(-2) = 0$$
$$2 \text{ (iron charge)} = +6$$
$$\text{iron charge} = +3$$

c. Both the positive and negative ions in this compound are polyatomic—the ammonium ion (NH_4^+) and the carbonate ion (CO_3^{2-}). The name of the compound is simply the combination of the names of the two polyatomic ions: *ammonium carbonate*.

EXERCISES AND PROBLEMS

Valence Electrons

6.1 How many valence electrons do atoms with the following electron configurations have?
a. $1s^2 2s^2$
b. $1s^2 2s^2 2p^2$
c. $1s^2 2s^2 2p^5$
d. $1s^2 2s^2 2p^6 3s^2$
e. $1s^2 2s^2 2p^6 3s^2 3p^6 4s^2 3d^{10} 4p^1$
f. $1s^2 2s^2 2p^6 3s^2 3p^6 4s^2 3d^{10} 4p^6$

6.2 How many valence electrons do atoms of each of the following elements have?
a. $_3Li$ c. $_{10}Ne$ e. $_{53}I$
b. $_8O$ d. $_{20}Ca$ f. $_{55}Cs$

6.3 If Q is the symbol of a representative element nonmetal and it has three valence electrons, in what group is Q in the periodic table?

Electron-Dot Structures for Atoms

6.4 Draw electron-dot structures for atoms of the following elements.
a. $_{12}Mg$ c. $_{19}K$ e. $_1H$
b. $_{15}P$ d. $_{18}Ar$ f. $_4Be$

6.5 Each of the following electron-dot structures represents a period 2 element. Determine the element's identity in each case.
a. ·X·   b.  ·Ẋ·   c.  ·Ẍ: d. 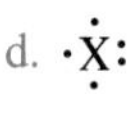 ·Ẋ:

Octet Rule

6.6 What is unique about the electron configurations of the noble gases?

6.7 State the octet rule.

Notation for Ions

6.8 Why does an atom that loses electrons become positively charged?

6.9 What would be the charge, if any, on particles with the following subatomic makeups?
a. 15 protons, 17 neutrons, 18 electrons
b. 7 protons, 7 neutrons, 10 electrons
c. 12 protons, 14 neutrons, 10 electrons
d. 3 protons, 4 neutrons, 2 electrons

6.10 What is the difference in meaning associated with the notations K and K^+?

6.11 Calculate the number of protons and electrons in each of the following ions.
a. $_{20}Ca^{2+}$ c. $_{17}Cl^-$ e. $_{16}S^{2-}$
b. $_8O^{2-}$ d. $_{11}Na^+$ f. $_{13}Al^{3+}$

Ionic Charge

6.12 Predict the charge on the monoatomic ion formed by each of the following elements.
a. $_{12}Mg$ c. $_7N$ e. $_{19}K$
b. $_9F$ d. $_3Li$ f. $_{15}P$

6.13 Indicate the number of electrons lost or gained when each of the following atoms forms an ion.
a. $_4Be$ c. $_{35}Br$ e. $_{38}Sr$
b. $_{34}Se$ d. $_{37}Rb$ f. $_{53}I$

Ionic Compound Formation and Electron-dot Structures

6.14 Show the formation of the following ionic compounds using electron-dot structures.
a. K_2O b. CaO   c. Na_3P d. $AlCl_3$

6.15 Show the formation of the following ionic compounds using electron-dot structures.
a. LiI   b. Be_3N_2 c. Al_2O_3 d. AlN

6.16 Using electron-dot structures, show how ionic compounds are formed by atoms of
a. Be and O   c. K and N
b. Mg and S   d. Ca and F

Formulas for Binary Ionic Compounds

6.17 Write the formula for an ionic compound formed from Ba^{2+} ion and each of the following ions.
a. Cl^- c. Br^- e. N^{3-}
b. O^{2-} d. S^{2-} f. P^{3-}

6.18 Write the formula for an ionic compound formed from F^- and each of the following ions.
a. magnesium   d. lithium
b. potassium   e. aluminum
c. sodium   f. calcium

6.19 Write the formulas for the compounds formed from the following ions.
a. Na^+ and S^{2-} d. Al^{3+} and P^{3-}
b. Be^{2+} and F^- e. Li^+ and Cl^-
c. K^+ and O^{2-} f. Mg^{2+} and Br^-

Structure of Ionic Compounds

6.20 Describe in words the general structure of an ionic solid.

6.21 What does the formula for an ionic compound actually represent?

6.22 Explain why it is inappropriate to talk about molecules of an ionic compound.

Nomenclature for Binary Ionic Compounds

6.23 Name the following binary ionic compounds.
a. KI   c. BeO   e. AlF_3
b. Na_3P d. $CaCl_2$ f. Ca_2C

6.24 Indicate, with a yes or no, whether a Roman numeral is required in the name of each of the following binary ionic compounds.
a. $AuCl_3$ c. NaCl   e. ZnF_2
b. FeO   d. AgF   f. Cu_2S

6.25 Calculate the charge on the metal ion in each of the following binary ionic compounds.
a. Au_2O c. CuO   e. SnO_2
b. Fe_2O_3 d. $SnCl_4$ f. SnO

6.26 Name the following binary ionic compounds.
a. FeO   c. Au_2O_3 e. CuS
b. Co_2O_3 d. PbO_2 f. $FeCl_3$

6.27 Name the following binary ionic compounds.
a. AuCl   c. KCl   e. CaO
b. AgCl   d. NiO   f. FeN

6.28 Write formulas for the following binary ionic compounds.
a. potassium bromide   d. barium phosphide
b. silver oxide   e. gallium nitride
c. beryllium fluoride   f. iron(II) iodide

6.29 Write formulas for the following binary ionic compounds.
a. cobalt(II) sulfide   d. gold(III) sulfide
b. cobalt(III) sulfide   e. aluminum sulfide
c. tin(IV) sulfide   f. silver sulfide

Single-Covalent Bonds

6.30 Draw electron-dot structures to illustrate the covalent bonding found in each of the following molecules.
a. I_2 c. HBr   e. OF_2
b. ClF   d. NCl_3 f. CCl_4

6.31 Write electron-dot structures for the compounds most likely to be formed between these pairs of elements.
a. phosphorus and hydrogen
b. bromine and chlorine
c. hydrogen and iodine
d. carbon and bromine
e. carbon and iodine
f. sulfur and fluorine

Multiple-Covalent Bonds

6.32 What is the difference between a single-covalent bond, a double-covalent bond, and a triple-covalent bond?

6.33 Draw the electron-dot structures for the following

molecules, each of which contains at least one double bond. (The skeletal arrangement of atoms in each molecule is given underneath its formula.)

a. C_3H_4

```
  H     H
  C  C  C
  H     H
```

b. N_2F_2

```
F  N  N  F
```

c. H_2CO

```
  H
  C  O
  H
```

d. $C_2H_2Br_2$

```
Br    Br
 C    C
 H    H
```

6.34 Draw the electron-dot structures for the following molecules, each of which contains at least one triple bond. (The skeletal arrangement of atoms in each molecule is given underneath its formula.)

a. C_2H_3N

```
  H
H C  C  N
  H
```

b. C_3H_4

```
    H     H
H   C  C  C  H
    H     H
```

c. C_2N_2

```
N  C  C  N
```

d. CH_2N_2

```
  H
  N  C  N
  H
```

Coordinate-Covalent Bonds

6.35 What is a coordinate-covalent bond?

6.36 Once formed, how (if at all) does a coordinate-covalent bond differ from an ordinary covalent bond?

6.37 Draw an electron-dot structure for the molecule S_2O (S S O), which contains a coordinate-covalent bond. Indicate which bond is involved with coordinate covalency.

Electronegativity and Polarity of Bonds

6.38 In each of the following pairs of elements, indicate which element is the more electronegative element.

a. H and F
b. C and O
c. N and S
d. Na and Mg
e. N and P
f. Cl and Br

6.39 Arrange each of the following sets of bonds in order of increasing polarity.

a. H—Cl, H—O, H—Br
b. O—F, P—O, Al—O
c. H—Cl, Br—Br, B—N
d. P—N, S—O, Br—F

6.40 Place a δ^+ above the atom that is relatively positive and a δ^- above the atom that is relatively negative in each of the following bonds.

a. B—N
b. F—O
c. C—Cl
d. Al—S
e. N—C
f. Br—P

6.41 Classify each of the following bonds as covalent, polar covalent, or ionic.

a. carbon–oxygen
b. beryllium–chlorine
c. nitrogen–oxygen
d. phosphorus–phosphorus
e. cesium–oxygen
f. sodium–fluorine

Molecular Polarity

6.42 Is it possible for a molecule to be nonpolar when it contains polar bonds? Explain.

6.43 For each of the following hypothetical triatomic molecules, indicate whether the bonds are polar or nonpolar and whether the molecule is polar or nonpolar. Assume A, X, and Y have different electronegativities.

a. X—A—X

b. A—X—X

c.
```
      X
     / \
    X   X
```

d.
```
      A
     / \
    X   X
```

e. Y—A—X

6.44 Indicate whether each of the following molecules is polar or nonpolar. The geometry of each molecule is given in parentheses.

a. NCl_3 (trigonal pyramid)—The nitrogen atom is the apex.
b. H_2S (bent)—The sulfur atom is in the middle.
c. CS_2 (linear)—The carbon atom is in the middle.
d. $CHCl_3$ (tetrahedral)—The carbon atom is in the center.

Nomenclature for Binary Covalent Compounds

6.45 Write the number that corresponds to each of the following prefixes.

a. octa
b. tri
c. tetra
d. hepta
e. deca
f. penta
g. hexa
h. di

6.46 Name the following binary molecular compounds.

a. SF_4
b. Cl_2O
c. P_4O_6
d. CO
e. ClO_2
f. PI_3
g. S_4N_2
h. As_2O_3

6.47 Write formulas for the following binary covalent compounds.

a. iodine monochloride
b. nitrogen trichloride
c. sulfur hexafluoride
d. silicon tetrafluoride
e. dinitrogen monoxide
f. tetraphosphorus decoxide
g. hydrogen peroxide
h. ammonia

Compounds Containing Polyatomic Ions

6.48 Give names for the following polyatomic ions.

a. NO_3^-
b. SO_4^{2-}
c. NH_4^+
d. OH^-
e. CN^-
f. CO_3^{2-}
g. PO_4^{3-}
h. HCO_3^-

6.49 Write formulas (including charge) for the following polyatomic ions.

a. ammonium
b. hydronium
c. chlorate
d. chlorite
e. hydrogen sulfate
f. hydrogen phosphate
g. dichromate
h. permanganate

6.50 Indicate which of the following compounds contain polyatomic ions and identify the polyatomic ion by name if it is present.

a. Al_2S_3
b. $Ca_3(PO_4)_2$
c. $ZnSO_4$
d. KNO_3
e. NH_4Cl
f. $NaOH$
g. KCN
h. $KClO$

6.51 Write formulas for the compounds formed between these positive and negative ions.

a. Ba^{2+} and NO_3^-
b. Fe^{3+} and OH^-
c. Al^{3+} and CO_3^{2-}
d. Na^+ and ClO_4^-
e. K^+ and PO_4^{3-}
f. NH_4^+ and SO_4^{2-}
g. Co^{2+} and $H_2PO_4^-$
h. NH_4^+ and $C_2H_3O_2^-$

6.52 Write formulas for the compounds formed between the following positive and negative ions.

a. Ca^{2+} and acetate
b. K^+ and nitrate
c. Fe^{3+} and cyanide
d. Au^+ and sulfate
e. Cu^{2+} and phosphate
f. Be^{2+} and hydroxide
g. Mg^{2+} and carbonate
h. Li^+ and chlorate

6.53 Name the following compounds containing polyatomic ions.

a. Na_2SO_4
b. $Ca(OH)_2$
c. $Mg_3(PO_4)_2$
d. NH_4NO_3
e. $FeCO_3$
f. $AgNO_3$
g. Cu_2SO_4
h. $CuSO_4$

6.54 Write formulas for the following compounds containing polyatomic ions.

a. potassium bicarbonate
b. silver carbonate
c. gold(III) nitrate
d. beryllium cyanide
e. copper(II) hydroxide
f. sodium dihydrogenphosphate
g. aluminum phosphate
h. calcium hypochlorite

7 Gases, Liquids, and Solids

Objectives

After completing Chapter 7, you will be able to:

7.1 Compare and contrast the macroscopic distinguishing properties of gases, liquids, and solids.
7.2 List the five statements of kinetic molecular theory and understand the roles that disruptive forces (kinetic energy) and cohesive forces (potential energy) play in determining the physical state of a system.
7.3 Use kinetic molecular theory to explain the characteristic properties of solids.
7.4 Use kinetic molecular theory to explain the characteristic properties of liquids.
7.5 Use kinetic molecular theory to explain the characteristic properties of gases.
7.6 Contrast, on a molecular level, the differences among solids, liquids, and gases.
7.7 State the units commonly used for each of the three gas law variables: pressure, temperature, and volume.
7.8 State Boyle's law in words and as a mathematical equation; know how to use the law in problem solving; and be familiar with common applications of the law.
7.9 State Charles' law in words and as a mathematical equation; know how to use the law in problem solving; and be familiar with common applications of the law.
7.10 State the mathematical form of the combined-gas law and use it in problem solving.
7.11 Use Dalton's law of partial pressures to calculate the partial or total pressure of mixtures of gaseous substances.
7.12 Understand the terminology used to describe various changes of state and distinguish between exothermic and endothermic changes of state.
7.13 Explain, on a molecular basis, what happens during the process of evaporation and list the factors that affect evaporation rates.
7.14 Understand the relationship between vapor pressure and an equilibrium state and know the factors that affect the magnitude of a given liquid's vapor pressure.
7.15 Understand the process of boiling from a molecular viewpoint; understand what is meant by the term *normal boiling point*; and know the general relationship between the boiling point of a liquid and external pressure.
7.16 Understand the origins of dipole–dipole interactions and the special significance of hydrogen bonds.

7.17 Understand the differences, on a molecular level, between the processes of sublimation and melting.
7.18 Explain, on a molecular level, what happens when a substance undergoes decomposition.

INTRODUCTION

In Chapters 3 and 5, we considered the structure of matter from a submicroscopic point of view—in terms of molecules, atoms, protons, neutrons, and electrons. In this chapter, we will be concerned with the macroscopic characteristics of matter as represented by the physical states solid, liquid, and gas. Of particular concern will be the properties that are exhibited by matter in the various physical states and a theory that correlates these properties with molecular behavior.

7.1 PROPERTY DIFFERENCES BETWEEN PHYSICAL STATES

A comparison of four macroscopically observable properties of each of the physical states of matter will serve as our starting point for discussion about these states. The four properties are: (1) volume and shape, (2) density, (3) compressibility, and (4) thermal expansion. We discussed the property of density in Section 1.12. *Compressibility* involves the change in volume resulting from a pressure change. *Thermal expansion* involves the volume change resulting from a temperature change. Table 7–1 contrasts the previously listed properties for the three states of matter.

7.2 THE KINETIC MOLECULAR THEORY OF MATTER

The physical characteristics of the solid, liquid, and gaseous states listed in Table 7–1 can be explained by *kinetic molecular theory*, which is one of the fundamental theories of chemistry. The **kinetic molecular theory of matter** *is a set of five state-*

kinetic molecular theory of matter

TABLE 7–1 Distinguishing Properties of Solids, Liquids, and Gases

Property	Solid State	Liquid State	Gaseous State
volume and shape	definite volume and definite shape	definite volume and indefinite shape; takes the shape of container to the extent it is filled	indefinite volume and indefinite shape; takes the volume and shape of container that it fills
density	high	high, but usually lower than corresponding solid	low
compressibility	small	small, but usually greater than corresponding solid	large
thermal expansion	very small: about 0.01% per °C	small: about 0.10% per °C	moderate: about 0.30% per °C

ments that are used to explain the physical behavior of the three states of matter (*solids, liquids, and gases*). The basic idea of this theory is that the particles (atoms, molecules, or ions) present in a substance, independent of the physical state of the substance, have *motion* associated with them. The word *kinetic* comes from the Greek word *kinesis*, which means movement; hence, the name *kinetic molecular theory*.

The specific statements that are part of the kinetic molecular theory of matter are:

1. Matter is ultimately composed of tiny particles (atoms, molecules, or ions) that have definite and characteristic sizes that do not change.
2. The particles are in constant random motion and therefore possess kinetic energy.
3. The particles interact with each other through attractions and repulsions and therefore possess potential energy.
4. The velocity of the particles increases as the temperature is increased. The average kinetic energy of all particles in a system depends on the temperature, and it increases as the temperature increases.
5. The particles in a system transfer energy to each other during collisions in which no net energy is lost from the system. The energy of any given particle is thus continually changing.

Both kinetic energy and potential energy are mentioned in the kinetic molecular theory of matter. We considered definitions for these two forms of energy in Section 1.14. Recall that kinetic energy is energy associated with motion and potential energy is stored energy that results from an object's position, condition, or composition.

The amount of kinetic energy a particle possesses depends on both its mass and its velocity. The exact mathematical relationship between kinetic energy and the mass and velocity of a particle is

$$\text{kinetic energy} = \frac{1}{2} mv^2$$

where m is the mass of the particle and v is its velocity. We can see from this expression that any differences in kinetic energy between particles of the same mass must be caused by differences in their velocities. Similarly, differences in kinetic energy between particles moving at the same velocity must be caused by mass differences.

Unlike kinetic energy, there is no single, simple equation that can always be used to calculate the amount of potential energy that an object has. This is because of the many diverse origins of potential energy (Section 1.14).

The potential energy of greatest importance when considering the three states of matter is that originating from electrostatic interactions between particles. **Electrostatic interactions** *are attractions and repulsions that occur between charged particles; particles of opposite charge (one positive and one negative) attract each other and particles of identical charge (both positive or both negative) repel each other.* The magnitude of electrostatic interactions depends on the sizes of the charges associated with the particles ($+1, +2, -1, -2$, and so on) and the distance of separation between particles.

When using kinetic molecular theory to explain the differences among the solid, liquid, and gaseous states of matter, the relative magnitudes of kinetic energy and potential energy (electrostatic attractions) for the system under consideration are of key importance. Kinetic energy can be considered to be a *disruptive force* within the chemical system that tends to make the particles of the system increasingly independent of each other. As a result of the energy of motion, the particles tend to move away from each other. Potential energy of attraction can be considered to be a *cohesive force* that tends to cause order and stability among the particles of the system.

The kinetic energy of a chemical system depends on its temperature. Kinetic energy increases as temperature increases (Statement 4 of the kinetic molecular theory of matter). Thus, the higher the temperature is, the greater the magnitude of disruptive influences within a chemical system is. Potential energy magnitude, or cohesive force magnitude, is essentially independent of temperature. The fact that one of the types of forces depends on temperature (disruptive forces) and the other does not (cohesive forces) causes temperature to be the factor that determines in which of the three physical states a given sample is found. We will discuss the reason for this in later sections.

Sections 7.3, 7.4, and 7.5 deal with kinetic molecular theory explanations in terms of disruptive forces, cohesive forces, and temperature magnitude for the general properties of the solid, liquid, and gaseous states.

7.3 THE SOLID STATE

The solid state is characterized by a dominance of potential energy (cohesive forces) over kinetic energy (disruptive forces). The particles in a solid are drawn close together in a regular pattern by the strong cohesive forces present. Each particle occupies a fixed position, about which it vibrates because of disruptive kinetic energy. Using this model, the characteristic properties of solids (Table 7–1) can be obtained as follows:

1. *Definite volume and definite shape.* The strong, cohesive forces hold the particles in essentially fixed positions, resulting in definite volume and definite shape.
2. *High density.* The constituent particles of solids are located as close together as possible (essentially touching each other). Therefore, a given volume contains large numbers of particles, resulting in a high density.
3. *Small compressibility.* Because there is very little space between particles, increased pressure cannot push the particles any closer together; therefore, it has little effect on the solid's volume.
4. *Very small thermal expansion.* An increased temperature increases the kinetic energy (disruptive forces), thereby causing more vibrational motion of the particles. Each particle occupies a slightly larger volume and the result is a slight expansion of the solid. The strong, cohesive forces prevent this effect from becoming very large.

7.4 THE LIQUID STATE

The liquid state consists of particles that are randomly packed but relatively near to each other. The molecules are in constant, random motion, freely sliding over one another, but they do not have sufficient energy to separate from each other. *Thus, the liquid state is one in which neither potential energy (cohesive forces) nor kinetic energy (disruptive forces) dominates.* The fact that the particles freely slide over each other indicates the influence of disruptive forces; however, the fact that the particles do not separate indicates fairly strong influences from cohesive forces. Using this model, the characteristic properties of liquids (Table 7–1) can be explained as follows:

1. *Definite volume and indefinite shape.* The attractive forces are strong enough to restrict particles to movement within a definite volume. They are not strong enough, however, to prevent the particles from moving over each other in a random manner that is limited only by the container walls. Thus, liquids have no definite shape, with the exception that they maintain a horizontal upper surface in containers that are not completely filled.

2. *High density.* The particles in a liquid are not widely separated; they are still touching each other. Therefore, there will be a large number of particles in a given volume and a resultant high density.
3. *Small compressibility.* Because the particles in a liquid are still touching each other, there is very little empty space. Therefore, a pressure increase cannot squeeze the particles much closer together.
4. *Small thermal expansion.* Most of the particle movement in a liquid is vibrational because a particle can move only a short distance before colliding with a neighbor. The increased particle velocity that accompanies a temperature increase results only in increased vibrational amplitudes. The net effect is an increase in the effective volume a particle occupies, which causes a slight volume increase in the liquid.

7.5 THE GASEOUS STATE

The gaseous state is characterized by a complete dominance of kinetic energy (*disruptive forces*) *over potential energy* (*cohesive forces*). As a result, the particles of a gas move essentially independently of each another, in a totally random manner. Under ordinary pressure, the particles are relatively far apart, except when they collide with each other. Between collisions with each other or the container walls, gas particles travel in straight lines.

The kinetic theory explanation of gaseous-state properties follows the same pattern that we saw earlier for solids and liquids.

1. *Indefinite volume and indefinite shape.* The attractive (cohesive) forces between particles have been overcome by high kinetic energy and the particles are free to travel in all directions. Therefore, gas particles completely fill their container and the shape of the gas is that of the container.
2. *Low density.* The particles of a gas are widely separated. There are relatively few particles in a given volume (compared to liquids and solids), which means little mass per volume (a low density).
3. *Large compressibility.* Particles in a gas are widely separated; essentially, a gas is mostly empty space. When pressure is applied, the particles are easily pushed closer together, decreasing the amount of empty space and the volume of the gas (see Figure 7–1).
4. *Moderate thermal expansion.* An increase in temperature means an increase in particle velocity. The increased kinetic energy of the particles enables them to push back whatever barrier is confining them into a given volume and the volume increases. Note that the size of the particles is not changed during expansion or compression of gases, solids, or liquids; they merely move either farther apart or closer together. It is the space between the particles that changes.

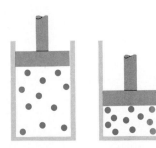

Gas at low pressure   Gas at higher pressure

FIGURE 7–1 The compression of a gas—decreasing the amount of empty space in the container.

7.6 A COMPARISON OF SOLIDS, LIQUIDS, AND GASES

The average distance between particles is only slightly different in the solid and liquid states, but it is markedly different in the gaseous state. Roughly speaking, at ordinary temperatures and pressures, particles in a liquid are about 10% farther apart than those in the solid state. However, particles in a gas are about 1000% farther apart than solid state particles. The distance ratio between particles in the three states (solid to liquid to gas) is thus 1:1.1:10. The similarities and differences in the states of matter are illustrated in Figure 7–2.

7.7 GAS LAWS AND GAS LAW VARIABLES

We will now consider the gaseous state from a quantitative (numerical) viewpoint. Behavior of matter in the gaseous state can be described by some very simple

FIGURE 7–2 Similarities and differences in the states of matter.

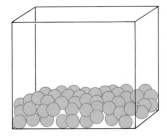
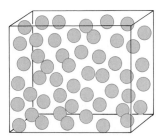

SOLID
Definite volume, definite shape
High density, small compressibility
Very small thermal expansion
Dominance of cohesive forces

LIQUID
Definite volume, indefinite shape
High density, small compressibility
Small thermal expansion
Equally strong cohesive and disruptive forces

GAS
Indefinite volume, indefinite shape
Low density, large compressibility
Moderate thermal expansion
Dominance of disruptive forces

---

quantitative relationships called *gas laws*. **Gas laws** *are generalizations that describe the relationships among the pressure, temperature, and volume of a specific quantity of gas in mathematical terms.*

gas laws

Only the gaseous state is able to be described by simple mathematical relationships. Laws describing liquid- and solid-state behavior are mathematically extremely complex. Consequently, quantitative treatments of liquid-state and solid-state behavior will not be given in this text.

Prior to discussing the mathematical form of the various gas laws, some comments concerning the major variables involved in gas-law calculations—volume, temperature, and pressure—are in order. Two of these variables, volume and temperature, have been discussed previously in Sections 1.10 and 1.13. The units *liter* and *milliliter* are generally used to specify gas volume. Only one of the three temperature scales discussed in Section 1.13, the *Kelvin scale*, can be used in gas law calculations if the results are to be valid. Therefore, you should be thoroughly familiar with the conversion of Celsius and Fahrenheit scale readings to the Kelvin scale (Section 1.13).

Pressure *is the force applied per unit area; that is, the total force on a surface divided by the area of that surface.*

pressure

$$P \text{ (pressure)} = \frac{F \text{ (force)}}{A \text{ (area)}}$$

For a gas, the force involved in pressure is that which is exerted by the gas molecules or atoms as they constantly collide with the walls of their container. Barometers, manometers, and gauges are the instruments most commonly used to measure gas pressures.

Millimeters of mercury, atmospheres, and pounds-per-square-inch are the three kinds of pressure units in most common use. Barometers and manometers measure pressure in terms of the height of a column of mercury in millimeters. Gauges are usually calibrated in terms of force per area, as in pounds-per-square-inch.

The air that surrounds the Earth exerts a pressure on all the objects it is in contact with. A **barometer** *is a device used to measure atmospheric pressure.* The essential components of a simple barometer are shown in Figure 7–3 (p. 152). A

barometer

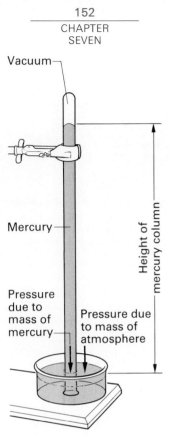

FIGURE 7–3 The essential components of a mercury barometer.

barometer can be constructed by taking a long glass tube that is sealed at one end and filling it all the way to the top with mercury; then the tube should be inverted into a dish of mercury, without letting any air into it. The mercury in the tube falls until the pressure from the mass of the mercury in the tube is just balanced by the pressure of the atmosphere on the mercury in the dish. The pressure of the atmosphere is expressed in terms of the height of the supported column of mercury. Mercury is the liquid of choice in a barometer for two reasons: (1) it is a very dense liquid, so only a short tube is needed; and (2) it does not readily vaporize (evaporate) so the pressure reading does not have to be corrected for the presence of mercury vapor.

The pressure unit *millimeters of mercury* is usually written in the abbreviated form *mmHg*. An alternate name for this unit is *torr*; this name honors the Italian physicist Evangelista Torricelli, the inventor of the barometer.

$$1 \text{ mmHg} = 1 \text{ torr}$$

The pressure of the atmosphere varies with altitude, decreasing at the rate of approximately 25 mmHg per 1000 feet increase in altitude. It also fluctuates with weather conditions. Think about the terminology used in a weather report—high pressure ridge, low pressure front, and so on. At sea level, barometric pressure fluctuates between 740 and 770 mmHg and averages about 760 mmHg, depending on weather conditions. The pressure unit *atmospheres*, abbreviated *atm*, is defined in terms of this average sea level pressure. By definition, we have

$$1 \text{ atm} = 760 \text{ mmHg} = 760 \text{ torr}$$

Because of its size, which is 760 times larger than an mmHg, the atmosphere unit is used whenever high pressures are encountered.

The relationship between the pressure units pounds-per-square-inch and atmospheres is

$$1 \text{ atmosphere} = 14.7 \text{ pounds-per-square-inch}$$

Two abbreviations exist for the pounds-per-square-inch pressure unit, lb/in^2 and psi.

In clinical work in allied health fields, the pressure unit millimeters-of-mercury is the unit most often encountered and used. For example, when considering the process of respiration, oxygen and carbon dioxide pressures are almost always specified using mmHg. (Additional information concerning the magnitude of oxygen and carbon dioxide pressures at various stages of the respiration cycle will be given in Section 7.11.)

Often the fact that a pressure measurement associated with the functioning of the human body is in mmHg is not explicitly stated. It is just assumed that you know that this is the case. For example, a normal blood pressure reading is written as 120/80 or is stated as "120 over 80". What this really means is that the maximum pressure of the blood in the arteries is 120 mmHg greater than atmospheric pressure and the minimum pressure when the heart is relaxed between beats is 80 mmHg greater than atmospheric pressure. The units mmHg associated with blood pressure readings are never specifically mentioned.

7.8 BOYLE'S LAW: A PRESSURE–VOLUME RELATIONSHIP

Of the several relationships that exist between gas-law variables, the first to be discovered was the one that relates gas pressure to gas volume. It was formulated over 300 years ago, in 1662, by the British chemist and physicist Robert Boyle

FIGURE 7-4 A pressure–volume graph illustrating Boyle's law data.

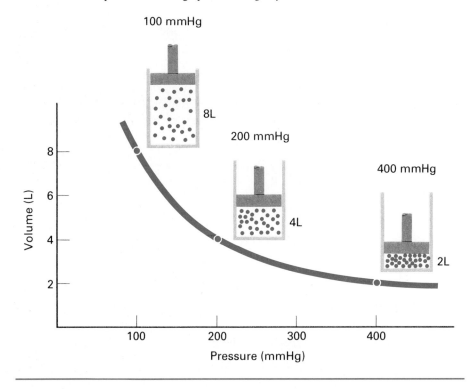

and is known as *Boyle's Law*. **Boyle's law** states that *the volume of a sample of a gas is inversely proportional to the pressure applied to the gas if the temperature is kept constant*. This means that if the pressure on the gas increases, the volume decreases proportionally, and conversely, if the pressure is decreased, the volume will increase. Doubling the pressure cuts the volume in half; tripling the pressure cuts the volume to one-third its original value; quadrupling the pressure cuts the volume to one-fourth, and so on. Any time two quantities are *inversely proportional*, which is the case with pressure and volume of a gas (Boyle's law), one increases as the other decreases. Figure 7–4 illustrates Boyle's law.

Boyle's law may be stated mathematically as

$$P_1 \times V_1 = P_2 \times V_2$$

The P_1 and V_1 are the pressure and volume of a gas at an initial set of conditions. The P_2 and V_2 are the pressure and volume of the same sample of gas under a new set of conditions, with the temperature remaining constant. The Boyle's law equation is only valid if the temperature remains constant. When we know any three of the four quantities in the Boyle's Law equation we can calculate the fourth, which will usually be the final pressure, P_2, or the final volume, V_2.

Boyle's law

A sample of O_2 gas occupies a volume of 1.50 L at a pressure of 735 mmHg and a temperature of 25°C. What volume will it occupy if the pressure is increased to 770 mmHg, with no change in temperature?

Example 7.1

Solution
A suggested first step in working gas-law problems involving two sets of

conditions is to analyze the given data in terms of initial and final conditions. Doing this, we find that

$$P_1 = 735 \text{ mmHg} \qquad P_2 = 770 \text{ mmHg}$$
$$V_1 = 1.50 \text{ L} \qquad V_2 = ? \text{ L}$$

We know three of the four variables in the Boyle's Law equation, so we can calculate the fourth, V_2. We will rearrange Boyle's law to isolate V_2 (the quantity to be calculated) by itself on one side of the equation. This is accomplished by dividing both sides of the Boyle's law equation by P_2.

$$P_1 V_1 = P_2 V_2 \quad \text{(Boyle's law)}$$

$$\frac{P_1 V_1}{P_2} = \frac{\cancel{P_2} V_2}{\cancel{P_2}} \quad \text{(divide each side of the equation by } P_2\text{)}$$

$$V_2 = V_1 \times \frac{P_1}{P_2}$$

Substituting the given data into the rearranged equation and doing the arithmetic gives

$$V_2 = 1.50 \text{ L} \times \left(\frac{735 \text{ mmHg}}{770 \text{ mmHg}}\right)$$
$$= 1.43 \text{ L}$$

Our answer is reasonable. Increasing the pressure should decrease the volume; the volume went from 1.50 L to 1.43 L.

Boyle's law is consistent with kinetic molecular theory (Section 7.2). The pressure that a gas exerts results from collisions of the gas molecules with the sides of the container. The number of collisions within a given area on the container wall during a given time is proportional to the pressure of the gas at a given temperature. If the volume of a container holding a specific number of gas molecules is increased, the total wall area of the container will also increase and the number of collisions in a given area (the pressure) will decrease because of the greater wall

FIGURE 7–5 The average number of times a molecule hits container walls is doubled when the volume of the gas, at constant temperature, decreases by one-half.

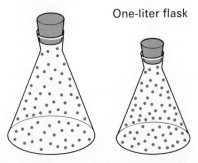
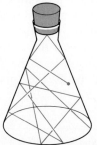

(a) The volume is decreased by one-half.

(b) A given molecule hits container wall twice as often.

area. Conversely, if the volume of the container is decreased, the wall area will be smaller and there will be more collisions within a given wall area. Figure 7–5 illustrates the situation that occurs when the volume is decreased.

The process of filling a medical syringe with liquid also involves Boyle's law. As the plunger is drawn out of the syringe (see Figure 7–6), the increase in volume inside the syringe chamber results in decreased pressure there. The liquid, which is under the influence of atmospheric pressure, flows into this reduced pressure area. The liquid is then expelled from the chamber by pushing the plunger back in. This ejection of the liquid does not involve Boyle's law; a liquid is incompressible and mechanical force pushes it out.

A direct application of Boyle's law is found in the process of breathing, which is illustrated in Figure 7–7. Human lungs, which are elastic (expandable) structures open to the atmosphere, are located within an airtight region of the body called the thoracic cavity. The diaphragm, which is a muscle that is able to move up and down (contract and relax), forms the floor of the thoracic cavity. Breathing in (inhaling or inspiration) occurs when the diaphragm flattens out (contracts). This contraction causes the volume of the thoracic cavity to increase. At the same time, the pressure within the cavity drops (Boyle's law) below atmospheric (external) pressure. Air flows into the lungs and expands them because the pressure is greater outside the lungs than within them. Breathing out (exhaling or expiration) occurs when the diaphragm relaxes (moves up). This action decreases the volume of the thoracic cavity and simultaneously increases the pressure (Boyle's law) within the

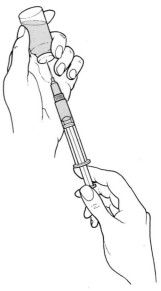

FIGURE 7–6 Filling a syringe with liquid is an application of Boyle's law.

FIGURE 7–7 The mechanics of breathing.

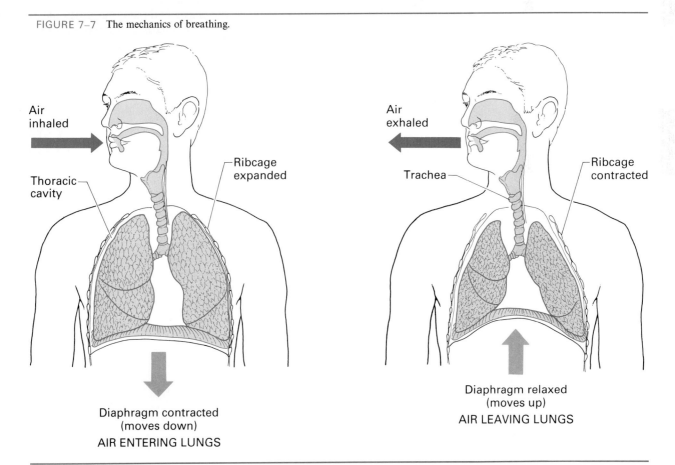

cavity to a value greater than the external pressure. Air flows out of the lungs. The air flow direction is always from a high-pressure region to a low-pressure region.

A respirator is designed to simulate the sequence of pressure changes that are experienced in the thoracic cavity during normal breathing. A respirator contains a movable diaphragm that works in opposition to a patient's lungs. When the diaphragm is moved so that the volume inside the respirator increases, the lower pressure in the respirator allows air to expand out of the patient's lungs. When the diaphragm is moved in the opposite direction, the higher pressure inside the respirator compresses the air into the lungs and causes them to increase in volume.

The Heimlich maneuver, which is an emergency procedure used to help someone who is choking, also works on the principle of Boyle's law. The procedure involves wrapping one's arms below the ribcage of the choking person and giving the person an upward bearhug. The thoracic cavity volume decreases, the pressure within the cavity increases (Boyle's law), and often material that is lodged in a person's trachea is expelled.

7.9 CHARLES' LAW: A TEMPERATURE–VOLUME RELATIONSHIP

The relationship between the temperature and volume of a gas at constant pressure is called *Charles' law*, after the French scientist Jacques Charles. This law was discovered in 1787, nearly one hundred years after the discovery of Boyle's law. **Charles' law** states that *the volume of a sample of gas is directly proportional to its Kelvin temperature if the pressure is kept constant.* Contained within the wording of this law is the phrase *directly proportional*; this contrasts with Boyle's law, which contains the phrase *inversely proportional*. Whenever a *direct* proportion exists between two quantities, one increases when the other increases and one decreases when the other decreases. Thus, a direct proportion and an inverse proportion portray opposite behaviors. The direct proportion relationship of Charles' law means that if the temperature increases, the volume will also increase and if the

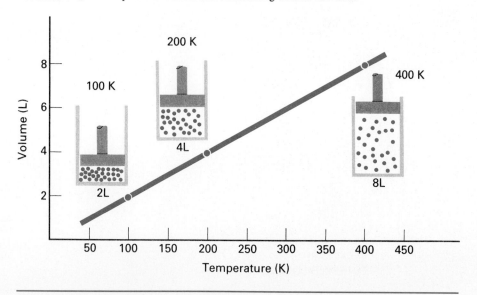

FIGURE 7–8 A temperature–volume curve illustrating Charles' law data.

temperature decreases, the volume will also decrease. Figure 7–8 illustrates Charles' law.

A balloon filled with air provides a qualitative illustration of Charles' law. If the balloon is placed near a heat source such as a light bulb that has been on for some time, the heat will cause the balloon to increase in size (volume). The change in volume is visually noticeable. Putting the same balloon in the refrigerator will cause it to shrink.

Charles' law, stated in mathematical form, is

$$\frac{V_1}{T_1} = \frac{V_2}{T_2}$$

In this expression, V_1 is the volume of a gas at a given pressure and T_1 is the Kelvin temperature of the gas. V_2 and T_2 are the volume and Kelvin temperature of the gas under a new set of conditions, with the pressure remaining constant. When using the Charles' law expression, temperatures must always be Kelvin scale temperatures. When we know any three of the four quantities in the Charles' law equation we can calculate the fourth, which will usually be the final volume, V_2, or the final temperature, T_2.

Example 7.2

A sample of the gaseous anesthetic cyclopropane, with a volume of 425 mL at a temperature of 27°C, is cooled, at constant pressure, to 20°C. What is the new volume of the sample?

Solution
First, we will analyze the data in terms of initial and final conditions. Doing this, we find that

$V_1 = 425$ mL $\qquad V_2 = ?$ mL
$T_1 = 27°C + 273 = 300$ K $\qquad T_2 = 20°C + 273 = 293$ K

Note that both of the given temperatures have been converted to Kelvin scale readings. This change is accomplished by simply adding 273 to the Celsius scale value (Section 1.13).

We know three of the four variables in the Charles' law equation, so we can calculate the fourth, V_2. We will rearrange Charles' law to isolate V_2 (the quantity desired) by multiplying each side of the equation by T_2.

$$\frac{V_1}{T_1} = \frac{V_2}{T_2} \quad \text{(Charles' law)}$$

$$\frac{V_1 T_2}{T_1} = \frac{V_2 \cancel{T_2}}{\cancel{T_2}} \quad \text{(multiply each side by } T_2\text{)}$$

$$V_2 = V_1 \times \frac{T_2}{T_1}$$

Substituting the given data into the equation and doing the arithmetic gives

$$V_2 = 425 \text{ mL} \times \left(\frac{293 \cancel{K}}{300 \cancel{K}}\right)$$

$$= 415 \text{ mL}$$

Our answer is consistent with what reasoning says it should be. A decrease in the temperature of a gas at constant pressure should result in a volume decrease.

Charles' law is easy to understand in terms of the kinetic molecular theory. The theory states that when the temperature of a gas increases, the velocity (kinetic energy) of the gas molecules increases. The speedier particles hit the container walls harder and more often. In order for the pressure of the gas to remain constant, it is necessary for the container volume to increase. In a larger volume, the particles will hit the container walls less often and the pressure can remain the same. A similar argument applies if the temperature of the gas is lowered; this time, the velocity of the molecules decreases and the wall area (volume) must also decrease in order to increase the number of collisions in a given area in a given time.

From Charles' law we have learned that heating a gas at constant pressure causes it to expand, or increase in volume. As a result of this expansion, the density of the gas decreases; the same amount of gas is now in a larger volume.

The density changes associated with Charles' law behavior are important in a number of situations. Hot-air balloon operation is based on density change. The hot air of the balloon is of lower density than the surrounding air. Therefore, the balloon rises in the denser, cooler air.

Charles' law is the principle involved in the operation of an incubator. The air in contact with the heating element expands and becomes less dense. This warm, less-dense air rises, causing continuous circulation of warm air. This same principle explains what happens in closed rooms with ineffective air circulation; the warmer and less-dense air stays near the top of the room. This may be desirable in the summer, but it certainly is not in the winter.

7.10 THE COMBINED-GAS LAW

combined-gas law

The **combined-gas law** is an expression obtained by mathematically combining Boyle's and Charles' laws. Its mathematical form is

$$\frac{P_1 V_1}{T_1} = \frac{P_2 V_2}{T_2}$$

This combined-gas law is a much more versatile equation than either of the laws from which it is derived. A change in pressure, temperature, or volume that is brought about by changes in both of the other two variables can be calculated using this equation. Both Boyle's and Charles' laws require that one of the three variables be held constant.

Example 7.3

A sample of O_2 gas occupies a volume of 1.62 L at 755 mmHg pressure and has a temperature of 0°C. What volume will this gas sample occupy at 725 mmHg pressure and 50°C?

Solution
Students often have problems rearranging the standard form of the combined-gas law into the format needed in a particular problem. For example, in this problem we need to rearrange things so that V_2 is by itself on one side of the equation.

The following rule from algebra is useful in this situation. If two fractions are equal,

$$\frac{a}{b} = \frac{c}{d}$$

then the numerator of the first fraction (*a*) times the denominator of the second fraction (d) is equal to the numerator of the second fraction (c) times the denominator of the first fraction (b).

$$\text{If } \frac{a}{b} = \frac{c}{d}, \text{ then } a \times d = c \times b$$

Applying this rule to the standard form of the combined-gas law gives

$$\frac{P_2 V_2}{T_2} = \frac{P_1 V_1}{T_1} \text{ and } P_2 V_2 T_1 = P_1 V_1 T_2$$

With the combined-gas law in the form $P_1 V_1 T_2 = P_2 V_2 T_1$, any of the six variables can be isolated by a simple division. In order to isolate V_2, we divide both sides of the equation by $P_2 T_1$.

$$\frac{P_2 V_2 T_1}{P_2 T_1} = \frac{P_1 V_1 T_2}{P_2 T_1} \text{ or } V_2 = \frac{V_1 P_1 T_2}{P_2 T_1}$$

Prior to inserting values into this equation, let us analyze the given data in terms of initial and final conditions.

$P_1 = 755$ mmHg $\qquad P_2 = 725$ mmHg
$V_1 = 1.62$ L $\qquad V_2 = ?$ L
$T_1 = 0°C + 273 = 273$ K $\qquad T_2 = 50°C + 273 = 323$ K

Substituting these numerical values into the combined-gas law, which has been rearranged so V_2 is on the left side by itself, gives

$$V_2 = 1.62 \text{ L} \times \frac{755 \text{ mmHg}}{725 \text{ mmHg}} \times \frac{323 \text{ K}}{273 \text{ K}}$$
$$= 2.00 \text{ L}$$

7.11 DALTON'S LAW OF PARTIAL PRESSURES

In a mixture of gases that do not react with each other, each type of molecule moves around in the container as if the other kinds were not there. This type of behavior is possible because a gas is mostly empty space (Section 7.5) and attractions between molecules in the gaseous state are negligible at most temperatures and pressures. Each gas in the mixture occupies the entire volume of the container; that is, it distributes itself uniformly throughout the container. The molecules of each type strike the walls of the container as frequently and with the same energy as if they were the only gas in the mixture. Consequently, the pressure exerted by a gas in a mixture is the same as it would be if the gas were alone in the same container under the same conditions.

English scientist John Dalton was the first to notice the independent behavior of gases in mixtures. In 1803, he published a summary statement concerning this behavior, which is now known as *Dalton's law of partial pressures*. **Dalton's law of partial pressures** states that *the total pressure exerted by a mixture of gases is the sum of the partial pressures of the individual gases.* A new term, *partial pressure*, is used in stating Dalton's law. A **partial pressure** is *the pressure that a gas in a mixture would exert if it were present alone under the same conditions.*

Expressed mathematically, Dalton's law states that

$$P_T = P_1 + P_2 + P_3 + \ldots$$

FIGURE 7-9 Dalton's Law of Partial Pressure.

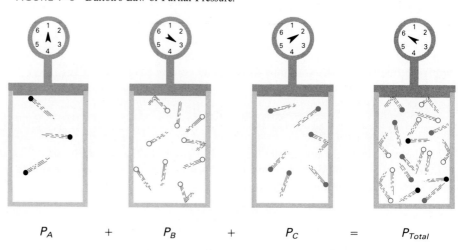

$$P_A \quad + \quad P_B \quad + \quad P_C \quad = \quad P_{Total}$$

where P_T is the total pressure of a gaseous mixture and P_1, P_2, P_3, and so on are the partial pressures of the individual gaseous components of the mixture.

As an illustration of Dalton's law, consider the four identical gas containers shown in Figure 7-9. Suppose we place amounts of three different gases (represented by A, B, and C) into three of the containers and measure the pressure exerted by each sample. We then place all three samples in the fourth container and measure the pressure exerted by this mixture of gases. We find that

$$P_{Total} = P_A + P_B + P_C$$

Using the actual gauge pressure values given in Figure 7-11, we see that

$$P_{Total} = 1 + 3 + 2 = 6$$

Example 7.4

The total pressure exerted by a mixture of the three gases oxygen, nitrogen, and water vapor is 742 mmHg. The partial pressures of the nitrogen and oxygen in the sample are 581 mmHg and 143 mmHg, respectively. What is the partial pressure of the water vapor present in the mixture?

Solution

Dalton's law says that

$$P_{Total} = P_{N_2} + P_{O_2} + P_{H_2O}$$

The known values for variables in this equation are

$$P_{Total} = 742 \text{ mmHg}$$
$$P_{N_2} = 581 \text{ mmHg}$$
$$P_{O_2} = 143 \text{ mmHg}$$

Rearranging Dalton's law to isolate P_{H_2O} on the left side of the equation gives

$$P_{H_2O} = P_{Total} - P_{N_2} - P_{O_2}$$

Substituting the known numerical values into this equation and doing the arithmetic gives

$$P_{H_2O} = 742 \text{ mmHg} - 581 \text{ mmHg} - 143 \text{ mmHg} = 18 \text{ mmHg}$$

Dalton's law of partial pressures is important when considering the air of our atmosphere, which is a mixture of numerous gases. The composition of dry air, from which all water vapor has been removed, is virtually constant throughout the world. Table 7–2 gives the composition of clean, dry air by percentage volume and lists all components that make up at least 0.001% of the total volume. Atmospheric pressure is the sum of the partial pressures of the gaseous components present in air. Table 7–2 also gives the partial pressure of each component of air in a situation where total atmospheric pressure is 760 mmHg.

The composition of air is not absolutely constant. The variability in composition is caused predominantly by the presence of water vapor, which is not listed in Table 7–2 because that table deals with *dry* air. The amount of water vapor in air varies between values of a few tenths of 1% and 5–6%, depending on weather and temperature.

At higher altitudes, the total pressure of air decreases, as do the partial pressures of the individual components of air. For example, at an altitude where total atmospheric pressure is 640 mmHg (4000–5000 feet), the partial pressures of nitrogen and oxygen would be 499.7 mmHg and 134.1 mmHg, respectively. An individual going from sea level to a higher altitude usually experiences some tiredness because his body is not functioning as efficiently at the higher altitude. The red blood cells in his body absorb a smaller amount of oxygen because the oxygen partial pressure at the higher altitude is lower. A person's body acclimates itself to the higher altitude after a period of time.

The concept of partial pressures is an integral part of the explanation of how the process of respiration occurs in the human body. In general, the act of breathing replenishes oxygen, which has been consumed by cellular processes, in the blood and removes carbon dioxide, the waste product from cellular processes, from the blood. The basics of this process are shown in Figure 7–10 (p. 162).

Pressure gradients resulting from partial-pressure differences govern the movement of oxygen and carbon dioxide throughout the human body. Both gases always move from areas of higher partial pressure to areas of lower partial pressure. Blood returning to the lungs from tissues (shown in the upper part of Figure 7–10b) has been depleted of its oxygen supply (low P_{O_2}) and is carrying waste product carbon dioxide (high P_{CO_2}). Within the lungs (shown on the left side of Figure 7–10b), carbon dioxide leaves the blood and oxygen enters the blood. The partial pressure of carbon dioxide is greater in the blood than within the lungs; for oxygen, the exact opposite is true. Blood leaving the lungs (shown in the bottom part of Figure 7–10b), which has been replenished in oxygen (high P_{O_2}), travels to the tissues (low P_{O_2}). The oxygen from the blood diffuses into the tissues (shown on the right side of Figure 7–10b). At the same time, carbon dioxide produced in the cells (high P_{CO_2}) diffuses into the bloodstream (low P_{CO_2}). Note the actual values of the partial pressures for oxygen and carbon dioxide given in Figure 7–10b at

TABLE 7–2 The Major Components of Clean, Dry Air

Gaseous Component	Formula	Volume Percentage	Partial Pressure (in mmHg)
nitrogen	N_2	78.084	593.4
oxygen	O_2	20.948	159.2
argon	Ar	0.934	7.1
carbon dioxide	CO_2	0.031	0.2
neon	Ne	0.002	—
helium	He	0.001	—

FIGURE 7-10 The movement of oxygen and carbon dioxide in the blood and tissues depends upon the partial pressure of each gas.

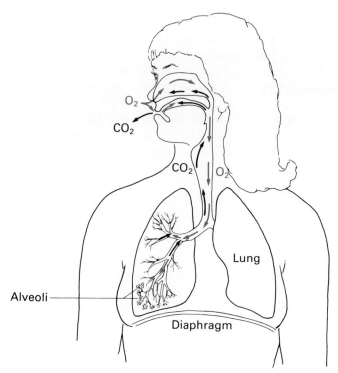

(a) Schematic summary of external movement of CO_2 and O_2

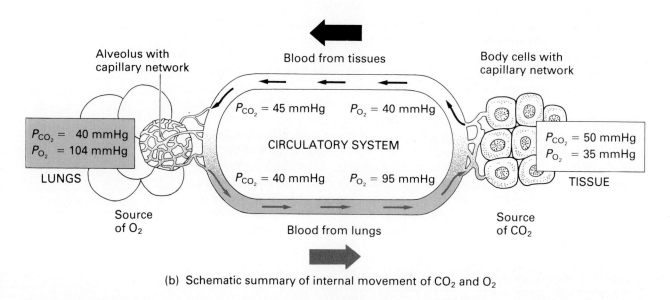

(b) Schematic summary of internal movement of CO_2 and O_2

various locations in the respiration cycle. For carbon dioxide, the partial pressure varies between 40 and 50 mmHg; this is a small pressure gradient. For oxygen, a much larger pressure gradient exists, with the partial pressure being 104 mmHg in the lungs and 35 mmHg in the tissues.

FIGURE 7-11 The six changes of state and the terms used to describe them.

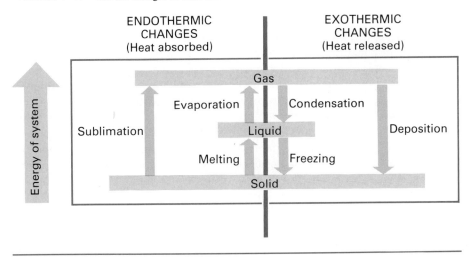

7.12 CHANGES OF STATE

A **change of state** *is a process in which a substance is transformed from one physical state to another*. Changes of state are usually accomplished through heating or cooling a substance. Pressure change is also a factor in some systems. As noted previously (Section 1.4), changes of state are examples of physical changes; that is, changes in which chemical composition remains constant. No new substances are ever formed as a result of state change.

There are six possible changes of state. Figure 7-11 identifies each of these changes and gives the terminology used to describe the changes. Four of the six terms used in describing state changes are familiar: freezing, melting, evaporation, and condensation. The other two terms—sublimation and deposition—describe changes involving the solid and gaseous states; these changes are not as common as the other types.

Changes of state can be classified into two categories, based on whether heat (thermal energy) is given up or absorbed during the change process. An **endothermic change of state** *is a change that requires the input (absorption) of heat*. The endothermic changes of state are melting, sublimation, and evaporation (the processes in Figure 7-11 in which the arrows point up). An **exothermic change of state** *is a change that requires heat to be given up (released)*. Exothermic changes of state are the reverse of endothermic changes of state and include freezing, condensation, and deposition (the processes in Figure 7-11 in which the arrows point down).

change of state

endothermic change of state

exothermic change of state

7.13 EVAPORATION OF LIQUIDS

Evaporation *is the process by which molecules escape from the liquid phase to the gas phase*. It is a familiar process to us. We are all aware that water left in an open container at room temperature slowly disappears by evaporation.

The phenomenon of evaporation can readily be explained using kinetic molecular theory. Statement 5 of this theory (Section 7.2) indicates that the molecules in a liquid (or solid or gas) do not all possess the same kinetic energy. At any given instant, some molecules will have above-average kinetic energies and others will have below-average kinetic energies as a result of collisions between molecules.

evaporation

A given molecule's energy constantly changes as a result of collisions with neighboring molecules. Molecules that happen to be considerably above average in kinetic energy at a given moment can overcome the attractive forces (potential energy) holding them in the liquid and escape if they are on the liquid surface and are moving in a favorable direction relative to the surface.

Evaporation is a surface phenomenon. Molecules within the interior of a liquid are surrounded on all sides by other molecules, which makes escape very improbable. Surface molecules are subject to fewer attractive forces because they are not completely surrounded by other molecules; thus, escape is much more probable. Liquid surface area is an important factor when determining the rate at which evaporation occurs. Increased surface area results in an increased evaporation rate because a greater fraction of molecules occupy surface locations.

Water evaporates faster from a glass of hot water than from a glass of cold water. Why is this so? A certain minimum kinetic energy is required for molecules to escape from the attractions of neighboring molecules. As the temperature of a liquid increases, a larger fraction of the molecules present acquire this needed minimum kinetic energy. Consequently, the rate of evaporation always increases as liquid temperature increases. Figure 7–12 contrasts the fraction of molecules possessing the needed minimum kinetic energy for escape at two temperatures. Note that at both the lower and higher temperatures, a broad distribution of kinetic energies is present and that some molecules possess the needed minimum kinetic energy at each temperature. However, at the higher temperature, a larger fraction of the molecules present have the requisite kinetic energy.

The escape of high-energy molecules from a liquid during evaporation affects the liquid in two ways: the amount of liquid decreases and the liquid temperature is lowered. The lower temperature reflects the fact that the average kinetic energy of the remaining molecules is lower than the pre-evaporation value because of the loss of the most energetic molecules. (Analogously, if all the tall people are removed from a classroom of students, the average height of the remaining students decreases.) A lower average kinetic energy corresponds to a lower temperature (Statement 4 of kinetic molecular theory); hence, a cooling effect is produced.

Evaporative cooling is important in many processes. Our own bodies use evaporation in order to maintain a constant temperature. We perspire in hot weather

FIGURE 7–12 The kinetic energy distributions of molecules of a liquid at two different temperatures. The dashed line represents the minimum kinetic energy required for molecules of the liquid to overcome attractive forces and escape into the gas phase. Molecules in the shaded area have the energy that is necessary to overcome attractions.

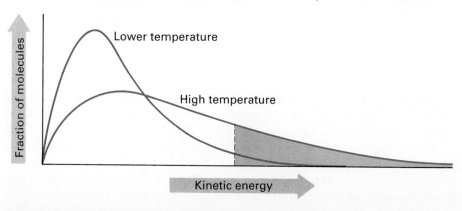

because evaporation of the perspiration cools our skin. The cooling effect of evaporation is quite noticeable when someone first comes out of a swimming pool on a hot day (especially if a breeze is blowing). A humid atmosphere necessarily retards evaporation and therefore lessens the cooling effect derived from it; this explains the fact that the same degree of temperature seems hotter in humid climates than it does in dry ones.

Certain alcohols are used externally as rubbing agents to reduce fevers in small children. The alcohol, which evaporates more readily than water, lowers the body temperature of the feverish child.

The molecules that escape from an evaporating liquid are often collectively referred to as *vapor*, rather than gas. The term **vapor** describes *gaseous molecules of a substance at a temperature and pressure in which we ordinarily would think of the substance as a liquid or solid.* For example, at room temperature and atmospheric pressure, the normal state for water is the liquid state. Molecules that escape (evaporate) from liquid water at these conditions are frequently called *water vapor*.

vapor

7.14 VAPOR PRESSURE OF LIQUIDS

The evaporative behavior of a liquid in a closed container is quite different from its behavior in an open container. In a closed container, we observe that some liquid evaporation occurs; this is indicated by a drop in liquid level. However, unlike the open-container system, with time, the liquid level ceases to drop (becomes constant); this is an indication that not all of the liquid will evaporate.

Kinetic molecular theory explains these observations in the following way. The molecules that evaporate in the closed container are unable to move completely away from the liquid, as they did in the open container. They find themselves confined in a fixed space immediately above the liquid (see Figure 7–13a). These trapped vapor molecules undergo many random collisions with the container walls, other vapor molecules, and the liquid surface. Molecules that collide with the liquid surface are recaptured by the liquid. Thus, two processes, evaporation (escape) and condensation (recapture), take place in a closed container.

For a short time, the rate of evaporation in a closed container exceeds the rate of condensation and the liquid level drops. However, as more of the liquid evaporates, the number of vapor molecules increases; the chance of their recapture through striking the liquid surface also increases. Eventually, the rate of condensation becomes equal to the rate of evaporation and the liquid level stops dropping

FIGURE 7–13 Evaporation of a liquid in a closed container.

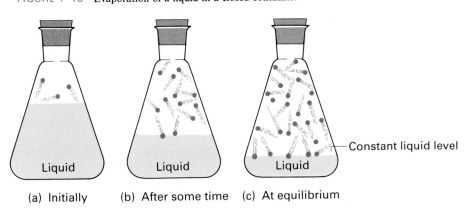

TABLE 7-3 Vapor Pressure of Water at Various Temperatures

Temperature (°C)	Vapor Pressure (mmHg)	Temperature (°C)	Vapor Pressure (mmHg)
0	4.6	60	149.4
10	9.2	70	233.7
20	17.5	80	355.1
30	31.8	90	525.8
40	55.3	100	760.0
50	92.5		

(see Figure 7–13c). At this point, the number of molecules that escape in a given time is the same as the number recaptured; a steady-state situation has been reached. The amount of liquid and vapor in the container does not change, even though both evaporation and condensation are still occurring.

This steady-state situation, which will continue as long as the temperature of the system remains constant, is an example of a *state of equilibrium*. A **state of equilibrium** *is a situation in which two opposite processes take place at equal rates.* For systems in a state of equilibrium, no net macroscopic changes can be detected. However, the system is dynamic; the forward and reverse processes are occurring at equal rates.

When there is a liquid–vapor equilibrium in a closed container, the vapor in the fixed space immediately above the liquid exerts a constant pressure on both the liquid surface and the walls of the container. This pressure is called the *vapor pressure* of the liquid. **Vapor pressure** *is the pressure exerted by a vapor above a liquid when the liquid and vapor are in equilibrium.*

The magnitude of a vapor pressure depends on the nature and temperature of a liquid. Liquids that have strong attractive forces between molecules will have lower vapor pressures than liquids that have weak attractive forces between particles. Substances that have high vapor pressures at room temperature are said to be *volatile*.

The vapor pressure of all liquids increases with temperature. Why? An increase in temperature results in more molecules having the minimum kinetic energy required for evaporation. Hence, the pressure of the vapor is greater at equilibrium. Table 7–3 shows the variation in vapor pressure, for increasing temperature, of water.

The size (volume) of the space that the vapor occupies does not affect the magnitude of the vapor pressure. A larger fixed space will enable more molecules to be present in the vapor at equilibrium. However, the larger number of molecules spread over a larger volume results in the same pressure as a small number of molecules in a small volume.

7.15 BOILING AND THE BOILING POINT

In order for a molecule to escape from the liquid state, it usually must be on the surface of the liquid. **Boiling** *is a special form of evaporation where conversion from the liquid state to the vapor state occurs within the body of a liquid through bubble formation.* This phenomenon begins to occur when the vapor pressure of a liquid, which steadily increases as the liquid is heated, reaches a value equal to that of the prevailing external pressure on the liquid; for liquids in open containers, this value is atmospheric pressure. When these two pressures become equal, bubbles of vapor form around any speck of dust or around any irregularity associated

TABLE 7-4 Variation of the Boiling Point of Water with Elevation

Location	Elevation (feet above sea level)	Boiling Point of Water (°C)
San Francisco, Calif.	sea level	100.0
Salt Lake City, Utah	4,390	95.6
Denver, Colo.	5,280	95.0
La Paz, Bolivia	12,795	91.4
Mount Everest	28,028	76.5

with the container surface. These vapor bubbles quickly rise to the surface and escape because they are less dense than the liquid itself. The quick ascent of the bubbles causes the agitation associated with a boiling liquid. At this point we say the liquid is boiling.

The **boiling point** of a liquid is defined as *the temperature at which the vapor pressure of a liquid becomes equal to the external (atmospheric) pressure exerted on the liquid.* Because the atmospheric pressure fluctuates from day to day, the boiling point of a liquid does also. Thus, in order to compare the boiling points of different liquids, the external pressure must be the same. The boiling point of a liquid that is most often used for comparison and tabulation purposes is called the *normal* boiling point. A liquid's **normal boiling point** is *the temperature at which a liquid boils under a pressure of 760 mmHg.*

At any given location, the changes in the boiling point of a liquid caused by *natural* variations in atmospheric pressure seldom exceeds a few degrees; in the case of water, the maximum is about 2°C. However, variations in boiling points *between* locations at different elevations can be quite striking, as is shown in the data in Table 7-4.

The boiling point of a liquid can be increased by increasing the external pressure. This principle is used in the operation of a pressure cooker. Foods cook faster in pressure cookers because the elevated pressure causes water to boil above 100°C. An increase in temperature of only 10°C will cause food to cook in approximately one-half the normal time. (Cooking food involves chemical reactions and the rate of a chemical reaction generally doubles with every 10°C increase in temperature.) Table 7-5 gives the boiling temperatures reached by water under normal household pressure-cooker conditions. Hospitals use this same principle to sterilize instruments and laundry in autoclaves; there, sufficiently high temperatures are reached to destroy bacteria.

Liquids that have high normal boiling points or that undergo undesirable chemical reactions at elevated temperatures can be made to boil at low temperatures

boiling point

normal boiling point

TABLE 7-5 Boiling Point of Water at Various Pressure-Cooker Pressures

Pressure above Atmospheric		Boiling Point of Water (°C)
lb/in.²	mmHg	
5	259	108
10	517	116
15	776	121

by reducing the external pressure. This principle is used in the preparation of numerous food products, including frozen fruit juice concentrates. Some of the water in a fruit juice is boiled away at a reduced pressure, thus concentrating the juice without having to heat it to a high temperature. Heating juices to high temperatures causes changes that spoil the taste of the juice and reduce its nutritional value.

7.16 INTERMOLECULAR FORCES IN LIQUIDS

Boiling points among substances vary greatly. The boiling points of some substances are well below zero; for example, oxygen has a boiling point of −183°C. As we know, numerous other substances do not boil until the temperature is much higher. An explanation of this variation in boiling points involves a consideration of the nature of the *intermolecular* forces that must be overcome in order for molecules to escape from the liquid state into the vapor state. **Intermolecular forces** *are forces that act between a molecule or ion and another molecule or ion.*

Intermolecular forces are similar in one way to the previously discussed *intramolecular* forces (*within* molecules) that are involved in covalent bonding (Sections 6.8 and 6.9); they are electrostatic in origin. A major difference between inter- and intramolecular forces is their magnitude; the former are much weaker. However, despite their relative weakness, intermolecular forces are sufficiently strong to influence the behavior of liquids and often do so in a very dramatic way.

A major type of intermolecular force is the dipole–dipole interaction. **Dipole–dipole interactions** *are electrostatic attractions between polar molecules.* Polar molecules, which are often called *dipoles* (Section 6.12), are electrically unsymmetrical. Therefore, when polar molecules approach each other, they tend to line up so that the relatively positive end of one molecule is directed toward the relatively negative end of the other molecule. As a result, there is an electrostatic attraction between the molecules. The greater the polarity of the molecules, the greater the strength of the dipole–dipole interactions; the greater the strength of the dipole–dipole interactions is, the higher the boiling point of the liquid is. Figure 7–14 shows the many dipole–dipole interactions that are possible for a random arrangement of polar ClF molecules.

Particularly strong dipole–dipole interactions occur between polar molecules that contain hydrogen bonded to a highly electronegative element such as fluorine, oxygen, and nitrogen. Two factors account for the extra strength of these dipole–dipole interactions: (1) the great polarity of the (hydrogen)–(highly-electronegative-element) bond, and (2) the close approach of the dipoles, which is allowed by the very small size of the hydrogen atom. Dipole–dipole interactions of this type are sufficiently unique to be given a special name—hydrogen bonds. A **hydrogen bond** *is an extra-strong dipole–dipole interaction between a hydrogen atom that is bonded to a nitrogen, oxygen, or fluorine atom and another nitrogen, oxygen, or fluorine atom.*

Water (H_2O) is the most commonly encountered substance that shows hydrogen bonding. Figure 7–15 depicts the process of hydrogen bonding among water molecules. Because of the polarity of water molecules (Section 6.12), each hydrogen atom carries a partial positive charge (denoted as δ^+) and each oxygen atom carries a partial negative charge (denoted as $2\delta^-$). Hydrogen bonds result from the close approach of the very small, partially charged hydrogen atoms to nonbonding pairs of electrons on the partially charged oxygen atoms of other water molecules.

Hydrogen bonds are much weaker than ionic or covalent bonds. The strongest hydrogen bonds are about one-tenth as strong as a covalent bond. In spite of the

FIGURE 7–14 Dipole–dipole interactions between randomly arranged ClF molecules. The positive end of one molecule is attracted to the negative end of another neighboring ClF molecule.

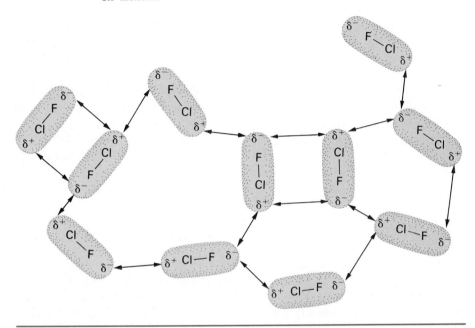

fact that they are weak, hydrogen bonds profoundly affect the properties of substances in which they occur. The effects of hydrogen bonding on the properties of water are considered in Chapter 8 (Sections 8.2 through 8.5).

Many molecules of biological importance, such as DNA and proteins, contain O—H and N—H bonds, and hydrogen bonding plays a role in the behavior of these substances. Certain bonds in these compounds must be capable of breaking and reforming with relative ease, and only hydrogen bonds have the right energies to permit this. It is not an overstatement to say that hydrogen bonding makes life possible. The importance of hydrogen bonding in molecules of biological significance will be explored further in the biochemistry chapters of this text.

FIGURE 7–15 Hydrogen bonds (dashed lines) among water molecules.

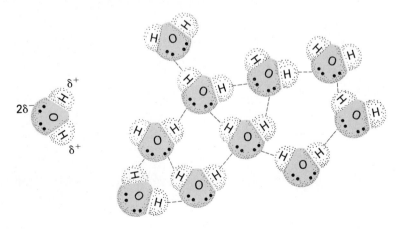

7.17 SUBLIMATION AND MELTING

Solids, like liquids, have vapor pressures. The odors associated with mothballs, spices, and instant coffee are evidence of this. The kinetic molecular theory applies to solids in much the same way as it does to liquids. Particles in the solid state have a range of energies, and when these particles are on a solid's surface, they can escape into the gas phase if they acquire sufficiently high energies. **Sublimation** *is the name of the process whereby a solid changes directly to a gas (vapor).*

The vapor pressures for solids are generally much lower than those of liquids because the cohesive forces in solids are not overcome by opposing disruptive forces to the extent that they are in liquids (Section 7.3). As expected, the vapor pressure of a solid increases with increasing temperature. At room temperature and pressure, there are very few solids for which the vapor pressure is sufficiently high that the transition from solid to vapor is noticeable.

Two sublimation processes that most people have encountered in everyday life involve solid carbon dioxide (dry ice) and naphthalene (mothballs). Sublimation of solid water (ice) is also a common process during winter months. Ice disappears (sublimes) from sidewalks and driveways in the coldest part of the winter even though the temperature does not rise above freezing. Wet laundry that is hung out to dry in freezing weather eventually dries as the frozen water sublimes. A commercial sublimation process called *freeze drying* is used to dry biological materials or foodstuffs that would be damaged by heating (see Figure 7–16). A reduced pressure is usually a necessary part of such commercial freeze-drying processes. Freeze drying can also be used to preserve tissue cultures and bacteria. These freeze-dried specimens can be stored for an extended period of time.

Although all solids have vapor pressures and the potential to sublime, most of them do not sublime. Instead, most solids melt when heated. **Melting** *is the process in which a solid is converted to a liquid.* As the temperature of a solid is increased, the motion of the particles about their fixed positions becomes more vigorous and the particles are forced farther apart. Eventually, the particles gain sufficient kinetic energy to break down (collapse) the rigid structure associated with the solid state. When this happens, we say that the substance has melted.

FIGURE 7–16 Freeze drying is used to prepare cultures for analysis. Water is removed from the cells through sublimation.

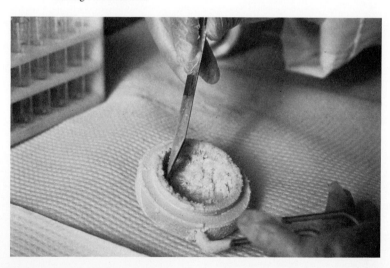

7.18 DECOMPOSITION

Nearly every gas becomes a liquid and then a solid if its constituent particles are slowed (cooled) enough to allow the cohesive forces that are present in all matter to dominate over the disruptive forces. It is not true, however, that heating changes all solids into liquids or all liquids into gases. In some cases, the atoms that make up the molecules of a solid or liquid can acquire enough energy and vibrate so violently that bonds within the molecules are broken before the solid or liquid can change state. When this occurs, we say that the substance has decomposed. **Decomposition** *is the process in which chemical bonds (intramolecular forces) are overcome (broken) before intermolecular forces are overcome.* For example, paper and cotton do not melt when heated, but char or decompose.

decomposition

EXERCISES AND PROBLEMS

States of Matter

7.1 List the macroscopic characteristics that distinguish solids from liquids and liquids from gases.

7.2 The following statements relate to the terms *solid state*, *liquid state*, and *gaseous state*. Match the terms to the appropriate statements.

a. This state is characterized by the lowest density of the three states.
b. This state is characterized by an indefinite shape and a high density.
c. Temperature changes significantly influence the volume of this state.
d. Pressure changes influence the volume of this state more than the other two states.
e. In this state, constituent particles are less free to move around than in other states.

Kinetic Molecular Theory of Matter

7.3 According to the statements of the kinetic molecular theory of matter:
a. What two types of energy do particles possess?
b. How do molecules transfer energy to each other?
c. What is the relationship between temperature and the average velocity with which the particles move?

7.4 Contrast the effects of cohesive and disruptive forces on a system of particles. What type of energy is related to cohesive forces and disruptive forces?

7.5 What effect does temperature have on the magnitude of disruptive forces and cohesive forces?

7.6 Distinguish among the gaseous, liquid, and solid states of a substance from the point of view of the relative magnitude of the kinetic and potential energies of the constituent particles.

7.7 Explain each of the following observations using the kinetic molecular theory of matter.
a. Gases have a low density.
b. Solids maintain characteristic shapes.
c. Liquids show little change in volume with changes in temperature.
d. Both liquids and solids are practically incompressible.
e. A gas always exerts a pressure on the object or container with which it is in contact.

7.8 Using kinetic molecular theory, explain why it is dangerous to throw an aerosol can into an open fire.

7.9 Illustrate the differences among the three states of matter on the microscopic (molecular) level by drawing three different arrangements of 15 to 20 spheres.

Gas Law Variables

7.10 What are the volume units that are most often used when doing gas-law calculations?

7.11 What temperature scale must always be used to specify temperature when the temperature is used in a gas-law equation?

7.12 What are the three most common units for specifying the pressure a gas exerts and what are the interrelationships among these three units?

7.13 Carry out the following pressure unit conversions using the dimensional-analysis method of problem solving.
a. 735 mmHg to atmospheres
b. 0.53 atm to mmHg
c. 0.53 atm to torr
d. 12.0 psi to atmospheres
e. 777 mmHg to pounds-per-square-inch

7.14 The pressure unit mmHg can be used to characterize fluid pressure as well as gas pressure. What is the actual range of fluid pressure, expressed in mmHg, if a blood pressure is reported as 130/80?

Boyle's Law

7.15 Give a written statement and a mathematical equation for Boyle's law.

7.16 When we say that two quantities are inversely proportional, what do we mean?

7.17 A balloon is inflated to a volume of 12.6 L on a day when the atmospheric pressure is 652 mmHg. The next day, as a storm front arrives, the atmospheric pressure drops to 620 mmHg. Assuming the temperature remains constant, what is the new volume for the balloon, in liters?

7.18 A sample of carbon monoxide gas (CO) in an expandable container exerts a pressure of 740 mmHg when the volume is set at 4.00 L. Determine the volume settings that would produce the following gas pressures, assuming the temperature remains constant.
a. 760 mmHg c. 450 mmHg
b. 900 mmHg d. 200 mmHg

7.19 A sample of N_2 gas occupies a volume of 2.00 L at 27°C and 760 torr. Determine the pressure that this sample will exert at the same temperature if the volume of the gas is changed to each of the following values.
a. 1.00 L   c. 3.70 L   e. 1.20 L
b. 1.43 L   d. 5.00 L

7.20 Explain Boyle's law using the kinetic molecular theory of matter.

Charles' Law

7.21 Give a written statement and a mathematical equation for Charles' law.

7.22 When we say that two quantities are directly proportional, what do we mean?

7.23 A sample of H_2 gas has a volume of 3.03 L at 27°C and 1.2 atm pressure. What volume will the H_2 gas occupy at each of the following temperatures if the pressure is held constant?
a. 227°C   c. −73°C   e. 1200 K
b. 327°C   d. 1200°C

7.24 A sample of O_2 gas occupies a volume of 350 mL at 25°C and a pressure of 1.00 atm. At the same pressure; determine the temperature, in °C, at which the volume of the gas would be equal to each of the following.
a. 385 mL   c. 2500 mL   e. 1.23 L
b. 12.0 mL d. 700 mL

7.25 An adult human breathes in approximately 0.500 L of air at 36°C and 1.00 atm pressure with every breath. What would this breath volume be at the same pressure if the temperature drops to 26°C?

7.26 Explain Charles' law using the kinetic molecular theory of matter.

Combined-Gas Law

7.27 Give a written statement and a mathematical equation for the combined-gas law.

7.28 Rearrange the standard form of the combined-gas law equation so that each of the following variables is by itself on the left side of the equation.
a. T_1 b. V_1 c. P_2 d. T_2

7.29 A sample of Cl_2 gas has a volume of 25.0 L at a pressure of 4.00 atm and a temperature of 27°C. Determine the volume that this gas will occupy, in liters, after it undergoes the following temperature and pressure changes.
a. 35°C and 2.43 atm   d. 600°C and 3.00 atm
b. 382°C and 25.0 atm   e. 1000°C and 100 atm
c. −63°C and 0.532 atm

Dalton's Law of Partial Pressures

7.30 Give a written statement and a mathematical equation for Dalton's law of partial pressures.

7.31 The total pressure exerted by a mixture of oxygen, nitrogen, and carbon dioxide gas is 2200 mmHg. The partial pressures of the nitrogen and oxygen are 520 mmHg and 650 mmHg, respectively. What is the partial pressure of the carbon dioxide?

7.32 Exhaled breath is a mixture of nitrogen, oxygen, carbon dioxide, and water vapor. What is the partial pressure of water vapor in exhaled breath at body temperature (37°C) on a day when the atmospheric pressure is 760 mmHg if the partial pressure of oxygen is 116 mmHg, the partial pressure of nitrogen is 569 mmHg, and the partial pressure of carbon dioxide is 28 mmHg?

7.33 Indicate whether the partial pressure of each of the following gases is higher in the lungs or in active cells of body tissue.
a. oxygen   b. carbon dioxide

Changes of State

7.34 Indicate whether each of the following is an exothermic or endothermic change of state.
a. sublimation   d. evaporation
b. freezing   e. condensation
c. melting   f. deposition

7.35 Match the following statements to the appropriate term: *evaporation, vapor pressure, boiling point, deposition,* and *decomposition.*

a. This is a temperature at which the liquid vapor pressure is equal to the external pressure on a liquid.
b. This process corresponds to a direct change from the gaseous state to the solid state.
c. This process takes place when a liquid changes to a vapor.
d. In this process, intramolecular forces rather than intermolecular forces are overcome.
e. This property can be measured by allowing a liquid to evaporate in a closed container.
f. At this temperature, bubbles of vapor form within a liquid.
g. This temperature changes appreciably with changes in atmospheric pressure.

7.36 Offer a clear explanation for each of the following observations.
a. All liquids do not have the same vapor pressure at a given temperature.

b. Changing the volume of a container in which there is a liquid–vapor equilibrium will not change the magnitude of the vapor pressure.

c. Increasing the temperature of a liquid–vapor equilibrium system causes an increase in the magnitude of the vapor pressure.

7.37 Offer a clear explanation for each of the following observations.

a. The boiling point of a liquid varies with atmospheric pressure.

b. It takes more time to cook an egg in water on a mountain top than at sea level.

c. Food will cook just as fast in boiling water with the stove set at low heat as in boiling water at high heat.

d. Food cooks faster in a pressure cooker than in an open pan.

7.38 Offer a clear explanation for each of the following observations.

a. A person feels more uncomfortable in humid air at 90°F than in dry air at 90°F.

b. A person emerging from an outdoor swimming pool on a breezy day gets the shivers.

c. During the cold winter months, snow often slowly disappears without melting.

7.39 Distinguish between the terms *boiling point* and *normal boiling point*.

7.40 What two factors affect the rate at which a substance evaporates?

7.41 What factors affect the magnitude of the vapor pressure a liquid exerts?

7.42 Criticize the statement: "All solids will melt if heated to a high enough temperature".

Intermolecular Forces in Liquids

7.43 What is the difference in meaning between the terms *intermolecular force* and *intramolecular force*?

7.44 In liquids, what is the relationship between boiling point and the strength of intermolecular forces?

7.45 Describe in writing what a dipole–dipole attraction is.

7.46 What are hydrogen bonds?

7.47 Which is harder to break, an ordinary covalent bond or a hydrogen bond?

7.48 Which three nonmetals, other than hydrogen, are the elements most often involved in hydrogen bonding?

8 Water, Solutions, and Colloids

Objectives

After completing Chapter 8, you will be able to:

8.1 Discuss the distribution of water on the Earth and the ways in which water enters and leaves the human body.
8.2 List the unique properties of the substance water.
8.3 Discuss the effect that hydrogen bonding has on water's boiling and freezing points.
8.4 Discuss the important consequences of water having thermal properties with higher than normal values.
8.5 Discuss the abnormal density behavior of water and the important consequences of this behavior.
8.6 Define the terms *solution*, *solute*, and *solvent*.
8.7 List and give an example of the nine types of two-component solutions.
8.8 Define and distinguish between the terms *saturated* and *unsaturated*, *dilute* and *concentrated*, and *miscible* and *immiscible*.
8.9 Describe the solution process, at a molecular level, as an ionic solute dissolves in water.
8.10 Explain the rule "likes dissolve likes" and know the solubility guidelines for ionic solutes in water.
8.11 Explain what is meant by the term *concentration*.
8.12 Define and work problems involving the concentration units *percent-by-mass*, *mass-volume percent*, *percent-by-volume*, and *milligram percent*.
8.13 Define and work problems involving the concentration unit *molarity*.
8.14 Define and work problems involving the concentration unit *milliequivalents-per-liter*.
8.15 Calculate the concentration of a solution that is obtained by diluting a solution of known concentration.
8.16 Explain how the presence of solute in a solution causes vapor pressure lowering, boiling point elevation, and freezing point depression and discuss the important consequences of these colligative properties.
8.17 Discuss the process of osmosis, osmotic pressure, osmolarity, and the biological importance of osmotic processes.
8.18 Distinguish among the terms *isotonic*, *hypertonic*, and *hypotonic*, and between the processes of hemolysis and crenation.
8.19 List the differences between colloidal dispersions and true solutions.
8.20 Understand the differences and similarities between the processes of dialysis and osmosis.

INTRODUCTION

All samples of matter can be classified into the two categories of pure substances and mixtures. We discussed this concept in Section 2.1. We have tended to limit our discussion to pure substances since that early discussion of matter classification. Now we will turn our attention to mixtures, with particular emphasis on homogenous mixtures (Section 2.2).

Another name for a homogeneous mixture is a *solution*; we will use this terminology in this chapter. Solutions are common in nature and they represent an abundant form of matter. Solutions carry nutrients to the cells of our bodies and carry away waste products. The ocean is a solution of water, sodium chloride, and many other substances (even gold). A large percentage of chemical reactions take place in solution, including most of those that will be discussed in later chapters of this book.

Most of the solutions we encounter are water-based solutions. A good starting point for learning about solutions is to consider the substance *water*. In both the liquid and solid states, water is an unusual compound when compared to other liquids and solids. We will discuss the reasons for this unusual behavior and the important ramifications of it.

8.1 ABUNDANCE AND DISTRIBUTION OF WATER

Water is the most abundant compound on the face of the Earth. We encounter it everywhere we go: as water vapor in air, as a liquid in rivers, lakes, and oceans, and as a solid (ice and snow) both on land and in the oceans.

Approximately 75% of the Earth's surface is covered by water; the oceans are the major repositories for water. Table 8-1 lists the major reservoirs for water on Earth.

Water is also the most abundant compound in the human body. It constitutes more than 65% of the total body mass. A human being can survive for only a few days without taking in water. Water serves a variety of functions in the body. It serves as the fluid medium in which the chemical reactions of the cells take place. In the blood, water is the major transport medium for distributing oxygen and nutrients to the body cells, as well as for carrying away their waste products. Water plays an important role in the excretion of wastes through the kidneys, intestines, and sweat glands. The water that is eliminated through the sweat glands also plays a key role in the body's temperature-control mechanisms.

TABLE 8-1 Distribution of Water on the Earth

Location	Percent
oceans	97.2
glaciers and polar caps	2.16
subsurface water	0.62
lakes and rivers	0.019
atmospheric water	0.001

TABLE 8–2 Water Content of Selected Foods

Food	Percent water
apples	85
broccoli	90
cheddar cheese	35
eggs	75
grapes	80
milk	87
potatoes	78
steak	73
tuna fish	60

Water enters the body in three different ways. Two major sources of water are the liquids we ingest and the foods we eat. Table 8–2 gives the water content of selected foods. The third source of water is one that is less universally known. The cells in our bodies produce water as a byproduct of the reactions in which food is broken down into simpler components, and this water enters the bloodstream. (Some desert insects are capable of using the water generated metabolically as their sole source of water.)

Water normally leaves the body by four exits: kidneys (urine), lungs (water in expired air), skin (by diffusion and sweat), and intestines (feces). In order for a body to function normally, a fluid balance must exist within it. This means that the total volume of water entering the body must normally equal the total volume of water leaving the body. The body's water content is continuously monitored in the brain. A sensation of thirst is triggered if more water is needed to meet the body's needs. Too much water in the tissues triggers control mechanisms that restore fluid balance by increasing fluid elimination.

Approximately 2400 mL of water enter and leave the body daily. Table 8–3 lists typical normal values for each portal of water entry and exit. These values can vary considerably and still be considered normal. Much of the variance relates to the fat content of the body. Fat people have a lower water content per kilogram of body mass than slender people. As a person grows older, the amount of water in his or her body also decreases. A newborn infant's body contains about 77% water and an older adult man's body only contains about 60% water.

TABLE 8–3 Typical Normal Values for Each Portal of Water Entry and Exit in the Human Body

Intake	
ingested liquids	1500 mL
water in foods	700 mL
water formed as a byproduct in cell breakdown of food	200 mL
	2400 mL
Output	
kidneys (urine)	1400 mL
lungs (water in expired air)	350 mL
skin	
by diffusion	350 mL
by sweat	100 mL
intestines (in feces)	200 mL
	2400 mL

8.2 PROPERTIES OF WATER

From a chemical viewpoint, the most fascinating thing about water is that it is unique in many of its properties. Numerous properties of water have values that fall outside the normal range exhibited by compounds similar in structure. It is the unusual properties of water that make it the key substance in the forms of life on this planet.

Most of water's unusual behavior is a consequence of the extensive hydrogen bonding (Section 7.16) that occurs between water molecules, both in the liquid state and in the solid state. Three important hydrogen-bonding-influenced properties of water are: (1) higher than expected boiling and freezing points, (2) higher than expected thermal properties, and (3) an unusual temperature–density relationship. The next three sections of the text consider these three properties.

8.3 BOILING AND FREEZING POINTS OF WATER

The vapor pressures (Section 7.14) of liquids that have significant hydrogen bonding are significantly lower than those of similar liquids where little or no hydrogen bonding occurs. The presence of hydrogen bonds makes it more difficult for molecules to escape from the condensed state; additional energy is needed to overcome the hydrogen bonds. The greater the hydrogen bond strength is, the lower the vapor pressure is at any given temperature.

The boiling point of a liquid depends upon vapor pressure (Section 7.15), so liquids with low vapor pressures have to be heated to higher temperatures in order to bring their vapor pressures up to the point where boiling occurs (atmospheric pressure). As a result, boiling points are much higher for liquids in which hydrogen bonding occurs.

The effect that hydrogen bonding has on water's boiling point can be seen by comparing it with the boiling points of other hydrogen compounds of Group VIA elements—H_2S, H_2Se, and H_2Te. In this series of compounds—H_2O, H_2S, H_2Se, and H_2Te—water is the only compound where significant hydrogen bonding occurs. Normally, the boiling points of a series of compounds that contain elements

FIGURE 8-1 Boiling points of the hydrogen compounds of Group VIA elements. Water is the only one of the compounds where significant hydrogen bonding is present.

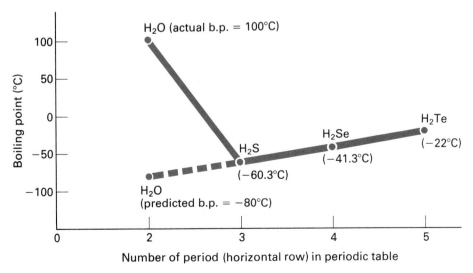

in the same periodic table group increase with increasing formula weight. Thus, in the hydrogen–Group VIA series, we would expect that H_2Te, the heaviest member of the series, would have the highest boiling point and that water, the compound of lowest formula weight, would have the lowest boiling point. Contrary to expectation, H_2O has the highest boiling point, as can be seen from the boiling point data shown in Figure 8–1 (p. 177). The data in Figure 8–1 indicate that water should have a boiling point of approximately $-80°C$; this value is obtained by extrapolation (extension of the line connecting the three heavier compounds. The actual boiling point of water, $100°C$, is nearly $200°$ higher than predicted. Indeed, in the absence of hydrogen bonding, water would be a gas at room temperature and life as we know it on Earth would not be possible.

A higher-than-expected freezing point is also characteristic of water, and a plot of freezing points for similar compounds would have the same general shape as the boiling point plot of Figure 8–1.

8.4 THERMAL PROPERTIES OF WATER

Three important thermal properties for any substance are specific heat, heat of vaporization, and heat of fusion. The property *specific heat* was defined and discussed in Section 1.15. Liquid water has a specific heat of 1.00 cal/g°C. This specific heat value is at least double that of most other substances (Table 1.6).

Heat energy is absorbed or evolved in changes of state (Section 7.12). The **heat of fusion** *is the amount of energy required to convert one gram of a solid to a liquid at its melting point.* This same amount of energy is released to the surroundings when one gram of the liquid is changed back to solid during a freezing process. The **heat of vaporization** *is the amount of energy required to convert one gram of a liquid to a gas at its boiling point.* This same amount of energy is released to the surroundings when one gram of the gas condenses to a liquid.

The numerical values of the heats of fusion and vaporization for water are 79.8 cal/g and 540 cal/g, respectively. Both of these values are higher than those for most other substances. By comparison, liquid CH_4 (a non-hydrogen-bonded liquid) has a heat of vaporization of 122 cal/g, which is less than one-fourth that of water.

The presence of extensive hydrogen bonding among water molecules significantly increases the ability of water to absorb heat energy when evaporating (heat of vaporization) and to release heat energy when freezing (heat of fusion). Additional heat energy is required in the evaporation process in order to overcome (break) the hydrogen bonds present.

Water's thermal properties and its abundance account for its widespread use as a coolant. Large bodies of water exert a temperature-moderating effect on their surroundings, primarily because of water's ability to absorb and release large amounts of heat energy. In the heat of a summer day, extensive water evaporation occurs, and in the process, energy is absorbed from the surroundings. The net effect of the evaporation is a lowering of the temperature of the surroundings. In the cool of the evening, some of this water vapor condenses back to the liquid state, releasing heat, which raises the temperature of the surroundings. In this manner, the temperature variation between night and day is reduced. A similar process occurs in the winter; water freezes on cold days and releases heat energy to the surroundings. The hottest and coldest regions on Earth are all inland regions that are farthest away from the moderating effects of large bodies of water.

Water's ability to absorb large amounts of heat during the evaporation process is a factor in the cooling of the human body. We noted in Section 7.13 that

FIGURE 8–2 Water's high heat of vaporization reduces the amount of water (perspiration) needed to carry away body heat during strenuous exercise.

evaporation of perspiration from the skin reduces body temperature (see Figure 8–2). Because water's heat of vaporization is so high, the amount of water lost from the body is minimized; this makes it easier to maintain fluid balance (Section 8.1) within the body.

8.5 WATER'S TEMPERATURE–DENSITY RELATIONSHIP

An unusual behavior pattern of water is the variation of its density with temperature. For most liquids, density increases with decreasing temperature and reaches a maximum for the liquid at its freezing point. The density pattern for water is different. Water's maximum density is reached at a temperature that is a few degrees higher than its freezing point. As shown in Figure 8–3 (p. 180), the maximum density for liquid water occurs at 4°C. This abnormality, that water at its freezing point is less dense than water at slightly higher temperatures, has tremendous ecological significance. Furthermore, at 0°C, solid water (ice) is significantly less

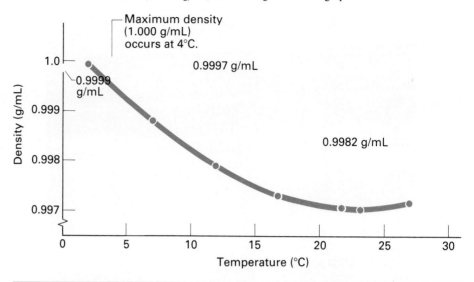

FIGURE 8–3 A plot of water density versus temperature for liquid water. The density of solid water at 0°C, 0.9170 g/mL, does not register on this graph.

dense than liquid water—0.9170 g/mL versus 0.9999 g/mL. Water's unusual density behavior is directly related to hydrogen bonding.

The fact that hydrogen bonding among water molecules is directional in nature is of prime importance when explaining water's peculiar density behavior. Hydrogen bonds can form only between molecules that are at certain angles. These angles are dictated by the location of the nonbonding pairs of electrons of water's oxygen atom. The net result is that when water molecules are hydrogen bonded, they are farther apart than when they are not hydrogen bonded. Figure 8–4 shows the hydrogen-bonding pattern characteristic of ice.

On a molecular scale, let us now consider what happens to water molecules when the temperature is lowered. At high temperatures such as 80°C, the kinetic energy of the water molecules is large enough to prevent hydrogen bonding from having much of an orientation effect on the molecules. Hydrogen bonds are rapidly and continually being formed and broken. As the temperature is lowered, the accompanying decrease in kinetic energy decreases molecular motion and the molecules move closer together. This results in an increase in density. The kinetic energy is still sufficient to negate most of the orientation effects of hydrogen bonding. When the temperature is lowered still further, the kinetic energy finally becomes insufficient to prevent hydrogen bonding from orienting molecules into definite patterns that require open spaces between molecules. At 4°C, the temperature at which water has its maximum density, the trade-off between random motion from kinetic energy and orientation from hydrogen bonding is such that the molecules are as close together as they will ever be. Cooling below 4°C causes a decrease in density because hydrogen bonding causes more and more open spaces to be present in the liquid. Density decrease continues down to the freezing point, at which temperature the hydrogen bonding causes molecular orientation to the maximum degree, then the solid crystal lattice of ice is formed. This solid crystal lattice of ice (Figure 8–4) is an extremely open type of structure; this results in solid ice having a lower density than liquid water. Water is one of only a few known substances where the solid phase is less dense than the liquid phase.

FIGURE 8-4 The crystalline structure of ice. Normal covalent bonds between oxygen and hydrogen, which hold the water molecules together, are shown by solid short lines. The weaker hydrogen bonds between molecules are shown by dashed longer lines.

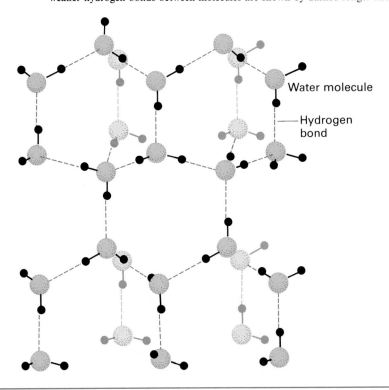

Two important consequences of ice being less dense than water are the facts that ice floats in liquid water and liquid water expands upon freezing.

If a container is filled with water and sealed, the force generated from the expansion of the water upon freezing will break the container. This is the reason that antifreeze is used in car radiators in the winter. Water in the radiator will freeze and crack the engine block; sufficient force is generated from expansion upon freezing to burst even iron or copper parts. During the winter season, the weathering of rocks and concrete and the formation of "potholes" in streets are hastened by the expansion of freezing water in cracks.

Water's density pattern also explains why lakes freeze from top to bottom and not vice versa, and why aquatic life can continue to exist for extended periods of time in bodies of water that are frozen over.

8.6 COMPONENTS OF A SOLUTION

In previous sections of this chapter we considered the importance of water and some of its unique properties. We now turn our attention to solutions, with particular emphasis on water-based solutions. We very seldom find water in its pure state. Instead, we encounter water in which other substances have been dissolved, water solutions. For example, in the human body, pure water is not the fluid medium for the chemical reactions of cells. These reactions always take place in water-based solutions.

solution A **solution** *is a homogeneous (uniform) mixture of two or more substances.* The requirement of homogeneity for a solution means that the intermingling of solution components must be on the molecular level; that is, the particles present must be of atomic or molecular size.

solvent When discussing solutions, it is often convenient to call one component the *solvent* and the others *solutes*. The **solvent** *is the component of the solution that is present in the greatest amount.* The solvent can be thought of as the medium in which the other substances present are dissolved. A **solute** *is a solution component that is present in a small amount relative to that of solvent.* More than one solute can be present in the same solution. For example, both sugar and salt (two solutes) can be dissolved in a container of water to give salty sugar water.

solute

In most of the situations we will encounter, the solutes present in a solution will be of more interest to us than the solvent. The solutes are the active ingredients in the solution. They are the substances that undergo reaction when solutions are mixed.

The solutions used in laboratories and clinical settings are usually liquids, and the solvent is almost always water. However, as we shall see shortly, gaseous solutions and solid solutions of numerous types do exist.

Because a solution is homogeneous, it will have the same properties throughout. No matter where we take a sample from a solution, we will obtain material with the same composition as that of any other sample from the same solution. The composition of a solution can be varied, usually within certain limits, by changing the relative amounts of solvent and solute present. If the composition limits are transgressed, a heterogeneous mixture is formed.

8.7 TYPES OF SOLUTIONS

Nine types of two-component solutions can exist, according to a classification scheme based on the physical states of the solvent and solute before mixing. Table 8–4 lists these types of solutions and gives an example of each type. The type of solution which will be emphasized in this book is that in which the final state of the solution components is a liquid.

The physical state of a solute becomes that of the solvent when a solution is

TABLE 8–4 Examples of Various Types of Solutions

Solution Type (solute listed first)	*Example*
Gaseous Solutions	
gas dissolved in gas	dry air (oxygen and other gases dissolved in nitrogen)
liquid dissolved in gas[a]	wet air (water vapor in air)
solid dissolved in gas[a]	moth repellent (or moth balls) sublimed into air
Liquid Solutions	
gas dissolved in liquid	carbonated beverage (CO_2 in water)
liquid dissolved in liquid	vinegar (acetic acid dissolved in water)
solid dissolved in liquid	salt water
Solid Solutions	
gas dissolved in solid	hydrogen in platinum
liquid dissolved in solid	dental filling (mercury dissolved in silver)
solid dissolved in solid	sterling silver (copper dissolved in silver)

[a] An alternate viewpoint is that liquid-in-gas and solid-in-gas solutions do not actually exist as true solutions. From this viewpoint, water vapor or moth repellent in air is considered to be a gas-in-gas solution because the water or moth repellent must evaporate or sublime first in order to enter the air.

formed. For example, solid naphthalene (moth repellent) must be sublimed in order for it to dissolve in air. Finely pulverizing a solid and dispersing it in air does not produce a solution. (Dust particles in air are an example of this.) The particles of the solid must be subdivided to the molecular level; that is, the solid must be sublimed. Similarly, fog is a suspension of water droplets in air; the droplets are large enough to reflect light, a fact that becomes evident when we drive an automobile on a foggy night. Thus, fog is not a solution. However, water vapor is present in solution form in air. When hydrogen gas dissolves in platinum metal (a gas in solid solution), the gas molecules take up fixed positions in the structure of the metal; the gas becomes solidified.

8.8 TERMINOLOGY USED IN DESCRIBING SOLUTIONS

In addition to *solvent* and *solute*, several other terms are used to describe characteristics of solutions. The **solubility** *of a solute is the amount of solute that will dissolve in a given amount of solvent*. Many factors affect the numerical value of a solute's solubility in a given solvent, including the nature of the solvent itself, the temperature, and, in some cases, the pressure and presence of other solutes.

solubility

Common units for expressing solubility are grams of solute per 100 grams of solvent. The temperature of the solvent must also be specified. Table 8–5 gives the solubilities of selected solutes in the solvent water at three different temperatures.

A **saturated solution** *is a solution that contains the maximum amount of solute that can be dissolved under the conditions at which the solution exists*. A saturated solution containing excess undissolved solute is an equilibrium situation where the rate of dissolution of undissolved solute is equal to the rate of crystallization of dissolved solute. Consider the process of adding table sugar (sucrose) to a container of water. Initially, the added sugar dissolves as the solution is stirred. Finally, as we add more sugar, we reach a point where no amount of stirring will cause the added sugar to dissolve. The last-added sugar remains as a solid on the bottom of the container; the solution is saturated. Although it appears to the eye that nothing is happening once the saturation point is reached, this is not the case on the molecular level. Solid sugar from the bottom of the container is continuously dissolving in the water and an equal amount of sugar is coming out of solution. Accordingly, the net number of sugar molecules in the liquid remains the same. The equilibrium situation in the saturated solution is somewhat similar to the previously discussed evaporation of a liquid in a closed container (Section 7.14). Figure 8–5 illustrates the dynamic equilibrium process occurring in a saturated solution that contains undissolved excess solute.

saturated solution

TABLE 8–5 Solubilities of Various Compounds in Water at 0°C, 50°C, and 100°C.

	Solubility (g solute/100 g H_2O)		
Solute	0°C	50°C	100°C
lead(II) bromide ($PbBr_2$)	0.455	1.94	4.75
silver sulfate (Ag_2SO_4)	0.573	1.08	1.41
copper(II) sulfate ($CuSO_4$)	14.3	33.3	75.4
sodium chloride (NaCl)	35.7	37.0	39.8
silver nitrate ($AgNO_3$)	122	455	952
cesium chloride (CsCl)	161.4	218.5	270.5

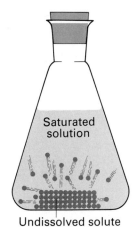

FIGURE 8–5 The dynamic equilibrium process occurring in a saturated solution that contains undissolved excess solute.

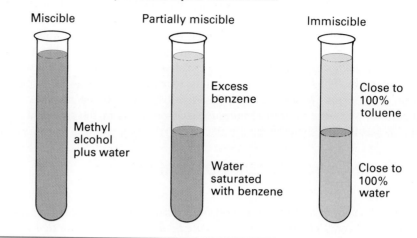

FIGURE 8-6 Miscibility of selected liquids with each other.

unsaturated solution

An **unsaturated solution** *is a solution where less solute than the maximum amount possible is dissolved in the solution.* Most solutions we encounter fall into this category.

The terms *dilute* and *concentrated* are also used to convey qualitative information about the degree of saturation of a solution. A **dilute solution** *is a solution that contains a small amount of solute relative to the amount that could dissolve.* On the other hand, a **concentrated solution** *is a solution that contains a large amount of solute relative to the amount that could dissolve.* A concentrated solution does not have to be a saturated solution.

dilute solution

concentrated solution

When dealing with liquid-in-liquid solutions, the terms *miscible, partially miscible,* and *immiscible* are frequently used to describe solubility characteristics associated with the liquids. **Miscible** *substances dissolve in any amount in each other.* For example, methyl alcohol (CH_3OH) is completely miscible with water; that is, they completely mix with each other in any and all proportions. After these two liquids are mixed, only one phase is present. **Partially miscible** *substances have limited solubility in each other.* Benzene (C_6H_6) and water are partially miscible. If benzene is slowly added to water, a small amount of benzene initially dissolves and a single phase results. However, as soon as the benzene solubility limit is reached, the excess benzene forms a separate layer on top of the water because it is less dense. **Immiscible** *substances do not dissolve in each other.* When these substances are mixed, two layers (phases) immediately form. Very few liquids are totally immiscible in each other; however, toluene (C_7H_8) and water approach this limiting case. Figure 8–6 illustrates the results obtained by mixing liquids of various miscibilities with each other.

miscible

partially miscible

immiscible

Another term commonly encountered in solution discussions is *aqueous solution.* An **aqueous solution** *is a solution in which water is the solvent.* Aqueous solutions are the most common type of solutions.

aqueous solution

8.9 SOLUTION FORMATION

In a solution, solute particles are uniformly dispersed throughout the solvent. Considering what happens at the molecular level during the solution process will help us to understand how this is achieved.

In order for a solute to dissolve in a solvent, two types of interparticle attractions must be overcome: (1) attractions between solute particles (solute–solute attractions), and (2) attractions between solvent particles (solvent–solvent attractions). Only when these attractions are overcome can particles in both pure solute and pure solvent separate from each other and begin to intermingle. A new type of interaction, which does not exist prior to solution formation, arises as the result of the mixing of solute and solvent. This new interaction is the attraction between solute and solvent particles (solute–solvent attractions). These attractions are the primary driving force for solution formation. The extent to which a substance dissolves depends on the degree to which the newly formed solute–solvent attractions are able to compensate for the energy needed to overcome the solute–solute and solvent–solvent interactions. A solute will not dissolve in a solvent if either solute–solute or solvent–solvent interactions are too strong to be compensated for by the formation of the new solute–solvent interactions.

An important type of solution process is one in which an ionic solid dissolves in water. Let us consider in detail the process of dissolving sodium chloride, a typical ionic solid, in water. We will consider this process in steps. The fact that water molecules are polar (Section 6.12) is very important in our considerations.

Figure 8–7 shows what is thought to happen when sodium chloride is placed in water. The polar water molecules become oriented in such a way that the negative oxygen portion points toward positive sodium ions and the positive hydrogen portion points toward negative chloride ions. As the polar water mol-

FIGURE 8–7 The solution process for an ionic solid in water.

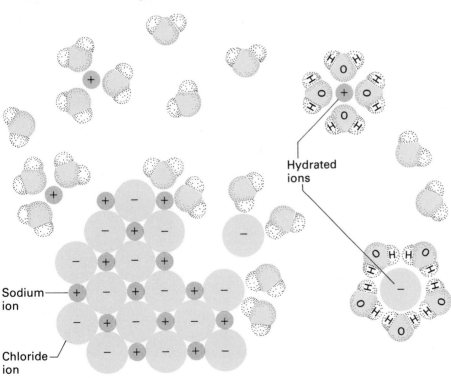

ecules begin to surround ions on the crystal surface, they exert sufficient attraction to cause these ions to break away from the crystal surface. After leaving the crystal, an ion retains its surrounding group of water molecules; it has become a *hydrated ion*. As each hydrated ion leaves the surface, other ions are exposed to the water, and the crystal is picked apart ion by ion. Once in solution, the hydrated ions are uniformly distributed either by stirring or by random collisions with other molecules or ions.

The random motion of solute ions in solutions causes them to collide with each other, with solvent molecules, and occasionally with the surface of any undissolved solute. Ions undergoing the latter type of collision occasionally stick to the solid surface and thus leave the solution. When the number of ions in solution is low, the chances for collision with the undissolved solute are low. However, as the number of ions in solution increases, so do the chances for collisions, and more ions are recaptured by the undissolved solute. Eventually the number of ions in solution reaches a level where ions return to the undissolved solute at the same rate as other ions leave. At this point, the solution is saturated, and the equilibrium process discussed in the last section is in operation.

8.10 SOLUBILITY RULES

In this section, we will present some rules for qualitatively predicting solute solubilities. These rules summarize in a concise form the results of thousands of experimental solute–solvent solubility determinations.

A very useful generalization that relates polarity to solubility is "*Substances of like polarity tend to be more soluble in each other than substances that differ in polarity.*" This conclusion is often expressed as the simple phrase "*likes dissolve likes*". Polar substances, in general, are good solvents for other polar substances, but not for nonpolar substances. Similarly, nonpolar substances exhibit greater solubility in nonpolar solvents than they do in polar solvents.

The generalization "likes dissolve likes" is a useful tool for predicting solubility behavior in many, but not all, solute–solvent situations. Results that agree with this generalization are almost always obtained in the cases of gas-in-liquid and liquid-in-liquid solutions and for solid-in-liquid solutions in which the solute is not an ionic compound. For example, NH_3 gas (a polar gas) is much more soluble in H_2O (a polar liquid) than O_2 gas (a nonpolar gas) is. (The actual solubilities of NH_3 and O_2 in water at 20°C are 51.8 g/100 g H_2O and 0.0043 g/100 g H_2O respectively.)

TABLE 8–6 Solubility Guidelines for Ionic Compounds in Water.

Ion Contained in the Compound	Solubility	Exceptions
Group IA (Li^+, Na^+, K^+, etc.)	soluble	
ammonium (NH_4^+)	soluble	
acetates ($C_2H_3O_2^-$)	soluble	
nitrates (NO_3^-)	soluble	
chlorides (Cl^-), bromides (Br^-) and iodides (I^-)	soluble	Ag^+, Pb^{2+}, Hg_2^{2+}
sulfates (SO_4^{2-})	soluble	Ca^{2+}, Sr^{2+}, Ba^{2+}, Pb^{2+}
carbonates (CO_3^{2-})	insoluble	Group IA and NH_4^+
phosphates (PO_4^{3-})	insoluble	Group IA and NH_4^+
sulfides (S^{2-})	insoluble	Groups IA and IIA and NH_4^+
hydroxides (OH^-)	insoluble	Group IA, Ba^{2+}, Sr^{2+}

In the common case of solid-in-liquid solutions in which the solute is an ionic compound, the rule "likes dissolve likes" is not adequate. Because of their polar nature, one would predict that all ionic compounds are soluble in a polar solvent such as water. This is not the case. The failure of the generalization for ionic compounds is related to the complexity of the factors involved in determining the magnitude of the solute–solute (ion–ion) and solvent–solute (ion–solvent) interactions. Among other things, both the charge and size of the ions in the solute must be considered. Changes in these factors affect both types of interactions, but not to the same extent.

Some guidelines concerning the solubility of ionic compounds in water, which should be used in place of "likes dissolve likes", are given in Table 8–6.

Predict the solubility of the following solutes in the solvent indicated.

a. CH_4 (a nonpolar gas) in water
b. Ethyl alcohol (polar) in chloroform (polar)
c. AgCl in water
d. Na_2SO_4 in water
e. $AgNO_3$ in water

Example 8.1

Solution

a. Insoluble. They are of unlike polarity because water is polar.
b. Soluble. Both substances are polar, so they should be relatively soluble in each other—likes dissolve likes.
c. Insoluble. Table 8–6 indicates that all chlorides except those of silver, lead, and mercury(I) are soluble. Thus, AgCl is one of the exceptions.
d. Soluble. Table 8–6 indicates that all ionic sodium-containing compounds are soluble.
e. Soluble. Table 8–6 indicates that all compounds containing the nitrate ion (NO_3^-) are soluble.

All ionic compounds, even the most insoluble ones, dissolve to some slight extent in water. Thus, the "insoluble" classification used in Table 8–6 really means ionic compounds that have very limited solubility in water. You should become thoroughly familiar with the rules in Table 8–6. They are extensively used in chemical discussions.

Polarity plays an important role in the solubility of many substances in the fluids and tissues of the human body. For example, let us briefly consider vitamin solubilities. Some vitamins are water-soluble and others are fat-soluble. Vitamins that are water-soluble have polar molecular structures like water does. On the other hand, vitamins that are fat-soluble have nonpolar molecular structures that are compatible with the nonpolar nature of fats.

Vitamin C is an example of a water-soluble vitamin. Because of its water-solubility, Vitamin C is not stored in the body and must be taken in as part of our daily diet. Unneeded (excess) Vitamin C is eliminated rapidly from the body via body fluids. Vitamin A is a fat-soluble vitamin. It can be and is stored by the body in fat tissue for later use. If Vitamin A is consumed in quantities that are too large (from excessive vitamin supplements), illness can result. Because of its limited water-solubility, Vitamin A cannot be rapidly eliminated from the body by body fluids.

8.11 SOLUTION CONCENTRATIONS

concentration

The amount of solute present in a solution is specified by stating the concentration of the solution. For the concentration units we will discuss, the **concentration** *of a solution is the amount of solute present in a specified amount of solution.* Thus, concentration is a ratio of two quantities.

$$\frac{\text{amount of solute}}{\text{amount of solution}}$$

A variety of types of units are used to express concentration. Each has been chosen for convenience under some particular set of circumstances. We will discuss a commonly encountered type of concentration unit in each of the next three sections. The concentration expressions we will discuss are percentage of solute, molarity, and equivalents and milliequivalents per liter.

People who work with solutions, particularly in biomedical fields, are often required to make dosage or dilution calculations. The starting point for these calculations is a knowledge of the meaning of the solution concentration designations we will now discuss.

8.12 CONCENTRATION: PERCENTAGE OF SOLUTE

The concentration of a solution is often specified in terms of the percentage of solute in the total amount of solution. Different types of percentage units exist because the amounts of solute and solution present can be stated in terms of either mass or volume. The three most common are

1. Percent-by-mass (or mass-mass percent).
2. Percent-by-volume (or volume-volume percent).
3. Mass-volume percent.

percent-by-mass

Percent-by-mass (or mass-mass percent) is the percentage unit most often used in chemical laboratories. **Percent-by-mass** *is equal to the mass of solute divided by the total mass of solution multiplied by 100 (to put the value in terms of percentage).*

$$\text{percent-by-mass} = \frac{\text{mass of solute}}{\text{mass of solution}} \times 100$$

The solute and solution masses must be measured in the same unit, which is usually grams. The mass of the solution is equal to the mass of the solute plus the mass of the solvent.

$$\text{mass of solution} = \text{mass of solute} + \text{mass of solvent}$$

A solution whose mass percent concentration is 5.0% would contain 5.0 g of solute per 100.0 g of solution (5.0 g of solute and 95.0 g of solvent). Thus, percent-by-mass directly gives the number of grams of solute in 100 g of solution. The percent-by-mass concentration unit is often abbreviated as %(m/m).

Example 8.2

What is the percent-by-mass, % (m/m), concentration of sucrose (table sugar) in a solution made by dissolving 7.6 g of sucrose in 83.4 g of water?

Solution
Both the mass of solute and mass of solvent are known. Substituting these numbers into the equation

$$\% \text{ (m/m)} = \frac{\text{mass of solute}}{\text{mass of solution}} \times 100$$

gives

$$\% \text{ (m/m)} = \frac{7.6 \text{ g sucrose}}{7.6 \text{ g sucrose} + 83.4 \text{ g water}} \times 100$$

Note that the denominator of the preceding equation is mass of solution, which is the combined mass of solute and solvent.

Doing the mathematics gives

$$\% \text{ (m/m)} = \frac{7.6 \text{ g}}{91.0 \text{ g}} \times 100 = 8.4\%$$

Example 8.3

How many grams of sucrose must be added to 375 g of water to prepare a 2.75% by mass solution of sucrose?

Solution

Often, when a solution concentration is given as part of a problem statement, the concentration information is used in the form of a conversion factor when solving the problem. That will be the case in this problem.

The given quantity is 375 g of H_2O (grams of solvent) and the desired quantity is grams of sucrose (grams of solute).

$$375 \text{ g } H_2O = ? \text{ g sucrose}$$

The conversion factor relating these two quantities (solvent and solute) is obtained from the given concentration. In a 2.75% by mass sucrose solution, there are 2.75 g of sucrose per every 97.25 g of water.

$$100.0 \text{ g solution} - 2.75 \text{ g sucrose} = 97.25 \text{ g } H_2O$$

The relationship between grams of solute and grams of solvent (2.75 to 97.25) gives us the needed conversion factor.

$$\frac{2.75 \text{ g sucrose}}{97.25 \text{ g } H_2O}$$

The problem is set up and solved, using dimensional analysis, as follows.

$$375 \text{ g } H_2O \times \left(\frac{2.75 \text{ g sucrose}}{97.25 \text{ g } H_2O}\right) = 10.6 \text{ g sucrose}$$

Percent-by-volume (or volume-volume percent), which is abbreviated % (v/v), is used as a concentration unit in situations where the solute and solvent are both liquids or both gases. In these cases, it is more convenient to measure volumes than masses. **Percent-by-volume** *is equal to the volume of solute divided by the total volume of solution multiplied by 100.*

$$\text{percent-by-volume} = \frac{\text{volume of solute}}{\text{volume of solution}} \times 100$$

Solute and solution volumes must always be expressed in the same units when using percent-by-volume.

When the numerical value of a concentration is expressed as a percent-by-volume, it directly gives the number of milliliters of solute in 100 mL of solution.

Thus, a 100 mL sample of a 5.0% (v/v) alcohol in water solution contains 5.0 mL of alcohol dissolved in enough water to give 100 mL of solution. Note that such a 5.0% (v/v) solution could not be made by adding 5 mL of alcohol to 95 mL of water because the volumes of two liquids are not usually additive. Differences in the way molecules are packed, as well as differences in distances between molecules, almost always result in the volume of the solution being different from the sum of the volumes of solute and solvent. For example, the final volume resulting from the addition of 50.0 mL of ethyl alcohol to 50.0 mL of water is 96.5 mL of solution.

Example 8.4

When 80.0 mL of methyl alcohol and 80.0 mL of water are mixed, they make a solution that has a final volume of 154 mL. What is the concentration of the solution expressed as percent-by-volume methyl alcohol?

Solution
To calculate percent-by-volume, the volumes of solute and solution are needed. Both are given in the problem statement.

$$\text{solute volume} = 80.0 \text{ mL}$$
$$\text{solution volume} = 154 \text{ mL}$$

Note that the solution volume is not the sum of the solute and solvent volumes. As mentioned previously, the volumes of two different liquids are not usually additive.

Substituting the given quantities into the equation

$$\text{percent-by-volume} = \frac{\text{volume solute}}{\text{volume solution}} \times 100$$

gives

$$\% \text{ (v/v)} = \left(\frac{80.0 \text{ mL}}{154 \text{ mL}}\right) \times 100 = 51.9\%$$

The third type of percentage unit in common use is mass–volume percentage. This unit, which is often encountered in clinical and hospital settings, is particularly convenient to use when working with a solid solute, which is easily weighed, and a liquid solvent. Solutions of drugs for internal and external use, intravenous and intramuscular injectables, and reagent solutions for testing are usually labeled in percent mass–volume.

mass–volume percent

Mass–volume percent, which is abbreviated %(m/v), *is equal to the mass of solute (in grams) divided by the total volume of solution (in milliliters) multiplied by* 100.

$$\text{mass-volume percent} = \frac{\text{mass of solute (g)}}{\text{volume of solution (mL)}} \times 100$$

Note that in defining mass–volume percent, specific mass and volume units are given. This is necessary because the units do not cancel, as was the case with mass percent and volume percent.

Mass–volume percent indicates the number of grams of solute dissolved in each 100 mL of solution. Thus, a 2.3%(m/v) solution of any solute will contain 2.3 g of solute in each 100 mL of solution and a 5.4%(m/v) solution will contain 5.4 g of solute in each 100 mL of solution.

Example 8.5

Normal saline solution that is used to dissolve drugs for intravenous use is 0.92%(m/v) NaCl in water. How many grams of NaCl are required to prepare 35.0 mL of normal saline solution?

Solution
The given quantity is 35.0 mL of solution and the desired quantity is grams of NaCl.

$$35.0 \text{ mL solution} = ? \text{ g NaCl}$$

The given concentration, 0.92%(m/v), which means 0.92 g NaCl per 100 mL of solution, is used as a conversion factor to go from milliliters of solution to grams of NaCl. The setup for the conversion is

$$35.0 \text{ mL solution} \times \left(\frac{0.92 \text{ g NaCl}}{100 \text{ mL solution}}\right)$$

Doing the arithmetic after canceling the units gives

$$\left(\frac{35.0 \times 0.92}{100}\right) \text{ g NaCl} = 0.32 \text{ g NaCl}$$

A variation of the concentration unit %(m/v), which is called *milligram percent* (mg%), is extensively used when dealing with biological solutions such as blood plasma and urine. A characteristic of these fluids is very low solute concentrations. **Milligram percent** is equal to the number of milligrams of solute divided by the total volume of solution (in milliliters) multiplied by 100.

milligram percent

$$\text{milligram percent} = \frac{\text{mass of solute (mg)}}{\text{volume of solution (mL)}} \times 100$$

Example 8.6

The concentration of sodium ions in blood plasma is usually reported in milligram percent. If a 3.00 mL sample of blood plasma contains 10.2 mg of sodium ions, what is the mg% Na^+ ion in the blood plasma?

Solution
The defining equation for mg% is

$$\text{mg\%} = \frac{\text{mg of solute}}{\text{mL of solution}} \times 100$$

Both of the quantities called for in this equation are known.

$$\text{mg of solute} = 10.2$$
$$\text{mL of solution} = 3.00$$

Substituting these values into the equation and doing the mathematics gives

$$\text{mg\%} = \left(\frac{10.2 \text{ mg}}{3.00 \text{ mL}}\right) \times 100 = 340\%$$

In Example 8.6, the answer to the problem was 340 mg%, a value greater than 100%. Such large percentage values are typical when dealing with blood chemistry.

Normal blood levels for cholesterol fall in the range of 120 to 220 mg%. The normal range for fasting glucose levels is 70 to 100 mg%.

8.13 CONCENTRATION: MOLARITY

molarity

The **molarity** of a solution, which is abbreviated M, is a ratio giving the number of moles of solute per liter of solution.

$$\text{molarity (M)} = \frac{\text{moles of solute}}{\text{liters of solutions}}$$

This concentration unit is often used in laboratories where chemical reactions are being studied. Because chemical reactions occur between molecules and atoms, the mole, a unit that counts particles, is desirable. Two solutions of the same molarity contain the same number of solute molecules. By contrast, two solutions of equal mass percent will not necessarily contain the same number of solute molecules. These solutions contain equal masses of solute. However, if the formula weights of the solutes differ, which will usually be the case, the number of moles (and molecules) will differ.

In order to find the molarity of a solution, we need to know the solution volume in liters and the number of moles of solute present. An alternative to knowing the number of moles of solute is knowing the number of grams of solute present and the solute's formula weight. The number of moles can be calculated using these two quantities.

Example 8.7

Determine the molarities of the following solutions.

a. 4.35 moles of $KMnO_4$ are dissolved in enough water to give 750 mL of solution.

b. 20.0 g of NaOH are dissolved in enough water to give 1.50 L of solution.

Solution

a. The number of moles of solute is given in the problem statement.

$$\text{moles of solute (KMnO}_4) = 4.35$$

The volume of the solution is also given in the problem statement, but not in the right units. Molarity requires liters for the volume units and we are given milliliters of solution. Making the unit change gives

$$750 \text{ mL} \times \left(\frac{1 \text{ L}}{1000 \text{ mL}}\right) = 0.750 \text{ L}$$

The molarity of the solution is obtained by substituting the known quantities into the equation

$$M = \frac{\text{moles of solute}}{\text{liters of solution}}$$

which gives

$$M = \frac{4.35 \text{ moles KMnO}_4}{0.750 \text{ L solution}}$$

$$= 5.80 \frac{\text{moles KMnO}_4}{\text{L solution}}$$

Note that the units for molarity are always moles per liter.

b. This time, the volume of solution is given in liters.

$$\text{volume of solution} = 1.50 \text{ L}$$

The moles of solute must be calculated from the grams of solute (given) and the solute's formula weight, which is 40.0 amu (calculated from a table of atomic weights).

$$20.0 \text{ g NaOH} \times \left(\frac{1 \text{ mole NaOH}}{40.0 \text{ g NaOH}}\right) = 0.500 \text{ mole NaOH}$$

Substituting the known quantities into the defining equation for molarity gives

$$M = \frac{0.500 \text{ mole NaOH}}{1.50 \text{ L solution}} = 0.333 \frac{\text{mole NaOH}}{\text{L solution}}$$

The mass of solute present in a known volume of solution is an easily calculable quantity if the molarity of the solution is known. When doing such a calculation, molarity serves as a conversion factor that relates liters of solution to moles of solute. In a similar manner, the volume of solution needed to supply a given amount of solute can be calculated using the solution's molarity as a conversion factor.

Example 8.8

How many grams of sucrose (table sugar, $C_{12}H_{22}O_{11}$) are present in 175 mL of a 2.50 M sucrose solution?

Solution
The given quantity is 175 mL of solution and the desired quantity is grams of $C_{12}H_{22}O_{11}$.

$$175 \text{ mL of solution} = ? \text{ g } C_{12}H_{22}O_{11}$$

The pathway used to solve this problem is

mL solution $\longrightarrow$ L solution $\longrightarrow$ moles $C_{12}H_{22}O_{11}$ $\longrightarrow$ g $C_{12}H_{22}O_{11}$

The given molarity (2.50 M) serves as the conversion factor for the second unit change; the formula weight of sucrose (which is not given and must be calculated) is used to accomplish the third unit change.

The dimensional-analysis setup for this pathway is

$$175 \text{ mL solution} \times \left(\frac{1 \text{ L solution}}{1000 \text{ mL solution}}\right) \times \left(\frac{2.50 \text{ moles } C_{12}H_{22}O_{11}}{1 \text{ L solution}}\right) \times \left(\frac{342 \text{ g } C_{12}H_{22}O_{11}}{1 \text{ mole } C_{12}H_{22}O_{11}}\right)$$

Canceling the units and doing the arithmetic gives

$$\left(\frac{175 \times 1 \times 2.50 \times 342}{1000 \times 1 \times 1}\right) \text{ g } C_{12}H_{22}O_{11} = 150 \text{ g } C_{12}H_{22}O_{11}$$

Example 8.9

A typical dose of iron(II) sulfate ($FeSO_4$) in the treatment of iron-deficiency anemia is 0.30 g. How many milliliters of a 0.10 M iron(II) sulfate solution would be needed to supply this dose?

Solution
The given quantity is 0.30 g of $FeSO_4$ and the desired quantity is milliliters of $FeSO_4$ solution.

$$0.30 \text{ g FeSO}_4 = ? \text{ mL FeSO}_4 \text{ solution}$$

The pathway used to solve this problem will involve the following steps.

$$\text{g FeSO}_4 \longrightarrow \text{moles FeSO}_4 \longrightarrow \text{L FeSO}_4 \text{ solution} \longrightarrow \text{mL FeSO}_4 \text{ solution}$$

We accomplish the first unit conversion by using the formula weight of $FeSO_4$ (which must be calculated) as a conversion factor. The second unit conversion involves the use of the given molarity as a conversion factor.

$$0.30 \text{ g FeSO}_4 \times \left(\frac{1 \text{ mole FeSO}_4}{152 \text{ g FeSO}_4}\right) \times \left(\frac{1 \text{ L solution}}{0.10 \text{ mole FeSO}_4}\right) \times \left(\frac{1000 \text{ mL solution}}{1 \text{ L solution}}\right)$$

Canceling units and doing the arithmetic gives

$$\left(\frac{0.30 \times 1 \times 1 \times 1000}{152 \times 0.10 \times 1}\right) \text{ mL solution} = 20 \text{ mL solution}$$

Molarity and mass–volume percent are probably the two most frequently used concentration units. Because they are both so common, we often need to convert from one to the other. Example 8.10 illustrates this simple conversion.

Example 8.10

A 5.0%(m/v) solution of glucose ($C_6H_{12}O_6$) in water is used to feed newborn infants. What is the molarity of this solution?

Solution

In order to determine molarity, we will need to know the moles of solute and liters of solution present in a sample of 5%(m/v) glucose solution. Solution concentration is independent of sample size, so we can use any size sample for our calculation. To simplify the mathematics, we will choose a 100.0 mL sample of solution.

Moles of solute. By definition, a 100.0 mL sample of 5%(m/v) glucose solution will contain 5.0 g of glucose. Only one step will be required to change this gram amount to moles.

$$5.0 \text{ g } C_6H_{12}O_6 \times \left(\frac{1 \text{ mole } C_6H_{12}O_6}{180 \text{ g } C_6H_{12}O_6}\right) = 0.028 \text{ mole } C_6H_{12}O_6$$

Liters of solution. We decided to use a sample volume of 100.0 mL, which is equivalent to 0.1000 liters.

$$100.0 \text{ mL solution} \times \left(\frac{1 \text{ L solution}}{1000 \text{ mL solution}}\right) = 0.1000 \text{ L solution}$$

Molarity. The molarity is obtained by substituting the moles of solute and liters of solution into the defining equation for molarity.

$$M = \frac{\text{moles solute}}{\text{liters solution}}$$

$$= \frac{0.028 \text{ mole } C_6H_{12}O_6}{0.1000 \text{ L solution}}$$

$$= 0.28 \frac{\text{mole } C_6H_{12}O_6}{\text{L solution}}$$

8.14 CONCENTRATION: MILLIEQUIVALENTS PER LITER

The concentration unit *milliequivalents-per-liter* (mEq/L) is used extensively in clinical laboratory work. It is the preferred unit for reporting the concentration of various ions present in blood, serum, urine, and other body fluids. The solute content of solutions to be intraveneously injected into the human body is often specified using these units.

In order to understand what a milliequivalent is, we will first consider the parent unit, an equivalent. An equivalent, like a mole, is a specific quantity. Equivalents can be defined in several ways. We will define it for *ionic solutes* here and for acids and bases in Section 10.10. The definition of an equivalent is tied to the behavior of ionic solutes as they dissolve in a solvent. Upon dissolving in water, ionic solutes dissociate (break up) into their constituent ions. The dissociation process for the ionic solutes KNO_3 and $Ca(NO_3)_2$ is shown in the following equations.

$$KNO_3 \longrightarrow K^+ + NO_3^-$$
$$Ca(NO_3)_2 \longrightarrow Ca^{2+} + 2\,NO_3^-$$

Considering each of these equations in further detail will give us the insights we need to understand what an equivalent is.

Interpreting the KNO_3 equation in terms of moles leads to the following statement.

$$1 \text{ mole } KNO_3 \longrightarrow 1 \text{ mole } K^+ \text{ ions} + 1 \text{ mole } NO_3^- \text{ ions}$$

Each K^+ ion carries one positive charge, and each NO_3^- ion carries one negative charge, so we can also write

$$1 \text{ mole } KNO_3 \longrightarrow 1 \text{ mole of positive charge} + 1 \text{ mole of negative charge}$$

We are now ready to define an equivalent. An **equivalent** of ionic solute is the *quantity of the substance that will supply one mole of positive charge or one mole of negative charge upon complete dissociation*. For KNO_3, one mole of compound produces one mole of positive or negative charge. Therefore, one mole is equal to one equivalent for this compound.

equivalent

$$1 \text{ mole } KNO_3 = 1 \text{ equivalent } KNO_3$$

For KNO_3, an equivalent and a mole are equal. This equality is not always the case, as we will see when $Ca(NO_3)_2$ is the solute.

Interpreting the $Ca(NO_3)_2$ equation in terms of moles leads to the following statement.

$$1 \text{ mole } Ca(NO_3)_2 \longrightarrow 1 \text{ mole } Ca^{2+} \text{ ions} + 2 \text{ moles } NO_3^- \text{ ions}$$

The one mole of Ca^{2+} ion produced from the dissociation is equivalent to two moles of positive charge because each ion carries a $+2$ charge. Two moles of negative charge are also produced; although each NO_3^- carries only a -1 charge, two moles of nitrate ions are produced. Thus, we can write

$$1 \text{ mole } Ca(NO_3)_2 \longrightarrow 2 \text{ moles of positive charge} + 2 \text{ moles of negative charge}$$

Therefore, because an equivalent is associated with one mole of charge, we can write

$$1 \text{ mole } Ca(NO_3)_2 = 2 \text{ equivalents } Ca(NO_3)_2$$

Dividing each side of this equation by two gives us

TABLE 8–7 Relationship Between an Equivalent and a Mole for an Ionic Solute

Solute	Dissociation Equation	Moles of Positive or Negative Charge Produced	Number of Equivalents in a Mole
MgO	MgO $\longrightarrow$ Mg^{2+} + O^{2-}	2	2
AlCl$_3$	AlCl$_3$ $\longrightarrow$ Al^{3+} + 3 Cl$^-$	3	3
CaCl$_2$	CaCl$_2$ $\longrightarrow$ Ca^{2+} + 2 Cl$^-$	2	2
Al$_2$(SO$_4$)$_3$	Al$_2$(SO$_4$)$_3$ $\longrightarrow$ 2 Al^{3+} + 3 SO$_4^{2-}$	6	6
KF	KF $\longrightarrow$ K$^+$ + F$^-$	1	1

$$1/2 \text{ mole Ca(NO}_3)_2 = 1 \text{ equivalent Ca(NO}_3)_2$$

Generalizing the two cases just considered gives the equation

$$1 \text{ equivalent of ionic solute} = \frac{1 \text{ mole of ionic solute}}{n}$$

where n *is equal to the number of moles of positive or negative charge produced upon dissociation of one mole of solute into ions.* Table 8–7 gives additional examples of the relationship between an equivalent and a mole of an ionic solute.

Dissociation of an ionic solute produces both positive and negative ions. It is often convenient to talk about equivalents of a particular ion rather than equivalents of the solute as a whole. When this is done, the equation

$$1 \text{ equivalent} = \frac{1 \text{ mole}}{n}$$

still holds, with the value of n being the magnitude of the charge on the ion of concern; $n = 1$ for Na$^+$, $n = 3$ for Al^{3+}, and $n = 2$ for SO$_4^{2-}$.

How much does an equivalent weigh? This question is answered in the same way as the question "How much does a mole weigh?" (Section 4.3). The mass of an equivalent is dependent upon the identity of the substance. For a substance where one equivalent is equal to one-third of a mole, the mass of an equivalent is simply one-third the mass of a mole of that substance. Equivalent masses are derived from formula weights.

As noted previously, equivalents are used extensively to specify the concentration of ions in body fluids. There must be an electrical balance between the

TABLE 8–8 Electrolytical Balance Associated with Extracellular Fluid

Ion	Normal concentration in mEq/L
Positive Ions	
sodium (Na$^+$)	142 ⎫
potassium (K$^+$)	5 ⎬ total of
calcium (Ca^{2+})	5 ⎨ 155 mEq/L
magnesium (Mg^{2+})	3 ⎭
Negative Ions	
chloride (Cl$^-$)	104 ⎫
bicarbonate (HCO$_3^-$)	27 ⎬ total of
hydrogen phosphate (HPO$_4^{2-}$)	2 ⎨ 155 mEq/L
sulfate (SO$_4^{2-}$)	1 ⎪
organic ions	21 ⎭

the charge carried by the positive and negative ions present in body fluids; otherwise the fluid would carry an electrical charge. Equivalents aid in keeping track of this electrical balance.

Table 8–8 shows the electrical charge balance associated with extracellular fluid (the fluid outside body cells), which constitutes 20% of body mass for young adults.

All of the values in Table 8–8 are given in milliequivalents-per-liter (mEq/L). These units are normally used to report concentrations of ions in body fluids because of the low concentrations of the solutions.

Example 8.11

The normal range for Na^+ ions in human blood is 135 to 148 mEq/L. How many grams of Na^+ ions are there in one liter of human blood if the Na^+ concentration is 141 mEq/L?

Solution
The given quantity is 141 mEq/L and the desired quantity is grams per liter of Na^+.

$$141 \text{ mEq/L} = ? \text{ g/L}$$

The pathway used to solve this problem involves the following sequence of unit changes.

$$\text{mEq } Na^+ \longrightarrow \text{Eq } Na^+ \longrightarrow \text{moles } Na^+ \longrightarrow \text{g } Na^+$$

The conversion factors needed to accomplish these changes are

$$\frac{141 \text{ mEq } Na^+}{1 \text{ L solution}} \times \left(\frac{1 \text{ Eq } Na^+}{1000 \text{ mEq } Na^+}\right) \times \left(\frac{1 \text{ mole } Na^+}{1 \text{ Eq } Na^+}\right) \times \left(\frac{23.0 \text{ g } Na^+}{1 \text{ mole } Na^+}\right)$$

Because the charge on Na^+ ions is +1, an equivalent and a mole of Na^+ ions will be the same ($n = 1$). This equality is the basis for the conversion factor in the second unit change. The final unit change, from moles to grams, is accomplished using the atomic weight of sodium, which is 23.0. (The atomic weights of Na atoms and Na^+ ions are considered to have the same value even though a Na^+ ion has one less electron than a Na atom; the masses of electrons are considered negligible when dealing with atomic weights.)

Canceling the units and doing the arithmetic gives us our answer.

$$\left(\frac{141 \times 1 \times 1 \times 23.0}{1 \times 1000 \times 1 \times 1}\right) \frac{\text{g } Na^+}{\text{L solution}} = 3.24 \frac{\text{g } Na^+}{\text{L solution}}$$

Example 8.12

How much does one milliequivalent of each of the following ions weigh? (Give each answer in milligrams.)

a. Ca^{2+} b. SO_4^{2-} c. HCO_3^-

Solution
Each part of this problem will have the same general setup.

$$1 \text{ mEq} = ? \text{ mg}$$

The dimensional-analysis pathway will be

$$\text{mEq ion} \longrightarrow \text{Eq ion} \longrightarrow \text{moles ion} \longrightarrow \text{g ion} \longrightarrow \text{mg ion}$$

a. For Ca^{2+} ions, two equivalents are equal to one mole because the charge on a Ca^{2+} ion is 2. We will need this fact to construct the conversion factor for going from equivalents to moles. The dimensional-analysis setup is

$$1 \text{ mEq Ca}^{2+} \times \left(\frac{1 \text{ Eq Ca}^{2+}}{1000 \text{ mEq Ca}^{2+}}\right) \times \left(\frac{1 \text{ mole Ca}^{2+}}{2 \text{ Eq Ca}^{2+}}\right) \times \left(\frac{40 \text{ g Ca}^{2+}}{1 \text{ mole Ca}^{2+}}\right)$$

$$\times \left(\frac{1000 \text{ mg Ca}^{2+}}{1 \text{ g Ca}^{2+}}\right) = 20 \text{ mg Ca}^{2+}$$

b. For SO_4^{2-}, as for Ca^{2+}, two equivalents are equal to one mole because the magnitude of the charge is 2. The setup for this problem differs from that in part (a) in only one way: the sulfate ion has a different formula weight than the calcium ion.

$$1 \text{ mEq SO}_4^{2-} \times \left(\frac{1 \text{ Eq SO}_4^{2-}}{1000 \text{ mEq SO}_4^{2-}}\right) \times \left(\frac{1 \text{ mole SO}_4^{2-}}{2 \text{ Eq SO}_4^{2-}}\right)$$

$$\times \left(\frac{96 \text{ g SO}_4^{2-}}{1 \text{ mole SO}_4^{2-}}\right) \times \left(\frac{1000 \text{ mg SO}_4^{2-}}{1 \text{ g SO}_4^{2-}}\right) = 48 \text{ mg SO}_4^{2-}$$

c. Similarly, we have

$$1 \text{ mEq HCO}_3^- \times \left(\frac{1 \text{ Eq HCO}_3^-}{1000 \text{ mEq CO}_3^-}\right) \times \left(\frac{1 \text{ mole HCO}_3^-}{1 \text{ Eq HCO}_3^-}\right)$$

$$\times \left(\frac{61 \text{ g HCO}_3^-}{1 \text{ mole HCO}_3^-}\right) \times \left(\frac{1000 \text{ mg HCO}_3^-}{1 \text{ g HCO}_3^-}\right) = 61 \text{ mg HCO}_3^-$$

Notice that in this part, one equivalent is equal to one mole.

In clinical methods textbooks the simple equation

$$1 \text{ mEq} = \frac{\text{formula weight of ion (mg)}}{\text{charge on the ion}}$$

is given for use when calculating the mass of a milliequivalent of an ion in milligrams. Use of this short-cut equation for problems like Example 8.12 will give the same answers that we obtained using dimensional analysis. We don't recommend that you use this equation until you thoroughly understand the dimensional-analysis method of solving these problems. The basis for this short-cut equation is the fact that the numerical values of 1000 that are found in the second and fifth conversion factors in each of the setups in Example 8.12 cancel each other. This will always be the case when going from one milli- prefixed unit to another milli- prefixed unit (from mEq to mg in our case).

8.15 DILUTION

dilution

A common problem encountered when working with solutions is that of diluting a solution of known concentration (usually called a stock solution) to a lower concentration. **Dilution** *is the process in which more solvent is added to a solution in order to lower its concentration.* Dilution always lowers the concentration of a solution. The same amount of solute is present, but it is now distributed in a larger amount of solvent (the original solvent plus the added solvent).

Because most of the solutions we deal with are liquids, dilution is normally a volumetric procedure. Often, we prepare a solution of a specific concentration by adding a predetermined volume of solvent to a specific volume of stock solution.

A simple relationship exists between the volumes and concentrations of the diluted and stock solutions. It is

$$\begin{pmatrix}\text{concentration of}\\\text{stock solution}\end{pmatrix} \times \begin{pmatrix}\text{volume of}\\\text{stock solution}\end{pmatrix} = \begin{pmatrix}\text{concentration of}\\\text{diluted solution}\end{pmatrix} \times \begin{pmatrix}\text{volume of}\\\text{diluted solution}\end{pmatrix}$$

or

$$C_s \times V_s = C_d \times V_d$$

When using this equation, you must use the same units of concentration and units of volume on both sides of the equation. A more general form of the preceding equation is written as follows

$$\begin{pmatrix}\text{initial}\\\text{concentration}\end{pmatrix} \times \begin{pmatrix}\text{initial}\\\text{volume}\end{pmatrix} = \begin{pmatrix}\text{final}\\\text{concentration}\end{pmatrix} \times \begin{pmatrix}\text{final}\\\text{volume}\end{pmatrix}$$

or

$$C_i \times V_i = C_f \times V_f$$

If the volume of the solution is tripled when you make a dilution, you have made a 1:3 dilution. Similarly, if the volume is increased by a factor of five, you have made a 1:5 dilution. When reconstituting frozen grape juice, the directions tell you to add three cans of water. You increase the total volume to four cans (three cans of water and one can of grape juice concentrate) by doing this. This dilution process is thus a 1:4 dilution. Note that when making a 1:4 dilution, you do not add four volumes of solvent, you just add three volumes.

Example 8.13

A 25%(v/v) isopropyl alcohol in water solution is needed for use as a cooling sponge bath. How much 70%(v/v) stock solution and how much water are needed to prepare 6.0 L of the desired solution?

Solution

We know three of the four variables in the equation

$$C_s \times V_s = C_d \times V_d$$

$C_s = 70\%(v/v)$ $C_d = 25\%(v/v)$
$V_s = ?$ $V_d = 6.0\ L$

Rearranging the equation to isolate V_s on the left side and substituting the known variables into it gives

$$V_s = \frac{C_d \times V_d}{C_s}$$

$$= \frac{25\%(v/v) \times 6.0\ L}{70\%(v/v)}$$

$$= 2.1\ L$$

In order to make up our desired solution, we will need 2.1 L of stock solution and 3.9 L (6.0 L − 2.1 L) of water.

Example 8.14

A nurse wants to prepare a 1.0%(m/v) silver nitrate solution from 20 mL of a 4.0%(m/v) stock solution of silver nitrate. How much water should be added to the 20 mL of stock solution?

Solution

The volume of water to be added will be equal to the difference between the final and initial volumes. The initial volume is known (20 mL). The final volume can be calculated using the equation

$$C_i \times V_i = C_f \times V_f$$

Once the final volume is known, the difference between the two volumes can be obtained.

Substituting the known quantities into the dilution equation, which has been rearranged to isolate V_f on the left side, gives

$$V_f = \frac{C_i \times V_i}{C_f}$$

$$= \frac{4.0\% \cancel{(m/v)} \times 20 \text{ mL}}{1.0\% \cancel{(m/v)}}$$

$$= 80 \text{ mL}$$

The solvent added is

$$V_f - V_i = (80 - 20) \text{ mL}$$
$$= 60 \text{ mL}$$

8.16 COLLIGATIVE PROPERTIES OF SOLUTIONS

The physical properties that a solvent possesses when it is pure undergo changes when a solute is added. A special group of such properties that change when a solute is added are called *colligative properties*. **Colligative properties** *are the physical properties of a solution that depend only on the number (concentration) of solute particles (molecules or ions) in a given quantity of solvent, and not on their chemical identities.* Vapor pressure, boiling point, freezing point, and osmotic pressure are all examples of colligative properties. We will briefly discuss the first three of these colligative properties and then consider the fourth, osmotic pressure, in detail.

colligative properties

Vapor Pressure

The vapor pressure of a solution containing a nonvolatile solute is always less than that of pure solvent at the same temperature. (A nonvolatile solute is one that has a low vapor pressure and therefore a low tendency to vaporize.) This lowering of vapor pressure is a direct consequence of some of the solute molecules or ions occupying positions on the surface of the liquid. Their presence decreases the probability of solvent molecules escaping; that is, the number of surface-occupying solvent molecules has been decreased. Figure 8–8 illustrates the decrease in surface concentration of solvent molecules when a solute is added. As the *number* of solute particles increases, the reduction in vapor pressure increases also; thus, vapor pressure is a colligative property. What is important is not the identity of the solute molecules, but the fact that they take up room on the surface of the liquid.

Boiling Point

The addition of a nonvolatile solute to a solvent *raises* the boiling point of the resulting solution above that of the pure solvent. This is logical when we remember that the vapor pressure of the solution is lower than that of pure solvent, and that the boiling point is dependent on vapor pressure (Section 7.15). A higher temperature will be required to raise the depressed vapor pressure of the solution to atmospheric pressure; this is the condition necessary for boiling.

A common application of the boiling point elevation phenomenon involves automobiles. The coolant *ethylene glycol* (a nonvolatile solute) is added to car radiators to prevent boilover in hot weather. The engine may not run any cooler, but the coolant–water mixture will not boil until a temperature well above the normal boiling point of water is reached.

Freezing Point

The addition of a nonvolatile solute to a solvent *lowers* the freezing point of the resulting solution. This freezing-point depression, like boiling-point elevation, is a direct consequence of the decrease in vapor pressure associated with the presence of solute. Solute added to a liquid (solvent) that is in equilibrium with its pure solid will decrease the rate at which liquid (solvent) molecules can crystallize as solid. The net result of this is a decrease in freezing point for the liquid.

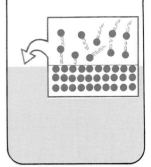

(a) Pure solvent—100% of surface positions are occupied by solvent molecules.

Applications of freezing-point depression are even more numerous than those for boiling-point elevation. In climates where the temperature drops below 0°C in the winter, it is necessary to protect water-cooled automobile engines from freezing. This is done by adding antifreeze (usually ethylene glycol) to the radiator. The addition of this nonvolatile material causes the vapor pressure and freezing point of the resulting solution to be much lower than that of pure water. Also in the winter, salt, usually NaCl or $CaCl_2$, is spread on roads and sidewalks to melt ice or prevent it from forming. The salt dissolves in the water to form a solution that will not freeze until the temperature drops much lower than 0°C, the normal freezing point of water. However, if the temperature of the ice is below the freezing point of water when its is depressed by the addition of salt, this method is ineffective.

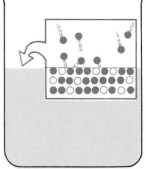

(b) Solvent and solute—some surface positions are occupied by solute molecules.

FIGURE 8–8 Close-up of the surface of a liquid solvent before and after solute has been added.

8.17 OSMOSIS AND OSMOTIC PRESSURE

The process of osmosis and the colligative property of osmotic pressure are extremely important phenomena when considering biological solutions. These phenomena govern many of the processes that occur in a functioning human body.

Osmosis *involves the passage of a solvent from a dilute solution (or pure solvent) through a semipermeable membrane into a more concentrated solution.* The simple apparatus shown in Figure 8–9a (p. 202) is helpful in explaining at the molecular level what actually occurs during the osmotic process. The apparatus consists of a tube containing a concentrated salt–water solution that has been immersed in a dilute salt–water solution. The immersed end of the tube is covered with a semipermeable membrane. A **semipermeable membrane** *is a thin layer of material that allows certain types of molecules to pass through, but prohibits the passage of others.* The selectivity of the membrane is based on size differences between molecules. The particles that are allowed to pass through (usually just solvent molecules like water) are relatively small. Thus, the membrane functions somewhat like a sieve. Using the experimental setup of Figure 8–9a, a net flow of solvent from the dilute to the concentrated solution can be observed over the course of time. This is indicated by a rise in the level of the solution in the tube and a drop in the level of the dilute solution, as shown in Figure 8–9b (p. 202).

osmosis

semipermeable membrane

What is actually happening on a molecular level as the process of osmosis occurs? Water is flowing in both directions through the membrane. However, the rate of flow into the concentrated solution is greater than the rate of flow in the

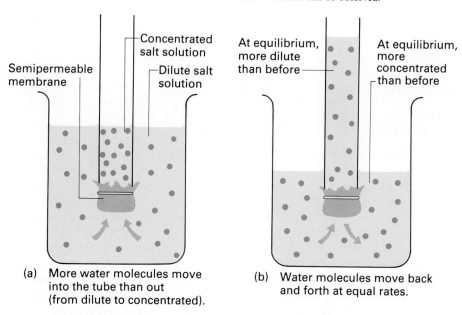

FIGURE 8-9 An apparatus in which osmosis, the flow of solvent through a semipermeable membrane from a dilute to a concentrated solution can be observed.

(a) More water molecules move into the tube than out (from dilute to concentrated).

(b) Water molecules move back and forth at equal rates.

other direction (see Figure 8–10). Why? The presence of solute molecules diminishes the ability of water molecules to cross the membrane. The solute molecules literally get in the way; they occupy some of the surface positions next to the membrane. As there is a greater concentration of solute molecules on one side of the membrane than on the other, the flow rates will differ. The flow rate will be diminished to a greater extent on that side of the membrane where the greater concentration of solute is present.

The net transfer of solvent across the membrane continues at a diminishing rate until the pressure of the solution on the concentrated side of the membrane becomes sufficient to counterbalance the greater escaping tendency of molecules from the dilute side. (This pressure results from the increased mass and volume of the concentrated solution.)

Osmotic pressure *is the amount of pressure that must be applied to prevent the flow of solvent through a semipermeable membrane from a solution of lower solute concentration to a solution of higher solute concentration.* In terms of Figure 8–9, osmotic pressure is the pressure required to prevent water from rising in the tube. Figure 8–11 shows how this pressure can be measured.

The greater the original concentration difference between the separated solutions, then the greater the magnitude of the osmotic pressure. Note that the concentrations of the two solutions involved in osmosis never become equal, even at equilibrium. However, the more concentrated solution is less concentrated than it originally was because of its increased volume. Of course, the amount of solute stays the same.

Cell membranes in both plants and animals are semipermeable in nature. The selective passage of fluid materials through these membranes governs the balance of fluids in living systems. Thus, osmotic-type phenomena are of prime importance for life. We say "osmotic-type phenomena" instead of "osmosis" because the semipermeable membranes found in living cells usually permit the passage of

FIGURE 8–10 An enlarged view of a semipermeable membrane separating (a) pure water and a salt solution and (b) a dilute salt–water solution and a concentrated salt–water solution. Because the solute molecules (large dots) interfere with the movement of water molecules (small dots), water moves more readily from an area of lower solute concentration to an area of greater solute concentration.

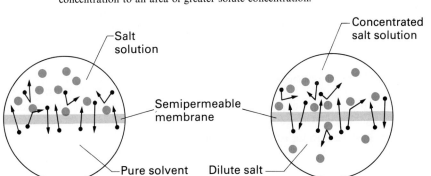

(a) More water molecules move into the salt solution from the pure solvent than vice versa.

(b) More water molecules move into the concentrated salt solution from dilute salt solution than vice versa.

FIGURE 8–11 Osmotic pressure is the amount of pressure needed to prevent the solution in the tube from rising as the result of the process of osmosis.

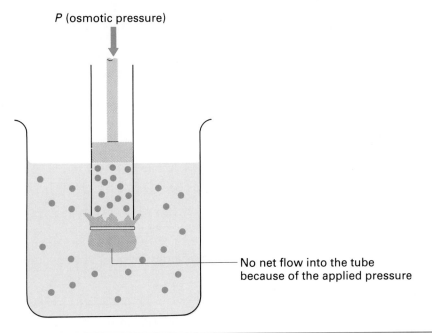

small solute molecules (nutrients and waste products) in addition to solvent. The term *osmosis* implies the passage of solvent only. The substances prohibited from passing through the membrane in osmotic-type processes are large molecules and insoluble suspended materials.

Plants will die if they are watered with salt water because of an osmotic-type process. The salt solution outside the root membranes is more concentrated than the solution in the root, so water flows out of the roots; then the plant becomes dehydrated and dies. This same principle is behind the reason for not drinking excessive amounts of salt water, even if you are stranded on a raft in the middle of the ocean. When salt water is taken into the stomach, water flows out of the stomach wall membranes and into the stomach; then the tissues become dehydrated. Drinking sea water will cause greater thirst because the body will lose water rather than absorb it.

The osmotic pressure of a solution depends upon the number of solute particles present. This in turn depends upon the solute concentration and whether the solute dissociates into ions once it is in solution. Notice that two factors are involved in determining osmotic pressure.

The fact that some solutes dissociate in solution is of utmost importance in osmotic pressure considerations. For example, the osmotic pressure of a 1 M NaCl solution is twice that of a 1 M glucose solution, despite the fact that both solutions have equal concentration (1 M). Sodium chloride is an ionic solute and it dissociates in solution to give two particles (a Na^+ and a Cl^- ion) per formula unit; however, glucose is a molecular solute and does not dissociate. It is the number of particles present that determines osmotic pressure.

The concentration unit *osmolarity* is used to compare the osmotic pressures of solutions. The **osmolarity** *of a solution is the product of its molarity and the number of particles produced per formula unit when the solute dissociates.* The equation for osmolarity is

$$\text{osmolarity} = \text{molarity} \times n$$

where n is the number of particles produced from the dissociation of one formula unit of solute.

Solutions of equal osmolarity will have equal osmotic pressures. If the osmolarity of one solution is three times that of another, the osmotic pressure of the first solution will be three times that of the second solution. A solution with high osmotic pressure will take up more water than a solution of lower osmotic pressure; thus, more pressure must be applied to prevent osmosis.

Example 8.15

What is the osmolarity for each of the following solutions?

a. 2 M NaCl   b. 2 M $CaCl_2$ c. 2 M glucose
d. 2 M in both NaCl and glucose
e. 1 M in glucose and 2 M in NaCl

Solution

The general equation for osmolarity that was previously given will be applicable in each of the parts of the problem.

$$\text{osmolarity} = \text{molarity} \times n$$

a. Two particles per dissociation are produced when NaCl dissociates in solution.

$$NaCl \longrightarrow Na^+ + Cl^-$$

The value of n is two and the osmolarity is twice the molarity.

$$\text{osmolarity} = 2\text{ M} \times 2$$
$$= 4$$

b. For $CaCl_2$, the value of n is three because three ions are produced from the dissociation of one $CaCl_2$ unit.

$$CaCl_2 \longrightarrow Ca^{2+} + 2\,Cl^-$$

The osmolarity will therefore be triple the molarity

$$\text{osmolarity} = 2\,M \times 3$$
$$= 6$$

c. Glucose is a nondissociating solute. Thus, the value of n is one and the molarity and osmolarity will be the same—two molar and two osmolar.

d. With two solutes present, we must consider the collective effects of both solutes. For NaCl, $n = 2$, and for glucose, $n = 1$. The osmolarity is calculated as follows

$$\text{osmolarity} = \underbrace{2\,M \times 2}_{\text{NaCl}} + \underbrace{2\,M \times 1}_{\text{glucose}}$$
$$= 6$$

e. This problem differs from the previous one in that the two solutes are not present in equal concentrations. This does not change the way we work the problem. The n values are the same as before and the osmolarity is

$$\text{osmolarity} = \underbrace{2\,M \times 2}_{\text{NaCl}} + \underbrace{1\,M \times 1}_{\text{glucose}}$$
$$= 5$$

8.18 OSMOSIS AND THE HUMAN BODY

The terms *isotonic solution*, *hypertonic solution*, and *hypotonic solution* are used repeatedly when dealing with osmotic-type phenomena in the human body. A consideration of what happens to erythrocytes (red blood cells) when they are placed in three different solution mediums will help us to understand the differences in the meanings of these three terms. The solution mediums are distilled water, concentrated sodium chloride solution, and physiological saline solution.

If red blood cells are placed in pure water, you can observe (using a microscope) that they swell up (enlarge in size) and finally rupture (burst); this process is called *hemolysis*. Hemolysis is caused by an increase in the amount of water entering the cells as compared to the amount of water leaving the cells. This is the result of cell fluid having a greater osmotic pressure than pure water.

If red blood cells are placed in a concentrated sodium chloride solution, a process opposite to that of hemolysis occurs. This time, water moves from the cells to the solution, causing the cells to shrivel (shrink in size); this process is called *crenation*. Crenation occurs because the osmotic pressure of the concentrated salt solution surrounding the red cells is greater than that of the fluid within the cells. Water always moves in the direction of greater osmotic pressure.

Finally, if red blood cells are placed in physiological saline solution, a 0.9%(m/v) sodium chloride solution, water flow is balanced and neither hemolysis nor crenation occurs. The osmotic pressure of physiological saline solution is the same as that of red blood cell fluid. Thus, the rate of water flow into and out of the red blood cells is the same. Figure 8–12 (p. 206) shows diagrammatically the three environments for red blood cells that we have just discussed.

FIGURE 8–12 The effects of bathing red blood cells in various types of solutions.

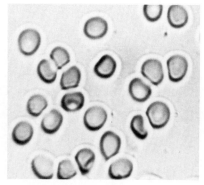
(a) Hypotonic Solution

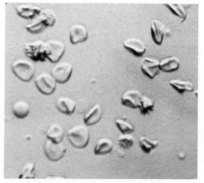
(b) Hypertonic Solution

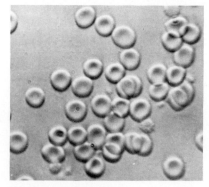
(c) Isotonic Solution

isotonic solution

We will now define the terms *isotonic*, *hypotonic*, and *hypertonic*. An **isotonic solution** *is a solution whose osmotic pressure is equal to that within cells.* Red blood cell fluid, physiological saline solution, and 5%(m/v) glucose water are all isotonic with respect to each other. The processes of replacing body fluids and supplying nutrients to the body intravenously require the use of isotonic solutions such as physiological saline and glucose water. If these isotonic solutions were not used, the damaging effects of hemolysis and crenation would occur.

hypotonic solution

hypertonic solution

A **hypotonic solution** *is a solution with a lower osmotic pressure than that within cells.* The prefix *hypo-* means "under" or "less than normal". Distilled water is hypotonic with respect to red blood cell fluid, and these cells will hemolyze when placed in it (Figure 8–12a). A **hypertonic solution** *is a solution with a higher osmotic pressure than that within cells.* The prefix *hyper-* means "over" or "more than normal". Concentrated sodium chloride solution is hypertonic with respect to red blood cell fluid and these cells undergo crenation when placed in it (Figure 8–12b).

It is sometimes necessary to introduce a hypertonic or hypotonic solution, under controlled conditions, into the body to correct an improper "water balance" situation in a patient. A hypertonic solution will cause the net transfer of water from tissues to blood; then the kidneys will remove the water. Some laxatives, such as Epsom salts, act by forming hypertonic solutions in the intestines. A hypotonic solution can be used to cause water to flow from the blood into surrounding tissue; blood pressure can be decreased in this manner. Table 8–9 summarizes the differences in meaning among the terms isotonic, hypotonic, and hypertonic.

TABLE 8–9 Characteristics of Isotonic, Hypertonic, and Hypotonic Solutions

	Type of Solution		
	Isotonic	Hypertonic	Hypotonic
osmolarity relative to body fluids	equal	greater than	less than
osmotic pressure relative to body fluids	equal	greater than	less than
osmotic effect on cells	equal water flow into and out of cells	net flow of water out of cells	net flow of water into cells

8.19 COLLOIDAL DISPERSIONS

Colloidal dispersions (or colloids) are heterogeneous mixtures that have many properties similar to those of solutions, although they are not true solutions. In a broad sense, colloids may be thought of as solutions in which a material is *suspended* rather than *dissolved*. A **colloidal dispersion** is *a dispersion (suspension) of small particles of one substance in another substance*. The terms *solute* and *solvent* are not used to indicate the components of a colloidal dispersion. Instead, the particles suspended in a colloidal dispersion are called the *dispersed phase*, and the material in which they are suspended is called the *dispersing medium*.

Suspended colloidal material is small enough in size that: (1) it is usually not discernible by the naked eye, (2) it does not settle out under the influence of gravity, and (3) it cannot be filtered out with filter paper that has relatively large pores. In these aspects, the suspended material behaves similarly to the solute of a solution. However, the suspended material is sufficiently large to make the dispersion nonhomogeneous.

A beam of light detects the nonhomogeneity of a colloidal dispersion. If we shine a beam of light through a true solution, we cannot see the track of the light. However, a beam of light passing through a colloid can be observed because the light is scattered by the colloidal particles. This scattered light is reflected into our eyes. Figure 8-13 contrasts the different ways a true solution and a colloidal dispersion behave with a beam of light.

The diameters of the suspended particles in a colloidal dispersion are in the range of 10^{-7} and 10^{-5} centimeters. This compares to diameters of less than 10^{-7} centimeters for particles such as ions, atoms, and molecules. Thus, colloidal particles are up to 1000 times larger than those present in a true solution. The suspended particles are usually aggregates of molecules, but this is not always the case. Some protein molecules are sufficiently large to form colloidal dispersions that contain single molecules in suspension. Colloidal dispersions containing particles with diameters larger than 10^{-5} centimeters are usually not encountered. Suspended particles of this size usually settle out under the influence of gravity.

Types of Colloidal Dispersions

Colloids can be classified into categories by using the system that was previously employed for classifying solutions (Section 8.9). This system is based on the physical states of the pure constituents before they interact. Only eight types of colloidal dispersions are possible, as contrasted to nine types of solutions. A gas-

colloidal dispersion

FIGURE 8-13 A beam of light travels through a true solution without being scattered. This is not the case for a colloidal dispersion.

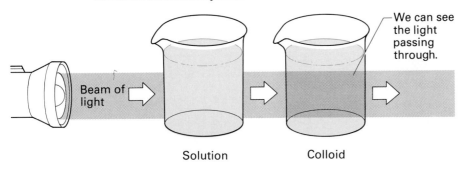

TABLE 8–10 Types of Colloidal Dispersions

Type			
Dispersing medium	Dispersed phase	Name	Examples
gas	liquid	aerosol	fog, mist, aerosol sprays, some types of air pollutants
gas	solid	aerosol	smoke, some types of air pollutants
liquid	gas	foam	whipped cream, shaving cream
liquid	liquid	emulsion	milk, mayonnaise, cosmetic creams
liquid	solid	sol	paint, ink, blood, gelatin, hot chocolate
solid	gas	solid foam	marshmallow, foam rubber
solid	liquid	solid emulsion	butter, cheese
solid	solid	solid sol	pearls, opals, colored glass

in-gas system cannot produce a colloid because gas molecules are not large enough to be colloidal. The eight types of colloidal systems are listed in Table 8–10, along with examples and the special names by which the general types are known.

Note the wide variety of substances that are colloidal in nature. They range from natural items such as blood and milk to manufactured foods like cheese and butter, shaving cream, paint, and ink.

Gelatin desserts and jellies belong to a special subclass of sols called *gels*. In these colloids, the solid dispersed phase has a very high affinity for the dispersing medium. The gel sets by forming a three-dimensional network of particles and water. Another example of a gel is canned heat (jellied alcohol), which is made from a mixture of ethyl alcohol and a saturated solution of calcium acetate.

Many different biochemical colloidal dispersions occur within the human body. Foremost among them is blood, which has numerous components that are colloidal in size. The colloidal nature of bile salts and bile protein help keep slightly soluble cholesterol in suspension in the blood. Fat is transported in the blood and lymph systems as colloidal-sized particles.

8.20 DIALYSIS

dialysis

In Section 8.18, we discussed semipermeable membranes, which selectively allow solvent molecules, but not solutes, to pass through during osmosis. Another membrane process, called *dialysis*, is also important in living organisms. **Dialysis** *is a*

FIGURE 8–14 In artifical kidney machines, blood is pumped through tubing made of a dialyzing membrane. Waste products dialyze out into the washing solution.

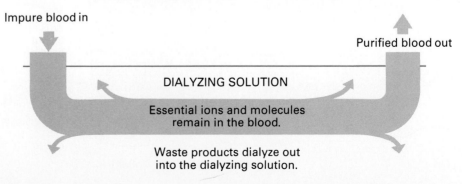

process in which a semipermeable membrane permits the passage of solvent, dissolved ions, and small molecules, but blocks the passage of colloidal particles and large molecules. Thus, dialysis allows for the separation of small particles from colloids. Many plant and animal membranes function as dialyzing membranes.

The human kidneys are a complex dialyzing system that is responsible for removing waste products from blood. The removed products are then eliminated by urine. When the kidneys fail, these waste products build up and eventually poison the body.

Individuals who once would have died because of kidney failure can now be helped through the use of artifical kidney machines, which clean the blood of toxic waste products. In these machines, the blood is pumped through tubing made of a dialyzing membrane. The tubing passes through a water bath that collects the impurities from the blood. Blood proteins and other important large molecules remain in the blood (see Figure 8–14).

EXERCISES AND PROBLEMS

Water

8.1 What are the major reservoirs for water on Earth and how do they compare to each other in size?

8.2 What are the major sources of fluid intake and fluid output for the human body?

8.3 If water were a "normal" compound, what would be its expected boiling point?

8.4 Explain why the maximum density of water occurs at a temperature that is higher than its freezing point.

8.5 Explain why lakes freeze from the top to the bottom rather than from the bottom up.

8.6 Explain, in terms of hydrogen bonding, why water has higher than expected thermal properties.

Solution Terminology

8.7 Indicate which substance is the *solvent* in each of the following solutions.
a. A solution containing 10.0 g of NaCl and 100.0 g of water.
b. A solution containing 20.0 mL of ethyl alcohol and 15.0 mL of water.
c. A solution containing 0.50 g of $AgNO_3$ and 10 mL of water.

8.8 Use Table 8–5 to determine whether each of the following solutions is saturated or unsaturated.
a. 1.94 g of $PbBr_2$ in 100 g of H_2O at 50°C.
b. 34.0 g of NaCl in 100 g of H_2O at 0°C.
c. 75.4 g of $CuSO_4$ in 200 g of H_2O at 100°C.
d. 0.540 g of Ag_2SO_4 in 50 g of H_2O at 50°C.

8.9 Based on the solubilities given in Table 8–5, characterize each of the following solutions as being a dilute or a concentrated solution.
a. 0.20 g of $CuSO_4$ dissolved in 100 g of H_2O at 100°C.
b. 1.50 g of $PbBr_2$ dissolved in 100 g of H_2O at 50°C.
c. 60 g of $AgNO_3$ dissolved in 100 g of H_2O at 50°C.
d. 0.50 g of Ag_2SO_4 dissolved in 100 g of H_2O at 0°C.

8.10 Using the concept of equilibrium, define the term *saturated solution*.

8.11 Must a saturated solution be a concentrated solution? Explain.

Solubility Rules

8.12 Predict whether the following solutes are very soluble or slightly soluble in water.
a. NH_3 (a polar gas)
b. C_2H_5OH (a polar liquid)
c. CCl_4 (a nonpolar liquid)
d. Na_2SO_4 (an ionic solid)
e. $AlPO_4$ (an ionic solid)

8.13 Using Table 8–6, predict whether each of the following ionic compounds is soluble or insoluble in water.
a. NaI
b. AgCl
c. $KC_2H_3O_2$
d. $Ba(NO_3)_2$
e. $CaSO_4$
f. $CuCO_3$
g. K_3PO_4
h. Al_2S_3

Percentage Concentration Units

8.14 Calculate the mass percent of solute in the following solutions.
a. 6.50 g of NaCl in 85.0 g of H_2O
b. 2.31 g of LiBr in 25.0 g of H_2O
c. 12.5 g of KNO_3 in 125 g of H_2O
d. 0.0032 g of NaOH in 1.2 g of H_2O

8.15 Would the same quantity of solute be present, in grams, in 200 mL of 5%(m/m) formic acid and in 200 mL of 5%(m/m) acetic acid?

8.16 Calculate the mass, in grams, of solute needed to prepare each of the following amounts of solution.
a. 250 g of a 4.00% by mass glucose ($C_6H_{12}O_6$) solution
b. 22 g of a 13.5% by mass sucrose ($C_{12}H_{22}O_{11}$) solution
c. 1000 g of a 1.20% by mass NaCl solution
d. a 5.00% by mass $NaNO_3$ solution containing 150 g of H_2O

8.17 Calculate the volume percent of solute in the following solutions.
a. 20.0 mL of methyl alcohol in enough water to give 500 mL of solution.
b. 4.00 mL of bromine in enough carbon tetrachloride to produce 87.0 mL of solution.
c. 75.0 mL of ethylene glycol in enough water to give 200 mL of solution.

8.18 What is the percent-by-volume of isopropyl alcohol in an aqueous solution made by diluting 20 mL of pure isopropyl alcohol to a volume of 125 mL?

8.19 A person has a 0.30%(v/v) alcohol level in the blood. Calculate the volume, in milliliters, of alcohol in the blood, assuming that the person's total blood volume is 5.8 L.

8.20 Determine the number of grams of solute that are needed to prepare each of the following solutions.
a. 500 mL of 5.00%(m/v) aqueous sodium bicarbonate ($NaHCO_3$).
b. 20 mL of 0.90%(m/v) NaCl (physiological saline solution).
c. 5 L of 5.00%(m/v) glucose ($C_6H_{12}O_6$) water.
d. 125 mL of 10.4%(m/v) NaOH solution.

8.21 Calculate the mass–volume percent of magnesium chloride in each of the following solutions.
a. 5.0 g of $MgCl_2$ in enough water to give 250 mL of solution.
b. 85.0 g of $MgCl_2$ in enough water to give 5.0 L of solution.
c. 1.0 g of $MgCl_2$ in enough water to give 35 mL of solution.

8.22 A solution of the drug digoxin, which is a powerful heart stimulant, contains 2.5 mg of drug per 10 mL of solution. What is this concentration in milligram percent?

8.23 Express a blood cholesterol level of 145 mg% as %(m/v).

Molarity

8.24 What is the difference in meaning between the phrases "1 mole KCl" and "1 molar KCl"?

8.25 Calculate the molarity of the following solutions.
a. 3.0 moles of potassium nitrate (KNO_3) in 0.50 L of solution.
b. 12.5 g of sucrose ($C_{12}H_{22}O_{11}$) in 80.0 mL of solution.
c. 25.0 g of sodium chloride (NaCl) in 1250 mL of solution.
d. 0.00125 mole of baking soda ($NaHCO_3$) in 2.50 mL of solution.

8.26 Calculate the number of milliliters of solution required to provide the following.

a. 1.00 g of sodium chloride (NaCl) from a 0.200 M sodium chloride solution.
b. 200.0 g of glucose ($C_6H_{12}O_6$) from a 4.2 M glucose solution.
c. 3.67 moles of silver nitrate ($AgNO_3$) from a 0.0045 M $AgNO_3$ solution.
d. 0.0001 mole of sucrose ($C_{12}H_{22}O_{11}$) from a 8.7 M sucrose solution.

8.27 Calculate the number of grams of solute in each of the following solutions.
a. 3.00 L of 0.01 M boric acid (H_3BO_3) solution.
b. 3.00 mL of 0.01 M boric acid (H_3BO_3) solution.
c. 375 mL of 7.5 M $CaCl_2$ solution.
d. 3.00 L of 0.0075 M $CaCl_2$ solution.

8.28 A 3.00%(m/v) solution of sodium bicarbonate ($NaHCO_3$) is often used to irrigate human eyes. What is the molarity of such a solution?

8.29 A 3.00 mL sample of blood is found to contain 0.740 mg of calcium ions. Calculate the molarity of calcium ions.

Equivalents and Milliequivalents

8.30 For each of the following ionic solutes, determine how many equivalents there are in one mole of solute.
a. KCl
b. Na_3PO_4
c. $MgCl_2$
d. Na_3N
e. $Al(NO_3)_3$
f. $Ca_3(PO_4)_2$

8.31 Determine the number of milligrams of Mg^{2+} that there are in a 10 mL sample of interstitial fluid if the Mg^{2+} concentration is 2.0 mEq/L.

8.32 The normal range of Na^+ in blood is 135–148 mEq/L. Does a person need Na^+ ion added to his blood if a blood analysis shows the sodium content to be 7.0 mg in a 50 mL sample of blood?

8.33 Express each of the following ionic concentrations in mEq/L.
a. 7.0 mg SO_4^{2-} in 50 mL solution
b. 1.0 mg Ca^{2+} in 2.0 mL solution
c. 3.0 g Na^+ in 500 mL solution
d. 1.0 mole Cl^- in 1500 mL solution

Dilution

8.34 Determine how much "stock solution" is needed to prepare each of the following solutions using dilution techniques.
a. 3.0 L of 2.0%(m/v) glucose solution from 5.0%(m/v) glucose solution.
b. 25 mL of 5.0%(m/v) NaCl solution from 7.0%(m/v) NaCl solution.
c. 500 mL of 6.0 M HCl from 12.0 M HCl.

8.35 For each of the following solutions, how many milliliters of water should be added to obtain a solution that has a concentration of 0.100 M?

a. 50.0 mL of 3.00 M NaCl
b. 2.00 mL of 1.00 M NaCl
c. 1.45 L of 6.00 M NaCl
d. 75.0 mL of 3.75 M NaCl

8.36 Determine the final concentration of each of the following solutions after 20.0 mL of water has been added.

a. 30.0 mL of 5.0%(m/v) glucose water
b. 30.0 mL of 5.0 mg% glucose water
c. 30.0 mL of 5.0 M glucose water
d. 30.0 mL of 5 mEq/L glucose water

8.37 Determine the final concentration of each of the following solutions if they are diluted in the specified manner.

a. 1:5 dilution of 6.0 M NaOH solution
b. 1:2 dilution of 5.0%(m/v) glucose water
c. 1:3 dilution of 4.0 mg% Na^+ ion solution
d. 1:10 dilution of 1.25 M NaCl solution

Colligative Properties of Solutions

8.38 What is a colligative property of a solution?

8.39 Explain why the vapor pressure of a solution containing a nonvolatile solute is always less than that of a pure solvent.

8.40 How are the boiling points and freezing points of water affected by the concentration of solute?

8.41 Will the freezing point of a 0.30 M NaCl solution be less than, the same as, or greater than that of a 0.20 M NaCl solution? Explain your answer.

8.42 The freezing point of a 0.10 M KCl solution is lower than that of a 0.10 M glucose solution. Explain why.

Osmosis and Osmotic Pressure

8.43 Explain what would happen if a 3.0%(m/v) glucose solution were separated by a semipermeable membrane from a 5.0%(m/v) glucose solution.

8.44 Humans must drink fresh water, not sea water. Explain why this is so.

8.45 Determine whether the osmotic pressure of a 0.1 M NaCl solution will be less than, the same as, or greater than that of the following solutions.

a. 0.1 M NaBr
b. 0.05 M $MgCl_2$
c. 0.1 M $MgCl_2$
d. 0.1 M glucose

8.46 Calculate the osmolarity of each of the following solutions.

a. 0.1 M NaCl
b. 0.1 M glucose
c. 0.1 M $CaCl_2$
d. 0.1 M $Al(NO_3)_3$
e. 0.1 M in both $CaCl_2$ and $Al(NO_3)_3$
f. 0.1 M in both NaCl and glucose

8.47 Two glucose solutions of unequal osmolarity are separated by a semipermeable membrane. Which solution will lose water, the one with higher or lower osmolarity? Explain your answer.

8.48 What is the ratio of the osmotic pressures of 0.30 M NaCl and 0.30 M $CaCl_2$?

8.49 Explain why cut flowers wilt quickly when they are placed in a sugar solution, and why they will quickly revive when transferred to pure water.

8.50 Explain what would happen if red blood cells were placed in each of the following solutions.

a. 0.9%(m/v) glucose solution
b. 0.9%(m/v) NaCl solution
c. 2.3%(m/v) glucose solution
d. 2.3%(m/v) NaCl solution
e. 5.0%(m/v) glucose solution
f. 5.0%(m/v) NaCl solution
g. distilled water

8.51 Classify each of the solutions in Problem 8-50 as isotonic, hypertonic, or hypotonic.

8.52 What is the relationship between hypotonic and hypertonic solutions and the processes of hemolysis and crenation?

Colloidal Dispersions

8.53 What is the difference between a colloidal dispersion and a true solution?

8.54 What terms are used in place of solute and solvent when describing the make-up of a colloidal dispersion?

8.55 Describe the difference in the passage of light through a colloidal dispersion and a solution.

8.56 Identify the dispersed phase (solid, liquid, or gas) and the dispersing medium, and also give an example of each of the following types of colloidal dispersions.

a. emulsion
b. aerosol
c. foam
d. gel
e. solid foam
f. solid emulsion

Dialysis

8.57 Explain the difference between an osmotic membrane and a dialyzing membrane.

8.58 A solution containing colloidal-sized protein, KCl, and glucose is placed inside a dialyzing bag and the bag is immersed in distilled water. Which compounds dialyze through the bag into the distilled water?

9 Reaction Rates and Chemical Equilibrium

Objectives

After completing Chapter 9, you will be able to:

9.1 Explain the conditions necessary for a reaction to take place in terms of collision theory.
9.2 Explain the relationship between reactant energies and product energies for exothermic and endothermic chemical reactions.
9.3 List four factors that affect reaction rates and explain how they operate in terms of collision theory.
9.4 Explain what is meant by *chemical equilibrium* and discuss the conditions necessary for the attainment of this state.
9.5 Understand the terminology used to denote the position of a chemical equilibrium.
9.6 Use Le Châtelier's principle to predict the effect that concentration, temperature, and pressure changes will have on an equilibrium system.

INTRODUCTION

In the previous two chapters, we considered the properties of matter in various pure and mixed states. Almost all of the subject matter dealt with interactions and changes of a physical nature. We will now concern ourselves with the *chemical* changes that can occur when matter interacts. We have not discussed the nature of chemical reactions in any detail, except for mass relationships that are associated with chemical changes (Section 4.8). In this chapter we will consider the fundamentals common to all chemical reactions. Of particular concern to us will be how fast chemical reactions go (reaction rates) and how far chemical reactions go (chemical equilibrium).

9.1 CONDITIONS NECESSARY FOR A CHEMICAL REACTION

The starting point for a general discussion of chemical reactions is a consideration of the conditions necessary before a chemical reaction will take place. **Collision theory** *is a set of statements that give the conditions required before a chemical reaction will occur.* Collision theory contains three fundamental concepts that were developed from the study of reaction rate information for many different reactions.

collision theory

1. Reactant particles must collide with each other in order for a reaction to occur.
2. Colliding particles must possess a certain minimum total amount of energy between them if the collision is to be effective; that is, it will result in reaction.
3. In some cases, reactants must be oriented in a specific way upon collision if reaction is to occur.

Let us consider each concept separately.

Molecular Collisions

When reactions involve two or more reactants, collision theory assumes (Statement 1) that the reactant molecules, ions, or atoms must come in contact (collide) with each other in order for any chemical change to occur. The validity of this statement is fairly obvious. Reactants cannot react with each other if they are miles apart.

Most reactions are carried out either in liquid solution or in the gaseous phase. The reason for this is directly related to the concept of molecular collisions. In both liquid solution and the gaseous phase, reacting particles are more free to move around; thus, it is easier for the reactants to come in contact with each other. Reactions in which reactants are solids can and do occur; however, the conditions for molecular collisions are not as favorable as they are for liquids and gases. Reactions of solids usually take place only on the solid surface and thus include only a small fraction of the total particles present in the solid. As the reaction proceeds and products dissolve, diffuse, or fall from the surface, fresh solid is exposed. In this way, the reaction can eventually consume all of the solid. The rusting of iron is an example of this type of process.

Activation Energy

The collisions between reactant particles do not always result in the formation of reaction products. Sometimes reactant particles rebound unchanged from a collision. Statement 2 of the collision theory indicates that in order for a reaction to occur, colliding particles must impact with a certain minimum energy; that is, the sum of the kinetic energies of the colliding particles must add up to a certain minimum value. **Activation energy** *is the minimum combined kinetic energy reactant particles must possess in order for their collision to result in a reaction.* Every chemical reaction has a different activation energy. In a slow reaction, the activation energy is far above the average energy content of the reacting particles. Only those few particles with above-average energy will undergo collisions that result in reaction; this is the reason for the slowness of the reaction.

It is sometimes possible to start a reaction by providing activation energy and then have the reaction continue on its own. Once the reaction is started, enough energy is released to activate other molecules and keep the reaction going. The striking of a kitchen match is an example of such a situation. Activation energy is initially provided by rubbing the match head against a rough surface; heat is generated by friction. Once the reaction is started, the match continues to burn.

Collision Orientation

Reaction rates are sometimes very slow because reactant molecules must be oriented in a certain way in order for collisions to lead successfully to products. How can this be? Statement 3 of the collision theory, which deals with the orientation of colliding particles at the moment of collision, relates to this situation. For nonspherical molecules and polyatomic ions, orientation relative to each other at the moment of collision is a factor that determines whether a collision is effective.

Consider the following hypothetical reaction with the diatomic molecules AB and CD as reactants.

$$A—B + C—D \longrightarrow A—C + B—D$$

In this reaction, B and C exchange places. The most favorable orientation during reactant molecule collisions would be one that simultaneously puts A and C in close proximity to each other (to form the molecule A—C) and B and D near each other (to form the molecule B—D). A possible orientation in which this situation exists is shown in Figure 9–1a. The possibility for a reaction resulting from this orientation is much greater than if the molecules were to collide while oriented as shown in Figures 9–1b or 9–1c. In Figure 9–1b, A is not near C and B is not near D. In Figure 9–1c, B is near D, but A and C are far removed from each other. Thus, certain collision orientations are preferred over others. The undesirable collision orientations of Figures 9–1b and 9–1c, however, could still result in a reaction if the molecules collided with abnormally high energies.

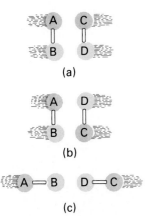

FIGURE 9–1 Different collision orientations for the reacting molecules AB and CD.

9.2 EXOTHERMIC AND ENDOTHERMIC CHEMICAL REACTIONS

In Section 7.12, the terms *exothermic* and *endothermic* were used to classify changes of state. Melting, sublimation, and evaporation are endothermic changes of state and freezing, condensation, and deposition are exothermic changes of state. The terms exothermic and endothermic are also used to classify chemical reactions. An **exothermic chemical reaction** *is one in which energy is released as the reaction* *occurs.* The burning of a fuel (reaction of the fuel with oxygen) is an exothermic process. An **endothermic chemical reaction** *is one that requires the continuous input of energy as the reaction occurs.* The photosynthesis process that occurs in plants is an example of an endothermic reaction. Light is the energy source for photosynthesis. Light energy must be continuously supplied in order for photosynthesis to occur; a green plant that is kept in the dark will die.

endothermic chemical reaction

What determines whether a chemical reaction is exothermic or endothermic? The answer to this question is related to the strength of chemical bonds, that is, the energy stored in chemical bonds. Different types of bonds, such as oxygen–hydrogen bonds and fluorine–nitrogen bonds, have different energies associated with them. In a chemical reaction, bonds are broken within reactant molecules and new bonds are formed within product molecules. The energy balance between this bond-breaking–bond-forming process determines whether there is a net loss or net gain of energy.

An exothermic reaction (release of energy) occurs when the energy required to break bonds in the reactants is less than the energy released by bond formation in the products. The opposite situation applies for an endothermic reaction. There is more energy stored in product molecule bonds than in reactant molecule bonds. The necessary additional energy must be supplied from external sources as the reaction proceeds.

Figure 9–2 shows schematically the energy relationships associated with exothermic and endothermic chemical reactions. As shown in Figure 9–2a, the products are at a lower potential energy than the reactants in an exothermic reaction; energy has been lost (released) as the reaction occurred. Conversely, as shown in Figure 9–2b, the products are at a higher potential energy than the reactants in an endothermic reaction; thus, the system required an input (absorption) of energy.

Note also that both of these diagrams contain a "hill" or "hump". The height of this "hill" corresponds to the activation energy needed in order for reaction

FIGURE 9–2 Energy diagrams for (a) an exothermic chemical reaction and (b) an endothermic chemical reaction.

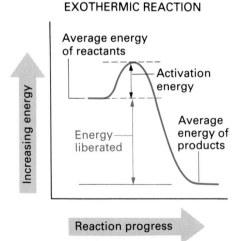

(a) In an exothermic reaction the average energy of the reactants is higher than that of the products, indicating energy has been released in the reaction.

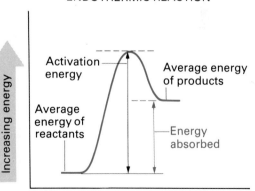

(b) In an endothermic reaction the average energy of the reactants is less than that of the products, indicating energy has been absorbed in the reaction.

between molecules to occur. This activation energy is independent of whether a given reaction is exothermic or endothermic.

9.3 FACTORS THAT INFLUENCE REACTION RATES

rate of a chemical reaction

The **rate of a chemical reaction** *is the rate or speed at which reactants are consumed or products are produced.* A number of variables affect the rate of a reaction; we routinely encounter some of them. One variable is temperature. Food is stored in a refrigerator to reduce the rate at which spoiling (a chemical reaction) occurs. State of subdivision of solids is another reaction rate variable. Sawdust and kindling wood burn (a chemical reaction) much faster than larger logs do.

In this section we will consider four different factors that affect reaction rate: (1) the physical nature of the reactants, (2) reactant concentrations, (3) reactant temperature, and (4) the presence of catalysts.

Physical Nature of Reactants

The physical nature of reactants refers not only to the physical state of each reactant (solid, liquid, or gas), but also to the state of subdivision, or particle size. In reactions where reactants are all in the same physical state, the reaction rate is generally faster between liquid state reactants than between solid reactants and is the fastest between gaseous reactants. Of the three states of matter, the gaseous state is the one where there is the most freedom of movement; hence, there is a greater frequency of collision between reactants in this state.

In the solid state, reactions occur at the boundary surface between reactants. The reaction rate increases as the amount of boundary surface area increases. Subdividing a solid into smaller particles increases surface area and thus increases reaction rate. A crushed aspirin tablet will exert its pain-relieving effect much faster than a whole tablet because of its increased surface area.

FIGURE 9-3 Extremely rapid combustion (reaction with oxygen) of grain dust produced the explosive effect that destroyed this grain elevator.

When the particle size is extremely small, reaction rates can be so fast that an explosion results. A lump of coal is difficult to ignite; coal dust ignites explosively. The spontaneous ignition of coal dust is a real threat to underground coal-mining operations. Grain dust (very finely divided grain particles) is a problem in grain storage elevators; the possibility of an explosive ignition of the dust from an accidental spark is always present. Figure 9–3 shows the destruction that can result from the accidental ignition of grain dust in a storage elevator.

Reactant Concentration

An increase in the concentration of a reactant causes an increase in the rate of the reaction. Combustible substances burn much more rapidly in pure oxygen than they do in air (21% oxygen). A person with respiratory problems such as pneumonia or emphysema is often given air enriched with oxygen because an increased partial pressure of oxygen facilitates the absorption of oxygen in the alveoli of the lungs and thus expedites all subsequent steps in respiration (Section 7.11).

Increasing the concentration of a reactant means that there are more molecules of that reactant present in the reaction mixture; thus, there is a greater possibility for collisions between this reactant and other reactant particles. An analogy to the reaction rate–reactant concentration relationship can be drawn from the game of billiards. The more billiard balls there are on the table, the greater the probability is of a moving cue ball striking one of them.

When the concentration of reactants is increased, the actual quantitative change in reaction rate is determined by the specific reaction. The rate always increases, but not to the same extent in all cases. You will not be able to determine how changes in concentration will affect the reaction rate by simply looking at the balanced equation for a reaction. This must be determined by actual experimentation. Sometimes the rate doubles with a doubling of concentration; however, this is not always the case.

Reaction Temperature

The effect of temperature on reaction rates can also be explained by using the molecular-collision concept. An increase in the temperature of a system results in an increase in the average kinetic energy of the reacting molecules. The increased molecular speed causes more collisions to take place within a given time. Also, because the average energy of the colliding molecules is greater, a larger fraction of the collisions will result in reaction from the point of view of activation energy.

As a rough rule of thumb, chemists have found that the rate of a chemical reaction doubles for every 10°C increase (a difference of about 18°F) in temperature for the temperature ranges we normally encounter. The chemical reactions involved in cooking foods take place faster in a pressure cooker because of a higher cooking temperature (Section 7.15). On the other hand, foods are cooled or frozen to slow down the chemical reactions that result in the spoiling of food, souring of milk, and ripening of fruit.

When a person has a fever, the rate of many chemical reactions in the body proceed at a faster rate than normal. The fever's effect translates into an increased pulse rate and increased breathing rate. For every 1°C increase in body temperature, 13% more oxygen is required by body tissues.

A decrease in body temperature has the opposite effect: cells use less oxygen than they normally do. This knowledge is applied clinically in certain medical situations such as open-heart surgery. During open-heart surgery, a patient's body temperature is usually lowered two to three Celsius degrees; both the metabolism rate and oxygen requirements of the patient decrease correspondingly.

Presence of Catalysts

catalyst

A **catalyst** *is a substance that increases a reaction rate without being consumed in the reaction.* In most cases only extremely small amounts of catalyst are needed. Catalysts enhance reaction rates by providing alternate reaction pathways that have lower activation energies than the original uncatalyzed pathway. This lowering of activation energy is diagrammatically shown in Figure 9–4.

Catalysts exert their effects in varying ways. Some catalysts provide a lower-energy pathway by entering into a reaction and forming an "intermediate", which then reacts further to produce the desired products and regenerate the catalyst. The following equations, where C is the catalyst, illustrate this concept.

Uncatalyzed Reaction: $X + Y \longrightarrow XY$

Catalyzed Reaction: *Step 1:* $X + C \longrightarrow XC$

                                     Step 2: $XC + Y \longrightarrow XY + C$

In Step 1 of the catalyzed reaction, the intermediate XC is formed from the interaction of the catalyst with one of the reactants (X). The activation energy for this reaction is different from (lower than) that of the uncatalyzed reaction. The intermediate from Step 1, XC, serves as a reactant for Step 2; this produces the desired product XY and regenerates the catalyst C.

Solid-state catalysts often act by providing a surface on which impacting reactant molecules are physically attracted and held with a particular orientation. These reactants are sufficiently close and favorably oriented toward each other that the reaction takes place. The products of the reaction then leave the surface and make it available to catalyze other reactants.

Catalysts are a key element in the functioning of automobile emission control systems. In these systems, solid catalysts speed up reactions that convert air

FIGURE 9–4   A catalyst lowers the activation energy for a chemical reaction.

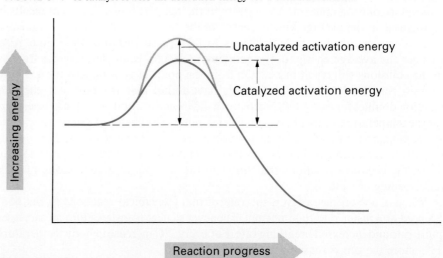

pollutants present in the exhaust to less harmful products. For example, carbon monoxide is converted to carbon dioxide through reaction with oxygen.

Catalysts are of extreme importance for the proper functioning of the human body and other biological systems. In the human body, special protein compounds called *enzymes* function as catalysts. They cause many reactions to take place rapidly under mild conditions and at body temperature; without these enzymes, the reactions would proceed very slowly and then only under harsher conditions. One of the dangers of an extremely high fever during illness is the possibility that some enzymes will become deactivated at the higher temperature. Enzymes are very sensitive to temperature changes. Chapter 25 will be devoted to the subject of enzymes.

9.4 CHEMICAL EQUILIBRIUM

In our discussions about chemical reactions up to this point, we have assumed that chemical reactions go to completion; that is, reactions continue until one or more of the reactants is used up. This is not true for many reactions. Experiments show that in numerous chemical reactions, the complete conversion of reactants to products does not occur, regardless of the time allowed for the reactions to take place. The reason for this is that product molecules (provided they are not allowed to escape from the reaction mixture) begin to react with each other to reform reactants. With time, a steady-state situation results where the rate of formation of products and the rate of reformation of reactants are equal. At this point, the concentrations of all reactants and all products remain constant and a state of *chemical equilibrium* is reached. **Chemical equilibrium** *is the process where two opposing chemical reactions occur simultaneously at the same rate.* We have discussed equilibrium situations in Sections 7.14 (vapor pressure) and 8.8 (saturated solutions); however, these previous examples involved physical equilibrium rather than chemical equilibrium.

chemical equilibrium

The conditions that exist in a system in a state of chemical equilibrium can best be seen by considering an actual chemical reaction. Suppose equal molar amounts of gaseous H_2 and I_2 are mixed together in a closed container and allowed to react.

$$H_2 + I_2 \longrightarrow 2\,HI$$

Initially, no HI is present, so the only possible reaction that can occur is that between H_2 and I_2. However, with time, as the HI concentration increases, some HI molecules collide with each other in a way that causes a reverse reaction to occur.

$$2\,HI \longrightarrow H_2 + I_2$$

The initially low concentration of HI makes this reverse reaction slow at first, but as the concentration of HI increases, the reaction rate does also. At the same time that the reverse-reaction rate is increasing, the forward-reaction rate (production of HI) is decreasing as the reactants are used up. Eventually, the concentrations of H_2, I_2, and HI in the reaction mixture reach a level at which the rates of the forward and reverse reactions become equal. At this point, a state of chemical equilibrium has been reached.

Figure 9–5 (p. 220) graphically shows the behavior of reaction rates and reaction concentrations over time for both the forward and reverse reactions in the H_2–I_2–HI system. Figure 9–5a shows that the forward- and reverse-reaction rates become equal as a result of the forward-reaction rate decreasing as reactants are

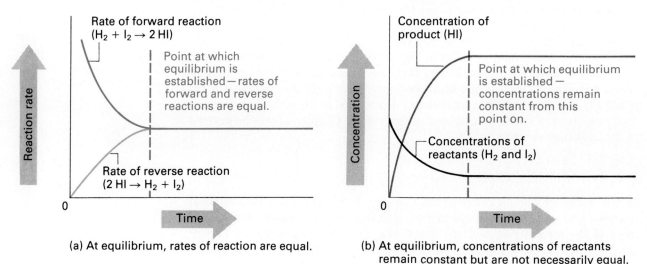

FIGURE 9-5 Variation of reaction rates and reactant concentrations over time in the chemical system H_2–I_2–HI.

(a) At equilibrium, rates of reaction are equal.

(b) At equilibrium, concentrations of reactants remain constant but are not necessarily equal.

used up and the reverse-reaction rate increasing as product concentration increases. Figure 9-5b illustrates the important point that the reactant and product concentrations are usually not equal at the point at which equilibrium is reached. The rates are equal, but not the concentrations. For the H_2–I_2–HI system, much more product HI is present than reactants H_2 and I_2 at equilibrium. Note in Figure 9-5b that the point at which equilibrium is established is the point where the two curves become straight lines.

The equilibrium involving H_2, I_2, and HI could have been established just as easily by starting with pure HI and allowing it to change into H_2 and I_2 (the reverse reaction). The final position of equilibrium does not depend on the direction from which equilibrium is approached.

Instead of writing separate equations for both the forward and reverse reactions for a system at equilibrium, it is normal procedure to represent the equilibrium by using a single equation and two half-headed arrows pointing in opposite directions. Thus, the reaction between H_2 and I_2 at equilibrium is written as

$$H_2 + I_2 \rightleftharpoons 2\,HI$$

The half-headed arrows denote a chemical system at equilibrium.

reversible reaction

The term *reversible* is often used to describe a reaction like the one we have just discussed. A **reversible reaction** is *a chemical reaction in which the products formed can react to yield the original reactants*. When the half-headed arrow notation is used in an equation, it means that a reaction is reversible.

Many reactions in living organisms are reversible; that is, they can proceed in both forward and reverse directions. An example of a reversible reaction that occurs in the human body is

$$\underset{\text{carbon dioxide}}{CO_2} + H_2O \rightleftharpoons \underset{\text{carbonic acid}}{H_2CO_3}$$

Carbon dioxide, which is formed as a waste product in the tissues, dissolves in the blood, where it is converted to carbonic acid. When the blood reaches the lungs, the carbonic acid breaks apart to regenerate the carbon dioxide, which is then exhaled.

The method by which oxygen is distributed throughout the body also involves a reversible reaction.

$$O_2 + Hb \rightleftharpoons HbO_2$$

In the lungs, inhaled O_2 reacts with hemoglobin, Hb, to form oxyhemoglobin, HbO_2. The oxyhemoglobin is transported by the blood to cells that require oxygen; there, the reverse reaction occurs.

9.5 POSITION OF EQUILIBRIUM

The **position of equilibrium** *specifies, in a qualitative way, the relative amounts of reactants and products at equilibrium.* The terms *far to the right, to the right, to the left,* and *far to the left* are commonly used to describe equilibrium position. In equilibrium situations where the concentrations of products are greater than those of reactants, the equilibrium position is said to lie *to the right* because products are always listed on the right side of an equation. If product concentrations are smaller than reactant concentrations at equilibrium, the equilibrium position lies *to the left. Far to the right* and *far to the left* denote equilibrium positions approaching 100% products and 100% reactants, respectively. Table 9–1 summarizes the equilibrium position terminology that is in common use.

position of equilibrium

Position of equilibrium can also be indicated by varying the length of the arrows in the half-headed arrow notation for a reversible reaction. The longer arrow indicates the direction of the predominant reaction. For example, the arrow notation in the equation

$$CO_2 + H_2O \rightleftharpoons H_2CO_3$$

indicates that the equilibrium position lies far to the left.

9.6 LE CHÂTELIER'S PRINCIPLE

The balance associated with a chemical system at equilibrium is very susceptible to disruption from outside forces. A change in temperature or a change in pressure can upset the balance within the equilibrium system. Changes in the concentrations of reactants or products upset an equilibrium also.

Disturbing an equilibrium has one of two results: either the forward reaction speeds up (to produce more products) or the reverse reaction speeds up (to produce additional reactants). Over time, the forward and reverse reactions again become equal, and a new equilibrium, which is not identical to the previous one, is established. If more products have been produced as a result of the disruption,

TABLE 9–1 Position of Equilibrium and Relative Amounts of Products and Reactants Present

Description of Equilibrium Position	Relative Amounts of Products and Reactants
far to the right	essentially all products
to the right	more products than reactants
to the left	more reactants than products
far to left	essentially all reactants

Le Châtelier's principle

the equilibrium is said to have *shifted to the right*. Similarly, when disruption causes more reactants to form, the equilibrium has *shifted to the left*.

An equilibrium system's response to disrupting influences can be predicted by using a guideline (principle) introduced by the French chemist Henry Louis Le Châtelier (1850–1926). **Le Châtelier's principle** *states that if a stress (change of conditions) is applied to a system in equilibrium, the system will readjust (change the position of equilibrium) in the direction that best reduces the stress imposed on it.* We will use this principle to consider how four types of changes affect equilibrium position. The changes are: (1) concentration changes, (2) temperature changes, (3) pressure changes, and (4) addition of catalysts.

Concentration Changes

Adding or removing a reactant or product from a reaction mixture at equilibrium always upsets the equilibrium. Le Châtelier's principle predicts that the reaction will shift in the direction that minimizes the change in concentration caused by the addition or removal. If an additional amount of any reactant or product has been *added* to the system, the stress is relieved by shifting the equilibrium in the direction that *consumes* (uses up) some of the added reactant or product. Conversely, if a reactant or product is *removed* from an equilibrium system, the equilibrium shifts in a direction that will *produce* more of the substance that was removed. The following two equations summarize the behavior of an equilibrium system when the stress put upon it is a concentration change.

1. The position of the equilibrium will shift to the right if a reactant concentration is increased or a product concentration is decreased.

$$\underset{A + B}{\text{addition of}} \rightleftharpoons \underset{C + D}{\text{removal of}}$$

2. The position of the equilibrium will shift to the left if a reactant concentration is decreased or a product concentration is increased.

$$\underset{A + B}{\text{removal of}} \rightleftharpoons \underset{C + D}{\text{addition of}}$$

Within the human body, numerous equilibrium situations exist that change (shift) in response to concentration change. Consider, for example, the equilibrium between glucose in the blood and stored glucose (glycogen) in the liver.

$$\text{glucose in blood} \rightleftharpoons \text{stored glucose} + H_2O$$

Strenuous exercise or hard work causes our blood glucose level to decrease. Our bodies respond to this stress (not enough glucose in the blood) by having the liver convert glycogen into glucose. Conversely, when an excess of glucose is present in the blood (after eating certain food items), the liver converts the excess glucose in the blood to its storage form (glycogen).

Temperature Changes

Le Châtelier's principle can be used to predict the influence of temperature changes on an equilibrium, provided you know whether the reaction of concern is exothermic or endothermic.

Consider the following general *exothermic* reaction.

$$A + B \rightleftharpoons C + D + \text{heat}$$

Heat is produced when the reaction proceeds to the right. Thus, if we add heat to an exothermic system at equilibrium (by raising the temperature), the system will shift to the left in an attempt to use up the added heat. When equilibrium is reestablished, the concentrations of A and B will be higher and the concentrations of C and D will have decreased. Lowering the temperature of an exothermic reaction mixture causes the reaction to shift to the right as the system attempts to replace the lost heat.

The behavior, with temperature change, of an equilibrium system involving an *endothermic* reaction

$$E + F + \text{heat} \rightleftharpoons G + H$$

is opposite that of an exothermic reaction because a shift to the left produces heat. Consequently, an increase in temperature will cause the equilibrium to shift to the right (to use up the added heat) and a decrease in temperature will produce a shift to the left (to generate more heat).

Pressure Changes

Changes in pressure do not significantly affect the concentrations of solids and liquids, but they do significantly alter the concentrations of gases. Thus, pressure changes affect systems at equilibrium only when gases are involved, and then only in cases where the chemical reaction is such that a change in the total number of moles in the gaseous state occurs. This latter point can be illustrated by considering the following two gas phase reactions.

$$2\,H_2(g) + O_2(g) \longrightarrow 2\,H_2O(g)$$
$$H_2(g) + Cl_2(g) \longrightarrow 2\,HCl(g)$$

In the first reaction, the total number of moles of gaseous reactants and products decreases as the reaction proceeds to the right because three moles of reactants combine to give only two moles of products. In the second reaction, there is no change in the total number of moles of gaseous substances present as the reaction proceeds because two moles of reactants combine to give two moles of products. Thus, a pressure change will shift the position of the equilibrium in the first reaction, but not in the second reaction.

Pressure changes are usually brought about through volume changes. A pressure increase results from a volume decrease and a pressure decrease results from a volume increase (Section 7.8). The use of Le Châtelier's principle correctly predicts the direction of the equilibrium position shift resulting from a pressure change only when the pressure change is caused by volume change. It does not apply to pressure increases caused by the addition of a nonreactive (inert) gas to the reaction mixture. This addition has no effect on the equilibrium position. The partial pressures (Section 7.11) of each of the gases involved in the reaction remain the same.

According to Le Châtelier's principle, the stress of increased pressure is relieved by decreasing the number of moles of gaseous substances in the system. This is accomplished by the reaction shifting in the direction of the fewer number of moles; that is, it shifts to the side of the equation that contains the fewer number of moles of gaseous substances. For the reaction

$$2\,NO_2(g) + 7\,H_2(g) \rightleftharpoons 2\,NH_3(g) + 4\,H_2O(g)$$

an increase in pressure would shift the equilibrium position to the right because there are nine moles of gaseous reactants and only six moles of gaseous products. A stress of decreased pressure results in an equilibrium system reacting to produce more moles of gaseous substances.

Addition of Catalysts

Catalysts cannot change the position of an equilibrium. Why this is so becomes clear when we remember that a catalyst functions by lowering the activation energy for a reaction. A catalyst speeds up both the forward and reverse reactions, and therefore it has no net effect on the position of the equilibrium. However, the lowered activation energy allows equilibrium to be established more quickly than if the catalyst were absent.

Example 9.1

How will the gaseous phase equilibrium

$$CH_4(g) + 2\ H_2S(g) + heat \rightleftharpoons CS_2(g) + 4\ H_2(g)$$

be affected by
a. the removal of $H_2(g)$?
b. the addition of $CS_2(g)$?
c. an increase in the temperature?
d. an increase in the volume of the container (a decrease in pressure)?

Solution
a. The equilibrium will *shift to the right*, according to Le Châtelier's principle, in an attempt to replenish the H_2 that was removed.
b. The equilibrium will *shift to the left* in an attempt to use up the extra CS_2 that has been placed in the system.
c. Raising the temperature means that heat energy has been added. In order to minimize the effect of this extra heat, the position of the equilibrium will *shift to the right*, which is a direction that consumes heat; heat is one of the reactants in an endothermic reaction.
d. The system will *shift to the right* in an attempt to produce more moles of gaseous reactants that will increase the pressure. In this way, the reaction produces five moles of gaseous products for every three moles of gaseous reactants consumed.

EXERCISES AND PROBLEMS

Collision Theory

9.1 What is collision theory and what are the main concepts of this theory?

9.2 Define the term *activation energy*.

9.3 What two factors determine whether a collision between two reactant molecules will result in a reaction?

Exothermic and Endothermic Reactions

9.4 Which of the following reactions are endothermic and which are exothermic?

a. $C_2H_4 + 3\ O_2 \longrightarrow 2\ CO_2 + 2\ H_2O + heat$
b. $N_2 + O_2 + heat \longrightarrow 2\ NO_2$
c. $2\ H_2O + heat \longrightarrow 2\ H_2 + O_2$
d. $2\ KClO_3 \longrightarrow 2\ KCl + 3\ O_2 + heat$

9.5 Draw energy diagrams to represent exothermic and endothermic chemical reactions. Label the following items in each diagram: (1) average energy of the reactants, (2) average energy of the products, (3) activation energy, and (4) energy liberated or absorbed.

9.6 One reaction occurs at room temperature and liberates 200 kilocalories of energy per mole of reactant. Another reaction does not take place until the reaction mixture is heated to 150°C. However, it also liberates 200 kilocalories of energy per mole of reactant. Draw an energy diagram for each reaction and indicate the similarities and differences between the two diagrams.

Factors That Influence Reaction Rates

9.7 List four factors that influence the rate of a reaction and describe how each effect can be explained in terms of collision theory.

9.8 Define the term *reaction rate*.

9.9 Substances burn more rapidly in pure oxygen than in air. Explain why.

9.10 Milk will sour in a couple of days when left at room temperature; however, it will remain unspoiled for two weeks when refrigerated. Explain why this is so.

9.11 What is a catalyst?

9.12 Draw an energy diagram for an exothermic reaction where no catalyst is present. Then draw an energy diagram for the same reaction when a catalyst is present. Indicate the similarities and differences between the two diagrams.

9.13 The characteristics of six reactions, each of which involves only two reactants, are as follows:

Reaction	Activation Energy	Temperature	Concentration of Reactants
1	low	low	1 mole/L of each
2	high	low	1 mole/L of each
3	low	high	1 mole/L of each
4	high	high	1 mole/L of each
5	low	low	1 mole/L of 1st reactant, 4 mole/L of 2nd reactant
6	low	low	4 mole/L of each

For each of the following pairs of the preceding reactions, compare the reaction rates when the two reactants are first mixed. Indicate which reaction is faster.

a. 1 or 2
b. 1 or 3
c. 1 or 5
d. 2 or 3
e. 2 or 4
f. 3 or 4
g. 5 or 6

Chemical Equilibrium

9.14 What is *equal* in a chemical reaction at equilibrium?

9.15 The concentrations of all species in an equilibrium mixture are necessarily constant over time; however, reactions are still occurring. Explain how this can be so.

9.16 Sketch a graph that shows how the concentrations of the reactants and products of a typical reaction vary with time.

9.17 What is a reversible reaction?

9.18 What notation is used to denote both a reversible reaction and a state of equilibrium?

9.19 List the four phrases that are used to denote, in a qualitative manner, the position of an equilibrium and indicate the meaning associated with each phrase's use.

9.20 Indicate how half-arrow notation can be used to specify the position of an equilibrium.

Le Châtelier's Principle

9.21 State Le Châtelier's principle.

9.22 What is meant by the phrase "an equilibrium shifts to the right"? Explain in terms of the forward and reverse reactions.

9.23 List the three changes in conditions that can affect the position of an equilibrium.

9.24 For the reaction

$$2\ Cl_2(g) + 2\ H_2O(g) \rightleftharpoons 4\ HCl(g) + O_2(g)$$

determine the direction that the equilibrium will be shifted by the following changes.

a. increasing the concentration of Cl_2
b. increasing the concentration of HCl
c. increasing the concentration of O_2
d. decreasing the concentration of Cl_2
e. decreasing the concentration of H_2O
f. decreasing the concentration of HCl

9.25 For the reaction

$$C_6H_6(g) + 3\ H_2(g) \rightleftharpoons C_6H_{12}(g) + heat$$

determine the direction that the equilibrium will be shifted by the following changes.

a. increasing the concentration of H_2
b. decreasing the concentration of C_6H_{12}
c. increasing the temperature
d. decreasing the temperature
e. increasing the pressure by decreasing the volume
f. decreasing the pressure by increasing the volume

9.26 Consider the following chemical system at equilibrium.

$$CO(g) + H_2O(g) + heat \rightleftharpoons CO_2(g) + H_2(g)$$

Determine the direction that the equilibrium will be shifted by the following changes.

a. heating the equilibrium mixture
b. adding CO to the equilibrium mixture
c. adding H_2 to the equilibrium mixture
d. increasing the pressure on the equilibrium mixture by adding an inert gas
e. increasing the pressure on the equilibrium mixture by decreasing the size of the container for the mixture
f. removing CO_2 from the equilibrium mixture
g. adding a catalyst to the equilibrium mixture
h. increasing the size of the container for the mixture

10 Acids, Bases, and Salts

Objectives

After completing Chapter 10, you will be able to:

10.1 State the definitions of acids and bases using (1) the Arrhenius concept, and (2) the Brønsted–Lowry concept.
10.2 Differentiate between and give examples of strong and weak Arrhenius acids and bases.
10.3 Differentiate among and give examples of mono-, di-, and triprotic acids, and write equations for the stepwise dissociation of di- and triprotic acids.
10.4 Recognize the compounds that are classified as salts.
10.5 List the classes of substances that are strong electrolytes and weak electrolytes.
10.6 Define the term *neutralization* and write equations for various acid–base neutralization processes.
10.7 Understand the relationship between $[H^+]$ and $[OH^-]$ in pure water and in aqueous solutions and tell whether a solution is acidic, basic, or neutral, when given its $[H^+]$ or $[OH^-]$.
10.8 Calculate the $[H^+]$ and $[OH^-]$ of a solution given the pH and tell whether the solution is acidic, basic, or neutral.
10.9 List the steps involved in the titration of an acid or base.
10.10 Express the concentration of an acidic or basic solution using the normality concentration unit and carry out titration calculations using this unit.
10.11 Explain the process of salt hydrolysis and predict which salts will hydrolyze and whether the hydrolysis produces an acidic or basic solution.
10.12 Describe how a buffer system causes a solution to be pH-change resistant.
10.13 Describe the operation of the carbonic acid/hydrogen carbonate and dihydrogen phosphate/hydrogen phosphate buffer systems in the human body.

INTRODUCTION

Acids, bases, and salts are among the most common and important compounds that are known; they are encountered in every walk of life. In the form of aqueous solutions, these compounds are key materials in both biological systems and the chemical industry. A major ingredient of gastric juice in the stomach is hydrochloric acid. Large quantities of lactic acid are produced when the human body is subjected to strenuous exercise. The lye used in making soap contains the base sodium hydroxide. Bases are ingredients in many stomach antacid formulations. The white crystals you sprinkle on your food to make it taste better represent only one of many hundreds of salts that exist. In this chapter we will define acids, bases, and salts from a chemical standpoint and consider many of the reactions that make them such important compounds.

10.1 ACID–BASE DEFINITIONS

Several definitions for acids and bases are now in use. We will consider two sets: the Arrhenius definitions and the Brønsted–Lowry definitions.

Arrhenius Definitions

In 1884, the Swedish chemist Svante August Arrhenius (1859–1927) proposed that acids and bases be defined in terms of the species they form upon dissolution in water. His definitions are the simplest and most commonly used definitions in use today. An **Arrhenius acid** is *a substance that releases hydrogen ions* (H^+) *in aqueous solution.* An **Arrhenius base** is *a substance that releases hydroxide ions* (OH^-) *in aqueous solution.*

Arrhenius acid

Arrhenius base

Some common examples of acids, according to the Arrhenius definition, are the substances HNO_3 and HCl.

$$HNO_3(l) \xrightarrow{H_2O} H^+(aq) + NO_3^-(aq)$$

$$HCl(g) \xrightarrow{H_2O} H^+(aq) + Cl^-(aq)$$

When Arrhenius acids are in the pure state (not in solution), they are covalent compounds; that is, they do not contain H^+ ions. These ions are formed through a chemical reaction when the acid is mixed with water. The chemical reaction between water and the acid molecules results in the removal of H^+ ions from acid molecules.

Two common examples of Arrhenius bases are $NaOH$ and KOH.

$$NaOH(s) \xrightarrow{H_2O} Na^+(aq) + OH^-(aq)$$

$$KOH(s) \xrightarrow{H_2O} K^+(aq) + OH^-(aq)$$

Arrhenius bases are usually ionic compounds in the pure state, in direct contrast to acids. When these compounds dissolve in water, the ions separate to yield the OH^- ions.

At an introductory level, the Arrhenius definitions for acids and bases adequately explain the behaviors of acids and bases in aqueous solution. These definitions are the only definitions that are found in many elementary textbooks. However, the definitions have two drawbacks: (1) they are adequate only in the case of aqueous solutions, and (2) the identity of the acidic species present in aqueous solutions is oversimplified. With regard to statement 2, unknown to Arrhenius, *free* hydrogen ions (H^+) cannot exist in water. The attraction between a hydrogen ion and polar water molecules is sufficiently strong to bond the hydrogen ion with a water molecule to form a hydronium ion (H_3O^+).*

$$H^+ + \overset{..}{\underset{H}{\overset{|}{O}}}-H \longrightarrow \left[H:\overset{..}{\underset{H}{\overset{|}{O}}}-H\right]^+$$

$$\text{hydronium ion}$$

The bond holding the hydrogen ion to the water molecule is a coordinate-covalent bond (Section 6.10) because both electrons are furnished by the oxygen atom.

Brønsted–Lowry Definitions

In 1923, Johannes Nicholas Brønsted (1879–1947), a Danish scientist, and Thomas Martin Lowry (1874–1936), a British scientist, independently and almost simultaneously proposed definitions for acids and bases that expanded upon the ideas of Arrhenius. Their definitions (1) extend the number of substances that can be considered to be acids and bases; (2) are not restricted to aqueous solutions; and (3) account for the fact that the acidic species in aqueous solution is the hydronium ion.

Brønsted–Lowry acid

Brønsted–Lowry base

A **Brønsted–Lowry acid** *is any substance that can donate a proton (H^+) to some other substance.* (Remember, an H^+ ion is a hydrogen atom (proton plus electron) that has lost its electron; hence, it is a bare proton.) A **Brønsted–Lowry base** *is any substance that can accept a proton from some other substance.* In short, a Brønsted–Lowry acid is a *proton donor* and a Brønsted–Lowry base is a *proton acceptor*. The Brønsted–Lowry definitions change the focus on acids and bases from that of the production of specific chemical species (H^+ and OH^-) to one dealing with chemical reactions involving proton exchange.

The Brønsted–Lowry acid–base definitions can best be illustrated by example. Let us consider the formation reaction for hydrochloric acid, which involves the dissolving of hydrogen chloride gas in water.

$$HCl(g) + H_2O(l) \longrightarrow H_3O^+(aq) + Cl^-(aq)$$

The hydrogen chloride behaves as a Brønsted–Lowry acid by donating a proton to a water molecule. Note that a hydronium ion is formed as a result. The base in this reaction is water because it has accepted a proton; no hydroxide ions are involved. The Brønsted–Lowry definition of a base includes any species capable of accepting a proton; hydroxide ions can do this, but so can many other substances.

It is not necessary that a water molecule be one of the reactants in a Brønsted–Lowry acid–base reaction; the reaction does not have to take place in the liquid state. Brønsted–Lowry acid–base theory can be used to describe gas-phase reactions. The white solid haze that often covers glassware in a chemistry laboratory results from the gas-phase reaction between HCl and NH_3.

---

* To keep our ions straight, let us summarize: a hydrogen ion is H^+, a hydroxide ion is OH^-, and a hydronium ion is H_3O^+.

$$HCl(g) + NH_3(g) \longrightarrow NH_4^+(g) + Cl^-(g) \longrightarrow NH_4Cl(s)$$

This is a Brønsted–Lowry acid–base reaction because the HCl molecules donate protons to the NH_3, forming NH_4^+ and Cl^- ions. These ions instantaneously combine to form the white solid NH_4Cl.

All Arrhenius acids are Brønsted–Lowry acids, and all Arrhenius bases are Brønsted–Lowry bases. However, the converse of this statement is not true. Brønsted–Lowry theory includes Arrhenius theory and much more.

Occasionally, it will be advantageous to discuss acid–base chemistry using the concepts of Brønsted and Lowry; however, the Arrhenius acid–base definitions will be adequate for most of our discussions. The notation H^+, instead of H_3O^+, for the acidic species in aqueous solution will also be adequate most of the time.

10.2 STRENGTHS OF ACIDS AND BASES

Acids can be classified as strong or weak based upon the percent of acid molecules that dissociate when dissolved in water. A **strong acid** *dissociates 100%, or very nearly 100%, in aqueous solution.* Thus, if an acid is strong, almost all of the acid molecules present dissociate into ions. This extensive dissociation causes many hydrogen ions to be present in the solution of a strong acid. A **weak acid** *dissociates only slightly, usually less than 5%, in aqueous solution.* Most acid molecules are present in undissociated form in a solution of a weak acid. Relatively speaking, there are only a few hydrogen ions present in the solution of a weak acid.

strong acid

weak acid

The extent to which an acid dissociates in solution depends on the molecular structure of the acid; molecular polarity and the strength and polarity of individual bonds are particularly important factors in determining whether an acid is strong or weak.

Only a few strong acids exist. The formulas and structures of the six most commonly encountered strong acids are listed in Table 10–1.

The vast majority of the acids that exist are weak acids. The formulas for numerous weak acids are listed in Table 10–3 (p. 232). The difference between a strong acid and a weak acid can also be stated in terms of the position of equilibrium (Section 9.5) for the reaction,

$$HA \rightleftharpoons H^+ + A^-$$

TABLE 10–1 Commonly Encountered Strong Acids

Name	Molecular Formula	Molecular Structure
nitric acid	HNO_3	H—O—N—O, ‖O
sulfuric acid	H_2SO_4	H—O—S(=O)—O—H, with O below
perchloric acid	$HClO_4$	H—O—Cl(=O)—O, with O below
hydrochloric acid	HCl	H—Cl
hydrobromic acid	HBr	H—Br
hydroiodic acid	HI	H—I

TABLE 10-2 Common Strong Bases

Group IA Hydroxides	Group IIA Hydroxides
LiOH	
NaOH	
KOH	
RbOH	Sr(OH)$_2$
CsOH	Ba(OH)$_2$

where HA represents the acid and H$^+$ and A$^-$ are the products from the dissociation of the acid. For strong acids, the equilibrium lies far to the right (100% or almost 100%).

$$HA \rightleftharpoons H^+ + A^-$$

For weak acids, the equilibrium position lies far to the left.

$$HA \rightleftharpoons H^+ + A^-$$

Thus, in solutions of strong acids, the predominant species are H$^+$ and A$^-$. In solutions of weak acids, the opposite situation applies. The predominant species is HA, the undissociated acid. A graphical representation of the differences between strong and weak acids, in terms of species present in solution is given in Figure 10-1.

Just as there are strong acids and weak acids, there are also strong bases and weak bases. As with acids, there are only a few strong bases. Strong bases are limited to the hydroxides of Groups IA and IIA that are listed in Table 10-2. Of the strong bases, only NaOH and KOH are commonly encountered in a chemical laboratory. The limited solubility in water of the listed Group IIA hydroxides limits their use. However, despite their limited solubility, these hydroxides are still considered to be strong bases; that which dissolves dissociates into ions 100%.

Only one of the many weak bases that exist is fairly common—aqueous ammonia. In this solution of ammonia gas (NH$_3$) in water, small amounts of OH$^-$ ions are produced through the reaction of NH$_3$ molecules with water.

$$NH_3(g) + H_2O(l) \rightleftharpoons NH_4^+(aq) + OH^-(aq)$$

A solution of aqueous ammonia is sometimes erroneously called *ammonium hydroxide*. Aqueous ammonia is the preferred designation because most of the NH$_3$ present has not reacted with water; the equilibrium position lies far to the left. Only a few ammonium ions (NH$_4^+$) and hydroxide ions (OH$^-$) are present.

It is important to remember that the terms *strong* and *weak* apply to the extent of dissociation and not to the concentrations of acid or base. For example, stomach acid (gastric juice) is a dilute (not weak) solution of a strong acid; it is 5% by mass hydrochloric acid. On the other hand, a 36% by mass solution of hydrochloric acid is considered to be a concentrated (not strong) solution of a strong acid.

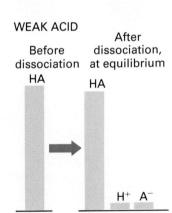

FIGURE 10-1 A strong acid is completely dissociated in solution, but in a weak acid only a small fraction of the molecules are dissociated in solution.

10.3 POLYPROTIC ACIDS

Acids can be classified according to the number of hydrogen ions they produce *per molecule* upon dissociation in solution. A **monoprotic acid** *is an acid that yields one H$^+$ ion (proton) per molecule upon dissociation.* Hydrochloric acid (HCl) and nitric acid (HNO$_3$) are both monoprotic acids.

A **diprotic acid** *is an acid that yields two H$^+$ ions (two protons) per molecule upon dissociation.* Carbonic acid (H$_2$CO$_3$) is a diprotic acid. The dissociation process for a diprotic acid always occurs in steps. For H$_2$CO$_3$, the two steps are

monoprotic acid

diprotic acid

$$H_2CO_3 \rightleftharpoons H^+ + HCO_3^-$$
$$HCO_3^- \rightleftharpoons H^+ + CO_3^{2-}$$

The second proton is not as easily removed as the first because it must be pulled away from a negatively charged particle, HCO_3^-. (Remember, particles with opposite charge attract each other.) Accordingly, HCO_3^- is a weaker acid than H_2CO_3. In general, each successive step in the dissociation of an acid occurs to a lesser extent than the previous step.

A few triprotic acids exist. A **triprotic acid** *is an acid that yields three H^+ ions (three protons) per molecule upon dissociation.* Phosphoric acid (H_3PO_4) is the most common triprotic acid. The three dissociation steps involved in the removal of H^+ ions from this molecule are

triprotic acid

$$H_3PO_4 \rightleftharpoons H^+ + H_2PO_4^-$$
$$H_2PO_4^- \rightleftharpoons H^+ + HPO_4^{2-}$$
$$HPO_4^{2-} \rightleftharpoons H^+ + PO_4^{3-}$$

The general term **polyprotic acid** describes *acids that are capable of producing two or more H^+ ions (protons) per molecule upon dissociation.*

polyprotic acid

The number of hydrogen atoms present in one molecule of an acid cannot always be used to classify the acid as mono-, di-, or triprotic. For example, a molecule of acetic acid contains four hydrogen atoms and yet it is a monoprotic acid. Only one of the hydrogen atoms in acetic acid is *acidic*; that is, only one of the hydrogen atoms leaves the molecule when it is in solution.

Whether or not a hydrogen atom is acidic is related to its location in a molecule; that is, to which other atom it is bonded. Let us consider our acetic acid example in more detail by looking at the structures of the species involved in the dissociation process. From a structural viewpoint, the dissociation of acetic acid can be represented by the equation

$$\begin{array}{c} H\ \ O \\ |\ \ \ \| \\ H-C-C-O-H \\ | \\ H \end{array} \rightleftharpoons H^+ + \left[\begin{array}{c} H\ \ O \\ |\ \ \ \| \\ H-C-C-O \\ | \\ H \end{array}\right]^-$$

Note that one hydrogen atom is bonded to an oxygen atom; the other three hydrogen atoms are each bonded to a carbon atom. The hydrogen atom bonded to the oxygen atom is the acidic one. Why? The hydrogen atoms that are bonded to carbon atoms are too tightly held to be removed by reaction with water molecules. Water has very little effect on a carbon–hydrogen bond because it is essentially nonpolar. On the other hand, the hydrogen bonded to oxygen is involved in a very polar bond because of oxygen's large electronegativity (Section 6.11). Water, which is a polar molecule, readily attacks this bond.

The formula for acetic acid is always written as $HC_2H_3O_2$ instead of $C_2H_4O_2$. We now understand why; one of the hydrogen atoms is acidic and the other three are not. In situations such as this, where some hydrogen atoms are acidic and others are not, it is accepted procedure to always write the acidic hydrogens first, thus separating them from the other hydrogen atoms in the formula. Citric acid, the principal acid in citrus fruits, is another example of an acid that contains both acidic and nonacidic hydrogens. Its formula, $H_3C_6H_5O_6$, indicates that three of the eight hydrogen atoms present in a molecule are acidic. Table 10–3 (p. 232) contains information about the use and occurrence of selected common mono-, di-, and triprotic acids, many of which contain nonacidic hydrogen atoms.

We have focused our attention on acids in the preceding discussion. It should be noted that, in a similar manner, molecules of Arrhenius bases can be the source

TABLE 10-3 Selected Common Mono-, Di, and Triprotic Acids

Name	Formula	Classification	Common Occurrence
acetic	$HC_2H_3O_2$	monoprotic; weak	vinegar
lactic	$HC_3H_5O_3$	monoprotic; weak	sour milk, cheese; produced during muscle contraction
salicylic	$HC_7H_5O_3$	monoprotic; weak	present in chemically combined form in aspirin
hydrochloric	HCl	monoprotic; strong	constituent of gastric juice; industrial cleaning agent
nitric	HNO_3	monoprotic; strong	used in urinalysis test for protein; used in manufacture of dyes and explosives
tartaric	$H_2C_4H_4O_6$	diprotic; weak	grapes
carbonic	H_2CO_3	diprotic; weak	carbonated beverages; produced in the body from carbon dioxide
sulfuric	H_2SO_4	diprotic; strong	storage batteries; manufacture of fertilizer
citric	$H_3C_6H_5O_7$	triprotic; weak	citrus fruits
boric	H_3BO_3	triprotic; weak	antiseptic eyewash
phosphoric	H_3PO_4	triprotic; weak	found in dissociated form (HPO_4^{-2}, $H_2PO_4^-$) in intracellular fluid; component of DNA; fertilizer manufacture

of more than one hydroxide ion. For example, calcium hydroxide, $Ca(OH)_2$, is a base that yields two hydroxide ions per molecule when it dissociates in solution. Base dissociation differs from acid dissociation in that dissociation occurs in one step rather than stepwise.

10.4 SALTS

To a nonscientist, the term *salt* connotes a white granular substance that is used as a seasoning for food. To the chemist, the term *salt* has a much broader meaning; sodium chloride (table salt) is only one of thousands of salts known to a chemist. "Pass the salt" is a very ambiguous request to a chemist.

salts

Salts *are ionic compounds that contain any negative ion except hydroxide ion and any positive ion except hydrogen ion.* Salts, as a class of compounds, differ from acids and bases in that they contain no common ion or species that characterizes the class. The way that an acid, a base, and a salt are related is that a salt is one of the products resulting from the reaction of an acid with a base.

$$\underset{\text{base}}{Na}OH + H\underset{\text{acid}}{Cl} \longrightarrow \underset{\text{salt}}{NaCl} + H_2O$$

This particular type of reaction will be discussed in further detail in Section 10.6.
Many salts occur in nature and numerous others have been prepared in the laboratory. The wide variety of uses found for salts can be seen from Table 10-4. Much information concerning salts has been presented in previous chapters, although the term *salt* was not explicitly used in these discussions. Formula writing and nonmenclature for binary ionic compounds (salts) was covered in Sections 6.5 and 6.7. As can be seen from the examples in Table 10-4, many salts contain polyatomic ions such as nitrate and sulfate. These ions were discussed in Section 6.14. The solubility of ionic compounds (salts) in water was the topic of Section 8.10.

TABLE 10-4 Some Common Salts and Their Uses

Name	Formula	Uses
ammonium nitrate	NH_4NO_3	fertilizer; explosive manufacture
barium sulfate	$BaSO_4$	X-rays of gastrointestinal tract
calcium carbonate	$CaCO_3$	chalk; limestone
calcium chloride	$CaCl_2$	drying agent for removal of small amounts of water
iron(II) sulfate	$FeSO_4$	treatment for anemia
potassium chloride	KCl	"salt" substitute for low sodium diets
sodium chloride	NaCl	table salt; used as a de-icer (to melt ice)
sodium bicarbonate	$NaHCO_3$	ingredient in baking powder; stomach antacid
sodium hypochlorite	NaClO	bleaching agent
silver bromide	AgBr	light-sensitive material in photographic film
tin(II) fluoride	SnF_2	toothpaste additive

All common soluble salts are dissociated into ions in solution (Section 8.9). Even if a salt is only slightly soluble, the small amount that does dissolve completely dissociates. Thus, the terms *weak* and *strong*, which are used to qualitatively denote the percent dissociation of acids and bases, are not applicable to salts. We do not use the term *weak salt*.

10.5 ELECTROLYTES

Aqueous solutions in which ions are present are good conductors of electricity. In general, the greater the number of ions present, the better the solution conducts electricity. Acids, bases, and soluble salts all have the common characteristic of producing ions in solution; thus, they all produce solutions that conduct electricity. All three types of compounds are said to be electrolytes. An **electrolyte** *is a substance that dissolves in water to form a solution that conducts electricity.* The presence of ions (charged particles) causes the electrical conductivity.

electrolyte

Some substances, such as table sugar (sucrose), glucose, or isopropyl alcohol, do not produce ions in solution. These substances are called *nonelectrolytes*. A

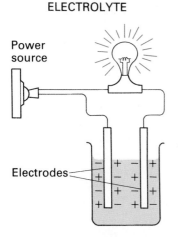
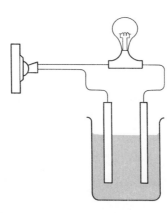

FIGURE 10-2 A simple device that can be used to distinguish between electrolyte and nonelectrolyte solutions. The light bulb glows for electrolyte solutions, but not for nonelectrolyte solutions.

nonelectrolyte

strong electrolyte

weak electrolyte

nonelectrolyte *is a substance that dissolves in water to form a solution that does not conduct electricity.*

Electrolytes can be divided into two groups—strong electrolytes and weak electrolytes. A **strong electrolyte** *is a substance that completely (or nearly completely) dissociates into ions in solution.* These electrolytes give strongly conducting solutions. All strong acids and strong bases and all soluble salts are strong electrolytes. A **weak electrolyte** *is a substance that only partially dissociates into ions in solution.* These electrolytes give solutions that are intermediate between those containing strong electrolytes and those containing nonelectrolytes in their ability to conduct an electric current. Weak acids, weak bases, and slightly soluble salts constitute the weak electrolytes.

You can determine whether or not a substance is an electrolyte in solution by testing the ability of the solution to conduct an electric current. A device such as that shown in Figure 10–2 (p. 233) can be used to distinguish between strong, weak and nonelectrolytes. If the medium between the electrodes (the solution) is a conductor of electricity, the light bulb glows. A strong glow indicates a strong electrolyte. A faint glow of the light bulb occurs for a weak electrolyte and there is no glow for a nonelectrolyte.

The fact that the presence of ions causes a solution to conduct electricity is of extreme biological significance. For example, messages to and from the brain are sent in the form of electrical signals. Ions in cellular and intercellular fluids are often the carriers of these signals. The presence of electrolytes (ions) is essential to the proper functioning of a human body.

The principal ions derived from electrolytes that are present in body fluids are shown in Figure 10–3. Within cells, K^+ is the predominant positive ion and HPO_4^{2-} is the principal negative ion. Outside of cells (interstitial fluid and blood plasma), Na^+ is the most abundant positive ion and Cl^- is the most abundant negative ion.

FIGURE 10–3 The principal electrolyte-derived ions present in body fluids.

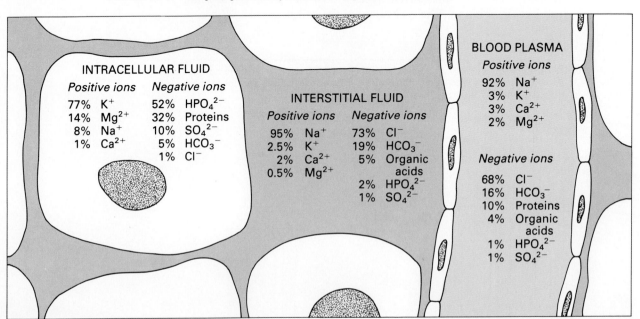

10.6 NEUTRALIZATION

When acids and bases are mixed, they react with each other; their acidic and basic properties disappear, and we say that they have *neutralized* each other. **Neutralization** *is the reaction between an acid and a base to form a salt and water.* The hydrogen ions from the acid combine with the hydroxide ions from the base to form water. The salt that is formed contains the negative ion from the acid and the positive ion from the base.

neutralization

Any time an acid is completely reacted with a base, neutralization occurs. It does not matter whether the acid and base are strong or weak. Sodium hydroxide (a strong base) and nitric acid (a strong acid) react as follows.

$$HNO_3 + NaOH \longrightarrow NaNO_3 + H_2O$$

The equations for the reaction of potassium hydroxide (a strong base) with hydrocyanic acid (a weak acid) are

$$HCN + KOH \longrightarrow KCN + H_2O$$

Note that in both reactions, the products are a salt ($NaNO_3$ in the first reaction and KCN in the second) and water.

10.7 ION FORMATION IN PURE WATER

Although we usually think of water as a covalent, nonionizing substance, experiments show that in reality, an *extremely small* percentage of water molecules in pure water dissociate to form ions. This dissociation reaction can be thought of as the transfer of a proton from one water molecule to another (Brønsted-Lowry theory, Section 10.1)

$$H_2O + H_2O \rightleftharpoons H_3O^+ + OH^-$$

or simply as the dissociation of a single water molecule (Arrhenius theory, Section 10.1).

$$H_2O \text{ (HOH)} \rightleftharpoons H^+ + OH^-$$

From either viewpoint, the net result is the formation of *equal amounts* of hydrogen (hydronium) ion and hydroxide ion.

At any given time, the number of H^+ and OH^- ions present in a sample of pure water is always extremely small. At equilibrium, at 25°C, the H^+ and OH^- concentrations are 1.00×10^{-7} M (0.000000100 M).

The constant concentration of H^+ and OH^- ions present in pure water at 25°C can be used to calculate a very useful relationship called the *ion product for water*. The **ion product for water** *is the numerical value* 1.00×10^{-14} *obtained by multiplying together the molar concentrations of* H^+ *and* OH^- *ion present in pure water.* We have the following equations for the ion product of water.

ion product for water

$$\text{ion product of water} = [H^+] \times [OH^-]$$
$$= (1.00 \times 10^{-7}) \times (1.00 \times 10^{-7})$$
$$= 1.00 \times 10^{-14}$$

Whenever square brackets are used in an equation, as in the equation for the ion product of water, they should be interpreted to mean concentration in moles per liter (molarity).

The ion-product expression for water is valid not only in pure water but also when solutes are present in water. At all times, the product of the hydrogen and

FIGURE 10-4 The relationship between H^+ and OH^- concentrations in an aqueous solution.

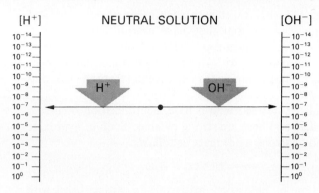

(a) In pure water the concentrations of hydrogen ions, $[H^+]$, and hydroxide ions, $[OH^-]$, are equal. Both are 1.00×10^{-7} M at 25°C.

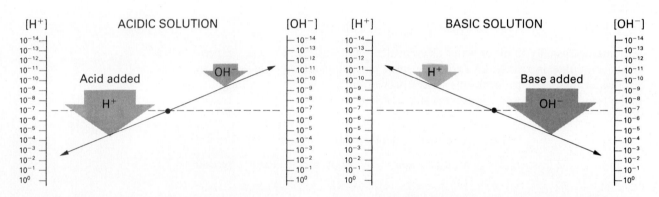

(b) If the $[H^+]$ is increased by a factor of 10^5 (from 10^{-7} M to 10^{-2} M), the $[OH^-]$ is decreased by a factor of 10^5 (from 10^{-7} M to 10^{-12} M).

(c) If the $[OH^-]$ is increased by a factor of 10^5 (from 10^{-7} M to 10^{-2} M), the $[H^+]$ is decreased by a factor of 10^5 (from 10^{-7} M to 10^{-12} M).

hydroxide ion molarities in an aqueous solution at 25°C must equal 1.00×10^{-14}. Thus, if the $[H^+]$ is increased by the addition of an acidic solute, the $[OH^-]$ must decrease so that their product will still be 1.00×10^{-14}. Similarly, if additional OH^- ions are added to the water, the $[H^+]$ must correspondingly decrease.

If we know the concentration of the other ion, we can easily calculate the con-

TABLE 10-5 Relationships Between $[H^+]$ and $[OH^-]$ in Aqueous Solutions at 25°C.

neutral solution:	$[H^+] = [OH^-] = 1.00 \times 10^{-7}$
acidic solution: $[H^+] > [OH^-]$	$[H^+]$ is greater than 1.00×10^{-7} $[OH^-]$ is less than 1.00×10^{-7}
basic solution: $[OH^-] > [H^+]$	$[H^+]$ is less than 1.00×10^{-7} $[OH^-]$ is greater than 1.00×10^{-7}

centration of either H^+ or OH^- ion present in an aqueous solution by simply rearranging the ion-product expression.

$$[H^+] = \frac{1.00 \times 10^{-14}}{[OH^-]}$$

$$[OH^-] = \frac{1.00 \times 10^{-14}}{[H^+]}$$

Example 10.1

Sufficient acidic solute is added to a quantity of water to produce a $[H^+] = 4.0 \times 10^{-3}$. What is the $[OH^-]$ in this solution?

Solution
The $[OH^-]$ can be calculated using the ion-product expression for water, rearranged in the form

$$[OH^-] = \frac{1.00 \times 10^{-14}}{[H^+]}$$

Substituting into this expression the known $[H^+]$ and doing the arithmetic gives

$$[OH^-] = \frac{1.00 \times 10^{-14}}{4.0 \times 10^{-3}} = 2.5 \times 10^{-12}$$

The interdependence of $[H^+]$ and $[OH^-]$—if one increases the other decreases—is illustrated diagrammatically in Figure 10-4. Note that the increase–decrease relationship is a linear one. If the $[H^+]$ increases by a factor of 10^2, then the $[OH^-]$ decreases by the same factor, 10^2.

A small amount of H^+ ion and OH^- ion are present in all aqueous solutions, whether they are basic or acidic. What then determines whether a solution is acidic or basic? It is the relative amounts of these two ions present. An **acidic solution** *is one in which the concentration of* H^+ *ion is higher than that of* OH^- *ion.* Similarly, a **basic solution** *is one in which the concentration of the* OH^- *ion is higher than that of the* H^+ *ion.* A basic solution is also often referred to as an *alkaline solution*. It is possible to have an aqueous solution that is neither acidic nor basic, but is a neutral solution. A **neutral solution** *is one in which the concentrations of* H^+ *and* OH^- *ions are equal.* Table 10-5 summarizes the relationships between $[H^+]$ and $[OH^-]$ that we have just considered.

acidic solution

basic solution

neutral solution

10.8 THE pH SCALE

Hydrogen-ion concentrations in aqueous solution range from relatively high values (10 M) to extremely small ones (10^{-14} M). It is inconvenient to work with numbers that extend over such a wide range; a hydrogen-ion concentration of 10 M is 1000 trillion times larger than a hydrogen-ion concentration of 10^{-14} M. The pH scale is a more practical way to handle such a wide range of numbers. The **pH scale** *is a scale of small numbers that are used to specify molar hydrogen-ion concentration in an aqueous solution.*

pH scale

The pH value for a solution whose molar hydrogen-ion concentration is an exact power of ten, such as 1.0×10^{-4}, 1.0×10^{-8}, or 1.0×10^{-11}, is very easy to calculate. In this situation, the pH is given directly by the negative of the exponent value on the power of ten.

$$[H^+] = 1.0 \times 10^{-x}$$
$$pH = x$$

Thus, if the hydrogen-ion concentration is 1.0×10^{-9}, then the pH will be 9.00. Note that the preceding simple relationship between pH and $[H^+]$ is valid only when the coefficient in the exponential expression for the hydrogen-ion concentration is 1.0. We will comment on how to calculate the pH when the coefficient is not 1.0 later in this section.

Example 10.2

Calculate the pH of each of the following solutions.
a. $[H^+] = 1.0 \times 10^{-10}$ M
b. $[H^+] = 0.001$ M

Solution
a. The power of ten associated with the molar hydrogen-ion concentration is -10. Thus, the pH will be 10.0, which is the negative of the exponential power.
b. We must first express the given molar hydrogen-ion concentration in scientific notation.

$$0.001 = 1 \times 10^{-3}$$

The power of ten associated with the molar hydrogen-ion concentration is -3. Thus, the pH will be 3.0, which is the negative of the exponential power.

Acidic, neutral, and basic solutions can be identified by their pH values because pH is simply another way of expressing hydrogen-ion concentration. At 25°C, a neutral solution has a pH value of 7. Values of pH that are less than 7 correspond to acidic solutions and values of pH that are greater than 7 are associated with basic solutions. The relationships between $[H^+]$, $[OH^-]$, and pH are

TABLE 10–6 Relationships Between pH Values and $[H^+]$ and $[OH^-]$ at 25°C.

pH	$[H^+]$	$[OH^-]$	
0	1	10^{-14}	
1	10^{-1}	10^{-13}	
2	10^{-2}	10^{-12}	
3	10^{-3}	10^{-11}	Acidic
4	10^{-4}	10^{-10}	
5	10^{-5}	10^{-9}	
6	10^{-6}	10^{-8}	
7	10^{-7}	10^{-7}	NEUTRAL
8	10^{-8}	10^{-6}	
9	10^{-9}	10^{-5}	
10	10^{-10}	10^{-4}	
11	10^{-11}	10^{-3}	Basic
12	10^{-12}	10^{-2}	
13	10^{-13}	10^{-1}	
14	10^{-14}	1	

TABLE 10–7 The Normal pH Range of Some Selected Body Fluids

Type of fluid	pH value
bile	6.8–7.0
blood plasma	7.3–7.5
gastric juices	1.0–3.0
milk	6.6–7.6
saliva	6.5–7.5
spinal fluid	7.3–7.5
urine	4.8–8.4

summarized in Table 10–6. Note the following trends from the information in Table 10–6:

1. The higher the concentration of hydrogen ion, the smaller the pH value. Another statement of this same trend is that lowering the pH always corresponds to increasing the hydrogen-ion concentration.
2. A change of one unit in pH always corresponds to a tenfold change in hydrogen-ion concentration. For example,

$$\text{difference of 1} \begin{cases} \text{pH} = 1, \text{then } [H^+] = 0.1 \text{ M} \\ \text{pH} = 2, \text{then } [H^+] = 0.01 \text{ M} \end{cases} \text{tenfold difference}$$

In a laboratory, solutions corresponding to any pH can be created. The range of pH values that are displayed by natural solutions is more limited than that of prepared solutions, but solutions corresponding to most pH values can be found. Some natural solutions have nearly constant pH values. For example, human blood plasma has a pH value between 7.3 and 7.5. Any significant deviation from this range—a few tenths of a pH unit—can cause severe physiological problems.

The pH values of several human body fluids are given in Table 10–7. Most human body fluids have pH values in a narrow range centered around neutrality, except for gastric juices. Both blood plasma and spinal fluid are always slightly basic fluids.

Most of the pH values found in Table 10–7 are nonintegral; that is, they are not whole numbers. Obtaining these pH values from measured hydrogen-ion concentrations involves the use of logarithms, a mathematical concept we will not deal with in this text. However, when given nonintegral pH values, we can still interpret them in a qualitative way without resorting to logarithm use. For example, a pH of 4.75 is between a pH of 4 and a pH of 5. We know that a pH of 4 corresponds to $[H^+] = 1 \times 10^{-4}$ and that a pH of 5 corresponds to $[H^+] = 1 \times 10^{-5}$. Hence, a pH of 4.75 will correspond to a hydrogen-ion concentration that is between 1×10^{-4} and 1×10^{-5}.

10.9 ACID–BASE TITRATIONS

The concentration of an acid or base in a solution and the solution's pH are two different entities. The pH of a solution gives information about the *concentration* of hydrogen (hydronium) ions in solution. Only dissociated molecules influence the pH value. The concentration of an acid or base gives information about the *total number* of acid or base molecules present; both dissociated and undissociated molecules are counted. Thus, acid or base concentration is a measure of total acidity or total basicity.

acid–base titration

The procedure most frequently used to determine the concentration of an acid or base solution is an acid–base titration. In an **acid–base titration**, *a measured volume of an acid or a base of known concentration is exactly reacted with a measured volume of a base or acid of unknown concentration.*

Suppose we want to determine the concentration of an acid solution by means of titration. We first measure out a known volume of the acid solution into a beaker or flask. Then, we slowly add a solution of base of known concentration to the flask or beaker by means of a buret (see Figure 10–5). We continue to add base until all the acid has completely reacted with the added base. The volume of base needed to reach this point is obtained from the buret readings. When we know the original volume of acid, the concentration of the base, and the volume of added base, we can calculate the concentration of the acid (Section 10.10).

In order to complete a titration successfully, we must be able to detect when the reaction between acid and base is complete. Neither the acid nor the base gives any outward sign that the reaction is complete. Thus, an indicator is always added to the reaction mixture. An **indicator** *is a compound that exhibits different colors depending on the* pH *of its surroundings.* Typically, an indicator is one color in basic solutions and another color in acidic solutions. An indicator is selected that will

indicator

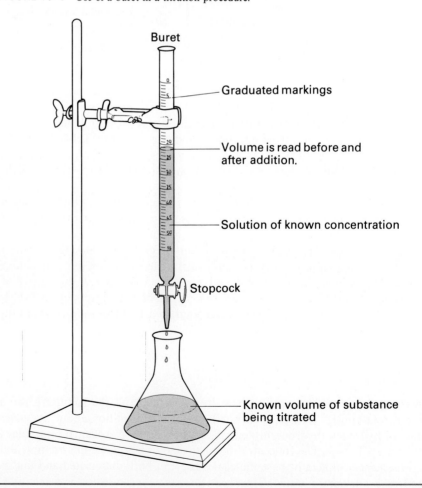

FIGURE 10–5 Use of a buret in a titration procedure.

FIGURE 10–6 Colors exhibited by various indicators as a function of pH value.

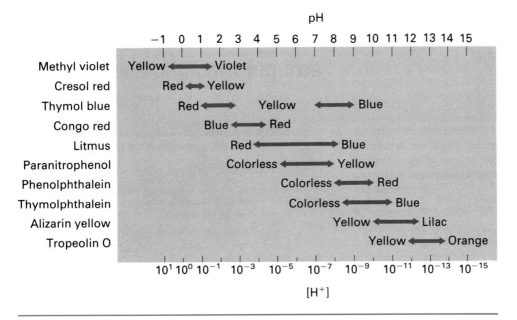

change color at a pH that corresponds as nearly as possible to the pH of the solution when the titration is complete. If the acid and base are both strong, the pH at that point is 7. However, for reasons to be discussed in Section 10.11, the pH is not 7 if a weak acid or weak base is part of the titration system. The colors shown by a number of indicators over various pH ranges are given in Figure 10–6.

10.10 ACID–BASE TITRATION CALCULATIONS

The concentration unit *normality* is the most convenient unit to work with when doing acid–base titration calculations. This is a concentration unit that is very closely related to molarity. The difference between the two units involves the expression of the amount of solute present. Moles of solute are specified for molarity and equivalents of solute are specified for normality. The **normality** *of a solution, which is abbreviated N, is a ratio giving the number of equivalents of solute per liter of solution.*

normality

$$\text{normality (N)} = \frac{\text{equivalents of solute}}{\text{liters of solution}}$$

Normality takes into account the fact that all acids and bases do not contain the same number of hydrogen or hydroxide ions per molecule. Even though their molar concentrations are equal, 0.10 M H_2SO_4 (which can supply two H^+ ions per molecule) will neutralize twice as much base as the same volume of 0.10 M HNO_3 (which can supply only one H^+ ion per molecule). However, equal volumes of 0.10 N H_2SO_4 and 0.10 N HNO_3 will neutralize exactly the same amount of base.

The key to comprehending normality concentration units is an understanding of the concept of equivalents. The term *equivalents* has been previously encountered in the text. In Section 8.14, an equivalent was defined in a way that was applicable for dealing with ionic salts. Definitions of *equivalent* that are applicable for acid–base work are different. An **equivalent of acid** is the quantity of acid that will supply one mole of H^+ ion in a chemical reaction. Similarly, an **equivalent of base** is the quantity of base that will react with one mole of H^+ ions. Examples 10.3 and 10.4 illustrate the use of these two definitions in problem-solving contexts.

equivalent of acid

equivalent of base

Example 10.3

Determine the number of equivalents of acid or base in each of the following samples.

a. 1 mole of H_3PO_4 b. 2 moles of $Ba(OH)_2$ c. 3 moles of HNO_3

Solution

a. The equation for the complete dissociation of H_3PO_4 is

$$H_3PO_4 \longrightarrow 3H^+ + PO_4^{3-}$$

Thus, one mole of H_3PO_4 gives up three moles of H^+ ion on complete dissociation. Three moles of H^+ ion is equal to three equivalents of H^+ ion by definition. Therefore,

$$1 \text{ mole } H_3PO_4 \times \left(\frac{3 \text{ equiv } H_3PO_4}{1 \text{ mole } H_3PO_4}\right) = 3 \text{ equiv } H_3PO_4$$

b. The equation for the complete dissociation of $Ba(OH)_2$ is

$$Ba(OH)_2 \longrightarrow Ba^{2+} + 2\,OH^-$$

Thus, one mole of $Ba(OH)_2$ yields two moles of OH^-. An equivalent of OH^- ion is defined in terms of reaction with H^+ ion. Because one mole of OH^- ion will always react with one mole of H^+ ion, the number of moles of OH^- ion and the number of equivalents of OH^- ion will be the same. One mole of $Ba(OH)_2$ will produce two equivalents of OH^- ion upon complete dissociation. Therefore,

$$2 \text{ moles } Ba(OH)_2 \times \left(\frac{2 \text{ equiv } Ba(OH)_2}{1 \text{ mole } Ba(OH)_2}\right) = 4 \text{ equiv } Ba(OH)_2$$

c. The equation for the complete dissociation of HNO_3 is

$$HNO_3 \longrightarrow H^+ + NO_3^-$$

One mole of acid yields only one mole (or equiv) of H^+ ion. Therefore,

$$3 \text{ moles } HNO_3 \times \left(\frac{1 \text{ equiv } HNO_3}{1 \text{ mole } HNO_3}\right) = 3 \text{ equiv } HNO_3$$

Example 10.4

Calculate the normality of a solution made by dissolving 25.0 g of H_3BO_3 in enough water to give 500.0 mL of solution.

Solution

In order to calculate normality, we need to know the number of equivalents of solute present and the solution volume in liters.

For H_3BO_3

$$1 \text{ mole} = 3 \text{ equivalents}$$

because one mole of H_3BO_3 yields three moles of H^+ ions.

$$H_3BO_3 \longrightarrow 3H^+ + BO_3^{3-}$$

The number of equivalents of H_3BO_3 in 25.0 g of H_3BO_3 is obtained by converting the grams of H_3BO_3 to moles of H_3BO_3 and then using the preceding mole–equivalent relationship as a conversion factor to obtain equivalents.

$$25.0 \text{ g } H_3BO_3 \times \left(\frac{1 \text{ mole } H_3BO_3}{61.8 \text{ g } H_3BO_3}\right) \times \left(\frac{3 \text{ equiv } H_3BO_3}{1 \text{ mole } H_3BO_3}\right) = 1.21 \text{ equiv } H_3BO_3$$

The solution volume of 500.0 mL, changed to liter units, is 0.5000 L.

The quantities called for in the defining equation for normality, which are equivalents of solute and liters of solution, are now known. Therefore, substituting into the defining equation, we get

$$N = \frac{1.21 \text{ equiv } H_3BO_3}{0.5000 \text{ L solution}} = \frac{2.42 \text{ equiv } H_3BO_3}{1 \text{ L solution}}$$

$$= 2.42 \text{ N}$$

In the laboratory, solution concentrations are usually specified in terms of molarity rather than normality. Conversion of acid or base molarities to normalities (or vice versa) is a very simple operation. Only the conversion factor relating moles to equivalents is needed.

Express the following molar concentrations as normalities.

a. 2.00 M H_2SO_4 b. 0.30 M $Ba(OH)_2$ c. 6.0 M HNO_3

Example 10.5

Solution

a. A 2.00 M H_2SO_4 solution contains 2.00 moles of H_2SO_4 per liter of solution. Using this fact and the fact that

$$1 \text{ mole } H_2SO_4 = 2 \text{ equiv } H_2SO_4$$

we calculate the normality as follows.

$$\left(\frac{2.00 \text{ mole } H_2SO_4}{1 \text{ L solution}}\right) \times \left(\frac{2 \text{ equiv } H_2SO_4}{1 \text{ mole } H_2SO_4}\right) = \frac{4.00 \text{ equiv } H_2SO_4}{1 \text{ L solution}}$$

$$= 4.00 \text{ N}$$

b. A 0.30 M $Ba(OH)_2$ solution contains 0.30 mole of $Ba(OH)_2$ per liter of solution. This information, coupled with the fact that

$$1 \text{ mole } Ba(OH)_2 = 2 \text{ equiv } Ba(OH)_2$$

enables us to calculate the normality as follows.

$$\left(\frac{0.30 \text{ mole } Ba(OH)_2}{1 \text{ L solution}}\right) \times \left(\frac{2 \text{ equiv } Ba(OH)_2}{1 \text{ mole } Ba(OH)_2}\right) = \frac{0.60 \text{ equiv } Ba(OH)_2}{1 \text{ L solution}}$$

$$= 0.60 \text{ N}$$

c. For 6.0 M HNO_3, we calculate the normality, using reasoning similar to that in parts (a) and (b), as follows.

$$\left(\frac{6.0 \text{ moles } HNO_3}{1 \text{ L solution}}\right) \times \left(\frac{1 \text{ equiv } HNO_3}{1 \text{ mole } HNO_3}\right) = \frac{6.0 \text{ equiv } HNO_3}{1 \text{ L solution}}$$

$$= 6.0 \text{ N}$$

Molarities can also be converted to normalities without involvement with dimensional analysis by using the following equations.

$$N_{acid} = M_{acid} \times \text{number of acidic hydrogens in the acid}$$
$$N_{base} = M_{base} \times \text{number of hydroxide ions in the base}$$

Both of these equations are based on the Arrhenius system for defining acids and bases. It is not recommended that you use these short-cut equations until you thoroughly understand the dimensional-analysis approach to obtaining normalities (Example 10.5).

In an acid–base titration, the chemical reaction of neutralization takes place (Section 10.6). The H^+ ions from the acid react with the OH^- ions from the base to produce water.

$$H^+ + OH^- \longrightarrow H_2O$$

These two ions always react in a one-to-one ratio; thus, one mole of H^+ ions always reacts with one mole of OH^- ions. Because one mole of each of these ions is an equivalent, it follows that at the end point of an acid–base titration, an equal number of equivalents of acid and base have been consumed.

$$\text{equivalents of reacted acid} = \text{equivalents of reacted base}$$

A simple equation, whose derivation is based on the fact that equal numbers of equivalents of acid and base are consumed during an acid–base titration, is employed in titration calculations.

$$N_{acid} \times V_{acid} = N_{base} \times V_{base}$$

where N stands for normality and V stands for volume.

The derivation of this equation is as follows. The defining equations for the normality of the acid and base are

$$N_{acid} = \frac{E_{acid}}{V_{acid}} \quad \text{and} \quad N_{base} = \frac{E_{base}}{V_{base}}$$

where E is the equivalents of solute and V is the volume of solution (in liters). Each of these equations can be rearranged as follows

$$E_{acid} = N_{acid} \times V_{acid}$$
$$E_{base} = N_{base} \times V_{base}$$

At the endpoint of the titration, where $E_{acid} = E_{base}$, we can write, by substitution

$$N_{acid} \times V_{acid} = N_{base} \times V_{base}$$

When using this equation, the volumes V_{acid} and V_{base} can be expressed in liters or milliliters, provided the same unit is used for both. Burets are calibrated in milliliters, so milliliters is the unit most often used in titration calculations.

The data obtained from an acid–base titration (Section 10.9) are (1) the volume of base used, (2) the volume of acid used, and (3) the concentration of the acid or base used to titrate the solution of unknown concentration. These quantities comprise three of the four variables in the equation

$$N_{acid} \times V_{acid} = N_{base} \times V_{base}$$

The fourth variable, the concentration of the acid or base that was titrated, can easily be calculated.

Example 10.6

In an acid–base titration, 17.3 mL of 0.126 M NaOH is required to neutralize 25.0 mL of H_2SO_4. What are the normality and the molarity of the H_2SO_4 solution?

Solution

We will use the equation

$$N_{acid} \times V_{acid} = N_{base} \times V_{base}$$

to solve this problem. Three of the four variables in this equation are known. The volumes of acid and base are 25.0 and 17.3 mL, respectively. The normality of the base is not directly given, but it can easily be calculated from the molarity of the base. One mole of NaOH contains one equivalent of OH^- ions, so the normality of the base will be the same as the molarity; that is, the base is 0.126 M and also 0.126 N.

Substituting these known values into the equation and rearranging so that N_{acid} is isolated on the left side, gives

$$N_{acid} = \frac{N_{base} \times V_{base}}{V_{acid}}$$

$$= 0.126 \text{ N} \times \left(\frac{17.3 \text{ mL}}{25.0 \text{ mL}}\right) = 0.0872 \text{ N}$$

The molarity of H_2SO_4 can be calculated from the normality. For H_2SO_4

$$1 \text{ mole } H_2SO_4 = 2 \text{ equiv } H_2SO_4$$

$$\left(\frac{0.0872 \text{ equiv } H_2SO_4}{1 \text{ L solution}}\right) \times \left(\frac{1 \text{ mole } H_2SO_4}{2 \text{ equiv } H_2SO_4}\right) = \frac{0.0436 \text{ mole } H_2SO_4}{1 \text{ L solution}}$$

$$= 0.0436 \text{ M}$$

10.11 HYDROLYSIS OF SALTS

Salts contain neither H^+ ions nor OH^- ions (Section 10.4). Consequently, we might expect that aqueous solutions of salts would be neutral (pH = 7). Surprisingly, many salt solutions are not neutral; they have pH values greater than 7 (basic) or less than 7 (acidic). How can this be? In certain solutions, a phenomenon called *salt hydrolysis* causes pH changes. **Salt hydrolysis** *is a reaction in which a salt interacts with water to produce an acidic or basic solution.* Salt hydrolysis reactions are important in numerous settings, including many involving the human body.

salt hydrolysis

In order for a solution to be acidic or basic, an excess of H^+ ion or OH^- ion must be present (Section 10.7). An explanation of why the addition of a salt to a solution can cause an excess of H^+ or OH^- ions involves a consideration of the following points.

1. In any aqueous salt solution, there are four, rather than two ions present. Positive and negative ions are produced by the salt as it dissolves (Section 10.4). Also present are a small number (1×10^{-7} M) of H^+ and OH^- ions, which are produced by the dissociation of water (Section 10.7). The number of H^+ and OH^- ions are equal because dissociation of a water molecule always produces one ion of each type.

$$H_2O \longrightarrow H^+ + OH^-$$

2. The positive ion from the salt reacts with the OH^- ions from the water under certain conditions. The stimulus (or driving force) for such a reaction is the production of a weak base. These substances exist predominantly in molecular form. Weak base formation reduces the amount of free OH^- ion present; the OH^- ions are tied up in molecules. The H^+ ions present in the solution are unaffected by this weak base formation and outnumber the OH^- ions; this situation is characteristic of an acidic solution.
3. The negative ion from the salt reacts with the H^+ ions from the water under certain conditions. The driving force for such a reaction is the production of a weak acid. Like weak bases, weak acids are substances that exist predominantly in molecular form. Thus, weak acid formation reduces the amount of free H^+ ion present; the H^+ ions are tied up in molecules. The OH^- ions present in the solution are unaffected by this weak acid formation and outnumber the H^+ ions; this situation is characteristic of a basic solution.
4. The trace amounts of H^+ and OH^- ion that are initially present in an aqueous salt solution are part of an equilibrium system.

$$H_2O \rightleftharpoons H^+ + OH^-$$

Reaction of either of these ions with the dissolved salt (hydrolysis) upsets the equilibrium. The stress is the removal (decrease in concentration) of one or both of these ions through weak acid and/or weak base formation. According to Le Châtelier's principle (Section 9.6), the equilibrium will shift to the right. This response to the stress produces more H^+ and OH^- ions that are available for continuation of the hydrolysis process. Hydrolysis continues until the weak acid or weak base being formed reaches an equilibrium condition.

Using the preceding considerations, let us now consider the hydrolysis of a specific salt, NaCN. Will the hydrolysis of this salt produce an acidic or basic solution? We will first write the formulas of the four ions present in the solution.

$$Na^+ \quad CN^-$$
$$OH^- \quad H^+$$

Will Na^+ ions tie up OH^- ions? The answer is "no" because NaOH is a strong base, and strong bases exist in ionic form in solution (Section 10.2). Will CN^- ions tie up H^+ ions? The answer is "yes" because HCN is a weak acid. (Recall the strong and weak acids and bases that were discussed in Section 10.2). After hydrolysis, the predominant species present in solution will be

$$Na^+$$
$$OH^- \quad HCN$$

After hydrolysis, the solution will be basic. There are free OH^- ions present, while most of the H^+ ions have been converted to HCN molecules.

Example 10.7

Predict whether solutions of each of the following soluble salts will be acidic, basic, or neutral.

a. sodium acetate, $NaC_2H_3O_2$
b. ammonium chloride, NH_4Cl
c. potassium chloride, KCl
d. ammonium fluoride, NH_4F

Solution
A familarity with the strengths of common acids and bases (weak or strong) is the key to solving hydrolysis problems. The strengths of acids and bases was the topic of Section 10.2.

a. The four ions present in solution are

$$Na^+ \quad C_2H_3O_2^-$$
$$OH^- \quad H^+$$

The Na^+ ion will not hydrolyze because the base that would be formed, NaOH, is a strong base. The $C_2H_3O_2^-$ ion does hydrolyze, forming the weak acid $HC_2H_3O_2$.

$$Na^+$$
$$OH^- \quad HC_2H_3O_2$$

The solution will be basic because of the presence of the free OH^- ions.

b. This time, the four ions present in solution are

$$NH_4^+ \quad Cl^-$$
$$OH^- \quad H^+$$

The NH_4^+ ion will hydrolyze because this hydrolysis produces the weak base aqueous ammonia. The Cl^- ion does not hydrolyze because HCl is a strong acid. The solution will be acidic. The OH^- ions have been tied up as the result of aqueous ammonia formation, but the H^+ ions are still "free".

c. Potassium chloride is a salt that does not hydrolyze. The four ions initially present are

$$K^+ \quad Cl^-$$
$$OH^- \quad H^+$$

The possible products from the hydrolysis are KOH and HCl; KOH is a strong base and HCl is a strong acid. There is no driving force for hydrolysis and the solution remains neutral.

d. It is possible for both the positive and negative ions from a salt to undergo hydrolysis at the same time. This is the situation for NH_4F. The four ions initially present are

$$NH_4^+ \quad F^-$$
$$OH^- \quad H^+$$

The hydrolysis products are aqueous ammonia (a weak base) and hydrofluoric acid (a weak acid). In a double hydrolysis situation such as this, it is not easy to predict whether the resulting solution will be acidic or basic. It depends on the relative strengths of the weak acid and weak base produced. This type of strength comparison is beyond the scope of this text, and you will not be expected to predict acidity or basicity in double hydrolysis situations. However, you should be able to identify double hydrolysis situations.

A set of simple rules for determining whether or not a salt will hydrolyze are easily formulated if salts are visualized as being the products of the neutralization reaction between an acid and a base (Section 10.6).

$$\text{acid} + \text{base} \longrightarrow \text{salt} + \text{water}$$

From this viewpoint, four categories of salts exist. They are formed from the reaction of:

1. a strong acid and a strong base
2. a strong acid and a weak base

3. a weak acid and a strong base
4. a weak acid and a weak base

Salts that will hydrolyze are obtained from a neutralization that involves either a weak acid, a weak base, or both. Table 10–8 gives further information about the hydrolytic behavior of the various types of salts. Note from the table that in all cases, the ions that hydrolyze are those that were originally part of a weak acid or weak base. Category 2 salts always produce acidic solutions; weak base formation is the driving force for the hydrolysis. Category 3 salts always produce basic solutions; weak acid formation is the driving force for the hydrolysis. Category 4 salts always give a double hydrolysis situation; both a weak acid and a weak base are formed.

Generally, hydrolysis reactions take place only to a small extent, about 2–5%. However, in some cases, this is sufficient to cause very significant changes in pH, as can be seen from the pH values in Table 10–9.

The fact that some salts hydrolyze is of extreme physiological importance. Very slight changes in pH can alter the activity of enzymes, which are the catalysts for many body reactions.

Blood plasma has a slightly basic pH (7.3–7.5)—as shown in Table 10–7. The reason for this is related to hydrolysis and becomes apparent on inspection of the plasma electrolyte composition that was previously given in Figure 10–3.

The most abundant positive ion (Na^+) comes from a strong base. The predominant negative ion (Cl^-) comes from a strong acid. Together, these two ions would produce a neutral solution because neither hydrolyzes. The third most abundant ion is HCO_3^-, which comes from the weak acid H_2CO_3. Hydrolysis of this ion will tie up hydrogen ions and leave hydroxide in excess. Thus, the plasma fluid has a slightly basic character. Other negative ions such as HPO_4^{2-} will also hydrolyze and add to the basic character. However, their effect on the pH will not be as great as that of HCO_3^- because of their lower concentrations.

10.12 BUFFER SOLUTIONS

As we discussed in the last section, certain salts that hydrolyze can change the pH of water. In this section, we will consider a second phenomenon involving salts and their effect on solution pH. Certain *combinations* of compounds can protect the pH of a solution from change. These combinations, which always involve

TABLE 10–8 Hydrolytic Characteristics of Various Types of Salts

Type of Salt	Hydrolysis Characteristics	pH of Solution
1. negative ion from strong acid / positive ion from strong base	neither ion hydrolyzes	neutral
2. negative ion from strong acid / positive ion from weak base	positive ion hydrolyzes	acidic
3. negative ion from weak acid / positive ion from strong base	negative ion hydrolyzes	basic
4. negative ion from weak acid / positive ion from weak base	both positive and negative ions hydrolyze	acidic, basic, or neutral depending on the relative strengths of the weak acid and base

TABLE 10–9 Approximate pH of Selected 0.1 M Aqueous Salt Solutions

Name of Salt	Formula of Salt		Category of Salt
ammonium nitrate	NH_4NO_3	5.1	strong acid–weak base
ammonium nitrite	NH_4NO_2	6.3	weak acid–weak base
ammonium acetate	$NH_4C_2H_3O_2$	7.0	weak acid–weak base
sodium chloride	NaCl	7.0	strong acid–strong base
sodium fluoride	NaF	8.1	weak acid–strong base
sodium acetate	$NaC_2H_3O_2$	8.9	weak acid–strong base
ammonium cyanide	NH_4CN	9.3	weak acid–weak base
sodium cyanide	NaCN	11.1	weak acid–strong base

at least one salt, are called *buffers* and the solutions containing them are called *buffer solutions*. A **buffer solution** *is a solution that resists a change in pH when small amounts of acid or base are added to it.* A **buffer** *is the solute (or solutes) present in a buffer solution that cause it to be pH-change resistant.*

buffer solution

buffer

A buffer solution is simply a solution containing substances that have the ability to react with either H^+ or OH^- and remove them from solution. Many common buffers are a combination of a weak acid and one of its salts or a weak base and one of its salts.

As an illustration of buffer action, consider a buffer solution containing approximately equal concentrations of acetic acid (a weak acid) and sodium acetate (a salt of this weak acid). This solution resists pH change by the following mechanisms.

1. When a small amount of a strong acid such as HCl is added to the solution, the newly added H^+ ions react with the acetate ions from the sodium acetate to give acetic acid.

$$H^+ + C_2H_3O_2^- \longrightarrow HC_2H_3O_2$$

   Most of the added H^+ ions are tied up in acetic acid molecules, and the pH changes very little.

2. When a small amount of a strong base such as NaOH is added to the solution, the newly added OH^- ions react with the acetic acid (neutralization) to give acetate ions and water.

$$OH^- + HC_2H_3O_2 \longrightarrow C_2H_3O_2^- + H_2O$$

   Most of the added OH^- ions are converted to water and the pH changes only slightly.

The reactions that are responsible for the buffering action in the acetic acid/acetate ion system can be summarized as follows:

$$C_2H_3O_2^- \underset{OH^-}{\overset{H^+}{\rightleftharpoons}} HC_2H_3O_2$$

Note that one member of the buffer pair removes excess H^+ ion and the other removes excess OH^- ion. The buffering action always results in the active species being converted to its partner species.

Buffer systems have their limits. If large amounts of H^+ or OH^- are added to a buffer, the buffer capacity can be exceeded; then the buffer system is overwhelmed and the pH changes. For example, if large amounts of H^+ were added to the acetate/acetic acid buffer just discussed, the H^+ ion would react with acetate ion until the acetate was depleted. Then, the pH would begin to drop as free H^+ ions build up in the solution.

10.13 BUFFERS IN THE HUMAN BODY

Buffer solutions play an important role in the functioning of the human body. All body fluids have definite pH values that must be maintained within very narrow ranges because living cells are extremely sensitive to even slight changes in pH. The protection against pH change is provided by buffers, which can be referred to as "chemical shock absorbers" or "chemical sponges" because of their key protection role.

Blood is a vital buffer solution. Even small departures from the normal pH range of blood (7.35–7.45) can cause serious illness, and death can result from variations that exceed a few tenths of a pH unit. This situation results because many of the key reactions that take place in blood are enzyme-catalyzed (Section 25.1) and reach optimum conditions only within the narrow normal pH range. Altering the pH slows down or stops the action of the enzymes.

The major buffer system in blood is composed of carbonic acid, H_2CO_3, and bicarbonate salts such as sodium bicarbonate, $NaHCO_3$. Certain proteins (Section 24.1) and, to a small extent, hydrogen phosphate ions, also help buffer blood. The carbonic acid/bicarbonate buffering system in blood operates in the following manner. Any acid formed in the blood reacts with bicarbonate ion to give carbonic acid.

$$H^+ + HCO_3^- \rightleftharpoons H_2CO_3$$

Carbonic acid is an unstable acid that readily decomposes to give carbon dioxide and water.

$$H_2CO_3 \rightleftharpoons H_2O + CO_2$$

Carbon dioxide in the blood, which is formed from this decomposition, is removed from the blood by the lungs and is exhaled.

Formation of OH^- in the blood is not a common occurrence. If it does occur, the carbonic acid/bicarbonate buffer adjusts for its presence through the reaction

$$H_2CO_3 + OH^- \rightleftharpoons HCO_3^- + H_2O$$

Excessive HCO_3^- ions can be eliminated from the body through the kidneys. Figure 10–7 summarizes the workings of the carbonic acid/bicarbonate ion buffer system in the blood. The capacity of the carbonic acid/bicarbonate buffer system to handle OH^- ion increases is not as great as its capacity to handle H^+ ion increases. This is because the ratio of HCO_3^- ion to H_2CO_3 in blood is about 20:1.

A very important buffer system within cells is the dihydrogen phosphate ($H_2PO_4^-$)/hydrogen phosphate (HPO_4^{2-}) system. The control of pH within cellular fluids is maintained by the reaction of OH^- ion with $H_2PO_4^-$ and the reaction of H^+ with HPO_4^{2-}.

FIGURE 10–7 Workings of the carbonic acid/bicarbonate ion buffer system in human blood.

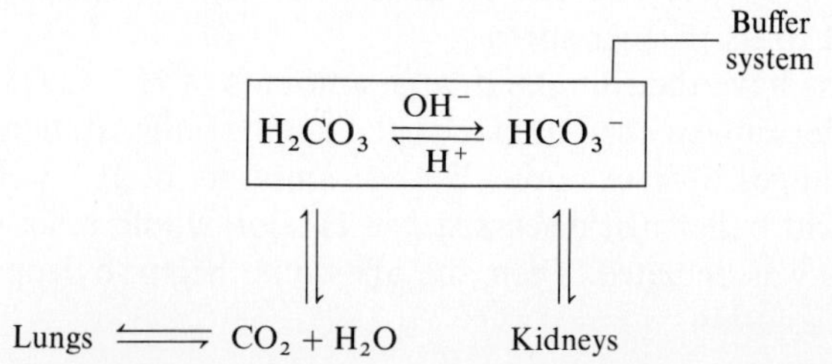

$$\underbrace{H_2CO_3 \underset{H^+}{\overset{OH^-}{\rightleftharpoons}} HCO_3^-}_{\text{Buffer system}}$$

Lungs $\rightleftharpoons CO_2 + H_2O \qquad$ Kidneys

$$H_2PO_4^- + OH^- \rightleftharpoons HPO_4^{2-} + H_2O$$
$$HPO_4^{2-} + H^+ \rightleftharpoons H_2PO_4^- + H_2O$$

The overall dihydrogen phosphate/hydrogen phosphate buffering action can be summarized as

$$H_2PO_4^- \underset{H^+}{\overset{OH^-}{\rightleftharpoons}} HPO_4^{2-}$$

The normal hydrogen phosphate/dihydrogen phosphate ion ratio in cellular fluids is about 4:1. Thus, the phosphate buffer system is better equipped to handle influxes of acid than influxes of base. Significant amounts of acids (up to 10 moles a day) are produced in a human body as the result of normal metabolic reactions. For example, lactic acid, $HC_3H_5O_2$, is produced in muscle tissue during exercise.

Under normal conditions, the body's carbonate and phosphate buffer systems are adequate to maintain normal-range pH's. However, under certain stress conditions, the buffer systems can be temporarily overwhelmed. When this happens, compensatory mechanisms involving the lungs and kidneys help to return the pH to normal. Both the lungs and kidneys play a role in pH control at all times, but this role is more important during stress periods.

Acidosis *is a body condition in which the* pH *of blood drops from its normal value of 7.4 to 7.1–7.2.* Various factors can cause acidosis, including hypoventilation (breathing too little) caused by emphysema, congestive heart failure, diabetes mellitus, excess loss of bicarbonate ion in severe diarrhea, or decreased excretion of hydrogen ions in kidney failure. A temporary condition of acidosis can result from prolonged, intensive exercise. The body reacts to alleviate this acidosis condition in two ways. Excess carbon dioxide (formed from the decomposition of carbonic acid) is expelled by increasing the rate of respiration. Also, kidney-system changes occur that increase excretion of H^+ and retention of HCO_3^-; this results in acidic urine.

Alkalosis *is a body condition in which the* pH *of blood increases from its normal value of 7.4 to a value of 7.5.* Alkalosis can result from hyperventilation (excess breathing) caused by anxiety or hysteria, extreme fevers, severe vomiting, or exposure to high altitudes (altitude sickness). The body's responses to alkalosis include a decrease in respiration rate (less expulsion of carbon dioxide by the lungs) and an increase in HCO_3^- excretion by the kidneys, resulting in alkaline urine. Alkalosis is not as common as acidosis.

acidosis

alkalosis

EXERCISES AND PROBLEMS

Acid-Base Definitions

10.1 What are the Arrhenius definitions for acids and bases?

10.2 What are the Brønsted–Lowry definitions for acids and bases, and what are their advantages over the Arrhenius definitions?

10.3 Write equations for the dissociation of the following Arrhenius acids and bases in water.
a. HI (hydroiodic acid)
b. HClO (hypochlorous acid)
c. LiOH (lithium hydroxide)
d. KOH (potassium hydroxide)

10.4 Identify the Brønsted–Lowry acid and base in the following reactions.
a. $HF + H_2O \longrightarrow H_3O^+ + F^-$
b. $H_2O + S^{2-} \longrightarrow HS^- + OH^-$
c. $H_2O + H_2CO_3 \longrightarrow H_3O^+ + HCO_3^-$
d. $HCO_3^- + H_2O \longrightarrow H_3O^+ + CO_3^{2-}$

10.5 Write equations to illustrate the acid–base reactions that can take place between the following Brønsted–Lowry acids and bases.
a. acid: HClO, base: H_2O
b. acid: $HClO_4$, base: NH_3
c. acid: H_3O^+, base: NH_2^-
d. acid: H_3O^+, base: OH^-

Strengths of Acids and Bases

10.6 What is the principal distinction between strong and weak acids? Is the distinction between strong and weak bases of a similar nature?

10.7 Classify each of the following as weak or strong acids.
a. H_2SO_4 c. H_3PO_4 e. H_3BO_3
b. HNO_3 d. HClO   f. HCl

10.8 Classify each of the following as weak or strong bases.
a. NaOH   b. $Ba(OH)_2$ c. NH_3 d. KOH

10.9 Make a listing of the six common strong acids and seven common strong bases.

10.10 What is the distinction between the terms *weak acid* and *dilute acid*?

Polyprotic Acids

10.11 Identify the following acids as monoprotic, diprotic, or triprotic.
a. $HClO_4$ (perchloric acid)
b. $H_2C_2O_4$ (oxalic acid)
c. $HC_2H_3O_2$ (acetic acid)
d. $HC_4H_7O_2$ (butyric acid)
e. H_3PO_4 (phosphoric acid)
f. H_2SO_4 (sulfuric acid)

10.12 Write equations showing all steps in the dissociation of the following acids.
a. H_2CO_3 (carbonic acid)
b. $H_3C_6H_5O_7$ (citric acid)
c. $HC_4H_7O_2$ (butyric acid)

10.13 The formula for lactic acid is preferably written as $HC_3H_5O_3$, rather than as $C_3H_6O_3$. Explain why this is so.

10.14 Pyruvic acid, which is produced in metabolic reactions, has the following structure.

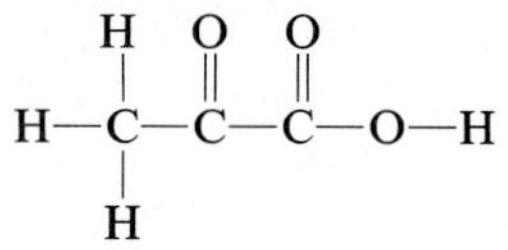

$$\begin{array}{c} HOO \\ |\|\| \\ H-C-C-C-O-H \\ | \\ H \end{array}$$

Would you predict that this acid is a mono-, di-, tri-, or tetraprotic acid? Give your reasoning for arriving at your answer.

Salts

10.15 Give a formula and a name for the positive and negative ions present in each of the following salts. (You may have to refer back to Sections 6.7 and 6.13 for a quick review of nomenclature.)
a. NaF   d. K_3N g. Li_2CO_3
b. KNO_3 e. CaS   h. $BeSO_4$
c. $Ca(C_2H_3O_2)_2$ f. $Mg_3(PO_4)_2$

10.16 Which of the salts listed in Problem 10.15 are soluble in water? (The solubility rules for salts were given in Section 8.12.)

10.17 Write a balanced equation for the dissociation (breakup) in water of each of the following soluble salts.
a. $Ba(NO_3)_2$ d. K_2CO_3 g. $Mg(C_2H_3O_2)_2$
b. Na_2SO_4 e. $AlCl_3$ h. LiI
c. $CaBr_2$ f. CaS

10.18 Identify each of the following substances as an acid, base, or salt.
a. HBr   d. $AlPO_4$ g. HNO_3
b. NaI   e. $Ba(OH)_2$ h. $HC_2H_3O_2$
c. NH_4NO_3 f. KOH

10.19 The term *weak* is used to describe certain acids and bases, but it is never used to describe salts. Why?

Electrolytes

10.20 What is the difference between
a. an electrolyte and a nonelectrolyte?
b. a strong electrolyte and a weak electrolyte?

10.21 Classify each of the following types of compounds as a strong electrolyte or a weak electrolyte.
a. strong acid   c. strong base   e. soluble salt
b. weak acid   d. weak base   f. slightly soluble salt

10.22 Classify each of the following solutions as a strong electrolyte, a weak electrolyte, or a nonelectrolyte.
a. nitric acid
b. acetic acid
c. aqueous ammonia
d. sucrose
e. sodium hydroxide
f. sodium chloride
g. isopropyl alcohol
h. copper(II) sulfate

Neutralization

10.23 Which of the following equations represent neutralization reactions?
a. $2 HNO_3 + Ba(OH)_2 \longrightarrow Ba(NO_3)_2 + 2 H_2O$
b. $2 AgNO_3 + K_2CO_3 \longrightarrow Ag_2CO_3 + 2 KNO_3$
c. $H_2SO_4 + 2 NaNO_2 \longrightarrow 2 HNO_2 + Na_2SO_4$
d. $HF + KOH \longrightarrow KF + H_2O$
e. $2 H_2 + O_2 \longrightarrow 2 H_2O$
f. $HC_2H_3O_2 + NaOH \longrightarrow NaC_2H_3O_2 + H_2O$

10.24 Write a balanced molecular equation to represent each of the following acid–base neutralizations.
a. acid: HCl, base: NaOH
b. acid: HNO_3, base: KOH
c. acid: H_2SO_4, base: LiOH
d. acid: H_3PO_4, base: $Ba(OH)_2$

10.25 Write a balanced molecular equation for the preparation of each of the following salts using an acid–base neutralization reaction.
a. Li_2SO_4 (lithium sulfate)
b. NaCl (sodium chloride)
c. KNO_3 (potassium nitrate)
d. $Ba_3(PO_4)_2$ (barium phosphate)

Hydrogen Ion and Hydroxide Ion Concentrations

10.26 What is the concentration of hydrogen and hydroxide ion in pure water at 25°C?

10.27 Calculate the H^+ ion concentration of a solution if the OH^- ion concentration is
a. 3.0×10^{-3} M
b. 6.7×10^{-6} M
c. 9.1×10^{-8} M
d. 1.2×10^{-11} M

10.28 How are acidic, basic, and neutral aqueous solutions defined in terms of $[H^+]$ and $[OH^-]$?

10.29 Indicate whether the following solutions are acidic, basic, or neutral.
a. $[H^+] = 1.0 \times 10^{-3}$
b. $[H^+] = 1.0 \times 10^{-5}$
c. $[H^+] = 3.0 \times 10^{-11}$
d. $[OH^-] = 4.0 \times 10^{-6}$
e. $[OH^-] = 2.3 \times 10^{-10}$
f. $[OH^-] = 1.0 \times 10^{-7}$

pH Scale

10.30 Calculate the pH of the following solutions.
a. $[H^+] = 1.0 \times 10^{-4}$
b. $[H^+] = 1.0 \times 10^{-11}$
c. $[H^+] = 1.0 \times 10^{-8}$
d. $[OH^-] = 1.0 \times 10^{-3}$
e. $[OH^-] = 1.0 \times 10^{-7}$
f. $[OH^-] = 1.0 \times 10^{-10}$

10.31 How are acidic, basic, and neutral aqueous solutions defined in terms of pH values?

10.32 Determine the $[H^+]$ in a solution that has a pH of
a. 3   b. 7   c. 10   d. 13

10.33 A teacher once mistakenly told a class that the $[H^+]$ increased as the pH increased. Explain what is wrong with this statement.

10.34 Solution A has a pH of 2 and Solution B has a pH of 5. Which solution is more acidic? How many times more acidic is the one solution than the other?

10.35 How many grams of NaOH are necessary to make 275 mL of a solution with a pH of 8.00?

Normality of Acids and Bases

10.36 Determine how many equivalents of acid or base are present in each of the following samples.
a. 1 mole NaOH
b. 2 moles H_3PO_4
c. 0.50 mole HNO_3
d. 0.50 mole H_2CO_3
e. 2.0 g HCl
f. 2.0 g H_2SO_4
g. 2.0 g H_3PO_4
h. 2.0 g KOH

10.37 Determine the normalities of the following solutions.
a. 1.00 L of solution containing 0.500 equiv of NaOH
b. 275 mL of solution containing 0.250 equiv of NaOH
c. 1.00 L of solution containing 0.100 mole of HNO_3
d. 325 mL of solution containing 0.200 mole of H_3PO_4
e. 2.50 L of solution containing 25.0 g of H_2SO_4
f. 25.0 mL of solution containing 1.25 g of $Ba(OH)_2$

10.38 Express the following molarities as normalities.
a. 0.050 M H_3PO_4
b. 1.30 M HNO_3
c. 1.00 M KOH
d. 5.2 M H_2SO_4
e. 0.0020 M $HC_2H_3O_2$
f. 2.00 M $Ba(OH)_2$

10.39 Express the following normalities as molarities.
a. 0.500 N H_2SO_4
b. 2.50 N HNO_3
c. 1.33 N NaOH
d. 0.15 N H_3PO_4
e. 6.2 N H_2CO_3
f. 1.00 N $Ba(OH)_2$

Titration Calculations

10.40 Briefly describe the procedure involved in titrating an acid with a base. Make clear the importance of the endpoint and the use of an indicator.

10.41 Determine how many milliliters of 0.100 N NaOH solution would be needed to neutralize each of the following acid samples.
a. 5.00 mL of 0.250 N HNO_3
b. 20.00 mL of 0.500 N H_2SO_4
c. 25.00 mL of 0.250 N HCl
d. 15.00 mL of 0.100 N $HC_2H_3O_2$

10.42 How many equivalents of base would each of the following acid samples neutralize?
a. 100.0 mL of 0.500 N HNO_3
b. 100.0 mL of 0.500 N H_2SO_4
c. 100.0 mL of 0.500 N H_3PO_4
d. 100.0 mL of 0.500 M HNO_3
e. 100.0 mL of 0.500 M H_2SO_4
f. 100.0 mL of 0.500 M H_3PO_4

10.43 In an acid–base titration, 40.0 mL of 0.300 M NaOH is used to titrate 25.0 mL of citric acid ($H_3C_6H_5O_7$). What are the normality and molarity of the citric acid solution?

10.44 What volume, in milliliters, of 0.100 N NaOH is needed to titrate a 50.0 mL sample of a 0.0732 N HCL solution?

Salt Hydrolysis

10.45 It is useful in hydrolysis discussions to classify salts into four categories. What are these categories? Give an example of a salt that belongs to each category.

10.46 Explain why Category 1 salts do not hydrolyze.

10.47 Explain why, upon hydrolysis, Category 2 and Category 3 salts give acidic and basic solutions.

10.48 The salts NH_4NO_2, $NH_4C_2H_3O_2$, and NH_4CN are all Category 4 salts—salts that undergo double hydrolysis. Upon hydrolysis of the salt, the first listed salt gives an acidic aqueous solution, the second gives a neutral solution, and the third gives a basic solution. Explain how this is possible.

10.49 Identify the ion (or ions) present in each of the following salts that will undergo hydrolysis in aqueous solution.
a. NaF   d. $LiC_2H_3O_2$ g. Na_2CO_3
b. NH_4Cl e. Na_3BO_3 h. KNO_3
c. KCN   f. $(NH_4)_3PO_4$

10.50 Predict whether each of the following aqueous salt solutions will be acidic, basic, or neutral.
a. $NaNO_3$ d. LiCl   g. Na_2CO_3
b. NaCN   e. $KC_2H_3O_2$ h. NaF
c. NH_4Cl f. NH_4NO_3

10.51 Which ion is responsible for the principal hydrolysis reaction that causes blood plasma to be slightly basic?

Buffer Solutions

10.52 Explain the general concept associated with the operation of a chemical buffer system.

10.53 Predict whether each of the following pairs of substances could function as a buffer system in aqueous solution.
a. HCl and NaCl
b. HCl and HCN
c. HCN and KCN
d. NaOH and NaCl
e. $HC_2H_3O_2$ and $KC_2H_3O_2$
f. NaCl and NaCN

10.54 Write an equation for each of the following buffering actions.
a. The response of a $H_2PO_4^-/HPO_4^{2-}$ buffer to the addition of H^+ ions.
b. The response of a H_2CO_3/HCO_3^- buffer to the addition of OH^- ions.
c. The response of a H_2CO_3/HCO_3^- buffer to the addition of H^+ ions.
d. The response of a HCN/CN^- buffer to the addition of OH^- ions.
e. The response of a $H_2PO_4^-/HPO_4^{2-}$ buffer to the addition of OH^- ions.
f. The response of a HCN/CN^- buffer to the addition of H^+ ions.

10.55 What compensatory mechanisms does the body use to correct the conditions of (a) acidosis and (b) alkalosis?

Oxidation–Reduction Processes

Objectives

After completing Chapter 11, you will be able to:

11.1 Define the terms *oxidation, reduction, oxidizing agent,* and *reducing agent* in terms of loss and gain of electrons.

11.2 Determine the oxidation number of an element in a molecule or ion and define the terms *oxidation, reduction, oxidizing agent,* and *reducing agent* in terms of increase and decrease in oxidation number. You will also be able to identify, in a given redox equation, the substance reduced, the substance oxidized, the oxidizing agent, and the reducing agent.

11.3 Classify reactions as either redox or metathetical. You will also be able to classify reactions as synthesis, decomposition, single-displacement, or double-displacement.

11.4 Recognize important characteristics of the following redox processes: combustion of fossil fuels, oxidation of food within the human body, and photosynthesis.

INTRODUCTION

In Chapter 10, we discussed a major type of chemical reaction, the acid–base reaction. In this chapter, we will examine another important class of chemical reactions, oxidation–reduction reactions. Acid–base reactions can be regarded, in the Brønsted–Lowry sense (Section 10.1), as proton-transfer reactions. Oxidation–reduction reactions involve electron transfer between the reactants.

Oxidation–reduction reactions occur all around us and even inside us. The bulk of the energy needed for the functioning of all living organisms is obtained from food through oxidation–reduction processes. Diverse phenomena such as the electricity obtained from a battery to start a car, the use of natural gas to heat a home, iron rusting, illumination from a flashlight, and the functioning of antiseptic agents to kill or prevent the growth of bacteria all involve oxidation–reduction reactions. In short, a knowledge of this type of reaction is fundamental to understanding many biological and technological processes.

11.1 OXIDATION–REDUCTION TERMINOLOGY

The terms *oxidation* and *reduction*, like the terms *acid* and *base* (Section 10.1), have several definitions. Historically, the word *oxidation* was first used to describe the reaction of a substance with oxygen. According to this definition, each of the following reactions involves oxidation.

$$4\,Fe + 3\,O_2 \longrightarrow 2\,Fe_2O_3$$
$$S + O_2 \longrightarrow SO_2$$
$$CH_4 + 2\,O_2 \longrightarrow CO_2 + 2\,H_2O$$

The substance on the far left in each of these equations is said to have been oxidized.

Originally, the term *reduction* referred to processes where oxygen was removed from a compound. A common type of reduction reaction according to this original definition is the removal of oxygen from a metal oxide to produce the free metal.

$$CuO + H_2 \longrightarrow Cu + H_2O$$
$$2\,Fe_2O_3 + 3\,C \longrightarrow 4\,Fe + 3\,CO_2$$

The term *reduction* comes from the reduction in mass of the metal-containing species; the metal has a mass less than that of the metal oxide.

Today the words *oxidation* and *reduction* are used in a much broader sense. Current definitions include the previous examples and much more. Scientists now recognize that the changes brought about in a substance from reaction with oxygen can also be caused by reaction with numerous substances that do not contain oxygen. For example, consider the following reactions.

$$2\,Mg + O_2 \longrightarrow 2\,MgO$$
$$Mg + S \longrightarrow MgS$$
$$Mg + F_2 \longrightarrow MgF_2$$
$$3\,Mg + N_2 \longrightarrow Mg_3N_2$$

In each of these reactions, magnesium metal is converted to a magnesium com-

TABLE 11–1 Oxidation–Reduction Terminology in Terms of Electron Transfer

Terms Associated with the Loss of Electrons	Terms Associated with the Gain of Electrons
process of oxidation	process of reduction
substance oxidized	substance reduced
reducing agent	oxidizing agent

pound that contains Mg^{2+} ions. The process is the same—magnesium atoms change to magnesium ions by losing two electrons; the only difference is the identity of the substance that causes magnesium to undergo the change. All of these reactions are considered to involve oxidation when the modern definition of oxidation is applied. **Oxidation** *is the process whereby a substance in a chemical reaction loses one or more electrons.* The modern definition for reduction involves the use of similar terminology. **Reduction** *is the process whereby a substance in a chemical reaction gains one or more electrons.*

Oxidation and reduction are complementary processes, not isolated phenomena. They always occur together; you cannot have one without the other. If electrons are lost by one species, they cannot just disappear; they must be gained by another species. Electron transfer, then, is the basis for oxidation and reduction. The phrase **oxidation–reduction reaction** *is used to describe any reaction in which electrons are transferred from one reactant to another reactant.* This designation is often shortened to simply *redox reaction*.

There are two different ways of looking at the reactants in a redox reaction. First, the reactants can be viewed as being "acted upon". From this viewpoint, one reactant is *oxidized* (the one that loses electrons) and one is *reduced* (the one that gains electrons). Second, the reactants can be looked at as "bringing about" the reaction. In this approach, the terms *oxidizing agent* and *reducing agent* are used. An **oxidizing agent** *causes oxidation by accepting electrons from the other reactant.* This acceptance of electrons means that the oxidizing agent itself is reduced. Similarly, the **reducing agent** *causes reduction by providing electrons for the other reactant to accept.* Thus, the reducing agent and the substance oxidized are one and the same, as are the oxidizing agent and substance reduced.

$$\text{substance oxidized} = \text{reducing agent}$$
$$\text{substance reduced} = \text{oxidizing agent}$$

The terms *oxidizing agent* and *reducing agent* sometimes cause confusion because the oxidizing agent is not oxidized (it is reduced) and the reducing agent is not reduced (it is oxidized). A simple analogy is that a travel agent is not the one who takes a trip; he or she is the one who causes the trip to be taken. Table 11–1 summarizes the terminology presented in this section.

11.2 OXIDATION NUMBERS

Oxidation numbers are used to help determine whether oxidation and reduction have occurred in a reaction, and if they have, the identity of the oxidizing and reducing agents. An **oxidation number** *is the charge that an atom appears to have when the electrons in each bond it is participating in are assigned to the more electronegative of the two atoms involved in the bond.**

---

* In some textbooks, the term *oxidation state* is used in place of *oxidation number*. In other textbooks, the two terms are used interchangeably. We will use *oxidation number* in this textbook.

The following set of operational rules are used to determine oxidation numbers.

Rule 1: *The oxidation number of an atom in its elemental state is zero.* For example, the oxidation number of copper is zero, and the oxidation number of chlorine in Cl_2 is zero.

Rule 2: *The oxidation number of any monoatomic ion is equal to the charge on the ion.* For example, the Na^+ ion has an oxidation number of $+1$ and the S^{2-} ion has an oxidation number of -2.

Rule 3: *The oxidation numbers of Groups IA and IIA elements are always $+1$ and $+2$, respectively.*

Rule 4: *The oxidation number of fluorine is always -1 and that of the other Group VIIA elements (Cl, Br, and I) is usually -1.* The exception for these latter elements occurs when they are bonded to more electronegative elements. In this case, they exhibit positive oxidation numbers.

Rule 5: *The usual oxidation number for oxygen is -2.* The exceptions occur when oxygen is bonded to the more electronegative fluorine (O then has a positive oxidation number) or is found in compounds containing oxygen–oxygen bonds (peroxides). In peroxides, the oxidation number is -1 for oxygen. Peroxides form only between oxygen and hydrogen (H_2O_2), Group IA elements (Na_2O_2, etc.), and Group IIA elements (BaO_2, etc.).

Rule 6: *The usual oxidation number for hydrogen is $+1$.* The exception occurs in hydrides, compounds where hydrogen is bonded to a metal of lower electronegativity. In these compounds, hydrogen is assigned an oxidation number of -1. Examples of hydrides are NaH, CaH_2, and LiH.

Rule 7: *The algebraic sum of the oxidation numbers of all atoms in a neutral molecule must be zero.*

Rule 8: *The algebraic sum of the oxidation numbers of all atoms in a polyatomic ion is equal to the charge on the ion.*

Example 11.1 illustrates the use of these rules.

Example 11.1

Assign oxidation numbers to each element in the following compounds or polyatomic ions.

a. P_2O_5 b. $KMnO_4$ c. SnF_4 d. NO_3^-

Solution

a. The sum of the oxidation numbers of all of the atoms present must add to zero (Rule 7).

$$2 \text{ (oxid. no. P)} + 5 \text{ (oxid. no. O)} = 0$$

The oxidation number of oxygen is -2 (Rule 5). Substituting this value into the previous equation will enable us to calculate the oxidation number of phosphorus, an element for which a specific oxidation number rule does not exist.

$$2 \text{ (oxid. no. P)} + 5(-2) = 0$$
$$2 \text{ (oxid. no. P)} = +10$$
$$\text{(oxid. no. P)} = +5$$

Thus, the oxidation numbers for the elements involved in this compound are

$$P = +5 \quad \text{and} \quad O = -2$$

Note that the oxidation number of phosphorus is not $+10$; that is the calculated charge associated with two phosphorus atoms. Oxidation number is always specified on a *per atom* basis.

b. The sum of the oxidation numbers of all of the atoms present must add up to zero (Rule 7).

$$(\text{oxid. no. K}) + (\text{oxid. no. Mn}) + 4\,(\text{oxid. no. O}) = 0$$

The oxidation number of potassium, a Group IA element is $+1$ (Rule 3) and the oxidation number of oxygen is -2 (Rule 5). Substituting these two values into the Rule 7 equation will enable us to calculate the oxidation number of manganese.

$$(+1) + (\text{oxid. no. Mn}) + 4\,(-2) = 0$$
$$(\text{oxid. no. Mn}) = 8 - 1 = +7$$

Thus, the oxidation numbers for the elements involved in this compound are:

$$K = +1,\ Mn = +7,\ \text{and}\ O = -2$$

Note that all of the oxidation numbers add up to zero when it is taken into account that there are four oxygen atoms.

$$(+1) + (+7) + 4\,(-2) = 0$$

c. The sum of the oxidation numbers of all atoms in the compound must add up to zero (Rule 7).

$$(\text{oxid. no. Sn}) + 4\,(\text{oxid. no. F}) = 0$$

The oxidation number of fluorine is always -1 (Rule 4). Therefore, we have

$$(\text{oxid. no. Sn}) + 4\,(-1) = 0$$
$$(\text{oxid. no. Sn}) = +4$$

Thus, the oxidation numbers for the elements involved in this compound are:

$$Sn = +4 \quad \text{and} \quad F = -1$$

d. The species NO_3^- is a polyatomic ion rather than a neutral compound. Thus, Rule 8 rather than Rule 7 applies; the oxidation number sum must add up to the charge on the ion rather than to zero.

$$(\text{oxid. no. N}) + 3\,(\text{oxid. no. O}) = -1$$

The oxidation number of oxygen is -2 (Rule 5). Substituting this value into the sum equation gives

$$(\text{oxid. no. N}) + 3\,(-2) = -1$$
$$(\text{oxid. no. N}) = -1 + (+6) = +5$$

Thus, the oxidation numbers for the elements involved in this polyatomic ion are:

$$N = +5 \quad \text{and} \quad O = -2$$

Many elements display a range of oxidation numbers in their various compounds. For example, nitrogen exhibits oxidation numbers ranging from -3 to $+5$ in various compounds. Selected examples are

NH_3	N_2O	NO	N_2O_3	NO_2	HNO_3
-3	$+1$	$+2$	$+3$	$+4$	$+5$

As shown in this listing of nitrogen-containing compounds, the oxidation number

TABLE 11-2 Oxidation–Reduction Terminology in Terms of Oxidation Number Change

Terms Associated with an Increase in Oxidation Number	Terms Associated with a Decrease in Oxidation Number
process of oxidation	process of reduction
substance oxidized	substance reduced
reducing agent	oxidizing agent

of an atom is written *underneath* the atom in the formula. This convention is used to avoid confusion with the charge on an ion.

Oxidizing and reducing agents can be defined in terms of changes in oxidation numbers. The **oxidizing agent** *in a redox reaction is the substance that contains the atom that shows a decrease in oxidation number.* Because the oxidizing agent is the substance reduced in a reaction, reduction involves a decrease in oxidation number; the oxidation number is reduced (decreased) in a reduction. The **reducing agent** *in a redox reaction is the substance that contains the atom that shows an increase in oxidation number.* Oxidation involves an increase in oxidation number because the reducing agent is the substance oxidized in a reaction. Table 11-2 summarizes the relationships between oxidation–reduction terminology and oxidation-number changes. A comparison of Table 11-2 with Table 11-1 shows that the loss of electrons and oxidation number increases are synonymous, as are the gain of electrons and oxidation number decreases. The fact that the oxidation number becomes more positive (increases) as electrons are lost is consistent with our understanding of the proton–electron charge relationships in an atom.

oxidizing agent

reducing agent

Example 11.2

Determine oxidation numbers for each atom in the following reactions and identify the oxidizing and reducing agents.

a. $2 SO_2 + O_2 \longrightarrow 2 SO_3$
b. $S_2O_8^{2-} + 2 I^- \longrightarrow I_2 + 2 SO_4^{2-}$

Solution
The oxidation numbers are calculated by the methods illustrated in Example 11.1.

a.
$$2 SO_2 + O_2 \longrightarrow 2 SO_3$$
$$+4 \; -2 \quad\;\; 0 \quad\quad +6 \; -2$$
Rules 5, 7   Rule 1   Rules 5, 7

The oxidation number of sulfur has increased from +4 to +6. Therefore, the substance that contains sulfur, SO_2, has been oxidized and is the reducing agent.

The oxidation number of the oxygen in O_2 has decreased from 0 to −2. Therefore, the O_2 has been reduced and is the oxidizing agent.

b.
$$S_2O_8^{2-} + 2 I^- \longrightarrow I_2 + 2 SO_4^{2-}$$
$$+7 \; -2 \quad\;\; -1 \quad\quad 0 \quad\;\; +6 \; -2$$
Rules 5, 8   Rule 2   Rule 1   Rules 5, 7

The oxidation number of iodine has increased from −1 to 0. Thus, I^-, the iodine-containing reactant, has been oxidized and is the reducing agent.

The oxidation number of sulfur has decreased from +7 to +6. Thus, $S_2O_8^{2-}$, the sulfur containing reactant, has been reduced and is the oxidizing agent.

11.3 CLASSES OF CHEMICAL REACTIONS

An almost inconceivable number of chemical reactions is possible. The problems associated with organizing our knowledge of chemical reactions are diminished considerably by grouping the reactions into classes. Two classification systems are in common use.

1. A system based on oxidation-number change that groups reactions into two categories: oxidation–reduction (redox) and metathetical.
2. A system based on the form of the equation for the reaction that recognizes four types of reactions: synthesis, decomposition, single displacement, and double displacement.

These two systems are not mutually exclusive and are commonly used together. For example, a particular reaction can be characterized as a single-displacement redox reaction.

As we have just learned (Section 11.2), reactions in which oxidation numbers change are called oxidation–reduction reactions. A **metathetical reaction** *is one in which there is no oxidation number change*. The word *metathetical* comes from a Greek word that means "to transpose, or change positions". Many metathetical reactions can be viewed as involving the replacement of an atom or group of atoms by another atom or group of atoms.

In the second classification system, the form of the equation is the basis for classification. A **synthesis reaction** *is one in which a single product is produced from two (or more) reactants*. The general equation for a synthesis reaction is

$$X + Y \longrightarrow XY$$

This is a simple type of reaction, in which two substances join together to form a more complicated product. The reactants X and Y can be elements or compounds or an element and a compound. The product of the reaction, XY, is always a compound.

Some representative synthesis reactions that have elements as the reactants are

$$Ca + S \longrightarrow CaS$$
$$N_2 + 3\,H_2 \longrightarrow 2\,NH_3$$
$$2\,Na + O_2 \longrightarrow Na_2O_2$$

Synthesis reactions such as these, which only have elements as reactants, are always oxidation–reduction reactions. Oxidation-number changes must occur because all elements (the reactants) have an oxidation number of zero and all the constituent elements of a compound cannot have oxidation numbers of zero.

Some examples of synthesis reactions in which compounds are involved as reactants are

$$SO_3 + H_2O \longrightarrow H_2SO_4$$
$$K_2O + H_2O \longrightarrow 2\,KOH$$
$$2\,NO + O_2 \longrightarrow 2\,NO_2$$
$$2\,NO_2 + H_2O_2 \longrightarrow 2\,HNO_3$$

The first two of these reactions are metathetical and the latter two are redox. Most synthesis reactions in which both reactants are compounds are metathetical; redox synthesis reactions of this type are not common.

A **decomposition reaction** *is one in which a single reactant is converted into two or more simpler substances.* Thus, it is the exact opposite of a synthesis reaction. The general equation for a decomposition reaction is

$$XY \longrightarrow X + Y$$

decomposition reaction

Both redox and metathetical decomposition reactions are common. All compounds can be broken down (decomposed) into their constituent elements at sufficiently high temperatures. These reactions, which are always redox reactions, include

$$2\,CuO \longrightarrow 2\,Cu + O_2$$
$$2\,H_2O \longrightarrow 2\,H_2 + O_2$$

Examples of decomposition reactions that result in at least one compound as a product are

$$CaCO_3 \longrightarrow CaO + CO_2$$
$$2\,KClO_3 \longrightarrow 2\,KCl + 3\,O_2$$

The first of these two reactions is metathetical; the second is a redox reaction.

A **single-displacement reaction** *is one in which an atom, molecule, or ion displaces another atom or group of atoms from a compound.* There are always two reactants and two products in this type of reaction. The general equation for a single-displacement reaction is

single-displacement reaction

$$X + YZ \longrightarrow Y + XZ$$

A common type of single-displacement reaction is one in which an element and a compound are reactants and an element and a compound are products. Typical examples of this type of reaction include

$$Fe + CuSO_4 \longrightarrow Cu + FeSO_4$$
$$Mg + Ni(NO_3)_2 \longrightarrow Ni + Mg(NO_3)_2$$
$$Cl_2 + NiI_2 \longrightarrow I_2 + NiCl_2$$
$$F_2 + 2\,NaCl \longrightarrow Cl_2 + 2\,NaF$$

The first two equations illustrate one metal replacing another metal from its compound. The latter two equations illustrate one nonmetal replacing another nonmetal from its compound.

A more complicated example of a single-displacement reaction, in which all reactants and products are compounds, is

$$4\,PH_3 + Ni(CO)_4 \longrightarrow 4\,CO + Ni(PH_3)_4$$

A **double-displacement reaction** *is one in which two compounds exchange parts with each other and form two different compounds.* The general equation for such a reaction is

double-displacement reaction

$$AX + BY \longrightarrow AY + BX$$

Double-displacement reactions generally involve acids, bases, and salts in aqueous solution. Most often, the positive ion from one compound exchanges places with the positive ion of the other compound. The process can be thought of as partner swapping because each negative ion ends up paired with a new positive ion.

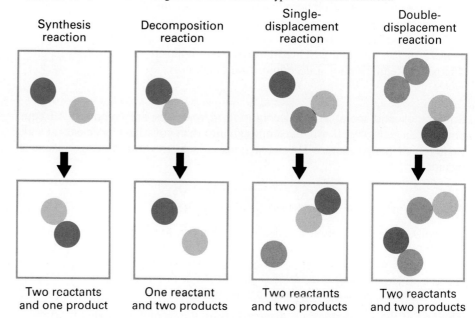

FIGURE 11–1 Schematic diagrams of four common types of chemical reactions.

Acid–base neutralizations (Section 10.6) are double-displacement reactions. For example, in the neutralization

$$NaOH + HCl \longrightarrow NaCl + HOH$$

the sodium and hydrogen have exchanged places. Many of the reactions of dissolved salts are also double-displacement reactions.

$$AgNO_3 + NaCl \longrightarrow NaNO_3 + AgCl$$

$$NaF + HCl \longrightarrow NaCl + HF$$

Figure 11–1 provides a pictorial summary of the four general types of chemical reactions.

Example 11.3

Classify each of the following reactions as redox or metathetical. Also classify them as synthesis, decomposition, single displacement, or double displacement.

a. $2\,C + O_2 \longrightarrow 2\,CO$
b. $2\,KNO_3 \longrightarrow 2\,KNO_2 + O_2$
c. $Zn + 2\,AgNO_3 \longrightarrow Zn(NO_3)_2 + 2\,Ag$
d. $Ni(NO_3)_2 + 2\,NaOH \longrightarrow Ni(OH)_2 + 2\,NaNO_3$

Solution

The oxidation numbers are calculated by the methods illustrated in Example 11.2.

a.

$$\begin{array}{ccccc} 2\,C & + & O_2 & \longrightarrow & 2\,CO \\ 0 & & 0 & & +2\ -2 \\ \text{Rule 1} & & \text{Rule 1} & & \text{Rules 5, 7} \end{array}$$

The oxidation numbers of both carbon and oxygen change; therefore, the reaction is a redox reaction. Two substances combine to form a single substance; hence, this reaction is classified as a synthesis reaction. We, thus, have a *redox synthesis reaction*.

b.
$$2\,KNO_3 \longrightarrow 2\,KNO_2 + O_2$$
$$+1\ +5\ -2 \qquad\qquad +1\ +3\ -2 \quad\ \ 0$$
$$\text{Rules 3, 5, 7} \qquad\qquad \text{Rules 3, 5, 7} \quad \text{Rule 1}$$

The oxidation number of nitrogen decreases from $+5$ to $+3$; the oxidation number for some oxygen atoms increases from -2 to 0. The reaction is a redox reaction. Two substances are produced from a single substance, and so it is also a decomposition reaction. Thus, we have a *redox decomposition reaction*.

c.
$$Zn + 2\,AgNO_3 \longrightarrow Zn(NO_3)_2 + 2\,Ag$$
$$\ 0 \qquad +1\ +5\ -2 \qquad\qquad +2\ +5\ -2 \qquad 0$$
$$\text{Rule 1}\quad\text{Rules 5, 7, 8} \qquad\ \ \text{Rules 5, 7, 8}\quad \text{Rule 1}$$

This is a redox reaction; zinc is oxidized and silver is reduced. Having an element and a compound as reactants and an element and compound as products is a characteristic of a single-displacement reaction. That is the type of reactant we have here: zinc and silver are exchanging places. Thus, we have a *redox single-displacement reaction*.

d.
$$Ni(NO_3)_2 + 2\,NaOH = Ni(OH)_2 + 2\,NaNO_3$$
$$+2\ +5\ -2 \qquad +1\ -2\ +1 \qquad +2\ -2\ +1 \qquad +1\ +5\ -2$$
$$\text{Rules 5, 7, 8} \quad\ \text{Rules 3, 5, 7}\quad\ \ \text{Rules 5, 7, 8}\quad\ \ \text{Rules 3, 5, 7}$$

This is a metathetical reaction; there are no oxidation-number changes. The reaction is also a double-displacement reaction; nickel and sodium are changing places. Thus, we have a *metathetical double-displacement reaction*.

11.4 SOME IMPORTANT OXIDATION–REDUCTION PROCESSES

As we noted in the introduction to this chapter, oxidation–reduction reactions are the basis for processes that are an absolute necessity for life itself, as well as for the high standard of living we enjoy. Let us now briefly consider some of these processes in general terms. Most of these processes will be discussed in depth in later chapters of the text.

Energy that is used to provide electrical power, heat our buildings, and operate our transportation systems is largely obtained from the combustion of fossil fuels (petroleum, coal, and natural gas) or the products obtained from them, such as gasoline. Fossil fuels and fossil fuel derivatives are all hydrogen–carbon compounds that burn rapidly in air (oxygen is the oxidizing agent) to produce carbon dioxide, water, and large amounts of energy.

$$\text{fossil fuel} + O_2 \longrightarrow CO_2 + H_2O + \text{energy}$$

Redox reactions, with oxygen as the oxidizing agent, are also the source of energy for a living organism. For humans, the substances oxidized are our foods: carbohydrates, fats, and proteins. These oxidations, which are multistep in nature, occur at a very slow, controlled rate, allowing cells to trap and use the released energy.

The oxygen needed to oxidize carbohydrates, fats, and proteins is supplied to cells by blood, which picks up oxygen in the lungs and distributes it throughout the body. Carbon dioxide and water are the products of these oxidations, as is shown in the following equations.

$$\text{carbohydrate} + O_2 \longrightarrow CO_2 + H_2O + \text{energy}$$
$$\text{fat} + O_2 \longrightarrow CO_2 + H_2O + \text{energy}$$
$$\text{protein} + O_2 \longrightarrow CO_2 + H_2O + \text{nitrogen-containing products} + \text{energy}$$

The blood returns the waste product carbon dioxide to the lungs to be exhaled; the water produced in the oxidation reaction either remains in the cells and tissues or is excreted through sweat or urine.

Another redox process that is of prime importance to humans is photosynthesis, which occurs in the plant world. Photosynthesis can be looked upon as the ultimate source of the food we eat. Photosynthesis, which consists of a complex series of redox reactions, enables plants to convert carbon dioxide (obtained from the air) into glucose, a simple carbohydrate. The glucose molecules are then combined in various ways within the plant to yield larger carbohydrates (the structural components of the plant). We use plants directly as food or indirectly by eating the meat of animals that have eaten the plants.

The overall reaction for photosynthesis can be written as

$$6\,CO_2 + 6\,H_2O + \text{solar energy} \longrightarrow \underset{\substack{\text{glucose—a}\\\text{building block}\\\text{of carbohydrates}}}{C_6H_{12}O_6} + 6\,O_2$$

Note that in addition to glucose, oxygen is also a product of photosynthesis. This pathway for the production of oxygen is nature's method for replenishing the Earth's oxygen supply. Photosynthesis compensates for processes such as combustion and human respiration (breathing) that continually remove oxygen from the atmosphere.

EXERCISES AND PROBLEMS

Oxidation-Reduction Terminology

11.1 Give definitions of *oxidation* and *reduction* in terms of
a. loss and gain of electrons
b. increase and decrease in oxidation number

11.2 Give definitions of *oxidizing agent* and *reducing agent* in terms of
a. loss and gain of electrons
b. increase and decrease in oxidation number
c. substance oxidized and substance reduced

11.3 In each of the following statements, choose the word in parentheses that best completes the statement.

a. An element that has lost electrons in a redox reaction is said to have been (oxidized, reduced).
b. Reduction always results in an (increase, decrease) in the oxidation number.
c. The substance oxidized in a redox reaction is the (oxidizing agent, reducing agent).

d. The reducing agent (gains, loses) electrons during a redox reaction.
e. The reducing agent causes an (increase, decrease) in the oxidation number of the oxidizing agent in a redox reaction.

Assignment of Oxidation Numbers

11.4 Determine the oxidation number of
a. S in SO_2
b. S in SO_3
c. S in SO_4^{2-}
d. Ba in Ba^{2+}
e. Zn in $ZnSO_4$
f. Cl in $HClO_3$
g. P in H_3PO_4
h. Cl in Cl_2

11.5 Determine the oxidation number of Cl in each of the following species.
a. ClF_4^+
b. $BeCl_2$
c. $Ba(ClO_2)_2$
d. Cl_2O_7
e. $AlCl_4^-$
f. NCl_3
g. ClF
h. ClO^-

11.6 Determine the oxidation number of Cr in each of the following species.

a. Cr_2O_3 c. Na_2CrO_4 e. $K_2Cr_2O_7$ g. CrO_2
b. $KCrO_2$ d. $BeCr_2O_7$ f. CrF_5 h. CrO_3

11.7 What are the oxidation numbers of all elements in the following species?

a. PF_3 c. H_2S e. $NaOH$ g. O_2^{2-}
b. H_2 d. Na_2SO_4 f. N^{3-} h. NH_4^+

11.8 Classify the following oxygen-containing compounds into the categories (1) oxygen has a −2 oxidation number, (2) oxygen has a −1 oxidation number (peroxide); or (3) oxygen has a positive oxidation number.

a. Na_2O c. OF_2 e. CaO_2
b. Na_2O_2 d. BaO f. FeO

11.9 Classify the following hydrogen-containing compounds into the categories (1) hydrogen has a +1 oxidation number or (2) hydrogen has a −1 oxidation number (hydrides).

a. CH_4 c. HCl e. CaH_2
b. NH_3 d. KH f. H_2Se

Oxidizing and Reducing Agents

11.10 In the following unbalanced equations, identify the oxidizing agent and the reducing agent.

a. $H_2 + Cl_2 \longrightarrow HCl$
b. $SO_2 + O_2 \longrightarrow SO_3$
c. $HBr + Mg \longrightarrow MgBr_2 + H_2$
d. $H_2 + FeCl_3 \longrightarrow HCl + FeCl_2$
e. $H_2S + HNO_3 \longrightarrow S + NO + H_2O$
f. $Zn + Cu^{2+} \longrightarrow Zn^{2+} + Cu$
g. $K_2S + I_2 + KOH \longrightarrow K_2SO_4 + KI + H_2O$
h. $MnO_4^- + H^+ + Hg \longrightarrow Mn^{2+} + H_2O + Hg_2^{2+}$

11.11 For each of the following reactions, indicate whether the underlined element has been (1) oxidized; (2) reduced; or (3) neither oxidized nor reduced.

a. $2 \underline{Mg} + O_2 \longrightarrow MgO$
b. $\underline{CuO} + H_2 \longrightarrow Cu + H_2O$
c. $2 \underline{Fe_2O_3} + 3 C \longrightarrow 4 Fe + 3 CO_2$
d. $\underline{Ag}^+ + Cl^- \longrightarrow AgCl$
e. $\underline{Ag}^+ + Fe^{2+} \longrightarrow Ag + Fe^{3+}$
f. $\underline{Zn} + 2 H^+ \longrightarrow Zn^{2+} + H_2$

g. $\underline{Ba}Cl_2 + H_2SO_4 \longrightarrow BaSO_4 + 2 HCl$
h. $2 K\underline{Cl}O_3 \longrightarrow 2 KCl + 3 O_2$

11.12 For each of the following reactions, identify the substance oxidized, the substance reduced, the oxidizing agent, and the reducing agent.

a. $2 Al + 3 Cl_2 \longrightarrow 2 AlCl_3$
b. $Fe_2O_3 + 2 Al \longrightarrow Al_2O_3 + 2 Fe$
c. $Zn + CuCl_2 \longrightarrow ZnCl_2 + Cu$
d. $4 NH_3 + 3 O_2 \longrightarrow 2 N_2 + 6 H_2O$
e. $K_2Cr_2O_7 + 14 HI \longrightarrow 2 CrI_3 + 2 KI + 3 I_2 + 7 H_2O$
f. $2 KCl + 3 O_2 \longrightarrow 2 KClO_3$
g. $Cu + 4 H^+ + 2 NO_3^- \longrightarrow Cu^{2+} + 2 NO_2 + 2 H_2O$
h. $SeO_3^{2-} + Cl_2 + 2 OH^- \longrightarrow SeO_4^{2-} + 2 Cl^- + H_2O$

Classes of Chemical Reactions

11.13 The following statements represent characteristics or examples of synthesis, decomposition, single-displacement, or double-displacement reactions. Match each statement to the type of reaction described.

a. Upon heating, a substance is changed into two new substances.
b. Iodine, I_2, and sodium chloride, NaCl, are produced when chlorine, Cl_2, is reacted with sodium iodide, NaI.
c. Carbon and oxygen react to form a single product, carbon dioxide.
d. Partners are swapped in this type of reaction.
e. Two substances react to become one substance.

11.14 Classify each of the following reactions as synthesis, decomposition, single displacement, or double displacement.

a. $SO_3 + H_2O \longrightarrow H_2SO_4$
b. $CuCO_3 \longrightarrow CuO + CO_2$
c. $2 AgNO_3 + K_2SO_4 \longrightarrow Ag_2SO_4 + 2 KNO_3$
d. $Mg + 2 HCl \longrightarrow MgCl_2 + H_2$
e. $2 Ag_2O \longrightarrow 4 Ag + O_2$
f. $Al(OH)_3 + 3 HCl \longrightarrow AlCl_3 + 3 H_2O$
g. $FeC_2O_4 \longrightarrow FeO + CO + CO_2$
h. $Fe_2O_3 + 3 C \longrightarrow 2 Fe + 3 CO$

11.15 Classify each of the reactions of Problem 11.14 as oxidation–reduction or metathetical.

12 Nuclear Chemistry and Radioactivity

Objectives

After completing Chapter 12, you will be able to:

12.1 Define the terms *nuclear reaction*, *nuclide*, and *radioactive nuclide*.
12.2 Name and write symbols that indicate the nature of the three types of radiation given off by naturally occurring radioactive materials.
12.3 Write equations, balanced for mass number and atomic number, to represent various alpha and beta decay processes. You will also be able to describe how the atomic number and mass number of a radionuclide changes as the result of alpha, beta, and gamma emission.
12.4 Understand how nuclear reactions can be brought about by bombarding nuclei with various particles and write balanced equations to represent these reactions.
12.5 Understand the half-life concept and determine the fraction of a radionuclide left after a given number of half-lives have elapsed, or vice versa.
12.6 State general methods of production for the synthetic elements and list their stability characteristics.
12.7 Describe the processes of ion-pair formation and free-radical formation that result from the interaction of radiation with matter.
12.8 Contrast the biological effects of exposure to alpha, beta, and gamma radiation.
12.9 State how radiation is detected using either film badges or Geiger counters.
12.10 Discuss the major sources of radiation exposure for human beings.
12.11 Discuss the basic principles behind the uses of radionuclides in diagnostic and therapeutic nuclear medicine.
12.12 Describe the general characteristics of nuclear fission and nuclear fusion.
12.13 Contrast the major differences between nuclear reactions and ordinary chemical reactions.

INTRODUCTION

In this chapter, we will examine the fundamentals of nuclear reactions and look at the numerous important applications of these reactions. Radioactivity, radiation exposure, nuclear weapons, nuclear power plants, and nuclear medicine all fall under the umbrella of nuclear reactions.

Historians talk about how we now live in a "nuclear age". What they really mean is that we now live in the age of nuclear reactions (nuclear change). Nuclear change has been a somewhat controversial subject in recent years. Problems with nuclear power plants and concerns about nuclear warfare have led to considerable negative press. However, there is also a positive side to the nuclear age—nuclear medicine. Today, nuclear change is routinely used in the diagnosis and treatment of numerous diseases. Diseases that were once regarded as incurable can now be treated effectively using nuclear medicine. When considering the pros and cons of nuclear reactions, it has been said that "it is far more likely that you will be saved by nuclear medicine than that you will be killed by nuclear weapons".

12.1 ATOMIC NUCLEI

A **nuclear reaction** *is a reaction in which changes occur in the nucleus of an atom.* Nuclear reactions are not considered to be ordinary chemical reactions. All nuclei remain unchanged in an ordinary chemical reaction. The governing principles for ordinary chemical reactions deal with the rearrangement of electrons; this rearrangement occurs as the result of electron transfer or electron sharing (Section 6.1). In nuclear reactions, it is nuclei rather than electrons that are the point of focus.

Atomic nuclei are the very dense, positively charged centers of atoms about which the electrons move. All nuclei of the atoms of a given element contain the same number of protons. It is this characteristic number of protons that determines the identity of the element. The *atomic number* of an atom gives the number of protons in the nucleus. The number of neutrons associated with the nuclei of a given element can vary within a small range. Atoms of a given element that differ in the number of nuclear neutrons they contain are called *isotopes*. The *mass number* of an atom is equal to the total number of protons and neutrons present in the nucleus. Isotopes of an element have different mass numbers, but the same atomic number. Refer back to Sections 3.5 and 3.6 for further details about atomic numbers, mass numbers, and isotopes.

The term *nuclide* is used extensively in discussions about nuclear reactions. A **nuclide** *is an atom of an element that has a specific number of protons and neutrons in its nucleus.* All atoms of a given nuclide must have the same number of protons and the same number of neutrons. The term *isotope* refers to different forms of the same element. The term *nuclide* is used to describe atomic forms of different elements. The species $^{12}_{6}C$ and $^{13}_{6}C$ are isotopes of the element carbon. The species $^{12}_{6}C$ and $^{16}_{8}O$ are nuclides of different elements.

In order to identify a nucleus or atom uniquely, both the atomic number and mass number must be specified. Two different notation systems exist for doing this. Consider a nuclide of nitrogen that has seven protons and eight neutrons.

nuclear reaction

nuclide

This nuclide can be denoted as $^{15}_{7}N$ or nitrogen-15. In the first notation, the superscript is the mass number and the subscript is the atomic number. In the second notation, the mass number is placed immediately after the name of the element. Both types of notation will be used in this chapter. Note that both notations give the mass number.

Some naturally occurring nuclides, as well as all synthetically produced nuclides, possess nuclei that are *unstable*. In order to achieve stability, these unstable nuclei spontaneously emit energy (radiation). The term **radioactive nuclide** *is used to describe atoms (nuclides) that possess unstable nuclei that spontaneously emit energy (radiation)*. The term *radioactive nuclide* is often shortened to *radionuclide*.*

12.2 THE NATURE OF EMISSIONS FROM RADIOACTIVE MATERIALS

The definition for a radioactive nuclide that was given in the previous section contains the two key concepts necessary for understanding the phenomenon of radioactivity. The two concepts are

1. Certain nuclides possess unstable nuclei.
2. Nuclides with unstable nuclei spontaneously emit energy (radiation).

Of the approximately 340 naturally occurring nuclides, approximately 70 possess unstable nuclei; that is, about 70 of them are radioactive. In addition, numerous unstable nuclides that are not found in nature have been produced in the laboratory.

In 1896, French physicist and engineer Antoine Henri Becquerel (1852–1908) observed the fact that an unstable nucleus spontaneously emits radiation. At that time, he was working with some uranium-containing materials. The first information concerning the nature of the radiation emanating from naturally radioactive materials was obtained by British scientist Ernest Rutherford (1871–1937) in 1898–99. Using an apparatus similar to that shown in Figure 12–1, he found that if radiation from uranium is passed between electrically charged plates, it is split into three components; this indicates the presence of three different types of emissions from naturally radioactive materials. A closer analysis of Rutherford's experiment reveals that one radiation component is positively charged (because it is attracted to the negative plate); a second component is negatively charged (because it is attracted to the positive plate); and the third component carries no charge (because it is unaffected by either charged plate). Rutherford chose to call the three radiation components alpha rays (α rays)—the positive component; beta rays (β rays)—the negative component; and gamma rays (γ rays)—the uncharged component. Alpha, beta, and gamma are the first three letters of the Greek alphabet. We mention Rutherford's nomenclature system because we still use the names for these radiation types today, even though we know much more about their identity. Additional research has substantiated Rutherford's conclusion that three distinct types of radiation are present in the emissions from naturally radioactive substances. This research has also supplied the necessary information for the complete characterization of each type of radiation; these complete identifications required many years of research. Early researchers in the field were hampered by

---

* In some textbooks, the term *radioactive isotope* (radioisotope) is used in place of *radioactive nuclide* (radionuclide). In other textbooks, the two terms are used interchangeably. We will use *nuclide* rather than *isotope* in this text.

FIGURE 12–1 Effect of an electric field on radiation emanating from a naturally radioactive substance.

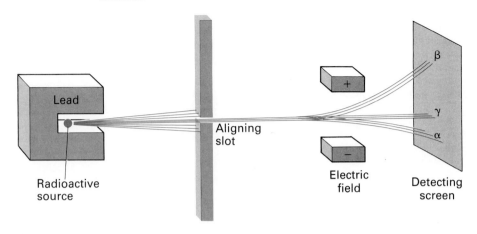

the fact that many of the details concerning atomic structure were not yet known; for example, the neutron was not identified until 1932.

Alpha rays *consist of a stream of positively charged particles (alpha particles),* *each of which is made up of two protons and two neutrons.* The notation used to represent an alpha particle is $^4_2\alpha$. The numerical subscript indicates that the charge on the particle is $+2$ (from the two protons). The numerical superscript indicates a mass of 4 amu. Alpha particles are identical with the nuclei of helium-4 (^4_2He) atoms.

alpha rays

Beta rays *consist of a stream of negatively charged particles (beta particles) whose* *charge and mass are identical to that of an electron.* However, beta particles are not extranuclear electrons; they are particles that have been produced inside the nucleus and then ejected. We will discuss this process in Section 12.3. The symbol used to represent a beta particle is $^{\ \ 0}_{-1}\beta$. The numerical subscript indicates that the charge on the beta particle is -1; it is the same as that of an electron. The use of the superscript zero for the mass of a beta particle should not be interpreted as meaning that a beta particle has no mass, but rather that the mass is very close to zero amu. The actual mass of a beta particle on the atomic weight scale is 0.00055 amu.

beta rays

Gamma rays *are not considered to be particles, but rather are pure energy without* *charge or mass.* They are very high energy radiation, somewhat like X-rays. The symbol for gamma rays is $^0_0\gamma$.

gamma rays

12.3 EQUATIONS FOR NUCLEAR REACTIONS

Alpha, beta, and gamma emissions come from the nucleus of an atom. These spontaneous emissions alter nuclei; obviously, if a nucleus loses an alpha particle (two protons and two neutrons), it will not be the same as it was before the departure of the particle. In the case of alpha and beta emissions, the nuclear alteration causes the identity of the atom to change; that is, a new element is formed. Thus, nuclear reactions differ dramatically from ordinary chemical reactions. In chemical reactions, the identity of the elements is always maintained. This is not the case for nuclear reactions.

The term *decay* (or *disintegration*) is used to describe a nuclear process where an element disappears; that is, the element changes into another element as a result radioactive decay — of radiation emission. **Radioactive decay** is *the process whereby a radionuclide is transformed into a nuclide of another element as a result of the emission of radiation.*

Alpha-Particle Decay

Alpha-particle decay, which is the emission of an alpha particle from a nucleus, always results in the formation of a nuclide of a different element. The product nucleus of this type of decay has an atomic number that is two less than that of the original nucleus and a mass number that is four less than the original nucleus. We can represent alpha particle decay in general terms by the equation

$$_{Z}^{A}X \longrightarrow {_{2}^{4}\alpha} + {_{Z-2}^{A-4}Y}$$

where X is the symbol for the nucleus of the original element undergoing decay and Y is the symbol of the element formed as a result of the decay.

In order to introduce us to actual nuclear equations, let us write equations for two alpha-particle-decay processes. Both $_{83}^{211}\text{Bi}$ and $_{92}^{238}\text{U}$ are radionuclides that undergo alpha-particle decay. The nuclear equations for these two decay processes are

$$_{83}^{211}\text{Bi} \longrightarrow {_{2}^{4}\alpha} + {_{81}^{207}\text{Tl}}$$

$$_{92}^{238}\text{U} \longrightarrow {_{2}^{4}\alpha} + {_{90}^{234}\text{Th}}$$

Let us contrast these two equations with those for ordinary chemical reactions. First of all, nuclear equations convey a different type of information than what is found in ordinary chemical equations. The symbols in nuclear equations stand for nuclei rather than atoms. (We do not worry about electrons when writing nuclear equations.) Second, mass numbers and atomic numbers (nuclear charge) are always used in conjunction with elemental symbols in nuclear equations. Third, the elemental symbols on both sides of the equation need not be (and usually are not) the same in nuclear equations.

balanced nuclear equation — The procedures for balancing nuclear equations are different than those used for ordinary chemical equations. In a **balanced nuclear equation**, *the sum of the subscripts (atomic numbers or particle charges) on each side of the equation are equal and the sum of the superscripts (mass numbers) on each side of the equation are equal.* Both of our examples are balanced. In the alpha decay of $_{83}^{211}\text{Bi}$, the subscripts on both sides total 83 and the superscripts total 211. For the alpha decay of $_{92}^{238}\text{U}$, the subscripts total 92 on both sides and the superscripts total 238 on both sides.

parent nuclide

daughter nuclide

The terms *parent nuclide* and *daughter nuclide* are often used to describe radioactive decay processes. The **parent nuclide** *is the nuclide that undergoes decay in a radioactive decay process.* The **daughter nuclide** *is the nuclide that is produced as a result of a radioactive decay process.* In our two previous equations, thallium-207 and thorium-234 are the daughter nuclides.

Beta-Particle Decay

Beta-particle decay always results in the formation of a nuclide of a different element. The mass number of the new nuclide is the same as that of the original atom. However, the atomic number has increased by one unit. The general equation for beta decay is

$$_{Z}^{A}X \longrightarrow {_{-1}^{0}\beta} + {_{Z+1}^{A}Y}$$

Specific examples of beta-particle decay are

$$^{10}_{4}\text{Be} \longrightarrow {}^{0}_{-1}\beta + {}^{10}_{5}\text{B}$$

$$^{234}_{90}\text{Th} \longrightarrow {}^{0}_{-1}\beta + {}^{234}_{91}\text{Pa}$$

Both of these nuclear equations are balanced; superscripts and subscripts add up to the same sums on each side of the equation.

At this point in the discussion, you may be wondering how a nucleus, which is composed only of neutrons and protons, ejects a negative particle (beta particle) when no such particle is present in the nucleus. The accepted explanation is that a neutron in the nucleus is transformed into a proton and a beta particle through a complex series of steps; that is,

$$^{1}_{0}\text{n} \longrightarrow {}^{1}_{1}\text{p} + {}^{0}_{-1}\beta$$

Once it is formed within the nucleus, the beta particle is ejected with a high velocity.

Gamma-Ray Emission

For naturally occurring radionuclides, gamma-ray emission almost always takes place in conjunction with an alpha- or beta-decay process; it never occurs independently. These gamma rays are often not included in the nuclear equation because they do not affect the balancing of the equation or the identity of the decay product. This can be seen from the following two nuclear equations.

balanced nuclear equation with gamma radiation included $\quad ^{226}_{88}\text{Ra} \longrightarrow {}^{222}_{86}\text{Rn} + {}^{4}_{2}\alpha + {}^{0}_{0}\gamma$

balanced nuclear equation with gamma radiation omitted $\quad ^{226}_{88}\text{Ra} \longrightarrow {}^{222}_{86}\text{Rn} + {}^{4}_{2}\alpha$

The fact that gamma rays are usually left out of nuclear equations does not imply that they are not important. On the contrary, gamma rays are more important than alpha and beta particles when the effects of external radiation exposure on living organisms are considered (Section 12.8). Note also that among *synthetically* produced radionuclides (Section 12.4), there are some pure "gamma emitters", radionuclides that give off gamma rays, but no alpha or beta radiations. These radionuclides are very important in diagnostic nuclear medicine (Section 12.11).

Example 12.1

Write a balanced nuclear equation for the decay of each of the following radioactive nuclides. The mode of decay is indicated in parentheses.

a. $^{70}_{31}\text{Ga}$ (beta emission)   c. $^{248}_{100}\text{Fm}$ (alpha emission)
b. $^{144}_{60}\text{Nd}$ (alpha emission) d. $^{113}_{47}\text{Ag}$ (beta emission)

Solution
In each case, the atomic and mass numbers of the daughter nucleus are obtained by writing the symbols of the parent nucleus and the particle emitted by the nucleus (alpha or beta). Then the equation is balanced.

a. Let X represent the product of the radioactive decay, the daughter nuclide. Then

$$^{70}_{31}\text{Ga} \longrightarrow {}^{0}_{-1}\beta + X$$

The sum of the superscripts on each side of the equation must be equal, so the superscript for X must be 70. In order for the sum of the subscripts on each side of the equation to be equal, the subscript for X must be 32. Then $31 = (-1) + (32)$. As soon as we determine the subscript of X, we can obtain the identity of X by looking at a periodic table. The element with an atomic number of 32 is Ge (germanium). Therefore,

$$^{70}_{31}\text{Ga} \longrightarrow \, ^{0}_{-1}\beta + ^{70}_{32}\text{Ge}$$

b. Letting X represent the product of the radioactive decay, we have for the alpha decay of $^{144}_{60}\text{Nd}$

$$^{144}_{60}\text{Nd} \longrightarrow \, ^{4}_{2}\alpha + X$$

We balance the equation by making the superscripts on each side of the equation total 144 and the subscripts total 60. We get

$$^{144}_{60}\text{Nd} \longrightarrow \, ^{4}_{2}\alpha + ^{140}_{58}\text{Ce}$$

c. Similarly, we write

$$^{248}_{100}\text{Fm} \longrightarrow \, ^{4}_{2}\alpha + X$$

Balancing superscripts and subscripts, we get

$$^{248}_{100}\text{Fm} \longrightarrow \, ^{4}_{2}\alpha + ^{244}_{98}\text{Cf}$$

In alpha emission, the atomic number of the daughter nuclide always decreases by two and the mass number of the daughter nuclide always decreases by four.

d. Finally, we write

$$^{113}_{47}\text{Ag} \longrightarrow \, ^{0}_{-1}\beta + X$$

In beta emission, the atomic number of the daughter nuclide always increases by one and the mass number does not change from that of the parent. The balancing procedure gives us the result

$$^{113}_{47}\text{Ag} \longrightarrow \, ^{0}_{-1}\beta + ^{113}_{48}\text{Cd}$$

12.4 BOMBARDMENT REACTIONS AND ARTIFICIAL RADIOACTIVITY

transmutation reaction

A **transmutation reaction** is *a nuclear reaction in which a nuclide of one element is changed into a nuclide of another element.* Radioactive decay, which was discussed in the last section, is an example of a natural transmutation process. There is also an artificial process that causes transmutation; it involves the use of bombardment reactions. A **bombardment reaction** is *a nuclear reaction in which small particles traveling at very high speeds are collided with stable nuclei; this causes them to undergo nuclear change.*

bombardment reaction

The first successful bombardment reaction was carried out in 1919, twenty-five years after the discovery of radioactive decay. Further research carried out by many investigators has shown that numerous nuclei experience change under the stress of small particle bombardment. In most cases, the new nuclide that is produced as the result of the transmutation is radioactive (unstable) rather than stable. **Artificial radioactivity** is *the name given to the radioactivity associated with radionuclides produced from stable nuclides through bombardment reactions.* Synthetically produced radionuclides undergo radioactive decay, like naturally occurring radionuclides. In many cases, the previously discussed alpha and beta particle

artificial radioactivity

modes of decay occur. Additional modes of decay, which we will not discuss in this text, are also encountered.

Over 1600 synthetically produced radionuclides that do not occur naturally are now known. Included in this total is at least one radionuclide of every naturally occurring element. In addition, nuclides of 21 elements that do not occur in nature have been produced in small quantities as the result of bombardment reactions. These synthetic elements will be discussed in Section 12.6.

The number of synthetically produced nuclides is approximately five times greater than the number of naturally occurring nuclides. There are significant uses for many synthetic radionuclides; they are not all idle laboratory curiosities. Most radioisotopes used in the field of medicine are synthetic radionuclides. For example, the synthetic radionuclides cobalt-60, yttrium-90, iodine-131, and gold-198 are used in radiotherapy treatments for cancer. Section 12.11 provides more information about the medical uses for radionuclides.

12.5 RATE OF RADIOACTIVE DECAY

All radioactive nuclides do not decay at the same rate. Some decay very rapidly; others undergo disintegration at extremely slow rates. This indicates that all radionuclides are not equally unstable. The faster the decay rate is, the lower the stability will be.

The concept of *half-life* is used to quantitatively express nuclear stability. The **half-life** *is the time required for one-half of any given quantity of a radioactive substance to undergo decay.* For example, if a radionuclide's half-life is 12 days and you have a 4.00 g sample of it, then after 12 days (one half-life), only 2.00 g of the sample (one-half the original amount) will remain undecayed; the other half will have decayed into some other substance.

half-life

Half-lives as long as billions of years and as short as a fraction of a second have been determined. Table 12–1 contains examples of the wide range of half-life

TABLE 12–1 Range of Half-lives Found for Naturally Occurring and Synthetic Radionuclides

Element	Half-life
Naturally Occurring Radionuclides	
vanadium-50	6×10^{15} yr
platinum-190	6.9×10^{11} yr
uranium-238	4.5×10^9 yr
uranium-235	7.1×10^8 yr
thorium-230	7.5×10^4 yr
lead-210	22 yr
bismuth-214	19.7 min
polonium-212	3.0×10^{-7} sec
Synthetic Radionuclides	
iodine-129	1.7×10^7 yr
nickel-63	92 yr
gold-195	200 days
lead-200	21 hr
silver-106	24 min
oxygen-19	29.4 sec
fermium-246	1.2 sec
beryllium-8	3×10^{-16} sec

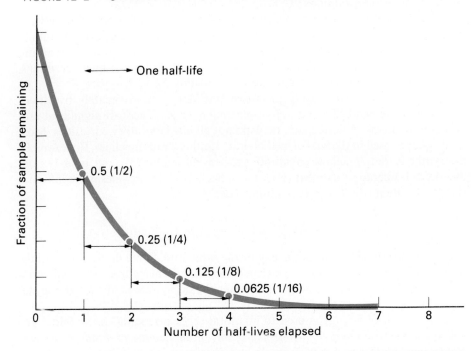

FIGURE 12–2 A general half-life decay curve for a radionuclide.

values that occur. Most naturally occurring radionuclides have long half-lives. However, some radionuclides with *short* half-lives are also found in nature. The decay rate (half-life) of a radionuclide is constant. It is independent of outward conditions such as temperatures, pressure, and state of chemical combination. It is dependent only on the identity of the radionuclide. For example, radioactive sodium-24 decays at the same rate, whether it is incorporated into NaCl, NaBr, Na_2SO_4, or $NaC_2H_3O_2$. Once something is radioactive, nothing will stop it from decaying and nothing will increase or decrease its decay rate.

Figure 12–2 shows graphically the meaning of the term *half-life*. After one half-life has passed, one-half of the original atoms have decayed, and half remain. During the next half-life, one-half of the remaining half will decay, and one-fourth of the original atoms remain undecayed. After three half-lives, $1/2 \times 1/2 \times 1/2 = 1/8$ of the original atoms remain undecayed, and so on. Note from Figure 12–2 that only a very small amount of original material (less than 1%) remains after seven half-lives have elapsed.

Calculations involving amounts of radioactive material decayed, amounts remaining undecayed, and time elapsed can be carried out by using the following equation.

$$\begin{pmatrix} \text{amount of radionuclide} \\ \text{undecayed after } n \text{ half-lives} \end{pmatrix} = \begin{pmatrix} \text{original amount} \\ \text{of radionuclide} \end{pmatrix} \times \begin{pmatrix} \frac{1}{2^n} \end{pmatrix}$$

Example 12.2

Iodine-131 is a radionuclide that is frequently used in nuclear medicine. Among other things, it is used to detect fluid buildup in the brain. The half-life of iodine-131 is 8.0 days. How much of a 0.50 g sample of iodine-131 will remain undecayed after a period of 32 days?

Solution
First, we must determine the number of half-lives that have elapsed.

$$32 \text{ days} \times \left(\frac{1 \text{ half-life}}{8.0 \text{ days}}\right) = 4 \text{ half-lives}$$

Knowing the number of elapsed half-lives and the original amount of radioactive iodine present, we can use the equation

$$\left(\begin{array}{c}\text{amount of radionuclide}\\ \text{undecayed after } n \text{ half-lives}\end{array}\right) = \left(\begin{array}{c}\text{original amount}\\ \text{of radionuclide}\end{array}\right) \times \left(\frac{1}{2^n}\right)$$

$$= 0.50 \text{ g} \times \frac{1}{2^4} \leftarrow \text{four half-lives}$$

$$= 0.50 \text{ g} \times \frac{1}{16}$$

$$= 0.031 \text{ g}$$

Example 12.3

Strontium-90 is a nuclide that is found in radioactive fallout from nuclear weapon explosions. Its half-life is 28.0 yr. How long will it take for 94% (15/16) of the strontium-90 atoms present in a sample of material to undergo decay?

Solution
If 15/16 of the sample has decayed, then 1/16 of the sample remains undecayed. In terms of $1/2^n$, 1/16 is equal to $1/2^4$; that is,

$$\frac{1}{2} \times \frac{1}{2} \times \frac{1}{2} \times \frac{1}{2} = \frac{1}{2^4} = \frac{1}{16}$$

Thus, four half-lives have elapsed in reducing the amount of strontium-90 to 1/16 of its original amount.

The half-life of strontium-90 is 28 years, so the total time elapsed will be

$$4 \text{ half-lives} \times \left(\frac{28.0 \text{ yr}}{1 \text{ half-life}}\right) = 112 \text{ yr}$$

In both Examples 12.2 and 12.3, the time elapsed was equivalent to a whole number of half-lives. In order to work problems involving a fractional number of half-lives, you must use more complicated equations with logarithms. These equations will not be presented in this text; you will be expected to be able to work only problems that involve a whole number of half-lives.

The subject matter areas for Examples 12.2 and 12.3 were nuclear medicine and nuclear fallout. These are two major areas where half-life is of prime importance. In nuclear medicine, almost all radionuclides in use today have relatively short half-lives. Thus, the radioactive effect of these nuclides within the body is limited to short periods of time. The effect that nuclear fallout has on the environment is directly related to the half-lives of the species present in the fallout. The longer the half-lives, the greater the time period that the environment is affected.

12.6 SYNTHETIC ELEMENTS

One of the most interesting facets of bombardment reaction research (Section 12.4) is the production of elements that do not occur in nature, or *synthetic elements*.

TABLE 12-2 Information About the Synthetic Transuranium Elements

Name	Symbol	Atomic Number	Atomic Weight of Most Stable Isotope	Half-life of Most Stable Isotope	Date of Discovery
neptunium	Np	93	237	2.14×10^6 yr	1940
plutonium	Pu	94	244	7.6×10^7 yr	1940
americium	Am	95	243	8.0×10^3 yr	1944
curium	Cm	96	247	1.6×10^7 yr	1944
berkelium	Bk	97	247	1400 yr	1950
californium	Cf	98	251	900 yr	1950
einsteinium	Es	99	252	472 days	1952
fermium	Fm	100	257	100 days	1953
mendelevium	Md	101	258	56 days	1955
nobelium	No	102	259	1 hr	1958
lawrencium	Lr	103	260	3 min	1961
unnilquadium	Unq	104	261	70 sec	1969
unnilpentium	Unp	105	262	40 sec	1970
unnilhexium	Unh	106	263	0.9 sec	1974
unnilseptium	Uns	107	262	0.005 sec	1980
unniloctium	Uno	108	265	0.002 sec	1984
unnilennium	Une	109	266	0.005 sec	1982

Four synthetic elements that were produced between 1937 and 1941 filled gaps in the periodic table for which no naturally occurring element had been found. These four are technicium (Tc, element 43), an element with numerous uses in nuclear medicine; promethium (Pm, element 61); astatine (At, element 85); and francium (Fr, element 87). The reactions for their production are

$$^{96}_{42}\text{Mo} + ^{2}_{1}\text{H} \longrightarrow ^{97}_{43}\text{Tc} + ^{1}_{0}\text{n} \quad \text{(half-life} = 2.6 \times 10^6 \text{ yr)}$$

$$^{142}_{60}\text{Nd} + ^{1}_{0}\text{n} \longrightarrow ^{143}_{61}\text{Pm} + ^{0}_{-1}\beta \quad \text{(half-life} = 265 \text{ days)}$$

$$^{209}_{83}\text{Bi} + ^{4}_{2}\alpha \longrightarrow ^{210}_{85}\text{At} + 3^{1}_{0}\text{n} \quad \text{(half-life} = 8.3 \text{ hr)}$$

$$^{230}_{90}\text{Th} + ^{1}_{1}\text{p} \longrightarrow ^{223}_{87}\text{Fr} + 2^{4}_{2}\alpha \quad \text{(half-life} = 22 \text{ min)}$$

Bombardment reactions have also been the source for elements 93 to 109. These elements are called the *transuranium elements* because they occur immediately following uranium in the periodic table. (Uranium is the highest atomic numbered, naturally occurring element.) All isotopes of all of the transuranium elements are radioactive. Table 12-2 gives information about the stability of the transuranium elements.

12.7 THE IONIZING EFFECTS OF RADIATION

The alpha, beta, and gamma radiations produced from radioactive decay travel outward from their nuclear sources into the material surrounding the radioactive substance. There, they interact with the atoms and molecules of the material and thus lose their energy. Let us consider in closer detail these interactions between radiation and atoms and molecules.

Because the nucleus of an atom occupies such a small portion of the total volume of an atom (Section 3.4), it is not surprising that in the great majority of radiation–atom interactions, the extranuclear electrons of an atom are more directly involved than the nucleus is. In many cases, energy transfer during radiation–atom interaction is sufficient to knock away electrons from atoms; that is, ionization occurs and ion pairs are formed. An **ion pair** *is the electron and positive ion that are*

ion pair

FIGURE 12–3   The interaction of radiation with atoms to form ion pairs.

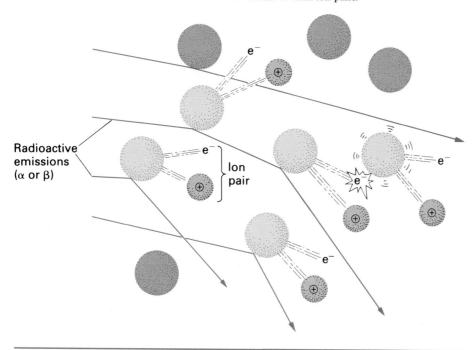

produced during an ionization collision between an atom and radiation. This ionization process is not the voluntary transfer of electrons that occurs during ionic compound formation, but rather a nonchemical, involuntary removal of electrons from atoms to form ions. Figure 12–3 diagrammatically shows ion-pair formation. Many ion pairs are produced by a single "particle" of radiation because such a particle must undergo many collisions before its energy is reduced to the level of the surrounding material. The electrons ejected from an atom frequently have enough energy to bombard neighboring molecules and cause additional ionization.

Free-radical formation is another effect caused by alpha, beta, and gamma radiation. A **free radical** *is a highly reactive uncharged molecular fragment; that is, an uncharged piece of a molecule.* Free radicals are very reactive entities because they contain an unpaired electron. (Recall from Section 6.8 that electrons occur in pairs in normal bonding situations.)

free radical

The ionizing effect of radiation (ion-pair formation and free-radical formation) is the factor that makes radiation harmful to both living matter and inert materials. In living matter, the formation of ions and free radicals disrupts cellular function.

12.8 THE BIOLOGICAL EFFECTS OF VARIOUS TYPES OF RADIATION

The three types of naturally occurring radioactive emissions—alpha particles, beta particles, and gamma rays—differ in their ability to penetrate matter and cause ionization. Consequently, the extent of the biological effects of radiation is dependent upon the type of radiation involved.

Alpha particles are the most massive and also the slowest particles involved in natural radioactive decay; they are emitted from nuclei at a velocity of about one-tenth of the speed of light. They have low penetrating power and cannot penetrate

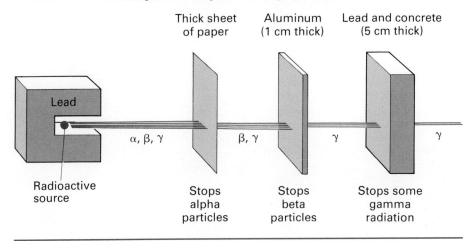

FIGURE 12-4 Penetrating abilities of alpha, beta, and gamma radiation.

the body's outer layers of skin. The major danger from alpha radiation occurs when alpha-emitting radionuclides are ingested, for example, in contaminated food. There are no protective layers of skin within the body.

Beta particles are emitted from nuclei at speeds of about nine-tenths that of light. With their greater velocity, they can penetrate much deeper than alpha particles and can cause severe skin burns if their source remains in contact with the skin for an appreciable time. They do not ionize molecules as readily as alpha particles do because of their much smaller size. An alpha particle is approximately 8000 times heavier than a beta particle. It is estimated that a typical alpha particle travels about 6 cm in air and produces 40,000 ion pairs and a typical beta particle travels 1000 cm in air and produces about 2000 ion pairs. Internal exposure to beta radiation is as serious as internal alpha exposure.

TABLE 12-3 The Effects of Short-Term Whole-Body Radiation Exposure on Humans

Dose (rems)[a]	Effects
0–25	No detectable clinical effects.
25–100	Slight short-term reduction in number of some blood cells; disabling sickness not common.
100–200	Nausea and fatigue, vomiting if dose is greater than 125 rems; longer-term reduction in number of some blood cells.
200–300	Nausea and vomiting first day of exposure; up to a two-week latent period followed by appetite loss, general malaise, sore throat, pallor, diarrhea, and moderate emaciation. Recovery in about three months unless complicated by infection or injury.
300–600	Nausea, vomiting, and diarrhea in first few hours. Up to a one-week latent period followed by loss of appetite, fever, and general malaise in the second week, followed by hemorrhage, inflammation of mouth and throat, diarrhea, and emaciation. Some deaths in two to six weeks. Eventual death for 50 percent if exposure is above 450 rems; others recover in about six months.
600 or more	Nausea, vomiting, and diarrhea in first few hours. Rapid emaciation and death as early as second week. Eventual death of nearly 100 percent.

[a] A rem is the quantity of ionizing radiation that must be absorbed by a human to produce the same biological effect as 1 roentgen of high penetration X-rays. A roentgen is the quantity of high penetration X-rays that produces approximately 2×10^9 ion pairs per cm^3 of dry air at 0°C and 1 atmosphere.

Gamma radiation is released at a velocity equal to that of the speed of light. Gamma rays readily penetrate deeply into organs, bone, and tissue.

Figure 12–4 contrasts the abilities of alpha, beta, and gamma radiations to penetrate paper, aluminum foil, and a thin layer of a lead–concrete mixture.

The minimum radiation dosage that causes human injury is unknown. However, the effects of larger doses have been studied and are listed in Table 12–3. As you can see, very serious damage or death can result from large doses of ionizing radiation. The dosages causing the various effects listed in Table 12–3 are given in terms of the radiation unit called a *rem*, which is defined in a footnote to the table.

12.9 DETECTION OF RADIATION

You cannot hear, feel, taste, see, or smell low levels of radiation. However, there are numerous methods for detecting the presence of radiation. Becquerel's initial discovery of radioactivity (Section 12.2) was the result of the effect of radiation on photographic plates. Radiation affects photographic film in the same way as ordinary light; the film is exposed. Technicians and others who work around radiation usually wear film badges (see Figure 12–5) to record the extent of their exposure to radiation. When the film from the badge is developed, the degree of darkening of the film negative indicates the extent of radiation exposure. By using different filters, various parts of the film register exposures to the types of radiations (alpha, beta, gamma and X-rays).

Radiation can also be detected by making use of the fact that radiation ionizes atoms and molecules (Section 12.7). The Geiger counter operates on this principle. The basic components of a Geiger counter are shown in Figure 12–6. The detection part of such a counter is a metal tube filled with a gas (usually argon). The tube has a thin-walled window made of a material that can be penetrated by alpha, beta, or gamma rays. In the center of the tube is a wire attached to the positive terminal of an electrical power source. The metal tube is attached to the negative terminal of the same source. Radiation entering the tube ionizes the gas, which

FIGURE 12–5 Film badges, shown here on the lapel of a lab coat, are worn by technicians and others whose work requires them to be around radiation. The extent of exposure of all personnel is carefully logged to help prevent overexposures.

FIGURE 12–6 The principle of operation of a Geiger counter. When radiation enters through the window, it ionizes one or more gas atoms, producing ion pairs. The electrons from the ion pairs are attracted to the central wire, and the positive ions are drawn to the metal tube. This constitutes a pulse of electric current, which is amplified and displayed on the meter or other readout.

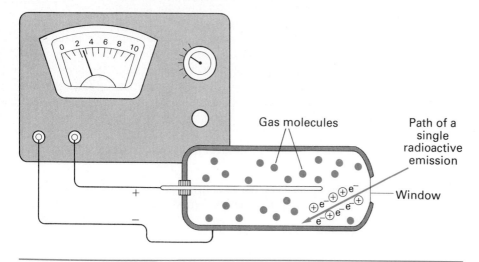

allows a pulse of electricity to flow. This pulse is amplified and displayed on a meter or some other type of readout display.

12.10 SOURCES OF RADIATION EXPOSURE

Most of us will never come in contact with the radiation dosage necessary to cause the effects listed in Table 12–3. Nevertheless, *low-level* exposure to ionizing radiation is something we constantly encounter. In fact, there is no way we can totally avoid this low-level exposure because much of it results from naturally occurring environmental processes.

Five major sources of low-level radiation exposure exist. They are (1) natural sources, (2) medical sources, (3) fallout from nuclear weapon testing, (4) nuclear power production, and (5) consumer products. Estimates of per capita radiation exposure from these five sources are given in Figure 12–7.

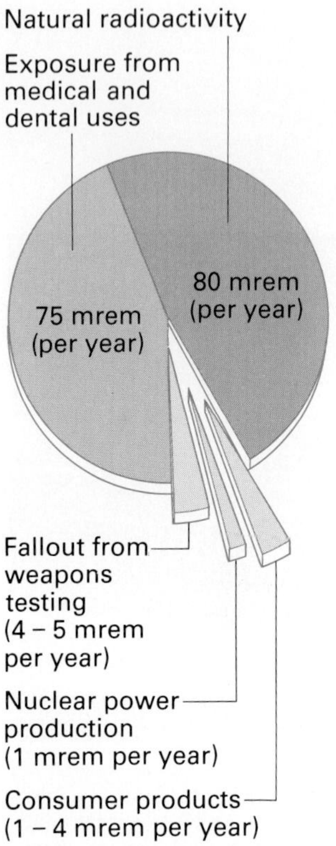

A comparison of the values of Figure 12–7 with those of Table 12–3 that takes into account the fact that the units in former are millirems and those in the latter are rems shows that the current dosage levels received by the general population are very small, compared to those known to cause serious radiation sickness. In the future, it will be important to monitor carefully the amount of radioactive materials we cause to enter the environment in order to ensure that radiation levels remain low.

With low-level radiation exposure, cell damage rather than cell death frequently occurs. If the damaged cells repair themselves improperly, which often occurs, new, abnormal cells are produced when the cells replicate. Thus, radiation may cause damage by altering genetic codes. Much still needs to be learned about the long-term effects of cell damage caused by low-level radiation exposure.

Cells that reproduce at a rapid rate, such as those in bone marrow, lymph nodes, and embryonic tissue, are the most sensitive to radiation damage. The sensitivity of embryonic tissue to radiation damage is the reason that pregnant women need to be protected from radiation exposure. One of the first signs of overexposure to radiation is a drop in red blood cell count. This directly relates to the sensitivity of bone marrow, the site of red cell formation, to radiation.

FIGURE 12–7 Estimates of per capita radiation dose from all sources of exposure.

- Natural radioactivity: 80 mrem (per year)
- Exposure from medical and dental uses: 75 mrem (per year)
- Fallout from weapons testing: 4–5 mrem per year
- Nuclear power production: 1 mrem per year
- Consumer products: 1–4 mrem per year

12.11 NUCLEAR MEDICINE

Radionuclides are used both diagnostically and therapeutically in medicine. In diagnostic applications, technicians use small amounts of radionuclides whose progress through the body or localization in specific organs can be followed. Larger quantities of radionuclides are used in therapeutic applications.

The fundamental chemical principle behind the use of radionuclides in diagnostic medical work is the fact that a radioactive isotope of an element has the same chemical properties as a nonradioactive isotope of the element. (Radioactive and nonradioactive isotopes of an element differ in nuclear properties, but not in chemical properties.) Thus, the body chemistry is not upset by the presence of a small amount of a radioactive substance that is already present in the body in a nonradioactive form.

The criteria used in selecting radionuclides for diagnostic procedures include the following.

1. At low concentrations (to minimize radiation damage), the radionuclide must be detectable by instrumentation placed outside the body. Almost all diagnostic radionuclides are gamma emitters because alpha and beta particles have penetrating power that is too low.

2. It must have a short half-life so that the intensity of the radiation is sufficiently great to be detected. A short half-life also limits the time period of radiation exposure.
3. It must have a known mechanism for elimination from the body so that the material does not remain in the body indefinitely.
4. Its chemical properties must be such that it is compatible with normal body chemistry. It must be able to be selectively transmitted to the part or system of the body that is under study.

The circulation of blood in the body can be followed by using radioactive sodium-24. A small amount of this is isotope is injected into the bloodstream in the form of a sodium chloride solution. The movement of this radionuclide through the circulation system can be followed easily with radiation-detection equipment. If it takes longer than normal for the nuclide to show up at a particular part of the body, this is an indication that the circulation is impaired at that spot. Sodium-24 can also be used to locate blood clots because the amount of radiation will be high on one side and low on the opposite side of the clot.

Radiologists evaluate the functioning of the thyroid gland by administering iodine-131 to a patient, usually in the form of a sodium iodide (NaI) solution. The radioactive iodine behaves in the same manner as ordinary iodine and is absorbed by the thyroid at a rate related to the activity of the gland. If a hypothyroid condition exists, the amount accumulated is less than normal, and if a hyperthyroid condition exists, a greater than average amount accumulates.

The size and shape of organs, as well as the presence of tumors, can be determined in some situations by scanning the organ in which a radionuclide tends to concentrate. Iodine-131 and technetium-99 are used to generate thyroid and brain scans. In the brain, technetium-99, in the form of a polyatomic ion (TcO_4^-), concentrates in brain tumors more than in normal brain tissue; this helps radiologists determine the presence, size, and location of brain tumors. Figure 12–8 shows a

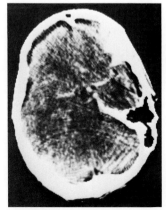

FIGURE 12–8 Brain scan obtained using radioactive technetium-99.

TABLE 12–4 Selected Radionuclides Used in Diagnostic Procedures

Isotope	Half-life	Part of body	Use in diagnosis
barium-131	11.6 d	bone	detection of bone tumors
chromium-51	27.8 d	blood	determination of blood volume and red blood cell lifetime
		kidney	assessment of kidney activity
iodine-131	8.05 d	brain	detection of fluid buildup in the brain
		kidney	location of cysts
		lung	location of blood clots
		thyroid	assessment of iodine uptake by thyroid
iron-59	45 d	blood	evaluation of iron metabolism in blood
phosphorus-32	14.3 d	blood	blood studies
		breast	assessment of breast carcinoma
potassium-42	12.4 h	tissue	determination of intercellular spaces in fluids
sodium-24	15.0 hr	blood	detection of circulatory problem; assessment of peripheral vascular disease
technetium-99	6.0 h	brain	detection of brain tumors, hemorrhages, or blood clots
		spleen	measurement of size and shape of spleen
		thyroid	measurement of size and shape of thyroid
		lung	location of blood clots

brain scan obtained using the radionuclide technetium-99. In this figure, the bright spot at the upper right indicates a tumor that has absorbed a greater amount of radioactive material than the normal brain tissue.

Table 12–4 (p. 283) lists a number of radionuclides that are used in diagnostic procedures. The half-life of the radionuclide, the parts of body it concentrates in, and its diagnostic value are also given in the table.

When radionuclides are used for therapeutic purposes, the objectives are entirely different than those for diagnostic procedures. The main objective for radionuclides in therapeutic use is to *selectively* destroy abnormal (usually cancerous) cells. The radionuclide is often, but not always, placed within the body. There is no need to monitor the radiation produced with an external detector. Therapeutic radionuclides implanted in the body are usually alpha or beta emitters because an intense dose of radiation in a small localized area is needed.

A commonly used implantation radionuclide that is effective in the localized treatment of tumors is yttrium-90, a beta emitter with a half-life of 64 hours. Yttrium-90 salts are implanted by using small hollow needles that are inserted into the tumor.

External, high-energy beams of gamma radiation are also extensively used in the treatment of certain cancers. Cobalt-60 is frequently used for this purpose; a beam of radiation is focused on the small area of the body where the tumor is located (see Figure 12–9). This therapy usually causes some radiation sickness because normal cells are also affected, but to a lesser extent. The operating principle

FIGURE 12–9 The use of a cobalt-60 source in radiation therapy.

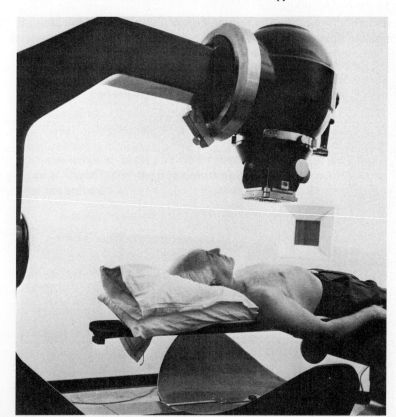

TABLE 12–5 Selected Radionuclides Used in Radiation Therapy

Isotope	Half-Life	Type of Emitter	Use in Therapy
cobalt-60	5.3 y	gamma	external source of radiation of treatment of cancer
iodine-131	8 d	beta, gamma	cancer of thyroid
phosphorus-32	14.3 d	beta, gamma	treatment of some types of leukemia and widespread carcinomas
radium-226	1620 yr	alpha, gamma	used in implantation cancer therapy
radon-222	3.8 d	alpha, gamma	used in treatment of uterine, cervical, oral, and bladder cancers
yttrium-90	64 hr	beta, gamma	implantation therapy

here is that abnormal cells are more susceptible to radiation damage than normal cells are (Section 12.10). Radiation sickness is the price paid for abnormal cell destruction. Table 12–5 lists selected radionuclides that are used in therapy.

12.12 NUCLEAR FISSION AND NUCLEAR FUSION

Our glimpse into the world of nuclear chemistry would not be complete without a brief mention of two additional types of nuclear reactions—nuclear fission and nuclear fusion. **Nuclear fission** *is the process in which a heavy element nucleus splits into two or more lighter nuclei as the result of nuclear bombardment.*

nuclear fission

The first fissionable nucleus to be discovered, which remains the most important one, was uranium-235. Bombardment of this nucleus with neutrons causes it to split into two fragments. Characteristics of the uranium-235 fission reaction include the following.

1. There is no unique way in which the uranium-235 nucleus splits. Thus, many different lighter elements are produced during uranium-235 fission reactions. The following are examples of the ways in which this fission process proceeds.

$$^{235}_{92}U + ^{1}_{0}n \longrightarrow \begin{cases} ^{135}_{53}I + ^{97}_{39}Y + 4\,^{1}_{0}n \\ ^{139}_{56}Ba + ^{94}_{36}Kr + 3\,^{1}_{0}n \\ ^{131}_{50}Sn + ^{103}_{42}Mo + 2\,^{1}_{0}n \\ ^{139}_{54}Xe + ^{95}_{38}Sr + 2\,^{1}_{0}n \end{cases}$$

2. Very large amounts of energy, which are many times greater than that released by ordinary radioactive decay, are emitted during the fission process. It is this large release of energy that makes nuclear fission of uranium-235 the important process that it is. In general, the term *nuclear energy* is used to refer to the energy released during a nuclear fission process. An older term for this energy is *atomic energy*.
3. Neutrons, which are reactants in the fission process, are also produced as products. The number of neutrons produced per fission depends on the way in which the nucleus splits, and ranges from two to four (as can be seen from the fission equations previously given). On the average, 2.4 neutrons are produced per fission. The significance of the produced neutrons is that they can cause the fission process to continue by colliding with further uranium-235 nuclei. Figure 12–10 (p. 286) shows diagrammatically the chain reaction that can occur once the fission process is started.

The process of nuclear fission, or "splitting the atom" as it is called in popularized science, can be carried out in both an uncontrolled manner and a controlled manner. The key question involves what happens to the neutrons produced during fission. Do they react further, causing further fission, or do they escape into the surroundings? If the majority of produced neutrons react further (Figure 12–10), an uncontrolled nuclear reaction (an atomic bomb) results. When only a few neutrons react further (on the average one per fission), the fission reaction self-propagates in a controlled manner.

The process of nuclear fission is the basis for the operation of nuclear power plants that are used to produce electricity. In this case, the fission process is carried out in a controlled manner. The reaction is controlled with rods that absorb excess neutrons (so that they cannot cause unwanted fissions) and with moderating substances that decrease the speed of the neutrons. The energy produced during the fission process, which appears as heat, is used to operate steam-powered, electricity generating equipment.

It is now known that there is another type of nuclear reaction that produces even more energy than nuclear fission. **Nuclear fusion** *is the process in which small nuclei are put together to make larger ones.* This process is essentially the opposite of nuclear fission. In order for fusion to occur, a very high temperature, of several hundred million degrees is required.

nuclear fusion

FIGURE 12–10 A fission chain reaction caused by further reaction of the neutrons produced during fission. (The "free" neutrons are shown here proportionally larger than those contained within the nuclei.)

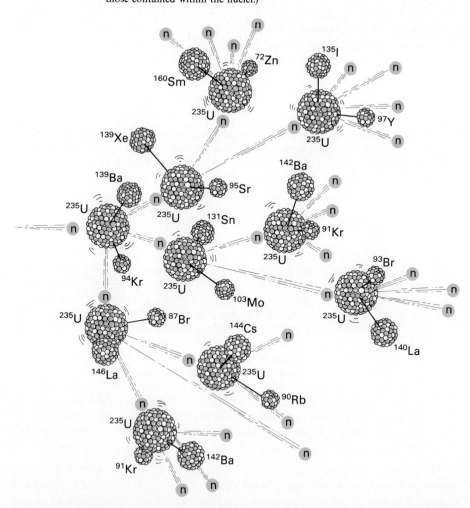

TABLE 12-6 Differences Between Nuclear and Chemical Reactions

Chemical Reaction	Nuclear Reaction
1. Different isotopes of an element have practically identical chemical properties.	1. Different isotopes of an element have different properties in nuclear processes.
2. The chemical reactivity of an element depends on the element's state of combination (free element, compound, etc.).	2. The nuclear reactivity of an element is independent of the state of chemical combination.
3. Elements retain their identity in chemical reactions.	3. Elements may be changed into other elements during nuclear reactions.
4. Energy changes that accompany chemical reactions are relatively small.	4. Nuclear reactions involve energy changes a number of orders of magnitude larger than those in chemical reactions.

One place that is hot enough for nuclear fusion to occur is the interior of the sun. Nuclear fusion is the process by which the sun generates its energy. Within the sun, in a three-step reaction, hydrogen-1 nuclei are converted to helium-4 nuclei with the release of extraordinarily large amounts of energy.

The use of nuclear fusion on Earth might seem impossible because of the high temperatures required. It has, however, been accomplished in a hydrogen bomb. In such a bomb, a *fission* device (an atomic bomb) is used to achieve the high temperatures needed to start the following fusion reaction.

$$^{3}_{1}H + ^{2}_{1}H \longrightarrow ^{4}_{2}He + ^{1}_{0}n$$

At the high temperature of fusion reactions, electrons completely separate from nuclei. Neutral atoms cannot exist. This high-temperature, gas-like mixture of nuclei and electrons that results is called a *plasma* and is considered by some scientists to represent a fourth state of matter (in addition to solids, liquids, and gases).

12.13 A COMPARISON OF NUCLEAR AND CHEMICAL REACTIONS

As can be seen from the discussion of the previous sections of this chapter, nuclear chemistry is a field that is quite different from ordinary chemistry. Many of the laws of chemistry must be modified when we consider nuclear reactions. The major differences between nuclear reactions and ordinary chemical reactions are listed in Table 12-6. This table serves as a summary of many of the concepts presented in this chapter.

EXERCISES AND PROBLEMS

Nuclear Symbols

12.1 Write an alternate nuclear symbol for each of the following nuclides.
a. nitrogen-14
b. chlorine-35
c. tin-121
d. gold-197
e. boron-10
f. sodium-24

12.2 Indicate the number of protons and neutrons contained in each of the following nuclides.
a. $^{24}_{12}Mg$
b. $^{27}_{13}Al$
c. $^{235}_{92}U$
d. $^{127}_{53}I$
e. $^{75}_{34}Se$
f. $^{133}_{54}Xe$

Types of Radiation

12.3 Three distinct types of radiation are present in the emissions given off by naturally occurring radioactive substances.
a. What are the names of these types of radiation?
b. What notation is used to represent these radiations?
c. What charge is associated with each type of radiation?

Equations for Radioactive Decay Processes

12.4 Write balanced nuclear equations for the alpha decay of each of the following radionuclides.
a. $^{200}_{84}Po$
b. curium-240
c. $^{244}_{96}Cm$
d. uranium-238
e. $^{229}_{90}Th$
f. bismuth-210

12.5 Write balanced nuclear equations for the beta decay of each of the following radionuclides.
a. $^{10}_{4}Be$
b. carbon-14
c. $^{21}_{9}F$
d. sodium-25
e. $^{77}_{32}Ge$
f. uranium-235

12.6 What is the relationship between the mass numbers and atomic numbers for parent and daughter nuclides when the following decay processes occur?
a. An alpha particle is emitted.
b. A beta particle is emitted.

12.7 Supply the missing symbol for each of the following radioactive decay processes.
a. $^{32}_{14}Si \longrightarrow ^{32}_{15}P + ?$
b. $? \longrightarrow ^{28}_{13}Al + ^{0}_{-1}\beta$
c. $^{252}_{99}Es \longrightarrow ^{248}_{97}Bk + ?$
d. $? \longrightarrow ^{230}_{92}U + ^{4}_{2}\alpha$
e. $^{84}_{35}Br \longrightarrow ? + ^{0}_{-1}\beta$
f. $^{204}_{82}Pb \longrightarrow ? + ^{4}_{2}\alpha$

12.8 Identify the mode of decay for each of the following radionuclides, given the identity of the daughter nuclide.
a. $^{190}_{78}Pt$ (daughter = $^{186}_{76}Os$)
b. $^{19}_{8}O$ (daughter = $^{19}_{9}F$)
c. uranium-238 (daughter = thorium-234)
d. rhodium-104 (daughter = palladium-104)

Equations for Bombardment Reactions

12.9 Identify the missing symbol in each of the following bombardment reactions.
a. $^{24}_{12}Mg + ? \longrightarrow ^{27}_{14}Si + ^{1}_{0}n$
b. $^{27}_{13}Al + ^{2}_{1}H \longrightarrow ? + ^{4}_{2}\alpha$
c. $? + ^{1}_{1}H \longrightarrow 2\,^{4}_{2}He$
d. $^{14}_{7}N + ^{4}_{2}\alpha \longrightarrow ? + ^{1}_{1}H$
e. $? + ^{11}_{5}B \longrightarrow ^{257}_{103}Lr + 4\,^{1}_{0}n$
f. $^{9}_{4}Be + ? \longrightarrow ^{12}_{6}C + ^{1}_{0}n$

12.10 Write equations for the following nuclear bombardment processes.
a. Beryllium-9 captures an alpha particle and emits a neutron.
b. Nickel-58 is bombarded with a proton and an alpha particle is emitted.
c. Bombardment of a radionuclide with an alpha particle results in the production of curium-242 and one neutron.
d. Bombardment of curium-246 with a small particle results in the production of $^{254}_{102}No$ and four neutrons.

12.11 How does the number of synthetically produced radionuclides compare with the number of naturally occurring nuclides?

Rate of Radioactive Decay

12.12 What is meant by the term *half-life* when it is applied to a radionuclide?

12.13 Explain the fallacy in the conclusion that the whole-life of a radionuclide is equal to twice the half-life.

12.14 Technetium-99 has a half-life of 6.0 hours. What fraction of the technetium-99 atoms in a sample will remain after
a. 12 hours
b. 36 hours
c. 3 days
d. three half-lives
e. six half-lives
f. ten half-lives

12.15 Sodium-24, which is used to locate blood clots in the human circulatory system, has a half-life of 15.0 hours. If you start with 10.0 g of sodium-24, how much will be left in 60.0 hours?

12.16 Some objections to nuclear power plants are based on the need to store the radioactive wastes that result. One radionuclide found in fission reactor wastes is strontium-90, which has a half-life of 28 years. How long would this nuclide have to be stored in order to reduce its amount to about 1/1000 of what was originally in the waste?

Ionizing Effects of Radiation

12.17 What are ion pairs and how are they produced?

12.18 What is a free radical and how is it produced?

Biological Effects of Radiation

12.19 Why is ionizing radiation harmful to living cells?

12.20 Contrast the penetrating power of alpha, beta, and gamma rays into matter. Which of these types of radiation produces the most ion pairs?

12.21 What would be the expected effects of each of the following short-term, whole-body radiation exposures?
- a. 10 rems
- b. 30 rems
- c. 150 rems
- d. 400 rems

Sources of Radiation Exposure

12.22 What are the five major sources of low-level radiation exposure for human beings? Contrast these sources in terms of mrem per year exposure that an average person receives from each source.

Nuclear Medicine

12.23 The radionuclides used for diagnostic procedures are almost always gamma emitters. Why is this so?

12.24 The radionuclides used in diagnostic procedures almost always have short half-lives. Why is this so?

12.25 Explain how each of the following radionuclides is used in diagnostic medicine.
- a. barium-131
- b. sodium-24
- c. iron-59
- d. potassium-42
- e. iodine-131
- f. technetium-99

12.26 How do the radionuclides used for therapeutic purposes differ from the radionuclides used for diagnostic purposes?

12.27 Contrast the different manners in which cobalt-60 and yttrium-90 are used in radiation therapy.

12.28 Exposure to ionizing radiation can cause cancer; yet radionuclides are used in the treatment of some cancers. Explain this apparent contradiction.

Nuclear Fission and Nuclear Fusion

12.29 Identify which of the following characteristics apply to the fission process, the fusion process, or both processes.
- a. An extremely high temperature is required to start the process.
- b. An example of the process occurs on the sun.
- c. Transmutation of elements occurs.
- d. Large amounts of energy are released in the process.
- e. Neutrons are needed to start the process.
- f. Energy released in the process is called *atomic energy*.
- g. The process is now used to generate some electrical power in the United States.

12.30 Identify the following nuclear reactions as fission, fusion, or neither.
- a. $^{3}_{2}He + ^{3}_{2}He \longrightarrow ^{4}_{2}He + 2\,^{1}_{1}H$
- b. $^{239}_{92}U \longrightarrow ^{239}_{93}Np + ^{0}_{-1}\beta$
- c. $^{235}_{92}U + ^{1}_{0}n \longrightarrow ^{144}_{55}Cs + ^{90}_{37}Rb + 2\,^{1}_{0}n$
- d. $^{230}_{90}Th + ^{1}_{1}H \longrightarrow ^{223}_{87}Fr + 2\,^{4}_{2}\alpha$
- e. $^{3}_{1}H + ^{2}_{1}H \longrightarrow ^{4}_{2}He + ^{1}_{0}n$

12.31 A fourth state of matter called a *plasma* is encountered when studying fusion reactions. What are the characteristics of matter in this state?

13

Chemistry of Several Common Elements and Compounds

Objectives

After completing Chapter 13, you will be able to:

13.1 List the most abundant elements in the Earth's crust and tell where and in what form they are found in the crust.
13.2 State the chemical characteristics of the lower regions of the atmosphere.
13.3 Be familiar with the chemistry and polluting effects of carbon monoxide.
13.4 Be familiar with the chemistry and polluting effects of sulfur dioxide, especially the phenomenon of acid rain.
13.5 Be familiar with the chemistry and effects of the primary pollutant nitrogen monoxide and the secondary pollutant nitrogen dioxide.
13.6 List the chemical characteristics of the hydrosphere.
13.7 Be familiar with the chemistry and effects of hard water.
13.8 Be familar with the chemistry of water-softening processes.
13.9 Be familiar with the chemistry of water fluoridation.

INTRODUCTION

This chapter may be looked upon as a "bridge" chapter, paving the way for a dramatic change that is about to occur in our focus on the subject of chemistry. Up to this point (Chapters 1 through 12), our focus has been on general chemical theory—chemical principles that apply in a broad spectrum of situations. We have been concerned with all of the elements and general guidelines, many of them mathematical in nature, concerning their behavior.

After this chapter, the remainder of the book (Chapters 14 through 33) is descriptive rather than theoretical in nature. It deals predominantly with the element carbon, rather than with all of the elements. Descriptive chemistry emphasizes the structures and reactions of specific compounds. The substances that are produced in a specific chemical reaction will be of particular concern to us.

This chapter contains both the descriptive aspect of the chapters to follow and the general aspect of previous chapters. We will look at many different substances, and we will be concerned with the reactions and properties of these substances.

The main reason most students take this chemistry course is because they want basic information about carbon chemistry—the fields of organic and biochemistry (Chapters 14 through 33). However, it is still important that students obtain some general knowledge about other elements. Many of these other elements are extremely important in the world in which we live. Any student who has had a year of chemistry should know some basics about a number of elements, regardless of what his or her chosen vocation is. Admittedly, a one-chapter exposure to descriptive chemistry of the non-carbon world must be limited in scope. Nevertheless, it is sufficient to expose many important facets of the everyday world.

13.1 ABUNDANCES OF THE ELEMENTS

In Section 2.6, we considered the abundances of the elements from various viewpoints. The viewpoints included the entire universe, the Earth (including its core), the Earth's crust, the atmosphere, the hydrosphere, the human body, and vegetation. Abundances differed markedly when going from one frame of reference to another. In the big picture (the universe), hydrogen is the dominant element. On Earth, oxygen is the dominant element. We will now consider further details relative to elemental abundances within the Earth's crust.

In terms of mass, a relatively small number of elements dominate the makeup of the Earth's crust, as can be seen from the numbers in Table 13-1 (p. 292). Using the numbers from Table 13-1 and an electronic hand calculator, we can quickly determine that:

1. Two elements account for 75.2% of the mass of the Earth's crust.
2. Five elements account for 90.8% of the mass of the Earth's crust.
3. Ten elements account for 99.2% of the mass of the Earth's crust.
4. Twenty elements account for over 99.9% of the mass of the Earth's crust.

Thus, most of the 88 naturally occurring elements (over three-fourths of them) are really quite rare; collectively, they account for less than 0.1% of the mass of the Earth's crust.

TABLE 13–1 The Most Abundant Elements Within the Earth's Crust on a Mass Basis

oxygen	O	49.5%		chlorine	Cl	0.19%	
silicon	Si	25.7		phosphorus	P	0.12	
aluminum	Al	7.5		manganese	Mn	0.09	
iron	Fe	4.7		carbon	C	0.08	
calcium	Ca	3.4	99.2%	sulfur	S	0.06	0.7%
sodium	Na	2.6		barium	Ba	0.04	
potassium	K	2.4		chromium	Cr	0.033	
magnesium	Mg	1.9		nitrogen	N	0.030	
hydrogen	H	0.87		fluorine	F	0.027	
titanium	Ti	0.58		zirconium	Zr	0.023	
				All others		0.1%	

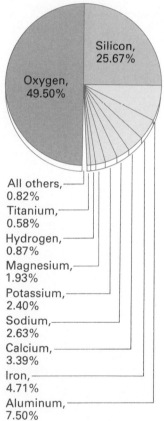

FIGURE 13–1 Mass percent of various elements in the Earth's crust.

Let us now focus our attention on the ten most abundant elements (on a mass basis) in the Earth's crust and the forms in which they occur in nature. Figure 13–1 gives us a better perspective on the relative amounts of these elements.

Oxygen

By far the most abundant element, oxygen accounts for almost one-half of the mass of the Earth's crust. It is the most abundant element in both the solid portion of the Earth's crust and the hydrosphere and is the second most abundant element in the atmosphere.

The majority of rocks and minerals are silicon–oxygen-based systems to which small amounts of metals such as iron, aluminum, and calcium are appended. The fundamental building block for these systems is the SiO_4^{4-} ion or species derived from it. Hence, oxygen accounts for a large percentage of the mass of rock materials.

The waters of this planet are 89% oxygen by mass because oxygen takes up 16 out of the 18 units of atomic weight in water.

Elemental oxygen (O_2) occurs in the atmosphere, although it is not the predominant element there; nitrogen is the most abundant element in the atmosphere (see Table 7–2). Very small amounts of oxygen in combined form are also present in the atmosphere in CO_2 and in air pollutants such as CO, SO_2, NO, and NO_2. The total mass of oxygen in the atmosphere is negligible compared to what is found in the liquid and solid portions of the Earth's crust. We have previously noted (Section 5.9) that oxygen is one of the four macronutrients needed for the functioning of the human body. Most of the compounds found within the human body contain oxygen.

Silicon

Most people (except those who have had a chemistry class) are surprised when they are told that silicon is the second most abundant element. It is an element most people have never encountered in pure form in everyday life. Where is all of the silicon on the Earth? Approximately 85% of the solid portion of the Earth's crust is silicon-based material. Sand and sandstone are impure forms of SiO_2 and quartz is a pure, crystalline form of the same substance. Silicates are additional silicon–oxygen compounds, as was mentioned in our previous discussion of oxygen (see Figure 13–2a).

FIGURE 13-2 (a) Silica, SiO_2 (on the left) is the mineral from which the element silicon (on the right) is obtained. (b) A tiny microcomputer chip (held on the end of a paperclip) is fabricated from a single piece of highly purified silicon.

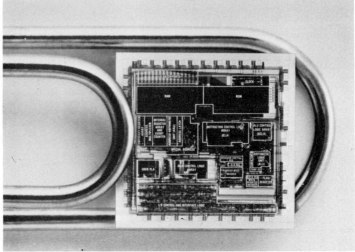

Silicon is to geology as carbon is to biochemistry. Just as almost all biologically significant molecules contain carbon, most of the rocks, sands, and soils of the Earth's crust contain silicon. Pure silicon is a shiny, silvery solid. Its electrical conductivity increases with increasing temperature. Substances that exhibit this type of relationship between these two properties are called *semiconductors*. Pure silicon is primarily used as a semiconductor in electronic devices such as transistors and microcomputer chips (see Figure 13-2b).

Aluminum

The two elements that outrank aluminum in abundance, oxygen and silicon, are nonmetals. Thus, aluminum is the most abundant metal. It is too reactive to be found in nature in the free state. The most useful ores of aluminum are Al_2O_3 in the mineral bauxite, and Na_3AlF_6 in the mineral cryolite.

Even though it is the most abundant metal, aluminum is not the most used metal. It ranks second, in both the United States and the world, to iron in amount produced.

Iron

Iron is the second most abundant and the most used of all metals. The principal iron-containing compounds in nature are oxides such as Fe_2O_3 (hematite) and Fe_3O_4 (magnetite). These oxides are usually associated with silicate materials.

Iron production is nine times greater than that of all other metals combined; 90% of all metal consumed is iron. Nearly all iron production goes into the manufacture of steel. There are many kinds of steel; all are alloys in which iron is the predominant metal.

Iron is one of the 15 trace elements (Section 5.9) known to be essential to life. It is the trace element that is required in the greatest amount. Iron atoms are a necessary component of hemoglobin and myoglobin molecules and of some enzymes.

Calcium

Calcium occurs in numerous solid state forms. Limestone, $CaCO_3$, contains some clay and other impurities and is an important natural source of calcium. Gypsum, which is a hydrated form of calcium sulfate ($CaSO_4 \cdot 2\,H_2O$), is another important calcium-containing mineral. Calcium is also part of some silicate minerals. Calcium is the micronutrient (Section 5.9) required by the human body in the greatest amount.

Sodium

Vast deposits of sodium salts occur in relatively pure form as a result of the evaporation of ancient seas. Large underground deposits of rock-salt (NaCl) occur in many regions. Unlimited supplies of NaCl are present in natural brines and oceanic water. Sodium, like calcium, is a micronutrient required by the human body.

Potassium

Potassium occurs principally as the salt KCl and in complex silicate structures. Sodium is 30 times more abundant than potassium in oceanic water partly because potassium salts are less soluble than sodium salts. In the human body, the micronutrient potassium is required in greater amounts than the micronutrient sodium.

Magnesium

Crustal rocks contain magnesium, mainly in the form of insoluble carbonates and sulfates. Substantial deposits of magnesium-containing evaporates that are associated with bodies of salt water also exist. Magnesium, like calcium, sodium, and potassium, is a micronutrient required by the human body.

Hydrogen

The amount of hydrogen associated with the solid part of the Earth's crust is very small. It is water (11.1% hydrogen) that causes hydrogen to be among the ten most abundant elements in the Earth's crust. Nearly all of the 0.87% hydrogen listed in Figure 13–1 originates with water.

The Earth's atmosphere does not contain elemental hydrogen, H_2, even though hydrogen is the most abundant element in outer space (Table 2–2). The reason for this is the high velocities at which gaseous H_2 molecules travel. Because of their small mass (smaller than that of any other molecule), H_2 molecules are able to acquire velocities that are greater than those of other molecules. (There is a direct relationship between molecular velocity and molecular weight—the smaller the mass, the greater the velocity, at any given temperature.) In the case of gaseous H_2, the molecular speed obtained is sufficient to enable H_2 molecules to overcome the gravitational attraction of the Earth and escape to outer space.

Hydrogen is one of the four macronutrients present in the human body (Section 5.9). On an atom basis, it is the most abundant element in the human body, and on a mass basis, it is the third most abundant element present (Table 5–5).

Titanium

Titanium is not a widely known element, despite its relatively large abundance (compared to most metals), because of the difficulties involved in extracting it from its naturally occurring sources. Many silicaceous materials contain titanium in the form of TiO_2 and $FeTiO_3$.

We have just briefly considered the ten most abundant elements in the Earth's crust. Ranking of the elements this way does not mean that these elements are the ten most important of the known elements. Abundance and importance are two entirely different things. Some elements that are found in minor quantities in the Earth's crust are more widely known and used than some of the top ten elements. Why is this so? It is because of the availability of their deposits. Some elements have been concentrated by natural processes into localized areas. Less abundant elements that are found in concentrated form are more useful than more abundant elements that have not been concentrated. For example, the elements copper, silver, and gold (less abundant elements) have more uses than titanium (an abundant element) does because of this concentration process. The difficulty and cost of isolating an element from its natural sources is also a factor in determining its importance. In the remaining sections of this chapter, we will consider other chemical aspects about the world in which we live.

13.2 THE ATMOSPHERE

In Section 7.11, while considering Dalton's law of partial pressures, we briefly discussed the major components of clean, dry air (Table 7-2). From that discussion, we recall that nitrogen (N_2—78.08 volume %) and oxygen (O_2—20.95 volume %) are the two major components of air. Only three other substances are present in volume percents greater than 0.1%—argon (Ar—0.93%), carbon dioxide (CO_2—0.03%), and water, with a percentage that varies from a few tenths of one percent up to five to six percent, depending on weather and temperature. We will now consider further details about the atmosphere and the chemical substances present in it.

The two most important regions of the atmosphere from a human activity viewpoint are the two regions closest to Earth—the troposphere and the stratosphere. The troposphere extends to an average height of about seven to ten miles and the stratosphere extends from this point to about 30 miles altitude. Naturally, the troposphere (the region closest to the Earth) is affected to a greater degree by human activities than the stratosphere is. Most commercial airline flights traverse the upper parts of the troposphere. Most air pollutants remain within the lower regions of the troposphere. Almost all weather-related phenomena occur within the troposphere. Very little human activity touches the stratosphere. Supersonic aircraft fly in the lower regions of the stratosphere. Some very unreactive (inert) air pollutants occasionally reach the stratosphere, as do materials from large volcanic eruptions. At the present time, there is concern about the extent of the interaction of stratospheric pollutants with the ozone naturally present in the stratosphere. We will look at this problem later on in this section.

Chemically, nitrogen (N_2) and oxygen (O_2) are the predominant gases in both the troposphere and the stratosphere. In the troposphere, carbon dioxide and water vapor are also critically important. Within the stratosphere, the species O_3 (ozone) has an important role to play relative to life on Earth.

The temperature of the troposphere decreases steadily with increasing altitude to the top of this region, where the temperature reaches about $-70°F$. Within the

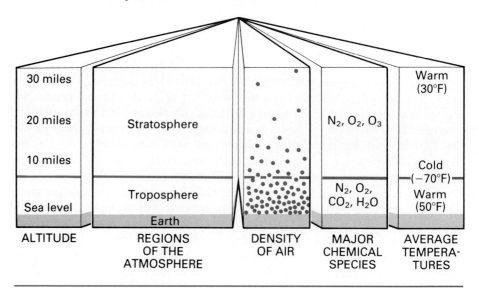

FIGURE 13-3 Various characteristics of the troposphere and stratosphere regions of the atmosphere.

stratosphere, the temperature increases with altitude, warming to about 30°F in the upper stratosphere.

Circulation (or lack of circulation) is a major difference between the troposphere and stratosphere. Much turbulence exists within the troposphere and there is strong vertical mixing of the various gases present in this region. The stratosphere, on the other hand, is relatively calm, with little vertical mixing. This distinction between the two regions will become important in later considerations in this section. Figure 13-3 summarizes some of the differences between the troposphere and the stratosphere.

Life, as we now know it on this planet, is very dependent on the chemical species found within the troposphere and stratosphere. The oxygen content of the lower atmosphere is an absolute necessity for normal respiration in human beings and animals. Carbon dioxide in the air is the basis for the process of photosynthesis that occurs within plants. Atmospheric water, which is deposited on Earth in the form of dew, rain, and snow, is the water source that sustains human, animal, and plant life. Ozone in the stratosphere screens us from the harmful effects of ultraviolet radiation from the sun. Atmospheric nitrogen, which is the predominant chemical species in lower atmospheric regions, is the ultimate source of the nitrogen fertilizers that assist in producing the food necessary to sustain the world's inhabitants.

The trace components of the atmosphere also affect our lives. No discussion of the atmosphere would be complete without an examination of those substances, which we usually refer to as *air pollutants*. Both natural and human sources exist for all air pollutants, with the amounts originating from natural sources usually exceeding those from human sources. However, it is the human sources that create the "problems" because they result in pollutants being released into the atmosphere in localized geographical areas in amounts that overwhelm naturally existing mechanisms for their removal, such as wind and rain.

Three air pollutants that are present in the atmosphere in almost all areas of the world are carbon monoxide, sulfur dioxide, and nitrogen dioxide. We will now briefly consider how these three pollutants form and what their adverse effects are.

13.3 CARBON MONOXIDE

Carbon monoxide (CO), which is a colorless, odorless, and tasteless gas that is not appreciably soluble in water, is the most abundant and widely distributed of all air pollutants found in the lower atmosphere. Its main human activity source is the incomplete combustion of carbon-containing fuels.

Fuels such as natural gas, gasoline, kerosene, heating oil, and diesel fuel all contain the element carbon. When these fuels are burned in a plentiful supply of oxygen, all of the carbon atoms present in the fuel are converted to carbon dioxide.

$$\text{complete combustion} \quad \text{C (in fuel)} + O_2 \longrightarrow CO_2$$

When less than an adequate amount of oxygen is present during fuel combustion, complete conversion of all carbon atoms to carbon dioxide is impossible. Instead, some of the carbon atoms end up as the monoxide instead of dioxide.

$$\text{incomplete combustion} \quad 2\,\text{C (in fuel)} + O_2 \longrightarrow 2\,CO$$

If a fuel–air mixture contains too much fuel and not enough air (oxygen), considerable amounts of carbon monoxide may be formed. (Note that carbon dioxide is the main product even for incomplete combustion.) Carbon monoxide may be an end product, even with sufficient oxygen present in a combustion mixture, if the fuel and air are poorly mixed. Poor mixing leads to localized areas of oxygen deficiency in the air–fuel mixture.

Approximately two-thirds of the carbon monoxide released into the atmosphere as a result of human activities comes from motor vehicle exhaust. It is impossible to completely prevent the formation of carbon monoxide in internal combustion engines such as those in automobiles because of imperfect mixing of air and fuel.

In principle, once carbon monoxide is released into the atmosphere it should be rapidly converted to carbon dioxide because there is plenty of oxygen present in the atmosphere.

$$2\,CO + O_2 \xrightarrow{\text{catalyst}} 2\,CO_2$$

However, this reaction does not appreciably decrease carbon monoxide concentration because the rate at which it occurs is extremely slow at normal atmospheric temperatures.

There are definite health effects associated with carbon monoxide pollution. When inhaled, carbon monoxide enters the bloodstream via the lungs and reduces the ability of the blood to transport oxygen through the body. The carbon monoxide does this by interacting with the hemoglobin of red blood cells. Normally, hemoglobin picks up oxygen in the lungs and distributes it to oxygen-deficient cells throughout the body

$$\text{hemoglobin} + O_2 \rightleftharpoons \text{oxyhemoglobin}$$

Carbon monoxide also combines with hemoglobin, but it does so much more strongly; it has an affinity for hemoglobin that is about 200 times greater than that of O_2.

TABLE 13–2 Carboxyhemoglobin Percentages in the Blood of Humans under Various Conditions

	Percentage Carboxyhemoglobin
continuous exposure, 10 ppm carbon monoxide	2.1
continuous exposure, 50 ppm carbon monoxide	8.5
nonsmokers, Detroit	1.6
smokers, Detroit	5.6
nonsmokers, Salt Lake City	1.2
smokers, Salt Lake City	5.1

$$\text{hemoglobin} + \text{CO} \rightleftharpoons \text{carboxyhemoglobin}$$

Once carboxyhemoglobin forms in a red blood cell, the cell loses its ability to carry oxygen. A relatively small quantity of carbon monoxide can inactivate a substantial fraction of the hemoglobin in the blood for oxygen transport. This reduces the amount of oxygen delivered to all tissues of the body.

The extent of the adverse effects on carbon monoxide on the oxygen-transport system of the body is determined by the amount of air breathed, the concentration of the carbon monoxide, and the length of exposure. Symptoms of carbon monoxide poisoning such as headache, fatigue, and dizziness are present when 10% of a person's hemoglobin is tied up as carboxyhemoglobin. When the fraction rises to 20%, death can result unless the victim is removed from the poisonous atmosphere. The tying up of hemoglobin by carbon monoxide is a reversible process. However, considerable time is needed (5–8 hours) in an unpolluted atmosphere in order for this reversal to occur.

It should be noted that cigarette smoke also contains carbon monoxide. Therefore, cigarette smokers have a portion of their hemoglobin inactivated by this source as well as by external air pollution. Numerous studies have shown that the concentration of carboxyhemoglobin in the blood of people who smoke is two to five times higher than that in nonsmokers. Table 13–2 shows the percentages of carboxyhemoglobin in blood that are typical of various groups of people. The background concentration of carbon monoxide in relatively clean, unpolluted air rarely exceeds 1.0 ppm. By comparison, in city traffic, concentrations of 50 ppm are often reached and carbon monoxide concentration may go as high as 140 ppm in a traffic jam. The inhaled smoke from cigarettes contains about 400 ppm carbon monoxide.

13.4 SULFUR DIOXIDE

Sulfur-containing compounds are naturally present in the atmosphere, even in "unpolluted" air. Natural sources of sulfur-containing compounds include volcanic eruptions and the bacterial decay of natural organisms. The major sulfur-containing compound that enters the air from these natural sources is the gas hydrogen sulfide (H_2S). Human activities also serve as a major source of the sulfur-containing compounds that enter the atmosphere. Usually the predominant compound from these sources is sulfur dioxide (SO_2).

As was the case for carbon monoxide, the combustion of fuel is the major source of sulfur dioxide.

$$\text{S (in fuel)} + O_2 \longrightarrow SO_2$$

Approximately 80% of the sulfur dioxide that enters the atmosphere as the result of human activities comes from coal combustion at power plant locations. Sulfur is present in all coals in amounts ranging between 0.2% and 7% of the coal's mass. Coal is thought to have originated from plants and marine organisms that died, decayed, and were transformed by heat and pressure into the "coal state". All living organisms contain sulfur; hence, all coals contain sulfur. Other fossil fuels (petroleum and natural gas) originated in ways that are similar to coal and also contain sulfur. However, these fuels do not present the same sulfur dioxide pollution problems that coal does because most of the offending sulfur is removed during natural gas processing and petroleum refining. The technology available for sulfur removal for these two fuels depends on the gaseous and liquid states of these fuels; this technology is not transferable to the solid state of coal. Metallurgical processes that involve the metals copper, zinc, mercury, and lead rank a distant second as an overall source of sulfur dioxide air pollution. However, in certain localized areas, these metallurgical operations are major sources of sulfur dioxide pollution.

Both plants and human beings are susceptible to sulfur dioxide, with plants being affected at lower levels than humans. Most of the effects of sulfur dioxide on human health are related to the irritation of the respiratory system and eyes. Of particular concern are the effects that sulfur-dioxide exposure has on individuals who suffer from chronic respiratory diseases such as bronchitis or asthma. These individuals exhibit diminished-lung functions at much lower sulfur dioxide levels than healthy adults do.

Much of the sulfur dioxide that enters the atmosphere is oxidized to sulfur trioxide (SO_3) in the presence of oxygen and sunlight and with time.

$$2\,SO_2 + O_2 \longrightarrow 2\,SO_3$$

Sulfur trioxide is an exceedingly reactive, colorless gas that has a high affinity for water; it rapidly dissolves in water to produce sulfuric acid (H_2SO_4).

$$SO_3 + H_2O \longrightarrow H_2SO_4$$

There is much concern about the possible adverse effects caused by the sulfuric acid produced in this way because it contributes to the phenomenon called *acid rain*. Even in unpolluted areas, rain water is naturally slightly acidic because of the presence of carbon dioxide in the atmosphere, which reacts with water to produce carbonic acid (a weak acid).

$$CO_2 + H_2O \longrightarrow H_2CO_3$$

This reaction produces rain water with a pH of approximately 5.5. (The pH scale for specifying acidity was discussed in Section 10.8.) In recent years, however, the rainfall in certain areas has been found to have much lower pH values, usually between 4 and 5, but sometimes as low as 2. This acid-rain phenomenon has been most often observed in the northeastern United States and in the maritime provinces of Canada (see Figure 13–4).

The most observable effect of acid rain is the corrosion of building materials; this phenomenon is sometimes called "stone leprosy". Materials made of limestone and marble, which are forms of calcium carbonate ($CaCO_3$) are the most severely affected. Sulfuric acid (acid rain) readily attacks carbonate-based building materials; the calcium carbonate present is slowly converted to calcium sulfate ($CaSO_4$).

$$CaCO_3(s) + H_2SO_4(aq) \longrightarrow CaSO_4(s) + CO_2(g) + H_2O(l)$$

The calcium sulfate formed is more soluble than calcium carbonate and is gradually washed away by the acid rain (see Figure 13–5, p. 300).

Highest acidity

Moderate to high acidity

FIGURE 13–4 The areas of the United States and Canada that are most affected by the "acid-rain" phenomenon.

FIGURE 13-5 Many limestone and marble monuments of Europe, such as this one in West Germany, are being damaged by acid rain. These photos contrast how this particular statue appeared in 1908 (208 years after it was put in place) and how it appears today.

13.5 NITROGEN MONOXIDE AND NITROGEN DIOXIDE

Two oxides of nitrogen must be dealt with in nitrogen oxide air pollution considerations: nitrogen monoxide (NO) and nitrogen dioxide (NO_2). Nitrogen monoxide is a *primary* pollutant; that is, it is discharged directly into the atmosphere. Nitrogen dioxide is a *secondary* pollutant; that is, it is produced within the atmosphere from reactions involving primary pollutants. In our case, it is the primary pollutant nitrogen monoxide that reacts further (once in the atmosphere) to produce nitrogen dioxide.

Nitrogen monoxide is a colorless, odorless, nonflammable gas. Nitrogen dioxide is a nonflammable, reddish-brown gas that possesses a strong choking odor. It is nitrogen dioxide's reddish-brown color that is primarily responsible for smog's yellow-brown color. Both nitrogen monoxide and nitrogen dioxide are toxic gases to humans.

Over 90% of the nitrogen oxide pollution that originates from human activities comes from fuel combustion reactions; almost equal amounts of pollution come from transportation and stationary fuel combustion sources. Fossil fuels (coal, oil, and natural gas) usually have only trace amounts of nitrogen-containing materials and they are not the sources of nitrogen oxide pollution. The source of this pollution is the other ingredient that is common to all fuel combustion processes—air. The temperatures associated with combustion are high enough to cause the two major components of air—nitrogen and oxygen—to become slightly reactive toward each other.

$$N_2 + O_2 \longrightarrow 2\,NO$$

Only small amounts of nitrogen monoxide are produced by this reaction, but they are significant in terms of atmospheric pollution concentrations. This nitrogen monoxide enters the atmosphere in combustion exhaust gases and it is rapidly converted to nitrogen dioxide.

$$2\,NO + O_2 \longrightarrow 2\,NO_2$$

Nitrogen dioxide plays a major role in the atmospheric reactions that produce photochemical smog. This type of air pollution was first noted in Los Angeles, but it is now common in many cities. The production of photochemical smog requires a number of reaction steps. Nitrogen dioxide is the sole reactant in the initial step that leads to photochemical smog production. Exposure to sunlight causes nitrogen dioxide to dissociate (break apart) into nitrogen monoxide and oxygen.

$$NO_2 \xrightarrow{\text{sunlight}} NO + O$$

These oxygen atoms, which are called *atomic oxygen*, are highly reactive species that initiate the second step in the smog formation process—the production of ozone (O_3). Atomic oxygen reacts with the normal diatomic oxygen (O_2) present in the atmosphere; this results in ozone formation.

$$O + O_2 \longrightarrow O_3$$

Ozone is considered to be the key chemical species in photochemical smog. Not only does it cause direct damage, but also it can react further to produce other damaging species. Ozone is a powerful oxidizing agent that can directly damage rubber, plastic materials, and plant and animal tissues. In addition, it reacts with other pollutants in the air such as unreacted gasoline vapor (always present in automobile exhaust) to form other, more complicated oxidizing agents. Collectively, ozone and its reactant products constitute the mixture of chemicals we call "smog". Figure 13-6 (p. 302) contrasts the appearance of downtown Los Angeles on smoggy and clear days.

Nitrogen oxides and ozone can irritate the lungs, cause bronchitis and pneumonia, and lower resistance to respiratory infections such as influenza.

Nitrogen dioxide present in the air that does not photochemically dissociate to produce atomic oxygen can react with water vapor in the air to form nitric

FIGURE 13–6 Photos showing downtown Los Angeles on a smoggy day and a clear day. Ozone and nitrogen dioxide are two of the main chemical species in the smog layer in the first photo.

acid. This nitric acid contributes to the problem of acid rain that was previously discussed in Section 13.4. Acid rainfall in the western United States, an area that is less dependent on coal combustion than the industrialized midwest and northeast, involves HNO_3 in addition to H_2SO_4.

13.6 THE HYDROSPHERE

The first ten sections of Chapter 8 dealt with the subject of water. We discussed water's physical and chemical properties in detail at that time. Because of the importance of this simple compound, H_2O, we will again return to the subject. This time, we will be particularly concerned with "natural waters", or water as it occurs in nature.

All water as it occurs in nature is impure in a chemical sense. These impurities include suspended matter, microbiological organisms, dissolved minerals, and dissolved gases. The diversity of dissolved substances present in natural waters illustrates the broad solvent properties of this compound. Water's "great" solvent ability stems from the highly polar nature of water molecules (Section 8.2). The solvent properties of water are both beneficial and detrimental from the viewpoint of humans. They are beneficial because water is a useful medium for carrying nutrients and oxygen throughout the human body and a useful material in washing and cleansing processes. They are detrimental because materials that are harmful to the health of humans tend to dissolve in water.

Natural waters can be classified into three categories on the basis of their dissolved mineral content: fresh water, inland brackish water, and sea water. Water containing less than 0.1% by mass dissolved solids is generally considered to be fresh, but the U.S. Public Health Service has set 0.05% mass percent as the recommended upper limit for drinking water. The inland underground and surface supply of brackish water contains dissolved solids in the range of 0.1% to 3.5% by mass, with an average of 0.6 mass percent. The dissolved mineral content of sea water varies from 3.3 to 3.7%, with an average of 3.5% by mass.

Sea water is often referred to as *saline water*; the salinity of the water is the percent by mass of dissolved solids present in the water. Over 70 different elements contribute to the salinity of sea water, but most of them are present only in very low concentrations.

Ninety-seven percent of the Earth's total water supply is sea water. The inland brackish supplies of water make up about 2.5 percent, leaving only 0.5 percent as usable fresh water. Fortunately, fresh water, unlike most natural resources, is renewable because it is part of a natural hydrologic cycle. Evaporation from oceans and lakes, as well as transpiration by plants (loss of water vapor by plants), places water in the atmosphere. Eventually, fresh water finds its way back to the Earth in the form of rain and snow. Figure 13–7 diagrammatically shows nature's hydrologic cycle.

It is interesting to compare ion concentrations in human body fluids with those of nature's seawater and fresh water. This is done in Table 13–3 (p. 304) for blood serum, the pale, yellowish liquid that is left after blood cells and clotting agents have been removed from whole blood.

Note the total concentration of ions in each solution that is given at the bottom of Table 13–3. On a relative scale, where the total concentration of ions in fresh water (the most dilute solution) is assigned a value of 1, blood plasma would have a value of approximately 100 (actually 115) and sea water would have a value of approximately 500 (actually 474). Thus, the concentration of ions in blood plasma is approximately 100 times greater than that of fresh water, but only approximately one-fifth that of the ions in sea water.

Note also the ions that dominate in each of the three solutions. Blood serum is more like sea water than fresh water in that Na^+ is the dominant positive ion and Cl^- is the dominant negative ion in both blood serum and sea water. This

FIGURE 13–7 Nature's hydrologic cycle causes fresh water to be a renewable natural resource.

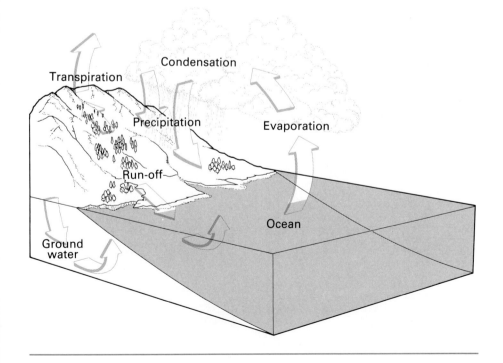

TABLE 13-3 A Comparison of Concentrations of the Major Ionic Constituents in Seawater, Fresh Water, and Blood Serum. (All concentrations are expressed in millimolar units, which is molarity/1000)*

	Seawater		Fresh Water		Blood Serum	
Positive Ions						
Na^+	460	(41%)	0.27	(11%)	145	(54%)
K^+	10	(0.90%)	0.06	(2.5%)	5.1	(1.9%)
Ca^{2+}	10	(0.90%)	0.38	(16%)	2.5	(0.92%)
Mg^{2+}	54	(4.9%)	0.34	(14%)	1.2	(0.44%)
Negative Ions						
Cl^-	550	(49%)	0.22	(8.5%)	103	(38%)
SO_4^{2-}	28	(2.5%)	0.12	(5.1%)	2.5	(0.92%)
HCO_3^-	2.3	(0.215%)	0.96	(41%)	12	(4.4%)
Total Ions	1114	(100%)	2.35	(100%)	271	(100%)

* (The percentages given in the table are the percents that the various ions contribute to the total ionic concentration of the solution.)

contrasts with fresh water, where Ca^{2+} and Mg^{2+} are the most abundant positive ions and HCO_3^- is the most abundant negative ion.

13.7 HARD WATER

hard water

Hard water *is water that contains dissolved metal ions (principally Ca^{2+}, Mg^{2+}, and Fe^{2+}) that form insoluble compounds (precipitates) either with soap or upon heating.* Almost all natural fresh waters contain these metal ions; thus, almost all natural fresh waters are hard.

Conventional water purification steps do not alter the hardness of water because they do not remove dissolved minerals. The focus in water purification is upon removing substances that are harmful to human health. At the concentrations

FIGURE 13-8 Boiler scale deposited in a section of a hot-water pipe over a two-year time period has nearly closed the pipe off.

normally present in natural fresh waters, dissolved minerals are not harmful to health. Indeed, some of the taste of water is caused by these ions; water without metal ions would taste unpleasant to most people.

Although water hardness does not affect its drinkability, it does affect other uses for water. Two very noticeable effects of water hardness are the formation of a sticky, curdy precipitate (scum) with soap and the deposition of a hard scale in steam boilers, tea kettles, and hot water pipes. The scum, among other things, produces a ring on bathtubs and imparts a dull, gray appearance to washed clothes. The hard, scaly deposit inside boilers and pipes forms an effective insulating layer, and heat is not transferred efficiently to the water inside. Ultimately, pipes can become completely clogged with these deposits (see Figure 13–8). Both the "scum" and "scale" effects of hard water are related to the insolubility of certain calcium, magnesium, and iron salts.

13.8 SOFT WATER

Soft water *is water in which the ions that cause hardness have been removed or tied up chemically.* Many different methods for softening water are in use today. One of the simplest ways to remove the offending metal ions from hard water is by adding chemicals that precipitate them; that is, to cause them to form insoluble compounds.

soft water

The most popular method for softening water in the home utilizes a process called ion exchange. In this process, the offending Ca^{2+}, Mg^{2+}, and Fe^{2+} ions are exchanged for sodium ions Na^+.

Ion exchange is accomplished by letting hard water percolate through a container (commercially, the water softener tank) that holds a finely divided substance capable of performing the ion exchange. Both naturally occurring materials called *zeolites*, which are complex sodium aluminum silicates, and synthetic materials are available to perform ion exchanges. The reaction that occurs in zeolite ion exchanges can be represented as follows (M^{2+} represents the hard water ions).

$$2\ NaAlSiO_4 + M^{2+}(aq) \longrightarrow M(AlSiO_4)_2 + 2\ Na^+(aq)$$
removed from solution   released into the water

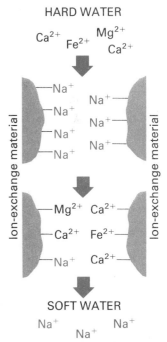

FIGURE 13–9 Pictorial representation of what happens when hard water is percolated through an ion-exchange material.

The preceding ion-exchange process is shown pictorially in Figure 13–9. Notice in Figure 13–9 that the hard water ions become attached to the ion-exchange material.

Over time, an ion-exchange material becomes saturated with hard-water ions and can no longer effectively soften water; all of the Na^+ ions have been used up. The ion-exchange material can be regenerated by washing it with a concentrated salt (NaCl) solution. This treatment, which "recharges" the water softener, reverses the original "discharging" process; the hard-water ions are replaced by Na^+ ions.

$$M(AlSiO_4)_2 + 2\ Na^+ \longrightarrow 2\ NaAlSiO_4 + M^{2+}$$
discarded in waste water from the recharging

Concentrated sodium chloride salt solutions are used in this process simply because sodium chloride is the cheapest source of Na^+ ion available.

It should be obvious to you that soft water produced by the preceding type of ion exchange has a much higher than normal Na^+ ion concentration. Often people with high blood pressure (hypertension) or kidney problems are advised to avoid drinking softened water. An increased intake of sodium can lead to water retention in the body's tissues and adds to the work of the kidneys. A good rule is: hard water for drinking and soft water for washing and heating.

Ion-exchange materials that are more expensive than those used in standard home water softener equipment are now available; these materials can produce extremely pure water called *deionized water*. **Deionized water** *is softened water in which dissolved ions (both positive and negative) have been removed*. In the production of deionized water, the hard water is first passed through an "acidic" ion-exchange material that replaces all positive ions with H^+ ions. Then the water is passed through a "basic" ion-exchange material that replaces all negative ions with OH^- ions. The H^+ ions and OH^- ions now present in the water immediately react with each other (neutralization—Section 10.6) to produce additional water.

$$H^+ + OH^- \longrightarrow H_2O$$

The net result is almost pure water. The gallon jugs of water that are available at grocery and variety stores for home use in steam irons is almost always deionized water.

The pure water used in many laboratory situations, where only small amounts of water are needed, is *distilled water*. This water is produced by distillation, the process in which water is boiled and the resulting steam, which is ion-free, is condensed back into water. At present, distillation is too expensive (because of the large amount of heat energy required) for large-scale use as a softening or purification process.

13.9 FLUORIDATED WATER

Many communities fluoridate their drinking water. In this process, about 1 part per million (m/m) F^- is added to drinking water. Salts such as NaF serve as the source of F^- ion. The purpose of fluoridation is to reduce dental caries. What is the chemistry involved here? The compound hydroxyapatite, $Ca_{10}(PO_4)_6(OH)_2$, is the major constituent of teeth. Ions associated with this compound, such as OH^- ion, are readily replaced by F^- ions. The resulting fluoride-containing material is less soluble and less reactive than the nonfluoride material is. This decreased solubility and decreased reactivity afford greater protection against caries. There is some opposition to fluoridation of water because many fluorine-containing compounds (at concentrations much greater than in fluoridated drinking water) exhibit toxic effects.

Fluoridated toothpaste is much less effective than fluoridated drinking water in reducing the incidence of tooth decay. In many areas in which drinking water is not fluoridated, dentists prescribe fluoride tablets or drops to children.

EXERCISES AND PROBLEMS

Abundances of the Elements

13.1 What is the definition of the "Earth's crust"?

13.2 How many of the ten most abundant elements in the Earth's crust are metals and how many are nonmetals? List the metals and the nonmetals.

13.3 What is the composition of the majority of the rocks and minerals found in the Earth's crust?

13.4 What is the chemical composition associated with the following substances?
a. sand
b. limestone
c. bauxite
d. hematite

13.5 The atmosphere contains little or no H_2 gas. Why is this so?

The Atmosphere

13.6 What are the five gaseous components of clean air that are present in volume percents greater than 0.1?

13.7 Contrast the properties of the stratosphere and troposphere regions of the atmosphere.

13.8 Discuss the major source of each of the following air pollutants.
a. CO   b. SO_2 c. NO

13.9 Write a chemical equation for the production of each of the following air pollutants.
a. CO   c. SO_3 e. NO_2
b. SO_2 d. NO

13.10 What is the distinction between a primary air pollutant and a secondary air pollutant? List two secondary air pollutants.

13.11 Discuss the chemistry of photochemical smog production.

13.12 Discuss the chemistry involved in the conversion of SO_2 to "acid rain" and describe the effects of acid rain.

The Hydrosphere

13.13 Discuss the basis for the classification of natural waters as fresh, brackish, or sea water.

13.14 What are the three most abundant metallic ionic species and the three most abundant nonmetallic ionic species in sea water?

13.15 What metal ions found in natural waters are responsible for the hardness of water?

13.16 List two common problems that arise from hard water use.

13.17 What is the chemistry involved in recharging a mechanical water softener unit by running a concentrated salt–water solution through it?

13.18 What is the difference between distilled water and deionized water?

13.19 What is the chemistry involved in using fluoridated water to help prevent dental caries?

14 Saturated Hydrocarbons

Objectives

After completing Chapter 14, you will be able to:

14.1 Understand the historical and modern definitions of organic and inorganic chemistry.
14.2 Define the term *hydrocarbon* and list the three major classes of hydrocarbons.
14.3 Define the terms *saturated hydrocarbon*, *alkane*, and *cycloalkane* and understand why carbon atoms in alkanes and cycloalkanes are always bonded to four other atoms.
14.4 Draw the structures of simple alkanes.
14.5 Understand the phenomenon of structural isomerism as it relates to alkanes.
14.6 Write expanded and condensed structural formulas for normal-chain and branched-chain alkanes and recognize structural formulas that simply represent different conformations of the same molecule.
14.7 Name alkanes using IUPAC rules, given the structural formulas, and vice versa.
14.8 Recognize by structure and name all three- and four-carbon branched alkyl groups.
14.9 Name cycloalkanes using IUPAC rules and draw structural formulas, including those of a geometrical figure type, for cycloalkanes.
14.10 List the major natural sources of alkanes.
14.11 List some physical properties of alkanes.
14.12 Write equations for alkane combustion and alkane halogenation and name halogenation products using IUPAC nomenclature rules.
14.13 Discuss the occurrence of alkanes in living organisms.

INTRODUCTION

This chapter is the first of a series that deals with the subject of organic chemistry and organic compounds. Organic compounds are the chemical basis for life itself, as well as the basis for our current standard of living. Proteins, carbohydrates, enzymes, and hormones are organic molecules; in addition, organic compounds include natural gas, petroleum, coal, gasoline, and many synthetic materials such as dyes, plastics, and nylon and dacron fibers.

14.1 ORGANIC CHEMISTRY—AN HISTORICAL PERSPECTIVE

During the latter part of the eighteenth century and early part of the nineteenth century, chemists began to distinguish between two types of compounds: organic compounds and inorganic compounds. The origin of the compound provided the distinction between the two types. Did a compound come from a "living" or "nonliving" source? The term *organic* was used for compounds that came from plants and animals (organisms), and the term *inorganic* was used for compounds obtained from the mineral constituents of the Earth.

During that time period, chemists also thought that organic compounds contained a special "vital force" that only a living organism could supply. They thought the only source of these compounds was nature itself. The fact that no scientists were able to successfully synthesize a known organic compound from inorganic starting materials gave credence to this "vital force" theory.

We now know that the "vital force" theory is incorrect. In 1828, the German chemist Friedrich Wöhler obtained unexpected results from a routine inorganic experiment he was conducting. While attempting to recrystallize the inorganic salt NH_4CNO from a solvent, he inadvertently produced the well-known organic compound *urea*, a component of urine. Wöhler's results were the stimulus for the renewed efforts of scientists to synthesize organic substances from inorganic starting materials. This time, after a century of negative results, there were numerous successful reactions. By 1860, the "vital force" theory was laid to rest, and Friedrich Wöhler became known as the father of organic chemistry. Despite the fall of the "vital force" theory, with its emphasis on the living and nonliving origins of substances, the terminology associated with the theory (the categories of *organic* and *inorganic*) is still in use. However, the original definitions for these terms have changed.

Today, **organic chemistry** *is defined as the study of hydrocarbons (binary compounds of hydrogen and carbon) and their derivatives.* Interestingly, almost all compounds found in living organisms still fall in the field of organic chemistry when this modern definition is applied. In addition, many compounds that are synthesized in the laboratory, which have never been found in nature or in a living organism, are considered to be organic compounds.

In a less rigorous manner, organic chemistry is often defined as the study of carbon-containing compounds. It is true that almost all carbon-containing compounds qualify as organic compounds. There are, however, some exceptions. The oxides of carbon, carbonates, cyanides, and metallic carbides are considered to

organic chemistry

FIGURE 14–1 Simple and complex chains and rings of carbon atoms.

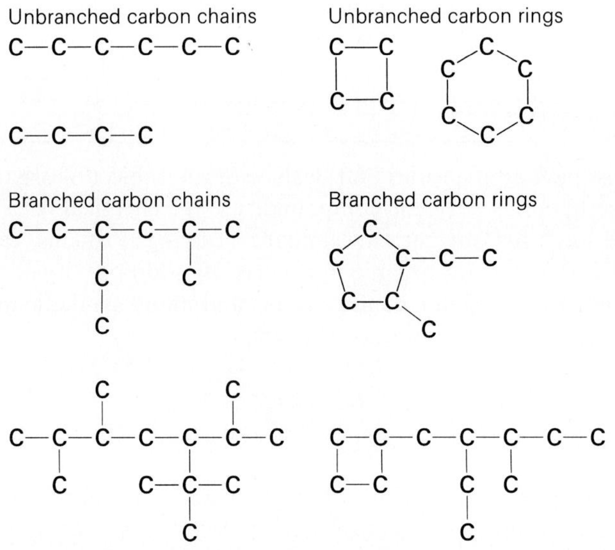

be inorganic compounds rather than organic compounds. The field of *inorganic chemistry* encompasses the study of all noncarbon-containing compounds (the other 108 elements) and the few carbon-containing compounds just mentioned.

In essence, organic chemistry is the study of one element (carbon) and inorganic chemistry the study of 108 elements. Why, relative to their study, is there such an unequal partitioning of the elements? The answer is simple. The chemistry of carbon is so much more extensive than the chemistry of the other elements that there is justification in making its study a field by itself. There are approximately 5 million known organic compounds. Fewer than 250,000 inorganic compounds exist. This implies an approximate twenty-to-one ratio between organic and inorganic compounds.

Why does carbon form twenty times as many compounds as all of the other elements combined? Carbon possesses the unique ability to form bonds with itself in long chains, rings, and complex combinations of chains and rings. Chains and rings of all lengths are possible. All of these chains and rings can contain carbon atom side chains as well.

The number of possible arrangements for carbon atoms bonded to each other is literally limitless. Scientists have calculated that there are 366,319 different ways of arranging 20 carbon atoms, based on a chain of atoms and allowing for side chains.

Figure 14–1 illustrates some of the possible ways of arranging carbon atoms to form organic molecules.

14.2 HYDROCARBONS

hydrocarbons

As the name implies, **hydrocarbons** *are compounds that contain only the two elements hydrogen and carbon.* There are two distinctly different families (or classes) of hydrocarbons. They are *saturated hydrocarbons* and *unsaturated hydrocarbons*. We will discuss saturated hydrocarbons for the remainder of this chapter. Unsaturated hydrocarbons are considered in Chapters 15 and 16.

14.3 SATURATED HYDROCARBONS

Saturated hydrocarbons *are compounds in which all of the carbon–carbon bonds are single bonds.* Saturated hydrocarbons can be divided into two groups—the alkanes and the cycloalkanes—based on the arrangement of carbon atoms. **Alkanes** *are saturated hydrocarbons in which the carbon atom arrangement is that of a unbranched or branched chain.* **Cycloalkanes** *are saturated hydrocarbons in which at least one cyclic arrangement of carbon atoms is present.*

saturated hydrocarbons

alkanes

cycloalkanes

Before considering the structures of specific alkanes and cycloalkanes, let us note one general bonding characteristic of all carbon atoms in any hydrocarbon or hydrocarbon derivative: *carbon atoms that are present in hydrocarbons or hydrocarbon derivatives must always have four bonds.*

Carbon can meet this four-bond requirement in three different ways:

1. By bonding to four other atoms. This situation requires the presence of four single bonds

$$-\overset{|}{\underset{|}{C}}-$$
four single bonds

2. By bonding to three other atoms. This situation requires the presence of two single bonds and one double bond.

$$-\overset{|}{C}=$$
two single bonds and one double bond

3. By bonding to two other atoms. This situation requires the presence of either two double bonds or a triple and a single bond.

$$=C=\qquad -C\equiv$$
two double bonds   one triple bond and one single bond

It is the first of these three possibilities, four single bonds, that is relevant to our discussion of saturated hydrocarbons. (The other possibilities will become important in Chapter 15 when we discuss unsaturated hydrocarbons.)

14.4 ALKANES

The simplest alkane hydrocarbon is *methane*, which has the formula CH_4. The methane molecule has a tetrahedral structure. The carbon atom is found at the center of the tetrahedron and the four hydrogen atoms bonded to the carbon are at the corners of the tetrahedron. Different representations of this tetrahedral structure for methane are shown in Figure 14–2 (p. 312). The hydrogen atoms are at the corners of the tetrahedron, so they are as far away from each other as possible. This minimizes the repulsions between the electrons associated with the various bonds in the molecule.

The next member of the alkane hydrocarbon series is *ethane*, which has a molecular formula of C_2H_6. This molecule can be thought of as a methane molecule with one hydrogen atom removed and a $-CH_3$ group put in its place. Perspective drawings of the ethane molecule shown in Figure 14–3 (p. 312) illustrate that the carbon bonds still exhibit a tetrahedral geometry about each carbon atom.

Propane, the third alkane, has the molecular formula C_3H_8. Once again, we can produce this formula by removing a hydrogen atom from the preceding compound

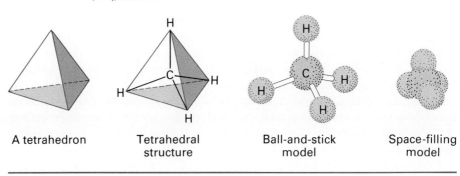

FIGURE 14-2 Different ways of showing the tetrahedral arrangement of atoms in the methane (CH_4) molecule.

A tetrahedron | Tetrahedral structure | Ball-and-stick model | Space-filling model

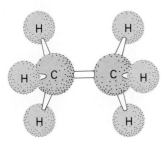

Ball-and-stick model

Space-filling model

FIGURE 14-3 Ball-and-stick and space-filling models of the hydrocarbon ethane (C_2H_6) molecule.

(ethane) and substituting a —CH_3 group in its place. All six hydrogen atoms of ethane are equivalent, so it makes no difference which one we choose to replace. Both ball-and-stick and space-filling models of the propane molecule are shown in Figure 14-4.

The formulas of all alkanes can be represented by a general formula, C_nH_{2n+2}, where n is the number of carbon atoms. Let's examine how this works. For each of the three molecules discussed so far (methane, ethane, and propane) and for longer chains, each carbon is attached to at least two hydrogens. The only carbons that have three hydrogens are the end carbons on the chain. Because they are on the end, they need an extra hydrogen to terminate the chain and give each carbon a total of four bonds. We can explain the formula C_nH_{2n+2} by this concept—each carbon in the chain has two hydrogens plus the two extra hydrogens, one on each end of the chain. For example, propane has three carbons, so the number of hydrogens is $2(3) + 2 = 8$. The mathematical formula predicts a molecular formula of C_3H_8, which agrees with the actual drawing of the molecule. This mathematical formula can be applied to any alkane, regardless of its length or size. For example, an alkane with 20 carbon atoms would have a molecular formula of $C_{20}H_{42}$. The number of hydrogen atoms present is always twice the number of carbon atoms plus two more.

14.5 STRUCTURAL ISOMERISM

We can also use the procedures outlined in the last section for establishing the structures of the ethane and propane molecules (replacement of a hydrogen with a —CH_3 group) to generate other members of the alkane series that contain more carbon atoms. However, a complication arises when four or more carbon atoms are present: different structures can be obtained depending on which hydrogen is replaced.

Two different structural arrangements of four carbons can be produced by removing a hydrogen atom from propane and replacing it with a —CH_3 group. This is the result of all the hydrogens in propane not being geometrically equivalent. The two hydrogens attached to the central carbon atom in propane (Figure 14-4) are equivalent to each other, but are distinct from the six hydrogens associated with the end carbons, which in turn are all equivalent to each other. Replacement of a hydrogen on an end carbon gives *butane*, the compound shown in Figure 14-5a. The compound in Figure 14-5b, *isobutane*, is the result of a —CH_3 group replacing a hydrogen on the central carbon atom.

FIGURE 14–4 Ball-and-stick and space-filling models of the hydrocarbon propane (C_3H_8).

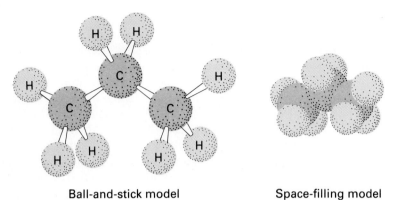

Ball-and-stick model   Space-filling model

Although butane and isobutane both have the same molecular formula of C_4H_{10}, they are separate compounds with different properties. The melting point of butane is $-138.3°C$ and that of isobutane is $-160°C$. The boiling points of the two compounds are $-0.5°C$ and $-12°C$, respectively. Their densities at 20°C also differ: 0.579 g/mL for butane and 0.557 g/mL for isobutane.

FIGURE 14–5 Ball-and-stick and space-filling models for (a) butane and (b) isobutane.

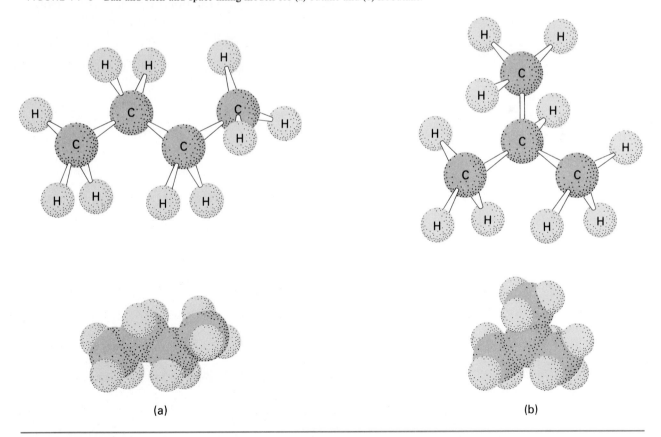

(a)   (b)

structural isomers

Compounds such as butane and isobutane are called *structural isomers*. **Structural isomers** *are compounds that have the same molecular formula, but different structural formulas; that is, they have different arrangements of atoms within the molecule.* Structural isomers are not rare in organic chemistry; in fact, they are the rule rather than the exception. The phenomenon of structural isomerism is one of the major reasons why there are so many organic compounds.

In Figure 14–5 (p. 313), three-dimensional representations were used to represent the structures of butane and isobutane. These representations give both the arrangement and spatial orientation of the atoms in a molecule. However, it is often difficult to draw these structures, especially if artistic talent is lacking, and it is also time consuming. Because of these drawbacks, an easier system for indicating structure has been developed. This alternate system involves structural formulas.

structural formulas

Structural formulas *are two-dimensional (planar) representations of the arrangement of the atoms in molecules.* These formulas give complete information about the arrangement of the atoms in a molecule, but not the spatial orientation of the atoms. The structural formulas for butane and isobutane are

$$CH_3-CH_2-CH_2-CH_3 \qquad CH_3-\underset{\underset{CH_3}{|}}{CH}-CH_3$$

butane   isobutane

Note that when writing structural formulas, each carbon atom is followed by the attached hydrogens. The number of attached hydrogens is determined by the number of other carbon atoms to which a given carbon atom is bonded. Because each carbon atom must have four bonds (Section 14.3), carbon atoms attached to only one other carbon will need three hydrogen atoms (CH_3). Carbons bonded to two other carbons will need two hydrogens (CH_2); those bonded to three other carbons need only one hydrogen (CH); and those bonded to four other carbons need no hydrogens.

When we consider alkanes with five carbon atoms, we find that there are three structural isomers.

$$CH_3-CH_2-CH_2-CH_2-CH_3 \qquad CH_3-\underset{\underset{CH_3}{|}}{CH}-CH_2-CH_3 \qquad CH_3-\underset{\underset{CH_3}{|}}{\overset{\overset{CH_3}{|}}{C}}-CH_3$$

pentane   isopentane   neopentane

TABLE 14–1 Possible Number of Structural Isomers for Selected Alkanes

Molecular Formula	Possible Number of Structural Isomers
CH_4	1
C_2H_6	1
C_3H_8	1
C_4H_{10}	2
C_5H_{12}	3
C_6H_{14}	5
C_7H_{16}	9
C_8H_{18}	18
C_9H_{20}	35
$C_{10}H_{22}$	75
$C_{15}H_{32}$	4,347
$C_{20}H_{42}$	336,319
$C_{30}H_{62}$	4,111,846,763

Like the two C_4H_{10} isomers, these C_5H_{12} isomers are distinctly different compounds with different properties.

The number of possible structural isomers increases rapidly with the number of carbon atoms in an alkane, as seen in Table 14–1.

14.6 STRUCTURAL FORMULAS

In Section 14.5, we used structural formulas to designate the various butane and pentane isomers. Formulas of this type are extensively used in organic chemistry.

Two types of structural formulas are commonly encountered. **Expanded structural formulas** *show, in two dimensions, all of the bonds within a molecule.* These formulas are used only when it is necessary to show all of the bonds within the molecule. **Condensed structural formulas** *use groupings of carbon and hydrogen atoms to indicate the structural arrangement of atoms within a molecule.* In a condensed structural formula, each carbon is written down, followed by the attached hydrogens. Table 14–2 gives both of these types of structural formulas for the alkanes with one to four carbons.

expanded structural formulas

condensed structural formulas

Even condensed structural formulas are cumbersome for molecules containing many carbon atoms. In these cases, the condensed structural formulas are condensed even more. For example, the formula

$$CH_3-CH_2-CH_2-CH_2-CH_2-CH_2-CH_2-CH_3$$

can be further abbreviated as

$$CH_3-(CH_2)_6-CH_3$$

TABLE 14–2 Expanded and Condensed Structural Formulas for Selected Alkanes

Number of Carbons in Alkane	Expanded Structural Formula	Condensed Structural Formula
1	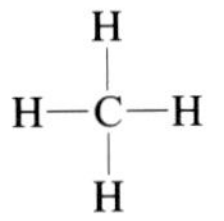 H—C—H with H above and H below	CH_4
2	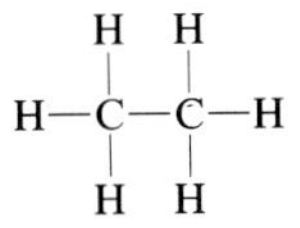 H—C—C—H with H's above and below each C	CH_3-CH_3
3	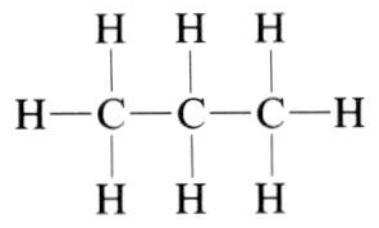 H—C—C—C—H with H's above and below each C	$CH_3-CH_2-CH_3$
4	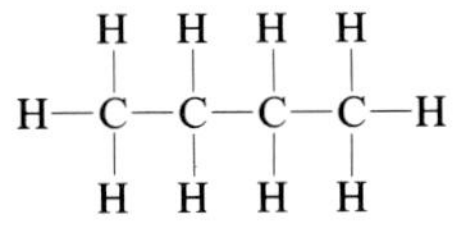 H—C—C—C—C—H with H's above and below each C	$CH_3-CH_2-CH_2-CH_3$
	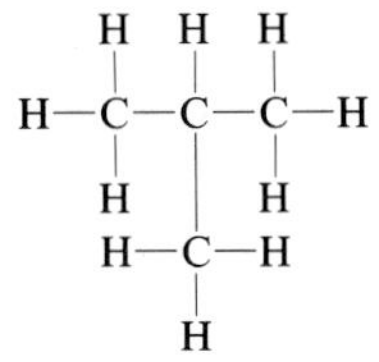 H—C—C—C—H with H's above and below, and H—C—H branching off middle C	$CH_3-CH-CH_3$ with CH_3 below middle C

FIGURE 14-6 Different orientations of atoms in a molecule are possible because of rotation around single bonds.

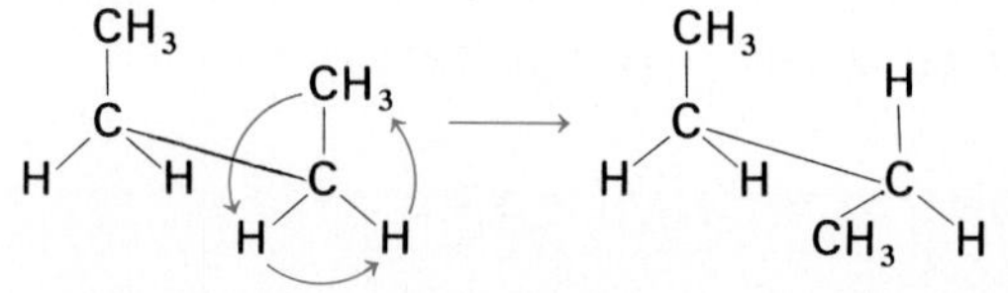

In this notation, parentheses and a numerical indicator are used to denote the number of —CH_2— groups in the chain.

In situations where the arrangement of carbon atoms is of utmost importance, you can draw condensed structure formulas that omit the hydrogen atoms present. For example, the following two notations are considered equivalent to each other.

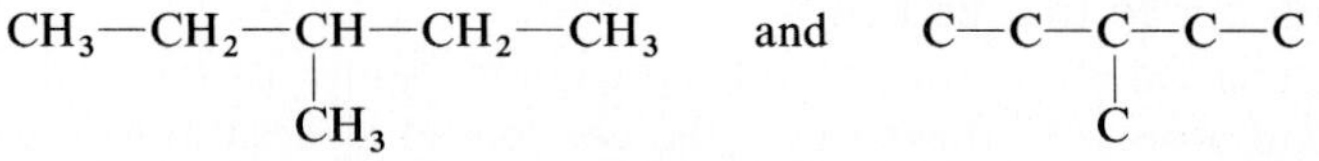

$$CH_3-CH_2-\underset{\underset{CH_3}{|}}{CH}-CH_2-CH_3 \quad \text{and} \quad C-C-\underset{\underset{C}{|}}{C}-C-C$$

The second notation of only carbon atoms still contains all of the information needed to uniquely identify the compound it represents because we know that each carbon atom shown must have enough hydrogen atoms attached to it to give it four bonds.

Structural formulas are necessary to indicate the identity of almost all organic compounds because of the phenomenon of isomerism. Only rarely will a molecular formula provide sufficient information for identification. We have already seen that the molecular formula C_4H_{10} is ambiguous because it corresponds to two different alkanes. We use a molecular formula in organic chemistry only when it represents one structure, such as CH_4 for methane. The necessity for structural formulas in organic chemistry is in direct contrast to inorganic chemistry, where molecular formulas are sufficient (CO_2, H_2O, NH_3, and so on).

A pitfall concerning structural formulas is that these formulas are not true three-dimensional representations of structures. Structural formulas incorrectly imply that carbon atoms are in a straight line for chains of three or more carbons; the ball-and-stick models of Figures 14-4 and 14-5a show that this is not the case.

Organic molecules are in constant motion, twisting, turning, vibrating, and bending. Groups of atoms connected by a single bond to another atom are capa-

FIGURE 14-7 Selected ways of representing a chain of five carbon atoms. All drawings represent the same structure.

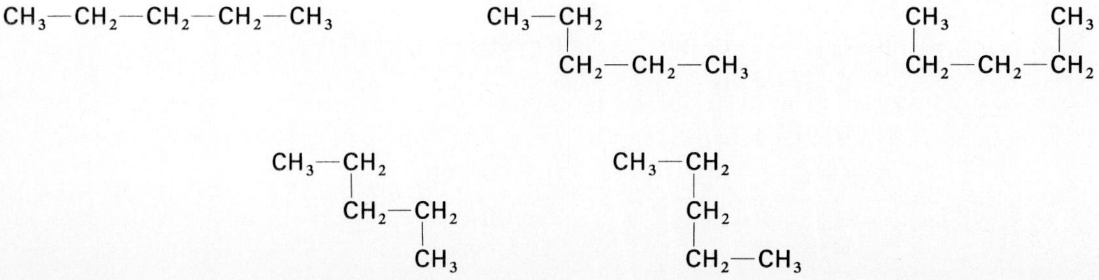

ble of rotating about that bond, much like a wheel rotates around an axle (see Figure 14–6).

As a result of rotation, molecules can exist in an infinite number of different orientations. For example, the molecules present in a sample of pentane (a chain of five carbon atoms) are present in numerous orientations at any given time, and each orientation would rapidly change into another. Simplified drawings (no attempt is made to show tetrahedral bonding angles) of some of these orientations are shown in Figure 14–7. Note that all of the drawings in Figure 14–7 represent the same molecule. In each case, five carbon atoms are bonded in a continuous (unbranched) chain. Note that it is "legal" to go around corners when determining that each of these drawings represents a chain of five carbon atoms. Two structures are isomers only if it is necessary to break bonds in order to convert one into the other. Structural isomers are permanent arrangements, but different orientations are transient in nature.

Example 14.1

Determine whether each of the following pairs of structures represent isomers or simply two different ways of drawing the same molecule.

a. $CH_3-CH_2-CH_2-CH_3$ and $\begin{array}{c} CH_2-CH_2 \\ | \quad\quad | \\ CH_3 \quad CH_3 \end{array}$

b. $\begin{array}{c} CH_2-CH_2-CH_3 \\ | \\ CH_3 \end{array}$ and $\begin{array}{c} CH_3-CH_2-CH_2 \\ \quad\quad\quad\quad | \\ \quad\quad\quad\quad CH_3 \end{array}$

c. $\begin{array}{c} CH_3-CH-CH_3 \\ | \\ CH_3 \end{array}$ and $\begin{array}{c} CH_3-CH_2-CH_2 \\ \quad\quad\quad\quad | \\ \quad\quad\quad\quad CH_3 \end{array}$

Solution

a. These structural formulas represent different ways of drawing the same molecule. In both cases, we have a chain of four carbon atoms. For the second structure, we need to go around two corners in order to get the chain of four carbon atoms.

$$C-C-C-C \quad\quad \begin{array}{c} C-C \\ | \quad | \\ C \quad C \end{array}$$

b. Again, we have two different ways for drawing the same molecule. Both structures are simply a chain of four carbon atoms.

$$\begin{array}{cc} C-C-C & C-C-C \\ | & \quad\quad | \\ C & \quad\quad C \end{array}$$

c. This time we have isomers. The longest chain of carbon atoms in the first structure has three carbons. For this structure, there is no way we can include the fourth carbon in the chain without back-tracking through the middle carbon; this procedure is not allowed. The second structure has a chain of four carbon atoms.

$$\begin{array}{cc} C-C-C & C-C-C \\ | & \quad\quad | \\ C & \quad\quad C \end{array}$$

14.7 IUPAC NOMENCLATURE FOR ALKANES

When relatively few organic compounds were known, chemists named them using what are today called *common names*. These names were arbitrarily selected, and include isopentane and neopentane (Section 14.5). However, as more hydrocarbons became known, this nonsystematic method of naming compounds became unwieldy.

The IUPAC (International Union of Pure and Applied Chemistry) nomenclature system is the solution to this problem. This system, which is now accepted worldwide, assigns to each organic compound a unique name that not only identifies it, but also gives enough information about it so that the structural formula for the compound can be drawn.

Only a few simple rules are needed to assign IUPAC names to alkanes. In order to simplify our presentation of these rules, we will classify alkanes into the categories of normal chain and branched chain. A **normal-chain** *organic compound is one in which all carbon atoms are connected in a continuous nonbranching chain.* In a **branched-chain** *organic compound, one or more side chains of carbon atoms are attached at some point to a continuous chain of carbon atoms.* The following four-carbon alkane isomers (Section 14.5) illustrate this classification system.

$$CH_3-CH_2-CH_2-CH_3 \qquad CH_3-CH-CH_3$$
$$\qquad\qquad\qquad\qquad\qquad\qquad\qquad\quad |$$
$$\qquad\qquad\qquad\qquad\qquad\qquad\qquad\quad CH_3$$

normal chain branched chain

In the branched-chain compound, the longest continuous chain is three carbons, to which a —CH_3 group (the branch) is attached (on the middle carbon of the three).

Normal-chain alkanes are structurally the simplest alkanes. The IUPAC names for normal-chain alkanes that have one through ten carbons are given in Table 14–3. Note that all of the names end in *-ane*, the characteristic ending for all alkanes. The first four names in Table 14–3 do not indicate the number of carbon atoms in the molecule; however, beginning with the five-carbon alkane, we use Greek numerical prefixes that directly give the number of carbon atoms present. It is important to memorize the names of these normal-chain alkanes because they are the basis for the entire IUPAC nomenclature system. (The IUPAC system also includes names for normal-chain alkanes that have more than ten carbon atoms, but we will not consider them. The names given in Table 14–3 will be sufficient for our purposes.)

TABLE 14–3 IUPAC Names for Normal-Chain Alkanes with One through Ten Carbon Atoms

Molecular Formula	IUPAC Prefix	IUPAC Name	Structural Formula
CH_4	meth-	methane	CH_4
C_2H_6	eth-	ethane	CH_3-CH_3
C_3H_8	prop-	propane	$CH_3-CH_2-CH_3$
C_4H_{10}	but-	*n*-butane	$CH_3-CH_2-CH_2-CH_3$
C_5H_{12}	pent-	*n*-pentane	$CH_3-CH_2-CH_2-CH_2-CH_3$
C_6H_{14}	hex-	*n*-hexane	$CH_3-CH_2-CH_2-CH_2-CH_2-CH_3$
C_7H_{16}	hept-	*n*-heptane	$CH_3-CH_2-CH_2-CH_2-CH_2-CH_2-CH_3$
C_8H_{18}	oct-	*n*-octane	$CH_3-CH_2-CH_2-CH_2-CH_2-CH_2-CH_2-CH_3$
C_9H_{20}	non-	*n*-nonane	$CH_3-CH_2-CH_2-CH_2-CH_2-CH_2-CH_2-CH_2-CH_3$
$C_{10}H_{22}$	dec-	*n*-decane	$CH_3-CH_2-CH_2-CH_2-CH_2-CH_2-CH_2-CH_2-CH_2-CH_3$

TABLE 14–4 Names for Normal-Chain Alkyl Groups that Contain from One to Six Carbon Atoms

Number of Carbons	Structural Formula	Stem of Alkane Name	Suffix	Alkyl Group Name
1	—CH$_3$	meth-	-yl	methyl
2	—CH$_2$—CH$_3$	eth-	-yl	ethyl
3	—CH$_2$—CH$_2$—CH$_3$	prop-	-yl	propyl
4	—CH$_2$—CH$_2$—CH$_2$—CH$_3$	but-	-yl	butyl
5	—CH$_2$—CH$_2$—CH$_2$—CH$_2$—CH$_3$	pent-	-yl	pentyl
6	—CH$_2$—CH$_2$—CH$_2$—CH$_2$—CH$_2$—CH$_3$	hex-	-yl	hexyl

The prefix *n-* (for normal) is attached to the name of a continuous-chain alkane when four or more carbon atoms are present. It is at four carbons that isomers become possible for alkanes. Thus, the correct IUPAC name for

$$CH_3-CH_2-CH_2-CH_2-CH_2-CH_3$$

is *n*-hexane, rather than simply hexane.

In order to name branched-chain alkanes, you must be able to recognize and name the branch or branches that are attached to the main carbon chain. These branches are formally known as *alkyl* groups. An **alkyl group** *is an alkane from which one hydrogen atom has been removed.* It is the removal of the one hydrogen atom that allows the branches (alkyl groups) to be attached to a main carbon chain. Alkyl groups have the general formula C_nH_{2n+1} (one less hydrogen atom than the corresponding alkane from which they may be considered to be derived).

alkyl group

The two most commonly encountered alkyl groups are the two simplest ones— the one-carbon and the two-carbon alkyl groups. Their formulas and names are

—CH$_3$ —CH$_2$—CH$_3$
methyl group   ethyl group

Alkyl groups are named by taking the stem of the name of the alkane containing the same number of carbon atoms and adding the ending *-yl*. Table 14–4 gives the names for normal-chain alkyl groups that contain from one to six carbon atoms. The names and structures for some more complicated alkyl groups will be presented in Section 14.8.

The long bond in these alkyl group structures (as well as those in Table 14–4) denotes the point of attachment to the carbon chain. Note that alkyl groups do not lead a stable, independent existence; they are always attached to other groups of carbon atoms.

We will now present the formal IUPAC rules for naming branched-chain alkanes; that is, alkanes where the carbon atoms are not arranged in a continuous chain. Two examples of the use of each rule will accompany each rule's presentation.

Rule 1: Select the longest continuous carbon-atom chain in the molecule as the base for the name. This longest carbon-atom chain is named as in Table 14–3.

Example:
$$CH_3-CH_2-\underset{\underset{CH_3}{|}}{CH}-CH_2-CH_2-CH_2-CH_3$$

In this example, the longest continuous carbon-atom chain (shown in color) is seven carbon atoms. Therefore, the base name (but not the complete name) for this compound is *heptane*.

Example:

$$CH_3-CH_2-CH-CH_2-CH-CH_3$$
$$||$$
$$CH_3CH_2$$
$$|$$
$$CH_2$$
$$|$$
$$CH_3$$

In this example, the longest continuous carbon chain (shown in color) possesses eight carbon atoms. Note that the carbon atoms in the longest continuous chain do not necessarily have to lie in a straight line. The base name (but not the complete name) for the alkane will be *octane*.

Rule 2: *The carbon atoms in the longest continuous chain of carbon atoms are numbered consecutively from the end that will give the lowest number(s) to any carbon(s) to which an alkyl group is attached.* If only one alkyl group is present, begin numbering the chain at the end closest to the alkyl group. When there are two or more alkyl groups, number the carbons so that the carbons with the alkyl groups are given numbers with the lowest possible sum.

Example:

$$CH_3 \longleftarrow \text{alkyl group}$$
$$|$$
$$CH_3-CH_2-CH_2-CH-CH_2-CH_3$$
$$123456 \text{(left to right)}$$
$$654321 \text{(right to left)}$$

carbon atom to which the alkyl group is attached

A chain always has two ends, so there are always two ways to number the chain: either from left to right or right to left. In this example, the left–right numbering system assigns the number 4 to the carbon atom bearing the alkyl group. If the chain is numbered the other way, the alkyl group is on carbon number 3. Because 3 is lower than 4, the right-to-left numbering system is the one used.

Example:

$$CH_3$$
$$|$$
$$CH_3-CH_2-C-CH-CH-CH_2-CH_3$$
$$|||$$
$$CH_3CH_3CH_3$$
$$1234567 \text{(left to right)}$$
$$7654321 \text{(right to left)}$$

In this example there are four alkyl groups (side chains) attached to the main chain. If the chain is numbered from left to right, the alkyl groups are found on carbons 3, 3, 4, and 5, the sum of which is 15. (Note that when a carbon atom carries two alkyl groups, the carbon's number must be counted twice in the sum.) Numbering from right to left, we generate the numbers 3, 4, 5, and 5, the sum of which is 17. The left-to-right numbering system is the one used for this compound, because 15 is smaller than 17.

Rule 3: *The complete name for an alkane contains the location by number and the name of each alkyl group and the name of the longest carbon chain.* The alkyl group names with their locations always precede the name of the base chain of carbon atoms.

Example:

$$\overset{1}{C}H_3-\overset{2}{C}H-\overset{3}{C}H_2-\overset{4}{C}H_2-\overset{5}{C}H_3 \quad \text{is} \quad \text{2-methylpentane}$$
$$\phantom{\overset{1}{C}H_3-}|$$
$$\phantom{\overset{1}{C}H_3-}CH_3$$

Example: $\overset{1}{C}H_3-\overset{2}{C}H_2-\overset{3}{C}H-\overset{4}{C}H_2-\overset{5}{C}H_3$ is 3-methylpentane
$\qquad\qquad\qquad\qquad\quad\;\;|$
$\qquad\qquad\qquad\qquad\; CH_3$

Note that the names are written as one word, with a hyphen between the number (location) and the name of the alkyl group.

Rule 4: *If two or more of the same kind of alkyl group (two methyl groups or two ethyl groups, and so forth) are present in a molecule, the number of them is indicated by the prefixes di-, tri-, tetra-, penta-, and so on. In addition, a number specifying the location of every identical group must be included. These position numbers, separated by commas, are put just ahead of the numerical prefix. Numbers are always separated from words by hyphens.*

Example: $\overset{1}{C}H_3-\overset{2}{C}H-\overset{3}{C}H_2-\overset{4}{C}H-\overset{5}{C}H_3$ is 2,4-dimethylpentane
$\qquad\qquad\qquad\quad\;\; |\qquad\qquad\;\; |$
$\qquad\qquad\qquad\; CH_3\qquad\;\; CH_3$

Example: $\overset{1}{C}H_3-\overset{2}{C}H_2-\overset{3}{\underset{|}{\overset{|}{C}}}-\overset{4}{C}H_2-\overset{5}{C}H_3$ is 3,3-dimethylpentane

with CH_3 groups above and below the central C.

In the case of 3,3-dimethylpentane, notice that *every* group attached to the parent chain needs a number describing its location, so we must include two 3's in the name to identify each of the methyl positions. If a molecule happened to have three methyl groups attached to a parent chain, we would need three numbers to specify position, even if two of them were the same. In addition, a "trimethyl" would also appear in the name.

Rule 5: *When two different kinds of alkyl groups are present on the same carbon chain, each group is numbered separately and the names of the alkyl groups are listed in alphabetical order. As before, all numbers and words must be separated by a hyphen, and all numbers must be separated from other numbers by commas.*

Example: $\overset{1}{C}H_3-\overset{2}{C}H-\overset{3}{C}H_2-\overset{4}{C}H-\overset{5}{C}H_2-\overset{6}{C}H_3$
$\qquad\qquad\qquad\quad\;\; |\qquad\qquad\;\; |$
$\qquad\qquad\qquad\; CH_3\qquad\;\; CH_2$
$\qquad\qquad\qquad\qquad\qquad\quad\;\; |$
$\qquad\qquad\qquad\qquad\qquad\; CH_3$

4-ethyl-2-methylhexane

Example: $\overset{1}{C}H_3-\overset{2}{C}H_2-\overset{3}{C}H-\overset{4}{C}H-\overset{5}{C}H-\overset{6}{C}H_2-\overset{7}{C}H_3$
$\qquad\qquad\qquad\qquad\quad\;\; |\quad\;\; |\quad\;\; |$
$\qquad\qquad\qquad\qquad CH_3\; CH_2\; CH_3$
$\qquad\qquad\qquad\qquad\qquad\quad\;\; |$
$\qquad\qquad\qquad\qquad\qquad\; CH_2$
$\qquad\qquad\qquad\qquad\qquad\quad\;\; |$
$\qquad\qquad\qquad\qquad\qquad\; CH_3$

3,5-dimethyl-4-propylheptane

The first example demonstrates that *alphabetical* order is more important than *numerical* order of substituents (*e*thyl precedes *m*ethyl, no matter what the numbers in front of these alkyl groups are). In the second example, *m*ethyl precedes *p*ropyl, so the methyl groups are listed first in the name. Notice that it is the *m* in di*m*ethyl that is important, and not the *d*. The numerical prefixes do not determine alphabetical order of names; only the alkyl name is considered when a choice must be made. For example, in

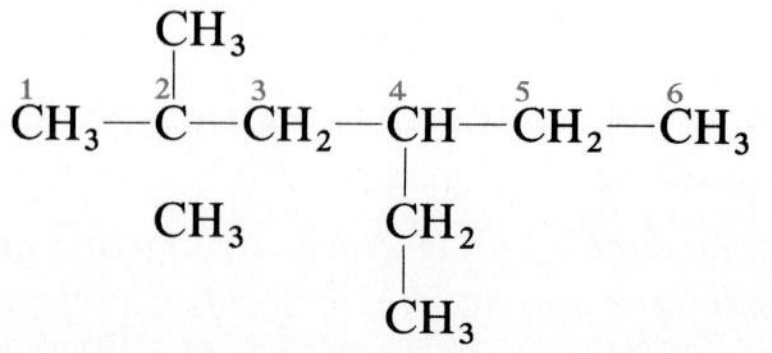

$$\overset{1}{C}H_3-\overset{\underset{|}{CH_3}}{\underset{\underset{|}{CH_3}}{C}}-\overset{3}{C}H_2-\overset{4}{\underset{\underset{\underset{|}{CH_3}}{CH_2}}{C}H}-\overset{5}{C}H_2-\overset{6}{C}H_3$$

4-ethyl-2,2-dimethylhexane

we see that *e*thyl precedes di*m*ethyl because the *e* and the *m* must be compared and not the *e* and the *d*.

Example 14.2

In Section 14.5, we used common names to name the four-carbon alkane isomers (butane and isobutane) and the five-carbon alkane isomers (pentane, isopentane and neopentane). What is the IUPAC name for each of these compounds?

Solution

a. *Butane*, $CH_3-CH_2-CH_2-CH_3$

The IUPAC name is butane or n-*butane*. Thus, the common and IUPAC names correspond.

b. *Isobutane*, $CH_3-\underset{\underset{|}{CH_3}}{CH}-CH_3$

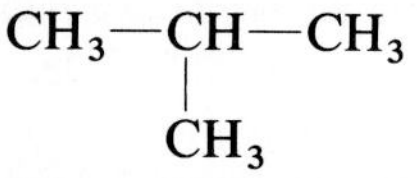

The longest chain contains three carbon atoms, and the attached methyl group is at carbon 2, so the name is *2-methylpropane*.

c. *Pentane*, $CH_3-CH_2-CH_2-CH_2-CH_3$

The IUPAC name is pentane or n-*pentane*. Thus, the common and IUPAC names correspond.

d. *Isopentane*, $CH_3-\underset{\underset{|}{CH_3}}{CH}-CH_2-CH_3$

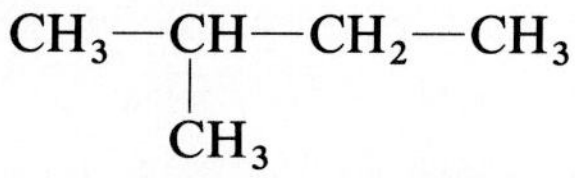

The longest chain contains four carbon atoms, and a methyl group is attached to the chain at carbon 2. The IUPAC name is *2-methylbutane*.

e. *Neopentane*, $CH_3-\overset{\underset{|}{CH_3}}{\underset{\underset{|}{CH_3}}{C}}-CH_3$

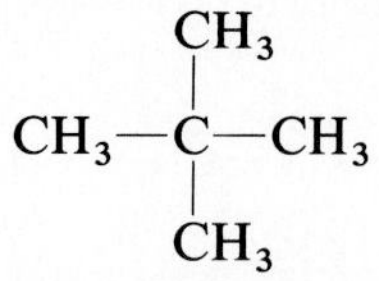

The longest chain contains only three carbons, and two alkyl groups are attached to it. Both alkyl groups are methyl groups and both are attached to carbon 2. The IUPAC name is *2,2-dimethylpropane*.

The ultimate goal in learning to name organic molecules such as alkanes is to accurately and concisely communicate descriptions of structures to others. After learning the rules for naming alkanes, it is a relatively easy process to reverse the procedure and translate the name of an alkane into a structural formula. Example 14.3 shows how this is done.

323
Saturated Hydrocarbons

Example 14.3

Draw the structural formula for 3-ethyl-2,3-dimethylpentane.

Solution
Step 1: This molecule is a pentane, so the longest continuous chain of carbon atoms is five. Draw this chain of five carbon atoms and number it.

$$\overset{1}{C}-\overset{2}{C}-\overset{3}{C}-\overset{4}{C}-\overset{5}{C}$$

Step 2: Complete the carbon skeleton by attaching alkyl groups as they are specified in the name. An ethyl group goes on carbon 3 and methyl groups are attached to both carbons 2 and 3.

```
              C
              |
    1   2   3|   4   5
    C — C — C — C — C
        |   |
        C   C
            |
            C
```

Step 3: Add hydrogen atoms to the carbon skeleton so that each carbon atom has four bonds. Note that some carbon atoms need three hydrogens and some need only two or one. (In the case of carbon 3, none are needed.)

```
                   CH₃
                   |
     1      2     3|     4       5
    CH₃ — CH — C — CH₂ — CH₃
           |   |
          CH₃ CH₂
               |
              CH₃
```

14.8 BRANCHED-ALKYL GROUPS

All of the alkyl groups encountered in Section 14.7 were straight-chain (unbranched) alkyl groups. Just as there are straight-chain and branched-chain alkanes, there are also straight-chain and branched-chain alkyl groups. Four of these branched alkyl groups are sufficiently common that you should know their names and structures. They are listed in Figure 14–8.

FIGURE 14–8 Names and structures of common branched alkyl groups.

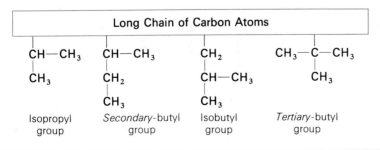

Note that the naming of these branched groups is based on a system of prefixes: iso-, *secondary* (abbreviated *sec-* or *s-*), and *tertiary-* (abbreviated *tert-* or *t-*). Two examples of alkanes containing branched alkyl groups (along with their names) are as follows

$$\overset{1}{C}H_3-\overset{2}{C}H_2-\overset{3}{C}H-\overset{4}{C}H_2-\overset{5}{C}H_2-\overset{6}{C}H-\overset{7}{C}H_2-\overset{8}{C}H_2-\overset{9}{C}H_3$$

with CH—CH$_3$ / CH$_3$ branch at C3 and CH$_2$—CH$_2$—CH$_3$ branch at C6

4-isopropyl-6-propylnonane

$$\overset{1}{C}H_3-\overset{2}{C}H_2-\overset{3}{C}H-\overset{4}{C}H_2-\overset{5}{C}H_2-\overset{6}{C}H_3$$

with CH$_3$—C—CH$_3$ / CH$_3$ branch at C3

3-*tert*-butylhexane

In the first example, note the difference between a propyl group and an isopropyl group. A propyl group involves three carbon atoms attached to a chain through one of the end carbons. An isopropyl group also involves three carbon atoms, but the attachment to the chain is through the middle of the three carbon atoms.

14.9 CYCLOALKANES

In Section 14.3, we noted that saturated hydrocarbons can be subclassified into two families: alkanes and cycloalkanes. Nothing has been said about cycloalkanes since that time. Recall from Section 14.3 that cycloalkanes contain cyclic arrangements of carbon atoms.

The simplest cycloalkane is cyclopropane, which is a cyclic arrangement of three carbon atoms; it takes a minimum of three carbon atoms to form a cyclic system. Figure 14–9 shows ball-and-stick models of cyclopropane, cyclobutane (a four-membered carbon ring), and cyclopentane (a five-membered carbon ring).

It has been experimentally found that cycloalkanes having four or more ring atoms are not planar (flat), as is implied by Figure 14–9. Rather, some puckering

FIGURE 14–9 Ball-and-stick models of cyclopropane, cyclobutane, and cyclopentane.

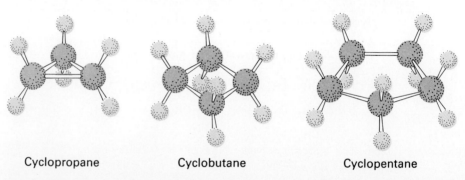

Cyclopropane   Cyclobutane   Cyclopentane

FIGURE 14–10 Nonplanar structure adopted by a cyclohexane molecule.

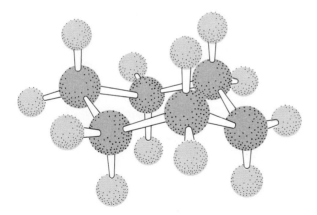

of the ring occurs, as is shown in Figure 14–10 for cyclohexane (a six-membered carbon ring).

Two sizes of cycloalkane rings, C_5 and C_6 rings, are especially stable and there are many important molecules, which will be discussed in later chapters, that contain these cyclic systems. Extremely large ring systems that contain up to 30 carbon atoms are known to exist. However, they are not as important as the smaller-sized ring systems.

The IUPAC nomenclature for cycloalkanes is very similar to that for alkanes (Section 14.7). There are only two modifications to the alkane nomenclature rules we have already discussed.

1. The prefix *cyclo-* is placed before the name that corresponds to the noncyclic chain having the same number of carbon atoms as the ring.
2. When alkyl groups are present, they are located by numbering the carbons in the ring according to a system that yields the lowest numbers for the carbons at which alkyl groups are attached.

Example 14.4 illustrates the use of these rule modifications in naming cycloalkanes.

Example 14.4

Assign IUPAC names to each of the following cycloalkanes.

a.
$$\begin{array}{c} CH_2 \\ H_2C \quad CH_2 \\ H_2C—CH_2 \end{array}$$

b.
$$\begin{array}{c} H_2C—CH_2 \\ |\quad\quad | \\ H_2C—CH \\ \quad\quad\quad \backslash CH_3 \end{array}$$

c.
$$\begin{array}{c} CH_2 \\ H_2C \quad CH—CH_3 \\ H_2C \quad CH—CH_3 \\ CH_2 \end{array}$$

d.
$$\begin{array}{c} CH_2 \\ H_2C \quad CH—CH_3 \\ H_2C \quad CH_2 \\ CH \\ | \\ CH_3 \end{array}$$

Solution

a. The five-carbon alkane is called *pentane*. Five carbon atoms in a cyclic arrangement is called *cyclopentane*.

b. The molecule is a cyclobutane (a four-membered carbon ring) to which a methyl group is attached. Its name is simply *methylcyclobutane*. The position of a single alkyl group on a cycloalkane ring does not need to be located with a number because all positions on the ring are equivalent to each other and thus, there is only one way to attach the methyl group.

c. This compound is a dimethyl cyclohexane. The positions of the two methyl groups, relative to each other, must be specified with numbers. The ring is numbered clockwise so that the methyl groups are on carbons 1 and 2. We want to keep the numbers as low as possible.

This compound's complete name is *1,2-dimethylcyclohexane*.

d. This molecule is isomeric with the previous molecule. The only difference between the two molecules is the positioning of the methyl groups on the ring. In part **c**, the methyl groups were on adjacent carbons (carbons 1 and 2). Here, the methyl groups are on carbons 1 and 3.

The name of this compound is *1,3-dimethylcyclohexane*. Note that when numbering a ring to which alkyl groups are attached, the number 1 is always assigned to one of the carbon atoms that bears an alkyl group.

For brevity and simplicity in drawing cycloalkane structures, geometrical figures are often used to represent the cyclic part of cycloalkane structures: a triangle is used for a three-carbon ring, a square is used for a four-carbon ring, and so on. When these geometrical figures are used, it is assumed that each corner of the figure represents a carbon atom, along with the number of hydrogens needed to give the carbon four bonds. In the following examples, this geometrical figure notation is used in drawing structural formulas for cyclohexane, 1,2-dimethylcyclopropane, and 1,3-dimethylcyclopentane.

cyclohexane   1,2-dimethylcyclopropane   1,3-dimethylcyclopentane

A somewhat similar notation is occasionally used to denote long chains of carbon atoms in alkanes. Here, a saw-tooth pattern is used to denote the chain, and a carbon atom is understood to be present at every point where two lines meet. Using this saw-tooth notation, the molecule *n*-octane

$$CH_3-CH_2-CH_2-CH_2-CH_2-CH_2-CH_2-CH_3$$

would be designated as

Saw-tooth notation is not used as frequently as the geometrical figure notation for cycloalkanes is. However, there will be occasions in later chapters where it will be advantageous to use saw-tooth notation.

The general formula for cycloalkanes is C_nH_{2n}. Thus, any given cycloalkane contains two fewer hydrogen atoms than an alkane with the same number of carbon atoms. (Recall, from Section 14.4, that the general formula for an alkane is C_nH_{2n+2}.) Thus, butane (C_4H_{10}) and cyclobutane (C_4H_8) are not isomers. Isomers must have the same molecular formula (Section 14.5).

It is easy to visualize why cycloalkanes contain two fewer hydrogen atoms than the corresponding alkanes. Consider what would have to occur if butane were to be converted to cyclobutane. Cyclization would not be possible without the removal of a hydrogen atom from each terminal carbon atom.

As with alkanes, structural isomers are possible for cycloalkanes. For example, there are four cycloalkane structural isomers that have the formula C_5H_{10}—one based on a five-membered ring, one based on a four-membered ring, and two based on a three-membered ring. These isomers are:

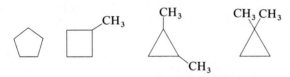

Note that the last isomer has both methyl groups attached to the same carbon atom. The name of this isomer would be 1,1-dimethylcyclopropane.

14.10 NATURAL SOURCES OF ALKANES

Natural gas and petroleum (oil) constitute the largest and most important natural sources of alkanes. Deposits of these substances are found in many locations throughout the world. These deposits are usually associated with dome-shaped rock formations (Figure 14–11) at a depth of several thousand feet. When a hole

FIGURE 14–11   Rock formations associated with petroleum and natural gas deposits.

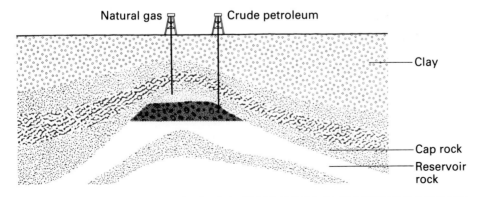

FIGURE 14–12 Much oil is now obtained from vast deposits under the ocean. This photo shows an offshore oil-drilling complex.

is drilled into such a rock formation, it is possible to recover some of the trapped hydrocarbons (see Figure 14–12). Note that petroleum and natural gas do not occur in the Earth in the form of liquid pools; this is a common misconception. They are dispersed throughout a porous rock formation. It is the structural features of the rock formation that allow the hydrocarbons to accumulate.

We do not completely understand the processes by which fossil fuels were produced in nature. Chemists have debated the question of their primary source for many years. Did they originate from animals, plants, or nonbiological minerals? It is now generally accepted that they were derived from the decomposition of marine organisms, both plants and (to a much smaller extent) animals, in the absence of oxygen in the environment of deep, ocean-bottom sediments.

Unprocessed natural gas contains 50–90% methane, 1–10% ethane, and up to 8% higher alkanes (predominantly propane and butanes). Most natural gas produced in the United States comes from the southwestern states. A vast network of pipelines carries the gas to other parts of the country. The higher alkanes found in crude natural gas are removed prior to releasing the gas into the pipeline distribution systems. Because the removed alkanes can be liquefied by the use of moderate pressure, they are stored as liquids under pressure in steel cylinders and are marketed as bottled gas. This material is used in rural areas and other locations not supplied by natural gas lines.

Crude petroleum is an extremely complex mixture of hydrocarbons. Some crudes consist chiefly of alkanes; others contain as much as 40% of other types of hydrocarbons—cycloalkanes and aromatics (Chapter 15).

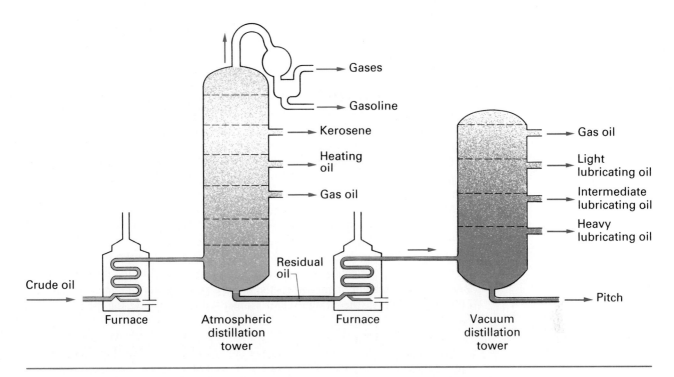

FIGURE 14-13 The fractionation of crude petroleum. The complex mixture of hydrocarbon compounds present in petroleum is separated into simpler mixtures (fractions) according to boiling-point differences.

In its natural state, crude petroleum has very few uses; however, this complex mixture can be separated into useful fractions through refining. The resulting fractions are still hydrocarbon mixtures, but each one is simpler (contains fewer compounds). During refining, the physical separation of the crude into component fractions is accomplished by fractional distillation, a process that takes advantage of boiling-point differences between the components of the crude petroleum. Each fraction contains hydrocarbons within a specific boiling-point range. The fractions obtained from a typical fractionation process are shown in Figure 14-13 and the properties and uses of each fraction are listed in Table 14-5. Note the many important and familiar products obtained from petroleum.

In addition to the uses listed in Table 14-5, a significant amount of the petroleum fractions produced annually is used as feedstock for the petrochemical in-

TABLE 14-5 Products Obtained From the Fractional Distillation of Petroleum

Fraction	Molecular Size Range	Boiling-point Range (°C)	Typical Uses
gas	C_1–C_4	−164–30	gaseous fuel
petroleum ether	C_5–C_7	30–90	solvent, dry cleaning,
straight-run gasoline	C_5–C_{12}	30–200	motor fuel
kerosene	C_{12}–C_{16}	175–275	fuel for stoves, diesel, and jet engines
gas oil or fuel oil	C_{15}–C_{18}	up to 375	furnace oil
lubricating oils	C_{16}–C_{20}	350 and up	lubrication, mineral oil
greases	C_{18}–up	semisolid	lubrication, petroleum jelly
paraffin (wax)	C_{20}–up	melts at 52–57	candles
pitch and tar	high	residue in boiler	roofing, paving

dustry. From this industry flow many consumer products that are considered to be necessities today. Examples of these products are plastics, synthetic fibers, adhesives, dyes, and medicines.

14.11 PHYSICAL PROPERTIES OF ALKANES

Alkanes have physical properties that are typical of nonpolar molecules. This is consistent with the fact that the electronegativities of carbon and hydrogen are very similar (Section 6.11).

The solubilities of alkanes in various solvents obey the polarity-based rule "likes dissolve likes" (Section 8.10). Alkanes are insoluble in water, but are soluble in nonpolar solvents such as carbon tetrachloride (CCl_4).

The densities of alkanes are usually less than 1.0 g/mL. This explains why water in gasoline sinks to the bottom of the container. Similarly, water is ineffective in fighting oil fires.

At ordinary temperatures and pressures, the lower-molecular-weight alkanes (C_1 to C_4) are gases. Straight-chain (unbranched) alkanes containing from 5 to 17 carbon atoms are liquids, and alkanes that are longer than this are solids at room temperature. As a general rule, the boiling point increases roughly 30°C for every carbon added to the chain. This concept can be seen graphically in Figure 14–14. Branching the carbon chain lowers the boiling point of the alkane. Thus, in a set of isomeric alkanes, the straight-chain isomer will always have the highest boiling point.

Alkanes that are solids at room temperature often have a waxy appearance, and the term *paraffins* is used to describe them collectively. Paraffin wax is a mixture of these compounds.

14.12 CHEMICAL PROPERTIES OF ALKANES

As a class, the alkanes are very unreactive compounds. The primary reason for this is the fact that each carbon atom in an alkane is already bonded to the maximum number of atoms possible—four. In order for alkanes to react with other substances, either carbon–carbon or carbon–hydrogen bonds must be broken.

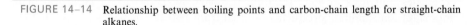
FIGURE 14–14 Relationship between boiling points and carbon-chain length for straight-chain alkanes.

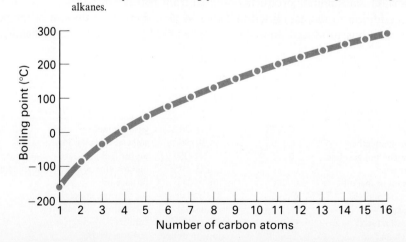

Because of their unreactive nature and their immisicibility in water (Section 14.11), alkanes are widely used as protective coatings. Paraffin wax is used to seal homemade jams. Lubricating oils help protect metal parts of cars and trucks from both rust and frictional heat. Petroleum jelly protects skin from moisture and mineral oil can act as a laxative by coating the intestines.

Alkanes do not react with aqueous solutions of acids or bases and they are not readily oxidized by most oxidizing agents, including oxygen. They do, however, react rapidly with oxygen in the presence of a catalyst or when they are ignited. This reaction with oxygen is called *combustion*. In the presence of an initiator (usually light energy), alkanes react with halogens (Group VIIA elements). This process is called *halogenation*. Combustion and halogenation are sufficiently important reactions that we need to consider both processes in detail.

Combustion

Combustion *is the rapid reaction of a substance with oxygen such that a flame, heat, and light are produced.* Alkanes readily undergo combustion when ignited. When sufficient oxygen is present to support total combustion, the products of combustion for all alkanes are carbon dioxide and water. Two representative alkane combustion reactions are

combustion

$$CH_4 + 2\,O_2 \longrightarrow CO_2 + 2\,H_2O + \text{energy}$$

$$2\,C_6H_{14} + 19\,O_2 \longrightarrow 12\,CO_2 + 14\,H_2O + \text{energy}$$

The exothermic nature of alkane combustion accounts for the extensive use of alkanes as fuels. Natural gas, which is predominantly methane, is used in heating homes and in cooking. Propane is used for home heating in rural areas and as a recreational fuel.

Incomplete combustion can occur if insufficient oxygen is present during the combustion process. When this is the case, some carbon monoxide (CO) or elemental carbon are reaction products along with carbon dioxide (CO_2). In a chemical laboratory setting, incomplete combustion is often observed during the process of heating beakers or other types of laboratory glassware with Bunsen burners. The appearance of deposits of carbon black (soot) on the bottom of the glassware is physical evidence that incomplete combustion is occurring. The problem is that the air–fuel ratio for the Bunsen burner is not correct. It is too rich; it contains too much fuel and not enough oxygen (air).

Halogenation

Halogens, the elements of Group VIIA of the periodic table (fluorine, chlorine, bromine, and iodine), are some of the few substances that are able to break carbon–hydrogen bonds. **Halogenation** *is the reaction of an alkane with a halogen (most commonly chlorine or bromine) to produce a hydrocarbon derivative in which one or more halogen atoms has been substituted for hydrogen atoms.* In the process, carbon–hydrogen bonds are broken and new carbon–halogen bonds are formed. Halogenation is an example of a general type of reaction known as a *substitution reaction*. A **substitution reaction** *is a reaction in which part of a small reacting molecule replaces a hydrogen atom on a hydrocarbon or a hydrocarbon group.*

halogenation

substitution reaction

For alkanes, halogenation reactions take place readily in sunlight, but not often in the absence of light. The usual result of halogenation is the formation of more than one type of product. This is because more than one hydrogen atom on the alkane can be replaced. For example, if methane, the simplest alkane, is reacted with Cl_2 over a period of time, varying amounts of four different products are formed: CH_3Cl, CH_2Cl_2, $CHCl_3$, and CCl_4.

Let us consider the chlorination of methane in further detail. The chlorination process that gives the four different products actually occurs in steps. The first step in the process involves the substitution of one chlorine atom for a hydrogen atom on the methane molecule.

$$CH_4 + Cl_2 \xrightarrow{light} CH_3Cl + HCl$$

Note that only half the chlorine molecule ends up in the organic product, which is called *chloromethane* or *methyl chloride*. The other chlorine atom becomes part of the inorganic molecule HCl. The hydrogen atom in HCl came from the methane molecule that lost one hydrogen atom in order to be able to bond to a chlorine atom. This substitution process can be diagrammed as follows.

$$\begin{array}{c} H \\ | \\ H-C-H + Cl-Cl \\ | \\ H \end{array} \longrightarrow \begin{array}{c} H \\ | \\ H-C-Cl + H-Cl \\ | \\ H \end{array}$$

In Step 2 of the methane chlorination process, the chloromethane formed in Step 1 reacts further to give dichloromethane (or methylene chloride).

$$CH_3Cl + Cl_2 \xrightarrow{light} CH_2Cl_2 + HCl$$

The dichloromethane from Step 2 then reacts further to give trichloromethane (or chloroform).

$$CH_2Cl_2 + Cl_2 \xrightarrow{light} CHCl_3 + HCl$$

Finally, the trichloromethane from Step 3 reacts with Cl_2 to give CCl_4 (tetrachloromethane or carbon tetrachloride).

$$CHCl_3 + Cl_2 \xrightarrow{light} CCl_4 + HCl$$

In actual practice, chlorination of methane gives all four of these products; the amounts of each depend on the time allowed for the reaction and the mole ratio of CH_4 and Cl_2.

Regarding the chlorination of methane, it is interesting to note how product properties change as chlorine is introduced into the methane molecule. Each time an additional Cl atom is added, the flammability of the product goes down, but its toxicity goes up. Carbon tetrachloride was once used in fire extinguishers and as a dry cleaning solvent because of its low flammability and ability to dissolve fats and oils. It is no longer used for either purpose because of its high toxicity. Chloroform was one of the first general anesthetics. It has the desired property of low flammability. However, it is now known that chloroform is toxic to the liver and may even be carcinogenic.

In many cases, even monohalogenation can result in the formation of more than one product. For example, in the bromination of propane, two monobromopropanes result.

$$2\,Br_2 + 2\,CH_3-CH_2-CH_3 \longrightarrow \underset{\substack{| \\ Br \\ \text{1-bromopropane} \\ (n\text{-propyl bromide})}}{CH_3-CH_2-CH_2} + \underset{\substack{| \\ Br \\ \text{2-bromopropane} \\ (\text{isopropyl bromide})}}{CH_3-CH-CH_3} + 2\,HBr$$

TABLE 14-6 Some Industrially Important and Useful Halogenated Alkanes

Name	Formula	Typical uses
dichlorodifluoromethane (Freon-12)	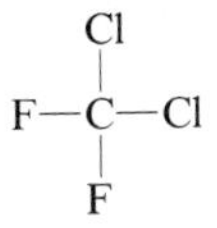 F—C(Cl)(Cl)—F	a refrigerant and aerosol propellant
1-bromo-1-chloro-2,2,2-trifluoroethane (halothane)	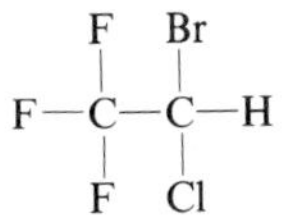 F—C(F)(F)—C(Br)(Cl)—H	a general anesthetic
1,2-dibromoethane	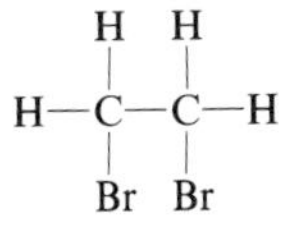 H—C(H)(H)—C(H)(H)—H with Br, Br on the two carbons	gasoline additive—removes lead deposits from engine
chloroethane (ethyl chloride)	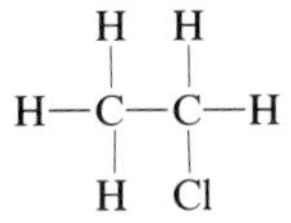 H—C(H)(H)—C(H)(Cl)—H	a topical freezing agent—used in minor surgery
1,2-dichloroethane (ethylene chloride)	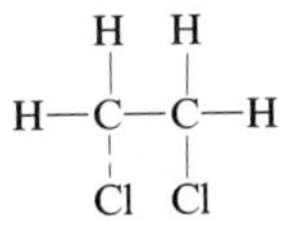 H—C(H)(Cl)—C(H)(Cl)—H	a dry cleaning solvent, chewing gum solvent

Further bromination of these monobrominated products would produce an even more complex mixture. The actual process is much more complicated and will not be discussed here. A large number of industrial solvents are mixtures of many halogenated hydrocarbons. Table 14–6 lists additional uses for selected halogenated hydrocarbons.

You may have noted that two different names have been given for each of the halogenated alkanes we have discussed. The first name is the IUPAC name and the second is the common name. Common names exist for almost all hydrocarbon derivatives that have commercial value. The reason for this is that, for the layperson, many IUPAC names are cumbersome and difficult to pronounce because of their length. The name 1-bromo-1-chloro-2,2,2-trifluoroethane is overwhelming; halothane is much better (see Table 14–6).

The IUPAC rules for naming halogen derivatives of alkanes are very similar to those used in naming branched alkanes. The halogen group or groups, which are referred to as *fluoro-*, *chloro-*, *bromo-*, and *iodo-*, are handled in the same way as alkyl groups; this is illustrated in Example 14.5.

Example 14.5

Name the following halogenated alkanes using IUPAC rules.

a. 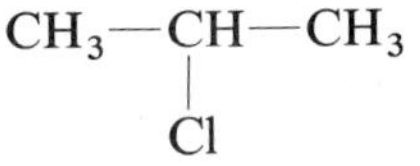 $CH_3—CH(Cl)—CH_3$

b. 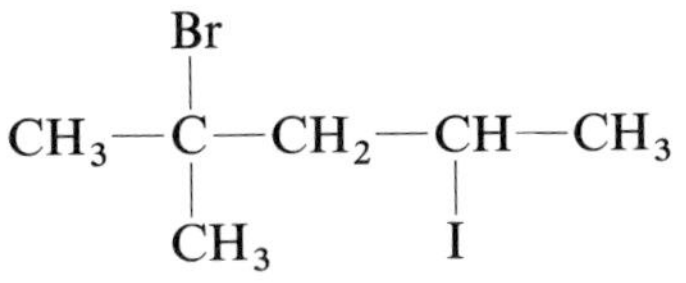 $CH_3—C(Br)(CH_3)—CH_2—CH(I)—CH_3$

Solution

a. The longest carbon chain contains three atoms, and a chlorine atom is attached to the middle carbon (carbon 2). Thus, the name is *2-chloropropane*.

b. The longest carbon chain contains five atoms that are numbered using a left-to-right numbering system. (We number from the chain end that gives the lowest set of numbers to the attachments.)

$$\overset{1}{C}H_3-\overset{2}{\underset{CH_3}{\overset{Br}{C}}}-\overset{3}{C}H_2-\overset{4}{\underset{I}{C}H}-\overset{5}{C}H_3$$

There are three attachments to the base chain—an iodo group on carbon 4, a bromo group on carbon 2, and a methyl group that is also on carbon 2. Listing these attachments in alphabetical order, the name of the compound becomes *2-bromo-4-iodo-2-methylpentane*.

14.13 ALKANES IN LIVING ORGANISMS

Living organisms contain a wide variety of organic molecules. Of these, comparatively few are simple alkanes. The functions of the few that have been found are quite interesting.

Some plants and trees produce higher-molecular-weight alkanes, such as the natural waxy coating of pears, which contains a 29-carbon alkane. Several kinds of pine trees produce *n*-heptane, and *n*-tetradecane ($C_{14}H_{30}$) is found in chrysanthemums. Microorganisms that evolve methane as byproducts of metabolism are being studied in an effort to use them to produce natural gas in large methane farms. This might provide an alternate energy source if the gas can be collected, concentrated, and sold economically.

Some alkanes help insects communicate with other members of their own species. These communication chemicals are known as *pheromones*, derived from the Greek *pherin*, "to carry", and *hormon*, "to excite or stimulate". Pheromones are excreted outside the insect's body, where they cause specific reactions from other insects of the same species.

Ants discharge a variety of pheromones to communicate with other members of the colony. Two of these compounds have been identified as the normal alkanes *n*-undecane ($C_{11}H_{24}$) and *n*-tridecane ($C_{13}H_{28}$). In tiger moths, 2-methylheptadecane functions as a sex pheromone. When small amounts of these sex pheromones are secreted by female insects, males of the same species can be attracted from great distances.

$$CH_3-(CH_2)_9-CH_3 \qquad CH_3-(CH_2)_{11}-CH_3 \qquad CH_3-\underset{CH_3}{\overset{|}{C}H}-(CH_2)_{14}-CH_3$$

undecane tridecane 2-methylheptadecane

EXERCISES AND PROBLEMS

Terminology

14.1 Contrast the modern day and historical definitions of *organic chemistry*.

14.2 Define the following terms.
a. hydrocarbon
b. saturated hydrocarbon
c. alkane
d. cycloalkane

Formulas for Alkanes

14.3 Draw three-dimensional ball-and-stick structures for the three simplest alkanes: methane, ethane, and propane.

14.4 Using the general formula for an alkane, calculate the number of hydrogen atoms that would be present in an alkane molecule containing
a. 8 carbon atoms   c. 22 carbon atoms
b. 14 carbon atoms   d. 35 carbon atoms

14.5 Convert each of the following expanded structural formulas into a condensed structural formula.

a.
```
    H   H   H   H
    |   |   |   |
H — C — C — C — C — H
    |   |   |   |
    H   H   H   H
```

b.
```
    H   H   H   H   H
    |   |   |   |   |
H — C — C — C — C — C — H
    |   |   |   |   |
    H   H   |   H   H
          H — C — H
            |
            H
```

c.
```
                H
                |
            H — C — H
    H   H   H   H   |   H
    |   |   |   |       |
H — C — C — C — C — C — C — H
    |   |   |   |   |   |
    H   H   |   H   H   H
          H — C — H
            |
            H
```

d.
```
    H   H   H   H   H
    |   |   |   |   |
H — C — C — C — C — C — H
    |   |   |   |   |
    H   H   |   H   H
          H — C — H
            |
          H — C — H
            |
            H
```

14.6 The following structural formulas for alkanes are incomplete, in that the hydrogen atoms attached to each carbon are not shown. Complete each of these formulas by writing in the correct number of hydrogen atoms attached to each carbon atom; that is, rewrite each of these formulas as a condensed structural formula such as $CH_3—CH_2—CH_3$.

a.
```
C — C — C — C
    |
    C
```

b.
```
C — C — C — C — C — C
    |   |   |
    C   C   C
```

c. C — C — C — C — C — C

d.
```
    C
    |
C — C — C — C
    |
    C
```

14.7 Draw the indicated type of formula for the following alkanes:

a. The expanded structural formula for a straight-chain alkane with the formula C_5H_{12}.
b. The expanded structural formula for $CH_3—(CH_2)_6—CH_3$.
c. The condensed structural formula, using parentheses for the $—CH_2—$ groups, for the straight-chain alkane $C_{10}H_{22}$.
d. The molecular formula for the alkane $CH_3—(CH_2)_4—CH_3$.

Nomenclature for Alkanes

14.8 The first step in naming an alkane is to identify the longest continuous carbon-atom chain. Give the number of carbon atoms in the longest continuous carbon-atom chain in each of the following carbon-atom arrangements.

a.
```
                    C
                    |
C — C — C — C — C — C — C
            |
            C
            |
            C
```

b.
```
C — C — C — C — C — C
    |
    C
    |
    C
```

c.
```
                C — C — C
                |
C — C — C — C — C
                |
                C — C — C — C
```

d.
```
    C—C—C
    |
C—C—C—C
|
C
|
C—C
```

e.
```
C—C—C—C—C—C—C
    |       |
    C       C
    |       |
    C       C
    |       |
    C       C
```

14.9 Give the proper IUPAC name for each of the following alkanes.

a.
```
    H  H  H  H  H  H
    |  |  |  |  |  |
H—C—C—C—C—C—C—H
    |  |  |  |  |  |
    H  H  H  H  H  H
```

b. CH_3—CH_2—CH_2—CH—CH_3
                          |
                          CH_3

c. CH_3—CH—CH—CH_2—CH—CH_3
           | | |
           CH_2 CH_3 CH_3
           |
           CH_3

d.
```
              CH₃
              |
CH₃—CH—C—CH₂—CH₃
      |  |
      CH₃ CH₂
          |
          CH₃
```

e.
```
       CH₃
       |
CH₃—C—CH₃
       |
       CH₃
```

f.
```
CH₂—CH₂—CH—CH₂—CH₂
|          |        |
CH₃        CH₂      CH₃
           |
           CH₂
           |
           CH₃
```

g.
```
     CH₃ CH₃
     |   |
CH₃—C—C—CH₃
     |   |
     CH₃ CH₃
```

h. CH_3—CH_2—CH_2—CH—CH_3
                          |
                          CH_3—CH_2—CH_3

i. CH_3—CH_2—CH—CH_2—CH_3
                |
                CH_2
                |
                CH_3

j.
```
          CH₃ CH₃
          |   |
    CH₃  CH₂ CH₂ CH₃
    |    |   |   |
    CH₂—C———C—CH₂
         |   |
         CH₃ CH₂
             |
             CH₃
```

14.10 Draw a structural formula for each of the following compounds.

a. 2-methylbutane
b. 3,4-dimethylhexane
c. 3-ethyl-3-methylpentane
d. 2,3,4,5-tetramethylheptane
e. 3,5-diethyloctane
f. 4-*n*-propylnonane

14.11 The following names are *incorrect*. Draw the carbon skeleton for each and explain why the name is incorrect.

a. 4-methylpentane
b. 2-ethyl-2-methylpropane
c. 2,3,3-trimethylbutane
d. 2,2,2-trimethylpentane
e. 4-ethyl-5-methylhexane

Structural Isomers for Alkanes

14.12 What are structural isomers?

14.13 For each of the following pairs of structures, determine if they are

1. different conformations of the same molecule,
2. structural isomers, or
3. neither different conformations of the same molecule nor structural isomers.

a. CH_3—CH_2—CH_2—CH_2—CH_3

   and CH_3—CH—CH_3
                |
                CH_2
                |
                CH_3

b. CH_3—CH_2—CH_2 and CH_3—CH_2
                    | |
                    CH_3 CH_2—CH_3

c.
```
       CH₃
       |
CH₃—C—CH₃      and   CH₃—CH—CH₂—CH₃
       |                    |
       CH₃                  CH₃
```

d. $CH_3-\underset{\underset{CH_2-CH_3}{|}}{CH}-CH_3$ and $CH_3-\underset{\underset{CH_3}{|}}{CH}-CH_2-CH_3$

e. $CH_3-\underset{\underset{\underset{CH_3}{|}}{CH_2}}{\overset{|}{CH}}-CH_2-CH_3$

and $CH_3-\underset{\underset{CH_3}{|}}{CH}-\underset{\underset{CH_3}{|}}{CH}-CH_3$

f. $CH_3-CH_2-CH_2-\underset{\underset{CH_3}{|}}{CH}-CH_3$

and $CH_3-\underset{\underset{CH_3}{|}}{CH}-CH_2-CH_3$

g. $\underset{CH_2-CH_2-CH_2}{\overset{CH_3 CH_3}{| |}}$

and $CH_3-CH_2-CH_2-\underset{\underset{CH_3}{|}}{CH}-CH_3$

14.14 For each of the following pairs of molecules, indicate whether they are structural isomers.

a. 2-methylpentane and 3-methylpentane
b. 2-methylpentane and 2-methylhexane
c. 2,3-dimethylbutane and 2,2-dimethylbutane
d. 2,3-dimethylbutane and 2-methylpentane
e. n-hexane and 2-methylhexane
f. n-hexane and 2,2-dimethylbutane
g. 3-methylheptane and 3-ethylheptane

14.15 Write structural formulas (showing only carbon atoms) for all the isomers for each of the following molecular formulas.

a. C_6H_{14} (five isomers are possible)
b. C_7H_{16} (nine isomers are possible)

Alkyl Groups

14.16 Indicate the total number of alkyl groups on each of the molecules listed in Problem 14.9.

14.17 Determine the length of the *shortest* carbon chain to which the following alkyl groups could be attached without lengthening the chain.

a. an *n*-propyl group c. an isopropyl group
b. an ethyl group d. a methyl group

14.18 For the following molecule, list all possible alkyl groups that "R" could represent while retaining only five carbon atoms in the longest carbon chain.

$CH_3-CH_2-\underset{\underset{R}{|}}{CH}-CH_2-CH_3$

14.19 Draw the structures of each of the following alkyl groups:

a. isopropyl-
b. *tert*-butyl-
c. *n*-butyl-
d. *sec*-butyl-
e. isobutyl-

14.20 Name the following alkanes, each of which contains at least one branched alkyl group.

a. $CH_3-CH_2-CH_2-\underset{\underset{\underset{CH_3}{|}}{CH-CH_3}}{\overset{|}{CH}}-CH_2-CH_3$

b. $CH_3-CH_2-\underset{\underset{\underset{CH_3}{|}}{CH_3-C-CH_3}}{\overset{|}{CH}}-CH_2-CH_3$

c. $CH_3-CH_2-CH_2-\underset{\underset{\underset{\underset{CH_3}{|}}{CH_2}}{\overset{|}{CH-CH_3}}}{\overset{|}{CH}}-\overset{\overset{CH_3}{|}}{CH}-CH_2-CH_3$

d. $CH_3-CH_2-\underset{\underset{\underset{CH_3}{|}}{CH-CH_3}}{\overset{\overset{\overset{CH_3}{|}}{CH_2}}{|}}{C}-CH_2-CH_2-\underset{\underset{CH_3}{|}}{CH}-CH_3$

Formulas for Cycloalkanes

14.21 What is the general formula for a cycloalkane?

14.22 All of the bonds in both *n*-butane (C_4H_{10}) and cyclobutane (C_4H_8) are single bonds, but the latter compound contains two less hydrogens than the former. Why?

14.23 How many hydrogen atoms are present in each of the following molecules?

a. [cyclohexane ring] c. [cyclopropane with CH_3]

b. [cyclobutane with two CH_3 groups] d. [cyclopentane with two CH_3 groups]

337

Nomenclature for Cycloalkanes

14.24 Assign IUPAC names to each of the compounds in Problem 14.23.

14.25 Write structural formulas for each of the following cycloalkanes, using geometrical figure notation.
a. 1,2,4-trimethylcyclohexane
b. 3-ethyl-1,1-dimethylcyclopentane
c. *n*-propylcyclobutane
d. isopropylcyclobutane

14.26 Draw and name the three possible dimethylcyclobutane isomers.

14.27 Determine what the general formula would be for cycloalkanes that have two cyclic portions to their structure. Write the molecular formula for the following compound and use it to help you.

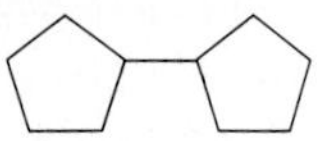

Sources and Physical Properties of Alkanes

14.28 Describe why the concept that petroleum occurs underground in liquid pools is a misconception.

14.29 What is the rough composition of unprocessed natural gas?

14.30 Describe the process of fractional distillation that is used to produce useful products from crude petroleum.

14.31 Consider the series of *n*-alkanes with 1 to 20 carbon atoms. Describe the changes in appearance, physical state, and boiling point of the compounds as carbon-chain length increases.

Chemical Reactions of Alkanes

14.32 Write equations representing the complete combustion of the following alkanes, and then balance each equation.

a. C_3H_8
b. *n*-butane
c. cyclobutane
d. *n*-octane

14.33 Write balanced chemical equations for the production of each of the four possible products that can be obtained from the bromination of methane.

14.34 Low-molecular-weight chlorofluorocarbons are often used as refrigerants or propellants in aerosol cans. These are marketed under the trade name "Freons". Draw the structures of the following freons.
a. trichlorofluoromethane
b. dichlorodifluoromethane
c. 1,2-dichloro-1,1,2,2-tetrafluoroethane

14.35 Name the following halogenated alkanes using IUPAC nomenclature.

a.
$$\begin{array}{c} \text{H} \quad \text{H} \\ | \quad\;\; | \\ \text{H}-\text{C}-\text{C}-\text{Br} \\ | \quad\;\; | \\ \text{H} \quad \text{Br} \end{array}$$

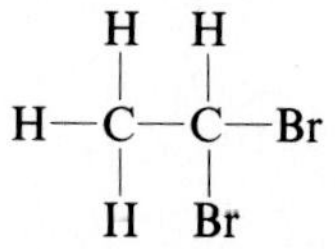

b.
$$\begin{array}{c} \text{CH}_3-\text{CH}-\text{CH}-\text{CH}_3 \\ \quad\quad\quad | \quad\;\; | \\ \quad\quad\quad \text{CH}_3 \;\; \text{I} \end{array}$$

c.
$$\begin{array}{c} \text{H} \quad \text{Cl} \quad \text{Cl} \\ | \quad\;\; | \quad\;\; | \\ \text{H}-\text{C}-\text{C}-\text{C}-\text{H} \\ | \quad\;\; | \quad\;\; | \\ \text{H} \quad \text{Br} \quad \text{H} \end{array}$$

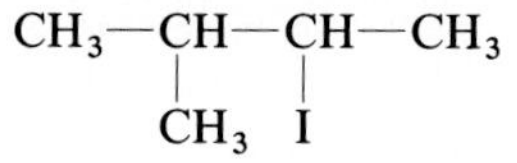

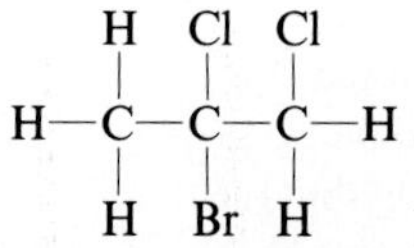

d. $F-CH_2-CH_2-F$

14.36 Assume the following reaction yields only *monobrominated* products.

$$n\text{-butane} + Br_2 \xrightarrow{\text{light}} \text{isomers of } C_4H_9Br + HBr$$

Draw structures for all possible isomeric products and name each one.

15 Unsaturated Hydrocarbons

Objectives

After completing Chapter 15, you will be able to:

15.1 Define the term *unsaturated hydrocarbon* and identify the functional groups present in these compounds.
15.2 Define the terms *alkene* and *cycloalkene* and give general formulas for these compounds that contain one double bond.
15.3 Name alkenes and cycloalkenes using IUPAC rules, given the structural formulas, and vice versa.
15.4 Discuss the nature of carbon–carbon multiple bonds in terms of sigma and pi bonds.
15.5 Understand the phenomenon of geometric isomerism (*cis–trans* isomerism) and recognize, identify, and name these isomers.
15.6 Describe the physical properties of alkenes and discuss their occurrence in nature.
15.7 Write equations and the structures of products for both symmetrical and unsymmetrical addition reactions of alkenes, and use Markovnikov's rule, when necessary, to predict major products of an addition.
15.8 Define the terms *oxidation* and *reduction* in an organic reaction context, using the concepts of loss or gain of hydrogen atoms and loss or gain of oxygen atoms.
15.9 List several common addition polymers and the alkene monomers from which they are synthesized.
15.10 Define the term *alkyne* and give the general formula for an alkyne with one triple bond.
15.11 Name alkynes using IUPAC rules, given the structural formulas, and vice versa.
15.12 Describe the physical properties of alkynes and discuss the chemical reactions alkynes undergo.

INTRODUCTION

In Chapter 14, we discussed saturated hydrocarbons; in these compounds, all carbon–hydrogen and carbon–carbon bonds were single bonds. A single bond is the only type of bond that is possible between carbon and hydrogen. No such restriction exists for bonds between two carbon atoms. With carbon atoms, single, double, and triple bond formation is possible. In this chapter, we will consider hydrocarbon molecules that contain one or more carbon–carbon double or triple bonds. The chemical characteristics of these molecules are very different from those of saturated hydrocarbons.

15.1 UNSATURATED HYDROCARBONS

An **unsaturated hydrocarbon** *is a hydrocarbon that contains one or more carbon–carbon double or triple bonds.* Compared to their saturated counterparts (alkanes and cycloalkanes), unsaturated hydrocarbons always contain fewer hydrogen atoms; that is, they are unsaturated with respect to hydrogen. By means of chemical reactions, these unsaturated molecules can be converted to saturated ones through the addition of hydrogen. The term *polyunsaturated* is often heard on television commercials dealing with cooking oils and margarines. This term refers to the presence of several double bonds within a molecule.

The term *multiple bond* is commonly used to collectively refer to both double and triple bonds. Carbon–carbon multiple bonds are less stable, that is, more reactive, than carbon–carbon single bonds are. The presence of these multiple bonds within a hydrocarbon molecule governs the chemistry of unsaturated hydrocarbons.

Whenever a certain portion of a molecule governs its chemical reactions, that portion of the molecule is called a *functional group.* A **functional group** *is an atom or group of atoms that serves as a basis for characterizing the chemical behavior of a family of organic compounds.* The one or more multiple bonds present in an unsaturated hydrocarbon comprise the functional group that characterizes this type of compound.

The ability to recognize and understand the behavior of functional groups is a powerful tool in organic chemistry. Although millions of organic compounds exist, they can be grouped into a relatively small number of categories, based upon the functional groups they contain. Once we understand how a particular functional group behaves, we can immediately understand how almost all molecules containing the same functional group behave.

Unsaturated hydrocarbons are divided into two groups: those that contain double bonds and those that contain triple bonds. We will discuss the hydrocarbons containing carbon–carbon double bonds first. The principles we learn about these compounds will then be applicable, with only slight modification, to hydrocarbons containing carbon–carbon triple bonds.

15.2 ALKENES

Alkenes *are hydrocarbons that contain one or more carbon–carbon double bonds.* Numerous noncyclic (open-chain) and cyclic alkenes are known.

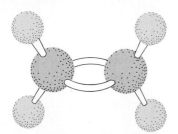

Ball-and-stick model

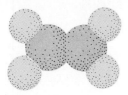

Space-filling model

FIGURE 15–1 Perspective drawings of the structure of ethene (C_2H_4), the simplest alkene.

The simplest alkene is ethene (common name, ethylene), which has the formula C_2H_4. Obviously, a one-carbon alkene cannot exist because two carbon atoms are required to have a carbon–carbon double bond. Ball-and-stick and space-filling models of ethene are shown in Figure 15–1. Note from the three-dimensional ball-and-stick model in Figure 15–1 that the arrangement of bonds about the carbon atoms is not tetrahedral as it is in alkanes; it is planar. The two carbon atoms and four hydrogen atoms all lie in a plane with an angle of 120° between bonds. Any carbon involved in multiple bonding will always have a nontetrahedral arrangement of bonds around it.

The following are expanded and condensed structural formulas for ethene.

$$\begin{array}{c} H \\ \diagdown \diagup H \\ C{=}C \\ \diagup \diagdown \\ H H \end{array} \qquad CH_2{=}CH_2$$

expanded structural formula condensed structural formula

The notation used for a double bond is the same as that used in inorganic chemistry (Section 6.9)—two parallel lines.

The general formula for a noncyclic alkene containing only one double bond is C_nH_{2n}. These alkenes contain two less hydrogen atoms than the maximum number possible in hydrocarbons (alkanes). The reduced number of hydrogen atoms is caused by the presence of the double bond.

In Section 14.9, we noted that the general formula for a cycloalkane is C_nH_{2n}. Thus, alkenes containing one double bond and cycloalkanes have the same general formula. This means that compounds from these two families that have the same number of carbon atoms are isomeric.

Cycloalkenes *are hydrocarbons that have a ring structure and also one or more double bonds within the ring structure.* Cycloalkenes containing one double bond have the general formula C_nH_{2n-2}. This general formula reflects the loss of four hydrogen atoms from maximum (C_nH_{2n+2}); two hydrogen atoms are lost because of the double bond and two hydrogen atoms are lost as a result of cyclization.

The simplest cycloalkene is the compound cyclopropene, C_3H_4, a three-membered carbon ring containing one double bond.

15.3 IUPAC NOMENCLATURE FOR ALKENES

The rules we previously developed for naming alkanes (Sections 14.7 through 14.9) can be used, with slight modifications, to name alkenes. The modifications needed to name alkenes containing only one double bond are

1. The *-ane* ending characteristic of alkanes is changed to *-ene* for alkenes. The *-ene* ending means at least one double bond is present.
2. The root name for the alkene is derived from the longest continuous chain (or ring) of carbon atoms *containing the double bond*.
3. For noncyclic molecules containing more than three carbon atoms, the position of the double bond must be specified because there is more than one position it may occupy. The position is given by a single number (the number of the lower numbered carbon involved in the double bond), which is placed immediately in front of the base chain name.

Examples: $\overset{1}{C}H_2=\overset{2}{C}H-\overset{3}{C}H_2-\overset{4}{C}H_3$ is 1-butene

and $\overset{1}{C}H_3-\overset{2}{C}H=\overset{3}{C}H-\overset{4}{C}H_3$ is 2-butene

The compounds 1-butene and 2-butene are isomers. Note that in the case of 1-butene, the double bond involves carbons 1 and 2, but we only write the lower of the two numbers in the name.

4. For noncyclic molecules where alkyl groups are present in addition to the double bond, the numbering system chosen is the one that assigns the lowest possible number to the carbon atom at which the double bond originates. Numbering alkyl groups is of lower priority than numbering the double bond.

Example: $\overset{4}{C}H_3-\overset{3}{C}H-\overset{2}{C}H=\overset{1}{C}H_2$ is 3-methyl-1-butene
$\quad\quad\quad\quad\quad\quad\;\;|$
$\quad\quad\quad\quad\quad\;\;CH_3$

The chain is always numbered in such a way as to give the double bond the lowest number possible, even if this means that alkyl groups must get higher numbers.

5. For cycloalkenes containing one double bond, the ring is numbered to give the double-bonded carbons the numbers 1 and 2. The direction of numbering is chosen so that any alkyl groups present receive the lowest numbers. In this case, the position of the double bond is not given because we know it is between the number 1 and 2 carbon atoms.

Example: is 4-methylcyclohexane

Example 15.1

Assign IUPAC names to each of the following alkenes.

a. $CH_2=CH-CH_2-CH_2-CH_2-CH_3$

b. $CH_3-CH-CH=CH-CH_3$
$\quad\quad\;\;|$
$\quad\;\;CH_3$

c. $CH_3-CH_2-C=CH_2$
$\quad\quad\quad\quad\;|$
$\quad\quad\quad\;CH_2$
$\quad\quad\quad\quad\;|$
$\quad\quad\quad\;CH_3$

d.

$\;\;CH_3$

Solution

a. The longest continuous chain containing the double bond is six carbons long. If this were an alkane, the base name would be *hexane*. In this case, the base name is *hexene* because the *-ane* ending must be changed to *-ene* to indicate the presence of the double bond. The double bond is located between carbon atoms 1 and 2. Therefore, the IUPAC name is *1-hexene*.

b. $\overset{5}{C}H_3-\overset{4}{C}H-\overset{3}{C}H=\overset{2}{C}H-\overset{1}{C}H_3$
$\quad\quad\quad\;|$
$\quad\quad\;CH_3$

The longest carbon chain containing the double bond has five carbons. This gives a base name of *pentene*. Numbering the chain from right to left so that the double bond receives the lowest number possible results in the name *4-methyl-2-pentene*. If we had numbered the chain in the opposite direction, from left to right, we would have obtained the name 2-methyl-4-pentene. This is an incorrect name even though the numbers are the same as in the correct name. The correct numbering system is the one that assigns the lowest number possible to the double bond.

c. $\overset{4}{CH_3}-\overset{3}{CH_2}-\overset{2}{C}=\overset{1}{CH_3}$
$|$
CH_2
$|$
CH_3

You may be tempted to identify the longest carbon chain, which is five carbons long, as the base for the name of this compound. That is incorrect. The base name of an alkene is derived from the longest carbon chain *that contains the double bond*. The five-carbon chain does not contain the double bond. The longest chain containing the double bond has four carbon atoms. Therefore, the proper base name is *butene*. The chain is numbered from right to left—from the end closest to the double bond. An ethyl group is attached on carbon number 2. The complete name for this alkene is *2-ethyl-1-butene*.

d.

The base name for a cyclic alkene is determined by the number of carbon atoms in the ring; in this case, it will be *cyclobutene*. The IUPAC rules state that we must number the ring so that the first carbon of the double bond is number 1. We then proceed to number the ring in the direction of the double bond, so that the other carbon of the double bond is number 2. This numbering system gives the number 3 to the carbon to which the methyl group is attached. The complete IUPAC name for this cycloalkene is *3-methylcyclobutene*. The location of the double bond is not explicitly given because it is understood that it will always involve carbon atoms 1 and 2.

There are many alkenes that contain more than one double bond. Compounds containing two double bonds are named *alkadienes*, and those with three double bonds are *alkatrienes*.

The general family names just mentioned give us the key to naming compounds containing more than one multiple bond. A prefix is added to the base chain ending that indicates the number of double bonds—*diene* for two double bonds, —*triene* for three double bonds, and so on. In addition, a number is used to locate each double bond. If there are three double bonds, three numbers are needed.

Example 15.2

Name the following compounds, each of which contains more than one double bond.

a. $CH_2=CH-CH=CH_2$

b. $CH_3-\underset{\underset{CH_3}{|}}{C}=CH-CH_2-\underset{\underset{CH_3}{|}}{CH}-CH=CH_2$

c. [cyclopentadiene structure]

d. [cycloheptatriene with CH$_3$ structure]

Solution

a. The longest carbon chain contains four carbons and two double bonds. This compound is a *butadiene*. Notice that we used *-diene* for the ending of the chain name, rather than simply *-ene*. Numbering the chain from either end gives the same results; the double-bond positions are 1 and 3. The correct name is *1,3-butadiene*.

b. There are seven carbon atoms and two double bonds in the longest carbon chain. The base name is *heptadiene*. Locating the positions of the two methyl groups and the two double bonds with numbers presents an apparent conflict. Numbering the chain from left to right gives the methyl groups the lowest numbers, but numbering from right to left gives the double bonds the lowest numbers. The IUPAC rules solve the problem—double bonds always take precedence over alkyl groups. Thus, we use the right-to-left numbering system.

$\overset{7}{C}H_3-\underset{\underset{CH_3}{|}}{\overset{6}{C}}=\overset{5}{C}H-\overset{4}{C}H_2-\underset{\underset{CH_3}{|}}{\overset{3}{C}H}-\overset{2}{C}H=\overset{1}{C}H_2$

The complete name for this compound is *3,6-dimethyl-1,5-heptadiene*.

c. A five-membered ring with two double bonds is a *cyclopentadiene*. In order to locate the double bonds, we number the ring beginning with the leading carbon of one double bond and in a direction to yield the lowest numbers possible. The correct numbering system in our case is

[numbered cyclopentadiene ring with positions 1-5]

The complete name of the compound is *1,3-cyclopentadiene*.

d. A ring system containing seven carbon atoms and three double bonds is called a *cycloheptatriene*. Two different numbering systems yield the same locations for the double bonds, but one direction yields a lower number for the methyl group.

[numbered cycloheptatriene ring with two numbering systems and CH$_3$]

The clockwise (outer) numbering system is the correct one. The full name of the compound is *3-methyl-1,3,5-cycloheptatriene*.

The two carbon atoms involved in a carbon–carbon double bond are referred to as vinyl carbons. Any atom or group of atoms attached to one of the carbon atoms of a double bond is said to be bonded to a *vinyl* carbon. The industrial name of the halogenated alkene *chloroethene* is *vinyl chloride*.

$$\begin{array}{c} H \\ \diagdown \\ C=C \\ \diagup\diagdown \\ HCl \end{array} \begin{array}{c} H \\ \diagup \\ \end{array}$$

vinyl chloride (chloroethene)

conjugated double bonds

The term *conjugation* is associated with the presence of double bonds in hydrocarbons. This term describes particular arrangements of double bonds. **Conjugated double bonds** *are double bonds separated from each other by one single bond.* The presence of conjugated double bonds in a molecule gives rise to special physical and chemical properties. Many molecules containing conjugated-double-bond systems absorb visible light. Such substances are used as dyes and pigments. In chemical reactions, conjugated-double-bond systems usually react differently than nonconjugated double bonds. Some examples of conjugated double bond systems include

$$CH_2=CH-CH \quad CH_2$$
1,3-butadiene

and $\quad CH_2=CH-CH=CH \quad CH-CH_2$
1,3,5-hexatriene

15.4 THE NATURE OF CARBON–CARBON MULTIPLE BONDS

In a carbon–carbon multiple bond, either a double bond or a triple bond, there are two types of bonds present. These bonding types are known as *sigma bonds* and *pi bonds*.

A quick review of two key concepts about chemical bonding that were covered in earlier chapters will help us to formulate the difference between a sigma and pi bond.

1. In Section 5.4, we learned that different types of atomic orbitals are available for electrons to occupy and that these different orbital types, which we called *s*, *p*, *d*, and *f*, differ in shape. An "*s*" orbital, for example, has a spherical shape. A dumbbell shape is associated with a "*p*" orbital.
2. In Section 6.8, we learned that electron-sharing (covalent-bond formation) results from the overlap of atomic orbitals. Because of orbital overlap, shared electrons are able to move around in the area between two covalently bonded atoms.

To these concepts we add a third concept, which was implied but not explicitly stated in the bonding discussions of Chapter 6.

3. Single-, double-, and triple-covalent bonds respectively involve the simultaneous overlap of one, two, and three *pairs* of atomic orbitals.

The manner in which two atomic orbitals overlap in the formation of a covalent bond is what determines whether a given bond is a sigma bond or a pi bond.

sigma bond

A **sigma bond** *is a covalent bond in which the overlap between atomic orbitals lies along the axis joining the two bonded atoms.* This definition can best be illustrated pictorially. Consider the following three examples of atomic orbital overlap—all of which give rise to sigma bonds.

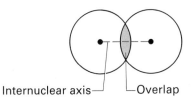 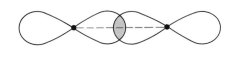 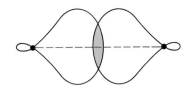

What is it that these three overlap diagrams have in common? It is not the shape of the orbitals; they are different in each case. The common thread is the location of the overlap region, shown in color in each diagram. In each case, this overlap area contains the internuclear axis (the dashed line). As our definition for a sigma bond states, the overlap between atomic orbitals must always lie along the axis joining the two bonded atoms.

Now let us contrast this sigma bonding situation with what occurs in pi bonding. A **pi bond** *is a covalent bond in which the overlap between atomic orbitals is above and below (but not on) the internuclear axis.* A pictorial representation of a pi bonding situation shows that two *p* orbitals (dumbbell-shaped orbitals) overlap sideways, rather than on their ends.

pi bond

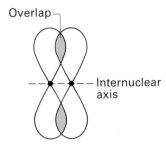

Note in this diagram that the overlap region (actually two regions) does not involve the internuclear axis; it is above and below the axis. This is the distinguishing feature of a pi bond. Note also that although two overlap regions result during pi-bond formation (above and below), the two regions collectively constitute one pi bond. The sideways overlap of two *p* orbitals that we have just discussed is the type of pi bond present in all carbon–carbon multiple bonds.

Now let us relate the concepts of sigma and pi bonds to the various types of bonding that can occur between carbon atoms.

1. When only one bond is present between two carbon atoms (a single bond), that bond is always a sigma bond.
2. When two bonds are present between carbon atoms (a double bond), that bond always consists of one sigma bond and one pi bond. Figure 15–2a (p. 348) shows diagrammatically the pi bond present in ethene (C_2H_4). The pi bond is formed from the overlap of one *p* orbital from each carbon atom.
3. When three bonds are present between carbon atoms (a triple bond)—a situation that we will discuss shortly (Section 15.10)—that bond always consists of one sigma bond and two pi bonds. Figure 15–2b (p. 348) shows the two pi bonds present in the molecule C_2H_2, which is the simplest hydrocarbon possessing a carbon–carbon triple bond. The two pi bonds present in this molecule are formed from the overlap of two *p* orbitals from each carbon atom.

Why is it important to know that pi bonding is present in carbon–carbon double bonds? *The presence of pi bonding causes a bond to be structurally rigid, which prevents free rotation of the carbon atoms involved in the bond.* Rotation is not compatible with a carbon–carbon multiple bond (see Figure 15–3, p. 349). If the

FIGURE 15-2 (a) A carbon–carbon double bond contains a sigma bond and a pi bond formed by the overlap of one *p* orbital from each carbon atom. (b) A carbon–carbon triple bond contains a sigma bond and two pi bonds formed by the overlap of two *p* orbitals from each carbon atom.

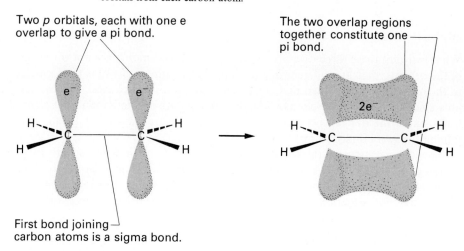

(a) C_2H_4 — a hydrocarbon with a carbon–carbon double bond.

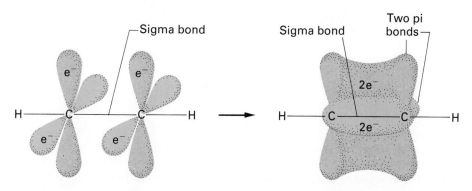

(b) C_2H_2 — a hydrocarbon with a carbon–carbon triple bond.

two carbon atoms in a double bond did rotate, the sigma bond would remain, but the pi bond would have to be broken. Remember that pi bonding occurs when orbitals overlap side-by-side and this is possible only when the atoms are in one particular alignment. This concept of restricted rotation is the key concept associated with the existence of sigma and pi bonds.

15.5 GEOMETRIC ISOMERS

geometric isomers

Lack of rotation around a chemical bond often gives rise to a special type of isomerism called *geometric isomerism*. **Geometric isomers** *are compounds that have the same molecular and structural formulas, but have different relative arrangements of atoms in space.*

FIGURE 15-3 Unlike (a) a carbon–carbon single bond, where rotation is allowed, (b) a carbon–carbon double bond is structurally rigid, preventing free rotation of the carbon atoms.

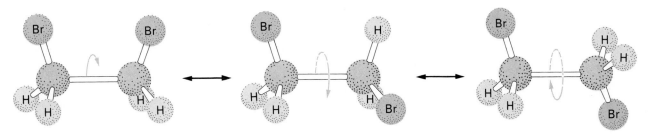

(a) The molecule 1,2-dibromoethane can rapidly interconvert between the three orientations shown because a carbon–carbon single bond is not structurally rigid.

(b) The molecule 1,2-dibromoethene cannot interconvert between the two forms shown because a carbon–carbon double bond is structurally rigid.

Two geometric isomers exist for the compound *1,2-dibromoethene*. They are

$$\begin{array}{c}\text{H}\\ \text{Br}\end{array}\!\!\!\!\!\!\!\!\!\text{C=C}\!\!\!\!\!\!\!\!\!\begin{array}{c}\text{H}\\ \text{Br}\end{array} \quad \text{and} \quad \begin{array}{c}\text{H}\\ \text{Br}\end{array}\!\!\!\!\!\!\!\!\!\text{C=C}\!\!\!\!\!\!\!\!\!\begin{array}{c}\text{Br}\\ \text{H}\end{array}$$

Because of the presence of the structurally rigid double bond, these two molecules cannot be converted into each other without breaking the pi bond and rotating about the sigma bond. Thus, despite their having the same structural formula—both carbon atoms in each molecule have a bromine atom and hydrogen atom attachment—these molecules are distinct. The two arrangements of atoms represent different compounds that have different boiling and melting points and other physical properties.

Geometrical isomers that result from the hindered rotation inherent in a double bond, such as the two 1,2-dibromoethenes just discussed, are often called *cis–trans* isomers. This is because the prefixes *cis-* and *trans-* are used when naming these isomers. The 1,2-dibromoethene molecule with both bromine atoms on the same side of the double bond (both below the double bond in our case) is called the *cis* isomer. Having the two bromine atoms on opposite sides of the double bond (one above and one below) results in the *trans* isomer.

$$\begin{array}{c}\text{H}\\ \text{Br}\end{array}\!\!\!\!\!\!\!\!\!\text{C=C}\!\!\!\!\!\!\!\!\!\begin{array}{c}\text{H}\\ \text{Br}\end{array} \qquad \begin{array}{c}\text{H}\\ \text{Br}\end{array}\!\!\!\!\!\!\!\!\!\text{C=C}\!\!\!\!\!\!\!\!\!\begin{array}{c}\text{Br}\\ \text{H}\end{array}$$

cis-1,2-dibromoethene   *trans*-1,2-dibromoethene

cis isomer

trans isomer

The ***cis* isomer** *is the geometric isomer in which the specified atoms or groups of atoms of concern are on the same side of the structurally rigid bond.* In our previous example, the bromine atoms were the atoms of concern. The ***trans* isomer** *is the geometric isomer in which the specified atoms or groups of atoms of concern are on opposite sides of the structurally rigid bond.* Looking back at Figure 15–3b, we can now see that the two molecules shown there are the *cis* and *trans* isomers of 1,2-dibromoethene.

Cis and *trans* isomers exist for the molecule 2-butene.

$$CH_3-CH=CH-CH_3$$

This is not readily apparent until the 2-butene structure is redrawn in a way that emphasizes the location of the atoms or groups of atoms attached to each of the double-bonded carbons.

    cis-2-butene   *trans*-2-butene

Not all alkenes exhibit geometric isomerism. Consider 2-methylpropene.

If the pi bond in this molecule were broken for a moment and rotation occurred, an identical molecule would result. This is because each of the double-bonded carbons has two identical groups attached to it. *Only alkenes that have two different groups attached to each of the double-bonded carbon atoms can exist in cis and trans forms.*

It is also possible to have *cis–trans* isomerism in certain cycloalkanes (Section 14.9). The cyclic structure of cycloalkanes hinders rotation about carbon–carbon single bonds in the ring. Thus, "up" and "down" positions exist for the attachments to the ring carbons, as can be seen in the case of 1,2-dichlorocyclobutane.

    cis-1,2-dichlorocyclobutane   *trans*-1,2-dichlorocyclobutane

Carbon atoms 1 and 2 of the ring—the carbons to which the chlorine atoms are attached—cannot rotate freely because of the ring; this traps the chlorine atoms on one or the other side of the ring.

In cycloalkane ring systems, we are not limited to *cis–trans* groups on adjacent carbon atoms. Groups separated by one or more carbons can also be *cis* or *trans* to each other, as is illustrated by the compound 1,3-diiodocyclobutane.

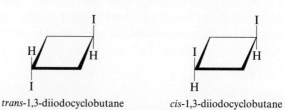

    trans-1,3-diiodocyclobutane   *cis*-1,3-diiodocyclobutane

A heavy, wedge-shaped bond to a ring structure is often used to indicate a bond is *above* the plane of the ring (coming out of the printed page). A broken line is used to indicate a bond *below* the plane of the ring (going into the printed page). The following example, involving 1,2-dimethylcyclopropane, illustrates the use of this notation.

351
Unsaturated Hydrocarbons

cis-1,2-dimethylcyclopropane   trans-1,2-dimethylcyclopropane

---

Example 15.3

Determine if geometric (*cis–trans*) isomers are possible for each of the following compounds.

a. CH=CH
   | |
   F   F

b.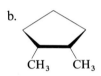
   CH₃ CH₃

c. 3-hexene

d. 1,1-dimethylcycloprane

Solution

a. Yes. Geometric isomers are possible because each carbon atom in the double bond has two different atoms attached to it.

   cis   and   trans

b. Yes. Geometric isomers are possible because the two methyl groups can be either on the same side or on opposite sides of the ring.

   cis   and   trans

c. Yes. *Cis–trans* isomerism is possible because the two carbons in the double bond each have a hydrogen and an ethyl group attached.

   cis   trans

d. No. Geometric isomers are not possible because all carbons in the ring have two identical atoms or groups attached. If we interchange the methyl groups on carbon number 1, we still have the same molecule.

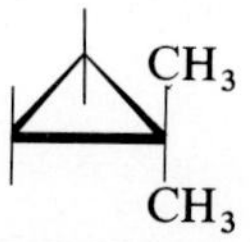

Cis–trans isomerism plays an important role in many biochemical processes, including the reception of light by the retina of the eye. The retina contains a compound called *retinal* that absorbs light and allows us to see. Retinal, which is attached to a protein called *opsin*, contains five conjugated carbon–carbon double bonds, one of which is in the *cis* configuration (see Figure 15–4a). When light is

FIGURE 15–4 The process of light reception by the retina of the eye involves converting *cis*-retinal to *trans*-retinal.

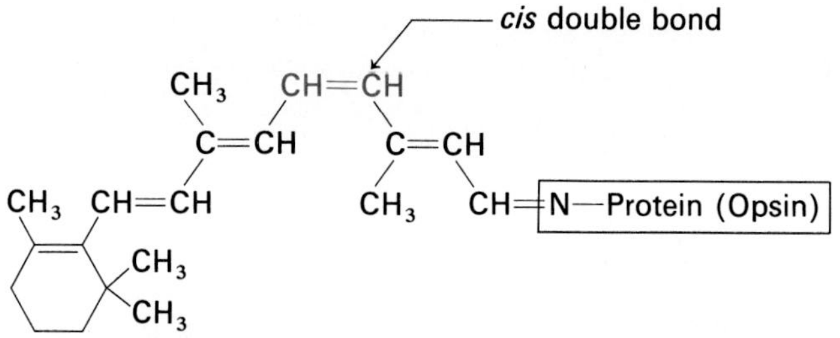

(a) Location of the *cis* double bond in *cis*-retinal.

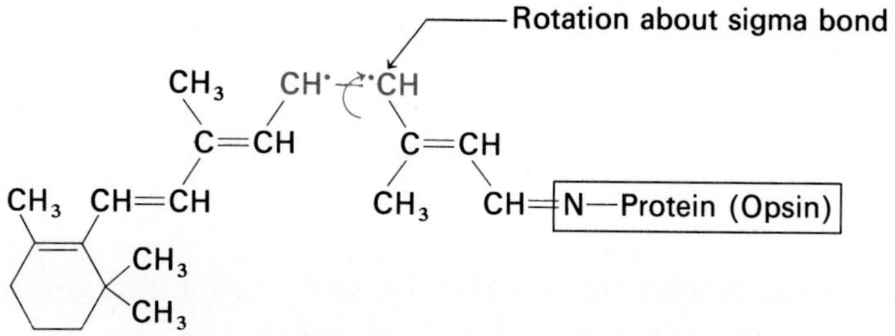

(b) Absorption of light momentarily breaks the pi bond.

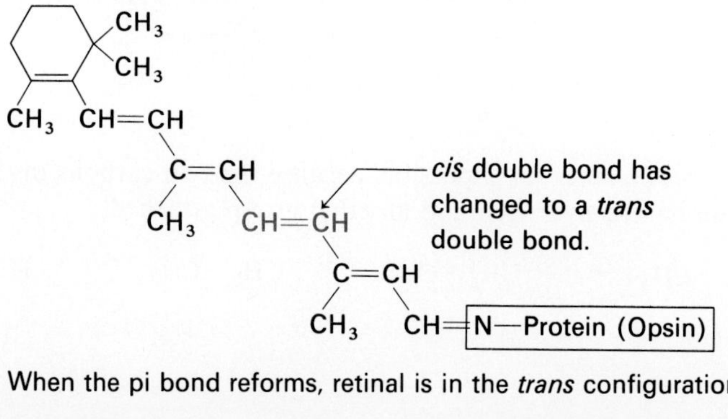

(c) When the pi bond reforms, retinal is in the *trans* configuration.

absorbed by retinal, the pi bond in this *cis* double bond is broken for a brief instant, allowing free-rotation about the sigma bond (see Figure 15–4b). The free-rotation switches the *cis* bond to a *trans* bond prior to the pi bond reforming (see Figure 15–4c). The *cis–trans* switch causes further chemical reactions in the cells of the retina in the eye and eventually sends a nerve impulse to the brain.

Trans-retinal must be converted back to *cis*-retinal before it can again trigger a nerve impulse. A large portion of this isomerization is done by enzymes in the liver. *Trans*-retinal is separated from the protein that holds it in the retina and is transported to the liver, where it is converted back to *cis*-retinal and transported back to the eye. This process takes a few moments, and so there is a time period required for the eye to adjust when a person goes from a brightly lit area to a darkened room.

15.6 PHYSICAL PROPERTIES AND NATURAL OCCURRENCE OF ALKANES

The physical properties of alkenes are similar to those of alkanes. Alkenes containing up to four carbons are gases under ordinary conditions and those containing five or more carbon atoms are liquids or solids.

Alkenes are important in many life processes. Consequently, they are found in abundance in natural substances. Ethene (ethylene), the simplest alkene, is synthesized in many species of trees, where it stimulates the normal ripening of fruit.

The deeply colored pigment found in yellow–orange vegetables such as carrots is beta-carotene, an alkene containing numerous carbon atoms and many conjugated double bonds.

Mammals and humans possess enzymes that can split beta-carotene in half, producing two molecules of vitamin A. The structure of vitamin A is

Vitamin A's structure is very similar to that of retinal, the visual pigment discussed in Section 15.5. A very simple biochemical reaction within the body transforms Vitamin A into retinal. Thus, it is true that eating carrots and taking Vitamin A can help improve your eyesight.

Hydrocarbons containing double bonds are responsible for the scents of numerous flowers and spices. The structures of four of these compounds are given in Figure 15–5 (p. 354).

The housefly sex-attractant pheromone *muscalure* is a C_{23}-alkene with a *cis* double bond.

cis-9-tricosene (muscalure)

FIGURE 15-5 The structures of unsaturated compounds found in nature that are associated with the scents of flowers and spices.

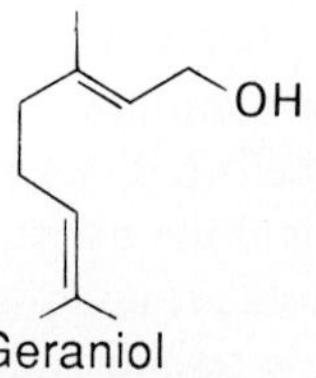

Geraniol
(from roses and other flowers)

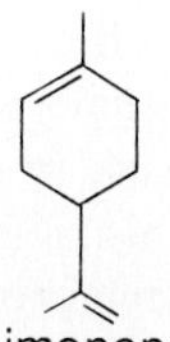

Limonene
(lemons)

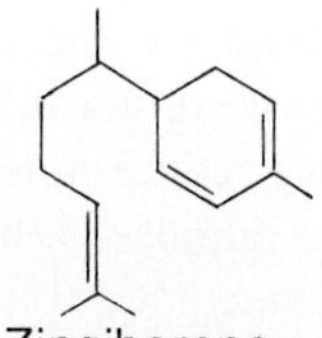

Zingiberene
(from oil of ginger)

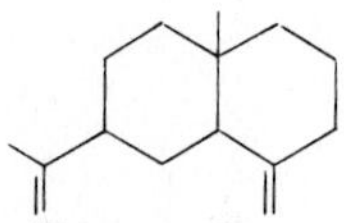

Beta-selinene
(from oil of celery)

Three particularly long-lasting and potent insecticides that are used to protect homes from termites are chlordane, heptachlor, and aldrin. All three compounds are chlorinated alkenes that have structures containing more than one ring of carbon atoms.

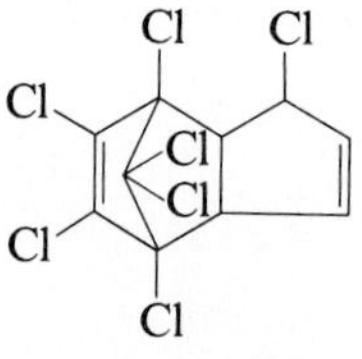

heptachlor 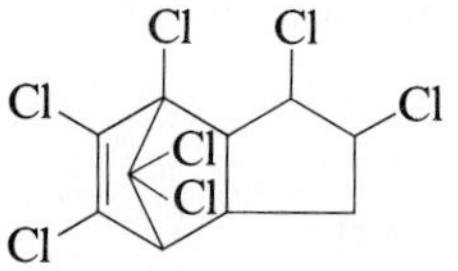   chlordane 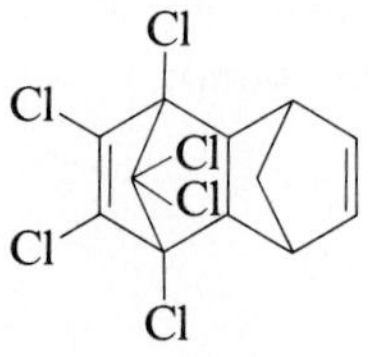   aldrin

These insecticides are the most effective, long-lasting and economical termite controls available. They are also effective in controlling many other insects. However, agricultural uses of these chemicals were canceled by the Environmental Protection Agency (EPA) in the late 1970s because of their persistence in the environment. Buildup of these chemicals in the environment led to concerns about their effects on forms of life other than insects.

15.7 CHEMICAL REACTIONS OF ALKENES

Alkenes, like alkanes (Section 14.12), are extremely flammable; that is, they readily undergo combustion. The combustion products are the same in each case—carbon dioxide and water.

$$2 C_2H_6 + 7 O_2 \longrightarrow 4 CO_2 + 6 H_2O$$
ethane

$$C_2H_4 + 3 O_2 \longrightarrow 2 CO_2 + 2 H_2O$$
ethene

As a group, alkenes are much more chemically reactive than alkanes because carbon–carbon double bonds are more reactive than carbon–carbon single bonds. Almost all of the important reactions of alkenes occur at double-bond sites; most of them occur in such a way that the carbon–carbon double bond is converted to a carbon–carbon single bond.

When a substance interacts with a carbon–carbon double bond and causes it to become a carbon–carbon single bond, an addition reaction has taken place. An **addition reaction** is *a reaction in which atoms or groups of atoms are added to each carbon atom of a carbon–carbon multiple bond.* The net result of an addition reaction is always a decrease in the multiplicity of the carbon–carbon multiple bond. In an alkene addition reaction, the pi bond is broken, but the sigma bond remains intact.

addition reaction

Addition reactions can be classified as *symmetrical* or *unsymmetrical*. The difference between these types of additions involves the identity of the new groups added to the carbon atoms of the carbon–carbon multiple bond. In **symmetrical addition**, *identical atoms or groups of atoms are added to each carbon of the multiple bond.* By contrast, in **unsymmetrical addition**, *different atoms or groups of atoms are added to the carbon atoms of the multiple bond.* We will now consider each type of addition in detail.

symmetrical addition

unsymmetrical addition

Symmetrical Addition

Hydrogenation and halogenation are the two most frequently encountered alkene symmetrical addition reactions. *Hydrogenation* involves the addition of a hydrogen atom to each carbon of the double bond. During *halogenation*, halogen atoms are added to each carbon of the double bond.

Hydrogenation is usually accomplished by heating the alkene and H_2 in a pressurized vessel in the presence of a nickel catalyst. The following two reactions are representative of the many alkene hydrogenation reactions that can occur.

$$CH_3-CH_2-CH=CH-CH_3 + H_2 \xrightarrow[\substack{25°C \\ 1-2 \text{ atm}}]{Pt} CH_3-CH_2-\underset{\underset{H}{|}}{CH}-\underset{\underset{H}{|}}{CH}-CH_3$$
2-pentene   n-pentane

$$CH_2=CH-CH_3 + H_2 \xrightarrow[\substack{150°C \\ 12-15 \text{ atm}}]{Ni} \underset{\underset{H}{|}}{CH_2}-\underset{\underset{H}{|}}{CH}-CH_3$$
propene   propane

The identity of the catalyst used in hydrogenation is specified by writing it above the arrow in the chemical equation. In general terms, hydrogenation of an alkene can be written as

$$\underset{\text{alkene}}{\ce{>C=C<}} + H_2 \xrightarrow{\text{catalyst}} \underset{\text{alkane}}{-\underset{\underset{|}{|}}{\overset{\overset{H}{|}}{C}}-\underset{\underset{|}{|}}{\overset{\overset{H}{|}}{C}}-}$$

The hydrogenation of vegetable oils is a very important commercial process today. Vegetable oils from sources such as soybeans and cottonseeds are composed of long-chain organic molecules that contain many double bonds. When these oils are hydrogenated, their melting points increase and they become low-melting solids. We recognize and use these hydrogenated vegetable products in the form of margarine and cooking shortening.

Hydrogenation reactions are often referred to as reduction processes because each carbon atom in the double bond gains more electron density (more electrons) than was present in the double bond. Each hydrogen atom contributes an additional electron to its carbon atom. Reduction is a process involving the gain of electrons (Section 11.1).

Halogenation of alkenes occurs readily for the halogens Cl_2, Br_2, and I_2. No catalyst or ultraviolet light is needed. The products are always dihalogenated species, as is shown by the following representative halogenation reactions.

$$CH_3-CH=CH-CH_3 + Cl_2 \longrightarrow CH_3-\overset{\overset{\displaystyle Cl}{|}}{CH}-\overset{\overset{\displaystyle Cl}{|}}{CH}-CH_3$$
2-butene   2,3-dichlorobutane

cyclohexene + Br_2 ⟶ *trans*-1,2-dibromocyclohexane

In general terms, halogenation of an alkene can be written as

$$\underset{\text{alkene}}{\diagdown\!\!\!\!\diagup_{C=C}\diagdown\!\!\!\!\diagup} + X_2 \longrightarrow \underset{\text{dihalogenated alkane}}{-\overset{\overset{\displaystyle X}{|}}{C}-\overset{\overset{\displaystyle X}{|}}{C}-} \qquad (X = Cl, Br)$$

Bromination is often used to test for the presence of carbon–carbon double bonds in organic substances. Bromine in water or carbon tetrachloride is reddish-brown in color. The dibromides formed from the symmetrical addition of bromine to an organic compound are colorless. Thus, the decolorization of a Br_2 solution indicates the presence of double bonds.

Halogenation reactions involving chlorine and bromine are often referred to as oxidation processes. Both chlorine and bromine are more electronegative than carbon. As a result, they draw electron density away from carbon atoms in the products of these reactions. This loss of electron density classifies halogenation as an oxidation reaction (Section 11.1).

Unsymmetrical Addition

Hydrohalogenation and hydration are the two most common unsymmetrical addition processes. *Hydrohalogenation* involves the reactants HCl, HBr, or HI. The reactant in *hydration* is H_2O. Hydrohalogenation reactions require no catalyst. However, hydration reactions require a small amount of H_2SO_4 (sulfuric acid) to catalyze the reaction, which otherwise would not occur.

For symmetrical alkenes such as ethene or 2-butene, only one product results from either hydrohalogenation or hydration.

$$CH_2=CH_2 + HCl \longrightarrow CH_2-CH_2$$
ethene   chloroethane

with H and Cl substituents

$$CH_3-CH=CH-CH_3 + HBr \longrightarrow CH_3-CH-CH-CH_3$$
2-butene   2-bromobutane

with H and Br substituents

$$CH_2=CH_2 + H-OH \xrightarrow{H_2SO_4} CH_2-CH_2$$
ethene   ethyl alcohol

with H and OH substituents

In the last equation, which is a hydration, the water (H_2O) is written as H—OH to emphasize how it adds to a double bond—the H goes to one carbon and the OH to the other one. Note also that the product from this hydration reaction contains an —OH group. This is our first encounter with hydrocarbon derivatives of this type; they are called *alcohols* and will be discussed in detail in Chapter 17.

More than one product is possible in unsymmetrical addition reactions involving alkenes that are themselves unsymmetrical. For example, the addition of HCl to propene could produce either 1-chloropropane or 2-chloropropane, depending on whether the Cl atom from the HCl attaches itself to carbon 1 or carbon 2.

$$CH_2=CH-CH_3 + HCl \longrightarrow \underset{\text{1-chloropropane}}{CH_2(Cl)-CH(H)-CH_3}$$
propene

or

$$CH_2=CH-CH_3 + HCl \longrightarrow \underset{\text{2-chloropropane}}{CH_2(H)-CH(Cl)-CH_3}$$
propene

A little more than 100 years ago, the Russian chemist Vladimir W. Markovnikov (1838–1904) studied many such reactions where two products were possible. He found that, in all of these reactions, one of the possible products was always formed in larger amounts than the other. From this research, he formulated a very useful rule that helps us identify which of the two possible products will be most prevalent. **Markovnikov's rule** states that *when an unsymmetrical molecule of the form H–Q adds to an unsymmetrical alkene, the hydrogen atom from the H–Q becomes attached to the unsaturated carbon atom that already has the most hydrogen atoms.* Thus, the hydrogen-rich atom gets an additional hydrogen. The major product in our previous example involving propene is 2-chloropropane.

Markovnikov's rule

Example 15.4

Using Markovnikov's rule, predict the predominant product in each of the following addition reactions.

a. $CH_3-CH_2-CH_2-CH=CH_2 + HBr \longrightarrow$

b. $CH_3-\underset{\underset{CH_3}{|}}{C}=CH_2 + H_2O \longrightarrow$

c. [cyclopentene with CH_3 substituent] $+ HCl \longrightarrow$

Solution

a. The hydrogen atom will add to carbon 1 rather than carbon 2 because carbon 1 already contains more hydrogen atoms than carbon 2. The predominant product of the addition will be 2-bromopentane.

$$CH_3-CH_2-CH_2-\underset{\underset{}{}}{\overset{Br}{\underset{|}{C}}}H-\overset{H}{\underset{|}{C}}H_2$$

b. Again, the hydrogen atom will add to carbon 1, which already contains more hydrogen atoms than carbon 2. The addition product that predominates is the alcohol with the —OH group on carbon 2.

$$\begin{array}{c} \text{OH} \ \ \text{H} \\ | \ \ \ \ \ | \\ CH_3-C-CH_2 \\ | \\ CH_3 \end{array}$$

c. Carbon 1 of the double bond (the carbon to which the methyl group is attached) does not have any H atoms directly attached to it. The other carbon of the double bond has one H atom (not shown in the structure) attached to it. The H atom from the HCl will add to this latter carbon, giving 1-chloro-1-methyl cyclopentane as the product.

(cyclopentane with CH₃, Cl on C1 and H on C2)

Unsymmetrical addition of H₂O to a double bond (hydration) results in a product containing one —OH group. It is also possible to symmetrically add two —OH groups to an alkene by treating the alkene with a dilute aqueous solution of potassium permanganate (KMnO₄).

$$CH_3-CH=CH-CH_3 + KMnO_4 + H_2O \longrightarrow CH_3-\underset{|}{\overset{OH}{CH}}-\underset{|}{\overset{OH}{CH}}-CH_3 + MnO_2 + KOH$$
(purple solution) (brown solid)

This reaction is sometimes used as a quick test for alkenes because the purple color of the permanganate solution disappears and the brown solid MnO₂ forms.

Hydrohalogenation and hydration represent processes in which an electronegative element is attached to one unsaturated carbon atom (resulting in loss of electron density) and a hydrogen atom is attached to the other unsaturated carbon atom (resulting in gain of electron density). Thus, one of the carbon atoms is oxidized and the other one is reduced. The overall process of hydrohalogenation or hydration is considered to be an oxidation process because the electron density withdrawal by the electronegative atom is always greater than the electron density gain from the hydrogen atom.

Example 15.5

Supply the structural formula of the missing substance in each of the following alkene addition-reaction equations.

a. $CH_3-CH_2-CH=CH_2 + H_2O \xrightarrow{H_2SO_4} \ ?$

b. $? + Br_2 \longrightarrow$ (cyclohexane with Br, Br, CH₃)

c. (cyclohexene with CH₃) $+ ? \longrightarrow$ (cyclohexane with Br, CH₃)

d. $CH_3-CH=CH_2 + KMnO_4 + H_2O \longrightarrow \ ? + MnO_2 + KOH$

Solution
a. This is a hydration reaction. Using Markovnikov's rule, we determine that the —OH group will be attached to carbon 2, which has fewer hydrogen atoms.

$$CH_3-CH_2-\underset{\underset{OH}{|}}{CH}-CH_3$$

b. The reactant alkene will have to have a double bond between the two carbon atoms that bromine atoms are attached to in the product.

(cyclohexene with CH₃ substituent)

c. The small reactant molecule that adds to the double bond is HBr. The added Br atom from the HBr is explicitly shown in the product structural formula, but the added H atom is not shown. (In geometrical figure cyclic structures, H atoms are usually not shown.)

d. Reaction of an alkene with potassium permanganate solution results in the symmetrical addition of —OH groups to both carbons of the double bond.

$$CH_3-\underset{\underset{OH}{|}}{CH}-\underset{\underset{OH}{|}}{CH_2}$$

15.8 OXIDATION AND REDUCTION IN ORGANIC REACTIONS

In general, oxidation numbers are seldom used in discussions about oxidation and reduction for organic chemical reactions. The focus is always on the increase or decrease in electron density caused by the addition or loss of hydrogen and oxygen (or other electronegative) atoms. The following guidelines are useful in deciphering whether a particular organic reaction is an oxidation or reduction process.

1. A carbon atom in an organic compound is *oxidized* if it gains oxygen (or other electronegative) atoms or loses hydrogen atoms in a chemical reaction.
2. A carbon atom in an organic compound is *reduced* if it loses oxygen (or other electronegative) atoms or gains hydrogen atoms in a chemical reaction.

These guidelines will be particularly helpful when we come to the biochemistry portions of this textbook.

15.9 POLYMERIZATION OF ALKENES

Certain low-molecular-weight alkenes readily undergo reactions that link many like alkene units together in long chains. The resulting product is called a *polymer*. A **polymer** *is a very large (high-molecular-weight) molecule made from the repetitious union of thousands of small (low-molecular-weight) molecules.* The small molecules used in forming a polymer are called *monomers*.

A multiple-addition reaction using alkene monomers produces a polymer. In this process, the double bond present in the alkene monomer is converted to a single bond (the pi bond breaks, but the sigma bond remains), and the new bonding capacity of each monomer is used to link the monomer units together.

The simplest alkene addition polymers have ethylene (ethene) as the monomer. Using appropriate catalysts, ethylene readily adds to itself.

$$\begin{array}{c}H\\H\end{array}\!\!>\!\!C\!=\!C\!\!<\!\!\begin{array}{c}H\\H\end{array} + \begin{array}{c}H\\H\end{array}\!\!>\!\!C\!=\!C\!\!<\!\!\begin{array}{c}H\\H\end{array} + \begin{array}{c}H\\H\end{array}\!\!>\!\!C\!=\!C\!\!<\!\!\begin{array}{c}H\\H\end{array} \longrightarrow \cdots-\underset{\underset{H}{|}}{\overset{\overset{H}{|}}{C}}-\underset{\underset{H}{|}}{\overset{\overset{H}{|}}{C}}-\underset{\underset{H}{|}}{\overset{\overset{H}{|}}{C}}-\underset{\underset{H}{|}}{\overset{\overset{H}{|}}{C}}-\underset{\underset{H}{|}}{\overset{\overset{H}{|}}{C}}-\underset{\underset{H}{|}}{\overset{\overset{H}{|}}{C}}-\cdots$$
polyethylene

The resulting polymer is called *polyethylene*—with *poly* meaning "many", and the name of the monomer used in forming the polymer. (Usually it is the industrial, rather than IUPAC, name of the monomer that is the base for the polymer name.)

An exact formula for a polymer cannot be written because the length of the carbon chain varies from polymer molecule to polymer molecule; at all times, there are many thousands of monomer units. In recognition of this, the notation used for denoting polymer formulas is independent of carbon chain length. We write the formula of the simplest repeating unit (the monomer with the double bond changed to a single bond) in parentheses and then add the subscript *n* after the parentheses, with *n* being understood to represent a very large number. Using this notation, we have for the formula of polyethylene

TABLE 15–1 Some Common Polymers Obtained from Ethene-Based Monomers

Polymer Formula and Name	Monomer Formula and Name	Uses of Polymer
Polyethylene $-\left(\!\!\begin{array}{cc}H&H\\|&|\\C-C\\|&|\\H&H\end{array}\!\!\right)_{\!n}\!\!-$	Ethylene $\begin{array}{cc}H&H\\|&|\\C=C\\|&|\\H&H\end{array}$	bottles, packaging, toys, electrical insulation
Polypropylene $-\left(\!\!\begin{array}{cc}H&H\\|&|\\C-C\\|&|\\H&CH_3\end{array}\!\!\right)_{\!n}\!\!-$	Propylene $\begin{array}{cc}H&H\\|&|\\C=C\\|&|\\H&CH_3\end{array}$	carpet fibers, molded parts (including heart valves)
Polyvinylchloride (PVC) $-\left(\!\!\begin{array}{cc}H&H\\|&|\\C-C\\|&|\\H&Cl\end{array}\!\!\right)_{\!n}\!\!-$	Vinyl chloride $\begin{array}{cc}H&H\\|&|\\C=C\\|&|\\H&Cl\end{array}$	floor coverings, records, garden hose, plastic pipe, synthetic leather
Teflon $-\left(\!\!\begin{array}{cc}F&F\\|&|\\C-C\\|&|\\F&F\end{array}\!\!\right)_{\!n}\!\!-$	Tetrafluoroethylene $\begin{array}{cc}F&F\\|&|\\C=C\\|&|\\F&F\end{array}$	cooking utensil coverings, electrical insulation, bearings
Polystyrene $-\left(\!\!\begin{array}{cc}H&H\\|&|\\C-C\\|&|\\H&C_6H_5\end{array}\!\!\right)_{\!n}\!\!-$	Styrene $\begin{array}{cc}H&H\\|&|\\C=C\\|&|\\H&C_6H_5\end{array}$	toys, styrofoam packaging, thermal insulation

$$\left(\begin{array}{cc} \overset{H}{\underset{H}{|}} & \overset{H}{\underset{H}{|}} \\ -C-C- \\ \overset{|}{H} & \overset{|}{H} \end{array}\right)_n$$

This notation clearly identifies the basic repeating unit found in the polymer.

Slight modification of the alkene monomer, which is accomplished by attaching various groups to the unsaturated carbon atoms, gives rise to a wide variety of polymers; each polymer has different characteristics and applications. Table 15–1 gives formulas, trade names, and uses for some of these polymers and Figure 15–6 illustrates some end-products made from these polymers.

When conjugated dienes like 1,3-butadiene are used as the monomers in addition-polymerization reactions, the resulting polymers are not saturated, but retain double bonds.

$$CH_2=CH-CH=CH_2 \xrightarrow{\text{polymerization}} -(CH_2-CH=CH-CH_2)_n-$$

1,3-butadiene polybutadiene

In general, these unsaturated polymers are much more flexible than the ethene-based saturated polymers whose formulas are given in Table 15–1.

Natural rubber, the best known natural organic polymer, is an addition polymer whose repeating unit (called isoprene) is 2-methyl-1,3-butadiene.

$$CH_2=\underset{\underset{CH_3}{|}}{C}-CH=CH_2 \xrightarrow{\text{polymerization}} -\left(CH_2-\underset{\underset{CH_3}{|}}{C}=CH-CH_2\right)_n-$$

isoprene polyisoprene
(2-methyl-1,3-butadiene) (natural rubber)

When sulfur is added to soft, tacky, natural rubber and the mixture is heated, the sulfur forms bonds between polymeric chains that cross-link the chains together.

FIGURE 15–6 Polymers obtained from ethene-based monomers are extensively used in many aspects of life.

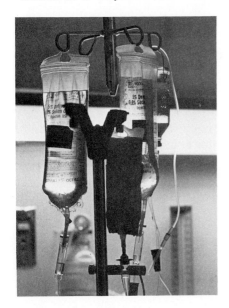

This process, called *vulcanization*, converts the naturally occurring rubber into a resilient and strong, yet flexible material that is useful in a variety of applications.

Neoprene, which is a synthetic rubber, has properties that are comparable with, or, in some cases, even superior to those of natural rubber. The monomer for this material is a chloro derivative of 1,3-butadiene—2-chloro-1,3-butadiene.

$$CH_2=C-CH=CH_2 \longrightarrow \left(CH_2-C=CH-CH_2 \right)_n$$
$$\overset{|}{Cl}\overset{|}{Cl}$$

chloroprene　　　　　　　　polychloroprene
(2-chloro-1,3-butadiene)　　　　(Neoprene)

Although we have focused on polymers made from a single monomer, we should note that there are many polymers that are synthesized from mixtures of monomers. When two or more different monomers are reacted to form a single polymer, the resulting polymer is called a *copolymer*.

The commercial product Saran Wrap is a copolymer of chloroethene (vinyl chloride) and 1,1-dichloroethene.

[Structures showing vinyl chloride + 1,1-dichloroethene → Saran Wrap copolymer, with 1st monomer and 2nd monomer labeled]

15.10 ALKYNES

alkynes

Alkynes *are hydrocarbons in which there is one carbon–carbon triple bond.* Alkynes have two less hydrogen atoms than the corresponding alkene and four less hydrogen atoms than the corresponding alkane.

[Structures of ethane (C_2H_6), ethene (C_2H_4), and ethyne (C_2H_2)]

The general formula for alkynes containing one triple bond is C_nH_{2n-2}. Thus, the simplest member of this series has the formula C_2H_2 and the next member, $n = 3$, the formula C_3H_4.

The presence of a carbon–carbon triple bond in a molecule always results in a linear arrangement for the two atoms attached to the carbons of the triple bond; that is, all four atoms will always lie in a straight line. Figure 15–7 gives perspective drawings of the straight-line geometry for C_2H_2. Because of the linearity (180° angles) about the triple bond, *cis–trans* isomers are not possible for alkynes such as $CH_3-C\equiv C-CH_3$, as was found for the 2-butenes ($CH_3-CH=CH-CH_3$). Cycloalkynes, molecules containing a triple bond as part of a ring structure, are known, but they are not common.

The simplest alkyne, ethyne (C_2H_2), is commonly called *acetylene*. (Unfortunately, the common name ends in *-ene*, which incorrectly suggests that a double bond is present.) Industrially, acetylene is the most important of all alkynes. This

FIGURE 15-7 Perspective drawing of C₂H₂, the simplest alkyne.

Ball-and-stick model

Space-filling model

colorless gas, which has a boiling point of $-84°C$, has traditionally been prepared by the reaction between calcium carbide and water.

$$CaC_2(s) + 2\ H_2O(l) \longrightarrow C_2H_2(g) + Ca(OH)_2(aq)$$

An alternate synthesis that is based on petroleum is slowly displacing the carbide process. This new process involves the controlled, high-temperature, partial oxidation of methane.

$$6\ CH_4(g) + O_2(g) \xrightarrow{1500°C} 2\ C_2H_2(g) + 2\ CO(g) + 10\ H_2(g)$$

When acetylene is burned with oxygen in an oxyacetylene welding torch, a very high temperature flame is produced. These torches are used extensively by welders for cutting and welding metals. Years ago, miners used helmet lanterns in which acetylene, produced by mixing small amounts of calcium carbide and water, was burned. These carbide lamps were used extensively until the development of today's safer and more efficient battery-powered lights.

15.11 NOMENCLATURE OF ALKYNES

The rules for naming alkynes are identical to those used to name alkenes (Section 15.3), except the ending *-yne* is used instead of *-ene*.

Example 15.6

Assign IUPAC names to each of the following alkynes.

a. CH₃—CH₂—CH₂—C≡C—CH₃

b. CH₃—CH—C≡C—CH₃
         |
         CH₃

c. CH₃   CH₃
   | |
   CH—C≡C—CH
   | |
   CH₃   CH₃

Solution

a. We use a right-to-left numbering system so that the triple bond will have the lowest number possible. This gives the IUPAC name *2-hexyne* for this molecule.

b. Again, we use a right-to-left numbering system that gives the triple bond low-number preference over the methyl group. The IUPAC name of this compound is *4-methyl-2-pentyne*.

c. The longest chain containing the triple bond is six carbons long. (We must go around two corners in order to have a chain length of six.) Methyl groups are bonded to carbons 2 and 5, and the triple bond begins at carbon number 3. (Both right-left and left-right numbering systems give the same results for this molecule because of its symmetrical nature.) The IUPAC name is *2,5-dimethyl-3-hexyne*.

15.12 PHYSICAL PROPERTIES AND CHEMICAL REACTIONS OF ALKYNES

The physical properties of alkynes are similar to those for alkenes and alkanes. In general, alkynes are insoluble in water but soluble in organic solvents, have low densities, and have boiling points that increase with molecular weight. Low-molecular-weight alkynes are gases at room temperature.

The chemical reactions of alkynes are similar to those for alkenes. The name of the game is again "addition". There is one difference between triple-bond additions and double-bond additions. Triple bonds can add two molecules of a specific reactant, but double bonds are limited to one molecule. This two-molecule addition proceeds stepwise. For example, propyne reacts with hydrogen to first form propene and then form propane.

$$CH\equiv C-CH_3 \xrightarrow{H_2}{Pt} CH_2=CH-CH_3 \xrightarrow{H_2}{Pt} CH_3-CH_2-CH_3$$
an alkyne → an alkene → an alkane

Halogenation and hydrohalogenation addition reactions are also possible for alkynes.

$$CH\equiv C-CH_3 \xrightarrow{Cl_2} \underset{Cl\ \ Cl}{CH=C-CH_3} \xrightarrow{Cl_2} \underset{Cl\ \ Cl}{\overset{Cl\ \ Cl}{CH-C-CH_3}}$$

$$CH\equiv C-CH_3 \xrightarrow{HBr} \underset{Br}{CH_2=C-CH_3} \xrightarrow{HBr} \underset{Br}{\overset{Br}{CH_3-C-CH_3}}$$

Note that in the last reaction sequence, which is a hydrohalogenation reaction, the products of the additions are those predicted by using Markovnikov's rule. It is possible to control alkyne two-step additions with special catalysts so that the addition stops after the first step; this gives an alkene as the product.

EXERCISES AND PROBLEMS

Unsaturated Hydrocarbons

15.1 What is an unsaturated hydrocarbon?

15.2 Classify each of the following compounds as saturated or unsaturated. In addition, classify each unsaturated compound as an alkene, alkyne, cycloalkene, or cycloalkyne.

a. CH$_3$—CH$_2$—CH$_3$
b. CH$_3$—CH=CH—CH$_3$
c. CH≡C—CH$_2$—CH$_3$
d.
e.
f. ☐
g. △—CH=CH$_2$
h.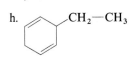

15.3 Write both the molecular formula and the generalized molecular formula (e.g., C$_n$H$_{2n-2}$) for each compound in Problem 15.2.

15.4 Write the molecular formula for compounds with each of the following structural features.

a. noncyclic, four carbon atoms, and no multiple bonds
b. noncyclic, five carbon atoms, and one double bond
c. noncyclic, six carbon atoms, and one triple bond
d. cyclic, five carbon atoms, and no multiple bonds
e. noncyclic, four carbon atoms, and two triple bonds
f. cyclic, seven carbon atoms, and two double bonds
g. cyclic, six carbon atoms, and three double bonds
h. cyclic with two rings, ten carbon atoms, and two double bonds

Nomenclature of Alkenes and Cycloalkenes

15.5 Assign an IUPAC name to each of the following compounds.

a. CH$_3$—CH=CH—CH$_3$
b. CH$_3$—C=CH—CH—CH$_3$
         | |
         CH$_3$ CH$_3$
c.
d. ⌂
e. CH$_2$=C—CH$_2$—CH$_2$—CH$_3$
         |
         CH$_3$—CH$_2$
f. ⬡—CH$_3$
g. ▱—CH—CH$_3$
         |
         CH$_3$
h. ⌇⌇⌇

15.6 Draw structural formulas (showing only carbon atoms) for each of the following unsaturated hydrocarbons.

a. 3-methyl-1-pentene
b. 3-methylcyclopentene
c. 1,3-butadiene
d. 3-ethyl-1,4-pentadiene
e. 4-propyl-2-heptene
f. 3,6-diethyl-1,4-cyclohexadiene
g. 1,3-cyclohexadiene
h. 4,4,5-trimethyl-2-heptene

15.7 Although the following names are incorrect according to IUPAC rules, they contain sufficient information to enable you to draw structural formulas. Draw the structural formula and then determine the correct IUPAC name for each compound.

a. 2-ethyl-2-pentene
b. 4,5-dimethyl-4-hexene
c. 3,5-cyclopentadiene
d. 1,2-dimethyl-4-cyclohexene

15.8 Draw structural formulas (showing only carbon atoms) for the five isomeric alkenes with the formula C$_5$H$_{10}$. (Do not consider cis–trans isomerism. There are five isomers without this consideration.)

15.9 Hydrocarbons with the formula C$_n$H$_{2n}$ can be either noncyclic alkenes or cyclic alkanes. Draw all of the possible structural isomers with the formula C$_4$H$_8$. (There is a combined total of five isomers without considering cis–trans isomerism.)

15.10 Which of the following compounds contain conjugated double bonds?

a. C=C—C=C—C=C—C=C
b. ⌇⌇⌇
c. ☐

365

d.

e.

f. C=C—C—C=C

Sigma and Pi Bonds

15.11 Explain the difference between a sigma bond and a pi bond.

15.12 Determine the number of carbon–carbon sigma bonds and the number of carbon–carbon pi bonds present in each of the following molecules.
a. $CH_3-CH_2-CH_2-CH_2-CH_3$
b. $CH_3-CH=CH-CH_3$
c. $CH_2=CH-CH=CH_2$
d. $CH\equiv CH$
e. $CH\equiv C-CH_2-CH_3$
f. cyclobutane with CH₃ substituent
g. cyclobutene
h. cyclohexyl—CH=CH—CH₃

Geometric Isomers

15.13 Define the term *geometric isomer* and indicate the structural prerequisites for the existence of geometric isomers.

15.14 Explain why it is not possible to have geometric isomerism in alkanes or alkynes.

15.15 For which of the following molecules does geometric (*cis–trans*) isomerism exist?
a. $CH_2=CH-CH_3$
b. $CH_3-CH_2-CH_2$ with Cl substituent
c. $CH_3-C=CH-CH_3$ with CH_3 substituent
d. $CH=CH$ with Br, Br substituents
e. 3-hexene
f. 1,2-dimethylcyclopentane
g. 4-methyl-2-pentene
h. methylcyclopropane

15.16 Assign IUPAC names to each of the following molecules. Be sure to include an indication of geometric isomerism where necessary.

a. CH_3, H on one carbon; CH_2-CH_3, H on other (C=C)
b. I, H on one carbon; H, Br on other (C=C)
c. F, F on one carbon; F, F on other (C=C)
d. cyclobutane with two CH_3 groups
e. cyclopentane with two CH_3 groups
f. cyclohexene with Br, Br substituents

15.17 Draw structural formulas for each of the following compounds.
a. *trans*-2-methyl-3-hexene
b. *cis*-2-pentene
c. *cis*-1,2-dichlorocyclobutane
d. *trans*-5-methyl-2-heptene
e. *cis*-1,4-dimethylcyclohexane
f. *trans*-1,3-pentadiene

15.18 How many geometric isomers are there for the compound 3,5-octadiene?

Reactions of Alkenes

15.19 Classify each of the following substances used in alkene-addition reactions as symmetrical or unsymmetrical.
a. H_2
b. H_2O
c. Cl_2
d. HBr
e. Br_2
f. HCl

15.20 Write chemical equations, showing reactants, products, and catalysts needed (if any), for the reaction of ethene (ethylene) with each of the following substances.
a. Cl_2
b. HCl
c. H_2
d. HBr
e. H_2O
f. Br_2

15.21 Write chemical equations, showing reactants, products, and catalysts needed (if any), for the reaction of propene (propylene) with each of the reactants listed in Problem 15.20. Be sure to utilize Markovnikov's rule where necessary.

15.22 Supply the structural formula of the missing substance in each of the following alkene-addition-reaction equations.

a. $CH_3-CH=CH-CH_3 + Cl_2 \longrightarrow$?

b. $CH_3-\underset{\underset{CH_3}{|}}{C}=CH_2 + HBr \longrightarrow$?

c. $CH_3-CH_2-CH=CH_2 + HCl \longrightarrow$?

d. ⬠ $+ H_2 \xrightarrow{Ni}$?

e. ⬠ $+ KMnO_4 \xrightarrow{\text{aqueous solution}}$?

f. ☐ $+ Br_2 \longrightarrow$?

g. $CH_3-CH=CH_2 + H_2O \xrightarrow{H_2SO_4}$?

h. 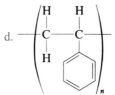—$CH_3 + HCl \longrightarrow$?

15.23 Reactions of alkenes with Br_2 or dilute $KMnO_4$ are often used as diagnostic tools to identify alkenes. Describe the characteristic color changes associated with each of these reactions.

15.24 Suggest a method for synthesizing each of the following substances, starting with unsaturated hydrocarbons.

a. $CH_3-\underset{\underset{OH}{|}}{CH}-CH_3$

b. $CH_3-CH_2-CH_2-\underset{\underset{Br}{|}}{CH}-\underset{\underset{CH_3}{|}}{CH}-CH_3$

c. $CH_3-\underset{\underset{Cl}{|}}{CH}-\underset{\underset{Cl}{|}}{CH}-CH_3$

d. $\underset{\underset{OH}{|}}{CH_2}-\underset{\underset{OH}{|}}{CH_2}$

e. $CH_3-CH_2-CH_3$

f. $CH_3-CH_2-CH_2-CH_2-CH_2-CH_3$

15.25 How many moles of H_2 gas will react with one mole of each of the following unsaturated compounds?

a. $CH_3-CH=CH-CH=CH-CH_3$

b. ☐—CH_3

c. ⬡—$CH=CH_2$

d. $CH_3-CH=C=\underset{\underset{CH_3}{|}}{C}-CH=CH_2$

Oxidation and Reduction

15.26 Classify each of the following types of alkene-addition-reactions as oxidation or reduction.
a. halogenation
b. hydrogenation
c. hydrohalogenation
d. hydration

15.27 Define the terms *oxidation* and *reduction* (in an organic reaction context) using the concepts of loss or gain of hydrogen and oxygen atoms.

Polymerization of Alkenes

15.28 Draw the structural formulas of the monomers from which each of the following polymers were made.

a. $+\!\!\left(\!\!\underset{\underset{F}{|}}{\overset{\overset{F}{|}}{C}}-\underset{\underset{F}{|}}{\overset{\overset{F}{|}}{C}}\!\!\right)\!\!_n$

b. $+\!\!\left(\!\!\underset{\underset{H}{|}}{\overset{\overset{H}{|}}{C}}-\underset{\underset{Cl}{|}}{\overset{\overset{H}{|}}{C}}\!\!\right)\!\!_n$

c. $+\!\!\left(\!\!\underset{\underset{H}{|}}{\overset{\overset{H}{|}}{C}}-\underset{\underset{Cl}{|}}{\overset{\overset{H}{|}}{C}}=\underset{\underset{H}{|}}{\overset{\overset{H}{|}}{C}}-\underset{\underset{H}{|}}{\overset{\overset{H}{|}}{C}}\!\!\right)\!\!_n$

d. $+\!\!\left(\!\!\underset{\underset{H}{|}}{\overset{\overset{H}{|}}{C}}-\underset{\underset{\text{C}_6\text{H}_5}{|}}{\overset{\overset{H}{|}}{C}}\!\!\right)\!\!_n$

15.29 Draw the "start" (the first five repeating units) of the structural formula for each of the following polymers.
a. polyethylene
b. natural rubber
c. teflon
d. polypropylene
e. PVC
f. Neoprene

Formulas and Nomenclature for Alkynes

15.30 What is the general formula for a(n)
a. alkyne with one triple bond
b. cycloalkyne with one triple bond
c. alkadiyne

367

15.31 There are two isomeric alkynes with the formula C_4H_6. Draw structural formulas for these two compounds.

15.32 Assign IUPAC names to each of the following compounds.

a. $CH_3-CH_2-CH_2-CH_2-C{\equiv}CH$

b. 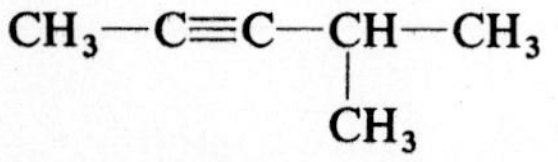 $CH_3-C{\equiv}C-\underset{\underset{\displaystyle CH_3}{|}}{CH}-CH_3$

c. $CH_3-\underset{\underset{\displaystyle CH_3}{|}}{\overset{\overset{\displaystyle CH_3}{|}}{C}}-C{\equiv}C-CH_2-CH_2-CH_3$

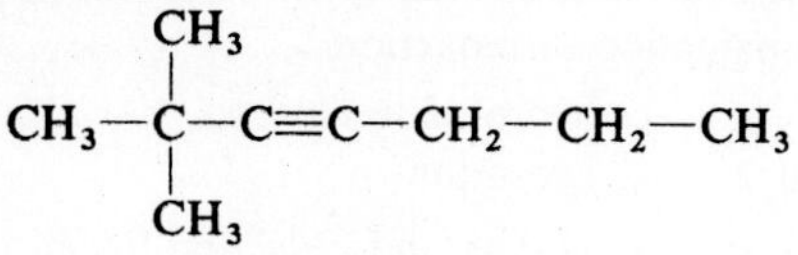

d. $CH_3-C{\equiv}C-CH_3$

e. 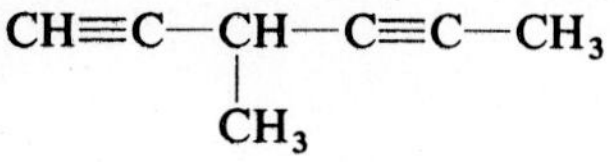 $CH{\equiv}C-\underset{\underset{\displaystyle CH_3}{|}}{CH}-C{\equiv}C-CH_3$

f. 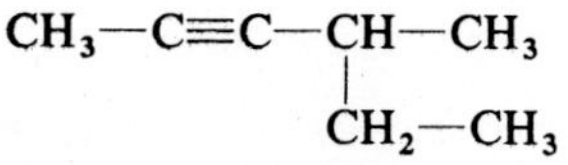 $CH_3-C{\equiv}C-\underset{\underset{\displaystyle CH_2-CH_3}{|}}{CH}-CH_3$

g. $CH{\equiv}C-C{\equiv}CH$

h. 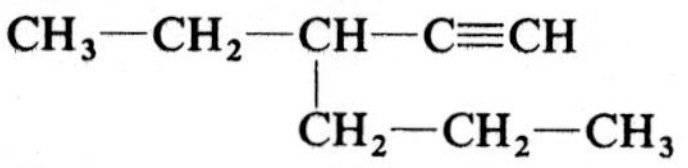 $CH_3-CH_2-\underset{\underset{\displaystyle CH_2-CH_2-CH_3}{|}}{CH}-C{\equiv}CH$

15.33 Draw structural formulas (showing only carbon atoms) for each of the following compounds.

a. 2-methyl-3-hexyne
b. 3-ethyl-1-pentyne
c. 4,4-dimethyl-2-hexyne
d. 4-isopropyl-2-hexyne
e. 1-cyclopropyl-2-butyne
f. 1,-3-hexadiyne

Chemical Reactions of Alkynes

15.34 What is the bonding angle between the atoms attached to an alkyne carbon? Compare this to the angles in alkanes and alkenes.

15.35 How many moles of H_2 will react with one mole of each of the compounds in Problem 15.22?

15.36 Supply the structural formula of the missing substance(s) in each of the following alkyne reaction equations.

a. $CH{\equiv}CH + 2 H_2 \xrightarrow{Ni}$?

b. $CH_3-C{\equiv}CH + 2 Br_2 \longrightarrow$?

c. $CH_3-C{\equiv}CH + 2 HBr \longrightarrow$?

d. $CH{\equiv}CH + 1 HCl \longrightarrow$?

e. 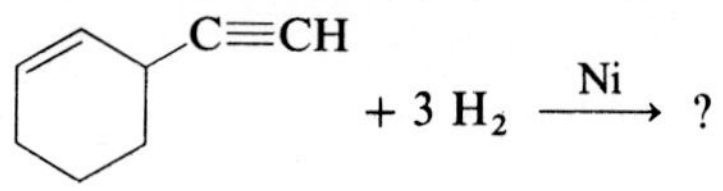 cyclohexene-$C{\equiv}CH + 3 H_2 \xrightarrow{Ni}$?

f. $CH_3-CH_2-C{\equiv}CH + 1 HBr \longrightarrow$?

15.37 Describe two methods of synthesizing acetylene.

16 Aromatic Hydrocarbons

Objectives

After completing Chapter 16, you will be able to:

16.1 Define the term *aromatic hydrocarbon*.
16.2 Describe how the bonding in benzene differs from that in nonaromatic hydrocarbons.
16.3 Name substituted benzene compounds using IUPAC or common names.
16.4 Recognize and write equations for aromatic substitution reactions involving alkylation, halogenation, and nitration.
16.5 List the uses of benzene and simple alkyl benzenes and the toxic effects associated with halogenated aromatic compounds.
16.6 Recognize the simpler polycylic aromatic hydrocarbons and understand the concept of "fused-ring" aromatic systems.
16.7 Describe the primary sources for aromatic hydrocarbons.
16.8 Identify the general structural feature associated with heterocyclic compounds.
16.9 Understand the interrelationships among the numerous families of hydrocarbons.

INTRODUCTION

In this chapter, we will discuss the last class of hydrocarbon molecules—the aromatic hydrocarbons. Many of the earliest known members of this family of hydrocarbons were found to possess pleasant odors; hence, the name *aromatic*. However, there are now many known aromatic hydrocarbons that have no odor associated with them. As a group, aromatic hydrocarbons are no more "odorous" than other hydrocarbon families.

Aromatic hydrocarbons are unsaturated cyclic compounds. However, their chemical properties are so different from those of the alkenes and alkynes discussed in Chapter 15 that they are classified separately from these other unsaturated compounds.

16.1 AROMATIC HYDROCARBONS

Unsaturated cyclic hydrocarbons can contain more than one multiple bond (Section 15.2). For example, introducing a second double bond into the molecule cyclohexene

produces either 1,3-cyclohexadiene or 1,4-cyclohexadiene

1,3-cyclohexadiene   1,4-cyclohexadiene

depending on where the additional double bond is positioned relative to the first one. These two compounds are isomers and differ only in the positions of the double bonds (1,3- versus 1,4-).

The introduction of a third double bond into 1,3-cyclohexadiene produces the compound

This compound contains a cyclic system of conjugated double bonds (Section 15.3). Using the IUPAC nomenclature rules for unsaturated compounds presented in the last chapter, we would expect to call this compound *1,3,5-cyclohexatriene*. However, this compound is not given that name; instead, its name is *benzene*. The reason for this "rule violation" is the fact that benzene does not possess the chemical properties normally associated with cyclic compounds containing multiple bonds. Benzene's properties are different enough from those of other unsaturated hydrocarbons that it is considered to be the first member of a new series of hydrocarbons—the aromatic hydrocarbons.

The **aromatic hydrocarbons** *are cyclic unsaturated compounds that possess a*   aromatic hydrocarbons *system of conjugated double bonds and do not readily undergo the addition reactions characteristic of other unsaturated compounds.*

16.2 BENZENE: STRUCTURE AND BONDING

A logical starting point for a discussion of aromatic compounds is a discussion of benzene, the prototype for these compounds. All of the aromatic compounds that we will discuss in this chapter (and later ones) will have at least one benzene ring as part of their structure.

Benzene, whose molecular formula is C_6H_6, is a colorless liquid at room temperature. It is a good solvent for numerous substances. Although first isolated in 1825, it was not until 1865 that a satisfactory structure was proposed for this extremely hydrogen-deficient compound. At that time, German chemist August Kekulé (1829–96) proposed the cyclic alternating single and double bond structure given in Section 16.1.

Kekulé also proposed that the double and single bonds in the carbon ring of benzene were in rapid oscillation around the ring and that this rapid oscillation somehow made the molecule resistant to addition reactions.

In order to indicate the rapid oscillation of bonds around the ring, Kekulé drew two equivalent structures for benzene that differed only in the location of the double bonds (1,3,5 positions versus 2,4,6 positions) and connected the two structures using a double-headed arrow.

This explanation of the lack of reactivity of benzene was used by chemists until the early 1900s, when a "better" explanation became available.

The current explanation for the structure of benzene takes Kekulé's two structures and combines them into one structure. The notation for this single structure is

delocalized bonding

The circle within the ring denotes a new type of bonding—delocalized bonding. In **delocalized bonding**, *three or more atoms are involved in the sharing of the same valence electrons.* (Up until now, all bonding discussions in the text have involved localized bonds; that is, bonds in which electrons are shared between two atoms.) Let us now consider how the delocalized bond concept helps explain the characteristics of benzene and other aromatic compounds.

Carbon atoms have four valence electrons for use in bonding. Three of the four electrons on each carbon atom in benzene will be used to form the sigma-bonding framework of a benzene molecule, as shown in the following diagram.

In this structure, each carbon has formed three bonds. Thus, each carbon atom still has one more electron available for bonding; this is shown as a dot in the following structure.

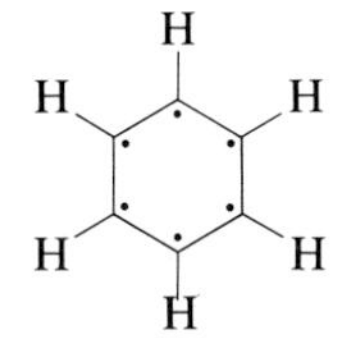

In each case, this additional electron is in a *p* orbital (Section 5.6) whose orientation is such that it can interact simultaneously with two other *p* orbitals—the *p* orbitals on each side of it in the carbon ring (see Figure 16–1a). This results in a continuous interaction around the ring and a delocalized doughnut-shaped pi bond that runs completely around the ring, as shown in Figure 16–1b. In this delocalized pi bond, the six electrons (one from each carbon atom) are considered to be shared equally by all six carbon atoms. All of the carbon–carbon bonds of the ring are equivalent; they consist of a sigma bond and a delocalized pi bond.

The delocalized pi bonding present in the benzene molecule imparts additional stability to it. This phenomenon, which is known as *aromaticity*, is what causes benzene and other aromatic molecules to be resistant to addition reactions. Addition reactions require a breaking up of the delocalized bonding within the ring so that additional sigma bonds can be formed.

The standard notation used to depict the delocalized bonding within a benzene molecule is a circle (almost as large as the benzene ring itself), which is placed within the ring. The presence of the circle is compulsory. Without the circle, we do not have the molecule benzene. A six-membered carbon ring without a circle drawn within it denotes the molecule cyclohexane.

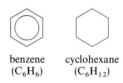

benzene (C_6H_6)   cyclohexane (C_6H_{12})

16.3 NOMENCLATURE OF BENZENE DERIVATIVES

The hydrogen atoms on the benzene ring can be replaced with other groups or atoms (Section 16.4). The resulting products are called derivatives of benzene. Frequently, only one or two hydrogen atoms are replaced, although there are

FIGURE 16–1   A representation of the bonding within the carbon ring of a benzene molecule. Six *p* orbitals (one on each carbon) interact (part a) to form a delocalized pi bond running completely around the ring (part b).

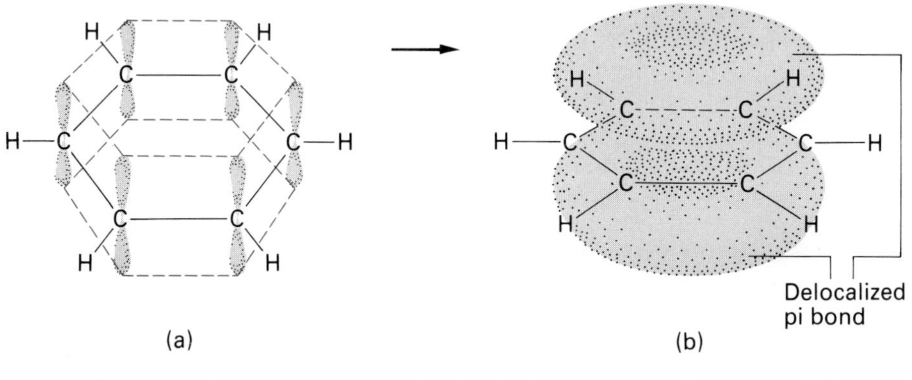

some derivatives in which all of the hydrogens have been replaced by other atoms or groups. Often the new group or groups attached to the ring are alkyl groups (methyl, ethyl, and so on) or halogens. Other commonly found groups are hydroxyl (—OH), amino (—NH_2), nitro (—NO_2), and carboxyl (—COOH).

The following IUPAC rules give the most important points concerning aromatic hydrocarbon nomenclature. Although these rules are insufficient to name all aromatic hydrocarbons, they allow us to name and recognize the aromatic compounds covered in this textbook.

Rule 1: When a single hydrogen atom attached to a benzene ring is replaced by a simple substituent, the compound is named by giving the name of the new substituent, followed by the name *benzene*.

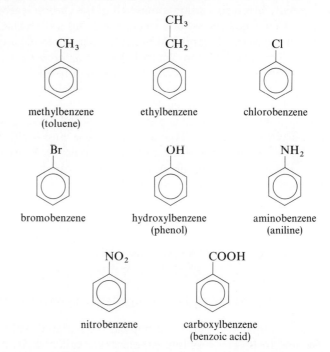

Note that some of these monosubstituted benzenes have common names (given in parentheses). You should learn these common names because they are often used as parent names for further substituted compounds.

As all of the hydrogen atoms in benzene are equivalent, it does not matter at which vertice of the ring the substituted group is located. Each one of the following formulas represents toluene (methylbenzene).

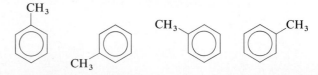

Rule 2: When two substituents are attached to a benzene ring, both their identity and their positions relative to each other must be specified. According to IUPAC rules, positioning on the ring is indicated by using numbers, as was the case for disubstituted cycloalkanes (Section 14.9) and substituted cycloalkenes (Section 15.3). Alternatively, nonnumeric prefixes can be used to denote the relative positions of the two attached groups (common nomenclature). The prefixes are ortho-, meta-, and para- (abbreviated *o*-,

m-, and *p*-). Ortho- means 1,2 disubstitution; the substituents are on adjacent carbon atoms. Meta- means 1,3 disubstitution; the substituents are one carbon removed from each other. Para- means 1,4 disubstitution; the substituents are on opposite sides of the ring.

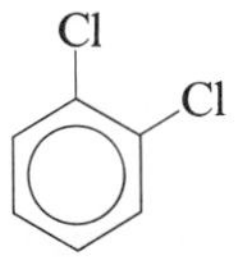

ortho-dichlorobenzene
(1,2-dichlorobenzene)

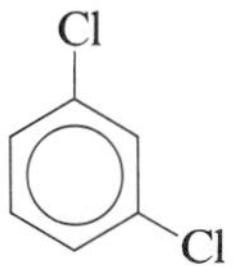

meta-dichlorobenzene
(1,3-dichlorobenzene)

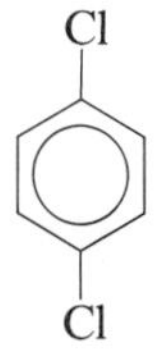

para-dichlorobenzene
(1,4-dichlorobenzene)

When the two substituents attached to the benzene ring are different, their names are given in alphabetical order.

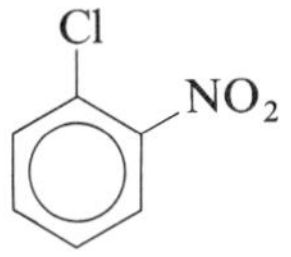

o-chloronitrobenzene

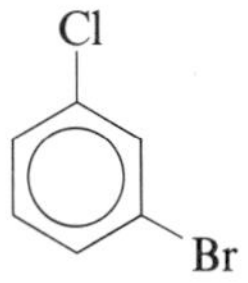

m-bromochlorobenzene

The dimethylbenzenes (three isomers) have the common name *xylene*. This term comes from *xylon*, which is Greek for wood. The dimethylbenzenes were first isolated from wood.

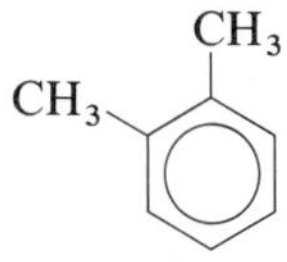

o-xylene
o-dimethylbenzene

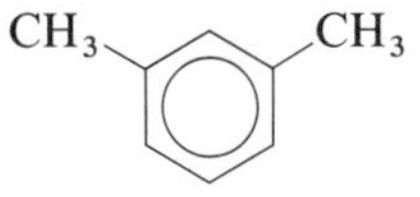

m-xylene
m-dimethylbenzene

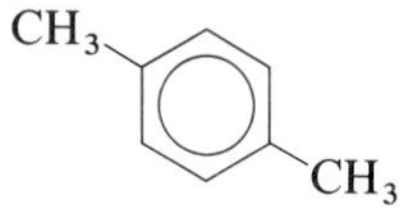

p-xylene
p-dimethylbenzene

When one of the substituents on a disubstituted benzene corresponds to a monosubstituted benzene that has a common name (Rule 1), the disubstituted compound can be named as a derivative of the monosubstituted compound.

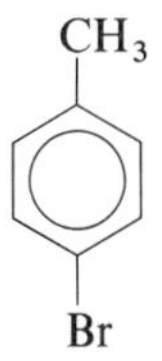

p-bromotoluene

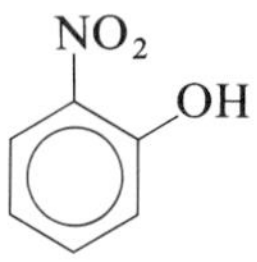

o-nitrophenol

Rule 3: If more than two substituted groups are present on a benzene ring, their positions are indicated using numbers rather than prefixes. The ring is numbered so as to obtain the lowest possible numbers for the carbons on which the groups are attached. If there is a choice for numbering systems (two systems give the same lowest total), the group that comes first alphabetically is given the lower number. If the compound is to be named as a derivative of a monosubstituted benzene (such as toluene or phenol), the group upon which the monosubstituted derivative is based will always be carbon 1.

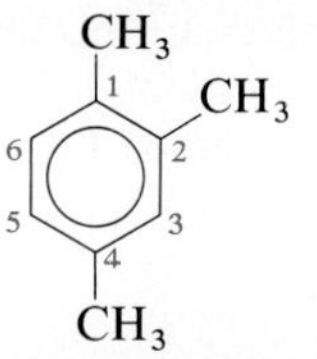

1,2,4-trimethylbenzene
or 2,4-dimethyltoluene

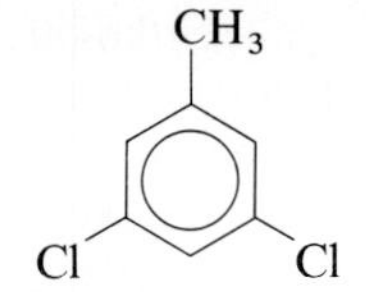

1,3-dichloro-5-methylbenzene
or 3,5-dichlorotoluene

Rule 4: For monosubstituted benzene rings that have a group attached that is not easily named as a substituent, the benzene ring is often treated as a group attached to this substituent. In this reversed approach, the benzene ring attachment is called a *phenyl* group, and the compound is named according to rules covered in previous chapters.

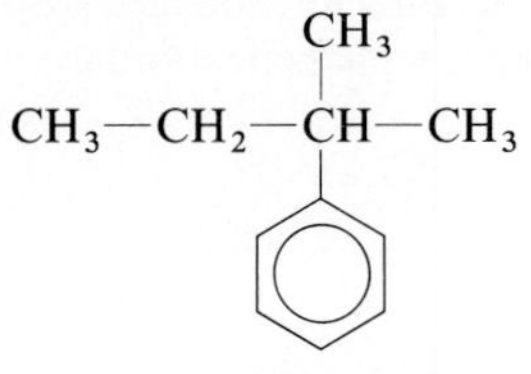

2-methyl-2-phenylbutane

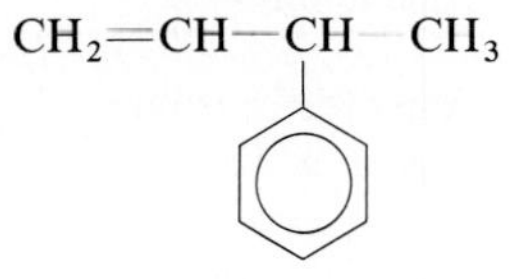

3-phenyl-1-butene

The term *phenyl* comes from "phene", a European term for benzene used during the 1800s. The presence of phenyl groups is often indicated by drawing out the ring structure for the group. It can, however, also be indicated by the notation C_6H_5. Thus, the following two formulas are equivalent ways to denote the molecule 2-phenylpropane.

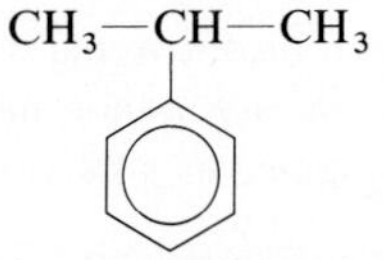

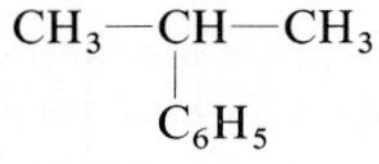

The substituted group derived from benzene is the phenyl group, as we have just seen. Another substituted group closely related to the phenyl group is the *benzyl* group, which is the substituted group derived from toluene.

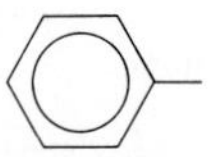

phenyl group

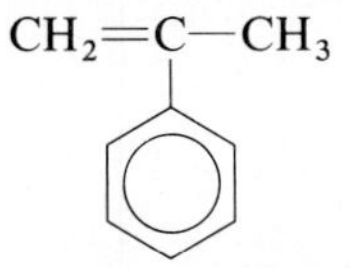

2-phenyl-1-propene

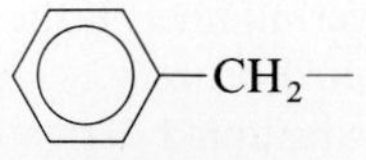

benzyl group

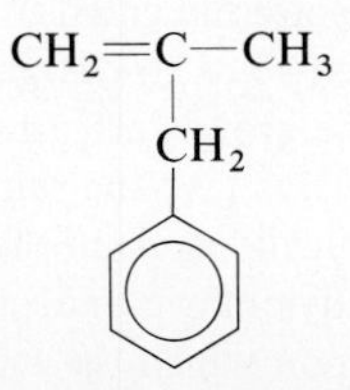

2-benzyl-1-propene

Name the following aromatic compounds.

Example 16.1

a. 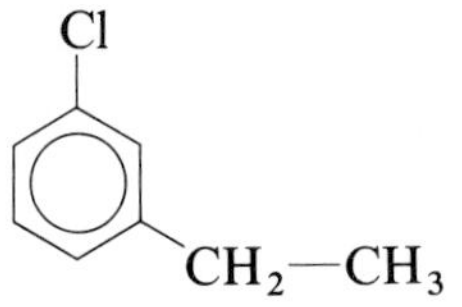 [structure: benzene ring with Cl and CH₂—CH₃ in meta positions]

b. 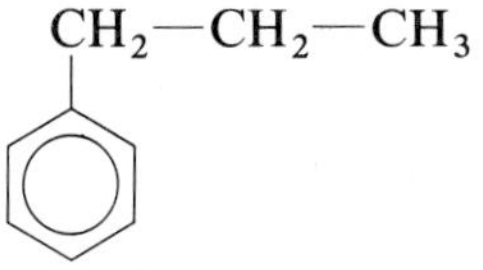 [structure: benzene ring with CH₂—CH₂—CH₃]

c. 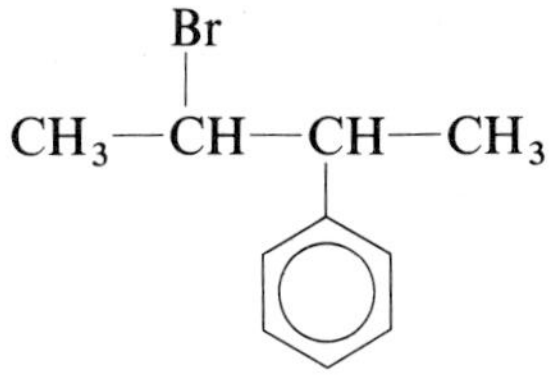 [structure: CH₃—CH(Br)—CH(phenyl)—CH₃]

d. 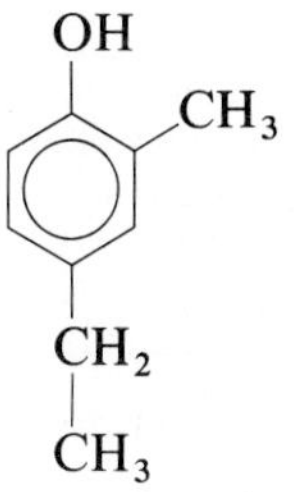 [structure: benzene ring with OH, CH₃ (ortho to OH), and CH₂—CH₃ (para to OH)]

Solution

a. This compound is named as a disubstituted benzene. The two substituents (chloro and ethyl) are listed alphabetically and either a prefix or numbers are used to locate the substituents. The compound is named either m-*chloroethylbenzene* or *1-chloro-3-ethylbenzene*.
b. This compound can be named either as a derivative of benzene: n-*propyl benzene*, or as a derivative of propane: *1-phenylpropane*.
c. This compound is most easily named as a derivative of butane. Numbering from left-to-right and arranging *bromo-* and *phenyl-* in alphabetical order gives the name *2-bromo-3-phenylbutane*.
d. Naming this compound as a trisubstituted benzene first requires that we obtain the correct numbering system for the substituents. The lowest-number-combination is achieved by beginning with the hydroxy group and numbering clockwise. Arranging the substituent names in alphabetical order, we obtain the name *4-ethyl-1-hydroxy-2-methylbenzene*. We could also name this compound as a substituted phenol. Using this approach, the name would be *4-ethyl-2-methylphenol*.

16.4 CHEMICAL REACTIONS OF BENZENE

We have already noted that aromatic compounds do not readily undergo the addition reactions characteristic of other unsaturated compounds. We now know the reason for this. Because of the aromaticity of the conjugated double bonds in the ring, pi electrons are free to move around the ring, and no one bond is a definite double bond. An addition reaction would require breaking up the aromaticity of the ring.

We can easily show the resistance of aromatic compounds to addition reactions by comparing their reactions to those of cycloalkenes. When we contrast the compounds cyclohexene and benzene, we note the following behaviors toward selected reactants.

Type of Reaction	Cyclohexene	Benzene
oxidation	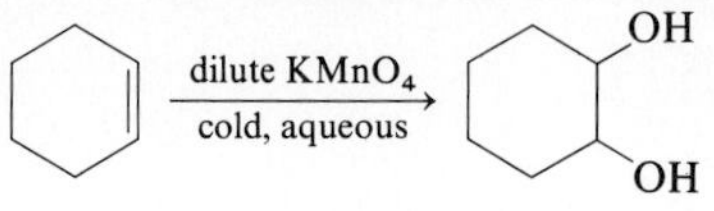 cyclohexene + dilute $KMnO_4$ (cold, aqueous) → cyclohexane-1,2-diol (OH, OH)	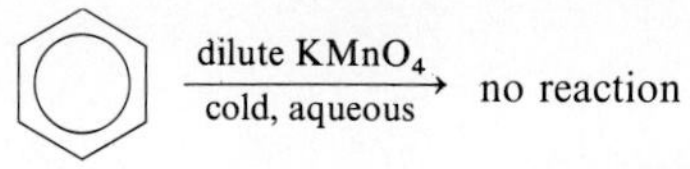 benzene + dilute $KMnO_4$ (cold, aqueous) → no reaction
symmetrical addition	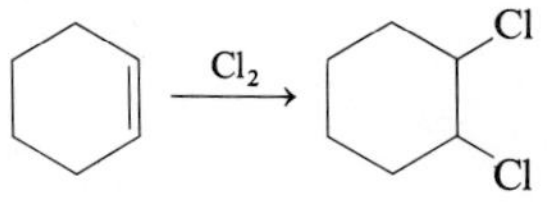 cyclohexene + Cl_2 → 1,2-dichlorocyclohexane	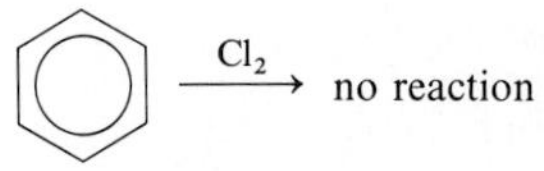 benzene + Cl_2 → no reaction
unsymmetrical addition	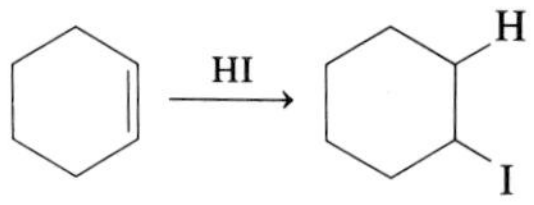 cyclohexene + HI → iodocyclohexane (H, I)	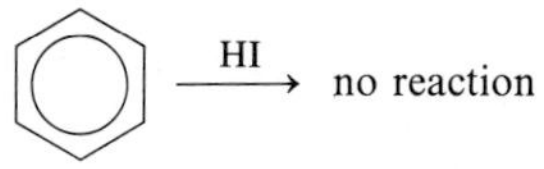 benzene + HI → no reaction
hydrogenation	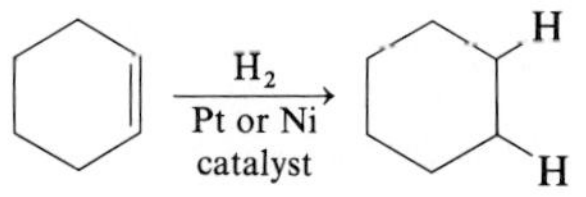 cyclohexene + H_2 (Pt or Ni catalyst) → cyclohexane (H, H)	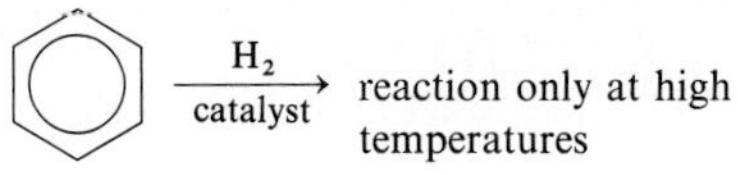 benzene + H_2 (catalyst) → reaction only at high temperatures
	very rapid reactions	*little or no reaction*

If benzene is so unresponsive to addition reactions, what type of reactions will it undergo? *Benzene undergoes substitution reactions, rather than addition reactions.* As you recall from Section 14.12, substitution reactions are characterized by different atoms or groups of atoms replacing hydrogen atoms in a hydrocarbon molecule. There are three important types of substitution reactions for benzene and other aromatic hydrocarbons: alkylation, halogenation, and nitration.

1. *Alkylation*: In an alkylation reaction, an alkyl group (R—) from an alkyl chloride (R—Cl) substitutes for a hydrogen atom on the benzene ring. A special catalyst, $AlCl_3$, is needed for alkylation. The following reaction is representative of the many benzene alkylation reactions that can occur.

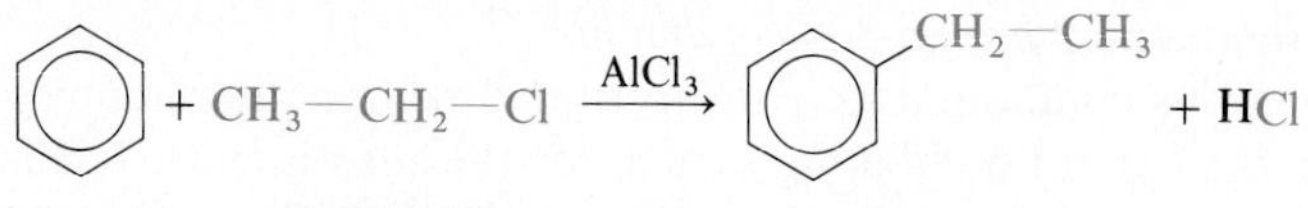

   benzene + CH_3—CH_2—Cl $\xrightarrow{AlCl_3}$ ethylbenzene (CH_2—CH_3) + HCl

   ethylchloride

   In general terms, the alkylation of benzene can be written as

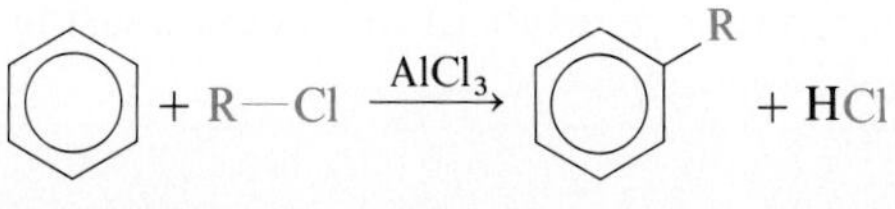

   benzene + R—Cl $\xrightarrow{AlCl_3}$ R-benzene + HCl

   Alkylation reactions are often called "Friedel–Crafts" reactions, after the team of chemists who first utilized this type of reaction. Alkylation is the most important industrial reaction of benzene.

2. *Halogenation* (bromination or chlorination): A hydrogen atom on a benzene ring can be replaced by bromine or chlorine if benzene is reacted with Br_2 or Cl_2 in the presence of a catalyst. The catalyst is usually $FeBr_3$ for bromination and $FeCl_3$ for chlorination.

$$\text{C}_6\text{H}_6 + \text{Br}_2 \xrightarrow{\text{FeBr}_3} \text{C}_6\text{H}_5\text{Br} + \text{HBr}$$

$$\text{C}_6\text{H}_6 + \text{Cl}_2 \xrightarrow{\text{FeCl}_3} \text{C}_6\text{H}_5\text{Cl} + \text{HCl}$$

Aromatic halogenation differs from alkane halogenation (Section 14.12) in that light is not required to initiate aromatic halogenation. Often aromatic halogenation reaction equations are written with "dark" or "no light" written on the reaction arrow. The following equations contrast the effect that light has on the halogenation of toluene; different products are obtained in the presence and absence of light.

$$\text{C}_6\text{H}_5\text{CH}_3 + \text{Cl}_2 \xrightarrow{\text{light}} \text{C}_6\text{H}_5\text{CH}_2\text{Cl} + \text{HCl}$$

$$\text{C}_6\text{H}_5\text{CH}_3 + \text{Cl}_2 \xrightarrow[\text{dark}]{\text{FeCl}_3} p\text{-ClC}_6\text{H}_4\text{CH}_3 + \text{HCl}$$

Note how substitution occurs in the alkane part (methyl group) of the aromatic compound in the presence of light.

3. *Nitration*: Nitric acid (HNO_3) reacts with benzene to form nitrobenzene (C_6H_5—NO_2). Nitric acid is often written as HO—NO_2 to more clearly emphasize that the —NO_2 group substitutes for hydrogen. The catalyst in this reaction is a small amount of concentrated H_2SO_4 (sulfuric acid), and the mixture must be heated. (This is indicated by a small triangle placed under the reaction equation arrow.)

$$\underset{\text{benzene}}{\text{C}_6\text{H}_6} + \underset{\text{nitric acid}}{\text{HO}-\text{NO}_2} \xrightarrow[\Delta]{\text{H}_2\text{SO}_4} \underset{\text{nitrobenzene}}{\text{C}_6\text{H}_5\text{NO}_2} + \text{H}_2\text{O}$$

Most of the reactions that benzene undergoes also work well with alkylated benzene rings such as toluene, ethylbenzene, and *n*-propylbenzene. In these reactions, a mixture of ortho-, meta-, and para- isomers is obtained, with the ortho- and para- isomers predominating. For example, in the nitration of toluene, the following isomer mix is obtained.

toluene $\xrightarrow[\text{H}_2\text{SO}_4]{\text{HNO}_3}$ *o*-nitrotoluene (59%) + *p*-nitrotoluene (37%) + *m*-nitrotoluene (4%)

At the beginning of this section, we noted that cold, dilute solutions of $KMnO_4$ will not oxidize benzene. Hot, concentrated $KMnO_4$ solutions also will not oxidize benzene. However, hot, concentrated KMnO₄ solutions will react with alkylbenzenes. The alkyl group is oxidized and all carbons are lost except the one attached to the ring. The resulting product is always benzoic acid. The reaction involving *n*-propylbenzene and hot, concentrated KMnO₄ solution illustrates alkylbenzene oxidation.

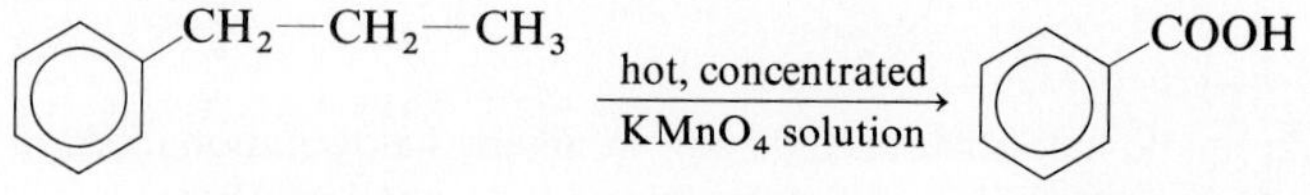

$$\text{C}_6\text{H}_5\text{—CH}_2\text{—CH}_2\text{—CH}_3 \xrightarrow[\text{KMnO}_4 \text{ solution}]{\text{hot, concentrated}} \text{C}_6\text{H}_5\text{—COOH}$$

n-propylbenzene → benzoic acid

Example 16.2

Identify the organic product (or products) in each of the following substitution reactions.

a. C₆H₆ + Br₂ —FeBr₃→ ?

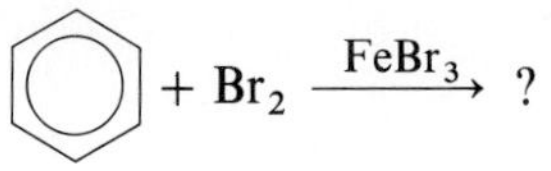

b. C₆H₆ + CH₃—Cl —AlCl₃→ ?

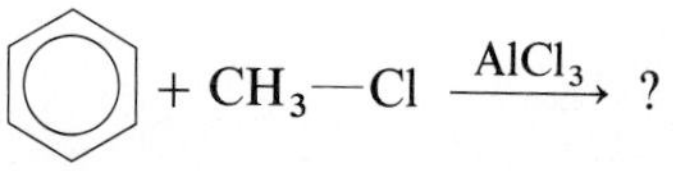

c. C₆H₅—CH₂—CH₃ + HO—NO₂ —H₂SO₄, Δ→ ?

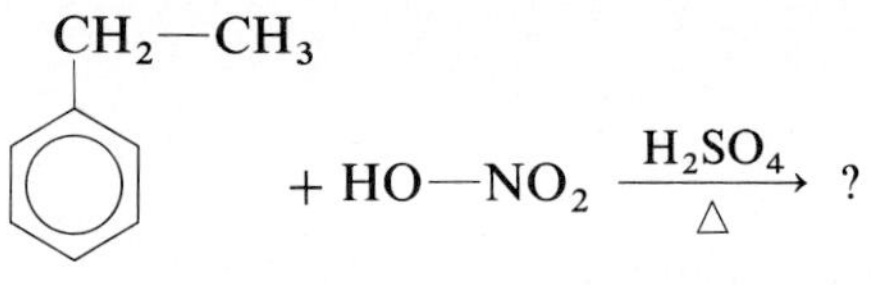

Solution

a. In the bromination of benzene, a bromine atom is substituted for a hydrogen atom on the benzene ring. The product is bromobenzene.

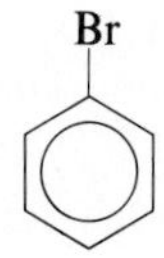

(bromobenzene structure)

b. This is an alkylation reaction, and the alkyl group is a methyl group. The reaction product will be methylbenzene.

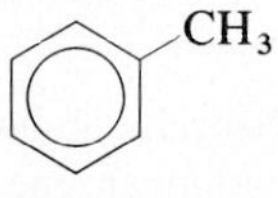

(methylbenzene structure)

c. Nitration of ethylbenzene will produce a mixture of *o*-, *m*-, and *p*-ethylnitrobenzene, with the *o*- and *p*- isomers predominating.

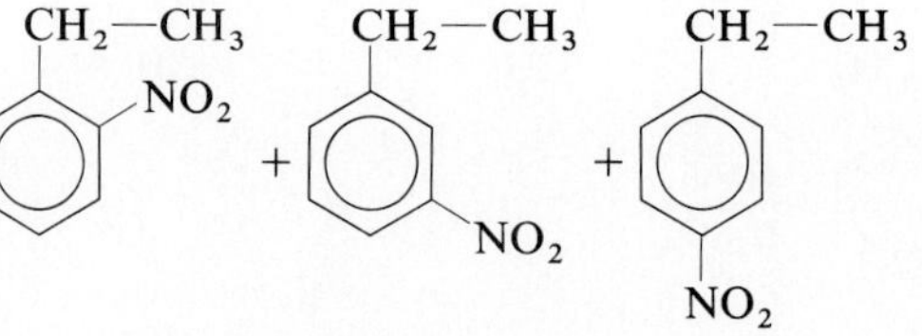

(o-, m-, and p-ethylnitrobenzene structures)

Example 16.3

Starting with benzene and using other chemical substances as needed, design a two-step method to synthesize the aromatic compound 4-chloro-1-ethylbenzene.

Solution

Two substitutions must be made in order to obtain the desired product. It is best to first attach the ethyl group to the benzene using a Friedel–Crafts alkylation.

Step 1: C_6H_6 + CH_3-CH_2-Cl $\xrightarrow{AlCl_3}$ ethylbenzene + HCl

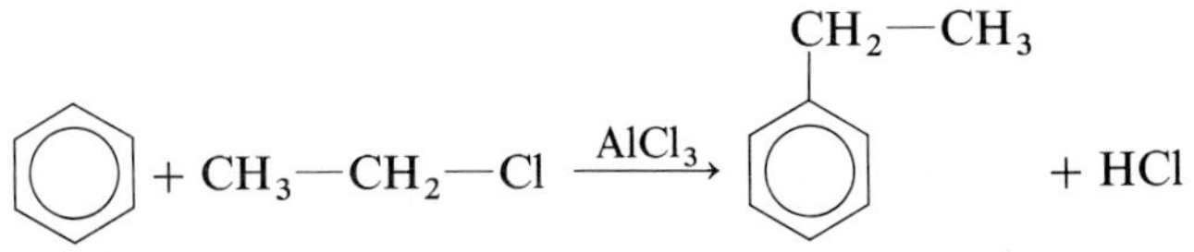

The ethylbenzene is then reacted with Cl_2 and a catalyst ($FeCl_3$) to substitute Cl in the number four (para) position.

Step 2: ethylbenzene + Cl_2 $\xrightarrow{FeCl_3}$ 4-chloroethylbenzene + HCl

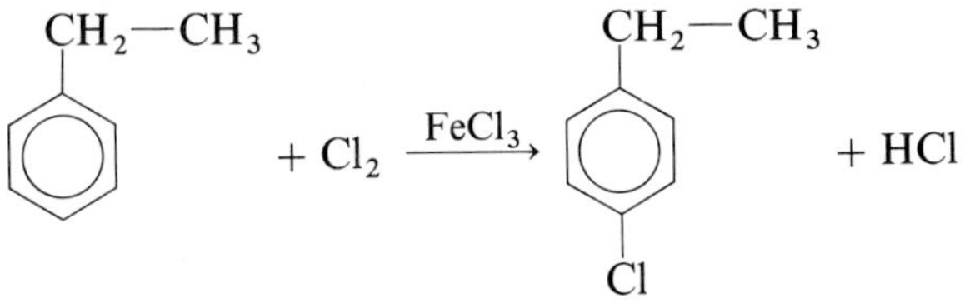

It is always best to first attach the alkyl group in these two-step reactions because this helps to make the ring slightly more reactive. It also helps direct the second substitution to the para (or ortho) position.

More than two substitutions (Example 16.3) are possible on benzene rings. Exhaustive substitutions under harsh conditions and long reaction periods can result in complete substitution. One such example is the complete chlorination of benzene to form hexachlorobenzene, which is a raw material used in the synthesis of pesticides.

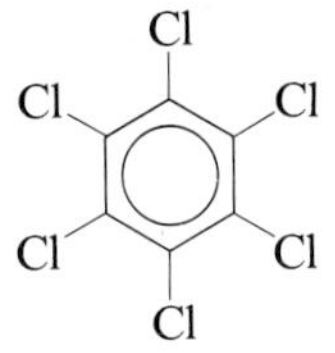

hexachlorobenzene

When toluene is heated with nitric acid, it is possible to substitute three nitro groups for hydrogens—at the para position and both ortho positions—forming the compound 2, 4, 6-*tri*ni*tro*toluene or "TNT", an explosive.

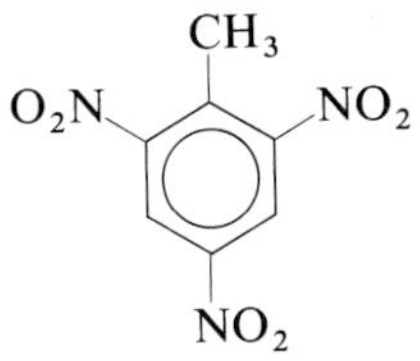

TNT (trinitrotoluene)

Any time numerous nitro groups are bonded to a molecule, the molecule tends to undergo vigorous, rapid combustion; this is a characteristic needed for explosives. When you dispose of chemicals in a laboratory, it is important to make sure that nitric acid is never mixed with organic waste. This prevents the inadvertent synthesis of nitro explosives in the waste materials.

16.5 USES OF BENZENE AND BENZENE DERIVATIVES

Benzene and most of the lower members of the alkyl benzene series (methylbenzene, ethylbenzene, dimethylbenzenes, and so on) are liquids under normal laboratory conditions. In addition to being good solvents, they also serve as the starting materials for the synthesis of hundreds of other valuable aromatic compounds. The wide range of commercial products made from benzene and its simple derivatives include explosives, numerous drugs, antiseptics, agricultural chemicals, and insecticides.

Until the late 1970s, benzene was widely used in laboratories as a solvent for other nonpolar organic compounds. This use of benzene has decreased markedly in recent years, based on findings that continued inhalation of benzene can cause a drop in white blood cell count and it has caused leukemia in laboratory test animals.

Chlorinated aromatic hydrocarbons have been used for over 40 years as pesticides. The most frequently used compound is DDT (*dichlorod*iphenyl*tri*chloroethane).

Discovered shortly before World War II, DDT was hailed as a miracle chemical. Its widespread and successful use during the 1940s and 1950s helped dramatically in controlling insect-borne malaria, as well as most agricultural crop pests. Because it was chemically inert, cheap to make, and posed little threat to humans, DDT seemed to be the ultimate answer to insect control.

We now know that there are problems associated with DDT use—its effects on nontarget organisms such as birds and fish. For example, DDT interferes with the calcium metabolism cycle in birds such as eagles, hawks, and falcons. The net result is eggs whose shells are thinner than normal—so thin that these eggs cannot withstand the rigors of incubation, and reproductive failure results (see Figure 16–2). Because of this, as well as other environmental considerations, DDT is no longer used as an insecticide. Other insecticides that have lower toxicities for mammals and birds, but are still extremely toxic for certain species of insects, have replaced DDT.

Polychlorinated biphenyls (PCBs) are another group of aromatic compounds that have made headlines in the last few years. These compounds are derived from the chlorination of biphenyl (phenylbenzene).

biphenyl   PCBs
(various numbers of Cl atoms in different positions)

A mixture of many different chlorinated biphenyls can be obtained, and the number of chlorine atoms on the product can vary from one to ten. Chemical inertness, thermal stability, desirable electrical properties, and low vapor pressure make PCBs useful in a variety of applications, including use as fire retardants, plasti-

FIGURE 16–2 This crushed egg (upper right) in the nest of a cinnamon teal had such a thin shell that the weight of the nesting parent's body destroyed it. High levels of DDT were associated with this egg.

cizers, insulation in electrical equipment, components of carbonless copy paper, and in some epoxy paints. It was not known until years after their introduction that PCBs act as a nerve poison and they accumulate in the fatty tissues of organisms over time with repeated exposure (see Figure 16–3). The use of PCBs has been

FIGURE 16–3 This photo shows the extreme precautions now taken during PCB cleanup operations. Here, workers clean up a spill that is the result of a fire that damaged a transformer.

16.6 FUSED-RING AROMATIC COMPOUNDS

Benzene and its substituted derivatives are not the only type of aromatic hydrocarbon that exists. Another large class of aromatic hydrocarbons is the fused-ring aromatic hydrocarbons. **Fused-ring aromatic hydrocarbons** *are aromatic compounds whose structures contain two or more carbon rings fused together.* Two carbon rings that share a pair of carbon atoms are said to be *fused*.

The simplest fused-ring aromatic hydrocarbon is naphthalene, a compound whose structure consists of two benzene rings that have two carbon atoms in common. The most common notation for depicting the structure and bonding in naphthalene is

This structure correctly emphasizes the fact that the delocalized bonding present in naphthalene involves both of the ring systems. A drawback to the use of this structure is that it incorrectly implies the presence of a delocalized system of six conjugated-double bonds because an aromatic circle drawn within a benzene ring represents a delocalized system of three conjugated-double bonds (Section 16.2). Actually, only five double bonds are present in the conjugated-double-bond system of naphthalene that becomes delocalized. Two structures for naphthalene that correctly indicate the presence of an original five conjugated-double-bond system are

In this text we will use structures of the latter type to describe the bonding in fused-ring systems. When using these structures, always remember that the delocalization of electrons involves more than one ring.

Figure 16–4 shows structures for additional fused-ring aromatic compounds. In unsubstituted fused-ring aromatic compounds like those in Figure 16–4, a single hydrogen atom is attached to each ring carbon atom except those that are at points of fusion which do not have any hydrogen attached. Most fused-ring aromatic hydrocarbons have fewer hydrogen atoms than carbon atoms. The molecular formulas for naphthalene, anthracene, and phenanthrene are $C_{10}H_8$, $C_{14}H_{10}$, and $C_{14}H_{10}$, respectively. (Anthracene and phenanthrene are isomers.)

FIGURE 16–4 Structures of selected fused-ring aromatic hydrocarbon compounds.

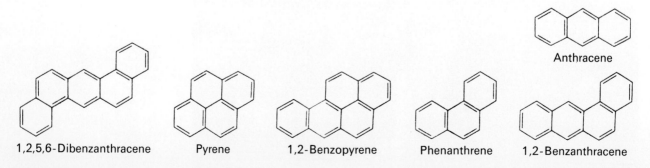

1,2,5,6-Dibenzanthracene   Pyrene   1,2-Benzopyrene   Phenanthrene   1,2-Benzanthracene   Anthracene

Naphthalene is a white solid that has a pungent odor and readily sublimes (Section 7.12) at room temperature. Napthalene is used in mothballs and moth-crystals (see Figure 16–5). Anthracene is an important starting material in the production of dyes. Phenanthrene is also used in dye production, as well as in making explosives and the synthesis of several drugs. Several fused-ring aromatic hydrocarbons have been found to be carcinogens (cancer-causing agents). One of these compounds, 1,2-benzopyrene, is present in cigarette smoke. This compound is believed to be at least partially responsible for the high incidence of lung cancer among cigarette smokers.

16.7 SOURCES OF AROMATIC HYDROCARBONS

When coal is heated in the absence of air, it produces a solid residue called *coke* and a liquid called *coal tar*. Although the coal tar obtained represents only a small percentage of the original coal, it amounts to many tons annually because of the large quantity of coal that is converted into coke. Coal tar serves as a major source of aromatic compounds. Many different aromatic compounds can be obtained from it by using distillation and other separation techniques.

Note that the aromatic compounds obtained from coal tar do not exist in coal as such; they are formed during the intense heating involved in coke production. The actual molecular structure of coal is an extremely large fused-ring system involving literally hundreds of rings. This multi-ring system breaks down into simpler substances at the high temperatures reached during coke production.

At one time, coal tar was the main source of aromatic hydrocarbons. This is no longer the case. Petroleum is now the number one source because it is cheaper. Also, the high demand for aromatic compounds for the production of drugs, plastics, explosives, rubber, and insecticides now far excedes the amount that coal tar can supply. This necessitates the conversion of saturated hydrocarbons, the main component of petroleum, into aromatic hydrocarbons. Using special catalysts, alkanes can be converted to aromatic hydrocarbons at high temperatures. The production of toluene from *n*-heptane is representative of such a conversion.

FIGURE 16–5 Naphthalene, the simplest fused-ring aromatic hydrocarbon, is a white solid that is used in moth repellant formations.

$$CH_3-CH_2-CH_2-CH_2-CH_2-CH_2-CH_3 \xrightarrow[500°C]{catalyst} C_6H_5CH_3 + 4H_2$$

16.8 HETEROCYCLIC COMPOUNDS

Before leaving the subject of aromatic hydrocarbons, we must note the existence of a large group of compounds that are closely related to aromatic hydrocarbons— the heterocyclic aromatic compounds. **Heterocyclic aromatic compounds** *are compounds in which one or more of the carbon atoms in an aromatic carbon ring have been replaced by other kinds of atoms.* The hetero atom or atoms are usually oxygen, nitrogen, or sulfur.

heterocyclic aromatic compounds

Many biologically important compounds contain aromatic heterocyclic structural features. Included among these are some proteins, some vitamins, DNA, and RNA, all of which will be discussed in subsequent chapters. Table 16–1 (p. 386) lists structures for some biologically important heterocyclic compounds where nitrogen is the hetero atom. The presence of hetero atoms in a carbon ring is not limited to aromatic systems. There are also saturated and nonaromatic unsaturated heterocyclic compounds.

TABLE 16-1 Nitrogen-containing Aromatic Heterocyclic Structures

Name	Structural Formula	Occurrence
pyridine	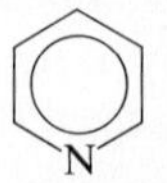	Found as part of the structure of niacin—one of the B vitamins—and of nicotine, a highly toxic material in tobacco.
pyrimidine		An important structural feature in DNA and RNA.
indole	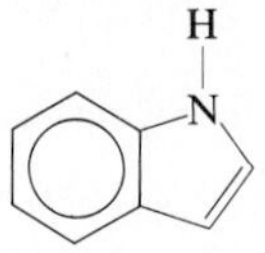	Important structural feature of tryptophan, an essential amino acid.
purine	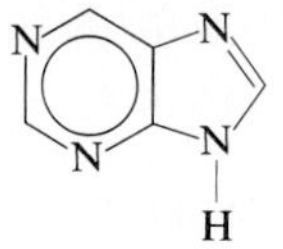	An important structural feature of DNA, RNA, ATP, and caffeine.

16.9 HYDROCARBONS: A SUMMARY

This chapter and the previous two chapters have focused on hydrocarbons, the simplest organic compounds. Because of the unique properties of carbon, almost limitless variations in hydrocarbon structure are possible. Beyond structural isomers for saturated hydrocarbons, the ability of carbon to form multiple bonds further extends the range of possible compounds.

FIGURE 16-6 Relationships among various families of hydrocarbon compounds.

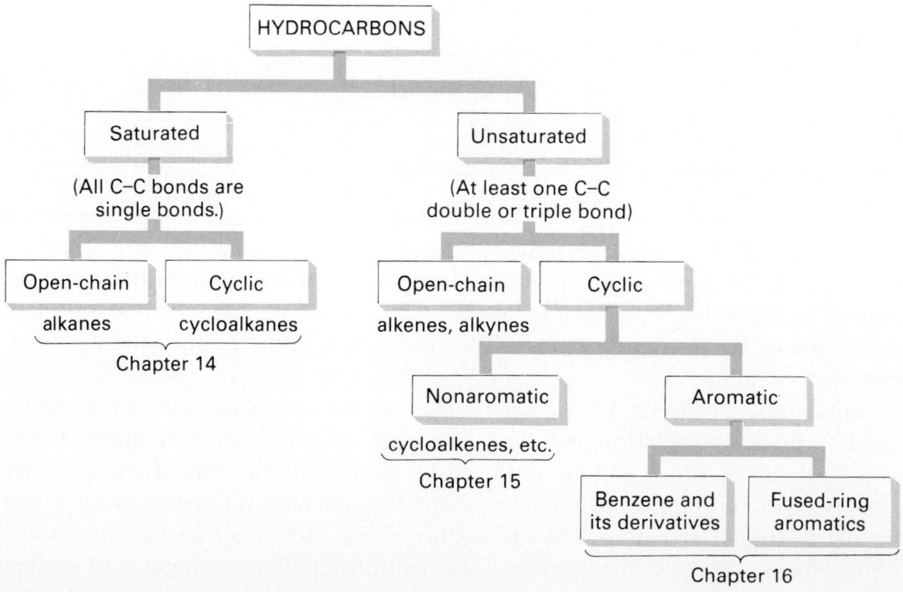

Classification of the many hydrocarbons into families (alkanes, alkenes, cycloalkenes, and so on) eases the task of dealing with their diversity. It is appropriate, as we end our consideration of hydrocarbons, to summarize the relationships among hydrocarbon families.

Hydrocarbons fall into two broad categories: saturated and unsaturated. Both saturated and unsaturated hydrocarbons are of two types: open-chain and cyclic. Cyclic unsaturated hydrocarbons can be further classified as aromatic or nonaromatic. Aromatic hydrocarbons are of two classifications: those based on a single ring of carbon atoms and those containing fused-carbon rings. Figure 16–6 shows diagrammatically the interrelationships among the hydrocarbon classifications.

EXERCISES AND PROBLEMS

Structure and Bonding in Benzene

16.1 Describe the structure of benzene. What evidence indicates conclusively that benzene has a symmetrical ring structure?

16.2 Describe the bonding in benzene. Be sure to make use of the concept of delocalized bonding in your description.

16.3 Both benzene and cyclohexane have a ring of six carbon atoms. Describe how the bonding in benzene differs from the bonding in cyclohexane.

16.4 Explain how the reactions of benzene provide evidence that its structure does not include double bonds like those found in alkenes.

Aromatic Hydrocarbons

16.5 Classify each of the following cyclic hydrocarbons into the categories aromatic, unsaturated nonaromatic, and saturated.

a. b. c. d. e. f.

16.6 How many hydrogen atoms are present in each of the molecules in Problem 16.5?

Nomenclature of Aromatic Compounds

16.7 Name each of the following compounds, using IUPAC names in which *benzene* is part of the name.

a. b. c. d. e. f.

16.8 What is the meaning associated with the prefixes ortho-, meta-, and para- when they are used to name di-substituted benzenes?

16.9 Name each of the following compounds, using IUPAC names in which ortho-, meta, or para- is part of the name.

a. b. c. d. e. f.

16.10 Name each of the following compounds, using IUPAC names in which the aromatic ring is treated as a substituent attached to a hydrocarbon chain.

a. $CH_3-CH-CH_2-CH_3$

b. $CH_3-CH_2-CH-CH_2$ with CH_2-CH_3

c. CH₃—CH—CH=CH₂
        |
        C₆H₅ (phenyl)

d. CH₃—CH—CH₂—CH—CH₃
        | |
        C₆H₅   C₆H₅

16.11 Name each of the following compounds, using IUPAC common names. Note that the word *benzene* should not appear in the name.

a. C₆H₅—OH
b. C₆H₅—NH₂
c. C₆H₅—CH₃
d. C₆H₅—COOH
e. 1,4-di-CH₃-C₆H₄
f. 1,3-di-CH₃-C₆H₄

16.12 Name each of the following compounds, using IUPAC-accepted common names. Note that the word *benzene* should not appear in the name.

a. 2-chlorophenol (OH, Cl on ring)
b. 3-ethyltoluene (CH₃ and CH₂—CH₃ on ring)
c. 4-chlorobenzoic acid (COOH, Cl)
d. 2-nitroaniline (NH₂, NO₂)
e. 3-bromoaniline (NH₂, Br)
f. 4-nitrophenol (NO₂, OH)

16.13 Assign an IUPAC name to each of the following compounds.

a. phenylcyclohexane
b. C₆H₅—CH₂—CH₂—C₆H₅

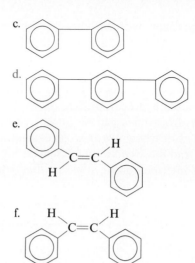

c. biphenyl structure
d. terphenyl structure
e. cis-stilbene (H, H on same side, C=C with two phenyls)
f. trans-stilbene

16.14 Write condensed structural formulas for each of the following compounds.

a. 1,3-diethylbenzene
b. *o*-xylene
c. *p*-ethylmethylbenzene
d. 1,4-diethyl-2,5-diiodobenzene
e. 3-methyl-3-phenylpentane
f. isopropylbenzene

16.15 Write condensed structural formulas for each of the following compounds.

a. triphenylmethane
b. 2,4-dinitrophenol
c. *p*-methylaniline
d. cyclopentylbenzene
e. 1,2,4-tribromobenzene
f. 2,4,5-tribromotoluene

Isomerism and Aromatic Hydrocarbons

16.16 Six different isomers can be produced by substituting two bromine atoms and one chlorine atom for hydrogens on a benzene ring. Draw structural formulas for the six isomers.

16.17 How many different compounds are there that fit each of the following descriptions?

a. monochlorobenzenes
b. dichlorobenzenes
c. trichlorobenzenes
d. tetrachlorobenzenes
e. bromotrichlorobenzenes
f. dibromodichlorobenzenes

16.18 There are eight isomeric substituted benzenes that have the formula C_9H_{12}. Draw structural formulas for these isomers and name them.

16.19 In Chapter 15 we learned that 1,2-dibromocyclohexane can exist in either *cis* or *trans* forms. Explain why no such geometric isomerism is possible for 1,2-dibromobenzene.

Chemical Reactions of Benzene and its Derivatives

16.20 For each of the following classes of compounds, indicate whether *addition* or *substitution* is the most characteristic reaction for the class.

a. alkanes   d. aromatic rings
b. alkenes   e. alkyl benzenes
c. alkynes   f. cycloalkenes

16.21 Complete the following reaction equations by giving the structure of the missing reactant or organic product. Do not balance the reactions or give inorganic products.

a. ⬡ + Br$_2$ $\xrightarrow{FeBr_3}$

b. ⬡ + HNO$_3$ $\xrightarrow{H_2SO_4}$

c. ⬡ + CH$_3$—CH$_2$—Cl $\xrightarrow{AlCl_3}$

d. ⬡ + Cl$_2$ $\xrightarrow{FeCl_3}$

e. ⬡ + ? ⟶ ⬡—CH$_2$—CH$_2$—CH$_3$ + HBr

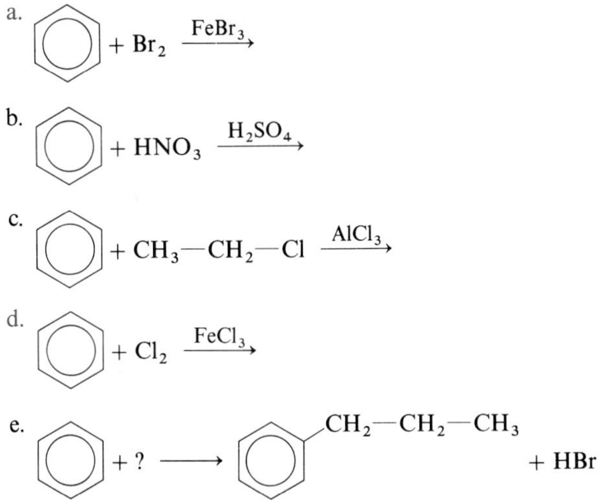

16.22 Complete the following reaction equations by giving the structure of the organic product. Do not balance the reactions or give inorganic products.

a. ⬡—CH$_3$ + conc. KMnO$_4$ $\xrightarrow{\Delta}$

b. ⬡—CH$_3$ + CH$_3$—C(CH$_3$)$_2$—Cl $\xrightarrow{AlCl_3}$

c. ⬡—CH$_3$ + Cl$_2$ $\xrightarrow{light}$

d. ⬡—CH$_3$ + Cl$_2$ $\xrightarrow[dark]{AlCl_3}$

e. ⬡—CH=CH$_2$ + Br$_2$ $\xrightarrow{no\ catalyst}$

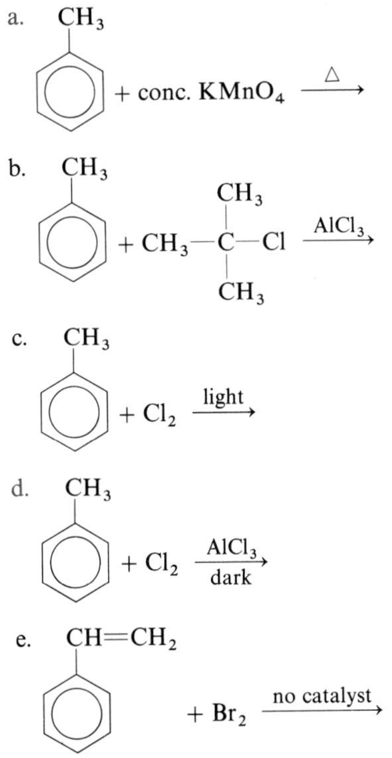

16.23 Often, in order to obtain a desired product, a multistep synthesis is required. In this situation, the product of one reaction is subjected to further reactions, and so on, until the desired product is obtained. Determine the final product in each of the following multistep syntheses.

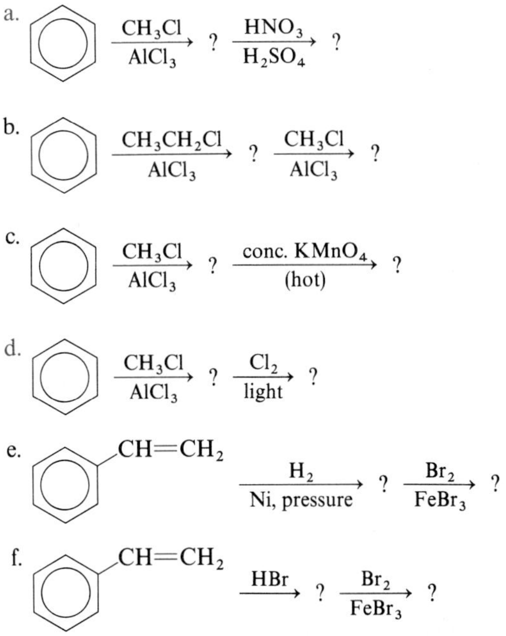

a. ⬡ $\xrightarrow{CH_3Cl}{AlCl_3}$? $\xrightarrow{HNO_3}{H_2SO_4}$?

b. ⬡ $\xrightarrow{CH_3CH_2Cl}{AlCl_3}$? $\xrightarrow{CH_3Cl}{AlCl_3}$?

c. ⬡ $\xrightarrow{CH_3Cl}{AlCl_3}$? $\xrightarrow{conc.\ KMnO_4}{(hot)}$?

d. ⬡ $\xrightarrow{CH_3Cl}{AlCl_3}$? $\xrightarrow{Cl_2}{light}$?

e. ⬡—CH=CH$_2$ $\xrightarrow{H_2}{Ni,\ pressure}$? $\xrightarrow{Br_2}{FeBr_3}$?

f. ⬡—CH=CH$_2$ $\xrightarrow{HBr}$? $\xrightarrow{Br_2}{FeBr_3}$?

16.24 Suppose we want to analyze an unknown organic liquid for functional groups. We add a few drops of red Br$_2$ solution, but no decolorization of the solution occurs. We then add a few shavings of iron metal, and within a few moments, decolorization occurs. Knowing that Fe and Br$_2$ form FeBr$_3$ in solution, explain what is happening to the unknown liquid. Is the unknown liquid an alkane, an alkene, or an aromatic compound?

16.25 Toluene and xylenes are routinely added to gasoline as "anti-knock" agents. This greatly increases the octane rating of the gasoline and makes the gasoline burn better. Write separate chemical equations for the complete combustion of both toluene and xylene. Balance each equation.

16.26 Terephthalic acid, which has the following structure, is one of the compounds used to synthesize Dacron polymers.

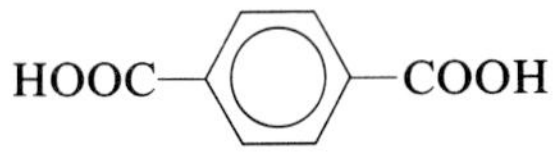

HOOC—⬡—COOH

What hydrocarbon, with the molecular formula C_8H_{10}, could be reacted with hot, concentrated KMnO$_4$ to form terephthalic acid?

Fused-ring Aromatic Hydrocarbons

16.27 Draw structural formulas for each of the following fused-ring aromatic hydrocarbons.

a. naphthalene   b. anthracene   c. phenanthrene

16.28 How many hydrogen atoms are present in each of the molecules in Problem 16.27?

16.29 How many different compounds (isomers) are there that fit each of the following descriptions?

a. monobrominated naphthalenes
b. monobrominated anthracenes
c. dibrominated napthalenes

17 Alcohols, Phenols, and Ethers

Objectives

After completing Chapter 17, you will be able to:

17.1 Recognize the functional groups and structural features of alcohols, phenols, and ethers.
17.2 Name alcohols and phenols using IUPAC rules, given the structural formulas, and vice versa, and know the common names for the simpler alcohols and phenols.
17.3 Classify alcohols as primary, secondary, or tertiary on the basis of their structures.
17.4 List general physical properties of alcohols and tell how hydrogen bonding influences them.
17.5 Give two general methods of preparation for alcohols.
17.6 Predict the organic reaction product(s) from alcohol dehydration, oxidation, and substitution reactions.
17.7 List common uses and properties of the alcohols methanol, ethanol, 2-propanol, 1,2-ethanediol, and 1,2,3-propanetriol.
17.8 List general physical properties and uses of phenols.
17.9 Discuss the occurrence of phenols in nature.
17.10 Assign ethers IUPAC or common names given the structural formulas, and vice versa.
17.11 List general physical properties and uses of ethers.
17.12 Recognize the epoxide structure and know the common uses for epoxides.
17.13 List the structural features and general reactions of thiols and disulfides.

INTRODUCTION

The element oxygen is necessary for life for mammals and almost all other organisms. Just as a fire needs oxygen to continue to burn, the human body needs oxygen to support its life processes. Most of the oxygen atoms in biological systems are found in water or in compounds where they are bonded to carbon atoms.

In this chapter, we will discuss three of the eight classes of organic compounds that contain carbon–oxygen bonds: alcohols, phenols, and ethers. These types of compounds all possess carbon–oxygen *single* bonds. Molecules containing carbon–oxygen *double* bonds are the subject of Chapters 18 and 19.

17.1 STRUCTURAL FEATURES OF ALCOHOLS, PHENOLS, AND ETHERS

Structurally, alcohols and phenols are closely related in that each type of compound possesses one or more hydroxyl groups (—OH) attached to carbon atoms. **Alcohols** *are compounds in which a hydroxyl group is attached to a saturated carbon atom; that is, a carbon atom that is also bonded to three other atoms.* In general terms, the designation R—OH is used to denote alcohols; R represents the carbon chain (alkyl group—Section 14.7) to which the —OH group is attached.

Phenols *are compounds in which a hydroxyl group is attached to an aromatic carbon ring.* The general formula for phenols is Ar—OH, where Ar— represents an *aryl* group. An **aryl group** *is an aromatic ring from which one hydrogen atom has been removed.*

Ethers *are compounds in which oxygen atoms are bonded to two carbon atoms by single bonds.* In an ether, the carbon atoms that are attached to an oxygen atom can be part of alkyl groups or aryl groups. Thus, three general types of ethers exist; they can be designated by the notations R—O—R′, Ar—O—Ar′, and R—O—Ar. In the first two notations, the prime is used to indicate that both R groups and

alcohols

phenols

aryl group

ethers

TABLE 17–1 Structural Characteristics of Organic Compounds Containing Carbon–Oxygen Single Bonds

	General Formula	Specific Examples
alcohols	R—OH	$CH_3—CH_2—CH_2—OH$
phenols	Ar—OH	phenol (benzene ring with OH)
ethers	R—O—R′	$CH_3—O—CH_2—CH_3$
	Ar—O—Ar′	diphenyl ether with CH₃ substituent
	R—O—Ar	$CH_3—O—$(benzene ring)

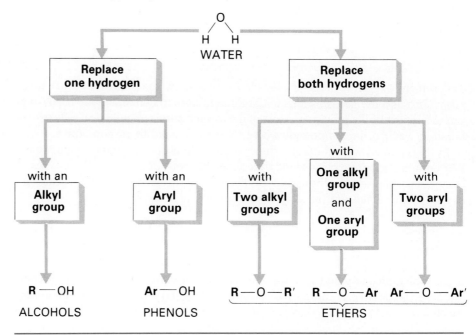

FIGURE 17-1 Structural relationships between water and organic compounds containing carbon–oxygen single bonds.

both Ar groups do not have to be identical. Note that, unlike alcohols and phenols, ethers do not possess hydroxyl (—OH) groups. Table 17-1 (p. 391) summarizes the basic structural characteristics of alcohols, phenols, and ethers.

These three types of compounds are structurally similar to water. When one or both of the hydrogen atoms of water are replaced by carbon atoms, the result is one of these types of molecules, as is illustrated in Figure 17-1.

Alcohols and phenols resemble water in many of their reactions, but ethers have very different properties. The different behavior of ethers is primarily caused by the lack of an —OH group in these compounds.

17.2 NOMENCLATURE OF ALCOHOLS AND PHENOLS

Simple alcohols, which contain only a few carbon atoms, are often named by identifying the alkyl group and then adding the word *alcohol*.

CH$_3$—OH   CH$_3$—CH$_2$—OH   CH$_3$—CH$_2$—CH$_2$—OH   CH$_3$—CH(CH$_3$)—OH
methyl alcohol   ethyl alcohol   n-propyl alcohol   isopropyl alcohol

Although this naming technique is useful, it is not sufficient for more complex alcohols. Formal IUPAC rules for naming alcohols can be used to name a much larger variety of compounds. The IUPAC rules for naming alcohols that contain a single hydroxyl group are:

Rule 1: Identify the longest carbon chain to which the —OH group is attached and use this as the name base.

Rule 2: Obtain the name of the longest chain by substituting the ending *-ol* for the *-e* of the corresponding alkane.

Rule 3: Number the longest chain, beginning with the end closest to the —OH group. The hydroxyl group should always receive the lowest number possible. Identify the numerical location of the —OH group by writing the number of the carbon atom to which it is attached in front of the chain name.

Rule 4: Locate and name any other groups attached to the chain. Include these in alphabetical order as prefixes to the chain name, as well as their numeric locations.

Rule 5: For nonaromatic cyclic alcohols, number the ring starting with the carbon atom bearing the hydroxyl group. Numbering continues to give the lowest numbers to carbon atoms bearing alkyl groups. The number 1 is not used in the name.

Name the following alcohols utilizing IUPAC nomenclature rules.

a. CH_3-OH

b.
$$CH_3-CH_2-\underset{\underset{OH}{|}}{\overset{\overset{CH_3}{|}}{C}}-CH_2-CH_2-CH_3$$

c.
$$CH_3-CH_2-\underset{\underset{CH_2-OH}{|}}{CH}-CH_2-CH_3$$

d.
[cyclohexane ring with CH_3 and CH_3 substituents and $-OH$ group]

Example 17.1

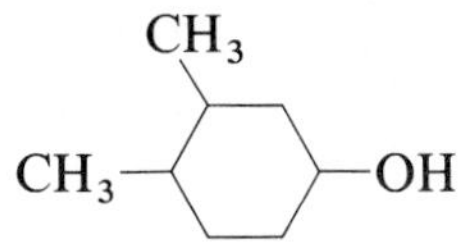

Solution

a. The longest carbon chain has only one carbon atom, so the root word is *methane*. Replacing *-e* with *-ol* yields the correct name, *methanol*. On such a short chain, no number is needed to locate the —OH group.

b. The longest carbon chain that contains the alcohol functional group has six carbons. Changing the *-e* to *-ol*, hexane becomes *hexanol*. Numbering the chain from left to right (from the end nearest the —OH group) identifies carbon number 3 as the location of the —OH group. Adding this to the name, with the usual hyphen between numbers and words, gives the name *3-hexanol*. There is also a methyl group attached to carbon number 3. The complete name is *3-methyl-3-hexanol*.

$$^1CH_3-{}^2CH_2-\underset{\underset{OH}{|}}{\overset{\overset{CH_3}{|}}{{}^3C}}-{}^4CH_2-{}^5CH_2-{}^6CH_3$$

c. The longest carbon chain in this molecule is five carbons long. However, the longest carbon chain *containing the* —OH *group* is only *four* carbons long. Beginning with the end of the chain closest to the —OH group, the chain is numbered as follows.

$$CH_3-CH_2-\underset{\underset{{}^1CH_2-OH}{|}}{\overset{2\quad 3\quad 4}{CH-CH_2-CH_3}}$$

The chain name for this alcohol is *1-butanol*. The ethyl group attached to carbon 2 is included to form the complete name *2-ethyl-1-butanol*.

d. We will need Rule 5 to name this compound. The base for this name is the ring system, cyclohexane, which is changed to *cyclohexanol*. The carbon to which the —OH group is attached is assigned the number 1. Numbering counterclockwise assigns the numbers 3 and 4 to the carbon atoms bearing methyl groups. The complete name for this alcohol is *3,4-dimethylcyclohexanol*.

$$CH_3-\underset{4}{\overset{CH_3}{\underset{}{\bigcirc}}}\overset{3\;2}{_{1}}-OH$$

Note that numbering the ring carbons clockwise would be incorrect because the methyl-bearing carbons would be 4 and 5, which are higher numbers than those obtained with the other numbering system.

Table 17–2 lists both IUPAC and common names for monohydroxyl alcohols containing four or less carbon atoms. Note that there are two isomeric three-carbon alcohols and four isomeric four-carbon alcohols.

Occasionally we encounter an alcohol that has more than one hydroxyl group. These alcohols, which are called *polyhydroxy alcohols*, can be named with only a slight modification of the IUPAC rules just discussed. A Greek numerical prefix is included in the name, preceding the *-ol* ending, to indicate how many hydroxyl groups are present (-diol, -triol, and so on). Numbers are used to designate the

TABLE 17–2 IUPAC and Common Names for Monohydroxyl Alcohols Containing Four or Less Carbon Atoms

Formula	IUPAC Name	Common Name
one carbon atom		
CH_3—OH	methanol	methyl alcohol
two carbon atoms		
CH_3—CH_2—OH	ethanol	ethyl alcohol
three carbon atoms		
CH_3—CH_2—CH_2—OH	1-propanol	*n*-propyl alcohol

CH_3—CH—CH_3
    |
    OH
	2-propanol	isopropyl alcohol

four carbon atoms		
CH_3—CH_2—CH_2—CH_2—OH	1-butanol	*n*-butyl alcohol

CH_3—CH—CH_2—OH
    |
    CH_3
	2-methyl-1-propanol	isobutyl alcohol

CH_3—CH_2—CH—CH_3
              |
              OH
	2-butanol	*sec*-butyl alcohol

CH_3 $\|$ CH_3—C—OH $\|$ CH_3	2-methyl-2-propanol	*t*-butyl alcohol

positions of each hydroxyl group. The -e at the end of the chain name is retained, in this case, for ease in pronunciation.

Example 17.2

Name the following polyhydroxy alcohols according to IUPAC rules.

a. 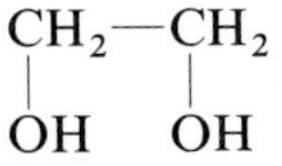 CH₂—CH₂
   | |
   OH   OH

b. 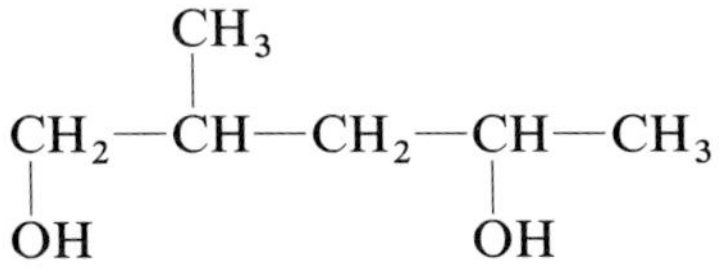   CH₃
            |
   CH₂—CH—CH₂—CH—CH₃
   | |
   OH   OH

c. 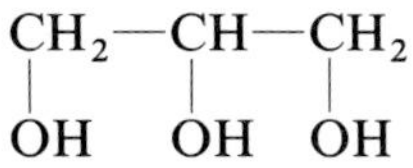 CH₂—CH—CH₂
   | | |
   OH OH OH

Solution

a. Two —OH groups are present; therefore, this compound will be a *-diol*. The longest carbon chain is two carbons long, so this is *ethanediol*. Specifying the positions of the —OH groups (carbons 1 and 2) yields the complete name: *1,2-ethanediol*.

b. The longest carbon chain contains five carbons and two hydroxyl groups; this is a *pentanediol*. Locating the —OH groups at carbons 1 and 4 yields *1,4-pentanediol*. Including the name and location of the methyl group completes the name: *2-methyl-1,4-pentanediol*.

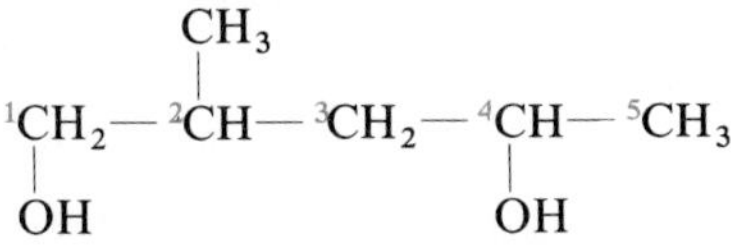

                CH₃
                |
   ¹CH₂—²CH—³CH₂—⁴CH—⁵CH₃
   | |
   OH   OH

c. Three —OH groups are present; this compound is a *-triol*. Three carbon atoms are present in the parent chain, and each carbon bears an —OH group. Numbering the chain in either direction yields the same name: *1,2,3-propanetriol*.

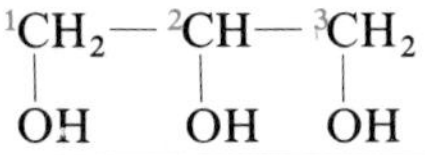

   ¹CH₂—²CH—³CH₂
   | | |
   OH OH OH

The simplest phenol (or aromatic alcohol) is the compound that involves a single —OH group and a benzene ring.

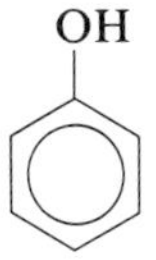

This compound, which can be considered the parent compound for all phenols, is itself called *phenol*. The name *phenol* is an IUPAC-acceptable name. Other phenol compounds can be named as derivatives of phenol, following the naming procedures previously discussed for aromatic compounds (Section 16.3). For example, consider the following compounds.

3-chlorophenol 4-ethyl-2-methylphenol 2,5-dibromophenol
(or meta-chlorophenol)

When determining names for phenol derivatives like the preceding examples, begin numbering the ring at the carbon where the —OH group is found.

We frequently encounter non-IUPAC names (common names) for methyl and hydroxy derivatives of phenol. Methylphenols are called *cresols*. The name *cresol* applies to all three isomeric methylphenols.

ortho-cresol meta-cresol para-cresol

For hydroxyphenols, each of the three isomers has a different common name.

catechol resorcinol hydroquinone

The chemical behavior of phenol and its derivatives is much different than that of alcohols. For this reason, they are named differently than alcohols.

17.3 CLASSIFICATIONS OF ALCOHOLS

In many cases, the characteristic chemistry of an alcohol depends on the number of carbon atoms bonded to the carbon atom that also bears the hydroxyl group. Alcohols can be classified as primary, secondary, or tertiary on this basis.

TABLE 17–3 Primary, Secondary, and Tertiary Alcohols

	Primary	Secondary	Tertiary
General formula	R—CH$_2$—OH	R—CH(R')—OH	R'—C(R)(R'')—OH
Specific example	CH$_3$—CH$_2$—OH	CH$_3$—CH(CH$_3$)—OH	CH$_3$—C(CH$_3$)(CH$_3$)—OH
	ethanol	2-propanol	2-methyl-2-propanol

A **primary alcohol** *is an alcohol in which the hydroxyl-bearing carbon atom is attached to one other carbon atom and two hydrogen atoms.* A **secondary alcohol** *is an alcohol in which the hydroxyl-bearing carbon is attached to two other carbon atoms and one hydrogen atom.* A **tertiary alcohol** *is an alcohol in which the hydroxyl-bearing carbon atom is attached to three other carbon atoms.* Table 17–3 gives specific examples of alcohols in these classifications, as well as the general structural characteristics for each classification. Methanol (CH_3OH), an alcohol in which the hydroxyl-bearing carbon atom is attached to three hydrogen atoms, is grouped with the primary alcohols.

primary alcohol

secondary alcohol

tertiary alcohol

17.4 PHYSICAL PROPERTIES OF ALCOHOLS

As a group, alcohols have higher boiling points than alkanes of similar molecular weight. However, like alkanes (Section 14.11), alcohols' boiling points increase as carbon-chain length increases. Table 17–4 illustrates both of these trends.

The differences in boiling-point behavior between alcohols and alkanes of similar molecular weight are directly related to the ability of alcohols to participate in hydrogen bonding. The hydrogen bonding that can occur between alcohol molecules is similar in nature to what occurs between water molecules (Section 7.16). Figure 17–2 (p. 398) contrasts the hydrogen bonding situations of water and methanol, the simplest alcohol. There is one major difference between water and an alcohol in hydrogen bonding situations: both ends of a water molecule can participate in hydrogen bonding, but only one end of an alcohol molecule can participate in hydrogen bonding.

Hydrogen bonding explains the higher boiling points of alcohols. In order to overcome the hydrogen bonding between alcohol molecules, more heat energy is needed to separate the molecules before they can enter the vapor phase; hence, alcohols in the liquid state must be heated to a higher temperature before they boil.

Low-molecular-weight alcohols are completely soluble in water, but their alkane counterparts have limited solubility in water. Hydrogen bonds that form between alcohol molecules and water molecules (Figure 17–3, p. 399) result in these alcohols having increased solubility in water.

Not all alcohols are soluble in water. As carbon-chain length increases, solubility decreases. There are two parts to an alcohol molecule: a polar part, the —OH group, and a nonpolar part, the alkyl group. The polar part of the alcohol

TABLE 17–4 Boiling Points of Some Alkanes and Alcohols (Alkanes and alcohols paired together have similar molecular weights.)

Name	Structure	Molecular Weight	Boiling Point (°C)
methyl alcohol	CH_3-OH	32	65
ethane	CH_3-CH_3	30	−88
ethyl alcohol	CH_3-CH_2-OH	46	78
propane	$CH_3-CH_2-CH_3$	44	−42
n-propyl alcohol	$CH_3-CH_2-CH_2-OH$	60	97
n-butane	$CH_3-CH_2-CH_2-CH_3$	58	−0.5
n-butyl alcohol	$CH_3-CH_2-CH_2-CH_2-OH$	74	118
n-pentane	$CH_3-CH_2-CH_2-CH_2-CH_3$	72	36

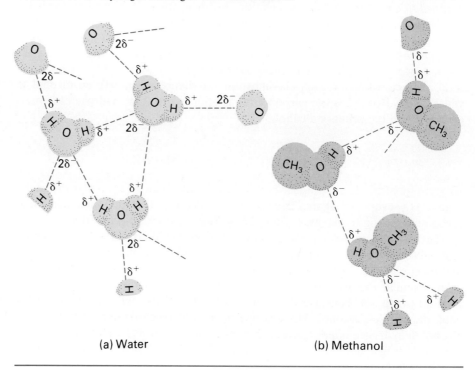

FIGURE 17-2 Hydrogen bonding in water and methanol.

(a) Water (b) Methanol

molecule increases solubility in water, but the nonpolar part decreases solubility. When the size of the nonpolar alkyl group is large, the solubility of the alcohol in water decreases dramatically.

methanol

CH$_3$—OH
nonpolar polar
very soluble in water

decanol

CH$_3$—CH$_2$—CH$_2$—CH$_2$—CH$_2$—CH$_2$—CH$_2$—CH$_2$—CH$_2$—CH$_2$—OH
nonpolar polar
insoluble in water

Table 17-5 gives data that quantitatively illustrates the relationship between carbon-chain length and the boiling point and solubility in water of normal-chain alcohols.

TABLE 17-5 Boiling Points and Solubilities for Normal-Chain Alcohols

Name	Formula	Boiling point (°C)	Solubility (g/100 mL in water)
methanol	CH$_3$—OH	65	miscible
ethanol	CH$_3$—CH$_2$—OH	78	miscible
1-propanol	CH$_3$—CH$_2$—CH$_2$—OH	97	miscible
1-butanol	CH$_3$—(CH$_2$)$_3$—OH	118	7.9
1-pentanol	CH$_3$—(CH$_2$)$_4$—OH	138	2.7
1-hexanol	CH$_3$—(CH$_2$)$_5$—OH	157	0.59
1-heptanol	CH$_3$—(CH$_2$)$_6$—OH	176	0.09
1-octanol	CH$_3$—(CH$_2$)$_7$—OH	195	insoluble
1-nonanol	CH$_3$—(CH$_2$)$_8$—OH	213	insoluble
1-decanol	CH$_3$—(CH$_2$)$_9$—OH	231	insoluble

FIGURE 17–3 Hydrogen bonding between water and methanol molecules in a water–methanol solution.

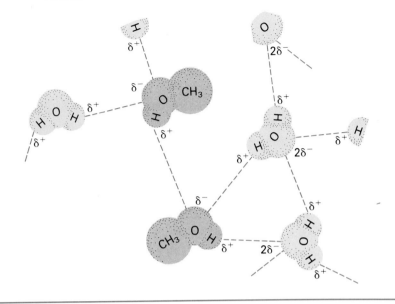

Alcohols containing more than one hydroxyl group per molecule provide more than one site for hydrogen bonding, and their properties reflect this. The simplest diol, 1,2-ethanediol, boils at 197°C. The lower diols are miscible with water, and diols containing as many as seven carbon atoms show appreciable solubility in water.

An excellent example of the relationship between number of hydroxyl groups and solubility involves the substance polyethylene glycol (PEG). In this polyethylene polymer (Section 15.9), every carbon atom in the long carbon backbone bears an —OH group.

$$\left(\begin{array}{cc} CH-CH \\ | \quad | \\ OH \quad OH \end{array}\right)_n \quad \text{polyethylene glycol (PEG)}$$

Aqueous solutions of this polymer are very viscous (thick) because of the great solubility of PEG in water. Polyethylene glycol is used as an additive in many shampoos. It contributes very little to the cleaning ability of the shampoo, but gives the shampoo texture or richness.

17.5 PREPARATION OF ALCOHOLS

One general method for preparing alcohols—the hydration of alkenes—was discussed in a previous chapter (Section 15.7). At that time, we learned that alkenes react with water (an unsymmetrical addition agent) in the presence of sulfuric acid (the catalyst) to form an alcohol. We used Markovnikov's rule to determine the predominant alcohol product of a hydration reaction.

Another method of synthesizing alcohols involves the addition of H_2 to a carbon–oxygen double bond (a carbonyl group, $\rangle C{=}O$). (The carbonyl group is a functional group that will be discussed in detail in Chapter 18.) A carbonyl group behaves very much like a carbon–carbon double bond when it reacts with

H_2 under proper conditions. As the result of H_2 addition, the oxygen of the carbonyl group is converted to an —OH group.

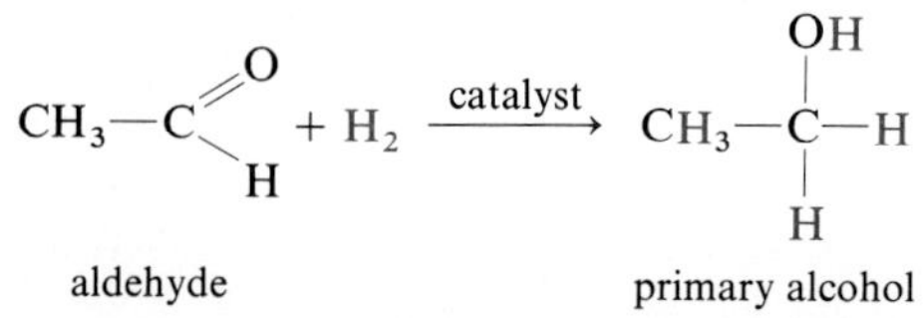

$$CH_3-C\underset{H}{\overset{O}{\diagup}} + H_2 \xrightarrow{catalyst} CH_3-\underset{H}{\overset{OH}{\underset{|}{C}}}-H$$

aldehyde   primary alcohol

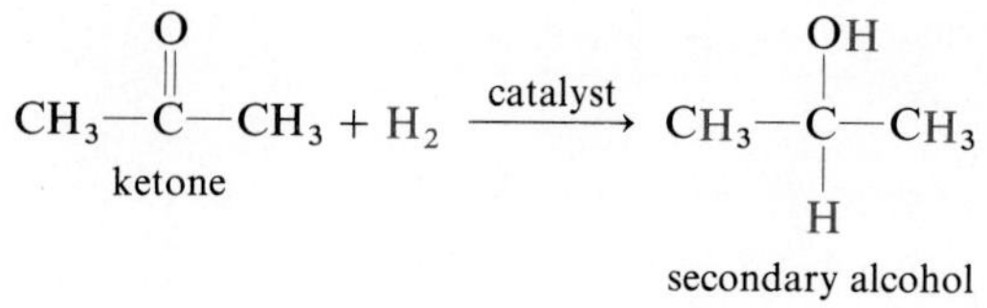

$$CH_3-\overset{O}{\overset{\|}{C}}-CH_3 + H_2 \xrightarrow{catalyst} CH_3-\underset{H}{\overset{OH}{\underset{|}{C}}}-CH_3$$

ketone   secondary alcohol

The hydration of an alkene to produce an alcohol is an oxidation process (Section 15.8). On the other hand, adding H_2 to a carbonyl group to produce an alcohol is a reduction process. Both of these processes are important biological methods of producing alcohols in living cells. Alcohol production plays an important role in the metabolism of both fats and carbohydrates.

17.6 REACTIONS OF ALCOHOLS

Alcohols are used to fuel some types of race cars, supply heat in "canned heat", and add a special touch to some gourmet recipes, such as flaming shish kebobs. Alcohols are used in these ways because they are very flammable substances that readily undergo combustion. The predominant products of alcohol combustion are carbon dioxide and water, as was the case with the combustion of hydrocarbons. The following reaction shows a typical combustion of an alcohol.

$$CH_3-CH_2-OH + 3\,O_2 \longrightarrow 2\,CO_2 + 3\,H_2O$$
ethanol

Additional important reactions of alcohols are dehydration, controlled oxidation and substitution. (Combustion is an uncontrolled oxidation process.)

Dehydration

Alcohol dehydration is the process in which a molecule of water is eliminated from an alcohol. Dehydration requires the presence of an acid (usually sulfuric acid) and the application of heat. The various classes of alcohols differ widely in their ease of dehydration. In general, tertiary alcohols are the easiest to dehydrate and primary alcohols are the most difficult.

The products that form as a result of alcohol dehydration depend on the temperature at which the dehydration is carried out. At high temperatures (180°C), the product is an alkene. At lower temperatures (140°C), the product is an ether. Let us look at these types of dehydration reactions in more detail.

High-temperature dehydration (180°C) of an alcohol can be characterized as an *intramolecular* dehydration process. The hydroxyl group and a hydrogen atom on a neighboring carbon are both lost (to form water) and an alkene is produced. In general terms, the equation for this process can be written as

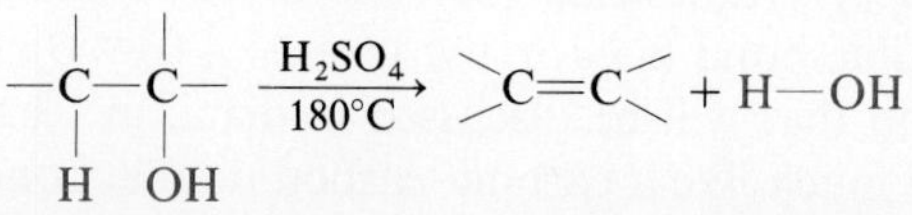

$$-\underset{H}{\overset{|}{C}}-\underset{OH}{\overset{|}{C}}- \xrightarrow[180°C]{H_2SO_4} \;\;>\!\!C\!\!=\!\!C\!\!<\; + \;H-OH$$

A specific example of intramolecular alcohol dehydration involves *n*-propanol, a primary alcohol.

$$CH_3-\underset{H}{\underset{|}{CH}}-\underset{OH}{\underset{|}{CH_2}} \xrightarrow[180°C]{H_2SO_4} CH_3-CH=CH_2 + H_2O$$

High-temperature dehydration of more complex alcohols can result in the production of more than one alkene product. This happens when there is more than one neighboring carbon atom from which hydrogen loss can occur. High-temperature dehydration of 2-butanol produces two alkenes.

$$\underset{\substack{H \\ \text{2-butanol}}}{\underset{|}{CH_2}}-\underset{OH}{\underset{|}{CH}}-\underset{H}{\underset{|}{CH}}-CH_3 \xrightarrow[180°C]{H_2SO_4} \underset{\substack{H \\ \text{2-butene}}}{\underset{|}{CH_2}}-CH=CH-CH_3 + \underset{\substack{H \\ \text{1-butene}}}{CH_2=CH-\underset{|}{CH}-CH_3}$$

In a reaction like this, the production of one of the alkenes is favored over the other one. The major product is the alkene that has the greatest number of alkyl groups attached to the carbons of the double bond. In our example, 2-butene (with 2 alkyl groups) is favored over 1-butene (with 1 alkyl group).

$$\underset{\text{2-butene}}{CH_3-CH=CH-CH_3} \quad \underset{\text{1-butene}}{CH_2=CH-CH_2-CH_3}$$

Note that high-temperature alcohol dehydration is the reverse of the hydration reaction for alkenes discussed in Section 15.7.

Low-temperature alcohol dehydration (140°C) can be characterized as an *intermolecular* process, rather than an intramolecular process. In this process, the —OH group from one alcohol and the hydrogen from the hydroxyl group of another alcohol are lost as water. The resulting "leftover" portions of the two alcohol molecules join to form an ether. The reaction for the low-temperature dehydration of ethanol illustrates this type of reaction.

$$\underset{\text{ethanol}}{CH_3-CH_2-OH} + \underset{\text{ethanol}}{H-O-CH_2-CH_3} \xrightarrow[140°C]{H_2SO_4} CH_3-CH_2-O-CH_2-CH_3 + H_2O$$

Note that if this reaction had been run at 180°C instead of 140°C, the product would have been ethene instead of an ether. As a general principle, temperature can be used in many organic reactions to alter the types of products formed.

Oxidation

Alcohols very readily undergo oxidation. In the human body, enzymes catalyze many different oxidation processes involving alcohols. In the laboratory, potassium permanganate ($KMnO_4$), potassium dichromate ($K_2Cr_2O_7$), and chromic acid (H_2CrO_4) are commonly used as oxidizing agents.

Several kinds of products can result from alcohol oxidation; the three most common are aldehydes, ketones, and carboxylic acids. All of these compounds contain carbon–oxygen double bonds.

Aldehyde production, which is the first step of primary alcohol oxidation, involves dehydrogenation—the removal of the elements of H_2 (H—H) from the alcohol. As shown in the following general reaction, one H atom is removed from the —OH group and the other H atom is removed from the C atom to which the —OH group is attached.

$$\text{one of these two hydrogens is removed} \quad R-\underset{\underset{H}{|}}{\overset{\overset{H}{|}}{C}}-OH \xrightarrow{\text{oxidizing agent}} R-C\overset{\nearrow O}{\underset{\searrow H}{}} + H_2O$$

this hydrogen is removed — an aldehyde

In addition to an aldehyde, water is also a product of this dehydrogenation. The two "extracted" H atoms combine with oxygen supplied by the oxidizing agent to give H_2O. The following example shows aldehyde production from a primary alcohol, 1-propanol.

$$CH_3-CH_2-\underset{\underset{OH}{|}}{CH_2} \xrightarrow{\text{oxidizing agent}} CH_3-CH_2-C\overset{\nearrow O}{\underset{\searrow H}{}} + H_2O$$

1-propanol an aldehyde

Aldehydes are very susceptible to further oxidation and can be isolated from oxidation reactions only when conditions are carefully controlled. When further oxidation occurs, a carboxylic acid is produced. The general reaction for the aldehyde–carboxylic acid transition is

$$R-C\overset{\nearrow O}{\underset{\searrow H}{}} \xrightarrow{\text{oxidizing agent}} R-C\overset{\nearrow O}{\underset{\searrow OH}{}}$$

an aldehyde a carboxylic acid

This second step of oxidation involves the gain of oxygen rather than the loss of hydrogen.

Laboratory oxidations of a primary alcohol, with $KMnO_4$ or $K_2Cr_2O_7$ as oxidizing agents, give an aldehyde as an intermediate product (which seldom can be isolated) and a carboxylic acid as the final product. The following example illustrates this two-step oxidation process, again using 1-propanol as the reactant.

$$CH_3-CH_2-\underset{\underset{OH}{|}}{CH_2} \xrightarrow{\text{loss of H atoms}} \left[CH_3-CH_2-C\overset{\nearrow O}{\underset{\searrow H}{}} \right] \xrightarrow{\text{gain of O atom}} CH_3-CH_2-C\overset{\nearrow O}{\underset{\searrow OH}{}}$$

primary alcohol an aldehyde a carboxylic acid

In the human body, where enzymes mediate the oxidation process, aldehydes are often the final product of primary alcohol oxidation.

In contrast to primary alcohols, secondary alcohols give ketones as products of oxidation. These compounds, which will be discussed in Chapter 18, are the final organic product because they do not easily undergo further oxidation. Secondary alcohol oxidation is a dehydrogenation reaction, as was the first step in primary alcohol oxidation. During the oxidation process the H of the —OH group and the H atom attached to the C bearing the —OH group are removed by the oxidizing agent. In general terms, we can represent the process as follows.

$$\text{both of these hydrogen atoms are removed} \quad R-\underset{\underset{H}{|}}{\overset{\overset{OH}{|}}{C}}-R \xrightarrow{\text{oxidizing agent}} R-\overset{\overset{O}{\|}}{C}-R + H_2O \quad \text{oxygen atom supplied by solvent or oxidizing agent}$$

secondary alcohol a ketone

A specific example of secondary alcohol oxidation involves 2-propanol.

FIGURE 17-4 A breath analyzer used in police departments to test drunken drivers.

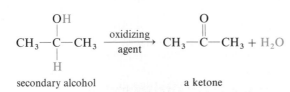

secondary alcohol → a ketone

Tertiary alcohols are not oxidized by the previously listed, common laboratory oxidizing agents. In a tertiary alcohol, the carbon atom bearing the —OH group has no hydrogen atoms attached to it; hence, no hydrogen can be removed from this carbon by an oxidizing agent.

$$R-\underset{\underset{R}{|}}{\overset{\overset{OH}{|}}{C}}-R \xrightarrow{\text{oxidizing agent}} \text{No reaction}$$

tertiary alcohol

Simple phenols, which have no hydrogen atoms on their hydroxy carbon, also cannot be oxidized in this manner.

The dichromate ion ($Cr_2O_7^{2-}$) exhibits a yellow-orange color in solution. As oxidation of the alcohol occurs, the dichromate ions are converted into Cr^{3+} ions, which are green in solution. This color change can be used to determine the amount of alcohol present in a person's breath when the person blows bubbles through a solution of potassium dichromate. Some breath analyzers that are used to test drunk driving suspects operate on this principle (see Figure 17–4).

Substitution Reactions

Alcohols also undergo substitution reactions. For example, the hydroxyl group of an alcohol can be replaced by a halogen atom. Phosphorus halides are commonly used as the source of halogen atoms in these processes. The following equations illustrate the use of PCl_3 as a halogenation agent.

$$CH_3-CH_2-\underset{OH}{CH_2} \xrightarrow{PCl_3} CH_3-CH_2-\underset{Cl}{CH_2}$$

$$CH_3-\underset{OH}{CH}-CH_3 \xrightarrow{PCl_3} CH_3-\underset{Cl}{CH}-CH_3$$

This method of making alkyl halides is more convenient than halogenation of an alkane (Section 14.12) because mixtures of products are not obtained. In the product, the halogen is found only where the —OH groups are originally located in the reactant.

17.7 IMPORTANT COMMONLY USED ALCOHOLS

In this section, we will focus our attention on five commonly used alcohols: methanol, ethanol, 2-propanol, 1,2-ethanediol, and 1,2,3-propanetriol. Structural formulas and common names, which are used more often than the IUPAC names, for these alcohols are given in Table 17–6.

Methanol

Methanol is the simplest alcohol and has one C atom and one OH group. It is a colorless liquid, with a characteristic odor similar to "duplicating fluid" (methanol is the primary component of duplicating fluid).

Methanol is sometimes called *wood alcohol* because it was originally obtained by heating wood to a high temperature in the absence of oxygen. Methanol is one of the compounds produced as the wood decomposes. Today, methanol is synthetically produced by reacting carbon monoxide with hydrogen gas.

$$CO + 2H_2 \xrightarrow[\text{catalysts}]{\text{heat, pressure}} CH_3OH$$

Methanol is extremely toxic. Once inside the body, methanol is quickly oxidized by the liver enzyme alcohol dehydrogenase to make formaldehyde.

$$\underset{\text{methanol}}{CH_3OH} \xrightarrow[\text{(oxidation)}]{\text{alcohol dehydrogenase}} \underset{\text{formaldehyde}}{H-\overset{\overset{\displaystyle O}{\|}}{C}-H}$$

TABLE 17–6 Structures and Common Names for Commonly Used Alcohols

IUPAC name	Structural Formula	Common Names
methanol	CH$_3$—OH	methyl alcohol, wood alcohol
ethanol	CH$_3$—CH$_2$—OH	ethyl alcohol, grain alcohol, drinking alcohol
2-propanol	CH$_3$—CH(OH)—CH$_3$	isopropyl alcohol, rubbing alcohol
1,2-ethanediol	CH$_2$(OH)—CH$_2$(OH)	ethylene glycol
1,2,3-propanetriol	CH$_2$(OH)—CH(OH)—CH$_2$(OH)	glycerol, glycerin

(The oxidation of a primary alcohol to an aldehyde was one of the reactions discussed in Section 17.6).

Formaldehyde is even more toxic than methanol. Within the body, formaldehyde disrupts protein function. Temporary blindness, permanent blindness, or death can result from the ingestion of only small amounts of methanol. As little as 1 oz (30 mL) of pure methanol can cause death. Prolonged breathing of methanol vapors is also a serious health hazard. Proteins within the eye are some of the first proteins affected by methanol (formaldehyde). In the lens of the eye, proteins are precipitated, which causes a clouding of the lens; with time, blindness can occur.

Ethanol

Ethanol is the compound that is commonly referred to as "alcohol" by the layperson. In addition to being the active ingredient in alcoholic beverages, it is also used in the pharmaceutical industry as a solvent (tinctures are ethanol solutions), a medicinal ingredient (cough syrups often contain as much as 20% by volume ethanol), and an industrial solvent. A 70% by volume ethanol solution is an excellent antiseptic and is used extensively for skin disinfection.

Like methanol, ethanol is toxic when taken internally. If rapidly ingested, as little as one pint of pure ethanol will kill most people. Nevertheless, many people still consume large quantities of alcoholic beverages. Because even the strongest alcoholic beverages seldom contain more than 45% ethanol and ethanol is less toxic than other simple alcohols, it is considered by some to be "safe" for human consumption. Longterm excessive use, however, is known to cause undesirable effects such as cirrhosis of the liver or loss of memory and can lead to strong physiological addiction. In recent years, links have been established between certain birth defects and the ingestion of ethanol by mothers during pregnancy.

Despite a knowledge of the harmful effects of ethanol, approximately two-thirds of the adult population of the United States drink alcoholic beverages occasionally. Why? At least a partial answer is the effects that ethanol initially has on the human body. Ethanol is a general central nervous system (CNS) depressant drug that slows down both physical and mental activity. For most people, CNS depression produces a euphoric state with an accompanying removal of inhibitions. Because of these effects, many people believe ethanol is a stimulant, which it is not.

The effects of ethanol on the human body that accompany alcoholic beverage consumption are summarized in Table 17-7. The numbers in Table 17-7 can be related to the problem of drunk driving by considering what constitutes a "drunk

TABLE 17-7 Approximate Relationship Between Drinks Consumed, Blood-Alcohol Level, and Behavior for a Moderate Drinker Weighing 150 lb

Number of Drinks*	Blood-Alcohol Level percent (m/v)**	Behavior
2	0.04	mild sedation; tranquility
4	0.08	lack of coordination
6	0.12	obvious intoxication
10	0.24	unconsciousness
20	0.40	possible death

* Rapidly consumed 1-oz shots of 90-proof whiskey or 12 oz of beer
** The numbers in this table are only average values and actual effects depend on many factors such as a person's weight, drinking history, and the amount of food in his or her stomach.

driver" in terms of blood-alcohol level. Drivers are considered to be DUI (driving under influence) in most states (46 of the 50) when the blood-alcohol level reaches 0.10%(m/v). Four states have more stringent guidelines—0.08%(m/v) in three states and 0.05%(m/v) in one state. A blood-alcohol level of 0.10%(m/v) is approximately equivalent to 1 drop of alcohol in 1000 drops of blood.

Once in the body, ethanol is broken down by the liver. The same liver enzymes that act on methanol act on ethanol. The product from ethanol oxidation is the two-carbon aldehyde, acetaldehyde.

$$CH_3-CH_2-OH \xrightarrow[\text{(oxidation)}]{\text{alcohol dehydrogenase}} CH_3-C\begin{subarray}{l}\diagup O \\ \diagdown H\end{subarray}$$

ethanol → acetaldehyde

Acetaldehyde, in turn, is oxidized to the two-carbon carboxylic acid *acetic acid*, which is a normal constituent of cells.

$$CH_3-C\begin{subarray}{l}\diagup O \\ \diagdown H\end{subarray} \xrightarrow{\text{oxidation}} CH_3-C\begin{subarray}{l}\diagup O \\ \diagdown OH\end{subarray}$$

acetaldehyde → acetic acid

(Recall from Section 17.6 that primary alcohol oxidation produces first an aldehyde and then a carboxylic acid.)

The intraveneous administration of ethanol has long been used in the treatment of methanol poisoning. The alcohol dehydrogenase enzymes in the liver prefer to oxidize ethanol. While they are tied up oxidizing ethanol to acetaldehyde, these enzymes cannot convert methanol to the more dangerous formaldehyde. The unoxidized methanol is later excreted from the body in the urine.

Drugs such as disulfiram (Antabuse) that are used to treat alcoholics prevent oxidation of acetaldehyde to acetic acid (the second step of primary alcohol oxidation). Buildup of acetaldehyde in the body causes dizziness, headaches, nausea, vomiting, and difficulty in breathing.

Ethanol was first produced by yeast fermentation of sugars found in plant extracts. *Fermentation* is an anaerobic process (no oxygen) in which sugars are converted to alcohol by a series of enzymatic reactions. The synthesis of ethanol from the fermentation of grains such as corn, rice, or barley led to its common name, *grain alcohol*. All alcohol destined for human consumption is still produced in this manner (see Figure 17–5). The maximum concentration of alcohol in fermentation processes is about 18% because enzymes and yeast cannot function above this level. Beverages with a higher concentration of alcohol than this are prepared either by distillation or fortification with alcohol that has been obtained by the distillation of another fermentation product. Table 17–8 lists the alcohol content of common alcoholic beverages.

The alcohol content of alcoholic beverages is sometimes stated using a system based on the word *proof*. The proof of a beverage is twice its alcohol percentage. An 86-proof brandy contains 43% alcohol. Pure ethanol (100% ethanol) is 200 proof. The origin of the proof system for concentration dates back to the seventeenth century. At that time, dealers of alcoholic beverages would pour some of the beverage over a small amount of gunpowder to prove to skeptical customers that their products had not been watered down. The mixture had to ignite as "proof" that the beverage was strong. When the concentration of alcohol is 50%(v/v), gunpowder ignites; this is why 100 proof is actually 50% ethanol.

FIGURE 17-5 The frothing seen in this fermentation vat is caused by the evolution of the carbon dioxide produced as ethanol forms.

Ethanol destined for human consumption in alcoholic beverages is heavily taxed by the United States government. However, ethanol that is used as a solvent in industry is exempt from taxes. Industrial ethanol is rendered unfit for human consumption by the addition of small amounts of toxic substances. Ethanol treated in this way is called *denatured alcohol*. Some denaturants that do not affect the solvent properties of the ethanol include methanol, benzene, and formaldehyde.

Absolute alcohol is ethanol from which all traces of water have been removed. Industrially produced ethanol, as well as ethanol from fermentation (after distillation), contains 5% by volume water.

Isopropyl Alcohol (2-propanol)

Isopropyl alcohol (2-propanol) is commonly known as *rubbing alcohol*. Drug store rubbing alcohol is a 70% by volume solution of isopropyl alcohol. Its rapid evaporation rate creates a dramatic cooling effect when it is applied to the skin; hence, alcohol rubs are often used to combat high body temperature. A 70% by volume isopropyl alcohol solution is as effective as ethanol in disinfection procedures (see Figure 17-6, p. 408).

TABLE 17-8 Alcohol Content of Common Alcoholic Beverages

Name	Source of Beverage	Amount of Ethyl Alcohol
beer	barley, wheat	3.2–9%
wine	grapes or other fruit	12% maximum, unless fortified
brandy	distilled wine	40–45%
whiskey	barley, rye, corn, etc.	45–55%
rum	molasses	~45%

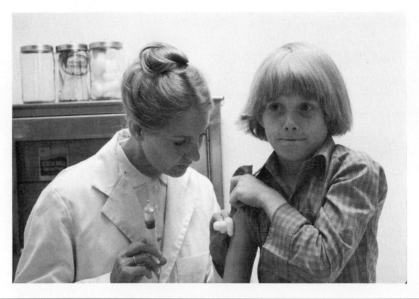

FIGURE 17-6 A 70% by volume aqueous solution of isopropyl alcohol is used as a disinfectant for the skin.

Isopropyl alcohol is for external use only. Ingestion induces vomiting, which eliminates the alcohol from the system. Other (longer carbon chain) alcohols also induce vomiting when taken internally.

Isopropyl alcohol has extensive use in industry as a solvent for products such as ink, cosmetics, paints, and some formulations of antifreeze. The synthetic source of this alcohol is the hydration of propene, and the synthesis is very similar to that of ethanol.

$$CH_2{=}CH{-}CH_3 + H_2O \xrightarrow[\text{catalyst}]{H_2SO_4} CH_3{-}\underset{\underset{\text{(2-propanol)}}{\text{isopropyl alcohol}}}{\overset{\overset{OH}{|}}{CH}}{-}CH_3$$

Note that 2-propanol, and not 1-propanol, is produced. This is in accordance with Markovnikov's rule (Section 15.7).

Ethylene Glycol (1,2-ethanediol)

Ethylene glycol (1,2-ethanediol) is the simplest alcohol containing two —OH groups. This compound is the main ingredient in most brands of permanent antifreeze for automobile radiators. It is nonvolatile (boiling point = 197°C), completely miscible with water, noncorrosive, and relatively inexpensive to produce. All of these characteristics are ideal properties for an antifreeze.

Ethylene glycol is extremely toxic when ingested. In the body, liver enzymes oxidize it to oxalic acid. (An aldehyde is produced as an intermediate.)

$$\underset{\text{ethylene glycol}}{HO{-}CH_2{-}CH_2{-}OH} \xrightarrow[\text{enzymes}]{\text{liver}} \underset{\text{oxalic acid}}{HO{-}\overset{\overset{O}{\|}}{C}{-}\overset{\overset{O}{\|}}{C}{-}OH}$$

Oxalic acid, as a calcium salt, crystalizes in the kidneys, which leads to renal problems. As with methanol poisoning, ethanol is used as an antidote. The liver enzymes oxidize ethanol, thus preventing the conversion of ethylene glycol to oxalic acid. Each year many children and even more household pets accidentally ingest ethylene glycol because partially used open antifreeze containers have been left in inappropriate places.

Ethylene glycol is also an important chemical raw material. Both Dacron and Mylar film are made from it (Section 19.12). Dacron is an important polyester fiber used for clothing and Mylar film is used in tapes for recorders and computers. An important industrial synthesis for ethylene glycol involves ethylene oxide (a cyclic ether), which in turn can be made from ethene.

$$CH_2{=}CH_2 + O_2 \xrightarrow[\text{heat}]{\text{Ag catalyst}} \underset{\text{ethylene oxide}}{\overset{O}{\underset{|\quad\;|}{CH_2{-}CH_2}}} \xrightarrow[\text{acid}]{H_2O} \underset{\text{ethylene glycol}}{HO{-}CH_2{-}CH_2{-}OH}$$

Glycerol (1,2,3-propanetriol)

Glycerol (glycerin or 1,2,3-propanetriol) is a clear, thick liquid that has the consistency of honey. It is a sweet-tasting liquid and is not toxic. It fact, as we shall see in Section 23.3, it plays an important role in the structure of fat molecules. Although glycerol is commercially synthesized from propene, it is also obtained as a by-product of the processing of animal and vegetable fats to make soap and other products. Glycerol has a great affinity for water vapor (moisture). Because of this, it is often added to pharmaceutical preparations such as skin lotions and soap. Florists sometimes use glycerol on cut flowers to help retain water and maintain freshness. Its lubricative properties also make it useful in shaving creams and applications such as glycerin suppositories for rectal administration of medicines.

17.8 PROPERTIES AND USES OF PHENOLS

Phenolic compounds are generally low-melting solids or oily liquids at room temperature. Most of them are only slightly soluble in water. Many of these compounds have antiseptic and disinfectant properties. (**Antiseptics** *are substances that kill microorganisms on living tissue* and **disinfectants** *are substances that kill microorganisms on inanimate objects.*)

antiseptics
disinfectants

The simplest phenolic compound is phenol, which is a colorless, low-melting solid (41°C) that exhibits slightly acidic properties in solution. The phenolic proton of phenol (the H atom of the OH group) can be removed in solution; however, it is difficult to do this because phenol is an extremely weak acid.

$$C_6H_5OH + H_2O \rightleftharpoons C_6H_5O^- + H_3O^+$$

Another name used for phenol, which draws attention to its very weak acidic properties, is *carbolic acid*. Other phenolic compounds are also weakly acidic. This contrasts with alcohols, which, as a group, do not exhibit acidic properties.

Dilute solutions of phenol (2%) have long been used as antiseptics. Joseph Lister, an English surgeon, introduced the use of phenol as a hospital antiseptic in the late 1800s. Before that time, antiseptics had not been used, and the specter of fatal infection confronted most surgical patients. Concentrated phenol solutions or pure phenol can cause severe skin burns. Today, phenol has largely been replaced by more effective antiseptics that are less irritating to the skin. Two such compounds are hexachlorophene and hexylresorcinol.

phenol
(Lister's original disinfectant)

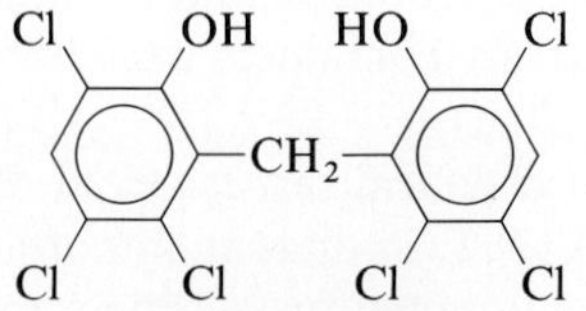

hexachlorophene
(used in hospitals as a disinfectant)

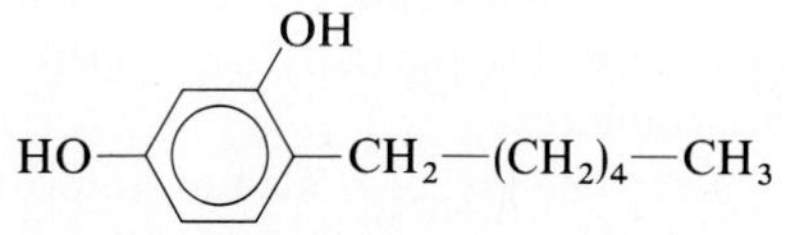

hexylresorcinol
(used in mouthwashes and throat lozenges)

17.9 PHENOLS IN NATURE

Many phenols and phenolic derivatives occur in nature, but we will mention only a few of them. Thymol, the flavoring constituent of thyme, has a pleasant flavor that makes it a favorite antiseptic ingredient in mouthwashes. Eugenol is responsible for the flavor of cloves. Dentists traditionally used clove oil as an antiseptic because it contains this phenol; to a limited extent they still do so today. Safrole, a cyclic compound closely related to eugenol, is a principal constituent of sassafras oil and has a strong odor characteristic of root beer. The structures of these three phenols are:

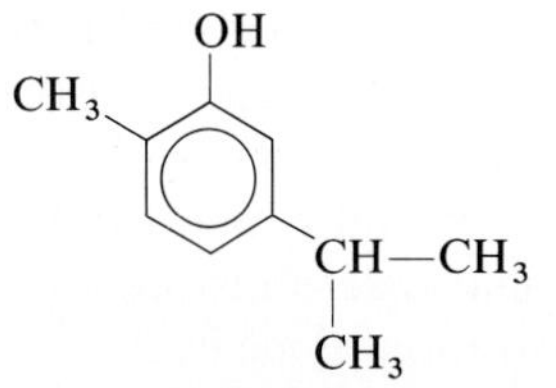

thymol

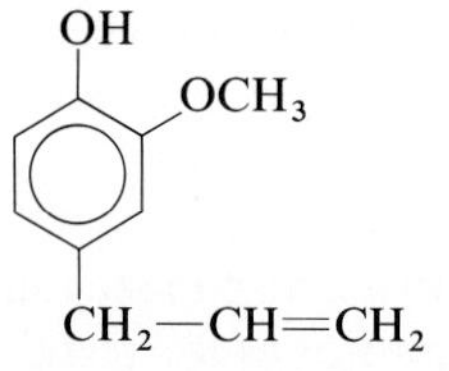

eugenol

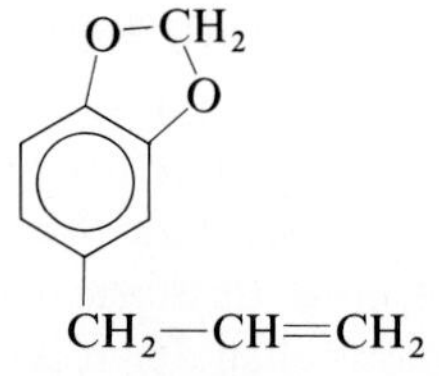

safrole

Certain phenols exert profound physiological effects. For example, the irritating constituents of poison ivy and poison oak are derivatives of catechol (ortho-hydroxyphenol). These skin irritants have a long alkyl chain, which has a variable degree of unsaturation, bonded to the number 3 position of the ring.

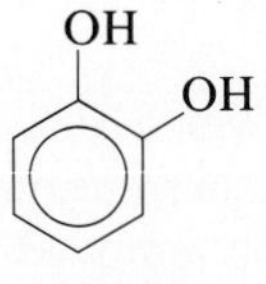

catechol

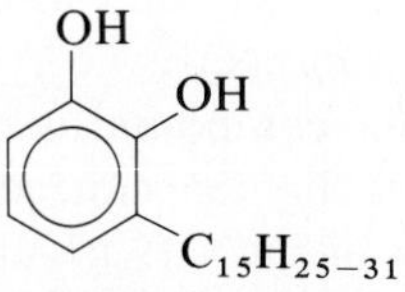

poison ivy irritants

The major psychological ingredient in marijuana is a complex phenol known as *te*tra*hy*dro*c*annabinol, which is abbreviated THC.

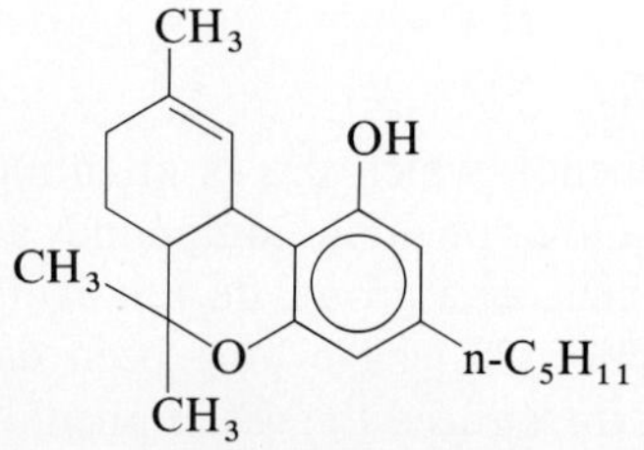

tetrahydrocannabinol

The term *marijuana* refers to a preparation made by gathering the leaves, flowers, seeds, and small stems of a hemp plant called *Cannabis sativa* (see Figure 17–7). These are generally dried and smoked. The THC content of marijuana varies considerably, depending on the genetic variety of plant. Most THC sold in the illegal drug market of North America has a THC content of about 1%.

One of the most popular indicators used for acid–base titrations (Section 10.9) is phenolphthalein, which is colorless in acidic solutions and red in basic solutions. This phenolic compound is also an excellent cathartic and is the active ingredient in some commercially available laxatives such as Ex-Lax.

phenolphthalein

FIGURE 17–7 The active ingredient in marijuana is the complex phenol tetrahydrocannabinol (THC).

Two "phenol-type" compounds associated with the human body are epinephrine and hydroquinone. Epinephrine (adrenalin) is a hormone secreted by the adrenal glands in response to emotional or physiological stress. It causes carbohydrate reserves to be quickly converted into glucose (blood sugar), thus providing the body with a quick burst of energy. Hydroquinone (parahydroxyphenol—Section 17.2) is an important compound in the process of respiration, where it serves as an electron-carrier. Under proper conditions, it can be reversibly oxidized and reduced (see Section 30.3).

adrenalin
(epinephrine)

hydroquinone
(*para*-hydroxyphenol)

Certain phenols are oxidized by enzymes in fruits and vegetables when they are exposed to air; this results in brown-colored products and causes the darkening of apple or potato slices. Chemicals that inhibit these enzymes prevent oxidation and are sold as fruit fresheners.

17.10 CLASSIFICATION AND NOMENCLATURE OF ETHERS

The general formula for an ether is R—O—R′, where the R groups can be any of a wide variety of alkyl or aromatic groups. A **simple** (or symmetrical) **ether** *is an ether in which the two R groups present are the same.* A **mixed** (or unsymmetrical) **ether** *is an ether in which the two R groups present are not the same.*

simple ether

mixed ether

Common names for open-chain ethers are easily derived by writing the names of the two hydrocarbon groups attached to oxygen, followed by the word *ether*. If the ether is symmetrical, the prefix *di-* is used as a prefix to the name for both substituents. Note that there is a space between words in the common names of ethers.

$$CH_3-O-CH_3 \qquad CH_3-O-CH_2-CH_3$$
dimethyl ether ethyl methyl ether diphenyl ether

IUPAC nomenclature for ethers allows more complex ethers to be named. The smaller R group and the oxygen atom are called an *alkoxy* group, as in methoxy ($-OCH_3$) or ethoxy ($-O-CH_2-CH_3$); this alkoxy group is regarded as a substituent attached to the larger R group.

$$CH_3-O-CH_2-CH_2-CH_3$$
methoxypropane

$$CH_3-CH-CH_2-CH_3$$
$$\quad\;\;|$$
$$\quad\;\;O-CH_2-CH_3$$
2-ethoxybutane

ethoxybenzene

phenoxybenzene

17.11 PROPERTIES AND USES OF ETHERS

Ethers generally have much lower boiling points than alcohols of the same molecular weight. Consider the difference between the boiling points of diethyl ether and *n*-butyl alcohol. Even though these two substances have the same molecular formula ($C_4H_{10}O$) and the same molecular weight, diethyl ether boils at 35°C and *n*-butyl alcohol boils at 117°C.

What is the reason for a difference of almost 100°C in the boiling points of these two isomers? The oxygen atom in an ether has no hydrogen atoms attached directly to it. Therefore, ether molecules cannot hydrogen bond to each other. This lack of hydrogen bonding causes ethers to have very low boiling points when compared to other substances where hydrogen bonding is occurring between molecules.

However, ethers are able to form hydrogen bonds with compounds such as water and they have solubilities in water that are similar to those of alcohols of the same molecular weight. For example, diethyl ether and *n*-butyl alcohol have the same solubility in water; it is approximately 8.0 g/100 mL at room temperature. The following is a general schematic for an ether–water hydrogen bond.

$$R-O\;\text{---}\;H$$
$$\;\;\;|\qquad\quad|$$
$$\;\;R'\qquad O$$
$$\qquad\quad\;|$$
$$\qquad\quad H$$

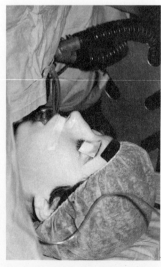

FIGURE 17–8 The use of ether general anesthetics during surgery has revolutionized medicine.

Diethyl ether is a popular laboratory solvent. Commonly called *ether*, diethyl ether has only limited solubility in water. When water and diethyl ether are poured together, two layers form. The ether floats on top of the water layer because it is less dense. Slightly polar organic compounds dissolved in water will transfer from the water layer into the ether layer. This transfer process, which is called *extraction*, can be used to selectively remove compounds from water. Extraction

TABLE 17-9 Some Common Inhalation Anesthetics That Are Ethers

Structure	Chemical name	Trade name
$CH_3-O-CH_2-CH_2-CH_3$	methyl propyl ether	Neothyl
$CH_2=CH-O-CH=CH_2$	divinyl ether	Vinethene
$\begin{array}{c}\text{Cl F} \quad\;\; \text{F}\\ \;\mid\;\; \mid \quad\;\; \mid\\ \text{H}-\text{C}-\text{C}-\text{O}-\text{C}-\text{H}\\ \;\mid\;\; \mid \quad\;\; \mid\\ \text{F F} \quad\;\; \text{F}\end{array}$	2-chloro-1,1,2-trifluoro-ethyldifluoromethyl ether	Enflurane or Ethrane
$\begin{array}{c}\text{Cl F} \quad\;\; \text{H}\\ \;\mid\;\; \mid \quad\;\; \mid\\ \text{H}-\text{C}-\text{C}-\text{O}-\text{C}-\text{H}\\ \;\mid\;\; \mid \quad\;\; \mid\\ \text{Cl F} \quad\;\; \text{H}\end{array}$	2,2-dichloro-1,1-difluoroethyl methyl ether	Methoxyflurane or Penthrane
$\begin{array}{c}\text{F H} \quad\;\; \text{H H}\\ \;\mid\;\; \mid \quad\;\; \mid\;\; \mid\\ \text{F}-\text{C}-\text{C}-\text{O}-\text{C}=\text{C}\\ \;\mid\;\; \mid \quad\quad\;\; \mid\\ \text{F H} \qquad\;\; \text{H}\end{array}$	2,2,2-trifluoroethyl vinyl ether	Fluoxene or Fluoromar

is a powerful tool used to isolate products of chemical reactions or naturally occurring substances from plants and other sources. After the ether is separated from the water, the ether is boiled away, leaving the extracted compound.

Ethers are generally unreactive. However, like alkanes, they burn readily when ignited. Diethyl ether is especially dangerous because of its high volatility. Special care must be used in laboratories so that ether flash fires do not occur.

Ethers are extensively used as general anesthetics. A **general anesthetic** *is a compound or mixture of compounds that acts on the brain to produce both unconsciousness and insensitivity to pain.* The availability of general anesthetics has revolutionized surgery and greatly reduced the risks of shock during and after surgery (see Figure 17–8). Table 17–9 lists the structures and names of some of the ethers now used as inhalation anesthetics.

general anesthetic

The mechanism by which a general anesthetic exerts its effects is related to its solubility in fat (a relatively nonpolar tissue). The anesthetic molecules dissolve in the fat-like membranes of nerve cells. This affects the permeability of the membranes to other substances and depresses the conductivity of the nerve cells.

The mixed ether, methyl *tert*-butyl ether (MTBE), can be used to dissolve gallstones. Gallstones are composed predominantly of cholesterol, which is highly soluble in MTBE. The MTBE is infused into the gallbladder using a thin catheter, the gallstones dissolve, and surgery is avoided.

$$\begin{array}{c}\text{CH}_3\\ \mid\\ \text{CH}_3-\text{O}-\text{C}-\text{CH}_3\\ \mid\\ \text{CH}_3\end{array}$$
methyl *tert*-butyl ether
(MTBE)

17.12 EPOXIDES

Epoxides *are cyclic ethers that contain a three-membered ring system involving one oxygen atom and two carbon atoms.* They are very reactive substances because of the strain present in the small three-membered ring.

epoxides

Ethylene oxide, the simplest epoxide, is produced by oxidizing ethylene under special conditions.

$$CH_2\!=\!CH_2 \xrightarrow[\text{catalyst, heat, pressure}]{\text{oxidation}} \underset{\text{ethylene oxide}}{CH_2\!\!-\!\!\overset{O}{\frown}\!\!-\!\!CH_2}$$

Ethylene oxide is highly reactive and is sometimes used for sterilization of medical equipment. In a pressurized container, ethylene oxide gas forces its way into tiny spaces and kills microorganisms by reacting in a variety of ways with biological compounds in the cells.

17.13 THIOLS AND DISULFIDES

Many inorganic and organic compounds containing oxygen have sulfur analogs, in which a sulfur atom has replaced an oxygen atom. This replacement should not be surprising because sulfur is in the same group in the periodic table as oxygen and thus, the two elements have similar electronic configurations. Rotten-egg gas is a prime example of such an analog. Hydrogen sulfide (H_2S), is the sulfur analog of water (H_2O). Hydrogen sulfide received its better-known name of "rotten-egg gas" because old, rotten eggs form this gas as a decomposition product of their sulfur-containing proteins. Hydrogen sulfide is a gas because sulfur is not as electronegative as oxygen and thus, intermolecular hydrogen bonding is reduced.

Sulfur analogs of alcohols are called *thiols*. These compounds contain —SH groups instead of —OH groups. The —SH moiety is known as the *sulfhydryl group*. An older, more general term used for thiols was *mercaptans*.

$$\underset{\text{alcohol}}{R\!-\!OH} \qquad \underset{\text{thiol (or mercaptan)}}{R\!-\!SH}$$

Thiols are named in the same way as alcohols in the IUPAC system except that the *-ol* becomes *-thiol*. The prefix *thio-* indicates the substitution of a sulfur atom for an oxygen atom in a compound. The following comparison shows the similarities in naming these two types of compounds:

$$\underset{\text{2-butanol}}{CH_3\!-\!\underset{\underset{OH}{|}}{CH}\!-\!CH_2\!-\!CH_3} \qquad \underset{\text{2-butanethiol}}{CH_3\!-\!\underset{\underset{SH}{|}}{CH}\!-\!CH_2\!-\!CH_3}$$

As in the case of diols and triols, the *-e* at the end of the alkane name is also retained for thiols.

Hydrogen sulfide and almost all thiols share two properties: (1) low boiling points because of reduced hydrogen bonding and (2) a strong, disagreeable odor. The familiar odor of natural gas is the result of the addition of thiols to the gas. The exceptionally low threshold of detection allows consumers to smell a gas leak long before the gas reaches dangerous levels. Thiols are responsible for the odor of skunks.

Many of the strong odors associated with foods are caused by the presence of thiols. The odors coming from freshly chopped onions, garlic, oysters, and cheddar cheese can be partially attributed to thiols. Structures for the sulfur-containing compounds present in these foods are listed in Table 17–10.

Thiols are easily oxidized, but yield different products than their alcohol analogs. Thiols form *disulfides* instead of aldehydes or ketones. Two thiol groups (usually

TABLE 17–10 Some Naturally Occurring Sulfur-containing Organic Compounds

Flavor or odor of	Structures of Compounds Involved
oysters	CH_3-SH
cheddar cheese	CH_3-SH
freshly chopped onions	$CH_3-CH_2-CH_2-SH$
garlic	$CH_2=CH-CH_2$ and $CH_2=CH-CH_2-S-S-CH_2-CH=CH_2$ $\quad\quad\quad\quad\;\;\,\mid$ $\quad\quad\quad\quad\;\,SH$

on different molecules) each lose a hydrogen atom, thus linking the two sulfur atoms together:

$$R-SH + HS-R \underset{\text{reduction}}{\overset{\text{oxidation}}{\rightleftarrows}} \underset{\text{a disulfide}}{R-S-S-R} + 2H$$

Mild reducing agents can reverse the reaction, yielding free thiols.

Cysteine, which contains a sulfhydryl group, is an important amino acid found in proteins. Oxidation of cysteines in various parts of proteins help define structures of proteins or link proteins together.

$$\underset{\text{cysteine}}{\begin{array}{c}COOH\\|\\HC-SH\\|\\NH_2\end{array}} + \underset{\text{cysteine}}{\begin{array}{c}COOH\\|\\HS-CH\\|\\NH_2\end{array}} \underset{\text{reduction}}{\overset{\text{oxidation}}{\rightleftarrows}} \begin{array}{cc}COOH & COOH\\|&|\\HC-S-S-CH\\|&|\\NH_2 & NH_2\end{array}$$

EXERCISES AND PROBLEMS

Structural Characteristics of Alcohols, Phenols, and Ethers

17.1 Classify each of the following compounds as an alcohol, phenol, or ether.

a. CH_3-CH_2-OH
$\quad\quad\;\;\;|$
$\quad\quad\;\;CH_3$

b. $CH_3-O-CH_2-CH_2-CH_3$

c. 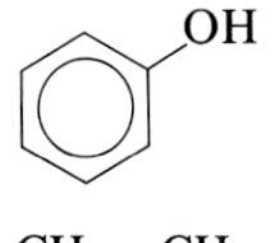 ⬡—OH

d. 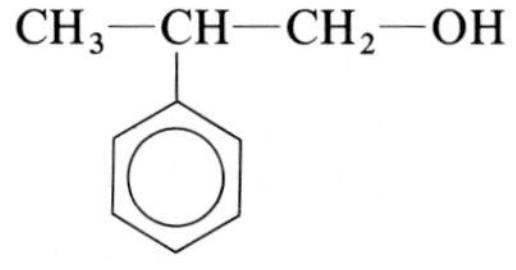 $CH_3-CH-CH_2-OH$
$\quad\quad\;\;\,|$
$\quad\quad\;\;\,$⬡

e. ⬡—O—⬡—CH_3

f. ⬡—O—CH_2—CH—CH_3
$\quad\quad\quad\quad\quad\quad\quad\;|$
$\quad\quad\quad\quad\quad\quad\;\,CH_3$

g. ⬡ with OH and $CH_2-CH_2-CH_3$

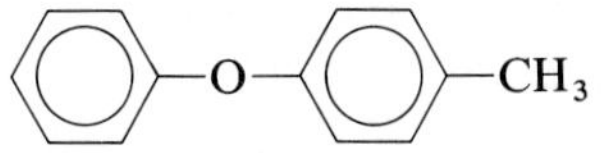

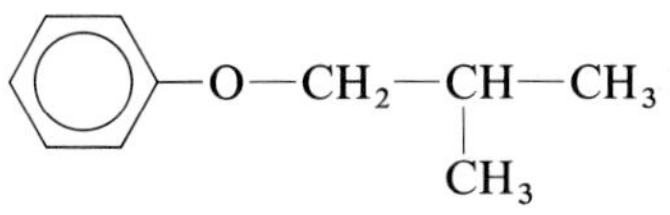

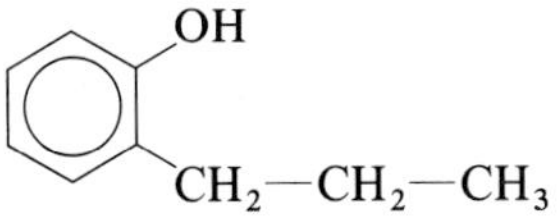

h. $\quad\quad\;\,OH$
$\quad\quad\;\;|$
CH_3-C-CH_3
$\quad\quad\;\;|$
$\quad\quad\;CH_3$

17.2 Explain the differences in structure between an alcohol and a phenol.

17.3 Explain how the structures of each of the following types of compounds are related to the structure of water.

a. alcohols
b. phenols
c. ethers

Nomenclature of Alcohols

17.4 Assign IUPAC names to each of the following alcohols.

a. 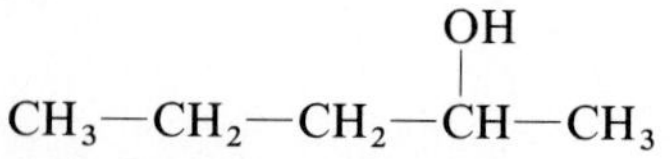 CH$_3$—CH$_2$—CH$_2$—CH(OH)—CH$_3$

b. CH$_3$—OH

c. 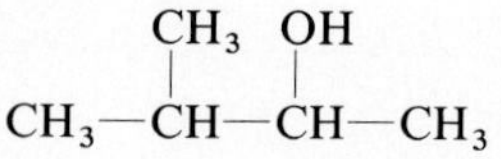 CH$_3$—CH(CH$_3$)—CH(OH)—CH$_3$

d. 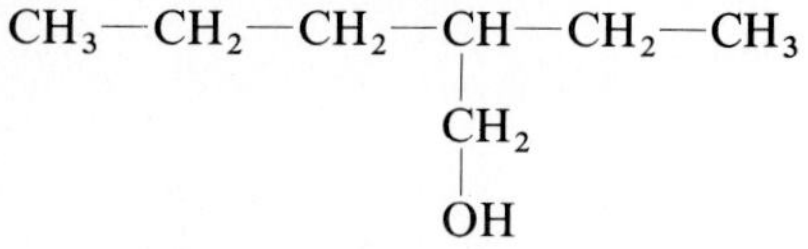 CH$_3$—CH$_2$—CH$_2$—CH(CH$_2$OH)—CH$_2$—CH$_3$

e. 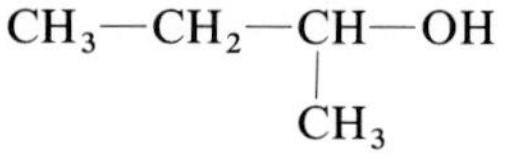 CH$_3$—CH$_2$—CH(OH)—CH$_3$ (with CH$_3$ branch)
   CH$_3$—CH$_2$—CH(CH$_3$)—OH

f. 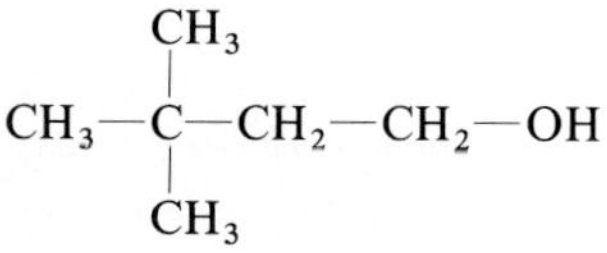 CH$_3$—C(CH$_3$)$_2$—CH$_2$—CH$_2$—OH

g. 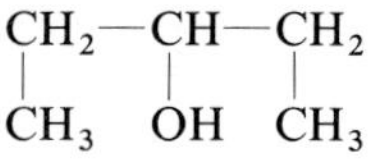 CH$_2$(CH$_3$)—CH(OH)—CH$_2$(CH$_3$)

h. 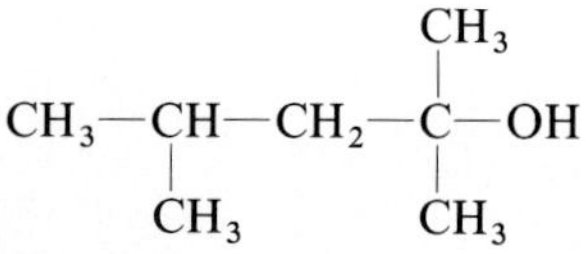 CH$_3$—CH(CH$_3$)—CH$_2$—C(CH$_3$)$_2$—OH

17.5 Write structural formulas for each of the following alcohols.

a. 3-pentanol
b. 3-ethyl-3-hexanol
c. 2-methyl-1-propanol
d. 4-methyl-2-pentanol
e. cyclohexanol
f. 2-phenyl-2-propanol
g. 2-methylcyclobutanol
h. 2-chloro-1-propanol

17.6 Write structural formulas and assign formal IUPAC names to each of the following alcohols.

a. ethyl alcohol
b. methyl alcohol
c. *n*-propyl-alcohol
d. isopropyl alcohol
e. *n*-butyl alcohol
f. isobutyl alcohol
g. sec-butyl alcohol
h. tert-butyl alcohol

17.7 Assign IUPAC names to each of the following polyhydroxy alcohols.

a. 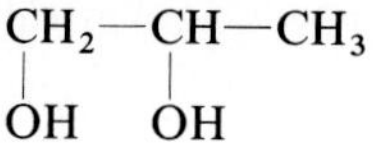 CH$_2$(OH)—CH(OH)—CH$_3$

b. 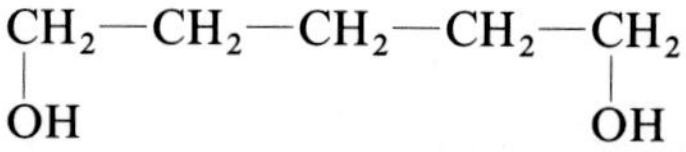 CH$_2$(OH)—CH$_2$—CH$_2$—CH$_2$—CH$_2$(OH)

c. 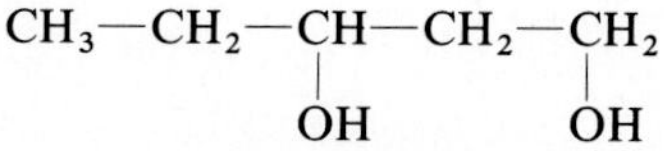 CH$_3$—CH$_2$—CH(OH)—CH$_2$—CH$_2$(OH)

d. 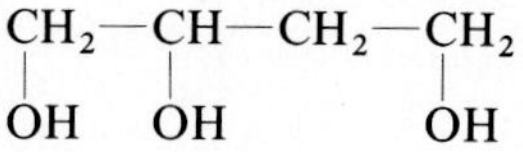 CH$_2$(OH)—CH(OH)—CH$_2$—CH$_2$(OH)

e. 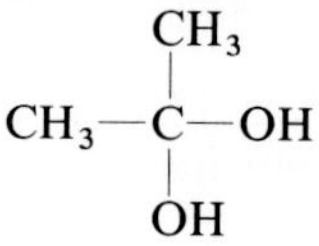 CH$_3$—C(CH$_3$)(OH)—OH

f. 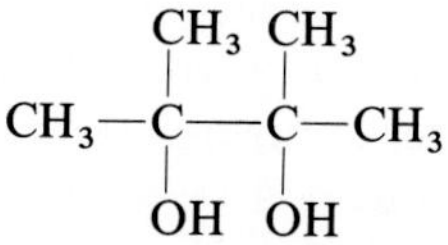 CH$_3$—C(CH$_3$)(OH)—C(CH$_3$)(OH)—CH$_3$

17.8 Write structural formulas for each of the following polyhydroxy alcohols.

a. 1,5-pentanediol
b. 1,2-propanediol
c. 1,2,4-butanetriol
d. 2,2-pentanediol
e. 1,4-cyclohexanediol
f. 1,2,3,4,5-pentanepentol

17.9 Utilizing IUPAC rules, name each of the following compounds. Don't forget to use *cis* and *trans* prefixes (Section 15.5) where needed.

a. 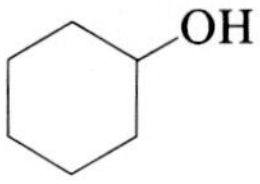 cyclohexane with OH
b. 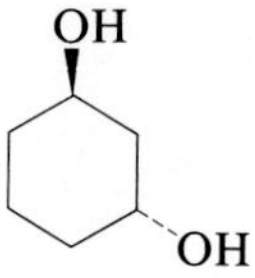 cyclohexane with two OH groups
c. 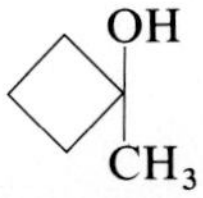 cyclobutane with OH and CH$_3$
d. 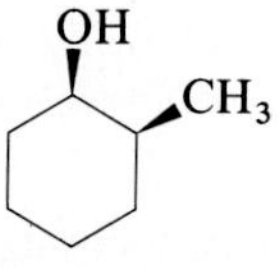 cyclohexane with OH and CH$_3$
e. 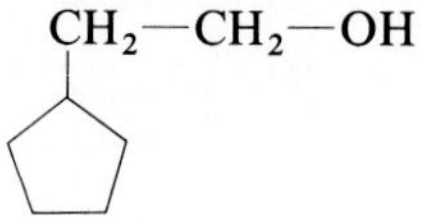 cyclopentane—CH$_2$—CH$_2$—OH
f. 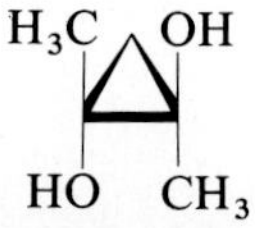 cyclopropane with H$_3$C, OH, HO, CH$_3$

17.10 Each of the following alcohols is named incorrectly. However, the names will give correct structural formulas. Draw structural formulas for the compounds and then write the *correct* IUPAC name for each alcohol.

a. 2-ethyl-1-propanol
b. 3,4-butanediol

c. 2-methyl-3-butanol
d. 1,4-cyclopentanediol
e. 3-methyl-3-butanol
f. 1,1-dimethyl-1-butanol

17.11 Isomerism is possible for alcohols. Draw all of the isomeric alcohols that have the formula C_4H_9OH. Also, assign an IUPAC name to each isomer.

Nomenclature of Phenols

17.12 Name the following compounds as phenol derivatives.

a. 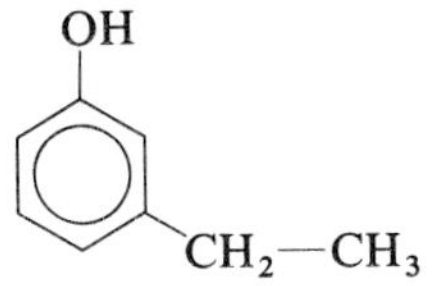 [OH on benzene with CH₂—CH₃]
b. 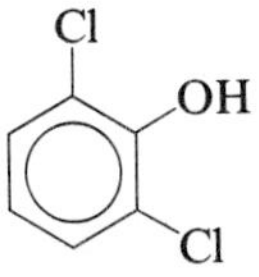 [benzene with Cl, OH, Cl]
c. 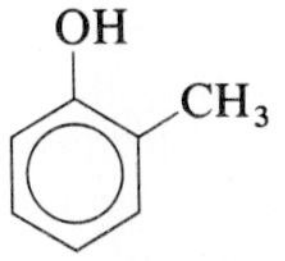 [OH on benzene with CH₃]
d. [benzene with OH and OH (para)]
e. [benzene with OH and Br]
f. 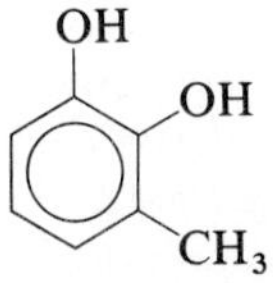 [benzene with OH, OH, CH₃]

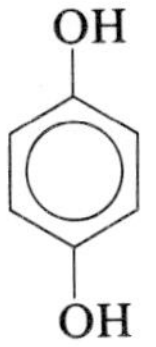

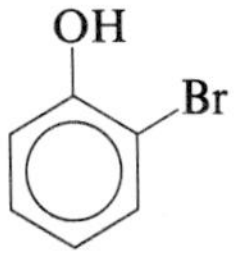

17.13 Draw structures for each of the following phenols.

a. 4-chlorophenol
b. 2-ethylphenol
c. 2,4-dinitrophenol
d. m-cresol
e. resorcinol
f. 2,6-diethyl-4-methylphenol

17.14 Catechol, resorcinol, and hydroquinone are structurally very similar. Describe the similarities in structure of these three phenolic compounds.

17.15 Draw structural formulas for all possible phenolic isomers that have the molecular formula C_8H_9OH, and name each one.

17.16 Assign IUPAC names to each of the following isomeric compounds. Explain why one is a phenol and the other is not.

a. 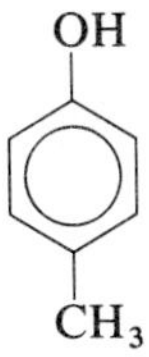 [benzene with OH and CH₃ para]
b. 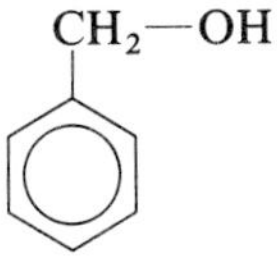 [benzene with CH₂—OH]

Classifications of Alcohols

17.17 Classify each of the alcohols in Problem 17.4 as primary, secondary, or tertiary alcohols.

17.18 Locate and classify *each* of the alcohol functional groups in the compounds in Problem 17.7 as a primary, secondary, or tertiary alcohol functional group.

17.19 Three amino acids found in proteins contain hydroxyl functional groups. Each contains a different "type" of alcohol group. Match the structures with their correct names, based on the type of alcohol described.

Names: a. serine (primary alcohol functional group)
b. threonine (secondary alcohol functional group)
c. tyrosine (phenolic functional group)

Structures:

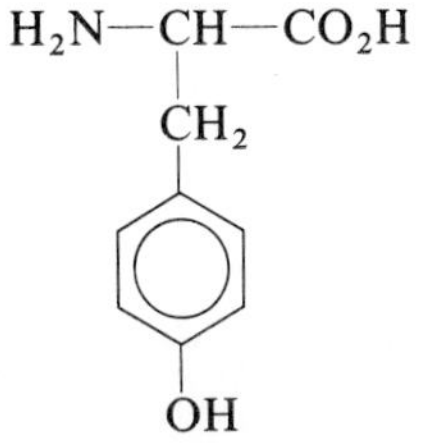

1. 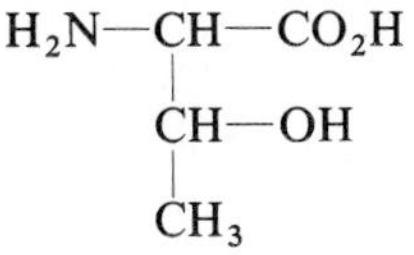 H_2N—CH—CO_2H
   $\quad\quad\quad\;\;|$
   $\quad\quad\quad\;$CH—OH
   $\quad\quad\quad\;\;|$
   $\quad\quad\quad\;\;CH_3$

2. H_2N—CH—CO_2H
   $\quad\quad\quad\;\;|$
   $\quad\quad\quad\;\;CH_2$
   $\quad\quad\quad\;\;|$
   [benzene ring with OH para]

3. H_2N—CH—CO_2H
   $\quad\quad\quad\;\;|$
   $\quad\quad\quad\;\;CH_2$
   $\quad\quad\quad\;\;|$
   $\quad\quad\quad\;\;OH$

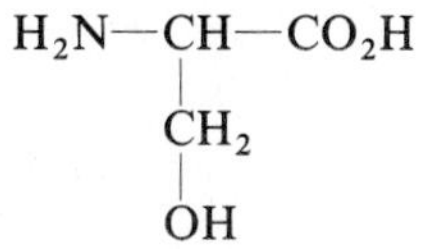

Physical Properties of Alcohols

17.20 Draw structures of the following molecules and show how hydrogen bonds can form between them, using dashed lines for the hydrogen bonds.

a. ethanol and water
b. ethanol and ethanol
c. ethanol and methanol

17.21 Explain why the boiling points of alcohols are much higher than the boiling points of alkanes that have similar molecular weights.

17.22 Select the member of each of the following pairs of alcohols that you would expect to be more soluble in water. Briefly explain your choice in each case.

a. n-pentane and 1-pentanol
b. 2-propanol and 2-pentanol
c. 2-butanol and 2,3-butanediol

Preparation of Alcohols

17.23 Write the structure of the expected predominant organic product from each of the following reactions.

a. $CH_2{=}CH_2 + H_2O \xrightarrow{H_2SO_4}$

b. 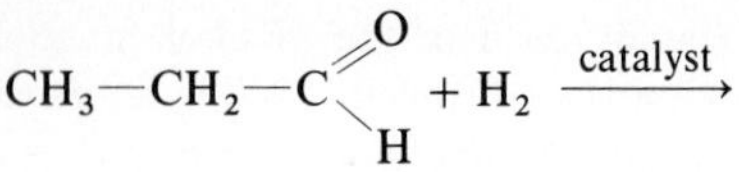 $CH_3-CH_2-C\overset{O}{\underset{H}{\diagdown}} + H_2 \xrightarrow{catalyst}$

c. $CH_3-CH_2-\underset{\underset{CH_3}{|}}{C}=CH_2 + H_2O \xrightarrow{H_2SO_4}$

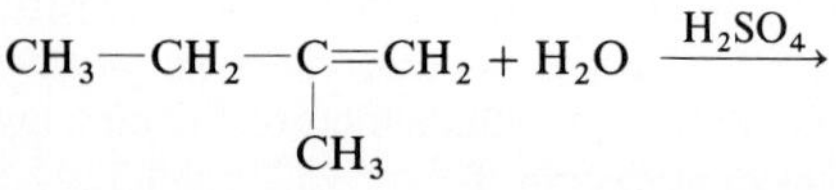

d. 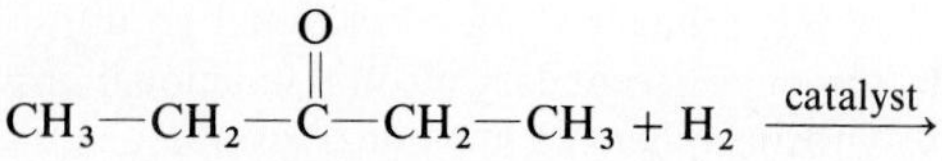 $CH_3-CH_2-\overset{O}{\overset{\|}{C}}-CH_2-CH_3 + H_2 \xrightarrow{catalyst}$

Reactions of Alcohols

17.24 Draw the structures of the expected predominant organic dehydration product when each of the following alcohols reacts with sulfuric acid at a temperature of 180°C.

a. 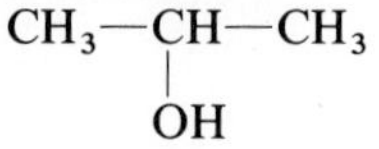 $CH_3-\underset{\underset{OH}{|}}{CH}-CH_3$

b. 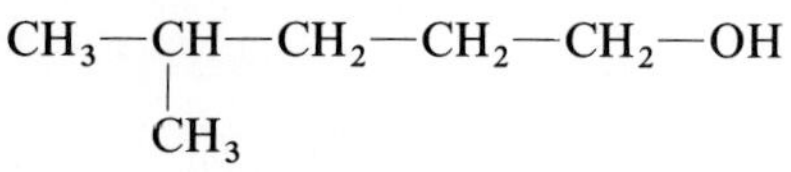 $CH_3-\underset{\underset{CH_3}{|}}{CH}-CH_2-CH_2-CH_2-OH$

c. 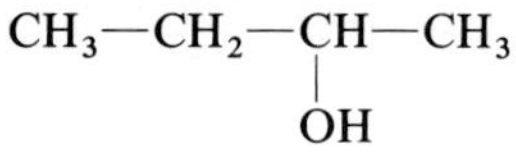 $CH_3-CH_2-\underset{\underset{OH}{|}}{CH}-CH_3$

d. 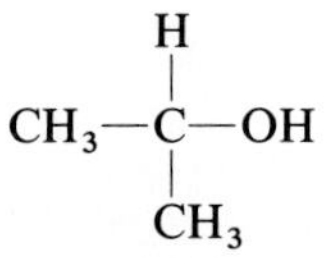 $CH_3-\underset{\underset{CH_3}{|}}{\overset{\overset{H}{|}}{C}}-OH$

e. 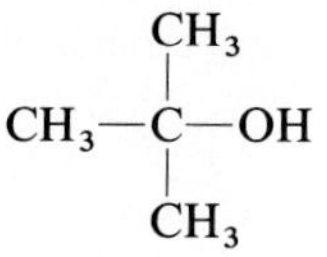 $CH_3-\underset{\underset{CH_3}{|}}{\overset{\overset{CH_3}{|}}{C}}-OH$

f. 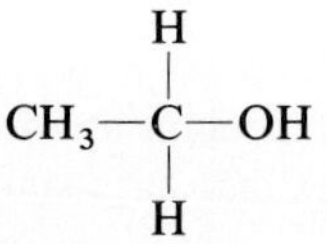 $CH_3-\underset{\underset{H}{|}}{\overset{\overset{H}{|}}{C}}-OH$

17.25 Draw the structures of the expected predominant organic dehydration product when each of the alcohols in Problem 17.24 reacts with sulfuric acid at a temperature of 140°C.

17.26 Which of the alcohols in Problem 17.24 will give a ketone when oxidized? Draw the structure of each ketone produced.

17.27 Which of the alcohols in Problem 17.24 will give a carboxylic acid in a two-step process when oxidized? Draw the structure of each carboxylic acid produced.

17.28 Draw the structure of the expected predominant organic product formed in each of the following reactions.

a. $CH_3-CH_2-CH_2-OH \xrightarrow{PCl_3}$

b. 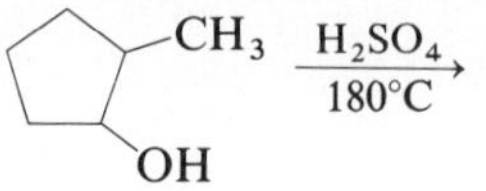 cyclopentane with $-CH_3$ and $-OH$ substituents $\xrightarrow[180°C]{H_2SO_4}$

c. $CH_3-\underset{\underset{OH}{|}}{CH}-CH_2-CH_3 \xrightarrow{K_2Cr_2O_7}$

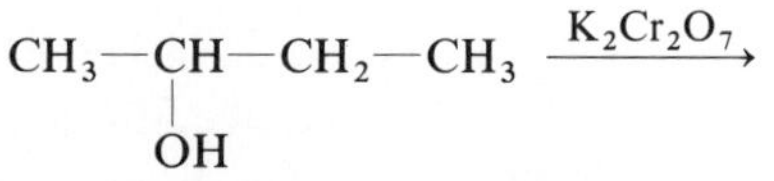

d. 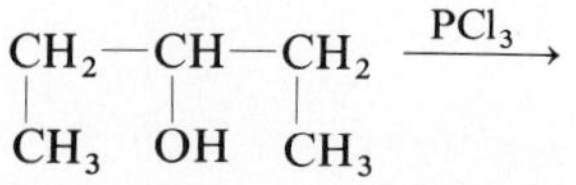 $CH_2-\underset{\underset{OH}{|}}{CH}-CH_2 \atop {\overset{|}{CH_3}\quad\overset{|}{}\quad\overset{|}{CH_3}} \xrightarrow{PCl_3}$

e. 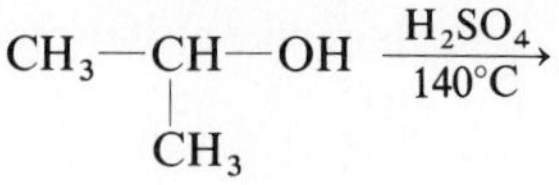 $CH_3-\underset{\underset{CH_3}{|}}{CH}-OH \xrightarrow[140°C]{H_2SO_4}$

f. 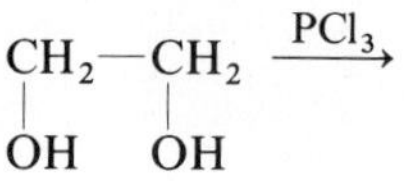 $CH_2-CH_2 \atop {\overset{|}{OH}\quad\overset{|}{OH}} \xrightarrow{PCl_3}$

g. 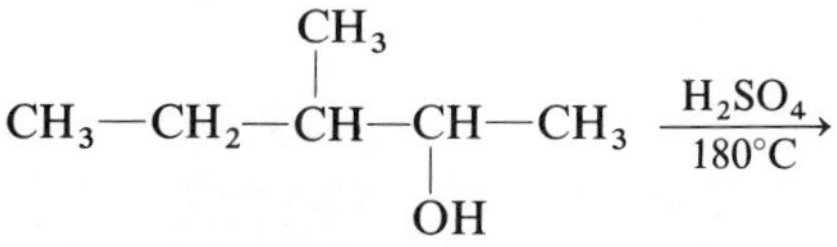 $CH_3-CH_2-\underset{}{CH}-\underset{\underset{OH}{|}}{\overset{\overset{CH_3}{|}}{CH}}-CH_3 \xrightarrow[180°C]{H_2SO_4}$

h. $CH_3-CH_2-CH_2-OH \xrightarrow{K_2Cr_2O_7}$

17.29 A series of reactions that converts one substance into another is often diagrammed by a series of reaction arrows. The needed reactants are written in above and below the arrows. Draw the structure of the final product of each of the following series of reactions.

a. $CH_3-CH=CH_2 \xrightarrow[H_2SO_4]{H_2O} \xrightarrow{PCl_3}$

b. $CH_3-CH=CH_2 \xrightarrow[H_2SO_4]{H_2O} \xrightarrow{K_2Cr_2O_7}$

c. 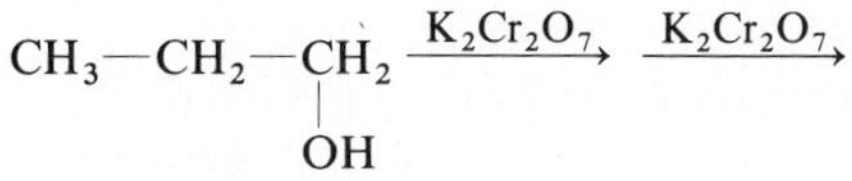 $CH_3-CH_2-\underset{\underset{OH}{|}}{CH_2} \xrightarrow{K_2Cr_2O_7} \xrightarrow{K_2Cr_2O_7}$

d. 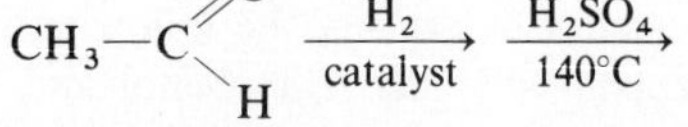 $CH_3-C\overset{O}{\underset{H}{\diagdown}} \xrightarrow[catalyst]{H_2} \xrightarrow[140°C]{H_2SO_4}$

17.30 The alcohol *4-methyl-3-heptanol* is a sex-attractant pheromone for certain kinds of bark beetles. If this alcohol is oxidized by $K_2Cr_2O_7$ in acidic solution, the resulting ketone is the alarm pheromone for the Texas leaf-cutter ant. Draw the structure of this ant-alarm pheromone.

17.31 Three isomeric pentanols with linear carbon chains exist. Which of these, upon dehydration at 180°C, yields only 2-pentene? Which isomer yields only 1-pentene?

17.32 A mixture of methanol and ethanol and H_2SO_4 is heated to 140°C. After reaction, the solution contains three different ethers. Draw a structural formula for each of the ethers. Why did three ethers form, rather than a single product?

17.33 An important series of reactions in the citric acid cycle (a biochemical pathway discussed in Chapter 29) converts fumaric acid into oxaloacetic acid. The steps involved in the reaction series involve hydration of a double bond to an alcohol, followed by the oxidation of the alcohol to a ketone. Given the structure of fumaric acid, draw the structures of the next two products in this reaction series.

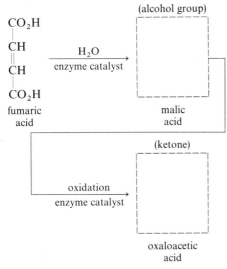

Important Alcohols

17.34 What is the meaning associated with the following terms?
a. absolute alcohol
b. grain alcohol
c. wood alcohol
d. denatured alcohol
e. rubbing alcohol
f. drinking alcohol
g. 70-proof alcohol

17.35 List an alcohol used in each of the following ways.
a. primary component of "duplicating fluid"
b. moistening agent in many cosmetics
c. "skin coolant" for the human body
d. antifreeze in car radiators
e. "active ingredient" in alcoholic beverages
f. component of polyester fibers
g. ingredient used in explosive manufacture

17.36 What is the relationship between the presence of methanol and formaldehyde in the human body?

17.37 What is the purpose for denaturing ethanol and what chemical substances are used as denaturants?

17.38 The maximum concentration of ethanol that is obtainable from fermentation is about 18%. Why is this so?

Phenols

17.39 List a phenol used in each of the following ways.

a. ingredient in mouthwashes and throat lozenges
b. "skin irritant" in poison ivy
c. hormone secreted in response to stress
d. responsible for the flavor of cloves
e. acid–base indicator
f. psychological ingredient in marijuana

17.40 Phenolic compounds are frequently used as antiseptics and disinfectants. What is the difference between an antiseptic and a disinfectant?

17.41 Phenol is sometimes called *carbolic acid*. Explain why phenol is sometimes referred to as an acid.

17.42 Phenolphthalein is a popular pH indicator used in acid–base titrations. It is colorless in acidic solution and in basic solution, it turns red. Based upon your understanding of Problem 17.41 and the structure of phenolphthalein, explain why the color change in phenolphthalein occurs. Draw the structure of the red-colored molecule.

Ethers

17.43 Assign IUPAC names to each of the following ethers.
a. $CH_3-O-CH_2-CH_2-CH_3$
b. $CH_3-CH_2-CH_2-O-CH_2-CH_3$
c. $CH_3-CH(O-CH_3)-CH_3$

d. phenyl–O–CH₃

e. cyclohexyl–O–cyclohexyl

f. CH_3-CH_2-O-cyclobutyl

17.44 Assign a non-IUPAC name to each of the ethers in Problem 17.43.

17.45 Classify each of the ethers in Problem 17.43 as a simple ether or a mixed ether.

17.46 Draw structural formulas for the six possible isomeric ethers that fit the molecular formula $C_5H_{12}O$.

17.47 Draw the structural formula of an alcohol that is isomeric with each of the following ethers.
a. $CH_3-O-CH_2-CH(CH_3)-CH_3$
b. $CH_3-CH_2-O-CH_2-CH_3$

17.48 Draw the structure of each of the following ethers.
a. propyl isopropyl ether
b. ethyl phenyl ether
c. methyl isobutyl ether
d. 2-ethoxypentane
e. ethoxycyclobutane
f. 1-methoxy-2, 2-dimethylpropane

17.49 Most ethers are not soluble in water. Explain why.

17.50 Dimethyl ether and ethanol are isomers. Dimethyl ether is a gas at room temperature and ethanol is a liquid at room temperature. Explain why.

17.51 After two laboratory periods of hard work, a handsome student identifies two unknown liquids as a primary alcohol and an ether. Unfortunately, he forgets to label the tubes and gets them mixed up. Noticing his predicament, an attractive coed seizes the opportunity to meet and impress him. Picking up some $K_2Cr_2O_7$ solution from the shelf in the lab, she approaches him and explains she can help him identify which tube is which with a quick test. What is her plan (as far as the alcohol and ether are concerned)?

17.52 What is an epoxide? Draw the structure of the simplest epoxide.

17.53 List an ether or epoxide that is used as:
a. a thickening agent in shampoos
b. a popular laboratory solvent
c. the first "ether" anesthetic
d. a halogenated inhalation anesthetic

Thiols and Disulfides

17.54 What is a sulfhydryl group? How does it differ from a disulfide group?

17.55 Describe the structural similarities and differences between
a. a disulfide and an ether
b. an alcohol and a thiol

17.56 According to law, thiols must be added to natural gas before the natural gas is distributed to the public. Why is this so?

17.57 Draw structures for the following thiols.
a. methanethiol
b. 2-propanethiol
c. 1-butanethiol
d. 3-methyl-1-pentanethiol
e. cyclopentanethiol
f. 1,2-ethanedithiol

17.58 Write formulas for the sulfur-containing organic products of the following reactions.

a. $2\ CH_3-CH_2-SH \xrightarrow{\text{oxidizing agent}}$

b. $CH_3-CH_2-S-S-CH_2-CH_3 \xrightarrow{\text{mild reducing agent}}$

17.59 Alcohols and thiols can both be oxidized in a controlled way. Discuss the differences in the products that result.

18 Aldehydes and Ketones

Objectives

After completing Chapter 18, you will be able to:

18.1 Understand the structure and bonding within the carbonyl functional group.
18.2 List the structural characteristics of aldehydes and ketones.
18.3 Assign IUPAC names to aldehydes and ketones.
18.4 List the general physical properties and uses of aldehydes and ketones.
18.5 Write equations for the preparation of aldehydes and ketones using oxidation reactions.
18.6 Predict the structure of alcohols formed by the reduction of aldehydes and ketones.
18.7 List the structural features present in hemiacetals and acetals and discuss how these compounds can be prepared.
18.8 Discuss the Tollens', Benedict's, and Fehling's tests used to differentiate between aldehydes and ketones, including the associated reagents and evidence for positive tests.

INTRODUCTION

In this chapter, we will continue our discussion of hydrocarbon derivatives that contain the element oxygen. In the previous chapter, the functional groups considered (alcohols, phenols, and ethers) had the common feature of the presence of carbon–oxygen *single* bonds. Carbon–oxygen *double* bonds are also possible in hydrocarbon derivatives. We will now consider the simplest types of compounds containing this structural feature: aldehydes and ketones.

18.1 THE CARBONYL GROUP

A **carbonyl group** *consists of a carbon atom bonded to an oxygen atom through a double bond.* Because any carbon atom in an organic molecule must form four bonds, the carbon atom of a carbonyl (pronounced "carbon-EEL") group must also be bonded to two other atoms.

carbonyl group

The double bond between carbon and oxygen in carbonyl groups consists of a sigma bond and a pi bond. Thus, it is similar to the carbon–carbon double bonds found in alkenes, which also consist of a sigma and a pi bond (Section 15.4). Electrons in the sigma bond are primarily found between the carbon and oxygen atoms, and electrons in the pi bond are found above and below the internuclear region, in the area where the two *p* orbitals overlap (see Figure 18–1). Both pairs of electrons are shared between the two atoms, but they are *not equally* shared. The higher electronegativity of the oxygen atom polarizes the bond, causing a slight negative charge around the oxygen atom and a slight positive charge at the carbon atom. Thus, the carbonyl bond is a *polar bond*, as contrasted to the carbon–carbon double bond, which is nonpolar. The carbonyl group is an extremely important functional group in both organic and biochemistry.

FIGURE 18–1 Schematic of the sigma and pi bonding present in the carbon–oxygen double bond of a carbonyl group.

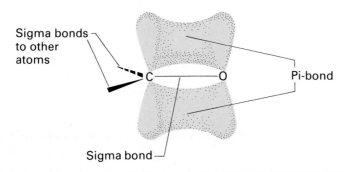

18.2 STRUCTURES OF ALDEHYDES AND KETONES

aldehydes

Aldehydes *are compounds that contain a carbonyl group in which the carbonyl carbon atom is bonded to at least one hydrogen atom.* The other group attached to the carbonyl carbon atom can be a hydrogen atom, an alkyl group (R), or an aromatic group (Ar).

$$H-\underset{H}{\overset{O}{\overset{\parallel}{C}}} \quad \text{or} \quad R-\underset{H}{\overset{O}{\overset{\parallel}{C}}} \quad \text{or} \quad Ar-\underset{H}{\overset{O}{\overset{\parallel}{C}}}$$

The aldehyde functional group, which is the structural feature common to each of the preceding structures is

$$-\underset{H}{\overset{O}{\overset{\parallel}{C}}}$$

In a linear expression, the aldehyde functional group is often written as —CHO. For example,

$$R-CHO \quad \text{is equivalent to} \quad R-\underset{H}{\overset{O}{\overset{\parallel}{C}}}$$

ketones

Ketones *are compounds that contain a carbonyl group in which the carbonyl carbon atom is bonded to two other carbon atoms.* The bonded groups can be any combination of alkyl or aromatic groups.

$$R-\overset{O}{\overset{\parallel}{C}}-R' \quad \text{or} \quad R-\overset{O}{\overset{\parallel}{C}}-Ar \quad \text{or} \quad Ar-\overset{O}{\overset{\parallel}{C}}-Ar'$$

When comparing the structural formulas for aldehydes and ketones, note that the distinguishing feature is the location of the carbonyl group. If a carbonyl group is located at the end of a hydrocarbon chain, the compound is an aldehyde. Ketones possess carbonyl groups that are not at the ends of the carbon chain.

There are a number of ways to spatially orient the various groups when writing structural formulas for either aldehydes or ketones. One way is not necessarily considered more correct than the others. Thus, the following three representations for an aldehyde, where the R group is an ethyl group, are equally correct.

$$CH_3-CH_2-\underset{H}{\overset{O}{\overset{\parallel}{C}}} \quad \text{or} \quad CH_3-CH_2-\overset{O}{\overset{\parallel}{C}}-H \quad \text{or} \quad \underset{H}{\overset{CH_3-CH_2}{\diagdown}}C=O$$

For the sake of consistency, the first of the preceding orientations for the aldehyde functional group will be the one most often used in this book.

There are also a number of different ways in which the structures of ketones can be drawn. Using as an example a ketone where both R groups are methyl groups, three possibilities are

$$CH_3-\underset{CH_3}{\overset{O}{\overset{\parallel}{C}}} \quad \text{or} \quad CH_3-\overset{O}{\overset{\parallel}{C}}-CH_3 \quad \text{or} \quad \underset{CH_3}{\overset{CH_3}{\diagdown}}C=O$$

Again, for the sake of consistency, we will favor one orientation; we will use the second of the preceding possibilities, the one where the R groups are in a linear arrangement.

There are also cyclic ketones in which the carbonyl carbon is part of the ring. Cyclic aldehydes are not possible because the carbonyl group is always the terminal carbon of a chain.

Example 18.1

Identify each of the following structures as an aldehyde or a ketone.

a. 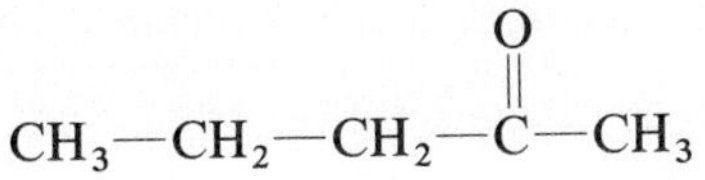 CH$_3$—CH$_2$—CH$_2$—C(=O)—CH$_3$

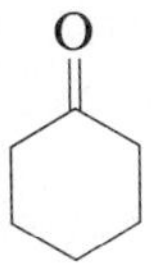

b. 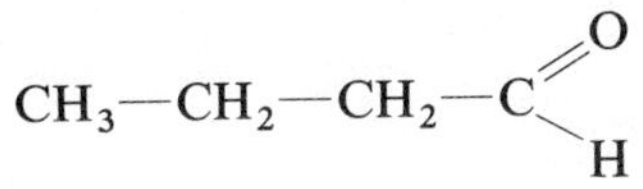 CH$_3$—CH$_2$—CH$_2$—CHO

c. cyclohexanone

d. 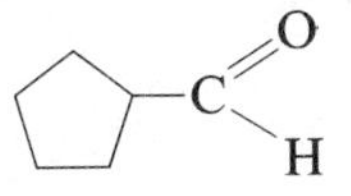 cyclopentyl-CHO

Solution
In all cases, ketones have carbonyl groups between two other carbon atoms. Aldehydes have at least one hydrogen atom bonded directly to the carbonyl group.

a. This is a ketone.
b. This is an aldehyde because the carbonyl carbon atom has a hydrogen attached to it.
c. This is a cyclic ketone.
d. This is an aldehyde where the R group is a cyclopentyl group.

18.3 NOMENCLATURE OF ALDEHYDES AND KETONES

According to the IUPAC system, *al*dehydes are named by dropping the final *-e* from the name of the hydrocarbon that has the same number of carbon atoms, and adding the ending *-al*. Ket*ones* are named in a similar way; their ending is *-one*. The position of the carbonyl group is indicated by the number of the carbonyl carbon atom as it is positioned in the carbon chain. The chain numbering is done from the end that results in the lowest number for the carbonyl group. Therefore, aldehydes will always have their carbonyl groups at number 1.

Example 18.2

Assign IUPAC names to each of the following aldehydes and ketones.

a. 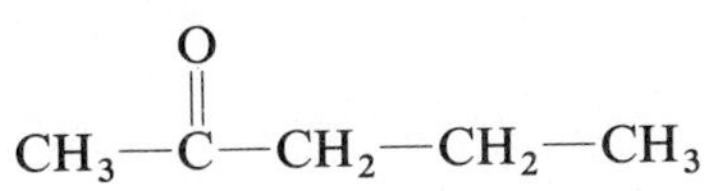 CH$_3$—C(=O)—CH$_2$—CH$_2$—CH$_3$

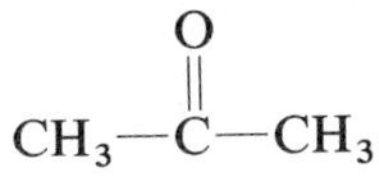

b. 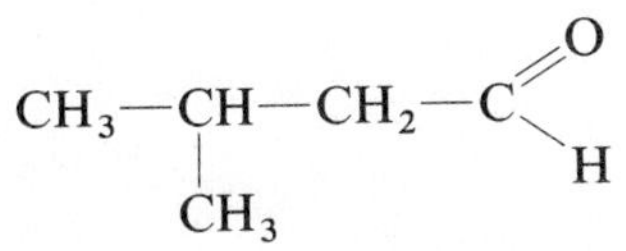 CH$_3$—CH(CH$_3$)—CH$_2$—CHO

c. CH$_3$—C(=O)—CH$_3$

d. 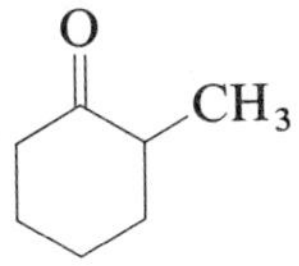 2-methylcyclohexanone

Solution
a. The longest continuous chain contains five carbons, so the root name is *pentanone*. Locating the carbonyl group at carbon number 2 completes the name: *2-pentanone*.
b. The longest chain contains four carbons, resulting in the root name of *butanal*. The methyl group is located at carbon number 3 because we must

always begin numbering at the carbonyl group in aldehydes. The final name is *3-methylbutanal*.
c. Three carbons form the chain, with the carbonyl group at number 2; the name is *2-propanone*.
d. This cyclic ketone has six carbons in the ring. Its root name is *cyclohexanone*. The methyl group is bonded to carbon number 2 because we begin numbering at the carbonyl carbon; the name is *2-methylcyclohexanone*.

You should pay particular attention to the *-al* ending associated with IUPAC aldehyde nomenclature because it closely resembles the *-ol* ending used to name alcohols (Section 17.2).

$$CH_3-CH_2-OH \qquad CH_3-C\underset{H}{\overset{O}{\lessgtr}}$$

ethan*ol*   ethan*al*

TABLE 18–1   IUPAC and Common Names for Selected Aldehydes and Ketones

Structure	IUPAC name	Common name
Aldehydes		
$H-CHO$	methanal	formaldehyde
CH_3-CHO	ethanal	acetaldehyde
CH_3-CH_2-CHO	propanal	propionaldehyde
$CH_3-CH_2-CH_2-CHO$	butanal	butyraldehyde
C_6H_5-CHO	benzaldehyde	benzaldehyde
Ketones		
$CH_3-CO-CH_3$	2-propanone	acetone or dimethyl ketone
$CH_3-CO-CH_2-CH_3$	2-butanone	ethyl methyl ketone
$C_6H_5-CO-CH_3$	1-phenylethanone	methyl phenyl ketone or acetophenone
cyclohexanone	cyclohexanone	cyclohexanone

The simpler aldehydes and ketones all have common names that are used in laboratory settings. The simplest aldehyde is methanal; in this aldehyde, two H atoms are attached to the carbonyl group. The common name of methanal is *formaldehyde*, which we mentioned (Section 17.7) when discussing the toxic effect of methanol. The common name of the two-carbon aldehyde (ethanal) is *acetaldehyde*.

One- and two-carbon ketones cannot exist. A minimum of three carbon atoms is required for a ketone—one C for the carbonyl group and one C for each R group. The common name for 2-propanone, the three-carbon ketone, is *acetone*. Frequently, the common names of ketones are derived by naming the two groups on either side of the carbonyl group (in alphabetical order), followed by the word *ketone*. An example of this nomenclature is *ethyl methyl ketone*. Table 18–1 gives a common name–IUPAC name comparison for selected common aldehydes and ketones.

18.4 PROPERTIES AND USES OF ALDEHYDES AND KETONES

The two simplest aldehydes, methanal (formaldehyde) and ethanal (acetaldehyde), are gases at room temperature. The C_3 through C_{11} straight-chain unsaturated aldehydes are liquids, and the higher aldehydes are solids. The presence of hydrocarbon branching tends to lower both boiling points and melting points, as does the presence of unsaturation in the carbon chain. Lower molecular weight ketones are colorless liquids at room temperature.

The boiling points of aldehydes and ketones are intermediate between those of alcohols and alkanes of similar molecular weight. Aldehydes and ketones have higher boiling points than alkanes, but lower boiling points than alcohols.

This trend in boiling points is due to the carbonyl functional group in aldehydes and ketones. The slightly polar carbon–oxygen double bond causes dipole–dipole interactions between these molecules, resulting in higher boiling points than is observed with alkanes. However, since no hydrogen bonding can occur between aldehyde or ketone molecules, their boiling points are lower than corresponding alcohols (see Table 18–2, p. 428).

Aldehydes and ketones can hydrogen bond to water because of the carbonyl group present in these compounds.

hydrogen bonds between water and carbonyl groups

This hydrogen bonding causes low-molecular-weight aldehydes and ketones to be water soluble. A striking example of water solubility caused by hydrogen bonding is exhibited by formaldehyde, the simplest aldehyde. When this colorless gas is bubbled through water, highly concentrated aqueous solutions, which are commonly called *formalin*, are produced. In this aqueous form, formaldehyde is used as a germicide for disinfecting surgical instruments and as a preservative that hardens tissues. Anyone who has experience in a biology laboratory is familiar with the pungent odor of the preservative formalin.

Low-molecular-weight ketones such as acetone are excellent solvents because they are miscible in both water and nonpolar solvents. Acetone is the main ingredient in gasoline treatments that are designed to solubilize water in the gas tank

TABLE 18-2 Relationship Between the Boiling Points of Aldehydes and Ketones and the Boiling Points of Other Organic Compounds

Type of Compound	Compound	Structure	Molecular Weight	Boiling Point (°C)
alkane	ethane	CH_3-CH_3	30	−89
alcohol	methanol	CH_3-OH	32	65
aldehyde	formaldehyde	$H-CHO$	30	−21
alkane	propane	$CH_3-CH_2-CH_3$	44	−42
alcohol	ethanol	CH_3-CH_2-OH	46	78
aldehyde	acetaldehyde	CH_3-CHO	44	20
alkane	butane	$CH_3-CH_2-CH_2-CH_3$	58	−1
alcohol	1-propanol	$CH_3-CH_2-CH_2-OH$	60	97
aldehyde	propionaldehyde	CH_3-CH_2-CHO	58	49
alkane	isobutane	$CH_3-CH(CH_3)-CH_3$	58	−12
alcohol	2-propanol	$CH_3-CH(OH)-CH_3$	60	83
ketone	acetone	$CH_3-CO-CH_3$	58	56

and allow it to pass through the engine in miscible form. Acetone can also be used to remove water from glassware in the laboratory or to strip old paint and varnish from furniture.

As the hydrocarbon portions of both aldehydes and ketones get larger, the water solubility of these compounds decreases. Simple aldehydes and ketones that have six or more carbons are usually insoluble in water.

Although low-molecular-weight aldehydes have pungent, penetrating, unpleasant odors, higher molecular-weight aldehydes (above C_8) and ketones are more fragrant, especially the aromatic ones. Table 18-3 gives the structures of selected naturally occurring aldehydes and ketones that are associated with various odors or flavors.

Numerous aldehydes and ketones are important in the functioning of the human body. Monosaccharides (simple sugars), which will be discussed in Section 22.3, are polyhydroxy aldehydes or polyhydroxy ketones. The following structures represent the compounds glucose, ribose, and fructose.

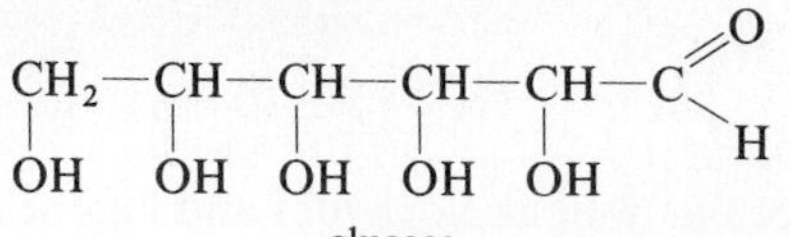

$$CH_2-CH-CH-CH-CH-C\overset{O}{\underset{H}{\diagup}}$$
$$OHOHOHOHOH$$
glucose

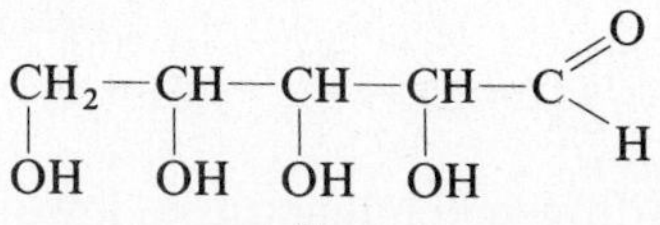

$$CH_2-CH-CH-CH-C\overset{O}{\underset{H}{\diagup}}$$
$$OHOHOHOH$$
ribose

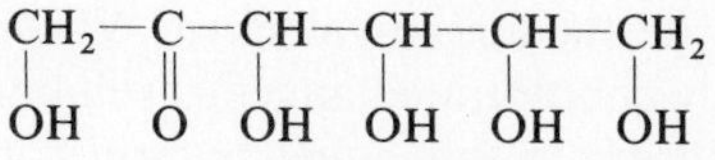

$$CH_2-\underset{O}{\overset{\|}{C}}-CH-CH-CH-CH_2$$
$$OHOHOHOHOH$$
fructose

Glucose is metabolized in our bodies to produce energy (Chapter 29). Ribose is a component in the structure of the hereditary material RNA (Chapter 26). Fructose, the sweetest simple sugar (Chapter 22), occurs in many fruits.

Many important steroid hormones (Chapter 23) are ketones, including testosterone, the hormone that controls the development of male sex characteristics;

TABLE 18-3 Selected Aldehydes and Ketones Whose Uses are Based on Their Odor or Flavor

Structural Formula	Name	Characteristics and Typical Uses
Aldehydes		
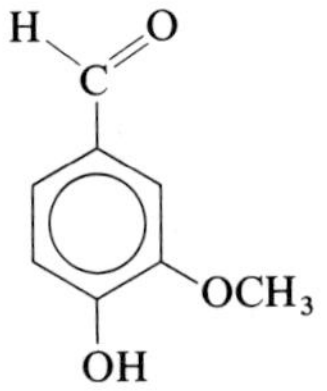 4-hydroxy-3-methoxybenzaldehyde structure	vanillin	vanilla flavoring
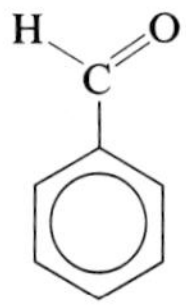 benzaldehyde structure	benzaldehyde	almond flavoring
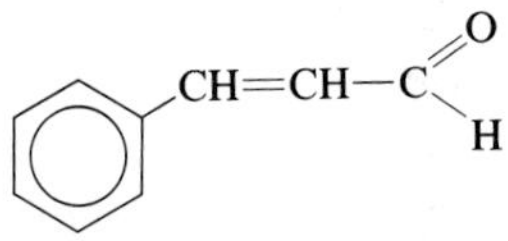 C₆H₅—CH=CH—C(=O)H	cinnamaldehyde	cinnamon flavoring
CH₃—C(CH₃)=CH—CH₂—CH₂—C(CH₃)=CH—C(=O)—H	citral	lemon oil
Ketones		
CH₃—C(=O)—CH₂—CH₃	2-butanone	used as a solvent and in fingernail polish remover
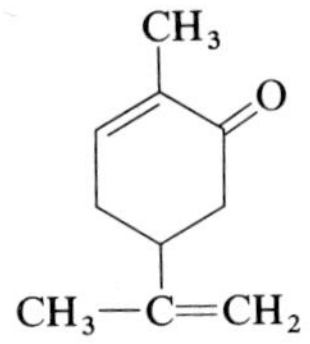 carvone structure	carvone	chief component of spearmint oil
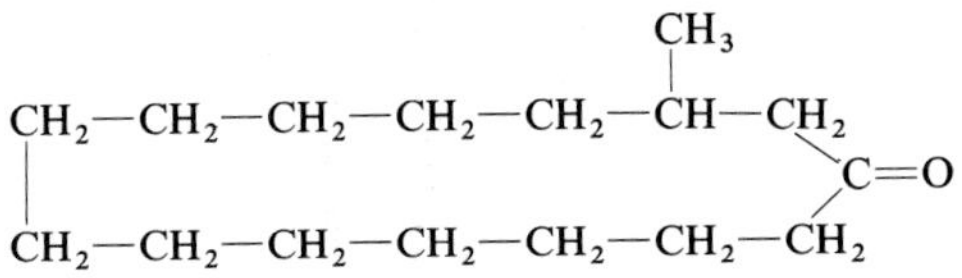 muscone structure (cyclopentadecane ring with methyl and ketone)	muscone	from gland of male musk deer, (used in perfume)

progesterone, the hormone secreted at the time of ovulation in females; and cortisone, a hormone from the adrenal glands that is used medicinally to relieve inflammation.

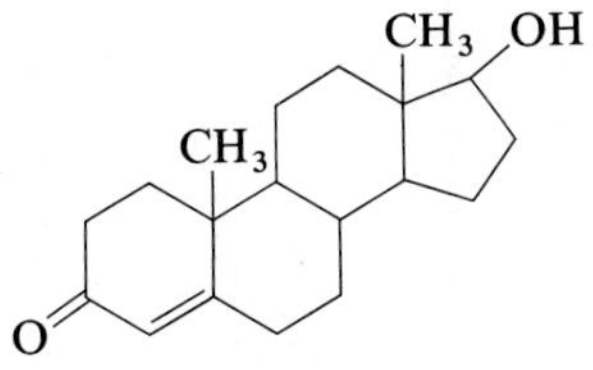

testosterone

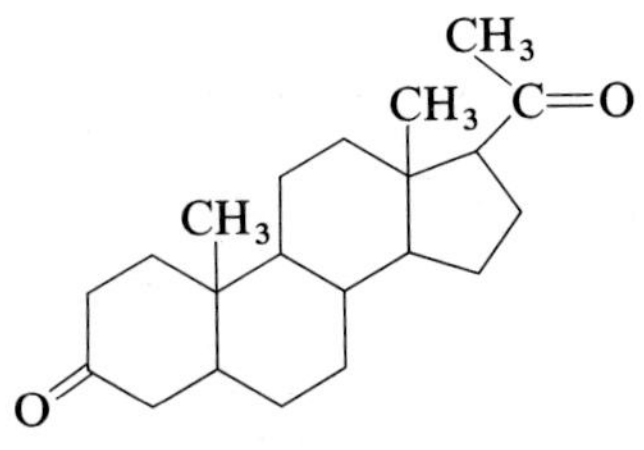

progesterone

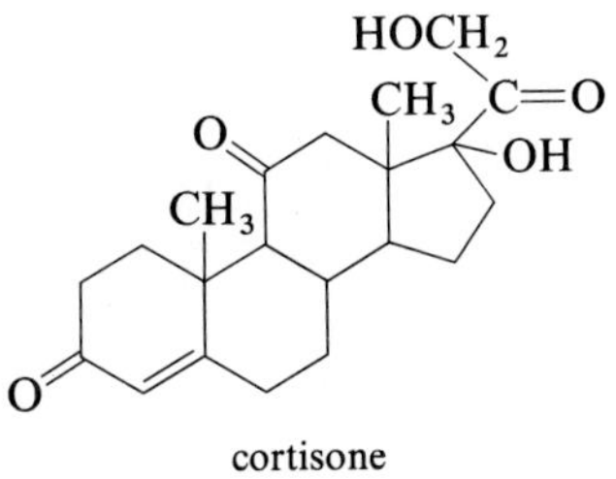

cortisone

18.5 PREPARATION OF ALDEHYDES AND KETONES

In Section 17.6, we saw that aldehydes and ketones can be produced by the oxidation of alcohols using oxidizing agents such as $K_2Cr_2O_7$ in acidic solution. Primary alcohols yield aldehydes, and secondary alcohols yield ketones.

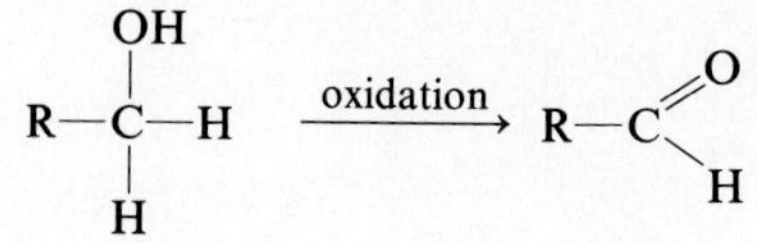

$$\underset{\text{primary alcohol}}{R-\underset{\underset{H}{|}}{\overset{\overset{OH}{|}}{C}}-H} \xrightarrow{\text{oxidation}} \underset{\text{aldehyde}}{R-C\overset{\nearrow O}{\underset{\searrow H}{}}}$$

$$\underset{\text{secondary alcohol}}{R-\underset{\underset{R'}{|}}{\overset{\overset{OH}{|}}{C}}-H} \xrightarrow{\text{oxidation}} \underset{\text{ketone}}{R-\overset{\overset{O}{\|}}{C}-R}$$

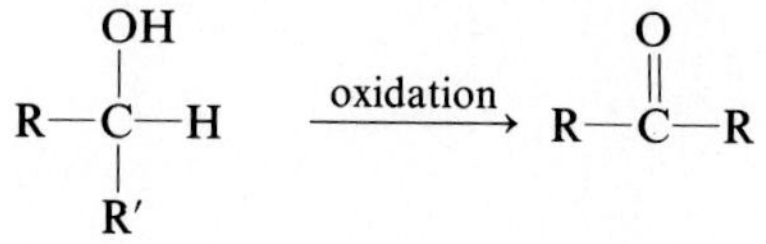

Oxidation reactions are often used to prepare aldehydes and ketones. However, reaction conditions must be sufficiently mild to avoid further oxidation of the aldehyde to a carboxylic acid, or the aldehyde must be removed from the reaction mixture before further oxidation can occur. Aldehydes can be removed by heating the reaction mixture to a temperature slightly lower than the boiling point of the alcohol. The lower boiling aldehyde then vaporizes and leaves the mixture as rapidly as it forms. The aldehyde is then condensed and collected.

Example 18.3

Predict the oxidation product of each of the following alcohols. Draw the structure and name each product.

a. $CH_3-CH_2-CH_2-OH$

b. 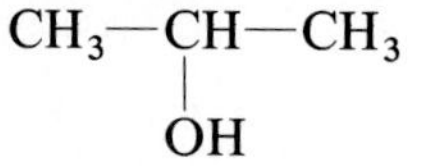 $CH_3-\underset{\underset{OH}{|}}{CH}-CH_3$

c. 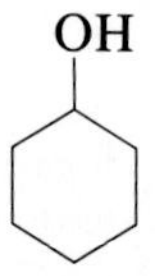 cyclohexanol (OH on cyclohexane ring)

Solution

a. This is a primary alcohol that will give the aldehyde *propanal* as the oxidation product.

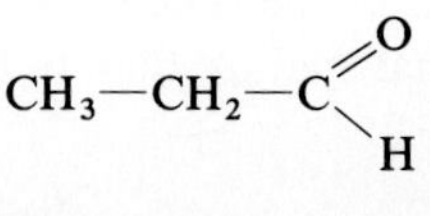

$$CH_3-CH_2-C\overset{\nearrow O}{\underset{\searrow H}{}} \quad \text{(propanal)}$$

b. This is a secondary alcohol. Upon oxidation, secondary alcohols are converted to ketones. The ketone produced here is *2-propanone* (acetone or dimethyl ketone).

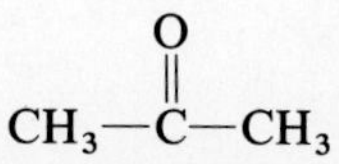

$$CH_3-\overset{\overset{O}{\|}}{C}-CH_3 \quad \text{(2-propanone)}$$

c. This cyclic alcohol is a secondary alcohol; hence, a ketone is the oxidation product. The ketone is *cyclohexanone*.

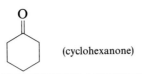
(cyclohexanone)

18.6 REDUCTION OF CARBONYL COMPOUNDS

Because alcohols can be oxidized to form aldehydes and ketones (Section 18.5), it follows that the reduction of aldehydes and ketones yields alcohols; oxidation and reduction are opposite processes. Carbonyl compounds can be reduced with H_2; this process is similar to the reduction of alkenes with H_2 to form alkanes (Section 15.7). Nickel or platinum metals catalyze the addition of H_2 across the carbonyl double bond.

$$R-\underset{H}{\overset{O}{C}} + H_2 \xrightarrow{Ni} R-\underset{H}{\overset{OH}{\underset{|}{C}}}-H$$

aldehyde → primary alcohol

$$R-\overset{O}{\underset{\|}{C}}-R + H_2 \xrightarrow{Ni} R-\underset{H}{\overset{OH}{\underset{|}{C}}}-R$$

ketone → secondary alcohol

Reduction of the aldehyde *acetaldehyde* and the ketone *acetone* yield ethanol and 2-propanol, respectively.

$$CH_3-\underset{H}{\overset{O}{C}} + H_2 \xrightarrow{Ni} CH_3-\underset{H}{\overset{OH}{\underset{|}{C}}}-H$$

ethanal → ethanol

$$CH_3-\overset{O}{\underset{\|}{C}}-CH_3 + H_2 \xrightarrow{Ni} CH_3-\overset{OH}{\underset{|}{CH}}-CH_3$$

acetone → 2-propanol

In general, the reduction of aldehydes yields primary alcohols, and the reduction of ketones yields secondary alcohols.

Reduction of carbonyl groups to alcohols is an important reaction in biological systems. Living systems use molecules other than H_2 gas as hydrogen donors, and enzymes as catalysts instead of platinum or nickel metals. A common hydrogen carrier is NADH (nicotinamide adenine dinucleotide). The final step in the fermentation of sugars by yeast to form alcohol is the reduction of acetaldehyde (ethanal) to form ethanol.

$$CH_3-\underset{H}{\overset{O}{C}} + NADH + H^+ \xrightarrow[\text{catalyst}]{\text{enzymatic}} CH_3-\overset{OH}{\underset{|}{CH_2}} + NAD^+$$

Other agents such as lithium aluminum hydride, $LiAlH_4$, or sodium borohydride, $NaBH_4$, are reducing agents that accomplish the same reaction in the laboratory without the need for H_2 gas.

Predict the reduction product of the following reaction.

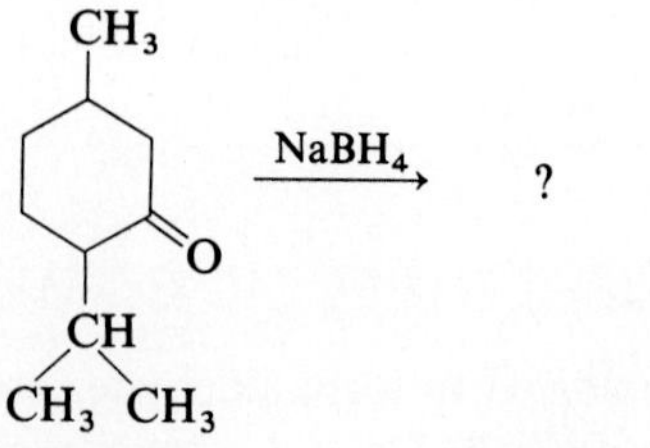

$$\text{menthone} \xrightarrow{\text{NaBH}_4} \text{?} \quad \text{(menthol)}$$

Solution

Reduction of this ketone yields the secondary alcohol *menthol*. Both menthone and menthol are partially responsible for the fragrance and flavor of mint plants.

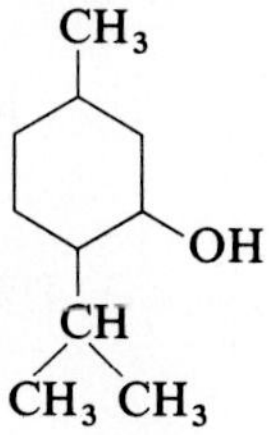

menthol

18.7 HEMIACETAL AND ACETAL FORMATION

hemiacetals

Like H_2, alcohols can add to the carbonyl group on aldehydes and ketones. The addition of alcohols to aldehydes yields hemiacetals. **Hemiacetals** *are compounds that contain an* —OH *group and an* —OR *group attached to the same carbon atom.*

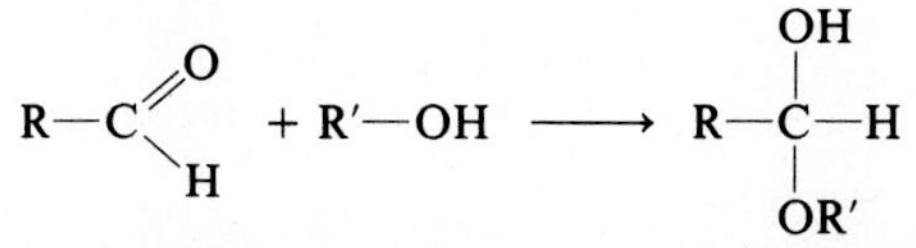

$$\underset{\text{an aldehedyde}}{R-\overset{O}{\underset{H}{C}}} + \underset{\text{an alcohol}}{R'-OH} \longrightarrow \underset{\text{a hemiacetal}}{R-\underset{OR'}{\overset{OH}{\underset{|}{C}}}-H}$$

A specific example is

$$CH_3-CH_2-\overset{O}{\underset{H}{C}} + CH_3-OH \longrightarrow CH_3-CH_2-\underset{OCH_3}{\overset{OH}{\underset{|}{CH}}}$$

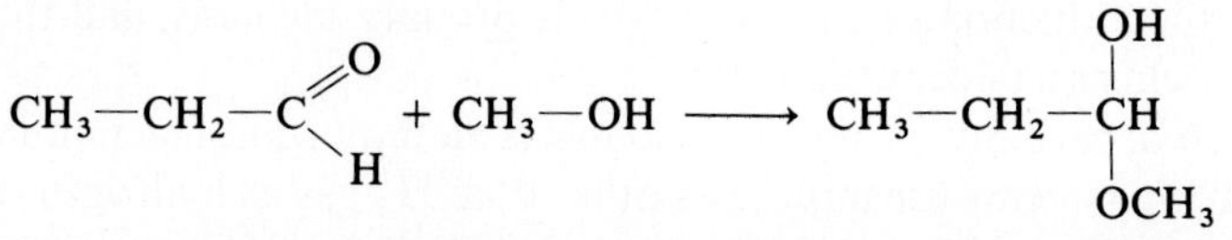

What is actually happening in this reaction is more easily understood if we visualize the alcohol molecule as being made up of two parts (circled in the following equation), with the two parts adding across the carbon–oxygen double bond of the aldehyde.

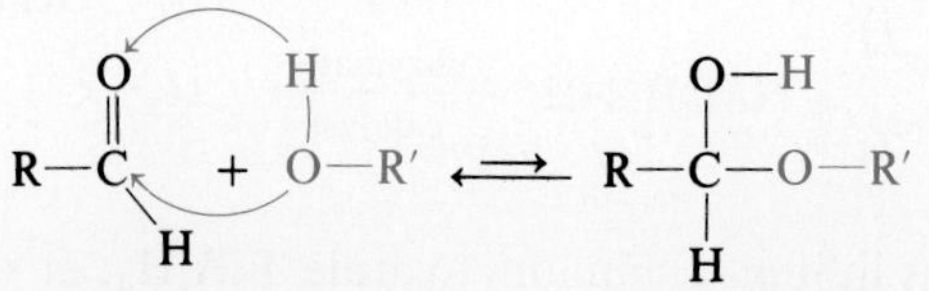

$$\underset{\text{aldehyde}}{R-\overset{O}{\underset{H}{C}}} + \underset{\text{alcohol}}{O-R'} \rightleftharpoons \underset{\text{hemiacetal}}{R-\underset{H}{\overset{O-H}{\underset{|}{C}}}-O-R'}$$

Most hemiacetals are unstable and break easily back down into the aldehyde and alcohol.

There are also cyclic hemiacetals. As a group, these compounds are much more stable than their noncyclic counterparts. They are also more important, from a biochemical viewpoint. In biochemistry, cyclic hemiacetal formation occurs in situations where both the aldehyde group and the alcohol group are part of the same molecule. The hemiacetal forms as the result of an intramolecular reaction.

As an illustration of this phenomenon, let us consider the sugar glucose and how it converts to a hemiacetal when in solution. Glucose is a six-carbon aldehyde in which every carbon atom other than the carbonyl carbon contains an alcohol group. As the carbon atoms in this molecule twist and rotate about their bonds, the alcohol group on carbon number 5 often comes into close proximity with the aldehyde group. Interaction between these two groups when they are in close proximity results in the formation of a cyclic hemiacetal.

In a cyclic hemiacetal such as this, the carbon atoms of the ring are the R portion of the —OR group. Glucose and many other sugars spend the majority of their time in solution as cyclic hemiacetals.

The prefix *hemi-* means half. Thus, a hemiacetal is half of an acetal. The reaction of a second alcohol molecule with a hemiacetal completes the formation of an acetal. **Acetals** are compounds that contain two —OR groups attached to the same carbon atom.

Some specific examples are

Acetals are stable in basic solutions, but tend to react with water (hydrolyze) in the presence of acids. The products of this reaction are the aldehyde and alcohols that originally reacted to form the acetal. The enzyme-catalyzed hydrolysis of acetal linkages is an important process in the digestion of carbohydrates. One simple way of identifying these functional groups is to look for a —C—O—C—O—C— group in an acetal and a —C—O—C—OH group in a hemiacetal.

When alcohols add to ketones, hemiketals and ketals are the products. They are analogous to hemiacetals and acetals in almost every way.

18.8 OXIDATION OF ALDEHYDES

Aldehydes differ from ketones in their ability to further oxidize to carboxylic acids. Mild oxidizing agents readily oxidize aldehydes to acids, but ketones generally remain unchanged.

$$R-\overset{O}{\underset{H}{C}} \xrightarrow{\text{mild oxidation}} R-\overset{O}{\underset{OH}{C}} \text{ carboxylic acid}$$

$$R-\overset{O}{\underset{}{C}}-R \xrightarrow{\text{mild oxidation}} \text{No Reaction}$$
ketone

As was mentioned earlier, under laboratory conditions, it is often difficult to stop the oxidation of a primary alcohol at the aldehyde because the aldehyde further oxidizes to the acid so quickly.

$$R-\underset{\underset{}{}}{CH_2}\text{—OH} \xrightarrow{\text{oxidation}} R-\overset{O}{\underset{H}{C}} \xrightarrow{\text{oxidation}} R-\overset{O}{\underset{OH}{C}}$$
primary alcohol   aldehyde (hard to isolate)   carboxylic acid

In biological systems, enzymatic catalysis carefully controls the oxidation of alcohols to aldehydes, and yields of aldehydes are typically 100% (no acid is formed). In the laboratory, very little aldehyde can be recovered because the acid is the major portion of the product mixture.

Example 18.5

Predict the products of the following oxidation reactions. (Assume mild oxidation conditions.)

a. $C_6H_5-\overset{O}{\underset{H}{C}} \xrightarrow{\text{oxidation}}$

b. $CH_3-\underset{\underset{CH_3}{|}}{CH}-\overset{O}{\underset{H}{C}} \xrightarrow{\text{oxidation}}$

c. $CH_3-\overset{O}{\underset{}{C}}-CH_2-CH_3 \xrightarrow{\text{oxidation}}$

Solution

a. The reactant is an aldehyde that will be oxidized to an acid.

$$C_6H_5-\overset{O}{\underset{OH}{C}}$$

b. This is also an aldehyde, so the oxidation product will be an acid.

$$CH_3-\underset{\underset{CH_3}{|}}{CH}-C\overset{O}{\underset{OH}{\diagup}}$$

c. The reactant is a ketone and ketones cannot be oxidized by a mild oxidizing agent. Therefore, no reaction occurs.

Harsh oxidation conditions such as combustion or hot $KMnO_4$ often break up the carbon backbone of the molecule and form CO_2. On the other hand, mild oxidizing agents can thus be used to distinguish between aldehydes and ketones. Three types of solutions are used expressly for this purpose: Tollens' reagent, Benedict's reagent, and Fehling's reagent.

Tollens' reagent is a mild oxidizing agent composed of a solution of silver nitrate in dilute aqueous ammonia. The active oxidizing agent is silver ion, which is reduced to metallic silver in the process. All aldehydes will react with the Ag^+, but ketones will not.

$$R-C\overset{O}{\underset{H}{\diagup}} + Ag^+ \xrightarrow[\text{warm}]{NH_3, H_2O} R-C\overset{O}{\underset{OH}{\diagup}} + Ag$$

an aldehyde   a carboxylic acid   silver metal

The metallic silver deposits on the glass and forms a mirror (see Figure 18–2).

Benedict's and Fehling's reagents are mild oxidizing agents composed of Cu^{2+} ions in alkaline solution. The Cu^{2+} ions can oxidize aldehydes, but not ketones. As the aldehyde is oxidized to an acid, the Cu^{2+} is reduced to Cu^+, which precipitates as Cu_2O (a brick-red solid). The Cu^{2+} ions are blue, and the reaction converts the blue color to a brick-red color.

$$R-C\overset{O}{\underset{H}{\diagup}} + Cu^{2+} \xrightarrow{NH_3} R-C\overset{O}{\underset{OH}{\diagup}} + Cu_2O$$

aldehyde   (blue solution)   acid   (red precipitate)

FIGURE 18–2 Tollens' reagent, a mild oxidizing agent composed of a solution of silver nitrate in dilute aqueous ammonia, is used to detect the presence of an aldehyde.

Both Benedict's and Fehling's reagents are used extensively to test urine for the blood sugar *glucose*, which contains an aldehyde group. Normal urine does not contain glucose. However, certain conditions or diseases such as diabetes cause glucose to appear in urine, and its rapid detection is a useful diagnostic tool. Positive results from such a test result in a color change ranging from yellow-green (0.5% glucose) to yellow-orange (1% glucose) to brick red (2% glucose).

EXERCISES AND PROBLEMS

The Carbonyl Group

18.1 Indicate which of the following compounds contain a carbonyl group.

a. $CH_3-CH_2-C(=O)OH$

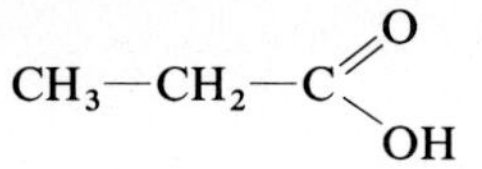

b. $CH_3-C(=O)-O-CH_3$

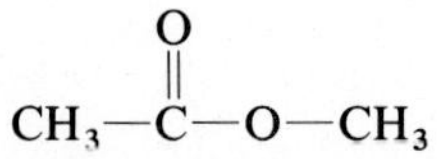

c. $CH_3-CH_2-O-CH_3$

d. $CH_3-CH_2-CH_2-OH$

e. $CH_3-CH_2-C(=O)OH$

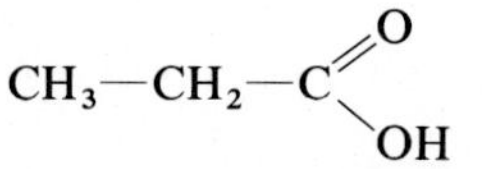

f. $CH_3-CH_2-C(=O)CH_3$

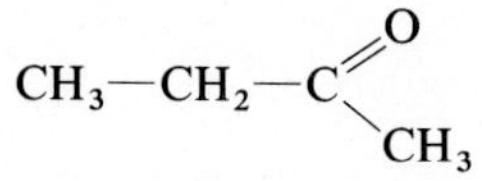

g. $CH_3-CH_2-C(=O)NH_2$

h. $CH_3-CH_2-C(OH)H$

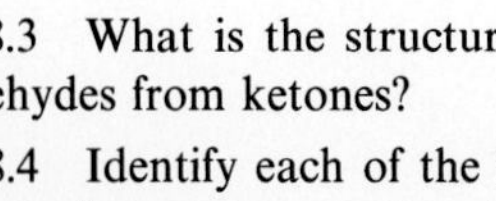

18.2 Discuss the similarities and differences between the bonding in a carbonyl group and a carbon–carbon double bond.

Structures of Aldehydes and Ketones

18.3 What is the structural feature that distinguishes aldehydes from ketones?

18.4 Identify each of the following structures as an aldehyde, ketone, or another class of compounds.

a. $CH_3-C(=O)-CH_2-CH_3$

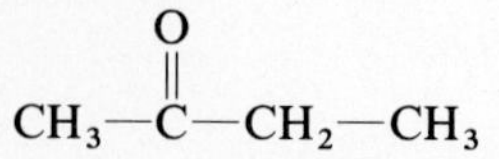

b. $CH_3-CH_2-CH_2-C(=O)O-CH_3$

c. $CH_3-CH_2-C(=O)CH_3$

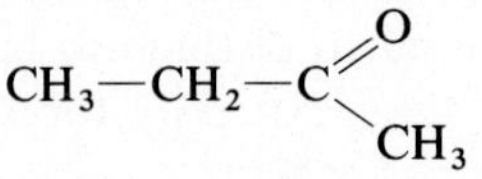

d. $CH_3-CH_2-C(=O)-H$

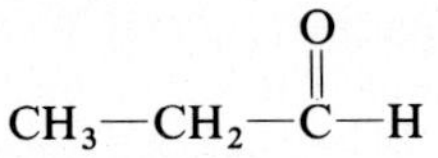

e. $CH_3-CH_2-CH(CH_3)-C(=O)H$

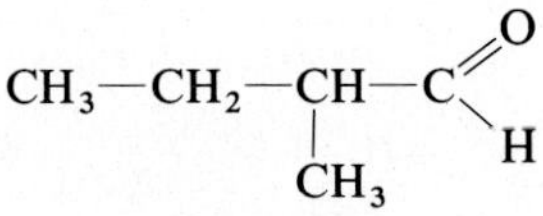

f. $CH_3-C(CH_3)(CH_3)-CH_2-CHO$

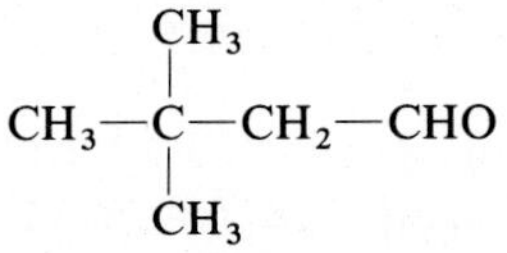

g. $CH_3-O-CH_2-CH_3$

h. $H-C(=O)-CH_3$

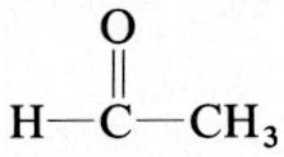

18.5 Identify each of the following structures as an aldehyde, ketone, or another class of compound.

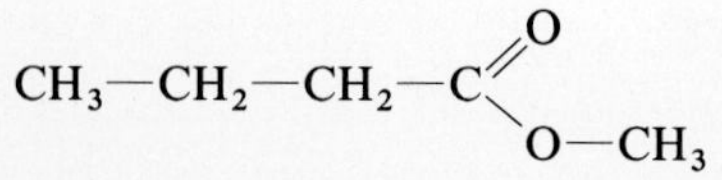

a. cyclohexane with OH and CH₃ substituents

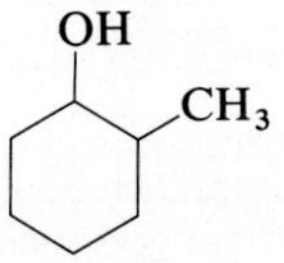

b. benzaldehyde (C₆H₅—CHO)

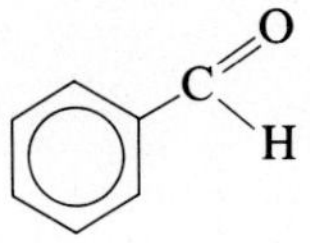

c. methyl benzoate (C₆H₅—C(=O)—O—CH₃)

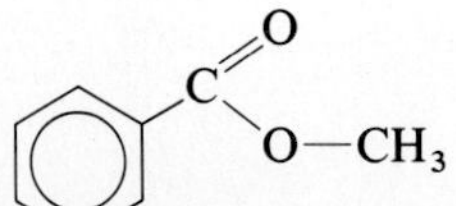

d. cyclohexanone

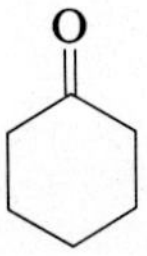

e. acetophenone (C₆H₅—C(=O)—CH₃)

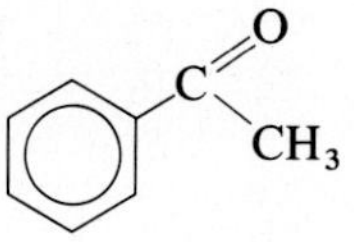

f. cyclohexane with C(=O)H and CH₃ substituents

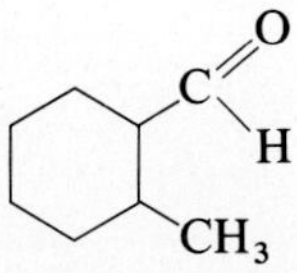

18.6 Draw the structures of the simplest aldehyde and the simplest ketone.

Nomenclature for Aldehydes and Ketones

18.7 Draw structural formulas for each of the following aldehydes.

a. 3-methylpentanal
b. formaldehyde
c. 2-ethylhexanal
d. 3,4-dimethylheptanal
e. propionaldehyde
f. 2,2-dichloropropanal
g. 2,4,5-trimethylheptanal
h. chloroacetaldehyde

18.8 Draw structural formulas for each of the following ketones.
a. 3-methyl-2-pentanone
b. 3-hexanone
c. cyclobutanone
d. 2,4-dimethyl-3-pentanone
e. ethyl phenyl ketone
f. acetone
g. dimethyl ketone
h. 2-propanone

18.9 Two ketones that act as alarm pheromones in several species of ants are 2-heptanone and 4-methyl-3-heptanone. Draw the structures of these pheromones.

18.10 Using IUPAC nomenclature, name each of the following compounds.

a. $CH_3-CH_2-CH_2-CH_2-C\overset{O}{\underset{H}{\diagup}}$

b. $CH_3-CH_2-CH-CH_2-C\overset{O}{\underset{H}{\diagup}}$
       $\quad\quad\quad\quad\;\; |$
       $\quad\quad\quad\quad CH_2-CH_3$

c. $CH_3-\underset{\underset{CH_3-CH_2}{|}}{CH}-\overset{O}{\overset{\|}{C}}-\underset{\underset{CH_2-CH_3}{|}}{CH}-CH_3$

d. Ph$-CH_2-\overset{O}{\overset{\|}{C}}-CH_3$

e. Ph$-CH_2-CH_2-\overset{O}{\overset{\|}{C}}-H$

f. CH_3-CH_2-CHO

g. $H-\overset{O}{\overset{\|}{C}}-CH_2-\underset{\underset{CH_3}{|}}{\overset{\overset{CH_3}{|}}{C}}-CH_3$

h. $Br-CH_2-CH_2-\overset{O}{\overset{\|}{C}}-CH_2-CH_3$

18.11 Using IUPAC nomenclature, name each of the following compounds.

a. cyclohexanone (cyclohexyl =O)

b. cyclohexyl–C(=O)–H

c. cyclohexyl–C(=O)–H

d. cyclohexyl–C(=O)–CH₃

e. cyclohexyl–C(=O)–cyclohexyl

f. cyclohexanone with –CH₃ substituent

g. cyclohexanone with CH₃ at 3-position

h. cyclohexanone with CH₃ at 2-position

18.12 Explain why the following names could never be correct names for any compound.
a. cyclopentanal
b. 1-pentanone

Physical Properties of Aldehydes and Ketones

18.13 Hydrogen bonds between aldehydes or ketones and water molecules occur readily. Draw a diagram showing the relevant interactions between water and acetone.

18.14 Explain, in terms of structure, why aldehydes and ketones have lower boiling points than alcohols that have similar molecular weights.

18.15 Why is hydrogen bonding between aldehyde molecules or ketone molecules in the pure state impossible?

Uses of Aldehydes and Ketones

18.16 Laboratory glassware often gets an oily, sticky film on the surface that does not wash off easily with water. Explain why acetone helps out so effectively in cleaning the glassware.

18.17 Name a ketone other than acetone that is used as a solvent.

18.18 What is "formalin" and what are two of its uses?

18.19 Give names and structures of three aldehydes that are used as flavoring agents.

Preparation of Aldehydes and Ketones

18.20 Draw the structure of the alcohol needed to prepare each of the following substances by oxidation of the alcohol.
a. ethanal
b. diethyl ketone
c. 1-phenyl-2-propanone
d. acetaldehyde
e. acetone
f. 2-ethylhexanal

Reduction of Carbonyl Compounds

18.21 Draw the structure of the major organic compound produced in the following reactions.

a. $CH_3-\overset{O}{\underset{\|}{C}}-CH_2-CH_3 \xrightarrow[\text{heat, pressure}]{H_2,\ Ni}$

b. 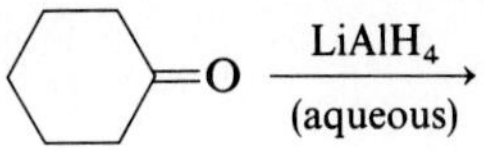 cyclohexanone =O $\xrightarrow[\text{(aqueous)}]{LiAlH_4}$

c. $CH_3-C\overset{O}{\underset{H}{\diagdown}} \xrightarrow[\text{(enzyme catalyst)}]{NADH,\ H^+}$

d. 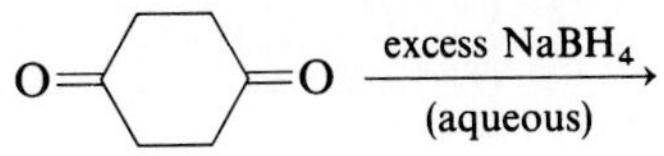 O= cyclohexane =O $\xrightarrow[\text{(aqueous)}]{\text{excess } NaBH_4}$

Hemiacetals and Acetals

18.22 Identify each of the following compounds as a hemiacetal, an acetal, or neither.
a. $CH_3-CH_2-O-CH_3$
b. CH_3-O
   $\quad\ |$
   $\ \ CH-OH$
   $\quad\ |$
   $\ \ CH_3$
c. $CH_3-O-CH_2-O-CH_3$
d. $CH_3-O-CH_2-CH_2-OH$
e. $\qquad O-CH_3$
   $\qquad\ \ |$
   CH_3-C-CH_3
   $\qquad\ \ |$
   $\qquad O-CH_3$
f. $CH_3-CH-OH$
   $\qquad\ \ |$
   $\ \ \ \ CH_2-OH$
g. 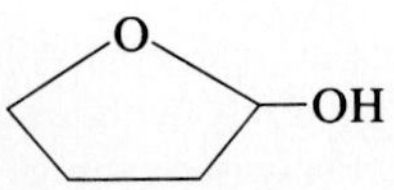 (cyclic ether with —OH)
h. 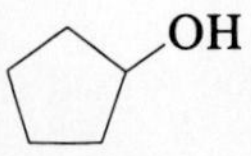 (cyclic with OH)

18.23 Write the structures of the hemiacetal and acetal that would form from each of the following pairs of reactants.

a. acetaldehyde and ethyl alcohol
b. ethanal and methanol
c. butanal and ethanol
d. formaldehyde and isopropyl alcohol

18.24 Draw the structure of the missing compound in each of the following reactions.

a. $CH_3-CH_2-CH_2-C\overset{O}{\underset{H}{\diagdown}}\ +$

$CH_3-CH_2-OH \xrightarrow{H^+} \ ?$

b. $?\ +\ CH_3OH \xrightarrow{H^+} CH_3-CH_2-\underset{\underset{O-CH_3}{|}}{\overset{\overset{OH}{|}}{CH}}$

c. $CH_3-\underset{\underset{CH_3}{|}}{CH}-CH_2-\underset{\underset{O-CH_3}{|}}{\overset{\overset{OH}{|}}{CH}}\ +$

$CH_3-CH_2-OH \xrightarrow{H^+} ?\ +\ H_2O$

d. 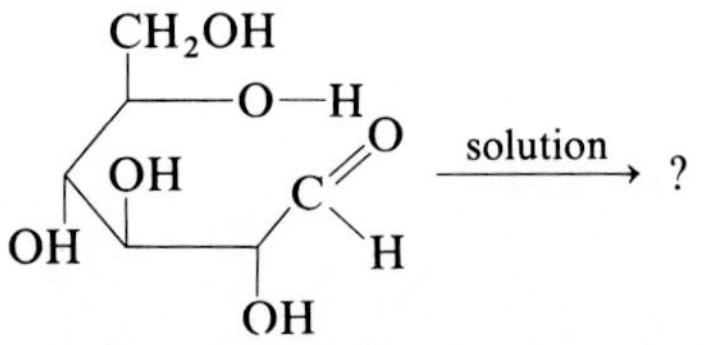 $\begin{array}{c} CH_2OH \\ \text{OH} \\ \text{OH} \\ \ \ \text{OH} \end{array}\ \overset{O-H}{\underset{H}{C}}\xrightarrow{\text{solution}} ?$

18.25 The compound 3-hydroxypropanal can form an intramolecular cyclic hemiacetal. Draw the structure of this hemiacetal.

Identification Tests for Aldehydes and Ketones

18.26 Tollens' test utilizes a redox reaction to test for a particular type of functional group.
a. What functional group gives a positive Tollens' test?
b. Describe the visible evidence for a positive Tollens' test.
c. What species is oxidized?
d. What species is reduced?

18.27 Benedict's and Fehling's reagents are also utilized in identifying whether a certain functional group is present. In addition, they can be used to determine the *quantity* present in solution.

a. Describe the color change for a positive Benedict's test.
b. What ion is responsible for the color change?
c. Is this ion oxidized or reduced?
d. What biological fluid can be readily tested with these tests?
e. A positive test means what kind of sugar is present?
f. Is this a good test for ketones? Explain.

18.28 An unknown Compound A ($C_4H_{10}O$) quickly reacts with $K_2Cr_2O_7$ in acidic solution to yield Compound B (C_4H_8O). Compound B does not give a positive result to Benedict's test, but it is easily reduced by $NaBH_4$ back to Compound A. What are the structures and names of Compounds A and B?

18.29 Two possible isomers exist for C_3H_6O that contain carbonyl groups. One gives a positive Tollens' test. Draw structures for both isomers, name each one, and identify the one that will form the silver mirror.

Multi-Step Syntheses Involving Aldehydes and Ketones

18.30 Consider the following series of reactions. Name the general type of reaction taking place at each step and draw structures of all compounds indicated by Roman numerals. (Some review of old reactions from previous chapters may be necessary.) Write NR for no reaction.

a. $CH_2=CH-CH_3 \xrightarrow{H_2O, H^+} I \xrightarrow[(H^+)]{K_2Cr_2O_7} II$

b. $Ph-CH=CH_2 \xrightarrow{H_2O, H^+} I \xrightarrow[(H^+)]{K_2Cr_2O_7} II$
$\qquad\qquad\qquad\searrow{CH_3OH, H^+} III$

c. 1,2-dimethylcyclohexene $\xrightarrow{H_2O, H^+} I \xrightarrow[(H^+)]{K_2Cr_2O_7} II$

d. 2-methylcyclopentanol $\xrightarrow[(180°)]{H_2SO_4} I \xrightarrow{H_2O, H^+} II \xrightarrow[(H^+)]{K_2Cr_2O_7} III$
$\qquad\qquad\qquad\searrow{K_2Cr_2O_7 \; (H^+)} IV$

19 Carboxylic Acids and Esters

Objectives

After completing Chapter 19, you will be able to:

19.1 List the structural features of the carboxyl group and carboxylic acids.
19.2 Determine both common and IUPAC nomenclature for carboxylic acids.
19.3 List the general physical properties of carboxylic acids.
19.4 List the types of chemical reactions utilized in the synthesis of carboxylic acids.
19.5 Relate the structures of carboxylic acids to their acidic behavior in solution.
19.6 Write equations for the production of carboxylic acid salts and discuss their properties, uses, and nomenclature.
19.7 List the structural features of the ester functional group.
19.8 State the general chemical reaction for the preparation of esters from carboxylic acids and alcohols.
19.9 Determine both common and IUPAC nomenclature for esters.
19.10 List the general physical properties of esters.
19.11 Write chemical equations for ester hydrolysis and saponification.
19.12 Discuss the formation of polyester condensation polymers.
19.13 Discuss the preparation and uses of esters of inorganic acids.
19.14 Draw the structures of phosphoric, pyrophosphoric, and triphosphoric acids and their esters.

INTRODUCTION

In Chapter 18, we discussed the carbonyl group and two families of compounds—aldehydes and ketones—that contain this group. In this chapter, we will discuss two more families of compounds in which the carbonyl group is present—carboxylic acids and esters. In these compounds, however, another oxygen-containing entity is present in addition to the carbonyl group.

Carboxylic acids and esters are among the most widely distributed substances in nature. Acetic acid, the substance that gives vinegar its sour taste, is a carboxylic acid. Other common carboxylic acids include citric acid, which gives citrus fruits their sour taste; lactic acid, which is present in sour milk; and salicylic acid, the substance from which aspirin is made. Many of the fragrances of flowers and fruits, as well as the flavors of many fruits, are caused by the presence of esters. Esters are also the chemical components of many of the artificial fruit flavors that are used in cakes, candies, ice cream, and soft drinks. High-formula-weight esters are found in fats and oils, important entities in the functioning of human beings.

19.1 THE STRUCTURE OF CARBOXYLIC ACIDS

Carboxylic acids *are compounds whose characteristic functional group is the carboxyl group.* These compounds were some of the first organic compounds studied in detail because of their wide distribution and abundance in natural products. A **carboxyl group** *is composed of a hydroxyl group bonded to a carbonyl carbon atom.* The carboxyl functional group is written

$$-\text{C} \begin{smallmatrix} \diagup\diagup \text{O} \\ \diagdown \text{OH} \end{smallmatrix}$$

carboxylic acids

carboxyl group

The carboxyl functional group can also be written in abbreviated linear form.

$$-\text{COOH} \quad \text{or} \quad -\text{CO}_2\text{H}.$$

Although a carboxyl group structurally appears to be a combination of a carbonyl group and a hydroxyl group (hence the name *carboxyl*), this group should not be thought of in terms of the properties of its component parts. Instead, we must think of it as a whole, a group containing two oxygen atoms directly bonded to the same carbon atom. The chemical properties of a carboxyl group are distinctly different from those of either an alcohol or a carbonyl group.

The carbon atom in a carboxyl group must form one more bond in order to have the required four bonds for a carbon atom. This fourth bond can be to a hydrogen atom, an alkyl group (R), or an aromatic group (Ar).

$$\text{H}-\text{C} \begin{smallmatrix} \diagup\diagup \text{O} \\ \diagdown \text{OH} \end{smallmatrix} \quad \text{or} \quad \text{R}-\text{C} \begin{smallmatrix} \diagup\diagup \text{O} \\ \diagdown \text{OH} \end{smallmatrix} \quad \text{or} \quad \text{Ar}-\text{C} \begin{smallmatrix} \diagup\diagup \text{O} \\ \diagdown \text{OH} \end{smallmatrix}$$

The case where the carboxyl group is bonded to a hydrogen atom produces the simplest carboxylic acid—the one-carbon carboxylic acid *formic acid*. Formic acid is a naturally occurring substance and is the irritant in the "sting" of red ants. (*Formica* is the Latin word for "ant".)

When the carboxyl carbon is bonded to a methyl group (—CH$_3$), the result is the two-carbon carboxylic acid *acetic acid*.

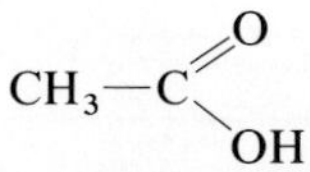

$$CH_3-C\begin{matrix}\nearrow O \\ \searrow OH\end{matrix}$$

acetic acid

Acetic acid is the most commonly encountered carboxylic acid. Vinegar is a dilute solution, approximately 5% (v/v), of acetic acid. (The Latin word for vinegar is *acetum*.) Pure acetic acid is called *glacial* acetic acid because it freezes into ice-like crystals in a cool room; its melting point is 17°C. A salt and vinegar solution is used as a preservation for pickled vegetables because of acetic acid's ability to inhibit microbial growth at moderate concentrations. Other low-molecular-weight carboxylic acids and their uses are listed in Table 19-2 on page 445.

The simplest aromatic carboxylic acid is *benzoic acid*; in this acid the carboxyl group is attached directly to a benzene ring

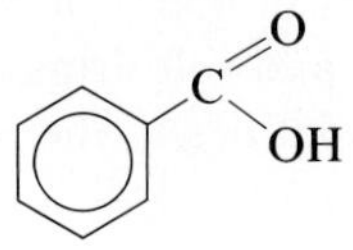

benzoic acid

Aspirin, the most widely used drug in America, can be considered a derivative of benzoic acid in which another functional group, in addition to the carboxyl group, is present on the benzene ring.

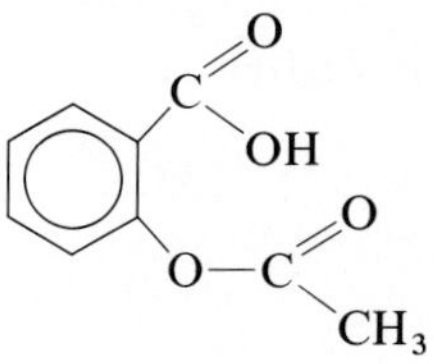

aspirin

The other functional group present in aspirin is an ester group; we will discuss this group later in this chapter. Another aromatic carboxylic acid used to relieve pain and inflammation is ibuprofen. Ibuprofen is the active ingredient in both Advil and Nuprin.

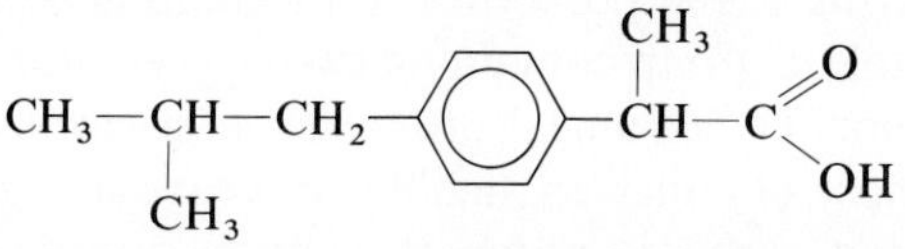

ibuprofen

There are also carboxylic acids that contain two or more carboxyl groups. Molecules that have two carboxyl groups are called *dicarboxylic acids* and those that have three carboxyl groups are called *tricarboxylic acids*.

The simplest dicarboxylic acid is *oxalic acid*.

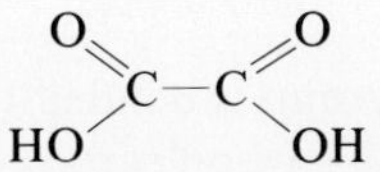

TABLE 19-1 Structures and Names of Selected Dicarboxylic Acids

Structure	Number of Carbons	Common Name
HOOC—COOH	2	oxalic acid
HOOC—CH$_2$—COOH	3	malonic acid
HOOC—CH$_2$—CH$_2$—COOH	4	*succinic acid
HOOC—CH$_2$—CH$_2$—CH$_2$—COOH	5	*glutaric acid
HOOC—CH$_2$—CH$_2$—CH$_2$—CH$_2$—COOH	6	adipic acid

*Biologically important

It is found in various plants including spinach, cabbage, and rhubarb. Commercially, oxalic acid is used to remove rust, bleach straw and leather, and remove ink stains. In high concentrations, oxalic acid is toxic to living organisms. The amount of oxalic acid present in spinach and cabbage is usually not harmful. However, rhubarb leaves should *not* be eaten because they contain much higher levels of oxalic acid. Table 19-1 lists the structures and common names of other simple dicarboxylic acids. Note that the acids with four and five carbon atoms are biologically important because they are intermediates in the citric acid cycle (Chapter 30).

There are many di- and tricarboxylic acids that contain one or more hydroxyl groups in addition to the acid groups. Many of these species are important in chemical cycles that occur within the human body. The dicarboxylic acid *malic acid* and the tricarboxylic acid *citric acid* are representative of these molecules. Both of these acids contain one —OH group in addition to the acid groups and both are important species in the metabolic reactions in which energy is obtained from sugars, fats, and proteins.

```
         COOH                    COOH
          |                       |
          |                      CH₂
     HO—CH                        |
          |                HO—C—COOH
         CH₂                      |
          |                      CH₂
         COOH                     |
                                 COOH
       malic acid              citric acid
```

19.2 NOMENCLATURE OF CARBOXYLIC ACIDS

The acid names used in the previous section were common names. Common-name usage is more prevalent for carboxylic acids than for any other family of organic compounds. Both the abundance of carboxylic acids in nature and the fact that these compounds were some of the earliest organic compounds to be studied (before IUPAC rules were in place) contribute to this situation. Nevertheless, IUPAC rules are available for naming carboxylic acids and you need to be acquainted with them.

As with previous IUPAC nomenclature, the longest carbon chain containing the functional group is identified and then it is numbered. In monocarboxylic acids, the carboxyl carbon atom is always the start of the carbon chain. Hence, the carboxyl carbon is the number 1 carbon and no number is required to locate the carboxyl group. The carboxylic acid name is formed by replacing the final *-e* of the hydrocarbon parent name with *-oic acid*. Thus, the following five-carbon unbranched carboxylic acid is named *pentanoic acid*.

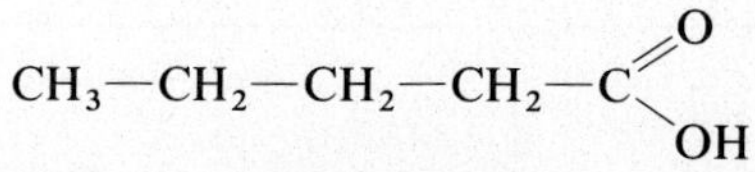

$$CH_3-CH_2-CH_2-CH_2-C\begin{smallmatrix}O\\\\OH\end{smallmatrix}$$

pentanoic acid

As in other families of compounds, the groups attached to the carbon chain are named and located numerically. Aromatic acids are named as derivatives of benzoic acid (the simplest aromatic carboxylic acid) in a manner similar to the way phenols were named (Section 17.2).

Example 19.1

Using IUPAC nomenclature, name the following carboxylic acids.

a. $CH_3-CH_2-CH_2-C(=O)OH$

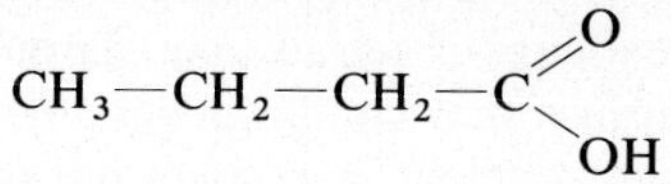

b. $CH_3-CH_2-CH_2-CH(CH_3)-CH_2-C(=O)OH$

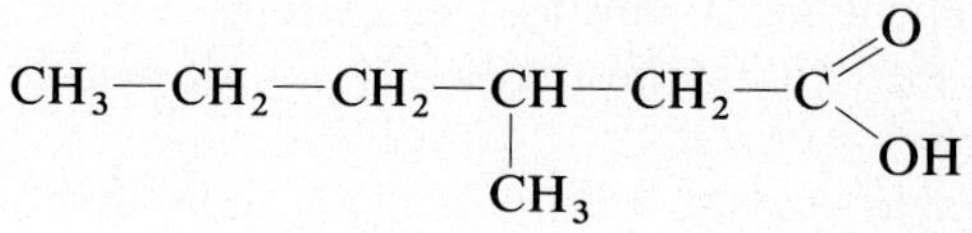

c. $Cl_3C-C(=O)OH$ (with three Cl on the first carbon)

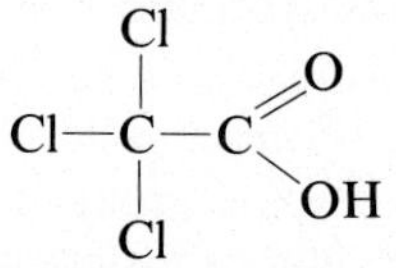

d. 3-ethyl substituted benzoic acid (benzene ring with COOH and CH₂CH₃ groups)

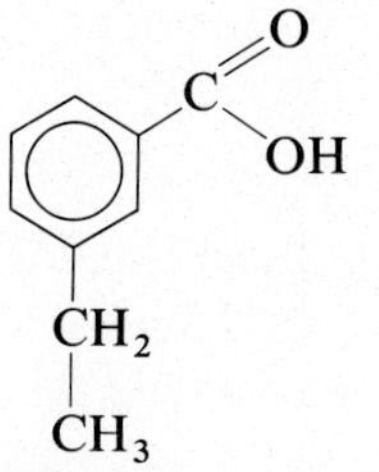

Solution

a. Four carbon atoms in the longest chain means that *butane* is the root name. Replacing the *-e* with *-oic acid* gives the IUPAC name *butanoic acid*.
b. The longest chain contains six carbons and a methyl group is attached to carbon number 3. The IUPAC name is *3-methylhexanoic acid*.
c. The formal IUPAC name here is *2,2,2-trichloroethanoic acid*. In the laboratory, this acid is often called by the common names "TCA" or trichloracetic acid. It is used to precipitate protein or DNA.
d. This is a derivative of benzoic acid and has an ethyl group attached to the ring at carbon number 3. The IUPAC name is *3-ethylbenzoic acid* (or *m*-ethylbenzoic acid).

Table 19–2 gives other examples of IUPAC and common names for certain carboxylic acids. Look particularly at the last two entries in the table. They show how IUPAC nomenclature is used to name carboxylic acids that contain more than one carboxyl group.

TABLE 19-2 Structure, Nomenclature, and Uses of Selected Carboxylic Acids

Common Name	IUPAC Name	Structural Formula	Characteristics and Typical Uses
formic acid	methanoic acid	H—C(=O)OH	stinging agents of red ants, bees, and nettles
acetic acid	ethanoic acid	CH_3—C(=O)OH	active ingredient in vinegar
propionic acid	propanoic acid	CH_3—CH_2—C(=O)OH	salts of this acid are used as mold inhibitors in breads and cereals
butyric acid	butanoic acid	CH_3—$(CH_2)_2$—C(=O)OH	odor-causing agent in rancid butter; present in human perspiration and vomit
caproic acid	hexanoic acid	CH_3—$(CH_2)_4$—C(=O)OH	characteristic odor of limburger cheese
lactic acid	2-hydroxypropanoic acid	CH_3—CH(OH)—C(=O)OH	found in sour milk and sauerkraut; formed in muscles during exercise
oxalic acid	ethanedioic acid	HO(O=)C—C(=O)OH	poisonous material in leaves of some plants such as rhubarb; used as cleaning agent for rust stains on fabric and porcelain
citric acid	3-hydroxy-3-carboxy-pentanedioic acid	HO(O=)C—CH_2—C(OH)(COOH)—CH_2—C(=O)OH	present in citrus fruits, used as a flavoring agent in foods

A number of unsaturated carboxylic acids that have one or more double bonds in the carbon chain are important components of fats and oils (Chapter 23). We name these compounds by replacing the *-ane* ending of the parent hydrocarbon's name with *-enoic*. An important example of this type of compound is *oleic acid* (common name), an 18-carbon monocarboxylic acid in which one double bond is present.

$$CH_3-(CH_2)_7-CH=CH-(CH_2)_7-C(=O)OH$$

The IUPAC name for this compound is *9-octadecenoic acid*. (The name for an 18-carbon chain is octadecane.)

Keto carboxylic acids, which contain a carbonyl group, are important substances in energy production cycles within the human body. These acids are named as follows.

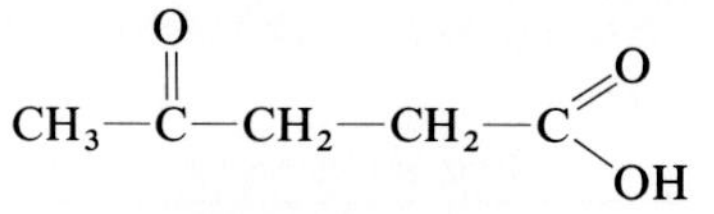

$$CH_3-\overset{\overset{O}{\|}}{C}-CH_2-CH_2-C\overset{\nearrow O}{\underset{\searrow OH}{}} \qquad CH_3-\overset{\overset{O}{\|}}{C}-C\overset{\nearrow O}{\underset{\searrow OH}{}}$$

4-ketopentanoic acid 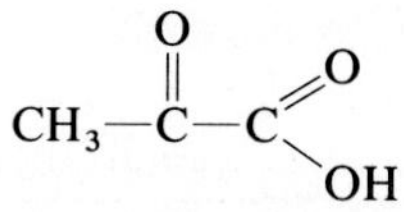   2-ketopropanoic acid

There is an alternate method for locating the positions of groups attached to the parent carbon chain in a carboxylic acid that is based on the Greek alphabet. The first four letters of the Greek alphabet are alpha (α), beta (β), gamma (γ), and delta (δ). The alpha carbon atom is carbon 2, the beta is carbon number 3, and so on. No group can be bonded to the carboxyl carbon, so the Greek letter system always starts with carbon number 2.

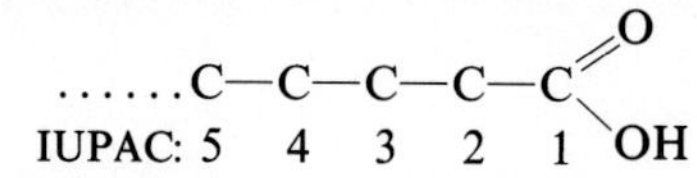

$$\ldots\ldots C-C-C-C-C\overset{\nearrow O}{\underset{\searrow OH}{}}$$

IUPAC: 5   4   3   2   1   OH
Greek letter: δ γ β α

Using the Greek letter system, the compound

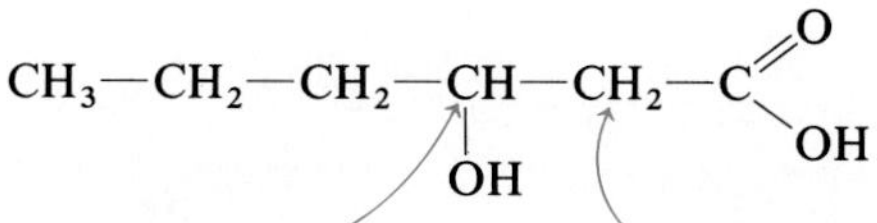

$$CH_3-CH_2-CH_2-\underset{OH}{CH}-CH_2-C\overset{\nearrow O}{\underset{\searrow OH}{}}$$

β carbon α carbon

would be named β-hydroxyhexanoic acid. β-Hydroxyacids are intermediates in the metabolism of fats (Section 31.3).

19.3 PHYSICAL PROPERTIES OF CARBOXYLIC ACIDS

Low-molecular-weight carboxylic acids are liquids at room temperature and have sharp or unpleasant odors; larger molecular-weight acids that have 10 or more carbons are waxlike solids and have little or no odor. Aromatic acids, as well as dicarboxylic acids, are odorless, crystalline solids at room temperature.

Because of the polarity of the carboxyl group and the presence of a proton on the singly bonded oxygen, carboxylic acids form strong hydrogen bonds among themselves, to water, and to other molecules (see Figure 19–1). In fact, acetic acid molecules form such strong hydrogen bonds among themselves that vapors es-

FIGURE 19–1   Hydrogen bonding situations involving carboxylic acids.

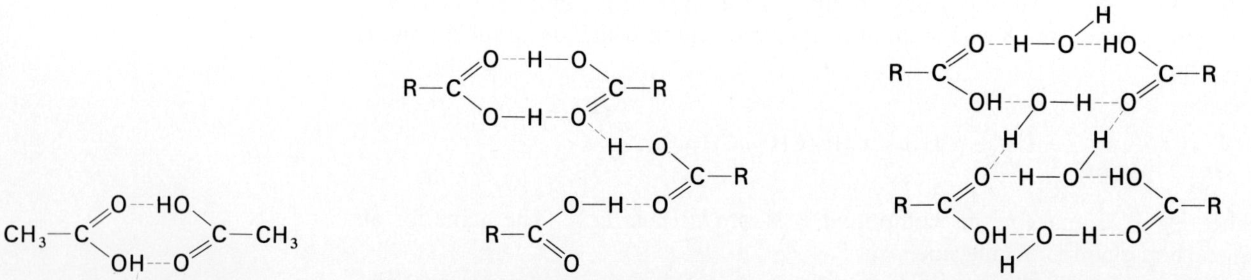

a. acetic acid "dimers"   b. between acid molecules   c. between water and acid molecules

caping from boiling acetic acid contain dimers (Figure 19–1) connected by hydrogen bonds! This characteristic explains the high boiling point of carboxylic acids when compared to compounds of similar molecular weight that do not exhibit hydrogen bonding. This also explains why low-molecular-weight carboxylic acids are soluble in water.

Certain acids exhibit profound biological effects because of the physical property of odor. One striking example is a carboxylic acid used by a queen bee to regulate the life of her entire honeybee colony. The queen secretes a mixture of pheromones, one of which is called the "queen substance", trans-9-keto-2-decenoic acid.

$$CH_3-\overset{\overset{O}{\|}}{C}-CH_2-CH_2-CH_2-CH_2-CH_2-CH=CH-C\overset{O}{\underset{OH}{\diagdown}}$$

This keto-acid attracts the workers and causes them to follow the queen during swarming.

In the hive, the workers (all female) constantly lick the queen, thereby obtaining some of the acid; this inhibits the growth of their ovaries. This acid, together with *trans*-9-hydroxy-2-decenoic acid (also a glandular secretion), prevents the rearing of new queens. The queen substance is also an effective sex attractant for the drones during the nuptial flight and is an aphrodisiac to all members of the colony. When the queen dies, the reproductive cycle of the workers is no longer inhibited and they begin to lay eggs. Some of these eggs will develop into queens, thus completing the cycle. Similarly, new queens are reared if the secretion of the queen bee substance diminishes or if the colony becomes too large for all the workers to obtain the limiting signal.

19.4 PREPARATION AND REACTIONS OF CARBOXYLIC ACIDS

As we discussed earlier (Sections 17.6 and 18.8), we can easily synthesize carboxylic acids through the oxidation of primary alcohols or aldehydes by strong oxidizing agents such as $KMnO_4$ or $K_2Cr_2O_7$.

$$\text{primary alcohol} \xrightarrow[\text{oxidation}]{\text{mild}} \text{[aldehyde]} \xrightarrow[\text{oxidation}]{\text{further}} \text{carboxylic acid}$$

The difficulty in synthesizing aldehydes by alcohol oxidation was that the resulting aldehyde often oxidized further to the corresponding acid. Acids, on the other hand, are quite stable to further oxidation by ordinary laboratory reagents.

The only way the carbon of the carboxyl group can be oxidized further is to form carbon dioxide (O=C=O). This results in the loss of the carboxyl group and is not often useful in laboratory syntheses. However, decarboxylation is an extremely useful biological reaction. For example, pyruvic acid is enzymatically decarboxylated to form acetaldehyde and carbon dioxide in microorganisms.

$$\underset{\text{pyruvic acid}}{CH_3-\overset{\overset{O}{\|}}{C}-C\overset{O}{\underset{OH}{\diagdown}}} \xrightarrow[\text{enzyme}]{\text{decarboxylase}} \underset{\text{acetaldehyde}}{CH_3-C\overset{O}{\underset{H}{\diagdown}}} + CO_2$$

Carbon dioxide is formed in the body by similar kinds of enzymatic decarboxylation reactions. Through the loss of carbon dioxide, sugars with *n* carbons can be converted into sugars with $n-1$ carbons. There are also enzymes that can add *carboxylate groups* to molecules; they add carbon dioxide and lengthen carbon chains by one carbon atom.

Derivatives of carboxylic acids are synthesized through substitution reactions in which the —OH group of the carboxylic acid is replaced by other groups. Carboxylic acids react with thionyl chloride ($SOCl_2$) to form acid chlorides; in these substances, a Cl atom is substituted for the —OH group of the carboxylic acid.

$$R-C(=O)OH + SOCl_2 \longrightarrow R-C(=O)Cl + SO_2 + HCl$$

thionyl chloride   acid chloride

Acid chlorides are extremely reactive substances. They must be kept away from moisture or they will hydrolyze back to the acid. Acid chlorides are much more reactive than their parent acids, and as a result, they are frequently used in the place of carboxylic acids in the synthesis of other acid derivatives (Section 19.8).

Carboxylic acids react with alcohols to form esters. The general equation for ester formation is

$$R-C(=O)OH + HO-R' \xrightarrow{\text{acid catalyst}} R-C(=O)O-R' + H_2O$$

alcohol   ester
(R can be H but R' cannot be H.)

Esters differ from carboxylic acids with respect to the group bonded to the non-carbonyl oxygen atom. In acids, a hydrogen atom is bonded to the noncarbonyl oxygen atom, and in esters, an alkyl or aryl group is bonded to that atom. Many biologically important substances are esters. Sections 19.7 through 19.13 of this chapter will discuss the ester functional group in detail.

When carboxylic acids are heated with ammonia (NH_3), the —OH group of the acid is replaced by a —NH_2 group to form an amide.

$$R-C(=O)OH + NH_3 \xrightarrow{\text{heat}} R-C(=O)NH_2 + H_2O$$

ammonia   amide

The nitrogen-containing amide functional group will be discussed in detail in Chapter 20.

19.5 ACIDITY OF CARBOXYLIC ACIDS

Acids are compounds that have the ability to donate protons (Section 10.1). In aqueous solution, carboxylic acids donate protons to water molecules.

$$R-C(=O)OH + H_2O \rightleftharpoons R-C(=O)O^- + H_3O^+$$

carboxylate ion

The portion of the acid molecule left after proton loss is called a carboxylate ion. **A carboxylate ion** *is the negative ion produced when a carboxylic acid loses one (or more) acidic protons.*

carboxylate ion

Carboxylate ions are named by dropping the *-ic acid* ending from the name of the parent acid and replacing it with *-ate*. For example, the carboxylate ion obtained from acetic acid is the *acetate ion* and benzoic acid produces the *benzoate ion*.

$$CH_3-C(=O)OH + H_2O \rightleftharpoons CH_3-C(=O)O^- + H_3O^+$$
$$\text{acetic acid} \qquad\qquad\qquad \text{acetate ion}$$

$$C_6H_5-C(=O)OH + H_2O \rightleftharpoons C_6H_5-C(=O)O^- + H_3O^+$$
$$\text{benzoic acid} \qquad\qquad\qquad \text{benzoate ion}$$

Although carboxylic acids are the most acidic organic compounds, they are weak acids when compared to inorganic acids such as HCl and HNO_3. Most carboxylic acids are less than 2 to 3% ionized in water. Hydrochloric acid and HNO_3 are virtually 100% ionized in water (Section 10.2). Note that the double arrow notation used in equations in this section conveys the message that carboxylic acids are weak acids. The arrow directed to the left is much longer than the arrow directed to the right; this indicates that the equilibrium position lies to the left.

As we have already seen for inorganic acids (Section 10.6), carboxylic acids are neutralized by bases to produce a salt and water. If benzoic acid, C_6H_5COOH, and sodium hydroxide, NaOH, are combined, they produce the salt sodium benzoate and water.

$$C_6H_5-C(=O)OH + NaOH \longrightarrow C_6H_5-C(=O)O^-Na^+ + H_2O$$
$$\text{benzoic acid} \qquad\qquad\qquad \text{sodium benzoate}$$

Another example of a neutralization reaction involving a carboxylic acid is the reaction between propanoic acid, C_2H_5COOH, and potassium hydroxide, KOH, to form the salt potassium propanoate and water.

$$CH_3-CH_2-C(=O)OH + KOH \longrightarrow CH_3-CH_2-C(=O)O^-K^+ + H_2O$$
$$\text{propanoic acid} \qquad\qquad\qquad \text{potassium propanoate}$$

Dicarboxylic acids require two moles of base for neutralization. Carboxylate ions in this case carry a -2 charge.

$$HO(O=)C-C(=O)OH + 2\,NaOH \longrightarrow Na^+{}^-O(O=)C-C(=O)O^-Na^+ + 2\,H_2O$$
$$\text{oxalic acid} \qquad\qquad\qquad \text{sodium oxalate}$$

Similarly, tricarboxylic acids require three moles of base for neutralization.

19.6 SALTS OF CARBOXYLIC ACIDS

Carboxylate salts are usually much more soluble in water than the carboxylic acid from which they were derived. Drugs and medicines that contain acid groups are often sold commercially as the sodium or potassium salt of the acid. This greatly

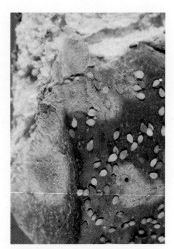

FIGURE 19–2 Bread that is not treated with calcium propionate will soon grow mold.

enhances the solubility of the medicine, and thus its effectiveness and ease of absorption by the body.

The salt *calcium oxalate* is an exception to the generalization that carboxylate salts are more soluble than the parent acid; it is rather insoluble in water. This is the reason for the toxicity of the oxalate ion; it depletes the body of soluble calcium. Some carboxylate salts are used as food preservatives (see Figure 19–2). Sodium benzoate is used in numerous foods including cider, ketchup, carbonated beverages, relishes, syrups, and dressings. Calcium propionate is added to bread as a preservative to prevent mold from growing.

Soap molecules are salts of long-chain acids that occur naturally in animal fat. When these "fatty acids" are treated with a base such as lye, an impure form of NaOH, they form carboxylate salts. These salts are the active ingredient in soap.

The cleansing action of soap is directly related to the structure of the carboxylate salt molecules present. Their structure is such that they exhibit dual polarity. The hydrocarbon portion of the molecule is nonpolar and the carboxylate portion of the molecule is polar. This dual polarity for the fatty acid salt *sodium stearate*, which is representative of all soap molecules, is as follows.

$$\underbrace{CH_3-(CH_2)_{16}}_{\text{nonpolar portion}}-\underbrace{C{\overset{\displaystyle O}{\underset{\displaystyle O^-Na^+}{\diagup}}}}_{\text{polar portion}}$$

sodium stearate

Soap solubilizes oily and greasy materials in the following manner: the nonpolar portion of the molecule dissolves in the oil or grease, while the polar portion maintains its solubility in water. Let us look closer at this solubilizing process.

FIGURE 19–3 The cleansing action of soap is caused by the ability of soap molecules to form micelles that encapsulate grease and carry it away.

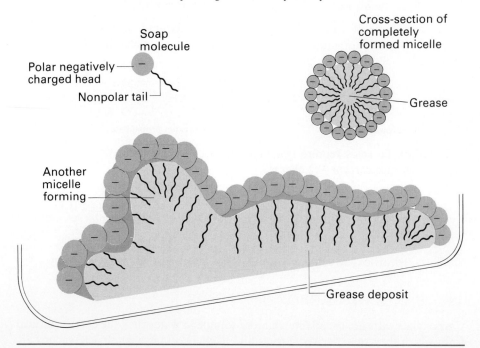

The penetration of the oil or grease by the nonpolar end of the soap molecules is followed by the formation of *micelles*. **Micelles** *are small, spherical grease-soap droplets that are soluble in water as a result of the polar groups on their surface* (the heads of the soap molecules—see Figure 19–3). The soap heads, carboxylate groups (COO^-), and water molecules are attracted to each other, causing the solubilizing of the micelle. The micelles do not combine into larger drops because their surfaces are all negatively charged; like charges repel each other. The water-soluble micelles are subsequently rinsed away, leaving a material void of oil and grease.

micelle

19.7 THE ESTER FUNCTIONAL GROUP

Esters *are organic compounds whose characteristic functional group is*

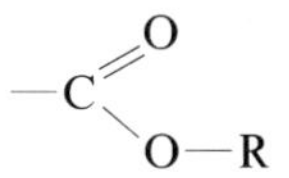

esters

$$-C\underset{O-R}{\overset{\parallel}{\overset{O}{}}}$$

The ester functional group is very similar to the acid functional group. The only difference is that the H atom of the —OH group of the acid has been replaced by a hydrocarbon group in the form of an alkyl group or an aromatic group.

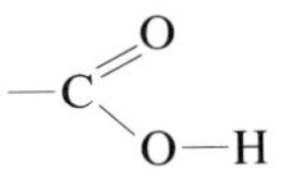

an acid functional group 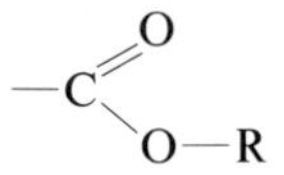   an ester functional group

In a linear form, the ester functional group can be represented as

—COOR   or   —CO$_2$R

Within the ester functional group, there are two carbon–oxygen single bonds. The carbon–oxygen single bond that involves the carbonyl carbon atom is called the *ester linkage*. This particular bond, which is highlighted in the following diagram, will be very important in our discussion of ester reactions.

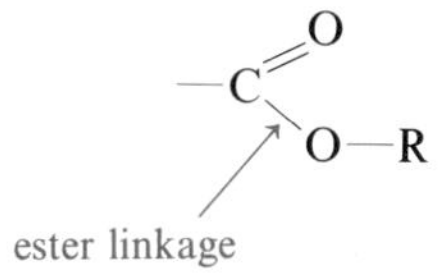

ester linkage

19.8 PREPARATION OF ESTERS

Esters are prepared from the chemical reaction between an alcohol and an acid chloride or a carboxylic acid.

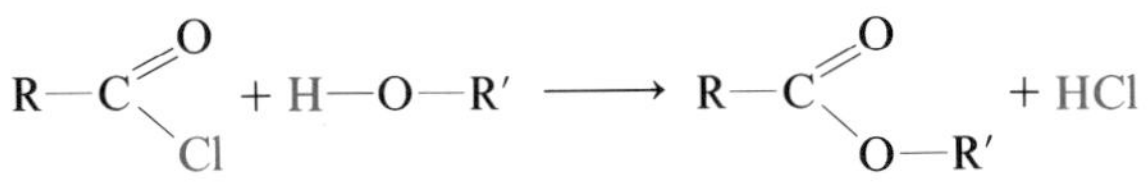

$$R-\underset{Cl}{\overset{O}{\overset{\parallel}{C}}} + H-O-R' \longrightarrow R-\underset{O-R'}{\overset{O}{\overset{\parallel}{C}}} + HCl$$

acid chloride   alcohol   ester

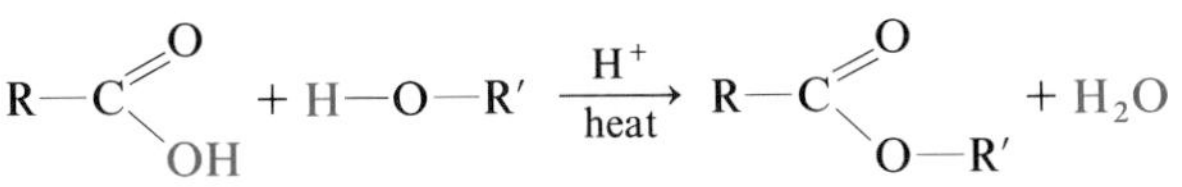

$$R-\underset{OH}{\overset{O}{\overset{\parallel}{C}}} + H-O-R' \xrightarrow[\text{heat}]{H^+} R-\underset{O-R'}{\overset{O}{\overset{\parallel}{C}}} + H_2O$$

carboxylic acid   alcohol   ester

In each of these reactions, the —OR' group from the alcohol bonds to the carbonyl carbon atom of the carboxyl group, which substitutes for a —Cl in the acid chloride and a —OH group in the carboxylic acid. The ester linkage (Section 19.7) is actually a newly formed bond that links the portion of the ester that came from the acid with the portion of the ester that came from the alcohol.

When a carboxylic acid is one of the reactants in esterification (ester formation), an acid catalyst and heat are needed. This is because carboxylic acids are less reactive than acid chlorides (Section 19.7).

Esterification with a carboxylic acid is a dehydration reaction; water is the inorganic product. This process is very similar to another dehydration process we have already discussed—one involving the reaction of two alcohol molecules to produce an ether (Section 17.6).

$$R-OH + H-O-R' \xrightarrow[140°C]{H_2SO_4} R-O-R' + H_2O$$

As a specific example of esterification, consider the following methods for producing the ester *ethyl acetate*. (The naming of esters will be covered in Section 19.9.)

$$CH_3-C(=O)Cl + HO-CH_2-CH_3 \longrightarrow CH_3-C(=O)-O-CH_2-CH_3 + HCl$$

acetyl chloride   ethyl alcohol   derived from the acid chloride   derived from the alcohol

$$CH_3-C(=O)OH + HO-CH_2-CH_3 \xrightarrow[heat]{H^+} CH_3-C(=O)-O-CH_2-CH_3 + H_2O$$

acetic acid   ethyl alcohol   derived from the acid   derived from the alcohol

The pain-reliever *aspirin* is produced from an esterification reaction. The acid involved is salicylic acid, a hydroxy derivative of benzoic acid.

salicylic acid

Salicylic acid was first isolated from plants of the genus *spirea*. For over 2000 years, herbal preparations containing this acid were used to relieve pain and fever, despite the side effects of mouth, throat, and stomach irritation. In the 1890s, esterification reactions led to the synthesis of two salicylic-acid based compounds that proved to be more effective and palatable pain-relievers than salicylic acid—aspirin (acetylsalicylic acid) and oil of wintergreen (methylsalicylate). *Aspirin* is a term coined from the "*a*" from *acetyl* and "*spir*" from the *spirea* family of plants.

Salicylic acid has both an acid group and an alcohol group, so it can form two different esters: one by reaction of the acid group with an alcohol and the other by reaction of the alcohol group with a carboxylic acid.

salicylic acid   acetic acid   aspirin

$$\underset{\text{salicylic acid}}{\underset{\text{OH}}{\underset{|}{\bigcirc}}\!\!-\!\!\overset{\overset{\text{O}}{\|}}{\text{C}}\!\!-\!\!\text{OH}} + \underset{\text{methanol}}{\text{H}\!-\!\text{O}\!-\!\text{CH}_3} \xrightarrow[\text{heat}]{\text{H}^+} \underset{\text{oil of wintergreen}}{\underset{\text{OH}}{\underset{|}{\bigcirc}}\!\!-\!\!\overset{\overset{\text{O}}{\|}}{\text{C}}\!\!-\!\!\text{O}\!-\!\text{CH}_3} + \text{H}_2\text{O}$$

Aspirin is probably the most widely used drug in the world. It has the ability to decrease pain (analgesic properties), lower body temperature (antipyretic properties), and reduce inflammation (anti-inflammatory properties). Aspirin is one of the few pain killers whose use does not result in physical addiction.

Aspirin is most frequently taken in tablet form. After ingestion, it is hydrolyzed (Section 19.11) to salicylic acid in the blood and liver. Salicylic acid is the active form of the drug. However, salicylic acid cannot be taken directly because it irritates the digestive tract. Despite aspirin's long use, only in the last 15–20 years have scientists begun to understand how it works in the body. We now know that aspirin inhibits prostaglandin synthesis. We will discuss prostaglandins in Chapter 23.

Oil of wintergreen, an ester also produced from salicylic acid, exhibits some of the same biological effects as aspirin. Because it is too toxic to be taken internally at therapeutic levels, it is used in skin rubs and liniments to help decrease the pain of sore muscles. It is absorbed through the skin, where it is hydrolyzed to produce salicylic acid; this gives local relief from pain.

Thioesters

In Section 17.13, we noted the existence of thiols, the sulfur analogs of alcohols. Thiols, like alcohols, will react with carboxylic acids to produce esters. These esters are called thioesters. **Thioesters** are esters in which the ester linkage involves a sulfur atom, rather than an oxygen atom.

thioester

$$\underset{\text{carboxylic acid}}{\text{CH}_3\!-\!\overset{\overset{\text{O}}{\|}}{\text{C}}\!-\!\text{OH}} + \underset{\text{thiol}}{\text{H}\!-\!\text{S}\!-\!\text{CH}_2\!-\!\text{CH}_3} \longrightarrow \underset{\text{thioester}}{\text{CH}_3\!-\!\overset{\overset{\text{O}}{\|}}{\text{C}}\!-\!\text{S}\!-\!\text{CH}_2\!-\!\text{CH}_3} + \text{H}_2\text{O}$$

Thioesters are very important biochemical compounds and are formed in the metabolism of both fats and sugars.

19.9 NOMENCLATURE OF ESTERS

The concept that an ester is an alcohol derivative of a carboxylic acid is the basis for naming esters in both the common and IUPAC systems of nomenclature. The name of the alcohol part of the ester is given first, and is followed by a separate word giving the name of the acid part of the ester. The name for the alcohol part of the ester is simply the name of the R group (alkyl or aromatic) present in the —OR portion of the ester. The name of the acid part of the ester is the name of the acid upon which the ester is based; this name is modified to end in *-ate*. Consider the ester derived from ethanoic acid (acetic acid) and methanol (methyl alcohol). Its name will be *methyl ethanoate* (IUPAC) or *methyl acetate* (common).

$$\underset{\substack{\text{ethanoic acid}\\\text{(acetic acid)}}}{\text{CH}_3\!-\!\overset{\overset{\text{O}}{\|}}{\text{C}}\!-\!\text{OH}} + \underset{\substack{\text{methanol}\\\text{(methyl alcohol)}}}{\text{HO}\!-\!\text{CH}_3} \longrightarrow \underset{\substack{\text{methyl ethanoate}\\\text{(methyl acetate)}}}{\text{CH}_3\!-\!\overset{\overset{\text{O}}{\|}}{\text{C}}\!-\!\text{O}\!-\!\text{CH}_3} + \text{H}_2\text{O}$$

Example 19.2 considers additional examples of ester nomenclature.

Example 19.2

Using IUPAC nomenclature, name the following esters.

a.
$$CH_3-CH_2-C{\overset{O}{\underset{O-CH_2-CH_3}{\diagdown}}}$$

b.
$$CH_3-CH(CH_3)-CH_2-C{\overset{O}{\underset{O-CH_2-CH_3}{\diagdown}}}$$

c.
$$H-C{\overset{O}{\underset{O-CH_3}{\diagdown}}}$$

d.
$$C_6H_5-C{\overset{O}{\underset{O-CH_2-CH_2-CH_3}{\diagdown}}}$$

Solution
a. This ester is derived from the alcohol *ethanol* and the acid *propanoic acid*. Its name is *ethyl propanoate*.
b. Here, the alcohol is the two-carbon alcohol *ethanol*, and the acid is the five-carbon acid, *3-methylbutanoic acid*. The name of the ester is *ethyl 3-methylbutanoate*.
c. The one-carbon alcohol *methanol*, and the one-carbon acid, *methanoic acid* (formic acid) are the source of this ester. The ester is named *methyl methanoate* (IUPAC) or *methyl formate* (common).
d. The source of the three-carbon —OR group is 1-propanol. The rest of the molecule is derived from benzoic acid. The name of the ester is *propyl benzoate*.

19.10 PROPERTIES OF ESTERS

Simple esters that are derived from monocarboxylic acids and monohydroxy alcohols are generally colorless liquids and solids. Ester molecules are polar, but are incapable of hydrogen bonding with each other. Consequently, their boiling points are lower than those of carboxylic acids and alcohols that have similar molecular weights. The following three compounds, which have the same molecular weight of 60.0 amu, are representative of this boiling point pattern.

$$CH_3-CH_2-CH_2-OH \qquad CH_3-C{\overset{O}{\underset{OH}{\diagdown}}} \qquad H-C{\overset{O}{\underset{O-CH_3}{\diagdown}}}$$

1-propanol   ethanoic acid   methyl methanoate
b.p. 97°C   b.p. 117°C   b.p. 32°C

Esters of low molecular weight are somewhat water soluble. These molecules are capable of hydrogen bonding to water molecules through the oxygen atom of the carbonyl group.

Low- and intermediate-molecular-weight esters usually possess pleasant odors. These compounds are responsible for the odor of many flowers and the odor and taste of many fruits. Many artificial flavoring agents contain esters. Table 19–3 lists the structures of selected esters that are used as flavoring agents. There are numerous cyclic esters that also have "odor". The cyclic ester *coumarin* is responsible for the pleasant odor of freshly cut hay. Cats of all types (from lions to house cats) are strongly attracted to nepetalactone, a cyclic ester found in the catnip plant.

coumarin   nepetalactone

Many esters exist throughout nature. Waxes are high-molecular-weight esters in which both the acid and alcohol portions contain long alkyl chains. A major component of beeswax is the ester involving a 16-carbon acid and a 30-carbon alcohol.

$$CH_3-(CH_2)_{14}-C\begin{matrix}\nearrow O \\ \searrow O-(CH_2)_{29}-CH_3\end{matrix}$$

an ester found in beeswax

TABLE 19–3 Fruit Odor or Flavor Associated with Selected Ester Flavoring Agents

Name	Structured Formula	Characteristic Flavor and Odor
isobutyl methanoate (isobutyl formate)	$H-C(=O)-O-CH_2-CH(CH_3)-CH_3$	raspberry
n-pentyl ethanoate (n-pentyl acetate)	$CH_3-C(=O)-O-(CH_2)_4-CH_3$	banana
n-octyl ethanoate (n-octyl acetate)	$CH_3-C(=O)-O-(CH_2)_7-CH_3$	orange
n-pentyl propanoate (n-pentyl propionate)	$CH_3-CH_2-C(=O)-O-(CH_2)_4-CH_3$	apricot
methyl butanoate (methyl butyrate)	$CH_3-(CH_2)_2-C(=O)-O-CH_3$	apple
ethyl butanoate (ethyl butyrate)	$CH_3-(CH_2)_2-C(=O)-O-CH_2-CH_3$	pineapple
methyl 2-aminobenzoate (methyl anthranilate)	2-aminobenzene-C(=O)-O-CH$_3$ (with NH$_2$ ortho)	grape

Animals and plants use waxes for protective coatings. The leaves and stems of most plants are coated with wax, which helps prevent both dehydration and the attack of microorganisms. The feathers of birds and the fur of animals are also coated with wax. The ears of humans are protected by ear wax. The natural function for waxes in nature is borrowed by humans when they use them in car polish, shoe polish, and cosmetics.

Esters are also found within the human body. One of the most important esters is *acetylcholine*, a substance involved in nerve impulse transmission.

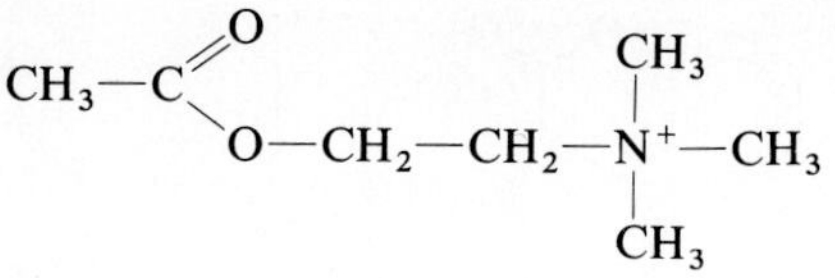

$$CH_3-C\underset{O-CH_2-CH_2-\underset{CH_3}{\overset{CH_3}{N^+}}-CH_3}{\overset{O}{\diagup}}$$

acetylcholine

Natural fats, which will be discussed in detail in Chapter 23, are triesters of the alcohol *glycerol*. Three long-chain acid molecules are bonded to the glycerol to give a triglyceride.

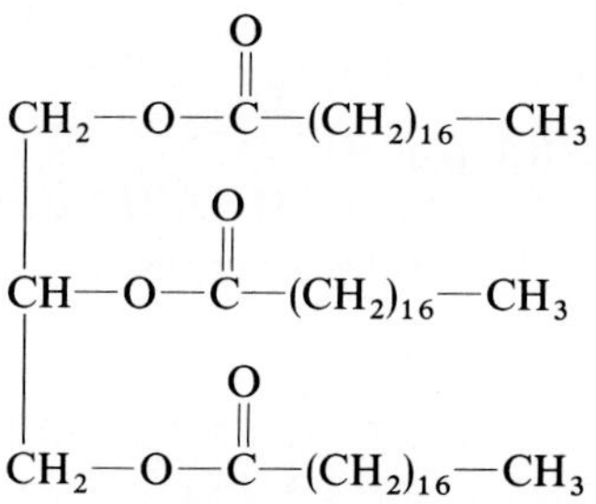

$$CH_2-O-\overset{O}{\overset{\|}{C}}-(CH_2)_{16}-CH_3$$
$$CH-O-\overset{O}{\overset{\|}{C}}-(CH_2)_{16}-CH_3$$
$$CH_2-O-\overset{O}{\overset{\|}{C}}-(CH_2)_{16}-CH_3$$

a triglyceride molecule

An important function of triglycerides is as a storage reserve of energy in living things.

Esters of *p*-hydroxybenzoic acid, commonly called *parabens*, prevent the growth of microorganisms such as molds and yeast. Parabens are extensively used as food and drug preservatives.

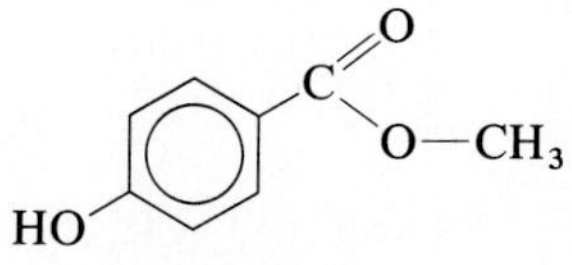

methyl paraben

19.11 REACTIONS OF ESTERS

The most important reaction of esters involves the breaking of the ester linkage, which results in an accompanying restoration of the acid and alcohol portions of the ester to their original form. This process is called *ester hydrolysis* because water is the substance that reacts with the ester. Ester hydrolysis is catalyzed by strong acids (H_2SO_4 or HCl) or by certain enzymes.

Hydrolysis of an ester is simply the reverse of esterification (the formation reaction for esters).

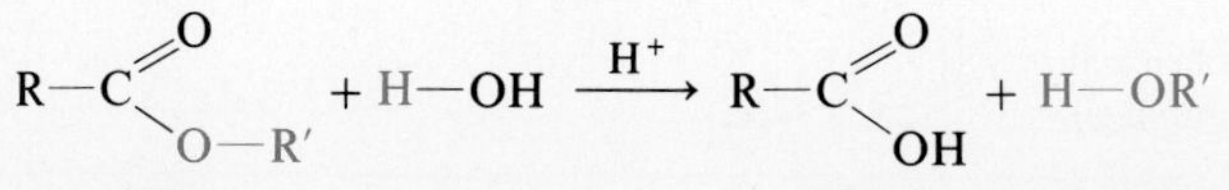

$$R-C\overset{O}{\underset{O-R'}{\diagup}} + H-OH \xrightarrow{H^+} R-C\overset{O}{\underset{OH}{\diagup}} + H-OR'$$

an ester   water   an acid   an alcohol

Notice that the H portion of the water molecule becomes part of the alcohol and the OH portion of the water molecule becomes part of the carboxylic acid.

Some specific examples of ester hydrolysis are

$$CH_3-C(=O)-O-CH_2-CH_3 + H-OH \xrightarrow{H^+} CH_3-C(=O)-OH + HO-CH_2-CH_3$$
ethyl acetate → acetic acid + ethyl alcohol

$$C_6H_5-C(=O)-O-CH_2-CH_2-CH_3 + H-OH \xrightarrow{H^+} C_6H_5-C(=O)-OH + HO-CH_2-CH_2-CH_3$$
n-propyl benzoate → benzoic acid + n-propyl alcohol

Animal fats and vegetable oils are esters of long-chain acids. Hydrolysis of these substances, assisted by enzymatic catalysts, is the first step in their digestion by the human body.

Saponification is the hydrolysis of an ester under basic conditions. The strong bases NaOH or KOH are usually the sources of the basic environment. The products of saponification are the alcohol and the salt of the carboxylic acid.

$$R-C(=O)-O-R' + NaOH \xrightarrow{H_2O} R-C(=O)-O^-Na^+ + HO-R'$$
an ester + a strong base → a carboxylate salt + an alcohol

A specific example of saponification is

$$C_6H_5-C(=O)-O-CH_3 + NaOH \xrightarrow{H_2O} C_6H_5-C(=O)-O^-Na^+ + HO-CH_3$$
methyl benzoate → sodium benzoate + methyl alcohol

Soaps are produced when fatty acid esters of glycerol, which are present in fats and oils, are saponified (Chapter 23).

19.12 POLYESTERS

Esterification is a condensation reaction. **A condensation reaction** *is a reaction in which two molecules are joined together to give a larger molecule, and a small molecule such as water is eliminated.*

condensation reaction

$$R-C(=O)-OH + HO-R' \longrightarrow R-C(=O)-O-R' + H_2O$$
carboxylic acid + alcohol → ester

Condensation reactions can be used to form polymers. In these reactions, the functional groups on two or more kinds of monomers react, giving both larger molecules and smaller molecules such as water. Polymers formed in this way are called *condensation polymers*. This polymerization process differs from addition polymerization (Section 15.9), where the entire monomer is included in the polymer.

The monomers used to form condensation polymers are always bifunctional molecules; that is, they are molecules that possess two functional groups. Dicarboxylic acids and dialcohols are two of many types of bifunctional molecules that exist.

a dicarboxylic acid   a dialcohol

polyesters

Polyesters *are condensation polymers that contain ester linkages between monomer units.* The monomers needed to form this type of polymer are a diacid and a dialcohol.

The best known of the many polyester polymers now marketed is the textile fiber Dacron. The monomers used to produce Dacron are terephthalic acid and ethylene glycol.

terephthalic acid   ethylene glycol

The reaction of one acid group of the diacid with one alcohol group of the dialcohol initially produces an ester molecule, with an acid group left over on one end and an alcohol group left over on the other end.

This species can react further. The remaining acid group can react with another alcohol group from another monomer, and the alcohol group can react with another acid group from another monomer. This process continues until an extremely large polymer molecule called a *polyester* is produced.

FIGURE 19–4 Dacron can be used to produce synthetic artery grafts that replace diseased arteries and also in the manufacture of heart-valve replacements.

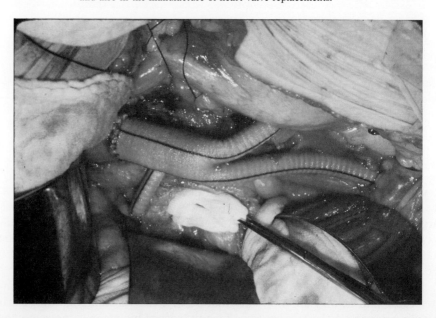

$$\cdots \overset{O}{\underset{\parallel}{C}} - \underset{}{\bigcirc} - \overset{O}{\underset{\parallel}{C}} \overset{\text{ester linkages}}{\overset{\curvearrowright}{-}} O - CH_2 - CH_2 - O - \overset{O}{\underset{\parallel}{C}} - \underset{}{\bigcirc} - \overset{O}{\underset{\parallel}{C}} \overset{\text{ester linkage}}{\overset{\curvearrowright}{-}} O - CH_2 - CH_2 - O \cdots$$

a portion of the polyester Dacron

Note that in a polyester polymer, the bonds that link the monomers are ester linkages; hence, the name *polyester*.

Dacron fibers are used in wash-and-wear clothing. When the same polymer is formed as a film, rather than a fiber, it is called *Mylar*. Magnetically coated Mylar tape is used in audio and video recording.

Dacron is also used in medicine. Because it is physiologically inert, Dacron is used in the form of a mesh to replace diseased sections of arteries. It has also been used in synthetic heart valves (see Figure 19-4).

19.13 ESTERS OF INORGANIC ACIDS

Alcohols can react with acids other than carboxylic acids to form esters. Mineral acids such as sulfuric, phosphoric, and nitric acids form esters. The structures of these acids resemble carboxylic acid's structure, allowing esterification reactions to occur that are very similar to those discussed for organic acids.

$$\underset{\text{carboxylic acid}}{R-\overset{O}{\underset{\parallel}{C}}-OH} \quad \underset{\text{sulfuric acid}}{HO-\overset{\overset{O}{\parallel}}{\underset{\underset{O}{\parallel}}{S}}-OH} \quad \underset{\text{phosphoric acid}}{HO-\overset{\overset{O}{\parallel}}{\underset{\underset{OH}{\mid}}{P}}-OH} \quad \underset{\text{nitric acid}}{\overset{O}{\underset{\underset{O}{\mid}}{N}}-OH}$$

Sulfate esters of long-chain alcohols are used as synthetic detergents. An example is lauryl hydrogen sulfate (used in shampoo).

$$\underset{\text{lauryl alcohol}}{CH_3-(CH_2)_{10}-CH_2-OH} + \underset{\text{sulfuric acid}}{HO-\overset{\overset{O}{\parallel}}{\underset{\underset{O}{\parallel}}{S}}-OH} \longrightarrow \underset{\text{lauryl hydrogen sulfate}}{CH_3-(CH_2)_{10}-CH_2-O-\overset{\overset{O}{\parallel}}{\underset{\underset{O}{\parallel}}{S}}-OH} + H_2O$$

Esters of nitric acid are formed easily from mixtures of nitric acids and alcohols.

$$\underset{\text{alcohol}}{R-OH} + \underset{\text{nitric acid}}{HO-NO_2} \longrightarrow \underset{\text{nitrate ester}}{R-O-NO_2} + H_2O$$

Nitroglycerin, a trinitrate ester of glycerol, is known for two properties: its explosive power and its biological action as a smooth-muscle relaxant and blood-vessel dilator. "Nitro" tablets placed in the mouth are quickly absorbed and dilate the blood vessels, allowing an easier flow of blood and an immediate drop in blood pressure. Nitric acid should never be mixed with organic waste materials in the laboratory because of the danger of nitrate-ester formation and the associated danger of explosion.

$$\underset{\text{glycerin}}{\begin{array}{c} CH_2-OH \\ | \\ CH-OH \\ | \\ CH_2-OH \end{array}} + 3\,HO-NO_2 \longrightarrow \underset{\text{nitroglycerin}}{\begin{array}{c} CH_2-O-NO_2 \\ | \\ CH-O-NO_2 \\ | \\ CH_2-O-NO_2 \end{array}} + 3\,H_2O$$

Phosphate esters, which are formed by reactions between phosphoric acid and alcohols, are very important in biological systems.

$$\text{R—OH} + \text{HO—}\underset{\underset{\text{OH}}{|}}{\overset{\overset{\text{O}}{\|}}{\text{P}}}\text{—OH} \longrightarrow \text{R—O—}\underset{\underset{\text{OH}}{|}}{\overset{\overset{\text{O}}{\|}}{\text{P}}}\text{—OH} + \text{H}_2\text{O}$$

an alcohol   phosphoric acid   a phosphate ester

Synthetic phosphate esters are used as insecticides. Many of these compounds, which are called *organophosphate insecticides,* are now in use. Organophosphate insecticides were developed as replacements for the persistent organochlorine insecticides such as DDT (Section 16.5). Because the organophosphate insecticides are esters, they break down rapidly through hydrolysis reactions (Section 19.11) and the problem of accumulation in the environment is avoided. Two organophosphate insecticides in common use are Dichlorvos and Parathion.

$$\underset{\text{Cl}}{\overset{\text{Cl}}{>}}\text{C=CH—O—}\underset{\underset{\text{O—CH}_3}{|}}{\overset{\overset{\text{O}}{\|}}{\text{P}}}\text{—O—CH}_3 \qquad \text{CH}_3\text{—CH}_2\text{—O—}\underset{\underset{\text{O—CH}_2\text{—CH}_3}{|}}{\overset{\overset{\text{S}}{\|}}{\text{P}}}\text{—O—}\langle\bigcirc\rangle\text{—NO}_2$$

dichlorvos
(Shell "No-Pest Strip")

parathion
(a potent insecticide)

19.14 ACID ANHYDRIDES

The structural feature of an oxygen atom bonded to a carbonyl carbon atom is common in both acids and esters. A third class of compounds also contains this feature—acid anhydrides. An **acid anhydride** *is an organic compound whose characteristic functional group is*

acid anhydride

$$\text{R—}\overset{\overset{\text{O}}{\|}}{\text{C}}\text{—O—}\overset{\overset{\text{O}}{\|}}{\text{C}}\text{—R}$$

In an acid anhydride, two carbonyl groups are bonded to the same oxygen atom. This linkage between the two carbonyl groups is called an *anhydride bond.*

$$\text{R—}\overset{\overset{\text{O}}{\|}}{\text{C}}\underbrace{\text{—O—}}_{\text{anhydride bond}}\overset{\overset{\text{O}}{\|}}{\text{C}}\text{—R}$$

Acid anhydrides can be prepared by reacting carboxylic acids with themselves in the presence of the dehydrating agent P_4O_{10}. The general equation for this process is

$$\text{R—C}\underset{\text{OH}}{\overset{\diagup\text{O}}{\diagdown}} + \underset{\text{HO}}{\overset{\text{O}\diagdown}{\diagup}}\text{C—R} \xrightarrow{P_4O_{10}} \text{R—}\overset{\overset{\text{O}}{\|}}{\text{C}}\text{—O—}\overset{\overset{\text{O}}{\|}}{\text{C}}\text{—R} + \text{H}_2\text{O}$$
acid anhydride

Note that both water and acid anhydride are products of this reaction. A specific example of anhydride formation, which involves acetic acid in the formation of acetic anhydride, is

$$\text{CH}_3\text{—C}\underset{\text{OH}}{\overset{\diagup\text{O}}{\diagdown}} + \underset{\text{HO}}{\overset{\text{O}\diagdown}{\diagup}}\text{C—CH}_3 \xrightarrow{P_4O_{10}} \text{CH}_3\text{—}\overset{\overset{\text{O}}{\|}}{\text{C}}\text{—O—}\overset{\overset{\text{O}}{\|}}{\text{C}}\text{—CH}_3 + \text{H}_2\text{O}$$
acetic anhydride

An important reaction of carboxylic acid anhydrides takes place with alcohols to form esters.

$$\underset{\text{acid anhydride}}{R-\overset{\overset{O}{\|}}{C}-O-\overset{\overset{O}{\|}}{C}-R} + \underset{\text{alcohol}}{R'-OH} \longrightarrow \underset{\text{ester}}{R-C\overset{\nearrow O}{\underset{\searrow O-R'}{}}} + \underset{\text{acid}}{R-C\overset{\nearrow O}{\underset{\searrow OH}{}}}$$

Both a carboxylic acid molecule and an ester molecule are products of this reaction.

In Section 19.4, we found that both acid chlorides and carboxylic acids react with alcohols to form esters. We can now list a third type of reactant that will effect esterification—acid anhydrides.

There are counterparts to acid anhydrides of carboxylic acids for inorganic acids. Of particular interest to us are the acid anhydrides of phosphoric acid. We can obtain phosphate esters, which are very important biological compounds, from these compounds.

Two phosphoric acid molecules can be converted to an anhydride in a manner analogous to that for carboxylic acids.

$$HO-\overset{\overset{O}{\|}}{\underset{OH}{P}}-OH + HO-\overset{\overset{O}{\|}}{\underset{OH}{P}}-OH \longrightarrow \underset{\text{pyrophosphoric acid}}{HO-\overset{\overset{O}{\|}}{\underset{OH}{P}}-O-\overset{\overset{O}{\|}}{\underset{OH}{P}}-OH} + H_2O$$

This anhydride, which involves two phosphoric acid units, is called *pyrophosphoric acid*.

Unlike carboxylic acid anhydrides, pyrophosphoric acid still possesses —OH groups that can react further to form higher anhydrides. The reaction of pyrophosphoric acid with another phosphoric acid molecule will produce an anhydride containing three phosphoric acid units. This anhydride is known as *triphosphoric acid*.

$$HO-\overset{\overset{O}{\|}}{\underset{OH}{P}}-O-\overset{\overset{O}{\|}}{\underset{OH}{P}}-OH + HO-\overset{\overset{O}{\|}}{\underset{OH}{P}}-OH \longrightarrow \underset{\text{triphosphoric acid}}{HO-\overset{\overset{O}{\|}}{\underset{OH}{P}}-O-\overset{\overset{O}{\|}}{\underset{OH}{P}}-O-\overset{\overset{O}{\|}}{\underset{OH}{P}}-OH} + H_2O$$

Note that triphosphoric acid contains two anhydride bonds.

$$HO-\overset{\overset{O}{\|}}{\underset{OH}{P}}-O-\overset{\overset{O}{\|}}{\underset{OH}{P}}-O-\overset{\overset{O}{\|}}{\underset{OH}{P}}-OH$$
— anhydride bonds

Just as alcohols react with carboxylic acids and their anhydrides to form esters, they also react with phosphoric acid and phosphoric acid anhydrides to produce phosphate esters. Esters of phosphoric acid, pyrophosphoric acid, and triphosphoric acid are all extremely important in the chemistry of living cells.

$$R-OH + HO-\overset{\overset{O}{\|}}{\underset{OH}{P}}-OH \longrightarrow \underset{\text{phosphate ester}}{R-O-\overset{\overset{O}{\|}}{\underset{OH}{P}}-OH} + H_2O$$

$$R-OH + HO-\underset{\underset{OH}{|}}{\overset{\overset{O}{\|}}{P}}-O-\underset{\underset{OH}{|}}{\overset{\overset{O}{\|}}{P}}-OH \longrightarrow R-O-\underset{\underset{OH}{|}}{\overset{\overset{O}{\|}}{P}}-O-\underset{\underset{OH}{|}}{\overset{\overset{O}{\|}}{P}}-OH + H_2O$$

diphosphate ester

$$R-OH + HO-\underset{\underset{OH}{|}}{\overset{\overset{O}{\|}}{P}}-O-\underset{\underset{OH}{|}}{\overset{\overset{O}{\|}}{P}}-O-\underset{\underset{OH}{|}}{\overset{\overset{O}{\|}}{P}}-OH \longrightarrow R-O-\underset{\underset{OH}{|}}{\overset{\overset{O}{\|}}{P}}-O-\underset{\underset{OH}{|}}{\overset{\overset{O}{\|}}{P}}-O-\underset{\underset{OH}{|}}{\overset{\overset{O}{\|}}{P}}-OH + H_2O$$

triphosphate ester

Within body fluids, phosphate esters exist as charged species because the H atoms in the —OH groups present are lost (ionized) at body fluid pH levels. The phosphate, diphosphate, and triphosphate species have charges of -2, -3, and -4, respectively. It is this ionization process that makes phosphate esters soluble in body fluids.

$$R-O-\underset{\underset{OH}{|}}{\overset{\overset{O}{\|}}{P}}-OH \longrightarrow R-O-\underset{\underset{O^-}{|}}{\overset{\overset{O}{\|}}{P}}-O^- + 2H^+$$

$$R-O-\underset{\underset{OH}{|}}{\overset{\overset{O}{\|}}{P}}-O-\underset{\underset{OH}{|}}{\overset{\overset{O}{\|}}{P}}-OH \longrightarrow R-O-\underset{\underset{O^-}{|}}{\overset{\overset{O}{\|}}{P}}-O-\underset{\underset{O^-}{|}}{\overset{\overset{O}{\|}}{P}}-O^- + 3H^+$$

$$R-O-\underset{\underset{OH}{|}}{\overset{\overset{O}{\|}}{P}}-O-\underset{\underset{OH}{|}}{\overset{\overset{O}{\|}}{P}}-O-\underset{\underset{OH}{|}}{\overset{\overset{O}{\|}}{P}}-OH \longrightarrow R-O-\underset{\underset{O^-}{|}}{\overset{\overset{O}{\|}}{P}}-O-\underset{\underset{O^-}{|}}{\overset{\overset{O}{\|}}{P}}-O-\underset{\underset{O^-}{|}}{\overset{\overset{O}{\|}}{P}}-O^- + 4H^+$$

Two key chemicals in the storage and transfer of energy in the human body are ATP (adenosine triphosphate) and ADP (adenosine diphosphate). Both of these compounds are phosphate esters. ATP is produced in the metabolism of fats and carbohydrates and stores some of the energy released in these reactions in the anhydride bonds present. When energy is needed, ATP is hydrolyzed (in the presence of enzymes) to produce ADP and 7.3 kcal of energy per mole of ADP produced. This energy-producing reaction and other similar reactions will be discussed in detail in Chapter 28.

$$\underbrace{\text{adenosine}-O-\underset{\underset{O^-}{|}}{\overset{\overset{O}{\|}}{P}}-O-\underset{\underset{O^-}{|}}{\overset{\overset{O}{\|}}{P}}-O-\underset{\underset{O^-}{|}}{\overset{\overset{O}{\|}}{P}}-O^-}_{\text{ATP}} + H_2O \xrightarrow{\text{enzyme}} \underbrace{\text{adenosine}-O-\underset{\underset{O^-}{|}}{\overset{\overset{O}{\|}}{P}}-O-\underset{\underset{O^-}{|}}{\overset{\overset{O}{\|}}{P}}-O^-}_{\text{ADP}} + H-O-\underset{\underset{OH}{|}}{\overset{\overset{O}{\|}}{P}}-O^-$$

EXERCISES AND PROBLEMS

The Carboxyl Group

19.1 Explain the structural differences among aldehydes, ketones, and carboxylic acids.

19.2 Indicate which of the following compounds contain a carboxyl group.

a. 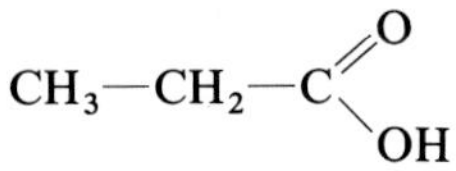 $CH_3-CH_2-C(=O)OH$

b. $CH_3-CH_2-CH_2-C(=O)-CH_3$

c. $CH_3-CH_2-CH(CH_3)-COOH$

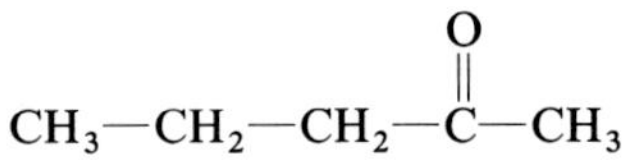

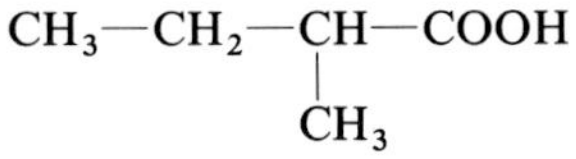

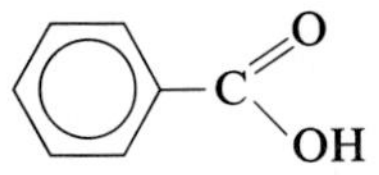

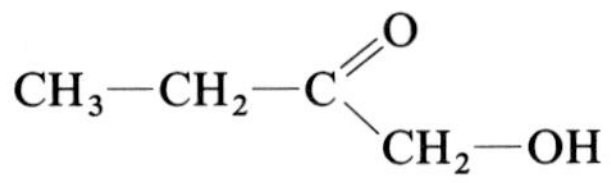

d. 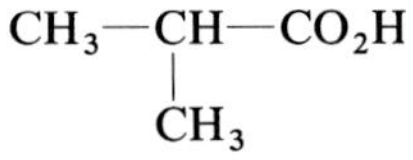 $CH_3-CH(CH_3)-CO_2H$

e. 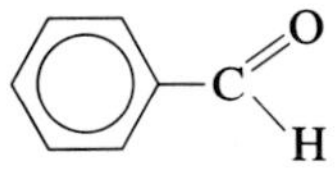 $C_6H_5-C(=O)H$

f. $C_6H_5-C(=O)OH$

g. $CH_3-CH_2-C(=O)-CH_2-OH$

h. $CH_3(HO)C=O$

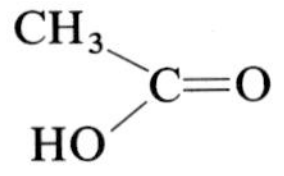

Nomenclature of Carboxylic Acids

19.3 Give the IUPAC name for each of the following carboxylic acids.

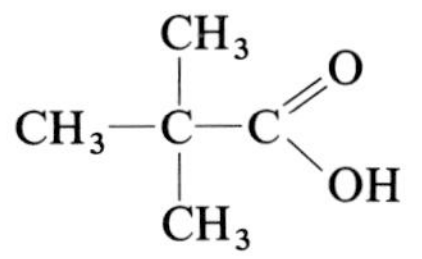

a. 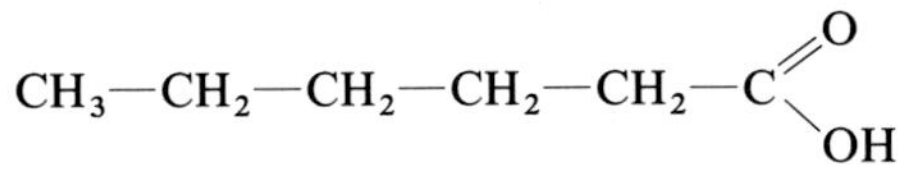 $CH_3-CH_2-CH_2-CH_2-CH_2-C(=O)OH$

b. $CH_3-C(CH_3)(CH_3)-C(=O)OH$

c. $CH_3-CH(Br)-CH_2-C(=O)OH$

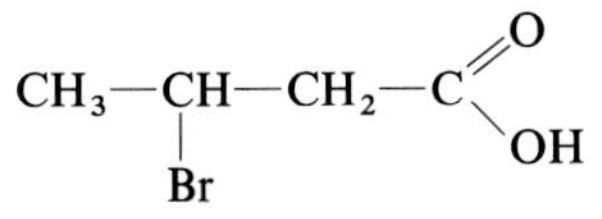

d. 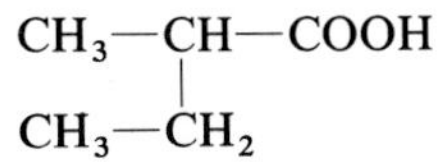 $CH_3-CH(CH_2CH_3)-COOH$

e. 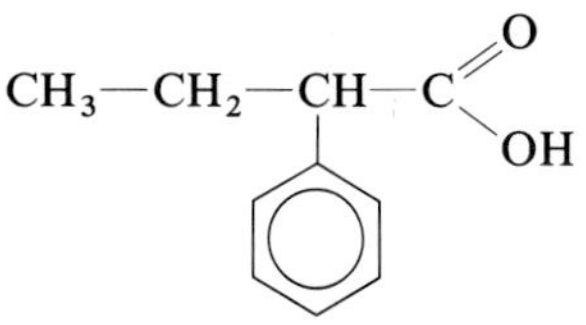 $CH_3-CH_2-CH(C_6H_5)-C(=O)OH$

f. 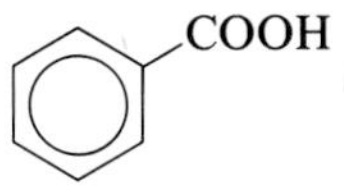 C_6H_5-COOH

g. 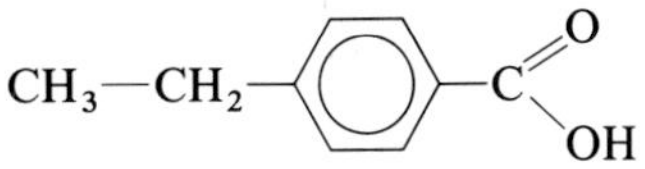 $CH_3-CH_2-C_6H_4-C(=O)OH$

h. 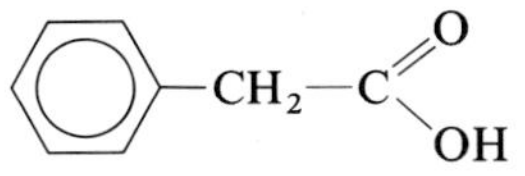 $C_6H_5-CH_2-C(=O)OH$

19.4 Draw structural formulas that correspond to each of the following carboxylic acids.

a. propanoic acid
b. 2-methylbutanoic acid
c. methanoic acid
d. 2-chloropropanoic acid
e. 4,4-dibromopentanoic acid
f. 2-chloro-4-ethylheptanoic acid
g. 2-methylbenzoic acid
h. o-chlorobenzoic acid

19.5 Draw structural formulas that correspond to each of the following carboxylic acids.

a. acetic acid
b. caproic acid
c. lactic acid
d. oxalic acid
e. malonic acid
f. succinic acid
g. malic acid
h. citric acid

19.6 Give the IUPAC name for each of the acids listed in Problem 19.5.

19.7 Draw structural formulas that correspond to each of the following carboxylic acids.

a. α-chloropropionic acid
b. β-bromopropionic acid
c. α-chloro-γ-methylcaproic acid
d. 3-pentenoic acid
e. 4-hexenoic acid
f. 3-ketopentanoic acid
g. 2-ketobutanoic acid
h. trans-9-keto-2-decenoic acid

19.8 Draw structural formulas to represent the four isomeric carboxylic acids that have the molecular formula $C_4H_9CO_2H$. Assign an IUPAC name to each isomer.

19.9 Gallic acid (3,4,5-trihydroxybenzoic acid) belongs to a group of naturally occurring hormones in plants that retard or inhibit certain physiological processes such as growth or seed germination. Draw the structure of gallic acid.

Physical Properties of Carboxylic Acids

19.10 Select the acid from each of the following pairs that you predict would be more soluble in water.

a.
$$CH_3-C\underset{OH}{\overset{O}{\diagdown}} \quad \text{and}$$

$$CH_3-CH_2-CH_2-CH_2-C\underset{OH}{\overset{O}{\diagdown}}$$

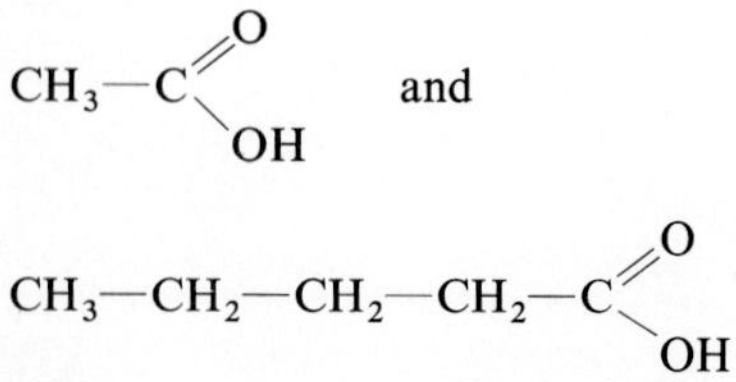

b. C_6H_5—COOH and CH_3-CH_2—COOH

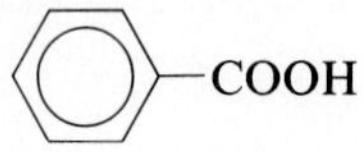

c. $CH_3-CH_2-CH_2$—COOH
and HOOC—CH_2-CH_2—COOH

19.11 Using structural formulas, show how hydrogen bonds can form between
a. two molecules of acetic acid
b. one molecule of acetic acid and two molecules of water

Preparation and Reactions of Carboxylic Acids

19.12 For each of the following reactions, predict the products and draw their structures.

a. $CH_3OH \xrightarrow[H_2O, H^+]{KMnO_4}$

b. $CH_3-C\underset{H}{\overset{O}{\diagdown}} \xrightarrow[H_2O, H^+]{K_2Cr_2O_7}$

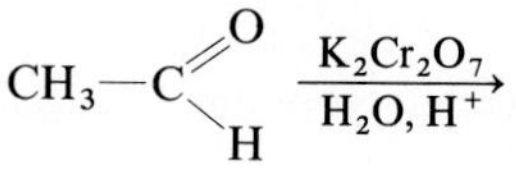

c. C_6H_5—$CH_2OH \xrightarrow[H_2O, H^+]{KMnO_4}$

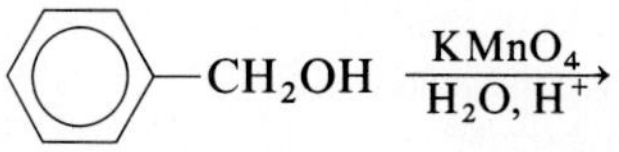

d. $CH_3-\underset{CH_3}{\underset{|}{CH}}-\underset{CH_3}{\underset{|}{CH}}-C\underset{H}{\overset{O}{\diagdown}} \xrightarrow[H_2O, H^+]{K_2Cr_2O_7}$

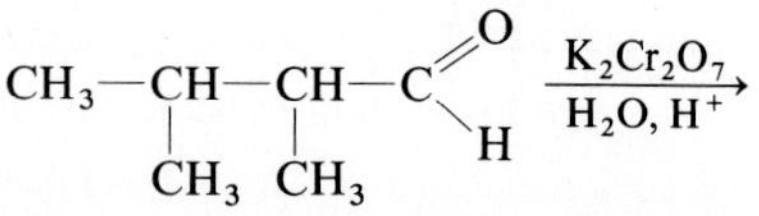

e. C_6H_5—$CH_2-CH_2OH \xrightarrow[H_2O, H^+]{K_2Cr_2O_7}$

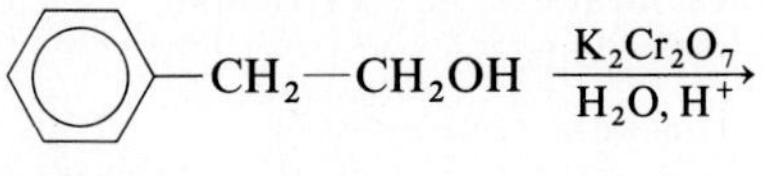

f. $\underset{CH_2OH}{\underset{|}{CH_2OH}} \xrightarrow[H_2O, H^+]{KMnO_4}$

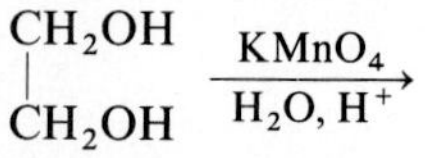

19.13 Explain how you could oxidize *trans*-9-hydroxy-2-decenoic acid (one queen bee secretion) to *trans*-9-keto-2-decenoic acid (the queen substance), using laboratory reagents. Draw structures for both pheromones in your chemical equation.

19.14 We often say carboxylic acids are resistant to further oxidation. However, certain enzymes can catalyze the oxidation of these acids. What are the general products of this type of oxidation in biological systems? Give a specific example.

19.15 Using a carboxylic acid as a reactant, write equations for the preparation of each of the following carboxylic acid derivatives.

a. $CH_3-CH_2-C\underset{Cl}{\overset{O}{\diagdown}}$

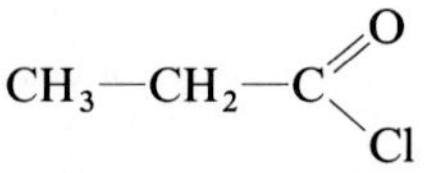

b. $\text{C}_6\text{H}_5-C\underset{O-CH_3}{\overset{O}{\diagdown}}$

c. $CH_3-\underset{CH_3}{\underset{|}{CH}}-C\underset{Cl}{\overset{O}{\diagdown}}$

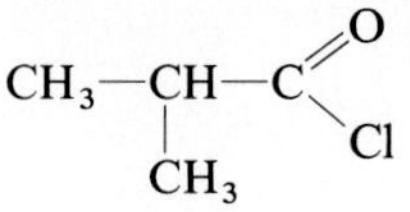

d. $CH_3-CH_2-CH_2-CH_2-CH_2-C\underset{NH_2}{\overset{O}{\diagdown}}$

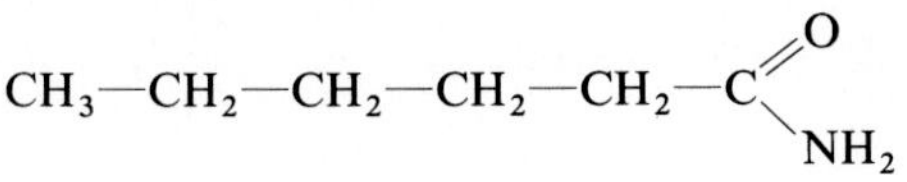

e. $CH_3-CH_2-C\underset{O-CH_2-CH_3}{\overset{O}{\diagdown}}$

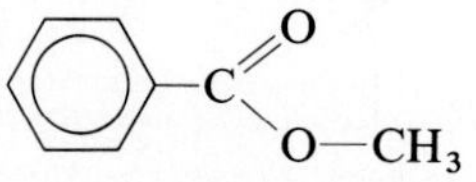

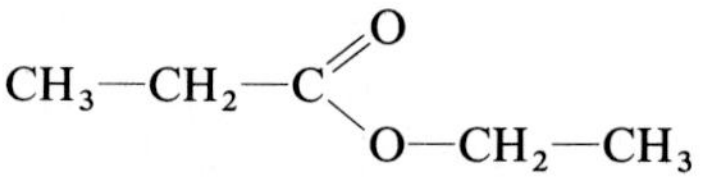

f. $CH_3-C\underset{NH_2}{\overset{O}{\diagdown}}$

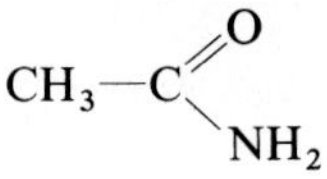

19.16 Utilizing any inorganic reagents you require, diagram a multi-step synthesis for each of the following conversions. (A review of reactions discussed in previous chapters may be necessary.)

a. $H_2C=CH-CH_2-CH_2-OH$ to $CH_3-\overset{O}{\overset{\|}{C}}-CH_2-\overset{O}{\overset{\|}{C}}-OH$

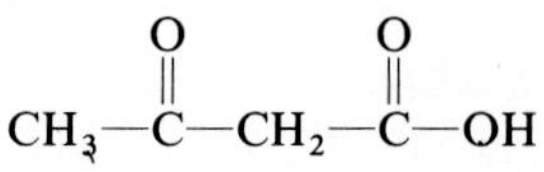

b. $H_2C=CH-C\underset{H}{\overset{O}{\diagdown}}$ to $\underset{Br}{\underset{|}{CH_2}}-\underset{Br}{\underset{|}{CH}}-C\underset{OH}{\overset{O}{\diagdown}}$

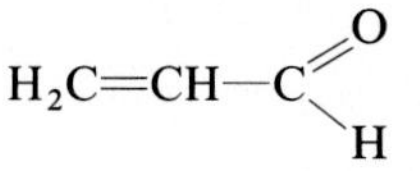

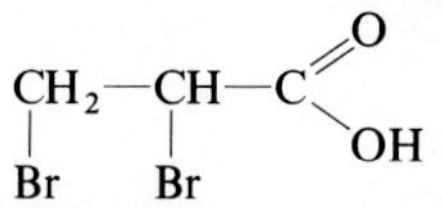

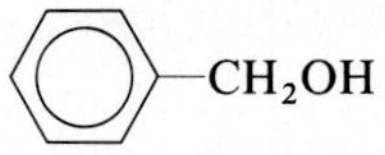

c. C_6H_5—CH_2OH to Cl—C_6H_4—COOH

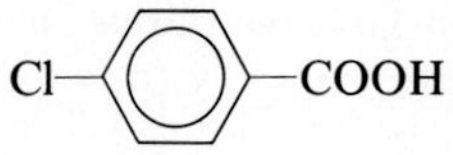

Acidity of Carboxylic Acids

19.17 Draw structures and give the names of carboxylate ions formed from each of the following carboxylic acids in aqueous solution.

a. $CH_3-C\underset{OH}{\overset{O}{\diagdown}}$

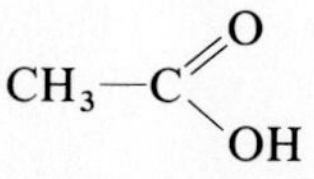

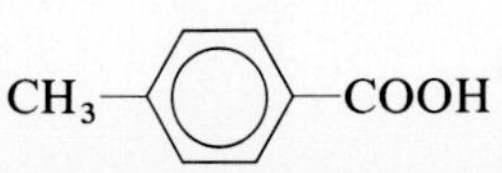

b. CH_3—C_6H_4—COOH

c. $H-C\underset{OH}{\overset{O}{\diagdown}}$

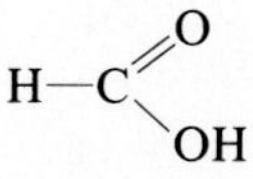

d. $CH_3-(CH_2)_{14}$—COOH

19.18 Acetic acid, like most carboxylic acids, is considered a weak acid. Explain what is meant by a "weak acid".

19.19 Most cooks realize that vinegar and baking soda (NaHCO$_3$) bubble vigorously when mixed together. Recalling that NaHCO$_3$ dissociates into Na$^+$ and HCO$_3^-$ ions in solution, and that the following equilibrium exists,

$$H^+ + HCO_3^- \rightleftharpoons H_2CO_3 \rightleftharpoons H_2O + CO_2,$$

explain why vinegar and soda mixtures yield so many bubbles.

19.20 Aspirin is not extremely soluble in pure water. Place an aspirin on a spoon and add a few drops of water. Observe the low solubility. Now sprinkle some baking soda onto the aspirin–water mixture and observe the results. Explain the enhanced solubility of aspirin. (Exasperated parents often use this trick to help children take an aspirin.)

Salts of Carboxylic Acids

19.21 Name each of the following salts.

a. 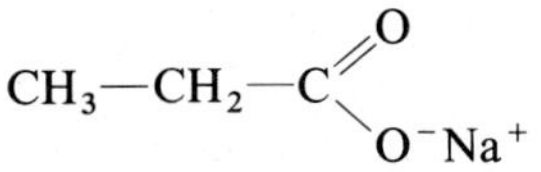 $CH_3-CH_2-COO^-Na^+$ (as structure with C=O and O$^-$Na$^+$)

b. $(CH_3-COO^-)_2 Ca^{2+}$

c. (phenyl)$-COO^-Na^+$

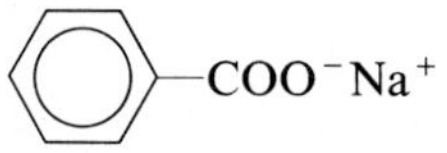

d. $K^+{}^-OOC-COO^-K^+$

19.22 Explain what happens when each of the salts in Problem 19.21 is dissolved in water.

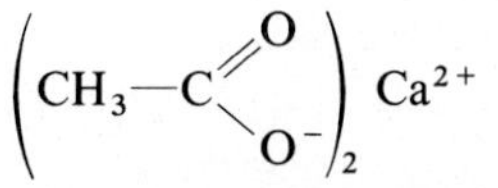

19.23 Which compound would you expect to be more soluble in water?
a. benzoic acid or sodium benzoate
b. potassium oxalate or oxalic acid
c. citric acid or trisodium citrate

19.24 In which solution would you predict that adipic acid would be more soluble: 0.1 N HCl or 0.1 N NaOH? Explain.

19.25 How many mL of 0.100 N NaOH would be required to titrate each of the following solutions? (Review Sections 10.8 and 10.9 if needed.)
a. 25 mL of 0.10 N acetic acid
b. 25 mL of 0.10 N oxalic acid
c. 25 mL of 0.10 N citric acid

19.26 How many milliequivalents are present in each of the following? (Review Section 8.16 if needed.)
a. 0.60 g of acetic acid   c. 1.82 g of citric acid
b. 0.60 g of oxalic acid   d. 0.83 g of sodium acetate

19.27 Explain the molecular basis for the action of soap. Compare the action of soap in acidic and basic aqueous solutions.

The Ester Functional Group

19.28 Which of the following structures represent esters?

a. 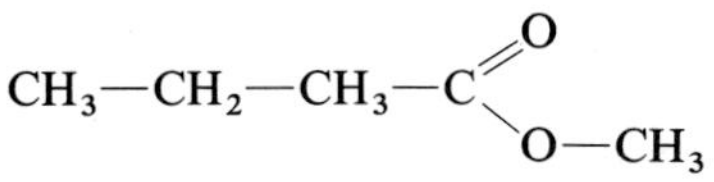 $CH_3-CH_2-CH_2-C(=O)-O-CH_3$

b. $CH_3-O-C(=O)-CH_3$

c. $CH_3-O-CH_2-C(=O)-CH_3$

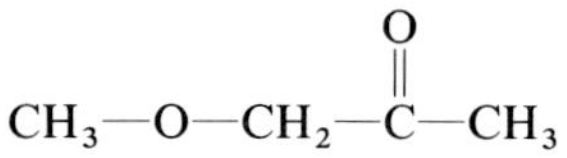

d. 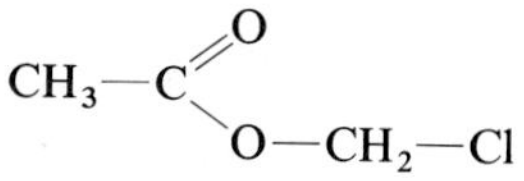 $CH_3-C(=O)-O-CH_2-Cl$

e. 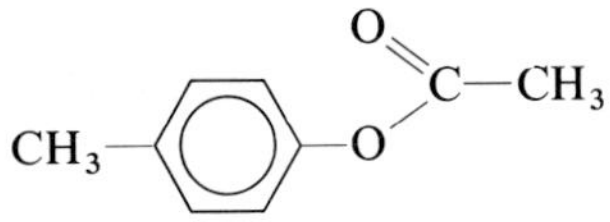 CH_3-(phenyl)$-O-C(=O)-CH_3$

f. 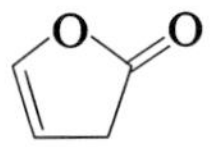 (cyclic structure with O-C(=O) in a five-membered ring containing a C=C)

g. 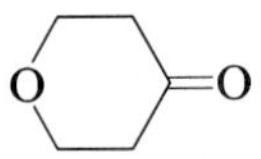 (six-membered ring with O and C=O)

h. 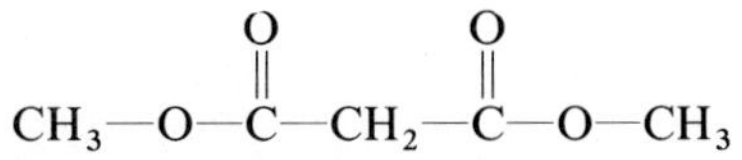 $CH_3-O-C(=O)-CH_2-C(=O)-O-CH_3$

19.29 For each of the following esters, identify the portion of the molecule that originally came from the acid by drawing a rectangle around it. Identify the alcohol portion with a circle. Draw a short arrow to indicate the location of the ester linkage.

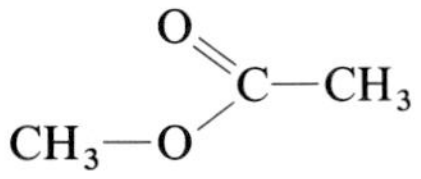

a. 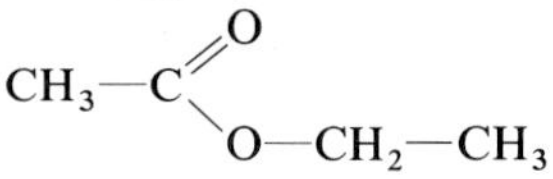 $CH_3-C(=O)-O-CH_2-CH_3$

b. $CH_3-CH_2-C(=O)-O-CH_3$

c. $CH_3-CH_2-CH_2-O-C(=O)-CH(CH_3)-CH_3$

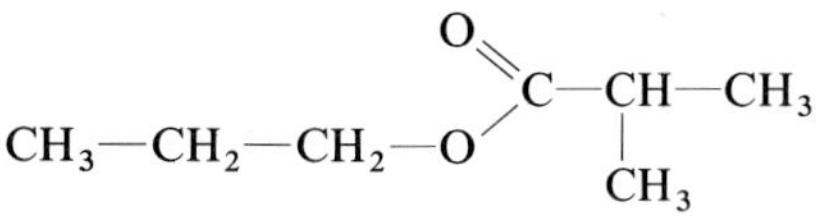

d. 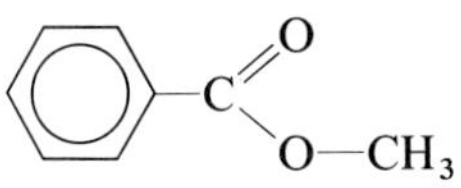 (phenyl)$-C(=O)-O-CH_3$

e. 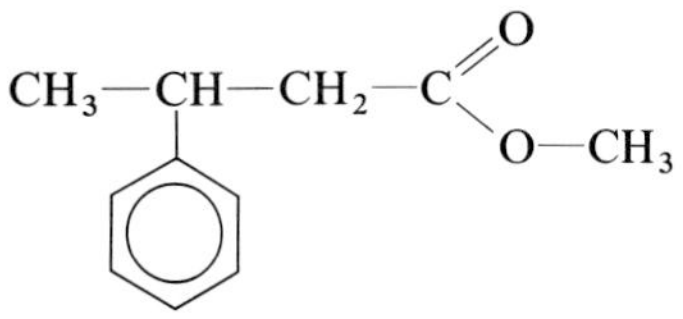 $CH_3-CH(C_6H_5)-CH_2-C(=O)-O-CH_3$

f. 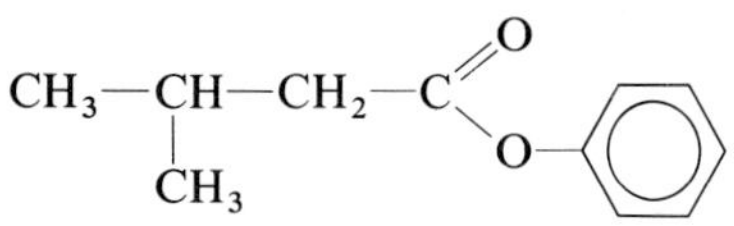 $CH_3-CH(CH_3)-CH_2-C(=O)-O-$(phenyl)

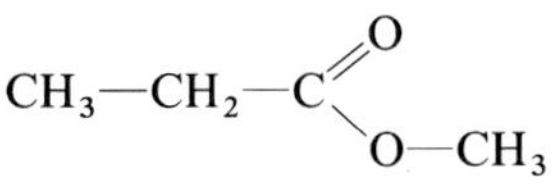

g.

$$CH_3-C\underset{O-CH_2-CH_2-CH_2-CH_2-CH_3}{\overset{O}{\diagup\!\!\!\!\diagdown}}$$

h.

(phenyl)—O—C(=O)—(phenyl)—Cl

Nomenclature of Esters

19.30 Assign an IUPAC name to each of the following esters.

a. $CH_3-CH_2-C(=O)-O-CH_2-CH_3$

b. $H-C(=O)-O-CH_3$

c. $CH_3-C(=O)-O-CH_3$

d. $CH_3-CH_2-CH_2-O-C(=O)-CH_3$

e. $CH_3-CH_2-C(=O)-O-CH_2-CH_2-CH_3$

f. $CH_3-CH_2-C(=O)-O-CH(CH_3)-CH_3$

g. (phenyl)—C(=O)—O—CH_3

h. $CH_3-C(=O)-O-CH_2-CH_3$

19.31 Assign a common name to each of the esters in Problem 19.30.

19.32 Draw structural formulas for each of the following esters.

a. methyl formate
b. *n*-propyl acetate
c. 4-bromophenyl butyrate
d. ethyl 2-phenylacetate
e. isopropyl acetate
f. *n*-octyl decanoate
g. 2-bromoethyl ethanoate
h. 2,2-dimethylpropyl formate

19.33 Assign IUPAC names to the simple esters that are produced from the reaction of the following carboxylic acids and alcohols.

a. acetic acid and ethanol
b. ethanoic acid and methanol
c. butyric acid and ethanol
d. lactic acid and *n*-propyl alcohol
e. ethyl alcohol and formic acid
f. methyl alcohol and acetic acid
g. isopropyl alcohol and formic acid
h. 1-pentanol and pentanoic acid

19.34 Draw and give IUPAC names to all possible esters that contain four carbon atoms. The general formula of these esters is $C_4H_8O_2$.

19.35 The pheromone *isopentyl acetate* is discharged with the venom when a bee stings an enemy, thus attracting other bees to sting the same spot. Draw the structure of isopentyl acetate. The isopentyl group is

$$CH_3-\underset{\underset{CH_3}{|}}{CH}-CH_2-CH_2-$$

Physical Properties of Esters

19.36 Explain why esters have lower boiling points than acids that have the same molecular weight.

19.37 List three esters that are used as flavoring agents. Draw the three structures and describe their odors.

19.38 Describe the general structural features of natural waxes.

Synthesis and Reactions of Esters

19.39 Write structural equations describing the formation of each of the following esters.

a. *n*-butyl acetate from an acid chloride and an alcohol
b. ethyl butyrate from a carboxylic acid and an alcohol
c. phenyl propanoate from an acid chloride and an alcohol
d. methyl benzoate from a carboxylic acid and an alcohol
e. isopropyl formate from an acid chloride and an alcohol
f. dimethyl oxalate from a carboxylic acid and an alcohol

19.40 Salicylic acid can be esterified in two different ways to form either aspirin or oil of wintergreen. Show the reactions that yield these esters. Identify the portions of the salicylic acid structure that undergo reaction in each case.

19.41 Explain the role of water in the formation and hydrolysis of esters.

19.42 What is the difference between the terms *ester hydrolysis* and *saponification*?

19.43 Draw structures of the reaction products in the following chemical reactions.

a. $CH_3-C(=O)-O-CH_2-CH_3 + H_2O \xrightarrow{H^+}$

b. (phenyl)—C(=O)—O—CH_3 + H_2O $\xrightarrow{H^+}$

c.
CH$_3$—CH(CH$_3$)—C(=O)—O—CH$_3$ + NaOH ⟶

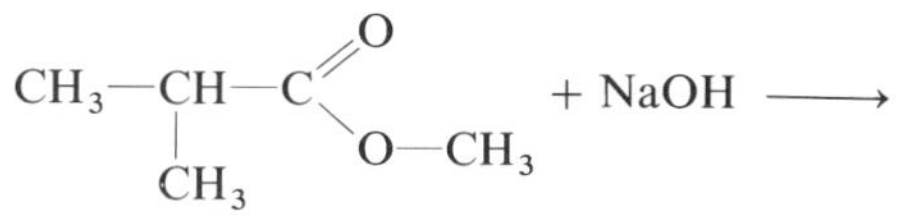

19.44 Some perfumes (and colognes) contain esters as ingredients. Individuals who have high acid concentrations in their skin complain that these perfumes smell awful after a few minutes or hours. Explain what is probably happening to cause this.

19.45 One hundred milliliters of ethyl alcohol is divided into two portions. Portion I is added to an aqueous solution of K$_2$Cr$_2$O$_7$ and allowed to react. The organic product of this reaction is mixed with Portion II of the alcohol. A trace of acid is added and the solution is heated. What is the final product of this reaction scheme?

19.46 Draw the structures of the thioesters formed as a result of a reaction between each of the following carboxylic acids and thiols.

a. CH$_3$—C(=O)OH and HS—CH$_2$—CH$_3$

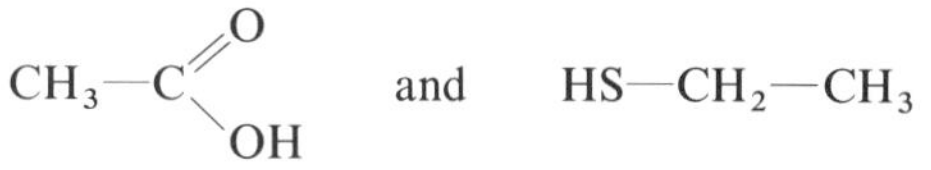

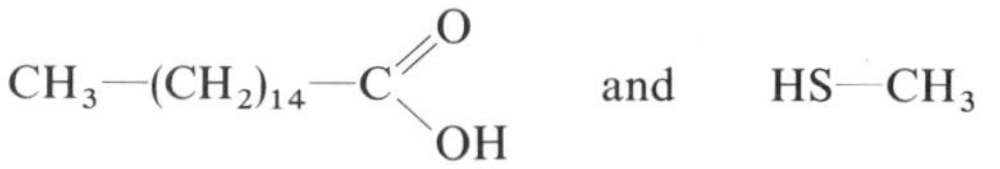

b. CH$_3$—(CH$_2$)$_{14}$—C(=O)OH and HS—CH$_3$

c. C$_6$H$_5$—C(=O)OH and HS—CH(CH$_3$)—CH$_3$

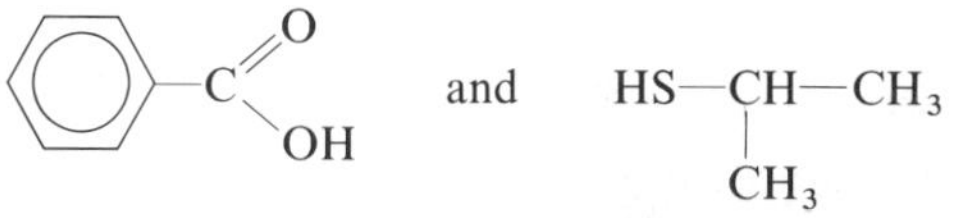

d. succinic acid and excess HS—CH$_2$—CH$_2$—CH$_3$

Polyesters

19.47 How do the acids and alcohols involved in polyester formation differ from those that commonly form simple esters?

19.48 Write a chemical reaction showing the synthesis of the polyester Dacron. Show the structures of both the reactants and the product.

19.49 Write the structure of the polyester polymer formed from oxalic acid and 1, 3-propanediol.

19.50 What are the differences between condensation polymers and addition polymers?

Esters of Inorganic Acids

19.51 Complete the following reactions.

a. CH$_3$—CH$_2$—CH$_2$—CH$_2$—CH$_2$—OH + HO—S(=O)$_2$—OH ⟶

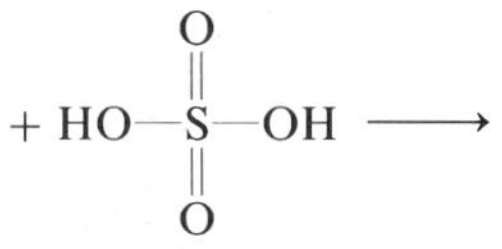

b. CH$_2$—OH, CH$_2$—OH + 2 HO—NO$_2$ ⟶

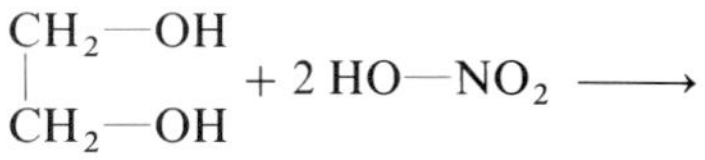

c. CH$_3$—CH$_2$—CH$_2$—OH + HO—P(=O)(OH)—OH ⟶

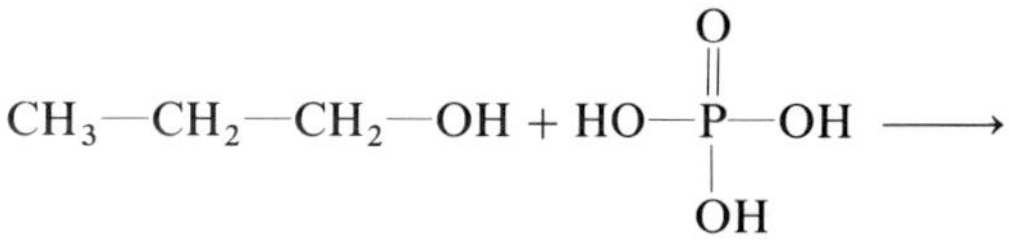

19.52 Glyceric acid (2,3-dihydroxypropanoic acid) is an important intermediate in the metabolism of sugars. A derivative of this acid, 2,3-diphosphoglycerate (2,3-DPG), plays an important role in the transport of oxygen by the blood. Draw the structure of 2,3-DPG.

19.53 Organic phosphate insecticides such as parathion are much more expensive than DDT. This increases the costs of crop production. During the 1960s many farmers were infuriated when DDT was banned. What explanation can you offer for the use of these more expensive pesticides?

Acid Anhydrides

19.54 Write the structure of the anhydride formed by reacting each of the following carboxylic acids with the dehydrating agent P$_4$O$_{10}$.

a. butyric acid
b. ethanoic acid
c. benzoic acid
d. hexanoic acid

19.55 The reaction of glutaric acid with P$_4$O$_{10}$ produces a compound with the molecular formula C$_5$H$_6$O$_3$. Draw the structure of this compound.

19.56 Write the structures of the organic products obtained from each of the following sets of reactants.

a. acetic anhydride and methanol
b. propanoic anhydride and methanol
c. acetic anhydride and ethanol
d. propanoic anhydride and ethanol

19.57 Write the structure of the organic reaction product obtained when one mole of methanol reacts with one mole of each of the following.

a. phosphoric acid
b. pyrophosphoric acid
c. triphosphoric acid

19.58 Show how triphosphoric acid is formed from three molecules of phosphoric acid. How many H$_2$O molecules are "split" out?

19.59 Write equations showing the hydrolysis of

a. methyl triphosphate ion to methyl diphosphate ion
b. methyl diphosphate ion to methyl monophosphate ion

20 Amines and Amides

Objectives

After completing Chapter 20, you will be able to:

20.1 List the structural features associated with the amine functional group and classify amines as primary, secondary, or tertiary amines.
20.2 Determine both common and IUPAC nomenclature for amines.
20.3 List the general physical properties of simple amines and the structures and uses (functions) of selected naturally occurring and synthetic β-phenylethylamines.
20.4 Write chemical equations for the synthesis of amines using a variety of techniques.
20.5 Explain why amines exhibit basic properties in aqueous solution and write chemical equations for the preparation of amine salts.
20.6 Write chemical equations for the conversion of amines to quaternary ammonium salts and amides.
20.7 List the general structural characteristics of the simple heterocyclic amines and the commonly occurring derivatives of these simple compounds.
20.8 List the structural features associated with the amide functional group and classify amides as primary, secondary, or tertiary amides.
20.9 Determine both common and IUPAC nomenclature for amides.
20.10 List the general physical properties of amides.
20.11 Write chemical equations for the synthesis of amides from amines and carboxylic acids.
20.12 Write chemical equations for amide hydrolysis reactions.
20.13 List the general structural features and uses of urea, acetanilides, barbiturates, and benzodiazepines.
20.14 State the general definition of an alkaloid and list the general structural features and uses of the alkaloidal narcotic analgesics.
20.15 Discuss the formation of polyamide and polyurethane polymers and give examples of their use.

INTRODUCTION

The four most abundant elements in living organisms are carbon, hydrogen, oxygen, and nitrogen (Section 2.6). In previous chapters of the "organic" portion of this text, we discussed compounds containing the first three of these elements. Alkanes, alkenes, alkynes, and aromatic hydrocarbons are all carbon–hydrogen compounds. The carbon–hydrogen–oxygen compounds we have discussed include alcohols, phenols, ethers, aldehydes, ketones, carboxylic acids, and esters. We will now extend our discussion to organic compounds that contain the element nitrogen.

Although we will mention a number of families of nitrogen-containing organic compounds in this chapter, our major emphasis will be on two types of compounds: amines and amides. Amines are carbon–hydrogen–nitrogen compounds, and amides contain oxygen in addition to these elements.

Amines and amides occur widely in nature in both plants and animals. Many of these naturally occurring compounds are highly active in a physiological sense. In addition, many drugs used for the treatment of mental illness, hay fever, heart problems, and other physical disorders are amines or amides. A thorough understanding of the properties of amines and amides is a prerequisite to our further study of biochemistry and its associated topics.

20.1 STRUCTURE AND CLASSIFICATION OF AMINES

Amines are *organic derivatives of ammonia (NH_3) in which one or more of the hydrogen atoms on the nitrogen has been replaced by an alkyl or aromatic group.* Thus, amines bear the same relationship to ammonia that alcohols and ethers do to water.

amines

Like alcohols, amines can also be classified as primary, secondary, or tertiary. However, the basis for these categorizations differs from that used for alcohols. Alcohols are classified by the number of nonhydrogen groups attached to the hydroxyl-bearing carbon atom, but amines are classified by the number of nonhydrogen groups attached directly to the nitrogen atom.

ammonia	H—N(H)—H	or NH_3
primary amine	R—N(H)—H	or $R-NH_2$
secondary amine	R—N(H)—R'	or R—NH—R'
tertiary amine	R—N(R'')—R'	

A comparison of the molecules *t*-butyl alcohol and *t*-butylamine illustrates the differences between the systems for classifying alcohols and amines.

$$\underset{\substack{\text{t-butyl alcohol}\\\text{(a tertiary alcohol)}}}{\underset{\underset{\text{CH}_3}{|}}{\overset{\overset{\text{CH}_3}{|}}{\text{CH}_3-\text{C}-\text{OH}}}} \qquad \underset{\substack{\text{t-butylamine}\\\text{(a primary amine)}}}{\underset{\underset{\text{CH}_3}{|}}{\overset{\overset{\text{CH}_3}{|}}{\text{CH}_3-\text{C}-\text{NH}_2}}}$$

tertiary carbon atom → (t-butyl alcohol)
tertiary carbon atom → (t-butylamine); primary nitrogen atom

Example 20.1

Classify each of the following amines as a primary, secondary, or tertiary amine.

a. $CH_3-CH_2-NH_2$

b. $CH_3-NH-\text{C}_6\text{H}_5$

c. $CH_3-NH-\underset{\underset{CH_3}{|}}{CH}-CH_3$

d. $CH_3-\underset{\underset{CH_3}{|}}{N}-CH_3$

e. $\text{C}_6\text{H}_5-\underset{\underset{CH_3}{|}}{N}-\text{C}_6\text{H}_5$

f. 3-methylpiperidine (cyclic amine with N—H)

Solution

The number of carbon atoms bonded to the nitrogen atom determines the amine classification. There is one carbon atom bonded to nitrogen for primary amines, two for secondary amines, and three for tertiary amines.

a. This is a primary amine because there is only one carbon–nitrogen bond.
b. This is a secondary amine because the nitrogen is bonded to both a methyl group and a phenyl group.
c. This is also a secondary amine. This time, the nitrogen atom is bonded to a methyl group and an isopropyl group.
d. Here we have a tertiary amine because the nitrogen atom is bonded to three alkyl groups.
e. This is also a tertiary amine; the nitrogen atom is bonded to two phenyl groups and a methyl group.
f. This is an example of a cyclic amine. Our classification system can also be used for these amines. The nitrogen atom is bonded to two carbon atoms (both within the ring), so it is a secondary amine.

20.2 NOMENCLATURE OF AMINES

Both common and IUPAC names are extensively used for amines. In the common system of nomenclature, simple amines are named by listing the alkyl group or groups attached to the nitrogen atom in alphabetical order, and following it with the suffix *-amine*; all of this appears as one word. Prefixes such as *di-* and *tri-* are added when identical groups are bonded to the nitrogen atom.

$$CH_3-CH_2-NH_2 \qquad CH_3-NH-CH_3$$
ethylamine dimethylamine

Ph−N(CH₃)−CH₂−CH₃
ethylmethylphenylamine

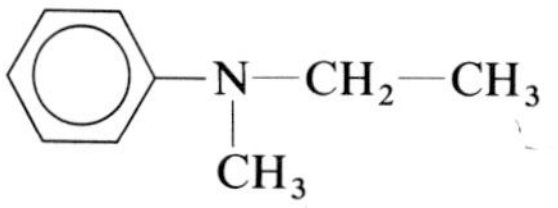

In the IUPAC system of nomenclature, amines are named as hydrocarbons and the location of the functional group is specified in the same way as for other hydrocarbon derivatives. Functional group naming proceeds as follows.

1. For primary amines, the —NH₂ group is called an *amino* group.
2. For secondary amines, the largest alkyl group is considered to be the base chain and the —NHR′ group is called an N—alkyl amino group. The capital N in the name of the —NHR′ group indicates that the alkyl group is attached to nitrogen and not to the base chain.
3. For tertiary amines, the largest alkyl group is again considered to be the base chain and the —NRR′ group is called a N,N—dialkyl amino group.

$$CH_3-CH_2-CH_2-NH_2 \qquad CH_3-CH_2-NH-CH_3$$
1-aminopropane N-methylaminoethane

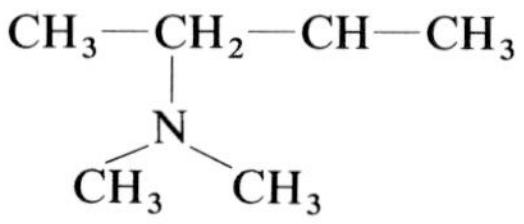

2-(N,N-dimethylamino)butane

The simplest aromatic amine is the compound *aniline* (aminobenzene).

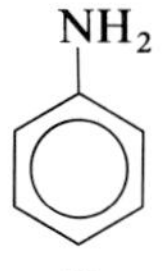

aniline

Aromatic amines are most frequently named as derivatives of aniline in a manner similar to phenols (aromatic alcohols) being named as derivatives of phenol (hydroxybenzene). In secondary and tertiary aromatic amines, groups attached to the nitrogen are located using a capital N—. In aniline rings that have multiple substituents, the numbering of the ring always begins with the carbon in the ring bearing the amino group. A benzene ring substituted with both an amino group and a methyl group is called *toluidine*. This represents a combination of the names *toluene* and *aniline*.

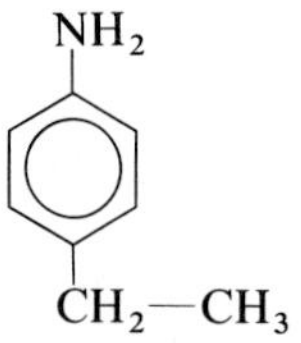

4-ethylaniline
(*p*-ethylaniline)

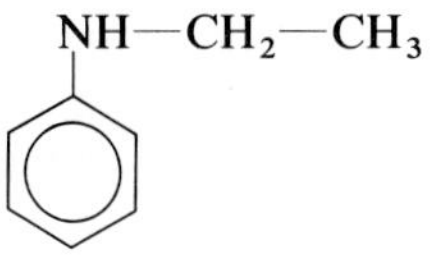

N-ethylaniline

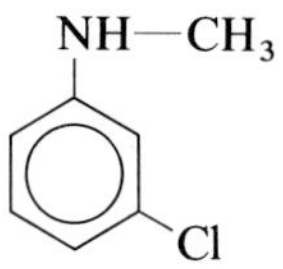

3-chloro-N-methylaniline
(*m*-chloro-N-methylaniline)

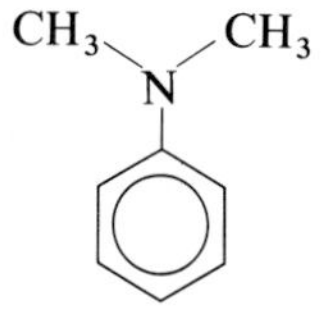

N,N-dimethylaniline

Example 20.2

Assign IUPAC names to each of the following amines.

a. 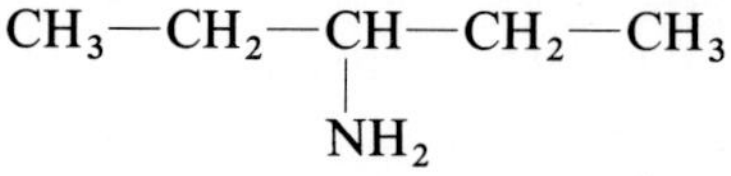 CH$_3$—CH$_2$—CH—CH$_2$—CH$_3$
              |
              NH$_2$

b. 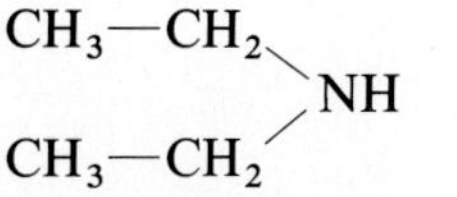 CH$_3$—CH$_2$
              \
               NH
              /
   CH$_3$—CH$_2$

c. 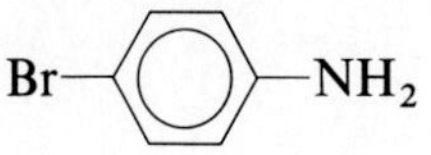 Br—⟨phenyl⟩—NH$_2$

d. 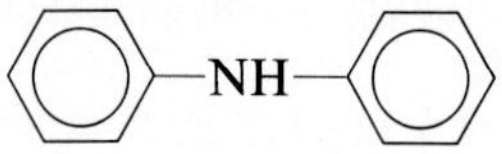 ⟨phenyl⟩—NH—⟨phenyl⟩

e. H$_2$N—CH$_2$—CH$_2$—NH$_2$

f. 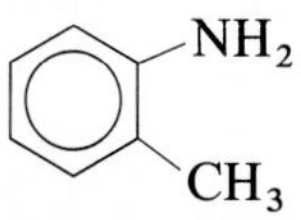 ⟨phenyl ring⟩ with NH$_2$ and CH$_3$ (ortho)

Solution

a. The name of this primary amine is *3-aminopentane*.
b. This is a secondary amine in which the longest alkyl group is an ethyl group. The name of the compound is *N-ethylaminoethane*.
c. This compound is named as a derivative of aniline—*4-bromoaniline* (or *p*-bromoaniline). The carbon in the ring to which the —NH$_2$ is attached is carbon number 1.
d. This is also named as a derivative of aniline. With a phenyl group attached to the nitrogen atom, we have *N-phenylaniline*.
e. Two —NH$_2$ groups are present in this molecule. The name is *1,2-diaminoethane*.
f. This compound is *o-toluidine*. An alternate name would be *2-methylaniline* (or *o*-methylaniline).

20.3 PROPERTIES, USES, AND OCCURRENCES OF AMINES

Low-molecular-weight alkylamines are gases or liquids at room temperature. Di- and triethylamine, as well as primary amines that have from three to ten carbons, are liquids; smaller amines are gases.

Amines with fewer than six carbon atoms are generally soluble in water because of intermolecular hydrogen bonding. Although nitrogen is not as electronegative as oxygen, it can polarize the N—H bond so that strong dipole-dipole attractions form between molecules (see Figure 20-1). Tertiary amines have no hydrogen atoms bonded directly to the nitrogen atom, so they do not form hydrogen bonds among themselves or with water. Consequently, their boiling points are lower than those of primary and secondary amines. As we shall find out later (Chapters 24 and 26) hydrogen bonds involving N—H bonds are important in maintaining the structures of various proteins and nucleic acids.

One of the most notable properties of low-molecular-weight amines is their foul smell. These volatile amines evaporate quickly and smell a lot like a mixture of

FIGURE 20–1 Hydrogen bonding situations involving primary amines.

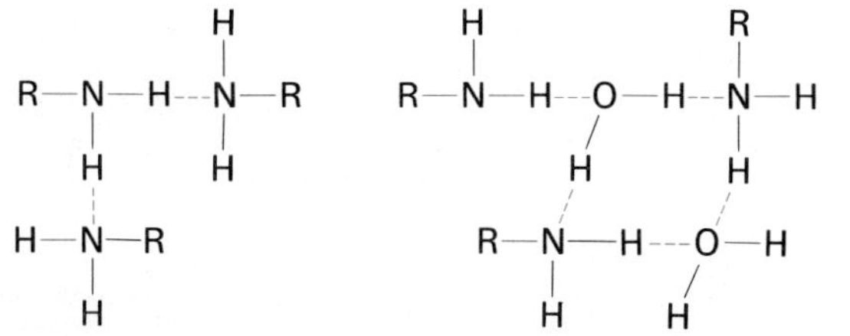

$$R-\underset{\underset{H}{|}}{N}-H\cdots\underset{\underset{H}{|}}{N}-R$$

$$\underset{\underset{H}{|}}{H-N-R}$$

Between primary
amine molecules

$$R-\underset{\underset{H}{|}}{N}-H\cdots O-H\cdots\underset{\underset{H}{|}}{N}-H$$

$$R-\underset{\underset{H}{|}}{N}-H\cdots O-H$$

Between primary amine
and water molecules

ammonia and decayed fish. Most decaying matter (especially tissues rich in protein) produces amines. Part of the odor of rendering plants, meat packing houses, and sewage treatment plants is caused by amines. Two naturally occurring amines that are partially responsible for the smell of dead, decaying animals are putrescine and cadaverine (whose names match their odors).

$H_2N-CH_2-CH_2-CH_2-CH_2-NH_2$ $HN_2-CH_2-CH_2-CH_2-CH_2-CH_2-NH_2$
putrescine   cadaverine
(1,4-diaminobutane)   (1,5-diaminopentane)

These compounds are formed by bacterial decomposition of proteins through enzymatic reactions.

Many low-molecular-weight amines are also toxic. Exposure to sufficiently high concentrations of some amines can cause irritations to mucous membranes, eyes, and skin. Although they are not extremely soluble in water, aromatic amines are rather soluble in fatty tissues and are easily absorbed through the skin. At least one aromatic amine, β-naphthylamine, has been shown to cause cancer, so these compounds should be handled carefully. Table 20–1 on page 474 gives the structure of β-naphthylamine, as well as selected other amines. Comments concerning the uses of these amines are also included in the table.

Two relatively simple amines are very important in the functioning of the human body. They are the adrenal gland secretions epinephrine (adrenalin) and norepinephrine.

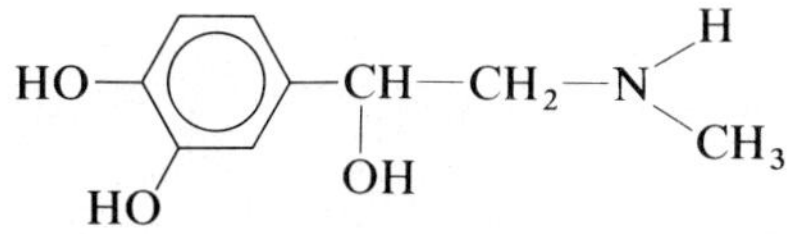

epinephrine

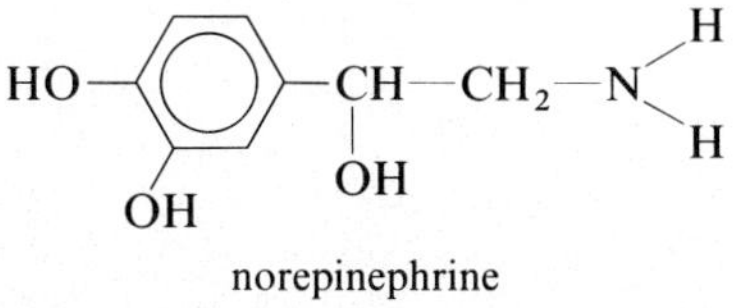

norepinephrine

When these two substances are secreted into the blood they increase cardiac output, raise the blood pressure, stimulate the central nervous system, increase heat production, and elevate the blood sugar concentration. Excitement or fear trigger the release of these substances into the blood.

Epinephrine and norepinephrine are members of a family of amines called β-phenylethylamines. β-Phenylethylamines have the general structure

TABLE 20–1 Structures and Uses of Selected Amines

Name	Structure	Uses
methylamine	CH_3-NH_2	synthesis of the insecticide Sevin and the rocket propellant monomethyl hydrazine
trimethylamine	$CH_3-N(CH_3)-CH_3$	production of choline chloride, a widely used vitamin supplement in animal feeds
diethylamine	$CH_3-CH_2-N(H)-CH_2-CH_3$	production of numerous rubber processing chemicals; production of chloroquine, an antimalaria agent
triethylamine	$CH_3-CH_2-N(CH_2-CH_3)-CH_2-CH_3$	solvent used in extraction and purification of antibiotics such as penicillin and tetracyclines
hexamethylenediamine	$H_2N-(CH_2)_6-NH_2$	raw material for the production of nylon
mechlorethamine	$Cl-CH_2-CH_2-N(CH_3)-CH_2-CH_2-Cl$	used in cancer therapy
aniline	(phenyl)–NH_2	used in manufacture of dyes, photographic chemicals, and prescription drugs
methadone	$CH_3-CH_2-C(=O)-C(phenyl)_2-CH_2-CH(CH_3)-N(CH_3)_2$	used in the treatment of heroin addiction
β-naphthylamine	(naphthyl)–NH_2	a carcinogen easily absorbed through the skin

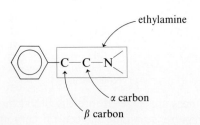

A phenyl group is attached to be the β carbon of an ethylamine.

A number of synthetic β-phenylethylamines are found in prescription medications. Some are used as nasal and lung decongestants, and others function as central nervous system stimulants.

Phenylephrine is used in many nasal sprays as a nasal decongestant. It is also used to treat low blood pressure resulting from shock and the effects of other drugs.

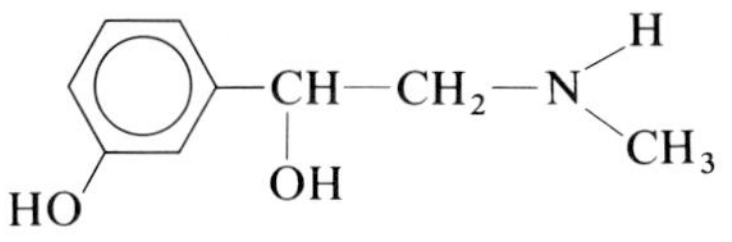

phenylephrine

These isomeric compounds metaproterenol and isoproterenol function as bronchodilators; they are used to decrease lung congestion.

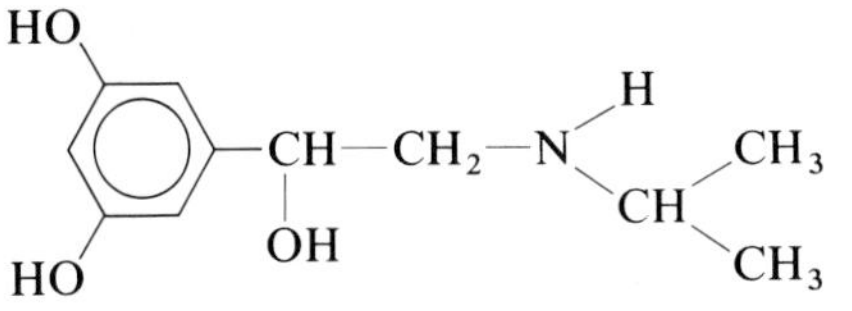

metaproterenol 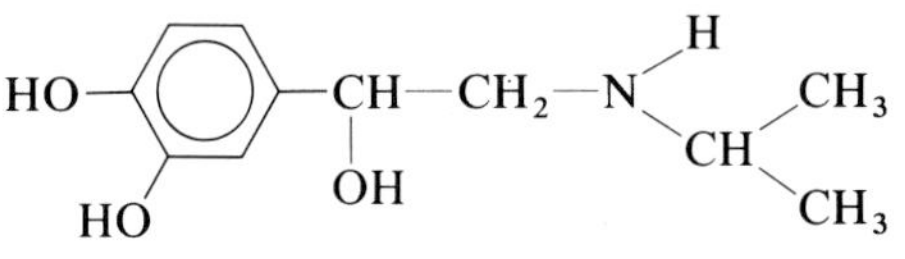   *isoproterenol*

These compounds differ from each other only in the positioning of the —OH groups on the aromatic ring.

Note that all three decongestant compounds are similar to epinephrine and norepinephrine in that both the phenyl group and the β carbon bear hydroxyl groups. Also note that norepinephrine is a primary amine and that epinephrine, phenylephrine, metaproterenol, and isoproterenol are all secondary amines.

A group of β-phenylethylamines that have a methyl group on the α carbon of the amine function as central nervous system stimulants. As a group, these compounds are called *amphetamines*, and are named after the parent compound.

The parent compound in this family, Benzedrine (amphetamine), was first synthesized in 1927 as a drug to stimulate the actions of the hormone adrenalin discussed earlier.

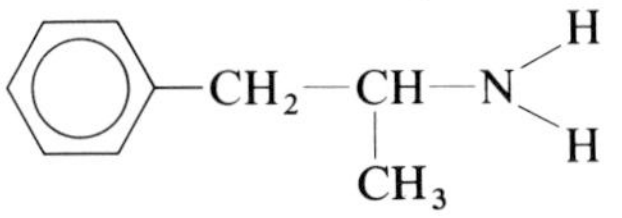

benzedrine (amphetamine)

Since the synthesis of the original Benzedrine, a number of other closely related compounds have been produced. Two very well known compounds are methamphetamine and methoxyamphetamine.

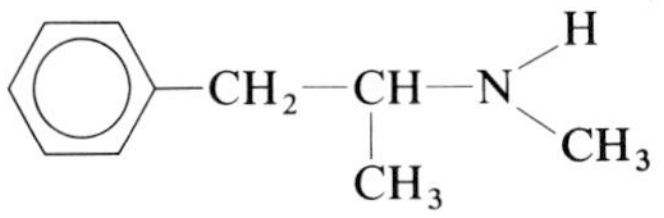

methamphetamine (methedrine) 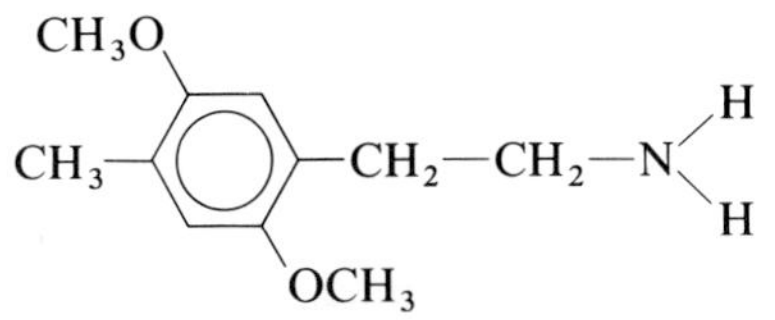   *methoxyamphetamine*

Amphetamines increase both heart and respiratory rates. They reduce fatigue and diminish hunger by raising the glucose level in the blood. Physicians use them to treat mild depression and narcolepsy, a rare form of sleeping sickness. At one time, they were widely used as appetite suppressants in the treatment of obesity, but because of many adverse effects, their use in weight control has diminished. Large quantities of these relatively inexpensive drugs have been diverted into the illegal drug market.

20.4 PREPARATION OF AMINES

A number of chemical processes are available to synthesize amines from starting materials such as alkyl halides, ketones, benzene, alcohols, acids, and nitriles. We will discuss four of these processes.

Alkyl Halide Substitution

A substitution reaction in which an amino group ($-NH_2$) from ammonia replaces the halide atom ($-X$) of an alkyl halide can be used to produce primary amines. The presence of a strong base such as NaOH is needed to carry out the reaction.

$$\underset{\text{an alkyl halide}}{R-X} + \underset{\text{ammonia}}{H-NH_2} + NaOH \longrightarrow \underset{\text{primary amine}}{R-NH_2} + H_2O + NaX$$

A specific example of this type of reaction is the preparation of ethylamine from ethyl bromide.

$$CH_3-CH_2-Br + H-NH_2 + NaOH \longrightarrow CH_3-CH_2-NH_2 + H_2O + NaBr$$

If the newly formed primary amine is not quickly removed from the reaction mixture, the nitrogen atom of the amine will react with further alkyl halide molecules, giving, in succession, secondary and tertiary amines.

$$NH_3 \xrightarrow{RX} \underset{\text{primary amine}}{RNH_2} \xrightarrow{RX} \underset{\text{secondary amine}}{R_2NH} \xrightarrow{RX} \underset{\text{tertiary amine}}{R_3N}$$

A drawback of this method of amine synthesis is that usually a mixture of different amines is obtained.

Reduction of Nitriles

A two-step synthesis, in which an alkyl halide is converted to a nitrile, which is then converted to an amine, is another method for producing primary amines. This preparation method has an advantage over the previous method in that a single product, rather than a mixture of products, is obtained.

A nitrile is a compound containing the $-C\equiv N$ (cyano) functional group. The nitrile group is easily substituted in the place of a halogen atom on an alkyl halide. Sodium cyanide (NaCN) is the salt frequently used to provide the CN^- ions for the substitution.

$$\underset{\text{an alkyl halide}}{R-X} + NaC\equiv N \longrightarrow \underset{\text{a nitrile}}{R-C\equiv N} + NaX$$

After the nitrile is formed, the carbon–nitrogen triple bond is reduced with H_2 gas under pressure, using a nickel catalyst; this process is similar to the hydrogenation of double and triple bonds in alkenes and alkynes.

$$\underset{\text{a nitrile}}{R-C\equiv N} \xrightarrow[\text{Ni}]{2H_2} \underset{\text{a primary amine}}{R-CH_2-NH_2}$$

Note that the resulting amine is one carbon longer than the original alkyl halide. You will recall from Chapter 17 that alcohols can be converted into alkyl halides by certain halogenation reagents. Now we have reactions at our command that, through a series of steps, convert alcohols into amines.

$$R-OH \xrightarrow{PCl_3} \underset{\text{an alkyl halide}}{R-X} \xrightarrow{NaCN} \underset{\text{a nitrile}}{R-C\equiv N} \xrightarrow[Ni]{2H_2} \underset{\text{an amine}}{R-CH_2-NH_2}$$

Reductive Amination

Amines can be easily synthesized from ketones or aldehydes by reductive amination. This is essentially the conversion of a carbonyl group into an amino group:

$$\underset{}{R-\overset{O}{\underset{\|}{C}}-R} + NH_3 \xrightarrow{H_2} R-\overset{NH_2}{\underset{|}{C}}-R + H_2O$$

The R-groups can be either alkyl or aromatic groups; this yields great versatility to this reaction. Biochemically, this type of reaction (with an enzyme catalyst) is important in the biosynthesis of amino acids used in the production of protein.

Reduction of Aromatic Nitro Groups

In Chapter 16, we learned that nitric acid was capable of nitrating a benzene ring in the presence of the catalyst sulfuric acid. The resulting product, nitrobenzene, can be reduced to aminobenzene (aniline) by either H_2 gas or a mixture of tin and HCl:

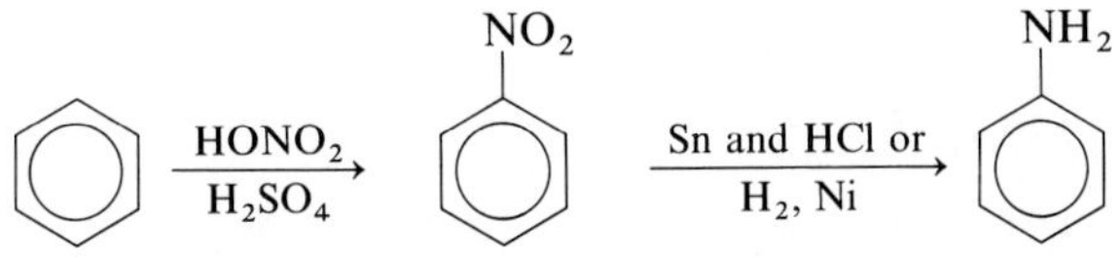

benzene $\xrightarrow[H_2SO_4]{HONO_2}$ nitrobenzene ($-NO_2$) $\xrightarrow[H_2, Ni]{\text{Sn and HCl or}}$ aniline ($-NH_2$)

Specific examples of the reduction step include the following:

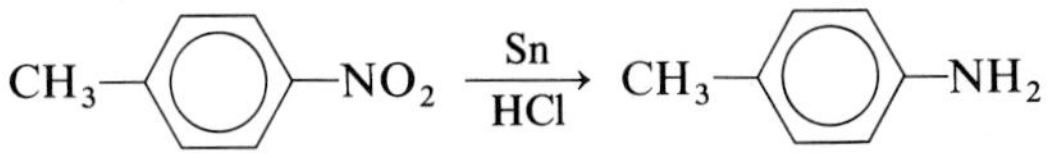

$$CH_3-\text{C}_6H_4-NO_2 \xrightarrow[HCl]{Sn} CH_3-\text{C}_6H_4-NH_2$$
p-nitrotoluene → p-toluidine

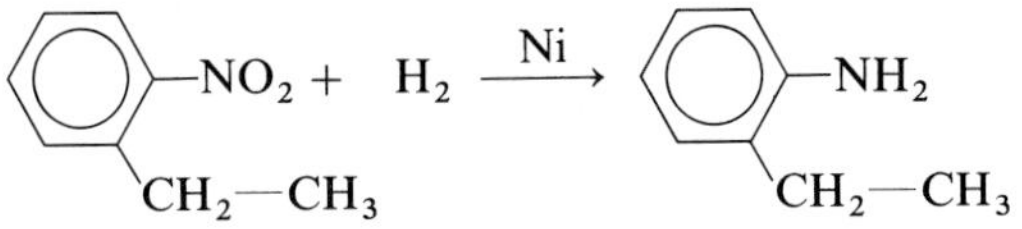

2-nitroethylbenzene + H_2 $\xrightarrow{Ni}$ o-ethylaniline

20.5 BASICITY OF AMINES

Solutions of ammonia in water are basic; they are called *aqueous ammonia* or *ammonium hydroxide* (Section 10.2). In addition to being bonded to three hydrogen atoms, the nitrogen atom in ammonia also possesses a nonbonding pair of electrons that are available for participation in an additional bond. In aqueous solution, water molecules donate a proton to ammonia molecules that results in the formation of ammonium ions and hydroxide ions.

$$\underset{\text{ammonia}}{:NH_3} + HOH \rightleftharpoons \underset{\text{ammonium ion}}{NH_4^+} + \underset{\text{hydroxide ion}}{OH^-}$$

Like ammonia, amines are also bases and they interact with water in an analogous manner. The result is a basic solution containing substituted ammonium ions (ammonium ions in which one or more H atoms have been substituted with alkyl or aromatic groups) and hydroxide ions. The following two reactions illustrate this process.

$$CH_3-\ddot{N}H_2 + HOH \rightleftharpoons [CH_3-NH_3]^+ + OH^-$$
$$\text{methylamine} \qquad\qquad \text{methylammonium} \quad \text{hydroxide}$$
$$\text{ion} \qquad\qquad \text{ion}$$

$$CH_3-CH_2-\underset{\underset{CH_3}{|}}{\ddot{N}H} + HOH \rightleftharpoons \left[CH_3-CH_2-\underset{\underset{CH_3}{|}}{NH_2}\right]^+ + OH^-$$
$$\text{ethylmethylammonium ion} \qquad \text{hydroxide ion}$$

Note that the substituted ammonium species formed by amines in water are charged species; that is, they are ions. There are a number of different notations for denoting these ions. Let us consider the notation system for these ions before proceeding further. In the first of the preceding equations, the methylammonium ion was produced. This ion can be denoted in the following three ways.

$$\left[CH_3-\underset{\underset{H}{|}}{\overset{\overset{H}{|}}{N}}-H\right]^+ \quad \text{or} \quad [CH_3-NH_3]^+ \quad \text{or} \quad CH_3-\overset{+}{N}H_3$$

In the first two notations, brackets were placed around the ion and the positive charge on the ion was placed outside the brackets. This notation correctly indicates that the charge associated with the ion is a property of the ion as a whole. In the last notation, no brackets are used and the positive charge sign is written above the nitrogen atom. This bracketless notation is used for the sake of brevity. It incorrectly suggests that all of the positive charge is associated with the nitrogen atom. Despite this drawback, bracketless notation is most often used to denote amine species where the nitrogen has four bonds. We will use this notation from now on.

Amines (and ammonia) will also accept protons from acids (either strong or weak). The product is a salt containing the substituted ammonium ion as the positive ion and the negative ion from the acid. These salts can be obtained in crystalline form by evaporating the water from the acidic solution. Amine salts are solids and have relatively high melting points. Like most salts, they are quite soluble in water. The following equations show the reaction of methylamine with hydrochloric acid (a strong acid) and acetic acid (a weak organic acid).

$$CH_3-NH_2(aq) + HCl(aq) \longrightarrow CH_3\overset{+}{N}H_3\ Cl^-$$
$$\text{methylammonium chloride}$$

$$CH_3-NH_2(aq) + CH_3-C\underset{OH}{\overset{\nearrow O}{}} \longrightarrow CH_3\overset{+}{N}H_3\ CH_3COO^-$$
$$\text{methylammonium acetate}$$

Many drugs that contain the amine functional group are administered to patients in the form of amine salts because of their increased solubility in water in this form.

Unknowingly, many people utilize acids to form amine salts when they put vinegar or lemon juice on fish. The often unpleasant odor of the evaporating amines in fish is avoided in acid solution because the protonated salt of the amine is no longer volatile. In addition, the tart taste of the excess acid adds flavor to the fish.

The process of forming amine salts with acids is easily reversed by a strong base such as NaOH. The amine is regenerated, as shown in the following equation.

$$CH_3-\overset{+}{N}H_3\;Cl^- + NaOH \longrightarrow CH_3-NH_2 + NaCl + H_2O$$
methylammonium chloride   methylamine

The ability of amines to pick up or lose protons depends on the pH of the solution. When the pH is low, there is an excess of protons and the amine is protonated. When the pH is high (as with addition of strong base), there are very few free protons in solution, and the amine therefore loses its proton:

$$R-NH_2 \underset{-H^+}{\overset{+H^+}{\rightleftharpoons}} R-\overset{+}{N}H_3$$
free amine at high pH   protonated amine at low pH

This phenomenon of protonation is identical to carboxylic acids that pick up protons at low pH and deprotonate at high pH.

$$R-COO^- \underset{-H^+}{\overset{+H^+}{\rightleftharpoons}} R-COOH$$
deprotonated acid at high pH   protonated acid at low pH

The difference between amines and acids becomes apparent at *neutral pH*, where the amine is positively charged and the acid bears a negative charge.

$$R-COO^- \qquad R-\overset{+}{N}H_3$$
carboxylic acid at neutral pH   an amine at neutral pH

Example 20.3

α-Amino acids, the building blocks of proteins, are carboxylic acids that contain an α-amino group. One of the simplest α-amino acids is *alanine*, whose structure is

$$CH_3-\underset{\underset{NH_2}{|}}{CH}-C\overset{\displaystyle O}{\underset{\displaystyle OH}{\diagup\!\!\!\diagdown}}$$

Assume you have three aqueous solutions of alanine, each at a different pH: low pH, neutral pH, and high pH. Draw structural formulas for alanine in each of these three solutions, showing the charges on the amino and carboxyl groups in each case.

Solution

The two groups act independently, like a free amine or free acid in solution. Using the principles discussed for each group at the different pH values, we obtain

$$CH_3-CH-COOH \atop {\underset{+}{NH_3}}$$
low pH
(excess H$^+$ in solution)
both groups protonated

$$CH_3-CH-COO^- \atop {\underset{+}{NH_3}}$$
neutral pH

$$CH_3-CH-COO^- \atop {NH_2}$$
high pH
(very few H$^+$ ions in solution)
both groups deprotonated

The key to solving this problem, of course, is knowing that when excess protons are present in solution, both groups are protonated, and at high pH (few protons in solution), both groups have lost their protons. The difference between the two is at neutral pH, where the acid has lost its proton, but the amine has not.

20.6 CHEMICAL REACTIONS OF AMINES

In this section, we will consider two additional types of amine reactions. They are:

1. Reaction of tertiary amines with alkyl halides to form quaternary ammonium salts, and
2. Reaction of amines with carboxylic acids to form amides.

Tertiary amines have three organic groups (alkyl or aromatic) attached to the nitrogen atom. It is possible to attach a fourth organic group to the nitrogen atom, thus producing a quaternary ammonium salt. A **quaternary ammonium salt** *is an ammonium salt in which all four groups attached to the nitrogen atom are organic groups.* Compounds containing quaternary ammonium groups are important in biological systems.

quaternary ammonium salt

Reaction of a tertiary amine with an alkyl halide in the presence of strong base produces a quaternary ammonium salt. This type of reaction is just a simple extension of the alkyl substitution reactions discussed in Section 20.5 in which primary, secondary, and tertiary amines are produced. In general terms, we have the following equation for this reaction.

$$R-\underset{R'}{\overset{R''}{N}}-R'' + R'''X \xrightarrow{OH^-} R-\underset{R''}{\overset{R'}{N^+}}-R''' + X^-$$

Like acid salts, these quaternary ammonium salts are colorless, odorless, crystalline solids that have high melting points and are usually water soluble.

Quaternary ammonium salts are named in a manner similar to that for acid salts of amines (Section 20.5). The positive organic ion is named as a substituted ammonium ion for alkyl amines and as a substituted anilinium ion for aromatic amines.

$$NH_3 \atop \text{ammonia} \qquad NH_4^+ \atop \text{ammonium ion}$$

$$CH_3-\underset{CH_3}{\overset{CH_3}{N^+}}-CH_3$$
tetramethylammonium ion

aniline — NH$_2$ attached to benzene ring

anilinium ion — $\overset{+}{N}H_3$ attached to benzene ring

trimethylanilinium ion:
$$CH_3-\overset{\overset{\displaystyle CH_3}{|}}{\underset{\underset{\displaystyle C_6H_5}{|}}{N^+}}-CH_3$$

The complete name for a quaternary ammonium salt has two parts: the name of the positive ion, followed by the name of the negative ion as a separate word.

ethyltrimethylammonium chloride:
$$CH_3-\overset{\overset{\displaystyle CH_3}{|}}{\underset{\underset{\displaystyle CH_3}{|}}{N^+}}-CH_2-CH_3 \quad Cl^-$$

diethylmethylanilinium hydrogen sulfate:
$$CH_3-CH_2-\overset{\overset{\displaystyle CH_3}{|}}{\underset{\underset{\displaystyle C_6H_5}{|}}{N^+}}-CH_2-CH_3 \quad HSO_4^-$$

Choline and acetylcholine are two important quaternary ammonium ions present in the human body. Choline has biologically important roles in both fat transport and growth regulation. Acetylcholine is involved in nerve impulse transmission.

choline:
$$CH_3-\overset{\overset{\displaystyle CH_3}{|}}{\underset{\underset{\displaystyle CH_3}{|}}{N^+}}-CH_2-CH_2-OH$$

acetylcholine:
$$CH_3-\overset{\overset{\displaystyle CH_3}{|}}{\underset{\underset{\displaystyle CH_3}{|}}{N^+}}-CH_2-CH_2-O-\overset{\overset{\displaystyle O}{\|}}{C}-CH_3$$

A quaternary ammonium compound with the common name *bephenium* is used to treat infections caused by both hookworm and roundworm parasites.

bephenium:
$$C_6H_5-O-CH_2-CH_2-\overset{\overset{\displaystyle CH_3}{|}}{\underset{\underset{\displaystyle CH_2}{|}}{N^+}}-CH_2-C_6H_5$$

Benzalkonium chloride (Zephiran chloride) solution is used as a skin disinfectant in cleaning injection and catheter sites (see Figure 20–2, p. 482). Solutions of these compounds are also used to sterilize medical instruments.

benzalkonium chloride
(R = mixture of alkyl groups from C_8H_{17} to $C_{18}H_{37}$)

$$C_6H_5-CH_2-\overset{\overset{\displaystyle CH_3}{|}}{\underset{\underset{\displaystyle CH_3}{|}}{N^+}}-R \quad Cl^-$$

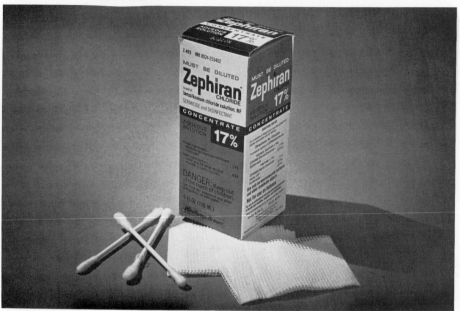

FIGURE 20-2 Benzalkonium chloride solution is used both to sterilize instruments and as a skin disinfectant.

When a carboxylic acid is heated with ammonia, a primary amine, or a secondary amine, a condensation reaction occurs in which the products are an *amide* (Section 20.8) and water.

$$R-C\underset{OH}{\overset{O}{\lessgtr}} + \underset{H}{\overset{H}{\gtrless}}N-H \xrightarrow{heat} R-C\underset{NH_2}{\overset{O}{\lessgtr}} + H_2O$$
ammonia

$$R-C\underset{OH}{\overset{O}{\lessgtr}} + \underset{H}{\overset{H}{\gtrless}}N-R' \xrightarrow{heat} R-C\underset{NHR'}{\overset{O}{\lessgtr}} + H_2O$$
primary amine

$$R-C\underset{OH}{\overset{O}{\lessgtr}} + \underset{H}{\overset{R'}{\gtrless}}N-R'' \xrightarrow{heat} R-C\underset{NR'R''}{\overset{O}{\lessgtr}} + H_2O$$
secondary amine

Tertiary amines do not form amides; the nitrogen atom of the amine has no hydrogen atoms attached to it.

The preceding condensation reactions for amide formation are very similar to the formation of an ester from a carboxylic acid and an alcohol (Section 19.8).

$$R-C\underset{OH}{\overset{O}{\lessgtr}} \quad HO-R' \longrightarrow R-C\underset{OR'}{\overset{O}{\lessgtr}} + H_2O$$

20.7 HETEROCYCLIC AMINES

In Section 16.8, the concept of heterocyclic organic compounds was first introduced. In these compounds, at least one atom present in a ring of atoms is not a carbon atom. Table 16–1 gave the structures of four compounds containing nitrogen as the hetero atom. All four of those compounds were actually heterocyclic amines.

A **heterocyclic amine** *is an organic compound in which the nitrogen atom of the amino group present is part of either an aromatic or nonaromatic ring system.* Heterocyclic amines will be an important part of many discussions in the biochemical chapters of this text. Much of the importance of heterocyclic amines lies in the fact that these compounds are components of the structures of more complicated molecules that are important to life processes.

In this section, we will first consider eight key, simple heterocyclic amines and then more complicated compounds in which each of these structures is present. Many heterocyclic amine derivatives are physiologically active compounds. Numerous drugs (both legal and illegal) contain a heterocyclic amine as part of their structure. Also, many compounds that are naturally produced within the human body, which are necessary for its proper functioning, have heterocyclic-amine-related structures.

Heterocyclic amines are conveniently discussed in categories according to ring size. We consider three 5-membered ring compounds, three 6-membered ring compounds, and two fused-ring compounds.

heterocyclic amine

Five-membered Ring Heterocyclic Amines

Three important five-membered ring heterocyclic amines are pyrrolidine, pyrrole, and imidazole.

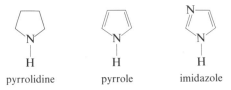

pyrrolidine   pyrrole   imidazole

The first two compounds are liquids at room temperature and the third is a solid. All of the compounds have limited solubility in water. Notice that in writing the structures of heterocyclic amines, we follow the usual notation of not writing the carbon atoms and their attached hydrogens. However, hydrogen atoms attached to hetero atoms, which are nitrogen in this case, are explicitly shown.

Pyrrolidine (C_4H_9N), which consists of a saturated ring containing one nitrogen atom and four carbon atoms, is the simplest five-membered heterocyclic amine. Two important compounds in which the pyrrolidine ring is found are the amino acid *proline* (Section 24.2) and *nicotine*, a central nervous system stimulant.

proline   nicotine

Nicotine also contains a six-membered heterocyclic amine (pyridine), which we will discuss later in this section.

Nicotine is the active ingredient in both smoking and chewing tobacco. Its mild effect on the central nervous system is transient in nature; after an initial response, depression follows. Most cigarettes contain about 2% nicotine, but only a small percent of the nicotine in a cigarette actually enters the body during smoking. Research on the effects of cigarette smoking indicates that nicotine is the principal compound that causes psychological dependence on cigarettes. In large doses, nicotine is a potent poison. Nicotine poisoning causes vomiting, diarrhea, nausea, and abdominal pain. Death occurs from respiratory paralysis. Nicotine was once used as an insectide.

Pyrrole (C_4H_5N) is an unsaturated five-atom ring containing a single nitrogen atom. A structure containing four pyrrole rings bonded together, called a *porphyrin ring*, is very important in the chemistry of life. The porphyrin ring structure is found in hemoglobin, myoglobin, vitamin B_{12}, chlorophyll, and cytochrome proteins. In each of these structures, there is a metal atom bonded to the four nitrogens of the pyrrole rings. This metal is iron in the heme unit present in hemoglobin and myoglobin.

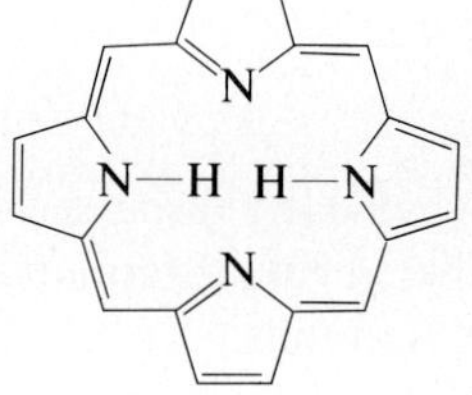

porphyrin ring

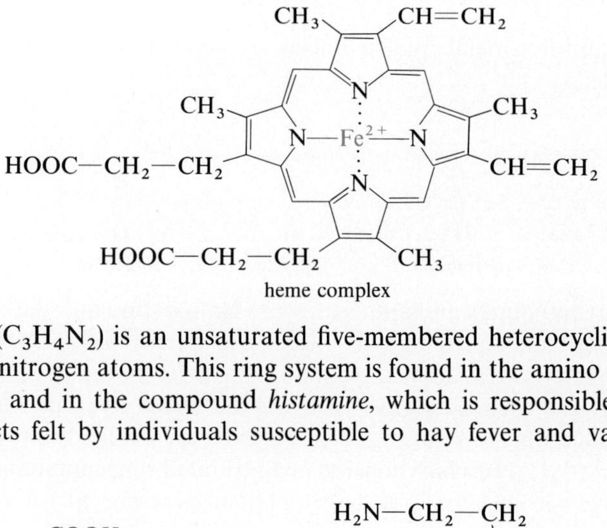

heme complex

Imidazole ($C_3H_4N_2$) is an unsaturated five-membered heterocyclic amine that contains two nitrogen atoms. This ring system is found in the amino acid *histidine* (Section 24.2) and in the compound *histamine*, which is responsible for the unpleasant effects felt by individuals susceptible to hay fever and various pollen allergies.

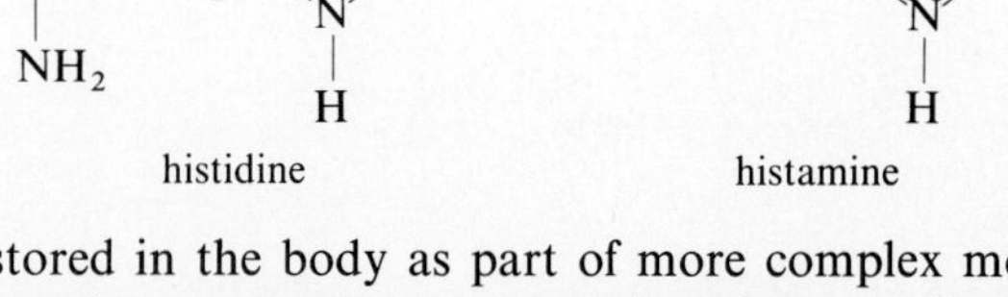

histidine

histamine

Histamine is stored in the body as part of more complex molecules. Various substances that people are allergic to trigger the release of histamine from these more complex molecules. The presence of free histamine in the body causes the allergy symptoms of headache, sinus congestion, and lung congestion.

A group of substances called *antihistamines* can be taken to counteract the ef-

FIGURE 20-3 The active ingredient in an antihistamine medication is usually an amine.

fects of histamines released in the body. Many of these antiallergy drugs (see Figure 20-3) are also amines. Examples of these antihistamines are brompheniramine and diphenhydramine (Benadryl).

brompheniramine   diphenhydramine (Benadryl)

Antihistamines do not provide a cure, but they do offer relief for the allergy victim.

Six-membered Ring Heterocyclic Amines

Three important six-membered ring heterocyclic amines are piperidine, pyridine, and pyrimidine.

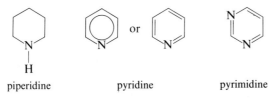

piperidine   pyridine   pyrimidine

All three compounds are liquids at room temperature and are miscible with water. The relationships among these compounds are the same as the relationships among the three 5-membered rings we have discussed. The first compound in each case has one nitrogen atom in a saturated ring; the second compound has one nitrogen atom in an unsaturated ring; and the third compound has two nitrogen atoms in an unsaturated ring.

Piperidine ($C_5H_{11}N$), the saturated six-membered ring system, is part of the structure of coniine, a very poisonous substance obtained from the hemlock plant and piperine, the active ingredient (6 to 10%) in black pepper.

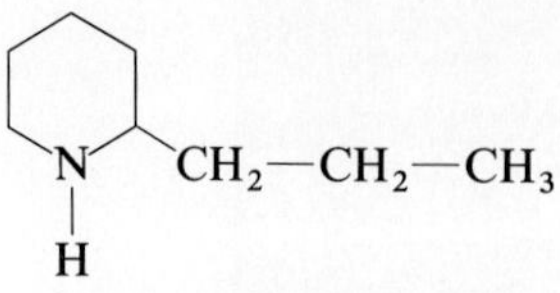

coniine

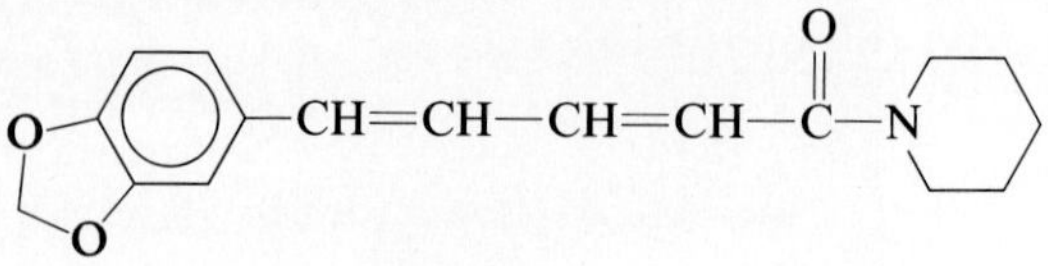

piperine

Two other substances that are also derived from plant extracts and contain the piperidine ring (in a more complex arrangement) are atropine and cocaine.

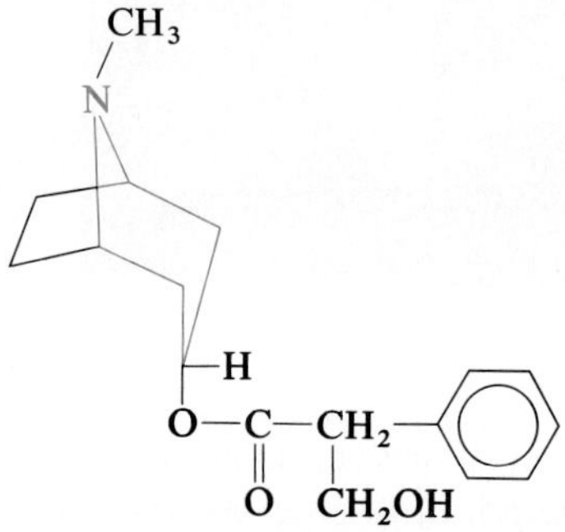

atropine

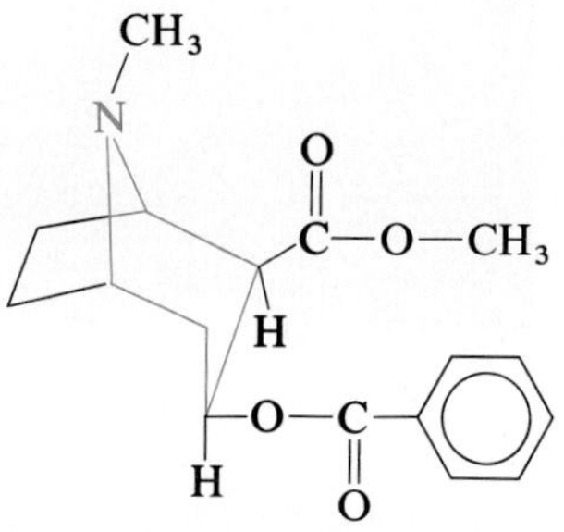

cocaine

Atropine is found in several poisonous plants, including henbane and belladonna. In dilute solution (0.05% to 1%), atropine is used to dilate the pupils of the eyes and as a smooth muscle relaxant. Cocaine is derived from the coca plant that grows on the eastern slopes of the Andes mountains in South America. It has the same heterocyclic ring structure as atropine and is a very powerful central nervous system stimulant. It increases mental awareness and decreases fatigue. As with many other stimulants, the stimulation effect is short-lived and is often followed by a period of deep depression. In the last decade, cocaine has become a dominant chemical in the illegal drug market. Recent studies indicate that cocaine is one of the most addictive known drugs; it is much more addictive than was previously thought.

Pyridine (C_5H_5N) has a structure similar to that of benzene except that a nitrogen atom replaces one of the carbon atoms. The three double bonds present are conjugated; thus, pyridine is an aromatic compound.

The substance nicotine contains a pyridine ring, along with a five-membered pyrrolidine ring. The pyridine ring is also part of the structure of vitamin B_3 (niacin) and vitamin B_6 (which is actually two substances).

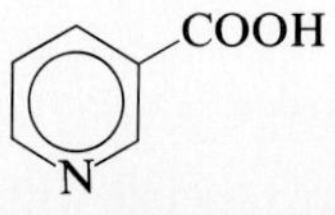

niacin

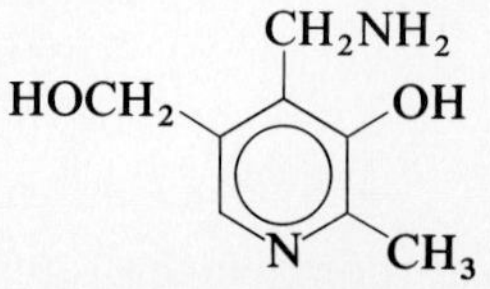

pyridoxamine

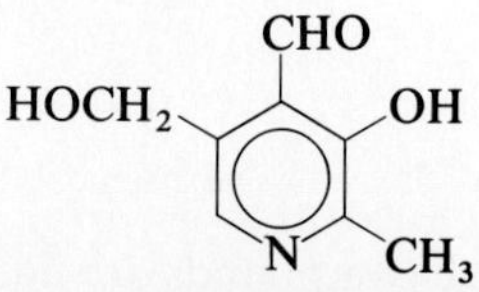

pyridoxal

the B_6 vitamins

More information about both of these vitamins is found in Chapter 27.

Pyrimidine ($C_4H_4N_2$), the dinitrogen heterocyclic amine, is present in another of the B vitamins; thiamine (vitamin B_1) contains this ring system, along with another ring system that has both nitrogen and sulfur present, in addition to carbon.

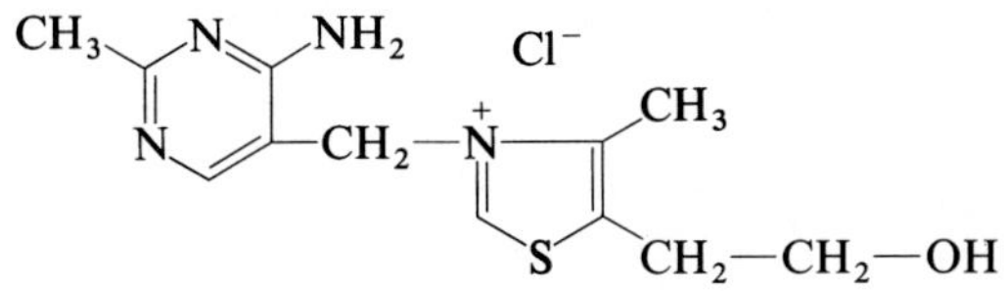

thiamine

Pyrimidine ring systems are a key part of the structure of the nucleic acids DNA and RNA (Chapter 26); these substances are responsible for the transmission of hereditary information and the control of protein synthesis. The chemical foundation for the genetic code (Chapter 26) involves five heterocyclic amines. Three of these amines have pyrimidine ring structures: cytosine, uracil, and thymine.

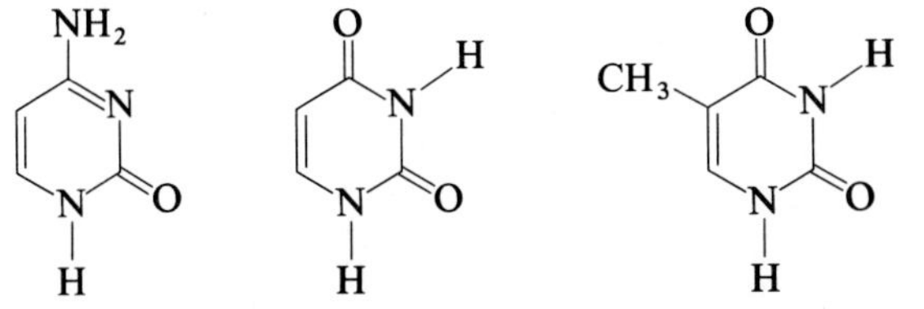

cytosine   uracil   thymine

Fused-ring Heterocyclic Amines

Fused-ring heterocyclic amines have one or more nitrogen atoms in a fused-ring system. Two important fused-ring cyclic amines are *purine* and *indole*; both have a six-membered ring fused to a five-membered ring.

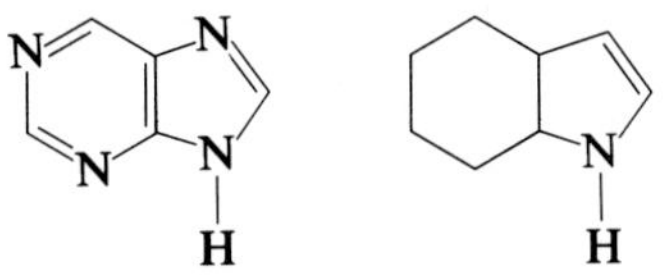

purine   indole

Both compounds are solids at room temperature.

Purine ($C_5N_4H_4$) has the greatest number of hetero nitrogen atoms (four) of any of the cyclic amines we have considered in this section. Both rings contain two nitrogen atoms. The six-membered ring is pyrimidine and the five-membered ring is imidazole.

The other two of the five heterocyclic amine structures that form the basis for the genetic code—adenine and guanine—are purine derivatives.

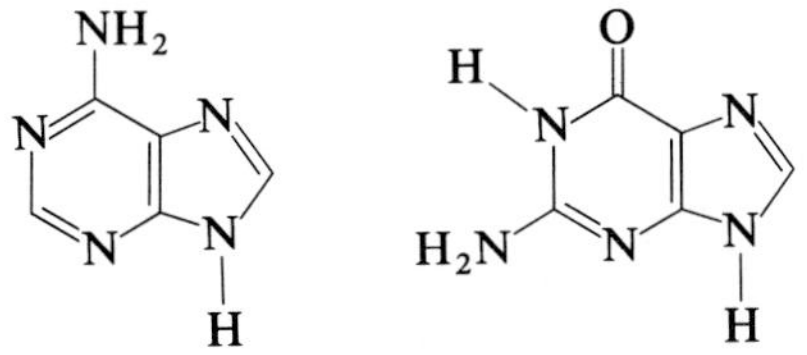

adenine   guanine

Adenine is also part of the structure of the energy storage molecule ATP (adenosine triphosphate), which was mentioned in Section 19.14.

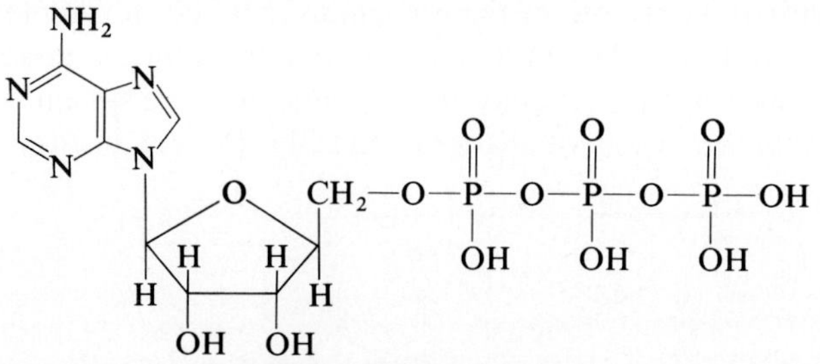

adenosine triphosphate (ATP)

Caffeine, which has a structure based on a purine ring system, is the most widely used, nonprescription, central nervous system stimulant. Its structure is

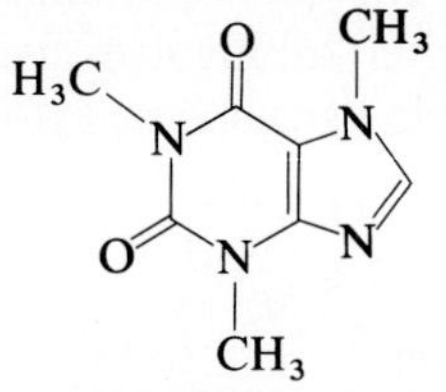

caffeine

In addition to stimulating the nervous system, caffeine can produce a variety of other effects, depending upon the amount consumed. It increases heartbeat and basal metabolic rate, promotes secretion of stomach acid, and steps up production of urine. The overall effect an individual experiences is usually interpreted as a "lift", a feeling of being wide-awake and able to focus on mental or manual tasks.

Excessive consumption of caffeine, which is known medically as *caffeine intoxication*, leads to restlessness and disturbed sleep, irritation of the stomach, and diarrhea. What constitutes an excessive intake varies among individuals; it is estimated to range from as low as 200 milligrams per day to 750 milligrams per day, depending upon the adult.

Caffeine is mildly addicting. People who ordinarily consume substantial amounts of caffeine-containing beverages or drugs experience withdrawal symptoms if caffeine is eliminated. These symptoms include headache and depression for a period of several days. Many people need a cup of coffee before they feel good each morning; this is a result of their dependence on caffeine.

Coffee beans and tea leaves are natural sources of caffeine. Although coffee is the major source of caffeine for most Americans, substantial amounts of caffeine can be consumed in soft drinks, tea, and numerous other products, including nonprescription medications. Table 20-2 gives caffeine amounts derived from several common sources.

Indole (C_8NH_7) has a structure with a benzene ring fused to a pyrrole ring system. *Serotonin*, a hormone that is involved in the normal functioning of the human brain, has an indole-based structure.

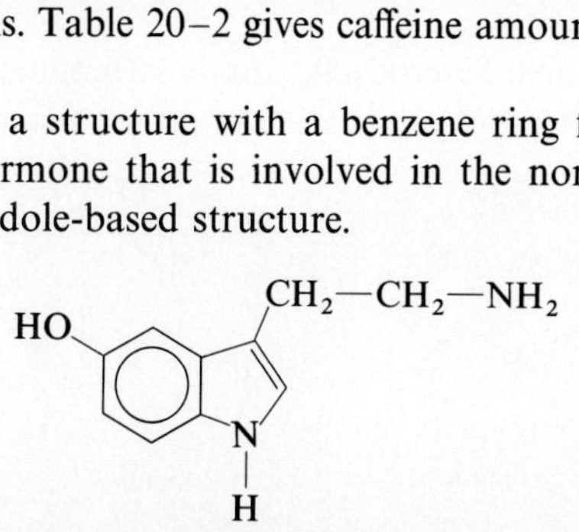

serotonin

TABLE 20-2 Common Sources of Caffeine

Product	Caffeine (in milligrams)
coffee	
drip (5 oz)	146
percolated (5 oz)	110
instant, regular (5 oz)	53
decaffeinated (5 oz)	2
tea	
one-minute brew (5 oz)	9 to 33
five-minute brew (5 oz)	20 to 46
canned iced tea (12 oz)	22 to 36
cola drinks	
Coca-Cola (12 oz)	40
Pepsi-Cola (12 oz)	43
Royal Crown Cola (12 oz)	40
cocoa and chocolate	
cocoa beverage (water mix, 6 oz)	10
milk chocolate (1 oz)	6
baking chocolate (1 oz)	35
nonprescription drugs (standard dose)	
stimulants	
Caffedrine capsules	200
NoDoz tablets	200
pain relievers	
Anacin	64
Excedrin	130
cold remedies	
Dristan	32
Triaminicin	30
diuretics	
Aqua-Ban	200
Permathene H_2Off	200

Serotonin is involved in the transmission of nerve impulses within the brain. This substance is released at the end of a nerve, travels a short distance (the synaptic gap) to the next nerve, and then triggers an impulse in this nerve.

Another substance that contains an indole ring system is the hallucinogenic drug LSD (*ly*sergic acid *d*iethylamide), one of the most potent known hallucinogens. This drug is a derivative of lysergic acid, which can be obtained from ergot, a fungus that grows mostly on rye, but also on wheat.

lysergic acid

lysergic acid diethylamide (LSD)

The drug LSD causes users to see, hear, feel, and smell things that either do not exist or are severe distortions of reality. Other than altered perceptions, the physical effects of LSD are increased body temperature, dilated pupils, and high blood pressure and pulse rate.

Researchers have established a link between serotonin and LSD. After serotonin has been released from a nerve ending, LSD blocks it from triggering an impulse in the next nerve. Scientists believe that this interference triggers the hallucinations associated with LSD use.

20.8 STRUCTURE AND CLASSIFICATION OF AMIDES

amides

Amides *are compounds in which an amino group or substituted amino group is bonded to a carbonyl carbon atom.* Amides can also be considered to be derivatives of carboxylic acids; in amides, the acid hydroxyl group is replaced by an amine or substituted amine group. Amides, like amines, can be classified as primary, secondary, or tertiary, depending on how many nonhydrogen atoms are attached to the nitrogen atom.

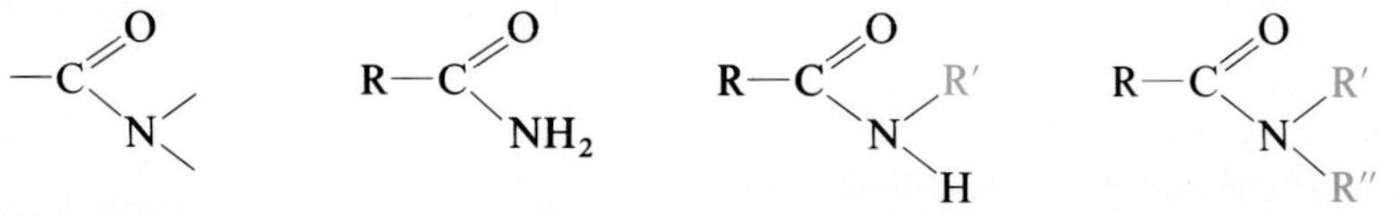

amide functional group   primary amide   secondary amide   tertiary amide

The carbon–nitrogen bond in the amide functional group is often called an *amide linkage*.

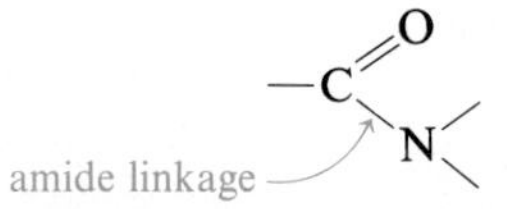

amide linkage

This terminology parallels that used for esters when there is an ester linkage (Section 19.7). Amide linkages in proteins are sometimes called *peptide bonds*. The latter term will be used extensively in Chapter 24 when we discuss proteins. Proteins are polymers of amino acids that are connected to each other by amide linkages (peptide bonds).

20.9 NOMENCLATURE OF AMIDES

The nomenclature for amides is similar to that of carboxylic acids. The root name for the acid from which the amide can be considered to be derived (either common name or IUPAC name) is changed by replacing the *-ic acid* ending (common system) or *-oic acid* ending (IUPAC system) with *-amide*. Alkyl groups attached to the nitrogen are included as prefixes, using a capital N– for each group to indicate location.

Example 20.4

Assign both common and IUPAC names to each of the following amides.

a. $CH_3-C(=O)-NH_2$

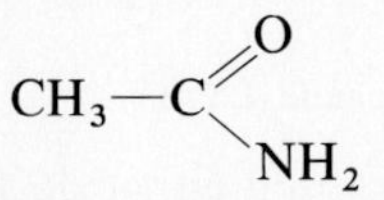

b. $CH_3-CH(Br)-C(=O)-NH-CH_3$

c. Ph$-C(=O)-N(Ph)_2$

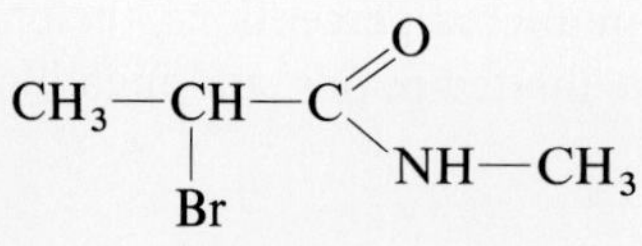

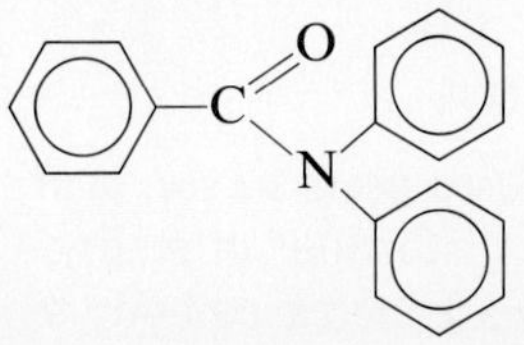

Solution
a. The parent acid for this amide is acetic acid (common) or ethanoic acid (IUPAC). The common name for this amide is *acetamide* and the IUPAC name is *ethanamide*.
b. The common and IUPAC names for this amide will be very similar because the common and IUPAC names of the acid are very similar; they are propionic acid and propanoic acid, respectively. The common name is *N-methyl-2-bromopropionamide* and the IUPAC name is *N-methyl-2-bromopropanamide*. The prefix *N-* must be used with the methyl group to indicate that it is attached to the nitrogen atom, rather than to the carbon chain.
c. In both the common and IUPAC systems of nomenclature, the name of the parent acid is the same: benzoic acid. The name of the amide is *N,N-diphenylbenzamide*.

20.10 PROPERTIES OF AMIDES

Unlike amines, amides are not bases; they are neutral compounds. This is caused in part by the electronegative oxygen atom in the carbonyl group drawing electrons away from nitrogen, leaving very little electron density on the nitrogen to bond to an incoming proton.

Although no acid–base equilibrium occurs with amides, they form excellent hydrogen bonds both among themselves and to water molecules. This is the reason that most unbranched amides are solids at room temperature and have correspondingly high boiling points. (In fact, many high-molecular-weight amides and amines decompose before they boil.) As alkyl groups are substituted on the nitrogen of an amide, the possibilities for intermolecular hydrogen bonding decrease, resulting in weaker cohesive forces between molecules. The water solubility of amides also depends on the ability of the amide to form hydrogen bonds to water molecules.

Example 20.5

Using acetamide and *N,N*-dimethylacetamide as representative amides, explain why primary amides form hydrogen bonds among themselves while tertiary amides do not. Predict which amide has the *higher* boiling point.

Solution

$$CH_3-\underset{\underset{H}{|}}{\overset{\overset{O}{\|}}{C}}-N-H\cdots O=\underset{\underset{H}{|}}{\overset{\overset{CH_3}{|}}{C}} \qquad CH_3-\overset{\overset{O}{\|}}{C}-\underset{\underset{CH_3}{|}}{N}-CH_3 \qquad O=\underset{\underset{CH_3}{|}}{\overset{\overset{CH_3}{|}}{C}}-CH_3$$

$$CH_3-\underset{\underset{H}{|}}{\overset{\overset{O}{\|}}{C}}-N-H \qquad CH_3-\overset{\overset{O}{\|}}{C}-\underset{\underset{CH_3}{|}}{N}-CH_3$$

acetamide
multiple possibilities
for hydrogen bonds

N,N-dimethylacetamide
no possibility for
hydrogen bonds

Because hydrogen bonding increases the cohesive forces between molecules, we predict acetamide would have the higher boiling point. This is the case; acetamide boils at 221°C and *N,N*-dimethylacetamide boils at 165°C.

Hydrogen bonding among amides is a critical factor in maintaining the delicate structure of proteins and enzymes.

20.11 PREPARATION OF AMIDES

The preparation method for synthetic amides has already been discussed. In Section 20.6, in our discussion of amine reactions, we noted that, upon heating, carboxylic acids react with ammonia, primary amines, and secondary amines to produce amides.

Specific examples of amide preparation using this method are

$$CH_3-C(=O)OH + H-NH_2 \longrightarrow CH_3-C(=O)NH_2 + H_2O$$

acetic acid   ammonia   acetamide (a primary amide)

$$CH_3-CH_2-CH_2-C(=O)OH + HN(H)CH_3 \longrightarrow CH_3-CH_2-CH_2-C(=O)NH-CH_3 + H_2O$$

butanoic acid   methylamine   N-methylbutanamide (a secondary amide)

$$C_6H_5-C(=O)OH + H-N(CH_2CH_3)_2 \longrightarrow C_6H_5-C(=O)N(CH_2CH_3)_2 + H_2O$$

benzoic acid   diethylamine   N,N-diethylbenzamide (a tertiary amide)

20.12 HYDROLYSIS OF AMIDES

Amides undergo a number of reactions; the most important biochemical amide reaction is *hydrolysis*. In the hydrolysis of an amide, the amide linkage is broken and free acid and amine are produced. Amide hydrolysis is essentially the reverse of amide formation. Amide hydrolysis is very similar to ester hydrolysis (Section 19.11). As with ester hydrolysis, amide hydrolysis is catalyzed by acids, bases, or certain enzymes; however, unlike esters, amide hydrolysis often requires sustained heating because the amide linkage is a stronger bond than the ester linkage.

$$R-C(=O)NR'R'' + H_2O \xrightarrow[\text{heat}]{\text{catalyst}} R-C(=O)OH + H-NR'R''$$

a disubstituted amide   water   a carboxylic acid   a disubstituted amine

$$C_6H_5-C(=O)NH-CH_3 + H_2O \xrightarrow[\text{heat}]{\text{catalyst}} C_6H_5-C(=O)OH + HNH-CH_3$$

N-methylbenzamide   benzoic acid   methyl amine

Note that in acidic or basic solutions, the free acid and amine will form their corresponding salts (Section 20.6). Occasionally the salts formed are insoluble and precipitate out of solution.

20.13 SELECTED AMIDES AND THEIR USES

Urea, the diamide of carbonic acid, is one of the simplest amides.

$$\text{HO}-\underset{\underset{\text{carbonic acid}}{}}{\overset{\overset{\text{O}}{\|}}{\text{C}}}-\text{OH} \qquad \text{H}_2\text{N}-\underset{\underset{\text{urea}}{}}{\overset{\overset{\text{O}}{\|}}{\text{C}}}-\text{NH}_2$$

Most excess nitrogen is eliminated from the human body in the form of urea. About 30 g of urea is excreted daily in the urine of an average adult. Urea is the endproduct of the breakdown of protein.

Urea is also an industrially important chemical; its most important industrial use is in fertilizer manufacture. Urea is synthesized industrially from CO_2 and NH_3.

$$CO_2 + 2\,NH_3 \xrightarrow[\text{high pressure}]{200°C} H_2N-\overset{\overset{O}{\|}}{C}-NH_2 + H_2O$$

When added to soil, urea is slowly hydrolyzed back to CO_2 and NH_3.

A number of synthetic amides exhibit physiological activity and are used as drugs in the human body. Three families of these compounds are the acetanilides, barbiturates, and benzodiazepines.

Acetanilide, phenacetin, and acetaminophen, which are structurally related aromatic amides, have found extensive use in over-the-counter medications.

Acetanilide was once used as an antipyretic (fever reducer). Because of its toxicity, it has been replaced by the less toxic phenacetin and acetaminophen. Both phenacetin and acetaminophen are comparable to aspirin (Section 19.8) as pain relievers. For many years, phenacetin was combined with aspirin and caffeine and sold as APC tablets. This use for phenacetin was discontinued in 1983 when it was implicated in problems involving kidney damage. Acetaminophen is the active ingredient in Tylenol and is frequently recommended for people who are allergic to aspirin. Like aspirin, acetaminophen acts as an analgesic and antipyretic, but it lacks anti-inflammatory action.

Barbiturates, which are cyclic amide compounds, are a heavily used group of prescription drugs that cause relaxation (tranquilizers), sleep (sedatives), and death (overdoses). All barbiturates are derivatives of barbituric acid, a cyclic amide that was first synthesized from urea and malonic acid.

(The researcher who first synthesized this compound named it after his girlfriend Barbara.)

TABLE 20-3 Identity of the R Groups Present in Common Barbiturates

Barbiturate Names	R_1	R_2
barbital	CH_3-CH_2-	CH_3-CH_2-
phenobarbital (luminal)	CH_3-CH_2-	C_6H_5-
amobarbital (amytal)	CH_3-CH_2-	$(CH_3)_2-CH-CH_2-CH_2-$
pentobarbital (nembutal)	CH_3-CH_2-	$CH_3-CH_2-CH_2-CH(CH_3)-$
secobarbital (seconal)	$CH_2=CH-CH_2-$	$CH_3-CH_2-CH_2-CH(CH_3)-$
aprobarbital	$CH_2=CH-CH_2-$	$(CH_3)_2CH-$

When various alkyl groups are substituted in place of the hydrogen atoms on the carbon atom between the two amide linkages, a large group of physiologically active compounds is obtained.

general formula of barbiturates

Table 20-3 lists common barbiturates and the R groups that are bonded to the barbituric acid structure of each.

Barbiturates depress the central nervous system in a manner similar to alcohol and general anesthetics. The duration of the effects varies with the barbiturate. Phenobarbital is long-lasting, and puts people to sleep for 10 to 12 hours. Pentobarbital and secobarbital are short-acting drugs and last from 3 to 4 hours. Amobarbital is an intermediate drug and lasts from 6 to 8 hours.

Barbiturates are one of the most abused classes of prescription drugs. Because they are sedatives, they are called "downers", as contrasted to amphetamines, which are "uppers". Barbiturates are especially dangerous when they are used with alcohol, which is often the case. The combined effect, which is called a *synergistic effect*, is greater than the sum of the effects when the two drugs are used separately.

The benzodiazepines are a group of tranquilizer drugs. The common feature of their structures is a seven-membered cyclic amide ring containing two nitrogen atoms. Valium, Librium, and Ativan are three popular benzodiazepines.

diazepam (Valium)   chlordiazepoxide (Librium)   lorazepam (Ativan)

Valium is one of the most frequently prescribed drugs in America. It is a mild, but addicting, tranquilizer.

20.14 ALKALOIDS

Primitive tribes in various parts of the world have known for centuries that physiological effects can be obtained by eating or chewing the leaves, roots, or bark of certain plants. By trial and error, these tribes have found that some plant extracts cure diseases and others are deadly poisons.

Today, thousands of different physiologically active compounds extracted from the leaves, bark, roots, flowers, and fruit of plants have been identified by chemists. Most of these compounds are water-insoluble, but they can be extracted from plant material by an acidic solution. Because they react with acid, these naturally occurring substances (most of which contain nitrogen—amides or amines) are called *alkaloids*, meaning "like a base" because bases react with acids (Section 10.6). An **alkaloid** *is a basic compound extracted from plant material.*

In Section 20.7, we encountered numerous alkaloid materials, although we did not call them alkaloids at that time. They included nicotine (tobacco plant), coniine (hemlock), piperine (black pepper), atropine (belladonna), cocaine (coca plant), and caffeine (coffee bean, tea leaf).

In this section, we will look at an important family of alkaloids—the narcotic analgesics. Narcotic analgesics are a class of drugs that are obtained or synthetically derived from alkaloids obtained from the resin of the oriental opium poppy plant. The most important drugs made from opium (the dried resin of the poppy plant) are morphine and its methyl ether derivative, codeine.

alkaloid

morphine

codeine

The nitrogen atom present in morphine and codeine is an amine nitrogen that is part of a six-membered heterocyclic ring.

Named after Morpheus, the ancient Greek god of dreams, morphine is one of the most effective painkillers known and is used extensively in modern medicine. Its analgesic properties are about 100 times greater than those of aspirin. Morphine acts by blocking the process in the brain that interprets pain signals coming from the peripheral nervous system. The major drawback of morphine is that it is addictive; hence, its classification as a narcotic.

Most often, morphine is administered to patients in the form of its hydrochloride or hydrogen sulfate salts because these salts are more soluble in water than morphine itself (Section 20.6). (Many medications are administered in the form of salts because of their enhanced water solubility.)

morphine hydrochloride

morphine hydrogen sulfate

Codeine, whose structure differs from morphine only with respect to one group bonded to the benzene ring, is a less potent painkiller than morphine. The most frequently used of all narcotic analgesics, codeine produces less sedation and depression of the central nervous system. It is also an effective cough suppressant and is used in cough syrup to decrease throat discomfort.

Heroin is a synthetic derivative of morphine; it is the diacetyl ester of morphine.

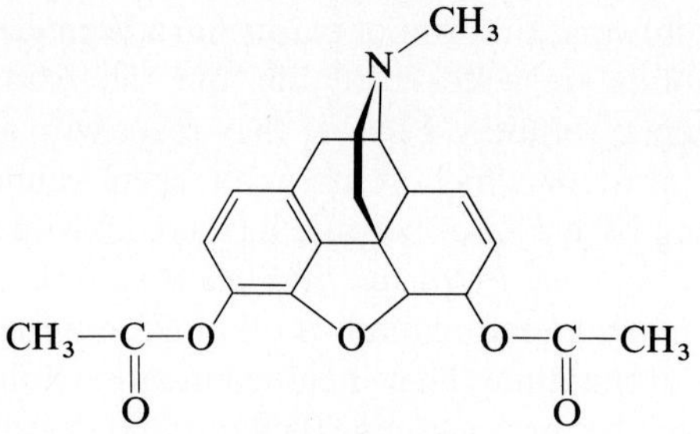

heroin

It was initially developed to help morphine addicts "kick the habit". Unfortunately, heroin is just as or even more addictive than morphine. Heroin addiction occurs more quickly and is also harder to cure. Heroin is not used for any medical reason and possession of it is illegal in the United States.

Many nonalkaloid drugs that have physiological actions similar to morphine have now been synthesized. The first totally synthetic analgesic drug is meperidine (Demerol). It is a more potent analgesic than codeine, but it is a significantly weaker analgesic than morphine.

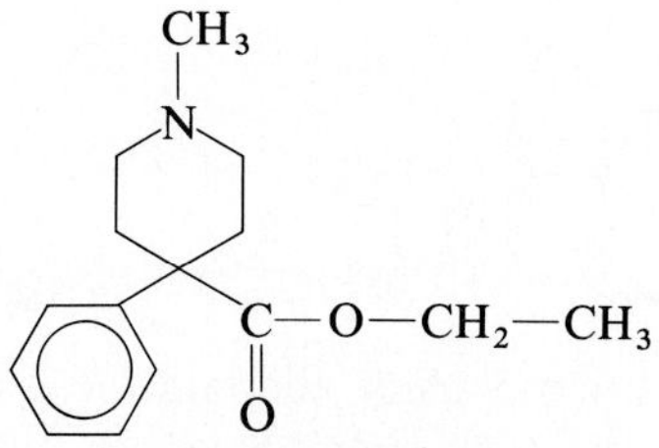

meperidine (Demerol)

20.15 POLYAMIDES AND POLYURETHANES

Amide polymers, *polyamides*, are synthesized by combining diamines and dicarboxylic acids in a condensation polymerization reaction (Section 19.12). A **polyamide** *is a condensation polymer that contains amide linkages between monomer units.*

polyamide

The most important synthetic polyamide is the substance *nylon*. Nylon is used in clothing and hosiery, as well as in carpets, tire cord, rope, and parachutes. It also has nonfabric uses; for example, it is used in paint brushes, electrical parts, valves, and fasteners. It is a tough, strong, nontoxic, nonflammable material that is resistant to chemicals. Surgical suture (Figure 20–4) is made of nylon because it is such a strong fiber.

There are actually many different types of nylons; however, all of them are polyamides. The most important nylon is Nylon 66, which is made using 1,6-diaminohexane and adipic acid as monomers.

1,6-diaminohexane

FIGURE 20-4 Nylon fibers are used to sew up surgical incisions.

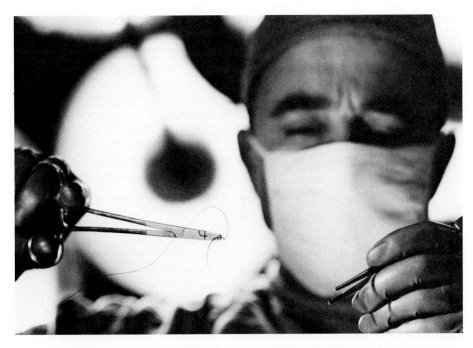

$$\underset{\text{HO}}{\overset{O}{\overset{\|}{C}}}-CH_2-CH_2-CH_2-CH_2-\overset{O}{\overset{\|}{C}}\underset{\text{OH}}{}$$

adipic acid

(The name Nylon 66 comes from the fact that each of the monomers has 6 carbon atoms.)

The formation of Nylon 66 (or any other polyamide) follows the same general principles that governed polyester formation (Section 19.12). The reaction of one acid group of the diacid with one amine group of the diamine initially produces an amide molecule; an acid group is left over on one end and an alcohol group is left over on the other end.

$$\underset{HO}{\overset{O}{\overset{\|}{C}}}-(CH_2)_4-\overset{O}{\overset{\|}{C}}\underset{OH}{} + \underset{H}{\overset{H}{N}}-(CH_2)_6-\underset{H}{\overset{H}{N}} \longrightarrow \underset{HO}{\overset{O}{\overset{\|}{C}}}-(CH_2)_4-\overset{O}{\overset{\|}{\underset{\text{leftover acid group which can react further}}{C}}}-\overset{H}{\underset{\text{amide linkage}}{N}}-(CH_2)_6-\underset{\underset{\text{leftover amine group which can react further}}{H}}{\overset{H}{N}} + H_2O$$

This species then reacts further, and the process continues until a long polymeric molecule, nylon, has been produced.

$$\cdots \overset{O}{\overset{\|}{C}}-(CH_2)_4-\overset{O}{\overset{\|}{C}}-\overset{H}{\overset{|}{N}}-(CH_2)_6-\overset{H}{\overset{|}{N}}-\overset{O}{\overset{\|}{C}}-(CH_2)_4-\overset{O}{\overset{\|}{C}}-\overset{H}{\overset{|}{N}}-(CH_2)_6-\overset{H}{\overset{|}{N}}-\overset{O}{\overset{\|}{C}}-(CH_2)_4-\overset{O}{\overset{\|}{C}}\cdots$$

a portion of the polyamide Nylon 66

As in the formation of polyesters, a water molecule is split out every time an amide linkage is formed. Note that in a polyamide polymer, the bonds that link the monomers are amide linkages (Section 20.8); hence, the name *polyamide*.

Silk and wool are naturally occurring polyamide fibers. Both of these substances are proteins and proteins are polyamides. Chapter 24 is devoted to the subject of proteins.

Polyurethanes are polymers related to polyesters and polyamides. The backbone (continuous chain of atoms) of a polyester contains carbon and oxygen atoms. The backbone of a polyamide contains carbon and nitrogen atoms. The backbone of a polyurethane contains both oxygen and nitrogen, in addition to carbon.

The following is a portion of the structure of a typical polyurethane.

$$\cdots\overset{O}{\underset{\|}{C}}-\overset{H}{\underset{|}{N}}-(CH_2)_6-\overset{H}{\underset{|}{N}}-\overset{O}{\underset{\|}{C}}-O-(CH_2)_4-O-\overset{O}{\underset{\|}{C}}-\overset{H}{\underset{|}{N}}-(CH_2)_6-\overset{H}{\underset{|}{N}}-\overset{O}{\underset{\|}{C}}-O-(CH_2)_4-O\cdots$$

Foam rubber in furniture upholstery, packaging materials, life preservers, elastic fibers, and many other products contain polyurethane polymers. A polyurethane fabric (trademark Biomer) is used as the diaphragm in the Jarvik-7 artificial heart (see Figure 20–5a) and a polyurethane material is used as a skin substitute in treating burn victims (Figure 20–5b). Membrane skin substitutes help patients recover more rapidly because they pass only oxygen and water.

Polyurethane polymer formation involves a rearrangement reaction. The functional groups involved are the isocyanate group (—N=C=O) and a functional group that has an active hydrogen site (such as —OH or —NH$_2$). The urethane linkage, which is

FIGURE 20-5 (a) The Jarvik-7-artificial heart; the base is aluminum and the valves are pyrolytic graphite and titanium. The base, valves, and valve-holding rings are coated with the polyurethane Biomer. Layers of this polymer serve as the diaphragm in the device. (b) Membranes made of polyurethane are being used as skin substitutes in treating severe burn victims. These membranes help patients recover more rapidly because they pass only oxygen and water.

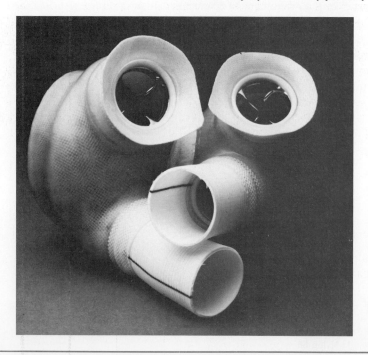
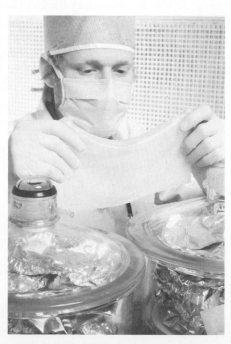

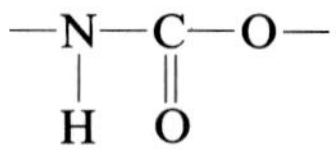

$$\begin{array}{c}-N-C-O-\\ |\|\\ HO\end{array}$$

is produced by a shift (rearrangement) of a hydrogen atom from the —OH or —NH$_2$ group to the nitrogen atom on the isocyanate group. For the dialcohol ethylene glycol and the diisocyanate *p*-phenylene diisocyanate, the rearrangement reaction is

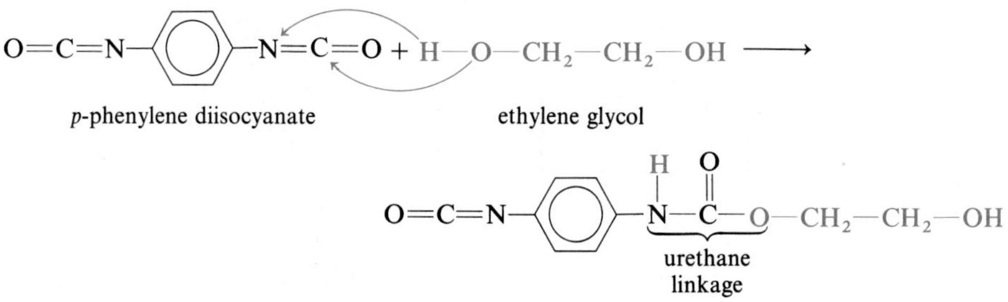

O=C=N—⟨⟩—N=C=O + H—O—CH$_2$—CH$_2$—OH ⟶

p-phenylene diisocyanate   ethylene glycol

O=C=N—⟨⟩—N—C—O—CH$_2$—CH$_2$—OH
              | ‖
              H O
         _____/
         urethane
          linkage

EXERCISES AND PROBLEMS

Structure and Classification of Amines

20.1 Indicate which of the following compounds contain an amine functional group.

a. CH$_3$—NH—CH$_2$—CH$_3$

b. 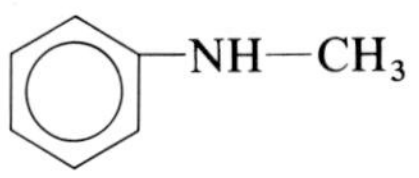 ⟨⟩—NH—CH$_3$

c. 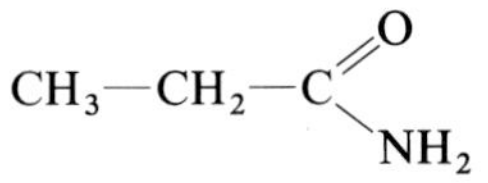 CH$_3$—CH$_2$—C(=O)NH$_2$

d. 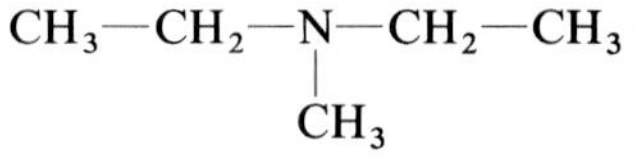 CH$_3$—CH$_2$—N(CH$_3$)—CH$_2$—CH$_3$

e. CH$_3$—CH$_2$—NH$_2$

f. 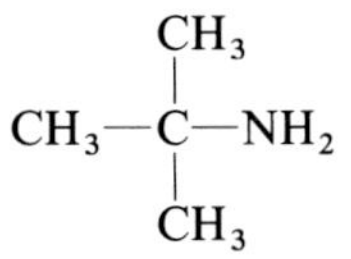 CH$_3$—C(CH$_3$)(CH$_3$)—NH$_2$

g. 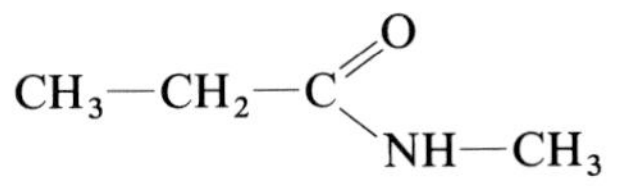 CH$_3$—CH$_2$—C(=O)NH—CH$_3$

h. 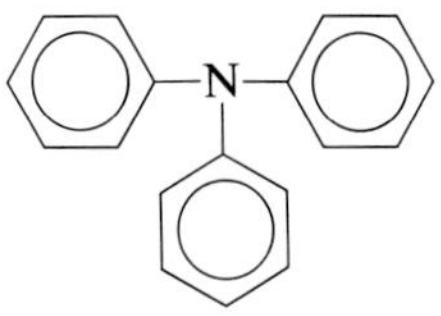 (C$_6$H$_5$)$_2$N—C$_6$H$_5$ (triphenylamine)

20.2 Explain the classification system of amines and contrast it with the classification system for alcohols.

20.3 Classify each of the amines in Problem 20.1 as a primary, secondary, or tertiary amine.

20.4 Classify each of the following amines as a primary, secondary, or tertiary amine.

a. 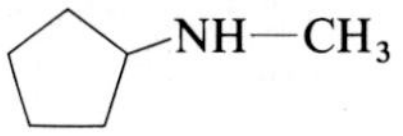 cyclopentyl—NH—CH$_3$

b. C$_6$H$_5$—NH$_2$

c. C$_6$H$_5$—N(CH$_3$)$_2$

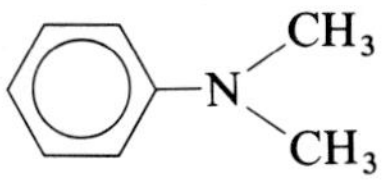

d. 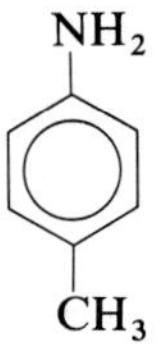 4-CH$_3$—C$_6$H$_4$—NH$_2$

e. 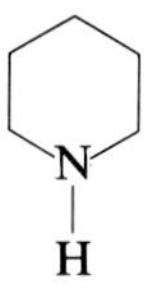 piperidine (N—H)

f. 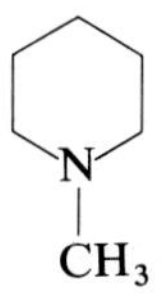 N-methylpiperidine (N—CH$_3$)

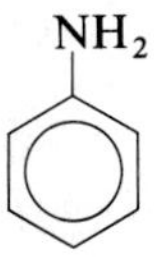

g.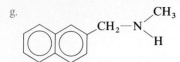

h.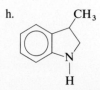

20.5 Each of the following molecules contains two or more amine sites. Classify each amine site within the molecules as a primary, secondary, or tertiary site.

a. nicotine, a toxin in tobacco

b. $H_2N-CH_2-CH_2-CH_2-CH_2-CH_2-CH_2-NH_2$
a monomer used in producing the polymer nylon

c. novocain, a local anesthetic

d. tryptophan, an amino acid

20.6 How many amine isomers are possible with the molecular formula

a. C_2H_7N b. C_3H_9N c. $C_4H_{11}N$

Nomenclature of Amines

20.7 Name each of the following amines by adding the name *amine* to alkyl or aromatic group names.

a. $CH_3-NH-CH_2-CH_3$

b. $CH_3-CH_2-CH_2-NH_2$

c. $CH_3-CH-CH_3$
        $|$
        NH_2

d. phenyl—NH—CH_3

e. diphenyl—N—CH_3

f. $CH_3-(CH_2)_4-CH_2-NH_2$

g. $CH_3-N-CH_2-CH_2-CH_3$
        $|$
        CH_3

h. phenyl—NH_2

20.8 Name each of the following amines by treating the amino group as a side-chain substituent.

a. $CH_3-CH-CH-CH-CH_3$ with NH_2 on middle C and CH_3 on two outer

b. $CH_3-CH-CH_2-CH_3$
        $|$
        NH_2

c. $CH_3-CH-CH-NH_2$
        $|$ $|$
        CH_3-CH_2 CH_2-CH_3

d. $CH_3-CH-CH=CH-CH_3$
        $|$
        NH_2

e. $CH_3-CH-CH_2-CH_2-NH_2$
        $|$
        NH_2

f. cyclobutane with CH_3, CH_2-CH_3, and NH_2

g. cyclopentane with two NH_2 groups

h. $CH_3-CH-CH-CH-CH_3$
        $|$ $|$ $|$
        Cl NH_2 OH

20.9 Name each of the following aromatic amines as derivatives of aniline.

a. Br—phenyl—NH_2 (para)

b. phenyl with NH_2 and $CH_2-CH_2-CH_3$

c. phenyl with CH_3 and NH_2

d. $CH_3-N-CH_2-CH_3$
        $|$
        phenyl

e. phenyl—NH—phenyl

f.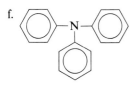

g.

![structure: p-chloroaniline with NH2 top and Cl bottom]

h.

![structure: 2,4-dinitroaniline with NO2 groups and NH2]

20.10 Draw structural formulas for each of the following amines.

a. ethylamine
b. triisopropylamine
c. o-methylaniline
d. dimethylamine
e. methylbutylamine
f. N,N-dimethylaniline
g. p-aminoaniline
h. α-aminopropanoic acid

20.11 Draw structural formulas for each of the following amines.

a. 1,3-diaminopropane
b. 3-amino-2-methyl-2-butanol
c. 2-ethyl-4-nitroaniline
d. 4-amino-2-chloropentanoic acid
e. 1-(N,N-dimethylamino)-butane
f. 3-(N-ethylamino)-1-butanol
g. o-chloro-N-methylaniline
h. toluidine

Properties and Uses of Amines

20.12 Indicate whether each of the following amines is a liquid or gas at room temperature.

a. n-butylamine
b. diethylamine
c. methylamine
d. n-heptylamine

20.13 Using structural formulas, show hydrogen bonding that is possible between a secondary amine and water.

20.14 Trimethylamine boils at 4°C and its isomer n-propylamine boils at 49°C. Explain this difference in boiling points.

20.15 List some uses for each of the following simple amines.

a. methylamine
b. trimethylamine
c. diethylamine
d. triethylamine
e. aniline

20.16 Which of the following compounds are beta-phenylethylamines?

a. epinephrine
b. norepinephrine
c. phenylephrine
d. beta-naphthylamine
e. methoxyamphetamine
f. benzedrine
g. isoproterenol

20.17 Identify a use or function for each of the compounds listed in Problem 20.16.

Preparation of Amines

20.18 Amines are often called derivatives of ammonia. This is not only true structurally but also chemically because we can actually transform ammonia into an amine. Write a chemical equation that would effect such a transformation.

20.19 Draw structural formulas for the products of each of the following reactions.

a. $CH_3-CH(CH_3)-CH_2-Cl + NH_3 \xrightarrow{NaOH}$

b. $CH_3-CH(CH_3)-NH_2 + CH_3-Br \xrightarrow{NaOH}$

c. $CH_3-CH(CH_3)-C\equiv N \xrightarrow{2 H_2}{Ni}$

d. p-chloronitrobenzene $\xrightarrow{Sn}{HCl}$

e. $(CH_3)_2C=O + NH_3 \xrightarrow{H_2}{Ni}$

f. o-nitrotoluene $\xrightarrow{H_2}{Ni}$

g. $CH_3-NH-CH_3 + CH_3-CH_2-CH_2-CH_2-Br \xrightarrow{NaOH}$

h. $CH_3-C\equiv N \xrightarrow{2 H_2}{Ni}$

Basicity of Amines

20.20 Explain why amines are considered to be bases in aqueous solution.

20.21 When diethylamine is dissolved in water, the pH increases. Explain this observation.

20.22 Show the structures of the missing compounds in the following acid–base equilibria.

a. $CH_3-CH_2-NH_2 + HCl \rightleftharpoons$?

b. $C_6H_5-NH_2 + HBr \rightleftharpoons$?

501

c.

? + HBr ⇌ CH$_3$—C(CH$_3$)(CH$_3$)—NH$_3^+$ Br$^-$

d.

? + HCl ⇌ CH$_3$—C(CH$_3$)(CH$_3$)—NH$_2^+$Cl$^-$ (with CH$_3$ groups)

e. norepinephrine + HCl ⇌ ?

20.23 Why are drugs that contain the amine functional group most often administered to patients in the form of amine chloride or sulfate salts?

20.24 Suppose you need to dissolve some *p*-toluamide in aqueous solution for a laboratory experiment, but it just won't dissolve, even after stirring and warming the solution. Suggest what you might do to rapidly solubilize the amine and explain your reasons for doing so.

20.25 List two reasons for putting lemon juice, which contains citric acid, on fish dishes.

Chemical Reactions Involving Amines

20.26 Draw the structure of the amide produced when each of the following carboxylic acids and amines react.
a. acetic acid and diethylamine
b. butanoic acid and ethylmethylamine
c. formic acid and methylamine
d. 2-methylbutanoic acid and dimethylamine

20.27 Draw the structure of the quaternary ammonium salt produced from each of the following sets of reactants.
a. trimethylamine and ethyl bromide
b. diisopropylmethylamine and methyl bromide
c. ethylmethylpropylamine and methylchloride
d. triethylamine and ethyl bromide

20.28 Name each of the quaternary ammonium salts produced in Problem 20.27.

20.29 Draw structures of all intermediates and final products as indicated for the following multi-step syntheses. (Some review of reactions in previous chapters may be necessary.)

a. CH$_3$—⟨ ⟩ $\xrightarrow{HONO_2}$ I $\xrightarrow{Sn}$ II
  (H$_2$SO$_4$)   HCl

b. CH$_3$—CH=CH$_2$ $\xrightarrow{HCl}$ III $\xrightarrow{NH_3}$ IV
  (NaOH)

c. CH$_3$—CH$_2$—CH(Br)—CH$_3$ $\xrightarrow{NaCN}$ V $\xrightarrow{2H_2}{Ni}$ VI

d. CH$_3$—CH=CH$_2$ $\xrightarrow{H_2O}{H^+}$ VII $\xrightarrow{K_2Cr_2O_7}$

  VIII $\xrightarrow{NH_3}{H_2}$ IX

e. CH$_3$—CH$_2$—OH $\xrightarrow{PCl_3}$ X $\xrightarrow{CH_3NH_2}{(NaOH)}$ XI

f. CH$_3$—CH$_2$—OH $\xrightarrow{PCl_3}$ $\xrightarrow{NaCN}$ $\xrightarrow{2H_2}{Ni}$ XII

Heterocyclic Amines

20.30 Indicate the size of the heterocyclic ring, the number of nitrogen atoms present, and the number of double bonds present in each of the following heterocyclic amines.
a. pyridine   e. piperidine
b. pyrimidine   f. purine
c. pyrrole   g. indole
d. pyrrolidine   h. imidazole

20.31 Give the name of the heterocyclic amine ring structure or structures present in each of the following compounds.
a. nicotine   e. niacin
b. histidine   f. pyrrolidine
c. cocaine   g. benadryl
d. caffeine   h. coniine

20.32 List three biologically important compounds that contain a porphyrin ring. Identify the heterocyclic amine that is repeated four times in a porphyrin ring.

20.33 Write the names and draw the structures of two amino acids that contain heterocyclic amines as part of their structure.

20.34 Write the names and draw the structures of three vitamins that have heterocyclic amines as part of their structure. Identify the heterocyclic amine present in each case.

20.35 Write the names and draw the structures of the five heterocyclic-amine-containing compounds that are the basis for the genetic code.

20.36 In each of the following pairs of caffeine sources, indicate which source would probably contain the most caffeine.
a. 1 cup of coffee or 1 cup of tea
b. 1 cup of coffee or 1 can of cola drink
c. 1 oz of milk chocolate or 1 oz of baking chocolate
d. 1 Anacin tablet or 1 can of cola drink

Classification and Structure of Amides

20.37 Which of the following compounds contain an amide functional group?

a. CH$_3$—CH$_2$—C(=O)—NH$_2$

b. ⟨ ⟩—C(=O)—N(CH$_3$)—CH$_2$—CH$_3$

c. 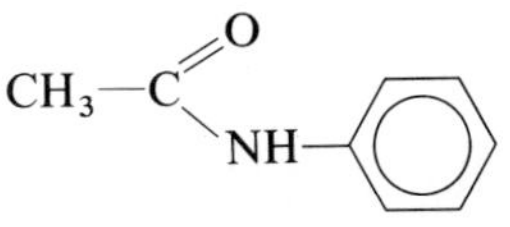
$$CH_3-C(=O)-NH-C_6H_5$$

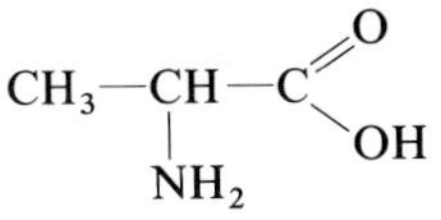

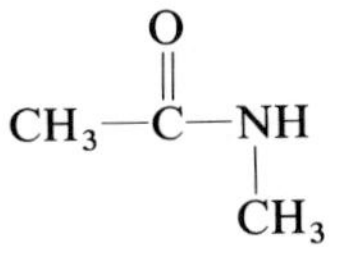

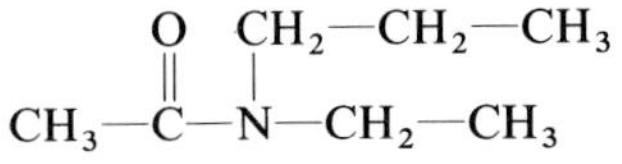

d. 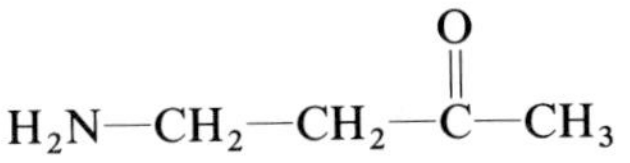
$$H_2N-CH_2-CH_2-C(=O)-CH_3$$

e.
$$CH_3-CH(NH_2)-C(=O)-OH$$

f.
$$CH_3-C(=O)-NH-CH_3$$

g.
$$CH_3-C(=O)-N(CH_2-CH_2-CH_3)(CH_2-CH_3)$$

h.
$$CH_3-CH(NH_2)-CH_2-C(=O)-NH_2$$

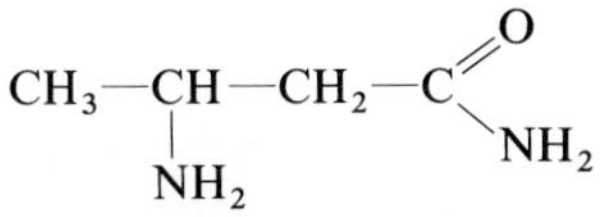

20.38 Classify each of the amides in Problem 20.37 as a primary, secondary, or tertiary amide.

20.39 Indicate whether each of the following compounds is an amine, an amide, both, or neither.

a. 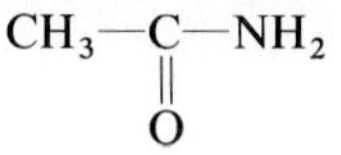 $CH_3-C(=O)-NH_2$

b. 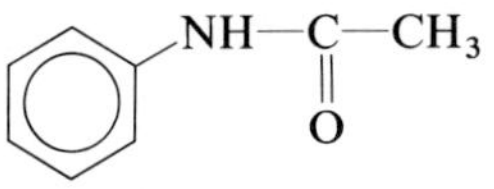 $C_6H_5-NH-C(=O)-CH_3$

c. $H_2N-CH_2-CH_2-NH_2$

d. 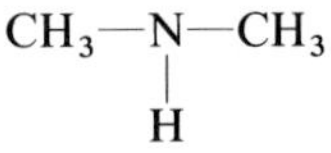 $CH_3-NH-CH_3$

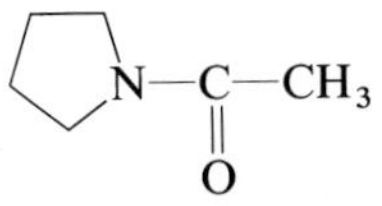

e. 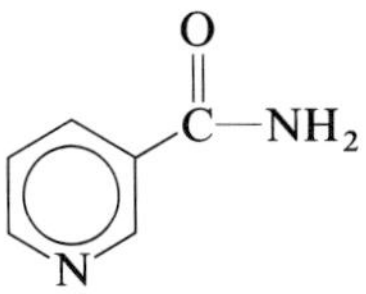 pyridine-3-C(=O)-NH_2

f. 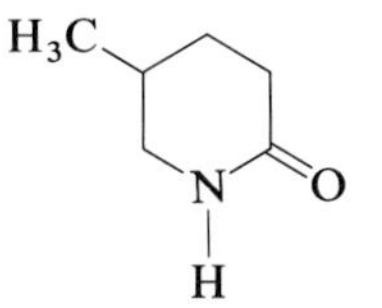 4-methylpiperidin-2-one (H_3C on piperidine ring with N-H and C=O)

g. pyrrolidine-N-C(=O)-CH_3

h. 2-(pyrrolidinyl)-C(=O)-CH_3

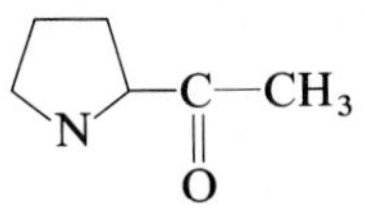

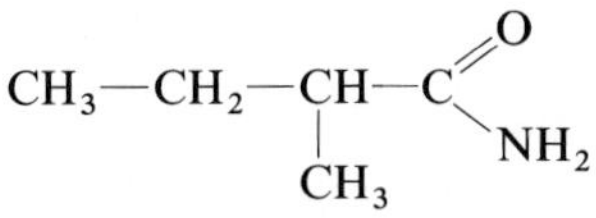

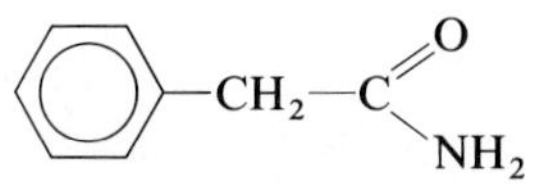

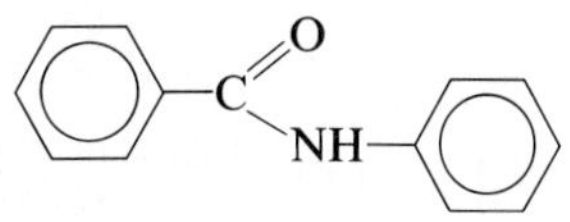

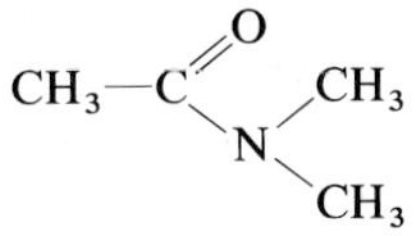

Nomenclature of Amides

20.40 Assign names to each of the following amides.

a. 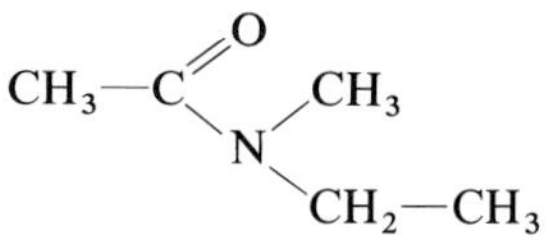
$$CH_3-C(=O)-N(CH_3)(CH_2-CH_3)$$

b.
$$CH_3-CH_2-CH(CH_3)-C(=O)-NH_2$$

c.
2-methylbenzamide (ortho-CH_3 on benzene with C(=O)-NH_2)

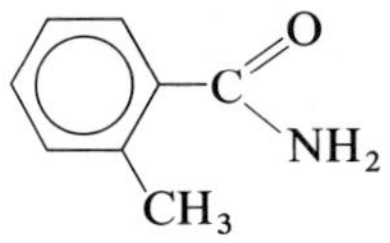

d. 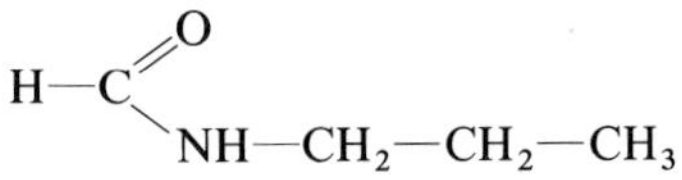
$$H-C(=O)-NH-CH_2-CH_2-CH_3$$

e.
$$C_6H_5-CH_2-C(=O)-NH_2$$

f.
$$C_6H_5-C(=O)-NH-C_6H_5$$

g.
$$CH_3-C(=O)-N(CH_3)(CH_3)$$

h.
$$CH_3-CH_2-CH_2-C(=O)-NH_2$$

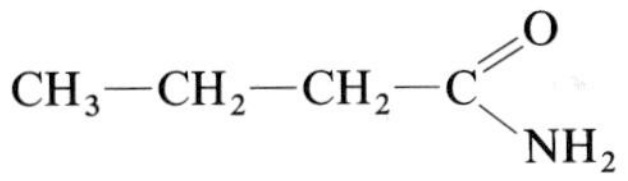

20.41 Write structural formulas for each of the following amides.
a. N,N-dimethylbutyramide
b. 2-methylbutyramide
c. N-methylbenzamide
d. N-methyl-3-methylbutanamide
e. N-phenylbenzamide
f. methanamide
g. propionamide
h. formamide

Properties of Amides

20.42 Although amides contain a nitrogen atom, they are not bases like amines. Explain why.

20.43 Explain why more sites for hydrogen bonding to water exist for toluamide than for N,N-dimethyltoluamide.

503

Preparation of Amides

20.44 Draw the structures of the products in each of the following reactions.

a. 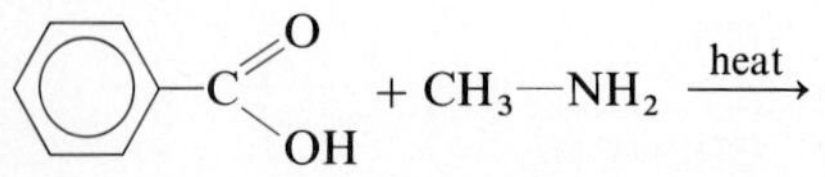 C₆H₅—C(=O)—OH + CH₃—NH₂ $\xrightarrow{\text{heat}}$

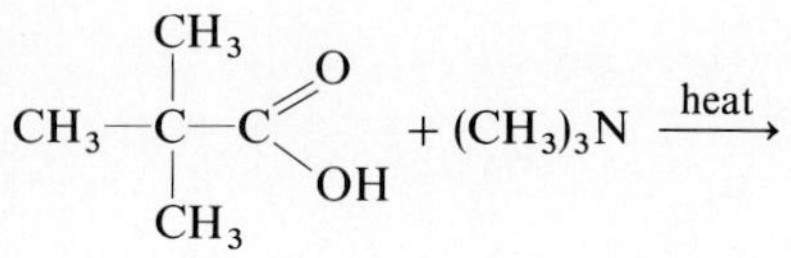

b. (CH₃)₃C—C(=O)—OH + (CH₃)₃N $\xrightarrow{\text{heat}}$

c. CH₃—CH₂—C(=O)—OH + H—N(piperidine) $\xrightarrow{\text{heat}}$

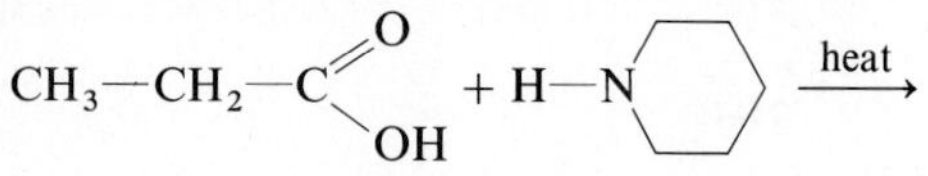

d. 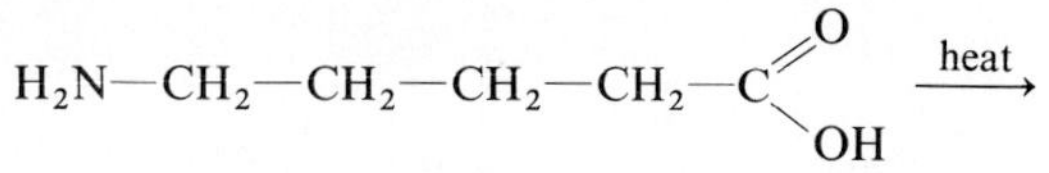 H₂N—CH₂—CH₂—CH₂—CH₂—C(=O)—OH $\xrightarrow{\text{heat}}$

Reactions of Amides

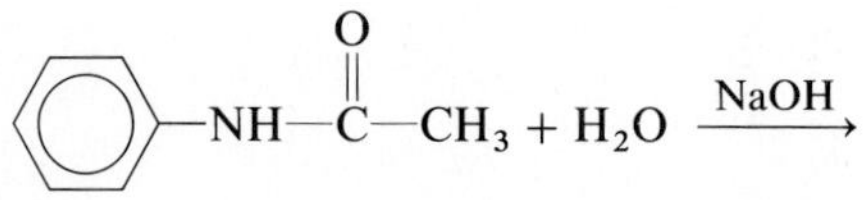

20.45 Draw the structures of the products in each of the following reactions.

a. 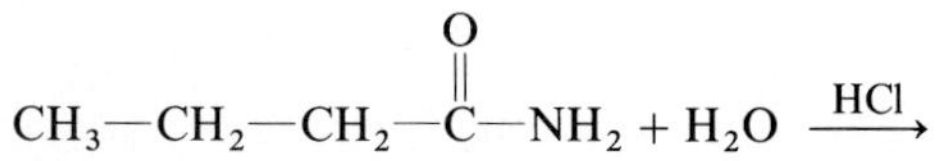 CH₃—CH₂—CH₂—C(=O)—NH₂ + H₂O $\xrightarrow{\text{HCl}}$

b. C₆H₅—NH—C(=O)—CH₃ + H₂O $\xrightarrow{\text{NaOH}}$

c. CH₃—CH(CH₃)—C(=O)—N(piperidine) + H₂O $\xrightarrow{\text{NaOH}}$

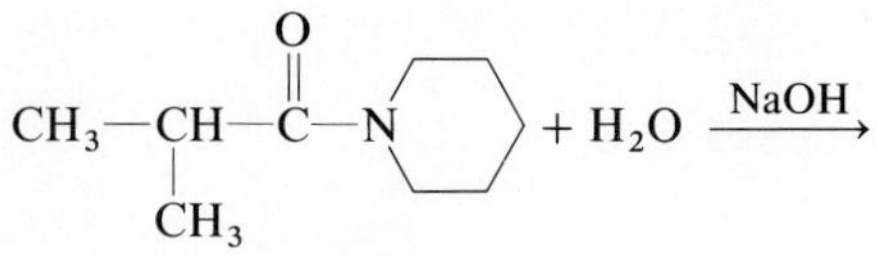

d. 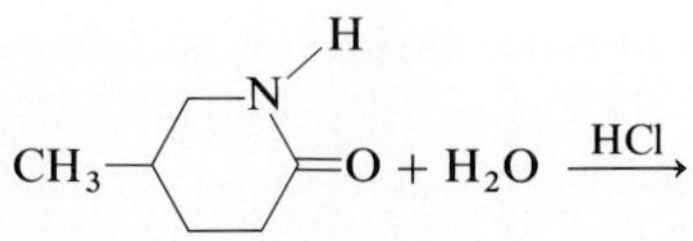 CH₃-(N-methylpiperidinone) + H₂O $\xrightarrow{\text{HCl}}$

Properties and Uses of Amides

20.46 Contrast the structures of dimethyl ketone, carbonic acid, and urea.

20.47 Write an equation for the industrial preparation of urea.

20.48 Identify the common features present in the structures of acetanilide, phenacetin, and acetaminophen.

20.49 What is the advantage and disadvantage of the use of acetaminophen over aspirin as a pain reliever?

20.50 Barbiturates are derivatives of urea. Identify the portion of the structure in barbital and secobarbital that contains the urea.

20.51 Identify the common features present in the structures of all barbiturates.

20.52 Indicate how the members of each of the following pairs of barbiturates differ structurally from each other.
a. barbital and amobarbital
b. pentobarbital and amobarbital
c. secobarbital and aprobarbital
d. pentobarbital and secobarbital

20.53 What is the difference between the physiological effects of barbiturates and amphetamines?

20.54 What is the structural feature common to the benzodiazepine tranquilizer drugs?

Alkaloids

20.55 From what plant are the narcotic analgesics obtained?

20.56 Compare the uses and biological effects of morphine and codeine.

20.57 What structural feature is common to morphine, codeine, and heroin? How do the three structures differ from each other?

20.58 Identify all of the functional groups present in the structures of the following substances.
a. morphine   c. heroin
b. codeine   d. demerol

20.59 Give an example of an alkaloid whose structure is based on each of the following types of ring systems.
a. piperidine   c. pyrrolidine
b. purine

Polyamides and Polyurethanes

20.60 List the general characteristics of the monomers needed to produce a
a. polyamide   b. polyurethane

20.61 How does an amide linkage differ from a urethane linkage?

20.62 Nylon is resistant to dilute acids or bases, but polyesters are damaged by acids or bases. Explain this difference.

20.63 Draw a structural representation of the polymer formed by each of the following pairs of monomers.
a. succinic acid and 1,4-diaminobutane
b. adipic acid and 1,2-diaminoethane
c. *p*-phenylene diisocyanate and propylene glycol
d. succinic acid and ethylene glycol

20.64 Characterize each of the polymers formed in Problem 20.63 as a polyamide, a polyurethane, or a polyester.

21 Stereoisomerism

Objectives

After completing Chapter 21, you will be able to:

21.1 Distinguish among the following types of isomerism: structural isomerism, stereoisomerism, geometric isomerism, and optical isomerism.

21.2 Understand the relationships that exist between enantiomers and draw the mirror image of a given structure.

21.3 Identify chiral carbon atoms in organic molecules, given their structural formulas.

21.4 Draw two-dimensional projection formulas when given three-dimensional perspective structures and know the conventions used in interpreting projection formulas.

21.5 Understand the meaning of D- and L- when they are used in naming enantiomers.

21.6 Describe the concept of optical activity, including the phenomenon of plane-polarized light, and understand the meaning of (+)- and (−)- relative to the optical activity of a compound.

21.7 Understand the difference between observed rotation obtained using a polarimeter and specific rotation, and calculate specific rotation from observed rotation.

21.8 Understand what is meant by the items *racemic mixture*, *racemization*, and *resolution of a racemic mixture*.

21.9 Draw projection formulas for all possible stereoisomers of a given compound and then label enantiomers, diastereomers, and meso structures. You will also be able to explain why meso compounds are optically inactive.

INTRODUCTION

The introduction to Chapter 13 noted that it could be viewed as a "bridge" chapter that provided the transition from inorganic chemistry to organic chemistry. This chapter can also be looked upon as a "bridge" chapter. Here, we make a transition from organic chemistry to biochemistry.

Additional information about isomerism will lead us to the subject of biochemistry. Isomerism is a topic that has occupied our attention numerous times in previous chapters. However, there is still much more to be learned about this subject.

All previous discussions of isomerism dealt with either structural isomers (Section 14.5) or geometrical isomers (Section 15.5). An additional type of isomerism, optical isomerism, will receive our attention in this chapter. Optical isomerism is of utmost importance when discussing how biological systems function. The metabolism of food (carbohydrates, lipids, and proteins) and the biosynthesis of molecules such as hormones, proteins, and DNA cannot be explained without taking into account the phenomenon of optical isomerism.

21.1 TYPES OF ISOMERISM

There are two general types of isomerism: structural isomerism and stereoisomerism. Section 14.5 defined structural isomers as compounds that have the same molecular formula, but different structural formulas; that is, they have different arrangements of the atoms within the molecule. Thus, structural isomerism is concerned with the order in which atoms are bonded to each other. The compounds n-butane and isobutane (2-methylpropane) are structural isomers.

$$CH_3—CH_2—CH_2—CH_3 \qquad CH_3—\underset{\underset{CH_3}{|}}{CH}—CH_3$$
$$\textit{n-butane} \qquad\qquad\qquad \textit{isobutane}$$

Stereoisomerism deals not only with the order in which atoms are bonded to each other, but also with the arrangement of these atoms in space. **Stereoisomers** *are compounds that have the same structural formula, but differ in the spatial arrangement of the atoms present in space; that is, they differ in molecular configuration.*

stereoisomers

Stereoisomers can be classified into a number of subtypes. We have already discussed the subtype known as *geometric isomerism*. We recall from Section 15.5 that geometric isomerism results from *hindered rotation* within a molecule; the hindered rotation is caused by either the presence of a double bond or a small ring structure. Geometric isomerism always involves hindered rotation associated with two or more carbon atoms.

A second subtype of stereoisomerism that is of extreme importance in biochemistry is *optical isomerism*. The majority of the remainder of this chapter deals with this topic.

Optical isomers *are isomers whose structures are the same, except that they are nonsuperimposable mirror images of each other.* Optical isomers differ from each other in the same way that your left and right hands differ from each other. Both

optical isomers

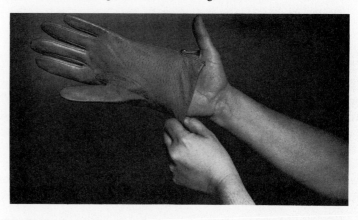

FIGURE 21-1 A left-handed glove will not fit the right hand.

of your hands are similar in that they have four fingers and a thumb. They are also very different, as is evidenced by what happens when you try to put your right hand into a left-handed glove (see Figure 21-1). The key to understanding optical isomerism lies in comprehending the concepts of *mirror images* and *nonsuperimposability*.

21.2 OPTICAL ISOMERISM

Let us begin our discussion of optical isomerism by exploring the relationships that exist between a person's left and right hands. The concepts developed using this familiar example will then be applied to molecules.

The fact that your two hands are mirror images of each other—they are structurally related to each other as are an object and its image in a mirror—is easily shown. Place your left hand in front of a mirror and compare what you see in the mirror to your right hand. What you see in the mirror has characteristics that are identical to those of your right hand, as is illustrated in Figure 21-2.

In addition to being mirror images of each other, your left hand and right hand also possess the property of nonsuperimposability. The act of trying to put your

FIGURE 21-2 The mirror image of the left hand looks like the right hand.

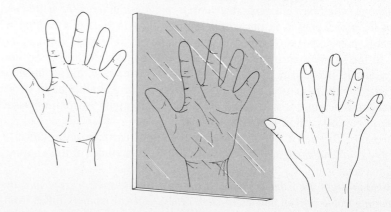

right hand into a left-handed glove (Figure 21-1) illustrates nonsuperimposability. In order for two objects to be superimposable, all parts of the objects must perfectly match when they are "imposed" on each other. Figure 21-3 shows what happens when one attempts to superimpose a pair of human hands on each other. Note that the thumbs point in opposite directions. Even the fingers do not impose correctly; the index fingers are on opposite sides.

Molecules that are optical isomers have the same relationship to each other that your right and left hand have. Both isomers have the same shape and the same order of attachment of atoms or groups of atoms. They differ in that they are nonsuperimposable mirror images of each other. Figure 21-4 illustrates this nonsuperimposable mirror image relationship for the simple organic molecule *bromochloroiodomethane* and the simple biochemical molecule *alanine* (an amino acid).

The mirror images shown in Figure 21-4 are not superimposable on each other. Figure 21-5 (p. 510) shows an attempt to superimpose the bromochloroiodomethane mirror images on each other. The hydrogen and chlorine atoms match, but the bromine and iodine atoms mismatch.

Optical isomers are also called *enantiomers*. The two terms are used interchangeably. Thus **enantiomers** *are stereoisomers whose structures are nonsuperimposable mirror images of each other.* Enantiomers resemble each other closely in their physical properties. In most physical properties, such as melting point, boiling point, and solubility, enantiomers exhibit identical characteristics. There is only one physical property in which they differ: they behave differently toward plane-polarized light. This property enables chemists to distinguish one enantiomer from the other. Section 21.6 deals with the concept of plane-polarized light in detail.

Enantiomers can and often do exhibit different chemical behavior, particularly in physiological environments. This fact was first discovered many years ago by the French chemist Louis Pasteur (1822-95). He observed that the growth of a particular species of mold was inhibited by one enantiomer of lactic acid, but not by the other.

FIGURE 21-3 Your left hand and right hand are not superimposable on each other.

enantiomers

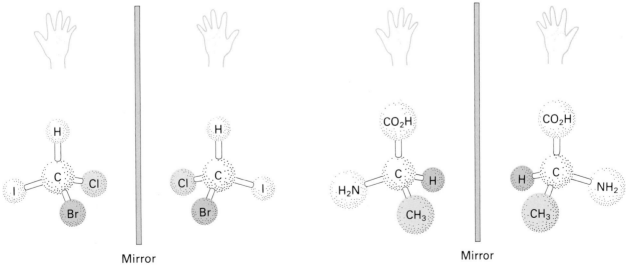

FIGURE 21-4 Mirror images of a hand, a simple organic molecule (bromochloroiodomethane), and a simple biochemical molecule (alanine).

FIGURE 21–5 (a) A molecule of bromochloroiodomethane and its mirror image. (b) Optical isomers (mirror images) of bromochloroiodomethane are not superimposable upon each other. Here there is a "mismatch" between the Cl and Br atoms.

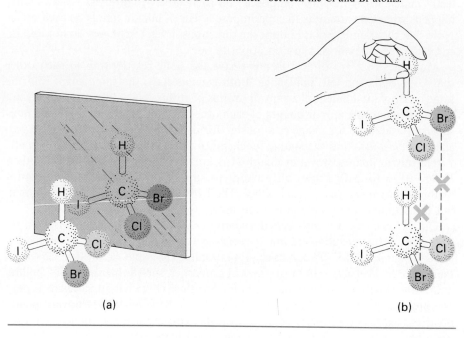

mold growth inhibited with this enantiomer (HO—C—H with COOH above and CH₃ below) | mold growth occurred with this enantiomer (H—C—OH with COOH above and CH₃ below)

A more modern example involving physiologically different behavior by enantiomers involves amphetamine (Section 20.3), which is one of the pep pills of the illegal drug culture. The two amphetamine enantiomers are

The isomer on the left is a much stronger stimulant than the other one; it is four times stronger than a mixture of the two isomers.

The structures just given for the lactic acid and amphetamine enantiomeric pairs were planar structures, rather than three-dimensional structures. More information concerning the interpretation of these structures from an optical isomerism viewpoint will be given in Section 21.3.

21.3 CHIRAL CARBON ATOMS

Not all organic compounds exist in two nonsuperimposable mirror image forms, that is, not all organic molecules exhibit optical isomerism. Which ones do and which ones do not? The key to making this determination involves the concept of chiral carbon atoms.

A **chiral carbon atom** *is a carbon atom in an organic molecule that has four different atoms or groups of atoms attached to it.* Any molecule containing a single chiral carbon atom exhibits optical isomerism. Molecules can contain more than one chiral center, in which case there can be more than two optical isomers.

Chiral carbon atoms within a molecule are often denoted by using a small asterisk. Note the chiral carbons in the following molecules.

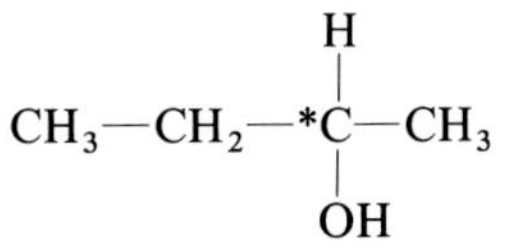

chiral carbon atom

$$CH_3-CH_2-\overset{H}{\underset{OH}{*C}}-CH_3 \qquad H-\overset{Cl}{\underset{I}{*C}}-CH_3 \qquad CH_3-CH_2-CH_2-\overset{CH_3}{\underset{H}{*C}}-CH_2-CH_3$$

2-butanol 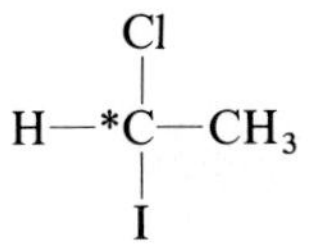   1-chloro-1-iodoethane 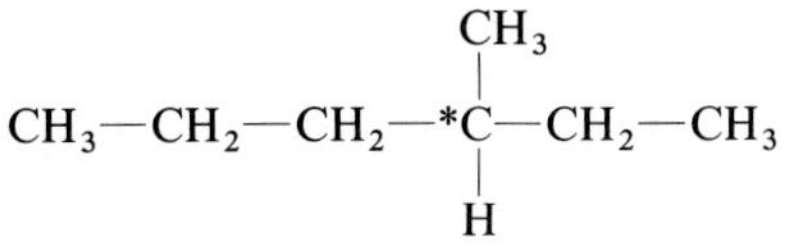   3-methyl hexane

A compound that contains one or more chiral carbon atoms is called a *chiral compound*. Compounds without chiral carbons are said to be *achiral*.

Identify with an asterisk the chiral carbon atoms in each of the following molecules. Draw a circle around each of the four different groups attached to the chiral carbon atoms. If a chiral carbon atom is not present, indicate this by writing *achiral*.

Example 21.1

a. 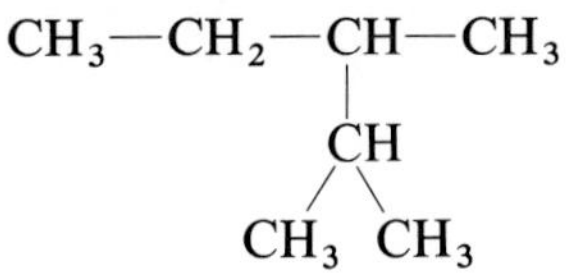 $CH_3-CH_2-\underset{\underset{CH_3 \quad CH_3}{CH}}{CH}-CH_3$

b. 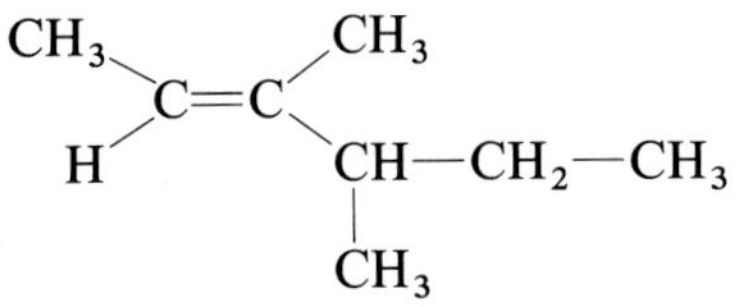 $\underset{H}{\overset{CH_3}{}}C=C\underset{\underset{CH_3}{CH-CH_2-CH_3}}{\overset{CH_3}{}}$

c. ⬡—$\underset{Br}{\overset{H}{\underset{|}{\overset{|}{C}}}}$—$\overset{O}{\overset{\|}{C}}$—OH

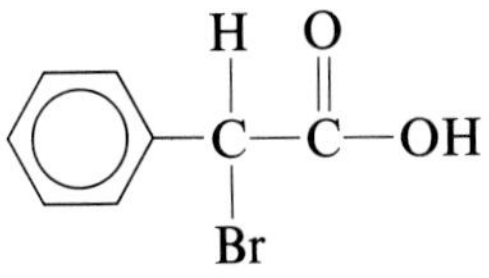

d. 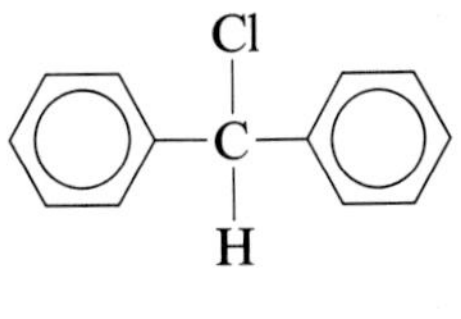 ⬡—$\underset{H}{\overset{Cl}{\underset{|}{\overset{|}{C}}}}$—⬡

e. 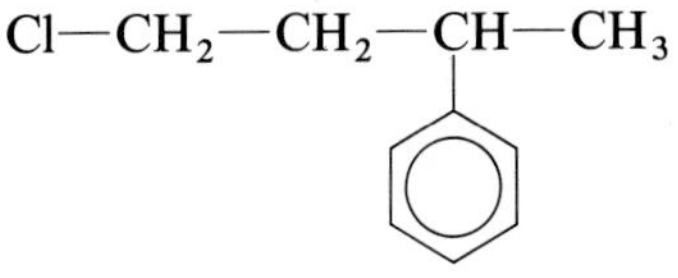 $Cl-CH_2-CH_2-\underset{\underset{\text{⬡}}{|}}{CH}-CH_3$

f. 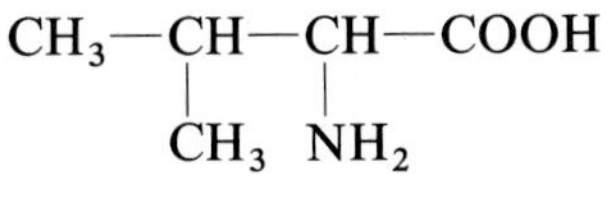 $CH_3-\underset{CH_3}{\underset{|}{CH}}-\underset{NH_2}{\underset{|}{CH}}-COOH$

Solution

Five of the six molecules contain a chiral carbon atom, a carbon atom to which four different groups are attached.

a. CH$_3$—CH$_2$—*C(H)—CH$_3$ with CH(CH$_3$)(CH$_3$) below

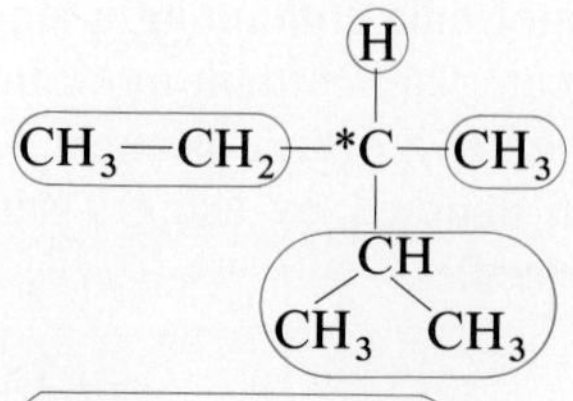

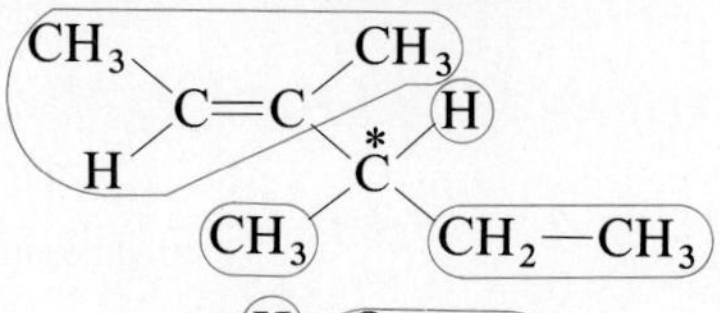

b. (CH$_3$)(H)C=C(CH$_3$)—*C(H)(CH$_2$—CH$_3$)

c. (phenyl)—*C(H)(Br)—C(=O)—OH

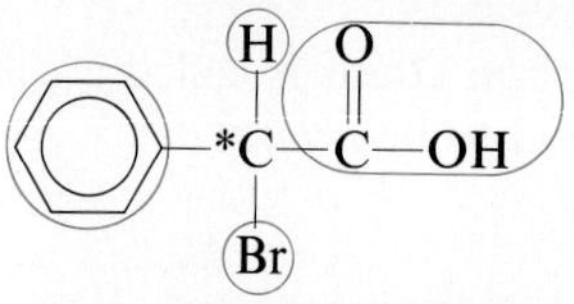

d. This molecule is achiral—two of the groups attached to the carbon atom are the same (two phenyl groups).

e. Cl—CH$_2$—CH$_2$—*C(H)(CH$_3$)(phenyl)

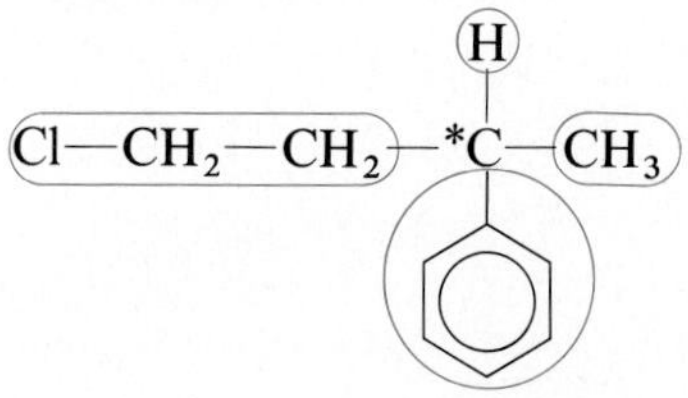

f. CH$_3$—CH(CH$_3$)—*C(H)(NH$_2$)—COOH

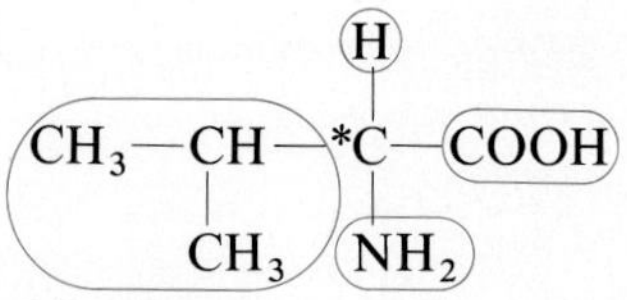

It is possible to have chiral carbon atoms in the rings of cyclic compounds. Two examples are 2-methylpiperidine and 2-methylcyclopentanone.

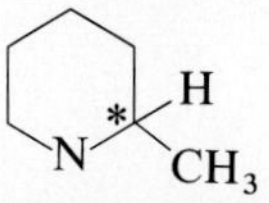

2-methylpiperidine 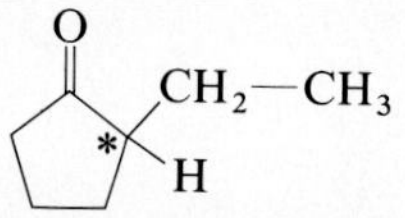   2-methylcyclopentanone

Many compounds have more than one chiral carbon, especially those found in biological systems. In order to identify each chiral carbon, you must treat each one separately. Consider threonine, an amino acid that contains two chiral carbons.

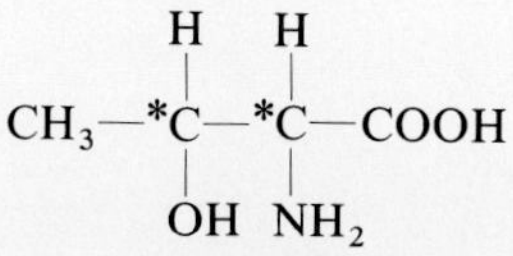

CH$_3$—*C(H)(OH)—*C(H)(NH$_2$)—COOH

threonine

The four different groups attached to each chiral carbon are as follows.

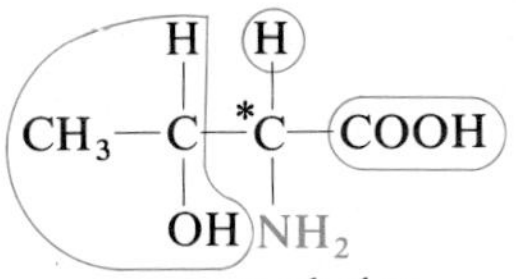

$$\text{CH}_3\text{—C}\overset{*}{\text{—}}\text{C—COOH}$$
with H, H on top and OH, NH$_2$ on bottom
groups attached to α-carbon

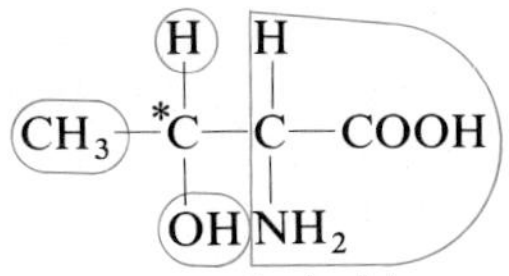

$$\text{CH}_3\overset{*}{\text{—}}\text{C—C—COOH}$$
with H, H on top and OH, NH$_2$ on bottom
groups attached to β-carbon

21.4 PROJECTION FORMULAS FOR ENANTIOMERS

A number of methods exist for denoting the structures of enantiomeric pairs of molecules. We will consider two methods in this section: three-dimensional perspective drawings, and two-dimensional projection formulas.

A three-dimensional perspective structure unambiguously gives the spatial relationships among the four different atoms or groups of atoms attached to a chiral carbon because it shows the tetrahedral arrangement of groups around the chiral atom. When only one chiral carbon atom is present, these structures are relatively easy to draw. Drawing difficulties increase in molecules containing more than one chiral atom.

Figure 21–6a shows perspective structures for the enantiomers of the one-chiral-carbon compound *glyceraldehyde*. The wedge and dashed line notation used in these drawings is not new; it was previously used in Section 15.5 during our discussion of geometric isomers. A wedge indicates a bond coming out of the printed page and a dashed line indicates a bond going into the printed page. The sphere in the center of these drawings represents the chiral carbon atom.

Parts (b) and (c) of Figure 21–6 show projection formulas for the glyceraldehyde enantiomers. These formulas are much easier and faster to create than perspective drawings. Because no attempt is made to indicate three-dimensional relationships in projection formulas, their use can present interpretation problems unless the following conventions of interpretation are understood.

FIGURE 21–6 Different ways of drawing the structures for the two enantiomeric glyceraldehyde molecules. Structures in part (a) are three-dimensional perspective drawings and those in parts (b) and (c) are two-dimensional projection formulas.

a.
CHO
H—●—OH   HO—●—H
CH$_2$OH   mirror   CH$_2$OH

b.
CHO   CHO
H—C—OH   HO—C—H
CH$_2$OH   CH$_2$OH

c.
CHO   CHO
H—|—OH   HO—|—H
CH$_2$OH   CH$_2$OH

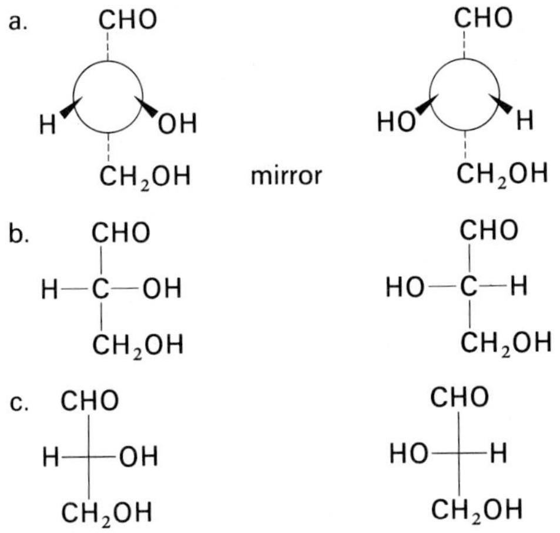

1. Groups attached to the vertical bonds (the CHO and CH₂OH groups in our case) extend away from the viewer into the paper.
2. Groups attached to the horizontal bonds (the —H and —OH groups in our case) project toward the viewer out of the paper.
3. If the chiral carbon atom is not explicitly shown, as in part (c), it is understood to be located where the bond lines intersect.

Example 21.2

Draw the three-dimensional perspective structure represented by the following projection formula.

$$\begin{array}{c} Cl \\ H{-}\!\!\!|{-}CH_3 \\ Br \end{array}$$

Solution

We will use a sphere to denote the chiral carbon atom. Projection formula convention dictates that horizontal bonds are understood to be coming out of the paper. We will use wedges to denote these bonds in the perspective structure. Projection formula convention states that vertical bonds are understood to be going into the paper. We will use broken lines to denote these bonds in the perspective structure. Our perspective structure is

$$\begin{array}{c} Cl \\ H \diagdown \bigcirc \diagup CH_3 \\ Br \end{array}$$

Projection formulas can also be used to represent the three-dimensional structure of molecules containing more than one chiral carbon atom. The usual convention with such molecules is to have the carbon chain in a vertical position, with the substituents to the right and left of the chain. A perspective drawing and projection formula for the molecule *2-bromo-3-chlorobutane* are as follows.

$$\begin{array}{cc} \begin{array}{c} CH_3 \\ H{-}C{-}Br \\ H{-}C{-}Cl \\ CH_3 \end{array} & \begin{array}{c} CH_3 \\ H{-}\!\!\!|{-}Br \\ H{-}\!\!\!|{-}Cl \\ CH_3 \end{array} \\ \text{perspective drawing} & \text{projection formula} \end{array}$$

When comparing different projection formulas to see if they represent the same molecule or are enantiomers, the following rules concerning movement of the projection formulas must be observed; otherwise, invalid interpretations of the three-dimensional relationships actually present in the molecules can result.

1. The single operation of rotating a projection formula by 180° within the plane of the paper (turning the formula upside down) is allowable.
2. The single operation of rotating a projection formula by 90° within the plane of the paper is *not* allowable.
3. The single operation of rotating a projection formula by 180° out of the plane of the paper (flipping it over) is *not* allowable.
4. The joint operations of rotating a projection formula by 90° within the plane

of the paper, followed by flipping it over is allowable. Neither one of these operations by itself (Rules 2 and 3) is allowed. However, when they are done jointly, they are allowed. (This is a case where two wrongs make a right.)

Example 21.3

Indicate whether each of the following pairs of projection formulas represent different structures or two ways of depicting the same structure.

a.
```
      CH₃                    CH₃
       |                      |
 Cl ——+—— H     and     H ——+—— Cl
       |                      |
      OH                     OH
```

b.
```
      CH₃                     Cl
       |                      |
 Cl ——+—— H     and    CH₃ ——+—— OH
       |                      |
      OH                      H
```

c.
```
      CH₃                    OH
       |                      |
 Cl ——+—— H     and     H ——+—— Cl
       |                      |
      OH                    CH₃
```

Solution

a. These are two different structures. The two structures are mirror images of each other. If the second structure is rotated 180° (Rule 1) in the plane of the paper, the Cl and H atoms would match, but the CH₃ and OH groups would mismatch.

b. Here, we have two ways of representing the same structure. When we apply Rule 4, it becomes apparent that they are the same structures. If we rotate the second structure by 90° (so that the CH₃ and OH groups are vertical) and then flip it over (interchanging the Cl and H atoms), we have a structure superimposable on the first structure.

c. Here, we have two identical structures. If the second structure is rotated 180° (Rule 1) in the plane of the paper, the two structures match exactly.

21.5 NOMENCLATURE FOR ENANTIOMERS

Enantiomers can be named by using the prefixes D- (*dextro* meaning "right") and L- (*levo*, meaning "left"). Which enantiomer is the D-isomer and which is the L-isomer? An arbitrary decision had to be made concerning this at the time the nomenclature system was set up. The standard for the decision was the molecule *glyceraldehyde*. The enantiomer of glyceraldehyde whose projection formula has the hydroxide group on the right side when the aldehyde group (CHO) is at the top is the D-isomer, and the enantiomer with the hydroxide on the left side is the L-isomer.

```
      CHO                    CHO
       |                      |
  H ——+—— OH           HO ——+—— H
       |                      |
     CH₂OH                  CH₂OH
 D-glyceraldehyde      L-glyceraldehyde
```

The D- and L- configurations for other chiral compounds is determined by chemically relating them to either D- or L-glyceraldehyde. For example, lactic acid, which differs from glyceraldehyde in that the CH_2OH group has been converted to CH_3 and the CHO group has been converted into a CO_2H group, would have D- and L- forms with the following structures.

$$\begin{array}{c} \text{CHO} \\ \text{H} - \text{OH} \\ \text{CH}_2\text{OH} \end{array} \xrightarrow{\text{chemical reactions}} \begin{array}{c} \text{CO}_2\text{H} \\ \text{H} - \text{OH} \\ \text{CH}_3 \end{array}$$
D-glyceraldehyde   D-lactic acid

$$\begin{array}{c} \text{CHO} \\ \text{HO} - \text{H} \\ \text{CH}_2\text{OH} \end{array} \xrightarrow{\text{chemical reactions}} \begin{array}{c} \text{CO}_2\text{H} \\ \text{HO} - \text{H} \\ \text{CH}_3 \end{array}$$
L-glyceraldehyde   L-lactic acid

D- and L- nomenclature for enantiomers will be used extensively in Chapters 22 (carbohydrates) and 24 (proteins). Chemists need D- and L- nomenclature in order to draw the correct structure of an enantiomer, when given its name.

21.6 OPTICAL ACTIVITY

In Section 21.2, we mentioned that the physical properties of enantiomers are identical except that they behave differently when interacting with plane-polarized light. The starting point for an understanding of this situation is a brief discussion concerning the nature of light.

Ordinary (unpolarized) light consists of electromagnetic radiation transmitted by wave motion in all directions (planes) perpendicular to the direction in which it is traveling. When this ordinary light passes through a polarizer, it emerges oscillating in only one plane and is known as *plane-polarized light*. Figure 21-7 contrasts the difference between ordinary light and plane-polarized light.

FIGURE 21-7 Diagram contrasting ordinary light, which vibrates in all directions, and plane-polarized light, which oscillates in only one direction.

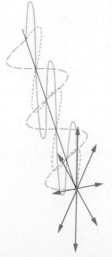

(a) Ordinary (unpolarized) light   (b) Plane-polarized light

Polarizers, the objects needed to transform ordinary light into plane-polarized light, are made from certain types of crystals. Usually properly oriented crystals are embedded in a transparent plastic to produce a polaroid filter.

When rays of plane-polarized light are passed through solutions of isolated enantiomers, the plane of the polarized light is rotated to the left or to the right, depending on which enantiomer is present. The extent of the rotation depends on the concentration of the enantiomer, as well as its identity. Furthermore, the two enantiomers rotate the plane-polarized light the same number of degrees, but in opposite directions. If a 0.5 M solution of one enantiomer rotates the light 30° to the right, a 0.5 M solution of the other one will rotate the light 30° to the left.

Instruments used to measure the degree of rotation of plane-polarized light by enantiomeric compounds are called *polarimeters*. The schematics shown in Figure 21–8 are the basis for these instruments.

Enantiomers, are said to be optically active because of the way they interact with plane-polarized light. An **optically active compound** *is a compound that rotates the plane of polarized light.*

Dextrorotatory compounds *are optically active compounds that rotate the plane of polarized light in a clockwise (to the right) direction.* **Levorotatory compounds** *are optically active compounds that rotate the plane of polarized light in a counterclockwise (to the left) direction.* If one member of an enantiomeric pair is dextrorotatory, then the other member must necessarily be levorotatory.

There are symbols that indicate the direction of rotation of plane-polarized light by optically active compounds. A plus or minus sign inside parentheses is used to denote the direction of rotation. A (+) notation means rotation to the right (clockwise) and a (−) notation means rotation to the left (counterclockwise). Using this notation, the enantiomer of lactic acid that is dextrorotatory would be designated (+)-lactic acid.

The D- and L- notation used to describe molecular configuration for enantiomers (Section 21–5) and the (+) and (−) notation used to describe the direction of rotation of plane-polarized light by enantiomers are not interconnected quantities. *There is no direct relationship between the two sets of quantities.* There is no way of knowing which way an enantiomer will rotate light until it is examined

optically active compound

dextrorotatory compounds

levorotatory compounds

FIGURE 21–8   The schematics upon which polarimeters are based.

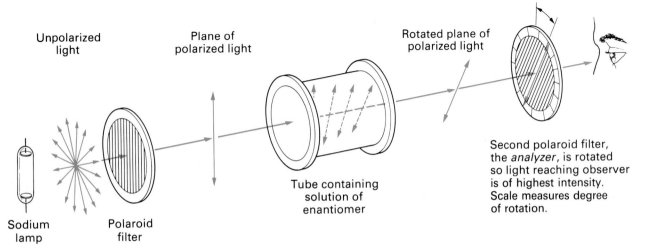

with a polarimeter. We only know that if a D-enantiomer rotates polarized light in one direction, then the L-enantiomer will equally rotate polarized light in the opposite direction.

When writing the name of an optically active molecule, you should include notation for both molecular configuration and direction of rotation to avoid ambiguity. Examples of complete names are D-(+)-glyceraldehyde and D-(−)-glyceric acid. Note how the D-forms of these two different substances rotate light in different directions.

In later chapters of this text, we will make important applications of these principles concerning enantiomerism and optical activity. For example, blood sugar (glucose) and amino acids (the building blocks for proteins) are specific non-interchangable enantiomers.

21.7 SPECIFIC ROTATION

The magnitude of the angle of rotation of plane-polarized light produced by an optically active compound depends on two parameters other than the identity of the compounds. They are the concentration of the substance and the length of the tube containing the sample. Because light interacts with molecules one by one the greater the number of molecules encountered, the greater the final angle of rotation is. Increased concentration and increased tube length both result in increased numbers of light–molecule interactions and increased angle of rotation.

Because different conditions produce different angles of rotation, a standard is needed for comparision purposes. This standard, which is called the *specific rotation*, is obtained by dividing the observed angle of rotation by the length of the tube and the concentration of the sample. The mathematical definition of **specific rotation**, is

$$\text{specific rotation} = \frac{\text{observed rotation (degrees)}}{\text{length (dm)} \times \text{concentration (g/cm}^3\text{)}}$$

This equation, in abbreviated form, is

$$[\alpha] = \frac{\alpha_{\text{obs}}}{l \times C}$$

where the symbol α (alpha) is used to denote angle of rotation. The units for length and concentration are somewhat different than those we normally use. Length is measured in decimeters because most sample tubes are 1.0 dm long. Concentration is measured in grams/cm³ of solution. These are the same units that are used for density, and for pure liquids, the density can be used. For dilute solutions, the concentration gives the grams of optically active substance dissolved in 1 cm³ of solution. The units on α_{obs} and $[\alpha]$ are both degrees of rotation.

The temperature and wavelength (color) of light used also affect specific rotation values. When specifying a specific rotation value, chemists use a numerical subscript to indicate the temperature (usually 20°C). The wavelength of light used is usually the yellow light obtained from a sodium lamp. This wavelength, which is 589 nm, is denoted using the symbol D as a subscript. A complete specific rotation notation would appear with the following format.

$$[\alpha]_D^{20} = \frac{\alpha_{\text{obs}}}{l \times C}$$

Example 21.4

Calculate the specific rotation, $[\alpha]_D^{20}$ of cholesterol, given that the observed angle of rotation was $-2.4°$ in a 1.00 dm tube for a solution of 0.061 g/cm³ in chloroform (using a sodium vapor lamp and temperature controlled at 20°C).

Solution
Checking the units, we see that they are all correct: the length is in dm and the concentration is in g/cm³. Dividing the observed angle by the length and concentration, we obtain

$$[\alpha]_D^{20} = \frac{-2.4°}{(1.00 \text{ cm})(0.061 \text{ g/cm}^3)} = -39°$$

Therefore, the specific rotation is $[\alpha]_D^{20} = -39°$. Notice how the small value for concentration substantially increases the angle of rotation after division. Also notice the negative angle. Light was rotated counterclockwise in the sample tube.

If the specific rotation for a substance is known, information obtained from polarimeter experiments can be used to calculate the concentration of that substance in a solution. Knowing $[\alpha]_D^{20}$, l, and the observed angle of rotation allows us to calculate the only remaining unknown in the specific rotation equation: concentration. To help matters out, $[\alpha]_D^{20}$ values for thousands of chiral compounds are listed in reference books such as *The Handbook of Chemistry and Physics*.

21.8 RACEMIC MIXTURES

A **racemic mixture** is a mixture containing equal amounts of a pair of enantiomers. This type of mixture is optically inactive and will show no rotation of plane-polarized light when tested in a polarimeter. As the plane of polarized light passes through a racemic mixture, the (+)-enantiomers still rotate the light to the right. However, an equal number of (−)-enantiomers rotate the plane of the light to the left. The net result of this is no change in the angle of rotation. Racemic mixtures of enantiomers are often represented by a (±)-prefix. For example, the notation for a racemic mixture of tartaric acid would be (±)-tartaric acid.

Racemization is the process of converting a solution of a single enantiomer into a racemic mixture. A number of enzymes (Chapter 25) within the human body have the ability to catalyze racemization processes. For example, the enzyme amino-acid-racemase can convert an optically active solution of one amino acid enantiomer into a racemic mixture of two enantiomers. Consider alanine (2-aminopropanoic acid) as an example.

racemic mixture

racemization

The progress of a racemization process such as this can be followed using a polarimeter by observing how fast the observed angle of rotation approaches 0°.

Chemical reactions in the laboratory most often result in racemic mixtures when a chiral carbon is present in the product. It is difficult to synthesize one enantiomer without having an equal amount of the other enantiomer formed simultaneously.

The **resolution of a racemic mixture** is *the process of separating enantiomers from each other*. This is a very challenging problem because the physical properties of the two enantiomers present are the same. They boil at the same temperature, freeze at the same temperature, have the same solubilities in common solvents, and often their crystals look identical.

resolution of a racemic mixture

Many enzymes within the human body have the ability to resolve racemic mixtures. One example is the enzyme L-*amino acid oxidase*. As the name of this enzyme implies, it specifically oxidizes only L-amino acids. When this enzyme is added to a racemic mixture of amino acid enantiomers, the L-amino acids are preferentially oxidized, leaving the D-amino acids in solution.

21.9 DIASTEREOMERS AND MESO COMPOUNDS

The number of stereoisomers (Section 21.1) possible for a given molecular structure increases as the number of chiral carbon atoms present increases. For a structure involving n chiral carbons, all of which are *nonequivalent*, 2^n stereoisomers are possible. In order for chiral carbon atoms to be nonequivalent, they cannot contain identical sets of substituents.

When a structure contains only one chiral carbon, the formula 2^n, with n equal to one, correctly predicts that only two stereoisomers (an enantiomeric pair) exist. All of our discussions of optical isomerism to this point have dealt with this n equals one situation.

When a structure contains two nonequivalent chiral carbon atoms, a total of four stereoisomers should exist ($2^2 = 4$). Let us consider the details of such a situation. Our example will be a compound with the following structural characteristics.

$$CH_3-\underset{OH}{CH}-\underset{Br}{CH}-CH_3$$

Projection formulas, which we have arbitrarily labeled A, B, C, and D for discussion purposes, for the four possible stereoisomers are as follows.

```
    CH3           CH3           CH3           CH3
 H──┼──OH     HO──┼──H      H──┼──OH     HO──┼──H
 H──┼──Br     Br──┼──H     Br──┼──H       H──┼──Br
    CH3           CH3           CH3           CH3
     A             B             C             D
```

Let us examine the relationships that exist among these four stereoisomers. In the process, we will identify a new type of stereoisomerism that we have not previously discussed.

First, we note that isomers A and B are nonsuperimposable mirror-images of each other and that a similar relationship exists between C and D. Thus, we have two pairs of enantiomers. The physical properties of the enantiomeric A and B will be similar, except for specific rotation. Likewise, the physical properties of C and D will be similar. However, the properties of A and B are *different* from those of C and D.

What is the relationship between A and C? They are not enantiomers because they are not mirror images of each other. However, they are still isomers because they contain the same number of atoms of each kind. Formulas A and C represent a new type of stereoisomerism called *diastereomerism*. **Diastereomers** are pairs of stereoisomers that are not enantiomers; that is, they are not nonsuperimposable mirror images of each other. There are four pairs of diastereomers for the compound now under discussion: A and C, A and D, B and C, and B and D. Diastereomers have different physical properties, but enantiomers do not.

diastereomers

If one or more chiral carbons in a structure are equivalent rather than nonequivalent, the number of stereoisomers possible is always less than that predicted by the 2^n formula. The 2^n formula assumes that all chiral carbons present are nonequivalent. Let us examine a structure where some equivalency exists.

$$HOOC-\underset{\underset{OH}{|}}{CH}-\underset{\underset{OH}{|}}{CH}-COOH$$

The formula 2^n predicts that four stereoisomers are possible. Projection formulas for these isomers should be as follows.

```
      COOH              COOH              COOH              COOH
   HO─┼─H            H─┼─OH            H─┼─OH            HO─┼─H
    H─┼─OH          HO─┼─H             H─┼─OH            HO─┼─H
      COOH              COOH              COOH              COOH
       A                 B                 C                 D
```

In a manner similar to our previous example, let us examine the relationships that exist among the four projection formulas.

Formulas A and B are nonsuperimposable mirror images and are enantiomers. Formulas C and D are also mirror images. However, if we rotate D 180° in the plane of the paper, we find it is superimposable on C. Thus, C and D are not isomers; rather, they are two representations of the same structure. Thus, only three stereoisomers exist for this compound.

Let us now look more closely at why C and D are one and the same. What caused the two structures to be superimposable on each other? There is a plane of symmetry that passes through the molecule.

```
         COOH
      H─┼─OH
   ─────┼────── plane of symmetry
      H─┼─OH
         COOH
```

The top half of the structure is the same as the bottom half. Structure C (or D) is an example of a *meso compound*. **A meso compound** is a compound whose structure contains chiral atoms and also a plane of symmetry. The mirror image of a meso structure is always superimposable on the original structure. The significance of meso compounds is that they are optically inactive even though they contain chiral carbons. Why would these compounds be optically inactive? The two halves of a meso structure are mirror images and mirror images affect plane-polarized light in opposite ways. Thus, the rotation of polarized light caused by the chiral carbon in the one half of the molecule is exactly compensated for by an opposite effect caused by the chiral atom in the other half of the molecule. We will have occasion to refer to meso structures in later chapters of the text.

meso compound

22 Carbohydrates

Objectives

After completing Chapter 22, you will be able to:

22.1 Define the term *biochemistry* and list the six major types of biochemicals.
22.2 Discuss the relationship between photosynthesis and carbohydrates.
22.3 Classify a carbohydrate as a monosaccharide, oligosaccharide, or polysaccharide.
22.4 Classify monosaccharides on the basis of the type of carbonyl group present and the number of carbon atoms present; and understand D- and L-nomenclature for monosaccharides.
22.5 List the names, structures, and occurrence of common monosaccharides.
22.6 Discuss the relationship between the open-chain and cyclic forms of a monosaccharide, recognize the structural characteristics of α and β anomers, and explain the process of mutarotation for cyclic monosaccharides in solution.
22.7 List the common chemical reactions that monosaccharides undergo, including acetal and ketal formation, oxidation, reduction, and esterification, and understand the concept of reducing and nonreducing sugars.
22.8 Draw the structures and list sources and uses of common disaccharides, and recognize and describe the various types of glycosidic bonds that exist.
22.9 Give the history of artificial sweeteners and the structures of the three most frequently used artificial sweeteners.
22.10 List the physical and chemical properties, uses, and structures of common polysaccharides and describe the biological significance of α and β glycosidic bonds within polysaccharides.

INTRODUCTION

In previous chapters, we have given numerous examples of the importance of chemistry in biological systems. Beginning with this chapter on carbohydrates, we will focus our attention almost exclusively on biochemistry, the chemistry of living systems. Like organic chemistry, biochemistry is a vast subject and we can discuss only a few of its facets. Our approach to biochemistry will be similar to our approach to organic chemistry. We will devote individual chapters to each of the major classes of biochemical compounds.

Carbohydrates are one of the classes of biochemical molecules. Other classes include proteins, lipids, and nucleic acids. We will begin this chapter with a general overview of the classes of biochemical substances. Then, we will focus specifically on carbohydrates, with an emphasis on the structures and reactions of these substances. This knowledge is necessary before discussing how our own bodies utilize carbohydrates (Chapter 29).

22.1 TYPES OF BIOCHEMICAL SUBSTANCES

Biochemistry *is the study of the chemistry of living systems.* It is a constantly changing field in which new discoveries are made almost daily about how life is maintained, how diseases occur, how diseases can be prevented, and how cells manufacture the molecules needed for life.

biochemistry

The substances found in living organisms, *biochemicals*, can be divided into two general groups: bioinorganic substances and bioorganic substances. Water and inorganic salts are the major bioinorganic substances. There are four major groups

FIGURE 22-1 Biochemical substances can be divided into two major groups, bioorganic substances and bioinorganic substances. There are four groups of bioorganic substances and two groups of bioinorganic substances.

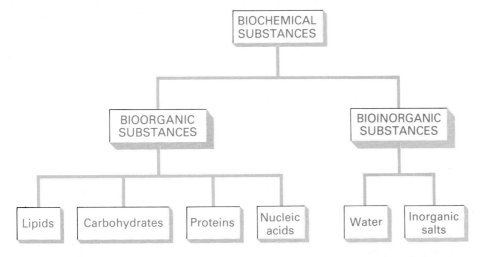

of bioorganic substances: carbohydrates, lipids, proteins, and nucleic acids. Figure 22–1 (p. 529) schematically shows the major types of biochemical substances present in living systems. The basic chemicals of life are similar, regardless of the type of living system (it could be a microorganism or an elephant).

The major types of biochemical substances are not equally abundant. As we pointed out in Section 8.1, water is the most abundant substance present in a living organism. Approximately two-thirds of the mass of a human body is water. Another 4–5% of human body mass is inorganic salts. The principal ions present in body fluids, which are derived from inorganic salts, were discussed in Section 10.5 when we considered the topic of electrolytes.

We tend to think of the human body as being made up of organic substances. Actually, bioorganic substances constitute only about one-fourth of the mass of a human body. Proteins are the most abundant type of bioorganic substance present in the human body ($\approx 15\%$). Lipids, a diverse group of compounds, are the second most abundant type of bioorganic substance present ($\approx 8\%$). The remaining two types of bioorganic substances, carbohydrates and nucleic acids, are found in relatively small amounts in most tissue ($\approx 2\%$). Composition data for the human body, in terms of the types of substances present, is summarized in Figure 22–2.

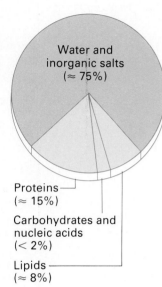

FIGURE 22–2 Composition of the human body in terms of major types of biochemical substances. (Percent by mass.)

22.2 OCCURRENCE OF CARBOHYDRATES

Carbohydrates, which include such familiar substances as glucose, table sugar (sucrose), starch, and cellulose, are of major importance to both plants and animals. Most of the matter in plants, with the exception of water, is made up of carbohydrates.

The name *carbohydrates* was given to this group of bioorganic substances by French scientists, who called them *hydrates de carbone* because their formulas approximated $(C \cdot H_2O)_n$. It has since been found that not all carbohydrates fit this general formula.

The process of photosynthesis, which is carried out by green plants (chlorophyll-containing plants), is the method by which almost all carbohydrates are produced. In this process, plants combine carbon dioxide from the air with water from the soil, with sunlight serving as the source of energy, to give simple carbohydrates.

$$6\,CO_2 + 6\,H_2O \xrightarrow[\text{chlorophyll}]{\text{sunlight}} \underset{\text{(a carbohydrate)}}{C_6H_{12}O_6} + 6\,O_2$$

The carbohydrates produced from photosynthesis serve two major functions within plants. Carbohydrates in the form of cellulose are structural elements, and carbohydrates in the form of starch provide needed nutritional reserves (energy) for the plant.

Animals and humans obtain carbohydrates by eating plants. Carbohydrates constitute 65% of the typical human diet. Carbohydrate food intake in humans serves two purposes: to provide energy, which is obtained through carbohydrate oxidation, and to supply carbon atoms for the synthesis of other biochemical substances (proteins, lipids, and nucleic acids).

22.3 DEFINITION AND CLASSIFICATION OF CARBOHYDRATES

carbohydrates

Carbohydrates *are polyhydroxy aldehydes or ketones, or substances that yield these compounds upon hydrolysis.* Compounds classified as carbohydrates vary in structure from those consisting of a few carbon atoms to gigantic (polymeric) molecules having molecular weights in the hundreds of thousands.

The general definition for a carbohydrate indicates that polyhydroxy aldehyde or polyhydroxy ketone units are present in these compounds. The general structures of polyhydroxy aldehydes and polyhydroxy ketones are as follows.

$$\begin{array}{c} \boxed{\underset{C}{H} \diagdown\!\!\!\diagup O} \leftarrow \text{aldehyde group} \\ | \\ (H-C-OH)_n \\ | \\ CH_2OH \end{array}$$
a polyhydroxy aldehyde

$$\begin{array}{c} CH_2OH \\ | \\ \boxed{C=O} \leftarrow \text{ketone group} \\ | \\ (H-C-OH)_m \\ | \\ CH_2OH \end{array}$$
a polyhydroxy ketone

A specific example of a polyhydroxy aldehyde is the compound *glucose*. An example of a polyhydroxy ketone is the compound *fructose*.

$$\begin{array}{c} \underset{C}{H} \diagdown\!\!\!\diagup O \\ | \\ H-C-OH \\ | \\ HO-C-H \\ | \\ H-C-OH \\ | \\ H-C-OH \\ | \\ CH_2OH \end{array}$$
glucose
(a polyhydroxy aldehyde)

$$\begin{array}{c} CH_2OH \\ | \\ C=O \\ | \\ HO-C-H \\ | \\ H-C-OH \\ | \\ H-C-OH \\ | \\ CH_2OH \end{array}$$
fructose
(a polyhydroxy ketone)

There are three major classes of carbohydrates: monosaccharides, oligosaccharides, and polysaccharides.

Monosaccharides *are carbohydrates that cannot be broken down into simpler units by hydrolysis reactions.* Monosaccharides consist of a single polyhydroxy aldehyde or polyhydroxy ketone unit. Both glucose and fructose are monosaccharides. Monosaccharides usually contain from three to six carbon atoms; five and six carbon atoms are especially common for naturally occurring monosaccharides. The most common monosaccharide is glucose.

monosaccharides

Oligosaccharides *are carbohydrates that contain from two to ten monosaccharide units.* Oligo comes from the Greek word *oligos*, which means "small" or "few". Oligosaccharides, like monosaccharides, are individual, identifiable, crystalline, water soluble compounds that have definite structures and molecular weights.

oligosaccharides

An oligosaccharide that contains two monosaccharide units is also called a *disaccharide*. Common table sugar, sucrose, is a disaccharide. (A number of monosaccharides and disaccharides taste sweet. For this reason, the term *sugar* is often used when referring to monosaccharides and disaccharides.) Similarly, oligosaccharides made up of three and four monosaccharide units could be (and often are) called *trisaccharides* and *tetrasaccharides*, respectively. The oligosaccharides of most concern to us in this text will be the disaccharides.

Polysaccharides *are carbohydrates in which there are more than ten monosaccharide units present.* Thousands of monosaccharide units are present in most polysaccharides. Polysaccharides are the most abundant carbohydrate form in plants. Both cellulose and starch are polysaccharides. We encounter various forms of these substances throughout our world. The paper on which this book is printed is mainly cellulose, as is the cotton in our clothes and the wood in our houses.

polysaccharides

Starch is a component of the flour used to make bread and of many other types of food such as potatoes, rice, corn, beans, and peas.

22.4 CLASSIFICATIONS AND STRUCTURES FOR MONOSACCHARIDES

Monosaccharides can be classified in two ways: on the basis of the number of carbon atoms present and by whether an aldehyde or ketone group is present. Most common monosaccharides contain from three to six carbon atoms. A three-carbon monosaccharide is called a *triose*. Similarly, monosaccharides containing four, five, and six carbon atoms are called *tetroses*, *pentoses*, and *hexoses*, respectively. Monosaccharides are classified as *aldoses* and *ketoses* based on the type of carbonyl group present. **Aldoses** *are monosaccharides containing an aldehyde group.* **Ketoses** *are monosaccharides containing a ketone group.*

The two monosaccharide classification systems are often combined. Using this combined system, glucose and fructose (whose structures were given in the previous section) are classified as an aldohexose and a ketohexose, respectively. An aldohexose is a six-carbon monosaccharide with an aldehyde group and a ketohexose is a six-carbon monosaccharide with a keto group. Figure 22–3 summarizes the terminology used in classifying monosaccharides.

FIGURE 22–3 All monosaccharides can be classified as aldoses or ketoses. Both aldoses and ketoses can be further classified as trioses, tetroses, pentoses, hexoses, and so on.

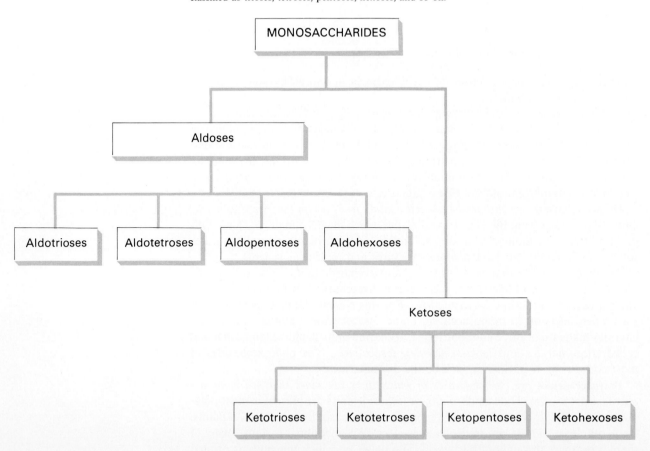

Aldoses

The simplest aldose is the triose *glyceraldehyde*. It exists in two optically active forms (enantiomeric forms—Section 21.2) because a chiral carbon atom (the middle carbon) is present in the molecule.

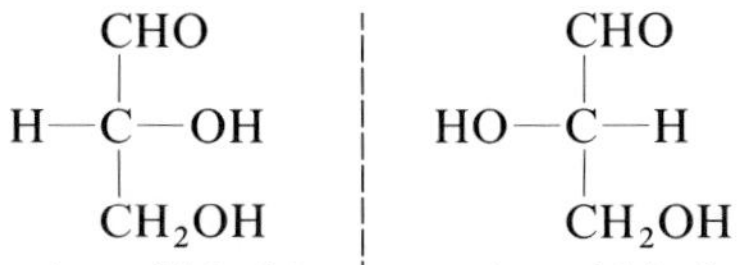

```
        CHO              CHO
         |                |
    H—C—OH           HO—C—H
         |                |
       CH₂OH            CH₂OH
    D-glyceraldehyde   L-glyceraldehyde
```

We noted in Section 21.5 that the D isomer of glyceraldehyde is the one whose projection formula has the —OH group on the middle carbon drawn to the right.

Other aldoses can be considered to be derived from glyceraldehyde by a lengthening of the carbon chain through insertion of additional H—C—OH groups between carbons 1 and 2. (The numbering of aldose carbon chains always begins with the aldehyde carbon atom.)

Insertion of an additional H—C—OH group into glyceraldehyde produces an aldotetrose. Four isomers are possible because an aldotetrose contains two chiral carbon atoms. (For molecules with n chiral carbon atoms, a maximum of 2^n different stereoisomers can exist—Section 21.9). The projection formulas for the four aldotetroses are:

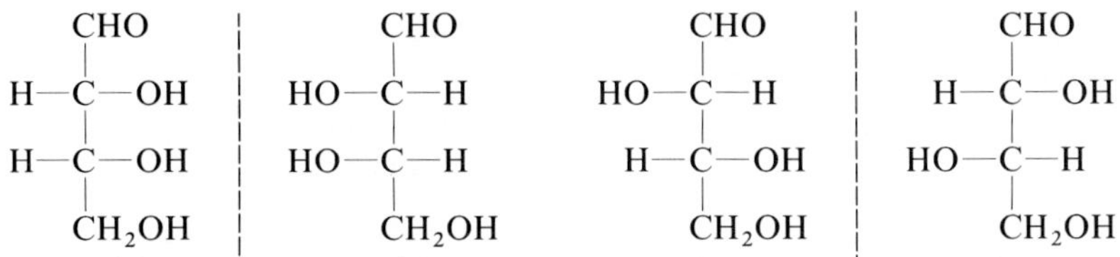

```
    CHO           CHO           CHO           CHO
     |             |             |             |
  H—C—OH       HO—C—H        HO—C—H         H—C—OH
     |             |             |             |
  H—C—OH       HO—C—H         H—C—OH        HO—C—H
     |             |             |             |
   CH₂OH         CH₂OH         CH₂OH         CH₂OH
  D-erythrose   L-erythrose    D-threose     L-threose
```

D-Erythrose and L-erythrose are enantiomers, as are D-threose and L-threose. The relationship between either of the erythroses and either of the threoses is that of diastereomers (Section 21.9).

In the erythroses, the two —OH groups bonded to the chiral carbon atoms point in the same direction in the projection formula; in the threoses, the two —OH groups point in opposite directions. In order to determine which is the D isomer and which is the L isomer in each enantiomeric pair, we look at the *penultimate carbon atom*, which is the chiral carbon atom that is most distant from the aldehyde group. (Penultimate means next to last.) If the —OH group on this carbon atom points to the right in the projection formula (as in D-glyceraldehyde), the enantiomer is the D form. If the —OH group on the penultimate carbon atom points to the left (as in D-glyceraldehyde), the molecule is in the L form.

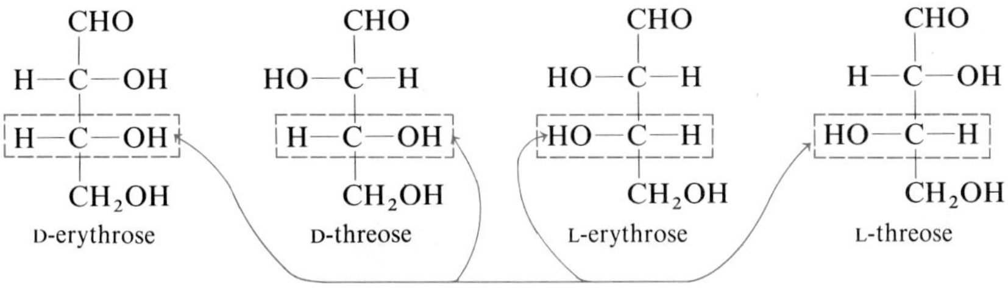

```
    CHO           CHO           CHO           CHO
     |             |             |             |
  H—C—OH       HO—C—H        HO—C—H         H—C—OH
     |             |             |             |
  [H—C—OH]     [H—C—OH]      [HO—C—H]      [HO—C—H]
     |             |             |             |
   CH₂OH         CH₂OH         CH₂OH         CH₂OH
  D-erythrose   D-threose     L-erythrose    L-threose
```

"Positioning" of the —OH group on the penultimate carbon atom (next to last carbon atom) determines D and L configuration.

Aldopentoses contain three chiral carbon atoms and thus, eight stereoisomeric compounds are possible ($2^3 = 8$). The eight stereoisomers will be four pairs of enantiomers. Projection formulas for one member of each enantiomeric pair, the D configuration, are given in Figure 22–4. The other four stereoisomers, the L configurations, will have projection formulas that are the mirror images of those given. Note that in each of the aldopentose structures in Figure 22–4, the —OH group on carbon number 4 (the penultimate carbon atom) is directed to the right; this is necessary in order for the structure to be a D configuration.

There are sixteen isomeric aldohexoses, or eight pairs of enantiomers. (There are four chiral carbon atoms present in an aldohexose and $2^4 = 16$.) Figure 22–4 gives projection formulas for the D form of each enantiomeric pair. The other eight compounds, the L forms, have structures that are the mirror images of the eight structures shown.

FIGURE 22–4 Projection formulas for aldoses with three through six carbon atoms. Only one member of each enantiomeric pair, the D form, is shown.

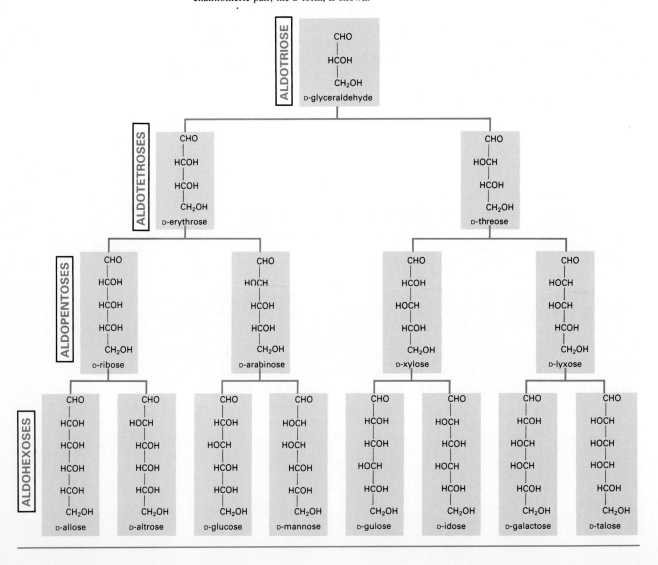

Example 22.1

Draw the projection formula for L-mannose.

Solution

L-Mannose is the enantiomer of D-mannose, whose projection formula is given in Figure 22–4. Enantiomers are mirror images of each other. Each hydroxyl group that is on the right in D-mannose is on the left in L-mannose. The hydroxyl groups on the left in D-mannose will be on the right in L-mannose.

```
        CHO                    CHO
         |                      |
    HO—C—H                 H—C—OH
         |                      |
    HO—C—H                 H—C—OH
         |                      |
     H—C—OH                HO—C—H
         |                      |
     H—C—OH                HO—C—H
         |                      |
       CH₂OH                  CH₂OH
      D-mannose    mirror    L-mannose
                   plane
```

Ketoses

The simplest ketose is the triose *dihydroxyacetone*.

$$\begin{array}{c} CH_2OH \\ | \\ C{=}O \\ | \\ CH_2OH \end{array}$$

dihydroxyacetone

A major difference between dihydroxyacetone and glyceraldehyde, the simplest aldose, is that the former does not possess a chiral carbon atom. Thus, D and L forms are not possible for dihydroxyacetone. This reduces by half (compared to aldoses) the number of stereoisomers possible for ketotetroses, ketopentoses, and ketohexoses. An aldohexose possesses four chiral carbon atoms, but a ketohexose has only three chiral carbon atoms. Figure 22–5 (p. 536) gives the projection formulas for the D forms of ketoses containing three, four, five, and six carbon atoms.

Note that the names of most aldoses and ketoses end in the suffix *-ose*. This is the characteristic ending for the common names of most monosaccharides as well as for the common names of other types of carbohydrates.

Most monosaccharides found in nature exist in the D form. Plants have the ability to produce only one form of an enantiomeric pair. Any monosaccharide synthesis in the laboratory will always result in the production of a racemic mixture (Section 21.8); that is, it will produce a mixture containing equal amounts of the D and L forms of the compounds.

22.5 IMPORTANT MONOSACCHARIDES

Figures 22–4 and 22–5 show that many different monosaccharides exist. Only a few of these substances appear to have important biological significance. These compounds include glucose, galactose, fructose, and ribose. Glucose and galactose are aldohexoses; fructose is a ketohexose; and ribose is an aldopentose. The D form

FIGURE 22-5 Projection formulas for ketoses with three through six carbon atoms. Only one member of each enantiomeric pair, the D form, is shown.

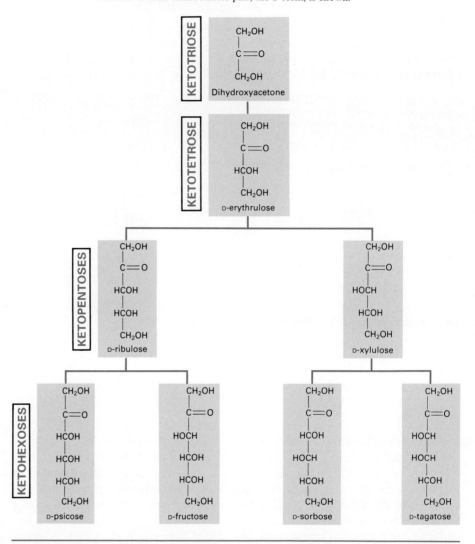

of each of these compounds is the biologically important form. You should know the structural characteristics of these four compounds in detail.

```
    CHO              CHO            CH₂OH
H—C—OH           H—C—OH            C=O             CHO
HO—C—H           HO—C—H          HO—C—H          H—C—OH
H—C—OH           HO—C—H           H—C—OH          H—C—OH
H—C—OH           H—C—OH           H—C—OH          H—C—OH
    CH₂OH            CH₂OH            CH₂OH            CH₂OH
 D-glucose       D-galactose       D-fructose       D-ribose
```

Glucose, galactose, and fructose all have the same molecular formula, $C_6H_{12}O_6$; the formula for ribose, which has one less carbon atom, is $C_5H_{10}O_5$.

D-Glucose

D-Glucose, a white crystalline solid with a high water-solubility, is the most abundant and most important monosaccharide. Natural sources for it are grapes, figs, and dates, as well as other fruits. Ripe grapes contain 20–30% glucose. Glucose is also found naturally as a component of the disaccharides sucrose, maltose, and lactose (Section 22.8) and is the monomer of the polysaccharides starch, cellulose, and glycogen (Section 22.10).

Two other names for D-glucose are dextrose and blood sugar. The name *dextrose* draws attention to the fact that the optically active D-glucose, in aqueous solution, rotates plane-polarized light to the right (Section 21.6). *Dextro* is the Latin word for "right". The term *blood sugar* draws attention to the fact that blood contains dissolved glucose. The concentration of glucose in human blood is fairly constant; it is in the range of 70–100 mg per 100 mL of blood. Cells use glucose as a primary energy source.

A 5% (m/v) solution of glucose in water is frequently used in hospitals as an intravenous (IV) source of nourishment for patients who cannot take food by mouth (see Figure 22–6). (A 5%(m/v) concentration is used so the IV solution will be isotonic with blood—Section 8.18). A 5% (m/v) glucose–water solution is also the source of nourishment that is given to newborn babies during their first few hours (and sometimes days) of life.

Certain body conditions, such as those caused by the disease *diabetes mellitus*, result in there being too much glucose in the blood. People with above-normal blood glucose levels are said to be *hyperglycemic*. (We will discuss hyperglycemia in Chapter 29). The opposite condition—lower than normal blood glucose levels—is called *hypoglycemia*. Symptoms of hypoglycemia include nausea and dizziness. Hypoglycemia is often treated by regulating the dietary intake of glucose.

FIGURE 22–6 A 5%(m/v) glucose solution is frequently used in a hospital as an intravenous (IV) source of nourishment for patients who cannot take food by mouth.

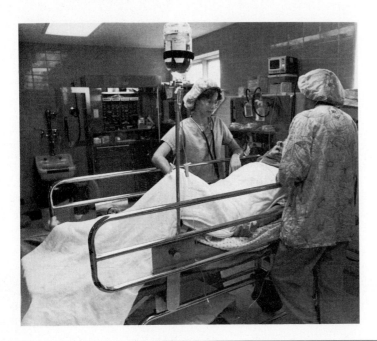

D-Galactose

Galactose does not frequently occur as a monosaccharide in nature. It is most often obtained by hydrolysis of the disaccharide *lactose*, which consists of a glucose unit and a galactose unit (Section 22.8). The structures of D-galactose and D-glucose differ only in the configuration of the —OH group on carbon number 4.

$$\begin{array}{cc}
^1\text{CHO} & ^1\text{CHO} \\
\text{H}-^2\text{C}-\text{OH} & \text{H}-^2\text{C}-\text{OH} \\
\text{HO}-^3\text{C}-\text{H} & \text{HO}-^3\text{C}-\text{H} \\
\boxed{\text{HO}-^4\text{C}-\text{H}} & \boxed{\text{H}-^4\text{C}-\text{OH}} \\
\text{H}-^5\text{C}-\text{OH} & \text{H}-^5\text{C}-\text{OH} \\
^6\text{CH}_2\text{OH} & ^6\text{CH}_2\text{OH} \\
\text{D-galactose} & \text{D-glucose}
\end{array}$$

epimers   D-Galactose and D-glucose are epimers. **Epimers** *are diastereomers that differ only in the configuration at one chiral carbon atom.* Galactose is synthesized from glucose in the mammary glands, where it is used to make the lactose present in milk. Galactose is also found in brain and nerve tissue as part of very complex molecules called *cerebrosides* and *gangliosides*.

D-Fructose

D-Fructose is the most important ketohexose. It is also known as *levulose* because aqueous solutions of naturally occurring D-fructose rotate plane-polarized light to the left (*levo* is Latin for "left"). From the third to the sixth carbon, the structure of D-fructose is identical to that of D-glucose.

$$\begin{array}{cc}
^1\text{CH}_2\text{OH} & ^1\text{CHO} \\
^2\text{C}=\text{O} & \text{H}-^2\text{C}-\text{OH} \\
\text{HO}-^3\text{C}-\text{H} & \text{HO}-^3\text{C}-\text{H} \\
\text{H}-^4\text{C}-\text{OH} & \text{H}-^4\text{C}-\text{OH} \\
\text{H}-^5\text{C}-\text{OH} & \text{H}-^5\text{C}-\text{OH} \\
^6\text{CH}_2\text{OH} & ^6\text{CH}_2\text{OH} \\
\text{D-fructose} & \text{D-glucose}
\end{array}$$

(same structure)

D-Fructose is the sweetest tasting sugar (Section 22.9). It is found in many fruits and is present in honey in equal amounts with glucose. D-Fructose is sometimes used as a dietary sugar, not because it has fewer calories per gram than other sugars, but because less is needed for the same amount of "sweetness".

D-Ribose

The three monosaccharides previously discussed in this section have all been hexoses. D-Ribose is a pentose. If carbon number 3 and its accompanying H and OH group were eliminated from the structure of D-glucose, the remaining structure would be that of D-ribose.

```
       ¹CHO                    ¹CHO
   H—²C—OH                H—²C—OH
  [HO—³C—H]                H—³C—OH
   H—⁴C—OH                H—⁴C—OH
   H—⁵C—OH                 ⁵CH₂OH
    ⁶CH₂OH
   D-glucose               D-ribose
```

D-Ribose is a component of a variety of complex molecules including ribonucleic acids (RNAs) and energy-rich compounds such as adenosine triphosphate (ATP).

A compound closely related to D-ribose, 2-deoxy-D-ribose, is also an important component in nucleic acid structures. As its name implies, this compound has one less oxygen atom than D-ribose. On the second carbon atom, deoxyribose has two hydrogen atoms, compared to a H and OH in D-ribose.

```
       ¹CHO                    ¹CHO
  [H—²C—OH]               [H—²C—H]
   H—³C—OH                H—³C—OH
   H—⁴C—OH                H—⁴C—OH
    ⁵CH₂OH                 ⁵CH₂OH
    D-ribose            2-deoxy-D-ribose
```

Deoxyribose is a component of the structures of deoxyribonucleic acid (DNA). DNA molecules control protein synthesis in cells and carry hereditary information (Chapter 26).

22.6 CYCLIC FORMS OF MONOSACCHARIDES

So far in this chapter we have used projection formulas to specify the structures of monosaccharides. These open-chain structures are oversimplifications for aldoses containing more than four carbon atoms and ketoses that have more than five carbon atoms. In these compounds, the open-chain structures exist in equilibrium with two cyclic structures; the cyclic structures are the predominant form in both aqueous solution and solid state.

Let us consider why and how cyclic structures originate for these compounds. Monosaccharides contain two types of reactive functional groups: a carbonyl group (aldehyde or ketone) and numerous —OH (hydroxyl) groups. In Section 18.7, we learned that aldehydes and ketones (carbonyl groups) react readily and reversibly with alcohols (hydroxyl groups) to form hemiacetals or hemiketals, depending on the type of carbonyl group present.

$$R-\underset{H}{\overset{O}{C}} + H-O-R' \longrightarrow R-\underset{H}{\overset{OH}{\underset{|}{C}}}-OR'$$

aldehyde   alcohol   hemiacetal

$$R-\underset{R}{\overset{O}{C}} + H-O-R' \longrightarrow R-\underset{R}{\overset{OH}{\underset{|}{C}}}-OR'$$

ketone   alcohol   hemiketal

(Remember, from Section 18.7, that the identifying characteristic for hemiacetals and hemiketals is an —OH group and an —OR group bonded to the same carbon atom.)

Monosaccharides with a sufficient number of carbon atoms to easily form a ring system form cyclic hemiacetals or cyclic hemiketals through an *intramolecular reaction* in which the carbonyl group and a hydroxyl group within the same molecule react with each other. Let us look at a specific example of this: the cyclization of D-glucose. Figure 22-7 will help us follow the process of intramolecular hemiacetal formation (cyclization) as it occurs in D-glucose.

Structure (a) at the top of Figure 22-7 is the projection formula for D-glucose.

FIGURE 22-7 Formation of the cyclic hemiacetal forms of D-glucose.

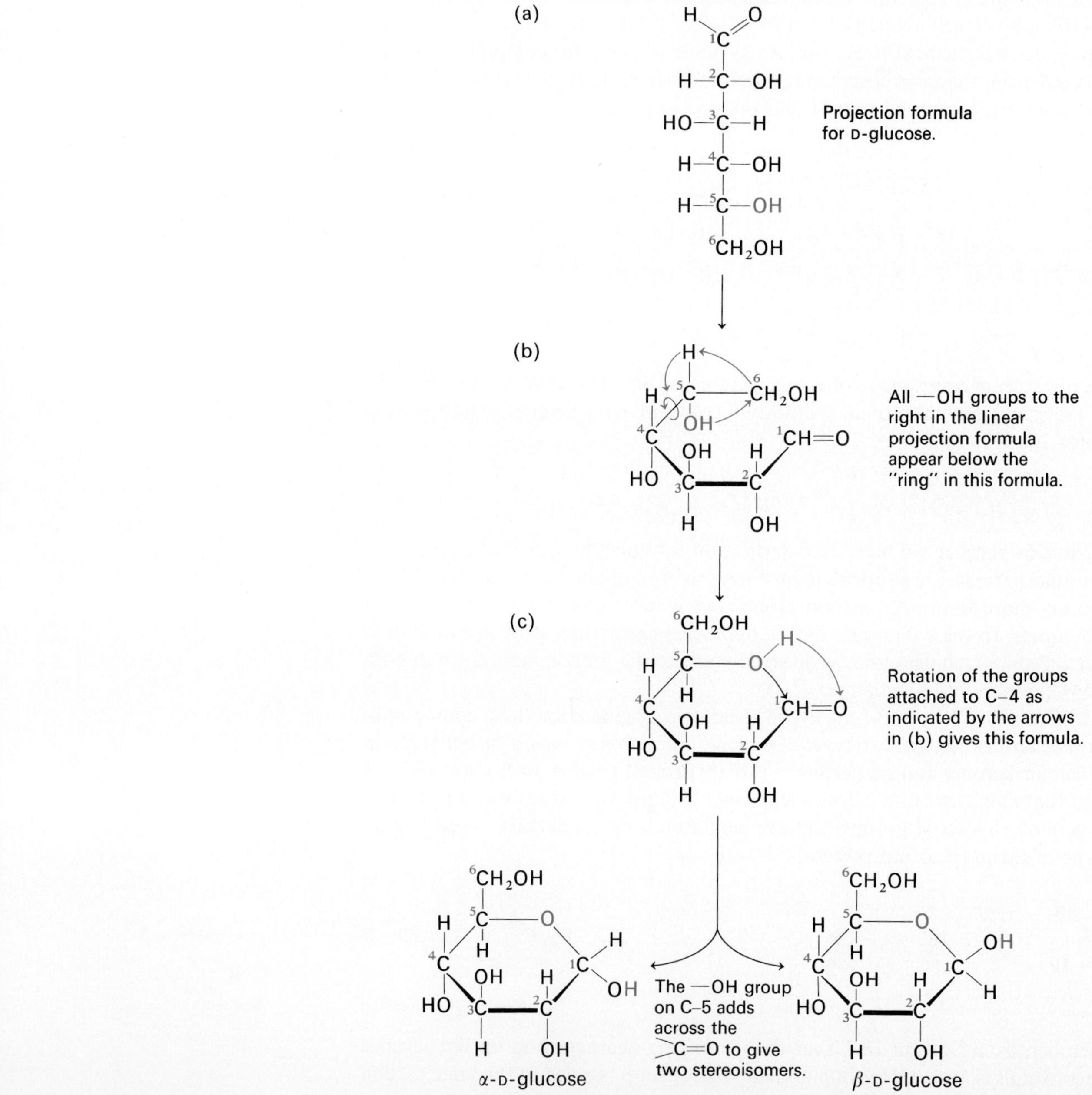

(a) Projection formula for D-glucose.

(b) All —OH groups to the right in the linear projection formula appear below the "ring" in this formula.

(c) Rotation of the groups attached to C-4 as indicated by the arrows in (b) gives this formula.

α-D-glucose ← The —OH group on C-5 adds across the C=O to give two stereoisomers. → β-D-glucose

Structure (b) is an alternate way of drawing this projection formula in which the carbon atoms have locations similar to those found in cyclic compounds. Note that in structure (b), all hydroxyl groups drawn to the right in the linear projection formula—structure (a)—appear below the "ring". Structure (c) is obtained by rotating the groups attached to carbon number four as indicated by the arrows in structure (b). Structure (c) has the atoms and groups in D-glucose in the positions where it is most easy to visualize how intramolecular hemiacetal formation occurs. The intramolecular reaction occurs between the —OH group on carbon number 5 and the carbonyl group (carbon number 1). The net result is addition of the —OH group across the carbon–oxygen double bond. The H of the —OH group bonds to the O atom of the double bond and the O of the —OH group bonds to the C atom of the double bond; the result is a cyclic hemiacetal. The hemiacetal carbon atom is carbon number 1. The ring system is a heterocyclic ring system containing five carbon atoms and one oxygen atom.

Two different cyclic forms (stereoisomers) of D-glucose result from the previously described reaction. They are shown at the bottom of Figure 22–7: α-D-glucose and β-D-glucose. These forms differ from each other in the orientation of the —OH group on the hemiacetal carbon atom. In the α form, the —OH group is on the opposite side of the ring from the CH$_2$OH group at carbon number 5; in the β isomer, both the CH$_2$OH group on carbon number 5 and the —OH group at carbon number 1 are on the same side of the ring.

In the open-chain form of D-glucose—structures (a), (b), and (c) in Figure 22–7—carbon number 1 is not a chiral carbon atom. In the hemiacetal forms of D-glucose this carbon atom is a chiral center; hence, two isomers are possible. Note that the two hemiacetal forms of D-glucose are not enantiomers; they are not mirror images of each other. They differ from each other only in the configuration of groups at carbon number 1. Diastereomers of this type are called *anomers*. **Anomers are diastereomers that differ only in the configuration at carbon number 1.** In cyclic monosaccharide structures, the hemiacetal carbon atom (carbon number 1) is often called the *anomeric carbon atom*.

anomers

Anomers are a subclass of epimers (Section 22.5). Epimers are any pair of diastereomers that differ only in the configuration at a single carbon atom; this can be any chiral carbon atom in the molecule. Anomers must have the difference at carbon number 1.

Both α-D-glucose and β-D-glucose can be isolated as pure compounds from glucose solutions. The properties of these two anomers are distinctly different. The melting point of α-D-glucose is 146°C, and β-D-glucose melts at 150°C. The specific rotation of α-D-glucose is +112° and that of β-D-glucose is +19°.

If either α-D-glucose or β-D-glucose is dissolved in water, the specific rotation gradually changes until a value of +52.7° is reached. This phenomenon is called *mutarotation*. **Mutarotation is the change in optical rotation shown by a solution of an optically active compound.** For D-glucose, mutarotation is the result of the cyclic hemiacetal (either α or β) opening up in solution to give the open-chain form, which then closes again, producing a mixture of the two anomers. The solution is a dynamic equilibrium situation with all three D-glucose forms present.

mutarotation

$$\alpha\text{-D-glucose} \rightleftharpoons \text{open chain} \rightleftharpoons \beta\text{-D-glucose}$$

The specific rotation for the equilibrium mixture of the two anomers is the observed 52.7°. Figure 22–8 (p. 542) shows diagrammatically the conversion process of α-D-glucose to β-D-glucose in solution (or vice versa).

When equilibrium is established, 63% of the molecules are β-D-glucose, 37% are α-D-glucose, and less than 0.01% are in the open-chain (aldehyde) form.

FIGURE 22-8 Equilibrium between anomeric forms of D-glucose.

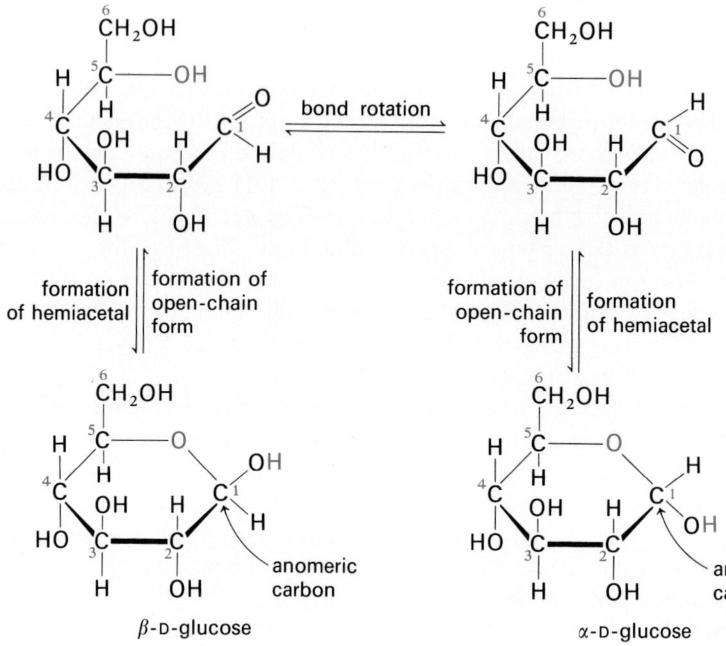

The difference between the α and β forms of D-glucose may seem small, but this difference is of utmost importance in glucose chemistry. In many reactions, only one form (α or β) of D-glucose is a suitable reactant to obtain a desired product.

In Section 22.4, we learned that carbon number 5 of the open-chain form of glucose determines D or L configuration. This carbon atom has the same function in the cyclic forms of glucose. The configuration at carbon number 5 determines whether carbon number 6 will be above the ring (D form) or below the plane (L form).

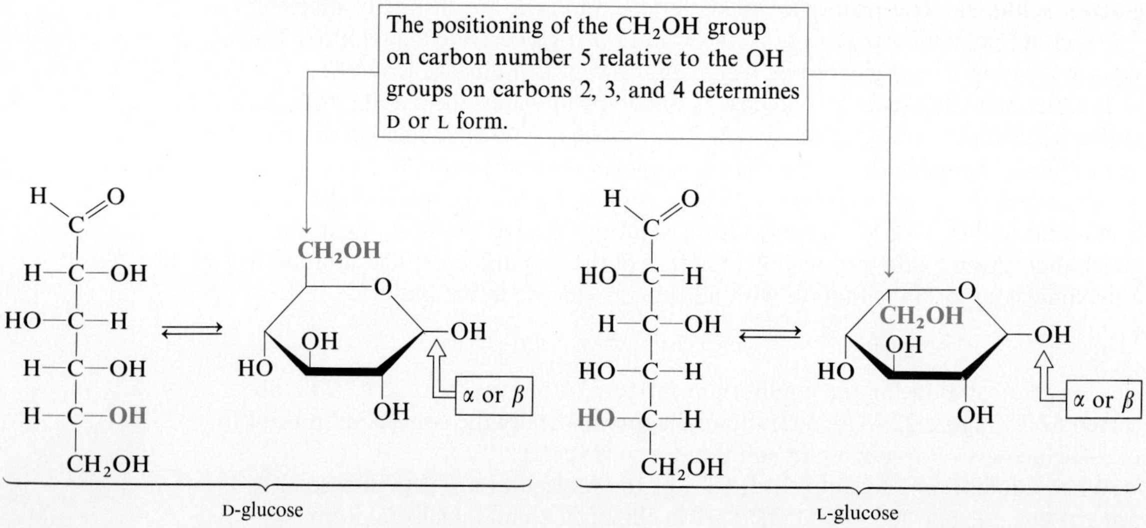

The positioning of the CH$_2$OH group on carbon number 5 relative to the OH groups on carbons 2, 3, and 4 determines D or L form.

In the preceding diagrams of D-glucose and L-glucose, note how the OH group on carbon number 1 has been drawn (it is in a horizontal position). If it is immaterial whether the molecule is in the α or β form (which is the case here), the OH group can be written as shown rather than in a "up" or "down" position. In summary, the positioning of groups around carbon number 1 of glucose determines α or β form and the positioning of groups around carbon number 5 determines D or L form.

So far, our discussion of cyclic forms of monosaccharides has been focused exclusively on glucose. This does not imply that glucose is the only monosaccharide that exists in cyclic hemiacetal forms. The principles we have discussed concerning the cyclic forms of glucose also apply to many other monosaccharides.

Like D-glucose, all other monosaccharides that undergo intramolecular hemiacetal or hemiketal formation show mutarotation when samples of the α or β isomer are dissolved in water; they form equilibrium mixtures of α, β, and open-chain forms in solution, with the cyclic forms predominating. The β isomer is not always the predominant isomer, as was the case for D-glucose. For example, D-mannose is a monosaccharide where the α isomer is the predominant form in solution.

22.7 REACTIONS OF MONOSACCHARIDES

In their open-chain form, monosaccharides contain two reactive functional groups, a carbonyl group and —OH groups. The presence of these functional groups leads to an extensive reaction chemistry for monosaccharides. In this section, we will consider four important types of reactions monosaccharides undergo. They are:

1. Reaction with alcohols to give acetals and ketals.
2. Oxidation with weak oxidizing agents to give carboxylic acids.
3. Reduction of either the carbonyl or hydroxyl group.
4. Reaction with acids to form esters.

Acetal and Ketal Formation

In Section 18.7, we discussed acetal and ketal formation. We recall that acetals and ketals contain two —OR groups attached to the same carbon atom. These compounds form when hemiacetals or hemiketals react with alcohols. The reaction is acid catalyzed, and the equilibrium can be shifted to the right either by addition of excess alcohol or removal of the water formed.

$$\underset{\text{hemiacetal}}{R-\underset{\underset{H}{|}}{\overset{\overset{OH}{|}}{C}}-OR' + \underset{}{\overset{\overset{H}{|}}{O}}-R'} \overset{H^+}{\rightleftharpoons} \underset{\text{acetal}}{R-\underset{\underset{H}{|}}{\overset{\overset{OR'}{|}}{C}}-OR' + H_2O}$$

$$\underset{\text{hemiketal}}{R-\underset{\underset{R}{|}}{\overset{\overset{OH}{|}}{C}}-OR' + \underset{}{\overset{\overset{H}{|}}{O}}-R'} \overset{H^+}{\rightleftharpoons} \underset{\text{ketal}}{R-\underset{\underset{R}{|}}{\overset{\overset{OR'}{|}}{C}}-OR' + H_2O}$$

Monosaccharides that have hemiacetal or hemiketal forms can also form acetals or ketals through reaction with an alcohol. As shown in Figure 22–9, (p. 544), methanol reacts with D-glucose to give an acetal that can exist in either an α or β form.

FIGURE 22–9 The formation of anomeric acetals from the reaction between D-glucose and methyl alcohol.

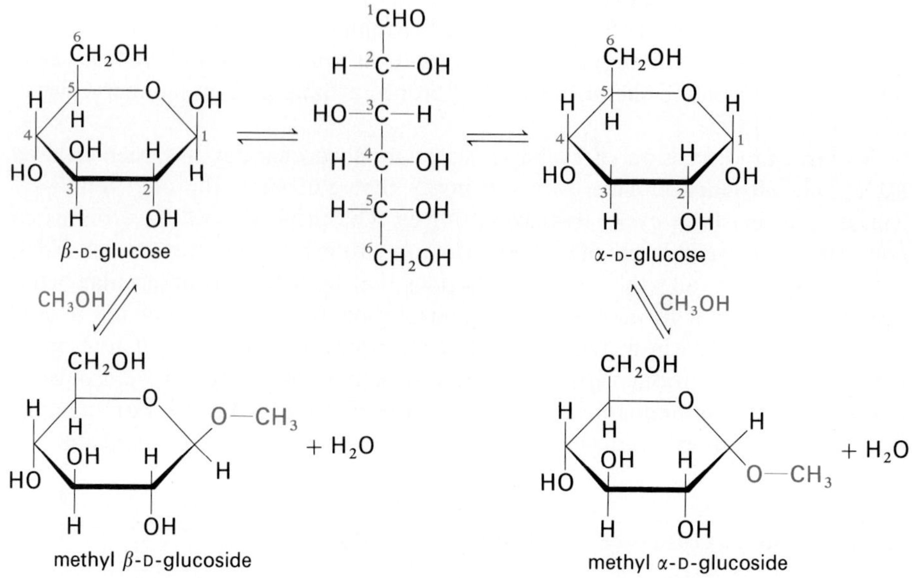

The acetals and ketals of monosaccharides (as well as other carbohydrates) are called *glycosides*. A **glycoside** *is an acetal or ketal that was formed from a cyclic form of a carbohydrate.* As shown in Figure 22–9, which illustrates acetal formation from the reaction of D-glucose and methyl alcohol, glycosides can exist either in an α form or a β form.

Glycosides are stable in neutral or basic solutions. This means that, unlike the rings of cyclic hemiacetals, the rings of glycosides cannot open and close. Thus, mutarotation does not occur in solution for glycosides. Glycosides are hydrolyzed in acid solution; this is the reverse of their formation reaction.

In later sections of this chapter, we will find that di- and polysaccharides are special kinds of glycosides.

Glycosides formed from glucose are often called *glucosides*, a name that combines the prefix of the name of the monosaccharide with the suffix of glycoside.

Oxidation

In Sections 18.5 and 18.8, we learned that aldehydes are easily oxidized to carboxylic acids. Aldoses, which have an aldehyde group in the open-chain form, are also easily oxidized to carboxylic acids. Tollens', Fehling's, and Benedict's solutions (Section 18.8), which are weak oxidizing agents, will oxidize aldoses.

Tollens', Fehling's, and Benedict's solutions also oxidize ketoses to carboxylic acids. This should be somewhat surprising to you because in Section 18.8, we emphasized that ketones do not undergo oxidation to carboxylic acids.

The explanation for this apparent inconsistency involves the concept of tautomers and the process of tautomerization. **Tautomers** *are isomers whose structures differ both in the point of attachment of a hydrogen atom and the location of a double bond. Tautomers exist in easy and rapid equilibrium with each other.*

Ketoses undergo tautomerization when treated with base (or basic reagents). Let us follow this process as it occurs for the ketohexose *fructose*. Three species are involved in the equilibrium process: a keto form, an enol form, and an aldo form.

```
 ¹CH₂OH              H                     H
  |                  |                     |
 ²C=O               ¹C—OH                 ¹C=O
  |                  ||                    |
HO—³C—H     ⇌      ²C—OH       ⇌       H—²C—OH
  |                  |                     |
 H—⁴C—OH           HO—³C—H               HO—³C—H
  |                  |                     |
 H—⁵C—OH           H—⁴C—OH               H—⁴C—OH
  |                  |                     |
 ⁶CH₂OH            H—⁵C—OH               H—⁵C—OH
                     |                     |
                   ⁶CH₂OH                 ⁶CH₂OH
 keto form         enol form             aldo form
```

These three species differ from each other only in the location of a hydrogen atom and the position of a double bond.

In going from the keto form to the enol form, a hydrogen atom on carbon number 1 is transferred to the carbonyl oxygen atom. In order for each carbon atom to maintain four bonds, the carbon–oxygen double bond becomes a single bond and the carbon–carbon single bond between carbons 1 and 2 becomes a double bond.

```
        H                         H
        |                         |
   H—¹C—OH                        C—OH
        |                         ||
       ²C=O           ⇌           C—OH
        |                         |
      HO—³C—H                   HO—C—H
        |                         |
      H—⁴C—OH                   H—C—OH
        |                         |
      H—⁵C—OH                   H—C—OH
        |                         |
       ⁶CH₂OH                    CH₂OH
     keto form                 enol form
```

Because the product of this rearrangement has structural features of both an alkene (-en) and an alcohol (-ol), it is called an *enol*.

In going from the enol form to the aldo form, the H atom of the —OH group on carbon number 1 of the enol form is transferred to carbon number 2 and the double bond moves to give a carbonyl group on carbon number 1.

```
        H                         H
        |                         |
       ¹C—OH                     ¹C=O
        ||                        |
       ²C—OH           ⇌       H—²C—OH
        |                         |
      HO—³C—H                   HO—C—H
        |                         |
      H—⁴C—OH                   H—C—OH
        |                         |
      H—⁵C—OH                   H—C—OH
        |                         |
       ⁶CH₂OH                    CH₂OH
     enol form                 aldo form
```

545

Carbohydrates

This process of tautomerization explains why fructose (a ketose) can be oxidized to a carboxylic acid. As soon as the ketose comes in contact with a basic solution, some aldose is formed through tautomerization. Although tautomerization occurs only to a very limited extent initially (less than 1% of the molecules are in the aldo form), this is sufficient for oxidation to occur. As the aldose is oxidized, more and more of the ketose converts to the aldo form via the enol intermediate (Le Châtelier's principle—Section 9.6) and eventually all of the ketose sample is oxidized.

Example 22.2

The following two isomeric compounds (a ketose and an aldose) can be interconverted. Draw the structure of their common intermediate, which is an enol.

$$
\begin{array}{c}
H \\
| \\
H-{}^1C-OH \\
| \\
{}^2C=O \\
| \\
H-{}^3C-OH \\
| \\
{}^4CH_2OH
\end{array}
\quad \text{and} \quad
\begin{array}{c}
H \\
| \\
{}^1C=O \\
| \\
H-{}^2C-OH \\
| \\
H-{}^3C-OH \\
| \\
{}^4CH_2OH
\end{array}
$$

Solution

The first compound is a ketose. The transfer of a hydrogen atom from carbon number 1 of the ketose to the carbonyl oxygen atom results in a change in the position of the double bond, giving the enol form.

$$
\begin{array}{c}
H \\
| \\
H-C-OH \\
| \\
C=O \\
| \\
H-C-OH \\
| \\
CH_2OH \\
\text{ketose}
\end{array}
\rightleftharpoons
\begin{array}{c}
H \\
| \\
C-OH \\
|| \\
C-OH \\
| \\
H-C-OH \\
| \\
CH_2OH \\
\text{enol form}
\end{array}
$$

The second compound is an aldose. Here, the transfer of the hydrogen atom from carbon number 2 to the carbonyl carbon atom results in a change in double bond position, giving the same enol form as obtained from the ketose.

$$
\begin{array}{c}
H \\
| \\
C=O \\
| \\
H-C-OH \\
| \\
H-C-OH \\
| \\
CH_2OH \\
\text{aldose}
\end{array}
\rightleftharpoons
\begin{array}{c}
H \\
| \\
C-OH \\
|| \\
C-OH \\
| \\
H-C-OH \\
| \\
CH_2OH \\
\text{enol form}
\end{array}
$$

We now come back to the reason we discussed the fact that ketoses tautomerize in basic solution. Tollens', Fehling's, and Benedict's solutions are all basic solutions. Thus, these solutions will oxidize not only aldoses, but also ketoses (because of tautomerization) to carboxylic acids.

Carbohydrates that react with Tollens', Fehling's, and Benedict's solutions are called *reducing sugars*. **Reducing sugars** *are carbohydrates that can be easily oxidized*. All monosaccharides are reducing sugars. The term *reducing* refers to the effect of the carbohydrate on the oxidizing agent; it reduces the oxidizing agent (Section 11.1). For example, the Cu^{2+} ions in Benedict's and Fehling's solutions are reduced to Cu^+ ions; in Tollens' solution, the Ag^+ ions are reduced to Ag (Section 18.8).

Clinically, oxidation can be used to detect the presence of glucose in urine. For example, using Benedict's solution, we observe that if no glucose is present in the urine (a normal condition), the Benedict's solution remains blue. The presence of glucose is indicated by the formation of a red precipitate (Section 18.8). Testing for the presence of glucose in urine is such a standard laboratory procedure that much effort has been put into the development of easy-to-use test methods. Plastic strips impregnated with all the necessary chemicals are now available and in common use. The strip is dipped into a urine sample and the color change that occurs on the strip can be compared to a standard to find the glucose concentration (see Figure 22–10).

reducing sugars

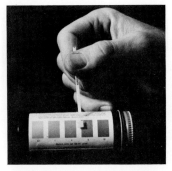

FIGURE 22–10 The glucose content of urine can be obtained by dipping a plastic strip treated with oxidizing agents in the sample and noting the color change of the strip. The color change is compared to a color chart that indicates glucose concentration.

Reduction

The carbonyl group present in a monosaccharide (either an aldose or a ketose) can be reduced to a hydroxyl group using hydrogen as the reducing agent. In Section 18.6, we discussed the fact that carbonyl groups can be reduced using hydrogen. For aldoses and ketoses, the product of the reduction is the corresponding polyhydroxyalcohol, which is sometimes called a sugar alcohol. For example, reduction of D-glucose gives D-sorbitol, which is also called D-glucitol.

$$
\begin{array}{c}
\text{H} \diagdown \!\!\!\! \diagup \text{O} \\
\text{C} \\
\text{H}-\text{C}-\text{OH} \\
\text{HO}-\text{C}-\text{H} \\
\text{H}-\text{C}-\text{OH} \\
\text{H}-\text{C}-\text{OH} \\
\text{CH}_2\text{OH} \\
\text{D-glucose}
\end{array}
\xrightarrow[\text{catalyst}]{H_2}
\begin{array}{c}
\text{CH}_2\text{OH} \\
\text{H}-\text{C}-\text{OH} \\
\text{HO}-\text{C}-\text{H} \\
\text{H}-\text{C}-\text{OH} \\
\text{H}-\text{C}-\text{OH} \\
\text{CH}_2\text{OH} \\
\text{D-sorbitol}
\end{array}
$$

Hexahydric alcohols such as D-sorbitol have properties similar to those of the trihydric alcohol *glycerol* (Section 17.7). These alcohols are used as moisturizing agents in foods and cosmetics because of their affinity for water.

D-Sorbitol is used as a sweetening agent in foods for diabetics who must restrict their intake of glucose. It is also one of the reactants in the commercial preparation of vitamin C. D-Sorbitol and D-xylitol, which is obtained from the reduction of the aldopentose *xylose* (Figure 22–4), are also used as sweetening agents in chewing gum. A major reason for their use in this context is the fact that bacteria that cause tooth decay cannot use them as food sources, as they can glucose and many other monosaccharides.

The reduction of D-glucose to D-sorbitol is an important process in the human body. The source of hydrogen for this reduction is the reducing agent NADPH (nicotinamide adeninedinucleotide phosphate), a compound that will be discussed in Section 28.4.

Within the human body, enzymes (biological catalysts—Chapter 25) can selectively direct the reduction of hydroxyl groups on a monosaccharide without affecting the carbonyl group. This process converts a hydroxyl group to a hydrogen atom, producing a deoxy-monosaccharide. For example, D-ribose can be converted to 2-deoxy-D-ribose in this manner.

$$\text{D-ribose} \xrightarrow{\text{enzyme-controlled reduction}} \text{2-deoxy-D-ribose}$$

Both D-ribose and 2-deoxy-D-ribose are essential components of nucleic acids (Chapter 26) as was previously mentioned in Section 22.5.

Ester Formation

The hydroxyl groups of a monosaccharide can also react with acids to form esters (Section 19.7). The first step in glucose metabolism within the human body involves ester formation. Cellular enzymes convert glucose to glucose-6-phosphate. Adenosinetriphosphate (ATP) is a highly energetic phosphorylation agent that provides the phosphate group in this reaction.

$$\text{glucose} + \text{ATP} \xrightarrow{\text{enzyme}} \text{glucose-6-phosphate} + \text{ADP}$$

Inorganic esters such as glucose-6-phosphate are quite stable in aqueous solution and play important roles in the metabolism of sugars (Chapter 29).

22.8 DISACCHARIDES

We saw in Section 22.7 that monosaccharides can be converted to glycosides through reaction with an alcohol, ROH. When the ROH molecule is a second monosaccharide, the glycoside product is a disaccharide. **Disaccharides** *are glycosides formed from two monosaccharides, in which one acts as the hemiacetal or hemiketal and the other acts as the alcohol.* The bond that links the two monosaccharide units together is called a *glycosidic bond*. A **glycosidic bond** *is the covalent bond between two monosaccharide units in a disaccharide (or polysaccharide).*

The glycosidic bond between the two monosaccharides in a disaccharide can be further characterized both according to the position numbers of the carbon atoms of the monosaccharides that are linked and the stereochemistry (α or β) of the linkage. Consider the following equation that illustrates the formation of a disaccharide containing an $\alpha(1 \rightarrow 4)$ glycosidic bond.

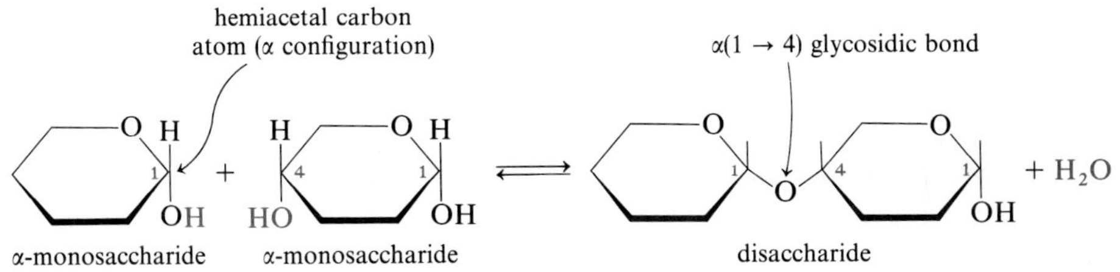

In this equation, all groups on the monosaccharides have been omitted except those involved in α or β designation and in the glycosidic bond formation. Note that a molecule of water is produced as a glycosidic bond forms. The (1 → 4) linkage that holds these two monosaccharides together is designated α because the —OH group of the hemiacetal was below the ring (α isomer). If this —OH group had been above the ring (β isomer), then a β(1 → 4) glycosidic bond would have resulted.

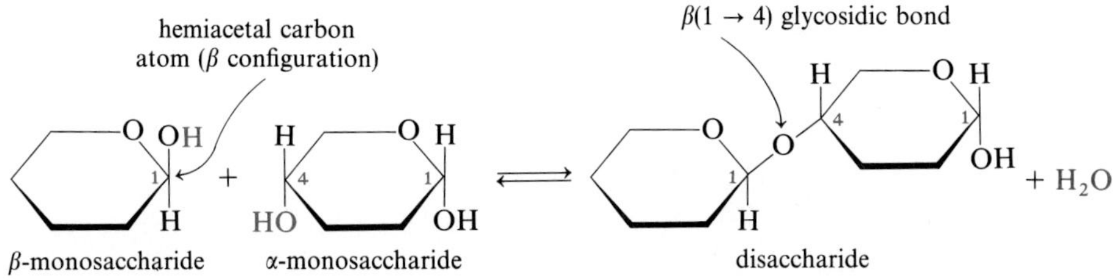

FIGURE 22–11 Some of the common types of glycosidic linkages found in disaccharides and polysaccharide molecules.

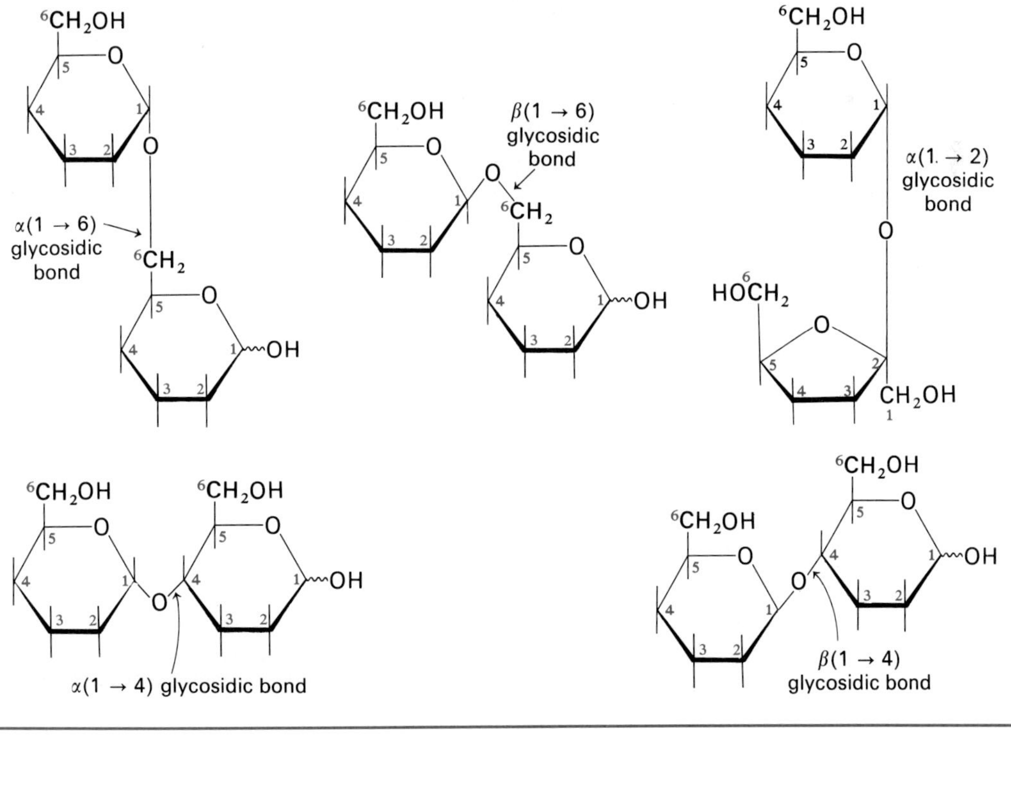

As we shall see later in this section, the distinction between an α(1 → 4) and a β(1 → 4) glycosidic linkage has important biological ramifications. Other common glycosidic linkages are α(1 → 6), β(1 → 6), and α(1 → 2), examples of which are shown in Figure 22–11 (p. 549).

We will now consider characteristics and properties of four specific disaccharides: maltose, cellobiose, lactose, and sucrose.

Maltose

Maltose is composed of two molecules of D-glucose that are bonded to each other by an α(1 → 4) glycosidic linkage.

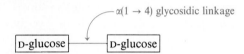

As shown in the following structural formulas for maltose, the D-glucose unit on the right still has a hemiacetal carbon atom; hence α and β isomers are possible.

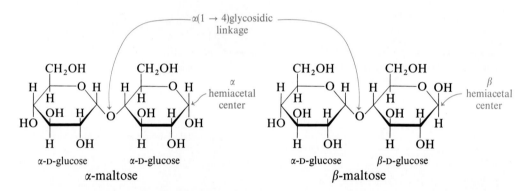

In solution, both forms of maltose exist in equilibrium with each other; mutarotation (Section 22.6) occurs. Note that the glycosidic linkage is α(1 → 4) for both α-maltose and β-maltose. The prefixes α and β refer to the configuration at the hemiacetal center; they say nothing about the type of glycosidic linkage present. The presence of the hemiacetal center in maltose makes it a reducing sugar (Section 22.7).

Maltose, which is also called *malt sugar*, does not occur abundantly in nature, except in germinating grain. Maltose is hydrolyzed by the enyzme *maltase* in the digestive tract to give two molecules of glucose. This reaction is the reverse of the formation reaction for maltose.

$$\text{maltose} + \text{H}_2\text{O} \xrightarrow{\text{maltase}} \text{glucose} + \text{glucose}$$

Maltose is an ingredient in formulas for feeding infants; it is an ingredient in corn syrup; and it is used in the production of beer.

Cellobiose

Cellobiose, like maltose, is composed of two molecules of D-glucose. The difference between cellobiose and maltose is in the glycosidic linkage, which is β(1 → 4), rather than α(1 → 4).

[Structural diagrams of α-cellobiose and β-cellobiose, showing β(1 → 4) glucosidic linkage between two D-glucose units]

Like maltose, cellobiose exists in solution as an equilibrium mixture of the α and β forms of the substance and it is also a reducing sugar.

Cellobiose is obtained from the hydrolysis of cellulose (a polysaccharide we will describe in Section 22.10). In contrast to maltose, the human digestive tract does not have enzymes that can catalyze the hydrolysis of cellobiose, so we cannot digest this substance. This apparently simple difference in configuration [α(1 → 4) versus β(1 → 4)] illustrates the remarkable difference in the specificity of enzymes.

Lactose

In maltose and cellobiose, the two units of the disaccharide were identical. This does not need to be the case for a disaccharide. The disaccharide *lactose* is made up of a D-galactose unit and a D-glucose unit linked by a β(1 → 4) glycosidic bond.

[Structural diagrams of α-lactose and β-lactose, showing β(1 → 4) glycosidic linkage between D-galactose and D-glucose]

Lactose is the major sugar found in milk. This accounts for its common name, *milk sugar*. Enzymes in mammalian mammary glands take glucose from the bloodstream and synthesize lactose in a two-step process. Epimerization of glucose yields galactose and then the $\beta(1 \rightarrow 4)$ linkage forms when carbon number 1 of galactose bonds to carbon number 4 of another glucose molecule.

The glucose hemiacetal center is unaffected when galactose bonds to glucose in the formation of lactose, and so lactose is a reducing sugar (the glucose ring can open to give an aldehyde). Lactose can also undergo mutarotation in solution.

Lactose is an important ingredient in commercially produced infant formulas that are designed to simulate mother's milk. When milk sours, it is caused by the conversion of lactose to lactic acid by bacteria in the milk. Pasteurization of milk is a quick-heating process that kills most of the bacteria and retards the souring process.

Like other disaccharides, lactose can be hydrolyzed by acid or enzyme catalysts, forming equimolar mixtures of galactose and glucose. The galactose so produced is then converted to glucose by other enzymes.

There are two methods or pathways by which free galactose in the human body is converted to glucose. Only one method is functional during infancy. Later in life, a second pathway also begins to function. Some infants are born with a genetic disease called *galactosemia*, which is an inability to convert galactose to glucose. If this disease is not detected, milk or other foods containing milk causes vomiting and diarrhea; in extreme cases it causes cataract formation and even mental retardation. Excluding milk from the diet by feeding the infant formulas based on other disaccharides usually eliminates the problem. In later years, as enzymes in the second pathway are produced, the problem usually disappears and milk can again be included in the diet.

Another type of lactose problem is *lactose intolerance*, a disease in which people lack the enzyme *lactase*, which is needed to hydrolyze lactose to galactose and glucose. Deficiency of lactase can be caused by a genetic defect, physiological decline with age, or injuries to the mucosa lining of the intestines. The consequences of an inability to hydrolyze lactose in the upper small intestine are an inability to absorb lactose and the bacterial fermentation of ingested lactose further along in the intestinal tract. Bacterial fermentation results in the production of gas (distension of the gut) and osmotically active solutes that draw water into the intestines, resulting in diarrhea.

The level of the enzyme lactase in humans varies with age and race. Most children have sufficient lactase during the early years of their life when milk is a much needed source of calcium in their diet. In adulthood, the enzyme level decreases and lactose intolerance results. This explains the change in milk-drinking habits of many adults. Some experts estimate that as many as one out of three adult Americans suffer a degree of lactose intolerance.

The amount of lactose in milk varies with species. Human mother's milk obtained by nursing infants (see Figure 22–12) contains 7–8% lactose; this is almost double the 4–5% lactose found in cow's milk.

FIGURE 22–12 Human mother's milk has a lactose content almost double that of cow's milk.

Sucrose

Sucrose is the carbohydrate (disaccharide) most people refer to simply as *sugar*; it is our common table sugar. Although sucrose is found in most fruits and vegetables, two plants are the dominant sources of sucrose: sugar cane and sugar beets (see Figure 22–13). Sugar cane contains up to 20% sucrose by weight and sugar beets contain up to 17% sucrose by weight.

Average per capita consumption of sucrose in the United States is approximately 150 g per day, which provides between one-fourth and one-third of the

FIGURE 22–13 Much of the sucrose (table sugar) used in the United States is obtained from sugar cane plants that grow in semitropical climates.

calories consumed daily. These values could be reduced as artificial low-calorie sweeteners (Section 22.9) become more popular. Much of the sucrose we consume is sugar that is added to foods rather than sugar that is naturally present. This can be seen from the numbers in Table 22–1, which lists the amount of sugar added to various foods.

The two monosaccharide units found within a sucrose molecule are D-glucose and D-fructose. Sucrose's structure is quite different from those of the three disaccharides previously discussed in this section because the two components of sucrose are linked by an $\alpha(1 \to 2)$ glycosidic bond. Maltose, cellobiose, and lactose all have $(1 \to 4)$ linkages in either α or β configurations.

TABLE 22–1 Amounts of Sugar Added to Selected Foods

Canned or Packaged Food	Grams of Sugar (serving size)	Percent of Total Mass	Percent of Calories
beverages			
cola type	37 (12 oz)	10	100
fruit juice drink	20 (6 oz)	12	85
Kool-Aid	25 (8 oz)	11	98
desserts			
peaches, light syrup	9 ($\frac{1}{2}$ cup)	7	48
heavy syrup	16 ($\frac{1}{2}$ cup)	12	61
pudding, starch type	25 (5 oz)	18	50
gelatin dessert	26 (5 oz)	18	97
milk chocolate candy	12 (1 oz)	44	32
brownies	12 (each)	50	30
coconut cream pie	24 ($\frac{1}{4}$ pie)	68	66
ready-to-eat cereals			
corn or wheat flakes	23 (1 oz)	7–11	11–15
presweetened, flavored	8–14 (1 oz)	29–44	26–45
100% natural	6 (1 oz)	19	15

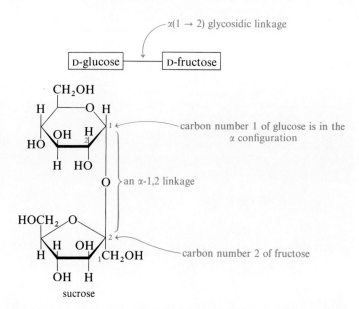

The linking of D-glucose (an aldose) with D-fructose (a ketose) produces both an acetal and a ketal functional group. No hemiacetal or hemiketal center is present in the molecule because the α(1 → 2) linkage involves the reducing end of both monosaccharides. Hence, sucrose is a nonreducing sugar (the first example we have encountered) and it does not undergo mutarotation in solution.

Sucrose can be hydrolyzed with acid or enzymatic catalysts to form an equimolar mixture of glucose and fructose. This mixture is called *invert sugar*.

The name *invert sugar* is appropriate because the initial positive value of specific rotation for sucrose (+65.5°) changes to a negative value (−19.9°) for a one-to-one mixture of glucose and fructose.

When sucrose is cooked with acid-containing foods such as fruits or berries partial hydrolysis takes place, forming some invert sugar. Jams and jellies prepared in this manner are actually sweeter than the pure sucrose added to the original mixture because invert sugar has a sweetness greater than that of sucrose (Section 22.9).

22.9 ARTIFICIAL SWEETENERS

Because of the high caloric value of sucrose, it is often difficult to satisfy a demanding "sweet tooth" without adding a few pounds or inches to the waistline. To combat this problem, scientists have sought to find artificial sweeteners that have little or no caloric value and offer sweet alternatives to sugar. Consumers for more than two decades have been in the middle of a seemingly unending controversy over three well-known artificial sweeteners: saccharin, sodium cyclamate, and aspartame. These artificial sweeteners have received wide publicity over the years and have been subjected to long review, scrutiny, and debate by industry, scientists, and the Federal Drug Administration (FDA). The structures of these three compounds are as follows.

saccharin sodium cyclamate aspartame

Saccharin, the sweetest of the three, contains no calories and has been around for the longest period of time. Discovered in 1879, saccharin was used initially as both an antiseptic and a food preservative. Soon diabetics began to use it, and, from the turn of the century until the 1950s, saccharin dominated the market.

Sodium cyclamate quickly became popular after it was approved for commercial use in 1951. Also having no caloric value, sodium cyclamate was widely used in canned fruits, chewing gum, toothpastes, and mouthwashes, as well as many other foods.

During the 1960s, sales of both saccharin and sodium cyclamate increased with the rising demand for diet soft drinks. Although saccharin is much sweeter, mixtures of the two were often sold because sodium cyclamate cut the aftertaste of saccharin. In 1970, sodium cyclamate was banned when questions about its cancer-causing potential came to light. By the late 1970s, the only available sweetener, saccharin, was being consumed at the rate of 6 million pounds annually (the majority of it in soda pop). During the late 1970s, warning labels about the potential health hazards of saccharin appeared in grocery stores because certain studies showed a possibility that saccharin might cause cancer. Although unproven by later studies, this possible link cast a shadow of doubt that clouded the economic future of saccharin.

Aspartame (Nutra-Sweet) came into the marketplace in 1981. It was accidentally discovered in 1965 by a research team working on ulcer drugs. It took 15 years for it to acquire all of the necessary approvals for marketing. Aspartame has the same food value (4 calories per gram) as sucrose, but it is 180 times sweeter than sucrose and it has no bitter aftertaste like saccharin. Aspartame has quickly found

TABLE 22–2 Comparative Sweetness of Selected Sugars and Artificial Sweeteners

Sugar or Artificial Sweetener	Sweetness Relative to Sucrose	Type
lactose	0.16	disaccharide
galactose	0.32	monosaccharide
maltose	0.33	disaccharide
glucose	0.74	monosaccharide
sucrose	1.00	disaccharide
invert sugar	1.25	mixture of glucose and fructose
fructose	1.73	monosaccharide
sodium cyclamate	30	artificial sweetener
aspartame	180	artificial sweetener
saccharin	450	artificial sweetener

its way into almost every diet food on the market today. So little aspartame is needed to sweeten foods that there is less than one calorie per serving.

The safety of aspartame lies with its hydrolysis products: the amino acids aspartic acid and phenylalanine. These amino acids are identical to those obtained from digestion of proteins (Chapter 24) and thus are considered safe. The only danger of aspartame is that it contains phenylalanine, an amino acid that can lead to mental retardation among young children suffering from PKU (phenylketonurea, which is discussed in Chapter 32). Warning labels on all products containing aspartame warn phenylketouretics of this potential danger.

The sweetness of sugars and artificial sweeteners is measured on a relative scale that compares their sweetness to that of sucrose, which is assigned a value of 1.00. Sweetness values for common sugars and artificial sweeteners are given in Table 22–2.

22.10 POLYSACCHARIDES

polysaccharide

A **polysaccharide** is *a polymer containing many monosaccharide units linked to each other by glycosidic bonds.* In some polysaccharides, the monosaccharide units are linked together to form a linear (unbranched) chain. In other polysaccharides, extensive chain branching occurs (see Figure 22–14). The number of monomer

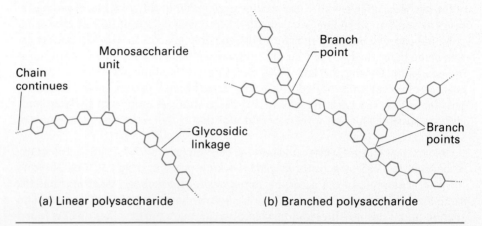

FIGURE 22–14 Monosaccharide units in a polysaccharide can form linear or branched-chain structures.

units in a polysaccharide chain varies from a few hundred up to 12–15 thousand, depending on the identity of the polysaccharide.

Unlike monosaccharides and most disaccharides, polysaccharides are not sweet and do not undergo mutarotation in solution. Polysaccharides do not give positive Tollens', Benedict's, or Fehling's tests (Section 22.7). They have limited water solubility because of their size. However, the —OH groups present can individually become hydrated by water molecules. The result is usually a thick colloidal suspension (Section 8.19) of the polysaccharide in water. Polysaccharides are often used as thickening agents in sauces, gelatins, desserts, and gravy.

Although many naturally occurring polysaccharides are known, we will focus our attention on three of them: cellulose, starch, and glycogen. All of these substances play vital roles in living systems; both cellulose and starch are important in plants and glycogen is a key substance in animals and humans.

Before going into details of structure for these substances, we note a common characteristic: when hydrolyzed (broken up into monosaccharide units), all three yield D-glucose as the *only* product. This means that these polysaccharides are polymers containing only glucose units.

Cellulose

Cellulose is the most abundant polysaccharide. It is the structural component of the cell walls of plants. It is estimated that approximately half of all the carbon atoms in the plant kingdom are contained in cellulose molecules. Structurally, cellulose is a linear (unbranched) D-glucose polymer in which the glucose units are linked by $\beta(1 \rightarrow 4)$ glycosidic bonds.

Typically, cellulose chains contain about 5,000 glucose units, which gives macromolecules with molecular weights of about 400,000 amu. The long chains usually line up side by side and are held together by hydrogen bonding between chains. This results in bundles of cellulose chains called *fibrils*. This structural feature makes cellulose fibrous, tough, and insoluble in water. Figure 22–15 is an electron micrograph picture of cellulose fibers. In addition to the hydrogen bonding present between cellulose chains, a non-carbohydrate, glue-like substance called *lignin* also helps hold the fibers together to give even greater strength.

Cotton is almost pure cellulose (95%) and wood is about 50% cellulose. Many important commercial products result from the chemical treatment of cellulose. The treatment of wood with various chemicals results in paper and the paper products industry. Rayon fiber is made from treating cellulose with sodium hydroxide (NaOH) and carbon disulfide (CS_2). Treatment with nitric acid produces cellulose nitrate, which is used in smokeless gun powder. Cellulose acetate is the raw material for the film used in the motion picture industry.

FIGURE 22–15 Electron micrograph of cellulose fibers.

Controlled laboratory hydrolysis of cellulose yields first the disaccharide cellobiose (Section 22.8) and finally the monosaccharide D-glucose.

$$\text{cellulose} \longrightarrow \text{cellobiose} \longrightarrow \text{D-glucose}$$
$$\text{(polysaccharide)} \quad \text{(disaccharide)} \quad \text{(monosaccharide)}$$

Even though it is a source of glucose, cellulose is not a source of nutrition for human beings. As we mentioned in Section 22.8, humans lack the enzymes capable of catalyzing the hydrolysis of $\beta(1 \rightarrow 4)$ linkages. Even grazing animals lack these necessary enzymes for cellulose digestion. However, the intestinal tracts of animals such as horses, cows, and sheep contain microorganisms (bacteria) that produce *cellulase*, an enzyme that can hydrolyze $\beta(1 \rightarrow 4)$ linkages and produce free glucose from cellulose. Thus, grasses and other plant materials are a source of nutrition for grazing animals. The intestinal tracts of termites contain the same microorganisms, enabling termites to use wood as their source of food. Microorganisms in the soil can also metabolize cellulose, thus allowing the biodegredation of dead plants.

Most food of plant origin contains some cellulose. Even though we get no nourishment from it, it still plays an important role in our diet. Cellulose provides the digestive tract with "bulk" or "fiber" that helps move food through the intestinal tract and facilitates the excretion of solid wastes. Cellulose absorbs a lot of water, leading to softer stools and frequent bowel action. Links have been found between the length of time stool spends in the colon and possible colon cancer, so many diet-conscious professionals recommend a high-fiber content in the foods we eat. Bran, celery, green vegetables, and fruits are rich in dietary fiber.

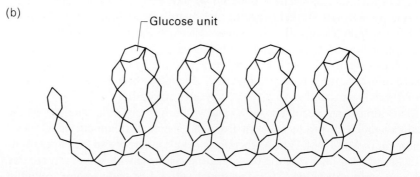

FIGURE 22-16 Two perspectives of the structure of amylose. (a) Glucose units are joined together via $\alpha(1 \rightarrow 4)$ linkages. (b) The amylose chain assumes a helical configuration.

(a) The $\alpha (1 \rightarrow 4)$ linkages of starch (amylose)

(b) Helical structure of the amylose component of starch

Starch

Starch, like cellulose, is a polymer containing only glucose units. It is the storage polysaccharide in plants. If excess glucose enters a plant cell, it is converted to starch and stored for later use. When the cell cannot get sufficient glucose from outside the cell, it hydrolyzes starch to release glucose.

Two different polyglucose polysaccharides can be isolated from most starches: amylose and amylopectin. *Amylose*, a straight-chain glucose polymer, usually accounts for 15–20% of the starch and *amylopectin*, a highly branched glucose polymer, accounts for the remaining 80–85% of the starch.

The straight-chain structure of amylose involves glucose units connected by $\alpha(1 \rightarrow 4)$ glycosidic bonds (Figure 22–16a). The number of glucose units present

FIGURE 22–17 The branched structure of the polysaccharide amylopectin. In (b) each circle represents a glucose unit.

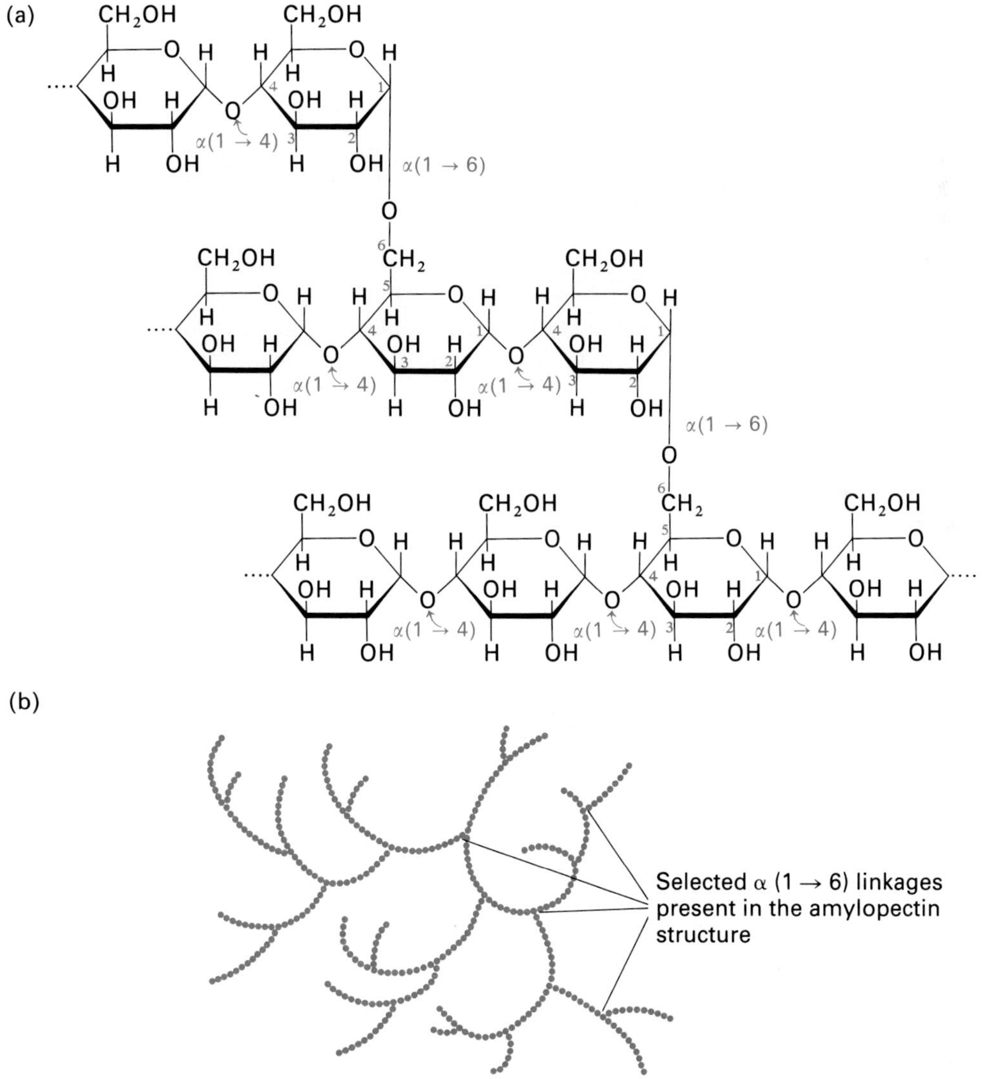

in a chain depends on the source of the starch, with 300–500 monomer units usually being present. Amylose polymers adopt a coiled structure called a *helix*. Helical structures resemble a stretched-out spring (Figure 22–16b).

Iodine is often used to test for the presence of starch in solution. Starch-containing solutions turn a dark blue-black when iodine is added. The dark blue-black results from the assembling, within the center of the helical amylose, of the I_2 molecules in long polyiodine chains. As starch is broken down through acid or enzymatic hydrolysis to glucose monomers, the helix is lost and the blue-black color disappears.

Amylopectin, the other polysaccharide in starch, is similar to amylose in that all linkages are α linkages. It is different in that there is a high degree of branching in the polymer. A branch occurs about once every 25 to 30 glucose units. The branch points involve $\alpha(1 \rightarrow 6)$ linkages (see Figure 22–17, p. 559). Because of the branching, amylopectin has a larger average molecular weight than the linear amylose. The average molecular weight of amylose is 50,000 amu or more, as compared to 300,000 or more for amylopectin.

Enzymatic hydrolysis of starch initially produces intermediate sized polysaccharides called *dextrins*. The dextrins are then further hydrolyzed to maltose and ultimately to D-glucose.

starch ⟶ dextrins ⟶ maltose ⟶ D-glucose
(polysaccharide) (polysaccharide) (disaccharide) (monosaccharide)

Dextrins have numerous commercial uses. Mucilage and some pastes contain dextrins; these adhesives are used on postage stamps and envelopes. A mixture of dextrins and maltose is used in baby foods and infant formulas.

Heat can also convert starch to dextrins. When bread is toasted, the change in texture as well as taste is the result of some starch being converted to dextrins.

Note that all of the glycosidic linkages in starch (both amylose and amylopectin) are of the α type. In amylose they are all $(1 \rightarrow 4)$, while in amylopectin both $(1 \rightarrow 4)$ and $(1 \rightarrow 6)$ linkages are present. Because α linkages can be broken through hydrolysis within the human digestive tract (with the help of the enzyme *amylase*), starch has nutritional value for humans. The starches present in potatoes, wheat, rice, and cereal grains account for approximately two-thirds of the world's food consumption.

Glycogen

Glycogen is the storage polysaccharide in animals and humans. Its function is similar to that of starch in plants, and sometimes it is referred to as *animal starch*. Although glycogen is found in many tissues in the human body, it is mainly concentrated in the liver and in muscles.

Glycogen has a structure similar to amylopectin (Figure 22–17); all linkages are the α type and both $(1 \rightarrow 4)$ and $(1 \rightarrow 6)$ linkages are present. The main difference between these two polymers is that glycogen is even more highly branched than amylopectin; there are only 8–12 D-glucose units between branches. Molecular weights of up to 3 million amu can be reached for glycogen polymers.

When excess glucose is present in the blood (normally from eating too much starch), the liver and muscle tissue convert the excess glucose to glycogen, which is then stored in these tissues. Whenever the glucose blood level drops (from exercise, fasting, or normal activities), some stored glycogen is hydrolyzed back to glucose. These two opposing processes are called *glycogenesis* and *glycogenolysis*, the formation and decomposition of glycogen.

FIGURE 22–18 Carbohydrate loading, the consumption of large amounts of starch, is practiced by many marathon runners the day prior to a race. This maximizes glycogen reserves, which can be drawn upon during the race.

$$\text{glucose} \underset{\text{glycogenolysis}}{\overset{\text{glycogenesis}}{\rightleftarrows}} \text{glycogen}$$

The processes of glycogenesis and glycogenolysis will be discussed in detail in Chapter 29.

In terms of actual numbers, the amount of stored glycogen in the human body is relatively small. Muscle tissue contains approximately 1% glycogen and liver tissue contains 2–3% glycogen. However, this amount is sufficient to take care of normal activity glucose demands for about 15 hours. During strenous exercise, glycogen supplies can be exhausted rapidly. At this point, the body begins using the oxidation of fat (Chapter 23) as a source of energy.

Many marathon runners (Figure 22–18) eat large quantities of starch foods such as spaghetti the day prior to a race. This practice, called *carbohydrate loading*, maximizes body glycogen reserves, which can then be drawn upon during the race.

Glycogen is an ideal storage form for glucose. The large size of these macromolecules prevents them from diffusing out of cells. Also, conversion of glucose to glycogen reduces osmotic pressure. Cells would burst if all of the glucose in

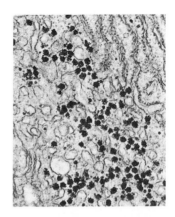

FIGURE 22–19 The small dense particles within this picture of a liver cell are glycogen granules. This picture was taken with the aid of an electron microscope.

EXERCISES AND PROBLEMS

Types of Biochemical Substances

22.1 What are the two general groups of biochemical substances and what are the major types of compounds found in each group?

22.2 Discuss the composition of the human body in terms of major types of biochemical substances present.

Classification of Carbohydrates

22.3 Give a general definition for *carbohydrates*.

22.4 What functional groups can be found in all carbohydrates?

22.5 Distinguish between the following types of carbohydrates.
a. a monosaccharide and an oligosaccharide
b. an oligosaccharide and a polysaccharide
c. a monosaccharide and a polysaccharide

Classification and Structure of Monosaccharides

22.6 Classify each of the following monosaccharides as an aldose or a ketose.

a.
```
      CHO
   H—C—OH
  HO—C—H
  HO—C—H
   H—C—OH
      CH₂OH
```

b.
```
      CH₂OH
      C=O
   H—C—OH
   H—C—OH
   H—C—OH
      CH₂OH
```

c.
```
      CHO
  HO—C—H
   H—C—OH
      CH₂OH
```

d.
```
      CHO
  HO—C—H
  HO—C—H
  HO—C—H
   H—C—OH
      CH₂OH
```

e.
```
      CH₂OH
      C=O
      CH₂OH
```

f.
```
      CH₂OH
      C=O
  HO—C—H
      CH₂OH
```

g.
```
      CHO
  HO—C—H
      CH₂OH
```

h.
```
      CH₂OH
      C=O
  HO—C—H
   H—C—OH
      CH₂OH
```

22.7 Assign a general name to each monosaccharide in Problem 22.6 that reflects the number of carbon atoms present (length of carbon chain) as well as the aldose or ketose nature of the compound.

22.8 For each of the monosaccharides in Problem 22.6, indicate how many chiral carbon atoms are present.

22.9 For each of the monosaccharides in Problem 22.6, indicate whether they belong to the D family, the L family, or neither.

22.10 Using the information in Figures 22–4 and 22–5, assign a name to each of the monosaccharides in Problem 22.6.

22.11 Explain the structural difference between an aldose and a ketose.

22.12 What is the structure (projection formula) and common name of
a. the simplest possible aldose?
b. the simplest possible ketose?

22.13 In an aldoheptose, configurational change around which carbon atom produces D and L isomerism?

22.14 Explain why the number of isomeric aldopentoses that exist (there are 8) is double the number of isomeric ketopentoses that exist (there are 4).

22.15 Projection formulas for the four possible aldotetroses are

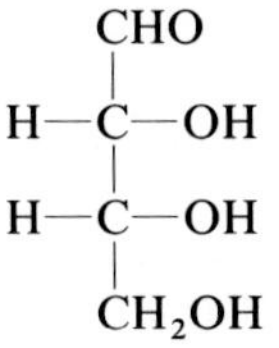

```
    CHO
H—C—OH
H—C—OH
   CH₂OH
      I

    CHO
H—C—OH
H—C—H
   CH₂OH
      II

    CHO
HO—C—H
HO—C—H
   CH₂OH
     III

    CHO
HO—C—H
H—C—OH
   CH₂OH
      IV
```

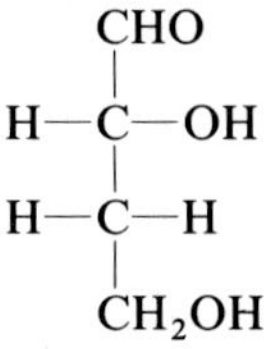

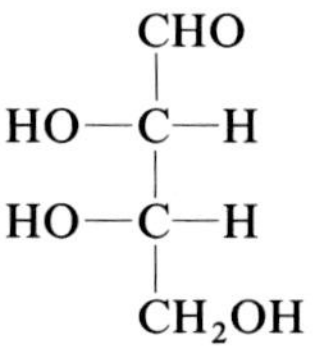

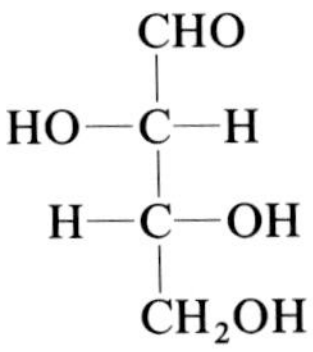

Classify each of the following pairs of the preceding compounds as enantiomers or diastereomers.
a. I and II   c. I and IV   e. II and IV
b. I and III   d. II and III   f. III and IV

22.16 What is the maximum number of stereoisomers that can exist that are
a. aldohexoses?   c. ketopentoses?
b. aldopentoses?   d. ketotetroses?

22.17 Each of the following monosaccharides is a hexose.

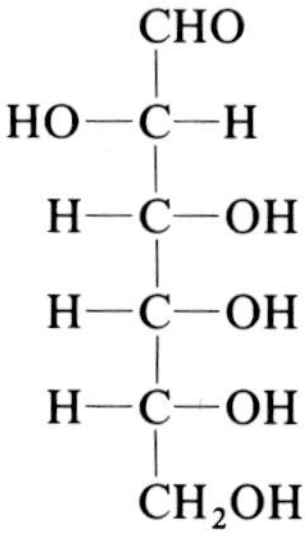

```
    CHO
HO—C—H
H—C—OH
H—C—OH
H—C—OH
   CH₂OH
      I

    CHO
H—C—OH
H—C—OH
HO—C—H
H—C—OH
   CH₂OH
      II

   CH₂OH
    C=O
H—C—OH
H—C—OH
H—C—OH
   CH₂OH
     III

   CH₂OH
    C=O
H—C—OH
HO—C—H
H—C—OH
   CH₂OH
      IV
```

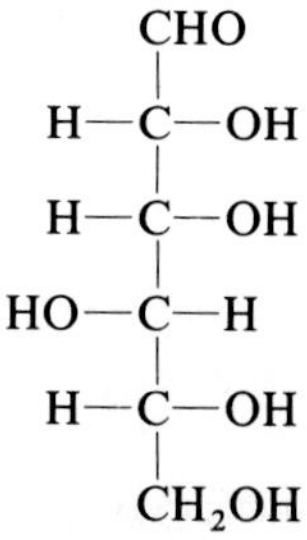

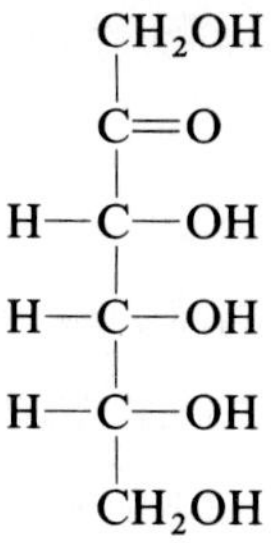

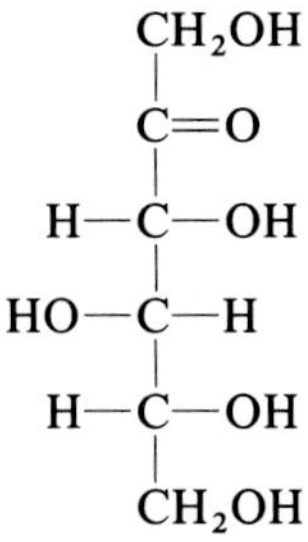

a. Which of the preceding compounds is a ketose?
b. Which of the preceding compounds is an aldose?
c. In which of the preceding compounds is the penultimate carbon atom carbon number 5?
d. Which of the preceding compounds belongs to the L family of monosaccharides?
e. Which of the preceding compounds has the molecular formula $C_6H_{12}O_6$?

Important Monosaccharides

22.18 Draw projection formulas for each of the following monosaccharides.
a. D-glucose   c. D-fructose
b. D-galactose   d. D-ribose

22.19 Explain the principal differences and similarities between the structures of the members of each of the following pairs of monosaccharides.
a. D-glucose and D-galactose
b. D-glucose and D-fructose
c. D-glucose and D-ribose
d. D-glucose and L-glucose

22.20 Identify the monosaccharide that has each of the following characteristics.
a. Is also known as *dextrose*.
b. Is also known as *levulose*.
c. Is an epimer of D-glucose.
d. Is also known as blood sugar.
e. Is the most abundant monosaccharide.
f. Is used in hospitals as an intravenous source of nourishment.
g. Has the sweetest taste of all monosaccharides.
h. Is a component of both RNA and ATP.

Cyclic Forms of Monosaccharides

22.21 Explain how to recognize a hemiacetal or hemiketal group.

22.22 The structure of glucose is sometimes written in an open-chain form and other times as a cyclic hemiacetal structure. Explain why either form is acceptable.

22.23 Write an equation, using structural formulas, to show the equilibrium between α, β, and open-chain forms of D-glucose.

22.24 Write the open-chain form of the following cyclic hemiacetal Write the open form with its chain coiled in the same way that it is coiled in the closed (cyclic) form.

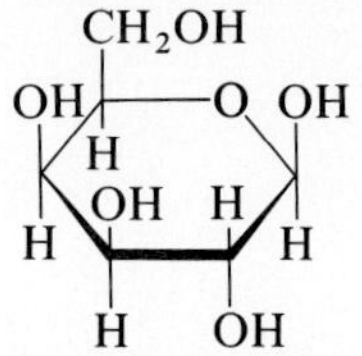

22.25 Which carbon atom determines
a. α or β configuration in aldohexoses?
b. α or β configuration in ketohexoses?
c. D or L configuration in aldohexoses?
d. D or L configuration in ketohexoses?

22.26 The monosaccharide D-allose is identical to D-glucose except that in its open-chain projection formula, the —OH group at carbon number 3 projects in the opposite direction. Draw structures for
a. α-D-allose.   c. α-L-allose.
b. β-D-allose.   d. β-L-allose.

22.27 Fructose and glucose both contain six carbon atoms. Why do the cyclic forms of fructose have a five-membered ring instead of the six-membered ring that is found in the cyclic forms of glucose?

22.28 Identify each of the following structures as being a hemiacetal or a hemiketal.

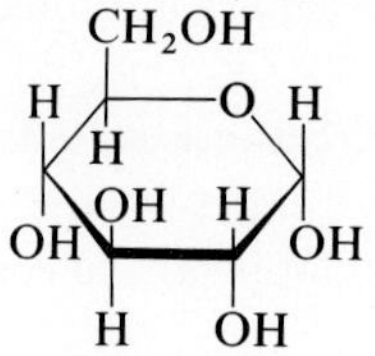

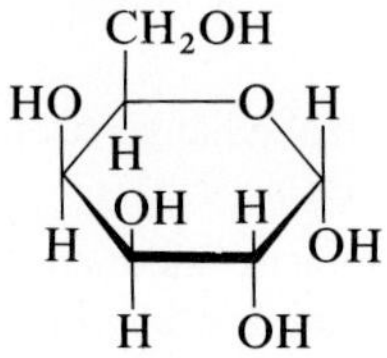

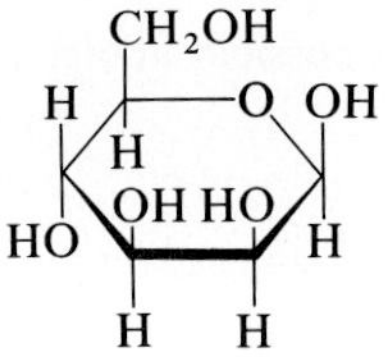

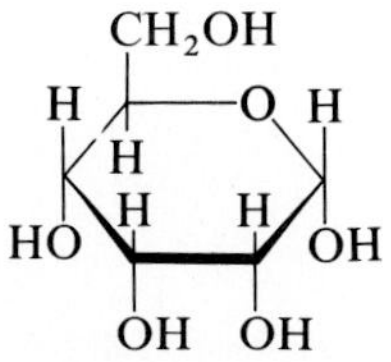

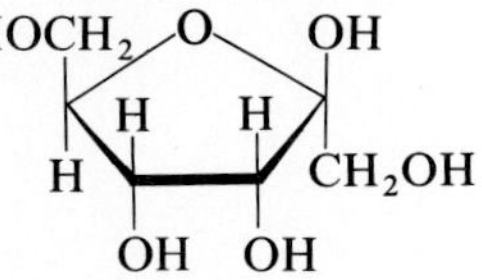

22.29 Identify each of the structures in Problem 22.28 as an α or β isomer.

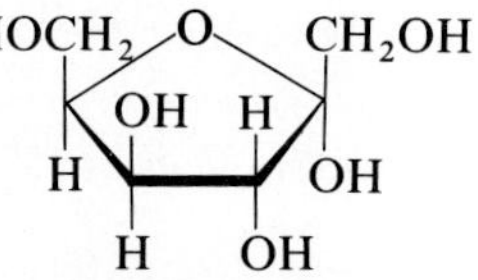

22.30 Assign a name to each of the compounds whose structures are given in Problem 22.28.

22.31 Sometimes a hemiacetal is referred to as a "potential aldehyde". Why is this?

22.32 What are anomers? How are they labeled?

22.33 Identify the anomeric carbon atom in each of the structures in Problem 22.28.

22.34 Write projection formulas for the two anomers of D-mannose.

22.35 Explain what happens during the process of mutarotation.

22.36 If pure α-glucose is dissolved in water, β-glucose is soon present. Explain how this is possible.

22.37 Which of the compounds whose structures are given in Problem 22.28 can undergo mutarotation in solution?

22.38 Why is a wavy line sometimes used for the bond between oxygen and carbon number 1 in the cyclic form of glucose?

Reactions of Monosaccharides

22.39 Identify each of the following compounds as being an acetal or ketal.

22.40 For each of the structures in Problem 22.39, identify the configuration at the acetal or ketal carbon atom as being α or β.

22.41 Define the term *glycoside*.

22.42 Draw structural formulas for the two glycosides formed when D-mannose reacts with ethyl alcohol.

22.43 Draw structures for the reactants and products of a chemical reaction that could be used to convert α-D-glucose to ethyl-α-D-glucoside.

22.44 What is the difference in meaning of the terms *glycoside* and *glucoside*?

22.45 Monosaccharides are sometimes called *reducing sugars*. Explain why.

22.46 Explain the process of tautomerization between ketose and aldose sugars, using fructose as an example.

22.47 Ketones do not react with Benedict's solution, but fructose (a ketohexose) does. Explain this observation.

22.48 In terms of oxidation and reduction, explain what occurs to both D-glucose and Tollens' solution when they react with each other.

22.49 Describe the chemical reaction used to detect glucose in urine that involves Benedict's solution.

22.50 Which of the compounds whose structures are given in Problem 22.28 are reducing sugars?

22.51 Explain the similarities and differences in the structures of D-glucose and D-sorbitol.

22.52 Give the name of the aldotetrose that you could reduce catalytically with H_2 to form the following polyhydroxy alcohol.

22.53 Give the name of the aldotetrose that you could oxidize with Benedict's solution to form the following compound.

22.54 Explain how the alcohol groups of monosaccharides can form esters. What important role do these esters play in living cells?

Disaccharides

22.55 Which monosaccharide is a component of the disaccharides sucrose, maltose, lactose, and cellobiose?

22.56 What are the names of the three nutritionally important disaccharides?

22.57 Identify a disaccharide that fits each of the following descriptions.

a. Is a household table sugar.
b. Is converted to lactic acid when milk sours.
c. Is an ingredient in infant formulas used as substitutes for mother's milk.
d. Is formed in germinating grain.
e. Hydrolyzes when cooked with acidic foods to give invert sugar.

22.58 What monosaccharides are produced from the hydrolysis of each of the following disaccharides?
a. sucrose
b. maltose
c. lactose
d. cellobiose

22.59 Match the following structures with the names sucrose, maltose, cellobiose, and lactose.

a. [structure]

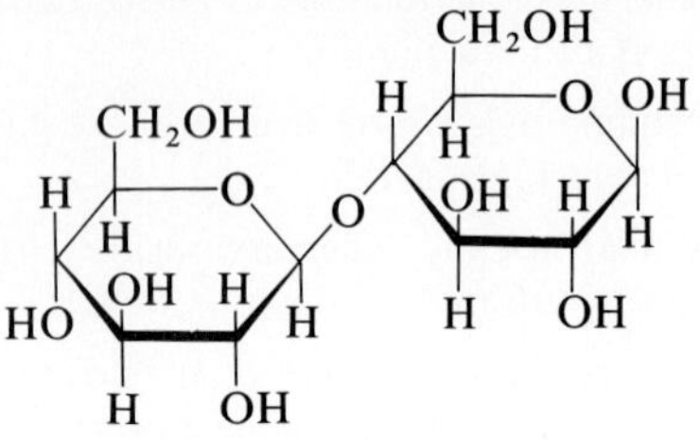

b. [structure]

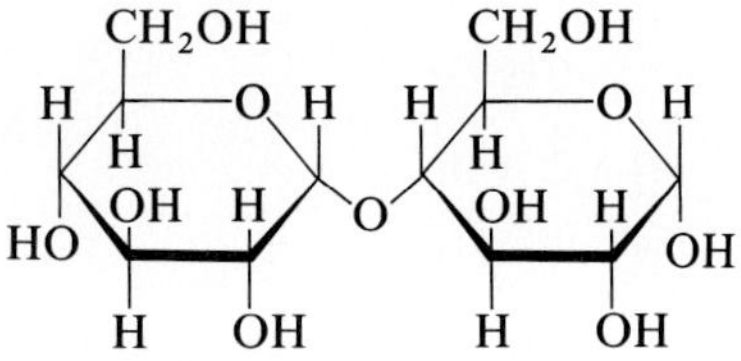

c. [structure]

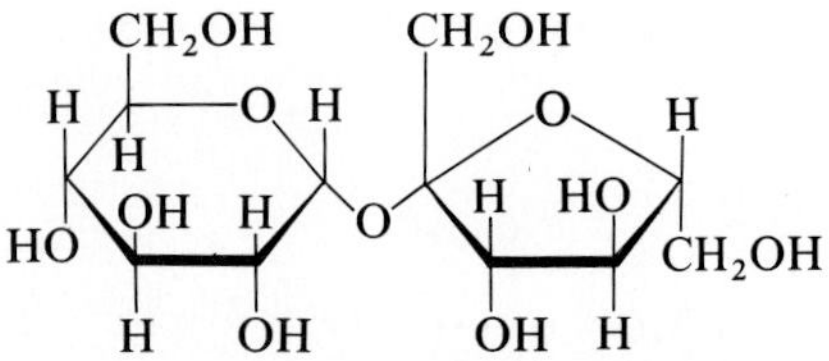

d. [structure]

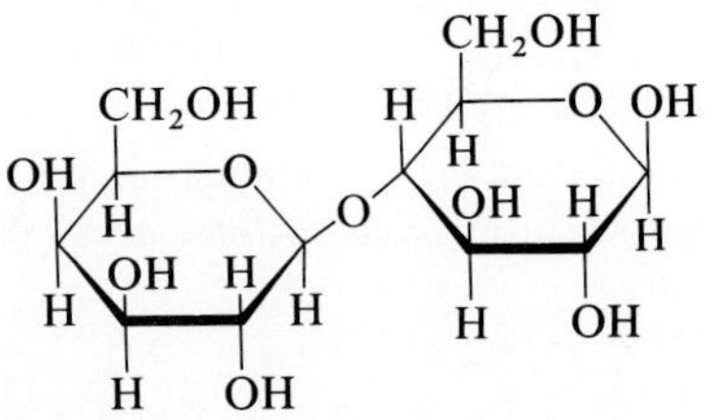

22.60 Characterize the type of glycosidic linkage present in each of the following disaccharides.

a. [structure]

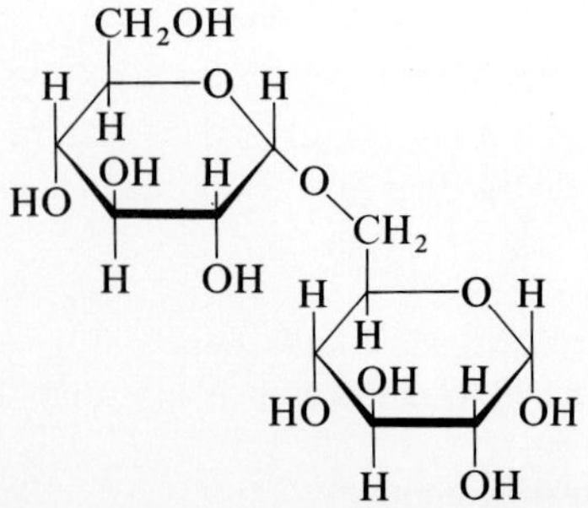

b. [structure]

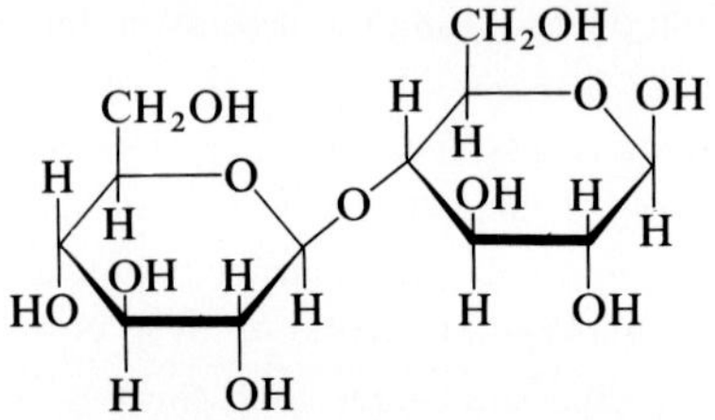

c. [structure]

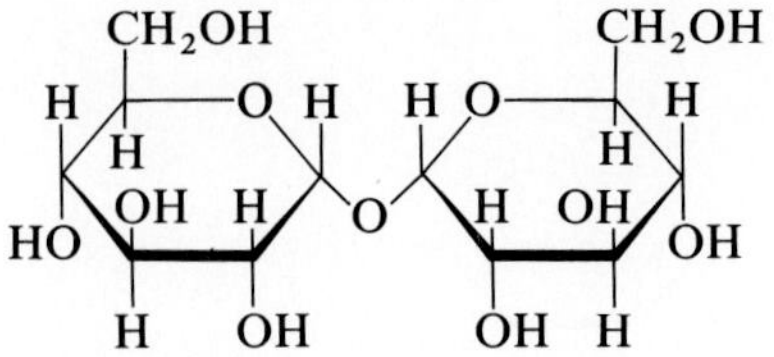

d. [structure]

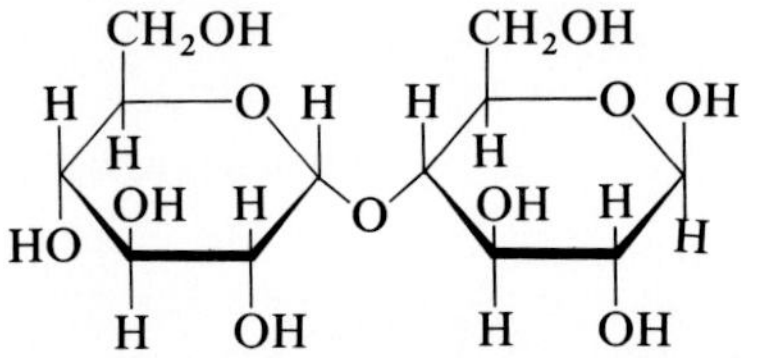

22.61 Identify all hemiacetal, hemiketal, ketal, or acetal carbon atoms present in each of the disaccharides listed in Problem 22.60.

22.62 Characterize each of the disaccharides listed in Problem 22.60 as a reducing sugar or a nonreducing sugar.

22.63 Write structures for disaccharides in which D-glucose units are linked together.
a. $\alpha(1 \rightarrow 4)$
b. $\beta(1 \rightarrow 4)$
c. $\alpha(1 \rightarrow 6)$
d. $\beta(1 \rightarrow 6)$

22.64 Explain why lactose and maltose, but not sucrose, are reducing sugars.

22.65 What is *invert sugar*? Describe how it received its name.

22.66 Maltose has a hemiacetal carbon center and is thus a reducing sugar. Draw the structure of maltose in which this group has changed to the open form.

22.67 Describe two types of allergies to milk that account for the fact that milk is not a good food for everybody.

Artificial Sweeteners

22.68 Why is invert sugar sweeter than the original sucrose from which it is obtained by hydrolysis?

22.69 Explain why aspartame is considered a low-calorie sweetener even though it has the same caloric content as sucrose (4 calories/gram).

Polysaccharides

22.70 Describe differences and similarities in the structures of each of the following pairs of polysaccharides.
a. glycogen and amylopectin
b. amylose and glycogen
c. amylose and cellulose
d. amylose and amylopectin

22.71 Match the following structural characteristics to the polysaccharides amylopectin, amylose, glycogen, and cellulose. A specific characteristic might apply to more than one polysaccharide.
a. Contains both $\alpha(1 \rightarrow 4)$ and $\alpha(1 \rightarrow 6)$ linkages.
b. Is composed of glucose monosaccharide units.
c. Contains acetal linkages between monosaccharide units.
d. Is composed of highly branched molecular chains.
e. Is composed of unbranched molecular chains.
f. Contains only $\beta(1 \rightarrow 4)$ linkages.

22.72 Name a polysaccharide that fits each of the following descriptions.
a. Is the unbranched polysaccharide in starch.
b. Is the most abundant polysaccharide in starch.
c. Is a storage form of carbohydrates in animals and humans.
d. Is a structural carbohydrate in plants.

22.73 Explain why human beings cannot digest cellulose, but mammals such as deer, sheep, and cows live quite comfortably on a diet high in cellulose content.

22.74 Although no caloric value is obtained from cellulose, it is still considered important in the diet of human beings. Explain why.

22.75 Why does I_2 turn blue-black in solutions of starch, but not in solutions of glucose?

22.76 What is the biological role of glycogen?

22.77 What is the difference between the processes of glycogenesis and glycogenolysis?

22.78 It has been determined that a polysaccharide very similar to amylopectin has an approximate molecular weight of 250,000 amu. What is the average number of monosaccharides in each molecule of the polysaccharide?

23 Lipids

Objectives

After completing Chapter 23, you will be able to:

23.1 Describe the property that distinguishes lipids from other biochemicals and understand the structural classification scheme for both saponifiable and nonsaponifiable lipids.
23.2 Recognize the general structure for fatty acids and understand the relationship between structure and melting points for fatty acids.
23.3 Recognize the general structure for fats and oils and distinguish between fats and oils on the basis of physical and structural properties.
23.4 Understand five types of reactions characteristic of triglycerides: hydrolysis, saponification, hydrogenation, rancidity, and iodination.
23.5 Recognize the general structure for phosphoglycerides and distinguish between lecithins and cephalins.
23.6 Recognize the general structure for waxes and list biochemical functions of waxes.
23.7 Recognize the general structure for sphingolipids and distinguish between sphingomyelins and glycosphingolipids.
23.8 Recognize the fused hydrocarbon ring system characteristic of steroid molecules and understand the function and general structure of various steroids.
23.9 Recognize the general structure for prostaglandins and describe the physiological effects regulated by prostaglandins.
23.10 Describe cell membrane structure in terms of the structural arrangement of lipid molecules and discuss the membrane processes of simple diffusion, facilitated diffusion, and active transport.

INTRODUCTION

In the last chapter, we noted that there are four major classes of bioorganic substances: carbohydrates, lipids, proteins, and nucleic acids. We now turn our attention to the second of the bioorganic classes, the compounds we call *lipids*.

Lipids, a large class of relatively water-insoluble compounds, are important in living systems. In humans and many animals, excess carbohydrates and other energy-yielding foods are converted to and stored in the body as lipids called *fats*. These fat reservoirs constitute a major storage form of chemical energy and carbon atoms in the body. Lipids surround and insulate numerous vital body organs. In this capacity, they provide protection from mechanical shock and help maintain the correct body temperature. Lipids also serve as basic structural components of all cell membranes. Many chemical messengers in the human body, substances we call *hormones*, are lipids.

According to some estimates, as much as 40% of the average American diet consists of edible lipids. Because the body cannot synthesize some lipids, certain lipids are necessary in the diet for good health.

The term *lipid* comes from the Greek "lipos" meaning *fat* or *lard*. The lipid family also includes oils, waxes, steroids, and some vitamins.

23.1 CLASSIFICATION OF LIPIDS

Lipids, in contrast to carbohydrates and most other classes of compounds, cannot be defined from a structural viewpoint. A variety of functional groups and structural features are found in molecules classified as *lipids*. What lipids share are their solubility properties. **Lipids** *are a structurally heterogeneous group of compounds that are water-insoluble but soluble in nonpolar organic solvents such as chloroform, acetone, diethyl ether, and carbon tetrachloride.* When biological material (animal or plant tissue) is homogenized in a blender and mixed with a nonpolar organic solvent, the substances that dissolve in the solvent will be the lipids.

lipids

Lipids can be divided into two major classes, *saponifiable* and *nonsaponifiable*, based on whether they do or do not undergo hydrolysis reactions in alkaline (basic) solution. **Saponifiable lipids** *can be hydrolyzed under alkaline conditions to yield salts of fatty acids.* **Nonsaponifiable lipids** *do not undergo hydrolysis reactions in alkaline solution.*

saponifiable lipids

nonsaponifiable lipids

We encountered the term *saponification* previously; recall, from Section 19.11, that saponification of an ester involves breaking of the ester linkage to produce an alcohol and a carboxylic acid salt. Amide linkages are also subject to saponification (Section 20.12).

Saponifiable and nonsaponifiable lipids can be divided into subclasses based on common structural features as shown in Figure 23–1 (p. 570).

23.2 FATTY ACIDS

We begin our discussion of saponifiable lipids by considering *fatty acids*, compounds that are a building block found in all saponifiable lipid structures.

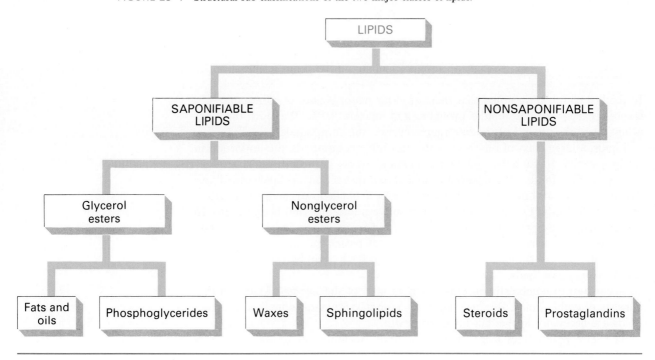

FIGURE 23–1 Structural sub-classifications of the two major classes of lipids.

fatty acids

Although they rarely exist in nature in the free state, fatty acids are lipids because they are insoluble in water but soluble in nonpolar organic solvents. **Fatty acids** *are monocarboxylic acids that contain long unbranched hydrocarbon chains.* The fatty acids of most interest in biochemistry involve carbon chain lengths in the range of 12–26 carbon atoms.

saturated fatty acids

Fatty acids are further classified as *saturated* or *unsaturated*. **Saturated fatty acids** *have a carbon chain in which all carbon–carbon bonds are single bonds.* An example is hexadecanoic acid, a 16-carbon acid whose common name is palmitic acid.

$$H-\underset{H}{\overset{H}{C}}-\underset{H}{\overset{H}{C}}-\underset{H}{\overset{H}{C}}-\underset{H}{\overset{H}{C}}-\underset{H}{\overset{H}{C}}-\underset{H}{\overset{H}{C}}-\underset{H}{\overset{H}{C}}-\underset{H}{\overset{H}{C}}-\underset{H}{\overset{H}{C}}-\underset{H}{\overset{H}{C}}-\underset{H}{\overset{H}{C}}-\underset{H}{\overset{H}{C}}-\underset{H}{\overset{H}{C}}-\underset{H}{\overset{H}{C}}-\underset{H}{\overset{H}{C}}-C\underset{OH}{\overset{O}{\diagup}}$$

palmitic acid

The structural formula of palmitic acid, or any other fatty acid, can be, and usually is, written in a more condensed form. The condensed structural formula for palmitic acid is

$$CH_3-(CH_2)_{14}-C\underset{OH}{\overset{O}{\diagup}}$$

unsaturated fatty acids

Unsaturated fatty acids *have a carbon chain with one or more carbon–carbon double bonds.* An example is oleic acid, an 18-carbon acid with one double bond, whose structural formula (expanded and condensed) is oleic acid

$$CH_3-(CH_2)_7-CH=CH-(CH_2)_7-C\overset{O}{\underset{OH}{\diagdown}}$$

TABLE 23–1 Structures, Melting Points, and Sources for Selected Saturated and Unsaturated Fatty Acids,

Name	Number of carbon atoms	Formula	Melting point (°C)	Common sources
Saturated fatty acids				
lauric acid	12	$CH_3-(CH_2)_{10}-COOH$	44	coconut oil
myristic acid	14	$CH_3-(CH_2)_{12}-COOH$	54	butterfat, coconut oil nutmeg oil
palmitic acid	16	$CH_3-(CH_2)_{14}-COOH$	63	lard, beef fat, butterfat, cottonseed oil
stearic acid	18	$CH_3-(CH_2)_{16}-COOH$	70	lard, beef fat, butterfat, cottonseed oil
arachidic acid	20	$CH_3-(CH_2)_{18}-COOH$	76	peanut oil
carotic acid	26	$CH_3-(CH_2)_{24}-COOH$	97	beeswax, wool fat
Unsaturated fatty acids				
oleic acid	18	$CH_3-(CH_2)_7CH=CH-(CH_2)_7-COOH$	16	lard, beef fat, olive oil, peanut oil
linoleic acid	18	$CH_3-(CH_2)_4-(CH=CH-CH_2)_2-(CH_2)_6-COOH$	−5	cottonseed oil, soybean oil, corn oil, linseed oil
linolenic acid	18	$CH_3-CH_2-(CH=CH-CH_2)_3-(CH_2)_6-COOH$	−11	linseed oil, corn oil
arachidonic acid	20	$CH_3-(CH_2)_4-(CH=CH-CH_2)_4-(CH_2)_2-COOH$	−50	corn oil, linseed oil, animal tissues

Note the angular arrangement of the hydrocarbon chain in oleic acid. This arrangement is caused by the double bond having a *cis* configuration. The configuration about double bonds in naturally occurring and biologically important fatty acids is almost always *cis*. The term *polyunsaturated fatty acid* is often used to describe fatty acids containing more than one double bond. Up to four double bonds are found in biologically important unsaturated fatty acids.

Table 23–1 (p. 571) gives formulas and information about the naturally occurring fatty acids most often encountered in lipid structures. Two features to note about the structures are: all hydrocarbon chains are straight rather than branched chains, and all chains have an even number of carbon atoms. The even number of carbon atoms relates to how these acids are synthesized.

The melting points of fatty acids depend on both the length of their hydrocarbon chains and their degree of unsaturation (number of double bonds per molecule). This fact can be seen by comparing the melting points of fatty acids in Table 23–1 (p. 571). Saturated fatty acids have higher melting points than unsaturated fatty acids because the added double bonds lower the melting points. The effect of unsaturation is clearly seen in the case of the 18-carbon molecules stearic acid (saturated), oleic acid (one double bond), and linoleic acid (two double bonds) which have, respectively, melting points of 70°C, 16°C, and −5°C.

These differences in melting points are further explained by comparing molecular attractions between carbon chains. In pure form, molecules of saturated fatty acids fit close together in a neat orderly fashion as the result of strong intermolecular attractions between carbon chains (see Figure 23–2a). Longer hydrocarbon chains allow for even more attractions between molecules and the melting point is pushed higher. The presence of *cis* double bonds, however, disrupts this orderly stacking pattern and weakens the intermolecular attractions (see Figure 23–2b). Hence, unsaturated fatty acids have lower melting points than saturated ones.

The long carbon chains in fatty acids make them and most of their derivatives water-insoluble.

Fatty acids are building blocks for the biosynthesis of many important lipids. The human body has the ability to synthesize most of these building blocks from other substances if needed. However, humans and other mammals cannot synthe-

FIGURE 23–2 Diagrammatic view of molecular stacking for saturated and unsaturated fatty acids.

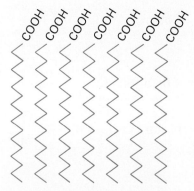

(a) Orderly stacking of saturated fatty acids results in high melting point.

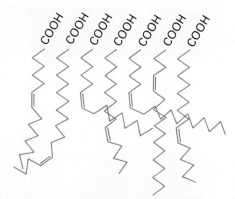

(b) Disorderly stacking of unsaturated fatty acids results in low melting point.

size fatty acids containing more than one double bond. Hence, polyunsaturated acids must be obtained from food. These dietary essential acids are often called *essential fatty acids*. The last three acids listed in Table 23–1 are of this type. Actually, our diet need contain only one of these three essential fatty acids because the body can convert any one of them into the other two.

Until recently, biochemists did not know why humans needed polyunsaturated fatty acids. It is now known that our body needs them to synthesize prostaglandins, a nonsaponifiable lipid we will discuss in Section 23.9.

We encountered fatty acids in Section 19.6, when we discussed sodium and potassium salts of fatty acids. These salts are the chemical basis for soaps. The mechanism by which soaps gain water solubility, or micelle formation, will be important later in this chapter.

23.3 FATS AND OILS

Fats and *oils* are the most abundant naturally occurring lipids. Fats are generally solids at room temperature and usually are obtained from animals. Oils are generally liquids at room temperature and usually are obtained from plants. We will discuss fats and oils together because they have identical structural building blocks. We will first cover the structural features common to fats and oils and then consider the fine points that distinguish fats from oils.

Fats and oils *contain triesters formed from the condensation reaction between three fatty acid molecules and a glycerol molecule.* Glycerol, a compound we discussed in Section 17.7, is an alcohol with three hydroxyl groups.

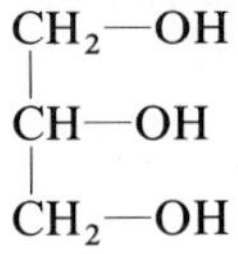

$$\begin{array}{c} CH_2-OH \\ | \\ CH-OH \\ | \\ CH_2-OH \end{array}$$
glycerol

Each hydroxyl group can react with the carboxyl group of a fatty acid molecule (an esterification reaction—Section 19.8) to give the triester structure of fats and oils.

The general formula and a block diagram showing structural components of these triesters are

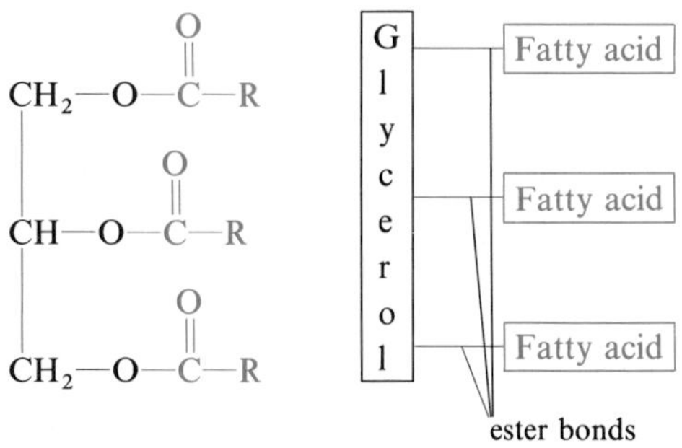

$$\begin{array}{c} CH_2-O-\overset{O}{\underset{\|}{C}}-R \\ | \\ CH-O-\overset{O}{\underset{\|}{C}}-R \\ | \\ CH_2-O-\overset{O}{\underset{\|}{C}}-R \end{array}$$

Triesters of this formula are commonly called *triglycerides*. The designation *triester* comes from the fact that each of the fatty acid molecules in a triglyceride is bonded to glycerol via an ester linkage (Section 19.7).

The reaction between glycerol and three molecules of stearic acid produces a triglyceride with the following structure.

$$
\begin{array}{l}
\text{CH}_2\text{OH} \quad \text{HO}-\overset{\overset{\text{O}}{\|}}{\text{C}}-(\text{CH}_2)_{16}-\text{CH}_3 \qquad \text{CH}_2\text{O}-\overset{\overset{\text{O}}{\|}}{\text{C}}-(\text{CH}_2)_{16}-\text{CH}_3 \\
| \\
\text{CHOH} \quad \text{HO}-\overset{\overset{\text{O}}{\|}}{\text{C}}-(\text{CH}_2)_{16}-\text{CH}_3 \longrightarrow \text{CHO}-\overset{\overset{\text{O}}{\|}}{\text{C}}-(\text{CH}_2)_{16}-\text{CH}_3 + 3\text{H}_2\text{O} \\
| \\
\text{CH}_2\text{OH} \quad \text{HO}-\overset{\overset{\text{O}}{\|}}{\text{C}}-(\text{CH}_2)_{16}-\text{CH}_3 \qquad \text{CH}_2\text{O}-\overset{\overset{\text{O}}{\|}}{\text{C}}-(\text{CH}_2)_{16}-\text{CH}_3
\end{array}
$$

Note that as each ester linkage forms, a molecule of water is also produced. The triester produced from glycerol and stearic acid is an example of a *simple triglyceride*. **Simple triglycerides** *are triesters formed from the reaction of glycerol and three molecules of one kind of fatty acid.*

simple triglycerides

The three fatty acid molecules of a triglyceride need not be identical. In the following reaction, three different fatty acid molecules are involved in esterification.

$$
\begin{array}{l}
\text{CH}_2\text{OH} \quad \text{HO}-\overset{\overset{\text{O}}{\|}}{\text{C}}-(\text{CH}_2)_{16}-\text{CH}_3 \qquad \text{CH}_2\text{O}-\overset{\overset{\text{O}}{\|}}{\text{C}}-(\text{CH}_2)_{16}-\text{CH}_3 \\
| \\
\text{CHOH} \quad \text{HO}-\overset{\overset{\text{O}}{\|}}{\text{C}}-(\text{CH}_2)_{14}-\text{CH}_3 \longrightarrow \text{CHO}-\overset{\overset{\text{O}}{\|}}{\text{C}}-(\text{CH}_2)_{14}-\text{CH}_3 + 3\text{H}_2\text{O} \\
| \\
\text{CH}_2\text{OH} \quad \text{HO}-\overset{\overset{\text{O}}{\|}}{\text{C}}-(\text{CH}_2)_{10}-\text{CH}_3 \qquad \text{CH}_2\text{O}-\overset{\overset{\text{O}}{\|}}{\text{C}}-(\text{CH}_2)_{10}-\text{CH}_3
\end{array}
$$

mixed triglycerides

Triesters of this type are called *mixed triglycerides*. **Mixed triclycerides** *are triesters formed from the reaction of glycerol with more than one kind of fatty acid molecule.* Figure 23–3 illustrates the difference between simple and mixed triglycerides using block diagrams.

In nature, simple triglycerides are rare. Most naturally occurring triglycerides are mixed triglycerides. Fats and oils that are biologically important are complex mixtures of mixed triglycerides. No single triglyceride structure adequately describes a naturally occurring fat or oil. The particular acid components in the triglycerides of a fat or oil depend on the source of the substance, as you can see from the composition data in Table 23–2 for selected naturally occurring fats and oils.

FIGURE 23–3 In simple triglycerides, all three fatty acid molecules are identical, while mixed triglycerides contain two or three different kinds.

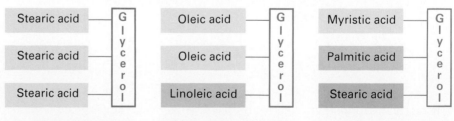

(a) A simple triglyceride — all three fatty acid molecules are the same.

(b) A mixed triglyceride — two different fatty acid molecules are present.

(c) A mixed triglyceride — three different fatty acid molecules are present.

TABLE 23-2 Fatty Acid Composition, by Mass Percent, of Selected Naturally Occurring Fats and Oils

| | Percent composition of fatty acids | | | | | | |
| | Saturated | | | | Unsaturated | | |
	Myristic	Palmitic	Stearic	Arachidic	Oleic	Linoleic	Linolenic
Animal fats							
butter	11	29	9	2	27	4	–
lard	1	28	12	–	48	6	–
human fat	3	24	8	–	47	10	–
beef tallow	5	28	23	–	40	3	–
Plant oils							
corn	1	10	3	–	50	34	–
cottonseed	1	23	1	1	23	48	–
linseed	–	6	2	1	19	24	47
olive	–	7	2	–	84	5	–
peanut	–	8	3	2	56	26	–
soybean	–	10	2	–	29	51	6

The percentages can vary widely for some fats and oils, Environmental factors affect fatty acid composition in plant materials and diet affects animal fat composition.

The percentages total less than 100% in some cases because of significant amounts of fatty acids above C_{18} or below C_{14}.

We can make an important generalization about fats and oils from composition data like that in Table 23-2. Triglyceride mixtures obtained from animals contain more saturated fatty acid components than do triglycerides from plants. This generalization leads to more specific definitions for fats and oils. **Fats** *are triglyceride mixtures having a relatively high percentage of saturated fatty acid residues.* **Oils** *are triglyceride mixtures having a relatively high percentage of unsaturated fatty*

fats

oils

FIGURE 23-4 Triglycerides containing unsaturated fatty acid components have lower melting points than those with only saturated fatty acid components because it is more difficult to pack the molecules together in the solid state. (Space-filling molecules are used here to emphasize packing.)

KEY
◯ Hydrogen ● Carbon ◯ Oxygen

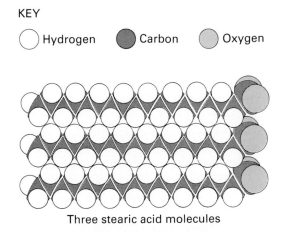

Three stearic acid molecules

(a) All fatty acid components are saturated.

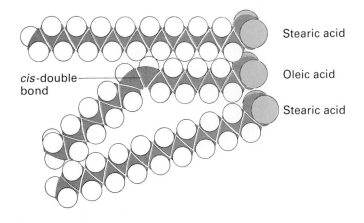

cis-double bond

Stearic acid
Oleic acid
Stearic acid

(b) Both saturated and unsaturated fatty acid components are present.

FIGURE 23-5 Plant oils such as corn oil and safflower oil are becoming increasingly popular as cooking oils because more unsaturated fatty acid components are present than in animal fats. Recent studies have produced evidence linking saturation in fats with arteriosclerosis and increased incidence of heart attacks.

acid components. These definitions are consistent with the fact that fats are solids and oils are liquids. Recall, from our discussion in Section 23.2, how degree of unsaturation affects packing of molecules, which in turn affects melting points for free fatty acids. This same principle applies to triglycerides. In Figure 23–4 (p. 575), note how unsaturation (*cis* double bonds present) affects the packing of the hydrocarbon chain portions of a triglyceride.

Several factors influence the degree of unsaturation among fatty acids in plants and animals. For example, linseed oil from flax seed grown in warm climates often contains up to twice as many double bonds as oil obtained from seed grown in cold climates. The degree of unsaturation in lard from hog fat depends on their diet; the fat of corn-fed hogs is more saturated than that of peanut-fed hogs.

The term *polyunsaturated* is a key word in many advertisements for cooking oils. Several studies suggest that triglycerides high in unsaturated fatty acids help prevent arterial deposits that restrict blood flow and can cause heart attacks and strokes. As a result, the popularity of unsaturated vegetable oils is increasing at the expense of saturated animal fats such as butter and lard (see Figure 23–5).

When a person consumes too much food, much of the excess energy (calories) is stored as fat. This fat is concentrated in special cells (adipocytes) that are nearly filled with large fat droplets. Adipose tissue contains these cells in various parts of the body—under the skin, in the abdominal cavity, in the mammary glands, and around various organs. Figure 23–6 shows an electron micrograph of fat cells. The bulging cells consist almost entirely of liquid fat. The deflated cells have lost some of their fat to meet energy needs.

Triglycerides store energy more efficiently than glycogen (Section 22.10). As an energy source, fats produce approximately twice as many kilocalories per gram as carbohydrates. The body can also store more fat than glycogen. Glycogen reserves

FIGURE 23-6 An electron micrograph of fat cells. Note the bulging spherical shape.

usually provide less than a day's supply of energy. Fat reserves contain sufficient energy for many days, and in the case of obese people, for several months.

23.4 CHEMICAL REACTIONS OF FATS AND OILS (TRIGLYCERIDES)

Fats and oils undergo many reactions; we will consider five: hydrolysis, saponification, hydrogenation, rancidity, and iodination.

Hydrolysis

The most important chemical reaction of fats and oils is *hydrolysis*, the breakdown of the fat or oil by water (with a catalyst). Hydrolysis occurs when the body breaks down fats during digestion in the small intestine. This reaction is the reverse of esterification by which triglycerides are formed, as you can see by the following hydrolysis equation for a triglyceride found in cottonseed oil.

$$\begin{array}{c} \text{CH}_3\text{—(CH}_2)_{14}\text{—}\overset{\overset{\displaystyle O}{\|}}{\text{C}}\text{—O—CH}_2 \\ \text{CH}_3\text{—(CH}_2)_7\text{—CH=CH—(CH}_2)_7\text{—}\overset{\overset{\displaystyle O}{\|}}{\text{C}}\text{—O—CH} + 3\text{H}_2\text{O} \xrightarrow{\Delta} \\ \text{CH}_3\text{—(CH}_2)_4\text{—(CH=CHCH}_2)_2\text{—(CH}_2)_6\text{—}\overset{\overset{\displaystyle O}{\|}}{\text{C}}\text{—O—CH}_2 \\ \text{a cottonseed oil} \end{array}$$

$$\begin{array}{ll} \text{CH}_3\text{—(CH}_2)_{14}\text{—}\overset{\overset{\displaystyle O}{\|}}{\text{C}}\text{—OH} & \text{HO—CH}_2 \\ \quad\text{palmitic acid} & \\ \text{CH}_3\text{—(CH}_2)_7\text{—CH=CH—(CH}_2)_7\text{—}\overset{\overset{\displaystyle O}{\|}}{\text{C}}\text{—OH} & +\ \text{HO—CH} \\ \quad\text{oleic acid} & \\ \text{CH}_3\text{—(CH}_2)_4\text{—(CH=CHCH}_2)_2\text{—(CH}_2)_6\text{—}\overset{\overset{\displaystyle O}{\|}}{\text{C}}\text{—OH} & \text{HO—CH}_2 \\ \quad\text{linoleic acid} & \quad\text{glycerol} \end{array}$$

Glycerol and three molecules of fatty acids are always the products of triglyceride hydrolysis. Within the human body hydrolysis occurs at body temperature with the help of enzymes. Commercially, hydrolysis of fats and oils is carried out using superheated steam.

Saponification

Saponification is *a hydrolysis reaction carried out in an alkaline (basic) solution.* For fats and oils, the products from saponification are glycerol and salts of the fatty acids. The active ingredients in soap are fatty acid salts produced during saponification.

saponification

$$\begin{array}{c}
\text{CH}_3\text{—(CH}_2)_{14}\text{—}\overset{\overset{\text{O}}{\|}}{\text{C}}\text{—O—CH}_2 \\
\text{CH}_3\text{—(CH}_2)_7\text{—CH=CH—(CH}_2)_7\text{—}\overset{\overset{\text{O}}{\|}}{\text{C}}\text{—O—CH} \quad + \quad 3\,\text{NaOH} \quad \longrightarrow \\
\text{CH}_3\text{—(CH}_2)_4\text{—(CH=CHCH}_2)_2\text{—(CH}_2)_6\text{—}\overset{\overset{\text{O}}{\|}}{\text{C}}\text{—O—CH}_2 \\
\text{a cottonseed oil} \hspace{6em} \text{strong base}
\end{array}$$

$$\begin{array}{cc}
\text{CH}_3\text{—(CH}_2)_{14}\text{—}\overset{\overset{\text{O}}{\|}}{\text{C}}\text{—O}^-\text{Na}^+ & \text{HO—CH}_2 \\
\text{sodium palmitate} & \\
\text{CH}_3\text{—(CH}_2)_7\text{—CH=CH—(CH}_2)_7\text{—}\overset{\overset{\text{O}}{\|}}{\text{C}}\text{—O}^-\text{Na}^+ & + \quad \text{HO—CH} \\
\text{sodium oleate} & \\
\text{CH}_3\text{—(CH}_2)_4\text{—(CH=CHCH}_2)_2\text{—(CH}_2)_6\text{—}\overset{\overset{\text{O}}{\|}}{\text{C}}\text{—O}^-\text{Na}^+ & \text{HO—CH}_2 \\
\text{sodium linoleate} & \text{glycerol} \\
\text{soaps} &
\end{array}$$

The overall reaction of saponification can be thought of as occurring in two steps. The first step is the hydrolysis of the ester linkages in the triglyceride to produce glycerol and three fatty acid molecules.

$$\text{fat or oil} + 3\,\text{H}_2\text{O} \longrightarrow 3\text{ fatty acids} + \text{glycerol}$$

The second step involves a reaction between the acid molecules and the base (usually NaOH) in the alkaline solution. This is an acid–base neutralization that produces water plus salts.

$$3\text{ fatty acids} + \text{NaOH} \longrightarrow 3\text{ fatty acid salts} + 3\,\text{H}_2\text{O}$$

The properties of the salts (soaps) obtained from saponification depend upon the base used. Sodium salts, known as "hard" soaps, are found in most cake soap used in the home. Potassium salts, or "soft" soaps, are used in some shaving creams and in liquid soap preparations.

Hydrogenation

Hydrogenation is a reaction we encountered in Section 15.7. Hydrogen is added across a carbon–carbon multiple bond converting carbon–carbon multiple bonds to carbon–carbon single bonds. Hydrogenation of fats and oils decreases the degree of unsaturation, as some double bonds are converted to single bonds. With this change there is a corresponding increase in the melting point of the fat or oil.

Hydrogenation is used to produce many food products. The peanut oil in many popular brands of peanut butter has been partially hydrogenated to convert the oil into a solid that does not separate out of the mixture as the oil would. Hydrogenation is used to produce solid cooking shortenings or margarines from liquid vegetable oils (see Figure 23–7). It is important not to complete the reaction and totally saturate all of the double bonds. If this is done, the product is hard and waxy (like beef tallow) instead of smooth and creamy.

Soft-spread margarines are partially hydrogenated oils. The extent of hydrogenation is carefully controlled to make the margarine soft at refrigerated temperatures (4°C). Consider the irony of those advertisements that promote these spreads as being "high in polyunsaturates"—many of the double bonds have purposely been changed to single bonds through controlled hydrogenation, increasing the degree of saturation.

$$\begin{array}{c}
CH_3-(CH_2)_{14}-\overset{O}{\underset{\|}{C}}-O-CH_2 \\
CH_3-(CH_2)_7-CH=CH-(CH_2)_7-\overset{O}{\underset{\|}{C}}-O-CH \quad + 2H_2 \xrightarrow[\text{pressure}]{\text{Ni catalyst}} \\
CH_3-(CH_2)_4-(CH=CHCH_2)_2-(CH_2)_6-\overset{O}{\underset{\|}{C}}-O-CH_2 \\
\text{a cottonseed oil}
\end{array}$$

$$\begin{array}{c}
CH_3-(CH_2)_{14}-\overset{O}{\underset{\|}{C}}-O-CH_2 \\
CH_3-(CH_2)_7-CH=CH-(CH_2)_7-\overset{O}{\underset{\|}{C}}-O-CH \\
CH_3-(CH_2)_{16}-\overset{O}{\underset{\|}{C}}-O-CH_2 \\
\text{solid shortening or margarine}
\end{array}$$

Rancidity

Fats and oils often develop unpleasant odors and/or flavors upon exposure to moist air at room temperature. These affected fats and oils have become *rancid*.

FIGURE 23–7 Hydrogenation is used in the production of many consumer food products from vegetable oil bases.

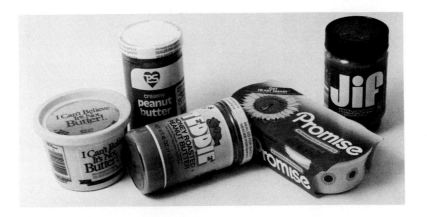

Rancidity results from two kinds of unwanted reactions: hydrolysis of triglyceride ester linkages, and oxidation of carbon–carbon double bonds in the fatty acid chains of triglycerides.

Hydrolytic rancidity results from the exposure of fats and oils to *moist* air. (All air contains some moisture.) Microorganisms in the air supply necessary enzymes to catalyze the hydrolysis. Volatile fatty acid molecules, especially those of low molecular weight, are freed as the result of triglyceride hydrolysis, and contribute to the disagreeable odors of rancid foods. For example, butyric acid is the source of the characteristic odor of rancid butter. Hydrolytic rancidity is prevented by storing fats and oils in closed containers (to minimize contact with microorganisms) and by refrigeration (all reactions go slower at lower temperatures). The unpleasant odor of sweat results from the hydrolysis of fats and oils on the skin.

Oxidative rancidity, in most cases, is a more important reaction than hydrolytic rancidity. In oxidative rancidity, double bonds in the unsaturated fatty acid components rupture and produce low molecular weight aldehydes. (This type of reaction was discussed in Section 18.5.) Many of the aldehydes have objectionable odors. In addition, these aldehydes can be further oxidized to give equally offensive low molecular weight carboxylic acids. Warmth and exposure to atmospheric oxygen induce oxidative rancidity. To avoid this unwanted oxidation, the food industry adds antioxidants to foods. Two naturally occurring antioxidants are vitamin C (ascorbic acid) and vitamin E (α-tocopherol). Two synthetic oxidation inhibitors are BHA and BHT, whose structures are

butylated hydroxy anisole (BHA)

butylated hydroxy toluene (BHT)

In the presence of air, antioxidants, rather than food, are oxidized.

Iodination and Iodine Numbers

iodination

Iodination *is an addition reaction in which an iodine-containing species* (I_2, *ICl, IBr*) *adds across the carbon–carbon double bond(s) of fatty acids.* Since I_2 is slow to react, reagents such as ICl or IBr are more commonly used in this reaction.

$$\underset{\text{fatty acid}}{-CH=CH-} + I_2 \longrightarrow -\underset{|}{\overset{I}{C}}H-\underset{|}{\overset{I}{C}}H- \quad \text{(slow)}$$

$$\underset{\text{fatty acid}}{-CH=CH-} + ICl \longrightarrow -\underset{|}{\overset{I}{C}}H-\underset{|}{\overset{Cl}{C}}H- \quad \text{(faster)}$$

Because the amount of I_2 or ICl that will react in this type of reaction is directly proportional to the number of double bonds in the fatty acids, this reaction can be used to determine the degree of unsaturation in fats and oils.

In practice, the iodine-containing compound in solution (a colored solution) is added dropwise to a solution of the triglyceride. As long as the added iodine compound is reacting with double bonds, the triglyceride solution remains color-

TABLE 23-3 Iodine Numbers of Selected Fats and Oils

Fat or Oil	Iodine Number
butterfat	32–35
beef tallow	40–42
lard	55–65
chicken fat	65–75
olive oil	80–88
peanut oil	93–95
corn oil	100–125
cottonseed oil	100–110
soybean oil	120–140
safflower oil	142–146

less. When all double bonds have added iodine (the end point), a very faint color appears in the triglyceride solution and persists with continued addition of iodine drops; the color results from unreacted iodine solution accumulating in the triglyceride solution.

An *iodine number* is used to specify the degree of unsaturation of a fat or oil. An **iodine number**, *for a fat or oil, is the number of grams of iodine absorbed per 100 grams of fat or oil.* (If another reagent such as ICl is used, the grams reacted is calculated and reported as if I_2 were used.) Table 23-3 lists the iodine numbers for selected fats and oils. Typically, animal fats have iodine numbers of less than 70 while the unsaturated vegetable oils have iodine numbers over 100.

iodine number

23.5 PHOSPHOGLYCERIDES

Referring to our classification scheme for saponifiable lipids (Figure 23-1a), we note a second group of glycerol esters besides fats and oils, the *phosphoglycerides*. **Phosphoglycerides** *are triesters of glycerol in which two* —OH *groups are esterified with fatty acids and one with phosphoric acid.*

phosphoglycerides

The block diagram and the general structure of a phosphoglyceride are

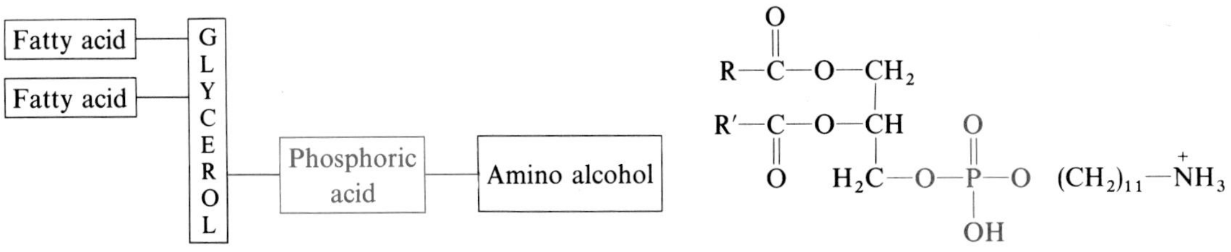

Fatty acid — GLYCEROL — Phosphoric acid — Amino alcohol

$$R-\overset{\underset{\parallel}{O}}{C}-O-CH_2$$
$$R'-\overset{\underset{\parallel}{O}}{C}-O-CH$$
$$H_2C-O-\overset{\underset{\parallel}{O}}{P}-O-(CH_2)_{11}-\overset{+}{N}H_3$$
$$OH$$

Note that the phosphoric acid molecule is also bonded to another entity besides glycerol, an amino alcohol. This bond between the phosphoric acid and the amino alcohol is also an ester linkage. Thus, phosphoglycerides can be alternatively viewed as phosphate diesters. The general structure for a phosphate diester is

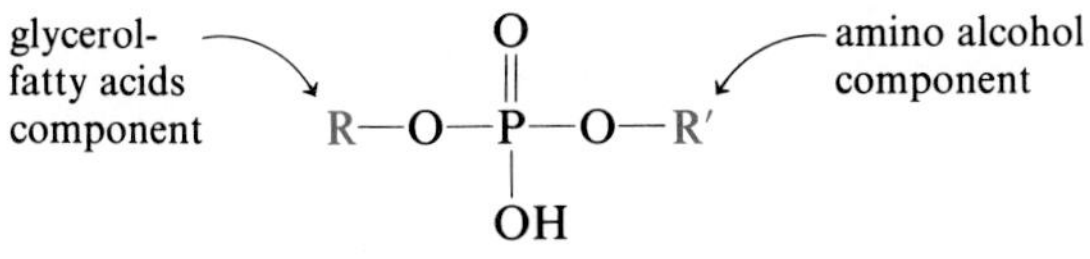

glycerol-fatty acids component → $R-O-\overset{\underset{\parallel}{O}}{P}-O-R'$ ← amino alcohol component
$$OH$$
phosphate diester

The nature of the amino alcohol attached to the phosphoric acid group determines the specific nature and function of the phosphoglyceride. Three of the most common amino alcohol substitutents found in biochemistry are choline, ethanolamine, and serine, whose structures are

$$HO-CH_2-CH_2-\overset{+}{N}(CH_3)_3 \qquad HO-CH_2-CH_2-NH_2 \qquad HO-CH_2-\underset{CO_2H}{\overset{H}{\underset{|}{\overset{|}{C}}}}-NH_2$$

choline   ethanolamine   serine

When the amino alcohol molecule is choline, the resulting phosphoglyceride is called a phosphatidyl choline, or, more commonly, a *lecithin*. The general structure for lecithins is

$$\begin{array}{c}
\overset{O}{\underset{\|}{}} \\
R-C-O-CH_2 \\
R'-C-O-CH \quad\quad \overset{O}{\underset{\|}{}} \\
\underset{O}{\overset{\|}{}} \quad H_2C-O-P-O-CH_2-CH_2-\overset{+}{N}(CH_3)_3 \\
\underset{O^-}{}
\end{array}$$

a phosphatidyl choline
(a lecithin)

Note that in a lecithin structure there are two charged sites. This is the form of lecithin in solution at a pH near 7 (the value for most body fluids). The —OH group of the phosphate has lost its hydrogen and the N atom has a positive charge because it is a quaternary ammonium ion (Section 20.6)

Phosphatidyl choline molecules are waxy solids that form colloidal suspensions in water. Egg yolks and soybeans are good dietary sources of these lipids. In the human body, lecithins are components of cell membranes.

Another group of phosphoglycerides are the *cephalins*. Ethanolamine and serine are the amino alcohols in the two most significant classes of cephalins. The general structures for these two types of phosphoglycerides are

$$\begin{array}{c}
R-C-O-CH_2 \\
R'-C-O-CH \quad\quad O \\
\underset{O}{\overset{\|}{}} \quad H_2C-O-P-O-CH_2-CH_2-\overset{+}{N}H_3 \\
\underset{O^-}{}
\end{array}$$

phosphatidyl ethanolamine

$$\begin{array}{c}
R-C-O-CH_2 \\
R'-C-O-CH \quad\quad O \quad\quad H \\
\underset{O}{\overset{\|}{}} \quad H_2C-O-P-O-CH_2-\underset{CO_2^-}{\overset{|}{\underset{|}{C}}}-\overset{+}{N}H_3 \\
\underset{O^-}{}
\end{array}$$

phosphatidyl serine

The phosphatidyl structure is common to all α phosphoglycerides.

Phosphatidyl ethanolamines have two positively charged sites; three charged sites are present in phosphatidyl serines. At a pH of 7, the amine group is protonated and both the —OH group of phosphoric acid and the carboxyl group (of serine) have lost their protons.

Note how phosphoglycerides are named. We write the name *phosphatidyl* followed by the name of the amino alcohol. A phosphatidyl group is the phosphoric acid—glycerol—fatty acid portion of the molecule.

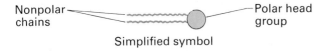

phosphatidyl group

Cephalins are found in the nerve tissue of the brain and the spinal cord. Much is yet to be learned about how these compounds function. Cephalins are also important in blood clotting.

The structures of phosphoglycerides, in general, can be represented by the simplified "head and two tail" structure shown in Figure 23–8a. In this structure the two nonpolar chains are the fatty acid chains attached to glycerol. The polar head represents the phosphate–aminoalcohol part of the molecule, or the part that contains the charged sites. We will use this "head and two tail" model for

FIGURE 23–8 The "head and two tail" representation for a phosphoglyceride.

(a) Symbolic representation of the head and two tails of a phosphoglyceride

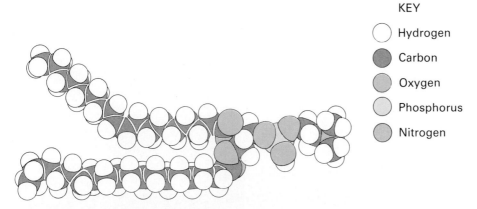

(b) Space-filling model of the phosphoglyceride phosphatidyl choline

phosphoglycerides when we consider the structure of cell membranes (Section 23.10). In Figure 23–8b we have, for comparison with the "head and two tail" model, a space-filling model of phosphatidyl choline. Note how easy it is to visualize "two tails" and a "head" for this structure.

23.6 WAXES

In our classification scheme for saponifiable lipids (Figure 23–1a), we noted two categories: glycerol esters and nonglycerol esters. We will now consider two categories of esters based on alcohols other than glycerols: *waxes* and *sphingolipids*.

Waxes *are monoesters formed from the reaction of a long-chain monohydroxy alcohol with a fatty acid molecule.* The block diagram for a wax is

| Fatty acid |—| Long-chain alcohol |

Many naturally occurring waxes are known. In most cases they function as protective coatings. Many plants that grow in dry regions have leaves coated with wax to minimize water loss. Human ears contain a wax secreted by glands. This wax acts as a protective barrier against infection by capturing air-borne particles. Bees secrete wax, which is used as a structural material for their hive. Commercially, waxes are marketed as protective coatings (automobile and floor waxes, and lotions) and for manufacturing candles.

Waxes coat the hair of many animals and the feathers of birds. For aquatic birds, this wax coating is a water-proofing agent (see Figure 23–9). Removing this wax from the feathers of aquatic birds has very serious consequences; the feathers become wet and the bird cannot maintain its buoyancy. Wax loss occurs when birds come in contact with an oil spill. Oil, a mixture of nonpolar hydro-

FIGURE 23–9 Waxes, esters of long chain alcohols, and fatty acids are essential to aquatic bird survival. Waxes coat feathers, water-proofing birds and insulating them from cold water.

TABLE 23-4 Structures, Sources, and Uses for Selected Important Waxes

Wax	Structure of a principal ester component	Source	Uses
beeswax	$CH_3-(CH_2)_{14}-C(=O)-O-(CH_2)_{29}-CH_3$	bees	candles, cosmetics, confections, medicinals, art preservation
carnauba wax	$(HO)_a-CH_2-(CH_2)_b-C(=O)-O-(CH_2)_c-CH_3$ $a = 0$ or 1 $b = 17-29$ $c = 31$ or 33	Brazilian palm trees	coatings for perishable products; polishing, candies, and pills; auto and floor waxes
rice bran wax	$CH_3-(CH_2)_m-C(=O)-O-(CH_2)_n-CH_3$ $m = 20$ or 22 $n = 21-35$	rice bran	lipstick base, plastics, processing aid
jojoba wax	$CH_3-(CH_2)_a-C(=O)-O-CH_2-(CH_2)_a-CH_3$ $a = 18$ or 20	jojoba beans	cosmetics and candles

carbons (Section 14.11), becomes a solvent for the waxes. Loss of feather wax also removes the insulative value of the feathers and in cold water some birds quickly die of exposure.

Most waxes are low-melting solids. They are not easily hydrolyzed. Saponification yields an alcohol and a fatty acid (as its carboxylate salt).

Table 23-4 gives structural formulas, sources, and uses for selected waxes. Natural waxes are always mixtures, like triglycerides, of numerous similarly structured esters. Sometimes natural waxes also contain other components besides esters. These components include free alcohols, free carboxylic acids, and long-chain hydrocarbons. Paraffin wax (Section 14.11) is a completely nonester material; it is a mixture of C_{20} and larger saturated hydrocarbons. Paraffin wax is used in home canning to seal jars.

23.7 SPHINGOLIPIDS

Sphingolipids *are saponifiable lipids derived from the amino dialcohol sphingosine.* The structure of sphingosine is

$$CH_3-(CH_2)_{12}-CH=CH-\underset{\underset{CH_2-OH}{|}}{\underset{H_2N-CH}{|}}{CH-OH}$$

sphingolipids

Note the long unsaturated nonpolar carbon chain in sphingosine.

All lipids derived from sphingosine have a fatty acid chain connected to the sphingosine through the amino group. In addition, the —OH group on the terminal carbon has a group attached to it. Various types of sphingolipids can be distinguished from each other according to the nature of this terminal —OH attachment. The block diagram for the general structure of a sphingolipid is

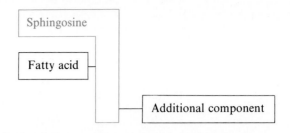

The fatty acid is linked to sphingosine by an amide group.

$$\cdots\text{C}-\overset{\overset{\text{O}}{\|}}{\text{C}}-\overset{\overset{\text{H}}{|}}{\text{N}}-\overset{|}{\text{C}}-\text{H}$$

$\underbrace{\phantom{\cdots\text{C}-\text{C}}}_{\text{fatty acid}}$ $\underbrace{\phantom{\text{N}-\text{C}}}_{\substack{\text{NH}_2 \text{ group} \\ \text{of sphingosine}}}$

This amide linkage contrasts with the ester linkages that connect fatty acids to glycerol in triglycerides. The amide linkage in sphingolipids can be hydrolyzed in basic solution to release the fatty acid. Hence, sphingolipids are saponifiable lipids.

We will now consider two classes of sphingolipids: sphingomyelins, which have a phosphate ester of choline attached to the terminal —OH group, and glycosphingolipids, which have a carbohydrate component attached to the terminal —OH group.

Sphingomyelins are found in all cell membranes and are an important component of myelin, the protective and insulating coating that surrounds nerves.

The block diagram for the structure of a sphingomyelin is

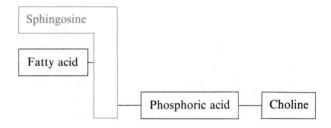

FIGURE 23-10 Space-filling model of a sphingomyelin molecule, a molecule with a "head and two tail" structure.

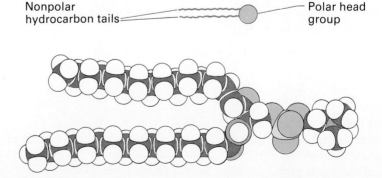

The following reaction shows the formation of a sphingomyelin from its components and gives a more detailed structural formula for a sphingomyelin.

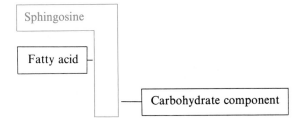

Note the formation of three water molecules as two ester linkages and one amide linkage form.

We should also add that sphingolipids, like phosphoglycerides, have structures that fit the "head and two tail" model discussed in Section 23.5. The space-filling model of a sphingomyelin molecule shown in Figure 23–10 illustrates this point. The fatty acid is one of the tails and the long carbon chain of sphingosine is the other one.

Glycosphingolipids are characterized by the presence of a carbohydrate component on the terminal —OH group in place of the phosphate fragment of sphingomyelins.

Cerebrosides and *gangliosides* are two important classes of glycosphingolipids. In *cerebrosides*, the carbohydrate component is either galactose or glucose. As the name suggests, cerebrosides are important constituents of brain tissue and are found in the white matter of the brain. They are also found in the myelin sheath around nerve fibers. The glucose or galactose in a cerebroside is attached to the terminal —OH group of sphingosine via a glycosidic linkage (Section 22.8). A general structure for a cerebroside, where galactose is the carbohydrate, is

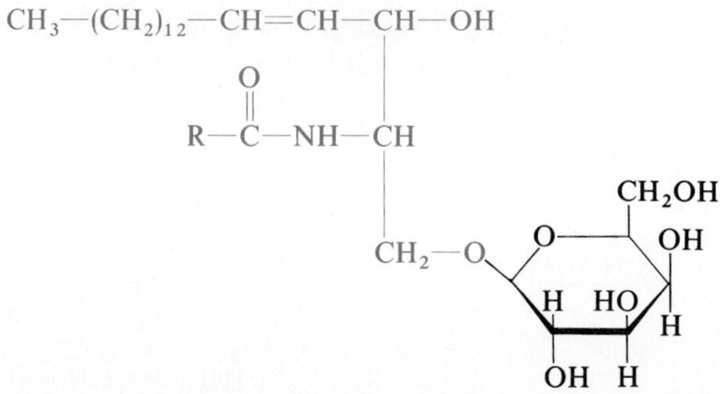

$$CH_3-(CH_2)_{12}-CH=CH-CH-OH$$
$$R-\overset{O}{\underset{\|}{C}}-NH-CH$$
$$CH_2-O-\text{(sugar ring with CH}_2\text{OH, OH, H, HO, OH, H substituents)}$$

Gangliosides contain an oligosaccharide instead of a monosaccharide as the carbohydrate component. The oligosaccharide can contain up to seven units. Gangliosides are found in the gray matter of the brain, in neural tissue, and are often parts of the receptor sites for neurotransmitters.

23.8 STEROIDS

This section is the first of two to deal with *nonsaponifiable* lipids. As shown in Figure 23–1b, *steroids* and *prostaglandins* (Section 23.9) are classes of nonsaponifiable lipids. (There are additional classes of these lipids, but we will not consider them in this text.)

steroids

Steroids *are lipids with structures that involve a fused ring system containing three six-membered rings and one five-membered ring.* The arrangement of rings within the fused system is as follows

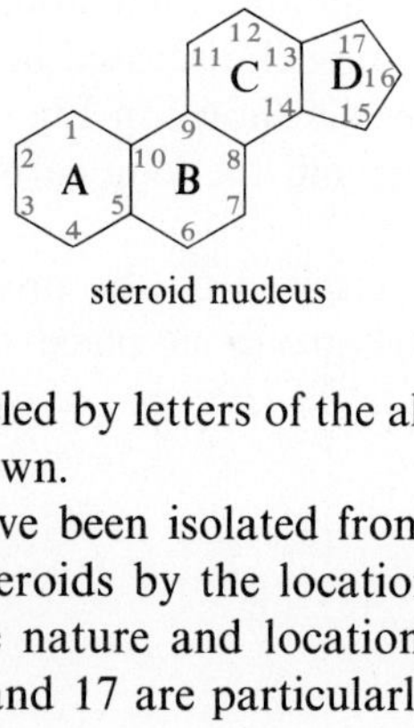

steroid nucleus

The rings are customarily labeled by letters of the alphabet and each carbon atom is labeled by a number as shown.

Many different steroids have been isolated from plants, animals, and human beings. We can distinguish steroids by the location of double bonds within the fused ring system and by the nature and location of substituents. Substituents located at carbons 3, 10, 13, and 17 are particularly common in biologically important steroids.

Steroids we will discuss in this section are cholesterol, bile salts, and steroidal hormones.

Cholesterol

Cholesterol is the most abundant steroid in the human body. It is found in cell membranes (up to 25% by mass), nerve tissue, and brain tissue (about 10% on a dry mass basis), and is the main component of gallstones. Human blood plasma contains about 50 mg of free cholesterol per 100 mL and about 170 mg of cholesterol esterified with various fatty acids.

cholesterol

Cholesterol *is an unsaturated steroid alcohol.* The presence of an —OH group on carbon number 3 of the steroid nucleus causes it to be an alcohol. The *-ol* ending in the name *cholesterol* reflects the presence of this —OH group. Unsaturation, in the form of a double bond, occurs between carbons 5 and 6 of the steroid nucleus. In addition, cholesterol has methyl groups bonded to carbon atoms 10 and 13, and a branched eight-carbon alkyl group is present on carbon 17.

cholesterol

Cholesterol has received much recent publicity because of a possible correlation between blood levels of cholesterol and the disease known as atherosclerosis. This condition occurs when lipid deposits form on the inner walls of blood vessels. These deposits, which contain large amounts of cholesterol, harden and can seriously obstruct the flow of blood.

Most body cholesterol is synthesized in the liver from carbohydrates, proteins, and fats in our food. Thus, eliminating cholesterol-containing foods from our diet will not necessarily lower cholesterol levels. We must also reduce the total caloric intake and change the types and amounts of fats we ingest. Certain unsaturated fish and vegetable oils, when substituted for saturated fats, lower the blood cholesterol level.

Plants do not synthesize cholesterol; therefore, foods such as margarine, sunflower oil and peanut butter contain no cholesterol. Animal fats do contain cholesterol, so that butter, milk, eggs, and meat are sources of cholesterol in our diet.

Cholesterol contributes to gall bladder attacks caused by gallstones. A large percentage of gallstones are almost pure cholesterol crystals that have precipitated from bile. Bile is a digestive juice excreted by the liver and stored by the gall bladder to help solubilize dietary fats within the small intestine.

Cholesterol is not totally bad. It occurs, in amounts up to 25%, in cell membranes, where its function involves membrane rigidity. There is a direct relationship between cholesterol concentration and rigidity.

Cholesterol plays a vital biological role in chemical synthesis within the human body. This substance is the starting material for the synthesis of numerous steroidal hormones, vitamin D, and bile acids.

Bile Salts

Bile salts *are emulsifying agents that make dietary lipids soluble in the aqueous environment of the digestive tract.* During digestion, bile salts are released into the intestine from the gall bladder, where they help digestion by emulsifying (solubilizing) fats and oils. Their mode of action is much like that of soap during washing (Section 19.6); micelles are formed.

bile salts

Bile salts are synthesized from cholesterol. Table 23–5 (p. 591) gives the specific structures for the major bile salts. All of the structures are similar. Note that except for sodium chenodeoxycholate, these salts have three —OH groups (carbons 3, 7, and 12) compared to one —OH group in cholesterol. In addition, the end of the carbon 17 side chain is a carboxylate anion which may or may not be coupled via an amide linkage to an additional group. The generalized structure for bile

salts may be represented as

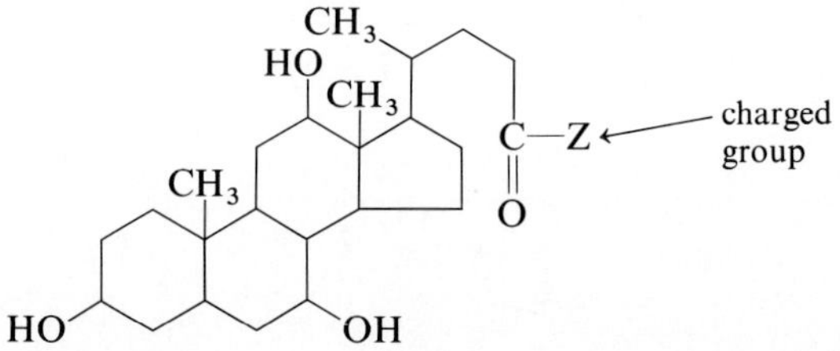

When bile is hydrolyzed, the most abundant steroid obtained is *cholic acid*. This compound can be considered the parent compound for all of the compounds shown in Table 23-5.

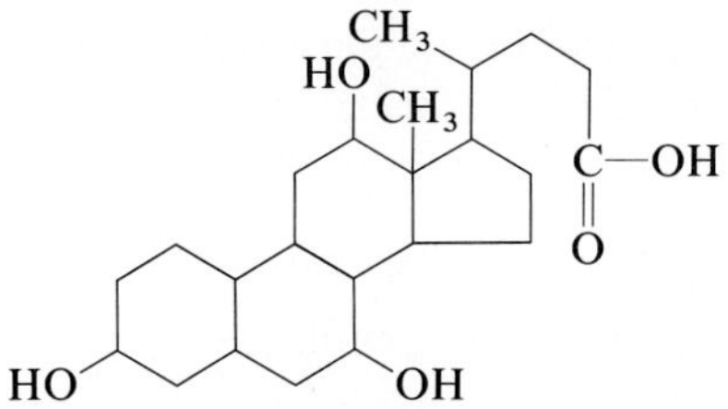

cholic acid

Fresh bile from the liver is golden yellow. On standing, bile darkens progressively from gold to green to blue and finally to brown as the bile pigments present gradually oxidize. Total excretion of bile salts varies from 0.5 to 2.0 grams daily and is responsible for the characteristic color of the feces.

Hormones

Hormones *are chemicals produced by ductless glands within the human body that provide a communication pathway between various tissues.* They are chemical messengers. Many, but not all, hormones in the human body are steroids. Such steroidal hormones are synthesized from cholesterol; hence they contain the four-ring fused system characteristic of that molecule.

Two areas of the human body are centers for steroidal hormone production: the adrenal glands and the gonads (the testes in males and the ovaries in females). The adrenal glands are found at the top of each kidney. More than 30 different hormones are produced by the adrenal glands.

Table 23-6 (p. 592) gives the structures of selected steroidal hormones. Small changes in structure for these compounds can alter function dramatically. For example, compare the structures and functions of progesterone and corticosterone (two of the entries in Table 23-6). Their structures differ by a single —OH group. Biological systems are extremely sensitive to minor changes in structure.

Section 27.6 will consider the subject of hormones in greater detail.

23.9 PROSTAGLANDINS

Prostaglandins *are twenty-carbon fatty acid derivatives that contain a cyclopentane ring.* Twenty-carbon fatty acids are converted into a prostaglandin structure when the eighth and twelfth carbon atoms of the fatty acid become connected to form a five-membered ring, as shown in the following equation.

arachidonic acid
(a 20-carbon fatty acid) → a prostaglandin

Note that the carbon chain of the fatty acid has been written in a "bent" form that approximates the carbon-chain configuration in prostaglandins in order to help visualize prostaglandin formation. Because prostaglandins do not contain an ester linkage, they are nonsaponifiable lipids.

Prostaglandins are named after the prostate gland, which was thought at first to be their only source. Today, more than 20 prostaglandins have been discovered in a variety of tissues. The physiological effects of prostaglandins are highly specific, acting primarily in the same tissue where they are synthesized.

Structures of six important prostaglandins are shown in Table 23–7 (p. 593). Prostaglandin structures are classified according to the number and arrangement of

TABLE 23–5 Structure for the Major Bile Salts that Help Solubilize the Dietary Lipids Fats and Oils

Name	Structure
sodium cholate	
sodium glycocholate	
sodium taurocholate	
sodium deoxycholate	

TABLE 23-6 Structures and Functions of Some Steroid Hormones

Structure	Name	Site of synthesis	Physiological function
	progesterone	corpus luteum	prepares uterus for pregnancy
	testosterone	testis	promotes development of secondary male sex characteristics
	estrone	ovary	promotes development of secondary female sex characteristics
	corticosterone	adrenal cortex	promotes formation of glycogen and stimulates degradation of fats and proteins
	aldosterone	adrenal cortex	increases blood pressure by stimulating reabsorption of sodium and chloride ions by kidneys

double bonds and hydroxyl and carbonyl groups. The E group prostaglandins have a carbonyl group at carbon number 9 while the F group compounds have a hydroxyl group at this position. In notations such as PGE_2 and PGF_3, the PG stands for prostaglandin, the third letter gives the group type, and the number indicates the number of double bonds present in the carbon chains.

Within the human body, prostaglandins are involved in many regulatory functions including raising body temperature, inhibiting the secretion of gastric juices, relaxing and contracting of smooth muscle, directing of water and electrolyte balance, intensifying pain, and enhancing inflammation responses.

Intravenous administration of PGE_1 and PGE_2 induces vasodilation, allowing a decrease in arterial blood pressure. PGE_2 also stimulates contractions of the

TABLE 23–7 Six Principal Naturally Occurring Prostaglandins and the Structure of Arachidonic Acid for Comparison

arachidonic acid

PGE$_1$ PGF$_{1\alpha}$

PGE$_2$ PGF$_{2\alpha}$

PGE$_3$ PGF$_{3\alpha}$

uterus during pregnancy. Infusion of PGE$_2$ at concentrations as low as only 4 micrograms per minute induces delivery within a few hours. Prostaglandins such as PGF$_2$ are being evaluated for potential use as contraceptives because they reduce the secretion of progesterone, a hormone needed for implantation of a fertilized ovum in the uterus.

Some prostaglandins raise body temperature (induce fever) and many are involved in the process of inflammation. Aspirin apparently reduces inflammation and fever because it can inactivate the enzyme needed for prostaglandin synthesis.

Natural prostaglandins cannot be taken orally because they are rapidly degraded in the digestive system. However, due to prostaglandins' potential as therapeutic drugs, researchers are trying to modify them so that they can be administered orally.

23.10 CELL MEMBRANES

All cells are surrounded by a membrane that confines their contents. Such membranes literally hold together the elements of life within cells. Figure 23–11 shows an electron micrograph of a cell membrane. We discuss cell membranes in this chapter because up to 80% of the mass of a cell membrane is lipid material.

The key to understanding cell membrane structure is the "head and two tail" model for the structure of certain types of lipids. We encountered this concept when discussing phosphoglycerides (Section 23.5) and again for sphingolipids

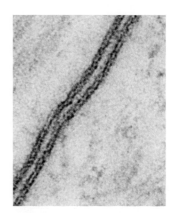

FIGURE 23–11 An electron micrograph showing a cell membrane.

FIGURE 23–12 Space-filling models of a phosphoglyceride and a sphingolipid molecule as well as the abbreviated symbol used to represent both.

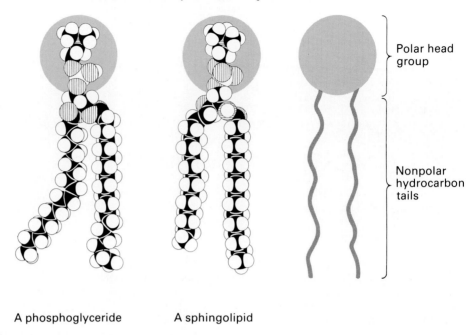

A phosphoglyceride   A sphingolipid

(Section 23.7). Figure 23–12 summarizes this concept by showing again the symbolism used to structurally represent these lipids.

The fundamental structure of a cell membrane is a *lipid bilayer* predominantly made up of various phosphoglyceride and sphingolipid molecules. A **lipid bilayer** *is a two-layer thick structure of lipid molecules in which nonpolar "tails" of the lipids*

lipid bilayer

FIGURE 23–13 The lipid bilayer model of cell membrane structure in which certain types of lipids form a double-layered sheet.

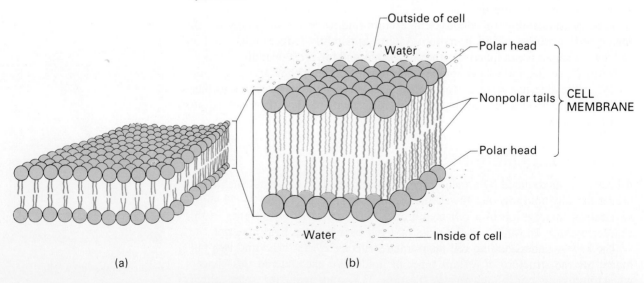

FIGURE 23–14   Cross-sectional views showing the comparison between structures of the lipid micelle and the lipid bilayer.

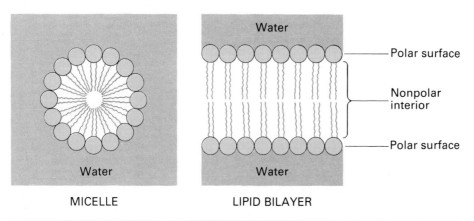

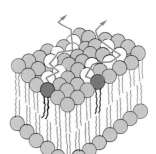

Lateral diffusion (rapid)

are in the middle and the polar "heads" are on the outside surface. Figure 23–13 shows a large section of a lipid bilayer and a closeup view of a smaller section of a bilayer.

Such a bilayer, as shown in Figure 23–13, is 6 to 9 billionths of a meter thick, that is, 6 to 9 nanometers thick. There are three distinct parts to such a layer: the exterior polar "heads", the interior polar "heads", and the central nonpolar "tail". These three distinct regions can be seen in the electron micrograph of a cell membrane (Figure 23–11).

A lipid bilayer is much like micelles (Section 19.6). Micelles, however, have only the one spherical surface, while the lipid bilayer has two flat parallel surfaces. Figure 23–14 shows the cross-section views of both a micelle and the sheet-like structure of the lipid bilayer.

As with the micelle, the lipid bilayer is held together by dipole–dipole interactions and not by covalent bonds. This means each phospholipid or sphingolipid is free to diffuse around the surface of a lipid bilayer. This diffusion is called *lateral diffusion* and can be very rapid. In fact, a lipid molecule can travel a distance greater than 50,000 times its own length across the surface of a lipid bilayer in only one second! On the other hand, the polar heads of a lipid bilayer seldom flip–flop from one side of the layer to the other. This process, called *transverse diffusion,* occurs at a rate only one billionth as fast as lateral diffusion. This rate is slower because the polar head group is repelled by the nonpolar interior of the bilayer. Figure 23–15 contrasts the processes of lateral and transverse diffusion.

Specific physical properties of cell membranes, such as flexibility, depend on the identity of the fatty acid chains present in the lipids. When phosphoglycerides containing a high percentage of *unsaturated* fatty acids dominate, the membrane is more flexible than when *saturated* fatty acids are present (see Figure 23–16 on p. 596). Recall, from Section 23.3, that unsaturated fatty acid chains are bent and prevent close packing of the fatty acids in the membrane. Just as this less organized packing decreases the melting point of unsaturated fatty acids, it also increases the fluidity of the membrane.

Proteins are also components of lipid bilayers. Overall, the proteins are responsible for moving substances such as nutrients and electrolytes across the membrane and they also act as receptors which bind hormones and neurotransmitters.

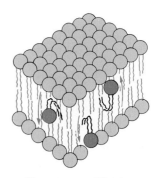

Transverse diffusion (extremely slow)

FIGURE 23–15   Diagram of lateral versus transverse diffusion.

FIGURE 23-16 Membranes rich in saturated fatty acid "tails" are more tightly packed and less flexible than those rich in unsaturated fatty acid "tails".

Saturated fatty acid chains fit tightly.

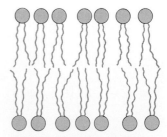

Unsaturated fatty acid chains are bent and fit loosely.

Two kinds of proteins—peripheral and integral—are present in cell membranes. **Peripheral proteins** *are associated only with the surface of the bilayer.* These proteins can be easily removed from the lipid bilayer. **Integral proteins** *are those that are within or extend through the lipid bilayer.* These proteins are essentially anchored in place. Figure 23-17 contrasts the two types of proteins found in a lipid bilayer.

The proteins in a lipid bilayer control the movement of other substances from the outside of the cell to the inside of the cell (or vice versa). Passage of molecules through the membrane occurs in three different ways: simple diffusion, facilitated diffusion, and active transport.

Small molecules and ions, such as H_2O, O_2, CO_2, and Cl^-, can enter the cell by *simple diffusion.* Here, integral protein sites serve as channels or pores through which the small ions and molecules can pass. No cellular energy is needed for this process, which depends only on the concentration gradient, as we discussed in connection with osmosis (Section 8.17). Although simple diffusion does occur,

FIGURE 23-17 Proteins are an important part of a cell membrane.

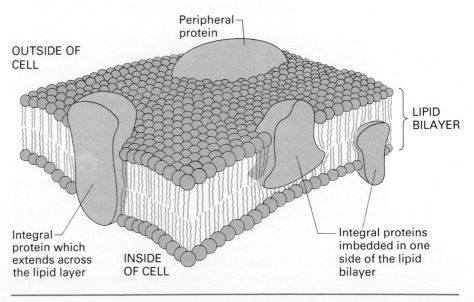

FIGURE 23-18 Diagram of membrane transport. Proteins mediate both facilitated diffusion and active transport. In the case of active transport, an expenditure of energy is needed to move the transported substance from an area of low concentration to one of higher concentration. In the case of facilitated diffusion, the protein merely acts to promote the natural tendency of substances to move from an area of high concentration to one of lower concentration.

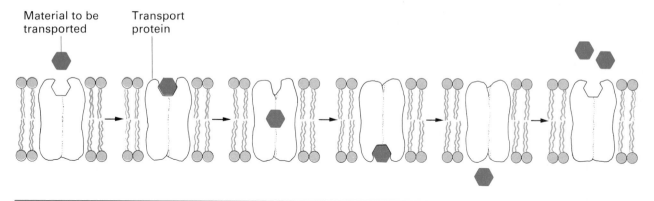

it is not the major pathway by which molecules cross cell membranes. The many larger molecules move across the cell membrane either by facilitated diffusion or active transport.

When substances are transported by proteins from an area of high concentration to an area of low concentration, the process is called *facilitated diffusion*. As with simple diffusion, no energy is required for such a process because the relative concentrations favor the movement. However, because of the size or chemical makeup of these molecules, they must enter the cell through a selective gate, which facilitates their entry.

There are times when a cell must bring in needed materials against a concentration gradient. For example, glucose and galactose, two sugars needed to provide the cell with energy, are almost always found in higher concentrations inside the cell. To move these nutrients across the cell membrane into the cytoplasm (the liquid inside the cell), the cell must expend energy. *Active transport* is a membrane transport process that requires the expenditure of energy. This energy is necessary because the material moves from a region of low concentration to an area of higher concentration.

Glucose and galactose are absorbed only by active transport. Fructose, on the other hand, is absorbed by cells of the small intestine by facilitated diffusion—concentrations of fructose are usually lower inside the cell than outside. Figure 23-18 shows how proteins act as intermediaries in moving such substances across the cell membrane.

Some compounds can be transported across the membrane only if another ion or compound is transported simultaneously. *Symport* is the process of simultaneous transport of two substances in the same direction across the membrane. Active transport of glucose into cells requires the symport of a Na^+ ion for every glucose molecule. Sodium ions are pumped out of a cell at one location and then diffuse to another site where they and glucose are simultaneously contratransported back into the cell. *Antiport* is a process in which two substances are carried in opposite directions across the membrane. This means one substance cannot enter the cell unless another substance leaves it at the same time. For example, Ca^{2+} ions are often expelled from the cell as Na^+ ions are brought in.

EXERCISES AND PROBLEMS

Classification of Lipids

23.1 Give a formal definition for the class of biochemicals called lipids.

23.2 In which of the following solvents would you expect lipids to be soluble? Explain your answers.
a. H_2O
b. $CH_3-CH_2-O-CH_2-CH_3$
c. CH_3-CH_2-OH
d. CCl_4

23.3 Which of the following types of lipids are nonsaponifiable lipids?
a. fats
b. oils
c. waxes
d. steroids
e. prostaglandins
f. sphingolipids

Fatty Acids

23.4 How does a fatty acid differ in structure from a simple carboxylic acid?

23.5 Name the saturated fatty acids that contain 16, 18, and 20 carbon atoms.

23.6 Compare the structures of stearic acid and oleic acid. How are they different? What do they have in common?

23.7 Draw the structures of four naturally occurring fatty acids having 18 carbon atoms.

23.8 Compare the melting points of arachidic acid and arachidonic acid. Suggest a reason for the dramatic difference between their melting points.

23.9 Place the following fatty acids in order of increasing melting point (lowest melting point first).
a. $CH_3-(CH_2)_{18}-COOH$
b. $CH_3-(CH_2)_6-CH=CH-(CH_2)_4-COOH$
c. $CH_3-(CH_2)_{12}-COOH$

23.10 What are essential fatty acids and why are they essential?

23.11 Elaidic acid and oleic acid are *trans* and *cis* isomers. Which of these two geometric isomers would you expect to have the lowest melting point? Explain.

Fats and Oils

23.12 What is the difference between a simple triglyceride molecule and a mixed triglyceride molecule?

23.13 Define in terms of structural features the class of compounds called triglycerides.

23.14 Draw the structure of a triglyceride formed from glycerol and
a. three molecules of palmitic acid.
b. two linoleic acid molecules and one myristic acid molecule.
c. one linoleic acid, one stearic acid, and one oleic acid molecule.

23.15 Identify the fatty acids present in the following triglyceride.

$$\begin{array}{l} CH_2-O-\overset{O}{\overset{\|}{C}}-(CH_2)_{14}-CH_3 \\ CH-O-\overset{O}{\overset{\|}{C}}-(CH_2)_{10}-CH_3 \\ CH_2-O-\overset{O}{\overset{\|}{C}}-(CH_2)_7-CH=CH-(CH_2)_7-CH_3 \end{array}$$

23.16 How does a fat differ from an oil in terms of
a. melting point
b. chemical structure

23.17 Explain the meaning of the term *polyunsaturated* when it is used in advertising for vegetable oils.

23.18 Using the information in Table 23–2, determine which one of each of the following pairs of naturally occurring triglyceride mixtures is more unsaturated.
a. lard or beef tallow
b. peanut oil or soybean oil
c. human fat or corn oil

23.19 Only one carbon atom in a triglyceride is capable of being a chiral carbon atom. Which carbon atom is this?

23.20 The hydrolysis of an optically active triglyceride yields one mole of glycerol, two moles of palmitic acid, and one mole of oleic acid. Draw the structural formula of this triglyceride.

Chemical Reactions of Fats and Oils

23.21 Name, in general terms, the products produced from
a. the hydrolysis of a triglyceride.
b. the saponification of a triglyceride.
c. the complete hydrogenation of a triglyceride.

23.22 Write two structural equations, one for the acid hydrolysis and one for the basic hydrolysis (NaOH) of the following triglyceride.

$$\begin{array}{l} CH_2-O-\overset{O}{\overset{\|}{C}}-(CH_2)_{14}-CH_3 \\ CH-O-\overset{O}{\overset{\|}{C}}-(CH_2)_{14}-CH_3 \\ CH_2-O-\overset{O}{\overset{\|}{C}}-(CH_2)_7-CH=CH-(CH_2)_7-CH_3 \end{array}$$

23.23 How many moles of H_2 will react with the following triglyceride?

$$\text{CH}_2-\text{O}-\overset{\overset{\text{O}}{\|}}{\text{C}}-(\text{CH}_2)_7-\text{CH}=\text{CH}-\text{CH}_2-\text{CH}=\text{CH}-(\text{CH}_2)_4-\text{CH}_3$$

$$\text{CH}-\text{O}-\overset{\overset{\text{O}}{\|}}{\text{C}}-(\text{CH}_2)_7-\text{CH}=\text{CH}-(\text{CH}_2)_7-\text{CH}_3$$

$$\text{CH}_2-\text{O}-\overset{\overset{\text{O}}{\|}}{\text{C}}-(\text{CH}_2)_7-(\text{CH}=\text{C}-\text{CH}_2)_2-\text{CH}=\text{CH}-\text{CH}_2-\text{CH}_3$$

23.24 Why can only unsaturated triglycerides undergo hydrogenation?

23.25 Describe how margarines are produced from vegetable oils.

23.26 Why do animal fats and vegetable oils become rancid when exposed to moist warm air?

23.27 Describe the differences between hydrolytic rancidity and oxidative rancidity.

23.28 Would you predict that fats or oils would have higher iodine numbers? Explain.

23.29 Describe the visual result obtained when I_2 solution or ICl solution is continuously added to a solution of an oil.

23.30 Which member of each of the following pairs has a lower iodine number?
a. olive oil or soybean oil
b. beef tallow or chicken fat
c. peanut oil or cottonseed oil

Phosphoglycerides

23.31 Draw the general structure of a phosphoglyceride.

23.32 How do phosphoglycerides differ structurally from triglycerides?

23.33 Draw the structures of three amino alcohols commonly esterified to the phosphate group in phosphoglycerides. How do the structures of the amino alcohols differ from each other?

23.34 Draw structural formulas for the products of complete hydrolysis of phosphatidyl choline.

23.35 Draw the structure of a cephalin that contains the following building blocks: phosphoric acid, glycerol, palmitic acid, linolenic acid, and serine.

23.36 How do the structures of cephalins differ from the structures of lecithins?

23.37 List a function in the human body for cephalins and for lecithins.

23.38 What is the structure of a phosphatidyl group?

23.39 How are phosphoglycerides similar to soaps and detergents?

23.40 What charges exist on the polar head of each of the following phosphoglycerides at physiological pH?
a. phosphatidyl choline
b. phosphatidyl ethanolamine
c. phosphatidyl serine

Waxes

23.41 Draw the general structure of a wax.

23.42 Name the two classes of organic compounds produced when a wax is hydrolyzed.

23.43 How does a wax differ from a triglyceride in structure and chemical properties?

23.44 Draw the structure of a wax formed from palmitic acid and cetyl alcohol ($\text{CH}_3-(\text{CH}_2)_{14}-\text{CH}_2-\text{OH}$).

23.45 How does beeswax differ from carnauba wax?

23.46 Describe a function that may be served by a wax
a. on the surface of a leaf.
b. within the human ear.
c. on the feathers of an aquatic bird.

Sphingolipids

23.47 What is the general structure for a sphingolipid?

23.48 What are the structural similarities and differences between sphingolipids and
a. triglycerides   b. phosphoglycerides

23.49 Glycerol is the structural backbone of all triglycerides. Give the name and draw the structure of the structural backbone of all sphingolipids.

23.50 Draw the structure of the sphingolipid that could form from sphingosine, stearic acid, phosphoric acid, and ethanolamine.

23.51 Sphingolipids are said to have two "nonpolar tails". One tail is the fatty acid residue. What is the identity of the other tail?

23.52 How do sphingomyelins and glycosphingolipids differ in structure?

23.53 How do cerebrosides and gangliosides differ in structure?

23.54 What type of bond joins the sugar unit to sphingosine in a glycosphingolipid?

23.55 What biological role is served by
a. sphingomyelins
b. cerebrosides
c. gangliosides

Steroids

23.56 Draw the fused hydrocarbon ring system characteristic of steroids.

23.57 Draw the structure of cholesterol.

23.58 Explain why steroids are classified as lipids although they do not contain any fatty acid residues.

23.59 Which organ in the human body synthesizes most of the cholesterol we need?

23.60 Explain why a diet rich in cholesterol is not necessarily the principal cause of high blood cholesterol.

23.61 Draw a generalized structure for a bile salt.

23.62 What role do bile salts play in the process of digestion in humans?

23.63 What are hormones and from what two sources are most steroidal hormones synthesized?

23.64 List all structural differences in each of the following pairs of steroids.
a. cholesterol and sodium cholate
b. esterone and testosterone
c. corticosterone and progesterone

Prostaglandins

23.65 What common structural features are present in all prostaglandin molecules?

23.66 What is the structure of the fatty acid from which most prostaglandins are synthesized?

23.67 Give at least three physiological effects that are regulated by prostaglandins.

23.68 Explain how aspirin appears to function in relieving inflammation.

23.69 Why must prostaglandins be administered intravenously?

Cell Membranes

23.70 What are the two major biochemical components of a cell membrane?

23.71 What are the structural characteristics common to lipids found in cell membranes?

23.72 Name and describe the two general types of membrane proteins.

23.73 Sketch a drawing of a membrane. Identify the polar and nonpolar regions of the membrane.

23.74 How would a cell membrane with a large percent of linoleic acid differ from one with a large percent of palmitic acid?

23.75 Explain the difference between lateral and transverse diffusion of lipids in a membrane. Which process is faster?

23.76 If a very small pin were used to pierce a membrane and then removed, would it leave a hole or would the membrane seal up the hole? Explain.

23.77 List three functions of proteins in cell membranes.

23.78 What is the difference between the protein-mediated processes of facilitated diffusion and active transport?

23.79 Compare the transport processes of symport and antiport.

23.80 Explain why the structural characteristics of a cell membrane are similar to those of a micelle.

24 Proteins

Objectives

After completing Chapter 24, you will be able to:

24.1 Understand the general functions in biological systems and the characteristics of protein molecules.
24.2 Recognize amino acids as the components of proteins, and be familiar with the distinguishing aspects of the common amino acids.
24.3 Explain why some amino acids are designated as essential.
24.4 Recognize a given structure of an amino acid enantiomer as a D or L form.
24.5 Write the structure of a given neutral amino acid in the zwitterionic, positively charged, and negatively charged form, and indicate the general pH conditions in which each is most likely to be found in solution.
24.6 Recognize peptide bonds between amino acids in polypeptides, write out all possible peptide sequences given the component amino acids, and name simple small peptides.
24.7 List the four levels of structural organization of proteins.
24.8 Understand that primary structure of proteins refers to the order of attachment of amino acids and the consequences of changes in this structure.
24.9 Understand the type of attractive interaction that contributes to the secondary structure of a protein and the three main types of secondary structure.
24.10 Recognize the four types of attractive interactions that contribute to the tertiary structure of a protein.
24.11 Understand that quaternary structure is the result of various subunit interactions in oligomeric proteins.
24.12 Realize that protein hydrolysis destroys the primary structure of proteins and that hydrolysis is the basis for protein digestion in the human body.
24.13 Describe on a molecular level how various physical and chemical changes can denature a protein.
24.14 Describe the structure and function of immunoglobulins.

INTRODUCTION

Proteins are components of all living cells. They are found in all forms of life, from the simplest (bacteria, algae, and other microorganisms) to the most complex (human beings). Proteins are required in the diets of animals and humans to synthesize tissues, enzymes, certain hormones, and some blood components. In addition, they are used to maintain and repair existing tissues and as a source of energy. This significant role of proteins in life processes is alluded to in the root of the word *protein*. Protein is derived from the Greek word "proteios" which means "of first importance".

Next to water, proteins are the most abundant substances in most cells—from 10 to 20% of the cell's mass (Section 22.1). All proteins contain the elements carbon, hydrogen, oxygen, nitrogen, and sulfur. Other elements such as phosphorus, iodine, and iron are essential constitutents of certain specialized proteins. Casein, the main protein of milk, contains phosphorus, an element very important in the diet of infants and children. Hemoglobin, the oxygen-transporting protein of blood, contains iron. The presence of the element nitrogen in proteins sets them apart from carbohydrates and lipids. The average nitrogen content of proteins is about 16% by mass.

Proteins are one of the three major categories of foods that humans need; the other two are carbohydrates and fats. High-protein foods include fish, beans, nuts, cheese, eggs, poultry, and meat. Because these foods are generally rather expensive, they are least available in underdeveloped countries. Protein deficiency is a significant problem among the world's undernourished peoples. A major area of scientific research addresses the problem of developing an adequate supply of high-quality protein for an ever-increasing population.

24.1 FUNCTIONS AND CHARACTERISTICS OF PROTEINS

An extraordinary number of different protein molecules, each with a different function, exist in the human body. A typical human cell contains about 9000 different kinds of protein molecules and it is estimated that the human body contains about 100,000.

The biological importance of proteins results from their wide variety of functions (Table 24–1, p. 604). *Structural proteins* provide mechanical support and are the chief constituent of skin, bones, hair, and fingernails. (The main structural material for plants is cellulose, Section 22.10; for animals and humans it is protein.) *Transport proteins* bind and transport specific molecules in cells and in the blood. (Recall, from Section 23.10, the role that proteins play in cell membranes.) *Catalytic proteins*, also called enzymes, catalyze virtually all reactions that take place in a living organism. We will discuss this most important type of protein in Chapter 25. *Nutrient proteins* serve as storage forms of nutrients for a developing organism. This situation is somewhat similar to the way starch and glycogen serve as storage forms for glucose (Section 22.10). *Regulatory proteins* serve as chemical messengers and help regulate cell processes. These proteins are often called protein hormones. *Contractile proteins*, found in muscle cells, are responsible for the property of

TABLE 24-1 The Biological Functions of Proteins

General function	Example	Specific Function
structural	collagen	forms connective tissue and cartilage
	keratin	forms hair, wool, skin, nails, feathers, and horns
	elastin	elastic connective tissue
transport	hemoglobin	transports oxygen in the blood to the cells
	lipoprotein	transports fatty acids in the blood
	transferrin	transports iron in the blood
catalytic (enzymes)	trypsin	hydrolyzes proteins
	sucrase	hydrolyzes sucrose
	lipase	hydrolyzes triglycerides
nutrient	ferritin	stores iron for reuse
	egg albumin	supplies amino acids to developing embryo
	zein	stores amino acids in corn
regulatory	insulin	regulates glucose levels in the blood and its metabolism
	somatotropin	growth hormone
	parathyroid hormone	regulates calcium in metabolism
contractile	actin	thin filaments in muscle
	myosin	thick filaments in muscle
	dynein	assists in movement of cilia and flagella
defense	antibodies	helps fight infectious diseases
	prothrombin	causes blood clotting
toxic	botulinus toxin	poison produced by *Clostridium botulinum*
	phosphodiesterase	one toxic enzyme in snake venom
	ricin	poisonous protein in caster beans
sensory	rhodopsin	absorbs light in the eye
	acetyl choline receptor	aids in nerve impulse transmission

motion. Muscle expansion and contraction are involved in all movements, from the swimming of sperm in the reproductive process to the blinking of an eye. *Defense proteins*, in vertebrates, serve as antibodies in the immune system. These proteins recognize and deactivate protein in invading organisms such as viruses and bacteria. *Toxic proteins*, such as snake venom, serve as protective devices for the organisms producing them. *Sensory proteins* aid in responses to stimuli such as nerve impulse or light (vision).

Proteins are extremely large molecules with molecular weights that vary from about 6000 to several million amu. Their immense size (in the molecular sense) can be appreciated by comparing glucose with hemoglobin, a relatively small protein. Glucose has a molecular weight of 180 amu, while the molecular weight of hemoglobin is 65,000 amu. The molecular formula of glucose is $C_6H_{12}O_6$, and that of hemoglobin is $C_{2952}H_{4664}O_{832}N_{812}S_8Fe_4$. The molecular weights of some representative proteins are given in Table 24-2.

With the great diversity of protein functions (Table 24-1) and the wide range in size (molecular weights) for proteins, what is the basic structural unit for such molecules? Proteins, like polysaccharides, are polymers. In polysaccharides, the monomeric units are monosaccharides (usually glucose)—Section 22.10. In proteins, the monomeric units are amino acids, substances that contain both an amino group and a carboxyl group. **Proteins** *are polymers of amino acids in which the amino acids are linked together by amide linkages formed between the carboxyl group of one amino acid and the amino group of another amino acid.*

TABLE 24–2 Molecular Weights of Some Common Proteins

Protein	Molecular Weight (amu)
insulin (human)	6,000
myoglobin (transports O_2)	17,000
pepsin (digestive enzyme)	24,000
ovalbumin	40,000
hemoglobin	65,000
serum albumin (human)	69,000
immunoglobulin G	150,000
catalase (cellular enzyme)	250,000
fibrinogen	400,000
pyruvate dehydrogenase complex (E. coli)	4,600,000

The starting point for a discussion of proteins is, thus, amino acids, the building blocks for proteins. Our knowledge of amines (Chapter 20), carboxylic acids (Chapter 19) and amides (Chapter 20) will all be important as we begin exploring amino acids and proteins.

24.2 AMINO ACIDS—BUILDING BLOCKS FOR PROTEINS

An **amino acid** is *an organic compound that contains both an amino (—NH_2) group and a carboxylic acid (—COOH) group.* In amino acids found in proteins, it is almost always true that the amino and carboxyl groups are separated from each other by a single carbon atom. Amino acids with this structural characteristic are called α-*amino acids*. The general formula for an α-amino acid is

$$H_2N-\underset{\underset{R}{|}}{CH}-\underset{\underset{}{\|}}{\overset{O}{C}}-OH$$

(α-carbon atom; amino group; acid group; characteristic side chain)

The R group in the general formula, called the *side chain*, may be hydrogen, an alkyl group, an aromatic ring, or a heterocyclic ring. The nature of this R group is responsible for the differences among the various α-amino acids.

The simplest α-amino acid is the one with hydrogen as the R group.

$$H_2N-\underset{\underset{H}{|}}{CH}-C\underset{OH}{\overset{O}{\diagup\!\!\!\diagdown}}$$

The common name of this amino acid is *glycine*. Next in complexity is the amino acid *alanine*, in which R is a methyl group.

$$H_2N-\underset{\underset{CH_3}{|}}{CH}-C\underset{OH}{\overset{O}{\diagup\!\!\!\diagdown}}$$

TABLE 24-3 The Common Amino Acids Found in Proteins

Nonpolar side chains

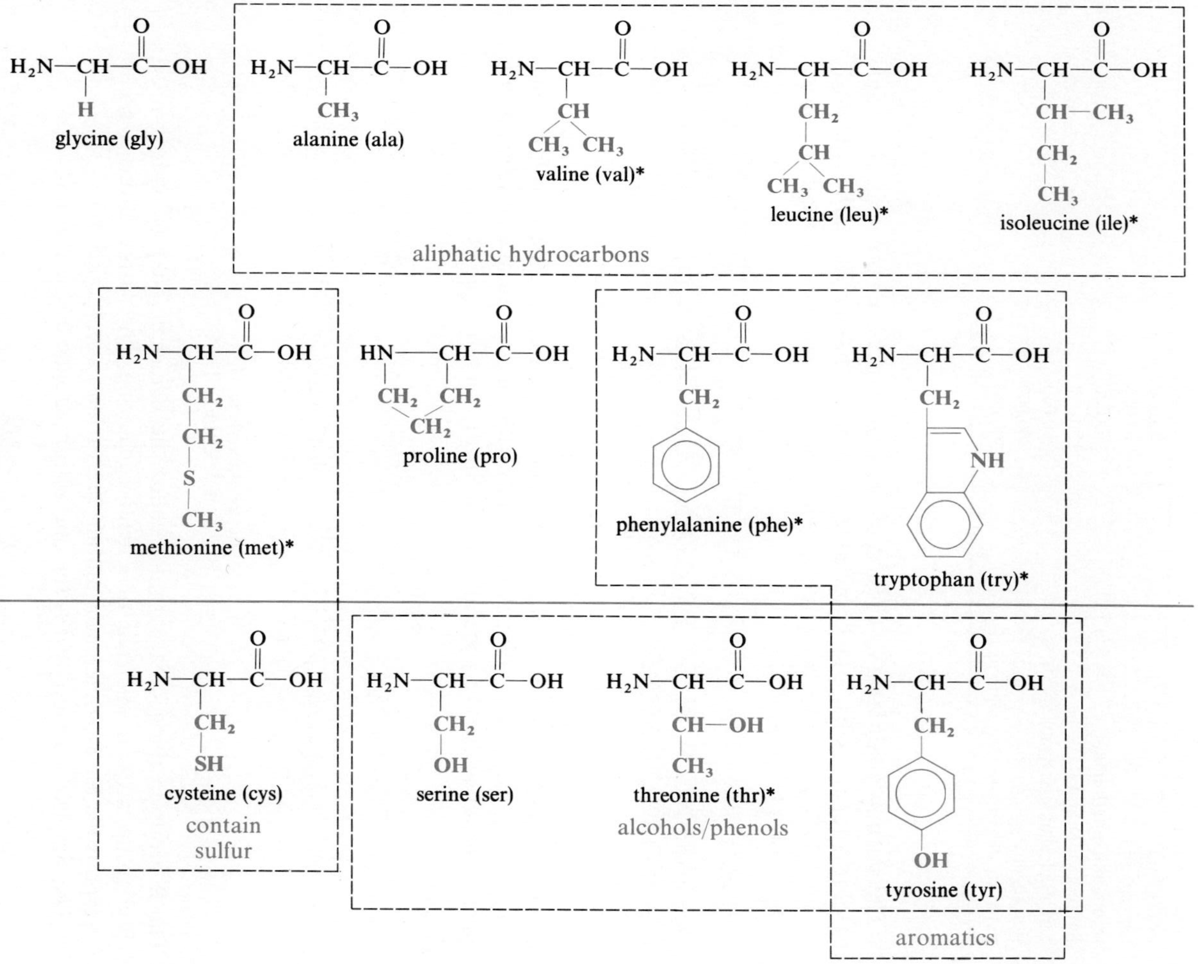

glycine (gly)

alanine (ala)
valine (val)*
leucine (leu)*
isoleucine (ile)*
— aliphatic hydrocarbons

methionine (met)*
proline (pro)
phenylalanine (phe)*
tryptophan (try)*

cysteine (cys) — contain sulfur
serine (ser)
threonine (thr)* — alcohols/phenols
tyrosine (tyr) — aromatics

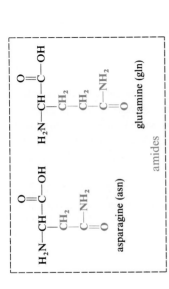

More than 30 different α-amino acids have been identified as constituents of protein. Only 20, however, are commonly encountered. Table 24–3 (pp. 606–607) lists the structures of these 20 most common α-amino acids. Because IUPAC names for these amino acids are quite cumbersome, common names, which are given in the table, are almost always used. The names of amino acids are often abbreviated by using the first three letters of the acid's common name. These name abbreviations are also given in Table 24–3.

Within Table 24–3, the amino acids are grouped according to the polarity properties of side chains. Note that there are three categories: nonpolar side chain, polar side chain, and very polar side chain.

The properties of specific proteins are mostly determined by the properties of the side chains on the amino acids. For example, insoluble, structural proteins (such as those in hair, wool, and tendons) contain high percentages of amino acids with nonpolar side chains. Membrane proteins also have large numbers of nonpolar side chains, thus providing attraction to the nonpolar interior of lipid bilayers (Section 23.10). The more water-soluble proteins (such as albumin and hemoglobin) contain large amounts of amino acids with polar side chains.

Because the amino acid composition of proteins determines their structure, function, and overall properties, it is worth our time to examine further the characteristics of the amino acids listed in Table 24–3.

Nine amino acids have nonpolar side chains. Four of these amino acids contain aliphatic hydrocarbon side chains: alanine, valine, leucine, and isoleucine. Proline has a hydrocarbon chain that is involved in a cyclic system. In essence, the hydrocarbon side chain wraps around and is bound to the amino group. Proline is the only member of the nonpolar side chain group that contains a secondary amino group.

Phenylalanine and tryptophan contain aromatic groups, as does tyrosine. Due to its phenolic group, tyrosine's side chain is classified as a polar side chain. Both methionine and cysteine have side chains that contain sulfur. Cysteine's sulfhydryl group (—SH) imparts enough polar character to the side chain to classify it as a polar side chain. As we will see shortly, cysteine side chains play an important role in protein structure.

The polar side chains contain functional groups we encountered in our study of organic chemistry. Serine and threonine contain primary and secondary alcohol

FIGURE 24–1 Charged forms of the acidic and basic amino acids in aqueous solution at physiological pH.

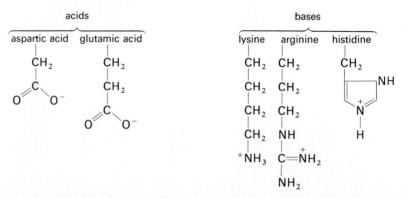

groups respectively. Asparagine and glutamine have side chains containing an amide functional group.

The term *very polar* side chain is reserved for amino acids whose side chains have acquired a charge at physiological pH. Aspartic acid and glutamic acid lose protons at neutral pH. Thus, their side chains are negatively charged. On the other hand, nitrogen atoms in the side chains of lysine, arginine, and histidine behave as bases at physiological pH by accepting protons from solution. These nitrogen atoms thus bear a positive charge under normal physiological conditions. Charged side chains, either positive or negative, are much more polar than the uncharged polar side chains of other amino acids. Figure 24–1 shows the charged forms of the two acids and three bases we have just discussed.

24.3 ESSENTIAL AMINO ACIDS

All of the amino acids in Table 24–3 are necessary constituents of human protein, but our body can synthesize only some of them from other materials. Evidence indicates that the human body is incapable of producing ten of these twenty amino acids fast enough or in sufficient quantities to sustain normal growth. These ten amino acids, called *essential amino acids*, must be obtained from our food. **Essential amino acids** *are amino acids that must be obtained from food because they cannot be biosynthesized in sufficient quantity.* Table 24–4 lists the ten essential amino acids. An adequate human diet must include foods that contain these amino acids.

A dietary protein that contains all ten of the essential amino acids is called a *complete protein*. Most animal proteins are complete proteins, including casein from milk and those found in meat, fish, and eggs. Casein contains all ten essential amino acids plus nine of the ten nonessential amino acids.

Gelatin, a substance obtained from a hydrolysis of collagen (found in meat) is an important animal protein that is not complete. It lacks tryptophan and is low in several other amino acids. We mention this because many liquid protein diets are gelatin based, and people on these diets will lack some essential amino acids if this is their only protein source.

Most plant proteins are incomplete. Rice protein lacks the amino acids lysine and threonine; wheat protein lacks lysine; corn protein lacks lysine and tryptophan; and soy protein is very low in methionine. Vegetarians must eat a variety of plant foods to obtain all of the essential amino acids.

Dietary proteins may be rated in terms of biological value on a percentage scale.

TABLE 24–4 The Ten Essential Amino Acids

isoleucine
leucine
lysine
methionine
phenylalanine
threonine
tryptophan
valine
arginine*
histidine*

* Probably essential

TABLE 24–5 Biological Value of Selected Dietary Proteins

Food	Biological Value (%)
whole hen's egg	94
whole cow's milk	84
fish	83
beef	73
soybeans	73
white potato	67
whole grain wheat	65
whole grain corn	59
dry beans	58

Such percentage values for selected proteins are given in Table 24–5. Note that, in general, animal proteins have higher biological value than plant proteins.

24.4 STEREOISOMERISM IN ALPHA-AMINO ACIDS

Again looking at the general formula for α-amino acids,

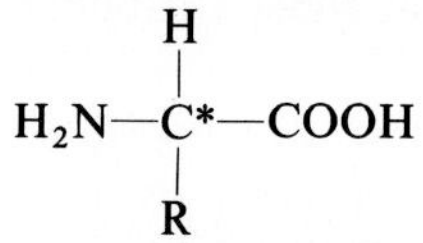

$$H_2N-\overset{\overset{\displaystyle H}{|}}{\underset{\underset{\displaystyle R}{|}}{C^*}}-COOH$$

we note that all of them, except for glycine, where R is equal to H, have a chiral carbon atom (which is starred in the general formula). Four different groups, R, H, COOH, and NH_2, are attached to this carbon atom. This means that there are D and L forms (two enantiomers—Section 21.2) for each of these amino acids (except glycine).

The D and L forms of alanine have the following structures (projection formulas).

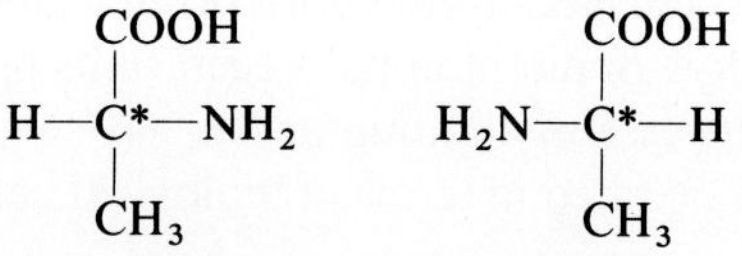

$$H-\overset{\overset{\displaystyle COOH}{|}}{\underset{\underset{\displaystyle CH_3}{|}}{C^*}}-NH_2 \qquad H_2N-\overset{\overset{\displaystyle COOH}{|}}{\underset{\underset{\displaystyle CH_3}{|}}{C^*}}-H$$

D-alanine   L-alanine

Enantiomeric amino acids are shown in projection formulas in the same way as in the configuration of D- and L-glyceraldehyde (Section 21.5). The COOH group is written at the top of the projection formula and the D configuration is denoted by writing the —NH_2 group to the right on the vertical carbon chain.

Both D and L α-amino acids are found in nature, but the amino acids found in all proteins are always from the L series. This is in direct contrast to the monosaccharide building blocks of polysaccharides, which always have the D configuration (Section 22.4).

24.5 ACID–BASE PROPERTIES OF AMINO ACIDS

In pure form, amino acids are white crystalline solids with relatively high melting points. Most of them are very soluble in water. High water solubility and relatively high melting points are properties associated with charged species (ionic compounds). Studies of amino acids confirm that they are charged species. Let us consider why.

Up to now, we have drawn the general structure of an amino acid as

$$H_2N-\underset{R}{\underset{|}{\overset{H}{\overset{|}{C}}}}-COOH$$

We now draw attention to the consequences of the presence of both an acidic group (—COOH) and a basic group (—NH$_2$) within an α-amino acid molecule.

In Chapter 19 we learned that acid groups, in neutral solution, have a tendency to lose protons (H$^+$ ions) to produce a negatively charged species.

$$-COOH \longrightarrow COO^- + H^+$$

In Chapter 20 we saw that amino groups (—NH$_2$) are basic, that is, in neutral solution they will accept protons to produce a positively charged species.

$$-NH_2 + H^+ \longrightarrow -\overset{+}{N}H_3$$

Consistent with the behavior of these groups, in neutral solution, the —COOH group of an amino acid donates a proton to the —NH$_2$ of the same amino acid. We can characterize this behavior as an *internal* acid–base reaction. The net result of this is, that in neutral solution, amino acid molecules have the structure

$$H_3\overset{+}{N}-\underset{R}{\underset{|}{\overset{H}{\overset{|}{C}}}}-COO^-$$

Such a molecule is known as a *zwitterion,* from the German term meaning "double ion". A **zwitterion** *is a molecule that has a positive charge on one atom and a negative charge on another atom.* Note that the net charge on a zwitterion is zero even though parts of the molecule carry charges. In solution and also in the solid state, α-amino acids are zwitterions.

zwitterion

Additional charge changes occur in zwitterions if the pH of a solution containing them is changed from neutral (or near neutral). In acidic solution (low pH), where additional H$^+$ ions are present, the acid group becomes reprotonated to give the species.

$$H^+ + H_3\overset{+}{N}-\underset{R}{\underset{|}{\overset{H}{\overset{|}{C}}}}-COO^- \longrightarrow H_3\overset{+}{N}-\underset{R}{\underset{|}{\overset{H}{\overset{|}{C}}}}-COOH$$

This resulting species is no longer neutral; it has a net charge of +1. In basic solution (high pH) where OH$^-$ ions predominate, the amino group is deprotonated to give the species

$$OH^- + H_3\overset{+}{N}-\underset{R}{\underset{|}{\overset{H}{\overset{|}{C}}}}-COO^- \longrightarrow H_2N-\underset{R}{\underset{|}{\overset{H}{\overset{|}{C}}}}-COO^- + H_2O$$

This resulting species has a net negative charge of −1.

In solution, the amounts of each of the three amino acid forms (zwitterion, negative ion, and positive ion) depend directly on pH. The three species are actually in equilibrium with each other and the equilibrium shifts with pH change. The overall equilibrium process can be represented as follows.

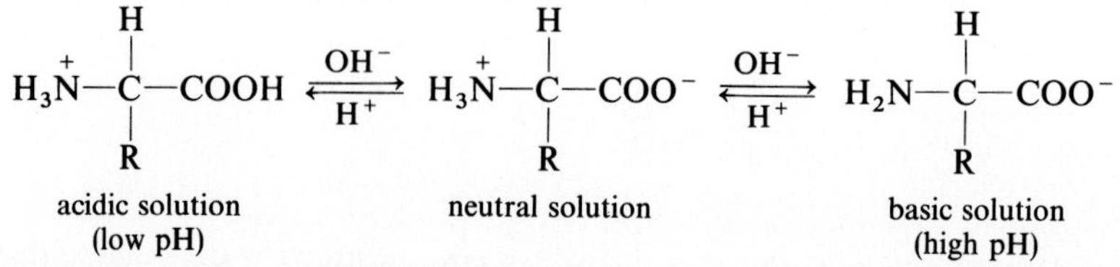

$$\underset{\substack{\text{acidic solution}\\\text{(low pH)}}}{\overset{H}{\underset{R}{H_3\overset{+}{N}-C-COOH}}} \underset{H^+}{\overset{OH^-}{\rightleftharpoons}} \underset{\substack{\text{neutral solution}}}{\overset{H}{\underset{R}{H_3\overset{+}{N}-C-COO^-}}} \underset{H^+}{\overset{OH^-}{\rightleftharpoons}} \underset{\substack{\text{basic solution}\\\text{(high pH)}}}{\overset{H}{\underset{R}{H_2N-C-COO^-}}}$$

In acidic solution, the positively charged species on the left predominates; near neutral solutions have the middle species (the zwitterion) as the dominant species; in basic solution, the negatively charged species on the right dominates.

Example 24.1

Draw an appropriate structural form for the amino acid alanine when in solution at each of the following pH values.

a. pH = 1 b. pH = 7 c. pH = 11

Solution

At low pH, both amino and carboxyl groups are protonated. At high pH, both groups have lost their protons. At neutral pH, the zwitterion is present.

a. 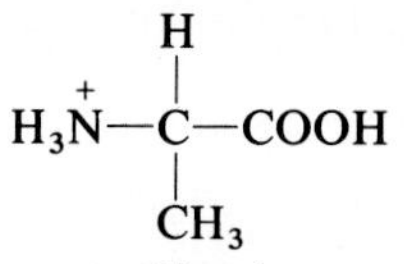
$$\overset{H}{\underset{CH_3}{H_3\overset{+}{N}-C-COOH}}$$
pH = 1
(net charge of +1)

b. 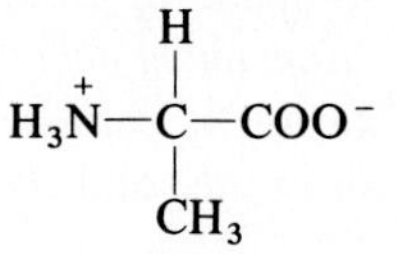
$$\overset{H}{\underset{CH_3}{H_3\overset{+}{N}-C-COO^-}}$$
pH = 7
(no net charge)

c. 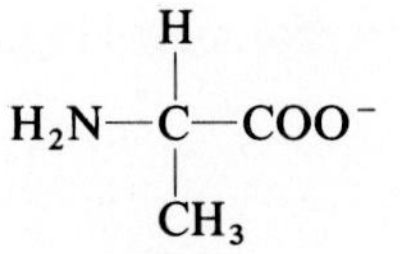
$$\overset{H}{\underset{CH_3}{H_2N-C-COO^-}}$$
pH = 11
(net charge of −1)

When two electrodes (one positively charged and one negatively charged) are immersed in an aqueous solution containing amino acids, those with a net positive charge are attracted to the negatively charged electrode. Negatively charged amino acids migrate toward the positively charged electrode. What would you predict the behavior of a zwitterion to be? Zwitterions, having both a positive and a negative charge, exhibit no net migration toward either electrode.

Since the charges on amino acids can be altered by changing the pH of their solutions, they can be made to migrate towards either electrode or not to migrate at all (see Figure 24–2). The pH value at which no net migration occurs is known

FIGURE 24–2 Migration of glycine towards positively or negatively charged electrodes in aqueous solutions of different pH.

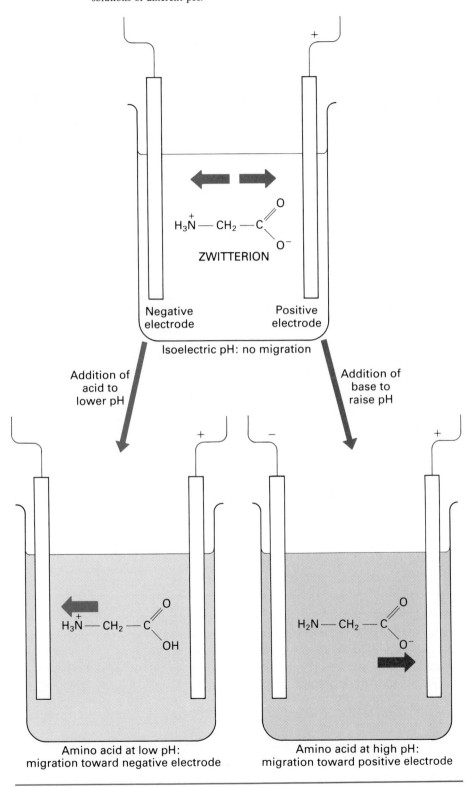

isoelectric point — as the isoelectric pH. The **isoelectric point** *for an amino acid solution is the pH value at which there is no migration of the amino acid in an electric field.* The zwitterion ion concentration for an amino acid is at a maximum at the isoelectric point.

In Section 24.2 we observed that aspartic acid and glutamic acid have negatively charged side chains at neutral pH. The bases lysine, arginine, and histidine have positively charged side chains in this same pH range. Hence, for these amino acids we must also consider the side chain charges in predicting their migration between electrodes and also the value of their isoelectric point.

Example 24.2

Predict the direction (if any) of migration toward the positively or negatively charged electrodes for the following amino acids in solutions of specified pH. Write "isoelectric" if no migration occurs.

a. lysine at pH = 7   b. glutamic acid at pH = 7   c. serine at pH = 1

Solution

lysine at pH = 7

$$H_3\overset{+}{N}-CH-\underset{\underset{\underset{\underset{{}^+NH_3}{|}}{\underset{CH_2}{|}}}{\underset{CH_2}{|}}}{\underset{CH_2}{|}}-\overset{O}{\overset{\|}{C}}-O^-$$

net positive charge (2 "+" and only 1 "−"); migrates toward negatively charged electrode

glutamic acid at pH = 7

$$H_3\overset{+}{N}-CH-\underset{\underset{\underset{O\overset{C}{\diagup}\hspace{-0.3em}\diagdown O^-}{|}}{\underset{CH_2}{|}}}{\underset{CH_2}{|}}-\overset{O}{\overset{\|}{C}}-O^-$$

net negative charge (1 "+" and 2 "−"); migrates toward positively charged electrode

serine at pH = 1

$$H_3\overset{+}{N}-CH-\underset{\underset{OH}{\underset{CH_2}{|}}}{|}-\overset{O}{\overset{\|}{C}}-OH$$

one positive charge; migrates toward negatively charged electrode

---

electrophoresis — Due to their different migration patterns at certain pH values, mixtures of amino acids in solution can be separated and identified using this concept. As an analytical procedure, this type of separation is called *electrophoresis*. **Electrophoresis** *is the process of separating charged molecules on the basis of their migration towards charged electrodes.* Because amino acids confer many of their own electrical properties to the proteins they form, proteins can also be separated by electrophoresis.

24.6 PEPTIDE FORMATION

In Section 20.6 we learned that a carboxylic acid and an amine can react with each other to give an amide. The general equation for such a reaction is

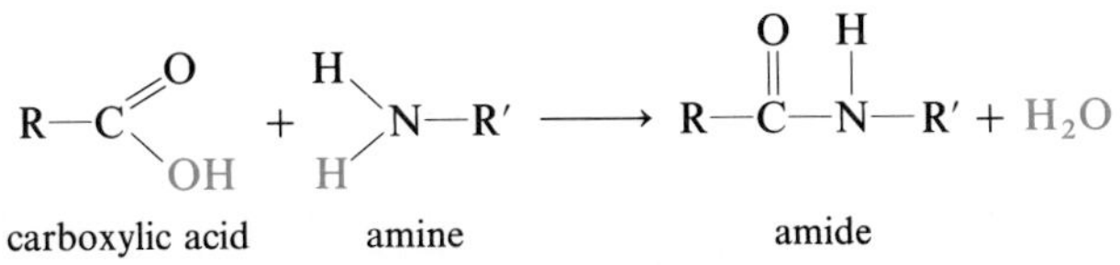

$$R-C{\overset{O}{\underset{OH}{\lessgtr}}} + {\overset{H}{\underset{H}{>}}}N-R' \longrightarrow R-{\overset{O}{\overset{\|}{C}}}-{\overset{H}{\underset{|}{N}}}-R' + H_2O$$

carboxylic acid   amine   amide

Two amino acids can react with each other in a similar way. The carboxyl group of one amino acid reacts with the amino group of the other amino acid. The product is a molecule containing two amino acids linked by an amide bond.

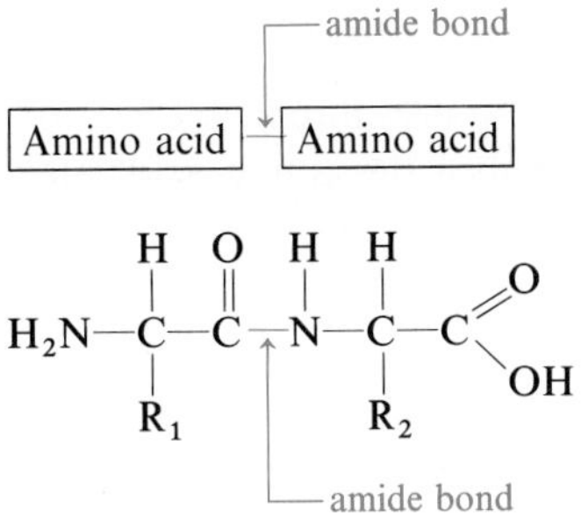

$$H_2N-\underset{R_1}{\overset{H}{\underset{|}{\overset{|}{C}}}}-\overset{O}{\overset{\|}{C}}-\underset{}{\overset{H}{\underset{|}{\overset{|}{N}}}}-\underset{R_2}{\overset{H}{\underset{|}{\overset{|}{C}}}}-C{\overset{O}{\underset{OH}{\lessgtr}}}$$

In amino acid chemistry, amide bonds that link amino acids together are given the special name of *peptide bond*. A **peptide bond** *is a bond between the carbonyl group of one amino acid and the amino group of another amino acid.*

Under proper conditions many amino acids can bond together to give chains of amino acids containing numerous peptide bonds. For example, four peptide bonds are present in a chain of five amino acids.

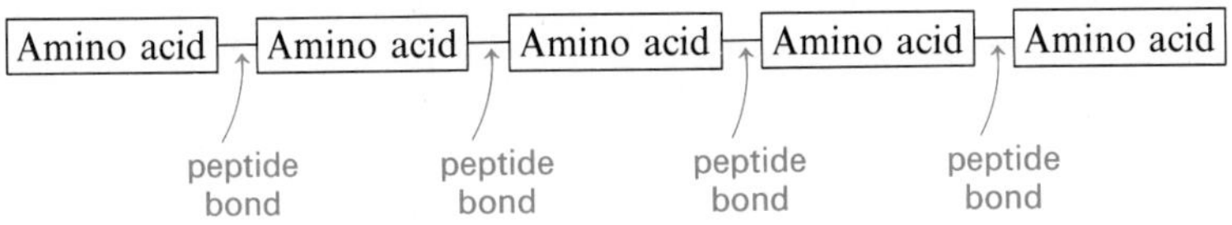

peptide bond

Short to medium-sized chains of amino acids are known as *peptides*. A **peptide** *is a sequence of amino acids, of up to 50 units, in which the amino acids are joined together through amide (peptide) bonds.* A compound containing two amino acids joined by a peptide bond is specifically called a *dipeptide*; three amino acids in a chain is a *tripeptide*, and so on. When the length of the chain becomes more than one wants to indicate in the name, it is merely called a *polypeptide*. For example, a chain containing ten amino acid units could be called a *decapeptide* or simply a polypeptide.

peptide

Example 24.3

Draw the general structure (showing atoms) of a hexapeptide. Draw a box around each amino acid. Use the symbol R with the subscripts 1 through 6 to indicate the various side chains.

Solution

All hexapeptides contain six amino acids, linked by peptide bonds. The amino acid at the beginning of the chain (on the left) will have a free amino group and the amino acid at the end of the chain (on the right) will have a free carboxyl group.

$$H_2N-\underset{R_1}{CH}-\overset{O}{\overset{\|}{C}}-NH-\underset{R_2}{CH}-\overset{O}{\overset{\|}{C}}-NH-\underset{R_3}{CH}-\overset{O}{\overset{\|}{C}}-NH-\underset{R_4}{CH}-\overset{O}{\overset{\|}{C}}-NH-\underset{R_5}{CH}-\overset{O}{\overset{\|}{C}}-NH-\underset{R_6}{CH}-\overset{O}{\overset{\|}{C}}-OH$$

free amino group ↑   ↑ free carboxyl group

Let us now look more closely at the formation of a dipeptide in order to gain insight into peptides that will be very important in later discussions about proteins.

Consider the reaction between the amino acids alanine and glycine to give the following dipeptide

$$H_2N-\underset{CH_3}{\overset{H}{\underset{|}{C}}}-\overset{O}{\overset{\nearrow}{C}}_{OH} + \underset{H}{\overset{H}{\underset{|}{N}}}-\underset{H}{\overset{H}{\underset{|}{C}}}-\overset{O}{\overset{\nearrow}{C}}_{OH} \longrightarrow H_2N-\underset{CH_3}{\overset{H}{\underset{|}{C}}}-\overset{O}{\overset{\|}{C}}-\underset{|}{\overset{H}{N}}-\underset{H}{\overset{H}{\underset{|}{C}}}-\overset{O}{\overset{\nearrow}{C}}_{OH} + H_2O$$

alanine   glycine   dipeptide

In this dipeptide there is an amino group at one end (at the left) and a carboxyl group at the other end (at the right). The amino end is called the *N-terminal end* and the carboxyl end the *C-terminal end*.

Using the abbreviations for amino acid names given in Table 24–3, we can abbreviate the structure of this dipeptide as

Ala–Gly

When using such notation, by convention, the N-terminal end of the peptide is always written on the left.

We note next that a second dipeptide of glycine and alanine exists which is isomeric with the first. It has the amino acid sequence

Gly–Ala

In this dipeptide, the glycine residue is at the N-terminal end and the alanine residue is at the C-terminal end.

Ala–Gly and Gly–Ala are different molecules (isomers). These isomers, which differ only in the order of attachment of the amino acids, have different chemical and physical properties. From this simple example we learn that the sequence of amino acids in a chain is very important. Changing the sequence of amino acids creates a new polypeptide with different properties from the original one.

The number of isomeric peptides possible increases rapidly as the length of the peptide chain increases. Let us consider the tripeptide Ala–Ser–Cys as another example.

$$H_2N-\underset{CH_3}{\overset{H}{\underset{|}{C}}}-\overset{O}{\overset{\|}{C}}-\overset{H}{\underset{|}{N}}-\underset{\underset{OH}{\overset{|}{CH_2}}}{\overset{H}{\underset{|}{C}}}-\overset{O}{\overset{\|}{C}}-\overset{H}{\underset{|}{N}}-\underset{\underset{SH}{\overset{|}{CH_2}}}{\overset{H}{\underset{|}{C}}}-\overset{O}{\overset{\nearrow}{C}}_{OH}$$

Ala–Ser–Cys

In addition to this sequence, five other arrangements of these three components

are possible, and each one represents an isomeric tripeptide. These sequences are:
Ala–Cys–Ser, Ser–Ala–Cys, Ser–Cys–Ala, Cys–Ala–Ser and Cys–Ser–Ala.

Example 24.4

List all possible sequences for the tripeptide containing one unit each of tyrosine (Tyr), glycine (Gly), and histidine (His).

Solution
There will be two isomers each with tyrosine, glycine, and histidine, as, respectively, the N-terminal end.

| Tyr–Gly–His | Gly–Tyr–His | His–Tyr–Gly |
| Tyr–His–Gly | Gly–His–Tyr | His–Gly–Tyr |

Formal names for peptides are derived from the names of the amino acid residues present. For dipeptides, the N-terminal amino acid is named as an alkyl group and the C-terminal amino acid retains its own name. The two isomeric dipeptides involving alanine and glycine are named as follows:

Ala–Gly   Alanylglycine
Gly–Ala   Glycylalanine

For larger peptides, all amino acid residues except the C-terminal one are named as alkyl groups. For example, we have the following

Ala–Ser–Cys   Alanylserylcysteine

We have noted that the sequence of amino acids in a polypeptide is vitally important. Consider the following example. The artificial sweetener aspartame is based on the dipeptide Asp–Phe. Aspartame's carboxyl end is esterified to a methyl group, but it is still the carboxyl end.

If the order of the amino acids is reversed, the dipeptide loses much of its sweet taste and cannot be used as an artificial sweetener (Section 22.9).

An analogy is often drawn between the structural similarities of polypeptides and words. Words, which convey information, are formed when the 26 letters of the English alphabet are properly sequenced. Polypeptides, which function biologically, are formed from the proper sequence of 20 amino acids. The proper sequence of letters in a word is necessary to make sense, just as the proper sequence of amino acids is necessary to make biologically active polypeptides. Furthermore, the letters that form a word are written from left to right, as are amino acids to form polypeptides. As any dictionary of the English language will document, a tremendous variety of words can be formed by different letter sequences. Imagine the possible number of sequences as longer polypeptides are considered. There are 1.55×10^{66} sequences possible for the 51 amino acids found in insulin! From these possibilities, the body reliably produces only *one*, illustrating the remarkable precision of life processes. From the simplest bacterium to the human brain cell, only those amino acid sequences needed by the cell are produced. The fascinating process of polypeptide biosynthesis and the way in which genes in DNA direct this process will be discussed further in Chapter 26.

24.7 PROTEINS

Proteins *are polypeptides that contain more than 50 amino acid units.* The dividing line between a polypeptide and a protein is arbitrary. The important point is that proteins are polymers containing a large number of amino acid units linked by peptide bonds. Polypeptides are short chains of amino acids. Some proteins have molecular weights in the millions. Some proteins also contain more than one polypeptide chain.

To aid us in describing protein structure, we will consider four levels of substructure: primary, secondary, tertiary, and quaternary. In considering these structure levels, one by one, remember that it is the total combination of all four levels of structure that control protein function.

24.8 PRIMARY STRUCTURE OF PROTEINS

The **primary structure** *of a protein is the sequence of amino acids present in the peptide chain or chains of the protein.* Knowledge of primary structures tells us what amino acids are present, the number of each, and the length and number of polypeptide chains. Every type of protein molecule in a biological organism has a different sequence of amino acids; that sequence allows the protein to carry out its function.

The first protein whose primary structure was determined was insulin, the hormone that regulates blood sugar level; a deficiency of insulin leads to diabetes. This project, which took over eight years, was completed in 1951. Today, hundreds of proteins have been sequenced; that is, researchers have determined the order of amino acids within the polypeptide chain or chains.

Figure 24–3 shows the primary structure of human insulin. The amino acids are parts of two polypeptide chains: Chain A contains 21 amino acids and Chain B contains 30 amino acids. The A and B chains are cross-linked by two disulfide bonds and a third disulfide linkage is also present between positions 6 and 11 of the A chain. Disulfide bonds are not considered part of the primary structure of the protein. They are however, very important in the overall structure of the protein and will be discussed in further detail in Section 24.10. For now, we note only that disulfide bonds are covalent bonds that result from the reaction of cysteine amino acid unit side chains with each other.

The primary structure of a specific protein is always the same, regardless of where the protein is found within an organism. The structure of certain proteins is even similar among different species of animals. For example, the primary structures of insulin in cows, pigs, sheep, and horses are very similar both to each other and to human insulin. This similarity is particularly important for diabetics who require supplemental injections of insulin. Since the amount of human insulin available for insulin injections cannot meet demand, insulin from cows, pigs, sheep, and horses can be used by humans with very few side effects. Table 24–6 compares the variations at positions 8, 9, and 10 on chain A and position 30 of chain B for insulin from humans and the four animals previously mentioned.

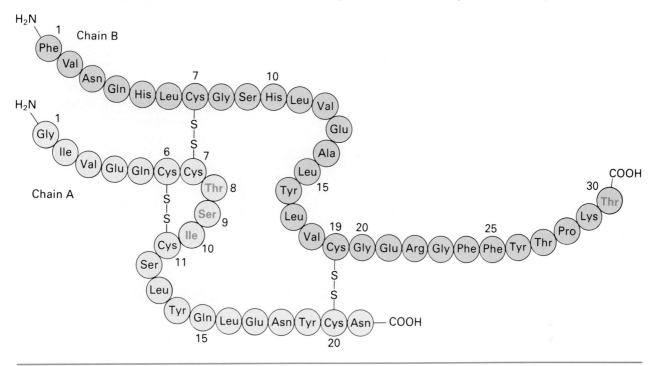

FIGURE 24-3 Diagram of human insulin showing primary structure. The locations where variance occurs between humans and certain other animals are shown in color type. (See Table 24-6 for comparisons of variance.)

The changes caused by the substitutions listed in Table 24-6 are not sufficient to alter significantly the effects of foreign insulin compared to human insulin. Both pig (ovine) and beef (bovine) insulins are commonly used in the treatment of human diabetics. Because of the differences in primary structure, some individuals will have an allergic response to the foreign insulin; however, the great majority of human diabetics can use the nonhuman insulin without complications. Due to genetic engineering, a topic to be discussed in Chapter 26, bacteria can now be made to produce human insulin. It is therefore projected that almost all diabetics will soon be able to have human insulin for injection, instead of that from another species.

In some proteins, unlike insulin, substitution of a single amino acid unit can sometimes significantly alter protein function. A case in point involves *hemoglobin*, the oxygen transport protein in the blood. Hemoglobin is a protein containing two different types of polypeptide chains. The hereditary disease sickle-cell anemia is a result of a change in primary structure involving only one amino acid among the 150 amino acids in one of the two different types of chains. The hemoglobin

TABLE 24-6 Variation of Amino Acids in positions 8, 9, and 10 of Chain A and position 30 of Chain B for various insulins

Species	Chain A			Chain B
	#8	#9	#10	#30
human	thr	ser	ile	thr
pig (ovine)	thr	ser	ile	ala
cow (bovine)	ala	ser	val	ala
sheep	ala	gly	val	ala
horse	thr	gly	ile	ala

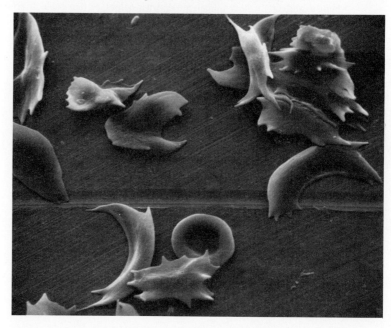

FIGURE 24-4 Blood cells from a person with sickle-cell anemia.

of an afflicted person contains valine in a position normally occupied by glutamic acid.

```
                              4    5    6    7    8    9
Normal Hemoglobin         -Thr-Pro-Glu-Glu-Lys-Ala-
Sickel-Cell Hemoglobin    -Thr-Pro-Val-Glu-Lys-Ala-
```

Substitution of the nonpolar amino acid valine at position 6 in place of the polar amino acid glutamic acid causes the new (sickle cell) hemoglobin to precipitate in the red blood cells. These cells become distorted and assume a characteristic crescent shape—the origin of the name *sickle-cell anemia* (see Figure 24-4). The distorted cells break apart more easily than healthy ones, clump together, and block capillaries. This, together with the impaired ability of hemoglobin itself to transport oxygen, results in sickle-cell anemia.

Clinical diagnosis of sickle-cell anemia is accomplished by electrophoresis. Hemoglobin S (from sickle cells) possesses less negative charge than hemoglobin A (normal adult hemoglobin). Hence, it does not migrate as quickly toward a positively-charged electrode in an electrical field. Many variant hemoglobins (besides hemoglobin S) can also be detected by electrophoresis.

24.9 SECONDARY STRUCTURE OF PROTEINS

If the only structural characteristic of proteins were the amino acid sequence, the molecules would consist of long chains coiled into random shapes. Experimental evidence shows that this is not the case. Most proteins fold into well-defined characteristic shapes because of interactions between peptide linkages at various points along the polypeptide chain. **Secondary structure** *is the three-dimensional shape of a protein in a relatively small, localized area of the protein.* Secondary structure is mostly a result of interactions between amino acids that are relatively close to one another along the polypeptide chain. Studies have found three major types of protein secondary structure: the *alpha-helix*, the *beta-pleated sheet*, and the *triple helix*. Let us examine each of them in detail.

Hydrogen bonding between carbonyl groups and the hydrogen–nitrogen bond in the amide groups along the polypeptide backbone of the protein is the major cause for all three types of secondary structure. This hydrogen bonding interaction may be diagrammed as follows

$$\diagup\!\!\!\!\!\!{>}\text{N}-\text{H}-----\text{O}=\text{C}{<}\!\!\!\!\!\!\diagup$$

The *alpha-helix* structure is tightly coiled and resembles a spring. It results from the coiling of the polypeptide into a spiral configuration, with the spiral configuration being maintained by hydrogen bonds between $>$N—H and $>$C=O groups of every fourth amino acid, as is shown diagrammatically in Figure 24–5 (p. 622).

The *beta-pleated sheet* structure, although not as common a secondary structure as is the α-helix, is found in such proteins as fibrin (the blood-clotting protein in blood), myosin (a protein of muscle), and keratin (the protein of hair). It occurs when two or more extended peptide chains (or two separate regions on the same extended chain) lie side by side. When separate chains are involved, the hydrogen bonding is between chains rather than between amino acids in the same chain, as was the case for the α-helix. A β-pleated sheet is described as *antiparallel* when neighboring chains (or two sections of a chain) run in opposite directions. The structure is said to be *parallel* when the chains run in the same direction. Drawings of the β-pleated sheet conformation are given in Figure 24–6 (p. 623). The "zig-zag" sheet arrangement minimizes unfavorable interactions between the amino acid side chains (R groups) that extend above and below the pleated sheet.

The β-pleated sheet is found extensively in the protein of silk. Since such proteins are already fully extended, silk fibers cannot be stretched. When wool protein (α-keratin) is steamed and then dried under tension, it maintains its stretched length because it has assumed the β-pleated sheet conformation.

Very few proteins have entirely α-helix or β-pleated sheet structures. Instead, most proteins have only certain portions of their molecules in these conformations. The rest of the molecule assumes a random coil. It is possible to have both α-helix and β-pleated sheet structures within the same protein. Figure 24–7 (p. 624) is a diagram of a protein that has more than one secondary structural feature.

Collagen, the most abundant protein in the human body, is the structural protein of connective tissue (bone, cartilage, tendon, and skin). Collagen's unique secondary structure is an excellent example of the *triple helix*. The triple helix involves three coiled helical polypeptide chains wound around each other about a common axis to give a ropelike arrangement of polypeptide chains (see Figure 24–8 (p. 624)). The unusually tight winding of the three chains is allowed because every third amino acid position in each chain is occupied by glycine. The majority of glycines are found on the interior region of the triple helix, where its unusually small side chain (a single hydrogen atom) takes up very little room and allows the close winding of the polypeptide chains.

Some cross-linking also occurs in the triple helix via covalent bonds between chains. The extent of cross-linking is a function of an animal's age. The older the animal, the greater the cross-linking and the tougher the tissue. The process of

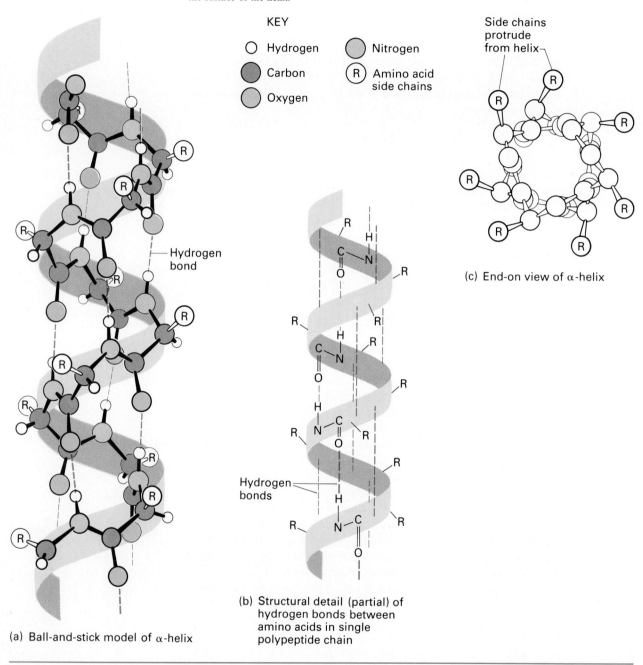

FIGURE 24-5 Three different drawings of the α-helix. The first two (a and b) show the helical nature of the coil and its hydrogen bonds. The end-on view (c) shows how side chains (represented by R) extend away from the surface of the helix.

tanning, which converts animal hides to leather, involves increasing the degree of cross-linking.

Collagen molecules are very long, thin, and rigid. Many such molecules, lined

FIGURE 24-6 Three different representations of the β-pleated sheet secondary structure adopted by some proteins. This structure is stabilized by hydrogen bonds between peptide chains that lie opposite each other.

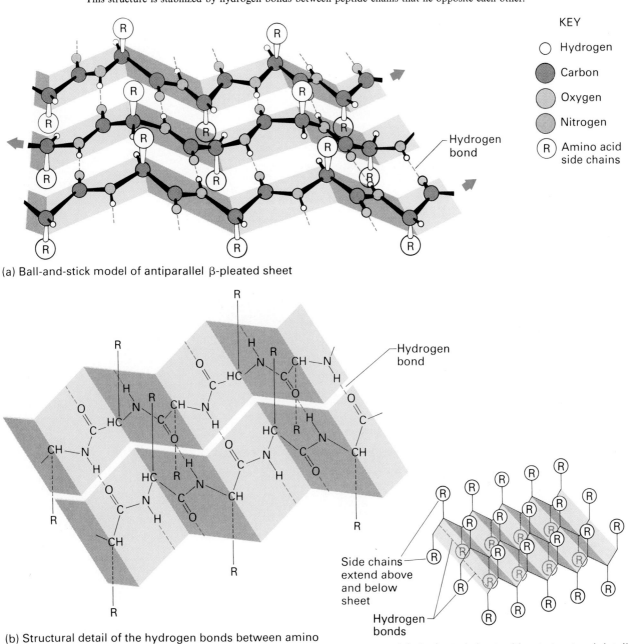

KEY
○ Hydrogen
● Carbon
○ Oxygen
○ Nitrogen
Ⓡ Amino acid side chains

(a) Ball-and-stick model of antiparallel β-pleated sheet

(b) Structural detail of the hydrogen bonds between amino acids of two polypeptide chains

(c) β-pleated sheet without structural detail

up alongside each other, combine to make collagen fibers. Cross-linking gives the fibers extra strength. Figure 24–9 (p. 624) is an electron micrograph showing how collagen fibers appear when viewed through an electron microscope.

623

FIGURE 24–7 Secondary structure of a single protein showing areas of α-helix, β-pleated sheet, and random coiling.

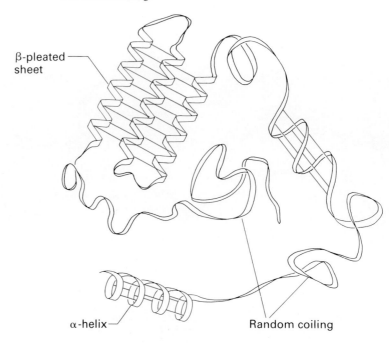

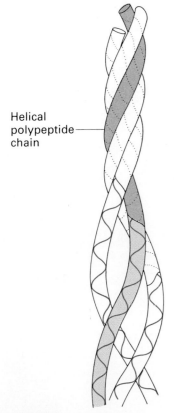

FIGURE 24–8 Three polypeptide chains interweave to form a triple helix. Here the chains are partially unwound and cut away to show their structure.

FIGURE 24–9 Collagen fibers, as seen through an electron microscope.

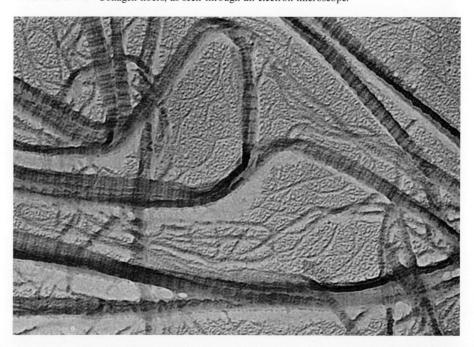

24.10 TERTIARY STRUCTURE OF PROTEINS

The **tertiary structure** *of a protein is the overall three-dimensional shape that results from the attractive forces among amino acid side chains (R groups) that are widely separated from each other within the chain.* The twists and folds in a protein chain relate to tertiary structure.

A good analogy for the relationships between the primary, secondary, and tertiary structures of a protein is that of a telephone cord (Figure 24-10, p. 626). The primary structure is the long straight cord. The coiling of the cord into a helical arrangement gives the secondary structure. The supercoiling arrangement the cord adopts after one hangs up the receiver is the tertiary structure.

If α-helix, β-pleated sheet, and triple helix describe secondary structural characteristics, what terms describe tertiary structures? The overall three-dimensional shape of proteins generally separates them into two categories: *globular proteins* and *fibrous proteins*. **Globular proteins** *are proteins whose overall shape is roughly spherical or globular.* Globular proteins can be spherical or more elongated, much like an ellipsoid or even an oblate spheroid (a squashed sphere). **Fibrous proteins** *are proteins that have long, thin, fibrous shapes.* Such proteins are composed of long strands or flat sheets.

Myoglobin was the first protein for which the complete three-dimensional shape was determined. The protein is relatively small (molecular weight of 17,500 amu) and stores and transports oxygen in muscle tissue. Figure 24-11 (p. 626) shows a diagram of the tertiary structure of myoglobin; only the backbone of the polypeptide chain is shown for clarity.

The structure of myoglobin also contains an extra feature, a *prosthetic group*. A **prosthetic group** *is a nonamino acid unit present in a protein.* The prosthetic group in myoglobin is an iron atom surrounded by a disk-like heme molecule.

tertiary structure

globular proteins

fibrous proteins

prosthetic group

heme

This group is an absolute necessity for the proper function of myoglobin. It assists in transporting oxygen; an O_2 molecule binds to the iron atom.

Other examples of globular proteins include insulin, egg albumin, ribonuclease, and carboxypeptidase A. The latter two of these proteins are enzymes (Chapter 25).

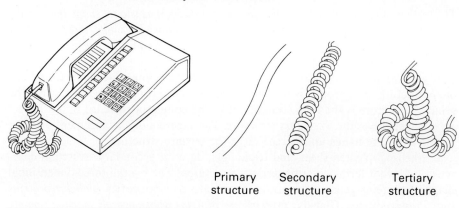

FIGURE 24–10 Three levels of telephone cord structure.

Primary structure   Secondary structure   Tertiary structure

The tertiary structure of fibrous proteins is heavily influenced by secondary structure features. Using collagen, a triple helix, as an example, we find that tertiary structure changes shape very little because tertiary interactions cause the intertwined peptide chains to be held even more strongly together.

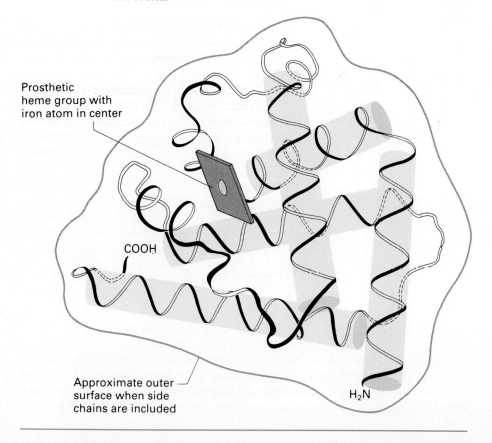

FIGURE 24–11 Diagram of the tertiary structure of myoglobin. Regions of secondary structure involving α helices are highlighted (rod-like structures). Several shorter helices can also be seen.

Four types of attractive interactions contribute to the tertiary structure, either globular or fibrous, of a protein: (1) disulfide bonds, (2) electrostatic attractions (salt linkages), (3) hydrogen bonds, and (4) hydrophobic attractions. All four of these interactions are interactions between amino acid R groups. This is a major distinction between tertiary structure interactions and secondary structure interactions. Tertiary structure interactions involve the R groups of amino acids while secondary structure interactions involve the peptide linkages between amino acid units.

Disulfide bonds form between the —SH groups of two cysteine amino acids. They are covalent bonds and differ in this respect from the other three types of interactions to be discussed.

Disulfide bonds may involve two cysteine units in the same chain or in different chains. Referring back to Figure 24–3, we note the presence of both types of sulfide bonds in the structure of insulin.

Electrostatic interactions, sometimes called salt bridges, always involve amino acids with ionized side chains. Such amino acids are those with very polar side chains (Table 24–3), that is, the acidic amino acids and the basic amino acids. The two R groups, one acidic and one basic, are held together by simple ion–ion attractions.

Hydrogen bonds can occur between amino acids with polar R groups. A variety of polar side chains can be involved, especially those that possess the following functional groups.

Hydrogen bonds are relatively weak and are easily disrupted by changes in pH and temperature.

Hydrophobic attractions result when two nonpolar side chains are close to each other. In aqueous solution, globular proteins usually have their polar R groups outward, toward the aqueous solvent (which is also polar) and their nonpolar R groups inward (away from the polar water molecules). The nonpolar R groups then interact with each other. Hydrophobic interactions are common between phenyl rings and alkyl side chains.

FIGURE 24-12 Types of interactions between amino acid R groups that lead to the tertiary structure of a protein.

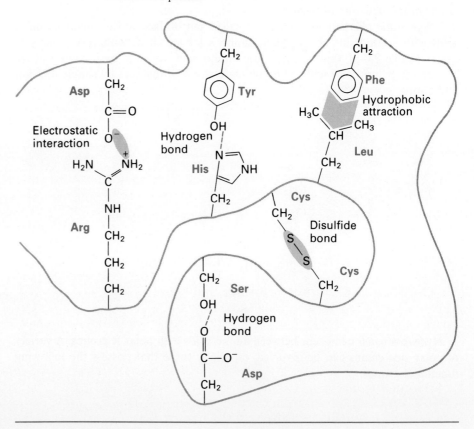

Although this type of interaction is weaker than hydrogen bonding or electrostatic interactions, it is a significant force because it occurs frequently.

Figure 24–12 summarizes the types of interactions that contribute to the tertiary structure of a protein.

24.11 QUATERNARY STRUCTURE OF PROTEINS

Quaternary structure is the highest level of protein organization. It is found only in proteins that have structures involving two or more polypeptide chains that are independent of each other, that is, not covalently bonded to each other. These multichain proteins are often called *oligomeric proteins*. The **quaternary structure** *of a protein involves the associations among the separate chains in an oligomeric protein.*

Most oligomeric proteins contain an even number of subunits (two subunits = a dimer, four subunits = a tetramer, and so on). Figure 24–13 diagrammatically represents the quaternary structure of a hypothetical tetramer.

Subunits of an oligomeric protein are held together by a variety of interactions between amino acids on the surface of the subunits. Most of these interactions are similar to the noncovalent tertiary interactions: electrostatic attractions, hydrogen bonding, and hydrophobic attractions.

The forces maintaining quaternary structure are more easily interrupted than those for tertiary structure. For example, only small changes in cellular conditions can cause a tetrameric enzyme to fall apart, dissociating into dimers or perhaps four separate subunits, with a resulting temporary loss of enzymatic activity (see Figure 24–14, p. 630). As conditions change back again, the oligomer automatically reforms and normal enzyme function is restored.

Phosphofructokinase (PFK), an important enzyme in sugar metabolism, is one example of an oligomeric protein whose function is disrupted through dissociation. A variety of different compounds in a cell can disrupt the tertiary structure of this tetramer, thus controlling the overall rate of sugar metabolism (Chapter 29).

Oligomeric proteins do not necessarily have to dissociate completely to exhibit a change in their normal function. Sometimes only a subtle change in quaternary structure is enough to affect protein function. Hemoglobin, the main oxygen transport protein of the circulatory system, is an excellent example of this phenomenon.

Adult hemoglobin is an oligometric protein with four subunits: two identical α chains and two identical β chains. Although differences occur in the primary structures of the different chains, both types of chains possess a secondary and tertiary structure similar to that of the myoglobin molecule (Figure 24–11). Each of the four subunits in hemoglobin contains a heme group, which serves as the binding site for oxygen. Since each site can transport one O_2 molecule, hemo-

FIGURE 24–13 A diagram of a tetrameric protein, showing the arrangement of subunits, which constitutes its quaternary structure.

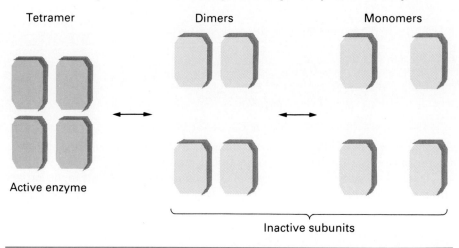

FIGURE 24–14 Diagram of an active tetrameric enzyme separating into inactive subunits. This process is the result of disruption of the quaternary structure of the protein.

globin can bind a total of four O_2 molecules simultaneously. Figure 24–15 shows the heme groups and the quaternary structure for hemoglobin.

As one individual subunit of hemoglobin binds an O_2 molecule, this subunit undergoes a change in its own tertiary structure. This change subsequently results in an alternation in its association with the other three subunits, that is, quaternary structure changes occur. These quaternary structure changes facilitate the binding

FIGURE 24–15 Quaternary structure. of a protein, represented by hemoglobin. There are four protein chains in this structure. The prosthetic heme units are also shown.

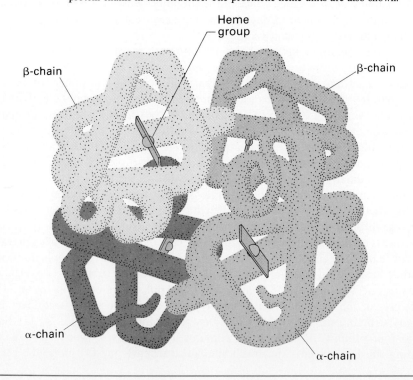

of three additional O_2 molecules. Conversely, similar changes occur, in reverse, when O_2 is released. The release of one O_2 molecule causes quaternary structure changes which facilitate the release of the remaining three oxygen molecules. Hence, quaternary structure plays an important role in oxygen transport by hemoglobin.

24.12 PROTEIN HYDROLYSIS

When a protein or polypeptide is heated to boiling in a solution of strong acid or strong base, the peptide bonds of the amino acid chain are hydrolyzed and free amino acids are produced. The hydrolysis reaction is the reverse of the formation reaction for a peptide bond. Amine and carboxylic acid functional groups are regenerated.

Let us consider the hydrolysis of the tripeptide Ala–Gly–Cys under acidic conditions. Hydrolysis produces one unit each of the amino acids alanine, glycine, and cysteine. The equation for the hydrolysis is

$$\underset{\text{Ala–Gly–Cys}}{H_2N-\underset{CH_3}{\underset{|}{\overset{H}{\overset{|}{C}}}}-\underset{}{\overset{O}{\overset{\|}{C}}}-\underset{H}{\underset{|}{\overset{H}{\overset{|}{N}}}}-\underset{H}{\underset{|}{\overset{H}{\overset{|}{C}}}}-\underset{}{\overset{O}{\overset{\|}{C}}}-\underset{}{\underset{|}{\overset{H}{\overset{|}{N}}}}-\underset{\underset{SH}{\underset{|}{CH_2}}}{\underset{|}{\overset{H}{\overset{|}{C}}}}-\overset{O}{\overset{\nearrow}{C}}_{OH}} \xrightarrow[\text{heat}]{H^+} \underset{\text{Ala}}{H_3\overset{+}{N}-\underset{CH_3}{\underset{|}{\overset{H}{\overset{|}{C}}}}-\overset{O}{\overset{\nearrow}{C}}_{OH}} + \underset{\text{Gly}}{H_3\overset{+}{N}-\underset{H}{\underset{|}{\overset{H}{\overset{|}{C}}}}-\overset{O}{\overset{\nearrow}{C}}_{OH}} + \underset{\text{Cys}}{H_3\overset{+}{N}-\underset{\underset{SH}{\underset{|}{CH_2}}}{\underset{|}{\overset{H}{\overset{|}{C}}}}-\overset{O}{\overset{\nearrow}{C}}_{OH}}$$

Note that the product amino acids in this reaction are written in positive ion form (Section 24.5) because of the acidic reaction conditions.

We should note also that complete hydrolysis destroys the primary structure of a protein. This result is in contrast to the denaturation reactions we will discuss in the next section; these reactions destroy quaternary, tertiary, and secondary structure, but leave primary structure intact.

Protein hydrolysis accounts for a number of the changes that are associated with the cooking of protein-rich foods. Collagen, an important protein in meat, as well as other proteins with similar triple helix structures, is not readily digestible. Cooking converts part of the collagen to gelatin, a water-soluble, digestible substance.

Within the human body, hydrolysis of protein takes place without heating, at body temperature, because of enzymes (which are also proteins—Chapter 25) which catalyze the hydrolysis reaction. Many different enzymes are involved. Some enzymes attack only N-terminal or C-terminal amino acids, whereas others break peptide bonds at specific internal amino acid locations.

Protein digestion is simply enzyme-catalyzed hydrolysis of ingested protein. The free amino acids produced from this process are absorbed through the intestinal wall into the bloodstream and transported to the liver. Here they become the raw materials needed for the synthesis of new protein tissue. This synthesis involves reassembling the amino acids into new and different orders. Excess amino acids are not stored in the body; hence, our diets should include a continuous and balanced intake of protein. This protein must contain all of the essential amino acids (Section 24.3). Chapter 32 is devoted to the subject of protein and amino acid metabolism.

24.13 PROTEIN DENATURATION

Denaturation *is the process in which the secondary, tertiary, and quaternary structure of a protein is disrupted, causing a loss in the protein's biological activity.* In sections

denaturation

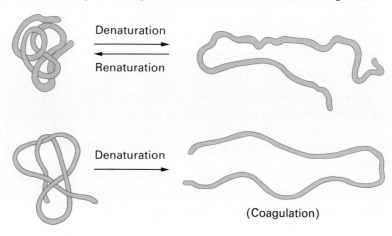

FIGURE 24–16 The processes of protein denaturation, renaturation, and coagulation.

24.9 through 24.11 we saw that the three-dimensional shape of a protein is determined by secondary, tertiary, and quaternary structure. Denaturation alters this shape, and biological activity is lost. In denaturation, the primary structure of a protein remains unchanged.

When denaturation occurs under mild conditions, the protein can often be restored to its original structure and effectiveness by carefully reversing the conditions of denaturation. This restoration is called *renaturation*. Denaturation under strong conditions is irreversible and usually such denatured proteins precipitate out of solution, a process called *coagulation*. Figure 24–16 contrasts the processes of denaturation, renaturation, and coagulation.

When quaternary structure is present in a protein, it is the first set of interactions to be upset by the denaturing agent. Next, the tertiary structural forces are disrupted and are followed by the secondary forces. The end product is an unfolded random coil protein that exhibits none of its original biological activity.

A number of physical and chemical changes will cause denaturation. They include heat, acids and bases, hydrogen-bonding solvents, salts of heavy metal ions, and reducing and oxidizing agents. We now consider these types of denaturing agents.

Heat

When proteins are heated, the added energy disrupts the weak interactions involved in quaternary, tertiary, and secondary protein structure. Gentle heating is usually reversible. Extensive heating almost always results in coagulation (irreversible denaturation) of the protein. Heating to just 50°C, only 13°C above normal body temperature, is usually sufficient to irreversibly denature a protein.

Cooking egg white (a concentrated solution of the protein albumin) changes it from a clear material with a jelly-like consistency to a white solid; this is due in large part to protein denaturation. This process is irreversible; you cannot uncook an egg. One of the reasons many foods are cooked is to denature protein so that it can be more easily digested.

Heat-induced denaturation is used to sterilize surgical instruments or for canning foods because bacteria are destroyed when the heat coagulates their protein. In surgery, heat is often used to seal small blood vessels. This process is called *cauterization*. Small wounds can also be sealed using cauterization.

Acids and Bases

Acids and bases affect pH. A pH change causes structural changes in selected amino acid R groups. These changes involve loss or gain of H^+ ion (Section 24.5). The net result is the disruption of electrostatic interactions involving these R groups.

The extent of denaturation by acids and bases depends on how much the pH is changed. The effects of a slight pH change can often be reversed; large pH changes produce irreversible denaturation.

A curdy precipitate of casein is formed in the stomach when the hydrochloric acid of gastric juice reacts with milk. The curdling of milk that takes place when milk sours or cheese is made results from the presence of lactic acid, a byproduct of bacterial growth. Serious eye damage can result from contact with acids or bases; irreversibly denatured and coagulated protein causes clouding of the cornea.

Hydrogen-bonding Solvents

Solvents with the ability to form hydrogen bonds can denature a protein by hydrogen bonding to it. This process disrupts the hydrogen bonds already existent within the secondary and tertiary structure of the protein.

Alcohols are an important type of hydrogen-bonding solvent involved in protein denaturation. Denaturation of bacterial protein occurs when isopropyl alcohol is used as a disinfectant. This denaturation accounts for the common practice of swabbing the skin with alcohol before giving an injection. Interestingly, pure isopropyl alcohol is less effective than the commonly used 70% solution because it quickly denatures and coagulates the bacterial surface, thus forming a barrier against further penetration by alcohol. This protective reaction does not take place when a 70% solution is used. The more dilute solution denatures more slowly and allows complete penetration to be achieved before coagulation of the surface proteins takes place.

Salts of Heavy Metal Ions

Ions such as Pb^{2+}, Hg^{2+}, and Ag^+, called heavy metal ions because of their large atomic weights, are extremely toxic to most living organisms. Their toxicity is caused by enzyme deactivation (see Section 25.7) as a result of reaction with sulfur-containing or acidic side chains of amino acids. The denaturation of a protein by Ag^+ ions through the formation of a carboxylate salt is illustrated by the following reaction.

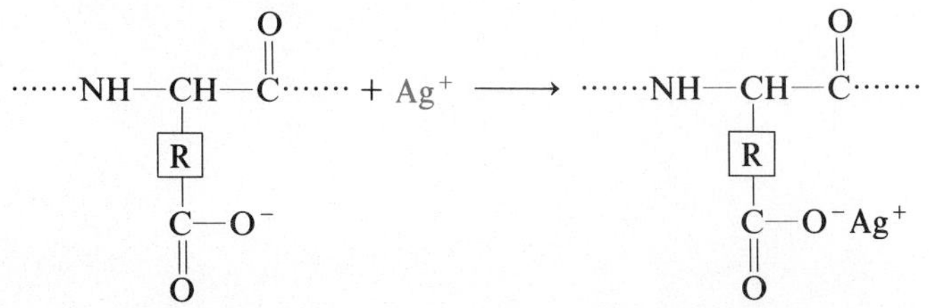

$$\cdots\cdots NH-\underset{\underset{\underset{O}{\overset{\|}{C}}}{\overset{\boxed{R}}{|}}}{CH}-\overset{O}{\overset{\|}{C}}\cdots\cdots + Ag^+ \longrightarrow \cdots\cdots NH-\underset{\underset{\underset{O}{\overset{\|}{C}}}{\overset{\boxed{R}}{|}}}{CH}-\overset{O}{\overset{\|}{C}}\cdots\cdots$$
$$C-O^-C-O^-Ag^+$$

$$\text{protein} \text{silver proteinate}$$

In poisonings resulting from heavy metal ingestion, egg whites and milk are often used as antidotes. The proteins in these substances are quickly denatured by combining with the toxic ions to form insoluble solids that prevent denaturation of other proteins in the body and absorption of the metal ions into the blood.

FIGURE 24-17 Reduction of disulfide bonds followed by oxidation may not regenerate the original protein conformation.

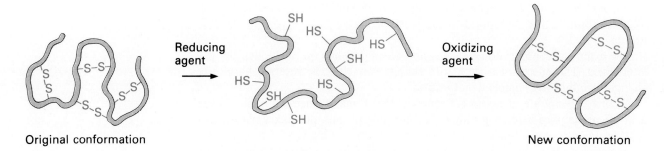

Sodium bicarbonate, a salt that raises the pH of a solution, is often administered along with the milk and egg whites to provide these proteins with excess negative charges. The negative charges enhance the ability of the proteins to attract the positive toxic metal ions. Vomiting is then induced so that the precipitated proteins are removed from the stomach before the harmful metal ions can be absorbed into the system.

Reducing and Oxidizing Agents

Mild reducing agents can convert disulfide bonds between cysteine amino acids (Section 24.10) to free sulfhydryl (—SH) groups. Since the disulfide bond is a covalent bond, this change seriously disrupts the tertiary structure of a protein.

Mild oxidizing agents can convert sulfhydryl groups back into disulfide bonds. However, the structure of the protein can be significantly different from the original

FIGURE 24-18 Hair protein is denaturated to give a permanent wave.

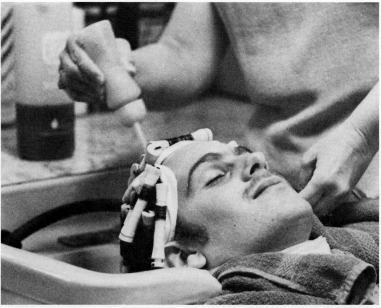

conformation because different combinations of cysteine units can be involved in the new bonds (see Figure 24–17).

Hair is a protein with a high content, about 16–18%, of the amino acid cysteine. This amino acid plays an important role in the hair structure. Hair salons, for instance, use reducing and oxidizing agents on disulfide linkages in hair to give a permanent (see Figure 24–18).

FIGURE 24–19 Structural changes occur in hair during treatment for a permanent wave because of protein denaturation.

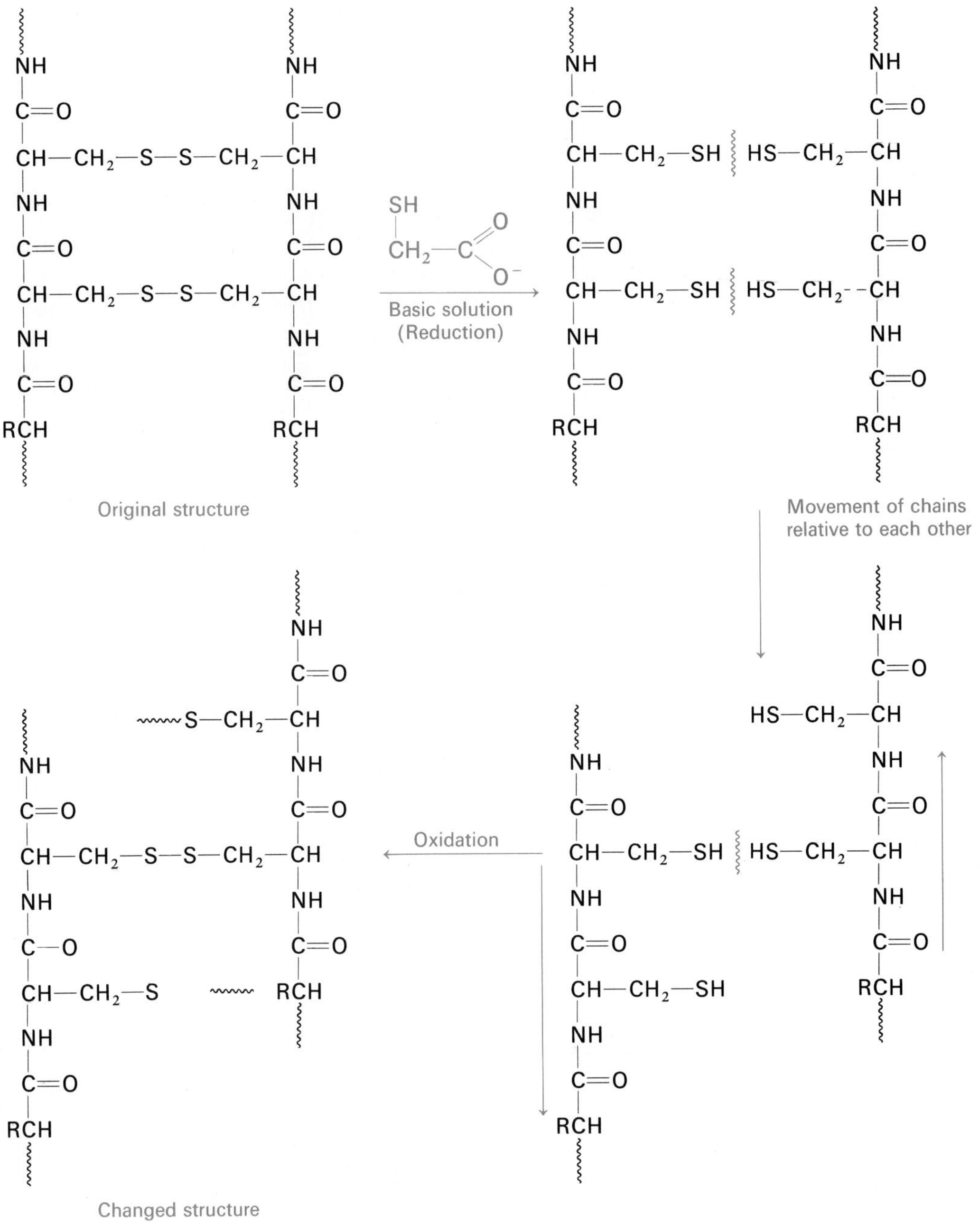

During a permanent, straight hair is reduced, using ammonium thioglycolate, to break disulfide bonds. The unraveled (denatured) hair, wound on rollers for shape, is then oxidized by a neutralizer, potassium bromate, to reform new disulfide bonds. The new shape and curl is maintained by the new disulfide bonds and resulting tertiary structure of the hair protein. Of course, as new hair grows out, the permanent will have to be repeated. Figure 24–19 (p. 635) shows, using chemical formulas, the principles involved in protein denaturation during a hair permanent.

24.14 IMMUNOGLOBULINS: AN EXAMPLE OF PROTEIN STRUCTURE AND FUNCTION

Immunoglobulins are among the most important and interesting of the soluble proteins in blood serum. Often called *antibodies* or *gamma-globulins*, **immunoglobulins** *are protein molecules produced by an organism in a protective response to the invasion of microorganisms or foreign molecules. The foreign molecules that elicit immunoglobulin production are known as* **antigens**. These antigens might be the foreign protein in bacteria or the nucleic acid polymer of a strain of virus.

It is the highly specialized structure of the immunoglobulin molecules that enables them to act in concert to inactivate the foreign substance. Each molecule has two identical "active sites" which can bind to two molecules of the antigen. The action of many such antibodies can create an *antigen-antibody complex* that will eventually destroy and eliminate the invading substance from the body (see Figure 24–20). Because viruses are not affected by antibiotics or most medications, the immunoglobulins represent our main line of defense against chronic infections.

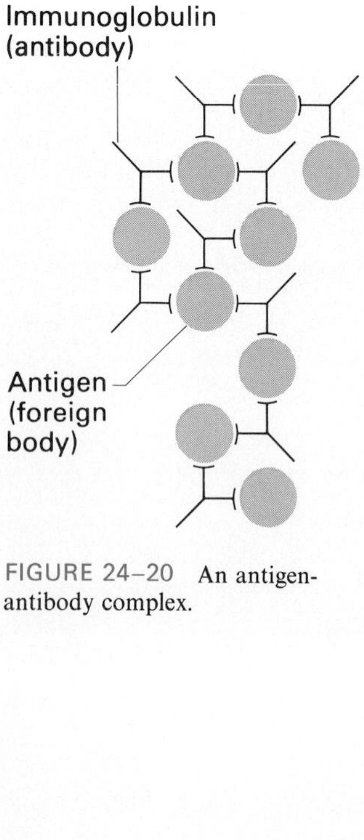

The human immune system has the capability to produce immunoglobulins that can respond to millions of different antigens. The presence of an antigen stimulates specialized cells in the blood called *lymphocytes* to produce the necessary immunoglobulin. Each lymphocyte produces a specific immunoglobulin in response to one and only one antigen. It is the immunoglobulin's exacting degree of specificity and the way in which it inactivates the antigen that make it an interesting study of protein form and function.

Despite the millions of different kinds of immunoglobulin molecules, they are classified by form and function into just five groups. Each category is labeled as an immunoglobulin (Ig) of a certain type (A, D, E, G, or M). We will concentrate on the most common type: immunoglobulin type G (IgG).

Each IgG molecule is composed of four polypeptide chains: two identical heavy (H) chains and two identical light (L) chains. The H chains are each 440 amino acids in length; the L chains are half as long, or about 220 amino acids in length. Numerous disulfide linkages link the four chains together (see Figure 24–21).

The first amino acid sequence of an IgG molecule was reported in 1969. Since then, other IgG molecules have been sequenced, revealing an important correlation between their primary structures: one region of both the H and L chains has relatively constant amino acid content from one molecule to another, while the other region varies in content from one IgG molecule to another. (The regions are shown in Figure 24–21.) It is in the variable region, in segments of the H and L chains that are called *hypervariable*, that each IgG molecule demonstrates its particular specificity for antigen binding.

The effective orchestration of protein form and function reveals itself in the secondary and tertiary structure of IgG. All IgG proteins have very similar secondary and tertiary structures. Upon folding into its active form, the molecule assumes

FIGURE 24–20 An antigen-antibody complex.

FIGURE 24-21 Diagram of the primary structure of IgG.

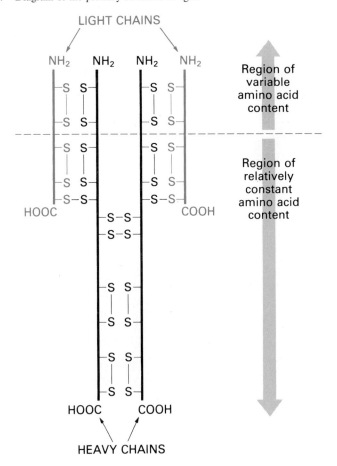

a Y-shaped conformation (see Figure 24-22, p. 638). The disulfide linkages between chains stabilize this structure, with one important exception. The S–S linkages between H chains at the branch are somewhat flexible, much like doors on a hinge. (This region is actually called the hinge region.) The antigen binding sites are located at the uppermost part of the Y's "arms".

Studies of secondary structure in the variable region of the H and L chains reveal extensive antiparallel β-pleated sheets. As can be seen in Figure 24-23 (p. 639), the rigid structure of the β-pleated sheets maintains the hypervariable regions at the very tips of the arms of the molecule. It is here that the antigen binds so specifically, and it is here that the amino acids differ significantly from one antibody to the next. The three-dimensional character of the binding site, in addition to the amino acid sequence, allows further selectivity. For this reason antibodies can even exhibit preferential binding to one stereoisomer over another. Furthermore, antibodies are very resistant to denaturation as compared to most other proteins.

The structure of the IgG molecule makes it perfectly suited to the antibody function it performs. Through changes in the hypervariable region alone, the antibodies provide the diversity needed to protect the body from millions of antigens.

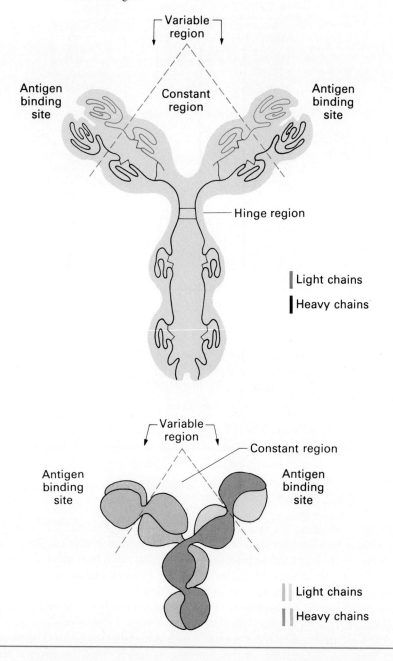

FIGURE 24-22 Diagramatic structure of IgG showing (a) local regions of folding and how they result in the overall tertiary structure of the molecule. (b) Three-dimensional representation of IgG.

There can be little doubt of the importance of immunoglobulins; this is amply and tragically demonstrated by the effects of AIDS (acquired immune deficiency syndrome). The AIDS virus upsets the body's normal production of immunoglobulins and leaves the body susceptible to what would otherwise not have been debilitating and deadly infections. Because some drugs suppress immunoglobulin synthesis, the result can be a loss of protection from infection; these drugs can be administered only under carefully controlled conditions. Individuals who receive

FIGURE 24-23 Schematic drawing of secondary and tertiary structure of one L chain of IgG. The hypervariable region is the antigen binding site. The arrows depict four β-pleated sheets (two in each region) and their direction from the NH₂ terminus.

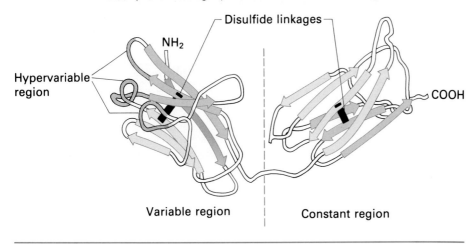

organ transplants must be given drugs to suppress the production of immunoglobulins against foreign proteins in the new organ, thus preventing rejection of the organ.

Immunoglobulins are only one example of the intricate manner in which the various levels of protein structure blend together to form functional molecules on which we depend so heavily for life processes. In the next chapter we examine another group of specialized proteins, namely enzymes, whose catalytic functions depend directly on their individual structures.

EXERCISES AND PROBLEMS

Functions and Characteristics of Proteins

24.1 Explain the origin of the word protein.

24.2 What elements are present in all proteins?

24.3 Match the protein names on the left with their functions on the right:

a. insulin
b. lipase
c. transferrin
d. somatotropin
e. collagen
f. ferritin
g. actin
h. rhodopsin

1. regulates skeletal growth
2. transports iron in the blood
3. stores iron for later use
4. contractile protein in muscle
5. catalyzes hydrolysis of triglycerides
6. controls blood glucose levels
7. absorbs light in the eye
8. helps form connective tissue

24.4 Classify each of the following proteins according to their function: regulatory, catalytic, transport, nutrient, contractile, structural, or defense.

a. myosin
b. hemoglobin
c. prothrombin
d. sucrase
e. insulin
f. egg albumin
g. elastin
h. zein

An experiment in the lab requires you to weigh out 5.0 mg (0.005 grams) of the protein enzyme catalase. Table 24-2 gives the molecular weight of catalase as 250,000 amu. How many moles of catalase are present in 5.0 mg of this enzyme?

Amino Acids

24.6 What functional groups are found in all amino acids?

24.7 What is an α-amino acid?

24.8 Which of the following structures do not represent α-amino acids?

a.
$$CH_3-CH-C\begin{matrix}\diagup O\\ \diagdown OH\end{matrix}$$
$$|$$
$$NH_2$$

b.
$$CH_3-CH-CH_2-C\begin{matrix}\diagup O\\ \diagdown OH\end{matrix}$$
$$|$$
$$NH_2$$

c.
$$CH_3-\underset{\underset{O}{\|}}{C}-CH_2-NH_2$$

d.
$$\underset{HO}{\overset{O}{\diagdown}}C-CH_2-NH_2$$

24.9 What distinguishes one α-amino acid from another?

24.10 Draw the structural formula for both leucine and serine. Clearly identify the following on both structures.
a. the amino group
b. the carboxyl group
c. the α-carbon atom
d. the side chain

24.11 Give the name and the structure for each of the following amino acids.
a. Glu   d. Phe
b. Gly   e. Tyr
c. Ala   f. Cys

24.12 Draw the structure of the amino acid in Table 24–3 that is a structural isomer of leucine.

24.13 Why is the structure of proline different from that of the other 19 common amino acids?

24.14 Give a definition for a protein in terms of amino acids.

24.15 Give the names of amino acids, from Table 24–3, that have each of the following side-chain characteristics:
a. two carbon atoms and an —OH functional group
b. a phenolic group
c. three carbon atoms and an S atom
d. an —SH group
e. a very polar side chain containing three N atoms
f. a polar side chain that is aromatic
g. an acidic side chain containing two C atoms
h. an aromatic nonpolar side chain that does not contain N.

24.16 Five of the 20 common amino acids have side chains that are very polar. Such side chains become charged in solution at physiological pH. Draw the structures (both the charged and uncharged forms) of
a. the two amino acids whose side chain acquires a negative charge.
b. the three amino acids whose side chain acquires a positive charge.

Essential Amino Acids

24.17 Why are some amino acids called essential while others are not?

24.18 Why should vegetarians carefully watch their intake of plant proteins?

24.19 Why should a proper diet include protein from a variety of sources?

24.20 How do the amino acids in milk, meat, and cereal compare in terms of biological value?

Stereoisomerism in Amino Acids

24.21 Which one of the 20 common amino acids is the only one that cannot exist in D and L forms?

24.22 Draw projection formulas for the D and L forms of valine.

24.23 Contrast the stereochemistry (D or L form) of the glucose units in polysaccharides with the amino acid units in proteins.

24.24 A number of D-amino acids are found in bacteria, one of which is D-glutamic acid. Draw the projection formula of D-glutamic acid.

Acid–Base Properties of Amino Acids

24.25 What is meant by the term *zwitterion*?

24.26 Draw the zwitterion structure for each of the following amino acids.
a. leucine   c. cysteine
b. tyrosine   d. serine

24.27 How will a zwitterion move if placed in an electric field?

24.28 List one evidence that amino acids exist as zwitterions in the solid state.

24.29 Draw the structure of the amino acid valine at each of the following pH values
a. 7   b. 1   c. 12

24.30 Consider the migration of alanine in an electric field.
a. At what pH will it move toward the positively-charged electrode?
b. At what pH will it move toward the negatively-charged electrode?
c. At what pH will it be stationary?

24.31 Predict the direction of movement of each of the following amino acids at the pH value specified. Indicate the direction as towards the positive electrode, towards the negative electrode, or as isoelectric (no net movement). Note: Do not forget the ionic contribution of the side chains when appropriate.
a. alanine at pH = 12
b. arginine at pH = 7
c. aspartic acid at pH = 1
d. serine at pH = 7
e. lysine at pH = 1
f. glutamic acid at pH = 12

24.32 A mixture of three amino acids, glycine, glutamic acid, and lysine, is to be separated using electrophoresis. Is this possible? If so, at what approximate pH would the separation be most efficient?

24.33 Most amino acids have isoelectric points between 5 and 6, but the isoelectric point of lysine is 9.7. Explain why lysine has such a high value for its isoelectric point.

24.34 The isoelectric point of hemoglobin is 6.8. What does this value indicate about the composition of hemoglobin?

Peptides and Polypeptides

24.35 How are the terms amide bond and peptide bond related?

24.36 How many amino acids are present in a hexapeptide? How many peptide bonds are present in a hexapeptide?

24.37 What is meant by the N-terminal end and the C-terminal end of a peptide?

24.38 Draw the complete structural formula for the tripeptide Try-Ala-Cys. Identify both the N-terminal end and the C-terminal end of the tripeptide.

24.39 Draw structural formulas for two different dipeptides that can be made from one glycine and one cysteine molecule.

24.40 Explain why the tetrapeptide Asp-Met-Arg-Val is not the same tetrapeptide as one written in reverse form, Val-Arg-Met-Asp.

24.41 There are a total of six different sequences for a tripeptide containing one molecule each of serine, valine, and glycine. Using three-letter abbreviations for the amino acids, draw the six possible sequences of amino acids.

24.42 How many different polypeptides containing two molecules of glycine and two molecules of leucine are possible?

24.43 How does the dipeptide glycylserine differ from the dipeptide serylglycine?

24.44 Draw the structural formulas of the following dipeptides.
a. serylhistidine
b. histidylglutamic acid
c. glycyllysine
d. methionylaspartic acid

24.45 Identify the amino acids contained in each of the following tripeptides.

a.
$$H_2N-CH-\overset{O}{\underset{\|}{C}}-\overset{H}{\underset{|}{N}}-CH-\overset{O}{\underset{\|}{C}}-\overset{H}{\underset{|}{N}}-CH-C\overset{O}{\underset{OH}{\diagdown}}$$
$$\underset{CH_2}{|} \quad \underset{CH_3}{|} \quad \underset{CH_2}{|}$$
$$\underset{OH}{|} \quad \quad \underset{SH}{|}$$

b.
$$H_2N-CH-\overset{O}{\underset{\|}{C}}-\overset{H}{\underset{|}{N}}-CH-\overset{O}{\underset{\|}{C}}-\overset{H}{\underset{|}{N}}-CH-C\overset{O}{\underset{OH}{\diagdown}}$$
$$\underset{CH_2}{|} \quad \underset{CH_2}{|} \quad \underset{CH}{|}$$
$$\underset{\underset{O}{\overset{\|}{C}}-OH}{} \quad \underset{\underset{O}{\overset{\|}{C}}-NH_2}{} \quad CH_3 \quad CH_3$$

c.
$$H_2N-CH_2-\overset{O}{\underset{\|}{C}}-\overset{H}{\underset{|}{N}}-CH-\overset{O}{\underset{\|}{C}}-\overset{H}{\underset{|}{N}}-CH_2-C\overset{O}{\underset{OH}{\diagdown}}$$
$$\underset{CH-OH}{|}$$
$$\underset{CH_3}{|}$$

24.46 The movement of a polypeptide in an electrical field depends on the charges of its free amino end, its free carboxyl end, and its side chains. Predict the direction of movement for each of the following polypeptides in an electrical field at neutral pH:
a. Angiotensin (produced by the blood to help control fluids and blood pressure):
Asp-Arg-Val-Tyr-Ile-His-Pro
b. Bradykinin (smooth muscle contractor)
Arg-Pro-Pro-Gly-Phe-Ser-Pro-Phe-Arg

Primary Structure of Proteins

24.47 Describe the difference in meaning of the terms polypeptide and protein.

24.48 What is meant by the primary structure of a protein?

24.49 Slight changes in the primary structure of a protein may or may not affect its ability to function. Give examples of cases where changes in amino acid sequences affect function to different degrees.

24.50 If you know the amino acid composition of a protein, can you predict its primary structure? Explain.

24.51 Draw a structural representation, containing four units, of the carbon-nitrogen backbone structure common to all polypeptides and proteins.

24.52 Compare the differences between the primary structure of human insulin and insulin from pigs and cows. Would pig or cow insulin be the best alternate source for human diabetics, based solely upon primary structure?

24.53 A valuable race horse develops diabetes. It is suggested that he be given daily supplemental shots of insulin, to keep him alive for breeding. Considering the types of insulin available (those in Table 24–6), which would be the best source of supplemental insulin? Which would be the second-best source?

Secondary Structure of Proteins

24.54 What is meant by the secondary structure of a protein?

24.55 Describe the three types of secondary structure commonly found in protein. Draw a sketch of each type.

25.56 What type of intramolecular force holds the "turns" in an α-helix in position?

24.57 What type of intramolecular force holds parallel protein chains together in the β-pleated sheet structure?

24.58 What two functional groups are involved in the hydrogen bonds associated with the secondary structure of a protein? Draw a diagram illustrating this type of hydrogen bond in
a. an α-helix.
b. a β-pleated sheet.

24.59 What common feature exists between the hydrogen bonding present in the α-helix and the β-pleated sheet?

24.60 What type of secondary structural feature gives rise to the high strength that the protein collagen exhibits?

24.61 Can more than one type of secondary structure be present in the same protein? Explain your answer.

24.62 What do we mean when we say that a section of a protein has a random coil arrangement?

24.63 Glycine occurs repeatedly in the amino acid sequence of collagen. Why is it necessary for glycine to be present in this connective tissue protein?

Tertiary Structure of Proteins

24.64 What is meant by the tertiary structure of a protein?

24.65 Two terms often associated with tertiary structure are "globular" and "fibrous". Describe these two types of tertiary structure and name a protein characteristic of each type.

24.66 State the four types of attractive interactions that give rise to tertiary structure for proteins.

24.67 Which class of amino acid side chains
a. are hydrophobic?
b. participate in hydrogen bonding?
c. participate in salt bridges?

24.68 Imagine that the folding of a protein (tertiary structure) brings the following pairs of amino acids into close proximity. Predict the most likely type of interaction between side chains.
a. phenylalanine and leucine.
b. arginine and glutamic acid
c. two cysteine residues
d. serine and tyrosine

24.69 Using structural formulas, sketch each of the side chain reactions predicted in Problem 24.68.

24.70 Distinguish between the origin of the attractive interactions in tertiary protein structure as compared with those in secondary protein structure.

24.71 What is a prosthetic group in a protein? Describe the prosthetic group in myoglobin. What is its function in myoglobin?

Quaternary Structure of Proteins

24.72 What is an oligomeric protein?

24.73 What is meant by the quaternary structure of a protein?

24.74 List the kinds of forces that hold the various units of an oligomeric protein in proper orientation to one another. Are these forces covalent bonds?

24.75 Can a protein with less than two subunits have a quaternary structure? Explain your answer.

24.76 In general terms, compare the structures and functions of myoglobin and hemoglobin.

Protein Hydrolysis

24.77 Write a balanced chemical equation with complete structures for the hydrolysis of the tripeptide Val-Gly-Ala.

24.78 Will hydrolysis of the dipeptides Ala-Val and Val-Ala give the same products? Explain.

24.79 A shampoo bottle lists "partially hydrolyzed protein" as one of its ingredients. What is the difference between partially and completely hydrolyzed protein?

24.80 Are proteins we absorb in our food the identical proteins we use in our body?

24.81 Drugs that are proteins, such as insulin, must always be injected rather than taken orally. Explain why.

24.82 Partial hydrolysis of various polypeptides yields fragments of the original molecules with the amino acid sequences listed below. Determine the primary structure of the original polypeptide in each case.
a. a nonapeptide: Ala-Ala-Trp-Gly-Lys
   Thr-Asn-Val-Lys
   Val-Lys-Ala-Ala-Trp
b. a hexapeptide: Val-Met-His
   His-Val-Arg
   Ala-Val
c. a decapeptide: Ile-Phe-Cys
   Arg-Gly-Gly
   Phe-His-Arg
   Gly-Ile
   Cys-Thr-Lys

Protein Denaturation

24.83 Which structural levels of a protein are affected by denaturation?

24.84 What is the difference on a molecular level between denaturation and hydrolysis of proteins?

24.85 In what way is the protein in a cooked egg the same as in a raw egg?

24.86 Why is 70% isopropyl alcohol more effective as a disinfectant than pure isopropyl alcohol?

24.87 What is the connection between sterilizing surgical instruments in an autoclave and the process of protein denaturation?

24.88 Why are milk and egg whites effective antidotes in treatment of heavy metal poisoning? Why should vomiting be induced after administration of these antidotes?

24.89 Why are two different chemical solutions used to give a hair permanent?

24.90 Why can strong acids or bases damage eyes?

Immunoglobulins

24.91 What are immunoglobulins? How do they function?

24.92 List two causes of abnormal immunoglobulin production.

24.93 What is the difference between an antigen and an antibody?

24.94 Describe the primary structure of IgG. Where do the regions of variable amino acid content of different IgGs occur? Where are the constant regions among different IgG molecules?

24.95 Describe the overall shape (tertiary structure) of IgG molecules. Indicate where antigens bind to IgG molecules.

24.96 Briefly describe the primary and secondary structure at the antigen binding site of IgG.

25

Enzymes: Biological Catalysts

Objectives

After completing Chapter 25, you will be able to:

25.1 Understand the highly specialized nature of enzymes and their importance in biological systems.
25.2 Recognize how enzymes are named and understand the relationship between substrate and active site as it pertains to enzymes.
25.3 Know the difference, in terms of structure, between simple enzymes and conjugated enzymes and be familiar with the terminology used in describing conjugated enzymes.
25.4 Understand the difference between a zymogen and the active enzyme that can be derived from it.
25.5 List some general properties of enzymes, including their specificity and categories of reactions they catalyze.
25.6 Describe the two common models used to explain enzyme action.
25.7 Explain how temperature, pH, substrate concentration, and enzyme concentration affect enzyme activity.
25.8 Understand how cells control their enzyme activity via competitive and noncompetitive inhibition, allosteric control, and covalent modification.
25.9 Describe how gastric and pancreatic zymogens are activated by covalent modification.
25.10 Using sulfanilamide and penicillin as specific examples, explain how some antibiotics inhibit specific bacterial enzymes.
25.11 Describe the unique structure of isoenzymes and how this affects their catalytic activity.
25.12 List a variety of uses and application of enzymes, emphasizing those in medically-related areas.

INTRODUCTION

Enzymes are among the most specialized of nature's proteins. With fascinating precision and selectivity, they catalyze biochemical reactions that store and release energy, make pigments in our hair and eyes, digest the food we eat, synthesize cellular building materials, and protect us by repairing damage or clotting our blood. These biocatalysts are sensitive to their environment, responding automatically to changes in the cell. The deficiency or excess of particular enzymes can cause a certain disease or help diagnose problems such as heart attacks and other organ damage. Applying our knowledge of protein structure will help us appreciate and better understand how enzymes function in living cells.

25.1 IMPORTANCE OF ENZYMES

Enzymes were used to make food long before they were understood. To make bread rise, sugar was oxidized to CO_2 by enzymes in yeast. Alcoholic beverages were made by the same yeast enzymes if sugars were partially oxidized in the absence of oxygen (fermentation). In fact, the word *enzyme* is of Greek origin, meaning "in yeast". A substance from stomachs of livestock (the enzyme rennin) curdled milk into cheese. Bacterial enzymes produced yogurt from milk and even apple cider could be fermented into vinegar. Not until recently have we understood the mysterious properties of these "magic" substances.

Enzymes *are highly specialized protein molecules that act as biological catalysts.* Enzymes are indispensable to maintain life processes in all organisms. Without enzymes, respiration, growth, muscle contraction, and other physical and mental activities could not occur. Enzyme-directed cellular reactions are more efficient than laboratory-run organic reactions; no undesired products are formed.

enzymes

Each cell in the body contains thousands of different enzymes because a different enzyme is needed to catalyze almost every chemical reaction. Enzymes accelerate reactions in the cell thousands or even millions of times faster than corresponding uncatalyzed reactions. As catalysts, they are neither created nor consumed during the reaction, but merely help the reaction occur more rapidly. For example, CO_2 is a waste product formed during respiration that must be transported by the blood to the lungs where it can be expelled from the body. Due to its limited solubility in aqueous solution, CO_2 must be transformed into carbonic acid (H_2CO_3) to transport it in sufficiently large quantities:

$$CO_2 + H_2O \xrightarrow{\text{carbonic anhydrase}} H_2CO_3$$

In the absence of the appropriate enzyme, carbonic anhydrase, the formation of carbonic acid takes place much too slowly to support the required exchange of carbon dioxide between the blood and the lungs. But in the presence of carbonic anhydrase, this vital reaction proceeds rapidly. Under optimal conditions, a single enzyme molecule can catalyze the formation of carbonic acid at the rate of 600,000 molecules per second. This is approximately 10^7 times faster than would occur without the enzyme!

Besides being fast, enzymes are specific. To hydrogenate an alkene in the laboratory, Pt or Ni can be used to catalyze the reaction for almost any alkene. This is not the case for enzymes. Almost every reaction in a cell requires its own specific enzyme. At first this may seem excessive or wasteful. Why not just a handful of enzymes to catalyze all the reactions in the cell? The answer lies in the unique ability of enzymatic catalysts to respond to the ever-changing conditions in the cell.

Many enzymes function only when needed, automatically turning on or off. Most laboratory catalysts need to be removed from the reaction to stop their catalytic action; not so with enzymes. If a certain chemical is needed in the cell, the enzyme responsible for its production senses the need and automatically "switches on". When a sufficient quantity has been produced, the enzyme "switches off". In other situations, the cell may produce more or less enzyme as required. By having individual enzymes for each reaction in the cell, certain reactions can be accelerated as needed, without affecting the rest of the cellular metabolism. Enzymatic activity is high for optimum, that is, "switched on", conditions. Low concentration of enzyme or its being "switched off" results in low enzymatic activity.

The advantages of these "high tech" catalysts are not achieved without associated drawbacks. Slight alterations in pH or temperature, as well as other protein denaturants, affect enzyme activity dramatically. Good cooks realize that overheating yeast kills the action of the yeast. Even a person suffering from a high fever (greater than 106°F) runs the risk of denaturing certain enzymes. The biochemist must exercise extreme caution in handling enzymes to avoid the loss of their activity. Even vigorous shaking of an enzyme solution can destroy enzyme activity.

Because of the delicate and elusive nature of enzymes, their protein character was not firmly established until the 1930s. Subsequent discoveries and studies on the thousands of enzymes now known give a clearer picture of how these specialized proteins function.

25.2 ENZYME NOMENCLATURE AND ASSOCIATED TERMINOLOGY

Historically, common names of enzymes were assigned by taking the name of the compound undergoing change and adding the ending -*ase*. For example, urease is an enzyme that catalyzes the hydrolysis of urea, while a peptidase enzyme catalyzes the hydrolysis of a peptide bond. Eventually, to avoid confusion, the type of reaction catalyzed by the enzyme also began to be included in the name. Glucose oxidase is an enzyme that catalyzes the oxidation of glucose; pyruvate decarboxylase catalyzes the decarboxylation of pyruvate; and alcohol dehydrogenase removes two hydrogen atoms from alcohol. A few enzymes received names that do not end in -*ase*. Most of these are enzymes involved in digestion or blood-clotting such as pepsin, trypsin, and thrombin. All of these are enzymes that cut or break the polypeptide backbone in other proteins. Most of older names are still used today.

In modern times, the International Enzyme Commission (EC) has developed a systematic, numerical nomenclature for enzymes. The technical name assigned to urease by the EC is EC 3.5.1.5 urea aminohydrolase and 4.1.1.1 2-oxo-acid carboxy-lyase for pyruvate decarboxylase. Each of the numerical digits in the name refers to a category of enzyme action and every enzyme has a unique set of numbers. Although extremely precise, the EC nomenclature is seldom used in routine discussion of enzymes. In this text we will refer to the enzymes by their common names.

Predict the function of the following enzymes: Example 25.1
a. cellulase   c. L-amino acid oxidase
b. sucrase   d. aspartate aminotransferase

Solution
a. cellulase: catalyzes the hydrolysis of cellulose
b. sucrase: catalyzes the hydrolysis of the disaccharide sucrose into the monosaccharides glucose and fructose
c. L-amino acid oxodase: catalyzes the oxidation of L-amino acids
d. aspartate aminotransferase: catalyzes the transfer of an amino group from aspartate to a different molecule

Since enzymes are usually quite specific for a particular reactant, a special term is given to the reactant involved in an enzymatic reaction. The **substrate** *in a biological reaction is the reactant on which the enzyme works.* For example, urea is the substrate for urease and cellulose is the substrate acted upon by cellulase. After the reaction has been completed, the altered substrate is called simply the *product* of the reaction.   substrate

The entire structure of an enzyme molecule is seldom involved in actual catalysis; only a small portion is actually used. The **active site** *on an enzyme is the relatively small portion of the enzyme molecule actually involved in the catalytic process.* It is the site where the substrate and enzyme interact (see Figure 25–1). The special properties of the active site are responsible for selective binding of the substrate and catalysis of the reaction. Any alteration of the active site affects enzyme activity, while changes in other regions of the enzyme may or may not affect its function.   active site

25.3 ENZYME COFACTORS

Enzymes can be divided into two general classes—*simple enzymes* and *conjugated enzymes*. **Simple enzymes** *are composed only of protein (amino acids).* In these enzymes, enzymatic function results from the proper three-dimensional arrangement of amino acids contained within the protein. **Conjugated enzymes** *have a nonprotein group in addition to a protein portion.* By themselves, neither the protein part of a conjugated protein nor the nonprotein portion possess catalytic properties. Only the combination of the two—the conjugated enzyme—has biological activity.   simple enzymes

conjugated enzymes

FIGURE 25–1 Diagram of the relatively small active site on an enzyme molecule.

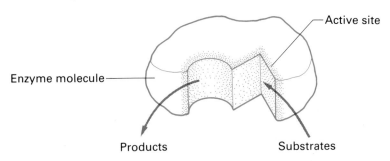

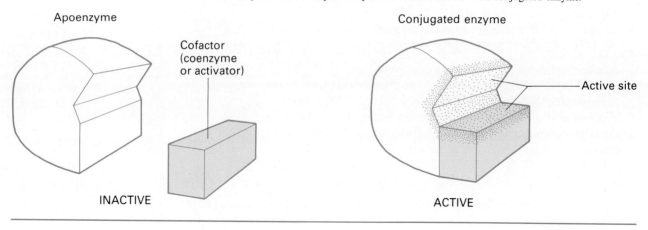

FIGURE 25–2 Schematic diagram of the components present at the active site of a conjugated enzyme.

apoenzyme

cofactor

coenzymes

activators

Two other terms, *apoenzyme* and *cofactor*, are associated with the concept of a conjugated protein. An **apoenzyme** *is the protein part of a conjugated enzyme without the nonprotein part that is required for it to function.* An apoenzyme is an inactive enzyme. A **cofactor** *is the nonprotein part of a conjugated protein.* It is the substance that combines with an apoenzyme to give a biologically active enzyme. Cofactors are generally located at the active site of a conjugated protein (see Figure 25–2).

Cofactors fall into two general categories: organic molecules and metal ions. **Coenzymes** *are organic molecules that function as cofactors.* Many of the vitamins we need in our diets are coenzymes (Chapter 27). Another typical cofactor is nicontinamide adenine dinucleotide (NADH). Alcohol dehydrogenase, for example, cannot dehydrogenate alcohol unless the coenzyme NAD^+ is present and it is concomitantly reduced to NADH during the reaction.

Coenzymes are sometimes covalently attached to the apoenzyme, but are usually held in place by weaker forces. Unlike the apoenzyme part of a conjugated enzyme, coenzymes are often chemically altered during the course of the reaction, as the previously mentioned example of the conversion of NAD^+ to NADH illustrates. If changed in any way, each coenzyme molecule must be replaced with another unchanged molecule before additional substrate can be acted upon.

Activators *are inorganic metal ions that function as cofactors.* Typical activators include such ions as Zn^{2+}, Mg^{2+}, Mn^{2+}, and Fe^{2+}. Even negative ions such as F^- are occasionally needed as activators. Aminopeptidases contain manganese

TABLE 25–1 Selected Conjugated Enzymes Containing a Trace Element Cofactor (Activator)

Trace Element	Conjugated Enzyme
copper	cytochrome oxidase
iron	heme enzymes
manganese	arginase, pyruvate oxidase
molybdenum	xanthine oxidase
nickel	urease
zinc	carbonic anhydrase, dehydrogenases, carboxypeptidase

or magnesium, whereas carboxypeptidases contain zinc. These activators are an important part of the minerals we consume daily in a well-balanced diet. Table 25–1 gives additional examples of conjugated enzymes where the cofactor is an activator (metal ion). Note that the metals involved are those identified in Section 5.9 as trace elements, elements known to be essential in humans and animals in small amounts.

25.4 ZYMOGENS OR PROENZYMES

In order to prevent their acting on (and destroying) the very glands or tissues that produce them, some enzymes, such as protein-digesting ones, are first generated in an inactive form that is then, upon need, converted to an active form. These inactive forms are called *zymogens* or *proenzymes*. A **zymogen** (*or proenzyme*) *is an inactive storage form of an enzyme.*

zymogen

Zymogens are activated by removing a small part of their polypeptide chain. Removing the peptide fragment not only shortens the chain, but also changes the three-dimensional structure (secondary and tertiary structure) of the peptide, allowing the enzyme to adopt its active configuration (see Figure 25–3). For example, the zymogen pepsinogen is secreted into the stomach, where hydrogen ions of the gastric juice convert it to pepsin, the active enzyme (Section 25.9).

25.5 GENERAL PROPERTIES OF ENZYMES

Enzymes are true catalysts. They are neither used up nor permanently changed during chemical reactions, but only speed up reactions that otherwise would take place very slowly.

As mentioned earlier, enzyme activity is extremely dependent upon the delicate structure of the protein. Since catalysis depends so much on the proper orientation of all amino acids in the active site, any slight changes at the active site can terminate normal activity. Mild conditions that might not denature the protein might, however, stop enzymatic activity. Examples of this effect include slight changes in pH, temperature, or solvent polarity, the chemical alteration of a single amino acid at the active site, and even changes in ionic strength of the solution.

Many enzymes are globular proteins that are soluble in water. The surfaces of these enzymes are covered with polar amino acid side chains and their interior

FIGURE 25–3 Conversion of a zymogen (or proenzyme) to an active enzyme involves removal of a short segment of the polypeptide chain.

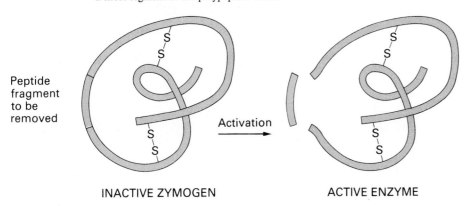

TABLE 25–2 Major Enzyme Classes

Category	Type of Reactions	Specific Example
oxidoreductases	oxidation/reduction	peroxidase
transferases	transfer of functional groups from one molecule to another	aspartate amino transferase
hydrolases	cleavage of covalent bonds by hydrolysis	carboxypeptidase
lyases	water, ammonia, or carbon dioxide is added to or removed from double bonds	pyruvate decarboxylase
isomerases	conversion of one isomer into another	glucose phosphate isomerase
ligases	two molecules are joined, usually requiring the input of energy from ATP	DNA ligase

portions contain the more hydrophobic residues. However, enzymes that function on water-insoluble substrates are usually found in the hydrophobic regions of membranes. These enzymes are rich in nonpolar amino acids and are not soluble in aqueous solution.

Enzymes can be grouped into categories that describe the types of reactions they catalyze. Six general categories are often used and are listed in Table 25–2. Many subcategories are found within each category. For example, peroxidases, dehydrogenases, oxidases, and reductases are all *oxidoreductases*, since they catalyze oxidation-reduction reactions.

Example 25.2

Assign each of the following enzymes to one of the categories in Table 25–2.

Enzyme | Reaction

a. succinate dehydrogenase

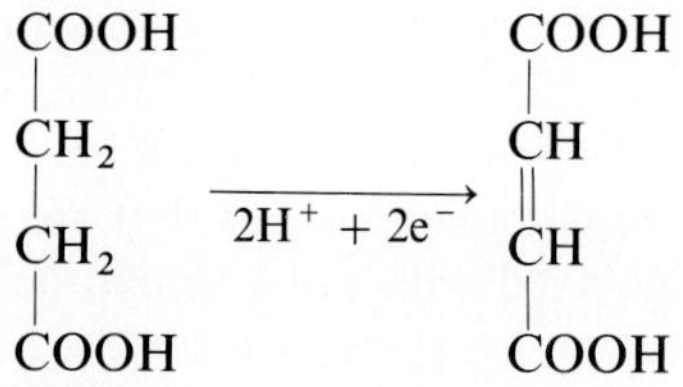

$$\begin{array}{c} \text{COOH} \\ | \\ \text{CH}_2 \\ | \\ \text{CH}_2 \\ | \\ \text{COOH} \end{array} \xrightarrow{2\text{H}^+ + 2e^-} \begin{array}{c} \text{COOH} \\ | \\ \text{CH} \\ \| \\ \text{CH} \\ | \\ \text{COOH} \end{array}$$

b. fumarase

$$\begin{array}{c} \text{COOH} \\ | \\ \text{CH} \\ \| \\ \text{CH} \\ | \\ \text{COOH} \end{array} \xrightarrow{+\text{H}_2\text{O}} \begin{array}{c} \text{COOH} \\ | \\ \text{HO—CH} \\ | \\ \text{CH}_2 \\ | \\ \text{COOH} \end{array}$$

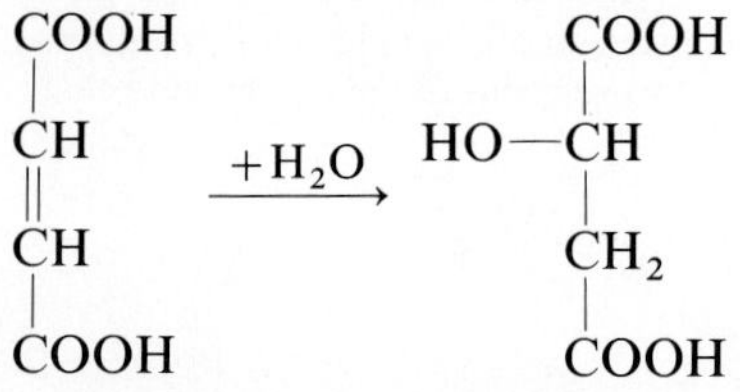

c. aconitase

$$\begin{array}{c} \text{COOH} \\ | \\ \text{CH}_2 \\ | \\ \text{HO—C—COOH} \\ | \\ \text{CH}_2 \\ | \\ \text{COOH} \end{array} \longrightarrow \begin{array}{c} \text{COOH} \\ | \\ \text{CH}_2 \\ | \\ \text{H—C—COOH} \\ | \\ \text{HO—CH} \\ | \\ \text{COOH} \end{array}$$

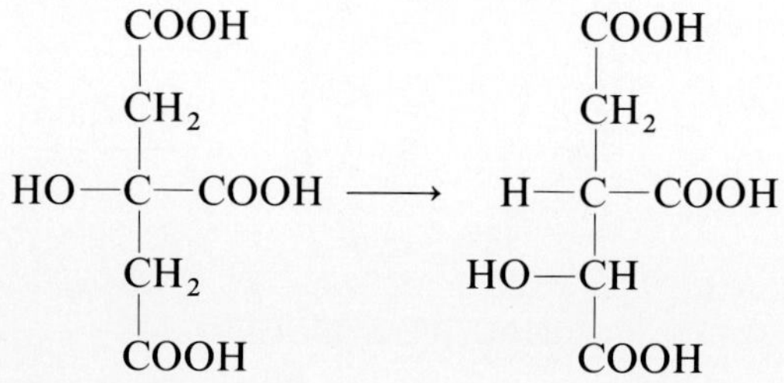

Solution
a. Succinate dehydrogenase catalyzes the oxidation of succinic acid, hence it is an oxidoreductase.
b. Fumarase assists in the addition of H_2O to a double bond, hence it is a lyase.
c. Aconitase catalyzes an isomerization reaction, hence it is an isomerase.

Enzymes exhibit different levels of selectivity or specificity for substrates. It is the active site of the enzyme that determines selectivity since it regulates what substrates can enter. Usually an enzyme accommodates only one substrate or a few closely related substrates.

Absolute specificity *means an enzyme accepts only one particular substrate and no others.* Urease exhibits absolute specificity for urea. Absolute specificity is the most limited case of enzyme specificity.

Other enzymes are only group specific. **Group specificity** *refers to enzymes that act on a set of closely related molecules, all of which contain the same functional group.* Phosphatases are group-specific enzymes that hydrolyze phosphate linkages in a variety of compounds possessing the phosphate functional group. Closely related to group specificity is linkage specificity. **Linkage specificity** *refers to enzymes that attack certain bonds within molecules.* Esterases catalyze the hydrolysis of ester linkages in a variety of types of molecules.

Stereochemical specificity *refers to enzymes that will accept substrates of only a particular stereochemical configuration.* L-Amino acid oxidase oxidizes only the amino acids containing an L configuration about the α carbon atom. Those with D configurations are ignored by the enzyme.

absolute specificity

group specificity

linkage specificity

stereochemical specificity

25.6 MODELS OF ENZYME ACTION

In Section 9.3, we learned the general concepts about how catalysts function. We can apply these concepts to explain how enzymes work as catalysts.

Catalysts, in general, offer an alternative pathway with lower activation energy (Section 9.3) through which a reaction can occur. This alternative pathway, in enzyme-controlled catalysis, involves the formation of an enzyme-substrate complex as an intermediate species in the reaction. An **enzyme-substrate complex** *is the intermediate reaction species formed when a substrate binds to the active site of an enzyme.* Within the enzyme-substrate complex, more reactions will occur, and products will be formed faster, than if the molecules were free. These better reaction conditions occur because the substrate is bound to the active site in a specific, rather than random, orientation.

enzyme substrate complex

To account for the highly specific way an enzyme selects a substrate and binds it to the active site, several models have been proposed. Two of the simplest models are the lock-and-key model, and the induced-fit model.

Lock-and-Key Model

The lock-and-key model is the simplest and the most often used to account for enzyme activity. The active site in the enzyme has a fixed, rigid geometrical conformation. Only substrates with a complementary geometry can be accommodated at such a site, much like a lock accepts only certain keys. Figure 25–4 (p. 652) illustrates the lock-and-key concept of substrate-enzyme interaction.

The lock-and-key model explains the action of many enzymes. It is, however, too restrictive for the action of other enzymes. Experimental evidence indicates

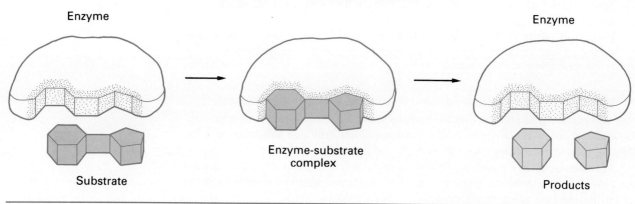

FIGURE 25-4 Lock-and-key model of substrate-enzyme interaction.

that many enzymes have flexibility in their shape. They are not rigid and static; there is constant change in their shape. The induced-fit model is used for this type of situation.

Induced-Fit Model

The induced-fit model allows for small changes in the change of the active site of an enzyme in order to accommodate a substrate. An anology would be the changes that occur in the shape of a glove when a hand is inserted into it. The induced fit is a result of the enzyme's flexibility which adapts to accept the incoming substrate. This model, shown pictorially in Figure 25-5, is a more complete concept of the active site properties of an enzyme because it includes the specificity of the lock-and-key model coupled with the flexibility of the enzyme protein.

An excellent example of the induced-fit model involves the enzyme hexokinase. Figure 25-6 shows, using actual models, the structural changes that occur when this enzyme binds to its glucose substrate. The surrounding of the substrate by the enzyme greatly facilitates the enzyme's function, which is to help in the formation of a bond between glucose and phosphate.

Lysozyme is an enzyme that breaks apart polysaccharide chains in bacterial

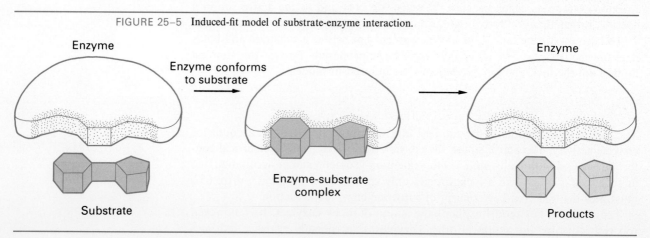

FIGURE 25-5 Induced-fit model of substrate-enzyme interaction.

FIGURE 25-6 Structural changes of hexokinase upon binding to glucose. Although slightly difficult to see in the drawing itself, the two lobes of the protein move in as much as 8 Å over an angle of 12° to surround the glucose when it binds. The enzyme is active only after it has folded up around the glucose substrate.

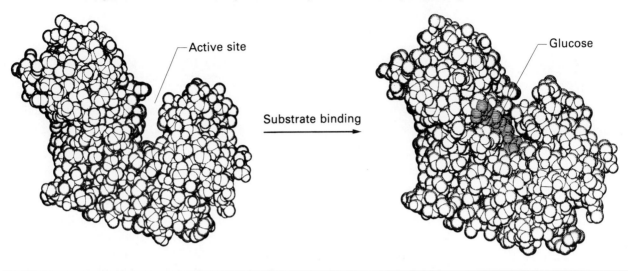

cell walls. The presence of lysozyme in tears and nasal secretions provides some protection against bacterial infection, caused by the weakening of the cell wall structure in susceptible bacteria. The way lysozyme catalyzes the hydrolytic cleavage destruction of these polysaccharides is in part due to how it binds them at its active site. A total of six sugar units in the polysaccharide bind to the enzyme. Five of these units fit nicely, but one of them (and its hemiacetal linkage) is strained almost to the breaking point as it binds. It is then easy for a water molecule to enter and hydrolyze the strained bond, as is shown in Figure 25-7.

What forces draw the substrate into the active site? Many of the same forces that maintain tertiary structure in the folding of polypeptide chains. Ionic attractions, hydrogen bonds, and hydrophobic attractions all help attract and bind substrate molecules. For example, a protonated (positively charged) amino group in

FIGURE 25-7 Diagram of lysozyme action, illustrating how formation of the enzyme-substrate complex strains the bond, making it easier for water to break the hemiacetal linkage between glucose units in the polysaccharide.

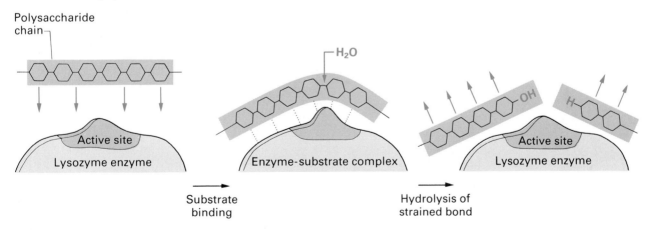

a substrate could be attracted and held at the active site by a negatively charged aspartate or glutamate residue. Alternatively, cofactors such as positively-charged metal ions often help bind substrate molecules with electron-dense groups.

25.7 FACTORS AFFECTING ENZYME ACTIVITY

enzyme activity

Enzyme activity *is a measure of the rate at which an enzyme converts substrate to products.* Four factors affect enzyme activity: (1) temperature, (2) pH change, (3) increasing substrate concentration, and (4) increasing enzyme concentration. In this section we will examine how each of these four factors affect enzyme activity.

Temperature Effect

Temperature is a measure of the kinetic energy (energy of motion) of molecules in a solution. Higher temperatures mean molecules are moving faster and colliding more frequently. This concept also applies to collisions between substrate molecules and enzymes. Therefore, as the temperature of an enzymatically catalyzed reaction increases, so does the rate (velocity) of the reaction. Consider, for example, how grass grows faster during the hot summer than during the cool spring or fall when temperatures are lower. Insects cannot move as fast in cold weather as they can on a hot day, and operating rooms are often cooled down to slow a patient's metabolism during surgery.

However, when the temperature increases beyond a certain point, the increased energy begins to cause disruptions in the tertiary structure of the enzyme; denaturation is occurring (Section 24.13). Loss of this structure at the active site impedes catalytic action, and the enzyme activity quickly decreases as the temperature climbs past this point (See Figure 25–8). Some enzymes are more resistant to denaturation than others. While most optimum temperatures are near body temperature (37°C), some enzymes resist denaturation at higher temperatures.

FIGURE 25–8 Effect of temperature on enzymatic activity.

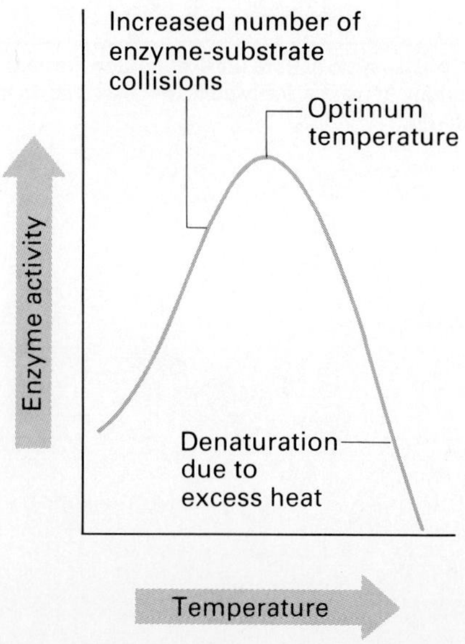

Organisms such as thermophillic bacteria maintain life processes at hot temperatures. Thermophillic bacteria have been found in natural hot pools with temperatures up to 60°C. However, this occurrence is the exception to the rule. Most proteins are irreversibly denatured at such high temperatures.

pH Effect

Most enzymes exhibit maximum activity at a certain pH. This is most often due to acidic and basic side chains of amino acids in the enzyme. These groups protonate at low pH (excess H^+) and lose their protons at higher pH. Consider an enzyme that requires a negatively charged glutamate at the active site to function properly. As the pH is lowered, the glutamate eventually protonates, the negative charge is lost, and catalytic action stops. Raising pH back up reforms the negative charge and restores enzyme activity. Because similar forces stabilize tertiary structure, pH can affect overall tertiary structure. If these changes alter the active site, the enzymatic activity is also affected. As with all proteins, high concentrations of strong acids and bases irreversibly denature enzymes.

Not all enzymes possess the same pH optimums. The pH often varies between different tissues and even among compartments within individual cells. For example, the pH in stomach fluids is much lower than that found in the small intestine. Pepsin, a protein-digesting enzyme in the stomach, functions best at lower pH, but another protein-digesting enzyme of the small intestine, carboxypeptidase A, functions best at neutral pH values (see Figure 25–9). Remember that enzymes function over a range of different pH values, but exhibit *maximum* activity only at their *optimum* pH.

Substrate Concentration Effect

If we keep the concentration of enzyme constant and increase the concentration of substrate, we obtain an enzyme activity dependence known as a *saturation curve*.

FIGURE 25–9 Effect of pH on enzyme activity.

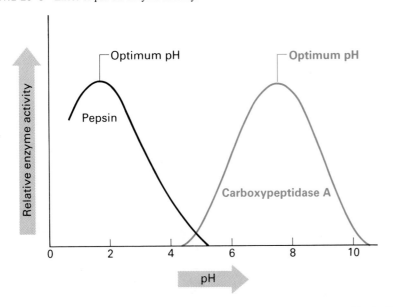

Enzyme activity increases to a certain substrate concentration, and thereafter remains constant. Figure 25–10 shows a saturation curve for enzymatic activity.

What limits enzymatic activity to a certain maximum value? If the amount of enzyme in each tube is constant as we increase substrate concentration, we eventually exceed the capacity of the enzymes to keep up with our expected reaction rate. Each substrate must diffuse in and bind to the enzyme and must occupy the active site for at least a short time, and then products must diffuse before the cycle can be repeated. When each enzyme is working at full capacity, the incoming substrate molecules must "wait their turn" at an active site. At this point the enzyme is said to be under saturating conditions.

We encounter this rate-limiting concept in our everyday lives. Consider a corporation that takes orders for merchandise over the phone. There are 20 telephone operators on duty at any given shift. During the summer months, business is slow and the operators often sit idle. However as Christmas approaches, the number of incoming calls increases daily. Eventually, every operator becomes saturated with calls and customers must often wait their turn to place an order. A graph of daily orders handled by operators for each of the last six months of the year might look a lot like Figure 25–10. If you were the boss of such a company and saw what was happening in November and December, what would you do to improve customer service? The answer is obvious: get more operators.

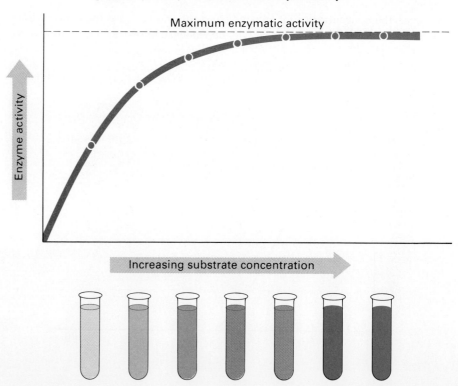

FIGURE 25–10 Plot of enzyme activity versus substrate concentration. To obtain a plot such as this, test tubes with identical amounts of enzyme and different concentrations of substrate are allowed to react for a certain period of time. The amount of product produced (in color) is a measure of the enzyme activity.

Enzyme Concentration Effect

What happens in the human body when the demands of a consistently high substrate concentration require more enzymes? Your body synthesizes them. Unique mechanisms inside each cell control the amounts of enzyme molecules on the job in the cell at any given time. Synthesis of new enzymes upon demand is called *enzyme induction*. Just as there is turnover of employees in the work force of a company, so is there a turnover of enzyme molecules in the cell. At some point, *enzyme degradation* occurs as they are hydrolyzed back down to their constituent amino acids. The number of enzymes in the cell at any given time is determined by these two processes, and is the most direct way the body controls enzyme activity in a cell.

Since enzymes are not consumed in the reactions they catalyze, the cell usually keeps the number of enzymes low compared to the number of their substrate molecules. This is for efficiency, because the cell avoids paying the energy costs of synthesizing and maintaining a large work force of enzymes. As the concentration of an enzyme changes, how does it affect the rate? Let us consider an experiment in which we keep the concentration of substrate constant and vary the amount of enzyme present. The results of this experiment are shown in Figure 25–11.

We can use a graph similar to Figure 25–11 to determine enzyme concentration in biological fluids. Let us look at an example. Increased concentrations of certain enzymes in the blood indicate that a heart attack has occurred. After obtaining a

FIGURE 25–11 Plot of enzyme activity versus enzyme concentration. Test tubes with identical amounts of substrate and different concentrations of enzyme are allowed to react for a certain period of time. The amount of product formed (in color) is a measure of the enzyme activity.

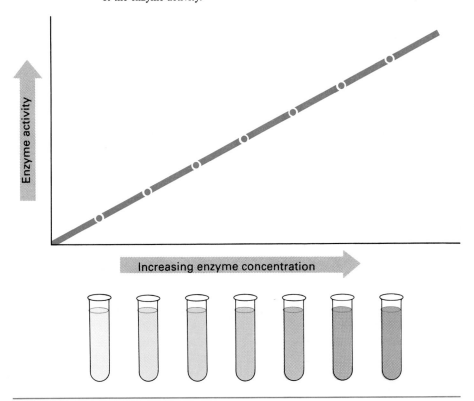

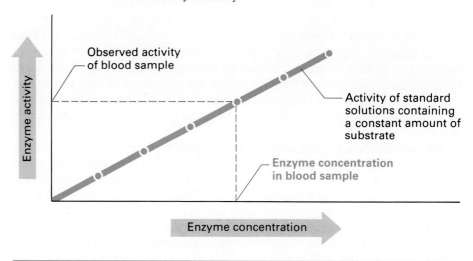

FIGURE 25-12 A graph of change in enzyme activity versus enzyme concentration at constant substrate concentration can be used to obtain the enzyme concentration of a sample with a known enzyme activity.

blood sample from a suspected heart attack victim, we can quickly measure the enzyme level for particular enzymes by comparing the enzymatic activity of the blood with that of reference solutions containing known amounts of the enzymes. To a test tube containing a premeasured amount of substrate (the amount on which our graph is based) for the enzyme of concern, we add a small known quantity of blood (our enzyme source) and then measure the enzymatic activity. We then obtain the enzyme concentration using the applicable graph, as shown in Figure 25-12.

25.8 CONTROL OF ENZYME ACTIVITY

As we mentioned in the beginning of this chapter, one of the most important advantages of enzymes over other types of catalysts is that their activity can be controlled. Responding to slowly changing conditions in the cell, enzyme synthesis and degradation keep the concentration of enzymes at specified levels. Furthermore, certain enzymes are synthesized only in compartments of the cell where they are needed. In this section we will consider further details about how the human body controls enzyme activity.

A variety of methods exist within the body to control enzyme activity. These methods can be categorized into a few general types of control. We will consider four: (1) competitive inhibition, (2) non-competitive inhibition, (3) allosteric control, and (4) covalent modification.

Competitive Inhibition

Enzyme function can be inhibited by certain chemicals called *enzyme inhibitors*. **Enzyme inhibitors** *are chemical substances that slow or stop the normal catalytic function of an enzyme.* Enzymes are quite specific about compounds which they accept at their active sites, but competitive inhibitors are so structurally similar to the intended substrate that they can also bind to the active site. If the enzyme accepts the intruder and is unable to catalyze a change in it, the active site is effectively blocked to other incoming substrates. Until the intruding chemical is released and diffuses away from the active site, the enzyme is inhibited and useless as a catalyst.

Competitive enzyme inhibitors *are enzyme inhibitors that compete with normal substrates for binding to the active site of an enzyme.* Malonate ion is a competitive inhibitor of succinate dehydrogenenase. This enzyme normally binds a succinate ion, catalyzes the removal of two hydrogen atoms from it in an oxidation reaction, and releases a fumarate ion as the product (see Figure 25–13a). Malonate ion acts as a competitive inhibitor because its structure is so similar to that of succinate ion, the normal substrate. When malonate ion binds to an active site, catalysis stops because there is no place for a double bond to form in a malonate ion (see Figure 25–13b). Until malonate is released from the active site, this particular enzyme molecule is non-functional as a catalyst. An inhibitor-enzyme

competitive enzyme inhibitors

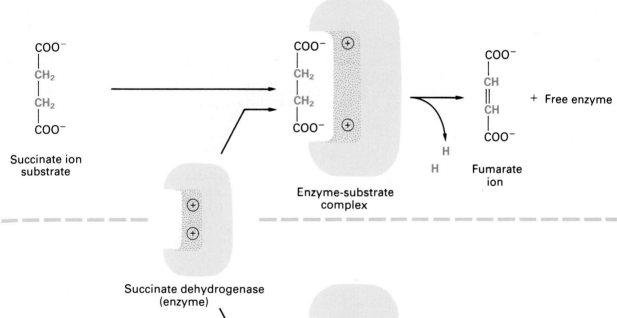

FIGURE 25–13 A contrast of the effect of succinate ion (normal substrate) and malonate ion (a competitive inhibitor) on the enzyme succinate dehydrogenase.

complex is often called a *dead-end complex* because the bound inhibitor cannot proceed to a product.

The success of a competitive inhibitor in binding to an active site depends on two factors: the affinity of the enzyme for the inhibitor compared to the substrate, and the concentrations of inhibitor and substrate. If an enzyme is selective enough to have only low affinity for the inhibitor, then the effect of that inhibitor is usually small at normal concentrations. High enzyme affinity for the inhibitor usually indicates it is a very potent inhibitor.

To illustrate concentration effects, let us examine a case for which the affinity of the enzyme for both substrate and competitive inhibitor is determined only by how much of each is in solution with the enzyme. If a solution of only substrate and enzyme is allowed to react, the enzyme activity can be measured. Now, if an equal amount of inhibitor is added to this solution, the activity slows down to about half of what it was originally. This change occurs because each bound enzyme site is occupied half of the time with competitive inhibitor and half of the time with normal substrate.

We can overcome competitive inhibition by simply increasing the concentration of the substrate. If we make its concentration greater than that of the inhibitor, then the substrate molecules have a much better chance of binding to the active site.

Non-Competitive Inhibition

non-competitive inhibitors

Non-competitive inhibitors *slow enzyme activity by binding to a site on an enzyme other than the active site.* In the process of binding to the enzyme, non-competitive inhibitors usually alter the overall tertiary structure of the enzyme, including the structure at the active site. Changing the active site lowers the affinity of the

FIGURE 25-14 A non-competitive inhibitor binds to an enzyme at a site other than the active site. This binding usually causes changes in structure at the active site.

(a) Normal enzyme activity

Substrate → Enzyme → Enzyme-substrate complex → Products   Free enzyme

(b) Non-competitive inhibition

Substrate
Non-competitive inhibitor
→ Enzyme-inhibitor complex (binding of substrate to enzyme more difficult) → Much slower reaction rate

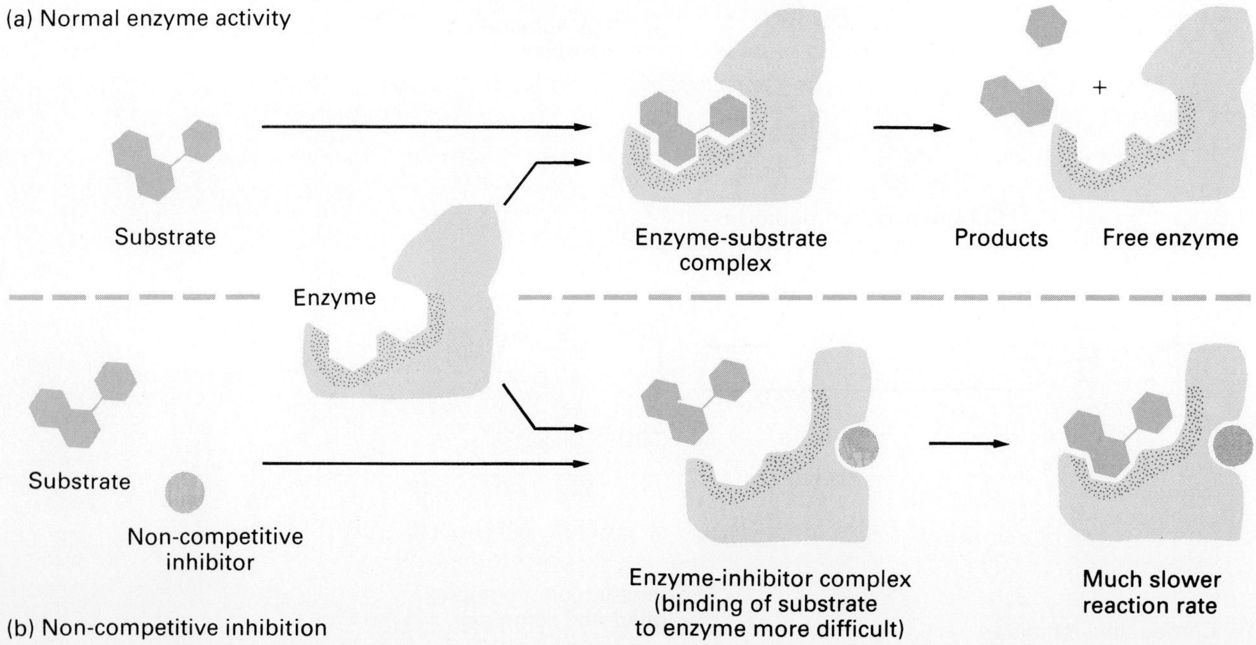

enzyme for its substrate and/or affects the enzyme's overall ability to function (see Figure 25–14).

Unlike competitive inhibition, increasing the concentration of substrate does not completely overcome the inhibitory effect in this case. Non-competitive inhibition alters the active site and therefore upsets the normal function of the enzyme. However, lowering the concentration of a noncompetitive inhibitor does free up many enzymes which then return to normal activity.

Allosteric Control

Many enzymes are composed of two or more subunits, thus allowing quaternary structure to play a role in enzyme activity. In Section 24.11 we learned that the relative ease of oxygen binding to each of the four sites of the hemoglobin tetramer depends on how many of the four sites are already bound to oxygen. The binding of oxygen to one site affects the other three sites so that they bind their oxygen molecules more readily than the first. We call this phenomenon *cooperativity*, a concept that applies equally well to oligomeric enzymes.

Consider an enzyme tetramer composed of four *identical* subunits, each with its own active site and each capable of catalysis. As one of the four active sites binds a substrate, a structural change occurs that activates the other three subunits through quaternary interactions. The result is a more active enzyme tetramer and higher enzyme activities than if four individual enzymes were working alone. This concept of one enzyme subunit affecting the activity of another is known as *allosteric control*. **Allosteric control** *involves regulation of enzyme activity through changes in quaternary structure of enzymes*. Recall, from Section 24.11, that quaternary structure for a protein involves how subunits in the protein associate with each other.

The effector of allosteric control is not always normal substrate (as in the previous example), nor does enzyme activity always increase (as in the previous example). Some allosteric control of enzyme activity is provided by the products of enzyme-catalyzed reactions. When these products bind to sites on the enzyme, they inhibit the enzyme. This is logical; higher concentrations of product mean that the need for enzyme activity is not as great as for low concentrations. The binding of product to the enzyme can, for example, cause the enzyme to dissociate into inactive subunits.

Many enzymes and reactions may be needed to convert a substrate into a final product. This series of reactions, which act one after another to convert substrate into an eventual product, is called a *reaction pathway*. By controlling the activity of the first enzyme of a reaction pathway, often called the *regulatory enzyme*, all of the reactions and enzymes in the pathway are also controlled. **Feedback inhibition** *is a special case of allosteric control of an enzyme, in which the final product of a reaction pathway returns to inhibit the first enzyme in the pathway*. Figure 25–15 shows the process of feedback inhibition for a reaction pathway involving three steps. When the end product concentration is high, the pathway does not function; when end product concentration is low, the first enzyme is activated and the entire pathway functions to produce more product.

Occasionally we encounter an enzyme composed of two types of subunits, catalytic subunits and modulator subunits. *Modulator subunits* are subunits of enzymes that do not catalyze reactions, but rather alter the activity or specificity of the catalytic subunits to which they are attached.

Lactose synthetase is a striking example of this concept. This enzyme is composed of two subunits: a catalytic subunit named galactosyl transferase and a mod-

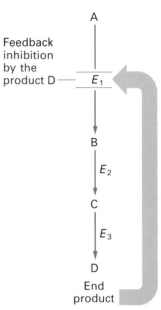

FIGURE 25–15 Feedback inhibition is a type of allosteric control in which the product of a reaction reversibly binds to the first enzyme in the reaction pathway to control the production of further product.

allosteric control

feedback inhibition

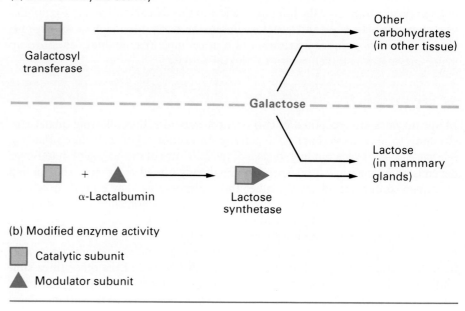

FIGURE 25-16 The concept of enzyme catalysis involving a modulator subunit.

ulator subunit named β-lactalbumin. Most mammalian tissues contain only the catalytic subunit, which normally catalyzes the transfer of activated galactose molecules during the synthesis of various carbohydrates. However it alone does *not* transfer galactose to glucose to make lactose. Only when the modulator subunit is bound to the enzyme does it produce lactose.

During pregnancy, this catalytic subunit is synthesized and stored in the mammary gland, but very little modulator subunit is present. At parturition, changes in the concentration of various hormones elicit the synthesis of large amounts of the modulator subunit which then attaches to the catalytic portion of the enzyme. Under allosteric direction, the resulting complex is now selective in its synthesis of lactose, the principal sugar in human milk. Figure 25-16 illustrates the concept of allosteric control through modulator subunits.

Covalent Modification

Covalent modification *of an enzyme alters the activity of the enzyme by the forming or breaking of covalent bonds in the enzyme molecule.* An example of how new covalent bonds change activity form is the way enzymes are phosphorylated inside cells to control their activity. In this process, the alcoholic side chain of the amino acid serine is often the acceptor of the phosphate group, as shown in the following equation.

The newly phosphorylated enzyme may be more or less active than the original. If serine was a critical residue needed at the active site for catalysis to occur, then the phosphorylated enzyme is deactivated. Phosphate groups can be removed by other enzymes called *phosphatases*. Thus, under normal cellular conditions, phosphorylation and dephosphorylation are used to control enzyme activity in a reversible fashion.

Chemicals that also attach to serine alcohol groups have been synthesized by chemists in the laboratory. These attachments are usually irreversible and permanently deactivate enzymes that require serine at their active sites. Most of these compounds are extremely toxic. One example is diisopropylphosphofluoridate (abbreviated DIPF or DFP), which reacts with serine groups in enzymes. Only 0.5 mg of DIPF per kilogram of body weight is considered lethal, and this amount can be readily absorbed through the skin! Chemicals such as these are active components of nerve gases and insecticides. Their poisonous action is due to the deactivation of acetylcholinesterase, an important enzyme involved in transmission of nerve impulses.

$$\text{(active)} \quad \text{Enzyme}-CH_2-OH + CH_3-CH(CH_3)-O-P(=O)(F)-O-CH(CH_3)-CH_3 \longrightarrow \text{(inactive)} \text{Enzyme}-CH_2-O-P(=O)-O-CH(CH_3)-CH_3 \text{ (with } CH_3-CH-\text{)} + HF$$

diisopropylphosphofluoridate (DIPF)

Another type of covalent modification used to control enzyme activity is the hydrolysis of specific peptide bonds in the protein chain structure of an enzyme. We described in Section 25.4 the type of compound in which this occurs—the zymogens, inactive storage forms of enzymes. When the primary structure of a zymogen is changed through hydrolysis of specific bonds, it refolds into a new, active conformation. If two or more bonds are broken, part of the polypeptide backbone may be lost in the process of activation. Other highly specific enzymes usually catalyze the hydrolysis of these peptide bonds in zymogens. The action of one enzyme on another is a very rapid and efficient method of enzyme activation. Digestive enzymes (Section 25.9) and blood-clotting enzymes (Chapter 32) are activated by this method of peptide hydrolysis.

25.9 PROTEIN-DIGESTING ENZYMES

A particularly interesting example of controlling enzyme activity by covalent modification is the activation of zymogens that help us digest food. This group of enzymes catalyzes hydrolysis of peptide bonds in proteins (which is required before individual amino acids can be absorbed by the small intestine). Enzymes that catalyze the hydrolysis of peptide bonds in proteins are called *proteolytic enzymes* or *peptidases*. To prevent proteolytic enzymes from destroying the proteins of the organism that synthesized them, they are produced as inactive zymogens. In mammals, zymogens are produced in the stomach and pancreas. After transport to the site where needed, these zymogens are activated by other enzyme molecules at that location.

Table 25–3 (p. 664) lists some important proteolytic enzymes as well as their

TABLE 25-3 Gastric and Pancreatic Proteolytic Enzymes

Site of Synthesis	Zymogen	Active Enzyme	Site of Activation
stomach (cell wall)	pepsinogen	pepsin	stomach
pancreas	trypsinogen	trypsin	small intestine
pancreas	chymotrypsinogen	chymotrypsin	small intestine
pancreas	proelastase	elastase	small intestine
pancreas	procarboxypeptidase	carboxypeptidase	small intestine

sites of precursor synthesis and activation. Your may recognize some of these enzymes as examples described in previous sections.

Once activated, some of these enzyme molecules can activate their own zymogen precursors. *Autoactivation* is the process by which active enzyme molecules catalyze the activation of their own zymogen precursors. Normally autoactivation only takes place at the location where needed because this is the only place active enzymes are found that can activate the zymogens. For example, before becoming fully activated, chymotrypsinogen must be acted upon by trypsin (which breaks one peptide bond) and also by another active chymotrypsin molecule (which breaks two peptide bonds). Active trypsin and chymotrypsin enzymes are normally found only in the fluids of the small intestine where activation occurs.

Beyond the synthesis of inactive zymogens, a healthy pancreas also protects itself from these enzymes by wrapping them up in a coat of lipids, forming storage compartments called *zymogen granules*. Furthermore, the pancreas synthesizes competitive inhibitors which bind to the active sites of these enzymes and block their normal proteolytic activity. Premature activation of zymogens such as trypsinogen and chymotrypsinogen occurs in hemorrhagic pancreatitis. In this disease, the patient's pancreas is destroyed by its own protein-digesting enzymes.

25.10 ANTIBIOTICS THAT INHIBIT ENZYME ACTIVITY

antibiotics

Antibiotics *are substances that kill or inhibit the growth of microorganisms.* A good antibiotic exerts its action selectively on bacteria and does not affect the normal metabolism of the host organism. The two best known families of antibiotics, the sulfa drugs and the pencillins, are inhibitors of specific enzymes essential to the life-processes of bacteria.

The antibiotic activity of a compound called sulfanilamide was first discovered in 1932 by a German chemist who was synthesizing sulfur-containing dyes, some of which were sulfanilamide derivatives. Since that time, scientists have synthesized many biologically active derivatives of sulfanilamide, collectively called "sulfa drugs".

Sulfanilamide inhibits bacterial growth because it is structurally similar to *p*-aminobenzoic acid (PABA).

sulfanilamide

p-aminobenzoic acid (PABA)

PABA is required by many bacteria to produce an important enzyme cofactor, folic acid. Sulfanilamide acts as a competitive inhibitor to enzymes in the bio-

FIGURE 25-17 Biosynthetic pathway from PABA to folic acid in bacteria, showing the competitive inhibition of sulfanilamide.

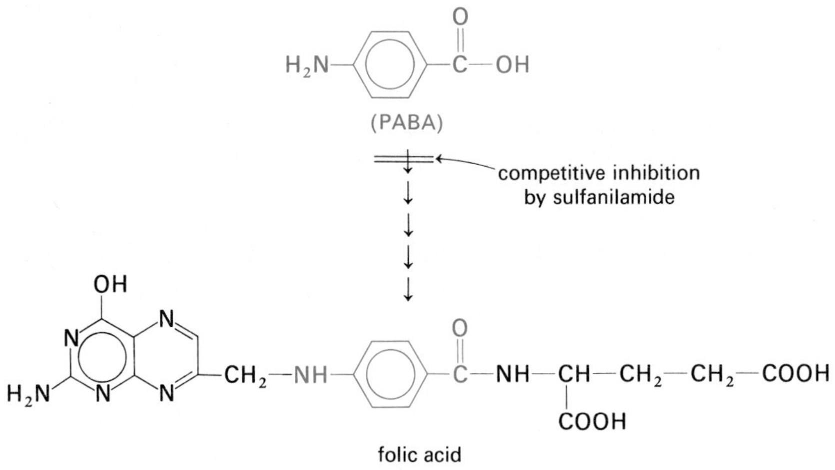

synthetic pathway that convert PABA into folic acid in these bacteria (see Figure 25-17). The depletion of folic acid retards growth of the bacteria and can eventually kill them.

Sulfa drugs selectively inhibit only bacteria metabolism and growth because mammals can absorb folic acid from their diets and thus do not need to use PABA for its synthesis. A few of the most common sulfa drugs and their structures are listed in Figure 25-18. The effectiveness of these drugs varies with the type of

FIGURE 25-18 Derivatives of sulfanilamide, showing names and effectiveness of each at inhibiting growth of *E. coli* bacteria.

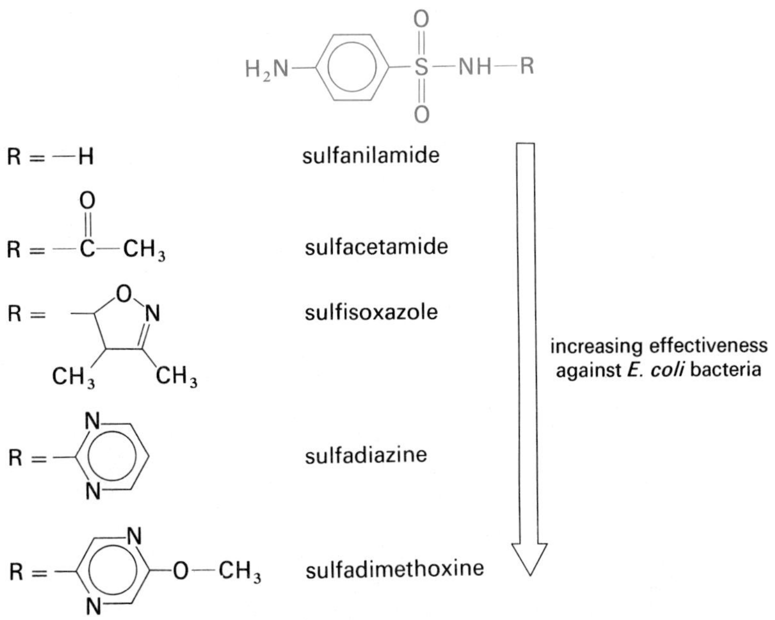

R–group attached to the sulfanilamide structure, as can be seen from the ordering of structures in Figure 25–18.

Sulfa drugs are not used as much as they used to be in treating bacterial infections; we now know of a side effect that involves the bone marrow and skin allergies. Today, physicians use sulfa drugs primarily to treat urinary tract infections because the drugs are efficiently absorbed and excreted in the urine.

Penicillin is an antibiotic that also inhibits a specific enzyme in bacteria. Discovered by Alexander Fleming in 1928, penicillin is produced and secreted by a mold for protection against bacteria. Fleming encountered difficulties trying to isolate the antibiotic, and it wasn't until 10 years later that the chemical structure of penicillin was deduced. Penicillin and its derivatives contain a thiazolidine ring fused to a β-lactam ring. This bicyclic ring structure (shown in Table 25–4) contains a highly reactive amide bond that plays an important role in its function. Like the sulfa drugs, variable R–groups can be attached to the basic penicillin molecule to make derivatives. Some of these derivatives are more effective as antibiotics than others and some are more able to survive the acidic digestive juices of the stomach.

TABLE 25–4 Penicillin Antibiotics

R-group	Name	Stability in Acid Solution
C₆H₅–CH₂–	Penicillin G (benzyl penicillin)	poor
C₆H₅–O–CH₂–	Penicillin V	good
2,6-dimethoxyphenyl (OCH₃, OCH₃)	Methicillin	poor
C₆H₅–CH(NH₂)–	Ampicillin	good
HO–C₆H₄–CH(NH₂)–	Amoxicillin	good
phenyl-methylisoxazolyl	Oxacillin	good

Penicillin inhibits *transpeptidase*, an enzyme that catalyzes the formation of peptide cross-links between polysaccharide strands in bacteria cell walls. These strands strengthen cell walls and are called *peptidoglycan strands* because they contain a few amino acids connected by peptide bonds as part of their polymeric structure. A strong cell wall is necessary to protect the bacterium from undergoing lysis (breaking open). By inhibiting transpeptidase, penicillin prevents the formation of a strong cell wall. Any osmotic or mechanical shock then causes lysis, killing the bacterium.

Penicillin's unique action depends on two aspects of enzyme deactivation we have discussed before: structural similarity to the enzyme's natural substrate, and covalent modification. Penicillin is *highly specific* in binding to the active site of transpeptidase because it so closely resembles the structure of the peptidoglycan strands normally cross-linked by the enzyme. In this sense, it acts as a very selective competitive inhibitor. However, unlike a normal competitive inhibitor, once bound to the active site, the β-lactam ring opens and the highly reactive amide bond forms a covalent linkage bond to a critical serine residue required for normal catalytic action. The result is a *permanently deactivated* transpeptidase enzyme (see Figure 25–19).

Penicillin does not interfere with normal metabolism in mammals because of its highly selective binding to bacterial transpeptidase, and because mammals have different cell wall structures than bacteria. This selectivity makes penicillin an extremely useful antibiotic.

However, some bacteria produce the enzyme *penicillinase*, which protects them from penicillin. Penicillinase selectively binds penicillin and catalyzes the opening of the β-lactam ring without forming a covalent bond to the enzyme, as is shown in Figure 25–20 (p. 668). Once the ring is opened, the penicillin is no longer capable of inactivating transpeptidase. When different R-groups are attached to the penicillin molecule, some derivatives become resistant to penicillinase action. Certain penicillins such as methicillin and oxacillin have been synthesized (Table 25–4); these resist penicillinase activity and are thus clinically important.

As we can see from the examples of both sulfa drugs and penicillins, research chemists devote much effort to using the functional portion of an antibiotic and then modifying other areas of the molecule to enhance its activity.

FIGURE 25–19 Diagram showing the selective binding of penicillin to the active site of transpeptidase and subsequent covalent modification.

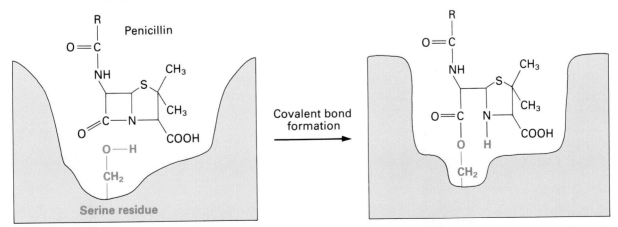

(a) Penicillin selectively binds to active site of transpeptidase.

(b) Covalent modification permanently blocks active site.

FIGURE 25–20 Penicillinase-catalyzed opening of the reactive amide bond in penicillin, rendering it useless as an antibiotic. The activity of penicillinase depends on the type of R-group attached to the β-lactam ring.

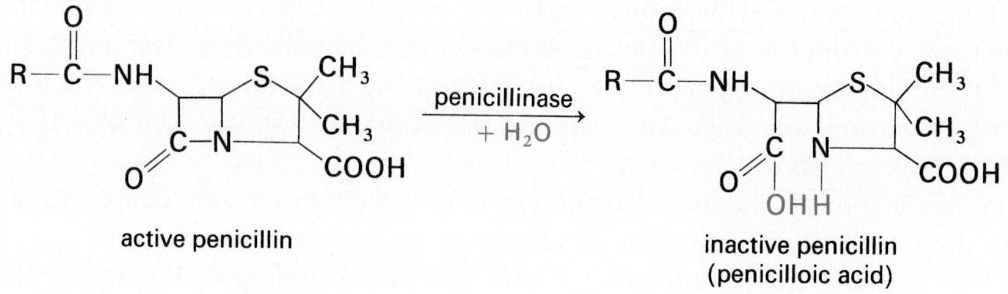

25.11 ISOENZYMES

isoenzymes

Isoenzymes *are groups of oligomeric enzymes (they contain subunits) with similar enzymatic functions, but with slightly different structures.* The five similar, but different, lactate dehydrogenase (LDH) enzymes found in human tissues represent the best known example of the phenomenon of isoenzymes.

All forms of LDH are tetramers and they can be made from two different kinds of subunits. All four subunits in each LDH isoenzyme from heart tissue are identical and are called heart (H) subunits. The four subunits that make up LDH from skeletal muscle are also identical to one another but are different from the H-subunits. They are identified as muscle (M) subunits. Intermediate forms of LDH, constructed from mixtures of H- and L-subunits, are found in other organs of the body, making a total of five different tetrameric forms of LDH. These multiple-molecular forms are shown in Figure 25–21. Each of the five different forms of LDH has different levels of enzymatic activity and thus controls rates of lactate metabolism in its own respective tissues. We will examine LDH action further as we discuss lactate metabolism in Chapter 29. In Section 25.12 we will see how the occurrence of different LDH isoenzymes in various locations of the body helps specialists diagnose disease and damage to various organs.

25.12 MEDICAL APPLICATIONS AND USES OF ENZYMES

By isolating and purifying enzymes from biological systems, we can use them outside their native cells. The specificity of enzymes, for example, allows pharma-

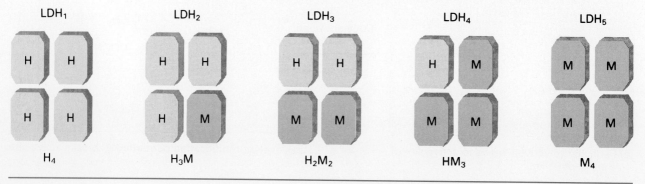

FIGURE 25–21 Five distinct forms of the lactate dehydrogenase isoenzyme. H_4 is found predominantly in heart tissue and M_4 in skeletal muscle. Intermediate forms are found in other organs and tissues.

ceutical companies to manufacture enantiomeric drugs. When used as catalysts, stereospecific enzymes help direct the formation of chiral carbon atoms in certain locations of the molecule being synthesized. Proteolytic enzymes have been incorporated into meat tenderizing products to help break down the tough, fibrous proteins in meat. At one time, enzymes were even added to detergents to help degrade food stains (a practice that has since been discontinued).

The medical applications of enzymes are especially useful in diagnosing disease. Many diseases appear to result from the absence of enzymes or from the presence of defective enzymes. These diseases are hereditary, and are often referred to as inborn errors of metabolism.

Phenylketonuria (PKU) is a disease in which a deficiency of the enzyme *phenylalanine hydroxylase* prevents the hydroxylation of phenylalanine to make tyrosine. This error in phenylalanine metabolism can lead to mental retardation in young children. A blood sample from newborn infants is routinely analyzed for phenylalanine hydroxylase activity. When low levels of enzyme activity are found, a diet low in phenylalanine is prescribed in an effort to prevent mental retardation. Other hereditary enzyme defects and their associated diseases are listed in Table 25–5. One of the long-range objectives of genetic engineering is to be able to transform proper cells and make them produce these necessary enzymes automatically, thus curing such hereditary diseases.

Enzymes can also be used to diagnose diseases that are not inborn errors of metabolism. Although the blood serum contains many enzymes, some enzymes

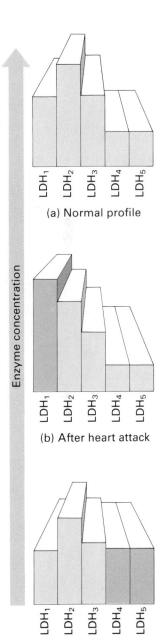

(a) Normal profile

(b) After heart attack

(c) In presence of liver disease

FIGURE 25–22 Concentrations of the various forms of the enzyme lactate dehydrogenase (LDH) in the blood can be used to diagnose various problems in the human body.

TABLE 25–5 Some Diseases Caused by Enzymic Defects

Disease	Enzyme
Gaucher's disease	β-glucosidase
Methemoglobinemia	methemoglobin reductase
Albinism	tyrosinase
Gout	hypoxanthine–guanine phosphoribosyltransferase
Niemann-Pick disease	sphingomyelinase
Tay-Sachs disease	hexosaminidase A
Galactosemia	phosphogalactose uridyltransferase

TABLE 25–6 Increased Serum Enzyme Levels Associated with Various Disease States

Organ	Enzyme	Disease State(s)
bone	alkaline phosphatase	Paget's disease (increased osteoblastic cellular activity), rickets
heart	lactate dehydrogenase (isoenzyme LDH_1)	myocardial infarction
	aspartate transaminase creatinine phosphokinase (CPK)	myocardial infarction
liver	aspartate transaminase (AST)	hepatocellular disease
	lactate dehydrogenase (isoenzymes LDH_4 and LDH_5)	hepatocellular disease
prostate	acid phosphatase	prostatic carcinoma
skeletal muscle	creatine kinase	muscular dystrophy
	aldolase	diseases involving muscle disintegration

are not normally found in the blood, but are produced only inside cells of certain organs and tissues. The appearance of these enzymes in the blood often indicates tissue damage in an organ and that cellular contents are spilling out (leaking) into the bloodstream. Assays of abnormal enzyme activity in blood serum can be used to diagnose many disease states, some of which are listed in Table 25–6 (p. 669).

Tests have been developed to differentiate between the various LDH isoenzymes in the blood. These tests are used as diagnostic tools in determining damage to specific organs. For example, an increase from "normal" in LDH_1 isoenzyme indicates heart muscle damage, while an increase in the presence of LDH_4 and LDH_5 points toward liver damage. Figure 25–22 (p. 669) gives quantitative information on the diagnostic use of LDH isoenzyme concentrations in the blood.

Scientists have recently attempted to use enzymes in the treatment of diseases. After the enzyme *urokinase* was shown to be capable of dissolving blood clots in the laboratory, subsequent clinical studies have now shown that it is extremely effective in treating massive internal blood clots. Other enzymes from bacteria have been used to treat the rash caused by poison ivy. When applied to the affected area of the skin, the enzymes rapidly catalyze the hydrolysis of the plant's toxic irritant to nontoxic products. Enzymes that dissolve cartilage have also been used to dissolve portions of disks in the spine that have slipped and are putting painful pressure on nerves. Injection of the enzymes and their subsequent action avoids painful and sometimes risky surgery. A recent advance in treating heart attacks is the use of tissue plasminogen activator (TPA), that activates the enzyme plasminogen. When so activated, this enzyme dissolves blood clots in the heart and often provides immediate relief.

EXERCISES AND PROBLEMS

Importance of Enzymes

25.1 Define the class of compounds called *enzymes*.

25.2 Why does the body need so many enzymes?

25.3 How do enzymes differ from ordinary (laboratory) catalysts?

25.4 Occasionally we refer to the "delicate" nature of enzymes. Explain why this adjective is appropriate.

Enzyme Nomenclature

25.5 What is the usual ending for the name of an enzyme?

25.6 Suggest a name for an enzyme that catalyzes the
a. hydrolysis of sucrose
b. decarboxylation of pyruvate
c. isomerization of glucose
d. oxidation of lactose (removal of hydrogens)

25.7 What is the difference between urea and urease?

25.8 Give the name of the reactant on which each of the following enzymes acts.
a. fructase
b. lactase
c. arginase
d. galactase

Enzyme Terminology

25.9 Identify the substrates for the enzymes lactase and cellobiase.

25.10 What is the fate of the substrate in an enzyme-catalyzed reaction?

25.11 What is the function of the active site of an enzyme?

25.12 Explain why an alteration at the active site of an enzyme affects enzymatic activity.

25.13 What is the difference between a simple enzyme and a conjugated enzyme?

25.14 What is the relationship between an apoenzyme and a cofactor?

25.15 What is the relationship between a coenzyme and a cofactor?

25.16 What is the relationship between a coenzyme and an activator?

25.17 What three names are used, in various situations, to describe the nonprotein component of a conjugated enzyme? Specify the situation in which each term is used.

25.18 What is a major role of minerals in our diet?

25.19 How does a zymogen differ from an enzyme in structure and biological activity?

25.20 What is the relationship between a zymogen and a proenzyme?

25.21 How does the body activate a zymogen?

General Properties of Enzymes

25.22 Six general categories are used to describe enzymes by the types of reactions they catalyze. Select the best category for an enzyme that catalyzes each of the reactions described.
a. conversion of a *cis* double bond to a *trans* double bond
b. dehydration of an alcohol to form a double bond
c. a carbonyl group is reduced to an alcohol
d. an amino group is transferred from one molecule to another
e. two strands of DNA are joined, requiring the input of extra energy.
f. water is used to cleave an ester linkage.

25.23 Identify the enzyme needed in each of the following reactions as: an isomerase, decarboxylase, dehydrogenase, lipase, or phosphatase.

a.
$$\begin{array}{c} CH_2-O-C(=O)-R \\ CH-O-C(=O)-R \\ CH_2-O-C(=O)-R \end{array} + 3H_2O \longrightarrow \begin{array}{c} CH_2-OH \\ CH-OH \\ CH_2-OH \end{array} + 3R-COOH$$

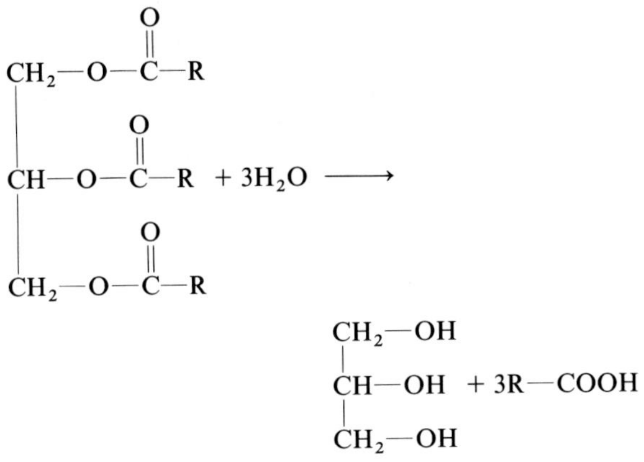

b. $CH_3-C(=O)-COOH \longrightarrow CH_3-C(=O)-H + CO_2$

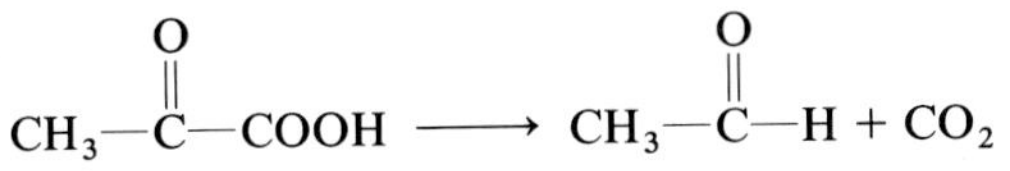

c. $H_2N-CH(CH_2-OPO_3^{2-})-COOH + H_2O \longrightarrow H_2N-CH(CH_2-OH)-COOH + HPO_4^{2-}$

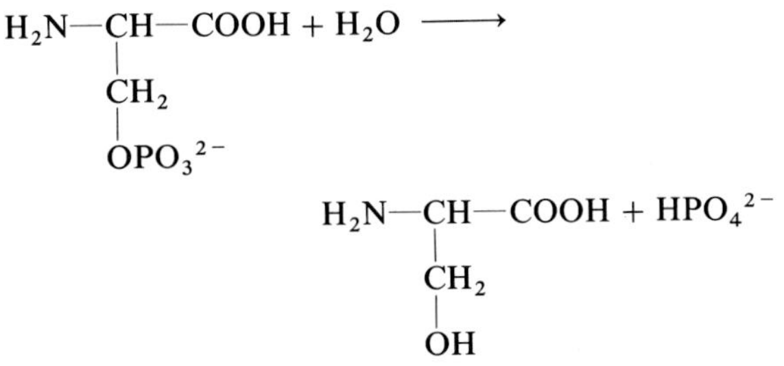

d. $CH_3-CH(OH)-COOH + NAD^+ \longrightarrow CH_3-C(=O)-COOH + NADH + H^+$

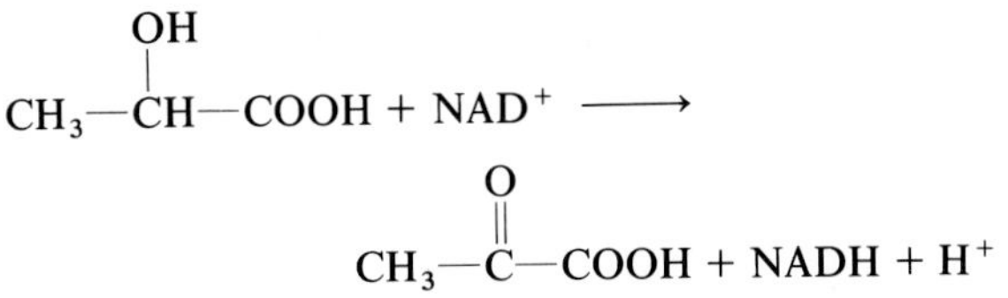

e.
$$\begin{array}{c} HC(=O)H \\ HC-OH \\ HO-CH \\ HC-OH \\ HO-CH \\ CH_2OPO_3^{2-} \end{array} \qquad \begin{array}{c} CH_2OH \\ C=O \\ HO-CH \\ HC-OH \\ HO-CH \\ CH_2OPO_3^{2-} \end{array}$$

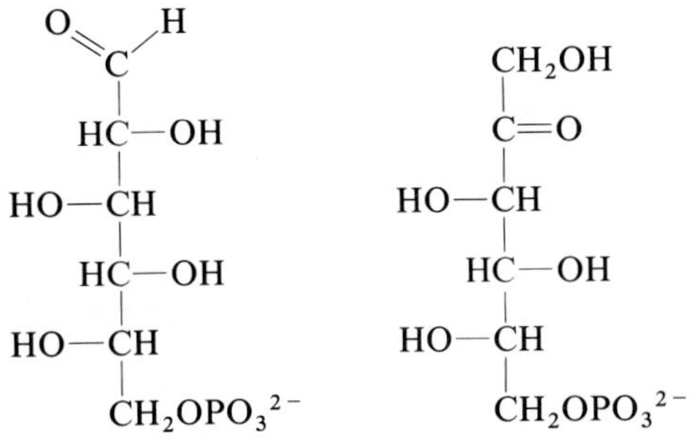

25.24 Enzymes exhibit different degrees and types of specificity. Explain the meaning of each of the following terms dealing with enzyme specificity.
a. absolute specificity
b. group specificity
c. linkage specificity
d. stereochemical specificity

25.25 Distinguish between absolute specificity and linkage specificity. Use an example to illustrate each type of specificity.

Models of Enzyme Action

25.26 What is an enzyme-substrate complex?

25.27 What is meant by complementary shape?

25.28 In what context do complementary shapes become extremely important?

25.29 How does the lock-and-key model explain enzyme specificity?

25.30 Contrast the differences between the lock-and-key and induced-fit models for enzyme specificity.

25.31 How does the induced-fit model for enzyme specificity explain the broad specificities of some enzymes?

25.32 What forces draw a substrate into an active site?

25.33 What type of amino acid side chains in an enzyme might assist in binding of a substrate containing a protonated amine group?

25.34 In what way does lysozyme weaken the bond to be hydrolyzed in its polysaccharide substrate?

Factors Affecting Enzyme Activity

25.35 List four factors that affect enzyme activity.

25.36 Temperature affects enzymatic reaction rates in two ways. An increase in temperature can accelerate the rate of a reaction or stop the reaction. Explain each of these effects.

25.37 Define the optimum temperature of an enzyme-catalyzed reaction.

25.38 Explain why a graph of rate-versus-temperature for most enzyme-controlled reactions is bell-shaped.

25.39 Explain why all enzymes do not possess the same optimum pH.

25.40 Explain why small pH changes, which do not denature the protein of an enzyme, can still cause a change in the rate of an enzyme-catalyzed reaction.

25.41 What does the phrase "saturating conditions" mean with respect to enzyme and substrate concentrations?

25.42 In an enzyme-catalyzed reaction, all of the enzyme active sites are saturated by substrate molecules at a certain substrate concentration. What happens to the rate of the reaction when the substrate concentration is doubled?

25.43 How does enzyme induction help control activity in a cell?

Control of Enzyme Activity

25.44 In competitive inhibition, can both the inhibitor and the substrate bind to an enzyme at the same time? Explain your answer.

25.45 Why is the binding of a competitive inhibitor to an enzyme overcome by a large increase in substrate concentration?

25.46 Compare the sites where competitive and noncompetitive inhibitors bind to enzymes.

25.47 Why is the binding of a noncompetitive inhibitor to an enzyme not overcome by an increase in substrate concentration?

25.48 Methanol is toxic because the enzyme alcohol dehydrogenase converts it to formaldehyde in the body. One method of treating methanol poisoning is to administer large amounts of ethanol until the methanol in the bloodstream can be harmlessly excreted. Explain why this treatment works if started immediately after methanol ingestion.

25.49 How is quaternary structure of oligomeric enzymes related to allosteric control of enzyme activity?

25.50 Describe what a reaction pathway is and how feedback inhibition controls all reactions in such a pathway.

25.51 What is a modulator subunit and how can it affect enzyme activity?

25.52 Give one example, not involving zymogens, of covalent modification used to control enzyme activity.

25.53 Why do small changes in primary structures of zymogens result in the formation of active enzymes?

Protein-digesting Enzymes

25.54 Why are the protein-digesting enzymes such as trypsin and carboxypeptidase synthesized as zymogens and not as active enzymes?

25.55 List two ways that the pancreas protects itself from protein-digesting enzymes, other than the synthesis of inactive enzyme precursors.

Antibiotics That Inhibit Enzyme Activity

25.56 Draw the chemical structure of sulfanilamide and the naturally-occurring compound it competitively inhibits.

25.57 Describe how sulfa drugs kill bacteria.

25.58 Draw the functionally-important portion of a penicillin molecule. Indicate the highly reactive bond which plays an important role in its function. Also identify the β-lactam ring.

25.59 What bacterial enzyme is inhibited by penicillin and why is penicillin so selective in binding to only this enzyme?

25.60 What amino acid in transpeptidase forms a covalent bond to penicillin?

25.61 Explain how some bacteria protect themselves from the antibiotic action of penicillin.

Isoenzymes

25.62 What are isoenzymes?

25.63 How can analysis of LDH in the blood stream often be used to diagnose damage to a particular organ in the body?

Medical Applications and Uses of Enzymes

25.64 How is the hereditary disease PKU diagnosed?

25.65 What peaks in the concentration of the LDH isoenzymes distinguish a myocardial infarction from a normal result?

25.66 List two ways that enzymes can be used to treat medical problems.

26 Nucleic Acids

Objectives

After completing Chapter 26, you will be able to:

26.1 Understand the major biological functions of nucleic acids.
26.2 Describe the composition and structure of nucleotides and nucleosides.
26.3 Discuss the formation of primary structure of nucleic acids.
26.4 Describe the secondary structure (double helix) of DNA and understand the relationships among the various kinds of bases present in DNA.
26.5 Describe the process of DNA replication, noting the important enzymes associated with various steps in the process.
26.6 List the major steps in the process of protein synthesis.
26.7 Indicate the characteristics and function of the three major types of RNA molecules.
26.8 Describe the RNA synthesis process (transcription) and understand the relationship among genes, chromosomes, and DNA molecules.
26.9 Understand the relationships among the genetic code, codons, nucleotide bases, and amino acid specification.
26.10 List the steps involved in protein synthesis (translation), be able to write the anticodon for a given codon, and be able to write the amino acid sequence of a peptide given the RNA or DNA sequence.
26.11 Describe how the substitution, addition, or deletion of a nucleotide base can result in a gene mutation.
26.12 Describe the action of viruses and how the human body combats them.
26.13 List the techniques used in forming recombinant DNA and give examples of practical applications of genetic engineering.

INTRODUCTION

One of the most remarkable properties of living cells is their ability to produce exact replicas of themselves throughout hundreds of generations. Furthermore, a single cell contains all of the instructions needed to make the millions of cells that form the human body with its variety of tissues and organs. This process requires that certain types of information be passed unchanged from one generation to the next. This storage and transfer of information is accomplished by nucleic acids.

Much like a book of master specifications describes the construction of all parts of a building, nucleic acids ultimately direct all life processes in the cell. Encoded within their polymeric structure are the directions for the synthesis of *all proteins* in the cell. The types of proteins and the number produced determine everything in a cell from its enzyme activity to its structural characteristics and function. Modern techniques of genetic engineering have allowed scientists to alter nucleic acid structure in cells, resulting in new research horizons for the diagnosis and treatment of disease.

26.1 NUCLEIC ACIDS

At the beginning of Chapter 22, we noted that there are four major classes of bioorganic substances—carbohydrates, lipids, proteins, and nucleic acids. We are now ready to consider the last of these four classes of compounds—nucleic acids.

Like carbohydrates (Chapter 22) and proteins (Chapter 24), nucleic acids are polymers. The repeating unit for the nucleic acid polymeric structure is a nucleotide. **Nucleic acids** *are polymeric molecules in which the repeating unit is a nucleotide.* Before discussing nucleotide structure, we will consider the discovery of nucleic acids and the major biological functions of these substances.

nucleic acids

Nucleic acid material was first isolated, in crude form, in 1869 by the Swiss biochemist Friedrich Miescher; he extracted, in basic solution, the contents of white blood cells. (His white blood cell source was pus washed from bandages of infected surgical patients of a nearby hospital.) Acidification of the materials in his basic extract gave a precipitate whose elemental analysis was much different from that known for carbohydrates, fats, and proteins. Not only were the elements carbon, hydrogen, and oxygen present but also nitrogen and phosphorus. The proportion of nitrogen and the presence of phosphorus made this material different from other biological materials. The name *nucleic acid* reflects the fact that this material was first isolated from cell nuclei and that it has acidic properties.

We now know that most cells contain two kinds of nucleic acids, *deoxyribo*nucleic *acids* (DNA) and *ribo*nucleic *acids* (RNA). (The structural differences between DNA and RNA will be considered in Section 26.2.) The majority of DNA is found in the nucleus of a cell. RNA is found both in the nucleus and outside the nucleus in the cytoplasm of the cell. **Cytoplasm** *is the material of the cell, exclusive of the nucleus,* as is shown in Figure 26–1 (p. 676). As a general rule, RNA molecules are shorter polymers and decompose much more readily than DNA.

cytoplasm

Two major biological functions exist for nucleic acids:

FIGURE 26-1 A simplified drawing of a cell showing its two major regions.

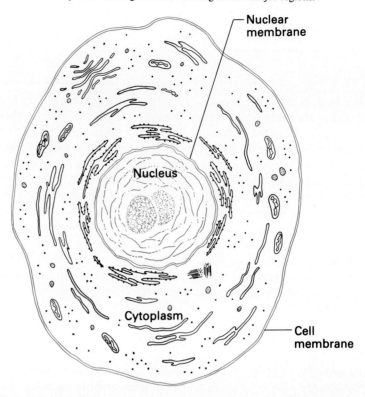

1. They store genetic information and transfer it between cells and transmit it from one generation to the next generation.
2. They are responsible for the control and direction of protein synthesis in cells.

In human cells, DNA is involved in the first of these major nucleic acid roles and both DNA and RNA are involved in the second of these roles.

26.2 NUCLEOTIDES

The monomers for nucleic acid polymers, nucleotides, are more complex in structure than carbohydrate monomers (monosaccharides—Section 22.4) or protein monomers (amino acids—Section 24.2). Within each nucleotide monomer are three subunits. A **nucleotide** is *a molecule composed of a pentose sugar bonded to both a phosphate group and a nitrogen-containing heterocyclic base.*

nucleotide

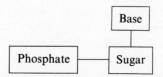

To further understand the details of nucleotide structure, let us examine each component (subunit) of a nucleotide in more detail.

One of two sugar molecules, ribose or deoxyribose, is always present in a nucleotide. Structurally, these two sugars, both pentoses, are very similar. The only

FIGURE 26–2 Purine and pyrimidine bases found in the nucleotides present in DNA and RNA.

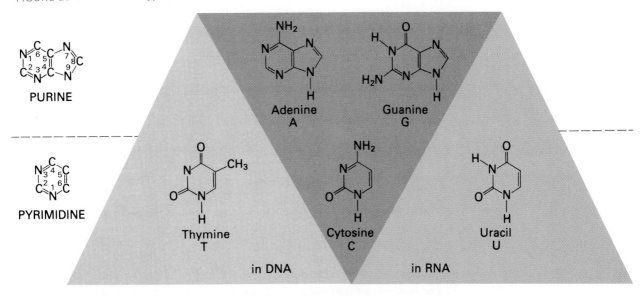

difference between them occurs at carbon number 2. In ribose, an —H and —OH group are bonded to this carbon atom, while in deoxyribose, two hydrogen atoms are present. The two general types of nucleic acids, RNAs and DNAs, can be distinguished from each other by the type of nucleotide-sugar-unit present. Nucleotides in RNA molecules always contain ribose—hence the R in RNA. Nucleotides present in DNA molecules always contain deoxyribose—hence the D in DNA.

β-D-ribose β-D-2-deoxyribose

Five nitrogeneous bases are found in nucleotides—adenine, guanine, cytosine, thymine, and uracil. These bases are the heterocyclic amines shown in Figure 26–2. As you may recall from Chapter 20, the extra (nonbonding) electrons on nitrogen atoms in these molecules cause them to attract protons, which is the characteristic of a basic molecule. Note from the structures in Figure 26–2 that the bicyclic compounds adenine and guanine are derivatives of purine, while the monocyclic rings in cytosine, thymine, and uracil are pyrimidine rings (Section 20.7). Adenine, guanine, and cytosine (usually abbreviated A, G, and C) are found in both DNA and RNA. Uracil (U) is normally found only in RNA while thymine (T) occurs only in DNA.

The nitrogeneous base portion of a nucleotide is connected to the sugar moiety by a nitrogen glycosidic bond (Section 22.8) formed in a condensation reaction involving carbon number 1 of the sugar and a nitrogen atom of the base. In purines, the nitrogen atom involved is nitrogen number 9 and in pyrimidines it is nitrogen number 1.

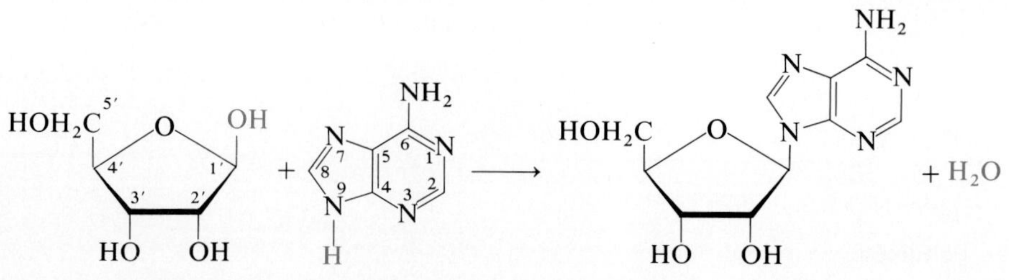

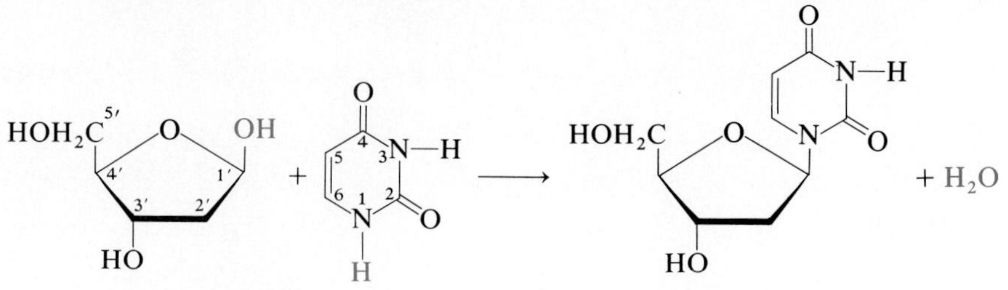

To avoid confusion between the carbon ring and nitrogen ring numbering systems, the numbers in the carbon ring are usually written with a prime: 1', 2', 3', 4', and 5'. The combined sugar-heterocyclic base portion of a nucleotide is often called a *nucleoside*. (The suffix *oside* reflects the fact that the bond that links these two subunits is a glycosidic bond.) A **nucleoside** *is a molecule that is a combination of a heterocyclic nitrogenous base and a pentose (ribose or deoxyribose)*. The structures and names of the four nucleosides that contribute to the structure of DNA are given in Figure 26–3. (Rules for naming nucleosides are presented later in this section.)

nucleoside

Example 26.1

The nucleoside uridine, which contains the sugar ribose and the base uracil, contributes to the structure of RNA molecules. Draw the structural formula of this nucleoside.

Solution

The base uracil is a pyrimidine base. Bases of this type bond to sugars, in nucleosides, through nitrogen number 1. Thus, the linkage between the two subunits of the nucleoside will involve the 1' carbon of ribose and the 1 nitrogen of uracil.

uridine:

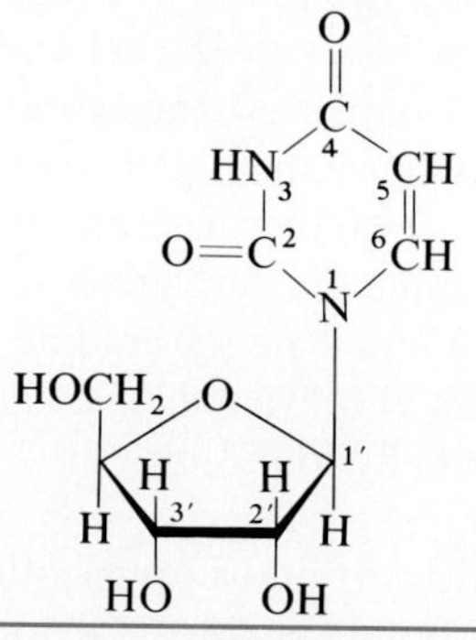

The formation of a phosphoester linkage between a phosphoric acid molecule and the —OH group on the 5'-carbon of the sugar converts nucleosides into nucleotides.

FIGURE 26-3 Structural formulas of nucleosides found in DNA.

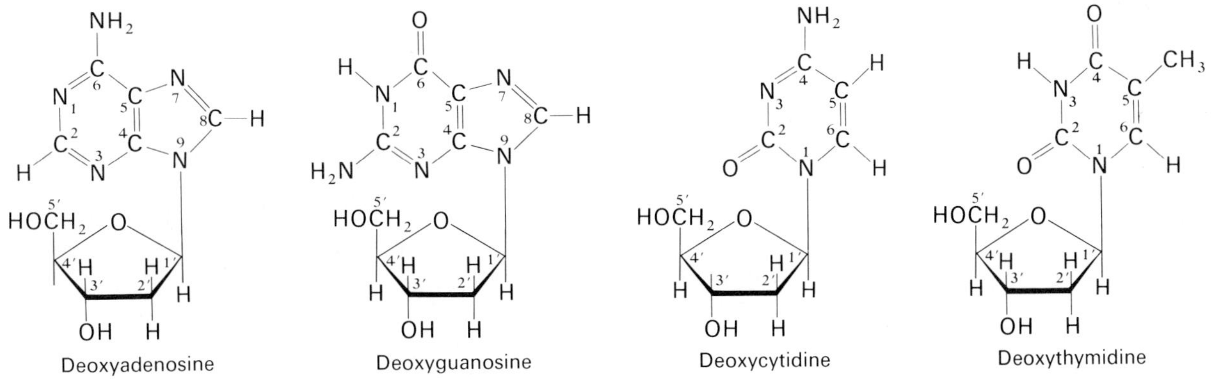

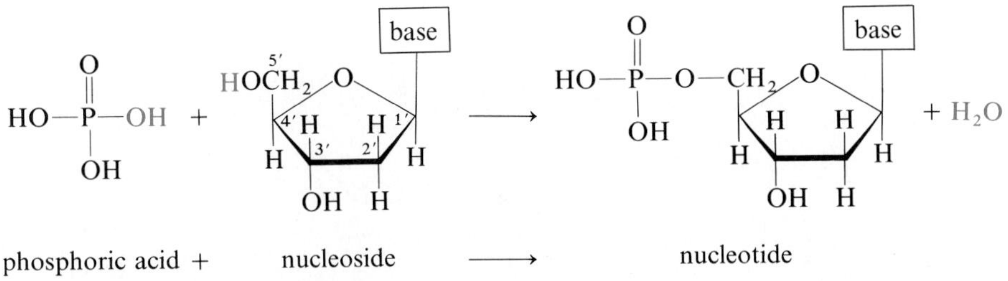

phosphoric acid + nucleoside ⟶ nucleotide

The names for common nucleosides and nucleotides associated with nucleic acids are given in Table 26–1. The name of a pyrimidine-containing ribonucleoside is obtained using the following rule.

Ribonucleoside name = pyrimidine name − ending + *idine*

For purine-containing ribonucleosides, the following rule applies.

Ribonucleoside name = purine name − *ine* + *osine*

For deoxyribonucleosides, the same rules as for ribonucleosides are followed, with the added feature that the prefix *deoxy* is added at the start of the name. Since nucleotides are monophosphate esters of nucleosides, they are named by writing the name of the nucleoside and then adding, as a separate word, the term 5'-

TABLE 26–1 Names of Bases, Nucleosides, and Nucleotides Found in RNA and DNA

Base	Nucleoside	Nucleotide
RNA		
adenine (A)	adenosine	adenosine 5'-monophosphate (AMP)
guanine (G)	guanosine	guanosine 5'-monophosphate (GMP)
uracil (U)	uridine	uridine 5'-monophosphate (UMP)
cytosine (C)	cytidine	cytidine 5'-monophosphate (CMP)
DNA		
adenine (A)	deoxyadenosine	deoxyadenosine 5'-monophosphate (dAMP)
guanine (G)	deoxyguanosine	deoxyguanosine 5'-monophosphate (dGMP)
thymine (T)	deoxythymidine	deoxythymidine 5'-monophosphate (dTMP)
cytosine (C)	deoxycytidine	deoxycytidine 5'-monophosphate (dCMP)

monophosphate. The number 5' indicates that the ester linkage involves the 5'-carbon of the sugar.

26.3 PRIMARY NUCLEIC ACID STRUCTURE

In Section 26.1 we noted that a nucleic acid is a polymer made up of nucleotide repeating units. Now that we know what nucleotides are, we are ready to consider the details of nucleic acid structure.

The arrangement of nucleotide units in nucleic acids is shown in Figure 26–4. The sugar unit of one nucleotide is bonded to the phosphate group of the next nucleotide in the sequence.

The sugar-phosphate linkage between nucleotides, a 3'-phosphoester linkage, results from a condensation reaction involving an —OH group of the phosphate and the 3'-carbon —OH group of the sugar, as shown in the following structural equation.

The complete chemical structure for a short segment of a nucleic acid (a tetranucleotide) is shown in Figure 26–5. We know that this segment represents a portion of a DNA molecule rather than an RNA molecule because the sugar is deoxyribose, not ribose.

Generalizations about the structures of nucleic acids, which can be drawn from figures like 26–4 and 26–5, include the following.

FIGURE 26–4 Diagram of the structure of a nucleic acid polymer showing the nucleotide repeating units.

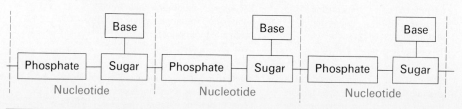

FIGURE 26–5 Chemical structure of a four-nucleotide long segment of a DNA molecule.

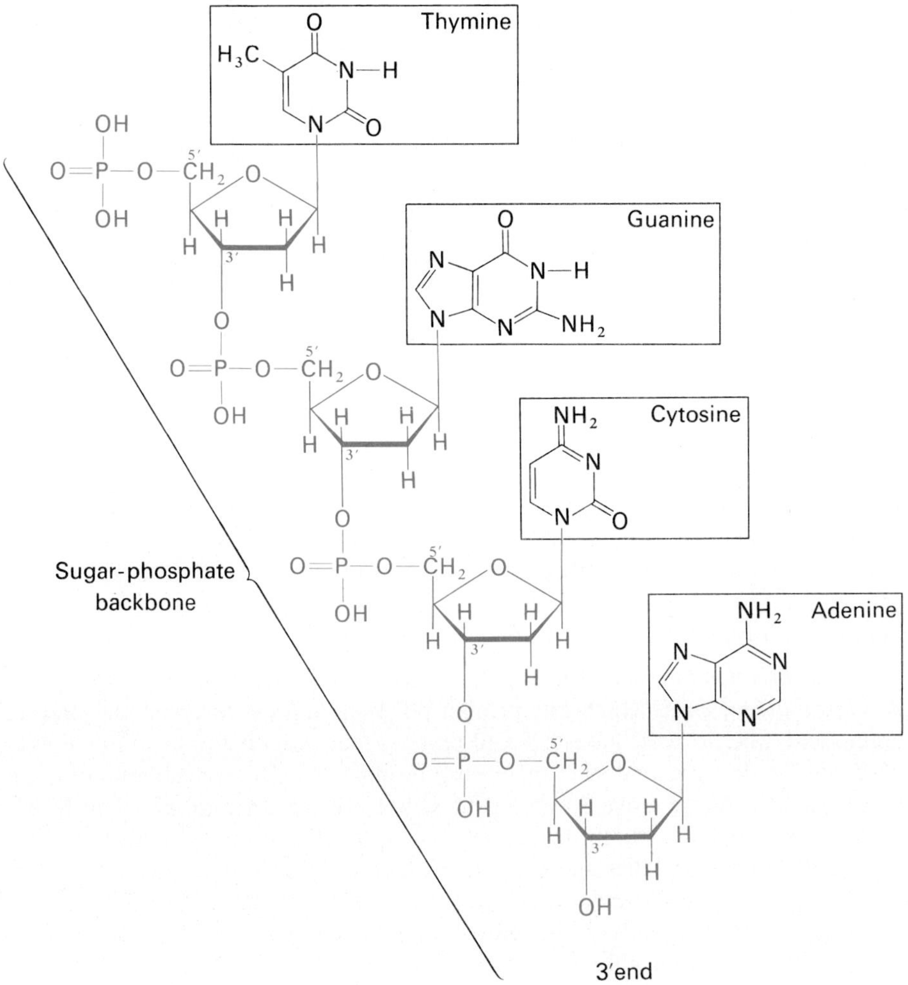

1. *The backbone of a nucleic acid molecule is a constant alternating sequence of sugar and phosphate groups.* For DNA molecules, this sequence involves deoxyribose and phosphate.

— Deoxyribose — Phosphate — Deoxyribose — Phosphate — Deoxyribose —

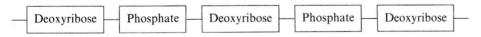

For RNA molecules, the alternation sequence involves ribose and phosphate.

— Ribose — Phosphate — Ribose — Phosphate — Ribose —

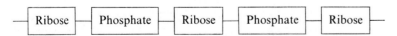

Sometimes the backbone of a nucleic acid molecule is described as sugar units (ribose or deoxyribose) linked by 3′,5′-phosphodiester bridges. The 3′-hydroxyl of the sugar unit of one nucleotide is joined to the 5′-hydroxyl of the next sugar unit by a phosphodiester bond.

$$\text{3',5'-phosphodiester linkage} \begin{cases} \\ \end{cases}$$

2. *The variable part of a nucleic acid molecule is the sequence of bases attached to the sugar units of the backbone.* The sequence of these base side chains distinguishes various DNAs from each other and various RNAs from each other. Only four types of base side chains are involved in any given nucleic acid structure. This situation is much simpler than that for proteins where 20 side chain entities (amino acids) are available (Section 24.2). In both RNAs and DNAs, adenine, guanine, and cytosine are encountered as side chain components; thymine is commonly found only in DNAs and uracil in RNAs (Section 26.2).

3. *The direction in which a DNA polymer is drawn is important because the polymer has different ends.* Much like protein polymers have a carboxyl end and an amino end, nucleic acids have a 5' end and a 3' end. Closer inspection of Figure 26–5 reveals that one end (the upper end) terminates with a phosphate attached to a 5'-carbon. As we move down the backbone, we can trace the covalent bonds across each sugar residue from its 5'-carbon to its 3'-carbon, through the phosphodiester linkage, and then across another sugar from its 5'-carbon to its 3'-carbon. Eventually, as we reach the other end of the polymer, we find a free 3' end.

If we are interested only in the sequence of bases along a nucleic acid polymer, we can write one-letter abbreviations for the bases. The sequence of bases in the tetranucleotide shown in Figure 26–5 can be written as 5'-TGCA-3'. The 5' and 3' indicate the appropriate ends of the strand. By convention these abbreviated representations of base sequence are always written beginning with the 5' end and continuing toward the 3' end. Oftentimes the 5' and 3' are omitted from the abbreviated representation for the base sequence.

4. *Each phosphoric acid molecule (phosphate group) involved in the backbone of the nucleic acid still possesses one of its three original —OH groups.* (The other two have become involved in the phosphodiester linkages.) This remaining —OH group is free to exhibit acidic behavior, that is, to lose its proton.

This behavior by many phosphate groups along the backbone gives nucleic acids their acidic properties.

nucleic acid

We can now define a nucleic acid in more chemical terms. A **nucleic acid** *is a polymeric molecule containing an alternating phosphate–pentose backbone with a nitrog-*

eneous base (heterocyclic amine) side chain attached to each pentose unit. This definition specifies the generalized primary structure of a nucleic acid.

As with proteins, nucleic acids have secondary as well as primary structure. The secondary structures of DNAs and RNAs differ, and we will discuss them separately.

26.4 DOUBLE HELIX STRUCTURE OF DNA

A key observation leading to the determination of DNA's complete structure involved measuring the amount of the side chain bases A, T, G, and C in DNA obtained from many different organisms. An interesting pattern emerged. The amount of A always equalled that of T, and the amounts of C and G were always equal.

The relative amounts of each of these base pairs in DNA varies depending upon the life form from which the DNA is obtained. (Each animal or plant has a unique base composition.) However, the relationships

$$\% A = \% T \quad \text{and} \quad \% C = \% G$$

always hold. For example, human DNA contains 30% adenine, 30% thymine, 20% guanine, and 20% cytosine. The percentages of A and T are equal and the percentages of G and C are equal. An explanation for this pattern of percentages came in 1953 when the complete structure for DNA was first proposed.*

A DNA molecule exists as two polynucleotide chains coiled around each other in a double helix. The bases (side chains) of each chain extend inward (within the interior of the double helix) toward the other chain, with each base in one chain hydrogen bonded to a base in the opposite chain.

The pairing of bases, through hydrogen bonding, is very selective. The base cytosine is always found opposite guanine, that is, C is paired with G; adenine is always found opposite thymine, that is, A and T are paired. No other pairing combinations are found. This arrangement of bases explains, very simply, why amounts of A and T are always equal, as are the amounts of C and G in any DNA molecule.

The pairing of bases, through hydrogen bonding, also explains why two strands of DNA polymer are attracted to each other. Although hydrogen bonds are relatively weak forces, each DNA contains so many base pairs that the hydrogen bonding attractions are sufficient in magnitude, collectively, to prevent the two DNA chains from spontaneously separating under normal physiological conditions.

Figure 26–6 (p. 684) shows diagrammatically the hydrogen bonding between adenine-thymine and cytosine-guanine base pairs. The A-T combination is held together by two hydrogen bonds and the C-G combination by three hydrogen bonds.

Each of the two observed pairings of bases involves a small pyrimidine base (monocyclic) and a large purine base (bicyclic). This is an important observation that sheds light on why other base pairing combinations are not observed. Two pyrimidine molecules would be too small to effectively hydrogen bond across the gap between chains in the interior of the DNA double helix. Hence, C-C, T-T, and C-T base pairing does not occur. Similarly, two purine bases would be too large to fit into the space in the interior of the double helix. Hence, A-A, G-G,

---

* Three scientists who contributed to this structural proposal—the English chemists Maurice Wilkins, Francis Crick, and James Watson—were later awarded a Nobel Prize for their achievement.

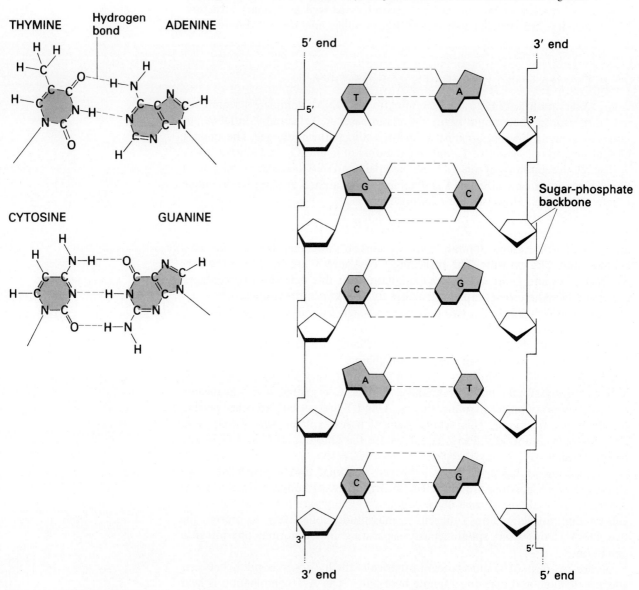

FIGURE 26-6 Hydrogen bonding between adenine-thymine and guanine-cytosine base pairs. This selective hydrogen bonding holds the two complementary, antiparallel strands of a DNA molecule together.

and A-G base pairing is not observed. To obtain a "proper fit", we must have one pyrimidine and one purine base.

We may ask why the pyrimidine-purine combinations A-C and G-T do not occur. The hydrogen bonding in these combinations is weaker than that in the observed A-T and C-G pairs.

Figure 26-7 shows three different representations of the double helix structure of a DNA molecule. At every ten nucleotides along the double helix, the helix completes a full turn.

Closer inspection of this DNA structure shows that the two strands of DNA are *antiparallel*, which means they run in opposite directions. If the backbone of one strand is oriented in the 5'-to-3' direction, the other strand is oriented in a 3'-to-5'

FIGURE 26–7 Three representations of the DNA double helix. (a) A drawing to show how the base pairs are stacked up like stairs in a circular staircase. (b) A picture of a space-filling model, similar to the ones used to determine the structure of DNA. (c) An abbreviated structure showing the complementary base pairing between chains. Note the antiparallel backbone structure (5′ and 3′ ends).

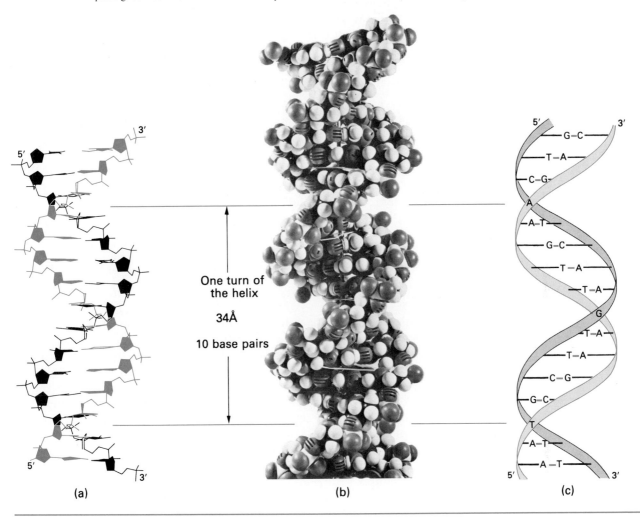

position (see Figure 26–8). This arrangement, as we shall see, has important ramifications in the way DNA is replicated.

FIGURE 26–8 Antiparallel orientation of double-stranded (duplex) DNA.

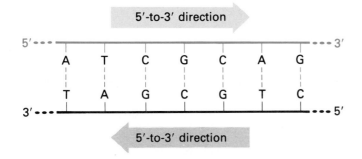

The complementary pairing of base pairs (A with T, and G with C) is one of the most important parts of the double helix model for DNA. By knowing the sequence of bases in one strand of helical DNA, we can immediately predict the sequence of bases on the complementary strand. Wherever A appears in one strand, T will appear in the complementary strand directly across from it. The same occurs with the G and C bases, as is seen in Figure 26–6.

Example 26.2

Predict the sequence of bases in and draw the structure of the complementary DNA strand for the single strand of DNA shown.

$$5'\text{-C-G-A-A-T-C-C-T-A-}3'$$

Solution

Since only A forms a complementary base pair with T and G with C, the complementary strand is as follows.

$$\begin{aligned} \text{given:} &\quad 5'\text{-C-G-A-A-T-C-C-T-A-}3' \\ \text{complementary strand:} &\quad 3'\text{-G-C-T-T-A-G-G-A-T-}5' \end{aligned}$$

DNA molecules also have tertiary structure. We seldom encounter straight double helixes. Instead they are coiled and twisted.

Some circular DNA molecules have been found. That is to say, the two ends of a linear DNA molecule, when covalently connected, form a circle of double-stranded DNA. Many bacterial and viral DNA molecules have circular shapes. Although these polymers have no ends, the two strands still retain their antiparallel configuration. Figure 26–9 shows possible twisting and coiling associated with the tertiary structure of a circular DNA molecule.

The length of even the shortest DNA is so long that determining the sequence of bases along one strand of DNA is a tremendous challenge. Scientists had to wait more than 25 years after the generalized structure of DNA was announced before methods were developed to quickly and efficiently sequence DNA. Today,

---

FIGURE 26–9 The tertiary structure (twisting and coiling) associated with a circular DNA molecule.

Circular DNA molecule

Coiled DNA molecule

laboratories can rapidly determine sequences of DNA by electrophoresis at a rate of around 100 to 200 bases per day.* Using such rapid techniques has allowed chemists to map large portions of the base sequence in DNAs obtained from many organisms.

26.5 DNA REPLICATION

We all know that physical and mental characteristics can be transmitted from parents to offspring and that this process continues generation after generation. For this to occur, genetic information must be passed on every time a cell divides. The carrier of genetic information within a cell is now known to be DNA. New cells obtain their genetic information through a process in which DNA reproduces itself, a process called *DNA* replication, which we now consider.

DNA replication *occurs when DNA molecules, in an enzyme catalyzed reaction, produce exact duplicates of themselves.* This reaction is unique among molecules. Let us now see why the double helix, with its complementary base pairing, is key to DNA replication.

To understand DNA replication, we must regard the two strands of the DNA double helix as a pair of *templates*. During replication, the templates separate. Each can then direct the synthesis of a new complementary strand. The result is two daughter DNA molecules with base sequences identical to those of the parent double helix. Let us consider details of this replication.

Under the influence of the enzyme DNA polymerase, an unwinding of the DNA double helix occurs. Associated with this process is the necessary cleavage of the hydrogen bonds between complementary bases. This unwinding process, as shown in Figure 26–10 (p. 688), is somewhat like opening a zipper.

The separated strands now have bases that are not involved in hydrogen bonding. They are free to form new base pairs with free nucleotides. As shown in Figure 26–10, the base pairing that results always involves C-G and A-T combinations. After the incoming nucleotides have formed hydrogen bonds with the template strand, DNA polymerase then verifies that the base-pairing is correct and catalyzes the formation of new phosphodiester linkages between nucleotides (represented by colored ribbons in Figure 26–10.)

Studies show that the energy needed to form the phosphodiester linkages between sugars, to give the backbone of the new DNA strand, is supplied by the individual nucleotides themselves because they carry two extra phosphate groups. The incoming nucleotides that base pair with the template are actually nucleoside-5′-triphosphates rather than nucleoside-5′-monophosphates. The general structure of such triphosphates is given in Figure 26–11 (p. 688).

As each incoming triphosphate nucleoside is added to the growing polymer, a diphosphate (pyrophosphate) unit is released. This diphosphate unit is quickly hydrolyzed by water to give two monophosphates. The energy released in breaking these high-energy phosphate linkages is sufficient to drive the formation of the new DNA strand. Figure 26–12 (p. 689) shows the triphosphate involvement in DNA replication.

The two daughter molecules of double-stranded DNA formed in the DNA rep-

---

* New technology currently under development will allow automated sequencing of thousands of bases per day.

FIGURE 26-10 (a) In DNA replication, the two strands of the DNA double helix must unwind, with the separated strands serving as templates. (b) Free nucleotides within a cell pair with the complementary bases on the separated strands of DNA. This process ultimately results in the complete replication of the DNA molecule.

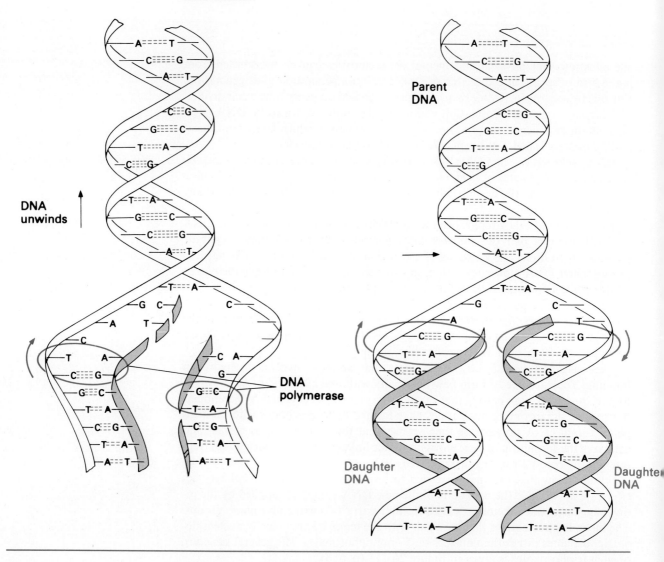

FIGURE 26-11 General structure of nucleoside-5'-triphosphates, used in the synthesis of DNA. The base may be A, T, G, or C. The corresponding abbreviations are dATP, dTTP, dGTP, and dCTP (For example, dATP is *deoxy*adenosine-5'-*tri*phosphate).

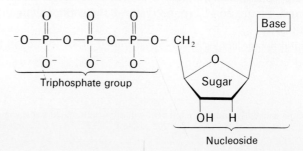

FIGURE 26-13 DNA polymerase catalyzes the addition of nucleotides to the growing chain of DNA. The loss of pyrophosphate and its subsequent hydrolysis provides the energy needed to drive the reaction. Only deoxynucleotides that have formed proper base pairs to the template strand are incorporated into the growing chain.

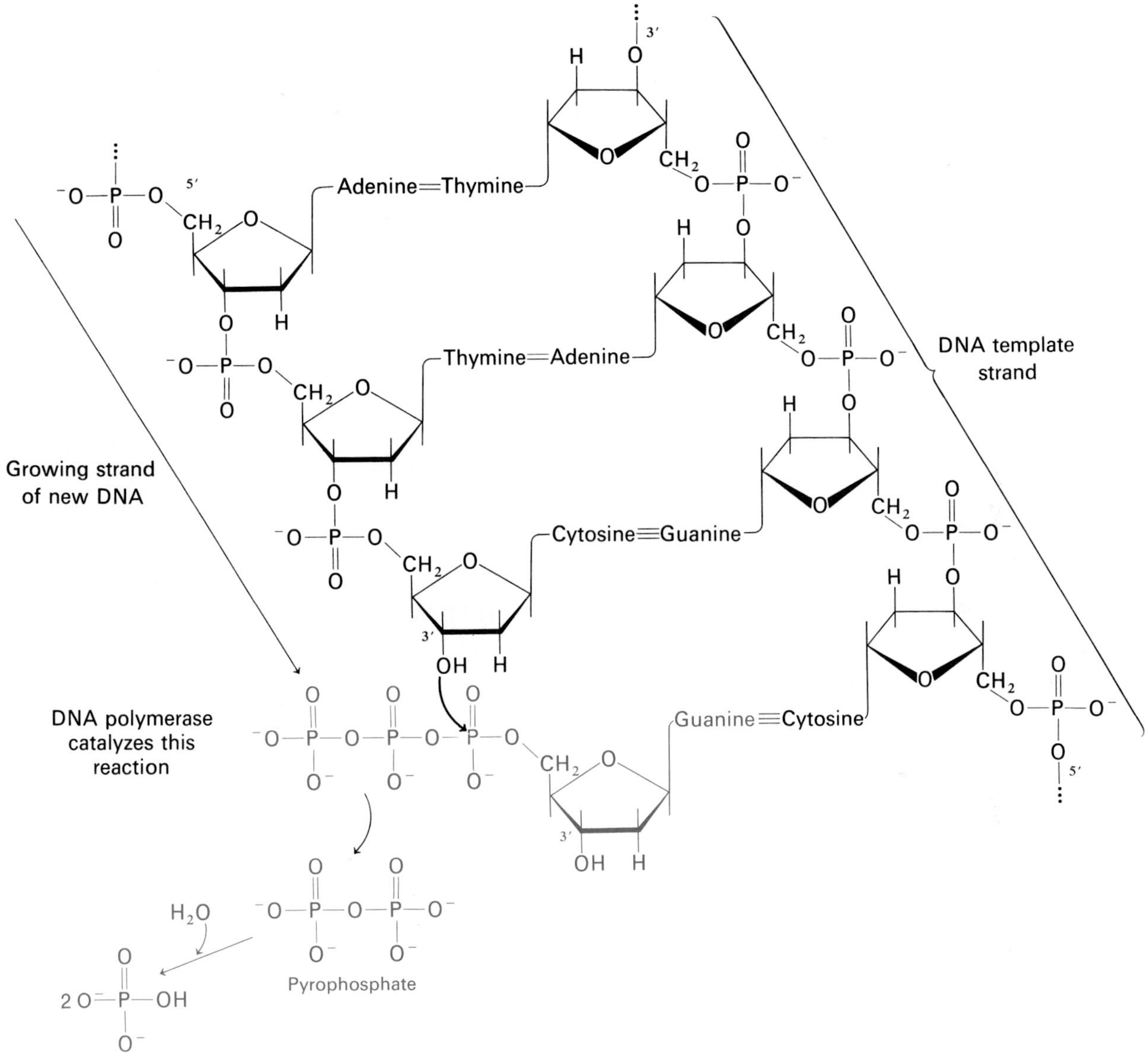

lication process each contain one strand from the original parent molecule and one newly formed strand.

Additional studies of the DNA replication process show that the enzyme DNA polymerase can catalyze the synthesis of DNA chains only in the 5′-to-3′ direction. Since the unwinding strands of parent DNA are antiparallel, only one strand can grow continuously in the 5′-to-3′ direction (see Figure 26–13, p. 690). The other strand, often called the *lagging* strand, must be formed in short segments as the DNA unwinds. These pieces of DNA are subsequently connected by action of the enzyme *DNA ligase*.

Circular DNA molecules also unwind and then replicate in a 5′-to-3′ direction.

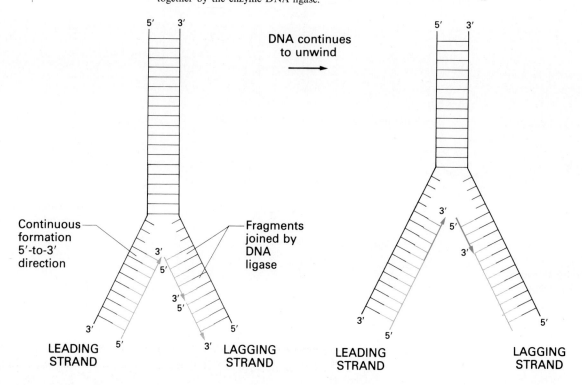

FIGURE 26-13 DNA is replicated in the 5'-to-3' direction. Since the strands of DNA run in opposite directions, only one strand (the leading strand) is synthesized in a continuous manner. The other strand (the lagging strand) is synthesized in short fragments. These short fragments are eventually joined together by the enzyme DNA ligase.

Unwinding begins in a small region of the molecule and then progresses around the circle. Under the electron microscope, this looks much like a bubble or loop between the strands of DNA as they separate (see Figure 26-14).

Once DNA has been completely replicated, it is organized into structural units, called *chromosomes*, that make an equal division of its genetic material possible. **Chromosomes** *are tightly packaged bundles of DNA and protein, which are visible only during cell division.* Different organisms have different numbers of chromosomes within a cell. The chromosome number is therefore a distinguishing characteristic of a species. A normal human has 46 chromosomes in each body cell; an onion, 16; a pine tree, 24; a corn plant, 20; a fruit fly, 8; a cat, 38; and a mosquito, only 6.

It is between periods of cell division, when DNA molecules are extended into a much less organized, less compact state throughout the nucleus, that replication occurs. As cell division approaches, the replicated DNA must fold and twist into the compact structure of a chromosome. To assist in the process of compression and packaging, the DNA winds around basic proteins called *histones*. *This complex structure of DNA and histone proteins is collectively called* **chromatin**. The replicated DNA molecules partially separate before the chromosomes become visible, so that they appear as duplicate pairs that are held together at only one point. As cell division takes place, these two chromosomes separate completely, with one member of each pair going to each daughter cell. Thus each daughter cell receives a full complement of genetic material.

FIGURE 26-14 (a) Schematic drawing of replication of circular DNA. (b) Electron micrograph shows actual replication.

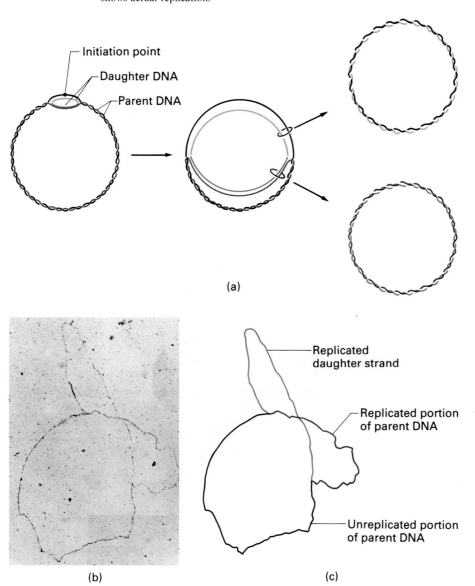

26.6 PROTEIN SYNTHESIS

We saw in the previous section how DNA can replicate so that a new cell contains the same genetic information as the one from which it was produced. We will now consider how the genetic information contained in a cell is expressed in cell operation. This brings us to the topic of protein synthesis. By participating in the synthesis of various proteins (skin, hair, hormones, enzymes, and so on), DNA establishes the similarities between parent and offspring that we regard as hereditary characteristics.

We can divide the overall process of protein synthesis into two steps. The first step is called *transcription* and the second *translation*. **Transcription** *is the process in which DNA directs the synthesis of RNA molecules that carry coded information needed for protein synthesis.* **Translation** *is the process in which the codes within RNA molecules are deciphered and a protein molecule is formed.* The following diagram summarizes the relationship between transcription and translation.

$$\text{DNA} \xrightarrow{\text{transcription}} \text{RNA} \xrightarrow{\text{translation}} \text{Protein}$$

Before discussing the details of transcription and translation, we need to learn more about RNA molecules. We will be particularly concerned with differences between RNA and DNA and various types of RNA molecules.

26.7 RIBONUCLEIC ACIDS

Four major differences exist between RNA molecules and DNA molecules. They are:

1. The sugar unit in the backbone of RNA is ribose rather than the deoxyribose found in DNA.
2. The base thymine found in DNA is replaced in RNA by uracil (Figure 26–2). Uracil forms base pairs (hydrogen bonds) with adenine in RNA, rather than thymine.
3. RNA is a single-stranded molecule rather than double-stranded (double helix) as is DNA. Consequently, base pairing is not required and correspondingly it is not necessary as with DNA, to have equal amounts of specific bases present.
4. RNA molecules are much smaller than long DNA molecules, ranging from as few as 75 nucleotides to a few thousand nucleotides.

We should note that the single-stranded nature of RNA does not prevent *portions* of an RNA molecule from folding back upon itself and forming double helical regions. If the base sequences along two portions of an RNA strand are complementary, a structure with a hairpin loop results, as shown in Figure 26–15.

Through transcription (Section 26.8), DNA produces three types of RNA distinguished by their function. The three types are: *ribosomal RNA* (rRNA), *messenger RNA* (mRNA), and *transfer RNA* (tRNA). Their characteristics are as follows.

Ribosomal RNA *combines with a series of proteins to form complex structures, called ribosomes, that serve as the physical sites for protein synthesis.* Ribosomes have molecular weights on the order of 3 million. The rRNA present in ribosomes has no informational function.

Messenger RNA *carries genetic information (instructions for protein synthesis)*

FIGURE 26–15 A hairpin loop in single-stranded RNA, caused by complementary base pairing between bases at different sites along the strand.

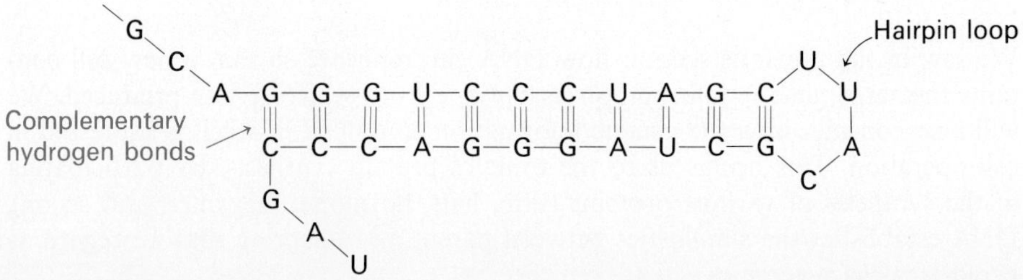

from DNA to the ribosomes. The size (molecular weight) of mRNA varies with the length of the protein whose synthesis it will direct.

Transfer DNA *delivers specific individual amino acids to the ribosomes, the site of protein synthesis.* These RNAs are the smallest of the RNAs, possessing only 75 to 90 nucleotide units.

More than 80% of the RNA in a cell is rRNA. Ribosomal RNA associates with protein in an approximate 1 to 2 weight ratio in forming ribosomes. It is at ribosomes that mRNA and tRNA interact to assemble amino acids into proteins. Less than 2% of the total RNA in a cell is mRNA.

If we understand the secondary structure of tRNA molecules, we can also better understand protein synthesis. We can talk about secondary structure in general terms because it is similar for all tRNAs. The representative secondary structure of a tRNA molecule, as shown in Figure 26–16, can be visualized as a cloverleaf resulting from extensive folding of the molecule so that complementary bases are hydrogen bonded to each other.

At the top of this cloverleaf structure are the 3′ and 5′ ends of the tRNA molecule. At the 3′ end is a free —OH group. An amino acid, carried by tRNA, binds to this site through an ester linkage (see Figure 26–17, p. 694). The remainder of the tRNA molecule contains three loops formed by base pairing along segments of the folded RNA chain. The loop opposite the 3′ and 5′ ends of the RNA has a site called an *anticodon* that can base pair with a complementary

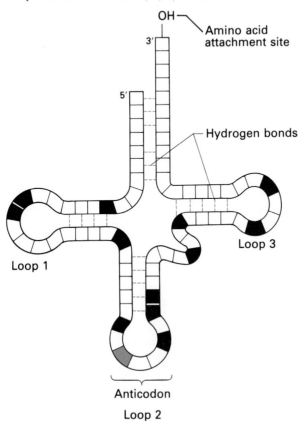

FIGURE 26–16 Idealized cloverleaf shape of a representative tRNA. The darkened segments represent bases other than A, C, G, and U.

FIGURE 26-17 The amino acid to be carried to the site of protein synthesis by a tRNA binds, via an ester linkage to the 3' end of the tRNA.

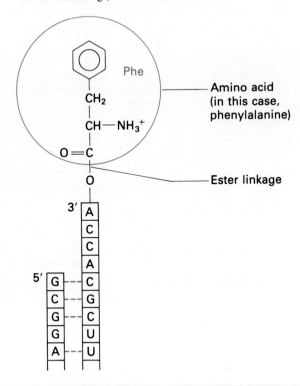

group of bases, called a *codon*, on an mRNA molecule. (The anticodon and codon concepts will be discussed further in Sections 26.9 and 26.10). The other loops (loops 1 and 3) function in recognizing a correct ribosome site and the specific enzyme associated with tRNA operation. We note also that a few rare heterocyclic bases, in addition to A, C, G, and U are found in tRNA. They are shown as dark spots in Figure 26-16.

The *actual* tertiary structure of a tRNA is more complicated than the idealized cloverleaf structure we've just discussed. Within this more complicated structure, however, we can find the major features of the cloverleaf, as we can see by comparing in Figure 26-18 the idealized and actual structures (determined by X-ray studies) of a tRNA that carries the amino acid phenylalanine.

26.8 TRANSCRIPTION: RNA SYNTHESIS

The mechanics of transcription are in many ways similar to what happens during DNA replication. Four steps are involved.

Step 1: A *portion* of the DNA double helix unwinds so as to expose nucleotide bases. The unwinding process is governed by the enzyme RNA polymerase rather than by DNA polymerase (replication enzyme). RNA polymerase recognizes certain base sequences as the starting point for the transcription.

Step 2: Free *ribo*nucleotides align along *one* of the exposed strands of DNA bases, forming new base pairs. In this process, U rather than T aligns with A in the base pairing process. Note that it is free ribonucleotides, not free deoxyribonucleotides; that are involved in forming the complementary RNA

FIGURE 26–18 Primary, secondary, and tertiary structure of a tRNA that carries the amino acid phenylalanine.

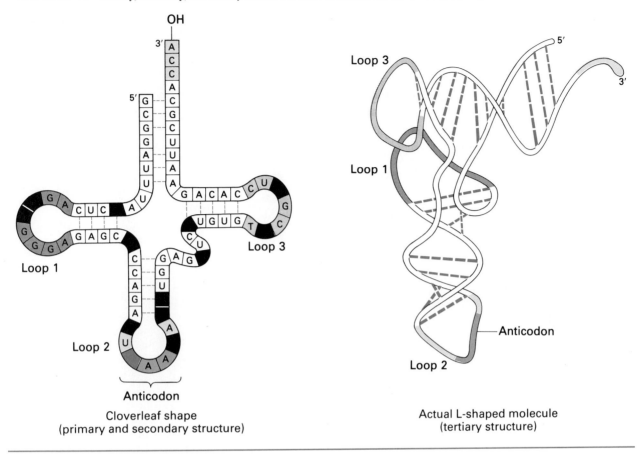

Cloverleaf shape
(primary and secondary structure)

Actual L-shaped molecule
(tertiary structure)

strand. Ribose, rather than deoxyribose, is incorporated into the backbone of the newly synthesized RNA strand.

Step 3: Linkage of the aligned ribonucleotides occurs. As with replication, this process actually involves triphosphate nucleotides (Section 26.5).

Step 4: Transcription ends when the RNA polymerase enzyme encounters a sequence of bases read as a stop signal. The newly formed RNA molecule is released as is the RNA polymerase enzyme and the DNA then rewinds to form the original double helix.

Figure 26–19, p. 696 shows the overall process of transcription of DNA to form RNA. Within a DNA chain are instructions for the synthesis of numerous specific mRNA molecules which in turn will direct the synthesis of many proteins. Short segments of DNA containing instructions for the formation of particular mRNAs are called *genes*. A **gene** *is a segment of a DNA molecule that contains the base sequence for the production of a single, specific RNA molecule.*

gene

In humans, most genes are composed of 1000 to 3500 nucleotide units. Hundreds of genes can exist along a DNA molecule strand. The DNA found in one human cell contains an estimated *5.5 billion nucleotide base pairs* that make up nearly one million genes.

In simpler life forms such as bacteria, the mRNA molecules produced at the DNA reach the ribosomes in an unaltered form to direct protein synthesis. How-

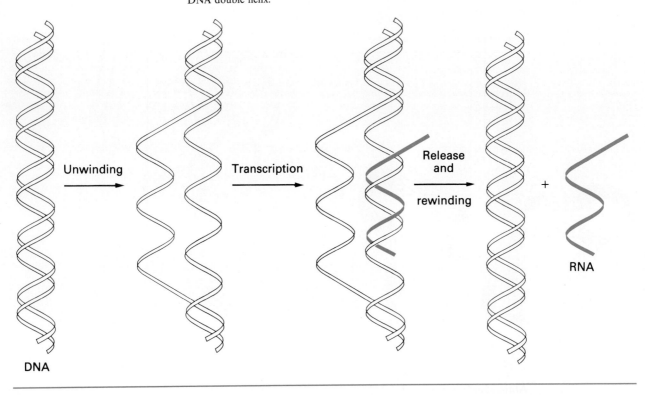

FIGURE 26–19 The process of transcription of DNA to form RNA involves the unwinding of a portion of the DNA double helix.

ever, in higher, more complex organisms, the mRNA is processed by specific enzymes before reaching the ribosomes. The initial mRNA molecule transcribed from the gene is called the *primary transcript*. Interior portions of the original mRNA are excised by processing enzymes, with the remaining portions being spliced together to yield the final mRNA. As much as 50% or more of the primary transcript may be cut during mRNA processing. Those portions of the RNA sequence that are expressed in the final mRNA are called *exons*; those sequences of the primary transcript that are removed during processing are called *introns*. The vast majority of genes in humans and other mammals are *split genes* made up of alternating exons and introns. The β-chain of hemoglobin is an excellent example of a split gene: it contains two introns that split the gene into three exons (see Figure 26–20). It is believed that split genes play an important role in cell growth and differentiation. Excision of introns and splicing of remaining exons represents only one type of RNA processing that occurs as these molecules move from the nucleus to the ribosomes in preparation for protein synthesis.

The gene content of a human cell is approximately 1 million genes, an extraordinary amount of information. To put this content in perspective, consider the following example. Let us assume you can see, with the naked eye, DNA molecules with their base pairs, and that you decided to write down the characteristics (base sequence) of all genes in a human cell. You use a lab notebook with one million pages and allow one page for each gene. Suppose you take five minutes (a very rapid speed) to record each gene's characteristics. When you finish the project, your notebook will be equivalent to 1250 eight hundred page textbooks like this one, and it would have taken you (working eight hours a day five days a week) 40 years to complete. All this information is found within one human cell.

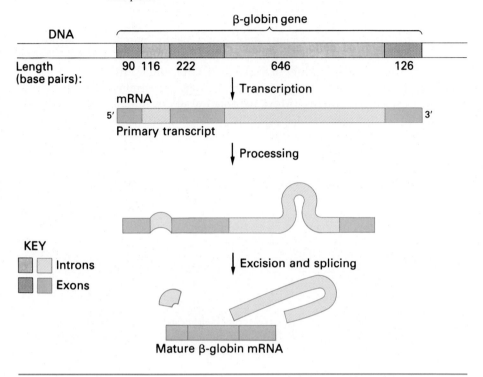

FIGURE 26-20 The β-globin gene, which codes for the β-chains present in hemoglobin, is a split gene. Two introns separate the three exons, both introns being excised from the mRNA primary transcript prior to translation. The primary mRNA transcript must be processed into the mature mRNA before translation at the ribosome can take place.

26.9 THE GENETIC CODE

We discussed the first step in protein synthesis, transcription (to form RNA), in the previous section. Before we discuss the second step of protein synthesis, translation, let us consider the *genetic code*, which will help you understand translation.

The base sequence along the backbone of an mRNA molecule is the informational part of this molecule. The base sequence must be decoded into the language of amino acids for protein synthesis to occur. In this section, we will consider the relationship between nucleotide base sequence and amino acid sequence.

How can the base sequence of an mRNA molecule (which involves only *four* different bases) encode enough information to direct proper sequencing of *twenty* amino acids in proteins? Obviously, if a single base codes for a single amino acid, we do not have enough bases—four bases versus twenty amino acids. A two-nucleotide code for amino acids will not suffice either. How many different two-letter "words" can be formed from an "alphabet" of four letters—the letters A, C, G, and U? The answer is sixteen. The "words" would be AA, AC, AG, AU, CA, CC, CG, CU, GA, GC, GG, GU, UA, UC, UG, and UU. Sixteen words are not enough to code for twenty amino acids. Resorting to a code involving three-letter words will give us more than enough combinations to code for the twenty amino acids. Sixty-four different three-letter "words" are possible using our A, C, G, and U base "alphabet."

Research has now shown that three-letter "words," that is, three-nucleotide base sequences, are the actual way mRNA molecules specify amino acids needed in protein synthesis. These three-nucleotide sequences (the three-letter "words,")

codon

are called *codons*. A **codon** *is a three-nucleotide sequence in mRNA that codes for a specific amino acid needed during protein synthesis.*

Among the 64 codons to choose from, which amino acid is specified by which codon? Researchers have deciphered codon-amino acid relationships by adding different *synthetic* mRNA molecules (whose base sequences are known) to cell extracts and then determining the structure of any newly formed protein. After a great number of these experiments, researchers finally matched all 64 possible codons with aspects of protein synthesis. They found that 61 of the 64 codons (formed by various combinations of the bases A, C, G, and U) are related to specific amino acids and that the other three combinations are termination codons (stop signals) for protein synthesis. (More information about termination codons will be given in the next section.) Collectively, these relationships between three-base sequences and amino acid identities are known as the *genetic code*. The **genetic code** *is a list of the* 20 *amino acids and the corresponding codons which signal their synthesis.* Table 26–2 is a listing of the genetic code. The determination of this code is one of the most remarkable twentieth century scientific achievements.

genetic code

TABLE 26–2 Codons in the Genetic Code Specific for Each Amino Acid

Amino Acid	Codons	No. of Codons	Amino Acid	Codons	No. of Codons
alanine	GCA	4	lysine	AAA	2
	GCC			AAG	
	GCG		methionine*	AUG	1
	GCU		phenylalanine	UUC	2
arginine	AGA	6		UUU	
	AGG		proline	CCA	4
	CGA			CCC	
	CGC			CCG	
	CGG			CCU	
	CGU		serine	UCA	6
asparagine	AAC	2		UCC	
	AAU			UCG	
aspartic acid	GAC	2		UCU	
	GAU			AGC	
cysteine	UGC	2		AGU	
	UGU		threonine	ACA	4
glutamic acid	GAA	2		ACC	
	GAG			ACG	
glutamine	CAA	2		ACU	
	CAG		tryptophan	UGG	1
glycine	GGA	4	tyrosine	UAC	2
	GGC			UAU	
	GGG		valine	GUA	4
	GGU			GUC	
histidine	CAC	2		GUG	
	CAU			GUU	
isoleucine	AUA	3	termination	UAA	3
	AUC			UAG	
	AUU			UGA	
leucine	CUA	6	Total number of codons		64
	CUC				
	CUG				
	CUU				
	UUA				
	UUG				

* AUG is also an *initiation* signal.

Table 26-3 General Features of the Genetic Code

Characteristic	Example
Codons are three-letter words.	GCA = alanine
The code is degenerate.	GCA, GCC, GCG, GCU all represent alanine
The code is nonoverlapping and chain requires no punctuation.	GCACUA = alanine, leucine, but not threonine (ACU)
Chain initiation is coded.	AUG = N-formylmethionine when used as first codon of a sequence
Chain termination is coded.	UAA, UAG, and UGA
The code is universal.	GCA = alanine, in almost all organisms

Let us look at this ingeniously efficient genetic code in some detail. Just a glance at Table 26-2 reveals that the code is *degenerate*, that is, there is more than one codon for most amino acids. Degeneracy does not imply chaos or confusion. It only means that there is more than one code word for certain amino acids. (By analogy, in the English language, there are often two or more words that have the same meaning.) It is important to point out that the reverse of degeneracy is not part of the code. No code specifies more than one amino acid. This specificity leads to order. [Many can specify one (degeneracy) but one cannot specify many (disorder)].

The genetic code is *nonoverlapping*. This means that parts of two adjacent codons cannot function as a new codon. For example, in the base sequence GCA CUA (codons for alanine and leucine), we cannot take the last part of the first codon, the A, and the first two parts of the second codon, CU, and combine them to form ACU (a codon for threonine).

allowed ⟶ alanine leucine
GC|A   CU|A
not allowed ⟶ threonine

Finally, studies of the genetic code in many organisms indicate that the code is the same in almost all of them, that is, the code is almost universal. The same codon specifies a given amino acid whether the cell is a bacterial cell, a corn plant cell, or a human cell. Table 26-3 summarizes the general features of the genetic code.

Example 26.3

Using Table 26-2, predict the sequence of amino acids coded by the mRNA sequence:

-GCC-AUG-GUA-AAA-UGC-GAC-CCA-

Solution
Matching the codons with the right amino acids yields:

mRNA: -GCC-AUG-GUA-AAA-UGC-GAC-CCA-
peptide: ala- met- val- lys- cys- asp- pro-

We are now ready to consider how a cell actually translates the language contained in mRNA into a sequence of amino acids that will combine to form a protein. We know who the "actors" will be—mRNA, rRNA, and tRNA (Section 26.7) and we know the "script"—the genetic code.

26.10 TRANSLATION: PROTEIN SYNTHESIS

In Section 26.6 we defined translation as the process in which the codons (Section 26.9) within an mRNA molecule are sequentially deciphered and that information is used to assemble a protein molecule. The substances needed for the translation process are the three types of RNA (mRNA, rRNA, and tRNA—Section 26.7), amino acids, and a number of enzymes.

The translation process can be conveniently discussed in terms of the following steps:

Step 1: A molecule of mRNA binds to a ribosome (rRNA and protein molecules) at a specific site.

Step 2: Molecules of tRNA carry amino acids to proper binding sites on the ribosome.

Step 3: Amino acids positioned in proper sequence at ribosome binding sites are joined together through peptide linkages.

Step 4: Protein synthesis is terminated by a "stop" codon on the mRNA molecule.

Further details, with accompanying diagrams, for each of these steps are as follows.

Step 1: Attachment of mRNA to Ribosome. A ribosome contains two subunits, a large subunit and a small subunit. The smaller unit is where the mRNA attaches itself and where the mRNA codons are read. The complete ribosome participates in the synthesis of the protein chain. In the following simplified diagrams, the ribosome and mRNA will be designated as

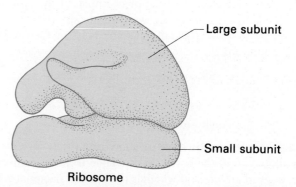

Ribosomes have two special sites for protein synthesis, a P site (peptidyl site) and an A site (aminoacyl site). Our ribosomal diagram with these sites labeled is

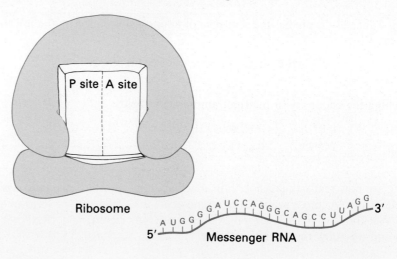

The messenger RNA attaches itself to the smaller ribosomal subunit so that its first codon (in the 5'-to-3' direction) is aligned at the P site and the second codon is at the A site.

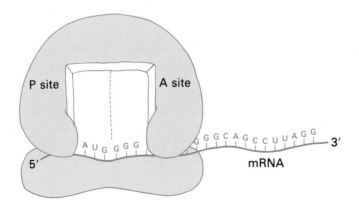

The first codon of mRNA, which lines up at the P site, is always the codon AUG, which initiates protein synthesis. When used in this capacity, AUG represents a derivative of the amino acid methionine, N-formylmethionine, which has the structure

The formyl group attached to the amino group of methionine blocks the N end of this amino acid so that protein chain growth can proceed in only one direction, toward the C-terminal end. Later the formyl group and sometimes the entire N-formyl methionine are removed from the finished protein by enzymatic cleavage.

Step 2: Amino Acid Transport by tRNA. The amino acids required for protein synthesis are brought to the ribosomal site for synthesis by tRNA molecules. The carrier tRNAs have the amino acids chemically bonded to them through an ester linkage. The ester bond forms from the interaction of the 3' end hydroxyl group of the tRNA molecule and the —OH of the acid functional group of the amino acid.

Note the representation used for the tRNA molecule. It is the cloverleaf-shaped structure discussed in Section 26.7. In most diagrams, we will use this notation (shown in the 3'-to-5' direction) rather than the more correct L-configuration (Figure 26–18) simply because it is easier to draw.

Specific tRNAs carry specific amino acids. Cells have at least one kind of tRNA that specifically binds each of the 20 common amino acid molecules. The tRNA molecules recognize their amino acids with help from enzymes that are simultaneously specific for the structures of the tRNA molecule and the amino acid molecule.

The amino acid that bonds to a given tRNA molecule is the one whose identity is consistent with the *anticodon* in the tRNA structure. Recall, from Figure 26–18 in Section 26.7, that the anticodon of a tRNA molecule is located within the loop opposite the 5' and 3' ends of the tRNA structure. An **anticodon** *is a three-base sequence in a tRNA molecule that serves as a complement to a codon in an mRNA molecule during protein synthesis.*

anticodon

The interaction between anticodon (tRNA) and codon (mRNA) leads to the proper placement of an amino acid into a growing peptide chain during protein synthesis. This interaction, which involves hydrogen bonding, can be diagrammed as follows.

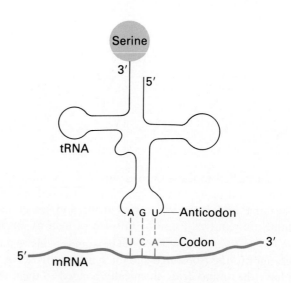

Only complementary codons and anticodons can interact in this manner.

The "landing point" for an amino-acid-carrying tRNA molecule is, thus, an mRNA codon that is the complement to its anticodon. All amino acids are brought into position at the ribosome by this principle: base pairing between a tRNA anticodon and an mRNA codon.

Initiation of protein synthesis begins when a tRNA, with the anticodon UAC, brings the amino acid derivative N-formyl methionine to the ribosomal P site. Another tRNA molecule, with an anticodon consistent with the codon at site A, brings its amino acid to the ribosome. In the diagram we are using, the second codon is GGG. Thus, the tRNA anticodon must be CCC, and this tRNA will be carrying the amino acid glycine. The codon GGG is one for glycine.

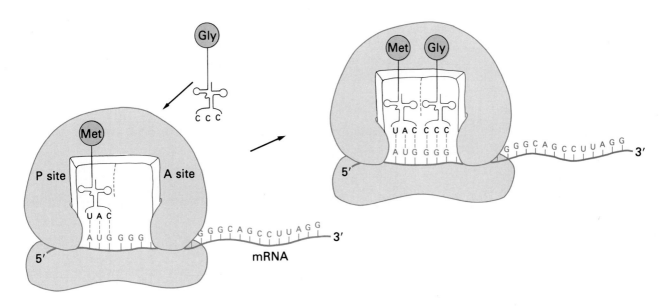

Step 3: Peptide Bond Formation Between Amino Acids. With amino acids in place at both the P site and A site of the ribosome, an enzyme catalyzes the formation of a peptide bond between the two amino acids. Note in the following diagram how the amino acid at the P site is detached from its tRNA and becomes attached to the amino acid at the A site, with the resulting formation of a dipeptide.

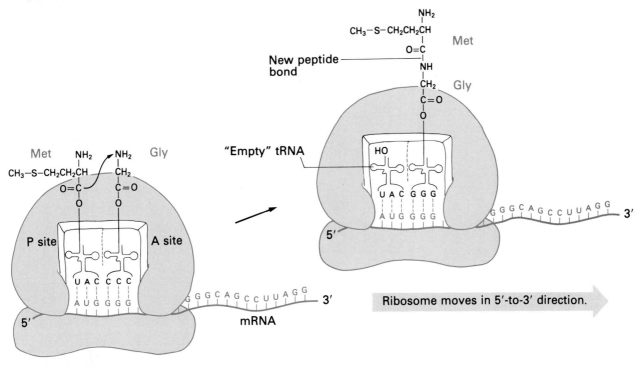

The "empty" tRNA at the P site now leaves that site to return to the cytoplasm to pick up another molecule of its specific amino acid and to be available for further use. Simultaneously with the tRNA release from the P site, the ribosome shifts, in the 3' direction, along the mRNA. This shift puts the newly formed dipeptide at the P site, and the third codon of mRNA is now available, at site A, to accept an aminoacyl tRNA molecule whose anticodon complements this codon.

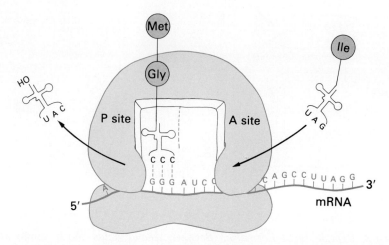

Now a repetitious process begins. The third codon, now at the A site, accepts an incoming tRNA with its accompanying amino acid, and then the entire dipeptide at the P site is transferred and bonded to the A site amino acid to give a tripeptide.

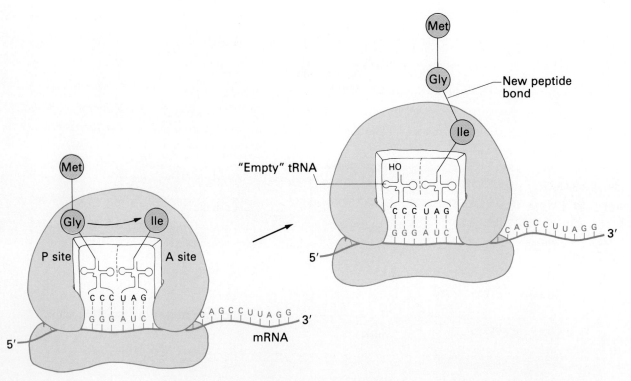

The "empty" tRNA at the P site is released, a ribosome shift along the mRNA occurs, and we are ready to begin formation of the tetrapeptide.

Note that except at the initiation of translation, where aminoacyl tRNAs occupy both the P and A sites, all incoming aminoacyl tRNAs base pair at the A site;

FIGURE 26-21 Simplified illustration of a polysome, showing peptide chains of various lengths protruding from the ribosomes as a single mRNA molecule directs their synthesis.

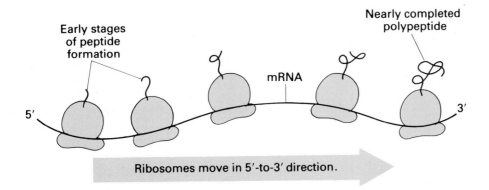

hence the designation A for aminoacyl. Also, the growing peptide is always transferred from the P site (peptidyl site) to the amino acid at the A site.

Step 4: Protein Chain Termination. The polypeptide continues to grow by way of translation (step 3) until all necessary amino acids are in place and bonded to each other. Appearance in the mRNA codon sequence of one of the three stop codons (UAA, UAG, or UGA) terminates the process. No tRNA has an anticodon that can base pair with these "stop" codons. The polypeptide is then cleaved from the tRNA through hydrolysis.

Electron microscopy studies have shown that one mRNA can be translated by many ribosomes at the same time. In fact, many ribosomes are usually attached to a single mRNA molecule, each simultaneously synthesizing a polypeptide chain (see Figure 26-21). This complex of ribosomes and mRNA is called a *polyribosome* or *polysome*.

How fast can cells produce proteins? Protein synthesis takes place very rapidly, considering the complexity of proteins. The speed of translation has been measured by injecting radioactive amino acids into animal cells. Within just a few minutes, these amino acids have been incorporated into newly synthesized proteins. Protein synthesis is rapid enough to respond to cellular demands for more proteins within just a few minutes.

Many antibiotics function by inhibiting bacterial protein synthesis. Most of

TABLE 26-4 Antibiotic Inhibitors of Protein Synthesis

Antibiotic	Biological Action
chloramphenicol	Inhibits an important enzyme (peptidyl transferase) in large ribosomal subunit.
erythromycin	Binds to the large subunit and stops the ribosome from moving along the mRNA from one codon to the next.
puromycin	Induces premature polypeptide chain termination.
streptomycin	Inhibits initiation of protein synthesis and also causes the mRNA codons to be read incorrectly.
tetracycline	Binds to the small ribosomal subunit and inhibits the binding of incoming tRNA molecules.

these inhibit one specific enzyme or another in the ribosome, thus halting protein synthesis. These antibiotics are not only useful in treating disease but also assist biochemical researchers in studying protein synthesis mechanisms in bacteria. The modes of action of a few representative antibiotics, whose structures are as follows, are given in Table 24–4 (p. 705).

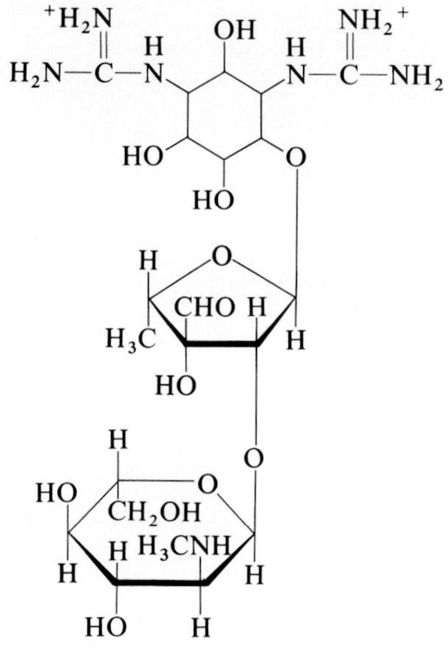

streptomycin

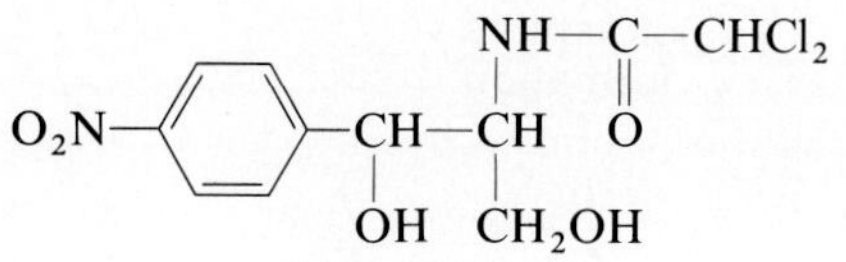

chloramphenicol

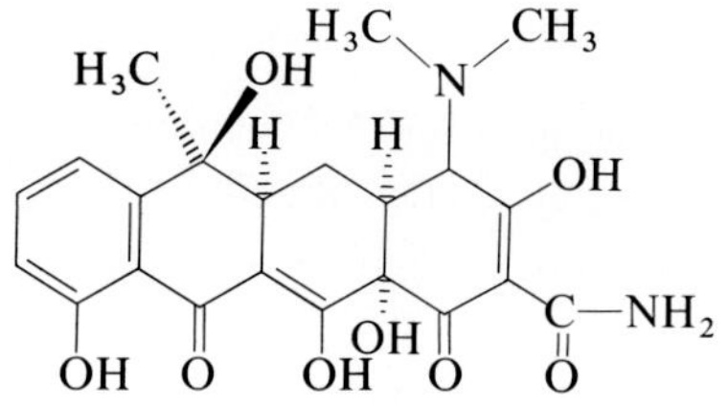

tetracycline

26.11 MUTATIONS

mutations

Mutations *are changes that alter the heterocyclic base sequence in DNA molecules.* These changes have two effects. First, they alter the genetic information that is passed on during transcription (Section 26.8) Second, changes in amino acid sequence during protein synthesis (Section 26.10) also occur. Sometimes such changes can have a profound effect on an organism.

mutagens

Mutagens *are the substances or agents that cause a change in the structure of a DNA molecule.* Radiation and chemical agents are two important types of mutagens. Radiation, in the form of ultraviolet light, X-rays, radioactivity (Chapter 12), and cosmic rays have the potential of being mutagenic. Ultraviolet light from the sun is the radiation that causes sunburn and can cause changes in the DNA of skin cells. Sustained exposure to ultraviolet light can lead to serious problems such as skin cancer.

Chemical agents can also cause mutagenic effects. Nitrous acid (HNO_2) is one such compound that causes deamination of heterocyclic nitrogen bases. For example, HNO_2 can convert cytosine to uracil.

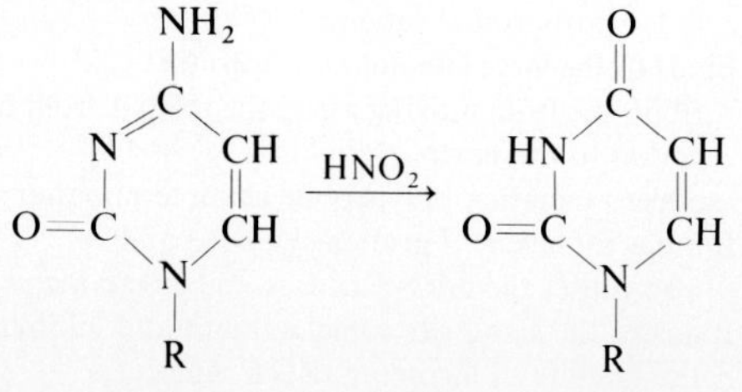

cytosine → uracil

A variety of chemicals—including nitrites, nitrates, and nitrosamines—can form nitrous acid in the body. The use of nitrates and nitrites as preservatives in foods such as bologna and hot dogs is a cause of concern because of their conversion to nitrous acid in the body and possible damage to DNA.

Fortunately, repair enzymes exist within the body that recognize and replace altered bases. The vast majority of altered DNA bases are normally repaired and mutations are avoided. However, occasionally the damage is not repaired and the mutation persists.

Mutations most often occur by changes in only one base in the sequence of bases found in a gene. These changes fall into three general types: *substitution* of one base for another, *insertion* of a new base into the sequence and *deletion* of a base from the sequence.

Let us consider, in general terms, the effects of these changes on the base sequence of a gene. To illustrate substitution, insertion, and deletion effects, let us use the analogy that the three-base codons are much like three-letter words used to construct a sentence. Any alterations in the spelling of these words can cause definite changes in meaning. Remembering that no punctuation or spaces are needed between codons, consider the following sentence:

SHECANNOWBUYTHENEWREDCAR

This sentence may seem a bit confusing at first glance, since we are used to reading words with spaces between words, so let's include some spaces between the three-letter words:

SHE CAN NOW BUY THE NEW RED CAR

What might happen to this sentence if it were subjected to mutations similar to those that occur in nucleic acids? *Substitution* of one letter for another often changes the entire meaning of the message:

SHE CAN NOT BUY THE NEW RED CAR

On the other hand, *deletion* of only one letter can result in total nonsense from the site of deletion on through to the end of the sentence. Consider the deletion of the letter Y from BUY. All letters after the deleted Y will move up one position. The result is

— deletion site

SHE CAN NOW BUT HEN EWR EDC AR

　　　　sense　　　　├── nonsense ──→

Similarly, *addition* of one letter can also cause confusion from the point of insertion, as shown by the following addition of the letter Z between the U and Y of BUY.

SHE CAN NOW BUZ YTH ENE WRE DCA R

　　　　sense　　　　├── nonsense ──→

Insertion and deletion mutations in DNA are called *frame-shift mutations*. The consequence of such a mutation is the disruption of the normal sequence of codons in the mRNA transcript and eventually in the amino acid sequence of the protein for which it codes. Consider the following frame-shift mutation, in which an extra adenosine nucleotide is incorporated into a DNA strand, and the effect that it has on RNA synthesis and then protein synthesis.

```
              Normal DNA                                  Mutant DNA
                                                         (addition of A)
                                                              ↓
DNA:      -TTA-GCG-TTT-ATG-TGG-TAG-           -TTA-GCG-TTT-A(A)T-GTG-GTA-
                         |                                  |   —frame shift→
                    transcription                      transcription
                         ↓                                  ↓
RNA:      -AAU-CGC-AAA-UAC-ACC-AUC-           -AAU-GCG-AAA-UUA-CAC-CAU-
                         |                                  |
                    translation                        translation
                         ↓                                  ↓
protein:   -asn- arg- lys- tyr- thr- ile-     -asn- arg- lys- leu- his- his-
                                                           —improper→
                                                              sequence
```

Substitution mutations, also called *point mutations,* are generally not as severe as insertion or deletion mutations because of the degeneracy of the genetic code. When multiple codons exist for one amino acid, it is usually the third base in the codon that varies between them. If a substitution occurs at this point in a codon, such that a different codon is formed that codes for the *same* amino acid, the effect on the cell is not even expressed. On the other hand, substitution involving just one base can produce serious effects. Sickle-cell anemia is the result of only one codon being altered in the gene that codes for the β-chain of hemoglobin. The sixth amino acid in this chain is normally glutamic acid, but is replaced by valine in sickle-cell hemoglobin.

Example 26.4

The codons for glutamic acid and valine are as follows.

Hemoglobin	*Amino Acid*	Codons
normal	glutamic acid	GAA, GAG
sickle-cell	valine	GUA, GUG, GUU, GUC

In sickle-cell anemia, which base in the codon has been altered, and what type of mutation occurred?

Solution
Since the entire sequence of hemoglobin's β-chain is unaltered *except* this one location, the mutation must be a substitution. The degeneracy of the codons for valine show that any base may be present at third position; however, the first two must be GU. Since the codon for glutamic acid also requires a G at the first position and varies at the third, *the substitution must have occurred at the second position,* with A being replaced by U. Either GUA or GUG might code for the valine in this case.

26.12 VIRUSES

viruses

Viruses are the lowest order of life. Indeed, their structure is so simple that some scientists do not consider them to be truly alive. **Viruses** *are tiny disease-causing packets of infectious nucleic acid surrounded by protective protein coats*; they come in various shapes and sizes (see Figure 26–22). The inner nucleic acid core contains DNA or RNA, but not both. This observation leads to the classification of

FIGURE 26–22 Electron micrographs of some viruses.

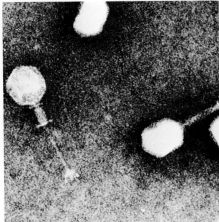

viruses into two major categories: DNA-containing viruses and RNA-containing viruses (see Table 26–5). Both types are important when considering the effects of viruses on other organisms.

There is no known form of life that is not subject to attack by viruses. Viruses attack bacteria, plants, animals, and humans. Many human diseases are of viral origin. They include the common cold, mumps, measles, small pox, rabies, influenza, infectious mononucleosis, hepatitis, and AIDS.

To reproduce, viruses must invade the cells of another organism and cause these host cells to carry out the reproduction tasks for them. Such invasion disrupts the normal operation of cells causing problems (diseases) within the host organism. In fact, the only function of a virus is reproduction; viruses do not generate energy or synthesize proteins of their own.

Viruses most often attach themselves to the outside of specific cells in a host organism. An enzyme within the protein overcoat of the virus catalyzes the break-

TABLE 26–5 Various Types of Nucleic Acids in Selected Viruses

Nucleic Acid	Representative Virus	Approximate Number of Genes
DNA		
single-stranded	$\varphi \times 174$ phage	5
double-stranded	polyoma virus	6
	adenovirus 2	30
	T4 phage	150
	vaccinia poxvirus	240
RNA		
single-stranded	$Q\beta$ phage	3
	Rous sarcoma virus	4
	tobacco mosaic virus	6
	poliovirus	8
	influenza virus	12
double-stranded	reovirus	22

Source: B.D. Davis, R. Dulbecco, H.N. Eisen, H.S. Ginsberg, and W.B. Wood, Jr *Microbiology* (Harper & Row, 1973).

down of the cell membrane, opening a hole in the membrane. The virus then injects its DNA or RNA into the cell. Once inside, this nucleic acid material is mistaken by the host cell as its own, whereupon it begins to translate and/or transcribe the viral nucleic acid. When all the virus components have been synthesized by the host cell, they assemble automatically to form many new virus particles. Within 20 to 30 minutes, after a single molecule of viral nucleic acid enters the host cell, hundreds of new virus particles have formed. So many are formed that they eventually burst the host cell and are free to infect other cells. Figure 26-23 shows the steps by which cells are killed by viral nucleic acids.

Viruses are highly specific in their selection of target cells. For example, a virus that attacks nerve cells in the spinal cord cannot attack liver cells or heart muscle cells. Furthermore, viruses that attack other species or plants cannot attack human cells.

RNA-containing viruses, once inside a host cell, reproduce unlike any other form of life. RNA must serve as the genetic material since no DNA is present. A single strand of viral mRNA, with the help of an enzyme called *reverse transcriptase*, is used to synthesize a complementary DNA strand (cDNA). This cDNA strand then directs the host cell to synthesize many molecules identical to the original viral mRNA.

A major line of defense of the human body against invading viruses is the production of antibodies (immunoglobulins, Chapter 24). Antibodies bind to viruses and prevent them from attaching to the host cell membrane. Thus, viral reproduction is stopped, and the virus is eventually eliminated from the body by the immune system.

Upon initial exposure, the immune system requires a number of days to produce antibodies that are specific to an invading virus. However, when the body is ex-

FIGURE 26-23 Steps in the mechanism by which a DNA-containing virus reproduces itself. Reproduction cannot occur without a host cell.

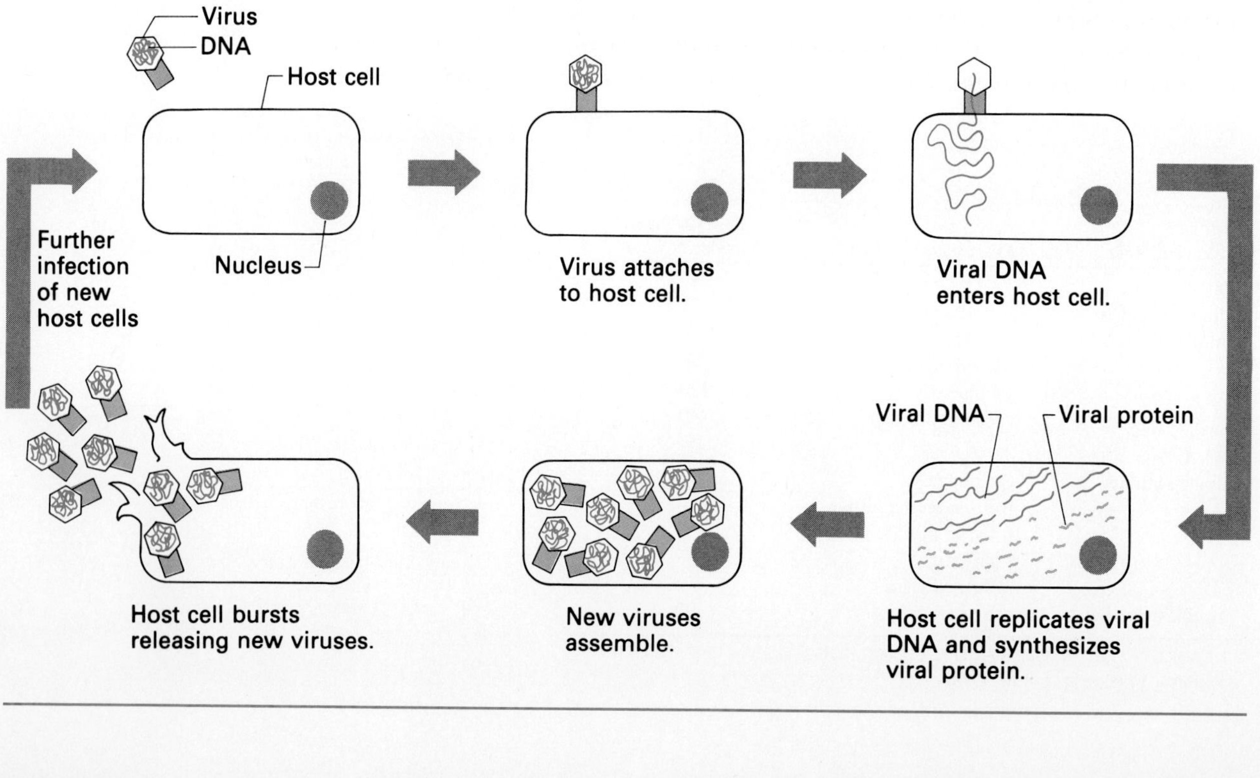

posed a second time to the same virus or bacteria, large quantities of the same antibodies are rapidly produced within a few hours, indicating that the body "remembers" the invaders or has kept a copy of the antibody on file for future use. Millions of different antibodies are found in the blood of a healthy person.

Preparation of antibodies ahead of time, before a particular virus invades, is the central concept of vaccination. **Vaccines** *contain inactive or attenuated forms of viruses.* The antibodies produced by the body against these specially modified viruses or bacteria have been found to effectively act against the naturally occurring forms as well. Many diseases, such as polio and mumps (caused by RNA-containing viruses) and smallpox and yellow fever (caused by DNA-containing viruses), are now seldom encountered because of vaccination programs.

Besides immunoglobulin production, cells also have more immediate response to an invading virus. They begin synthesizing and secreting a polypeptide called *interferon*. **Interferons** *are polypeptides that can inhibit viruses.* Interferons, produced by a cell under viral attack, bind to the membranes of neighboring cells, inducing an antiviral state in these cells. In this antiviral state, enzymes are produced that temporarily block protein synthesis in the cell and thereby protect against a broad spectrum of viruses.

Despite the existence of interferons, viral diseases still exist. Some viruses apparently manage to avoid stimulating interferon production or are very weak interferon stimulators. For such viruses, interferon concentrations never reach effective levels.

Researchers continue to study the use of interferons to treat viral diseases. Before the development of genetic engineering techniques, researchers could obtain only small amounts of interferons. Hundreds of thousands of pints of blood were required to obtain a few milligrams of interferon at a cost of millions of dollars. Today, researchers have relatively inexpensive interferon available, obtained from specially modified bacteria rather than from the human body. In 1986, FDA approval was given, for the first time, for clinical use of interferon. Its first use was a medication for treating particular forms of leukemia.

26.13 RECOMBINANT DNA AND GENETIC ENGINEERING

During the last decade, increased scientific knowledge about how DNA molecules behave under various chemical conditions has opened the door to a new and exciting field of research called *genetic engineering*. Human insulin, human growth hormone, and human interferons are now being manufactured in research laboratories and commercially (in small amounts) using genetic engineering techniques.

Commercially, many companies are producing human insulin. About 5% of the population of the United States suffer from diabetes, many of whom depend on injections of insulin obtained from the pancreas of slaughterhouse animals. Genetically transformed bacteria now produce human insulin that is available for individuals who have adverse side effects to insulin from normal animal sources. Pituitary growth hormone (used in treating human dwarfism) and human interferon can now also be made using genetically transformed bacteria, making larger quantities of these important proteins available.

Genetic engineering procedures involve a type of DNA called recombinant DNA. **Recombinant DNA** *molecules are DNA molecules that have been synthesized by splicing a segment of DNA, usually a gene, from one organism into DNA from another organism.* In 1978, the first report appeared that the gene (from rats) for producing insulin had successfully been incorporated into a strain of *E. coli* bacteria. As a result, these bacteria began to produce insulin. The scientific community immediately recognized exciting possibilities of this process and genetic engineering became a rapidly developing field. Let us examine the theory and procedures used in obtaining recombinant DNA through genetic engineering.

The bacterium *E. coli*, a one-celled organism, is the organism most used in experiments involving recombinant DNA. (Yeast cells are also now used, with increasing frequency, in this research.) *Escherichia coli*, or *E. coli*, are bacteria found in the intestinal tract of animals and humans. Important reasons for their extensive use include:

1. The genetics of this one-celled species is well known.
2. A number of strains of *E. coli* have been developed that cannot live outside of specially prepared culture media. There is, thus, little danger that these special strains could grow and reproduce outside the laboratory.
3. The genetic code for *E. coli* is the same as that for human cells. (Recall from Section 26.9 that the genetic code is almost universal, applying to most species.)

Structurally, one difference exists between *E. coli* cells and human cells. Not all DNA within *E. coli* is found in its single chromosome. Besides chromosomal DNA, *E. coli* also contain DNA in the form of small circular double stranded molecules called *plasmids*, with several plasmids existing within a cell. These plasmids, which carry only a few genes, replicate independent of the chromosome. Also they are transferred relatively easily from one cell to another. The plasmids of *E. coli* are used in recombinant DNA work.

Let us consider the overall steps to obtain *E. coli* cells that contain recombinant DNA. We will then consider details of specific steps.

Step 1: *E. coli* cells of a specific strain are placed in a solution that dissolves cell membranes, thus releasing the contents of the cells.

Step 2: The released cell components are separated into fractions with one fraction being the plasmids. The isolated plasmid fraction is the material used in further steps.

Step 3: A special enzyme, called a restriction enzyme, is used to cleave the double stranded DNA in a circular plasmid. The result is a linear (noncircular) DNA molecule.

Step 4: The same restriction enzyme is then used to remove a desired gene from a chromosome in another organism.

Step 5: The gene, from step 4, and the plasmid (linear DNA material), from step 3, are mixed in the presence of the enzyme DNA ligase (Section 26.5), which splices the two together. This splicing reaction takes place at both ends of the gene, resulting in a new circular plasmid (our recombinant DNA).

Step 6: The newly synthesized plasmids (recombinant DNA) are placed in a live *E. coli* culture where they are taken up by the *E. coli* bacteria. The *E. coli* culture into which the plasmids are placed need not be identical to that from which the plasmids were originally obtained.

Figure 26–24 summarizes the steps involved in obtaining recombinant DNA and incorporating it into a living cell.

In step 3 we noted that the conversion of a circular plasmid into a linear DNA molecule requires a restriction enzyme. *Restriction enzymes* are enzymes that can recognize specific base sequences in DNA and then cleave this base sequence at one particular site. To understand how a restriction enzyme works let us consider one that cleaves DNA between G and A bases in the 5′-to-3′ direction in the sequence GAATTC. This enzyme will cleave the double helix structure of a DNA molecule in the manner shown in Figure 26–25; p. 714.

Note that the double helix is not cut straight across; the individual strands

FIGURE 26-24 Recombinant DNA is made by inserting a gene obtained from a foreign cell into DNA obtained from another type of organism.

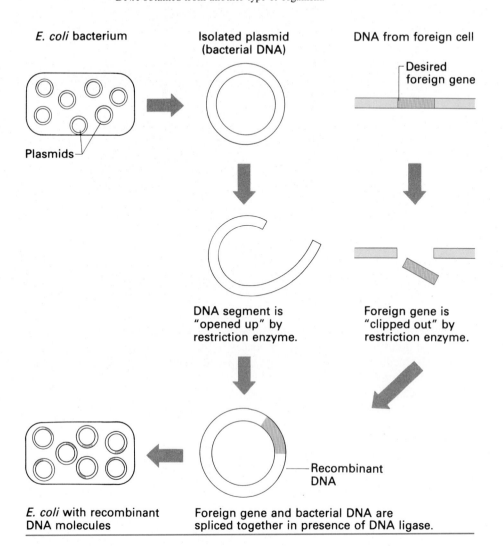

are cut at different points, giving a staircase cut. (Both cuts must be between G and A in the 5'-to-3' direction.) This staircase cut gives unpaired bases on each fragment produced from the cut. These unpaired base ends are called "sticky ends" because they are ready to "stick" (pair up) with a complementary section of DNA if they can find one.

If the same restriction enzyme used to cut a plasmid is also used to cut a gene from another DNA molecule, the sticky ends of the gene will be complementary to those of the plasmid. This enables the plasmid and gene to readily combine to give a new modified plasmid molecule. This modified plasmid molecule is called recombinant DNA. In addition to the newly spliced gene, the recombinant DNA plasmid contains all of the genes and characteristics of the original plasmid. Figure 26-26, p. 715 shows diagrammatically the match between sticky ends that occurs when plasmid and gene combine.

Step 6 involves inserting the recombinant DNA (modified plasmids) back into *E. coli* cells. The process is called *transformation*. **Transformation** *is the process of*

transformation

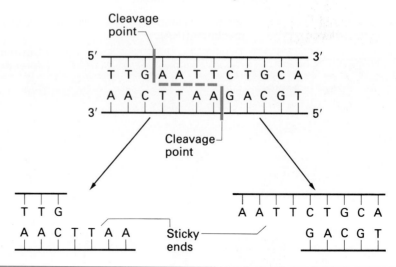

FIGURE 26–25 Cleavage pattern resulting from the use of a restriction enzyme that cleaves DNA between G and A bases in the 5'-to-3' direction in the sequence GAATTC. The double helix structure is not cut straight across.

incorporating recombinant DNA into a host cell. Usually only a few bacteria of the millions in a culture are transformed successfully. This makes the task of finding the transformed species difficult unless a specific label or marker is included in the recombinant DNA.

One powerful technique for selecting transformed bacteria is based upon antibiotic resistance. Not all bacteria are resistant to antibiotics such as penicillin. Those that are have DNA plasmids that contain penicillin-resistant genes. If a DNA plasmid containing a penicillin-resistant gene is used in the recombinant DNA, transformation of nonresistant bacteria will make them resistant to penicillin. Only the transformed cells will grow in media containing penicillin, allowing the selection of those containing the desired gene. The untransformed, nonresistant bacteria will not survive.

Once selection of transformed cells is accomplished, large numbers of identical cells called *clones* are obtained. **Clones** *are cells that have descended from a single cell and have identical properties.* Within a few hours, a single genetically altered cell can give rise to thousands of clones. Each clone has the capacity to synthesize the protein directed by the foreign gene it carries.

Many researchers who spend great effort and expense to develop new genetically altered strains of bacteria, actually patent their inventions. People who wish to use them are sometimes required to pay royalties to the inventors. To further protect their inventions, some genetic engineers also include their own trademark gene within their modified plasmids. Because bacteria are so easily passed around or stolen, the use of this marker, (a certain gene included in the original transformation) can protect the engineer's patent should a question arise as to the origin of this genetically altered strain.

Researchers are not limited to selection of naturally occurring genes for transforming bacteria. Chemists have developed nonenzymatic methods of linking nucleotides together, such that they can construct artificial genes of any sequence they so desire. In fact, bench-top instruments are now available that can be programmed by a microprocessor to synthesize any DNA base sequence *automatically.* The operator merely enters a sequence of desired bases, starts the instru-

FIGURE 26-26 The "sticky ends" of cut plasmid and cut gene are complementary and combine to form recombinant DNA.

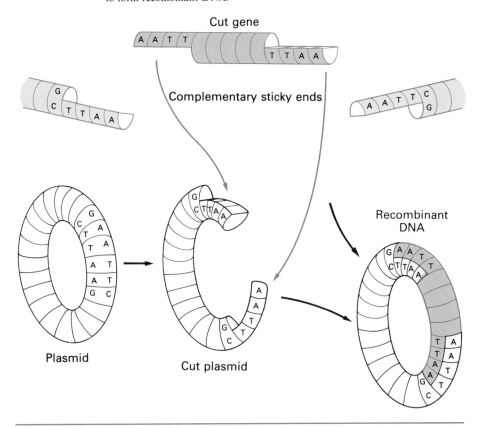

ment, and returns later to obtain the product. This synthesis requires only minutes per nucleotide (see Figure 26–27). This flexibility in manufacturing DNA has opened many doors, accelerated the pace of recombinant DNA research, and redefined the term "designer genes".

FIGURE 26-27 A fully automated bench-top DNA synthesizer.

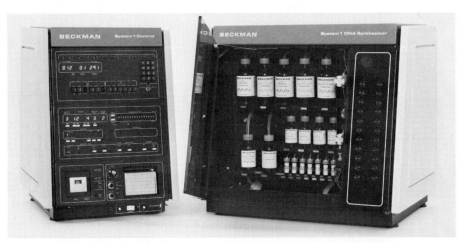

EXERCISES AND PROBLEMS

Nucleotides and Nucleosides

26.1 Describe the relationship of nucleotides to nucleic acids.

26.2 Identify the three chemical units that make up a nucleotide.

26.3 Describe the difference between the pentose sugars found in DNA and RNA.

26.4 Name the nitrogeneous bases found in DNA nucleotides. Identify each of them as a purine or pyrimidine derivative.

26.5 Name the nitrogeneous bases found in RNA nucleotides. Identify each of them as a purine or pyrimidine derivative.

26.6 Structurally, what is the difference between a purine base and a pyrimidine base?

26.7 From memory, give the names of the five bases whose symbols are C, T, U, A, and G.

26.8 Explain the difference between a nucleotide and a nucleoside.

26.9 Write a structural formula for a nucleoside containing
a. thymine and deoxyribose
b. adenine and ribose

26.10 Name each of the nucleosides from Problem 26.9.

26.11 Draw the structural formula for each of the following nucleotides.
a. one from DNA containing the base guanine
b. one from RNA containing the base uracil
c. cytidine-5'-monophosphate
d. deoxythymidine-5'-monophosphate
e. dGMP
f. UMP

26.12 Contrast the chemical differences in structure for nucleotides found in DNA and RNA.

Nucleic Acid Structure

26.13 Explain what is meant by the *backbone* of a nucleic acid molecule.

26.14 How does the backbone for a DNA molecule differ from that for a RNA molecule?

26.15 What distinguishes various DNA molecules from each other?

26.16 What is the difference between the 5' end and 3' end of a polynucleotide?

26.17 In what direction is the structure of a polynucleotide always read?

Double Helix Structure of DNA

26.18 Describe the secondary structural features of a DNA molecule.

26.19 What type of bonding holds a DNA double helix molecule together?

26.20 Where are the base pairs located in a DNA double helix?

26.21 The strands of DNA in the double helix are antiparallel. What does this mean?

26.22 What is meant by complementary base pair in DNA?

26.23 Using the concept of complementary base pairing, write complementary DNA strands for each of the following base sequences.
a. ACGTAT
b. TTACCGT
c. GCACTAAGTC

26.24 Draw the structures of and show with a diagram the two hydrogen bonds that stabilize base pairing between T and A in DNA.

26.25 Draw the structures of and show with a diagram the three hydrogen bonds that stabilize base-pairing between G and C in DNA.

26.26 The base content of a particular DNA molecule is 36% thymine. What is the percentage of each of the following bases in the molecule?
a. adenine
b. guanine
c. cystosine

DNA Replication

26.27 In general terms, describe the process of DNA replication.

26.28 What is the first conformational change that must occur in DNA before replication can take place?

26.29 What is the function of DNA polymerase in the replication process?

26.30 What principle insures that an exact duplicate copy of a DNA molecule results from replication?

26.31 In what direction are complementary bases added to a polynucleotide during replication?

26.32 Where are the DNA strands from a parent DNA molecule found within the daughter DNA molecules obtained from replication?

26.33 What is the role of triphosphate in DNA replication?

26.34 Explain why replication of the lagging strand of DNA occurs in a noncontinuous manner while that of the leading strand occurs in a continuous manner.

RNA Molecules

26.35 What are four major differences between RNA and DNA molecules?

26.36 What is the primary function of each of the following types of RNA?
a. tRNA   b. mRNA   c. rRNA

26.37 What is the most abundant type of RNA in a cell?

26.38 Describe the general structure, including secondary structural features, of a tRNA molecule.

Transcription: RNA Synthesis

26.39 Describe transcription by listing its four steps.

26.40 The following DNA base sequence is used as a template for formation of an RNA molecule

-T-A-C-G-G-G-T-A-A-C-T-C-G-

What will be the base sequence of the RNA formed from this template through transcription?

26.41 What is a gene?

26.42 What is the relationship between a gene and a chromosome?

The Genetic Code

26.43 What is a codon?

26.44 In what type of molecule are codons found?

26.45 What interrelationship is defined by the genetic code?

26.46 Which amino acid is coded for by each of the following codons?
a. CUU   d. GGG
b. AAU   e. GUA
c. AGU   f. CAC

26.47 Explain why the base sequence ATC could not be a codon.

26.48 The genetic code is *degenerate*. What does this mean?

26.49 The genetic code is *nonoverlapping*. What does this mean?

26.50 The genetic code is *universal*. What does this mean?

26.51 Using the genetic code, predict the sequence of amino acids coded by the RNA sequence

-AUG-AAA-GAA-GAC-CUA-UAC-CCC-

26.52 How many nucleotides must be present in mRNA to code for the placement of 125 amino acids in a single polypeptide chain?

Translation: Protein Synthesis

26.53 What are the four major steps in translation (protein synthesis)?

26.54 What is a ribosome, and what role do ribosomes play in protein synthesis?

26.55 Draw a sketch of a tRNA molecule and indicate
a. the site where an amino acid is attached during protein synthesis.
b. the site where the anticodon is located.

26.56 What is the purpose of the anticodon section of a tRNA molecule?

26.57 What anticodons in tRNA will base-pair with the codons in Problem 26.46?

26.58 In the growth step of protein synthesis, at which site in the ribosome does new peptide bond formation actually take place?

26.59 What two changes occur at a ribosome during protein synthesis immediately after peptide bond formation?

26.60 What occurs in protein synthesis immediately after the ribosome shifts in the 3′ direction along the mRNA molecule?

26.61 What is always the first codon that signals the beginning of protein synthesis? What special amino acid derivative is placed at the beginning of the new polypeptide chain as a result of this codon?

26.62 Describe the structure of a polysome. At which end (5′ or 3′) of the polysome are the growing peptide chains the longest?

26.63 Does every codon code for an amino acid in protein synthesis?

26.64 Why would two-base sequences in mRNA not be an appropriate code for amino acids?

Mutations

26.65 What is a mutation?

26.66 Describe three ways by which gene mutations might occur.

26.67 Why are substitution mutations generally less serious than deletion or addition mutations?

26.68 For the codon sequence

-GGC-UAU-AGU-AGC-CCC-

write the amino acid sequence produced
a. normally.
b. when a substitution mutation changes CCC to CCU.
c. when a substitution mutation changes CCC to ACC.
d. when an addition mutation changes GGC to GGCA.
e. when a deletion mutation changes GGC to GC.

Viruses

26.69 Describe the general structure of a virus.

26.70 Why are viruses sometimes considered inanimate?

26.71 What is the most common method by which viruses invade cells?

26.72 Why must a virus infect another organism in order to reproduce?

26.73 How does the reproduction of a DNA-containing virus differ from that of an RNA-containing virus?

26.74 What are antibodies and how do these substances effectively defend the body against viral attacks?

26.75 How do vaccines help in eradicating virus-caused diseases?

26.76 What is the relationship between viruses and interferons?

Recombinant DNA and Genetic Engineering

26.77 How does recombinant DNA differ from normal DNA?

26.78 What are *E. coli* plasmids and what is their role in genetic engineering experiments?

26.79 Describe what occurs when a particular restriction enzyme operates on a segment of double-stranded DNA.

26.80 Describe what occurs during transformation.

26.81 How are plasmids obtained from *E. coli* bacteria?

26.82 Recombinant DNA is made from the DNA of two different sources. What are, commonly, these two sources?

26.83 Only a small percentage of bacterial cells in a culture are transformed. Describe one way the transformed cells can be separated from the untransformed cells.

26.84 List two human proteins of commercial importance that can be produced by genetically altered bacteria.

26.85 A particular restriction enzyme will cleave DNA between A and A in the sequence AAGCTT in the 5'-to-3' direction. Assume the following DNA fragments are cleaved by this enzyme.

5'-C-C-A-A-G-C-T-T-G-3'
3'-G-G-T-T-C-G-A-A-C-5'   fragment 1

5'-G-G-A-A-G-C-T-T-A-3'
3'-C-C-T-T-C-G-A-A-T-5'   fragment 2

a. Draw a diagram showing the structural details of the "sticky ends" resulting from the cleavage of fragment 1.

b. Draw the structures of any new species that could result, through correct base pairing, when the cleaved pieces of the first fragment are mixed with the cleaved pieces of the second fragment in the presence of the enzyme DNA ligase.

27 Vitamins, Minerals, and Hormones

Objectives

After completing Chapter 27, you will be able to:

27.1 Understand some of the general characteristics of vitamins, important micronutrients for good health.
27.2 List the water-soluble vitamins, together with their roles in the body, amounts needed, and common foods that provide them.
27.3 Describe the functions, sources, and characteristics of the four fat-soluble vitamins, together with their associated overdose risks.
27.4 Recognize the important minerals needed by the body and some of their functions.
27.5 Outline the basic endocrine system and how the components of this glandular system communicate with each other.
27.6 Identify three classes of hormones based upon their structures, and give examples of each.
27.7 Describe two mechanisms of hormone action that are most often used in explaining how hormones function.
27.8 Understand how four principal hormones interact to direct the events of the female menstrual cycle and describe how some of the analogous, related chemical compounds are used as birth-control agents.

INTRODUCTION

Vitamins and *minerals* are necessary components of healthy diets and play important roles in cellular metabolism, but their role in body nutrition is often misunderstood. These substances do not provide energy directly, nor do they form "building blocks" of the body. In most cases, vitamins and minerals function as protective agents, enzyme cofactors, or carriers of functional groups during biosyntheses.

Although these substances occur in only very small amounts within cells, they are critically important. Absence of any of them is usually manifested as some deficiency disease.

Like vitamins and minerals, hormones are also present in the body in exceedingly small quantities. Synthesized within the body, they regulate many physiological processes and some are ultimately responsible for the different sexual characteristics exhibited by men and women. Abnormally low (or high) concentrations can result in dramatic consequences, causing hormone imbalances or diseases such as diabetes.

In this chapter, we will examine these three interesting classes of cellular microcomponents and their important roles in maintaining normal physiological function.

27.1 CHARACTERISTICS OF VITAMINS

Vitamins *are organic compounds necessary in small amounts for the normal growth and function of humans and some animals.* Adequate amounts of many of these compounds cannot be synthesized within the body and must therefore be obtained from other sources through the diet. The term *vitamine* was first used to describe the "vital amine," thiamine, which is needed to prevent *beriberi* (a once common disease among people who depended upon white rice for their main source of food). As other vitamins were discovered and characterized, not all of them were amines. The "e" was therefore dropped, but the generalized name "vitamin" survived.

vitamins

Researchers identified newly discovered vitamins by letters because the exact chemical structures were unknown. Later, what was thought to be one single vitamin often turned out to be many, and they added numerical subscripts to identify each different member of the group. One such group is the vitamin B complex, consisting of B_1, B_2, B_6, and B_{12}.

Some confusion also arose as to which vitamins were really necessary and which were not, resulting in gaps between numerical subscripts. For example, B_8 (adenylic acid), B_{13} (orotic acid), and B_{15} (parigamic acid) were removed from the list of essential vitamins. Others, originally designated as different, were later found to be the same compound. Vitamins H, M, S, W, and X were all eventually shown to be biotin; vitamin G became B_2 (riboflavin); and vitamin Y became B_6 (pyridoxine). At one time, vitamin M seems to have been used for three different vitamins: folic acid, pantothenic acid, and biotin. Today, scientists try to eliminate confusion by using chemical names of the now well-defined vitamins instead of letters.

Some vitamins are measured in **I.U.s (international units)**, *which is a measure of*

international unit (I.U.)

TABLE 27–1   United States Recommended Daily Allowances of Vitamins

	Unit	Infants (0–12 mo.)	Children under 4 yrs.	Adults and Children 4 or more yrs.	Pregnant or Lactating Women
vitamin A	IU	1500	2500	5000	8000
vitamin D	IU	400	400	400	400
vitamin E	IU	5	10	30	30
vitamin C	mg	35	40	60	60
folacin	mg	0.1	0.2	0.4	0.8
thiamine (B_1)	mg	0.5	0.7	1.5	1.7
riboflavin (B_2)	mg	0.6	0.8	1.7	2.0
niacin	mg	8	9	20	20
vitamin B_6	mg	0.4	0.7	2	2.5
vitamin B_{12}	mcg	2	3	6	8
biotin	mg	0.05	0.15	0.3	0.3
pantothenic acid	mg	3	5	10	10

IU = international unit
mg = milligram
mcg = microgram

biological activity. This measuring system is necessary because these vitamins have several natural forms that have different activities on an equal weight basis. Other vitamins are measured by weight in micrograms or milligrams. In the United States, the Federal Drug Administration (FDA) periodically issues a list of vitamins together with their **recommended daily allowances (RDA)**, *which are the highest amounts of vitamins that are needed by 95% of the population each day to maintain good health.* Table 27–1 lists the U.S. RDAs of vitamins.

To illustrate the small amount of vitamins needed by the human body, consider the RDA of vitamin B_{12}, which is 6 micrograms a day for an average adult. Just one gram of this vitamin could theoretically supply the daily needs of 166,666 people! Taking excess vitamins is a complete waste, both in money and effect since most of them can be stored only for a relatively short time. In fact, excess amounts of some vitamins can be harmful. A well-balanced diet will usually meet all the body's vitamin needs. Vitamin deficiency diseases are rarely seen in the United States. People who are *known* to have deficient diets require supplemental vitamins, as do those recovering from certain illnesses.

One of the most common myths associated with the nutritional aspects of vitamins is that vitamins from natural sources are superior to synthetic vitamins. In truth, synthetic vitamins, manufactured in the laboratory, are identical to the natural vitamins found in foods. The body cannot tell the difference and gets the same benefits from either source.

Vitamins are divided into two main classes: *water-soluble vitamins* and *fat-soluble vitamins*. The water-soluble vitamins are much better understood insofar as their enzyme cofactor roles are concerned, while the molecular aspects of fat-soluble vitamin functions are not as well understood. We will examine these two groups separately in the following sections.

27.2 WATER-SOLUBLE VITAMINS

Thiamine (Vitamin B_1)

Thiamine is a necessary component of the diet for most vertebrates and some microbial species. Soon after absorption into the body, thiamine is converted to *thiamine pyrophosphate* (*TPP*), the active cofactor for a number of enzymes involved in carbohydrate metabolism.

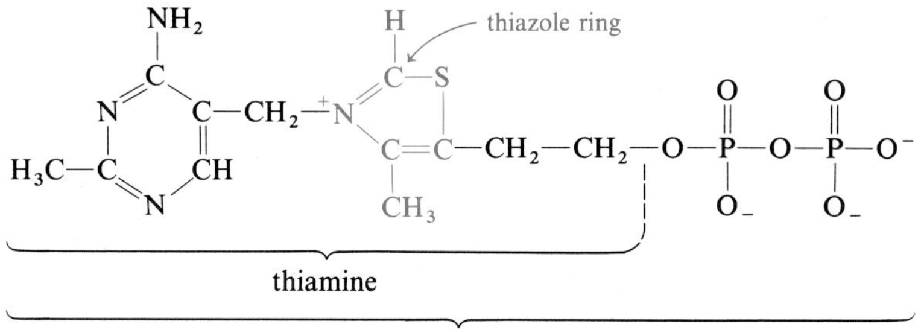

$$\underbrace{\underbrace{\begin{array}{c}NH_2\\|\\H_3C-C\overset{N}{\underset{N}{=}}\overset{C-CH_2}{\underset{CH}{C}}\end{array}\!\!\!-\!\!\!\overset{H}{\underset{}{\overset{|}{N}}}\!\!\overset{\text{— thiazole ring}}{\overset{C-S}{\underset{CH_3}{\overset{||}{C}=C}}}\!\!-CH_2-CH_2-}_{\text{thiamine}}O-\overset{O}{\underset{O_-}{\overset{||}{P}}}-O-\overset{O}{\underset{O_-}{\overset{||}{P}}}-O^-}_{\text{thiamine pyrophosphate}}$$

Two ring structures are found in thiamine: a pyrimidine ring and a five-membered thiazole ring. The carbon atom located between nitrogen and sulfur in the thiazole ring is the most important atom as far as function is concerned, since it readily forms bonds to the carbonyl carbons of α-keto acids. TPP is a cofactor for enzymes that decarboxylate α-keto acids or transfer aldehyde groups from one molecule to another. One such enzyme is pyruvate decarboxylase, which uses TPP to catalyze the following reaction.

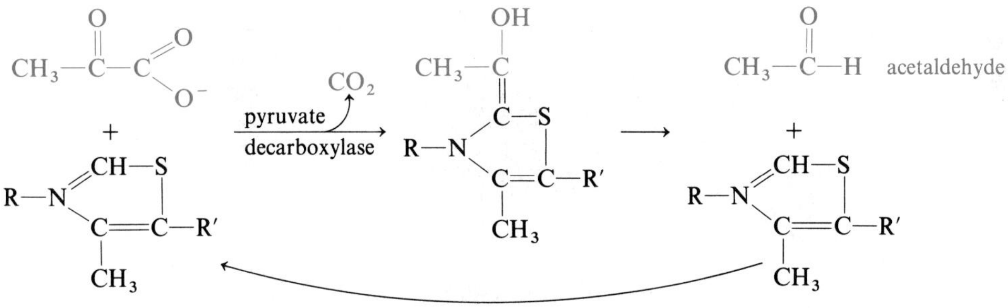

As a new bond forms to the α-keto carbon of pyruvate, the bond to the COO⁻ group breaks, resulting in the loss of CO_2 or decarboxylation of the acid. The remainder of the molecule (a modified aldehyde group) is bound to the TPP for a short time until it is removed to form the other product of this reaction, acetaldehyde.

As with most water-soluble vitamins, excess amounts beyond normal daily requirements (1.5 mg) are eliminated in the waste products of the body. Thiamine deficiency can result in depression, irritability, failure to concentrate, and altered reflexes. Extreme deficiencies lead to the disease beriberi, which is characterized by pain in the extremities, muscular weakness, and edema (swelling). Beriberi is often found in populations relying exclusively on polished white rice for food, because the rice husks, which are removed during refining, contain nearly all the thiamine in rice. Thiamine is also easily destroyed in foods that are baked, roasted, or fried for prolonged periods above 100°C.

Riboflavin (Vitamin B₂)

Riboflavin is a bright yellow, water-soluble vitamin. Foods rich in riboflavin include milk, meat, eggs, and cereal products. Its chemical structure, as well as those of its coenzyme forms, *flavin mononucleotide* (*FMN*) and *flavin adenine dinucleotide* (*FAD*), are shown in Figure 27–1 (p. 724).

These coenzymes take part in redox (reduction/oxidation) reactions catalyzed by flavin-linked dehydrogenase enzymes such as succinate dehydrogenase. Two

FIGURE 27-1 Riboflavin and its two coenzyme forms, flavin mononucleotide (FMN) and flavin adenine dinucleotide (FAD). The reactive group is in color.

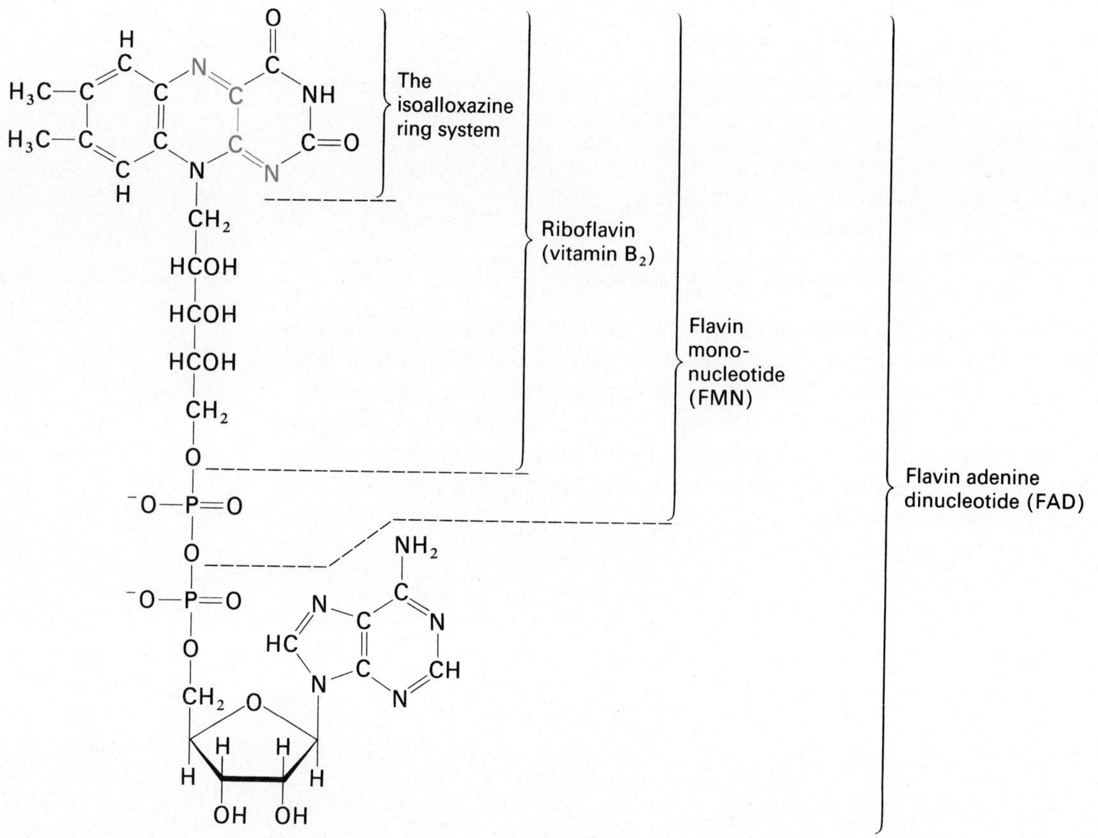

of the nitrogen atoms in the isoalloxazine (flavin) ring accept two hydrogen atoms with their two electrons in these reactions. The reduced riboflavin can reversibly return these electrons and hydrogen atoms to different substrates upon demand, making flavins excellent oxidizing or reducing agents in biological systems.

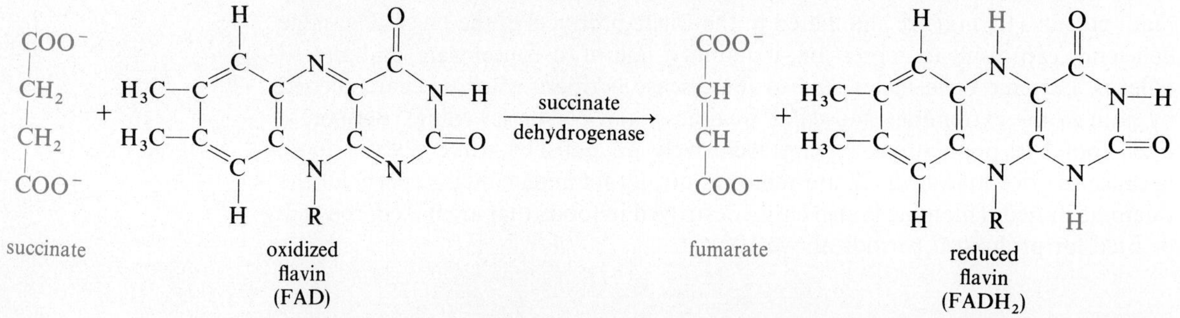

Excess riboflavin produces no known toxic effects, but a deficiency can produce symptoms of dermatitis, inflammation of the tongue, anemia, and even photophobia (intolerance to light). Riboflavin deficiencies are rare, but when they do occur, they are usually seen in chronic alcoholics. Riboflavin is stable at ordinary cooking temperatures. However, it is very sensitive to light. Milk exposed to direct sunlight in clear glass containers loses 50% of its riboflavin in two hours.

Niacin

Niacin is actually *nicotinic acid*. Both nicotinic acid and its amide derivitive, *niacinamide*, are active forms of this vitamin. The richest sources of niacin are similar to riboflavin: meats, peanuts and other legumes, eggs, and enriched cereals. Deficiency of niacin leads to the disease *pellagra* (Italian for "rough skin"), which is most often found among people who rely heavily upon corn as a primary food source.

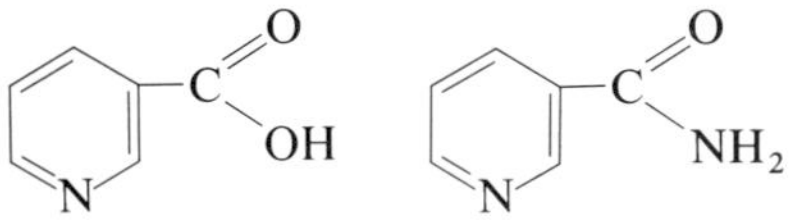

niacin
(nicotinic acid)

niacinamide

Pyridoxine (Vitamin B_6)

Pyridoxine (pyridoxol), pyridoxamine, and pyridoxal are all naturally occurring forms of vitamin B_6. Soon after ingestion, all three forms are efficiently converted by enzymes in the body to *pyridoxal phosphate* (often abbreviated PLP), the active coenzyme form of this vitamin (Figure 27–2).

Pyridoxal phosphate plays an important role in many phases of amino acid metabolism and is also required for synthesis of neurotransmitters such as serotonin and norepinephrine. These roles explain in part the symptoms of mild pyridoxine deficiencies, such as depression and irritability, as well as the more serious convulsive seizures associated with acute deficiency. Although present in most foods, the richest sources of vitamin B_6 are whole-grain cereals, egg yolks, meat, and vegetables.

Biotin

Once called vitamin H, biotin is now the sole descriptive term for this vitamin. The structure of biotin includes two five-membered heterocyclic rings fused

FIGURE 27–2 Various forms of vitamin B_6 and their conversion into pyridoxal phosphate, the active coenzyme form of this vitamin.

Pyridoxine
(Pyridoxol)

Pyridoxamine

Pyridoxal

Pyridoxal phosphate (PLP)

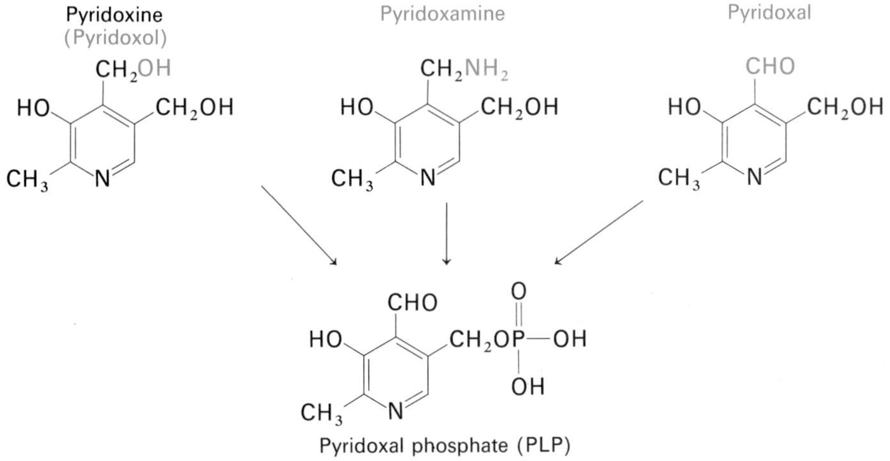

together. A nitrogen atom in one of these rings is capable of forming a bond to CO_2, carrying the CO_2 to a site where needed, and then transferring it to another molecule. Because of this unique ability, biotin functions as a cofactor for enzymes that catalyze carboxylation reactions. Red meats, poultry, fish, and milk are excellent sources of biotin.

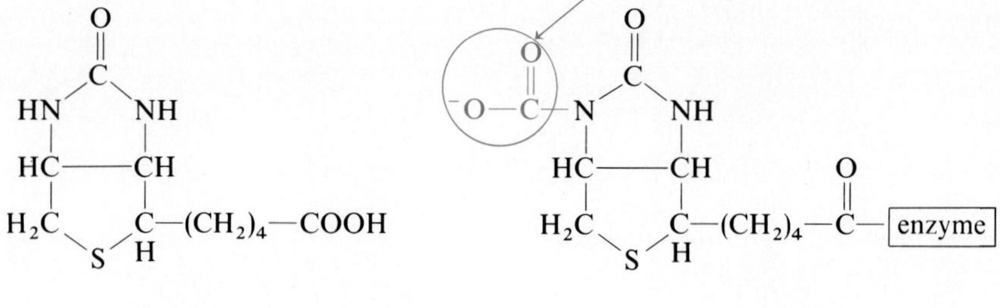

FIGURE 27-3 Structure of the vitamin, cobalamin (vitamin B_{12}). The commercially supplied vitamin often contains a cyano group in place of the 5'-deoxyadenosine and is called cyanocobalamin.

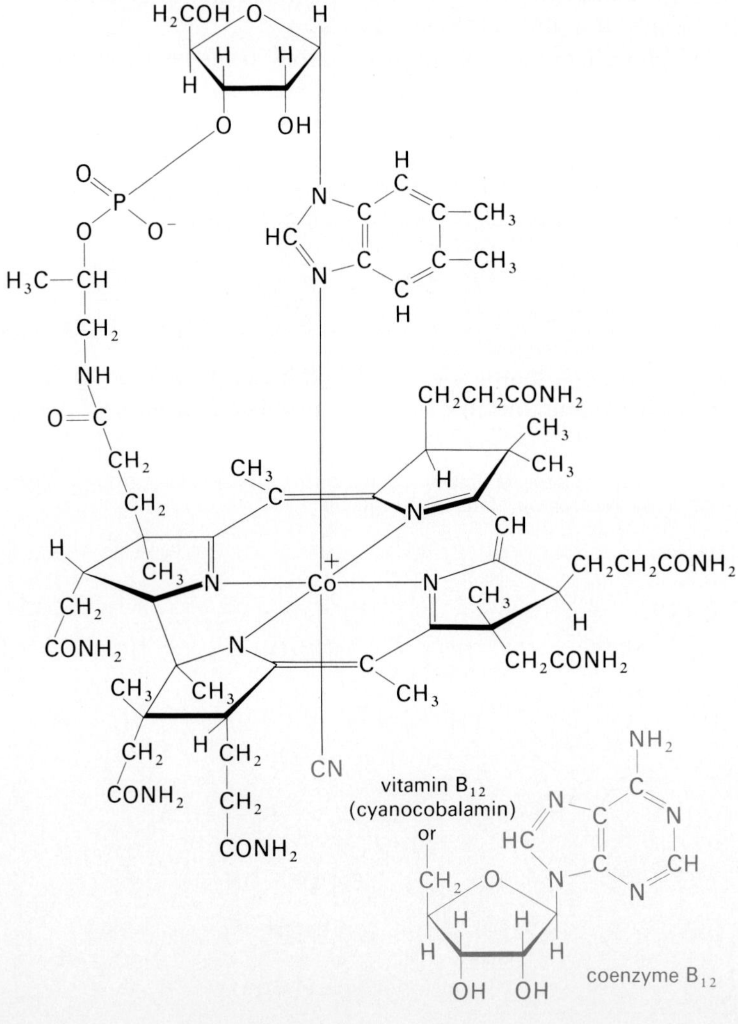

Cobalamin (Vitamin B_{12})

Structurally speaking, cobalamin is the most complex of all the vitamins. Not only does the vitamin have an extensive organic structure, but it also possesses a cobalt ion deep in the central cavity of a heme-like ring system (as can be seen in Figure 27–3).

Humans, as well as all animals and plants, are not capable of synthesizing cobalamin. Our only naturally occurring source is bacteria in normal human digestive tracts that supply the RDA of cobalamin (3–6 micrograms). The human liver is able to store enough cobalamin to last for years, and hence deficiencies of this vitamin are seldom seen today. However, until 1926, a serious and invariably fatal disease called *pernicious anemia* afflicted many people due to cobalamin deficiency. Pernicious anemia is a disease characterized by a decreased number of red blood cells and impairment of the central nervous system, which is caused by an inability to absorb vitamin B_{12}. People who suffer from this disease often consume sufficient amounts of the vitamin, but lack a specialized glycoprotein that assists in its absorption. Fortunately, extra amounts of vitamin B_{12} in the diet help overcome the problem of absorption. Due to storage of this vitamin in the liver, this organ is an excellent source of the vitamin, as are other foods such as milk products and seafoods (clams, oysters, lobsters, scallops, and haddock).

Folic Acid

The chemical structure of folic acid is composed of three parts: a bicyclic ring system (called pteridine) attached to *p*-aminobenzoic acid (PABA), which in turn is linked via an amide linkage to glutamic acid. Reduction of the pteridine ring with four hydrogen atoms and their electrons results in the formation of the active cofactor *tetrahydrofolate*, (the ionized form of tetrahydrofolic acid).

Since mammals are unable to synthesize pteridine ring systems, they obtain tetrahydrofolate from their diets or from microorganisms in their intestinal tracts. This cofactor functions in the transfer of one-carbon groups such as —CHO, —CH_3, —CH_2, and —CH=NH. It is essential in the formation of nucleic acids and therefore must be present for normal division of cells. As you may recall, the action of sulfa-type antibiotics selectively prevents the normal bacterial synthesis of folic acid, thus inhibiting bacterial growth and division.

Folic acid deficiency is believed to be the most common form of vitamin undernutrition. Deficiencies of folic acid among humans are quickly observed in the blood (which continually requires the new production of red blood cells), thus resulting in anemia. Weight loss and weakness are also symptoms of folic acid deficiency. It is estimated that the majority of people in underdeveloped countries have at least marginal folic acid deficiency. Pregnant women and infants are also particularly vulnerable, as well as some elderly people.

This vitamin owes its name to one of its sources, spinach leaves, from which it was first isolated (*folium*, Latin for "leaf"). Most green, leafy vegetables and meats are excellent sources of this vitamin. However, folic acid is destroyed by cooking. As much as 50% of the folic acid content of foods is lost during cooking.

Pantothenic Acid

Pantothenic acid (pantothenate in ionized form) is a medium-length acid that contains an amide bond (Figure 27–4). Humans are incapable of synthesizing this relatively simple molecule and must obtain it from outside sources. Once inside the body, it soon becomes an important part of *coenzyme A* (*CoA*), an important coenzyme for a large number of enzymes, most of which are involved in the transfer of acyl groups. Acyl groups are reversibly attached by a thioester linkage to the terminal sulfur of the long chain of coenzyme A. A pantothenic acid deficiency can produce gastrointestinal tract disturbances and depression; however, such deficiencies are seldom seen due to the ubiquitous occurrence of coenzyme A in almost all foods.

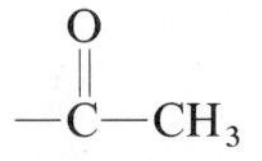

Ascorbic Acid (Vitamin C)

Ascorbic acid is a heterocyclic molecule containing an ester functional group in its five-membered ring. Vitamin C is an excellent reducing agent, using its vinyl hydroxyl groups as sources of hydrogen atoms and their associated electrons.

FIGURE 27–4 Pantothenic acid and its role as a component of coenzyme A, a carrier of acyl groups.

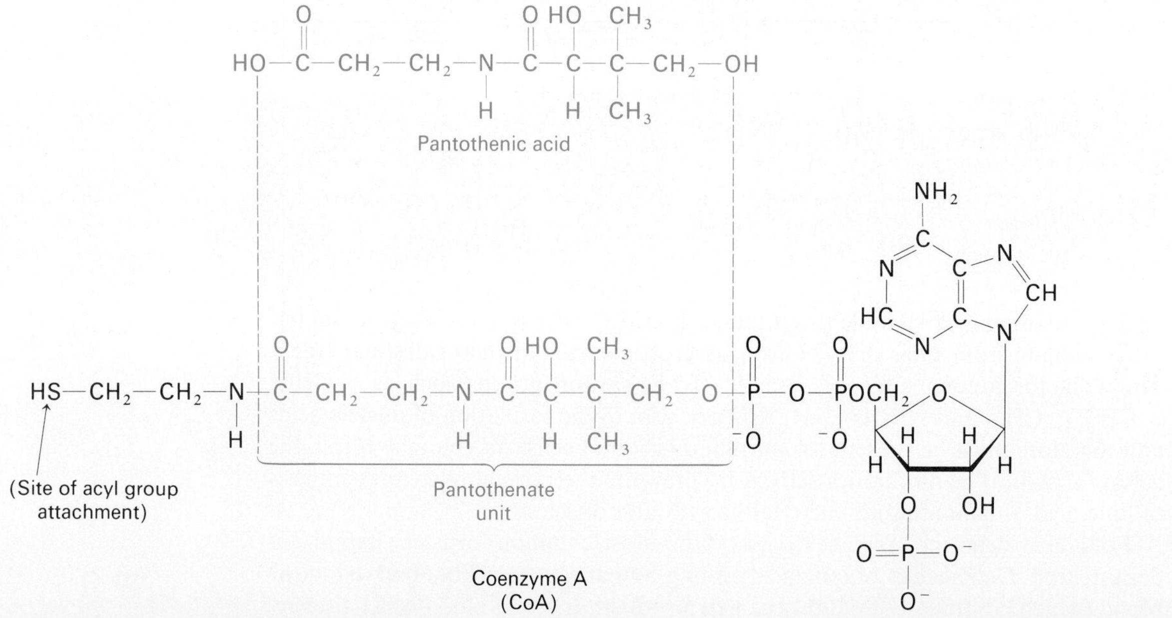

$$\text{ascorbic acid (reduced)} \underset{\text{reduction}}{\overset{\text{oxidation}}{\rightleftarrows}} \text{dehydroascorbic acid (oxidized)} + 2\text{H} \cdot$$

Ascorbic acid acts as a cofactor in redox (reduction/oxidation) reactions such as the processing of proline to hydroxyproline, a necessary step in the synthesis of collagen protein in connective tissue.

Hemorrhages of skin, gums, and joints were warnings to ancient sailors that death was near, due to the dreaded disease *scurvy*. Unknown to them, vitamin C is needed for proper maintenance of connective tissues. Around the 1700s, it became known that eating fresh fruits prevented or cured scurvy. British sailors were given the nickname, "limey", because they were always eating fresh limes. In fact, when researchers first characterized extracts of vitamin C from citrus fruits, the vitamin was named as an *anti-scurvy* or *a-scorbic* vitamin. When they eventually determined the acid nature of the molecule, the name became *ascorbic acid*.

Fresh fruits and vegetables are a necessary component of the diet to supply vitamin C. While 10 mg/day prevents scurvy, certain connective tissues and peripheral capillaries of the skin become fragile in due time at such low doses. The official U.S. RDA for vitamin C is 69 mg. In the past, some prominent scientists advocated that huge doses (250–1000 mg/day) help prevent or lessen the chances of catching the common cold. However, this claim has not been substantiated and some proof exists that these exceedingly high doses could even be harmful to a person's health, causing problems such as diarrhea, nausea, oxalic acid stones, or even loss of bone calcium.

27.3 FAT-SOLUBLE VITAMINS

There are four so-called fat-soluble vitamins, namely, A, D, E, and K. As the name implies and the chemical structures suggest, these vitamins are quite insoluble in water. This fact, and their small concentrations in cells hampered the study of their function for many years. Only recently have scientists begun to understand the exact role of these vitamins.

One interesting and clinically significant characteristic of these vitamins is that they are not readily excreted from the body like the water-soluble vitamins and can thus be stored in the body for months. As a result, complete lack of any of these vitamins in the diet may not be manifested physiologically for many weeks, or even months. Also, due to their water insolubility, they are not easily extracted into surrounding water during normal cooking of vegetables.

Retinol (Vitamin A)

Retinol is an isoprenoid alcohol (Figure 27–5, p. 730) that contains five conjugated double bonds that are responsible for its bright yellow color. Retinol, retinal, and retinoic acid are the active forms of this vitamin, with retinal playing an especially important role in vision. Other possible functions of this vitamin include regulation of Ca^{2+} transport, growth of bone-forming cells (osteoblasts), and various aspects of reproduction.

Most retinol is either absorbed directly from foods in the diet (such as liver, eggs, and milk) or is derived from β-carotene, a brightly colored pigment in vegetables (such as carrots). All three forms of vitamin A can be formed from β-

FIGURE 27-5 Formation of the different species of vitamin A, originating with the enzymatic cleavage of β-carotene.

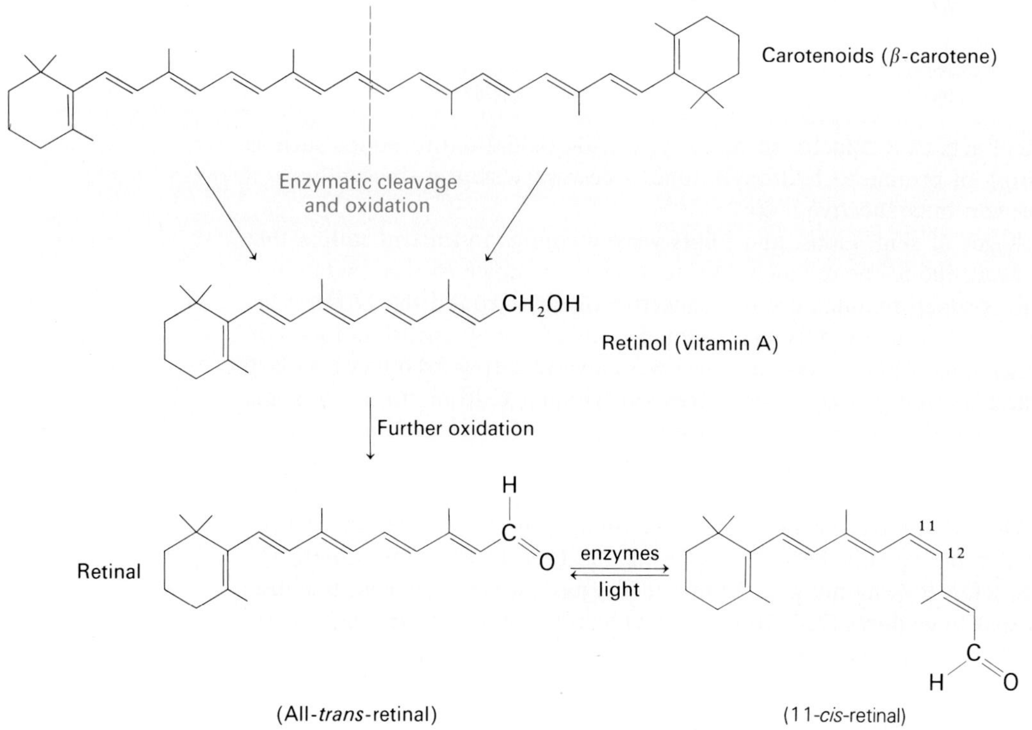

carotene, after it is cleaved enzymatically into two identical parts, as shown in Figure 27-5.

Excess vitamin A (hypervitaminosis A) can cause some serious problems, including liver and spleen enlargement, abnormal fetal development, fragile bones, skin problems, and loss of appetite. Some children and young people who have been given large doses of vitamin A have developed an increased pressure inside the skull that can mimic the *symptoms* of a brain tumor. Carotene, on the other hand, is practically nontoxic. Because of these detrimental effects, therapeutic doses of vitamin A require a doctor's prescription.

Cholecalciferol (Vitamin D)

Cholecalciferol is sometimes called the "sunshine vitamin" because it can be synthesized in the skin by sunlight irradiation of 7-dehydrocholesterol (a normal metabolite of cholesterol). In the polar regions of the Earth where sunshine is not always present during certain times of the year, humans must supplement their diets with vitamin D from sources such as fish oils, butter, or vitamin-D-fortified milk. Milk products are often fortified with vitamin D, obtained by irradiating yeast extracts with ultraviolet light.

The similarity of chemical structures between 7-dehydrocholesterol and vitamin D is seen in Figure 27-6, which illustrates the opening of the steroid ring system by the photochemical reaction. The biologically active form of vitamin D is 1,25-dihydroxycholecalciferol, formed from the addition of two hydroxyl groups in the liver and kidneys. This form of vitamin D acts as a hormone, promoting the

FIGURE 27-6 The structure of cholecalciferol. Vitamin D can be absorbed through the diet or synthesized in the skin by the action of ultraviolet rays in sunshine. Vitamin D is eventually converted into 1,25-dihydroxycholecalciferol, by the addition of two hydroxyl groups (one by enzymes in the liver and the other by enzymes in the kidneys).

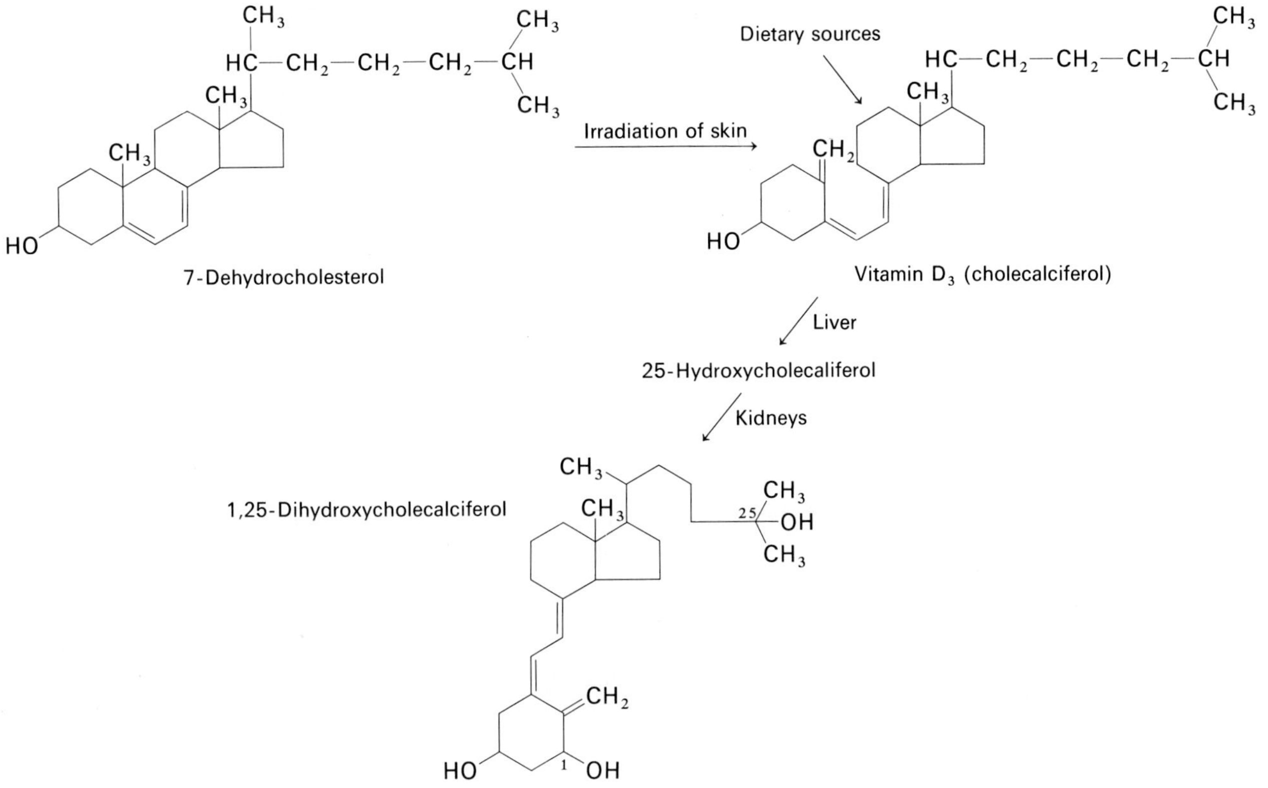

synthesis of calcium-binding protein in the intestinal mucosal cells and thereby regulating the uptake of calcium from the intestine.

Vitamin D doses of 6–10 times the RDA (60–100 micrograms) are toxic to small children, causing hypercalcemia (excess calcium in the blood) and calcification of soft tissues. On the other hand, vitamin D deficiency leads to *rickets* in children, a disease characterized by low levels of calcium, which results in soft and pliable bones, leading to bending and distortion.

Tocopherol (Vitamin E)

A number of tocopherols have been isolated and identified. Of these, α-tocopherol is the most abundant and biologically active.

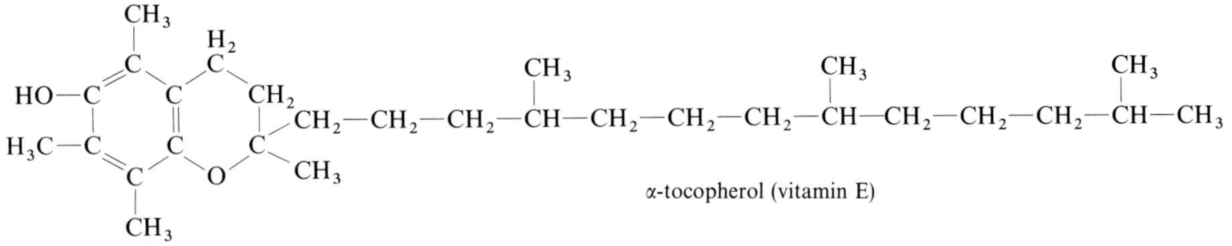

α-tocopherol (vitamin E)

Vitamin E was recognized in 1926 as a factor that helps prevent infertility in rats. Some researchers have extrapolated these findings to humans, but there is no proof that vitamin E might boost fertility in humans. Although the precise function of tocopherol is not known, some evidence suggests it acts as an antioxidant, preventing the oxidation of cell membrane phospholipids. This may not be totally true (or its only function), since other antioxidants do not prevent vitamin E deficiency symptoms from appearing in rats.

As yet, *symptoms of tocopherol deficiency in humans have not been fully identified.* While breast milk contains sufficient quantities of vitamin E, cow's milk does not. Premature infants fed on formulas low in vitamin E often develop a form of hemolytic anemia that can be corrected by vitamin E supplementation. Therefore, most manufacturers of infant formulas fortify their preparations with this vitamin. Vitamin E is destroyed when it is frozen or when fats containing it become rancid.

Coagulation Factor (Vitamin K)

Vitamin K was discovered by Henrik Dam in Denmark in the 1920s as a fat-soluble factor important in blood coagulation (K is for "koagulation" factor). Vitamin K is found naturally as K_1 in green vegetables and as K_2, which is synthesized by intestinal bacteria. The body is also able to convert synthetically prepared menadione (vitamin K_3) and a number of water-soluble analogs to the biologically active vitamin K_1.

Vitamin K is an important cofactor for an enzyme that participates in blood clotting. The only symptom of vitamin K deficiency is increased blood clotting time. Since vitamin K is relatively abundant in the diet and synthesized in the intestine, deficiencies are very rare among adults. However, some deficiencies can be caused by long-term antibiotic therapy, which may destroy the vitamin-producing organisms of the intestinal tract. The most common deficiency is observed in newborn infants (especially premature infants) because they have very low stores of vitamin K in their tissues, and breast milk is a relatively poor source of vitamin K. As soon as bacteria begin to form colonies in the intestines, the supply of vitamin K increases accordingly. At birth, as little as 1 mg is sufficient

---

FIGURE 27–7 Various forms of vitamin K, as well as dicoumerol, which is structurally similar and acts as an antagonist to vitamin K.

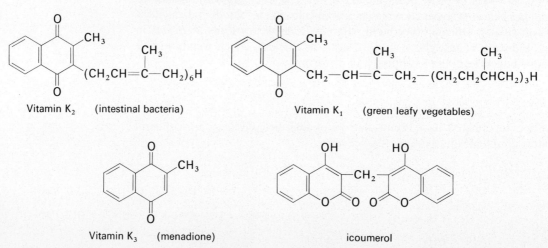

to prevent hemorrhagic disorders. Some patients are given vitamin K before surgery to ensure efficient blood clotting during and after an operation.

Dicoumerol is a naturally occurring anticoagulant that is structurally similar to vitamin K (Figure 27–7). In fact, dicoumerol and/or similar compounds are often used as anticoagulants to slow down clotting time in patients who have a high risk of thromboses (unwanted blood clots). These compounds are also used in rodent poisons because repeated high doses of them cause the animals to hemorrhage internally. These poisons are considered much safer than most because a single dose (as might be ingested by a child) is not fatal. Repeated doses over a period of time eventually result in death.

27.4 MINERALS

In Section 5.9 we noted that 26 elements are known to be essential to the proper functioning of living organisms and we classified these elements into three categories: macronutrients, micronutrients, and trace elements. Micronutrients and trace elements are also commonly called *minerals*. **Minerals** *are inorganic elements necessary (in small quantities) for normal growth, development, and health.* In Section 5.9, we briefly presented some information about the biological function of selected minerals. We will now consider in more detail the importance of minerals (both micronutrients and trace elements) in the human body.

Micronutrients

Calcium is the most abundant mineral in the body. Most calcium is found in the bones and teeth, but a small amount is also needed in biological fluids to function in processes such as blood clotting, muscle contraction, and activation of certain enzymes.

Dietary surveys have shown that one-third to one-half of the population in *developed countries* consume less than 50% of the RDA for calcium. Pregnant and lactating mothers, as well as their infants, are often lacking calcium, due to their increased needs for their mineral. The association of low calcium among older adults and the occurrence of *osteoporosis* (a bone disease characterized by demineralization of the bones) has led researchers to recommend supplemental calcium (1000–1200 mg/day) for this age group to help avoid the effects of this disease.

Phosphorus is a universal constituent of living cells and is always present in adequate amounts in foods. Phosphorus, like calcium, is an important mineral of bones and teeth. In addition, it is a structural component of nucleic acids and nucleotides, phosphoralated proteins, lipids, and sugars, and in the form of phosphate, it acts as an important buffer in the blood. One of the few deficiencies of phosphorous occurs following excessive use of antacids containing aluminum hydroxide or calcium carbonate, which form insoluble precipitates with phosphate and prevent its absorption.

Magnesium also occurs in all living cells. It is needed for optimum activities of many enzymes, especially those that involve adenosinetriphosphate (ATP) and other nucleotides. Low levels of magnesium impair kidney function, which can occur as a result of alcoholism or untreated diabetes. Some indications of hypomagnesemia (deficiency of magnesium) are insomnia, muscle cramps, or muscle weakness, while excess magnesium can upset the balances of the blood's electrolytes and interfere with normal muscular activity (such as the cardiac muscle). In plants, magnesium plays a vital role in photosynthesis, forming an important part of chlorophyll, the green pigment that absorbs sunlight in plants.

Sulfur is found in many proteins (chapters 24 and 25) in disulfide linkages, which stabilize protein structure. Other members of the micronutrients, potassium, sodium, and chloride, function as electrolytes in the blood to maintain osmotic balances and in membrane transport.

Trace Elements

Iron is one of the best known trace elements. About 70% of the iron in the body occurs in hemoglobin and myoglobin, where it functions in the transport of oxygen. Therefore, the daily supply of iron in the diet is one of the first considerations in diagnosing anemic conditions. Iron-poor blood has less oxygen-carrying capacity and hence results in characteristic symptoms of anemia. In addition to oxygen transport, iron acts as a cofactor for enzymes that contain heme groups such as cytochrome oxidase, catalase, and peroxidase (Figure 27–8). These enzymes participate in oxidation/reduction reactions, with the iron atoms acting as electron acceptors or donors as they alternate between ferrous (Fe^{2+}) and ferric (Fe^{3+}) states. Iron-sulfur enzymes are another important class of iron-containing enzymes that also function in electron-transfer reactions. These enzymes contain different types of active sites, most having equal numbers of iron and sulfur atoms; a typical structure is shown in Figure 27–9.

Copper is a cofactor for a number of enzymes. One example is cytochrome oxidase, an enzyme that requires both iron and copper ions to function properly. Another important enzyme is lysyl oxidase, an enzyme that catalyzes formation of crosslinks that strengthen collagen protein. Animals lacking copper in their diets suffer from weak arterial walls that tend to rupture easily. Very few cases of copper deficiency are seen in the population at large. However, an *excess* of copper in the tissues is associated with *Wilson's disease*, which can be treated with penicillamine, an agent promoting the excretion of copper.

Selenium is a cofactor for several enzymes, one of which functions to destroy toxic peroxides formed by some reactions in the cell. Too much selenium is toxic. In parts of Montana, the Dakotas, and other parts of the world where high con-

FIGURE 27–8 Iron acts as a cofactor for enzymes that contain heme groups.

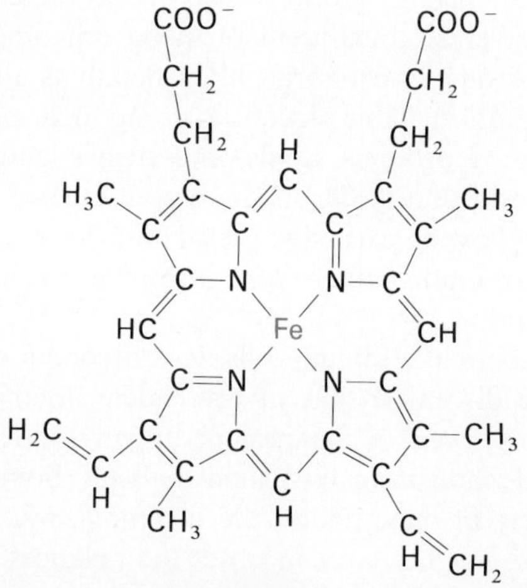

centrations of selenium salts in soil are taken up by range plants, cattle ingest the selenium while grazing and are poisoned.

Fluoride ions have proved effective in preventing tooth decay. This function is accomplished primarily by strengthening tooth enamel as fluoride ions replace some of the hydroxide ions in the main component of enamel, hydroxyapatite.

$Ca_5(PO_4)_3OH$ hydroxyapatite

$Ca_5(PO_4)_3F$ fluoride-strengthened hydroxyapatite

Fluoridated hydroxyapatite is even less soluble in saliva than normal hydroxyapatite and is therefore more resistant to decay. Periodic treatments with fluoride strengthen teeth, and it is therefore added to some municipal water supplies (in the form of NaF).

Zinc is an essential cofactor for many enzymes (over 100 discovered to date) such as carboxypeptidase, alcohol dehydrogenase, RNA and DNA polymerases, and carbonic anhydrase. Zinc also is required for the proper formation of taste buds and smell-sensing receptors of the nasal passages.

Iodine, an important element for the body, is used for the production of the hormone thyroxine. Each thyroxine molecule, whose structure is shown in Figure 27–16, p. 742, contains four iodine atoms and is one of the few molecules in the body that contains iodine. The best source of iodide ions is in iodized salt (which includes small amounts of KI or NaI). The clinical manifestation of iodine deficiency is *goiter*, characterized by an enlarged thyroid gland.

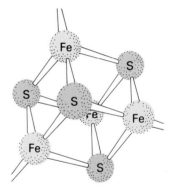

FIGURE 27–9 A typical iron-sulfur configuration found at the active site of some enzymes.

27.5 THE ENDOCRINE SYSTEM AND HORMONE FUNCTION

The **endocrine glands** *are a group of ductless organs that synthesize hormones and secrete them into the bloodstream.* The major endocrine glands are the pituitary gland, thyroid gland, parathyroid glands, adrenals, testes, ovaries, and the pancreas (see Figure 27–10, p. 736).

endocrine glands

Hormones *are chemical substances secreted by the endocrine glands and are transported to a "target tissue," where they evoke a particular biochemical or physiological response.* Some hormones have only one specific organ or target tissue, while others affect a wide variety of tissues. These extra-cellular chemical messengers received the name *hormone* from a similar Greek term that means to "excite or stimulate." Their dramatic effects on their target tissues are even more impressive when we consider the exceedingly small concentration in the blood (about 0.000001g/100 mL for the steroid hormones) that elicits a response.

hormones

Hormones regulate cellular processes by affecting the *rate of synthesis* of certain enzymes or proteins, the *rate of enzymatic catalysis*, or the *permeability of cell membranes* toward other substances. Generally, the observed physiological effects of hormone action are notably different in men and women. For the most part, men have a large bone structure, facial hair, and deep voice, while women have a high voice and small bone structure. These characteristics are a result of long-term effects by hormones on cellular processes such as those listed previously. In addition, hormones can elicit more immediate responses, such as the stimulation of the heart and muscle tissues during an emergency situation or the control of sugar metabolism.

Some hormones actually control the synthesis or release of other hormones. Because of this, any irregularity in the concentration of one hormone often affects

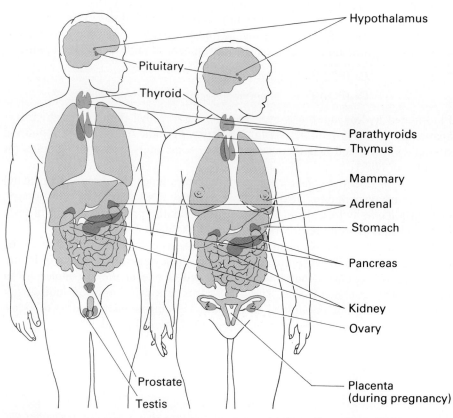

FIGURE 27-10 The major endocrine glands of the body.

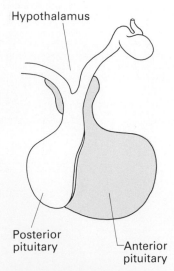

FIGURE 27-11 Diagram of the pituitary gland, showing the anterior and posterior lobes, as well as the hypothalamus.

the concentration of others. Needless to say, an intimate understanding of the overall endocrine system is necessary in treating any of its disorders or hormone imbalances.

The pituitary gland is the source of many hormones that stimulate other endocrine glands to produce or release their hormones. As such, for many years it was called the "master gland" of the endocrine system. Weighing only about 0.5 g, the pituitary is composed of two halves, the posterior and the anterior portions. Each of these sections has a slightly different function (Figure 27-11). Actually, the hypothalamus (which is controlled to a certain extent by the brain itself) is the master endocrine gland because it controls the pituitary gland. The hypothalamus secretes a number of small peptides, called *releasing factors*, that travel to the pituitary and direct the release of pituitary hormones. With the exception of oxytocin and vasopressin, which are synthesized by the hypothalamus and stored in the pituitary, the synthesis and release of each pituitary hormone is directed by its own individual releasing factor originating in the hypothalamus.

For example, when thyrotropin-releasing factor (TRF), is released by the hypothalamus, it stimulates the pituitary to release a thyroid-stimulating hormone (TSH). As the name implies, TSH travels through the bloodstream to the thyroid, where it stimulates this gland to synthesize and secrete thyroxine. Thyroxine eventually travels to a variety of target tissues, where it helps control the basic metabolic rate of the body (see Figure 27-12). Specialized receptor sites on the hypothalamus and pituitary sense the presence of sufficient thyroxine and respond

FIGURE 27-12 A schematic drawing of how the hypothalamus gland ultimately controls the release of hormones and factors that regulate the level of thyroxine in the body.

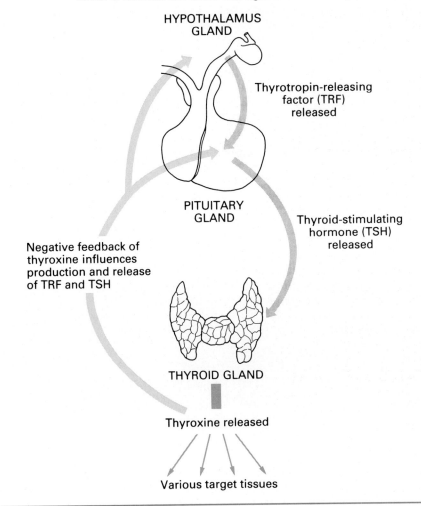

in a type of negative feedback-inhibition, so that fewer of the stimulating factors are released. Hence, we see in this relatively simple example the amazing complexities of chemical communication in the endocrine system. An outline of the main communication pathways for some of the endocrine system hormones and target tissues is given in Figure 27-13, p. 738.

thyrotropin-releasing factor (TRF)

FIGURE 27-13 The organization and sequence of hormone secretion controlled by the hypothalamic-pituitary system.

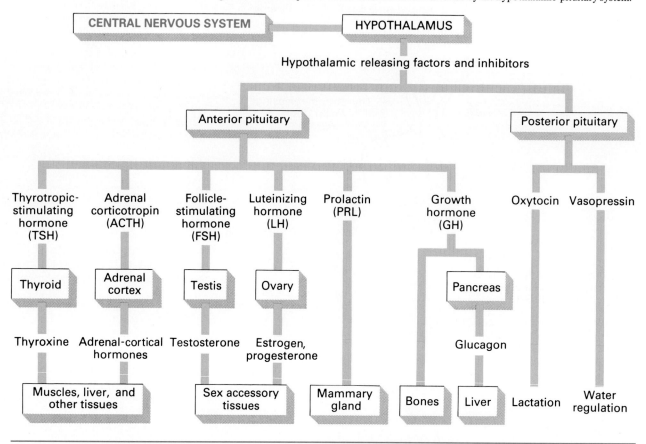

27.6 HORMONE STRUCTURE AND CLASSIFICATION

We encountered a number of hormones in past chapters, together with some of their functions. Most hormones can be classified into three general groups, according to their chemical structure: *amines*, which are often derivatives of amino acids, *polypeptides*, and *steroids*. Examples of each of these groups are shown in Figure 27–14.

Among the amines, three major hormones are *epinephrine*, *norepinephrine*, and *thyroxine*. Many polypeptide hormones range from 2–30 amino acids in length. However, some are much larger (such as insulin and its precursors) and are actually classified as proteins. Some polypeptide hormones contain sugar residues and are called *glycoprotein hormones*. The steroid hormones are derivatives of cholesterol and contain the characteristic four-ring system common to all steroids. Included among the steroids are the sex hormones: the female *estrogens* and the male *androgens*. A list of these and other important hormones is given in Table 27–2, p. 740.

Much as some enzymes are synthesized in an inactive form (e.g., the zymogens trypsinogen and chymotrypsinogen), some hormones are synthesized in an inactive

FIGURE 27-14 Examples of the three major classes of hormones.

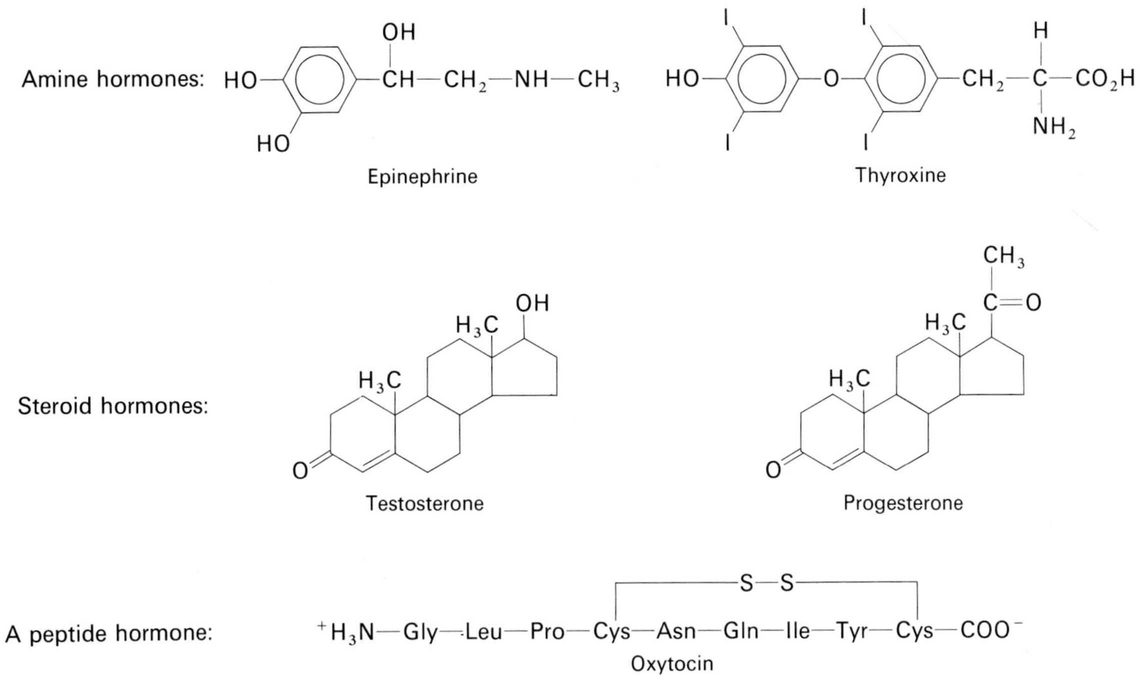

Amine hormones: Epinephrine, Thyroxine

Steroid hormones: Testosterone, Progesterone

A peptide hormone: ^+H_3N—Gly—Leu—Pro—Cys—Asn—Gln—Ile—Tyr—Cys—COO$^-$ (with S—S disulfide bridge between the two Cys residues)
Oxytocin

---

form. *Inactive precursors to active hormones are called* **prohormones**. Like zymogen enzymes, prohormones are inactive until activated by extracellular enzymes. The synthesis of insulin is an excellent example of this process.

The synthesis of insulin is a multi-step process. As you may recall, the active, mature form of insulin contains two polypeptide chains with a total of 51 amino acids. However, at the beginning of synthesis, insulin contains *102 amino acids* all in a *single polypeptide chain*. This precursor of insulin is called *preproinsulin*, a prohormone. When preproinsulin folds up into its characteristic three-dimensional structure, it forms the two disulfide linkages of the final insulin molecule. Some time after protein synthesis is complete, 16 amino acids are cleaved from the N-terminal end of preproinsulin, yielding proinsulin (86 residues in length). Proinsulin is subsequently broken into two chains, the 21-residue α-chain and a 32-residue β-chain, by enzymatic removal of a large interior portion of the polypeptide backbone that is 33 residues long. Upon removal of two arginines from the C-terminus of the β-chain, the active insulin molecule is completely formed (see Figure 27–15, p. 741).

Another sample of the prohormone conversion process is the conversion of the slightly active thyroxine (T_4) to the much more active 3,5,3'-triiodothyronine (T_3). This process is really quite simple; the liver enzyme, deiodinase, removes one of the four iodine atoms as shown in Figure 27–16, p. 742.

A number of polypeptide hormones are formed from a single prohormone parent, *pro-opiocortin*. This 29-kdal prohormone derives its name from the two immediate daughter peptide hormones it forms: *corticotropin* and the opiate hormone, *β-lipotropin*. Corticotropin is often referred to by its older, traditional

prohormones

TABLE 27-2 Major Hormones of the Human Body.

Chemical Nature	Gland and Tissue	Hormone	Effect
amino acid derivatives	adrenal gland medulla	epinephrine (adrenalin)	stimulates a variety of mechanisms to prepare the body for emergency action, including the conversion of glycogen to glucose
		norepinephrine (noradrenalin)	stimulates sympathetic nervous system, constricts blood vessels stimulates other glands
	thyroid	thyroxine	increases rate of cellular metabolism
peptides	hypothalamus	various releasing and inhibitory factors	triggers or inhibits release of pituitary hormones
proteins	pituitary, anterior lobe	human growth hormone (HGH)	controls the general body growth, and controls bone growth
		thyroid-stimulating hormone (TSH)	stimulates growth of the thyroid gland and production of thyroxin
		adrenal cortex-stimulating hormone (ACTH)	stimulates growth of the adrenal cortex and production of cortical hormones
		follicle-stimulating hormone (FSH)	stimulates growth of follicles in ovaries of females, sperm cells in testes of males
		luteinizing hormone (LH)	controls production and release of estrogens and progesterone from ovaries, testosterone from testes
		prolactin	maintains the production of estrogens and progesterone, stimulates the formation of milk
	pituitary, posterior lobe	vasopressin	stimulates contractions of smooth muscle, regulates water uptake by the kidneys
		oxytocin	stimulates contraction of the smooth muscle of the uterus, stimulates secretion of milk
	parathyroid	parathyroid	controls the metabolism of phosphorus and calcium
	pancreas, beta cells	insulin	increases cell usage of glucose, increases glycogen storage
	pancreas, alpha cells	glucagon	stimulates conversion of liver glycogen to glucose
steroids	adrenal gland, cortex	cortisol	stimulates conversion of proteins to carbohydrates
		aldosterone	regulates salt metabolism, stimulates kidneys to retain Na^+ and excrete K^+
	ovary, follicle	estradiol	stimulates female sex characteristics, regulates changes during menstrual cycle
	ovary, corpus luteum	progesterone	regulates menstrual cycle, maintains pregnancy
	testis	testosterone	stimulates and maintains male sex characteristics

name, *adreno*corticotropin *h*ormone, or "ACTH." Its target organ is the adrenal glands, where it stimulates glandular growth and the synthesis of a number of steroid hormones. The hormone β-lipotropin undergoes further enzymatic cleav-

FIGURE 27-15 The step-wise conversion of preproinsulin into active insulin.

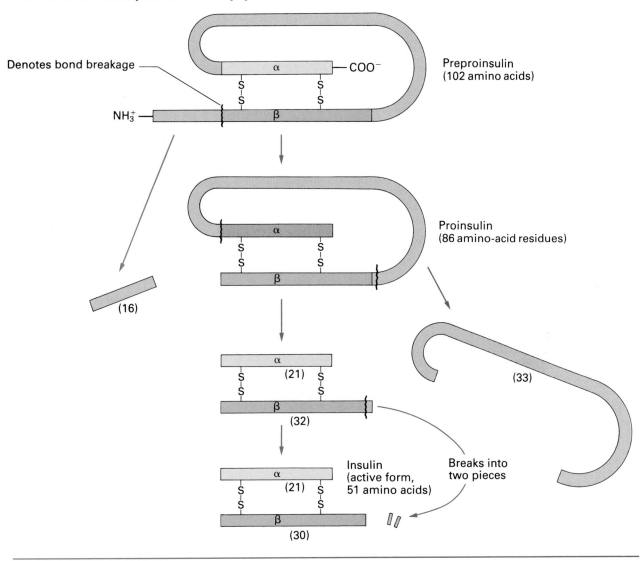

age to produce β-endorphin and γ-lipotropin. Both γ-lipotropin and corticotropin are further broken down by enzymatic reactions to two different melanocyte-stimulating hormones (MSH). This is illustrated in Figure 27-17, p. 742.

The discovery of endorphin hormones was an exciting event in biochemical research. Endorphins are potent pain relievers produced in the pituitary gland. These naturally occurring hormones bind to receptors throughout the body and in some (as yet not completely understood) fashion, interrupt the normal pain-sensing mechanism. Opiates such as morphine apparently bind to the same receptors and mimic the action of these hormones.

Naturally, if such compounds could be used to relieve pain without the undesirable side effects of addiction, they could revolutionize the treatment of pain. However, these polypeptides break down rapidly in the bloodstream due to proteolytic enzyme action. Unfortunately, like opiates, these natural pain relievers are also addictive. In fact, they have been blamed by some for the way in which joggers

FIGURE 27–16 Enzymatic conversion of the prohormone thyroxine (T_4) into 3,5,3′-triiodothyronine (T_3).

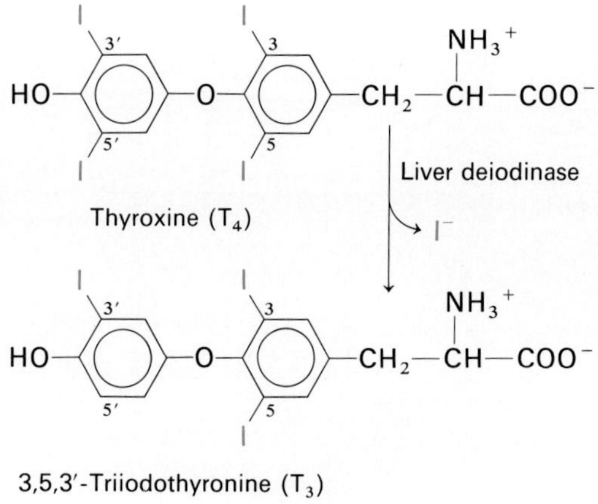

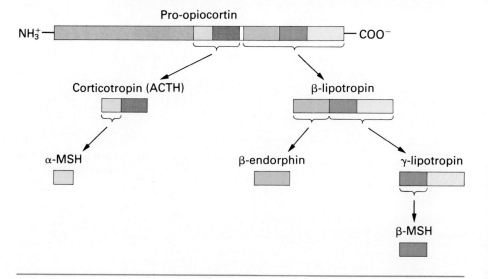

FIGURE 27–17 Enzymatic cleavage of pro-opiocortin produces a number of smaller active polypeptide hormones.

get "hooked" on running. It is hypothesized that the pain associated with consistent, intensive training results in higher levels of these natural pain relievers, which in turn cause addiction. Some cases have been reported in which avid runners actually suffer mild withdrawal symptoms when serious injuries prevent their daily workout or "fix."

27.7 MOLECULAR MECHANISMS OF HORMONE ACTION

Until the 1950s and 1960s, little was known about the molecular mechanisms associated with hormone control of cellular processes. Today, two general theories are used to explain hormone action. One theory involves binding the hormone to the outside of the cell at a specific receptor site; this causes events to occur inside the cell, without the hormone *ever actually entering the cell.* The second theory deals with the ability of some hormones to enter the cell and directly participate in the regulation of mRNA transcription in the nucleus. Let's examine each of these theories.

Earl Sutherland received the Nobel prize in 1971 for his *secondary messenger* hypothesis. His theory stated that, as certain hormones bind to the surface of their target cells, an enzyme in the membrane of the cell is positively modulated and begins to produce a secondary messenger on the inside surface of the membrane. Specifically, he showed that the membrane-bound enzyme *adenylate cyclase* is activated by the binding of certain hormones and begins to produce cyclic adenosine monophosphate (cAMP). As the concentration of cAMP increases inside the cell, it begins to affect many enzyme reaction rates (see Figure 27–18).

ATP

cyclic AMP

Epinephrine is often referred to as the "fight or flight" hormone because it is

FIGURE 27–18 The activation of membrane-bound adenylate cyclase and the subsequent synthesis of cAMP. The extracellular hormone binds to the receptor protein, activating adenylate cyclase to produce cAMP, which in turn acts as a secondary messenger inside the cell.

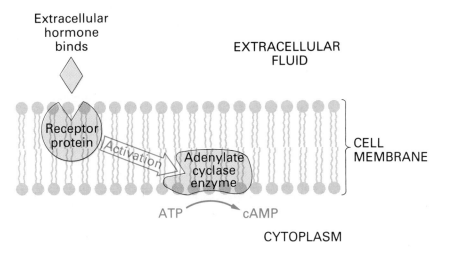

secreted by the adrenal gland during emergencies or times of stress and excitement. As it binds to receptor proteins at target tissues, it causes a dramatic and rapid increase in the turnover number of adenylate cyclase. The resulting higher concentration of cAMP in turn activates a series of enzymatic reactions that break down glycogen into free glucose monomers for immediate energy needs (a reaction pathway we will discuss in more detail in Chapter 29). As time goes on, and the emergency passes, normally occurring enzymes in the cytoplasm of the cell destroy the cAMP by eliminating its cyclic ester and converting it to non-cyclic AMP. This conversion removes the source of enzyme stimulation, and the breakdown of carbohydrate energy reserves (glycogen) slows down to normal rates.

cAMP → AMP

We should mention here that the enzyme reactions activated by cAMP are a *series of enzymatic reactions,* in which one enzyme activates another (mostly by phosphorylation of serine residues in the enzymes). Because of enzymes acting on each other, a *tremendous amplification of the hormone signal is achieved.* In each

FIGURE 27–19 Example of multiplier effect of biological amplification, in which one enzyme can activate many enzymes, each of which then activates others. After just a few such steps, a single molecule of cAMP can activate millions of enzymes in the cell.

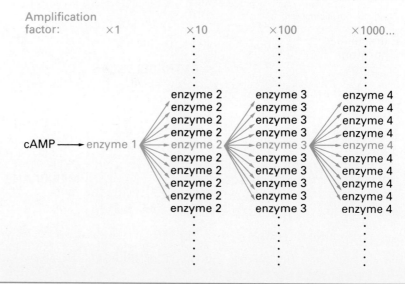

step of a series of enzyme reactions, each enzyme can activate many others. The activation process is like multiplying by tens (Figure 27–19). Just one hormone molecule binding to the outside of the cell can very quickly produce millions of active enzymes in just a few steps of the activation process. Consider the analogy of the spreading of a rumor. If each person who hears a rumor tells ten other people, it doesn't take too long to get lots of people involved—spreading the news at a terrific rate.

The active, 51-amino acid *insulin* hormone is one of the most interesting and most studied of hormones. Secreted by specialized cells in the pancreas, insulin is the most important regulator of energy stores in the body. It plays a role in controlling fat and protein metabolism, but its most observable effect is in the regulation of blood sugar. As the concentration of glucose in the blood rises, insulin is released into the bloodstream. Upon reaching target cells throughout the body, it stimulates them to both absorb sugars from the blood and convert free glucose molecules within the cell into glycogen. This conversion causes a rapid drop in blood sugar levels. Specific receptor proteins have been identified on the surface of target cells, and it is assumed that there is a secondary messenger inside the cell, whose production is stimulated by the binding of insulin to its receptor. However, *this secondary messenger for insulin has not yet been identified*, and finding it is the goal of many research groups.

Example 27.1

Why is cAMP definitely not the secondary messenger for insulin?

Solution
Insulin causes essentially the opposite effect, as does that of glucagon or epinephrin, upon binding to the exterior surface of the cell. If cAMP synthesis were stimulated by insulin binding, the cAMP would cause the same events to occur inside the cell as the other hormones. Since this is not the case, it is logical to assume that the secondary messenger for insulin is not cAMP.

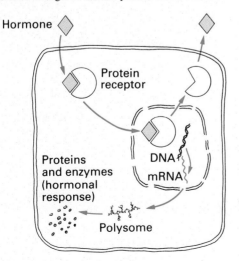

FIGURE 27–20 A proposed mechanism for hormone action whereby the hormone enters the cell and affects the transcription of mRNA. The effects of hormone action are eventually observed as changes in cellular protein concentrations.

A second model for hormone action is where the *hormone actually enters the cell and affects the rates of mRNA transcription.* For example, researchers believe that the active form of thyroxine, T_3, travels to the target tissue cells and is transported into the cell nucleus, where it stimulates the production of certain types of mRNA molecules. This production leads to higher levels of corresponding proteins and enzymes, stimulating the total metabolism of the cell and raising the basal metabolic rate.

Steroids are the most intensely studied hormones that function according to this second hypothesis. After entering the cell, steroid hormones such as estradiol bind to specific proteins in the cytoplasm. This hormone–protein complex then migrates to the nucleus, where it binds to DNA, affecting the subsequent transcription of mRNA. Eventually, as protein concentrations change because of altered levels of mRNA, the hormone response is observed (see Figure 27–20, p. 745). Thus, hormones that function by entering the cell and affecting transcription exhibit their effects more slowly than those that function like epinephrine, via a secondary messenger (cAMP).

27.8 SEX HORMONES AND THE MENSTRUAL CYCLE

Of all the hormones, sex hormones are usually the most interesting. Male sex hormones are collectively called *androgens* and are secreted by the testes. Androgens, the most important of which is testosterone, are responsible for the development of the male sex organs and sexual characteristics. About 6–10 mg of testosterone are produced daily by men, and smaller amounts (approximately 0.4 mg/day) are produced by women. A particularly striking dual effect of testosterone is stimulation of beard growth and the premature destruction of hair follicles of the head in genetically susceptible individuals. Because of this selective action of testosterone at its target tissues, a bald man can often grow a full, thick beard. Researchers don't know why or how this selectivity works but devote a great deal of research to this subject. If antagonists to androgen hormones are developed to prevent baldness, they must be extremely selective, acting only on the hair follicle cells of the scalp, so as to prevent thinning of the beard and loss of sexual drive.

Androgens also promote bone growth; the adolescent growth spurt in both males and females during puberty is believed to result from androgens. The predominantly greater height attained by men results in part from a higher concentration of androgen than occurs in women. Androgens also stimulate protein synthesis in the muscles (an *anabolic* effect). Many synthetic compounds have been developed in an effort to make "anabolic steroids" that will build muscles and strength, but not many have all of the other described effects. These efforts have been partially successful, and the use of anabolic hormones by athletes has become highly controversial. One such drug is norandrolone phenyl propionate, which exhibits five times as much anabolic activity as testosterone.

testosterone   norandrolone phenyl propionate

There are two important groups of female sex hormones: *estrogens* and *proges-*

tins. Estrogens are produced in the ovaries (as well as in the placenta during pregnancy) and contain an aromatic ring as part of their steroid ring system. They control female sexual functions such as the menstrual cycle, development of the breasts, and other sexual characteristics. Target tissues for estrogens include the uterus, mammary glands, and many other tissues throughout the body. Among other functions, estrogens are responsible for the overall higher fat content and smoother skin of females. Estradiol and estrone are two important estrogens.

estradiol   estrone

The most important progestin is progesterone, which prepares the uterus for pregnancy and prevents the further release of eggs from the ovaries during pregnancy. Progesterone is structurally very similar to testosterone, and is in fact synthesized from testosterone by a series of enzymatic reactions. A small amount of progesterone is even synthesized by males in their testes.

The female reproductive system is designed to secrete a fertilizable egg cell on a regular basis. The uterus must also be made ready to accept the fertilized egg cell (implantation) and must then be able to nourish the rapidly dividing cells during the early stages of pregnancy. A carefully balanced system of four essential hormones interacts in concerted fashion to regulate this cycle. The length of the menstrual cycle is variable, but averages 28 days. A graphical drawing in Figure 27–21 (p. 748) relates hormone levels, egg formation, and uterus behavior during the menstrual cycle.

At the beginning of the cycle, levels of both progesterone and estrogen are low. Estrogen synthesis increases as a result of release of *follicle-stimulating hormone* (*FSH*) from the pituitary. FSH stimulates growth of ovary follicles, which in turn produce estrogen. About midpoint in the cycle, secretion of another pituitary hormone, the *luteinizing hormone* (*LH*), causes the release of a mature egg from the ovary. After ovulation, progesterone secretion begins. *Progesterone is essential to maintenance of pregnancy*. If a fertilized egg is not implanted, production of this hormone decreases and menstruation occurs. However, if a fertilized egg is implanted, levels of progesterone stay high and prevent other eggs from being released.

When administered by injection, progesterone serves as an effective birth-control drug by establishing a false state of pregnancy and preventing the release of eggs from the ovary. The reason ovulation does not occur is because *progestins inhibit FSH and LH secretion from the pituitary by feedback inhibition*, thus preventing the normal stimulus for the ovaries to develop and release an egg. Because a woman does not ovulate, she cannot conceive.

The drawback of having to take injections to prevent pregnancy stimulated scientists to seek effective birth control drugs that could be taken orally. In the early 1950s, Carl Djerassi synthesized a compound very similar to progesterone that retained its birth-control properties when taken orally. This compound (sold under the trade name *Norlutin*) and others developed since then differ only slightly from progesterone (see Figure 27–22, p. 749).

Another synthetic compound that has estrogenic activity is *diethylstilbestrol* (*DES*). DES promotes rapid fat growth in livestock and was used for some years as a growth stimulant. Its application in this respect has been largely discontinued

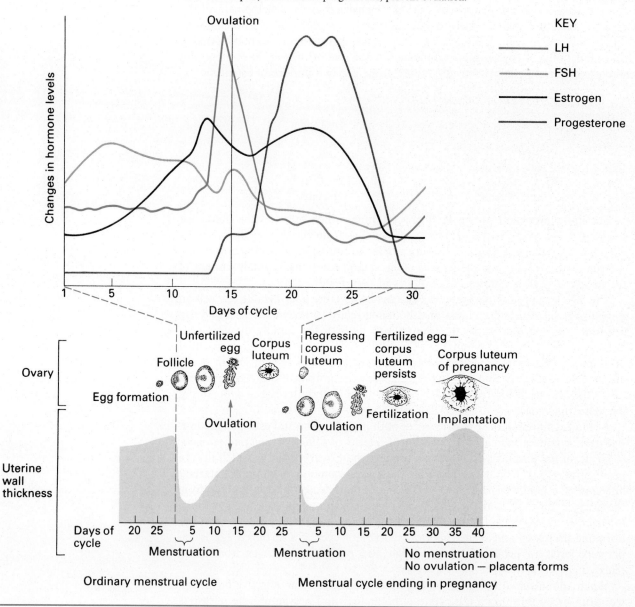

FIGURE 27–21 Changes in hormone levels, ovary, and uterus during the menstrual cycle. Pregnancy or synthetic birth control pills, which mimic progesterone, prevent ovulation.

because of demonstrated carcinogenic action in rats fed large amounts of the compound. Even though these potential dangers exist, another controversial use of DES is as a "morning-after pill" because it prevents the implantation of a fertilized egg in the uterine wall.

diethylstilbestrol (DES)

FIGURE 27-22 A few progesterone analogs that prevent ovulation and have been used as components of various contraceptive drugs.

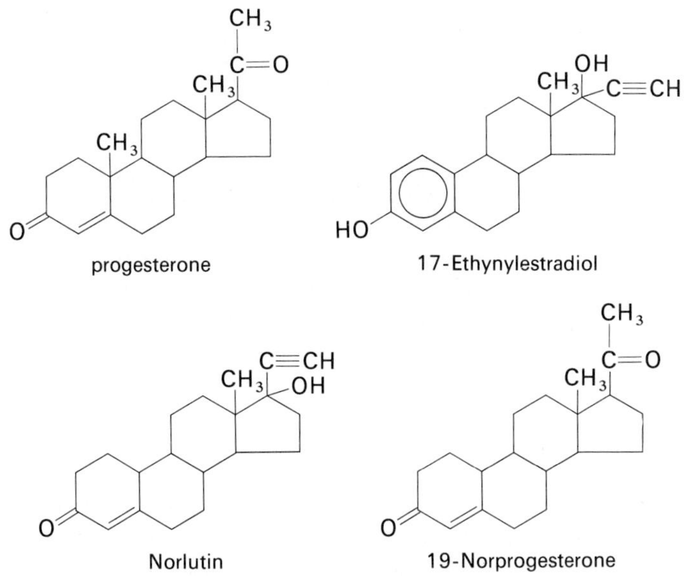

[Structures: progesterone, 17-Ethynylestradiol, Norlutin, 19-Norprogesterone]

The ultimate effects of prolonged tampering with the reproductive biochemistry of the human female remain to be seen. The "pill" has been used extensively by women in the United States since 1960. While many women experience no problems whatsoever, some of the undesirable side effects include hypertension, acne, and a tendency toward thromboses (undesirable blood clots that can lead to a heart attack or stroke). For this reason, women who have blood with an abnormal tendency to clot are normally advised not to take the "pill." The FDA requires warning labels on all drugs containing these estrogen- and progestin-like compounds, alerting consumers to possible harmful side effects.

Research efforts to develop a male contraceptive have been much less successful. One drug has been discovered that suppresses sperm production. This drug, testosterone enanthate (or Danazol), apparently has no side effects (except weight gain) if taken with supplemental testosterone. When it is discontinued, sperm are again produced. However, the high cost of Danazol prevents its current use as a popular and affordable contraceptive.

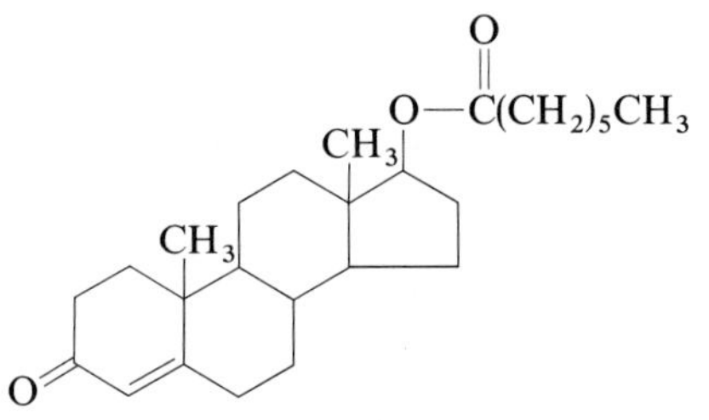

[Structure: testosterone enanthate (Danazol)]

EXERCISES AND PROBLEMS

Characteristics of Vitamins

27.1 Describe the origin of the term *vitamin*.

27.2 A label on a container of vitamins identifies the contents as "Vitamin B Complex." Explain what is meant by this complex vitamin.

27.3 Why were letters originally used to identify vitamins, instead of their actual chemical names?

27.4 Often the units "I.U." are used to describe a quantity of vitamins instead of micrograms or milligrams. What is meant by I.U.? Explain why I.U.s for some vitamins are different than for others.

27.5 What is the difference between the U.S. Recommended Daily Allowance for vitamins or minerals and the *minimum daily requirement*?

27.6 List some general advantages and disadvantages of taking excessive vitamins on a regular basis.

27.7 Many health-food advocates maintain that natural vitamins are much better for a person than synthetic vitamins. Is this true? Defend your answer.

27.8 Do *all* living organisms need *all* the vitamins discussed in this chapter in their diets? Explain.

Water-soluble Vitamins

27.9 After absorption into the body, thiamine is converted into an important enzyme cofactor. What is this cofactor and what purpose does it serve in general biosynthetic pathways?

27.10 Why is polished rice a poor source of thiamine? What disease often occurs as a result of thiamine deficiency among cultures heavily dependent on polished rice?

27.11 Riboflavin and niacin form cofactors that participate in what type of chemical reactions in the cell?

27.12 Which would contain more riboflavin: a mug of hot chocolate (made from whole milk) or a glass of cold milk that has been left on a picnic table for a couple of hours on a warm, sunny day?

27.13 Which vitamin contains a pteridine ring system?

27.14 Some vitamin bottle labels list vitamin B_6 under the names pyridoxol, pyridoxine, pyridoxamine, or pyridoxal. What happens to all of these different forms of vitamin B_6 after entering the system?

27.15 One vitamin was once identified by a variety of letters, including H, M, S, W, and X. What is the identity of this vitamin, which participates in carboxylation reactions in the cell?

27.16 What vitamin is structurally the most complex and contains an essential inorganic ion as part of its active form?

27.17 Which vitamin deficiency is believed to be the most common form of vitamin undernutrition? (Hint: This vitamin functions in the transfer of one-carbon groups in cellular reactions and is readily lost during cooking).

27.18 Which vitamin is required for the cellular production of coenzyme A, a very important carrier of acyl groups?

27.19 Vitamin C is required by the body in much higher amounts than any other vitamin. It acts as a reducing agent in redox reactions (especially important in the proper formation of collagen protein). Draw structures for both the oxidized and reduced forms of ascorbic acid in a chemical reaction that also shows the loss of two hydrogen atoms (with their associated electrons).

27.20 Describe how vitamin C got the name "ascorbic acid."

Fat-soluble Vitamins

27.21 Contrast the storage capabilities of fat-soluble versus water-soluble vitamins by the body.

27.22 What is hypervitaminosis? For which group of vitamins does this present potential problems (as compared to the other group)?

27.23 An old saying states that eating carrots helps a person see better. Is this true? Could this be said about other vegetables, too? Explain.

27.24 β-Carotene is added to some foods (such as margarine) as a food coloring. What vitamin might be provided to the body by ingesting such foods?

27.25 Sunshine destroys some vitamins. However, one vitamin is actually called the "sunshine vitamin." What vitamin is this? How is it formed in our bodies?

27.26 Vitamin D is often referred to as a *hormone*. Why? What biological process does it help control?

27.27 Fresh whole milk is an excellent source of vitamin D (as well as calcium). If a child does not drink milk, will he get rickets?

27.28 What are the symptoms of vitamin E deficiency in *humans*? How do these compare with the symptoms observed in laboratory animals?

27.29 Freezing is usually an excellent method of preserving foods. However, which vitamin usually does not survive frozen storage?

27.30 What is the source of vitamin K_1 as compared to vitamin K_2?

27.31 Why are some patients given extra vitamin K before major surgery?

27.32 Warfarin is structurally similar to dicoumerol. This poison is often included in rodent poisons. Describe how this toxic substance kills rodents and why it is considered "safer" than other types of poisons that could be used for pest control.

warfarin

27.33 Consider the U.S. RDA of the vitamins listed in Table 27–1. Rearrange these vitamins in descending order of the magnitude of their daily allowance.
a. Which category is generally higher on the list: water- or fat-soluble vitamins?
b. At which end of your list are International Units (I.U.s) used the most?
c. Toward which end of your list do you find the vitamins used as enzyme cofactors in *redox reactions*?

27.34 List the vitamins discussed in this chapter. Following each one, list the symptoms of their deficiencies and resulting known diseases (if any).

27.35 Prepare a simple table similar to the following example. In the left-hand column, list each vitamin; across the top, list some of the foods that are good sources of these vitamins. In the columns below the foods, mark an X for the vitamins they supply. Identify some of the best foods that supply many vitamins. Does any one food supply all needed vitamins?

Vitamin	Cereals	Dairy Products	Meats	Citrus ...
Vitamin C	—	—	—	X
⋮				

Minerals

27.36 What mineral deficiency appears to be a factor linked to the bone disease osteoporosis?

27.37 Name two chemicals that inhibit the absorption of phosphorus and describe why this happens.

27.38 What mineral is an important part of the chlorophyll pigment in plants?

27.39 Adenylate cyclase (the enzyme that converts ATP into cAMP in response to stimulation by various hormones) requires a certain mineral cofactor, just like most enzymes that act on ATP. What is this cofactor?

27.40 Wilson's disease is characterized by high levels of what mineral in body tissues?

27.41 Explain why selenium is both beneficial and toxic to mammals.

27.42 Explain the molecular reasons why fluoride helps prevent tooth decay.

27.43 What element is necessary for proper development of the taste buds?

27.44 In the event of a nuclear accident, many radioactive isotopes of iodine are released, such as ^{131}I (half life 8 days) and ^{125}I (half life 60 days). It is highly recommended that anyone exposed to these isotopes take large doses of non-radioactive KI to prevent damage to their thyroid gland. Why would this help? For how long would this therapy be necessary?

Endocrine System and Hormone Function

27.45 Two hormones, glucagon and epinephrine, have very different structures and different target tissues, yet both cause similar responses in their target tissues (glucagon in the liver and epinephrine in the muscles). What advantage might there be in having *two* hormones that stimulate the same process?

epinephrine

$^+$H$_3$N-His-Ser-Glu-Gly-Thr-Phe-Thr-Ser-Asp-Tyr-

-Ser-Lys-Tyr-Leu-Asp-Ser-Arg-Arg-Ala-Gln-

-Asp-Phe-Val-Gln-Trp-Leu-Met-Asn-Thr-COO$^-$

glucagon

27.46 Describe the roles of the hypothalamus and pituitary glands in the overall scheme of hormone production and secretion.

27.47 Explain how the increased levels of thyroxine in the bloodstream are sensed by the endocrine system and how it responds to these higher levels.

Hormone Structure and Classification

27.48 Identify three structural classes of hormones and give an example of each one.

27.49 Thyroxine is classified as an amine and is a derivative of an amino acid. Draw the structure of thyroxine and circle the portion that resembles the amino acid tyrosine.

27.50 Why are prostaglandins (Section 23.9) often called hormones? What is different about them as compared with hormones such as steroids or epinephrine (according to a classical definition)?

27.51 List three examples of prohormones and the active hormones that are eventually derived from them.

27.52 Early experiments with active insulin molecules in which the hormone was gently denatured and then allowed to renature or refold yielded confusing results. It was theorized that the hormone would again assume its active configuration upon renaturation. However, unlike many other proteins and enzymes, this was not the case. Propose an explanation as to why active insulin molecules will not refold into an active configuration after they are once denatured.

27.53 In Chapter 26, we described how active insulin hormones could be made by genetically altered bacteria. Using your knowledge of how insulin is produced by mammalian cells, what is the *minimum* length (in base pairs) of a gene that could be used to transform bacteria into insulin-producing bacteria?

27.54 What is the common parent peptide of the hormones β-lipotropin, β-endorphin, MSH, and ACTH? Draw a simple schematic diagram showing how all these hormones arise from a single parent molecule.

Molecular Mechanisms of Hormone Action

27.55 Explain how the enzyme adenylate cyclase is involved in Sutherland's model of hormone action involving a "secondary messenger."

27.56 Draw the chemical structure of cyclic AMP, and show how it differs from the structure of non-cyclic AMP.

27.57 Once cAMP is formed in the cell as a result of hormone action, why is the effect not permanent? Would such a permanent change be a good thing? Explain your reasoning.

27.58 What is meant by the term *biological amplification* of a hormone signal?

27.59 The secondary messenger for epinephrine is cAMP. Insulin causes essentially the opposite effect in many cells of the body. What is the secondary messenger for insulin?

27.60 List the events that must occur in a cell before a steroid response is observed.

Sex Hormones and the Menstrual Cycle

27.61 What is the main difference between the physiological roles of androgens and estrogens?

27.62 Some female athletes have been known to take anabolic steroids in an effort to build up their skeletal muscles. List a few possible side effects that may accompany the use of these testosterone-like hormones.

27.63 Give one reason why certain genetically susceptible bald men often have thick, full beards.

27.64 List four hormones that are intimately involved with the menstrual cycle. Also indicate their sites of production and what effect each has on the cycle.

27.65 What female hormone prevents the ovaries from producing and releasing mature, fertile eggs? Draw the structure of this hormone and two analogs used as active ingredients in birth-control pills.

27.66 A muscle-builder in your physics lecture approaches you after class one day to ask your advice (since he knows you are taking this chemistry course). His best friend has evidently told him that birth-control pills are the cheapest source of anabolic steroids, which he wants to start taking to help out in his weight-lifting class. He says this seems logical to him, but wants your opinion. Should he start taking the "pill?"

27.67 Why is DES (diethylstilbestrol) very possibly a dangerous drug for women who use it as a "morning-after pill?"

27.68 Why are male oral contraceptive drugs currently used less frequently than female ones?

28 Metabolism and Nutrition: An Overview

Objectives

After completing Chapter 28, you will be able to:

28.1 State the basic concepts of thermodynamics and recognize how they apply to chemical reactions in living cells.
28.2 Identify the free energy content of carbon atoms in different oxidation states and understand how nonspontaneous redox reactions can be driven by coupling them to spontaneous ones.
28.3 Explain the role of ATP as a storehouse of free energy in cells.
28.4 Recognize NADH, NADPH, and $FADH_2$ as universal electron carriers and understand their respective roles in redox reactions.
28.5 Compare the energy content of carbohydrates, lipids, and proteins and understand the nutritional considerations of each.
28.6 Understand how major components of foods are oxidized via three distinct phases of metabolism.
28.7 List a few of the important mechanisms that regulate metabolism.
28.8 Describe the role of transmembrane potentials in energy storage.
28.9 Outline the overall process of photosynthesis, including its role as vital link between solar energy and all living organisms.

INTRODUCTION

Human metabolism is quite remarkable. An average human adult who maintains constant weight for 40 years will process about 6 *tons* of solid food and 10,000 gallons of water, during which time the composition of the body and its weight remain essentially constant. Just as we must put gasoline in a car to make it go or plug in a kitchen appliance to make it run, we also need a source of energy to think, breathe, exercise, or work. As we have seen in previous chapters, even the simplest living cell is continually carrying on energy-demanding processes such as protein synthesis, DNA replication, RNA transcription, or membrane transport.

In this chapter, we will examine the laws that govern the extraction of energy from the foods we eat and we will discuss the overall energy-transducing strategy of metabolism. In addition, we will look at the ultimate source of all our energy—the sun—and how this energy reaches us.

28.1 CHEMICAL ENERGY

Until now, we have concentrated on molecular transformations without much emphasis on the associated energy changes that accompany these reactions. **Thermodynamics** *is the study of energy and its conversion from one form to another.* The heat released during the combustion of hydrocarbons is an easily observable energy change. Some of this released energy can be captured by machines and used to do work, such as moving a car or generating electricity. Other organic substances, such as carbohydrates and lipids, also contain stored energy that can be released through various reactions that eventually lead to the formation of carbon dioxide and water. Our cells, much like efficient machines, can harness much of this energy to do work.

A scientist named J. Willard Gibbs derived an expression (in 1878) that accurately describes the energy available from chemical reactions at constant temperature and pressure and is especially useful in describing biochemical reactions. We define Gibbs **free energy** *as the amount of energy that is actually free or available to use as a result of a chemical reaction; it is represented by the symbol* ΔG.

The molecules involved in a chemical reaction are called the *system*, and thus, ΔG is a measure of the energy content of the system. For any such molecular system, ΔG is negative when energy is released during a reaction. If energy must be put into a system to make a reaction occur, then the sign for ΔG is positive. *Reactions that liberate free energy (negative ΔG) are called* **exergonic** *reactions. Chemical reactions that require a net input of energy (positive ΔG) are called* **endergonic.** All living organisms obtain energy from exergonic reactions that release free energy from the foods they consume. Green plants also use energy from the sun to drive a series of endergonic reactions called *photosynthesis* (Section 28.9).

This concept is seen more easily from Figure 28–1, p. 756, a graphical representation of a reaction, in which we plot Gibbs free energy against the progress of a reaction.

Reactions such as those shown in Figure 28–1 are said to be spontaneous. **Spontaneous reactions** *have a negative value for ΔG and occur naturally.* Con-

FIGURE 28–1 A graph of ΔG during a reaction. Since the internal energy content of the products is lower than the reactants, energy is released during the reaction and ΔG is therefore negative.

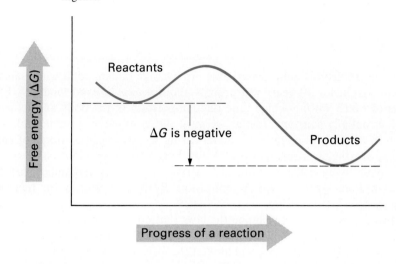

nonspontaneous reactions

versely, **nonspontaneous reactions** *have a positive ΔG value and do not occur of their own accord.* We should mention here that spontaneous does not necessarily mean immediately. Some spontaneous reactions occur very slowly.

Example 28.1

Draw a graph of a reaction like Figure 28–1, where ΔG is positive. Would this reaction be spontaneous? Would any energy be liberated that could be used for work?

Solution

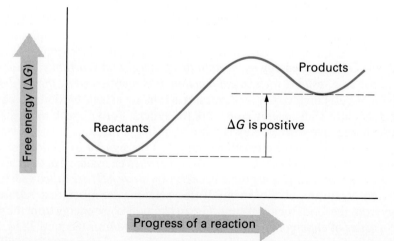

We can see from the graph that ΔG for this reaction is positive; therefore, this reaction will neither be spontaneous nor will it liberate energy. In fact, an input of external energy will be required to make the reaction occur at all.

The actual numerical value of Gibbs free energy is subject to change, according to the conditions of the reaction. For example, temperature, pH, reactant and product concentrations, and polarity of the solvent all affect ΔG. To help identify a standard value for ΔG, the symbol $\Delta G°$ (*standard free energy*) was adopted to indicate that this numerical value was measured under a standard set of conditions (such as all reactants and products at concentrations of 1.0 M). Although these conditions are not always present inside living cells, it makes an excellent comparison point for the $\Delta G°$ of many reactions, and we will use it to do so for many biochemical reactions. The units of energy associated with $\Delta G°$ are kilocalories (1 kcal = 1000 calories) or joules (1 cal = 4.184 joules). For this text, we will use the kilocalorie because of its widespread use in biochemistry and biology.

For any given reaction, $\Delta G°$ is always the same, whether the end product is obtained in one or several steps. Therefore, we can calculate the overall $\Delta G°$ for a reaction from the $\Delta G°$s of a series of other reactions that eventually lead to the overall product. Let us consider the hypothetical example A converting to C. In one case, A is converted directly into C. In another, A is first converted to B and then to C. We can measure $\Delta G°$ of this reaction as it occurs in a single step (A to C), or we can calculate $\Delta G°$ from a series of two reactions that lead to the same product.

Case 1 (Single Step)		*Case 2 (Multiple Steps)*	
Reaction	$\Delta G°$	Reaction	$\Delta G°$
A $\longrightarrow$ C	-3 kcal/mol	A $\longrightarrow$ B	$+2$ kcal/mol
		B $\longrightarrow$ C	-5 kcal/mol
		Sum: A $\longrightarrow$ C	-3 kcal/mol

Note that the same overall result for $\Delta G°$ is obtained from a stepwise approach or from a single-step reaction (see Figure 28–2). Many chemical reactions in

FIGURE 28–2 Graphic representation of the change in free energy during a single-step reaction as compared to a two-step reaction. The change in free energy from A to C is the same, regardless of the path taken.

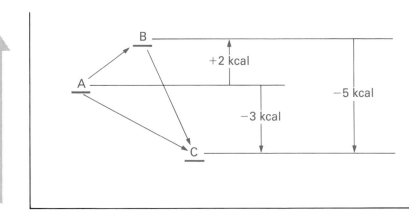

coupled reactions

biological systems are *coupled* to one another as in Case 2. **Coupled reactions** are *two or more reactions linked in such a way that one reaction cannot occur without the other*. Even though the first reaction (A → B) is endergonic and normally nonspontaneous, the second reaction (B → C) is quite exergonic. Therefore, a coupled overall reaction sequence of A → B → C *is spontaneous*, with an overall free energy change of −3 kcal/mol.

This concept is a powerful tool in analyzing biochemical reactions involved in metabolism. In living systems, energy-yielding foods such as fats and carbohydrates undergo a series of *many reactions* that eventually yield combustion products (CO_2 and H_2O). The overall release of energy is the same as in a one-step combustion process, but inside the cell, energy from each reaction step is harnessed and used in a most efficient way. Because of this, living systems use chemical energy *much more efficiently than mechanical machines.*

Machines such as steam engines or internal combustion engines do work by taking advantage of a heat-induced temperature change in which the expansion of hot gases pushes pistons and eventually turns wheels or gears. However, this is not practical for cells, since they do not tolerate significant temperature changes. Cells trap energy released by oxidation reactions in some chemical form before it is lost as heat. In the following sections, we will discuss just how this occurs and what chemicals are useful in trapping and storing this energy.

28.2 OXIDATION AND REDUCTION REACTIONS IN CELLS

As we discussed in Chapters 11 and 15, reactions involving the transfer of electrons are termed *redox reactions* (reduction/oxidation reactions). The loss of electrons is oxidation, while the opposite process, gaining electrons, is reduction. Electron transfer is quite easily observed with metal ions, as in the oxidation or reduction of iron.

$$\underset{\text{ferrous ion}}{Fe^{2+}} \underset{\text{reduction}}{\overset{\text{oxidation}}{\rightleftarrows}} \underset{\text{ferric ion}}{Fe^{3+}} + e^-$$

Redox reactions involving covalent compounds are not as obvious; to recognize oxidized or reduced atoms in these reactions, we must consider the difference in electronegativities between atoms involved in bonds broken or formed during a reaction. The partial oxidation of carbon in the conversion of methane to methanol is one example.

$$CH_4 + \tfrac{1}{2} O_2 \xrightarrow{\text{oxidation}} CH_3OH + \text{energy}$$

Because the oxygen atom in methanol is more electronegative than the corresponding hydrogen atom in methane, electrons are drawn more toward the oxygen atom in the product. Because of this chemical reaction, the carbon atom in methanol is more oxidized than it is in methane.

Oxidation of organic compounds in living cells is often accompanied by the loss of hydrogen atoms. Sometimes, hydrogen atoms act as electron carriers (as H· or as hydride ions, $H:^-$). In other reactions, electrons are lost to specialized carriers, and protons (H^+ ions) are lost separately. For example, the oxidation of quinone can occur in two different ways, both with the same overall result: loss of two electrons and two hydrogen atoms.

Hydrogen atoms carry away electrons:

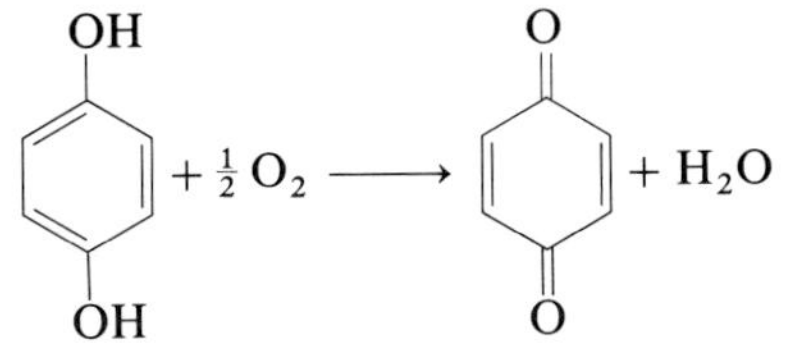

[structure: hydroquinone] + ½ O_2 ⟶ [structure: quinone] + H_2O

or

Electrons and H^+ ions lost independently (in two steps):

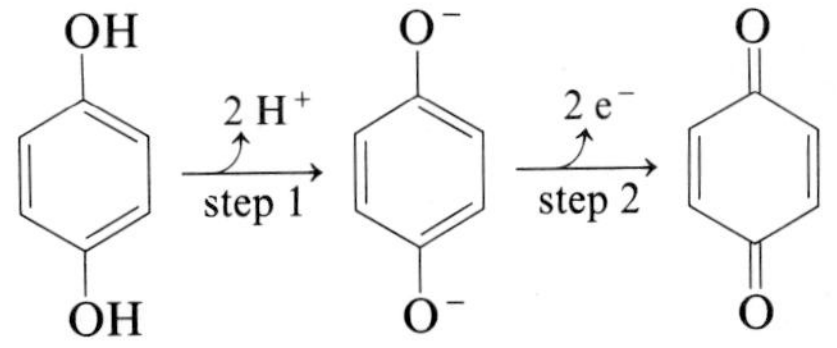

[hydroquinone] $\xrightarrow{2 H^+}_{\text{step 1}}$ [dianion] $\xrightarrow{2 e^-}_{\text{step 2}}$ [quinone]

Since oxidation states of carbon play such an important role in the storage of energy, we should review the degree of oxidation among various functional groups. Figure 28–3 shows the oxidation states of many of the organic functional groups we have studied. Note how more oxygens and fewer hydrogens result in highly oxidized carbon atoms, while more reduced carbons have fewer oxygen atoms and/or more hydrogens.

Almost all oxidation reactions yield free energy. Energy is produced because the chemical bonds in the more reduced reactants contain more energy than those in the oxidized products (Figure 28–4, p. 760). If an *oxidation* reaction were to run in the reverse direction, the result would be *reduction* and the value for $\Delta G°$ would be equal in magnitude, but of *opposite sign*. This is logical; if we trace the reverse reaction on the reaction graph in Figure 28–4, the same energy released in the forward reaction must be added back to reform the initial reactants, making $\Delta G°$ positive in this case. Since we have stated that $\Delta G°$ must be negative for a spontaneous reaction, the reverse reaction would require external energy and would not proceed spontaneously.

In the cells of our body, we can either oxidize sugars (a spontaneous process) to obtain energy, or we can synthesize sugars via a reductive process to store energy. In fact, many processes in the body are reductive ones, with positive $\Delta G°$ values. How is this possible? The solution to this paradox is relatively simple: Reactions can be coupled together, such that the free energy released from a spon-

FIGURE 28–3 Comparison of oxidation/reduction states of various functional groups, showing their relationship to potential (stored) energy.

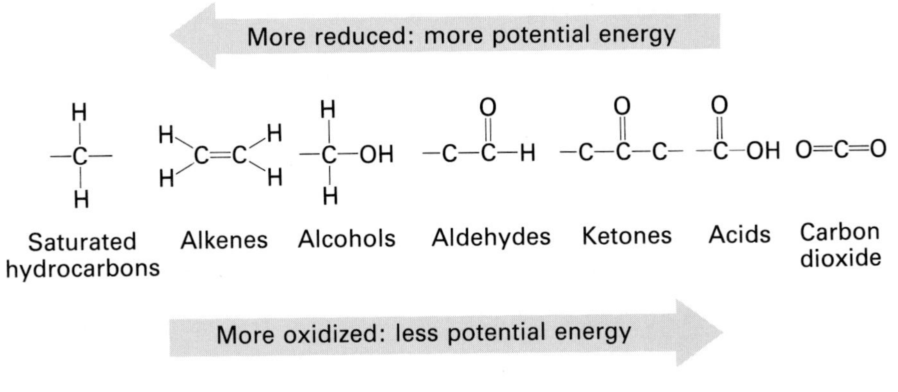

⬅ More reduced: more potential energy

| Saturated hydrocarbons | Alkenes | Alcohols | Aldehydes | Ketones | Acids | Carbon dioxide |

More oxidized: less potential energy ➡

FIGURE 28-4 A graph of ΔG during a reaction, showing the energies of bonds being broken in the reactants and new bonds forming in the products.

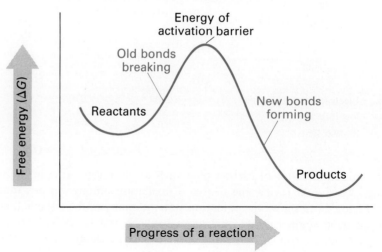

taneous reaction can be used to drive a nonspontaneous one. Thus, a thermodynamically unfavorable reaction can be driven by a thermodynamically favorable one.

In the next sections we will examine just how this coupling takes place in living cells and how both energy and electrons move from one molecule to another.

28.3 ATP: UNIVERSAL ENERGY CURRENCY OF THE CELL

In all living systems, energy obtained from the oxidation of foods is coupled to the formation of a single compound, *adenosine triphosphate* (*ATP*). This compound was introduced during our discussion of nucleic acids, in which ATP is one of four important nucleotides used in the synthesis of DNA. ATP is a high-energy compound, meaning that there are certain bonds within the compound that possess larger-than-normal amounts of potential energy. In this context, **potential energy** *refers to free energy that is "potentially" available to drive another reaction*. The unique energy-storage capacity of ATP is caused by two anhydride linkages between its phosphate groups. Formation of these two bonds requires a great deal of free energy, which in turn is released during hydrolysis.

The hydrolysis of ATP to ADP releases 7.3 kcal of free energy per mole.

$$\text{ATP} + \text{H}_2\text{O} \longrightarrow \text{ADP} + \text{P}_i \quad \Delta G° = -7.3 \text{ kcal}$$

(where P_i stands for inorganic phosphate, PO_4^{3-}). A second high-energy linkage liberates a similar amount of energy when ADP is hydrolyzed to AMP. Some enzymes, such as DNA polymerase, utilize the energy from both anhydride bonds by converting ATP directly to AMP. This releases pyrophosphate (PP_i), which in turn is rapidly hydrolyzed into two phosphates, releasing the energy stored in the pyrophosphate anhydride bond.

$$\text{ATP} \xrightarrow{\text{H}_2\text{O}} \text{AMP} + \underset{\substack{\text{pyrophosphate} \\ (PP_i)}}{{}^-\text{O}-\overset{\overset{\displaystyle O}{\|}}{\underset{\underset{\displaystyle O_-}{|}}{P}}-\text{O}-\overset{\overset{\displaystyle O}{\|}}{\underset{\underset{\displaystyle O_-}{|}}{P}}-\text{O}^-} \xrightarrow{\text{H}_2\text{O}} \underset{\substack{\text{phosphate} \\ (P_i)}}{2\ \text{HO}-\overset{\overset{\displaystyle O}{\|}}{\underset{\underset{\displaystyle O_-}{|}}{P}}-\text{O}^-}$$

One of the most beneficial aspects of energy storage in ATP is its stability in aqueous solution; hydrolysis of its anhydride linkages does not occur rapidly. Enzymes link the hydrolysis of ATP to reactions where energy is needed, avoiding the loss of stored energy by uncatalyzed hydrolysis. An obvious second benefit is that ADP and AMP molecules can be converted back into ATP, if coupled to oxidation reactions that supply at least 7.3 kcal of free energy.

Although we speak of "free" energy, we must remember that, like money, we must "earn" it before we spend it. We must work to earn money, which we can in turn trade or spend to acquire the things we need. When we get a little ahead, we can save it in the bank. For large investments, we can even borrow money. So it is with energy requirements in the cell, only instead of money, high-energy compounds such as ATP serve as our "energy currency." Only after "working" to obtain and store energy from oxidation of foods can we "spend" molecules such as ATP to power nonspontaneous reactions.

What is so special about an anhydride linkage between phosphates that allows it to store such large amounts of energy? In cellular fluids (around pH of 7), the three acidic phosphate groups lose four of their protons, generating four negative charges in close proximity to one another. Electrostatic repulsions of these negative charges strain the bonds between phosphates. Secondly, oxygen draws electron density away from the phosphorus atom, further straining the bonds. These strained bonds require a great deal of energy to form and, when broken, respond by releasing it back to the surroundings.

Phosphates from ATP are often used as a source of phosphate in reactions that phosphorylate other molecules such as proteins, sugars, and lipids.

$$\text{protein} + \text{ATP} \longrightarrow \text{protein-P} + \text{ADP}$$

Because both the phosphate and the energy to drive this type of reaction are supplied by ATP, we often refer to the phosphate in ATP as having a *high phosphate transfer potential*. Compounds with low phosphate transfer potentials cannot act as a source of phosphate in these reactions because insufficient energy is released to drive the reaction. Other nucleotides with high phosphate group transfer potentials include GTP, CTP, TTP, and UTP (Chapter 26). All of these are capable

FIGURE 28-5 High energy compounds that function in a similar manner to ATP in biological systems. Their $\Delta G°$ values for hydrolysis are listed for comparison.

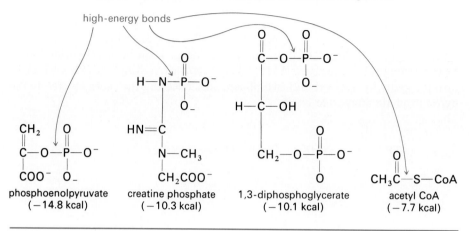

of transferring phosphates in phosphorylation reactions; however, ATP is used most often.

In addition to these nucleotides, other organic chemicals in cells carry phosphate groups with high transfer potentials. Much like ATP, these compounds also contain high-energy phosphate bonds that release their stored energy upon hydrolysis. Some of these compounds release even more energy than ATP, and are listed in Figure 28-5.

28.4 ELECTRON CARRIERS

Energy is not all that is transferred during a redox reaction. Electrons are also exchanged. In all redox reactions, one substance is oxidized, while another is concomitantly reduced. Although a wide variety of compounds are oxidized by cells, only a handful of specialized compounds are reduced. These specialized compounds are collectively called *electron carriers*. **Electron carriers** *accept electrons in redox reactions and "carry" them to sites where they can be used to reduce other substances.* Electron carriers are used for electron storage, much like ATP is used for energy storage. And, just as ATP possesses a high phosphate group transfer potential, *electron carriers have a high electron transfer potential.* This potential to drive the movement of electrons in chemical reactions is similar to the potential (voltage) needed to move electrons through a wire, and both are measured in volts. Higher (more positive) voltages indicate a greater reduction potential.

In aerobic organisms (those that require O_2), electrons eventually find their way to oxygen, the ultimate electron acceptor.

reduced fuel molecules $\xrightarrow{\text{oxidation}}$ electrons + oxidized products

$\downarrow$ electron carriers

$O_2 + 4\,H^+ + 4\,e^- \longrightarrow 2\,H_2O$

However, unlike combustion of organic compounds, electrons are not transferred from fuel molecules *directly* to O_2, but move through a series of electron carriers. For the most part, these specialized electron carriers are flavins or pyridine nucleotides that act as cofactors for enzymes involved in catalyzing redox reactions (Section 27.2). These cofactors transport electrons to different sites where electrons are needed for reduction of other compounds. Through a series of enzymatic reactions and electron carriers in the mitochondria called the *electron transport chain*, electrons are eventually transported to oxygen, resulting in its reduction and the subsequent formation of water.

Nicotinamide adenine dinucleotide (NAD^+) is a major electron acceptor in the oxidation of fuel molecules. The reactive part of the NAD^+ that actually carries electrons is its nicotinamide ring (derived from the vitamin niacin). Nicotinamide adenine dinucleotide phosphate ($NADP^+$) is very similar to NAD^+, differing only in a phosphate attached to the 2'-carbon in the ribose ring (see Figure 28–6). Both NAD^+ and $NADP^+$ accept one hydrogen ion and two electrons (equivalent to a hydride ion, $H:^-$) during their reduction. In their reduced states, these carriers are abbreviated NADH and NADPH, to distinguish them from their oxidized states.

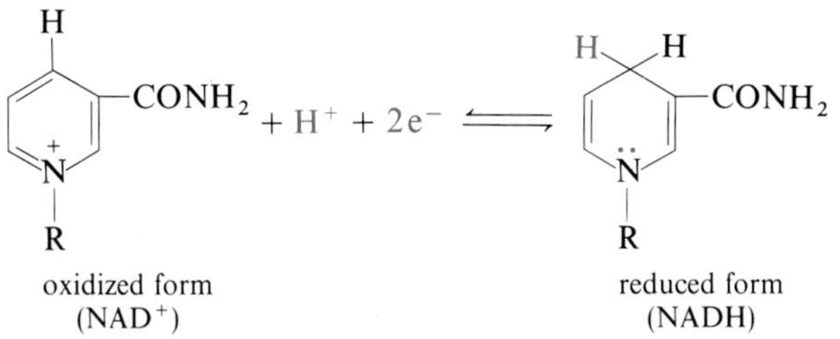

FIGURE 28–6 Structure of nicotinamide adenine dinucleotide (NAD^+) and the structurally similar nicotinamide adenine dinucleotide phosphate ($NADP^+$), which differs only in the attachment of a phosphate group at the 2'-position of the ribose ring.

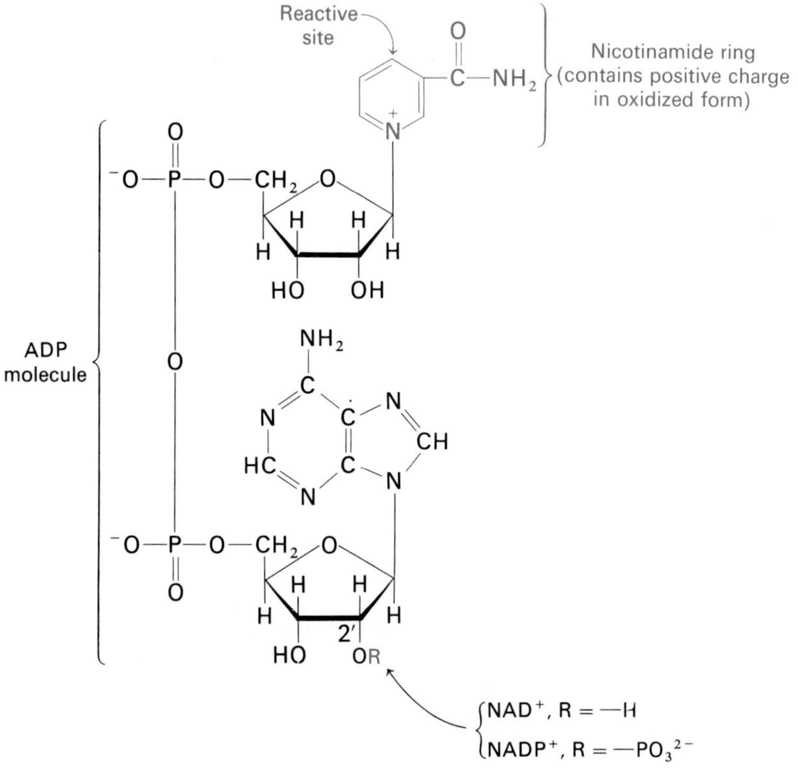

NAD$^+$ is an electron acceptor for most alcohol oxidations that yield ketones or aldehydes.

$$\underset{\underset{H}{|}}{\overset{\overset{OH}{|}}{R-C-R'}} + NAD^+ \rightleftharpoons R-\overset{\overset{O}{\|}}{C}-R' + NADH + H^+$$

Two hydrogen atoms and two electrons are lost in dehydrogenation (oxidation) of alcohols. Both electrons and one hydrogen are accepted by NAD$^+$ to form NADH, while the other hydrogen atom is lost to the surrounding solvent in the form of a proton (H$^+$).

The other major electron carrier in the oxidation of fuel molecules is *flavin adenine dinucleotide (FAD)*. (The complete structure of FAD is shown in Figure 27–1, p. 724.) Unlike NAD$^+$, FAD has no positive charge in neutral pH solutions. Its reduced form is FADH$_2$. Two of the four nitrogens in the flavin (isoalloxazine, Section 27.2) ring structure accept two hydrogen atoms and two electrons when this carrier is reduced.

oxidized form
(FAD)

$+ 2 H^+ + 2 e^- \rightleftharpoons$

reduced form
(FADH$_2$)

FAD functions as an electron acceptor for oxidation reactions in which saturated carbon–carbon single bonds are dehydrogenated to form carbon–carbon double bonds.

$$\underset{\underset{H}{|}\underset{H}{|}}{\overset{\overset{H}{|}\overset{H}{|}}{R-C-C-R'}} + FAD \longrightarrow \underset{\underset{H}{|}\underset{H}{|}}{R-C=C-R'} + FADH_2$$

While both NADH and FADH$_2$ function as electron carriers, *NADH has a higher electron transfer potential than FADH$_2$*. Therefore, NADH is a more powerful reducing agent than FADH$_2$.

In biosynthesis, NADPH plays a special role. It is used almost exclusively as a source of electrons for reduction reactions in cells. In most biosyntheses, redox reactions occur in which products are more reduced than reactants. NADPH represents an excellent source of *reductive power* for such reactions. Either NADPH or NADH could function as reducing agents, since their electron transfer potentials are almost identical, but most enzymes that catalyze reduction reactions are specific for NADPH.

In these two sections, we have examined two types of compounds essential in metabolism. ATP and similar compounds with high-energy bonds function as intermediate stores of free energy in the cell, supplying energy for numerous processes. At the same time, specialized electron carriers such as NADH, NADPH, and FADH$_2$ provide reductive power for biosyntheses. As we shall soon see, the reductive power in these electron carriers can also be converted into energy needed for ATP synthesis.

28.5 CALORIC CONTENT OF FOODS AND PRINCIPLES OF NUTRITION

Oxidation of foods containing carbohydrates, fats, and proteins yields energy. Once transformed and stored in the high-energy bonds of ATP, it can be used to perform work. Different foods yield different amounts of heat energy when completely metabolized by the body. For example, carbohydrates and proteins yield about 3–4 kcal/g, while fats yield more than twice as much, approximately 9 kcal/g.

Example 28.2

Recalling the general structure of lipids, as compared with carbohydrates and the oxidation states of carbon in each of these types of compounds, suggest two reasons why the caloric value of lipids is so much higher than carbohydrates.

Solution

One reason why more energy can be derived from lipids than from carbohydrates (or even proteins) is that the majority of carbon atoms in lipids are *more reduced* than carbon atoms in carbohydrates. More reduced substances have more potential energy, which is released during oxidation to yield CO_2 and H_2O.

Another reason is that lipids have larger molecular weights than monosaccharides. One mole of lipid contains many more carbon atoms than does a monosaccharide, thus yielding more CO_2 and more free energy during oxidation. Oxidation states of typical carbon atoms in carbohydrates and lipids are

carbohydrates: (partially oxidized)

$$-\underset{|}{\overset{OH}{\underset{|}{C}}}- \qquad -\overset{O}{\overset{\|}{C}}-H \qquad -\underset{|}{\overset{OH}{\underset{|}{C}}}-O-$$

lipids: (reduced)

$$-\underset{H}{\overset{H}{\underset{|}{C}}}-\underset{H}{\overset{H}{\underset{|}{C}}}-\underset{H}{\overset{H}{\underset{|}{C}}}-$$

Let us momentarily move from the molecular level of energy production to the overall dietary considerations of food consumption among humans. Purely from an energy standpoint, all energy obtained from foods is the same. However, as we observed in the previous chapter on vitamins and minerals, other considerations are important in planning a well-balanced diet (see Figure 28–7, p. 766).

Carbohydrates are the major source of energy for most of the world population. Cereals, sugars, starchy fruits (dates and figs), and roots and tubers provide most of the carbohydrates consumed. (Cereals alone account for nearly 40% of the total human intake of food worldwide.) Carbohydrates in the form of starch and sucrose supply 40–60% of the total number of calories in our diets. Before carbohydrates are absorbed, they are first hydrolyzed to monosaccharides such as glucose and galactose.

Although dietary fiber (cellulose) represents no useable energy source, it is receiving more and more attention as an important component in our diet. Even though bacteria in the intestines partially degrade cellulose, the products are not absorbed. Instead, these compounds absorb water and stimulate the bowel. This results in increased stool bulk, decreased transit time of intestinal contents through the gastrointestinal tract, and binding of harmful substances, thus preventing their

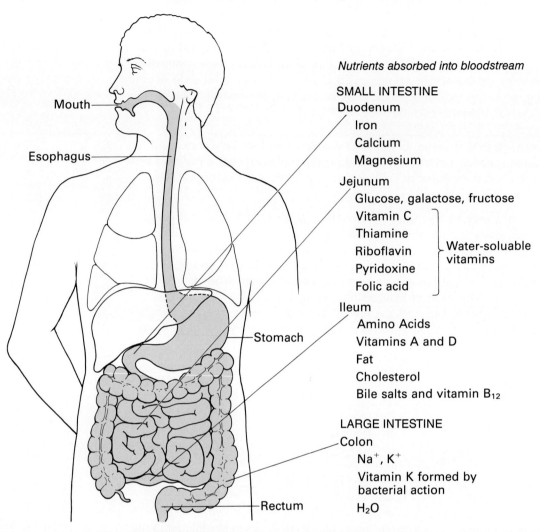

FIGURE 28-7 The absorption of important nutrients by the gastrointestinal tract.

absorption into the body. Experimental evidence with animals and studies of certain human populations suggest a correlation between the amount of dietary fiber ingested and the incidence of colon cancer. A decreased incidence of colon cancer was observed in test animals fed a diet high in fiber. Foods rich in fiber include bran cereals, whole wheat grains and breads, nuts, fresh fruits, carrots, celery, and other vegetables.

Fats (triglycerides) supply about 40% of the total calories in the average American diet. These concentrated energy stores provide free energy for the synthesis of many ATP molecules in the body (on a mole-to-mole basis). Many medical researchers recommend that the total percentage of daily calories contributed by fats be reduced to at least 30%, not because of energy considerations (unless you happen to be overweight), but because of studies showing that *excessive intake of fats* correlates with a high risk of heart disease. *Moderation* is the key to proper fat intake, since total elimination of fats from the diet is not recommended.

Proteins are first hydrolyzed by gastric and intestinal proteolytic enzymes into their constituent amino acids. These amino acids are absorbed and transported from the intestine by the bloodstream to other regions of the body. Amino acids are then used to make new proteins and to supply nitrogen for the synthesis of molecules such as heme and nucleic acids. Essential amino acids needed for protein synthesis must be supplied by the diet (Chapter 24). Alternatively, amino acids can be used as a source of energy after their carbon skeletons are converted into various sugars. Both vegetable protein and meat proteins can supply us with protein, but a person limited to vegetable protein must choose plant-derived protein wisely to avoid a deficiency in one or more essential amino acids. On the whole, meat proteins contain all essential amino acids, while plant proteins are often lacking in one or more essential amino acids.

Ingestion of food is soon followed by an increase in heat production above normal basal level (resting level), which is related to the digestion and absorption of foodstuffs. Later, as subsequent metabolism of these foods occurs, additional heat production may take place. Avid skiiers and mountain climbers know that protein-rich foods seem to provide a little extra body warmth and often snack on such foods while outdoors in cold weather. Why is this?

The ingestion of protein causes the greatest increase in heat production. This effect, termed **specific dynamic action**, *is the extra heat produced by the organism, over and above the basal heat production, as a direct result of eating.* For carbohydrates and lipids, the specific dynamic action is only 6% and 4%, respectively, of the overall energy content of the food eaten, but with protein it is approximately 30%. Therefore, eating 25 g of protein (at 4 kcal/g) represents 100 kcal of available energy, of which 30 kcal is released as extra heat production over the basal level. These 30 kcal are wasted as heat, and only the remaining 70 kcal can be used for energy demands of the body. It is essential, in calculating the caloric value of a diet, to make provisions for the specific dynamic effect of different foods.

specific dynamic action

28.6 MAJOR PATHWAYS OF OXIDATIVE METABOLISM

Metabolic pathways *are series of reactions that sequentially convert one chemical compound into another.* Some pathways are relatively short, while others involve many steps. Compounds formed between steps in a metabolic pathway are called *intermediates*, and intermediates from one pathway can enter other pathways, in a type of *branched pathway*. Indeed, the detailed charts of branched biochemical pathways resemble roadmaps, with exceedingly complex networks of interconnecting lines.

metabolic pathways

To help clarify the study of energy generation, a famous scientist named Hans Krebs described three stages in the metabolism of foods. In the first stage, large molecules are hydrolyzed or broken down into smaller units: proteins into amino acids, lipids into fatty acids and glycerol, and polysaccharides into monosaccharides. Very little if any energy is harvested from this stage, but these smaller compounds are more easily absorbed into the body and provide a common starting point for stage two.

In stage two, these numerous smaller molecules are degraded to a relatively few, simple units that play a central role in metabolism. The carbon skeletons in sugars, fatty acids, and many amino acids are converted into *acetyl-CoA*. CoA's role as a carrier of two-carbon acyl groups was discussed in Section 27.2. Although some ATP and reduced electron carriers (NADH) are formed during stage two, the real energy gain comes from stage three.

Stage three consists of two main pathways in the mitochondria: the *citric acid* (*Krebs'*) *cycle* and *oxidative phosphorylation* (often called *electron transport*). Acetyl-CoA acts as a connection between stages two and three, bringing acyl groups to the citric acid cycle. The citric acid cycle (discovered largely by Krebs)

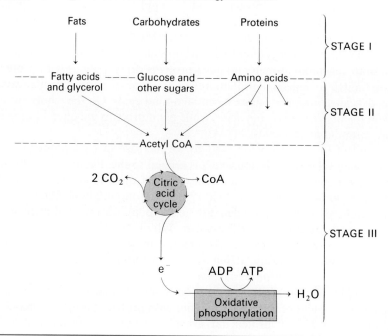

FIGURE 28-8 Stages in the extraction of free energy from food.

is a unique pathway, in that the product of the last step of the pathway becomes the first reactant at the beginning of the pathway (thus forming a cyclic, repetitious series of reactions). After entering the cycle, acyl groups are eventually oxidized to CO_2. Four redox reactions in this cyclic, oxidative pathway transfer electrons to NADH and $FADH_2$. These carriers, in turn, transport electrons to the oxidative phosphorylation pathway, a series of enzymes and proteins in the mitochondrial membrane. The relatively high electron transfer potential of NADH and $FADH_2$ is translated into energy that is used for ATP synthesis by this pathway. After moving from one carrier to the next, electrons eventually reach the end of this pathway and are accepted by O_2, which is reduced to water (see Figure 28-8).

Clearly, oxidation of foods to capture free energy is not as straightforward as simple combustion, but the products are the same. The carbon atoms of all three groups—carbohydrates, lipids, and amino acids—are oxidized to CO_2, with O_2 ultimately accepting electrons (and hydrogens) to form H_2O. Since the oxidation products are identical to combustion, thermodynamics dictates that the available free energy from both types of reactions is also the same.

28.7 MECHANISMS THAT REGULATE METABOLISM

Metabolic pathways are regulated by three different types of mechanisms, all of which we have discussed in previous chapters: the amount of a given enzyme present in the cell, the allosteric control of enzyme activity, and the entry of metabolites into cells.

The concentration of enzymes in cells is dictated by control sites on DNA that regulate mRNA transcription (Section 27.7). Higher concentrations of certain substances within the cell and certain types of hormones (particularly the steroid hormones) can stimulate the production of enzymes and proteins by controlling mRNA transcription (Figure 28-9).

FIGURE 28–9 Enzyme induction by hormone activation or by the presence of metabolites in the cell.

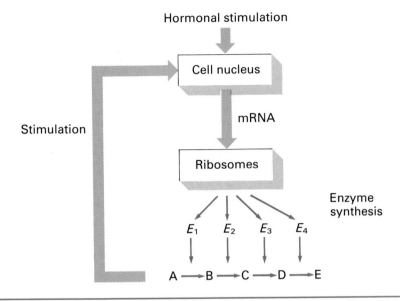

Allosteric control of enzyme activity is a more direct method of metabolic control and is the fastest and most responsive form of regulation. The binding of certain chemicals to allosteric enzymes at regulation sites can either increase or decrease the activity of the enzyme. Allosteric enzymes are most often located strategically at the beginning of metabolic pathways, where their controlling influence subsequently regulates the activity of the entire pathway (Section 25.8).

Controlling transport of substances into cells regulates their availability for metabolism. Enzymes and proteins involved in the transport of substances into cells respond to intracellular conditions automatically, keeping the concentration of metabolites at proper levels. We have seen one example of how insulin stimulates an increased absorption of sugar (Chapter 27), showing how certain extracellular factors can also affect transport.

An essential concept of metabolic regulation is that **catabolic pathways**, *which oxidize nutrients*, must be different from **anabolic pathways**, *which reduce or synthesize those same substances*. It would be counterproductive to simultaneously speed up both synthesis *and* degradation of the same compound. For this reason, such opposing pathways either are totally different or contain at least a few different enzymatic steps.

One example of separate pathways is the synthesis and breakdown of glycogen (see Figure 28–10, p. 770). Glycogen is a storage form of glucose in the body. It is synthesized by one pathway when the energy content of the cell is high and is broken back down to glucose by a different pathway when energy is needed. High concentrations of ATP simultaneously stimulate the formation of glycogen and slow down the opposite pathway. Insulin also stimulates the synthesis of glycogen and retards its hydrolysis. Just the opposite happens when the energy state of the cell drops (low ATP) or when hormones such as glucagon or epinephrin stimulate the production of their secondary messenger cAMP. Low ATP or high cAMP concentrations both act in similar fashion, stimulating the pathway that breaks down glycogen and effectively stopping the glycogen synthesis pathway.

catabolic pathways
anabolic pathways

FIGURE 28–10 Two different pathways control the synthesis and breakdown of glycogen. Agents that stimulate one pathway simultaneously impede the other.

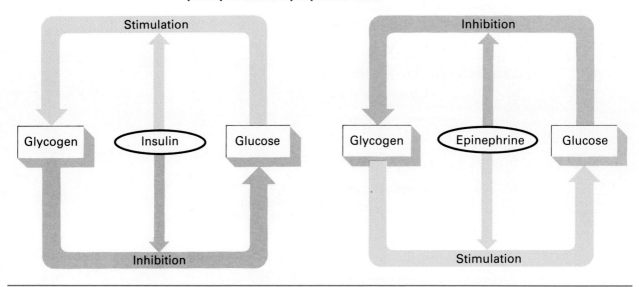

28.8 TRANSMEMBRANE POTENTIALS

transmembrane potentials

Some pathways or reactions cause changes in membrane potential. **Transmembrane potentials** are potential differences across a membrane caused by unequal concentrations of ions on either side of the membrane. The electrical charges of these ions cause an electrical potential or voltage difference across the membrane, which can be measured by a volt meter (less than 100 millivolts, or 0.1 volt).

FIGURE 28–11 By using microelectrodes on either side of a membrane, a potential difference can be measured. Membrane potentials are used by unequal concentrations of ions on one side compared to the other.

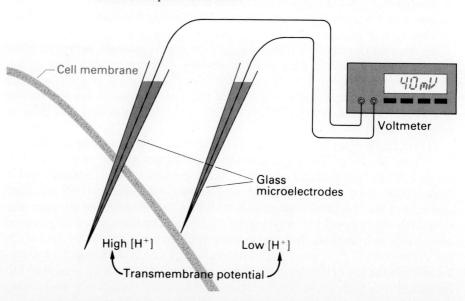

Placing small electrodes on either side of the membrane allows researchers to measure these voltage differences and observe changes as they occur (see Figure 28–11).

Energy can be stored in a transmembrane potential. The voltage associated with a transmembrane potential is directly proportional to its potential energy, since it is related to the differential concentrations of ions on one side compared to the other. Transmembrane potentials are caused by chemical reactions occurring within the membrane that cause ions to move vectorally across the membrane from one side to the other. A variety of ions (including H^+, Na^+, and K^+) are used to create transmembrane potentials.

As we will see in Chapter 30, energy from redox reactions in mitochondrial membranes is stored in transmembrane potentials and is then harnessed to produce ATP. Changes in membrane potential are also intimately associated with active transport across membranes, transmission of signals in nerve cells, and photosynthesis.

28.9 PHOTOSYNTHESIS

No discussion of biochemical energy would be complete without mention of *photosynthesis*. **Photosynthesis** *is the process in which plants absorb sunlight energy and use it to reduce carbon atoms.* Photosynthesis also produces ATP for energy needs in plants. Plants are classified as *phototrophs* ("light-eating" organisms). They provide the essential link between *chemotrophs* (chemical-eating organisms) and the ultimate source of energy for all life, the sun. Throughout this chapter, we have consistently mentioned the need among chemotrophs for reduced fuel molecules to supply free energy, which is released through oxidation of these substances to CO_2 and H_2O. This energy is then used for all aspects of metabolism. If we trace our food chain to its beginning, we eventually reach plants, whose photosynthetic capabilities trap solar energy and use it to synthesize reduced fuel molecules. Hence, *all living creatures depend on the sun for the energy of life.* In addition to plants, some bacteria are photosynthetic and can use sunlight directly for reducing their own fuel molecules.

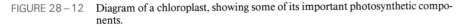

FIGURE 28–12 Diagram of a chloroplast, showing some of its important photosynthetic components.

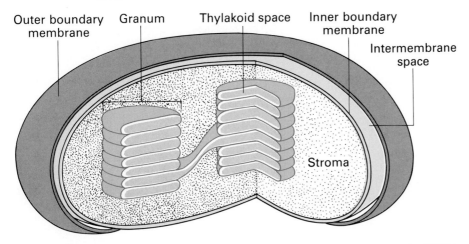

chloroplast   The site of photosynthesis in plants is a special organelle called the **chloroplast** (see Figure 28–12, p. 771). Chloroplasts contain all the necessary components needed for photosynthesis, including the green photosynthetic pigment, *chlorophyll*. This pigment makes it easy to identify chloroplasts under a light microscope because it is the only location in plant cells where chlorophyll is found.

The chloroplast has both an outer and an inner membrane. The space within the inner membrane, called the *stroma*, contains a solution of proteins and enzymes together with disk-shaped sacs called *thylakoids*. A stack of thylakoid disks is called a *granum*. Photosynthesis takes place in the membranes of individual thylakoid sacs, where the specialized enzymes of photosynthesis and chlorophyll molecules are located.

Structurally speaking, chlorophyll is similar to heme (Figure 28–13). Unlike heme, however, a *magnesium* atom is bound by attraction to nitrogens inside an aromatic, four-ring (tetra-pyrrole) structure. This part of the chlorophyll, with its extensive network of alternating double bonds, is responsible for its green color. Blue and red wavelengths of light are readily absorbed by chlorophyll, while green light is not. Many plants produce *auxillary pigments* other than chlorophyll to absorb green light that would otherwise be lost by the plant. In the fall of the year, the bright red and yellow colors of these pigments are more easily observed, after production of chlorophyll is discontinued for the winter.

FIGURE 28–13   The structure of chlorophyll and other auxillary pigments.

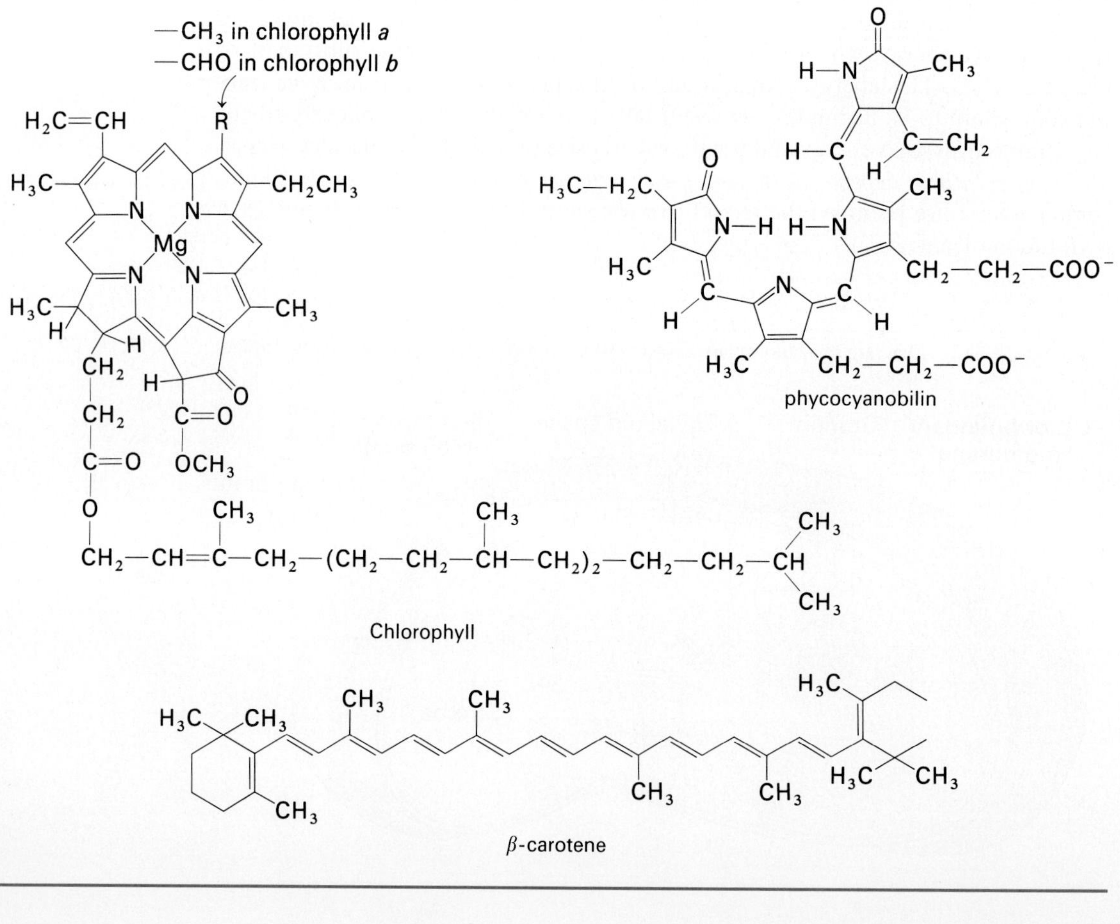

The main products of photosynthesis are carbohydrates. The overall reaction describing photosynthesis is the opposite of carbohydrate oxidation.

$$6\,CO_2 + 6\,H_2O \xrightarrow{\text{light}} 6\,O_2 + C_6H_{12}O_6$$

Just as oxidation processes in cells are more complex than a single step, photosynthesis also involves many steps. These steps are divided into two categories: the light reactions and the dark reactions.

After light energy is absorbed by either chlorophyll or auxiliary pigments, it is passed between many chlorophyll molecules until it funnels into a special reaction center complex where the first light reaction takes place. All the details of this first light-driven reaction are not known, but it involves many different membrane-bound proteins. Using the light energy, water is split apart.

$$2\,H_2O \xrightarrow[\substack{\text{(reaction center}\\\text{complex)}}]{\text{light energy}} O_2 + 4\,H^+ + 4\,e^-$$

The oxygen is released and diffuses out of the plant. The H^+ ions are pushed to one side of the thylakoid membrane, making a transmembrane potential that can be used by the plant to make ATP. The electrons are used to reduce a specialized electron carrier. After a series of more redox reactions, a second light-driven reaction takes place that eventually gives the electrons enough energy to reduce $NADP^+$ to NADPH. Only after enough NADPH is built up by the light reactions can the reduction of CO_2 occur.

In 1945, Melvin Calvin and his co-workers discovered that CO_2 could be converted into carbohydrates *in the dark*. Their experimental technique was to inject radioactive $^{14}CO_2$ into a suspension of photosynthetic algae. After a few seconds of shining a bright light on the suspension, the algae were quickly killed with alcohol and their sugars analyzed to see if radioactive sugars were present. As they suspected, the ^{14}C atoms rapidly became part of many sugar molecules inside the

FIGURE 28–14 A simplified summary of photosynthesis, showing both light and dark reactions and their products.

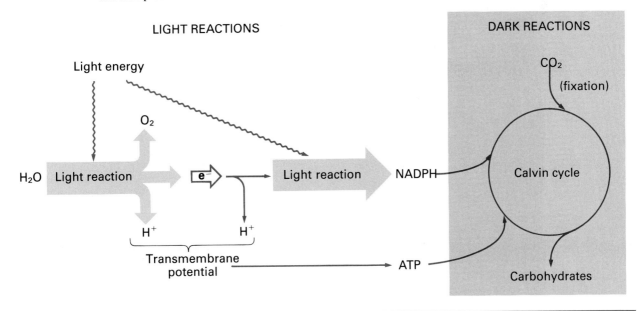

cells. However, one of the most startling results of these experiments was that under similar conditions, when radioactive $^{14}CO_2$ was injected *after the light was turned off*, it was still incorporated into newly synthesized carbohydrates. This meant that incorporation of CO_2 into sugar molecules (often called CO_2 *fixation*) actually occurred in the dark. Calvin and his group spent many years determining the actual way in which CO_2 was "fixed" into carbohydrates. Their results eventually yielded a series of reactions called the *Calvin cycle*.

To drive the reactions of the Calvin cycle, both ATP (a high energy source) and NADPH (reductive power) are required. After sufficient quantities of these have been supplied by light-driven reactions, CO_2 fixation can proceed in total darkness until their supply runs out.

Thus we see that the important process of photosynthesis is divided into two parts. A series of light-driven reactions provide both ATP, as a source of energy, and NADPH, a powerful reducing agent. In a subsequent, independent series of dark reactions, these energy-rich compounds are used to fix CO_2 and eventually produce carbohydrates, an essential source of the world's food supply (see Figure 28–14, p. 773).

EXERCISES AND PROBLEMS

Chemical Energy

28.1 Describe what is meant by free energy.

28.2 What is the sign of the free energy change ($\Delta G°$) for
a. a spontaneous reaction?
b. a nonspontaneous reaction?

28.3 Draw a graph of the free energy changes during a spontaneous reaction. Indicate the following on your graph:
a. the overall change in $\Delta G°$.
b. the part of the curve that corresponds to energy required to break bonds in the reactants.
c. the part of the curve that corresponds to energy released as new bonds form in the products.

Oxidation and Reduction Reactions in Cells

28.4 Why is carbon dioxide referred to as an energy-poor molecule?

28.5 Arrange each group in order of increasing oxidation state (from most reduced to most oxidized):
a. S^{4+}, S^{2-}, S^{6+}.
b. acids, alkenes, alcohols.
c. alkanes, esters, ketones.
d. aldehydes, alkenes, carbon dioxide.

28.6 Which of the following reactions is likely to yield more free energy? Justify your answer.
a. $CH_3-CH_2-CH_3 + 5\, O_2 \longrightarrow 3\, CO_2 + 4\, H_2O$
b. $HO-\underset{\underset{O}{\|}}{C}-CH_2-\underset{\underset{O}{\|}}{C}-OH + 2\, O_2 \longrightarrow 3\, CO_2 + 2\, H_2O$

28.7 How can nonspontaneous reactions be made to occur spontaneously?

ATP: The Universal Energy Currency of the Cell

28.8 How many high-energy bonds are present in ATP? Draw the chemical structure of ATP and identify these high-energy bonds with an arrow.

28.9 In some biochemical reactions, ATP is hydrolyzed directly to AMP with the loss of a single pyrophosphate group (PP_i). However, in such cases, the free energy change is the same as if *two* high-energy bonds are hydrolyzed. Why?

28.10 Suppose a new redox reaction has recently been discovered, for which $\Delta G° = -8.6$ kcal/mol. The scientist who discovered the reaction has written a new research grant proposal in which he requests more funds to prove the reaction is linked to the synthesis of ATP from ADP. In your opinion, is this theoretically possible? Explain.

28.11 A required step in metabolism of sugars is the phosphorylation of glucose to form glucose-6-phosphate. However, $\Delta G°$ for this reaction is $+3.3$ kcal/mol, making it nonspontaneous on its own. Coupling it to a reaction in which ATP is hydrolyzed to ADP provides the necessary energy to get the job done. Write two equations for each of these separate reactions and then the coupled reaction. Calculate $\Delta G°$ for the coupled reaction. ($\Delta G°$ for ATP hydrolysis to ADP is -7.3 kcal/mol.)

28.12 What is meant by "high phosphate transfer potential?" List two compounds other than ATP that have high phosphate transfer potentials.

28.13 During short bursts of energetic muscle movements (like high jumping), a fair amount of free energy is supplied by both ATP and creatine phosphate (Figure 28–5). Compare the $\Delta G°$ for the hydrolysis of ATP and creatine phosphate. Which yields the most energy on a mole-to-mole comparison?

Electron Carriers

28.14 How do electron carriers function in metabolic pathways?

28.15 Accompanying most NAD^+/NADH reductions, two hydrogen atoms are lost by the substrate being oxidized. However, only one hydrogen atom ends up on NADH. What happens to the other?

28.16 Why is NAD^+ written with a positive charge, while FAD is not? Why is NADH *not* written with a positive charge?

28.17 Draw the nicotinamide ring of NAD^+ and indicate the site where reduction occurs. What is the chemical species most often accepted by NAD^+ in its reduction? Show a simple chemical equation for the reduction of NAD^+ to NADH by this species.

28.18 Draw the isoalloxazine (flavin) ring structure of FAD in its oxidized form. Show the site(s) of reduction on this ring and draw the reduced form of this ring system ($FADH_2$).

28.19 What is the structural difference between NADH and NADPH? As compared to NADH, what is the *primary* role of NADPH in metabolism?

28.20 Select which electron carrier (NADH, NADPH or $FADH_2$) would most likely be used as a cofactor in each of the following reactions. Also label each reaction as oxidation or reduction.

a. ethanol $\longrightarrow$ ethanal

b. HOOC—CH_2—CH_2—COOH $\longrightarrow$
  HOOC—CH=CH—COOH

c. HOOC—CH(OH)—CH_2—COOH $\longrightarrow$
  HOOC—C(=O)—CH_2—COOH

d. CH_3—CH=CH—C(=O)—R $\longrightarrow$
  CH_3—CH_2—CH_2—C(=O)—R

28.21 The reductive potential of both NADH and $FADH_2$ can be converted by the oxidative phosphorylation pathway into free energy needed to synthesize ATP from ADP. Based upon the $\Delta G°$ of each reaction, what is the maximum number of ATP molecules that could *theoretically* be formed from ADP by a redox reaction between O_2 and NADH?

$$ADP + P_i + H^+ \longrightarrow ATP + H_2O$$
$$\Delta G° = +7.3 \text{ kcal/mol}$$
$$NADH + \tfrac{1}{2}O_2 + H^+ \longrightarrow NAD^+ + H_2O$$
$$\Delta G° = -52.6 \text{ kcal/mol}$$

Caloric Content of Foods and Principles of Nutrition

28.22 Among carbohydrates, lipids, and proteins, which
a. represents the most concentrated store of potential energy?
b. actually supplies the majority of daily calories for the general world population?
c. contains essential nutrients needed in the diet?
d. causes the greatest specific dynamic action (extra heat production immediately after eating)?
e. is associated with a higher risk of heart disease when eaten in *excess*?

28.23 Why is celluloid material (fiber) receiving more attention over the last few years as an important part of the diet, even though it supplies virtually no useable energy for humans?

Major Pathways of Oxidative Metabolism

28.24 Briefly outline the three phases for the oxidative metabolism of foods. In which phase is the majority of energy obtained by oxidative reactions?

28.25 Why is the citric acid cycle often called the Krebs cycle?

28.26 The ultimate products of oxidation, CO_2 and H_2O, are formed in totally different reactions during the third phase of metabolism. Identify the locations where they are formed in the general scheme of metabolism and explain how electrons make their way from their parent carbon atoms to oxygen.

28.27 What pathway converts the reductive potential of NADH and $FADH_2$ into free energy for ATP synthesis?

Mechanisms That Regulate Metabolism

28.28 List three ways that metabolic pathways are regulated by the body.

28.29 Controlling any one enzyme in a cyclic metabolic pathway tends to slow down the entire pathway. However, which steps in a linear pathway make the best control sites?

28.30 How can membrane transport processes be used to regulate metabolism inside the cell?

28.31 Explain the advantages of having totally (or even partially) different catabolic and anabolic pathways for a single compound.

28.32 High levels of ATP in the cell retard the oxidation of carbohydrates. What differential effect would you expect higher concentrations of ATP to have on lipid synthesis and lipid oxidation?

Transmembrane Potentials

28.33 What causes membrane potentials and how are they measured?

28.34 List three processes in the body that depend on transmembrane potentials for their energy requirements.

28.35 Certain toxic chemicals can dissolve into membranes that facilitate the diffusion of ions across membranes. One such chemical is 2,4-dinitrophenol (2,4-DNP). This compound can pick up a proton on one side of a membrane, diffuse across the membrane, and release it on the other side. What effect would you expect 2,4-DNP to have on the synthesis of ATP in mitochondria?

Photosynthesis

28.36 Why is sunlight so important to *all* living organisms?

28.37 Write an overall equation for photosynthesis, showing the reduction of carbon dioxide. (Be sure to balance the equation.)

28.38 Sketch a simple diagram of a chloroplast and identify some of its important structures. What is the site of photosynthesis in the chloroplast?

28.39 Why is chlorophyll green? Just because plants are green, does this mean that they have no way of trapping and using the energy in this region of the spectrum? Explain.

28.40 Does every chlorophyll molecule that absorbs light participate *directly* in photosynthesis? Explain.

28.41 Describe the products of both phases of photosynthesis: the light reactions and the dark reactions. Describe the experiment(s) done by Calvin and his co-workers which led to this division between light and dark reactions.

28.42 Under the following sets of conditions, predict the results asked for. Briefly explain your reasoning.

a. During irradiation by a bright light, could a plant make ATP for its needs, even if a powerful inhibitor of the Calvin cycle were present?
b. Under the same conditions, could the plant split H_2O and evolve O_2?
c. Again, under the same conditions, would radioactive carbon atoms (added as $^{14}CO_2$) be incorporated into newly synthesized sugar molecules?
d. Under different conditions, in which a plant is pre-irradiated with a bright light, could it continue to reduce CO_2 and incorporate it into sugar molecules after the light is turned off? If so, for how long?
e. Is *all NADPH* formed in photosynthesis used for CO_2 fixation all of the time? If not, what else is it used for?

29 Carbohydrate Metabolism

Objectives

After completing Chapter 29, you will be able to:

29.1 Understand how glucose is absorbed by cells of the body.
29.2 List the 10 steps of glycolysis from glucose to pyruvate, and describe the types of reactions that occur at each step.
29.3 Describe some energetically important metabolic fates of pyruvate.
29.4 Outline the gluconeogenic pathway, emphasizing the steps that are not simple reversals of glycolytic reactions.
29.5 Show the important products of the pentose phosphate pathway and how it allows for the metabolism of C_3, C_4, C_5, C_6, and C_7 sugars.
29.6 Explain why glycogenolysis and glycogenesis are not simple reverse pathways.
29.7 Describe the regulation of glycose metabolism, including both its storage and oxidation.
29.8 Understand free energy changes in carbohydrate metabolism.

INTRODUCTION

Glucose is the focal point of carbohydrate metabolism. Commonly called "blood sugar," *glucose* is supplied to the body via the circulatory system and, after being absorbed by the cell, can be either oxidized to yield energy or stored as glycogen for future use. When sufficient oxygen is present, glucose is totally oxidized to CO_2 and H_2O. However, in the absence of oxygen or necessary cellular machinery, glucose is only partially oxidized to lactic acid (or ethyl alcohol in microorganisms).

Besides energy needs, glucose and other six-carbon sugars can be converted into a variety of different sugars (C_2, C_3, C_4, C_5, and C_7) needed for biosyntheses. Some of the oxidative steps in carbohydrate metabolism also produce NADH and NADPH, sources of reductive power in cells.

29.1 GLUCOSE ABSORPTION

Glucose metabolism occurs inside cells; it must be absorbed or transported into the cell before it can be used. Polysaccharides consumed in the diet are quickly hydrolyzed by both salivary enzymes and enzymes in the small intestine (Chapter

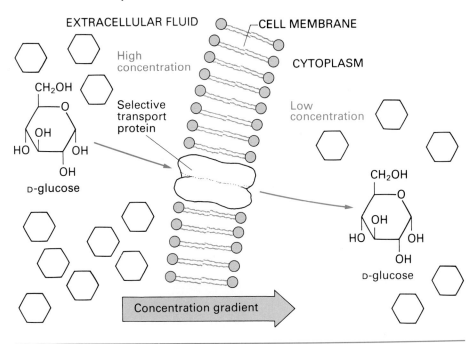

FIGURE 29–1 Transport of D-glucose into mammalian cells occurs down a concentration gradient. Membrane proteins act as selective gates for the passage of molecules in such transport mechanisms.

22), producing monosaccharides which are absorbed into the bloodstream for transport to tissues throughout the body.

Glucose is the primary sugar present in the blood, and is not readily permeable across cell membranes. Certain proteins in the cell membrane are specific carriers for glucose. They bind to the sugar and mediate its passage through the membrane like a selective gate. No extra energy is needed to bring glucose into the cell, because glucose transport takes place down a concentration gradient. *A concentration gradient is a condition in which different concentrations of a species exist on either side of a membrane* (Figure 29–1, p. 779). Glucose moves into the cell because the extracellular concentration of glucose is about 5 mM (5×10^{-3} M), while inside the cell the concentration is much lower. *A transport mechanism that allows the movement of a metabolite from an area of high concentration to one of lower concentration is called* **facilitated diffusion.**

The passive transport system for D-glucose also allows other D-sugars such as D-galactose and D-mannose to pass into the cell. However, L-isomers of glucose and these other sugars are not allowed to pass through. The transport of fructose into cells occurs by a totally different transport mechanism.

Glucose transport into cells of certain tissues is under strict hormone regulation. Insulin, for instance, regulates the passage of glucose into skeletal muscle and cardiac muscle. After binding of insulin to its own receptor, it not only accelerates the metabolism of glucose (by decreasing its intracellular concentration), but it also affects the membrane transport proteins by speeding up their transport activity. Glucose absorption in cells of other tissues, such as the brain, liver, and erythrocytes, is not affected by insulin, demonstrating the selectivity of hormones in their target tissues.

29.2 GLYCOLYSIS

Glycolysis *is a major metabolic pathway in all living cells that converts glucose (a C_6 sugar) into two molecules of pyruvate (a C_3 sugar)*. Usually included among the reactions of glycolysis are those that further metabolize pyruvate into acetyl-CoA in the mitochondria under aerobic conditions or to lactic acid in the cytoplasm under anaerobic conditions. On the other hand, certain microorganisms convert pyruvate to ethanol (an important reaction in making alcoholic beverages). (See Figure 29–2.) In this section, we will discuss the reactions of glycolysis leading to pyruvate; the fates of pyruvate are discussed in the following section.

FIGURE 29–2 Some metabolic fates of glucose after being absorbed by cells.

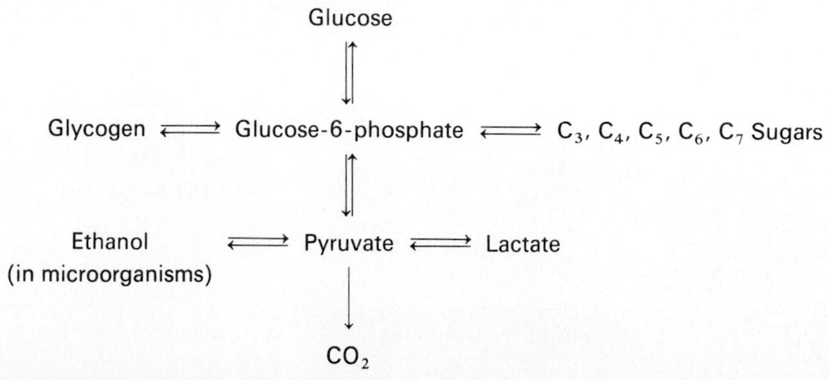

Glucose ⇌ Glycogen ⇌ Glucose-6-phosphate ⇌ C_3, C_4, C_5, C_6, C_7 Sugars ⇌ Ethanol (in microorganisms) ⇌ Pyruvate ⇌ Lactate → CO_2

TABLE 29–1 Balanced Chemical Equations for the Reactions of Glycolysis

Step	Reaction	Type	Enzyme
1	glucose + ATP $\longrightarrow$ glucose-6-phosphate + ADP + H^+	phosphorylation	hexokinase
2	glucose-6-phosphate $\rightleftharpoons$ fructose-6-phosphate	isomerization	phosphoglucose isomerase
3	fructose-6-phosphate + ATP $\longrightarrow$ fructose-1,6-diphosphate + ADP + H^+	phosphorylation	phosphofructokinase
4	fructose-1,6-diphosphate $\rightleftharpoons$ dihydroxyacetone phosphate + glyceraldehyde-3-phosphate	cleavage to aldehyde and alcohol	aldolase
5	dihydroxyacetone phosphate $\rightleftharpoons$ glyceraldehyde-3-phosphate	isomerization	triose phosphate isomerase
6	glyceraldehyde-3-phosphate + P_i + NAD^+ $\rightleftharpoons$ 1,3-diphosphoglycerate + NADH + H^+	oxidation	glyceraldehyde-3-phosphate dehydrogenase
7	1,3-diphosphoglycerate + ADP $\rightleftharpoons$ 3-phosphoglycerate + ATP	dephosphorylation	phosphoglycerate kinase
8	3-phosphoglycerate $\rightleftharpoons$ 2-phosphoglycerate	isomerization	phosphoglyceromutase
9	2-phosphoglycerate $\rightleftharpoons$ phosphoenolpyruvate + H_2O	dehydration	enolase
10	phosphoenolpyruvate + ADP + H^+ $\longrightarrow$ pyruvate + ATP	dephosphorylation	pyruvate kinase

Yeast cell extracts are used in baking to raise bread. When enzymes of the glycolysis pathway metabolize glucose (and other sugars) and produce CO_2, this CO_2 increases the volume of bread dough and makes it rise. Because yeast cultures are so easily grown and maintained, they became the main source of the cytoplasmic enzymes and cofactors in the study of glycolysis. Subsequent studies of many organisms show the same pathway functions in almost all cells.

Louis Pasteur stated in 1860 that glycolysis was absolutely impossible outside of living cells. Recognized for his other great discoveries, this incorrect hypothesis was readily accepted and remained a fundamental rule of biology until 1897, when a German scientist named Eduard Buchner discovered (quite by accident) that cellular extracts from broken yeast cells could also metabolize sugars via glycolysis. After more than 40 years of intensive effort by many researchers, the 10 enzymatic reactions of glycolysis and their individual products were finally pieced together. This pathway is often named the *Embden–Meyerhof pathway*, honoring two of these diligent workers, Gustav Embden and Otto Meyerhof.

Balanced equations for the ten reactions of glycolysis and their enzymes are listed in Table 29–1, while their position in the glycolytic pathway is shown in Figure 29–3 (p. 782).

As soon as glucose is absorbed by the cell, it is phosphorylated by the expenditure of one ATP to form glucose-6-phosphate. This first reaction of glycolysis effectively "traps" glucose inside the cell, preventing it from diffusing back out, since the phosphorylated glucose is rejected by the transport protein for glucose. In similar fashion, fructose is also phosphorylated to fructose-6-phosphate inside the cell to prevent its diffusion out of the cell. All enzymes that assist in the transfer of activated phosphate groups are named *kinases*. The enzyme that catalyzes phosphorylation of these C_6 sugars is called *hexokinase*.

$$\text{fructose} + \text{ATP} \xrightarrow{\text{hexokinase}} \text{fructose-6-phosphate} + \text{ADP} + \text{H}^+$$

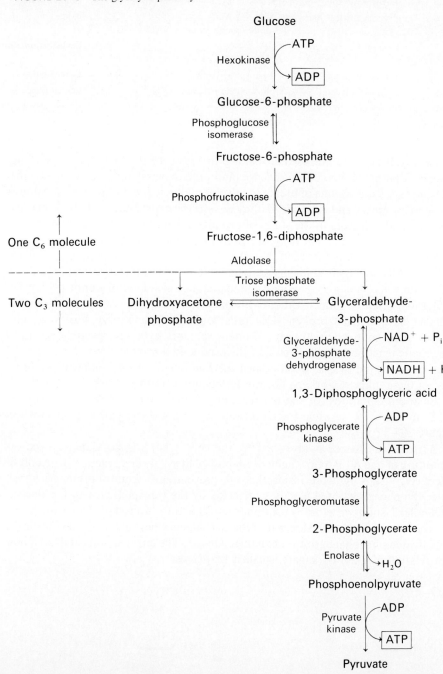

FIGURE 29-3 The glycolysis pathway.

In the second step of glycolysis, glucose-6-phosphate is isomerized to fructose-6-phosphate catalyzed by the enzyme phosphoglucoisomerase (or glucose phosphate isomerase). This is the same product produced by the direct phosphorylation of fructose. Hence, both fructose and glucose are converted to a common intermediate for further glycolytic reactions.

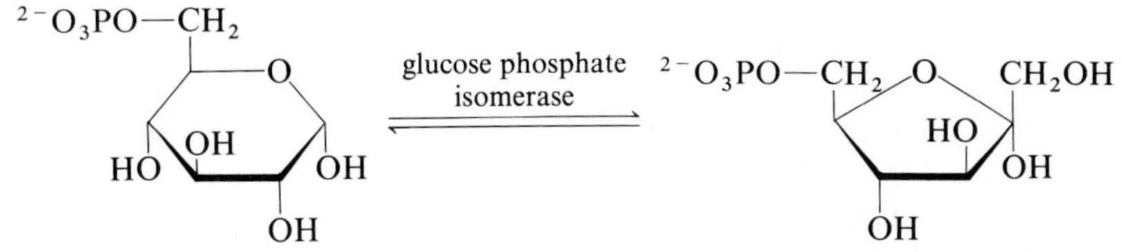

$$\text{glucose-6-phosphate} \xrightleftharpoons{\text{glucose phosphate isomerase}} \text{fructose-6-phosphate}$$

The next step in glycolysis expends another ATP to add a second phosphate to fructose-6-phosphate, forming fructose-1,6-diphosphate. The enzyme that catalyzes this reaction is *phosphofructokinase*, an exceedingly important enzyme in the control of glycolysis.

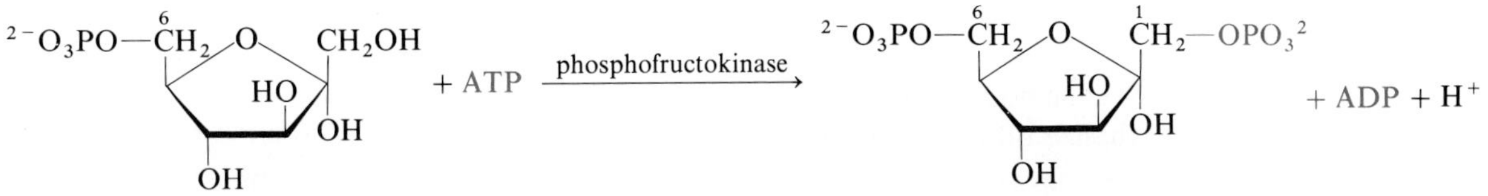

$$\text{fructose-6-phosphate} + \text{ATP} \xrightarrow{\text{phosphofructokinase}} \text{fructose-1,6-diphosphate} + \text{ADP} + \text{H}^+$$

The fourth step in glycolysis divides the six-carbon fructose-1,6-diphosphate into two smaller sugars, *each containing three carbons*: dihydroxyacetone phosphate and glyceraldehyde-3-phosphate. As can be seen from the structures given in the following reaction, one product is an aldehyde and the other is an alcohol. The name of the enzyme that catalyzes this reaction, *aldolase*, reflects the nature of these products.

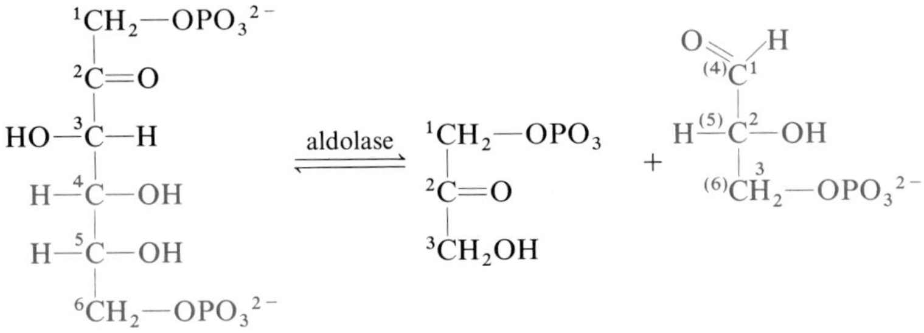

$$\text{fructose-1,6-diphosphate (open chain form)} \xrightleftharpoons{\text{aldolase}} \text{dihydroxyacetone phosphate} + \text{glyceraldehyde-3-phosphate}$$

The formation of two three-carbon sugars is significant, especially since in the next step, dihydroxyacetone phosphate is isomerized to glyceraldehyde-3-phosphate. Thus, through these two reactions, one molecule of fructose-1,6-diphosphate is converted into two molecules of glyceraldehyde-3-phosphate.

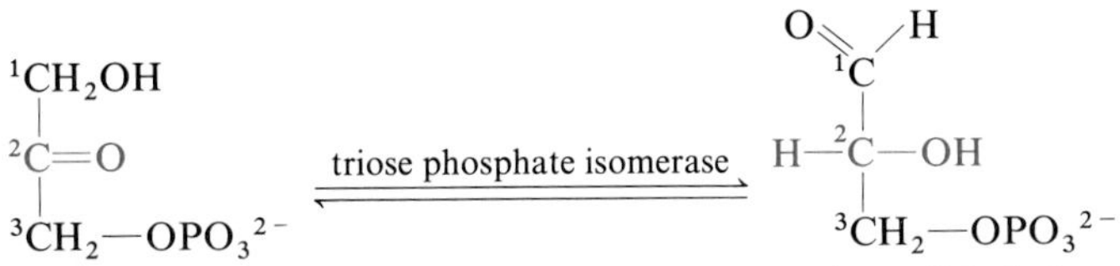

$$\text{dihydroxyacetone phosphate} \xrightleftharpoons{\text{triose phosphate isomerase}} \text{glyceraldehyde-3-phosphate}$$

From this point on in our discussion of glycolysis, we need to remember that what we describe for one molecule is actually happening to *two identical molecules*.

It is worth mentioning that until this point in glycolysis, we have harvested no free energy for the cell. In fact, one high-energy bond from each of two ATP molecules had to be expended to form fructose-1,6-diphosphate. Not until this step (the oxidation of glyceraldehyde-3-phosphate) does the cell actually benefit from the metabolism of glucose. The product of glyceraldehyde-3-phosphate oxidation is 1,3-diphosphoglyceric acid (1,3-DPG), a reaction catalyzed by glyceraldehyde-3-phosphate dehydrogenase (note the dehydrogenase classification, as with many enzymes that catalyze redox reactions). The free energy released from this oxidation is great enough to reduce NADH *and* to form a high-energy anhydride bond to phosphate.

$$\text{glyceraldehyde-3-phosphate} + NAD^+ + P_i \xrightleftharpoons{\text{glyceraldehyde-3-phosphate dehydrogenase}} \text{1,3-diphosphoglycerate} + NADH + H^+$$

The high phosphate transfer potential of 1,3-DPG is immediately used to synthesize ATP from ADP. The dephosphorylated sugar is simply 3-phosphoglycerate. Again we see that the enzyme that catalyzes high-energy phosphate transfer reactions is a kinase (phosphoglycerate kinase).

$$\text{1,3-diphosphoglycerate} + ADP \xrightleftharpoons{\text{phosphoglycerate kinase}} \text{3-phosphoglycerate} + ATP$$

Two reactions are now used to form another high-energy intermediate. In the first, 3-phosphoglycerate is isomerized to 2-phosphoglycerate.

$$\text{3-phosphoglycerate} \xrightleftharpoons{\text{phosphoglyceromutase}} \text{2-phosphoglycerate}$$

Dehydration of 2-phosphoglycerate now yields a compound with one of the highest phosphate transfer potentials of all cellular substances, phosphoenolpyruvate. The phosphorylated vinyl alcohol in this compound (an *enol* group) is responsible for its unusually high transfer potential and helped derive the name of the enzyme catalyzing this reaction, *enolase*.

$$\text{2-phosphoglycerate} \xrightleftharpoons{\text{enolase}} \text{phosphoenolpyruvate} + H_2O$$

The high phosphate transfer potential of phosphoenolpyruvate is used to form another ATP from ADP. In this last step of glycolysis, the loss of phosphate forms a short-lived intermediate, enolpyruvate, which rapidly rearranges to its more stable isomer, pyruvate. (This occurs so quickly that enolpyruvate is seldom shown in most pathway diagrams, but it helps you understand how phosphoenolpyruvate can form pyruvate upon loss of phosphate.) The kinase enzyme catalyzing this overall reaction is pyruvate kinase.

$$\underset{\text{phosphoenolpyruvate}}{\underset{|}{\overset{|}{\underset{CH_2}{C}}}-OPO_3^{2-}} + ADP + H^+ \xrightarrow[ATP]{\text{pyruvate kinase}} \left[\underset{\text{enolpyruvate}}{\underset{|}{\overset{|}{\underset{CH_2}{C}}}-OH}\right] \rightleftharpoons \underset{\text{pyruvate}}{\underset{|}{\overset{|}{\underset{CH_3}{C}}}=O}$$

Summarizing glycolysis up to this point, we can add up all the preceding reactions and write a single balanced equation showing the overall result of the ten steps involved.

$$\underbrace{1 \text{ glucose} + 2 \text{ NAD}^+ + 2 \text{ ADP} + 2 \text{ P}_i}_{\text{glycolysis} \downarrow}$$
$$\overbrace{2 \text{ pyruvate} + 2 \text{ NADH} + 2 \text{ ATP} + 2 \text{ H}^+ + 2 \text{ H}_2\text{O}}$$

Hence we see that the *net gain* in metabolizing glucose to pyruvate is only two moles of ATP and two moles of NADH.

Example 29.1

In the overall balanced equation for the oxidation of glucose to pyruvate, a net gain of only two ATP is shown. Yet, two molecules of 1,3-diphosphoglycerate yield two ATP and two molecules of phosphoenolpyruvate yield another two ATPs. Why is the net gain of ATP listed as two instead of four?

Solution
Four molecules of ATP are formed during glycolysis, but the *net* gain of ATP is only two. This is because two ATPs were expended early in the pathway to convert glucose into fructose-1,6-diphosphate.

$$\underset{\substack{\text{total ATP} \\ \text{formed}}}{4 \text{ ATP}} - \underset{\text{expended}}{2 \text{ ATP}} = \underset{\text{net gain}}{2 \text{ ATP}}$$

29.3 METABOLIC FATES OF PYRUVATE

Pyruvate has a variety of metabolic fates, depending upon the type of cell and its metabolic state. Figure 29-4 (p. 786) shows four major types of reactions in animal cells that can yield amino acids, oxaloacetate, acetyl-CoA, or lactate. With respect to energy-yielding metabolism, the oxidation of pyruvate to acetyl-CoA and its reduction to lactate are most important.

Oxidative decarboxylation of pyruvate to form acetyl-CoA is catalyzed by the pyruvate dehydrogenase complex, an integrated group of enzymes found only in mitochondria. The location of this enzyme complex requires pyruvate to diffuse into the mitochondria before this reaction can occur. In a multi-step mechanism requiring a variety of cofactors (Mg^{2+}, TPP, lipoic acid, CoA, FAD, and NAD^+), pyruvate is decarboxylated and the remaining two carbons (an acetyl group) are

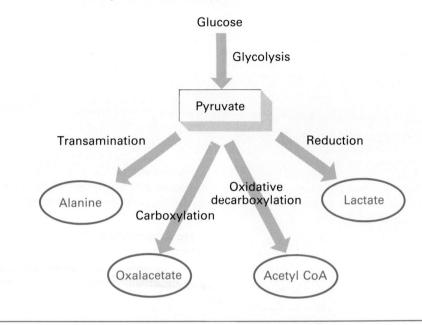

FIGURE 29-4 Four major metabolic fates of pyruvate.

covalently attached to CoA by a thioester linkage. (For a review of the structure and function of CoA, refer to Section 27.2.) This oxidative decarboxylation of pyruvate is linked to the reduction of NAD$^+$.

$$CH_3-\overset{O}{\underset{\|}{C}}-C\overset{O}{\underset{O^-}{\diagdown}} + CoA + NAD^+ \xrightarrow{\text{pyruvate dehydrogenase enzyme complex}} CH_3-\overset{O}{\underset{\|}{C}}-CoA + CO_2 + NADH$$

Glycolysis is of great importance in all tissues, but this is especially true for the brain. Approximately 120 g of glucose is used by an adult human brain each day in order to meet its exceedingly high demands for ATP. Almost all the pyruvate formed by glycolysis is further oxidized to CO_2 and H_2O in the brain mitochondria to provide high yields of ATP. In contrast to this, certain other cells in the body lack mitochondria and must rely upon glycolysis alone to provide ATP. Red blood cells (erythrocytes) lack mitochondria and therefore are unable to convert pyruvate to CO_2 and H_2O. The cornea and lens of the eye also lack mitochondria (which scatter light) and therefore depend on glycolysis as the major mechanism for ATP production.

Muscle tissues contain sufficient numbers of mitochondria to completely oxidize pyruvate, unless worked vigorously in the absence of sufficient oxygen (often called oxygen "debt" or an *anaerobic* condition). In situations like these, an excessive build-up of pyruvate in the cytoplasm of animal cells is avoided by a reaction that reduces pyruvate to lactic acid.

$$\underset{\text{pyruvate}}{CH_3-\overset{O}{\underset{\|}{C}}-CO_2^-} + NADH + H^+ \underset{}{\overset{LDH}{\rightleftharpoons}} \underset{\text{lactate}}{CH_3-\overset{OH}{\underset{|}{C}H}-CO_2^-} + NAD^+$$

The enzyme that catalyzes this reaction is lactate dehydrogenase (LDH), and as the name would indicate, it catalyzes the reaction in both directions. Lactate is a "dead end" in metabolism and must return to pyruvate before being metabolized further. If muscles are worked excessively under anaerobic conditions, lactic acid begins to accumulate in the cytoplasm of cells where it is produced. Some

FIGURE 29-5 The NAD^+ produced by the reduction of pyruvate can act as a source of NAD^+ in the oxidation of glyceraldehyde-3-phosphate. This results in a net gain of two ATP, but no NADH from glucose to lactate.

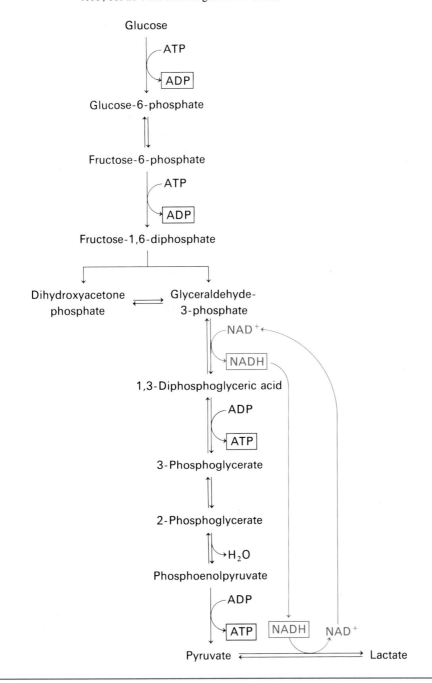

lactic acid diffuses out of cells into the blood, where it contributes to lower pH in the blood. During strenuous exercise, this lower pH triggers faster breathing to help supply needed oxygen. Lactic acid accumulation is the primary cause of stiff and sore muscles the day after strenuous exercise.

Recycling of NAD^+ produced in the reduction of pyruvate actually provides a source of NAD^+ for the oxidation of glyceraldehyde-3-phosphate. Hence, in the metabolic pathway from glucose all the way to lactate, there is *no net gain in NADH* (Figure 29–5, p. 787). A balanced equation that summarizes the overall conversion of glucose to lactate reveals this and a net gain of only two ATP molecules.

$$\text{glucose} + 2\,ADP + 2\,P_i \longrightarrow 2\,\text{lactate} + 2\,ATP + 2\,H_2O$$

fermentation

Anaerobic glycolysis in some microorganisms produces ethanol instead of lactate. **Fermentation** *is the biochemical process whereby sugar is converted into ethanol*. It is the key to production of alcoholic beverages. As in animal cells, pyruvate is formed from glycolysis, but instead of being reduced directly to lactic acid, it is first decarboxylated and then reduced to ethanol. The decarboxylation step forms acetaldehyde (ethanal), and is catalyzed by pyruvate decarboxylase. Acetaldehyde is reduced to ethanol by the enzyme alcohol dehydrogenase, using NADH as a reducing agent.

$$\underset{\text{pyruvate}}{CH_3-\underset{\underset{O}{\parallel}}{C}-CO_2^-} \xrightleftharpoons[\text{pyruvate decarboxylase}]{H^+ \quad CO_2} \underset{\text{acetaldehyde}}{CH_3-\underset{\underset{O}{\parallel}}{C}-H} \xrightleftharpoons[\text{alcohol dehydrogenase}]{NADH + H^+ \quad NAD^+} \underset{\text{ethanol}}{CH_3-CH_2-OH}$$

As in the case for lactate, ethanol must be converted back to pyruvate before further metabolism, a reaction catalyzed by alcohol dehydrogenase. (Alcohol dehydrogenase received its name from this reversible reaction.)

Animals and other organisms also synthesize the enzymes pyruvate decarboxylase and alcohol dehydrogenase. When alcohol enters the body, these enzymes convert it into pyruvate for subsequent metabolism through this pathway. Alcohol is reduced rather quickly into acetaldehyde after being absorbed. However, the carboxylation of acetaldehyde to pyruvate is much slower. This reaction results in the accumulation of acetaldehyde and is largely responsible for the symptoms of a "hangover" until its concentration can be decreased.

Alcohol is toxic. If microorganisms that produce alcohol under anaerobic conditions are left long enough without oxygen, ethanol concentrations build up to the point where the cells are killed (somewhere between 10–15% ethanol). Even among humans, excessive consumption of ethanol eventually leads to liver damage, the major site of its enzymatic conversion to pyruvate. When gulped down quickly, as little as *one pint* of pure ethanol will kill most people.

29.4 GLUCONEOGENESIS

gluconeogenesis

Demands for glucose by the brain and other tissues require a constant supply of this sugar in the blood. Sufficient supplies of glucose are present in the body (as either free sugar or glycogen) to supply its needs for about one day during average activity. When demands outstrip supply, the body must convert noncarbohydrate precursors into glucose via a pathway called *gluconeogenesis*. **Gluconeogenesis** *is the formation of glucose from a large variety of substances*. For example, much of the lactate formed during muscle activity is first converted back into pyruvate and subsequently to glucose by gluconeogenesis in the liver. This pathway is most active

FIGURE 29-6 The Cori cycle relates glycolysis and gluconeogenesis. Irreversible reactions of glycolysis are replaced by different reactions in gluconeogenesis.

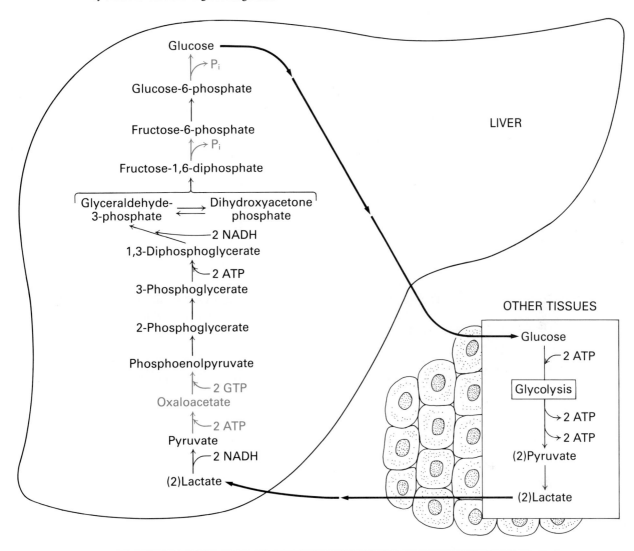

in the liver and kidneys, which produce glucose as a service for the more active tissues of the body. The cyclic nature of converting glucose into lactate in the muscles and lactate back to glucose in the liver was discovered by C. Cori and G. Cori in the 1930s and is often called the *Cori cycle*.

Gluconeogenesis is not a simple reversal of glycolysis. Although some of the reactions in glycolysis are reversible, *three of them (each involving ATP) are irreversible*: phosphorylation of glucose, phosphorylation of fructose-6-phosphate, and dephosphorylation of phosphoenolpyruvate to form pyruvate. These steps are bypassed by a separate set of reactions during the synthesis of glucose from pyruvate. In addition, gluconeogenesis requires more energy than is liberated by glycolysis. (See Figure 29-6.)

To circumvent the glycolytic degradation of phosphoenolpyruvate (PEP) to pyruvate, pyruvate is first carboxylated to form oxaloacetate. ATP supplies the free energy needed to drive this reaction, and the enzyme (pyruvate carboxylase) requires biotin as a cofactor. In the very next step, oxaloacetate gives up its newly acquired CO_2 group and GTP supplies an activated phosphate to form PEP.

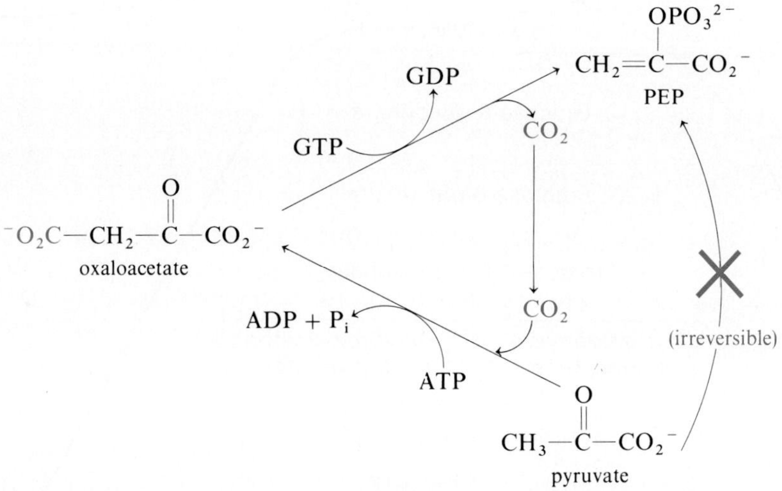

$$\begin{array}{c}
\text{oxaloacetate} \xrightarrow{\text{GTP} \to \text{GDP}, -CO_2} \text{PEP} \\
\text{pyruvate} \xrightarrow{\text{ATP} \to \text{ADP} + P_i, +CO_2} \text{oxaloacetate} \\
\text{pyruvate} \xrightarrow{\text{(irreversible)}} \text{PEP}
\end{array}$$

The other two irreversible reactions of glycolysis are those requiring ATP to phosphorylate glucose and fructose-6-phosphate. In gluconeogenesis, these steps are bypassed by phosphatase enzymes, which simply remove the phosphate by hydrolysis. This process in effect "wastes" a high-energy bond in each case, but is required to free glucose from its highly polar phosphate groups so that it may diffuse across the liver membrane into the bloodstream and be absorbed by other tissues.

By adding up all the individual reactions of gluconeogenesis, we can write an overall balanced reaction to summarize this pathway.

$$\underbrace{2 \text{ pyruvate} + 2 \text{ NADH} + 4 \text{ ATP} + 2 \text{ GTP} + 6 \text{ H}_2\text{O}}_{\downarrow \text{gluconeogenesis}}$$
$$1 \text{ glucose} + 2 \text{ NAD}^+ + 4 \text{ ADP} + 2 \text{ GDP} + 6 \text{ P}_i + 2 \text{ H}^+$$

This equation shows us that a total of six high-energy bonds are needed to drive gluconeogenesis, while glycolysis yields only two high-energy bonds (net as ATP). If we consider ATP equivalent to GTP as far as high-energy bonds are concerned, it requires a net input of four high-energy bonds to run one mole of glucose completely through the Cori cycle.

Example 29.2

Write a balanced equation for the conversion of two moles of lactate into one mole of glucose.

Solution

Only one extra step is needed to convert lactate to glucose (as compared to the conversion of pyruvate to glucose). This is an oxidation reaction and is coupled to the reduction of NAD^+ to NADH. Thus, the net change in NADH is zero, and the overall reaction is

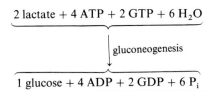

$$\underbrace{\text{2 lactate} + \text{4 ATP} + \text{2 GTP} + \text{6 H}_2\text{O}}$$
$$\Big\downarrow \text{gluconeogenesis}$$
$$\underbrace{\text{1 glucose} + \text{4 ADP} + \text{2 GDP} + \text{6 P}_i}$$

29.5 THE PENTOSE PHOSPHATE PATHWAY

Glucose is sometimes oxidized by an alternative route called the *pentose phosphate pathway* or *phosphate shunt*. The **pentose phosphate pathway** *provides for the interactive metabolism of sugars that contain 3, 4, 5, 6, and 7 carbon atoms.* Among these is the important C_5 sugar, ribose, needed for nucleic acid synthesis. In addition, this pathway is the primary source of NADPH in most cells, the main reducing agent needed for reductive biosyntheses.

pentose phosphate pathways

The pentose "shunt" actually begins as a branch in the glycolysis pathway. Glucose-6-phosphate is oxidized and decarboxylated in a series of four reactions, forming ribose-5-phosphate and its isomer, xylulose-5-phosphate (see Figure 29–7). The loss of CO_2 leaves only five carbons in these two sugars. Two NADPH are also formed by these reactions. Copious amounts of NADPH are needed for the reductive synthesis of fatty acids. Because this pentose phosphate pathway is an important supplier of NADPH, it is very active in adipose (fatty) tissue.

The next phase of the phosphate shunt involves the rearranging of carbon skeletons to form the C_3, C_4, C_5, C_6, and C_7 sugars. This is accomplished by moving

FIGURE 29–7 The oxidative decarboxylation of glucose-6-phosphate to form ribose-5-phosphate and xylulose-5-phosphate is the first phase of the pentose phosphate pathway.

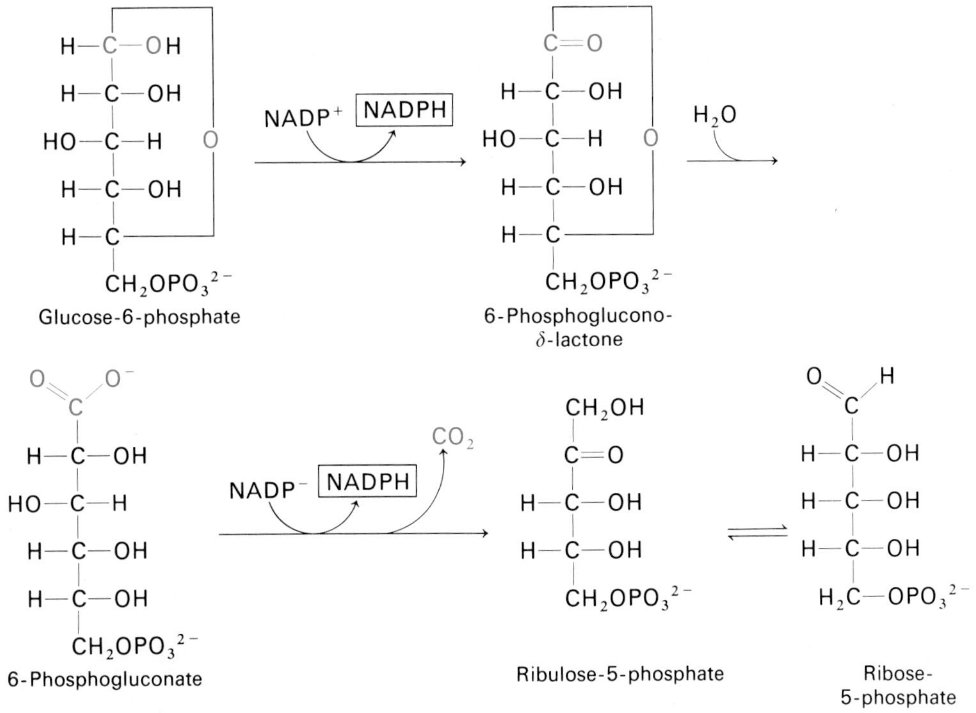

two-carbon or three-carbon groups between sugars. Two carbons from xylulose-5-phosphate are transferred to ribose-5-phosphate, forming glyceraldehyde-3-phosphate and sedoheptulose-7-phosphate. Simple math reveals that sedoheptulose is a C_7 sugar ($C_5 + C_2 = C_7$).

A different reaction then moves three carbons from sedoheptulose-7-phosphate back to glyceraldehyde-3-phosphate to form fructose-6-phosphate (a C_6 sugar) and erythrose-4-phosphate (a C_4 sugar).

While fructose-6-phosphate is free to enter back into the glycolytic pathway, erythrose-4-phosphate must receive two carbons from another xylulose-5-phosphate before being metabolized further. This transfer results in erythrose-4-phosphate becoming fructose-6-phosphate ($C_4 + C_2 = C_6$) and ribose-5-phosphate's loss converts it into glyceraldehyde-3-phosphate ($C_5 - C_2 = C_3$).

Two enzymes play an important role in this pathway. The C_2-transfers are catalyzed by the enzyme *transketolase* and the C_3-transfers by the enzyme *transaldolase*.

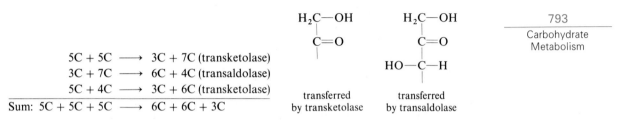

```
5C + 5C      ⟶  3C + 7C  (transketolase)
3C + 7C      ⟶  6C + 4C  (transaldolase)
5C + 4C      ⟶  3C + 6C  (transketolase)
─────────────────────────────────────────
Sum: 5C + 5C + 5C  ⟶  6C + 6C + 3C
```

Overall, three five-carbon isomers of ribose-5-phosphate are transformed into two fructose-6-phosphates and one glyceraldehyde-3-phosphate (Figure 29–8). The

---

FIGURE 29–8 The interrelationship of glycolysis and the pentose phosphate pathway.

end products of this pathway may enter back into glycolysis, while the intermediates are used in a variety of biosynthetic reactions.

29.6 GLYCOGEN METABOLISM

Glycogen is the polymeric storage form of glucose in animals (Chapter 22). Animals that receive their nutrition in pulses (meals), store extra glucose as glycogen. In well-nourished animals, glycogen is present in such high concentrations that it forms crystalline-like granules we can easily see under the microscope after appropriate staining (Figure 29–9).

As blood sugar levels drop, glycogen is broken back down to glucose monomers to provide a constant supply of carbohydrate energy throughout the system. To a certain extent, glycogen is stored in almost all tissues of the body, but the liver and muscles contain the vast majority of this polysaccharide (Figure 29–10).

Because of its important role in supplying the body with ample glucose, normal synthesis and degradation of glycogen are critical factors in good health. These opposite pathways of glycogen metabolism are described by phonetically similar names: **glycogenolysis** *is the degradation of glycogen*, while **glycogenesis** *is glycogen synthesis*. Though these names appear similar, their pathways are very different (Figure 29–11).

glycogenolysis
glycogenesis

The glucose units of glycogen are connected by α-1,4-linkages. Approximately every ten residues of an α-1,6-linkage cause a branch in the polymer, which results in many nonreducing ends and only one reducing end (Chapter 22). Individual glucose monomers are removed from the multiple nonreducing ends when needed by the enzyme *glycogen phosphorylase*. The large number of sites on glycogen available for degradation allow many glycogen phosphorylase enzymes to act simultaneously on the same molecule, releasing glucose much faster than if it were a linear polymer (Figure 29–12).

The enzyme glycogen phosphorylase received its name from the type of reaction it catalyzes. Instead of simple hydrolysis (water-lysis), this enzyme uses phosphate (P_i) to split the glycosidic linkage between glucose monomers (phosphorolysis). As a result, glucose-1-phosphate, rather than free glucose, is released. Glycogen phosphorylase continues along the glycogen chain, sequentially releasing one glucose monomer at a time. The direct formation of glucose-1-phosphate is beneficial because it cannot diffuse out of the cell and does not require

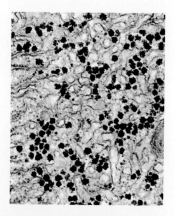

FIGURE 29–9 Electron micrographs showing glycogen granules (darkly stained material) in liver cells.

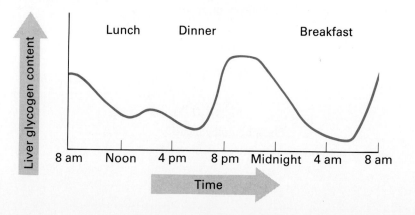

FIGURE 29–10 Variation of liver glycogen levels between meals.

FIGURE 29-11 The important pathways of glucose metabolism. Note that the glycogen degradation pathways end in *lysis*, while the glycogen synthesis pathways end with *genesis*.

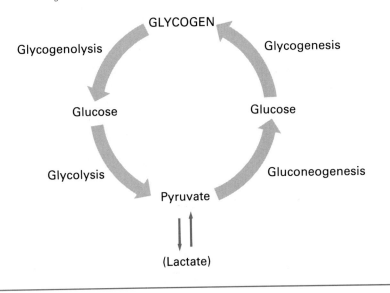

FIGURE 29-12 The branched nature of glycogen contains multiple nonreducing ends, each an enzymatic site for glycogen phosphorylase.

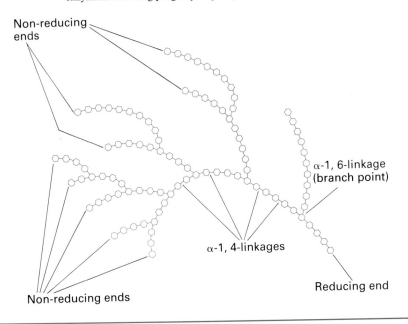

expenditure of ATP to prepare it for glycolysis. Another enzyme, phosphoglucomutase, isomerizes glucose-1-phosphate to glucose-6-phosphate without the need for extra ATP. In the liver (which needs to supply glucose to other organs), the phosphate is removed by a phosphatase enzyme, before its diffusion out of the cells into the bloodstream (see Figure 29-13, p. 796).

FIGURE 29-13 Glucose monomers are removed from the nonreducing ends of glycogen by the enzyme glycogen phosphorylase, which uses phosphate to split the glycosidic bond, rather than water. The resulting product, glucose-1-phosphate, is subsequently isomerized by phosphoglucomutase to glucose-6-phosphate or dephosphorylated in preparation for transport to other tissues.

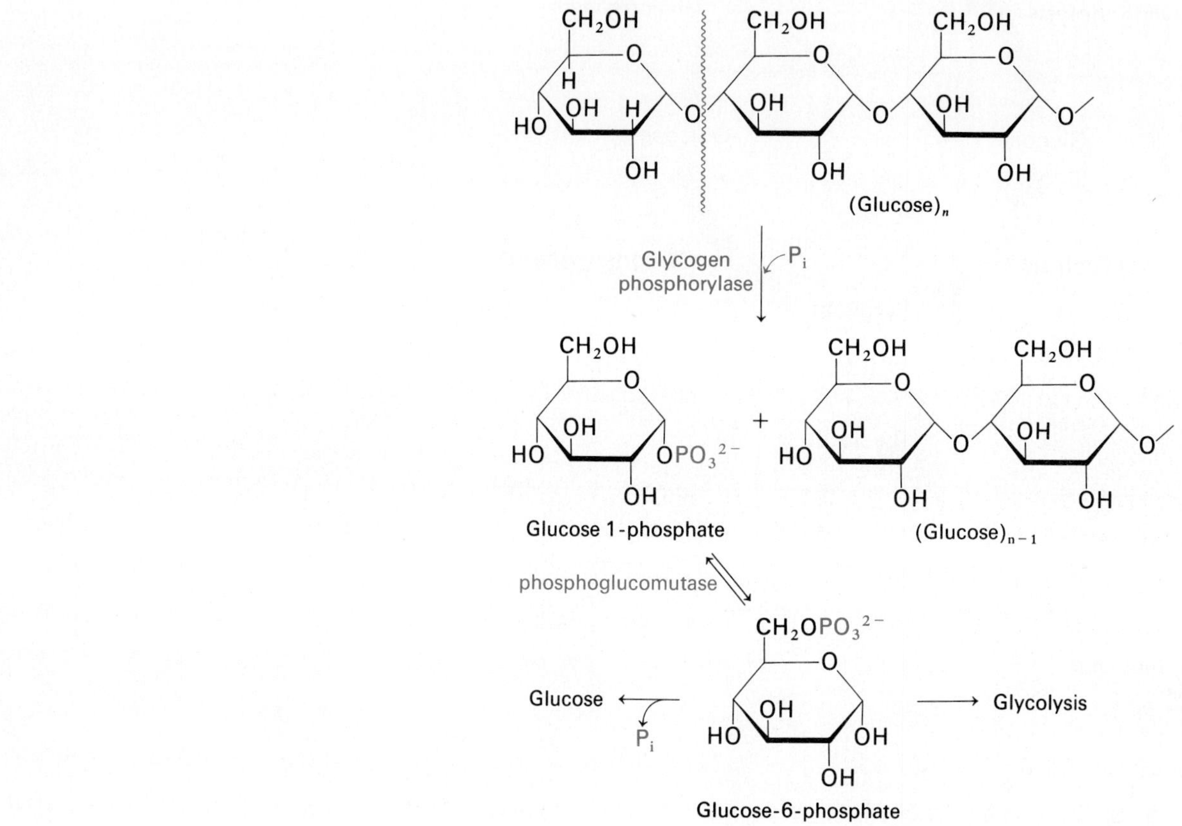

Although it is very efficient at cleaving α-1,4-linkages, glycogen phosphorylase cannot cleave α-1,6-linkages. The bulkiness of the enzyme near its active site prevents it from getting too close to branch points, and it stops cleaving α-1,4-linkages four residues from an α-1,6-branch point. Two other enzymes help out at this point. An enzyme named *transferase* shifts three glycosyl units from the branch to the end of the other main chain, attaching it by means of an α-1,4-linkage. The last remaining glucose unit (connected by its α-1,6-link) is cleaved by the *debranching enzyme*, α-1,6-*glucosidase*. The result is a linear polymer, free of its α-1,6-branch point. Glycogen phosphorylase can again proceed down the chain, releasing glucose-1-phosphate monomers until it nears another branch point, where the process will again be repeated (see Figure 29–14).

Glycogen is synthesized from glucose-1-phosphate. However, *the glycogen synthesis pathway in cells is distinctly different from its degradation pathway*. Glucose-1-phosphate molecules are first "activated" by coupling them to uridine diphosphate (UDP). Two high-energy bonds of UTP are sacrificed to activate glucose by forming UDP-glucose. One of the two phosphates in UDP-glucose comes from glucose-1-phosphate, while the other is the remaining group from UTP after the loss of its two high-energy phosphates.

FIGURE 29-14 Debranching steps in the degradation of glucose require two enzymes besides glycogen phosphorylase.

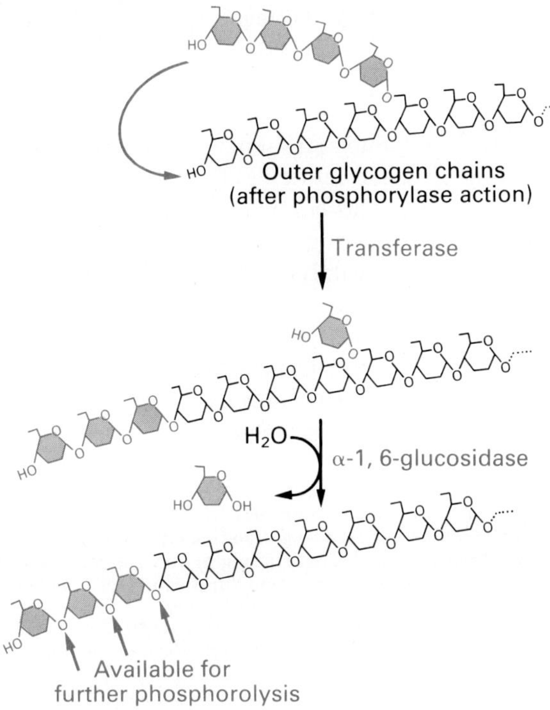

Outer glycogen chains
(after phosphorylase action)

Transferase

H_2O — α-1, 6-glucosidase

Available for further phosphorolysis

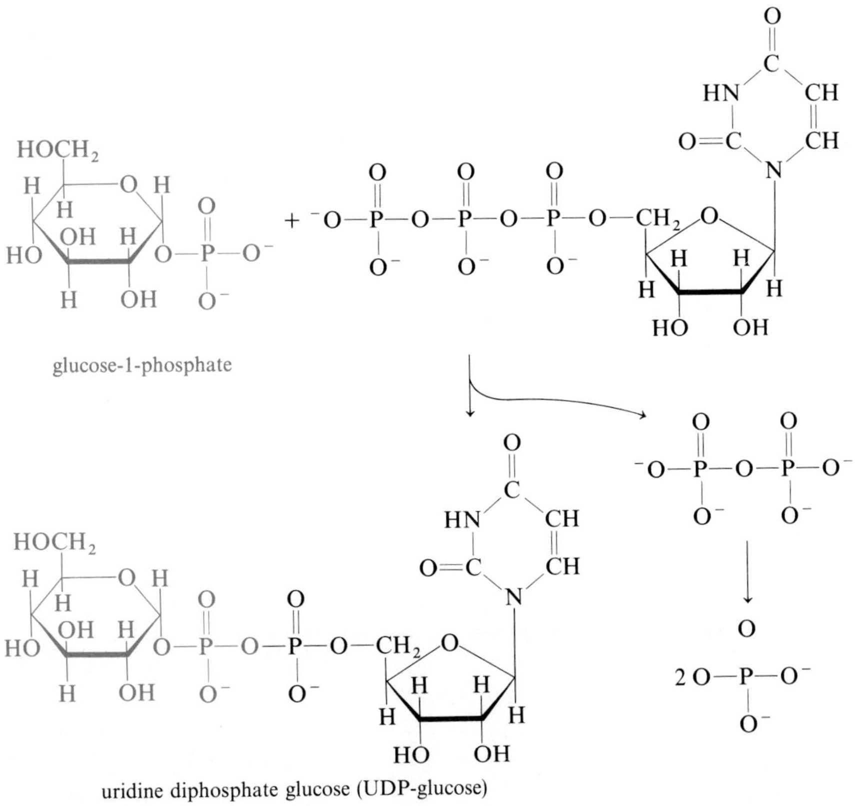

glucose-1-phosphate

uridine diphosphate glucose (UDP-glucose)

Once activated, UDP-glucose molecules undergo a reaction in which the glucose monomers are added to the nonreducing end of a growing glycogen molecule. UDP is released from the UDP-glucose as each glucose unit is added to the growing glycogen polymer. UTP can be regenerated from UDP by linking it to ATP hydrolysis.

$$\text{UDP} + \text{ATP} \qquad \text{UTP} + \text{ADP}$$

Glycogen synthetase forms only α-1,4-linkages and therefore does not form branches in the growing glycogen molecule. Branch points are formed later by a specialized *branching enzyme*. This enzyme breaks an α-1,4-linkage about seven residues from the end of the growing chain and moves the entire strand to a new branch point where it then catalyzes the formation of a new α-1,6-linkage (see Figure 29–15).

A number of genetically inherited glycogen-storage diseases are attributed to enzyme deficiencies. In fact, the first enzyme deficiency ever linked to a hereditary disease was the absence of glucose-6-phosphatase, which causes von Gierke's disease. This phosphatase removes phosphate from glucose-6-phosphate, so that

FIGURE 29–15 Glycogen synthesis is catalyzed by two different enzymes. Glycogen synthetase adds activated glucose molecules (in the form of UDP-glucose) to growing strands of glycogen. A second branching enzyme moves short strands (approximately seven residues in length) from the end of the growing strand and attaches them via α-1,6-linkages to the polymer strand "upstream," forming branch points.

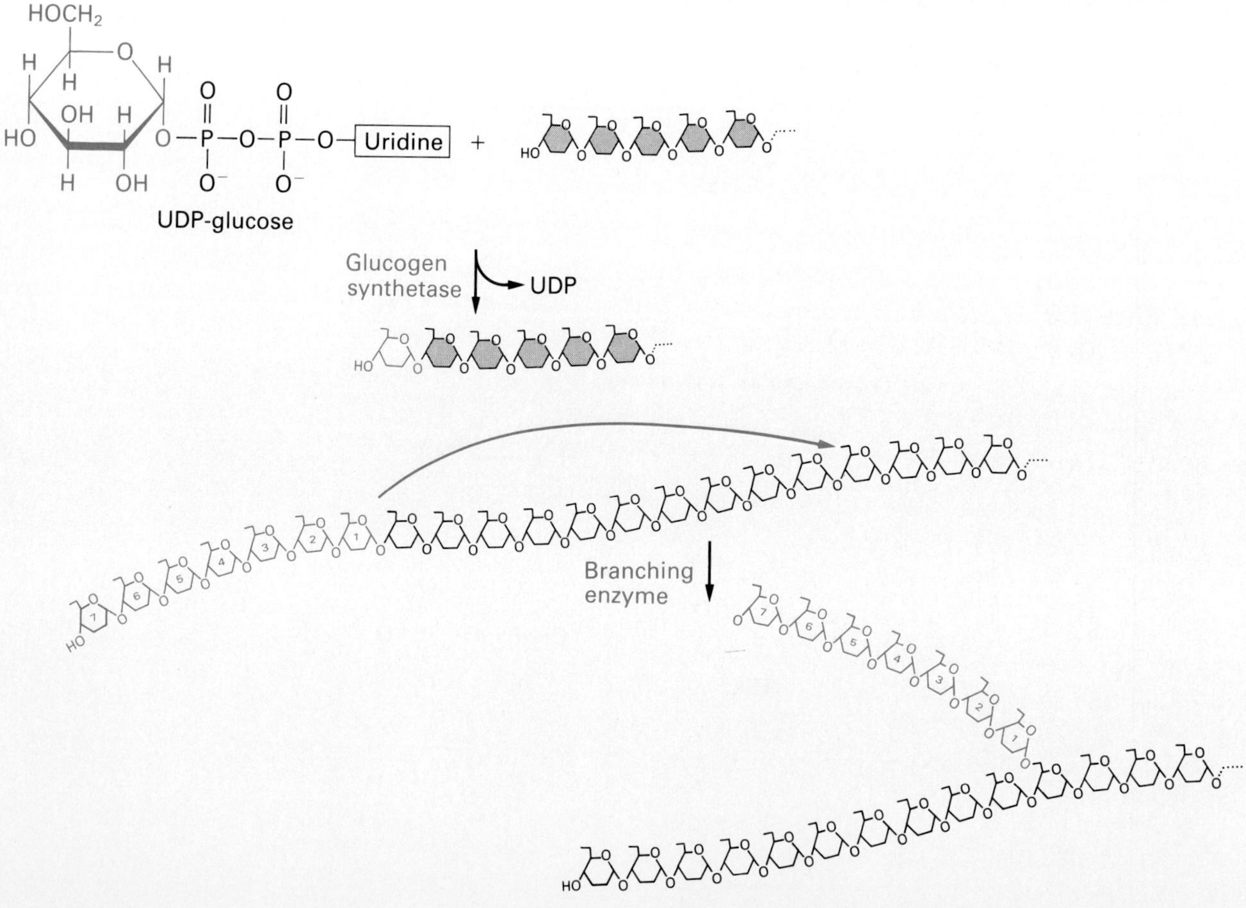

it can traverse membranes of the liver. Thus, when formed through glycogenolysis, glucose-6-phosphate can neither shed its phosphate nor leave the liver cells for transport throughout the body. Massive enlargement of the liver occurs and patients suffer from hypoglycemia and ketosis. The low levels of blood sugar can cause convulsions and failure to thrive among infants.

Defective glycogen phosphorylase in muscles (McArdle's disease) and in the liver (Hers' disease) limit strenuous physical activity because of painful muscle cramps. Anderson's disease (an absence of the branching enzyme) causes progressive cirrhosis of the liver and death before two years of age, while Cori's disease (an absence of the debranching enzyme) causes symptoms similar to von Gierke's disease.

29.7 REGULATION OF CARBOHYDRATE METABOLISM

Most people do not eat continually. As a component of the meals we eat, most carbohydrates are taken into the body during two or three meals each day. Soon after eating carbohydrates, relatively large amounts of glucose are absorbed by the body. The tissues respond rapidly and keep the blood glucose level at a constant level by absorbing the glucose and synthesizing glycogen. Between meals, as external glucose absorption decreases, these glycogen stores are broken down to keep blood glucose levels within tolerable limits. Let us explore some of the ways the body maintains this intricate balance between glucose storage and use.

The oxidation of glucose through the glycolytic pathway is controlled by three enzymes. The first is hexokinase. As the concentration of glucose-6-phosphate rises, it acts as a noncompetitive inhibitor of hexokinase. Glucose-6-phosphate increases because pathways that use it are not functioning, and its demand is low.

The other two enzymes that help control glycolysis are phosphofructokinase (PFK) and pyruvate kinase. High concentrations of ATP cause a reduction in activity of these enzymes. This is logical, since high levels of ATP mean that oxidation of glucose to supply ATP is not needed. On the other hand, high concentrations of ADP and AMP (low ATP) stimulate the activity of these two enzymes. We should note here, that if the ATP level of the cell is high and PFK activity is low, the pentose phosphate pathway can be activated to supply NADPH if needed.

During our discussion of insulin, epinephrine, and glucagon in previous chapters, we showed how these hormones control glucose levels of the blood by controlling enzyme activity. In response to high blood sugar, insulin increases the capacity of the liver to absorb glucose and store it as glycogen. Epinephrine and glucagon on the other hand, slow down the synthesis of glycogen and speed up glycogen degradation. Let us consider how these hormones function.

While epinephrine's primary target tissue is muscle and glucagon's is the liver, both stimulate the synthesis of cAMP in their respective tissues. In turn, higher cAMP levels cause a series of enzymatic activations that ultimately lead to the phosphorylation of glycogen phosphorylase. This occurs at a serine residue on this enzyme, and, once phosphorylated, the glycogen phosphorylase becomes *much more active* than before.

Increased concentrations of cAMP also stimulate the phosphorylation of glycogen synthetase. However, unlike the previous enzyme, glycogen synthetase becomes *much less active* when phosphorylated. Thus, cAMP stimulates the phosphorylation of two controlling enzymes in glycogen metabolism: one increases its activity while the other becomes quite inactive (see Figure 29–16, p. 800). Binding

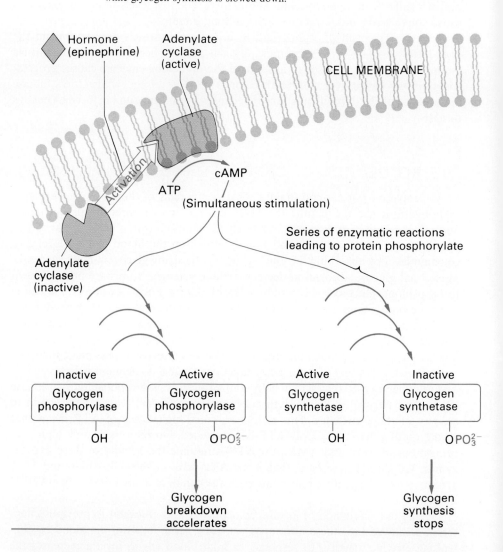

FIGURE 29-16 cAMP stimulates the phosphorylation of two important enzymes in glycogen metabolism. Glycogen phosphorylase becomes much more active, while the activity of glycogen synthetase decreases. As a result, glycogen degradation is accelerated while glycogen synthesis is slowed down.

of epinephrine or glucagon simultaneously stimulates glycogen breakdown and retards the pathway that synthesizes glycogen. As time goes on and the secondary messenger cAMP drops back to its normal concentration, phosphatase enzymes remove the phosphate groups from these enzymes and they return to normal activity.

Lack of enough insulin to control blood sugar results in hyperglycemia, as does an excess of glucagon or epinephrine. **Hyperglycemia** *is a state in which the concentration of blood sugar is higher than normal.* High concentrations of glucose in the blood often cause it to spill over into the urine, a symptom used to diagnose hyperglycemia. Another diagnostic technique for hyperglycemia is the glucose tolerance test. After ingestion of 100 g of glucose, blood sugar levels in a healthy adult should return to normal within two hours. In the diabetic, blood sugar remains elevated longer than this. **Diabetes mellitus** *is a condition of insufficient* *insulin production* that can often be treated by the intramuscular injection of extra insulin.

If too much insulin is released in response to elevated glucose, the tissues of the body overreact by absorbing too much glucose and result in an overproduction of glycogen. *Low blood sugar is called* **hypoglycemia**, and represents a potentially serious problem. Low blood sugar robs the brain and other tissues of needed nutrients. People who suffer from hypoglycemia must be careful not to eat too much sugar in a short period of time. Eating a candy bar for some extra energy can actually stimulate the release of too much insulin, resulting in a drastic reduction of glucose in the blood. Premature infants are particularly susceptible to hypoglycemia, not because of a hormone imbalance, but merely because their gluconeogenic pathway is not always functioning at full capacity and their supplies of glycogen are low.

29.8 FREE ENERGY CHANGES IN CARBOHYDRATE METABOLISM

Complete oxidation of glucose to CO_2 and H_2O yields far more energy (-686 kcal/mol) than the oxidation of glucose to lactate under anaerobic conditions (-47.0 kcal/mol). Hence, anaerobic glycolysis releases only 47/686 or 7% of the total available free energy in glucose. Reactions in the mitochondria that further oxidize pyruvate (discussed in Chapter 30) release much more of this free energy, but, for a variety of reasons discussed earlier, cells must occasionally depend completely on glycolysis for their energy needs.

Of the 47 kcal of free energy released by anaerobic glycolysis, what portion is trapped in the high-energy bonds of ATP? Recall that the overall balanced equation for anaerobic glycolysis (Section 29.2) shows a net gain of only two ATP. Since the free energy stored in one high-energy bond of ATP is 7.3 kcal, two moles of ATP mean that 14.6 kcal of energy are captured by this pathway. This represents an energy conversion efficiency of 31%.

$$\frac{\text{2 moles ATP formed}}{\text{1 mole glucose (partially oxidized to lactate)}} \longrightarrow \frac{2(-7.3 \text{ kcal/mol})}{-47.0 \text{ kcal/mol}} \times 100 = 31\%$$

Counting up ATP molecules formed by a pathway is an excellent way of analyzing the free energy actually harvested by the reactions within the pathway. As we will soon see, NADH formed in reaction pathways also represents harvested energy that can be used in reduction reactions *or to make ATP*.

Remember that lactate formation is only a short-term solution for energy needs in most cells. As soon as the cellular machinery can catch up or the liver can meet demands, lactate is oxidized back to pyruvate, where it can then be converted back to glucose (the Cori cycle) or be further oxidized in the mitochondria all the way to CO_2 and H_2O. This third and final stage of energy-harvesting metabolism inside mitochondria is our next topic (Chapter 30).

EXERCISES AND PROBLEMS

Glucose Absorption

29.1 In glucose absorption by animal cells, what is meant by facilitated diffusion? How does this process differ from active transport?

29.2 Sugars with an L-configuration (such as L-glucose) have been suggested for use as low-calorie dietary sweeteners. Experiments reveal that $\Delta G°$ of combustion is exactly the same for both D-isomers and L-isomers. Explain the rationale of using L-isomers as low-calorie sweeteners.

29.3 Briefly describe the role of insulin in regulating the absorption of glucose into cells of the body. Does insulin affect the brain to the same extent as cardiac muscle?

29.4 In an experiment using radioactively labeled glucose and thin tissue slices of liver, describe the effect on the *rate of glucose absorption* when each of the following is added to the medium surrounding the cells:

a. insulin
b. glucagon
c. increasing the concentration of labeled glucose

29.5 What organ uses glucose as its primary fuel source? What other organs supply the majority of this glucose?

Glycolysis

29.6 How is glycolysis involved in baking bread?

29.7 Why is glycolysis often called the Embden–Meyerhof pathway and not the Pasteur pathway?

29.8 Match the types of reactions in the left column with the type of glycolytic enzyme needed to catalyze each one in the right column.

Reaction Type	Enzyme Classification
a. phosphorylation	1. dehydrogenase
b. isomerization	2. aldolase
c. oxidation	3. kinase
d. alcohol & aldehyde formed	4. decarboxylase
e. loss of CO_2	5. isomerase

29.9 What is the first step in glycolysis and why is it important in retaining glucose inside the cell?

29.10 What is the first step in glycolysis that yields any energy for the cell?

29.11 The abbreviated name of phosphoenolpyruvate, PEP, is a very descriptive and appropriate one. Why can we say that PEP has a lot of "pep" and energy?

29.12 Why do we count *two* molecules of ATP being formed in the reaction of phosphoenolpyruvate (PEP) to pyruvate during glycolysis, when one mole PEP yields only one mole of ATP?

29.13 How many moles of ATP and NADH are formed (+) or consumed (−) in the following glycolytic conversions? (Assume one mole of each reactant on the left.)

Reaction Series		Net Gain in	
From	To	ATP	NADH
a. glucose	glucose-6-phosphate	−1	0
b. glyceraldehyde-3-phosphate	3-phosphoglycerate		
c. glucose	pyruvate		
d. glucose	lactate		
e. fructose-1,6-diphosphate	lactate		
f. 2-phosphoglycerate	pyruvate		

29.14 Human spermatozoa metabolize fructose anaerobically to form ATP, which provides necessary energy for their swimming motion. Within these cells, fructose is first phosphorylated to fructose-1-phosphate and then split in half, yielding two C_3 molecules: glyceraldehyde and dihydroxyacetone phosphate. One high-energy bond from ATP is used to phosphorylate glyceraldehyde, forming glyceraldehyde-3-phosphate. Draw the chemical structures of all these compounds in this short pathway, and show where these products enter the glycolytic pathway. Write a balanced chemical equation that shows the conversion of fructose to lactate through this pathway. Is this more or less efficient than normal glycolysis?

Metabolic Fates of Pyruvate

29.15 A sample of glucose has been labeled with ^{14}C at the number 1 carbon. Where would the label appear if this glucose were metabolized via glycolysis into

a. lactate?
b. ethanol and CO_2 via alcoholic fermentation?

29.16 Name two or three tissues that rely heavily upon anaerobic glycolysis of glucose to lactate. Explain why they do and under what conditions this occurs.

29.17 What important compound is recycled as a result of the pyruvate-to-lactate reduction in animal cells? What reaction uses this recycled substance?

29.18 Explain how fermentation allows yeast to survive with limited oxygen supplies.

29.19 Show a simple reaction pathway that demonstrates how ethanol is metabolized in animal cells.

29.20 An old treatment for hard-core alcoholics who were trying to "dry-out" and who were suffering traumatic withdrawal was to administer a *small amount* of acetaldehyde (or even formaldehyde). This helped some alcoholics whose systems had apparently become excessively dependent on acetaldehyde. What would cause the presence of acetaldehyde in the body tissues of a chronic alcoholic?

29.21 Why is it important to have an *airtight* container for the production of wine or other alcoholic beverages?

29.22 Home-made root beer can be carbonated by adding

a small amount of yeast and letting it metabolize the sugar in the root beer. If too much sugar is put in, or if it gets too warm, a sealed glass container can explode! Explain why.

29.23 Write a balanced equation for glucose fermentation. What is the maximum amount of ethanol (in grams) that could be produced from 10 pounds (1 lb = 454 g) of glucose?

Gluconeogenesis

29.24 In what tissues is gluconeogenesis most active?

29.25 Outline the overall Cori cycle, identifying the sites of glycolysis and gluconeogenesis.

29.26 List the three reactions of glycolysis that are essentially irreversible. Explain how gluconeogenesis gets around these reactions in synthesizing glucose from pyruvate.

29.27 What are the sources of high-energy bonds in gluconeogenesis?

29.28 What is the net energy cost of the Cori cycle (converting glucose to lactate and then back to glucose)?

The Pentose Phosphate Pathway

29.29 In what specialized capacity does NADPH function, as compared to NADH?

29.30 How many moles of NADPH are formed in the pentose phosphate pathway from glucose-6-phosphate to ribose-5-phosphate? How many CO_2 molecules are formed?

29.31 How are two C_5 sugars converted into C_3 and C_7 sugars, and how are C_4 sugars handled by the phosphate shunt?

29.32 What portion of a carbohydrate carbon skeleton is transferred by transketolase? What portion by transaldolase?

29.33 Under what circumstances might the pentose phosphate pathway be functioning at full capacity while glycolysis is nonfunctional? Inhibition of what key enzyme would cause this?

29.34 Explain how the pentose phosphate pathway could function as a *cyclic* pathway.

Glycogen Metabolism

29.35 Explain the difference between glycogenolysis and glycogenesis.

29.36 What structural aspect of glycogen contributes to the rapid release of glucose upon demand?

29.37 List three important enzymes involved in the degradation of glycogen.

29.38 Why is the enzyme glycogen phosphorylase not called glycogen hydrolase?

29.39 What nucleotide is needed to activate glucose-1-phosphate before it can be incorporated into glycogen?

29.40 Suppose that glucose is absorbed by an energy-rich cell that is in the resting state. List all the steps that must occur in converting this newly absorbed glucose molecule into glycogen for storage. (Hint: Four reactions are needed.)

29.41 Match each enzyme deficiency below with the proper disease.

Enzyme Deficiency	Disease
a. muscle glycogen phosphorylase	1. Anderson's disease
b. liver glycogen phosphorylase	2. McArdle's disease
c. glucose-6-phosphatase	3. Hers' disease
d. glycogen branching enzyme	4. von Gierke's disease
e. glycogen debranching enzyme	5. Cori's disease

Regulation of Carbohydrate Metabolism

29.42 List three enzymes that control the rate of glycolysis.

29.43 The concentration of ATP, ADP, and AMP all affect the rate of glycolysis. Explain the effect of each and what enzyme(s) are involved.

29.44 The secondary messenger, cAMP, eventually stimulates the phosphorylation of control enzymes in both glycogenolysis and glycogenesis. Which enzymes are these, and what effect does this have upon glycogen metabolism?

29.45 Describe the differences between hyperglycemia and hypoglycemia. How is the overproduction or underproduction of insulin involved with each of these conditions?

29.46 Why are the blood sugar levels of premature infants an important parameter in their care?

29.47 What conditions can be diagnosed by the glucose tolerance test? Explain the general results obtained in each case.

Free Energy Changes in Carbohydrate Metabolism

29.48 Explain why we say that only 7% of the available energy from glucose is released by its oxidation to lactate.

29.49 Considering the ATP synthesized from the anaerobic glycolysis of glucose to lactate, this process is said to be 31% efficient. Why?

29.50 What is the net gain or loss of high energy bonds in the following transformations?

Biochemical Transformation:		High-Energy Bonds (ATP) gain (+) or loss (−)
From	To	
a. glucose	glycogen	
b. 2 lactates	glucose	
c. 2 lactates	glycogen	
d. 1 glycogen unit	fructose-1,6-disphosphate	
e. 1 glycogen unit	2 lactates	

30

Mitochondrial Oxidation and Phosphorylation Pathways

Objectives

After completing Chapter 30, you will be able to:

30.1 Describe the structure of the mitochondrion and recognize it as the site of both the TCA cycle and oxidative phosphorylation.
30.2 List the reactions of the TCA cycle as well as the structural formulas of each chemical intermediate.
30.3 Outline the flow of electrons through the individual carriers of the electron transport chain.
30.4 Explain how transmembrane proton potentials form across the inner mitochondrial membrane as a result of electron transport and how energy is stored in such potentials.
30.5 Describe the structure of the ATP synthetase complex and how it uses energy from a transmembrane potential to synthesize ATP.
30.6 Accurately account for the number of ATP molecules produced by oxidation of glucose and other intermediates such as pyruvate or acetyl-CoA.
30.7 Explain how rates of the TCA cycle and electron transport are controlled by the cell.

INTRODUCTION

As we have seen in the preceding chapter, during aerobic glycolysis, glucose is first partially oxidized to pyruvate. Pyruvate is then transported from the cytoplasm into the mitochondrion, where it is completely oxidized to carbon dioxide and water. The total oxidation of pyruvate is linked to the production of ATP by a series of reactions in the tricarboxylic acid cycle and oxidative phosphorylation. Mitochondria are fascinating organelles which act as energy powerhouses in the cell. In this chapter, we will examine the unique structure of the mitochondrion and its enzymatic reactions that provide ATP by the total oxidation of pyruvate. Oxidation of fatty acids (another important role of the mitochondrion) is examined in Chapter 31.

30.1 THE MITOCHONDRION

All cells that utilize oxygen contain mitochondria distributed uniformly throughout the cytoplasm. In tissues where ATP use is high, one typically finds a higher concentration of mitochondria. For example, erythrocytes have no mitochondria and must rely totally upon glycolysis for their relatively small ATP needs. However, cardiac tissue is a highly aerobic tissue in which about *half* of the cytoplasmic volume of these cells is occupied by mitochondria. Liver is another tissue requiring a great deal of ATP, with the numbers of mitochondria in a single cell ranging between 800 and 2000.

Mitochondria are about the size of an average bacterium, about 0.5–1.0 μm

FIGURE 30–1 Electron micrograph of a typical mitochondrion. A sketch of the mitochondrion identifies many of the important structures and their locations within this organelle.

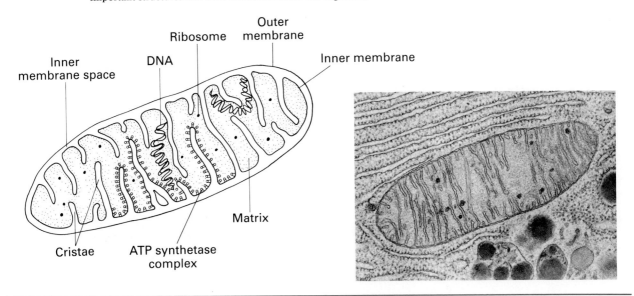

in diameter and several microns long. Depending on the tissue, some are round, some are branched, and some are extremely long. However, one distinguishing feature of these remarkable organelles is that *all mitochondria have both an outer and an inner membrane.* The outer membrane is about 50% lipid and 50% protein, while the inner membrane is only 20% lipid and 80% protein. The inner membrane contains most of the enzymes involved in electron transport, ATP synthesis, and many transport proteins. This inner membrane plays an important part in ATP synthesis by dividing the mitochondria into two separate compartments: *the interior region is called the* **matrix**, while *the region between inner and outer membranes is called the* **intermembrane space**. The inner membrane is highly folded. *The folds which protrude into the matrix of the mitochondrion are called* **cristae**. The threadlike appearance of the cristae in mitochondria are responsible for this organelle's name, *mitos,* Greek for "thread", and *chondrion,* Greek for "granule" (see Figure 30–1, p. 805).

The invention of high-resolution electron microscopes allowed researchers to see the interior structure of the mitochondrion more clearly and led to the discovery in 1962 of small spherical knobs attached to the cristae. *These small knobs appear only on the matrix side of the inner membrane and are called coupling factors or* **ATP synthetase complexes**. As their name would imply, these relatively small knobs are responsible for ATP synthesis, and their association with the inner membrane is critically important for this task (see Figure 30–2).

Most of the metabolic activity of mitochondria takes place within the matrix. As can be seen from Figure 30–3, most of the enzymes of the citric acid cycle are present in the matrix as are the mitochondrial DNA and enzymes for fatty acid oxidation.

30.2 THE TRICARBOXYLIC ACID CYCLE

As we discussed in Section 29.3, pyruvate has a variety of metabolic fates depending upon the type of cell and its present metabolic state. The most energetically favor-

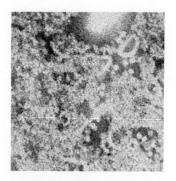

FIGURE 30–2 An electron micrograph of a single mitochondrial cristae, showing the ATP synthetase knobs extending into the matrix.

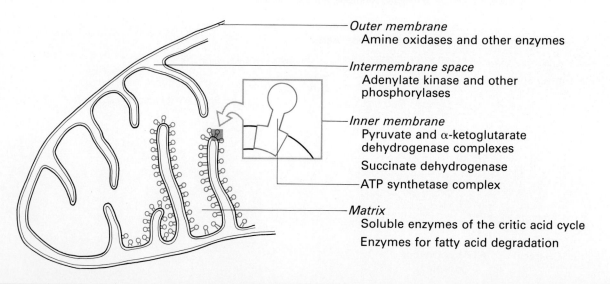

FIGURE 30–3 Localization of mitochondrial reactions.

Outer membrane
Amine oxidases and other enzymes

Intermembrane space
Adenylate kinase and other phosphorylases

Inner membrane
Pyruvate and α-ketoglutarate dehydrogenase complexes
Succinate dehydrogenase
ATP synthetase complex

Matrix
Soluble enzymes of the critic acid cycle
Enzymes for fatty acid degradation

able fate is the aerobic oxidation to carbon dioxide and water in the mitochondria. The first step in this process is the oxidative decarboxylation of pyruvate to form acetyl-CoA.

$$\underset{\text{(pyruvate)}}{CH_3-\overset{O}{\underset{\|}{C}}-COO^-} + CoA + NAD^+ \longrightarrow \underset{\text{(acetyl-CoA)}}{CH_3-\overset{O}{\underset{\|}{C}}-CoA} + CO_2 + NADH$$

This deceptively simple reaction is catalyzed by an integrated group of enzymes called the *pyruvate dehydrogenase complex*. Located only in the matrix of the mitochondrion, this multi-enzyme complex has an overall molecular weight of more than 6,000,000! This exceptional size allows it to be seen under the electron microscope. Figure 30–4 reveals the clustered nature of the many protein subunits that make up this enzyme complex.

The location of this enzyme complex requires the migration of pyruvate into the mitochondrial matrix before aerobic oxidation can occur. In a multi-step mechanism requiring a variety of cofactors (Mg^{2+}, TPP, lipoic acid, CoA, FAD, and NAD^+), pyruvate is decarboxylated and the remaining two carbons (an acetyl group) are covalently attached to CoA by a thioester linkage. NAD^+ is also reduced to NADH.

Acetyl-CoA is the entry point to the *tricarboxylic acid cycle*. Discovered to a great extent by Hans Krebs, this cyclic series of eight reactions is often called the "Krebs cycle". The combination of acetyl-CoA with oxaloacetate is the first step in this cycle, which forms citric acid. Citric acid is somewhat unique in that it contains *three* carboxyl (acid) groups. For this reason, this pathway is often called the *citric acid cycle* or the *tricarboxylic acid cycle* (abbreviated *TCA cycle*).

The primary function of the TCA cycle is twofold: (1) to completely oxidize acetyl-CoA to CO_2 and (2) to recover some of the free energy released during oxidation in the biologically useful form of electron carriers, NADH and $FADH_2$. The cyclic nature of this pathway is shown in Figure 30–5, p. 808, which shows all the intermediate compounds of the TCA cycle. During one turn of the cycle, two carbons enter (as acetyl-CoA) and two carbons are lost as CO_2. In addition to the CO_2 formed, three NADH, one $FADH_2$, and one GTP are produced. At first

FIGURE 30–4 Model of the pyruvate dehydrogenase complex.

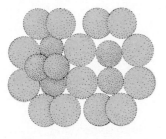

Side view   Top view

FIGURE 30-5 The tricarboxylic acid (TCA) cycle.

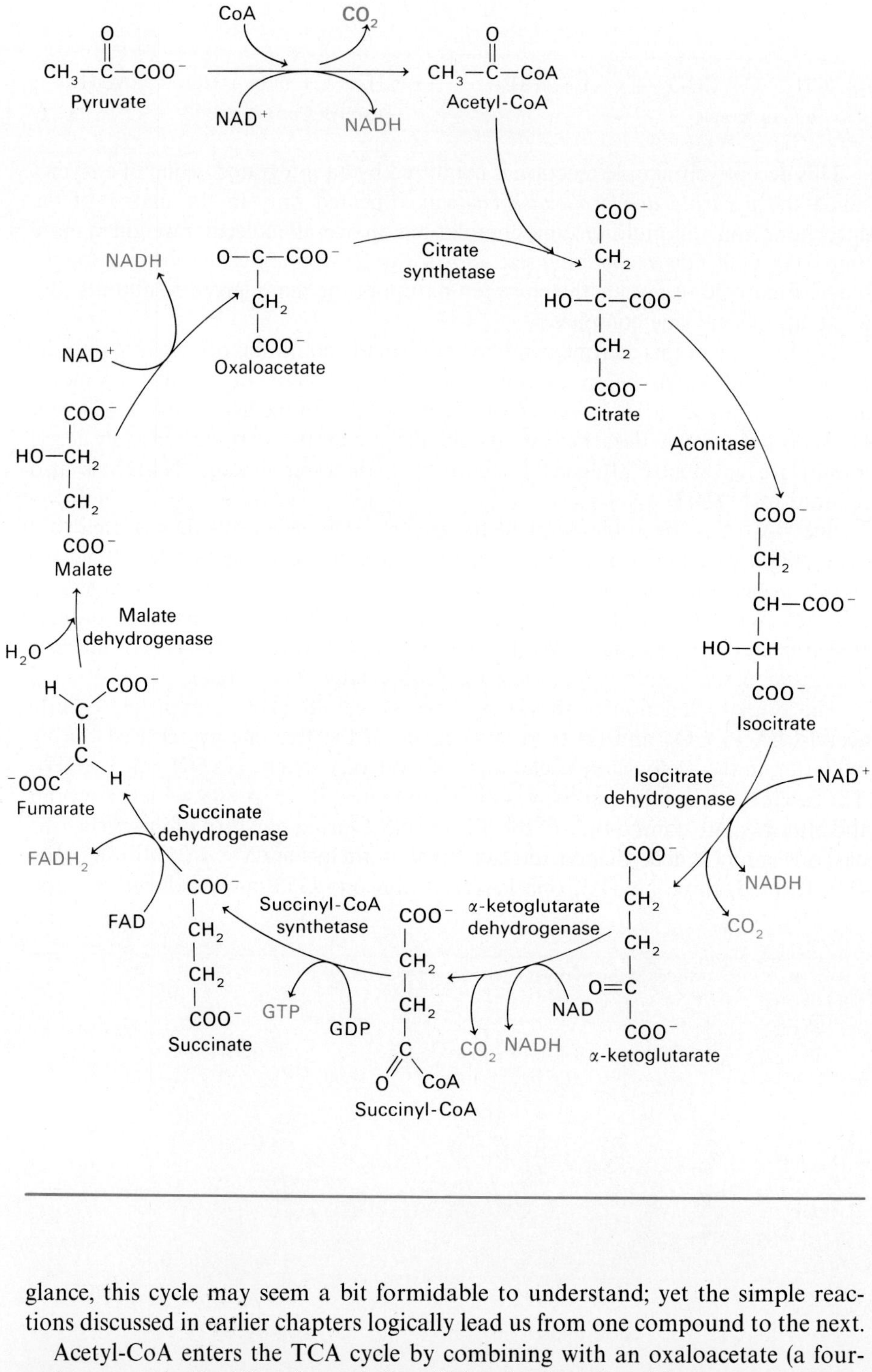

glance, this cycle may seem a bit formidable to understand; yet the simple reactions discussed in earlier chapters logically lead us from one compound to the next.

Acetyl-CoA enters the TCA cycle by combining with an oxaloacetate (a four-carbon organic diacid) to form citric acid (a tricarboxylic acid). The enzyme that catalyzes this condensation reaction is *citrate synthetase*, an important control point in the overall rate at which this pathway functions.

$$\underset{\text{oxaloacetate}}{\begin{array}{c} O=C-COO^- \\ | \\ CH_2 \\ | \\ COO^- \end{array}} + \underset{\text{acetyl-CoA}}{CH_3-\overset{O}{\underset{\|}{C}}-CoA} \xrightarrow{CoA} \underset{\text{citrate}}{\begin{array}{c} COO^- \\ | \\ CH_2 \\ | \\ HO-C-COO^- \\ | \\ CH_2 \\ | \\ COO^- \end{array}}$$

The second step of the TCA cycle is the isomerization of citrate in which the —OH group on carbon number three is moved down to carbon number four. The resulting compound is called by the logical name of *isocitrate*. *Aconitase*, the enzyme that catalyzes this isomerization, accomplishes this in two steps. First, citrate is dehydrated to form a double bond between C-3 and C-4 (*cis*-aconitate). Then the double bond is rehydrated with the —OH group adding to carbon number four.

$$\underset{\text{citrate}}{\begin{array}{c} {}^1COO^- \\ | \\ {}^2CH_2 \\ | \\ HO-{}^3C-COO^- \\ | \\ {}^4CH_2 \\ | \\ {}^5COO^- \end{array}} \xrightarrow[H_2O]{\text{aconitase}} \underset{\textit{cis}\text{-aconitate}}{\left[\begin{array}{c} COO^- \\ | \\ CH_2 \\ | \\ C-COO^- \\ \| \\ CH \\ | \\ COO^- \end{array}\right]} \xrightarrow[H_2O]{\text{aconitase}} \underset{\text{isocitrate}}{\begin{array}{c} {}^1COO^- \\ | \\ {}^2CH_2 \\ | \\ {}^3CH-COO^- \\ | \\ HO-{}^4CH \\ | \\ {}^5COO^- \end{array}}$$

Although citrate is a symmetrical molecule (having two identical groups attached to its central carbon), the aconitase enzyme can distinguish between the two seemingly identical —CH_2—COO^- groups. Radioactive labeling experiments have shown that the central —OH group (on carbon number three) is always moved to carbon number four (which came from oxaloacetate), and never to carbon number two (which originated with acetyl-CoA). This consistency is due to the three-dimensional structure of the aconitase active site and the way citrate binds to it.

Step three in the TCA cycle is the oxidation and decarboxylation of isocitrate to form *α-ketoglutarate* (note the keto group on the carbon in the alpha position, immediately next to the acid group). The carboxylate group on the central carbon in isocitrate is lost as CO_2, leaving only five carbons in α-ketoglutarate. The free energy released from the decarboxylation and the oxidation of an alcohol group to a keto group is used to reduce NAD^+ to NADH. The redox nature of this reaction is indicated by the name of its enzyme, *isocitrate dehydrogenase*.

$$\underset{\text{isocitrate}}{\begin{array}{c} COO^- \\ | \\ CH_2 \\ | \\ HC-COO^- \\ | \\ HO-CH \\ | \\ COO^- \end{array}} \xrightarrow[\substack{NAD^+ \\ \\ NADH + H^+}]{\substack{\text{isocitrate dehydrogenase} \\ \\ CO_2}} \underset{\text{α-ketoglutarate}}{\begin{array}{c} COO^- \\ | \\ CH_2 \\ | \\ CH_2 \\ | \\ O=C \\ | \\ COO^- \end{array}}$$

In the next step of the TCA cycle, another CO_2 is lost and another molecule of NADH is formed as α-ketoglutarate is transformed into *succinyl-CoA*. This reaction is very similar to the oxidative decarboxylation of pyruvate: pyruvate is an α-keto acid like α-ketoglutarate, CO_2 is lost, NAD^+ is reduced to NADH, and the product is attached to a molecule of CoA via a high-energy thioester linkage. Even the enzyme complex that catalyzes this reaction is similar, *α-ketoglutarate dehydrogenase*.

$$\text{α-ketoglutarate} \xrightarrow[\text{CO}_2 \quad \text{CoA} \quad \text{NAD}^+ \quad \text{NADH} + \text{H}^+]{\text{α-ketoglutarate dehydrogenase complex}} \text{succinyl-CoA}$$

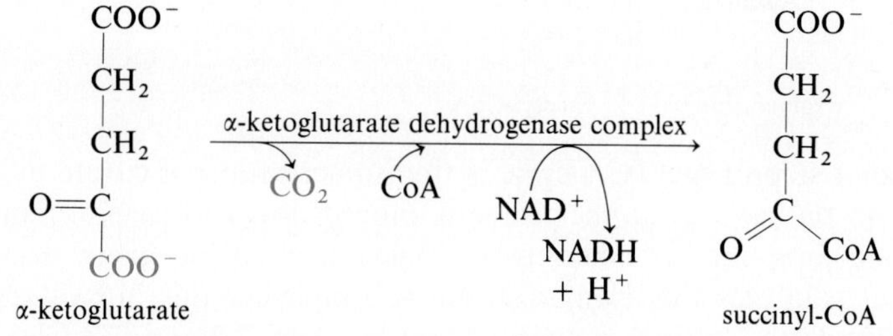

Some of the energy from the oxidation of α-ketoglutarate is conserved in the high-energy thioester linkage to CoA. In fact, during the next step of the TCA cycle, this thioester linkage is hydrolyzed. The free energy release is coupled to the direct formation of one high-energy bond in GTP. The product is merely the four-carbon diacid, succinate.

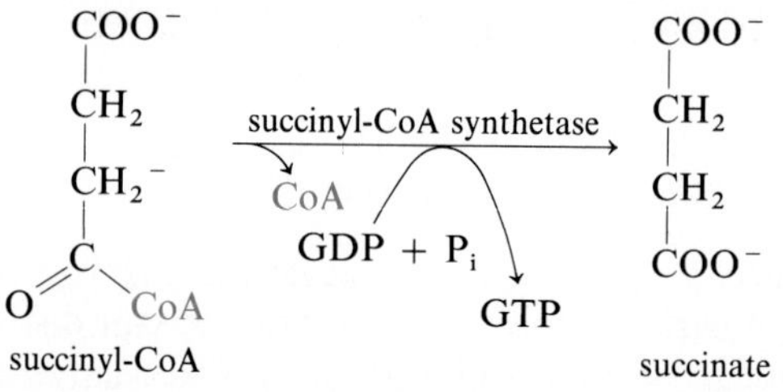

$$\text{succinyl-CoA} \xrightarrow[\text{CoA} \quad \text{GDP} + P_i \quad \text{GTP}]{\text{succinyl-CoA synthetase}} \text{succinate}$$

This reaction is readily reversible and the name of the enzyme was actually assigned on the basis of the reverse reaction: *succinyl-CoA synthetase*. The GTP formed in the forward reaction is equivalent to an ATP molecule (with respect to the high-energy bond), since a high-energy phosphate bond can be transferred via the following reaction.

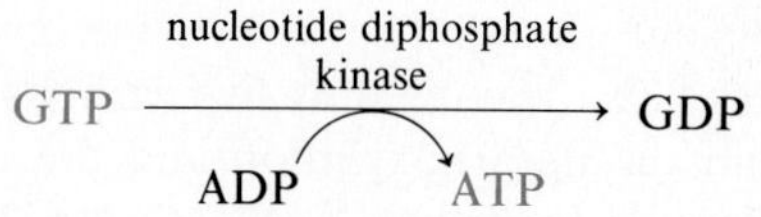

$$\text{GTP} \xrightarrow[\text{ADP} \quad \text{ATP}]{\text{nucleotide diphosphate kinase}} \text{GDP}$$

Until this point in the TCA cycle, oxidation reactions have involved the loss of CO_2. Such redox reactions release much more energy than simple dehydrogenation reactions. In the oxidation of succinate, a loss of two hydrogen atoms and their electrons (dehydrogenation) results in the formation of *fumarate*, a diacid with a *trans* double bond.

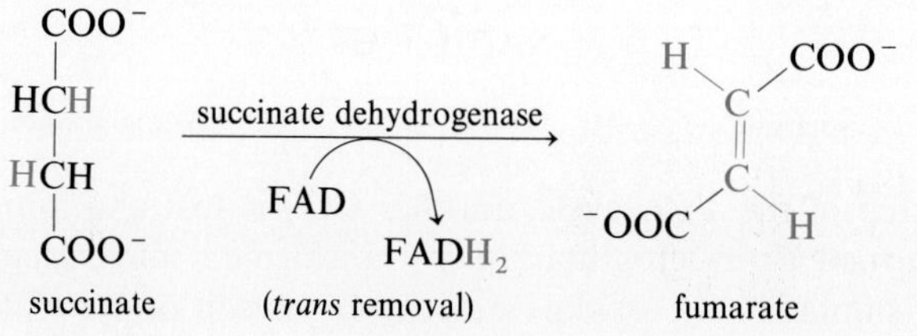

$$\text{succinate} \xrightarrow[\text{FAD} \quad \text{FADH}_2]{\text{succinate dehydrogenase}} \text{fumarate}$$
(*trans* removal)

Insufficient energy is liberated in this dehydrogenation to reduce NADH. Reduction of FAD to $FADH_2$ does not require as much energy (reductive potential) as the reduction of NADH, so the enzyme succinate dehydrogenase uses FAD as a cofactor, forming $FADH_2$ instead of NADH.

The seventh step in the TCA cycle is the hydration of the double bond in fumarate to yield *malate*, catalyzed by the enzyme *fumarase*.

$$\text{fumarate} \xrightarrow[H_2O]{\text{fumarase}} \text{malate}$$

The final reaction in the TCA cycle is another oxidation coupled to the reduction of NAD^+. Oxidation of the newly added alcohol group results in the formation of *oxaloacetate*, thus completing the cyclic pathway.

$$\text{malate} \xrightarrow{NAD^+ \rightarrow NADH + H^+} \text{oxaloacetate}$$

When cellular demands for energy are high, oxaloacetate is rapidly combined with acetyl-CoA to restart the cycle. However, if energy demands are low, oxaloacetate is not needed to continue the cycle. It can be transported out of the mitochondria and enter the gluconeogenic pathway, forming glucose and/or glycogen (Chapter 29).

Table 30–1 (p. 813) lists the eight reactions of the TCA cycle and their classifications. The overall TCA cycle is often easier to visualize by remembering the *type of reaction* that occurs at each step rather than committing each intermediate structure to memory.

Labeling studies, in which synthetic intermediates with radioactively labeled carbon atoms (^{14}C) are added to active mitochondria, allow each carbon atom to be traced throughout all steps of the cycle. For example, if we label the α-carbon of pyruvate with ^{14}C (often indicated with an asterisk, "*"), we can trace it throughout the TCA cycle after decarboxylation inside the mitochondria to form acetyl-CoA.

$$\text{pyruvate} \xrightarrow{CoA, \; CO_2} \text{acetyl-CoA} \longrightarrow \text{TCA cycle}$$

Other labeled intermediates of the cycle can also be introduced at various points in the pathway.

Although two carbons enter the cycle (in the form of acetyl-CoA) and two are lost during one turn of the cycle (as CO_2), you may have noted that these are not the same carbons. It is not until later turns of the cycle that these carbons are lost as CO_2.

Example 30.1

Explain why radioactive $^{14}CO_2$ is not evolved during the *first turn* of the TCA cycle when acetyl-CoA, labeled at its carbonyl carbon, is introduced to mitochondria.

Solution

Close examination of the structures of the TCA cycle reveals that the two CO_2 molecules lost during one turn of the TCA cycle come from the two acid groups of *oxaloacetate*. The radioactive label entering the cycle via acetyl-CoA appears in the acid group of citrate, which is not lost during this round of the cycle. This same labeled carbon is retained in the carbon skeleton of succinate as one of its terminal carbons. Because succinate is a symmetrical diacid, the enzymes that subsequently catalyze the formation of malate do not differentiate between one end or the other; 50% of the newly synthesized malate molecules are labeled at one end and 50% are labeled at the other. During the *second turn* of the cycle, both these groups are lost as $^{14}CO_2$.

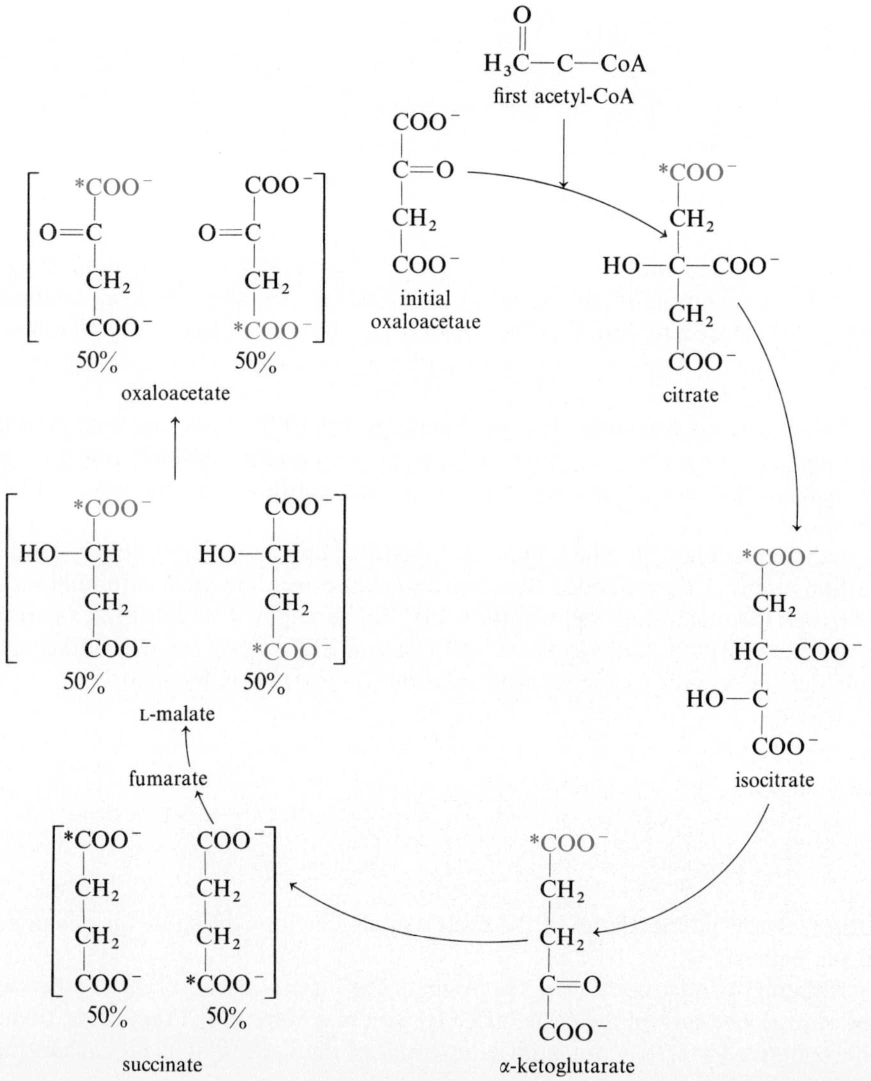

TABLE 30-1 Reactions of the Tricarboxylic Acid Cycle

Reaction	Type	Enzyme
1. acetyl-CoA + oxaloacetate + H_2O $\rightleftharpoons$ citrate + CoA + H^+	condensation	citrate synthetase
2. citrate $\rightleftharpoons$ isocitrate	isomerization	aconitase
3. isocitrate + NAD^+ $\rightleftharpoons$ α-ketoglutarate + CO_2 + NADH	oxidation and decarboxylation	isocitrate dehydrogenase
4. α-ketoglutarate + NAD^+ + CoA $\rightleftharpoons$ succinyl-CoA + CO_2 + NADH	oxidation and decarboxylation	α-ketoglutarate dehydrogenase
5. succinyl-CoA + P_i + GDP $\rightleftharpoons$ succinate + GTP + CoA	phosphorylation	succinyl-CoA synthetase
6. succinate + FAD $\rightleftharpoons$ fumarate + $FADH_2$	oxidation	succinate dehydrogenase
7. fumarate + H_2O $\rightleftharpoons$ malate	hydration	fumarase
8. malate + NAD^+ $\rightleftharpoons$ oxaloacetate + NADH + H^+	oxidation	malate dehydrogenase

Net Reaction:
acetyl-CoA + 3 NAD^+ + FAD + GDP + P_i + 2 H_2O $\longrightarrow$ 2 CO_2 + 3 NADH + $FADH_2$ + GTP + 2 H^+ + CoA

30.3 THE ELECTRON TRANSPORT CHAIN

The main energy objectives of the TCA cycle are to oxidize acetyl-CoA to CO_2 and to trap free energy in the electron carriers, NADH and $FADH_2$. As we know, energy is required to reduce NAD^+ and FAD to NADH and $FADH_2$. By using this stored energy (reduction potential), these carriers can transfer their electrons to various acceptors throughout the cell. One unique group of electron carriers in the mitochondria accepts these electrons and passes them from one to another along a specific pathway, with oxygen being the final acceptor. *This group of electron carriers, that transports electrons from NADH (or $FADH_2$) to O_2, is collectively referred to as the* **electron transport chain**.

electron transport chain

$$\text{NADH} \xrightarrow{(e^-)} \boxed{\text{electron transport chain}} \xrightarrow{(e^-)} O_2$$
(good reducing agent)   (easily reduced)

The majority of the electron transport chain components are located in the inner membrane of the mitochondria. Therefore, most of the electron transfers from one carrier to the next occur within this inner membrane. Some carriers require protons to accompany the electrons, while others do not. When needed, protons are taken from the aqueous solution on one side of the membrane. When they are not needed in a particular transfer, they are ejected into the aqueous solution on the other side of the membrane.

Many electron carriers make up the electron transport chain; most are cofactors associated with specialized enzymes in the membrane. Each is unique and different from the others, and each *has a different reduction potential*. As you recall, reduction potential is a measure of the tendency for a chemical compound to be reduced. Most often, reduction potentials are measured in volts, with higher (more positive) reduction potentials corresponding to an increased tendency to be reduced. Consider the reduction potentials for NADH and O_2.

Oxidized Form	Reduced Form	Reduction Potential
NAD^+	NADH	-0.32 volt
O_2	H_2O	$+0.82$ volt

Because O_2 has a higher reduction potential than NAD^+, if electrons are placed together under proper conditions, they will flow spontaneously from NADH to O_2. In fact, this difference in reduction potentials $[0.82 - (-0.32) = 1.14$ volts$]$ is equivalent to a free energy change of -52.6 kcal/mol.

$$NADH + \tfrac{1}{2} O_2 + H^+ \longrightarrow NAD^+ + H_2O \qquad \Delta G° = -52.6 \text{ kcal/mol}$$

The electron carriers of the electron transport chain all have slightly different reduction potentials, ranging from -0.32 volt (for NAD^+) to $+0.82$ volt (for O_2). The spontaneous flow of electrons from one carrier to the next is determined by the reduction potential of each carrier in the chain. Each reaction step *transfers electrons from carriers with lower potentials to carriers with higher reduction potentials.*

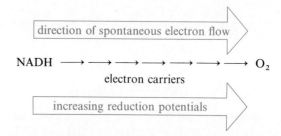

Theoretically, electrons can flow from NADH to O_2 directly, without passing through the electron transport chain. Indeed, this can be made to happen in certain laboratory experiments. However, in mitochondria, the only enzymatic pathway open to electrons from NADH to O_2 is through the electron transport chain.

Most of the membrane-bound carriers in the electron transport chain are clustered into three groups labeled *Site 1*, *Site 2*, and *Site 3*, with electrons passing through these three sites in numerical order. Each site contains a group of clustered enzymes, proteins, and cofactors that is so tightly bound together that the complex cluster comprising each site can be isolated as an intact group. The clustered groups at each site are independent from the other sites, allowing each site to diffuse freely about the inner membrane of the mitochondria.

NADH delivers its electrons to Site 1. After passing among the membrane-bound carriers at this site, a specialized carrier transports electrons to Site 2. This specialized carrier is **ubiquinone**, *a lipophilic quinone named for its "ubiquitous" occurrence in all living systems.* (An older name for ubiquinone is *coenzyme Q*.) A long aliphatic tail attached to the quinone ring of ubiquinone makes it lipophilic and it remains in the membrane, free to diffuse between sites 1 and 2. Since many ubiquinone molecules are present in the inner membrane of the mitochondria, they are sometimes called the ubiquinone "pool," a portion of which are reduced (abbreviated as QH_2) and a portion of which are oxidized (abbreviated as Q).

ubiquinone

oxidized ubiquinone (Q) $\xrightleftharpoons{2 e^- + 2 H^+}$ reduced ubiquinone (QH_2)

Upon delivery of electrons by ubiquinone to Site 2, electrons are passed between the various carriers clustered together at this location. Electrons are shuttled from

FIGURE 30-6 The heme group of cytochrome c (covalently bonded to the protein).

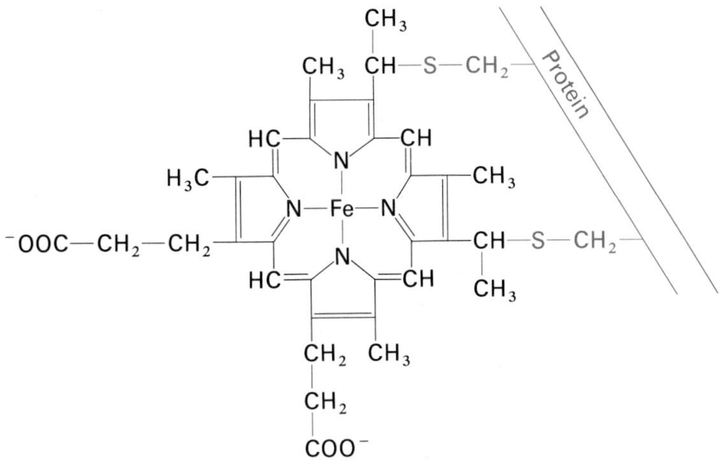

Site 2 to Site 3 by the water-soluble protein, *cytochrome c*. A great deal is known about cytochrome c, since it has been studied extensively. Its electron carrying site is an iron atom, located in the center of a covalently bonded heme group (see Figure 30-6). Because the iron atom is surrounded on all sides by the heme and protein side chains, no oxygen can bind to this heme (as was the case for hemoglobin and myoglobin), but electrons can reduce the ferric (Fe^{3+}) ion to a ferrous (Fe^{2+}) ion. Much as the membrane-bound ubiquinone pool delivers electrons to Site 2, a pool of water-soluble cytochrome c proteins delivers electrons to Site 3. However, unlike other carriers in the electron transport chain, this protein is water soluble, and it transports electrons to Site 3 through the aqueous solution of the mitochondria (see Figure 30-7).

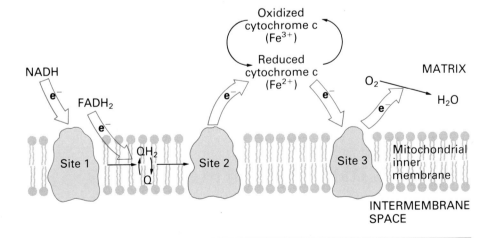

FIGURE 30-7 Schematic diagram of electron flow through the carriers of the mitochondrial electron transport chain, showing both the membrane-bound and water-soluble components. Electrons are transported between sites by the water-soluble protein, cytochrome c. After delivering its electron, it returns to transport more.

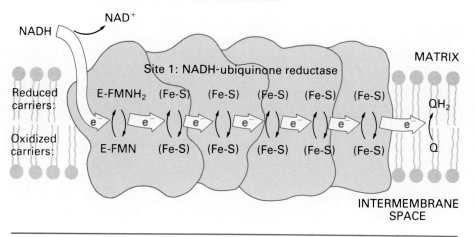

FIGURE 30–8 Schematic diagram of electron carriers at Site 1, all of which are embedded within the inner membrane of the mitochondria.

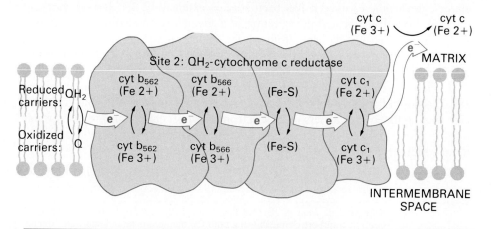

FIGURE 30–9 Schematic diagram of electron carriers at Site 2, all of which are embedded within the inner membrane of the mitochondria.

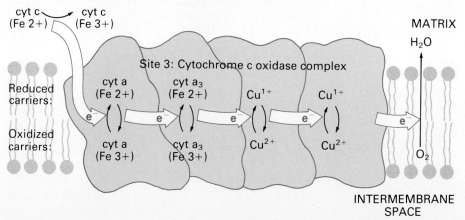

FIGURE 30–10 Schematic diagram of electron carriers at Site 3.

Site 3 consists of another group of electron carriers, the last of which transfers its electrons to O_2, reducing it to H_2O. The overall flow of electrons through the electron transport chain is summarized in Figure 30-7.

As we see in Figure 30-7, $FADH_2$ delivers its electrons to the electron transport chain by reducing Q to QH_2. Therefore, electrons from $FADH_2$ pass through sites 2 and 3 but *not Site* 1. This occurs partially because the reducing power of $FADH_2$ is less than that of NADH and it cannot reduce carriers at Site 1.

The technical name for Site 1 is *NADH-ubiquinone reductase*. Within this cluster of enzymes, proteins, and cofactors is a flavin-linked enzyme (a flavin mononucleotide cofactor, FMN) and at least five iron-sulfur proteins containing Fe_2S_2 and Fe_4S_4 centers (see Figure 30-8).

Reduced QH_2 molecules diffuse through the inner membrane and deliver electrons to Site 2. Site 2 is called the QH_2-*cytochrome c reductase complex*. It contains two cytochrome b proteins, cyt b_{562} and cyt b_{566}, identified by the different wavelengths of light they absorb (the subscripts are wavelengths in nanometers). Also present at Site 2 are another iron-sulfur protein and cytochrome c_1 (see Figure 30-9).

Site 3 is named the *cytochrome c oxidase complex*, containing two cytochromes (cyt a and cyt a_3). In some as yet unknown fashion, two copper atoms participate in the transfer of oxygen from cyt a_3 to oxygen (see Figure 30-10).

A number of chemical compounds are capable of specifically inhibiting electron flow through sites 1, 2, and 3. The fish poison *rotenone* and the barbiturate *amytal* (Figure 30-11) inhibit the electron transfer reactions at Site 1. As a result, enzymes at this site cannot accept electrons from NADH. The antibiotic *antimycin A* inhibits electron transfer at cytochrome b, thereby stopping electron flow through Site 2. Electron flow through Site 3 is stopped by *cyanide* (CN^-), *azide* (N_3^-), and *carbon monoxide* (CO). These relatively small species bind tightly to

FIGURE 30-11 Inhibitors of electron transfer and their sites of action on the electron transport chain.

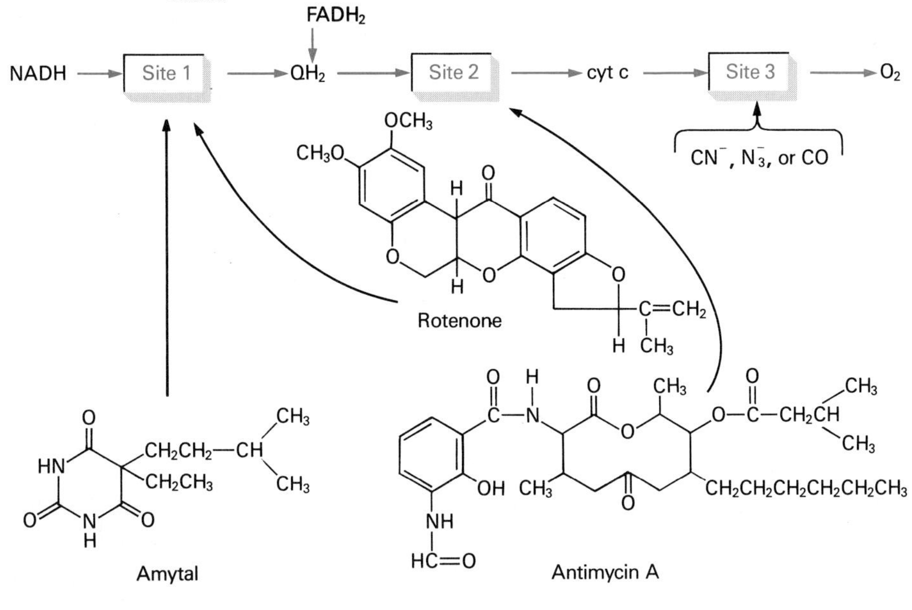

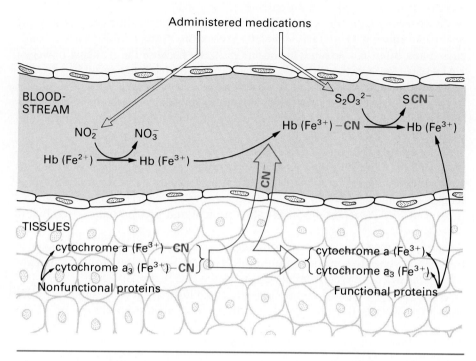

FIGURE 30–12 Treatment of cyanide poisoning by administrating nitrite (NO_2^-) and thiosulfate ($S_2O_3^{2-}$).

the oxidized heme iron (Fe^{3+}) in cytochromes a and a_3, blocking electron flow from cytochrome c to oxygen through this site.

Inhalation of hydrogen cyanide gas (HCN) or ingestion of potassium cyanide salt (KCN) rapidly inhibits the electron transport chain in all tissues, making cyanide one of the most potent and rapidly acting poisons known. Death is a result of tissue asphyxia, particularly in the central nervous system. Cyanide also blocks oxygen transport in the bloodstream by binding to the heme group in hemoglobin. One treatment for cyanide poisoning is administering various nitrites (NO_2^-), which oxidize the iron atoms of hemoglobin to Fe^{3+}. This form of hemoglobin helps draw CN^- back into the bloodstream where it can be converted to thiocyanate (SCN^-) by thiosulfate ($S_2O_3^{2-}$), which is also administered along with the nitrite (Figure 30–12).

30.4 TRANSMEMBRANE PROTON POTENTIALS: THE CHEMIOSMOTIC HYPOTHESIS

The most important aspect of the electron transfer reactions at sites 1, 2, and 3 is that as the electrons pass through these membrane-bound carriers, protons are picked up on one side of the inner membrane and deposited on the other. In 1961, Peter Mitchell proposed a unique hypothesis involving proton movement through the inner membrane of mitochondria. *Mitchell's detailed explanation of proton movement that accompanies electron transport became known as the* **chemiosmotic hypothesis**. His theory stated that each electron passing through each site literally forces one proton to be transported from the matrix to the intermembrane space. Since a pair of electrons is donated by each NADH (or

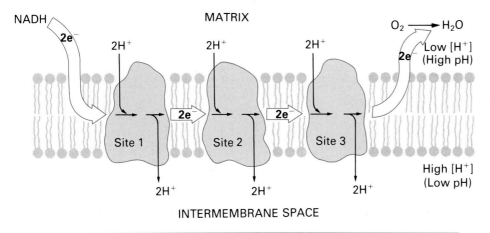

FIGURE 30–13 Diagram of proton "pumping," the basis of Peter Mitchell's chemiosmotic hypothesis.

FADH$_2$), two protons are "pumped" from one side of the membrane to the other as they pass through *each site* (see Figure 30–13).

Since protons are moved in only one direction across this membrane, they begin to "stack up" on the outside of the membrane. In time, this causes a differential concentration of protons on one side of the membrane compared to the other: high pH in the matrix (where H$^+$ concentration is lower) and low pH in the intermembrane space (where H$^+$ concentration is higher). Experimental evidence shows that during active electron transport, the intermembrane space is up to 25 *times more acidic* than the matrix.

A higher concentration of protons on one side of a membrane results in a potential energy difference called a **transmembrane proton potential** (Chapter 28). This potential energy difference exists because protons tend to diffuse back to the other side of the membrane in a process called *osmosis*, for which Mitchell's hypothesis is named. The insolubility of protons in the lipid membrane prevents their return and, as more and more are pumped out, the transmembrane potential continues to rise. Eventually, if the proton gradient is not dissipated, the transmembrane potential becomes so great that it begins to impede electron flow through the membrane. Much like pumping water from a lower reservoir to a higher one, the buildup of water eventually causes a back-pressure that slows down the pump.

Certain chemical compounds can dissipate transmembrane potentials. One such compound is 2,4-dinitrophenol (2,4-DNP). The phenolic proton in this compound comes on or off easily at physiological pH, and both the protonated and unprotonated forms are soluble in the membrane. When added to mitochondrial preparations that have built up a substantial proton potential difference, 2,4-DNP quickly dissipates the proton potential by transporting protons from the concentrated side through the membrane to the less concentrated matrix. This proton movement destroys the transmembrane potential and its accompanying resistance to electron flow, such that electron transport accelerates to a much faster pace than before the addition of 2,4-DNP. This chemical and others that function like it are poisonous because they rob mitochondria of energy stored in the transmembrane potential (see Figure 30–14, p. 820).

Transmembrane potentials are also found in bacteria and chloroplasts. Bac-

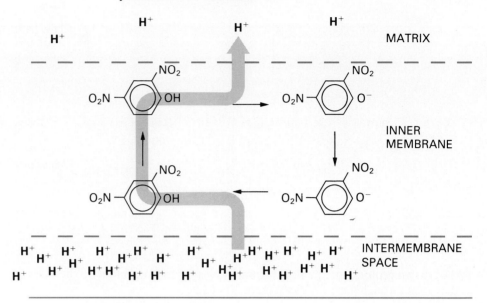

FIGURE 30-14 Action of 2,4-DNP. Picking up protons from the intermembrane space where their concentration is high, 2,4-DNP delivers them to the matrix and returns for other protons. Therefore, 2,4-DNP functions merely by facilitating diffusion of protons across the membrane.

teria pump protons in a similar fashion to create a transmembrane potential, which is used to power their flagella and thereby move through their medium. Chloroplasts pump protons across an inner membrane (of the granum) in a light-driven reaction. As we are about to see, both mitochondria and chloroplasts use transmembrane potentials to synthesize ATP.

30.5 ATP SYNTHESIS: OXIDATIVE PHOSPHORYLATION

The second aspect of Mitchell's hypothesis was that the energy stored in the form of a proton gradient can be coupled to the synthesis of ATP. As two electrons pass from one end of the electron transport chain (NADH) to the other (O_2), enough energy is generated by proton pumping to generate three ATP: *one ATP at each site.* **Oxidative phosphorylation** *describes phosphorylation of ADP to ATP, using the energy supplied by oxidation reactions of the TCA cycle and electron transport.* This was an excellent theory, but just how is the energy in a proton gradient coupled to the synthesis of ATP?

During the 1960s, elegant work by Ephriam Racker and his associates at Cornell University showed that the key to this paradox was the spherical nodules on the matrix side of the inner membrane, identified today as *the ATP synthetase complex.* Racker showed that the ATP synthetase complex is a group of proteins separate from the proteins of electron transport, extending through the inner membrane and forming a channel from one side to the other. When protons flow back through the ATP synthetase channel, the osmotically driven potential difference is used to synthesize ATP.

The ATP synthetase complex is composed of a number of protein subunits that are divided into two groups labeled F_0 and F_1 (F for coupling *factors*). The F_0 group of proteins is embedded inside the membrane and forms the channel through which protons can flow. The F_1 group forms the spherical nodules visible under

FIGURE 30-15 The transmembrane potential formed by electron transport is coupled to ATP synthesis at a remote site by the flow of protons back through the ATP synthetase complex.

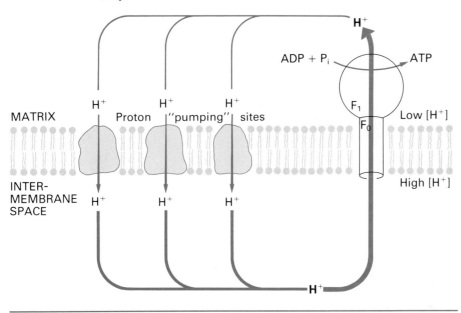

the electron microscope (Figure 30-2). Included in the F_1 group of proteins are the enzymes responsible for ATP synthesis. The driving force behind proton flow through the F_0 channel is harnessed by F_1 enzymes and used to form the high-energy anhydride linkage when ADP is converted to ATP.

Let us examine some of the experimental facts observed by Racker and others that verify this site as the one for ATP synthesis. First of all, if membranes of mitochondria are damaged so that protons can diffuse back through the inner membrane into the matrix, ATP synthesis stops, but electron transport continues, often at a faster rate than before the damage. When chemical treatments are used to remove F_1 from F_0, F_0 forms an open channel through the inner membrane and the protons stored in the gradient on one side of the membrane stream back through the open channel. Since the transmembrane proton potential is lost, ATP synthesis stops and electron transport accelerates (see Figure 30-15).

Example 30.2

Consider an experiment in which mitochondrial preparations are actively producing ATP by oxidative phosphorylation. Explain what happens to the rates of both electron transport and ATP synthesis upon the addition of 2,4-DNP.

Solution
As described earlier, 2,4-DNP dissipates the transmembrane proton potential produced by the electron transport chain. There is less resistance to proton pumping and electron transport from NADH to O_2 speeds up. On the other hand, the lack of a transmembrane potential completely stops ATP production. Under experimental conditions, addition of 2,4-DNP yields much the same effect as removing the F_1 complex from F_0.

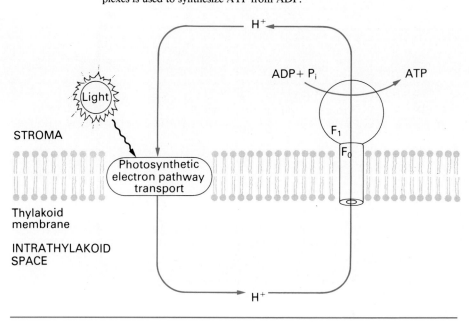

FIGURE 30–16 The proton cycle in photosynthesis. Photosynthetic reactions driven by light reactions pump protons across the thylakoid membrane to form a transmembrane proton potential. The flow of protons back through ATP synthetase complexes is used to synthesize ATP from ADP.

Thylakoid membranes in chloroplasts also have ATP synthetase complexes. A light-driven reaction and a subsequent electron transport series in this membrane pumps protons from the stroma into the intrathylakoid space. The resulting transmembrane potential is used by these ATP synthetase complexes to synthesize ATP as protons flow back through the complexes into the stroma (see Figure 30–16). Although the electron transport series in chloroplasts is different from that of mitochondria, the actual process of ATP synthesis seems almost identical.

30.6 ATP ACCOUNTING

We have now seen that NADH derived from the TCA cycle can diffuse to the inner membrane and deliver its electrons. As these electrons pass through the electron transport chain, a transmembrane potential is developed that is eventually used to synthesize ATP. An important question at this point is how many ATP molecules can be synthesized from the electrons donated by just one NADH molecule? To find the answer, we must remember that *two electrons are carried by each NADH molecule* and that *both enter the electron transport chain*. Each pair of electrons passes through the three proton "pumping" sites of electron transport, requiring the obligatory passage of two protons from the matrix into intermembrane space at each of these three sites. Enough energy is produced in the transmembrane potential for the production of one ATP molecule per site. For accounting purposes, we can say that *every NADH in the mitochondria supplies sufficient reductive energy to synthesize three ATP molecules from ADP.*

Example 30.3

$FADH_2$ formed by the TCA cycle does not deliver electrons to the electron transport chain at the same point as NADH. How many moles of ATP can be synthesized from one molecule of $FADH_2$?

Solution

FADH$_2$ delivers electrons to the electron transport chain via the ubiquinone pool, which in turn carries the electrons to Site 2 (bypassing Site 1). Since electrons from FADH$_2$ do not pass through Site 1, no protons are pumped at this site. However, electrons from FADH$_2$ do pass through sites 2 and 3, where protons are pumped, providing enough transmembrane potential to produce *two ATP molecules*.

ATP produced in the matrix of the mitochondria are transported to the cytoplasm of the cell by specific transport proteins. These require the antiport of one ADP and one inorganic phosphate (P$_i$) into the matrix as one ATP molecule leaves. This process keeps the ADP/ATP ratio carefully controlled inside the matrix.

Since we know that NADH oxidation produces three ATP molecules and oxidation of FADH$_2$ produces two molecules of ATP, we can count up the total ATP produced by the total oxidation of one molecule of pyruvate. Recall that pyruvate was the product of glycolysis in the cytoplasm and diffused into the mitochondria before further oxidation. Oxidative decarboxylation of pyruvate forms acetyl-CoA and one molecule of NADH. Three more molecules of NADH are formed by one turn of the TCA cycle, making a total of four NADH molecules produced by the oxidation of one molecule of pyruvate. One FADH$_2$ and one GTP are also formed. Counting GTP equivalent to ATP, three ATP for every NADH and two ATP for the FADH$_2$, adds up to 15 molecules of ATP produced for every molecule of pyruvate.

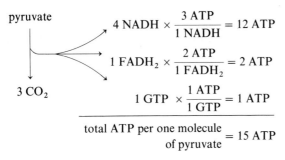

Let us now extend our ATP accounting to the oxidation of one molecule of glucose. Since two molecules of pyruvate are produced from glycolysis, we need to double the 15 ATP produced from one pyruvate. In addition to these, a net gain of two ATP molecules is achieved by glycolysis. This brings us to a total of (2 × 15) + 2, or 32 molecules of ATP. However, we must also consider the two molecules of NADH produced in the cytoplasm (from the oxidation of two molecules of glyceraldehyde-3-phosphate to 1,3-diphosphoglycerate).

These two molecules of NADH must enter the mitochondria before they can deliver their electrons to the electron transport chain. This presents somewhat of a problem, since *NADH cannot enter the mitochondria directly*. Electrons from cytoplasmic NADH are shuttled into the mitochondria by two different methods. One shuttle system (found in liver cells) is called the *malate shuttle*, since malate acts as an intermediate carrier between cytoplasmic NADH and NADH in the matrix. Electrons from NADH are used to reduce oxaloacetate to malate, which can traverse the mitochondrial inner membrane. Once inside, malate is oxidized back to oxaloacetate, giving up its electrons to form NADH in the matrix (see Figure 30–17, p. 824). In this fashion, matrix NADH is equivalent to cytoplasmic

FIGURE 30-17 Diagram of the malate shuttle. Although a number of other steps are involved in the return of oxaloacetate to the cytoplasm, malate actually carries electrons into the matrix.

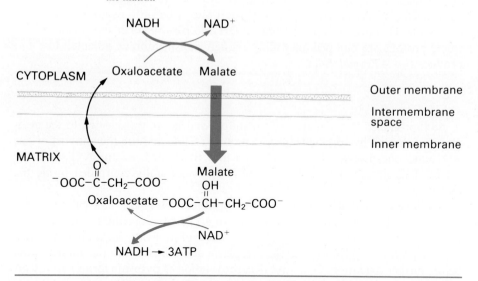

NADH, and three ATP molecules are produced. This results in the total production of 38 ATP molecules per molecule of glucose oxidized.

$$\text{glucose} \begin{cases} \text{net gain from glucose to pyruvate} = 2 \text{ ATP} \\ 2 \text{ NADH (transported via malate shuttle)} \times \dfrac{3 \text{ ATP}}{\text{NADH}} = 6 \text{ ATP} \\ 2 \text{ pyruvate} \times \dfrac{15 \text{ ATP}}{1 \text{ pyruvate}} = 30 \text{ ATP} \end{cases}$$

$\downarrow$ 6 CO_2

Net ATP gain from one molecule of glucose = 38 ATP
(via the malate shuttle)

Another shuttle system for NADH functions in some tissues (such as muscle), but it is not nearly as efficient. This system is called the *glycerol phosphate shuttle*. Electrons from NADH are used to reduce dihydroxyacetone phosphate to glycerol-3-phosphate. Glycerol-3-phosphate diffuses to the inner membrane where it reduces FAD to $FADH_2$. Therefore, when this shuttle is operational, electrons from cytoplasmic NADH must enter the electron transport pathway via $FADH_2$, resulting in only two ATP molecules being synthesized per cytoplasmic NADH instead of three as with the malate shuttle. This means only 36 total ATP can be generated from one molecule of glucose when the malate shuttle is functioning (see Figure 30-18).

Let us briefly consider the efficiency of total glucose oxidation from glycolysis through oxidative phosphorylation. If we select liver tissue, where the malate shuttle predominates, we gain 38 moles of ATP from the oxidation of one mole of glucose. With 7.3 kcal/mol of free energy stored in each ATP bond, this represents a total of 38 ATP × −7.3 kcal/mol (−277.4 kcal) of free energy captured for use. If we recall that the $\Delta G°$ for the direct combustion of glucose to CO_2 is −686 kcal/mol, we can calculate the efficiency in capturing this free energy.

FIGURE 30-18 The glycerol phosphate shuttle.

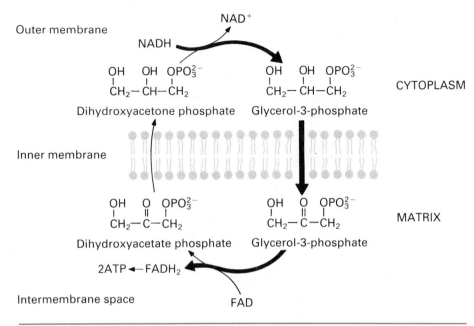

$$\frac{\text{energy stored in 38 ATP high-energy bonds} \longrightarrow -277 \text{ kcal}}{\text{free energy from combustion of glucose} \longrightarrow -686 \text{ kcal}} \times 100\% = 40\%$$

Although 40% may seem low, it actually represents an extremely efficient reclamation of energy from oxidation, far exceeding efficiencies of machines (such as internal combustion engines).

30.7 REGULATION OF THE TCA CYCLE AND OXIDATIVE PHOSPHORYLATION

The TCA cycle is strictly regulated by both its own metabolic product, NADH, and by ATP, the main product of oxidative phosphorylation. A number of enzymes in the TCA cycle act as control sites, responding automatically to the levels of these products.

The activity of *pyruvate dehydrogenase*, which produces acetyl-CoA for entry to the cycle, is inhibited by its own reaction products, NADH and acetyl-CoA (an excellent example of feedback inhibition). Furthermore, high levels of ATP also retard its activity (through phosphorylation of the enzyme). As levels of ATP, NADH, or acetyl-CoA drop in the mitochondria, inhibition of pyruvate dehydrogenase decreases and its activity increases accordingly.

Three main control points also exist within the cycle. Citrate synthetase and α-ketoglutarate dehydrogenase are also inhibited by high levels of either NADH or ATP. The isocitrate dehydrogenase enzyme is slightly inhibited by ATP or NADH, but unlike the other control enzymes, it is activated by ADP and NAD$^+$ (see Figure 30-19, p. 826).

Regulation of electron transport and oxidative phosphorylation are closely related, depending on the need for ATP. As the concentration of ATP in the mitochondria rises, it is increasingly difficult for the ATP synthetase complex to synthesize it. This effect is due primarily to Le Châtelier's principle: less ADP and more ATP stress the equilibrium, making it more difficult to convert ADP to

FIGURE 30–19 Critical control points in the TCA cycle.

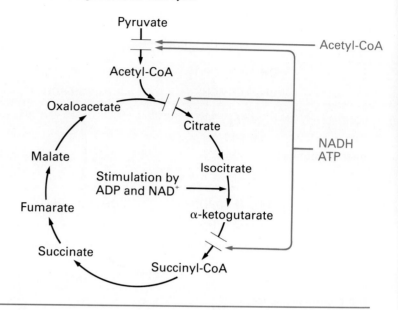

ATP. When the activity of the ATP synthetase complex decreases, the transmembrane potential builds up and slows the rate of electron transport. As we mentioned in Section 30.4, the obligatory proton movement associated with electron transport becomes increasingly difficult as the pH difference across the membrane increases. As a result, all electron carriers are gradually reduced and can no longer accept electrons from NADH or FADH$_2$. Unless the reduced forms of NADH and FADH$_2$ are needed for other redox reactions, their concentration also rises.

FIGURE 30–20 Diagram of effects of addition of antimycin A and its effect on electron flow, proton pumping, and ATP synthesis.

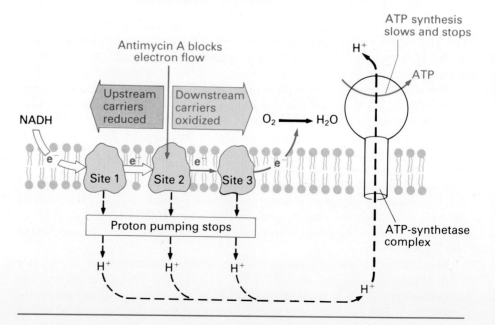

Slowing electron transport by increased levels of ATP resembles a major traffic jam. This massive backup of electrons and protons occurs because electrons are limited to a certain pathway through the carriers of the inner membrane, and exiting protons must return to the matrix only through the ATP synthetase complex. The backup continues until the ATP concentration drops and protons can again begin to move through the F_0 channel as ATP is synthesized.

Under laboratory conditions, experiments can be designed to artificially impede electron flow and ATP synthesis. Consider the effects of antimycin A. Addition of antimycin A blocks electron flow at Site 2, resulting in electron carriers "upstream" from Site 2 becoming reduced. Those "downstream" become oxidized as they pass along their electrons and none are available to replace them. The transmembrane proton potential decreases with time and eventually ATP synthesis slows and stops (see Figure 30–20).

EXERCISES AND PROBLEMS

The Mitochondrion

30.1 Sketch a simple drawing of a mitochondrion and identify each of the following parts:
a. inner and outer membranes
b. matrix
c. ATP synthetase complex
d. intermembrane space
e. region of higher pH during active oxidative phosphorylation
f. cristae
g. membrane where electron transport proteins are located

30.2 List and compare the lipid versus protein composition of the two mitochondrial membranes. Which is higher in protein concentration?

30.3 Describe the origin of the name *mitochondrion*.

The Tricarboxylic Acid Cycle

30.4 Explain why the tricarboxylic acid cycle is often called the Krebs cycle or the citric acid cycle.

30.5 Pyruvate is prepared for oxidation in the mitochondria by the enzyme pyruvate dehydrogenase. Write the structural formula for pyruvate and the product of this reaction. List the cofactors needed for this reaction sequence and explain why this enzyme is often called the pyruvate dehydrogenase *complex*.

30.6 What enzymatic reaction of the TCA cycle is *extremely* similar to that catalyzed by pyruvate dehydrogenase? Write the reaction and list some of its similarities.

30.7 How is oxaloacetate (containing four carbons) converted into citrate (containing six carbons)? Draw the structures for this reaction.

30.8 Consider the reactions that occur during *one turn* of the TCA cycle to answer each of the following:
a. How many carbons are gained?
b. How many carbons are lost?
c. How many molecules of NADH are formed?
d. How many times does water add to a carbon-carbon double bond?
e. How many times is a secondary alcohol oxidized?
f. How many molecules of $FADH_2$ are formed?
g. How many GTP molecules are formed?

30.9 If for some reason the level of oxaloacetate were too low to sustain normal TCA cycle reaction rates, how could the cell produce extra oxaloacetate other than by reactions of the TCA cycle? (Hint: Remember the gluconeogenic pathway.)

30.10 Which intermediate of the TCA cycle contains a high-energy linkage? Draw its structure and explain how the energy in this bond is used by the cycle.

30.11 Match the following enzymes with the appropriate type of reaction they catalyze.

Enzyme	Reaction Type
a. fumarase	1. isomerization
b. malate dehydrogenase	2. oxidation only
c. succinate dehydrogenase	3. hydration
d. succinyl-CoA synthetase	4. oxidation and decarboxylation
e. isocitrate dehydrogenase	
f. aconitase	5. phosphorylation

30.12 What is the fate of the radioactive carbon atom in each of the following compounds when they are added to cell-free extracts containing all necessary enzymes and cofactors of the TCA cycle and the glycolytic pathway?

a. $CH_3-\underset{\|}{\overset{O}{C}}-\overset{*}{C}OOH$

b. $\overset{*}{C}H_3-\underset{\|}{\overset{O}{C}}-COOH$

c.

HOOC—CH$_2$—CH$_2$—C(=O)—$\overset{*}{C}$OOH

d.

HOOC—$\overset{\underset{|}{OH}}{CH}$—CH$_2$—$\overset{*}{C}$OOH

e. glucose-6-phosphate labeled at C-6

The Electron Transport Chain

30.13 How do electrons from the TCA cycle reach the electron transport chain?

30.14 What is the location of the majority of electron carriers in the electron transport pathway?

30.15 What is the driving force for electrons to flow spontaneously from one carrier to the next in electron transport?

30.16 Considering the $\Delta G°$ for electron transfer between NADH and O$_2$ (-52.6 kcal/mol) and the energy needed to produce one ATP from ADP (-7.3 kcal/mol), calculate the efficiency of ATP synthesis by oxidative phosphorylation.

30.17 Suppose a new member of the electron transport chain were discovered, with a reduction potential of $+0.75$ volt. At what point in the electron transport chain would you predict it to function: (a) close to NADH, near the beginning of the chain, or (b) closer to O$_2$, near the end of the chain? Explain your reasoning.

30.18 What is meant by Site 1, Site 2, and Site 3 of the electron transport chain? What is the composition of a Site?

30.19 Why is the term ubiquinone *pool* used to describe this carrier of electrons? Where is this pool located?

30.20 What is the major difference between the oxidized and reduced forms of cytochrome c?

30.21 Where do electrons from NADH and FADH$_2$ enter the electron transport pathway? Why do these electrons not enter at the same location?

30.22 Sketch a simple diagram of the electron transport chain. Identify each of the following sites:
a. where electrons from NADH enter
b. where electrons from FADH$_2$ enter
c. the point at which each of the following blocks electron flow through the pathway:
  i. cyanide (CN$^-$)
  ii. amytal
  iii. azide (N$_3^-$)
  iv. antimycin A
  v. carbon monoxide (CO)
  vi. rotenone

30.23 Explain the chemical basis for treating cyanide poisoning with nitrite and thiosulfate.

Transmembrane Proton Potentials

30.24 Protons are needed in some electron transfer reactions at sites 1, 2, and 3. Where are these protons obtained from, and where are they deposited when no longer needed in subsequent reactions at these sites?

30.25 Why was the term *chemiosmotic* applied to Peter Mitchell's theory?

30.26 What is meant by *proton pumping*?

30.27 During active electron transport, protons move from one side of the inner membrane to the other. Identify the regions of the mitochondria that become more acidic and less acidic as a result of this proton movement.

30.28 Explain in your own words how energy can be stored in a transmembrane proton potential.

30.29 How does 2,4-dinitrophenol (2,4-DNP) destroy a transmembrane proton potential?

30.30 List two other locations in biological systems where proton-pumping mechanisms are found besides mitochondria, and explain the way energy from each of these potentials is used.

ATP Synthesis: Oxidative Phosphorylation

30.31 Define *oxidative phosphorylation*.

30.32 Draw a diagram of the ATP synthetase complex. Identify the F$_0$ and F$_1$ subunits and explain their function.

30.33 Why does electron transport slow down when levels of ATP are high in the matrix of the mitochondria?

30.34 Oligomycin is an antibiotic used in laboratory experiments that *blocks the F$_0$ channel* and prevents proton flow through this subunit. What effect would you predict oligomycin to have on ATP synthesis via oxidative phosphorylation? Explain.

30.35 Transmembrane potentials are also used to synthesize ATP in chloroplasts. What is the source of these proton potentials?

30.36 In a famous experiment reported in 1974, Ephriam Racker and Walther Stoeckenius formed artificial lipid vesicles (artificial cells) that contained only the ATP synthetase complex (from beef heart mitochondria) and a purple pigment/protein complex from *Halobacteria* (salt-loving bacteria). Racker's goal was to prove that ATP could be synthesized from any proton gradient and Stoeckenius wanted to prove that this purple pigment/protein complex pumped protons when illuminated with light. Both scientists were successful. Draw a diagram of these lipid vesicles with the F$_1$ subunit of the ATP synthetase complex extending to the outside of the vesicles and the purple pigment/protein complex at a different site. Using this diagram, explain what happened during their experiment. (For a detailed description, see their article in *Journal of Biological Chemistry*, **249**, 662–663.)

ATP Accounting

30.37 Explain why only two ATP are obtained from oxidation of FADH$_2$ by electron transport, while the oxidation of NADH under similar conditions yields three ATP.

30.38 Your frustrated friend, Susan, approaches you at lunch and asks for your help. Her botany textbook states that total oxidation of one molecule of glucose yields 36 ATP; yet her zoology textbook states that a total of 38 ATP molecules is produced for the *same oxidation* process. She has a test coming up next week in each class and wants to know which book is right. When you explain that they are both right, she becomes totally confused. Offer an explanation that will calm her down and help her pass both tests.

30.39 What is the ATP yield when each of the following intermediates is completely oxidized to CO_2 by cell-free extracts? (Assume that all oxidative pathways are completely functional: glycolysis, the TCA cycle, and oxidative phosphorylation. Also assume that the malate shuttle is functioning in each case.)

a. fructose
b. dihydroxyacetone phosphate
c. $FADH_2$
d. pyruvate
e. glucose-6-phosphate
f. phosphoenolpyruvate (PEP)
g. NADH

Regulation of the TCA Cycle and Oxidative Phosphorylation

30.40 List four enzymatic reactions that control the activity of the TCA cycle. Explain which compounds inhibit or activate these enzymes.

30.41 How do higher levels of ATP retard the flow of electrons through the electron transport chain?

30.42 At one location in northern California, the power company pumps water from a lower lake up to a higher lake during the night when demands for electrical power are low. During the day (when higher demands for electricity exist), the water is allowed to run back down to the lower lake. As it flows back down, the water is used to turn generators and provide electrical power. Compare this process to ATP synthesis in mitochondria. Be sure to indicate the similarities *and differences* between the two processes.

30.43 Predict the effects (if any) on *rates* of electron transport and ATP synthesis for each of the following experiments. In each case, assume tightly coupled mitochondrial preparations are used and that enough reduced NADH has been added to maintain a steady flow of electrons and constant rate of ATP synthesis initially.

Experimental Conditions	Changes in Rates of Electron Transport	ATP Synthesis
a. antimycin A is added		
b. 2,4-DNP is added		
c. cyanide is added		
d. both amytal and $FADH_2$ are added		
e. both azide and $FADH_2$ are added		
f. excess ATP is added		
g. excess ADP is added		

31 Lipid Metabolism

Objectives

After completing Chapter 31, you will be able to:

31.1 Identify metabolically important lipids used primarily for energy storage, together with their main sources and locations in the body.
31.2 Discuss the absorption and transport of dietary lipids.
31.3 Outline the steps of fatty acid mobilization and oxidation that result in energy production for the cell.
31.4 Explain how fatty acid synthesis occurs in cells, including the formation of triglycerides.
31.5 Determine accurately the amount of energy released by the complete oxidation of various length fatty acids as measured by net ATP synthesis.
31.6 Describe two important ways in which lipid metabolism is regulated in the body that involve both the mobilization of triglycerides and fatty acid oxidation.
31.7 Explain the origin of ketone bodies and their usefulness in the diagnosis of diabetes mellitus.

INTRODUCTION

Lipids represent a diverse group of biological compounds with a variety of physiological functions. In Chapter 23 we discussed many of these compounds and their roles. In this chapter, we will discuss how certain classes of lipids serve as storage and transport forms of metabolic fuel. These lipids play an extremely important role in cellular metabolism because they are an energy-rich fuel and can be stored in large amounts in adipose (fat) tissue. Between one-third and one-half of the calories consumed daily in the average American diet are supplied by lipids. Furthermore, excess energy derived from carbohydrates and proteins beyond normal daily needs is stored in the adipose tissue, later to be mobilized and used when needed.

In addition to serving energy needs, lipids help to insulate and protect the body. Accordingly, lipid deposits are often found in the peripheral tissues or surrounding organs. Analyzing the biochemical products of abnormal lipid metabolism assists in the diagnosis of various diseases and physiological disorders.

31.1 METABOLICALLY IMPORTANT SOURCES OF TRIGLYCEROL LIPIDS

In a typical mammal, at least 10 to 20% of the body weight is lipid, the bulk of which exists as *triglycerides*. **Triglycerides** *are triacyl-esters of glycerol in which each of the three alcohol groups of glycerol are esterified to a fatty acid*. These compounds are sometimes called *neutral fats, triacylglycerol,* or *triglycerols*. Most fatty acids in triglycerides contain an even number of carbon atoms usually ranging from 16 to 20 carbon atoms in length (see Figure 31–1). A triglyceride composed of glycerol and three palmitic acid molecules is given a specific name of *tripalmitin*, with similarly derived names for other triglycerides.

triglycerides

Triglycerides are distributed in all organs, particularly in adipose tissue, where droplets of this lipid represent more than 90% of the cytoplasm of some cells. Body lipid is a reservoir of potential chemical energy. About 100 times more energy is stored as mobilizable lipid than as mobilizable carbohydrate in the normal

FIGURE 31–1   A generalized triglyceride. Fatty acids vary in length and saturation from one triglycerol to another.

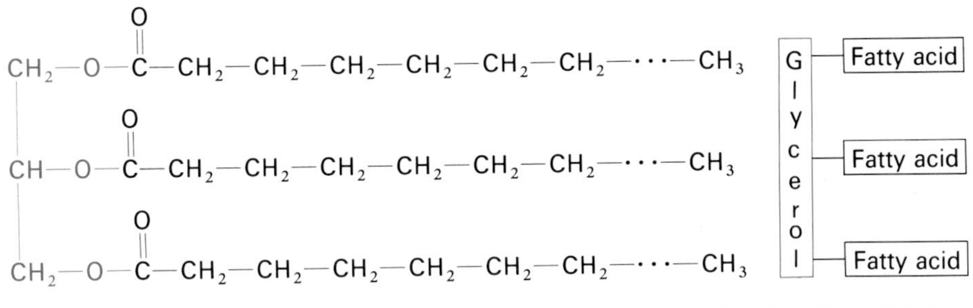

$$CH_2-O-\overset{\overset{O}{\|}}{C}-CH_2-CH_2-CH_2-CH_2-CH_2-CH_2-\cdots-CH_3$$

$$CH-O-\overset{\overset{O}{\|}}{C}-CH_2-CH_2-CH_2-CH_2-CH_2-CH_2-\cdots-CH_3$$

$$CH_2-O-\overset{\overset{O}{\|}}{C}-CH_2-CH_2-CH_2-CH_2-CH_2-CH_2-\cdots-CH_3$$

Glycerol — Fatty acid / Fatty acid / Fatty acid

human being. A normal man (70 kg) is estimated to have 15 kg of triglycerides scattered throughout his tissues and organs. Much of the lipid in mammals is located subcutaneously, where it serves as an insulator against excessive heat loss to the environment. Marine mammals such as whales and seals are excellent examples of animals that use fatty tissue for insulation.

More than 99% of the total lipid in human adipose tissue is triglyceride. These lipids are continuously being broken down and new lipid is synthesized to take their place. On the average, about 10% of the fatty acids in adipose tissue is replaced daily by new fatty acid molecules. This rate varies from one organ to another. (For example, the replacement of fatty acid molecules in the liver is five to seven times faster than in the brain.)

The fatty acid portions of triglyceride molecules are most important in terms of energy. Their highly reduced carbon atoms store chemical energy and their hydrophobic nature allows dense packing of the hydrocarbon chains.

Many of the fatty acids we use are supplied in our diet. Various animal and vegetable lipids are ingested, hydrolyzed to glycerol and free fatty acids, and then absorbed through the intestinal wall. They are distributed to the rest of the body through the lymphatic system and the bloodstream. After ingesting a meal containing triglycerides, the stomach contents enter the small intestine. Intestinal **peristalsis**, *the mechanical mixing of intestinal contents*, emulsifies the lipids, breaking up the large lipid droplets into smaller structures. Bile juice from the gall bladder further helps this emulsification. Pancreatic juice containing the enzyme *pancreatic lipase* enters the small intestine at this point, and catalyzes the decomposition of triglycerides into glycerol and free fatty acids. Fatty acids are then absorbed into the body through the intestinal mucosal cells.

$$\begin{array}{c} CH_2-O-\overset{O}{\underset{\|}{C}}-R \\ | \\ CH-O-\overset{O}{\underset{\|}{C}}-R \\ | \\ CH_2-O-\overset{O}{\underset{\|}{C}}-R \end{array} \xrightarrow[+3H_2O]{\text{pancreatic lipase}} \begin{array}{c} CH_2-OH \\ | \\ CH-OH \\ | \\ CH_2-OH \end{array} + 3\ HO-\overset{O}{\underset{\|}{C}}-R \longrightarrow \text{fatty acid absorption}$$

A genetically inherited deficiency of pancreatic lipase results in the impaired absorption of lipids.

Bile juice and its accompanying bile salts are absolutely necessary for normal lipid absorption. Without these detergent-like molecules, lipids are not efficiently emulsified, a prerequisite for lipase action and subsequent absorption. When bile is totally excluded from the intestinal tract as a result of severe liver dysfunction or biliary obstruction, lipid absorption decreases dramatically. This decrease causes an increased lipid content of the feces and results in characteristic clay-colored stools. The lipid-soluble vitamins are also poorly absorbed.

31.2 LIPID ABSORPTION AND TRANSPORT

A large portion of the lipid absorbed by the body is in the form of free fatty acids, following hydrolysis of triglycerides. However, a small portion of the triglycerides are absorbed in their intact form or as monoglycerides (molecules of glycerol with only one esterified fatty acid). Once free fatty acids are absorbed by the mucosal cells of the intestine, the fatty acids are reassembled into triglycerides for transport

TABLE 31-1 Characteristics of Lipoproteins

	Lipoprotein Type			
	Chylomicrons	Very Low Density Lipoproteins (VLDL)	Low Density Lipoproteins (LDL)	High Density Lipoproteins (HDL)
Average composition (%)				
triacylglycerols	86	55	8	5
phospholipids	8	20	20	30
cholesterol	2	10	10	5
cholesterol esters	2	6	37	15
protein	2	9	25	45
Density	0.92–0.96	0.95–1.006	1.006–1.063	1.063–1.21
Diameter (Å)	10^3-10^4	250–750	200–250	70–120

throughout the body. The reformed triglycerides (along with smaller amounts of phospholipids, cholesterol, and specific proteins) aggregate into spherical structures called *chylomicrons*. **Chylomicrons** *are similar to large micelles which contain triglycerides as a primary component, with phospholipids serving as a hydrophilic shell surrounding a hydrophobic core.* Once formed, chylomicrons enter the lymphatic system and finally reach the bloodstream. Soon after a meal heavily laden with fats has been ingested, the chylomicron content of both blood and lymph increases dramatically. If enough fatty acids and triglycerides are absorbed, the blood may even appear turbid and take on a yellow tint, due to the presence of so many chylomicrons. Usually within four hours after a meal, few if any chylomicrons remain in the blood, owing to their movement into adipose tissue cells or into the liver.

chylomicrons

In the postabsorptive state, when chylomicrons are virtually absent from the blood, some 95% of the lipids in blood are in the form of *lipoproteins*. **Lipoproteins** *are lipid transport proteins in the bloodstream composed of protein and lipids, such as triglycerides, cholesterol, and phospholipids.* Three main types of lipoproteins found in the blood are classified according to their densities (as determined by centrifugation in salt solutions or heavy water, ^2H_2O): *high density, low density,* and *very low density lipoproteins* (abbreviated *HDL, LDL,* and *VLDL,* respectively). The characteristics of these lipoproteins are summarized in Table 31-1.

lipoproteins

Lipoproteins have received considerable attention as indicators of a tendency toward heart disease and atherosclerosis. Diets high in saturated fats and cholesterol tend to increase LDL levels in the blood. In turn, increased LDL levels are associated with coronary artery disease and atherosclerosis. Table 31-1 reveals that LDL contains a comparatively large percentage of cholesterol. The resulting atherosclerosis is believed to be due in part to LDL binding to arterial cell membranes and subsequent deposition of cholesterol at the arterial wall surface. The result is a cholesterol plaque formation. On the other hand, HDL apparently interferes with LDL binding and cholesterol deposition at the arterial wall and is beneficial to human health. Some evidence indicates that exercise tends to elevate HDL concentration, which may account in part for the beneficial effects of exercise. Higher levels of HDL and lower levels of LDL are closely associated with good overall condition of the circulatory system.

31.3 FATTY ACID MOBILIZATION AND OXIDATION

Before the stored energy in lipids can be used, the fatty acids must be released from triglycerols in the adipose tissues and be transported to the peripheral tissues.

Within fat cells there are special lipase enzymes, capable of hydrolyzing the ester linkages of triglycerides. However, unlike pancreatic lipase, the intracellular lipase is *hormone sensitive*. Similar to the hormone regulation of glycogen metabolism, when the hormone epinephrine binds to the surface of a fat cell, it stimulates the synthesis of cAMP. Elevated levels of cAMP initiate a cascade of enzymatic reactions that eventually activates lipase (via phosphorylation). Within a short time, the fatty acid content in the blood rises after fat cells are stimulated by epinephrine (see Figure 31–2).

After absorption into the cell where needed, free fatty acids are coupled to CoA, forming *acyl-CoA*. (Note the difference between acyl-CoA and acetyl-CoA.) "Acyl" refers to a random length fatty acid which is covalently attached to CoA by a thioester linkage. This reaction requires *two high-energy ATP bonds* (from a single ATP molecule).

$$\underset{\text{free fatty acid}}{R-\overset{O}{\underset{\|}{C}}-O^-} + CoA + ATP \xrightarrow{\text{acyl-CoA synthetase}} \underset{\text{acyl-CoA}}{R-\overset{O}{\underset{\|}{C}}-CoA} + AMP + 2\,P_i$$

The site of fatty acid oxidation is the mitochondrial matrix. Fatty acids enter the mitochondria via a special transport mechanism involving a group of specialized transport proteins and a carrier molecule called *carnitine*.

$$\underset{\text{carnitine}}{H_3C-\overset{CH_3}{\underset{CH_3}{\overset{|}{\underset{|}{N^+}}}}-CH_2-\overset{H}{\underset{OH}{\overset{|}{\underset{|}{C}}}}-CH_2-C\overset{O}{\underset{O^-}{\diagup}}}$$

At this point fatty acids are ready for oxidation. A series of enzymatic reactions

FIGURE 31–2 Epinephrine stimulation of fatty acid mobilization. An enzymatic cascade of reactions similar to glycogen breakdown activates lipase by phosphorylation (the "-P" stands for phosphate).

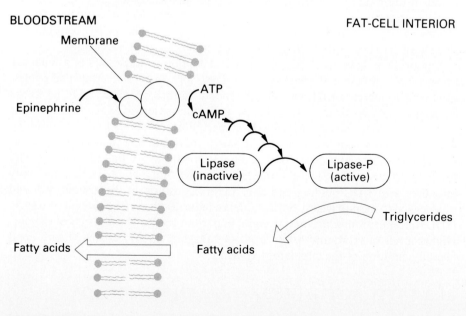

FIGURE 31-3 Greek letters used to indicate the position of carbon atoms in fatty acids.

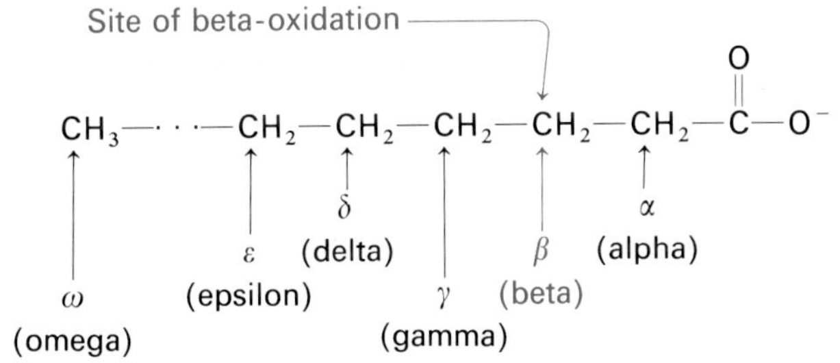

$$CH_3-\cdots-\underset{\omega\,(\text{omega})}{CH_2}-\underset{\varepsilon\,(\text{epsilon})}{CH_2}-\underset{\delta\,(\text{delta})}{CH_2}-\underset{\gamma\,(\text{gamma})}{CH_2}-\underset{\beta\,(\text{beta})}{CH_2}-\underset{\alpha\,(\text{alpha})}{CH_2}-\overset{O}{\underset{\|}{C}}-O^-$$

Site of beta-oxidation ↑

partially oxidize and remove carbon atoms (two at a time) from the fatty acids in a process called *β-oxidation of fatty acids*. The name of this pathway is derived from the fact that the carbon in the β position (see Figure 31–3) is oxidized to a carbonyl group, as the first two carbons of the chain are removed as acetyl-CoA. Let's examine the four reaction steps of β-oxidation in more detail.

The first step in β-oxidation of fatty acids is a dehydrogenation reaction in which a double bond is formed between carbons in the α and β positions. The enzyme catalyzing this reaction is fatty acyl-CoA dehydrogenase, which uses FAD as the hydrogen/electron acceptor.

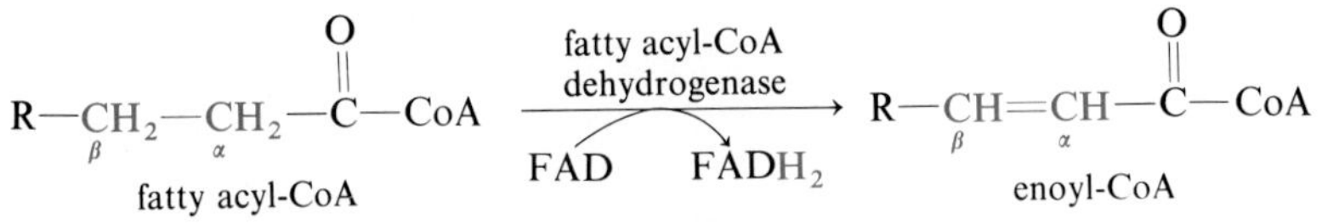

$$R-\underset{\beta}{CH_2}-\underset{\alpha}{CH_2}-\overset{O}{\underset{\|}{C}}-CoA \xrightarrow[\text{FAD} \quad \text{FADH}_2]{\text{fatty acyl-CoA dehydrogenase}} R-\underset{\beta}{CH}=\underset{\alpha}{CH}-\overset{O}{\underset{\|}{C}}-CoA$$

fatty acyl-CoA   enoyl-CoA

The second reaction in this pathway is hydration of the double bond, with the —OH functional group appearing on the β carbon.

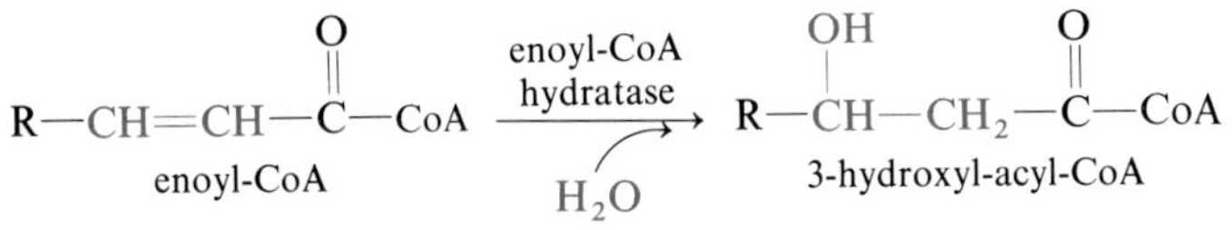

$$R-CH=CH-\overset{O}{\underset{\|}{C}}-CoA \xrightarrow[\text{H}_2\text{O}]{\text{enoyl-CoA hydratase}} R-\underset{|}{\overset{OH}{CH}}-CH_2-\overset{O}{\underset{\|}{C}}-CoA$$

enoyl-CoA   3-hydroxyl-acyl-CoA

Step three is an oxidation of the alcohol to a ketone functional group, resulting in the formation of a β-keto fatty acid. The dehydrogenase enzyme that catalyzes this oxidation is coupled to NAD$^+$, forming NADH as the reduced product.

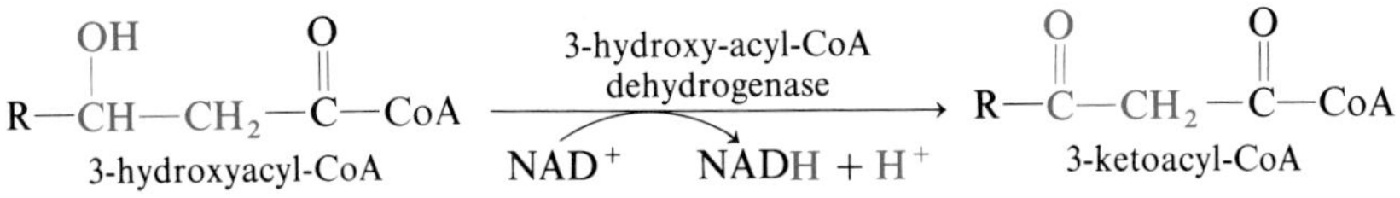

$$R-\underset{|}{\overset{OH}{CH}}-CH_2-\overset{O}{\underset{\|}{C}}-CoA \xrightarrow[\text{NAD}^+ \quad \text{NADH + H}^+]{\text{3-hydroxy-acyl-CoA dehydrogenase}} R-\overset{O}{\underset{\|}{C}}-CH_2-\overset{O}{\underset{\|}{C}}-CoA$$

3-hydroxyacyl-CoA   3-ketoacyl-CoA

It is now apparent why this series of reactions is called *β-oxidation*. The β-carbon has become substantially more oxidized as a ketone than as a totally reduced methylene (—CH$_2$—) group.

The fourth and final step in this pathway cleaves two carbons from the fatty acid; these are released as a molecule of acetyl-CoA. As the acetyl-CoA is released from the fatty acid, the β-keto-carbon becomes the first carbon in the shorter fatty acid and is coupled to another molecule of CoA.

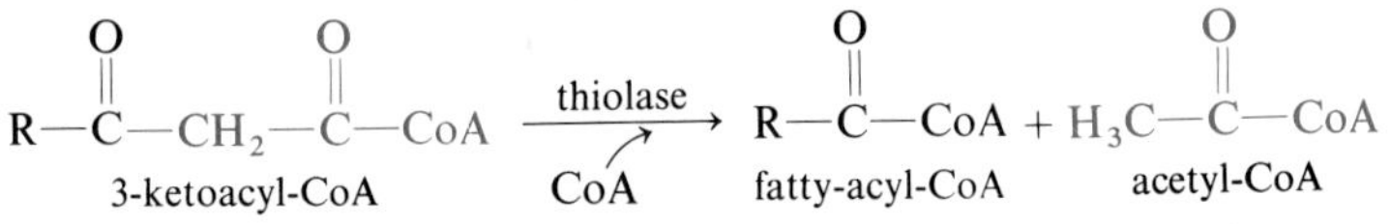

$$R-\overset{O}{\underset{\|}{C}}-CH_2-\overset{O}{\underset{\|}{C}}-CoA \xrightarrow[\text{CoA}]{\text{thiolase}} R-\overset{O}{\underset{\|}{C}}-CoA + H_3C-\overset{O}{\underset{\|}{C}}-CoA$$

3-ketoacyl-CoA   fatty-acyl-CoA   acetyl-CoA

835
Lipid Metabolism

The oxidative nature of this reaction keeps it spontaneous and supplies sufficient energy to attach CoA to the shorter fatty acid *without the expenditure of two high-energy bonds from ATP* (as was the case for the original fatty acid). Therefore, this remaining acyl-CoA molecule (now shorter by two carbons) is *recycled* through the same set of four reactions again. This yields another two-carbon acetyl-CoA molecule and more FADH and $NADH_2$. Recycling occurs again and again, until the entire fatty acid is converted to acetyl-CoA. Thus, the very long hydrocarbon

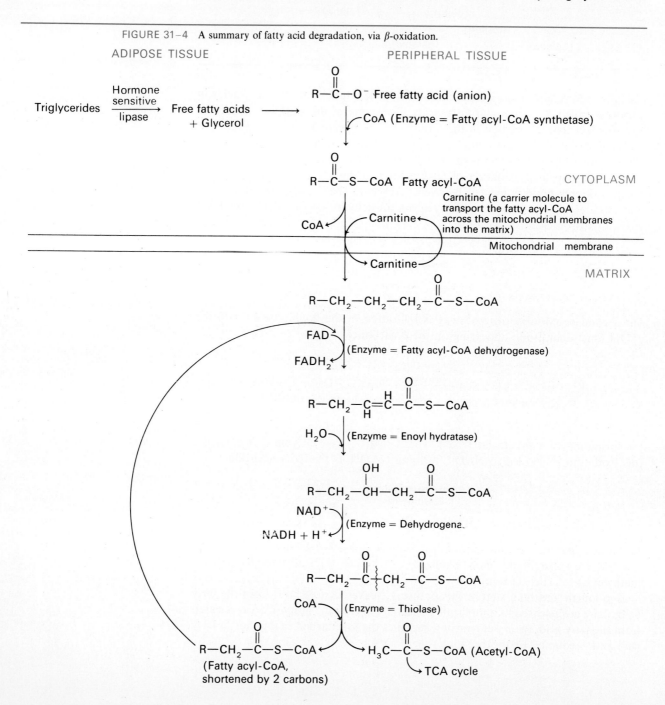

FIGURE 31–4 A summary of fatty acid degradation, via β-oxidation.

chain of a fatty acid molecule is *sequentially degraded,* two carbons at a time (see Figure 31–4). We should note that in the last round of reactions, a four-carbon fatty acid is simultaneously split into two molecules of acetyl-CoA, eliminating the need for another round of the pathway.

How many rounds of the β-oxidation pathway are needed to convert palmitic acid (a C_{16}-fatty acid) to acetyl-CoA?

Example 31.1

Solution
Palmitic acid contains *eight* two-carbon units. Only *seven* rounds of the β-oxidation pathway are needed to form eight acetyl-CoA molecules. In the last round, two molecules of acetyl-CoA are formed simultaneously.

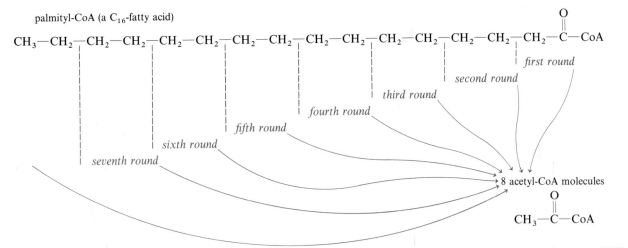

Note that *unsaturated fatty acids* can also be degraded to yield acetyl-CoA. An *isomerase* enzyme catalyzes the change in location of the double bond within the fatty acid molecule so that β-oxidation can proceed unhindered.

Each round of β-oxidation yields FADH and $NADH_2$. These can be used by the electron-transport chain to produce ATP. Acetyl-CoA, the primary product of this pathway, joins other acetyl-CoA molecules arising from other reactions in the cell (including those from glucose and amino acid oxidation). We have already examined how acetyl-CoA can be further oxidized via the TCA cycle to yield more ATP. In Section 31.5 we will explore just how many ATP molecules can be produced from the oxidation of various fatty acids.

31.4 FATTY ACID BIOSYNTHESIS

Fatty acid biosynthesis *(sometimes called lipid anabolism) is a series of enzymatic reactions that construct fatty acids from acetyl-CoA molecules.* Occurring primarily in the liver and adipose tissue, fatty acid biosynthesis occurs when *excess* carbohydrates and amino acids are present and when the energy from these substances is not needed for ATP production. Since acetyl-CoA is a central component in the metabolism of both carbohydrates and many amino acids, the body can use excess acetyl-CoA from these sources for fatty acid biosynthesis.

Fatty acid biosynthesis occurs through a series of reactions that at first glance *appear* to be a simple reversal of β-oxidation: acetyl-CoA is coupled to a growing

fatty acid biosynthesis

TABLE 31-2 Differences between Fatty Acid Biosynthesis and Fatty Acid Degradation

Property	Biosynthesis	Degradation
location	cytoplasm	mitochondrial matrix
enzymes	multienzyme complex	separate enzymes
acyl group carrier	acyl carrier protein	coenzyme A
cofactors	NADPH used	NADH (and $FADH_2$) made

chain of two-carbon units, followed by reduction of the β-keto group to an alcohol, dehydration, and reduction of a double bond to form a saturated hydrocarbon chain. However, *fatty acid oxidation and fatty acid synthesis occur by entirely different pathways*. Different enzyme systems and cofactors are used and the biosynthetic pathway is located in the cytoplasm of the cell, while oxidation occurs only in the mitochondrial matrix. Fatty acid degradation utilizes separate, independent enzymes, while the enzymes of fatty acid synthesis are arranged into a multienzyme complex. Table 31-2 summarizes some of the overall differences between fatty acid biosynthesis and degradation pathways.

Most fatty acids contain even numbers of carbon atoms because of the way fatty acids are synthesized. Two-carbon units from acetyl-CoA are linked together one at a time, until an acid of 14, 16, 18, or 20 carbon atoms is made. Let us examine the series of reactions that accomplish this synthesis.

The first step of fatty acid biosynthesis converts acetyl-CoA into malonyl-CoA, a reaction catalyzed by the enzyme *acetyl-CoA carboxylase*. This process involves the transfer of an activated CO_2 group in a *carboxylation reaction*. As we mentioned in Chapter 27, biotin is the cofactor used in this reaction to transfer CO_2 (usually in the form of bicarbonate, HCO_3^-). ATP is also required to supply the energy to carboxylate acetyl-CoA.

$$H_3C-\overset{O}{\overset{\|}{C}}-CoA + HCO_3^- \xrightarrow[\text{ATP} \quad \text{ADP} + P_i]{\text{acetyl-CoA carboxylase}} {}^-OOC-CH_2-\overset{O}{\overset{\|}{C}}-CoA$$

cytoplasmic acetyl-CoA   malonyl-CoA

The CO_2 group does not stay attached for long, however. As we will see, this CO_2 is lost during the next elongation step.

Both acetyl-CoA and malonyl-CoA are attached to *acyl carrier proteins* (abbreviated "ACP"). The **acyl carrier protein** *is a monomeric protein with relatively long side chains similar to acetyl-CoA's long phosphopantetheine group* (see Figure 31-5). The ACP forms an important part of the fatty acid synthetase complex with its long arm acting as a pendulum as it swings the acyl groups attached to it back and forth among the enzymes of the complex. Both malonyl-CoA and acetyl-CoA are converted to acetyl-ACP and malonyl-ACP in preparation for fatty acid synthesis.

acetyl-CoA + ACP $\rightleftharpoons$ acetyl-ACP + CoA
malonyl-CoA + ACP $\rightleftharpoons$ malonyl-ACP + CoA

After malonyl-ACP and acetyl-ACP have formed, a condensation reaction occurs linking them together and releasing the CO_2 acquired during malonyl-CoA synthesis. This process results in the formation of a four-carbon unit called *acetoacetyl-ACP*.

FIGURE 31-5 The ACP side chain and CoA are similar due to the presence of the phosphopantetheine group.

$$HS-CH_2-CH_2-\underset{\underset{O}{\|}}{N}-\overset{H}{\underset{}{C}}-CH_2-CH_2-\underset{\underset{O}{\|}}{N}-\overset{H}{\underset{H}{C}}-\overset{OH}{\underset{}{C}}-\overset{CH_3}{\underset{CH_3}{C}}-CH_2-O-\underset{\underset{O^-}{}}{\overset{O}{\|}}P-O-CH_2-Ser-$$

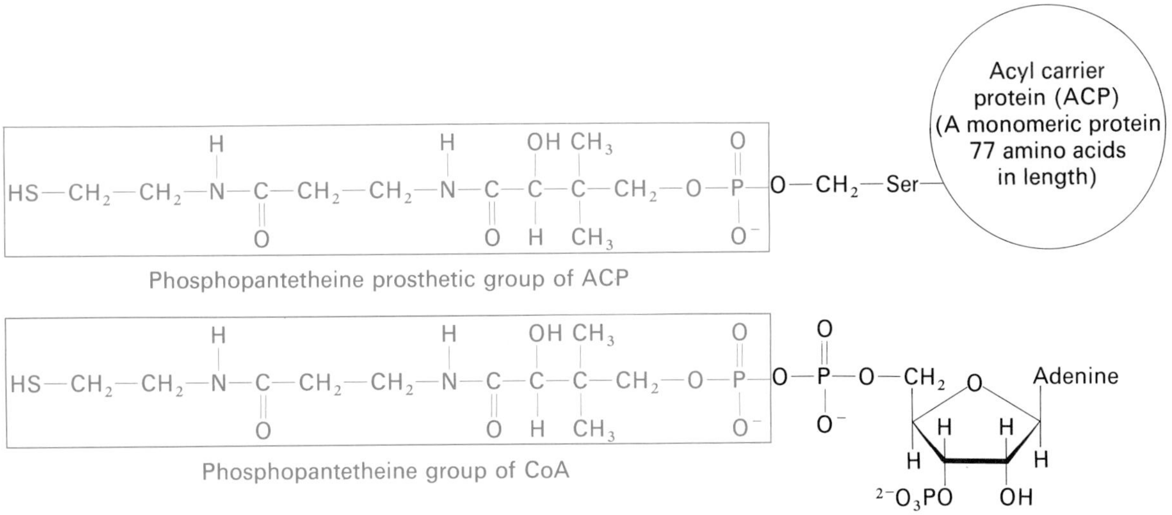

Acyl carrier protein (ACP) (A monomeric protein 77 amino acids in length)

Phosphopantetheine prosthetic group of ACP

$$HS-CH_2-CH_2-\underset{\underset{O}{\|}}{N}-\overset{H}{\underset{}{C}}-CH_2-CH_2-\underset{\underset{O}{\|}}{N}-\overset{H}{\underset{H}{C}}-\overset{OH}{\underset{}{C}}-\overset{CH_3}{\underset{CH_3}{C}}-CH_2-O-\underset{\underset{O^-}{}}{\overset{O}{\|}}P-O-\underset{\underset{O^-}{}}{\overset{O}{\|}}P-O-CH_2$$ Adenine

Phosphopantetheine group of CoA

$^{2-}O_3PO$ OH

$$CH_3-\overset{O}{\underset{\|}{C}}-ACP + {}^-OOC-CH_2-\overset{O}{\underset{\|}{C}}-ACP \longrightarrow CH_3-\overset{O}{\underset{\|}{C}}-CH_2-\overset{O}{\underset{\|}{C}}-ACP + ACP + CO_2$$

acetyl-ACP   malonyl-ACP   acetoacetyl-ACP

Now that a four-carbon segment has been synthesized from two smaller components, the reduction of the β-keto group to a methylene group (—CH$_2$) can be accomplished. This reaction happens in three steps, the first of which is reduction of the keto group to an alcohol. This reduction uses *NADPH as the reducing agent* (rather than NADH which forms during the reverse reaction in fatty acid degradation). Herein we see a general trend in biochemical reactions: *NADPH is often used as a reducing agent, while NADH is more often formed during energy-yielding oxidation reactions.*

During the next step of fatty acid biosynthesis, the alcohol group is removed in a *dehydration reaction*, which leaves a double bond between the α- and β-carbons. The final step in this reduction is a reaction that also uses NADPH to reduce the double bond to a single bond (hydrogenation), forming *butyryl-ACP*.

Once the four-carbon butyryl-ACP has been formed, the fatty acyl chain can be elongated further with repeated cycles of the fatty acid synthetase complex. Specifically, condensation with another malonyl-ACP occurs at the beginning of each cycle. Each two-carbon unit from malonyl-ACP binds to the end of the growing chain, just as in the first cycle when malonyl-ACP binds to acetyl-ACP. Once the complete chain has formed (usually 16 carbons long), the fatty acid is cleaved from the complex. A molecule of CoA is substituted for ACP, making a free molecule of *fatty acyl-CoA*.

$$CH_3-(CH_2)_{14}-\overset{O}{\underset{\|}{C}}-ACP + CoA \longrightarrow CH_3-(CH_2)_{14}-\overset{O}{\underset{\|}{C}}-CoA + ACP$$

fatty acyl-ACP   fatty acyl-CoA

A summary of the reactions that occur in fatty acid biosynthesis is shown in Figure 31-6 (p. 840).

FIGURE 31-6 Reactions of the fatty acid biosynthesis pathway. Chain elongation occurs as longer and longer acyl-ACP chains recycle through the pathway, adding two more carbons from malonyl-CoA during each round.

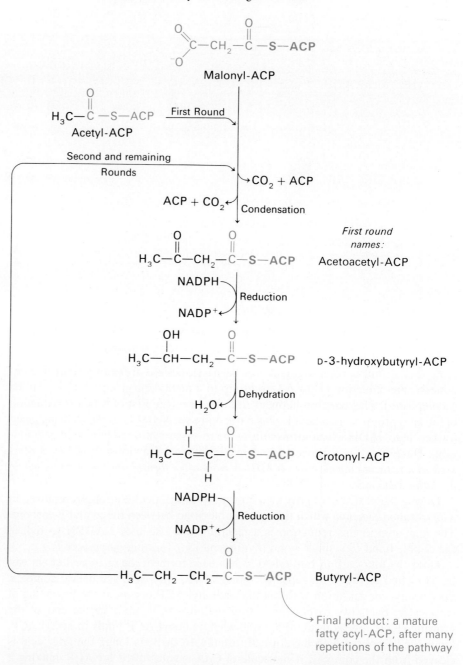

Example 31.2

How many cycles of the fatty acid biosynthetic pathway are needed to synthesize palmitoyl-CoA from acetyl-CoA and malonyl-CoA?

Solution

In the first cycle, malonyl-ACP and acetyl-ACP form the first four carbons of the growing fatty acid. Since 12 more carbons are needed (two being added in each round as malonyl-ACP condenses with the growing chain) only six more cycles are needed. This makes a total of only seven cycles to achieve the synthesis of palmitoyl-CoA.

Cycles	Reactions	Fatty acid length
1	malonyl-ACP + acetyl-ACP	⟶ C_4-chain
2, 3, 4, 5, 6, 7	malonyl-ACP + growing acyl-ACP	⟶ C_{16}-chain
Total 7 cycles		⟶ palmitoyl-CoA

Knowing that seven cycles of this reaction pathway are needed to synthesize palmitoyl-CoA, we can write a balanced equation for its synthesis:

$$8 \text{ acetyl-CoA} + 14 \text{ NADPH} + 7 \text{ ATP}$$

$$\downarrow \text{seven cycles of pathway}$$

$$\text{palmitoyl-CoA} + 14 \text{ NADP}^+ + 7 \text{ CoA} + 6 \text{ H}_2\text{O} + 7 \text{ ADP} + 7 \text{ P}_i$$

FIGURE 31-7   The stepwise formation of triglyceride biosynthesis, from fatty acyl-CoA molecules and glycerol phosphate.

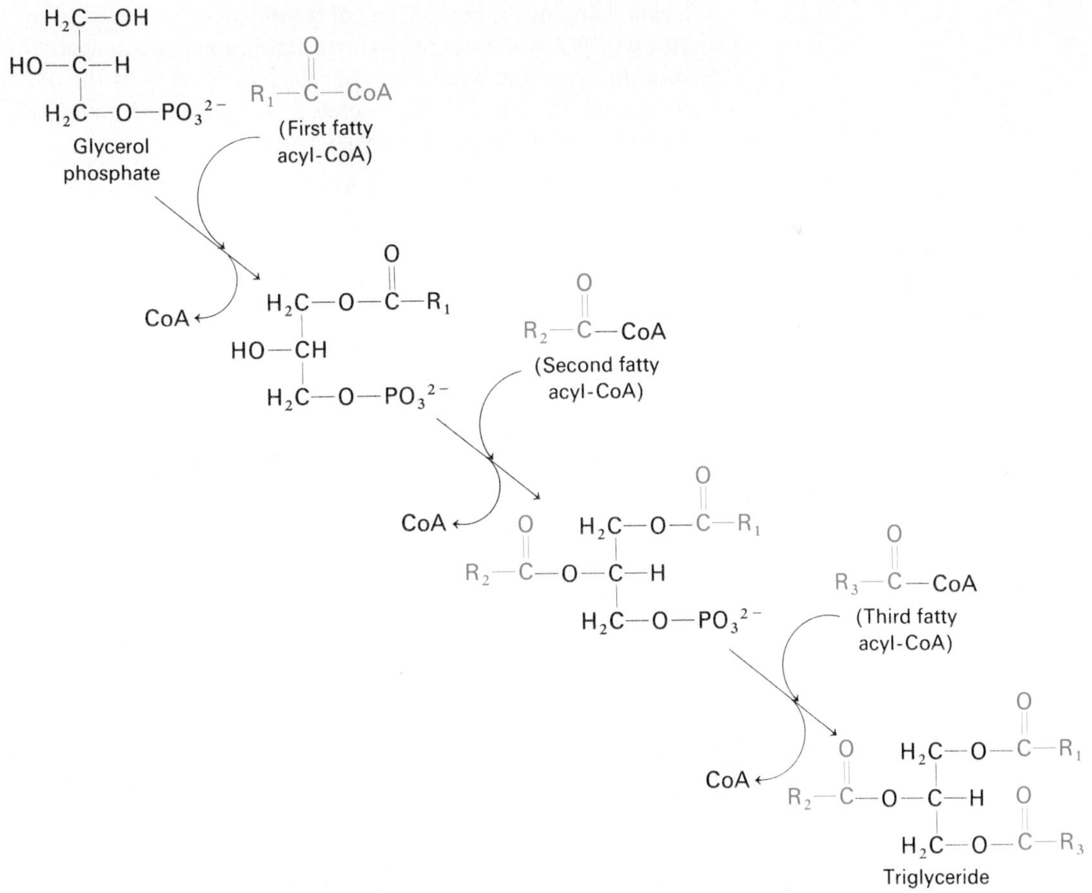

Humans (as well as other mammals) cannot synthesize certain unsaturated fatty acids, such as *linoleic acid* and *linolenic acid*. These substances are called *essential fatty acids* because they must be obtained in our diets. Although the body can synthesize a number of unsaturated fatty acids, humans lack the enzymes to introduce double bonds at carbon atoms beyond C-9 in the fatty acid chain.

linoleic acid: $H_3C(CH_2)_4CH=CH-CH_2-CH=CH-(CH_2)_7-COOH$

linolenic acid: $H_3C-CH_2-CH=CH-CH_2-CH=CH-CH_2-CH=CH(CH_2)_7COOH$

Once fatty acyl-CoA molecules have been formed and separated from the fatty acid synthetase complex, mono-, di-, and triglycerides can be formed. Enzymatic reactions use the potential energy stored in the ester linkage of acyl-CoA fatty acids by coupling the release of CoA to the formation of an ester linkage to glycerol. Each of the three fatty acids combine with glycerol phosphate in separate steps, yielding a triglyceride as the final product (see Figure 31–7, p. 841).

31.5 ATP ACCOUNTING IN LIPID METABOLISM

Oxidation of a fatty acid yields energy. In fact, fatty acids represent the *most concentrated form of energy storage in the body*. During β-oxidation, NADH, FADH$_2$, and acetyl-CoA are formed, all of which yield energy when further catabolized by the TCA cycle and oxidative phosphorylation. Let us examine exactly how much energy in terms of net production of ATP can be obtained from the total oxidation of one fatty acid molecule.

Recall that during catabolism of palmitoyl-CoA, only seven cycles of the β-oxidation pathway are needed to form eight molecules of acetyl-CoA (since during the seventh cycle, the remaining four-carbon portion of the chain is divided into two molecules of acetyl-CoA). Therefore, the balanced reaction for conversion of palmitoyl-CoA into acetyl-CoA is as follows.

$$\text{palmitoyl-CoA} + 7\text{ FAD} + 7\text{ NAD}^+ + 7\text{ CoA} + 7\text{ H}_2\text{O}$$

$$\downarrow \text{Seven cycles of } \beta\text{-oxidation}$$

$$8 \text{ acetyl-CoA} + 7 \text{ FADH}_2 + 7 \text{ NADH} + 7 \text{ H}^+$$

Further oxidation within the mitochondria yields the following numbers of ATP.

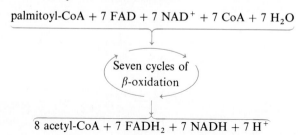

	TCA cycle and oxidative phosphorylation	
8 acetyl-CoA	$\times \dfrac{12 \text{ ATP}}{\text{acetyl-CoA}}$	= 96 ATP
7 FADH$_2$	$\times \dfrac{2 \text{ ATP}}{\text{FADH}_2}$	= 14 ATP
7 NADH	$\times \dfrac{3 \text{ ATP}}{\text{NADH}}$	= 21 ATP

Total (gross) = 131 ATP

Although the gross production of ATP from oxidation of palmitoyl-CoA is 131, we must also consider the activation of fatty acids, in which *two high-energy phosphate bonds of ATP are used* to form the acyl-CoA derivative of the fatty acid before oxidation. For accounting purposes, we can consider these two high-energy bonds equivalent to a single high-energy bond in two ATP molecules. Hence, the *net gain of ATP from total oxidation of palmitic acid is* 129 *ATP*.

We can calculate the efficiency of energy conversion from palmitic acid to CO_2 and H_2O in biological systems by comparing it to the free energy released during a one-step oxidation of palmitic acid in the laboratory. The standard free energy of hydrolysis of 129 moles of ATP is -940 kcal.

$$129 \text{ moles ATP} \times \frac{-7.3 \text{ kcal}}{\text{mole ATP}} = -940 \text{ kcal}$$

Since the standard free energy of palmitic acid oxidation is -2340 kcal, the fractional amount of energy received from 129 ATP yields approximately 40% efficiency. This percent efficiency is similar to that for oxidation of glucose (as we determined in Chapter 30).

$$\frac{\text{Energy stored in ATP}}{\text{Total energy available from palmitic acid}} \longrightarrow \frac{-940 \text{ kcal}}{-2340 \text{ kcal}} \times 100 = 40\%$$

Example 31.3

Calculate the *net gain of ATP* from the total oxidation of myristic acid (a saturated, C_{14}-fatty acid).

Solution

Since 14 carbons are in myristic acid, only six rounds of β-oxidation will be needed to form seven molecules of acetyl-CoA. During the six rounds of β-oxidation, six NADH and six $FADH_2$ will also be formed. Net ATP yield can be calculated as follows.

		TCA cycle and oxidative phosphorylation		
7 acetyl-CoA	×	$\dfrac{12 \text{ ATP}}{\text{acetyl-CoA}}$	=	84 ATP
6 $FADH_2$	×	$\dfrac{2 \text{ ATP}}{FADH_2}$	=	12 ATP
6 NADH	×	$\dfrac{3 \text{ ATP}}{\text{NADH}}$	=	18 ATP
		Subtotal	=	114 ATP
Less 2 ATP for activation of the free acid to its acyl-CoA form			−	2 ATP
Net ATP production from myristic acid	=			112 ATP

31.6 REGULATION OF FATTY ACID METABOLISM

Regulation of fatty acid metabolism occurs at several locations. The energy in stored fatty acids is not readily available because it is stored in the form of triglycerides. Triglycerides are mobilized when intracellular *hormone sensitive lipase*

TABLE 31–3 Regulatory Hormones in Triacylglycerol Metabolism

Enzyme	Hormones	Effect
hormone-sensitive lipase	"lipolytic hormones" such as epinephrine, glucagon, ACTH, and others	simulation by a cAMP-mediated mechanism that leads to phosphorylation and higher activity
	insulin	inhibition
	prostaglandins	inhibition
lipoprotein lipase	insulin	activation

hydrolyzes them into free fatty acids and glycerol. Some hormones activate this enzyme, while others inhibit it, as shown in Table 31–3.

Another important control site for fatty acid oxidation functions inside the cell. Once absorbed into cells, fatty acyl-CoA molecules have two major pathways open to them: transport into the mitochondria and subsequent oxidation or conversion into triglycerols inside the cytoplasm. Since the enzymes of β-oxidation are in the mitochondrial matrix and fatty acid biosynthesis enzymes are in the cytoplasm,

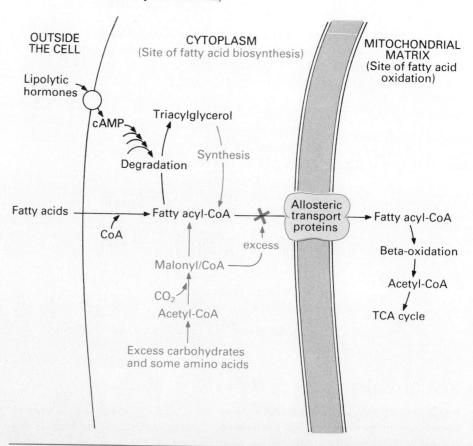

FIGURE 31–8 Summary of fatty acid storage and degradation, showing some critical aspects of regulation. (Fatty acid biosynthesis is shown in color to help differentiate between it and fatty acid oxidation.)

an acyl-CoA molecule is committed to oxidation when it leaves the cytoplasm and enters the matrix.

The fate of the fatty acyl-CoA molecules depends on the rate of transport into the mitochondria. Transport proteins carrying fatty acids across the mitochondrial membrane are allosterically controlled. Malonyl-CoA, the first intermediate in fatty acid biosynthesis, *inhibits transport of fatty acids into the mitochondria.* Whenever the cell is well supplied with carbohydrates and energy demands are low, the concentration of malonyl-CoA increases, blocking the transport of fatty acids into the mitochondria. This reaction effectively turns off fatty acid oxidation and keeps fatty acids in the cytoplasm where they can be stored as triacylglycerols (see Figure 31–8).

31.7 THE RELATIONSHIP BETWEEN CARBOHYDRATE AND LIPID METABOLISM

Lipid and carbohydrate metabolism are closely interrelated and under constant regulation in most organs. In a normal individual on a well-balanced diet, lipid and carbohydrate metabolism are in balance. Certain organs use both lipids and carbohydrates for their energy needs, but in varying amounts. For example, resting skeletal muscle primarily uses free fatty acids for ATP production. However, during muscular activity, carbohydrate oxidation (via aerobic or anaerobic glycolysis) also occurs to meet the increased energy demands. Heart muscle uses free fatty acids, carbohydrates, and even ketone bodies for its normal energy needs. The brain is unique in that it cannot use free fatty acids for its energy, but must depend on glucose as its main fuel. During severe metabolic stress (as in starvation), the brain can use ketone bodies as well as glucose for its energy demands.

Ketone bodies *are relatively small β-keto acids (or their derivatives), the most common of which are acetoacetate and β-hydroxybutyrate.* These small molecules are the most water-soluble form of lipid-based energy, differing only in the oxidation state of their β-carbons. Two molecules of acetyl-CoA are converted to acetoacetate by a series of reactions. Subsequent reduction of acetoacetate (with NADH) yields β-hydroxybutyrate.

ketone bodies

$$2\ CH_3-\overset{O}{\underset{\|}{C}}-CoA \longrightarrow \longrightarrow \longrightarrow CH_3-\overset{O}{\underset{\|}{C}}-CH_2-\overset{O}{\underset{\|}{C}}-O^-$$

acetyl-CoA   CoA   acetoacetate

$$CH_3-\overset{O}{\underset{\|}{C}}-CH_3$$
acetone

$$CH_3-\underset{\underset{OH}{|}}{CH}-CH_2-\overset{O}{\underset{\|}{C}}-CoA$$
β-hydroxybutyrate

(with NADH → NAD⁺)

Ketone bodies

The liver is the primary site for ketone body production (see Figure 31–9, p. 846); the kidney is also a site but to a lesser extent. In a nonenzymatic decarboxylation, a small portion of acetoacetate is spontaneously converted to acetone. Under normal conditions acetone formation is negligible, but when abnormally high

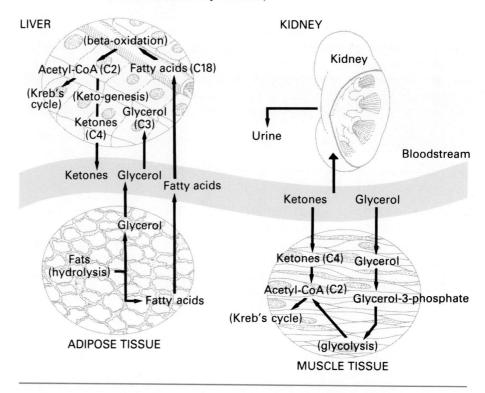

FIGURE 31-9 Fatty acid mobilization and catabolism. Note the role of the liver as the chief site of ketogenesis. (Numbers of carbon atoms forming the skeletal backbone of intermediates are shown in parentheses.)

concentrations of acetoacetate occur, the amount of acetone in blood can be sufficient to cause it to be detectable in a patient's breath.

A limited number of ketone bodies in the bloodstream is normal. However, *elevated concentrations of ketone bodies in the blood is a condition called* **ketosis** *and indicates abnormal fat or carbohydrate metabolism.* This can be caused by a number of pathological situations, two of the most common being fasting or starvation and uncontrolled diabetes mellitus.

During starvation carbohydrate intake is nonexistant, resulting in low glycogen stores and accelerated lipid mobilization and oxidation. When there is insufficient oxaloacetate to combine with the resulting acetyl-CoA, its concentration increases (primarily in the liver) and it begins to form excess ketone bodies which spill over into the bloodstream. These ketone bodies can be used by the peripheral tissues and the brain as an *alternative energy source* during this period. Excess ketone bodies not used by peripheral tissues are either excreted into the urine (a condition called *ketonuria*) or broken down to acetone, producing the characteristic "acetone breath" of a ketotic individual.

Ketone bodies are really organic acids; if their concentration rises above the capacity of buffers in the blood to maintain pH, blood pH decreases. *Low blood pH accompanied by ketone bodies is called* **ketoacidosis**. Usually, under starvation or fasting conditions, the pH of the blood can be controlled and ketoacidosis represents no threat. This is not the case with diabetes mellitus. As you recall, this form of diabetes is caused by low levels of insulin, which in turn results in low levels of glucose absorption by cells. In this abnormal condition, fatty acids are mobilized and used by the cells for energy demands, *even though an ample supply of glucose is present in the blood.* The liver produces excess ketone bodies and ketoacidosis occurs. The presence of relatively high glucose levels and ketone bodies in the blood is an important key in diagnosing diabetes mellitus.

EXERCISES AND PROBLEMS

Metabolically Important Sources of Triglycerol Lipids

31.1 Triglycerides are the most common form of lipid in the body. These molecules are often called neutral fats or triacylglycerol. Draw the generalized structure of a triglyceride molecule and explain the origin of these names.

31.2 List two important functions of lipids in warm-blooded animals.

31.3 How much more energy in our bodies is stored in lipids as compared to carbohydrates?

31.4 It is often difficult to lose extra, unwanted pounds of fat. After fat deposits have been formed, does each fat molecule in a deposit remain there until we can lose the weight? Explain.

31.5 Why is pancreatic lipase important in the initial digestion and absorption of dietary lipids? How does bile juice assist these processes?

Lipid Absorption and Transport

31.6 What is the most common form of lipids absorbed by the body? Are triglycerides ever absorbed directly? If so, what percentage of total absorbed lipid do they represent?

31.7 What are chylomicrons? Are they continually present in the blood?

31.8 List two types of lipoproteins and explain how they are classified.

31.9 A blood test during a routine examination reveals you have higher than normal levels of LDL. Is this good or bad? Explain why and list what steps could be taken to change LDL levels in the bloodstream.

Fatty Acid Oxidation

31.10 The stimulation of lipid mobilization by epinephrine is similar to the stimulation of glycogen breakdown. List the similarities between these two activation processes.

31.11 Where in the cell does fatty acid oxidation occur?

31.12 How are fatty acids activated before oxidation?

31.13 Only one molecule of ATP is used to activate fatty acids before oxidation occurs, yet we count this expenditure as *two* ATP for accounting purposes. Explain this apparent inconsistency.

31.14 Why is fatty acid oxidation often called β-oxidation?

31.15 Draw chemical structures for the two intermediate compounds formed as butanoyl-CoA is converted to β-ketobutanoyl-CoA by the β-oxidation pathway. Name the type of reaction that occurs at each step and show the cofactors used in each reaction.

$$CH_3-CH_2-CH_2-\overset{O}{\underset{\|}{C}}-CoA \longrightarrow \; ? \; \longrightarrow \; ?$$
butanoyl-CoA

$$\longrightarrow CH_3-\overset{O}{\underset{\|}{C}}-CH_2-\overset{O}{\underset{\|}{C}}-CoA$$
β-ketobutanoyl-CoA

31.16 Why is an ATP *not* needed between rounds of the β-oxidation pathway when longer fatty acids are processed?

31.17 Consider the case where capric acid (a saturated C_{10}-fatty acid) is converted entirely to acetyl-CoA.
a. How many rounds of the β-oxidation pathway are needed?
b. What is the yield of acetyl-CoA?
c. What is the yield of NADH?
d. What is the yield of $FADH_2$?
e. How many high-energy ATP bonds are consumed?

31.18 Consider the β-oxidation of the following unsaturated fatty acids. Which one of these fatty acids will require an isomerase enzyme for complete oxidation?

A: $CH_3-(CH_2)_4=(CH_2)-\overset{O}{\underset{\|}{C}}-CoA$

B: $CH_3-(CH_2)_3=(CH_2)-\overset{O}{\underset{\|}{C}}-CoA$

31.19 Which yields more $FADH_2$, saturated or unsaturated fatty acids? Do these differ in their yield of NADH? Explain.

Fatty Acid Biosynthesis

31.20 Where are the enzymes of fatty acid biosynthesis located? How does this location compare with the location of fatty acid degradation?

31.21 Coenzyme A (CoA) plays an important role in β-

oxidation of fatty acids. What is its counterpart in fatty acid biosynthesis and how does it differ from CoA?

31.22 Why do almost all fatty acids in the human body contain an even number of carbon atoms?

31.23 What is the difference between acetyl-ACP and malonyl-ACP? Write a chemical equation for the conversion of one to the other.

31.24 Select a fatty acid and show the four enzymatic reactions of fatty acid biosynthesis needed to lengthen it by two carbon atoms. Name the type of reaction occurring at each point and include the required cofactors (including ACP).

31.25 What molecule releases CO_2 during fatty acid biosynthesis? Does this happen every round or only once?

31.26 Consider the biosynthesis of palmitoyl-CoA from acetyl-CoA molecules.
a. How many rounds of the fatty acid biosynthetic pathway are needed?
b. How many molecules of malonyl-ACP must be formed?
c. How many high-energy ATP bonds are consumed?
d. How many NADPH molecules are needed?

31.27 From your knowledge of carbohydrate metabolism, what pathway produces the majority of NADPH needed for fatty acid biosynthesis?

31.28 Name the two essential fatty acids and explain why they are required in the diet.

31.29 What is the source of energy needed to form ester linkages between glycerol and fatty acids in triacylglycerols?

ATP Accounting in Lipid Metabolism

31.30 For the following fatty acids, what is the maximum number of ATP molecules that can be produced from their complete oxidation to CO_2 and H_2O?
a. capric acid (saturated C_{10}-fatty acid)
b. stearic acid (saturated C_{18}-fatty acid)
c. oleic acid (unsaturated C_{18}-fatty acid with only one double bond)

31.31 In Section 31.5, we determined that 129 ATP (net) could be produced from the complete oxidation of palmitic acid. How many ATP could be produced from three palmitic acid molecules? How many ATP from one tripalmitin fat molecule? (Careful!)

Regulation of Fatty Acid Metabolism

31.32 In what way is cAMP involved in regulating fatty acid mobilization?

31.33 Carnitine forms an *ester linkage* with fatty acids during transport into the mitochondria. Given the structure of carnitine, draw the structure of the carnitine-fatty acid molecule that is transported into mitochondria.

$$\text{carnitine:} \quad H_3C-\overset{\overset{\displaystyle CH_3}{|}}{\underset{\underset{\displaystyle CH_3}{|}}{N^+}}-CH_2-\underset{\underset{\displaystyle OH}{|}}{CH}-CH_2-COO^-$$

31.34 Explain how higher levels of malonyl-CoA help control the rate of fatty acid oxidation.

31.35 In diphtheria, bacteria produce a toxin that causes fatty acid accumulation and infiltration of the heart muscles, with eventual weakening of the cardiac tissues. Diphtheria toxin reduces the carnitine level in the tissues. Why would lower levels of carnitine encourage the synthesis of unusually large amounts of fatty acids?

The Relationship Between Carbohydrates and Lipid Metabolism

31.36 What are ketone bodies? Describe their origin in the body.

31.37 Ever since your friend has gone on her new diet, she complains about having bad breath. Offer an explanation that will make her feel like she is making progress with her diet. You should also suggest she include more carbohydrates in her diet. Why would this change be better for her?

31.38 An extremely thin coed checks into the emergency room late one night. She complains of continually being tired and losing weight, even though she claims to get enough rest and that she eats well. Even though she drinks lots of water, she always seems to be thirsty and frequently has to urinate. Her roommates are concerned that she has anorexia, but the doctor wants to check for diabetes. What lab tests should be ordered? How can the results of these tests be used to make a correct diagnosis?

31.39 The carbon skeletons from carbohydrates can be used to synthesize fatty acids. Can humans reverse this process, converting stored fatty acids into sugars? (Hint: Consider the net numbers of carbons taken in and lost during one round of the TCA cycle.)

32 Amino Acid Metabolism

Objectives

After completing Chapter 32, you will be able to:

32.1 Describe what is meant by the amino acid pool, and list various ways this pool is supplied and depleted by metabolic reactions.
32.2 Explain why the carbon skeleton of amino acids plays an important role in amino acid metabolism.
32.3 List two different types of reactions that remove amino groups from amino acids prior to oxidation.
32.4 Know the series of reactions that comprise the urea cycle.
32.5 Understand the overall biosynthetic pathways of amino acid production.
32.6 Understand that amino acid catabolism and biosynthesis occur by totally different pathways, regulated by different hormone control mechanisms.

INTRODUCTION

From an energy-production standpoint, amino acids supply only a small portion of the body's needs. Carbohydrates and fats supply 90% of the energy used in a normal diet, with only 10% of our daily calories coming from amino acid catabolism. However, amino acid metabolism plays an important role in maintaining good health beyond its minor role in energy production. An ample supply of amino acids is required for both protein synthesis and synthesis of other nitrogen-containing compounds in the cell.

In this chapter we will examine primarily the oxidative degradation of amino acids and briefly discuss their biosynthesis. Understanding these processes gives us a unique view of nitrogen metabolism in the body and several important parameters in the diagnosis of metabolic disorders.

32.1 OVERVIEW OF AMINO ACID METABOLISM

Amino acid synthesis and degradation is linked to protein turnover. Recall from Chapter 24 that proteins are continually being degraded and replaced by new ones. Some proteins, such as collagens, exhibit very slow turnover (replaced every three years or so), while hormones and enzymes have much shorter lifetimes (on the order of a few minutes to a few days). Dietary protein, protein turnover, and amino acid synthesis provide an **amino acid pool** *within cells that consists of moderate amounts of each of the 20 free amino acids commonly found in proteins.* The amino acid pool provides substrates for synthesis of proteins and other nitrogen compounds in the body. Many metabolic pathways contribute amino acids to or take amino acids from the pool. Figure 32-1 illustrates how some of the major cellular processes discussed in this chapter use the amino acid pool.

amino acid pool

FIGURE 32-1 Metabolic pathways that utilize the amino acid pool.

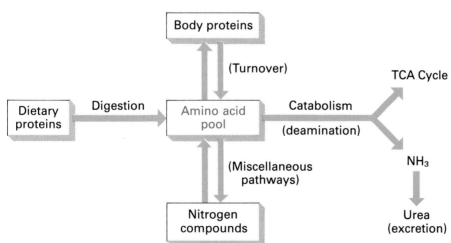

Usually a state of protein balance exists in the normal healthy adult body, that is, the rate of protein synthesis equals or balances the rate of protein degradation. **Nitrogen balance** *is the state when nitrogen taken into the body* (*as protein*) *equals the amount of nitrogen excreted from the body.* Two types of nitrogen imbalance can occur. When protein degradation exceeds protein synthesis, the amount of nitrogen in the urine exceeds the amount of nitrogen ingested (dietary protein). This condition is called *negative nitrogen balance* and accompanies a state of "tissue wasting" because more tissue proteins are being catabolized than are being replaced by protein synthesis. Protein-poor diets, starvation, and wasting illnesses, for example, produce a negative nitrogen balance.

A *positive nitrogen balance* (nitrogen intake exceeds nitrogen output) indicates that the rate of protein anabolism (synthesis) exceeds that of protein catabolism. This state indicates that large amounts of tissue are being synthesized, such as during growth, pregnancy, and convalescence from an emaciating illness.

While the overall nitrogen balance in the body often varies, the relative concentrations of amino acids within the amino acid pool remain essentially constant. No specialized storage forms for amino acids exist in the body as is the case for glucose (glycogen) and fatty acids (triglycerides). Therefore, the body needs a relatively constant source of amino acids to maintain normal metabolism. During negative nitrogen balance, the body must resort to degradation of proteins that were synthesized for specific functions other than amino acid storage. As we have seen throughout our discussion of metabolism, a well-balanced diet consisting of carbohydrates, lipids, and proteins is essential for good health and normal metabolism.

32.2 CARBON SKELETONS OF AMINO ACIDS

The first step in amino acid catabolism takes place in liver cells. It consists of removing the amino group and producing α-keto acids such as pyruvate. These keto acids can be oxidized by the TCA cycle or converted to fats or sugars. Most of the nitrogen removed from amino acids is converted to urea by liver cells (via the urea cycle, Section 32.4) and is later excreted in the urine (Figure 32–2).

FIGURE 32–2 Removal of nitrogen from amino acids yields keto acids, which can be further oxidized or converted to sugars or lipids for storage.

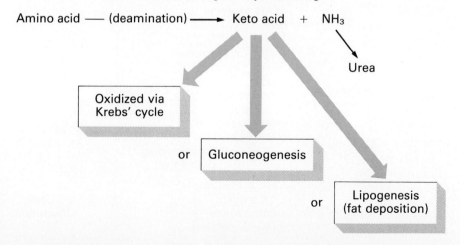

Due to the diversity of amino acids among carbon skeletons, their oxidation demands many different metabolic pathways. The carbon skeleton of amino acids resembles the carbon skeleton of many intermediates in metabolic pathways we have already encountered. Once the amino functional group is removed, many of the residual carbon skeletons can be metabolized by these pathways. Although the individual steps of all these pathways are beyond the scope of this text, let us examine the overall results of these pathways and how carbon skeletons of amino acids can be metabolized by pathways we covered in previous chapters.

The many steps of amino acid catabolism eventually result in the production of compounds that fuel the TCA cycle, as is shown in Figure 32–3.

Amino acids that are degraded into pyruvate or a TCA cycle intermediate can be used to make glucose. *Amino acids capable of being converted into glucose are classified as* **glucogenic**, while *amino acids that are converted into acetyl-CoA and acetoacetyl-CoA (keto acids) are called* **ketogenic**.

Of the five ketogenic amino acids, two can also be converted to glucose. This property makes them both ketogenic and glucogenic, depending on metabolic conditions. The remaining 15 amino acids are solely glucogenic.

During starvation, amino acid catabolism is increased dramatically. Proteins from the digestive tract and from the skeletal muscle system are degraded into amino acids for energy production. A significant portion of the glucogenic amino acids are used for gluconeogenesis to maintain adequate blood glucose levels for the brain. In addition to the ketone bodies (Section 31.7) formed by lipid mobilization under such circumstances, ketogenic amino acids produce more ketone bodies. Ketosis can also result from a diet high in protein and low in carbohydrates. For this and other reasons, high-protein diets (void of carbohydrates) can be dangerous to a person's health.

glucogenic
ketogenic

FIGURE 32–3   Amino acid catabolism forms many intermediates of the TCA cycle. Note that some provide carbon skeletons that can enter at different points (e.g., tyrosine).

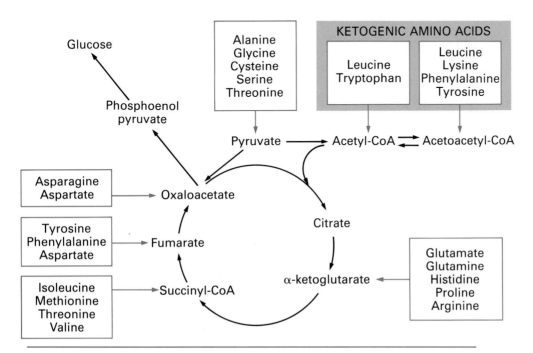

FIGURE 32-4 The catabolism of phenylalanine to acetoacetyl-CoA. Enzymatic defects at different steps along this pathway can lead to various diseases, of which PKU is most common.

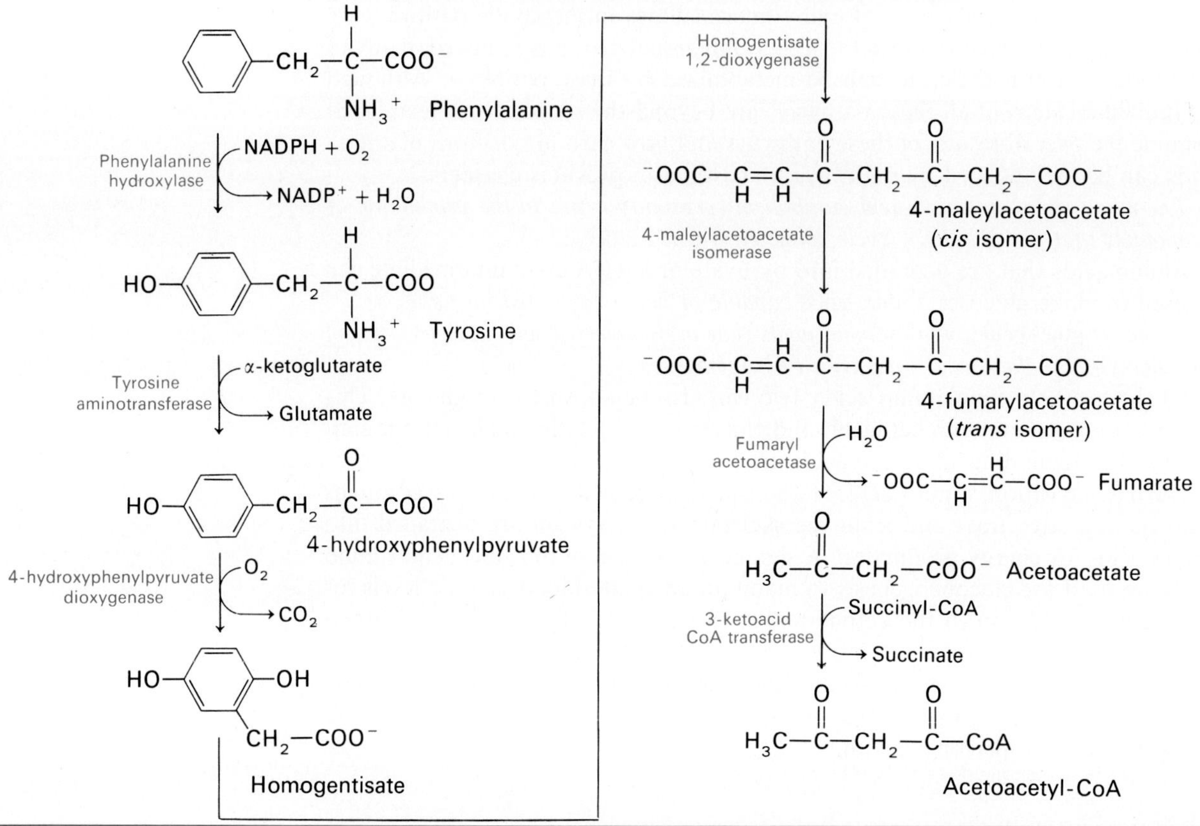

The catabolism of phenylalanine demonstrates the complexity of one amino acid oxidative pathway. In a series of seven steps, phenylalanine is converted into acetoacetyl-CoA (Figure 32-4). In the first of these steps, phenylalanine is hydroxylated to form tyrosine, a reaction catalyzed by *phenylalanine hydroxylase*. If the activity of this enzyme is low (due to deficiency or genetic defect), phenylalanine cannot be catabolized in a normal fashion, but is converted to phenylpyruvate (a phenyl-keto acid). Phenylpyruvate competitively inhibits pyruvate dehydrogenase, dramatically slowing down the TCA cycle. The brain's ability to use energy is severely handicapped, and mental retardation in young children can result. As we pointed out in Chapter 24, this disease is called PKU (phenylketonuria) and can be detected by the presence of phenylpyruvic acid in the urine or by a lack of phenylalanine hydroxylase activity in the blood.

TABLE 32-1 Genetically Linked Diseases and the Enzymatic Defects That Cause Them

Name	Defective Enzyme or Process
albinism	tyrosine-3-monooxygenase
alkaptonuria	homogentisate-1,2-dioxygenase
argininosuccinic acidemia	argininosuccinatelyase
homocystinuria	cystathionine β-synthase
maple syrup urine disease	branched-chain α-keto acid dehydrogenase
phenylketonuria	phenylalanine-4-monooxygenase
hypervalinemia	valine transaminase

A number of other enzymatic defects in amino acid metabolism that cause human genetically linked diseases are listed in Table 32–1.

32.3 DEAMINATION AND TRANSAMINATION

Removing the amino groups from amino acids (which yields carbon skeletons for further metabolism) most often occurs by two general types of reactions. Both reactions form α-keto acids because the amino group is replaced by a carbonyl group at the α position.

$$\text{R}-\underset{\underset{\alpha\text{-amino acid}}{}}{\overset{\overset{NH_3^+}{|}}{CH}}-COO^- \xrightarrow[\text{(oxidation)}]{\text{loss of amino group} \nearrow NH_4^+} \text{R}-\underset{\alpha\text{-keto acid}}{\overset{\overset{O}{\|}}{C}}-COO^-$$

Oxidative deamination *is a reaction in which an amino acid is converted into a keto acid accompanied by the release of a free ammonium ion.* These reactions are catalyzed by *amino acid oxidase* enzymes, and use flavin cofactors as the oxidizing agent. The oxidative deamination of glutamate to form α-ketoglutarate is an example of such a reaction.

oxidative deamination

$$^-OOC-(CH_2)_2-\underset{\underset{\text{L-glutamic acid}}{}}{\overset{\overset{NH_3^+}{|}}{CH}}-COO^- \xrightarrow[\text{NAD}^+ \downarrow \text{NADH + H}^+]{\nearrow NH_4^+} {}^-OOC-(CH_2)_2-\underset{\alpha\text{-ketoglutarate}}{\overset{\overset{O}{\|}}{C}}-COO^-$$

In a similar but slightly different reaction, serine and threonine are deaminated without producing a reduced cofactor. For example, water is formed in a dehydration reaction as serine is converted to pyruvate.

$$HO-CH_2-\underset{\underset{\text{serine}}{}}{\overset{\overset{NH_3^+}{|}}{CH}}-COO^- \xrightarrow[\downarrow H_2O]{\nearrow NH_4^+} H_3C-\underset{\text{pyruvate}}{\overset{\overset{O}{\|}}{C}}-COO^-$$

Transamination *also forms α-keto acids, but through an entirely different mechanism that transfers amino groups from one molecule to another.* In particular, the molecule that accepts the transferred amino group is initially an α-keto acid. Hence, the reactants of transamination reactions are an amino acid and an α-keto acid. The products are also an amino acid and an α-keto acid. The only difference between reactants and products is which carbon skeleton bears the amino functional group.

transamination

$$\overset{\alpha\text{-amino group transfer}}{\overset{\frown}{R_1-\underset{\underset{\alpha\text{-amino acid (1)}}{}}{\overset{\overset{NH_3^+}{|}}{CH}}-COO^- + R_2-\underset{\alpha\text{-keto acid (2)}}{\overset{\overset{O}{\|}}{C}}-COO^-}} \underset{}{\overset{\text{transaminase}}{\rightleftharpoons}} R_1-\underset{\alpha\text{-keto acid (1)}}{\overset{\overset{O}{\|}}{C}}-COO^- + R_2-\underset{\alpha\text{-amino acid (2)}}{\overset{\overset{NH_3^+}{|}}{CH}}-COO^-$$

Several transaminase (often called aminotransferase) enzymes are present in various tissues. Aspartate aminotransferase (AST)—also known as glutamate-oxaloacetate transaminase (GOT)—and alanine aminotransferase (ALT) are important transaminases. Aspartate aminotransferase catalyzes the transfer of the α-amino group of aspartic acid to α-ketoglutaric acid (a TCA cycle intermediate). The products of this reaction are glutamic acid and oxaloacetate (another TCA cycle intermediate).

TABLE 32–2 Clinical Conditions and Serum Transaminase Levels

Clinical Condition	Transaminase Level
1. myocardial infarction	elevated AST
2. infectious hepatitis (viral)	elevated AST and ALT
3. renal infarction	elevated AST
4. progressive muscular dystrophy	elevated AST
5. other muscle disease states (and injuries affecting the muscles—crushing of skeletal muscles in automobile and industrial accidents	elevated AST

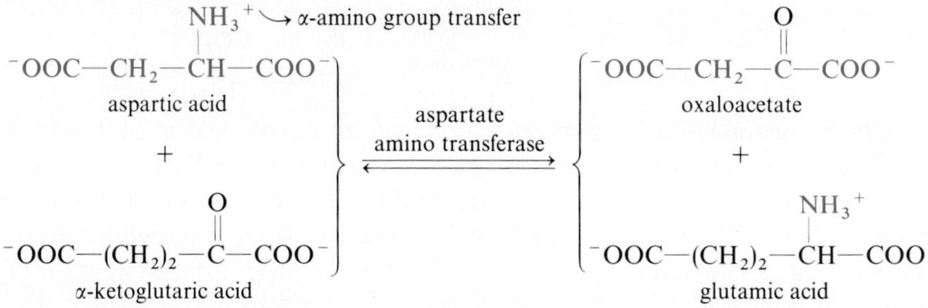

$$\text{aspartic acid} + \alpha\text{-ketoglutaric acid} \xrightleftharpoons[]{\text{aspartate amino transferase}} \text{oxaloacetate} + \text{glutamic acid}$$

(α-amino group transfer: NH_3^+ from $^-OOC-CH_2-CH(NH_3^+)-COO^-$ (aspartic acid) transferred, giving $^-OOC-CH_2-C(=O)-COO^-$ (oxaloacetate); $^-OOC-(CH_2)_2-C(=O)-COO^-$ (α-ketoglutaric acid) becomes $^-OOC-(CH_2)_2-CH(NH_3^+)-COO^-$ (glutamic acid).)

Although not often shown in transaminase reactions such as Figure 32–5, all amino-transferase enzymes utilize pyridoxal phosphate (PLP, derived from vitamin B_6) as an enzyme cofactor.

The enzymes AST and ALT are of particular clinical importance, since organs contain different amounts of these enzymes. Injury to any of these organs can be detected by the release of AST and ALT into the bloodstream, as we see in Table 32–2.

32.4 THE UREA CYCLE

Urea is the principal waste product whereby nitrogen is eliminated from the human body. This extremely water-soluble compound is an efficient carrier of nitrogen, with its two nitrogen atoms representing almost 47% of its total molecular weight.

$$\text{urea:} \quad NH_2-\overset{\overset{\displaystyle O}{\|}}{C}-NH_2$$

urea cycle

The **urea cycle** *is the metabolic pathway that converts ammonia into urea.* This important series of reactions was first elucidated by H. A. Krebs and K. Henseleit in 1932 (the same Krebs who discovered the TCA cycle). Occurring primarily in the liver, this cycle combines free ammonium ions (NH_4^+), carbon dioxide (CO_2), and an amino group from aspartic acid to eventually form urea. ATP is also required to drive the formation of urea, a balanced equation of which is shown here.

$$CO_2 + NH_4^+ + \text{aspartic acid} + 2\,H_2O + 3\,ATP \xrightarrow{\text{urea cycle}} \text{urea} + \text{fumaric acid} + 2\,ADP + AMP + 3\,P_i$$

Ammonia is transported from peripheral tissues to the liver by both glutamine and alanine. Ammonia is connected to glutamic acid by forming an amide linkage. Once glutamine has reached the liver, a different reaction releases free ammonia and regenerates glutamic acid. Alanine transports amino groups from other amino acids by using transaminase enzymes at both locations (see Figure 32–5).

The initial reaction of the urea cycle is catalyzed by the enzyme *carbamoyl phosphate synthetase* and occurs within the mitochondrial matrix. This reaction combines NH_3, CO_2, and H_2O to form an unstable compound called *carbamoyl phosphate*. This irreversible reaction requires the expenditure of two ATP molecules and is summarized in the following equation.

$$NH_3 + CO_2 + H_2O \xrightarrow[\text{2 ATP} \quad \text{2 ADP} + 2P_i]{} \underset{\text{carbamoyl phosphate}}{\overset{\overset{\text{carbamoyl group}}{\downarrow}}{H_2N-\overset{O}{\underset{\|}{C}}-O-\overset{O}{\underset{\underset{O^-}{|}}{\overset{\|}{P}}}-O^-}}$$

The carbamoyl phosphate group, which will eventually form part of the urea molecule, now enters the cyclic portion of the urea cycle. Condensing with ornithine (an amino acid similar to leucine), carbamoyl phosphate now forms citrulline. These first two reactions of the urea cycle occur in the mitochondria matrix, while the remainder of this metabolic pathway occurs in the cytoplasm. This requires citrulline to be transported into the cytoplasm before further reactions can occur.

A reaction between aspartate and citrulline now forms one of the largest metabolic intermediates we have discussed to this point, argininosuccinate. Two high-energy bonds from one ATP molecule are consumed in this reaction. The sole purpose of forming argininosuccinate appears to be *to enable aspartate to attach its amino group to the slowly forming urea molecule* (now still attached to argininosuc-

FIGURE 32–5 Transport of ammonia to the liver via glutamine.

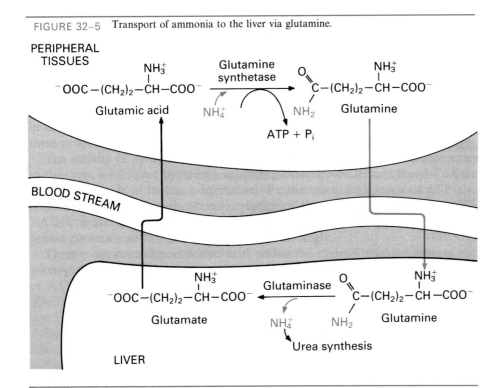

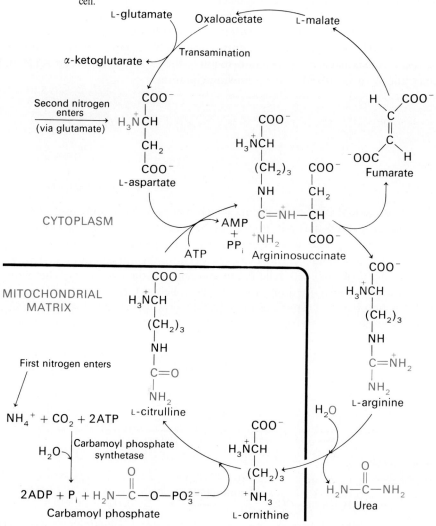

FIGURE 32-6 The urea cycle, showing reaction intermediates and their locations within the cell.

cinate). In the very next reaction, the four-carbon skeleton (resembling succinate) is cleaved off, yielding fumarate and a newly formed molecule of arginine. Fumarate is eventually recycled through malate and oxaloacetate, which picks up another amino group via transamination to again form aspartate (see Figure 32-6).

In the final reaction of the urea cycle, urea is hydrolyzed from arginine to reform ornithine. After reentering the mitochondria, the regenerated ornithine can again begin another cycle. A total of four high-energy ATP bonds are needed to synthesize one molecule of urea. Two ATP molecules are converted to ADP, while a third loses two high-energy bonds to form AMP.

Example 32.1

From your knowledge of other metabolic pathways, identify the intermediates in Figure 32-6 that are part of the TCA cycle.

Solution

Fumarate, malate, and oxaloacetate are all members of the TCA cycle. Hence we say that the urea cycle and the TCA cycle are linked. In fact, Krebs' work on this cycle aided him greatly in elucidating the TCA cycle.

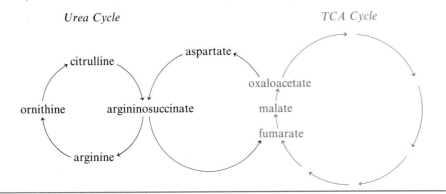

Urea is removed from the bloodstream and excreted by the kidneys. A steady state concentration of urea is normally present in the blood (15–39 mg urea/100 mL blood). This corresponds to about 7–18 mg of nitrogen/100 mL of blood (about 47% of the weight of urea). The term **blood urea nitrogen (BUN)** is used to express the actual amount of nitrogen present in the blood.

blood urea nitrogen

Increased BUN values accompany several physiological states. These are: increased protein catabolism—leading to increased urea synthesis and excretion by the liver, dehydration, kidney malfunction, or obstruction of the urinary tract (e.g., by stones). Decreased BUN levels are observed during protein malnutrition, pregnancy, or rehydration (massive water infusion).

32.5 AMINO ACID BIOSYNTHESIS

Living organisms differ considerably in their ability to synthesize the 20 amino acids commonly found in proteins. Some plants and different strains of bacteria can synthesize all 20 amino acids, while human adults can synthesize only 10. Those not synthesized by the body are called *nutritionally essential* and must be obtained from the diet, while those synthesized by the body are called *nonessential*. These categories of amino acids are presented in Table 32–3.

Most of the nonessential amino acids are easily synthesized via relatively short pathways in the body. Synthetic routes for the essential amino acids are usually longer and more complex. Some animals are lacking one or more enzymes in these pathways or the enzymes exhibit low activity. This condition impedes synthesis of these amino acids and makes essential their intake through foods in the diet.

In organisms that can synthesize them, amino acid biosynthesis is rather complex, but we have already discussed precursors of amino acids in earlier chapters. Figure 32–7 (p. 860) shows where pathways begin that lead to amino acid production. Note how some amino acids share a common biosynthetic pathway, increasing the efficiency of the cell. Amino acids are not usually synthesized by a simple reversal of the catabolic pathways that degrade them.

TABLE 32–3 Essential and Nonessential Amino Acids for Humans

Nonessential	Essential
glutamate	isoleucine
glutamine	leucine
proline	lysine
aspartate	methionine
asparagine	phenylalanine
alanine	threonine
glycine	tryptophan
serine	valine
tyrosine	arginine*
cysteine	histidine

* Essential in young children but not in adults.

The biosynthetic pathways for two or three amino acids are only one step long. Consider the conversion of pyruvate to alanine. Draw the structures for pyruvate and alanine and show the reaction that accomplishes this conversion.

Example 32.2

FIGURE 32-7   A schematic drawing of the initial precursors of amino acid biosynthesis. Many steps are often required in the synthesis of a single amino acid.

```
         α-ketoglutarate                    Oxaloacetate                  Ribose 5-phosphate
                ↓                                ↓                               ↓
             Glutamate                        Aspartate                       Histidine
          ↙     ↓     ↘                  ↙    ↓    ↓    ↘
   Glutamine  Proline  Arginine   Asparagine Methionine Threonine Lysine
                                                         ↓
                                                     Isoleucine

      Phosphoenolpyruvate              3-Phosphoglycerate              Pyruvate
              +                                ↓                    ↙    ↓    ↘
      Erythrose 4-phosphate                 Serine              Alanine Valine Leucine
          ↙     ↓     ↘                  ↙       ↘
  Phenylalanine→Tryosine Tryptophan  Cysteine  Glycine
```

Solution

Transamination of pyruvate forms alanine, a reaction catalyzed by alanine aminotransferase.

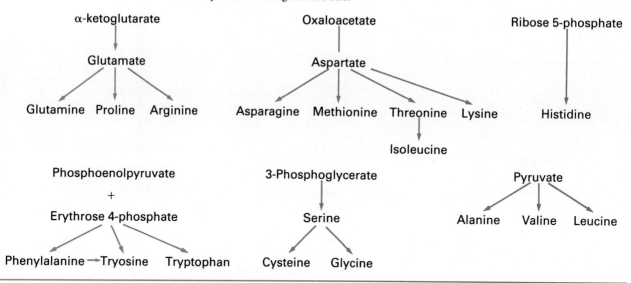

Interestingly, some ruminant animals, such as cows, secrete urea from the blood into the rumen. Microorganisms in the rumen use the urea as a source of NH_3 to manufacture amino acids, which are then absorbed and used by the cow. This efficient use of urea waste helps reclaim some of the energy lost during its synthesis.

32.6 CONTROL OF AMINO ACID METABOLISM

The amino acid pool contains very low concentrations of the 20 amino acids needed for protein synthesis, as compared to the intermediates and products of carbohydrate and fatty acid metabolism. Their use and flow through metabolic pathways is also relatively small. However, when needed, they must be made in correct ratios and at the right time to support protein synthesis. Therefore, their metabolism must be strictly controlled. As we have seen for other pathways, the catabolic pathways and biosynthetic pathways are separate and distinctly different. This allows independent regulation, which occurs primarily through the action of specific hormones.

Growth hormone and testosterone both have a stimulating effect on protein synthesis (anabolism). For this reason, they are called *anabolic hormones*. Anabolic steroids are sometimes taken by body builders to stimulate muscle production (Section 27.8).

Thyroid hormone is a unique hormone in the way it helps regulate amino acid metabolism. Low concentrations of thyroid hormone are necessary for and tend to promote protein anabolism, thereby stimulating growth when plenty of carbohydrates and fats are available for energy production. On the other hand, under different conditions (such as excessive thyroid hormone or low caloric intake), this hormone can promote protein mobilization and catabolism.

EXERCISES AND PROBLEMS

Overview of Amino Acid Metabolism

32.1 What is *protein turnover*? Compare the turnover rates between insulin and proteins in connective tissue.

32.2 What is the *amino acid pool*?

32.3 Compare the amounts of energy provided by amino acids with that produced by carbohydrates and lipids for an average adult on a well-balanced diet.

32.4 Why is a positive nitrogen balance necessary to sustain healthy growth? How can an expectant mother ensure a positive nitrogen balance for her unborn child?

32.5 What cellular processes deplete the amino acid pool?

Carbon Skeletons of Amino Acids

32.6 Describe why carbon skeletons play an important part in the metabolism of amino acids.

32.7 Can proteins be converted into glycogen or fats? Explain.

32.8 What is the difference between glucogenic and ketogenic amino acids?

32.9 Lab tests from a newly admitted hospital patient reveal elevated levels of ketone bodies in both the blood and urine. The patient is obviously not diabetic and insists he is eating well, yet complains of feeling weak and never having enough energy. What line of questioning should be pursued as to his eating habits? Explain.

32.10 What aromatic amino acid is produced as a degradation product of another?

32.11 Explain why the term phenylketonuria (PKU) is given to a disease caused by low levels of phenylalanine hydroxylase activity.

32.12 A transaminase enzyme is responsible for the conversion of phenylalanine to phenylpyruvic acid. Write a chemical equation for this reaction, showing structures of both reactants and products.

32.13 In many forms of albinism (light pigmentation of skin and other tissues) the enzymatic activity of tyrosinase is very low. This enzymatic reaction is the first in a series of reactions that convert tyrosine into melanin pigments in the skin. Many children suffering from PKU also have fair hair and blue eyes. This is not necessarily because their parents had light complexions; rather it is a direct result of the PKU disease. Suggest a biochemical reason for this mild albinism.

Deamination and Transamination

32.14 What are the general products of amino acid deamination and transamination?

32.15 How does deamination differ from transamination?

32.16 Simple deamination of aspartic acid yields what TCA cycle intermediate?

32.17 The following α-keto acid can be used as a substitute for a particular essential amino acid in the diet. Explain how this is possible and draw its structure.

$$CH_3-CH-CH_2-\overset{O}{\underset{\|}{C}}-COOH$$
$$\underset{CH_3}{|}$$

32.18 For each of the following transamination reactions, draw chemical structures for reactants or products as indicated.

a. oxaloacetate $\longrightarrow$?
b. ? $\longrightarrow$ α-ketoglutarate
c. alanine $\longrightarrow$?

32.19 What coenzyme is used by transaminase enzymes?

32.20 A patient is admitted to the hospital with severe kidney pain. Explain how lab test results of AST, ALT, and BUN levels in the blood can be used to diagnose either kidney stones or renal infarction.

The Urea Cycle

32.21 Draw the structure of urea. Identify the source of both nitrogens, the carbon, and the oxygen atom in this structure when urea is synthesized by the urea cycle.

32.22 Why is it often stated that the urea cycle is "linked" to the TCA cycle?

32.23 The site of carbamoyl phosphate formation is in the mitochondrial matrix; yet reactions of the TCA cycle and ultimate urea formation occur in the cytoplasm. Explain how these reactions can be linked.

32.24 Urea is an excellent way to excrete nitrogen; yet some organisms excrete nitrogen waste in other ways. Fish excrete

NH$_3$, while animals such as birds, lizards, and snakes excrete uric acid. Calculate the weight–percent of nitrogen in each of these excreted molecules, and compare these values to urea.

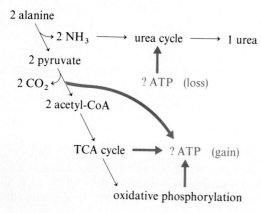

ammonia   urea

uric acid

32.25 Consider the transamination of an amino acid, in which its amino group is transferred to aspartic acid. Trace the fate of this nitrogen from this point until it reaches urea and is excreted.

32.26 Suppose two molecules of alanine are degraded according to the following pathway.

a. How many ATP molecules are generated when alanine is oxidized via this pathway?
b. How many ATP molecules are required to convert the two nitrogen atoms from two molecules of alanine into urea?
c. Considering both the loss and gain of ATP from these reactions, what is the *net* production of ATP?

Amino Acid Biosynthesis

32.27 Explain how transamination can be used to synthesize glutamic acid from α-ketoglutarate.

32.28 Why are some amino acids considered "essential" and others "nonessential?"

32.29 What amino acid functions in the transport of nitrogen to the liver from peripheral tissues? Draw the structure of α-ketoglutarate and show how it can be converted into this amino acid in only two steps.

32.30 From your study of carbohydrate metabolism of 3-, 4-, 5-, 6-, and 7-carbon sugars, what pathway provides the starting material for synthesis of the three aromatic acids?

Control of Amino Acid Metabolism

32.31 One of your classmates argues that biochemistry would be a lot simpler if amino acid biosynthesis and catabolism occurred as simple reversals of the same metabolic pathways. Explain why even though this might be easier to learn, it would *not* be advantageous for the cell.

32.32 Give an example of an anabolic *steroid* that stimulates protein biosynthesis.

32.33 Explain the cellular conditions under which thyroid hormone can stimulate either protein synthesis or catabolism.

Extracellular Fluids

Objectives

After completing Chapter 33, you will be able to:

33.1 Understand some of the main differences between extracellular and intracellular fluids, including their chemical composition.
33.2 Show that you know the major constituents and functions of blood.
33.3 Describe the functions of the lymphatic system in disease control and fluid maintenance.
33.4 Describe the chemistry of urine formation in the body, together with the role played by each of the important parts of the urinary tract.
33.5 Understand the basic mechanisms of fluid and electrolyte balance and state why they are so closely related.
33.6 Describe three ways body fluid pH is regulated.
33.7 Be familiar with some of the parameters that affect drug absorption and excretion from the extracellular fluids of the body.

INTRODUCTION

Throughout our study of metabolism, we have studied how important the supply of nutrients and oxygen is for proper cell maintenance. Beyond the supply of nutrients, waste products such as carbon dioxide and water must be removed and excreted from the system. Small, single-celled organisms merely absorb what they need directly from the immediate surroundings and excrete their waste products in similar fashion. The complex, integrated nature of the human body demands transport of these substances throughout our systems. Extracellular fluids provide this transport and also allow us to be independent of many changes in our environments.

Normal metabolic activities require the stringent regulation of the volume, ionic composition, and pH of these fluids. Of the several mechanisms involved in this regulation, kidney function is most important. Furthermore, abnormal constituents of extracellular fluids such as blood and urine are critically important parameters in diagnosing disease. Almost any drug administered to an individual soon shows up in the urine; urine analysis will indicate the drug's current presence in the body.

33.1 EXTRACELLULAR AND INTRACELLULAR BODY FLUIDS

Water represents approximately 60% of the adult male body weight. In females this value is less, about 50%, due to their slightly higher percentage of adipose tissue which contains less water per unit of weight. Of all the water in humans, about two-thirds is located within cells as the major part of the *intracellular fluids*. **Intracellular fluids** *are the fluids present within cells.* **Extracellular fluids** *are those fluids not present in cells* and account for the remaining one-third of body water. About 7% of all body fluid is in the *liquid part of the blood (called* **plasma***)*. Another 20–25% is found in the **interstitial fluid**, *the fluid in tissues between and around most cells.* The remaining extracellular fluids are found in specialized fluids such as lymph, urine, cerebrospinal fluid, synovial fluid, aqueous humor and others.

There are several differences between the composition of extracellular fluids and intracellular fluids. Inside cells, potassium, magnesium, phosphate, and sulfate are the principal ions, while the extracellular fluids predominate in sodium, chloride, and bicarbonate ions. These *ions are often called* **electrolytes**, since ions conduct electricity in aqueous solution. (Differences in electrolyte composition are shown graphically in Figure 33–1, p. 866.) These fluids also differ in their concentrations of nonelectrolytes such as urea and glucose. A third difference between these fluids is their protein content. The protein content of intracellular fluid is approximately four times the amount in blood plasma, which in turn contains roughly 10 times the amount of protein found in interstitial fluid. Hence, we see that considerable differences exist between extracellular and intracellular fluids.

intracellular fluids

extracellular fluids

plasma

interstitial fluid

electrolytes

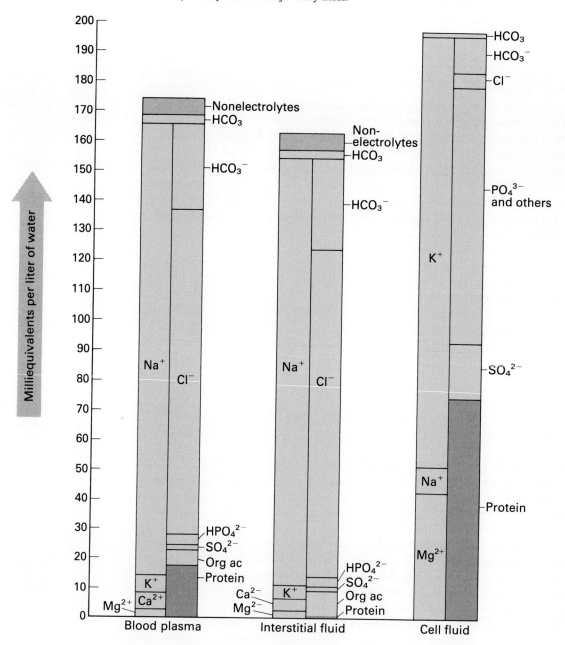

FIGURE 33-1 Electrolyte composition of major body fluids.

33.2 BLOOD

Blood is a complete transport medium that performs vital services for the body. It picks up nutrients such as carbohydrates, lipids, and amino acids, and oxygen from the digestive and respiratory systems, delivering them throughout the body. It also picks up waste products from cells and delivers them to excretory organs.

It transports hormones from endocrine glands to target tissues and contains important enzymes and buffers. Furthermore, it assists in maintaining temperature.

How much blood does an adult body contain? Extremely accurate determinations of total blood volume are accomplished by dilutions of radioisotopes (such as ^{32}P or ^{51}Cr) that are injected into the blood. After a relatively brief period of mixing throughout the circulatory system, the dilution factor is calculated by analyzing a small volume of blood containing the diluted isotope. Approximate blood volume is more often estimated from body weight; there is approximately 70–75 mL blood per kg of body weight.

Assuming an average of 75 mL blood/kg of body weight, calculate the total volume of blood in a 70-kg person.

Example 33.1

Solution
Simple multiplication yields the answer:

$$70 \text{ kg body weight} \times \frac{0.075 \text{ L}}{1 \text{ kg body weight}} = 5.3 \text{ L}$$

Loss of one pint of blood (about 0.5 liter), as when donating blood, usually does not have serious consequences.

About 55% of the total blood volume is plasma, with the rest being taken up by blood cells. Blood cells occur in three main types, as outlined in Table 33–1.

The term **blood type**, *refers to the type of antigens (often called "agglutinogens") in red blood cell membranes.* (Antigens are foreign proteins that stimulate production of antibodies, Chapter 24). Antigens A, B, and Rh are the most clinically important blood antigens as far as transfusion and newborn survival are concerned. These antigens cause antibody formation if injected into the bloodstream of a person who does not normally have them in his own red blood cell membranes. These antibodies are called *anti-A, anti-B,* and *anti-Rh,* and reside in the plasma.

blood type

TABLE 33–1 Types and Characteristics of Different Blood Cells

	Red Blood Cells	White Blood Cells	Platelets
Actual names	erythrocytes	leukocytes	thrombocytes
Subclasses		neutrophils eosinophils basophils lymphocytes monocytes	
Functions	O_2 & CO_2 transport	defense against infection	blood clotting & hemostasis
Diameter (μmeters)	7	15–20	2–4
Average number per cubic millimeter of whole blood			
men	5,500,000	5000–9000	250,000
women	4,800,000	5000–9000	250,000
Average cell lifetimes	105–120 days	3–200 days	10 days

agglutinins

They rapidly bind to and inactivate any foreign blood cell containing their target antigens. Thus, it is important during transfusion to match blood types carefully to avoid possible sensitization and eventually serious consequences. Table 33–2 compares different blood types, based upon these antigens.

Anti-A, anti-B, and anti-Rh antibodies are agglutinins. **Agglutinins** *are antibodies that agglutinate cells, causing them to stick together in clumps.* Such clumps of agglutinated red blood cells can be lethal for people receiving a mismatched blood transfusion, since these agglutinated cells block flow of the blood in small capillary blood vessels of the brain or heart.

One potentially serious problem can arise with Rh-negative mothers who give birth to Rh-positive babies. Some fetal red blood cells may find their way into the mother's blood at birth or cross the placenta membrane during pregnancy, causing the production of anti-Rh antibodies. These antibodies are produced throughout the mother's lifetime, causing potential risks to both the mother and her future Rh-positive children during pregnancy. Usually, mothers in this situation are given shots of other proteins to confuse the immune system during and after pregnancy, thus avoiding sensitization by the baby's blood factors.

Centrifugation of whole blood separates blood cells from the plasma. Blood plasma is roughly 90% water and 10% solutes. Proteins are by far the most important component of plasma, constituting about 6–8% of its total weight. Among these proteins are the *albumins* (55%), *globulins* (38%), and *fibrinogen* (7%). These proteins function in tasks such as transport, protection, controlling blood viscosity, and pH. Fibrinogen and prothrombin (an albumin) function in blood clotting (Chapter 24), while many of the globulins are part of the immune system.

Once nutrients in the blood have reached the peripheral tissues, what makes them pass out of the bloodstream into the surrounding interstitial fluids and cells? This movement of both water and dissolved materials through the capillary walls is governed by two factors: the pressure of blood against the capillary walls and osmosis. Due to pumping action of the heart, blood pressure is higher in the arteries than in veins. This gradient generates a tendency for water, along with dissolved electrolytes and some nonelectrolytes, to move out of the capillary into the interstitial fluid. As the blood flows through the capillaries, the blood pressure drops and the relatively constant osmotic pressure becomes the major force, drawing waste products from cells and interstitial fluids into the blood (see Figure 33–2).

Oxygen transport is accomplished by hemoglobin, which is located inside the red blood cells (thus giving them their red color). The process of oxygen (and carbon dioxide) binding to hemoglobin, together with the factors that promote their release were discussed in detail in Chapter 24. In Sections 33.5 and 33.6 we will study further the role of CO_2 in the control of electrolyte balance and pH within the bloodstream.

TABLE 33–2 Blood Types

Type	Antigen Present in Red Blood Cell Membranes	Can Accept Blood From	Can Donate Blood To
A	A	A, O	A, AB
B	B	B, O	B, AB
AB	A & B	AB, O, A, or B	only AB
O	neither A nor B	only O	O, AB, A, or B
Rh^+	Rh	Rh^+	Rh^+
Rh^-	no Rh antigens	Rh^-	Rh^-

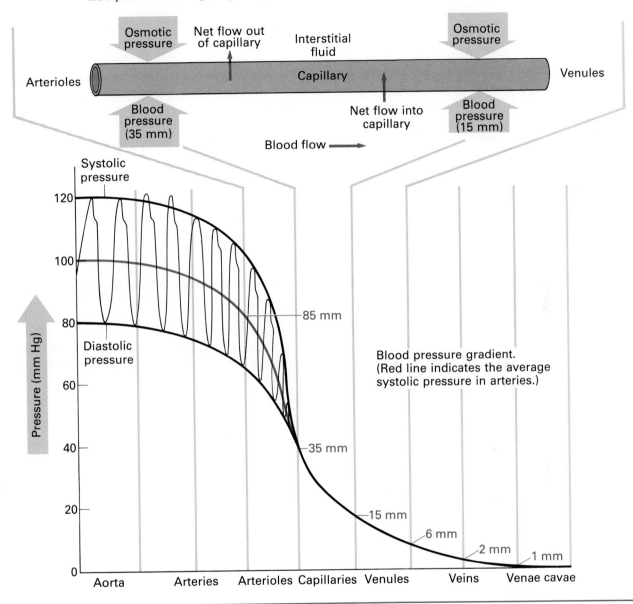

FIGURE 33-2 Movement of substances through capillary walls is directed by both blood pressure and osmotic pressures. Osmotic pressure is relatively constant, but blood pressure is much higher at the entrance to the capillaries than after passing through.

33.3 THE LYMPHATIC SYSTEM

The lymphatic system is actually a specialized component of the circulatory system because it consists of a moving liquid derived from blood and tissue fluids. Lymphatic vessels drain the peripheral areas of the body and return substances such as proteins and fats to the bloodstream for general circulation.

Lymph *is the clear, water-like fluid found in the lymphatic vessels.* About 2500 to 2800 mL of lymph fluid flows through the lymphatic system daily. Lymph somewhat resembles blood plasma and interstitial fluid in composition. The main difference is that the *protein concentration* in lymph (about 4 g/mL) is twice that of

lymph

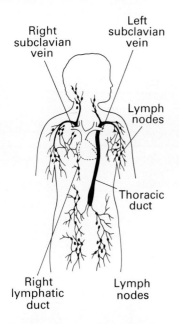

FIGURE 33-3 The lymphatic system, showing some of the major vessels and lymph glands.

other fluids. Water and solutes continually filter out of capillary blood into the interstitial fluid. In fact, each day about 50% of the total blood proteins "leak" out of the capillaries into the tissue fluid and return to the blood by way of the lymphatic vessels.

The lymphatic system not only returns proteins to the blood; it helps fight disease and removes diseased cells from the body. Lymph glands or "nodes" are oval-shaped structures along the lymphatic vessels. As the lymph passes through these glands, it is filtered to remove injurious particles such as microorganisms and diseased cells (i.e., cancer cells). These components are routinely destroyed and removed from the system unless large numbers of microorganisms are present in the lymph. This condition can cause swelling of the lymph gland(s), a fact physicians sometimes use to detect an active infection (see Figure 33-3).

33.4 THE URINARY SYSTEM

Under normal environmental conditions, there is a daily loss of approximately 1500 mL of water by human adults. Of this, about 600 mL is lost through the skin as perspiration, 400 mL in expired air, and 500 mL in the urine. Oxidation of glucose and lipids provides about 200 mL of water, thus requiring an intake of at least 1200 mL of water per day.

urinary system

*The **urinary system** consists of those organs that produce urine and eliminate it from the body.* Formed in the kidneys, urine flows through the ureters, is stored in the bladder, and is expelled through the urethra. The selective filtration of waste products from the blood and excretion via the urine is one of the most important processes in the body. The composition of substances in the blood depends not only on what we eat, but also what is kept by the kidneys. Concentrations of sodium, potassium, chloride, and nitrogen waste in the blood are all controlled by the kidneys, to mention only a few.

nephrons

The unique filtration process in the kidneys occurs in the microscopic structures called *nephrons*. **Nephrons** *are small groups of blood vessels and tubules in close association where water and solutes extracted from the blood are combined to form urine.* The characteristic structure and components of a nephron are shown in Figure 33-4. About 1.25 million nephrons are present in each human kidney and make up the bulk of the kidney tissue.

One-fifth of all blood pumped by the heart flows through the kidneys (about 1200 mL every minute). This blood passes through two main components of the nephrons: the *glomerulus* and the *tubule* (Figures 33-4 and 33-5). Urine is formed by three processes at these sites: *filtration*, *reabsorption*, and *secretion*.

Non-selective filtration occurs at the glomerulus, where electrolytes (such as Na^+ ions), glucose, water, and many other substances flow from the bloodstream into the beginning of the tubule (Bowman's capsule). This is not the only exchange of substances in the kidney, however. Blood continues its flow through small (peritubular) capillaries that are intertwined with the tubule leading from Bowman's capsule. Much of the water lost during filtration, and certain substances such as glucose, are reabsorbed from the tubules back into the blood. Glucose must be reclaimed by active transport in a mechanism that requires the co-transport of sodium ions. This reclamation is so efficient that glucose normally is never found in the urine (if blood glucose levels do not exceed 150 mg/100 mL). Chloride and bicarbonate ions are also absorbed back into the tubules at this point.

In addition to reabsorption, tubules also secrete certain substances. The distal tubules (Figure 33-5) secrete potassium, hydrogen, and ammonium ions into the urine, exchanging them for sodium ions, which are returned to the blood. Secretion of hydrogen ions by the kidneys is one method used by the body to rid

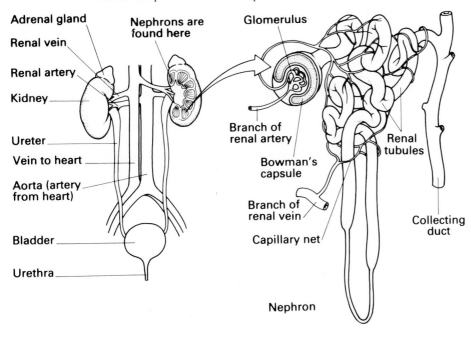

FIGURE 33-4 The principal components of the urinary system. The wedge-shaped section shows the microscopic structure of the nephron unit.

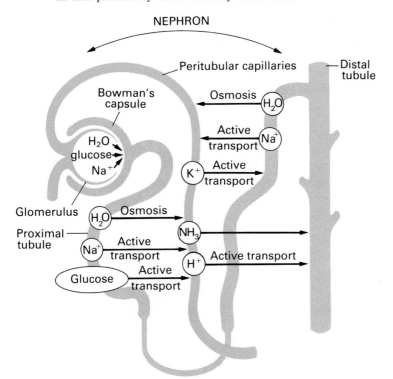

FIGURE 33-5 Diagram showing glomerular filtration, tubular reabsorption, and tubular secretion, the three processes by which the kidneys excrete urine.

itself of unwanted protons (pH maintainance). Tubule cells also secrete certain drugs (e.g., penicillin and sulfa-based drugs).

Initial filtration passes a large volume of water out of the bloodstream. In an average day, approximately 190 *liters* of water are filtered out of the glomerular blood. This is many times the total volume of the blood in the body. However, only a small portion of this is actually excreted in the urine. More than 99% is reclaimed by reabsorption back into the blood. About 270 g of glucose also crosses the glomerular membranes each day, with very little (if any) escaping through the urine.

Non-selective filtration of these *and* foreign substances, with selective reabsorption of only the things the body needs, provides a unique way of excreting almost any water-soluble substance from the body. For this reason, urine is an important tool for the doctor in diagnosing disease and detecting foreign materials present in the body (such as illegal drugs). Super-sensitive techniques for drug and steroid detection in the urine, for example, have been used to screen athletes at Olympic Games. Similar urine tests are often used as evidence for drug abuse. Table 33–3 outlines several abnormal constituents of urine and possible causes of their presence.

Urine volume is directly proportional to the total amount of solutes excreted in the urine. Elevated solute concentrations osmotically draw more water into renal tubules during the reabsorption and secretion stage, and thus tend to increase urine volume. The best known example of this control of urine volume occurs in untreated diabetes mellitus. Voiding of abnormally large volumes of urine is one of the first noticeable symptoms of a new diabetic. Excess glucose "spills over" into the urine and cannot be completely reabsorbed. This increase in solute concentration leads to **diuresis**, *an increase in the volume of discharged urine.*

diuresis

The body controls urine volume by the use of two hormones: *vasopressin* and *aldosterone.* Vasopressin (also known as the antidiuretic hormone, ADH) increases permeability of the distal tubules to water. This increases water reabsorption and decreases urine volume. Aldosterone stimulates the distal tubule absorption of sodium and other ions, leading to increased water reabsorption by osmosis and correspondingly lower urine volumes.

vasopressin

aldosterone

TABLE 33–3 A Few Abnormal Constituents of Urine

Substance Present	An Indication of
glucose	diabetes mellitus, hyperthyroidism
xylulose	lack of the enzyme, xylulose dehydrogenase (no harmful effects)
lactose	moderate excretion during lactation in women
ketone bodies	ketosis
bilirubin	obstructive jaundice
cystine & other amino acids	cystinuria and other renal diseases
Bence Jones protein	multiple myeloma
general protein	kidney damage

Diuretics *are substances that cause an increase in the quantity of urine produced* (*diuresis*). Diuretics function primarily by increasing filtration volumes in the kidneys or by decreasing the reabsorption from distal tubules. Caffeine, alcohol, and certain (often spicy) foods tend to cause diuresis. Diuretic drugs are administered for the relief or prevention of **edema** (*swollen tissues due to increased interstitial fluid volume*), for speeding up the excretion of ingested poisons, or for diluting urine to prevent precipitation of drugs in the kidneys.

The most common test used to evaluate kidney function is probably measuring the level of serum creatinine. Normally, no significant changes occur in the amount of creatinine in blood serum because it is determined by skeletal muscle mass, a factor that does not readily change. Elevation of the serum creatinine (above 1.5 mg/100 mL) indicates that the kidneys are not removing it efficiently, or that renal function is depressed.

creatinine

The kidneys also function in helping control blood pressure. In the case of rapid blood pressure drop (as occurs during hemorrhaging), the kidneys secrete an enzyme named *renin* into the blood. This enzyme brings about the rapid formation of **angiotensin II**, *a small nonapeptide hormone that is a powerful vasoconstrictor*. Due to constriction of peripheral blood vessels, bleeding slows and blood pressure increases.

33.5 FLUID AND ELECTROLYTE BALANCE

Total fluid volume and electrolyte balance are inseparably connected in the body, with the kidneys playing a major role in their regulation. As you recall, the term *electrolytes* refers to charged species (ions) in solution capable of conducting electricity. Differential concentrations of electrolytes on either side of a membrane cause osmosis, which in turn regulates fluid levels in various tissues and interstitial spaces. Hormones that control fluid balances in the kidneys also indirectly control fluid balances throughout the body. Usually, vasopressin levels in the body are sufficient to maintain the proper amount of water in tissues under various levels of water intake. However, severe dehydration upsets the fluid balance beyond the capacity of vasopressin to control it effectively (such as excessive sweating, low water intake, or diarrhea). In such cases, aldosterone also helps to maintain fluid levels. This hormone not only promotes the reabsorption of Na^+ ions in the kidneys, but thereby controls Na^+ levels in the blood. When Na^+ concentration rises, it is accompanied by an increase in Cl^- ions. Together they increase blood volume by increasing osmotic pressures that draw water into the blood.

Since total blood volume is an important parameter in determining blood pressure, higher blood volumes mean higher blood pressures. Hence, we can see why people suffering from high blood pressure (hypertension) are urged to minimize the amount of sodium in their diets.

The majority of body potassium is found inside the cells. For this reason, it is difficult to determine potassium levels merely from serum potassium levels. The body may lose one-third to one-half of its intracellular potassium reserves before the loss is reflected in lower potassium levels. **Hypopotassemia** *is the condition of potassium deficit* and occurs whenever there is cell breakdown as in starvation, trauma, or dehydration. Plasma potassium is rapidly excreted because it is not efficiently reabsorbed by the kidney.

33.6 REGULATION OF pH

Within the extracellular fluids, pH is normally maintained with phenomenal consistency at an average value of 7.4. Variation of just a few tenths of a pH unit from normal "neutral" values can have drastic results for the body: pH values greater than 7.8 or below 7.0 are generally not compatible with life.

acidosis

alkalosis

Acidosis *is a condition in which the* pH *value of the arterial blood is below* 7.4. The opposite situation is termed **alkalosis**, *when* pH *is greater than* 7.4. Even slight changes in pH usually mean very different acid levels; the pH scale is a logarithmic scale in which each whole unit of change corresponds to a tenfold change in proton concentration.

Like other equilibria in the body, the acid–base balance is a dynamic equilibrium. Acids and bases are taken in with the food and acids are produced during metabolic processes (mostly H_2CO_3). Three main mechanisms control the pH in body fluids: *buffer systems* in the body fluids resist pH change, *respiration* removes CO_2 (and with it, H_2CO_3), and *kidneys excrete* acids and bases (see Figure 33–6).

The blood has unusually strong buffers that resist pH changes. In Section 10.15 we studied the carbonic acid/bicarbonate (H_2CO_3/HCO_3^-) buffer system. This is the most prevalent buffer in the blood and exhibits unique traits that make it an outstanding buffer for all interstitial fluids. When CO_2 dissolves in water, carbonic acid forms.

$$CO_2 + H_2O \rightleftharpoons \underset{\text{carbonic acid}}{H_2CO_3} \rightleftharpoons H^+ + \underset{\text{bicarbonate}}{HCO_3^-}$$

There are considerably more HCO_3^- ions in extracellular fluid than any other buffer component. CO_2 is continually produced by cellular respiration, which in turn forms H_2CO_3. Therefore, both components of this buffer system are present in relatively high quantities. Le Châtelier's principle shifts the equilibrium as protons are added to or taken away from solution, thus maintaining pH at 7.4. To a lesser extent, phosphate buffers and protein buffers also help maintain blood pH at this value. However, due to the solubility of CO_2 and HCO_3^- across many cell membranes, the bicarbonate buffers are critically important in the interstitial fluids.

The buffer efficiency of the bicarbonate buffer system in the blood is further enhanced by the presence of erythrocytes. As CO_2 dissolves into blood serum, the pH drops accordingly. However, a great deal of the CO_2 diffuses into the red blood cells and binds to hemoglobin. While bound to hemoglobin, CO_2 cannot form carbonic acid and decrease the pH of the serum. In our study of the binding of CO_2 and O_2 to hemoglobin, we saw that CO_2 is given off in the lungs when CO_2 concentrations drop as these molecules diffuse into the lungs and are exhaled (Chapter 24).

This beneficial effect can work against someone who hyperventilates (breathes in and out very deeply and rapidly). The rapid, deep breaths drive off larger than normal amounts of CO_2 in the expired air. This increases pH, causing alkalosis. When the rapid breathing slows or stops, normal CO_2 levels are again resumed.

The pH of the urine varies from day to day, ranging between 4.5 and 8.0. The pH of urine is controlled primarily by what we eat and our general health. Ketone bodies (Chapter 31) in the urine lower its pH, as do other acids filtered out of the blood. In a unique mechanism, protons in the blood are shuttled into the nephron tubules via bicarbonate and into the urine via ammonium ions or simple diffusion. These protons are captured and held by salts in the urine. Sodium ions return to maintain positive charge balance (H^+ exchanges with Na^+), and are shuttled back into the blood via carbonate. Here, another cycle can begin again. Figure 33–7 shows diagrammatically how this entire process occurs.

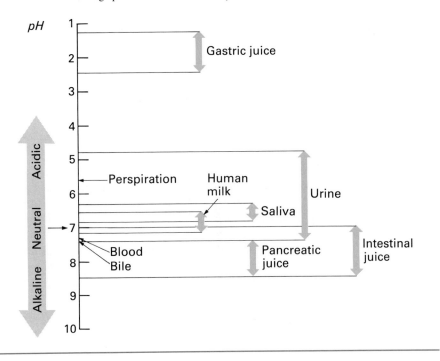

FIGURE 33–6 Average pH values of various body fluids.

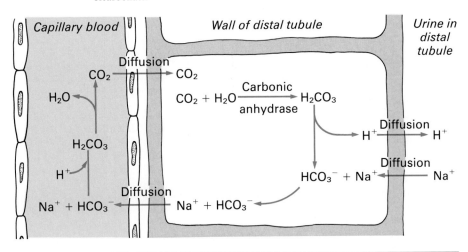

FIGURE 33–7 Kidney control of pH in the blood by shuttling protons out into the urine via bicarbonate.

33.7 DRUG ABSORPTION AND EXCRETION

Upon entering the body, drugs generally distribute themselves throughout the body fluids. Drug action is based upon a sequence of chemical events starting when the active agent enters the body, reaches the target tissue, and eventually is excreted from the body. **Pharmacokinetics** *is the area of pharmacology that studies rates of drug absorption, transport, and excretion.*

Many parameters affect the rate of drug absorption. For example, in the case of oral administration, some tablets are designed to dissolve quickly and allow all their active ingredients to be absorbed. Others are designed to dissolve slowly, in a time-release fashion. Some drugs are absorbed faster than others, depending on the nature of the drug, food contents in the stomach, pH optimum for absorption, and general metabolic state of the individual.

Once in the bloodstream, a drug is rapidly distributed throughout the body. Within a few minutes after absorption, it will have reached every organ in the body. While in the blood, a drug can bind to proteins in the blood or be exposed to metabolic attack (especially in the liver). In fact, some drugs must be activated by enzymatic reactions in the body before assuming their "active" form. Due to dilution throughout the circulatory system, only a small fraction of the administered dose reaches the intended organ and causes the desired response.

Because of the massive flow of blood through the kidneys and efficient filtration, the drug begins to be excreted and appears in the urine within minutes. In general, the more water-soluble the drug is, the more rapidly it is eliminated. Figure 33–8 summarizes some of these processes that increase and decrease the concentration of drugs in extracellular fluids.

Because so many parameters affect the concentrations and lengths of time that drugs are present in the body, scientists must study all of these processes before they can determine therapeutically effective and safe doses. The correct dose and

FIGURE 33–8 Some of the principal processes that take place in drug absorption, transport, and excretion.

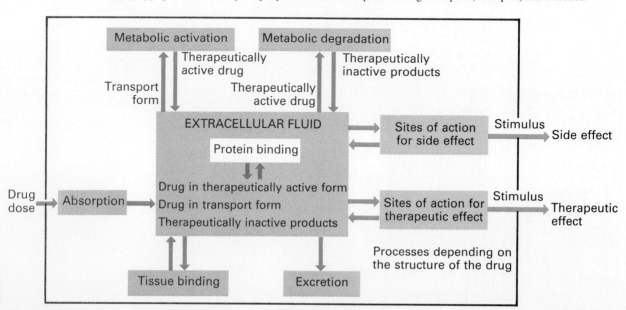

the time between doses must also be considered because many drugs are toxic at higher levels and ineffective at concentrations that are too low.

Figure 33–9 is a graph showing the concentration of a typical drug in the blood as a function of time after the initial dose was taken orally.

As we see in Figure 33–9, a therapeutically effective concentration in the bloodstream is often maintained for only a short time after a single dose. To maintain an effective concentration, repeated doses can be administered periodically. An alternative is to administer a time-release preparation, which would gradually release medication over a period of time. Both these possibilities are graphed in Figure 33–10.

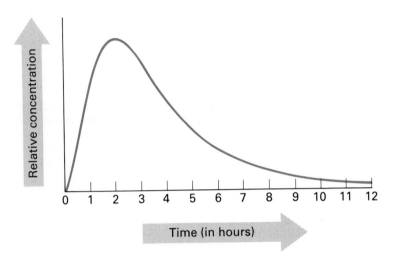

FIGURE 33–9 Concentration of a typical drug in the blood at various times after a single oral dose.

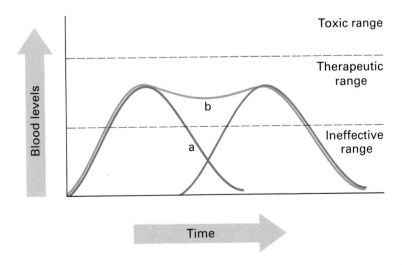

FIGURE 33–10 Concentration of a typical drug (identical to Figure 33–9) in the blood at various times after (a) two oral doses at different times and (b) a single dose of a time-release preparation.

TABLE 33-4 Characteristics of Common Drug Administration Routes

Route	Absorption Pattern	Special Usefulness	Limitations
(a) via gastro-intestinal tract, oral (p.o.)	variable	most convenient, safe and economical	requires patient cooperation; absorption potentially erratic and incomplete for poorly soluble drugs; can cause gastric irritation
rectal	good	useful for medications which cause gastric distress	poor patient-acceptability
sub-lingual	good	rapid onset; especially useful for materials absorbed erratically from gastro-intestinal tract	taste
(b) parenteral intravenous injection (i.v.)	absorption circumvented	potentially immediate effects; valuable for emergency use; allows titration of dose; can be used with large volumes and dilute solutions of irritating substances	increased risk of adverse effects; slow injections required; needs trained personnel; not suitable for oily or insoluble substances
intramuscular injections (i.m.)	prompt from aqueous solution; slow and sustained from depot preparations	suitable for moderate volumes, oily vehicles, and some irritating substances	often causes irritation around injection site; needs trained personnel
subcutaneous injection (s.c.)	as for i.m.	suitable for some insoluble suspensions and implantation of solid pellets	not suitable for large volumes; possible slough from irritating substances; can be done by untrained personnel
(c) lung: aerosol or spin-haler	good	drug acting on lung or broncho-respiratory tract	systemic absorption erratic; spin-haler may be difficult for small children to use
(d) topical: cream and ointment	erratic into systemic circulation	applying high levels of drug to required site	

Excretion of drugs in the urine closely parallels their concentration in the blood. Some drugs (such as steroids) have long lifetimes in the body, while others (such as water-soluble vitamins or aspirin) are quickly removed from the bloodstream, lasting in the body only a few hours.

Some drugs cannot survive the harsh digestive juices of the digestive tract. Table 33–4 lists a variety of different drug administration routes, together with the strengths and weaknesses of each method. Some of these administration techniques prolong the lifetime of the drug in the body, while others speed up absorption and possibly excretion.

EXERCISES AND PROBLEMS

Intracellular and Extracellular Body Fluids

33.1 Why does the relative percent of water in the human body differ slightly between men and women?

33.2 List three differences between the compositions of intracellular and extracellular body fluids.

33.3 What fluids are rich in sodium? In potassium? In chloride?

33.4 Compare the protein content of interstitial fluid vs. blood plasma.

33.5 You overhear a physician in the hospital request the laboratory test results for a patient's "gas and lytes". What is he requesting?

33.6 What important cation is lost from the intracellular fluid compartment as a result of cell breakdown in starvation or trauma?

Blood

33.7 Describe two different ways blood volumes can be determined.

33.8 As a service project, a campus fraternity and sorority go to the hospital, where each member donates one pint of blood. Compare the *percentage* of the total blood supply that a 91-kg fraternity brother and a 50-kg sorority sister donate. (Which person makes the greatest sacrifice?)

33.9 An accurate determination of a patient's blood supply is needed. Exactly 4.00 mL of radioactive ^{32}P-labeled tracer is injected. After allowing for dilution, exactly 4.00 mL of blood is drawn and checked for its radioactivity. Calculate the total blood volume, if the concentration of the original radioactive solution was 5.25×10^{-6} M and the final, diluted solution was determined to be 4.11×10^{-9} M.

33.10 List the three types of blood cells. Compare their numbers in normal whole blood and their average cell lifetimes.

33.11 Describe the determining factors in blood types, and what happens when transfusions of incompatible blood types occur.

33.12 Explain why people with type "O" blood are sometimes called universal donors, while people with type "AB" blood are considered universal acceptors.

33.13 How is plasma obtained from whole blood?

33.14 Explain why solutes in the blood pass *out* of the circulatory system as they approach the capillaries and why solutes pass *into* the blood on the other side.

The Lymphatic System

33.15 Explain the role of the lymphatic system in returning proteins to the bloodstream.

33.16 How does lymphatic fluid compare with other body fluids with respect to protein concentration?

33.17 Why are swollen lymph nodes often an indication of infection?

The Urinary System

33.18 Sketch and name the anatomical structure in the kidneys responsible for extraction of waste materials from the bloodstream.

33.19 Name the three processes whereby kidneys produce urine. Identify the general location for each of these on your sketch for Exercise 33.18.

33.20 Identify the location where each of the following pass between the capillaries and the nephron tubules. (Hint: Some may pass in different directions and at more than one point.)
a. water
b. K^+ ions
c. glucose
d. Na^+ ions
e. penicillin

33.21 Calculate about how much blood passes through the kidneys each day. Based upon this value, calculate each of the following (on a volume/volume basis).
a. How much water passes out of the blood and into the kidney at Bowman's capsule each day? Calculate the percentage of water in the blood volume passing through the kidneys.
b. How much water is actually excreted as urine each day? Calculate the percentage of this water compared to the blood volume passing through the kidneys.

33.22 Why can synthetic steroids be detected in the urine many days after an athlete has discontinued their use?

33.23 How do solutes secreted into the nephron tubules help control urine volume?

33.24 Why does diuresis often accompany untreated diabetes?

33.25 Describe the mechanisms whereby vasopressin and aldosterone control urinary output.

33.26 Outline the mechanism of how secretions of the kidney can help increase blood pressure following substantial decrease in blood volume.

Fluid and Electrolyte Balance

33.27 Why are people with heart problems or high blood pressure encouraged to limit their sodium intake?

33.28 What is meant by the term *electrolytes*?

33.29 Why is it difficult to estimate total potassium levels in the body merely by analyzing potassium in the blood?

Regulation of pH

33.30 What is the normal range for pH of body fluids?

33.31 Define *acidosis* and *alkalosis*.

33.32 What can be said about the relative concentrations of protons [H^+] in a fluid that differs by one pH unit? By two pH units?

33.33 List the three main mechanisms regulating pH of blood.

33.34 What chemical species are represented by the question marks in the drawing? Using the drawing, explain how bicarbonate helps resist pH changes. (Review Section 10.13 if needed.)

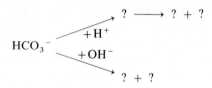

33.35 How does breathing into a paper bag help overcome the ill-effects of hyperventilation?

33.36 Do protons from the bloodstream move *directly* into the urine in the kidneys? (Hint: Are all the protons moving into the urine the *same* protons that leave the bloodstream?) Explain.

33.37 Consider a patient who has suffered a severe head injury. Her breathing is slow and depressed. Which, if either, is a potential danger: acidosis or alkalosis? Why?

33.38 A patient comes in for a routine check-up and complains of not feeling "up-to-par" lately. Urine analysis reveals slightly elevated levels of glucose, sodium, and protein. Creatine levels in the blood are also elevated. What do these results suggest could be wrong with the patient?

Drug Absorption and Excretion

33.39 List three factors influencing the rates of drug absorption in the body.

33.40 A certain new drug exhibits an excellent effect when administered by injection. However, initial studies show it does not work nearly so well when administered orally. Give two possible reasons.

33.41 Draw a graph of the concentration of a water-soluble drug in the blood after a single oral dose. Superimpose another curve of a lipid-soluble drug after a single oral dose. Explain your reasoning for how you drew these curves.

33.42 Compare the advantages and disadvantages of fast-dissolving pills vs. time-release preparations. Suggest two different conditions when each would be the best choice.

33.43 The doctor prescribes an anti-inflammatory agent for you. He tells you to be patient, since you won't be able to tell if it is helping for at least a day or two. Describe one possible reason for this lengthy response time. Should you help things out (on your own) by taking a larger dose? Explain.

33.44 Caffeine increases glomerular filtration rates of many substances. How might a cup of coffee affect blood levels of a medication you are taking?

33.45 Water-soluble vitamins such as B_2 often reach peak concentrations in the blood within two hours after a single dose. To treat a B_2 deficiency, would it be better for a person to take one extra-large dose once a day, or two smaller doses at 12-hour intervals? Explain your logic with a graph similar to the one in Exercise 33.41. Would this logic apply for a fat-soluble vitamin?

Appendix: Significant Figures in Calculations

To count the number of significant figures in a measurement, we observe the following rules.

1. *In numbers not containing any zeros all digits are significant.*

14.233	five significant figures
7.345	four significant figures
233	three significant figures

2. *All zeros between significant digits are significant.*

1.066	four significant figures
4044	four significant figures
3.0001	five significant figures

3. *In numbers less than one, zeros to the left of the first nonzero digit are not significant.*

0.000*34*	two significant figures
0.0*203*	three significant figures
0.*11*	two significant figures

4. *When a number has a decimal point, zeros to the right of the last nonzero digit are significant.*

16.30	four significant figures
120.0	four significant figures
0.*2300*	four significant figures

5. *When a number without a decimal point explicitly shown ends in one or more zeros, the zeros that end the number are not considered significant unless it is explicitly stated that such is the case.*

*32*00	two significant figures
93,000,000	two significant figures
*83,42*0	four significant figures

When two or more numbers representing measurements (expressed to their proper number of significant figures) are used in a calculation, the correct number of significant figures in the answer depends on the number of significant figures in the data used and also on the types of mathematical operations used in obtaining the result. The following two rules can be used to determine the correct number of significant figures for the answer to a calculation involving measurements.

1. *In multiplication or division of numbers, an answer should not have any more significant figures than the least number present in any of the numbers that were multiplied or divided.*

   Suppose we want to multiply the measurements 3.67 and 4.321. Performing this operation on an electronic calculator gives the number 15.85807 as the product. This product, however, is not the correct answer for the calculation because it is much more precise than the input numbers 3.67 and 4.321. The final answer can only contain three significant figures, the number of significant figures in the input number 3.67. Thus, the calculated answer 15.85807 must be rounded to three significant figures, giving 15.9 as the final answer.

2. *In addition or subtraction of numbers, an answer should not have digits beyond the last digit position common to all numbers being added or subtracted.*

   Suppose we want to add the measurements 1.32 and 4.4. On an electronic calculator our answer would be 5.72. This answer must be adjusted, however, because of significant figure considerations. Only the tenths digit is common to both input numbers. The hundredths digit is not known for the input number 4.4. Thus the final answer is reported only to the tenths digit. The answer 5.72 becomes 5.7. Using similar reasoning, the difference between the numbers 5.71 and 4.5 is 2.2 and not 2.21.

Answers to Selected Problems

CHAPTER ONE

1.2. (a) matter (d) nonmatter **1.8.** (a) physical (d) chemical (g) chemical **1.9.** (a) chemical (d) physical (g) physical **1.11.** (a) chemical (d) physical (g) physical **1.12.** (a) chemical (d) physical (g) physical **1.13.** (a) qualitative (d) quantitative **1.17.** (a) 6.34×10^2 (d) 2.1001×10^1 (g) 3.513×10^9 **1.18.** (a) 1.2×10^{-2} (d) 2.304×10^{-1} (g) 1.8×10^{-10} **1.19.** (a) 23,000 (d) 333,333.34 (g) 63,200,000,000 **1.20.** (a) 6,000 (d) 43,000 (g) 300.03 **1.21.** (a) 10^5 (d) 10^{-9} (g) 10^1 **1.22.** (a) 8×10^{-2} (d) 5.2×10^{-18} (g) 8.0×10^{12} **1.23.** (a) 10^2 (d) 10^{-2} (g) 10^8 **1.24.** (a) 2.0×10^{-10} (d) 4.0×10^{-11} (g) 5.0×10^0 **1.25.** (a) 10^2 (d) 10^{-17} **1.26.** (a) 2.0×10^9 (d) 2.2×10^{-11} **1.27.** (a) kilo (d) centi (g) nano **1.28.** (a) centimeter (d) gigagram (g) nanogram **1.31.** (a) 4.75×10^{-1} g (d) 4.75×10^1 cg **1.32.** (a) 1.6×10^5 cm (d) 3×10^7 cm **1.34.** 9×10^{-5} cm **1.37.** 3 tablets **1.40.** 3.41 lbs **1.42.** 6.6×10^{-6} lbs **1.44.** 183 lbs and 6.3 ft **1.46.** 266 qts/hr **1.48.** 89 km/hr **1.50.** 0.0022 lb **1.52.** 1380 g **1.54.** 0.535 lb **1.56.** 10.40 **1.58.** 0.534 **1.60.** $-38.0\,°C$ **1.62.** $-269\,°C$ and $-452\,°F$ **1.64.** $7.2\,°F$ **1.67.** (a) light and heat energy (d) mechanical energy **1.69.** (a) potential (d) potential **1.71.** 1100 kcal **1.73.** 0.155 cal/g $°C$ **1.74.** (a) 1000 cal (d) 31 cal **1.76.** 6040 cal

CHAPTER TWO

2.2. (a) heterogeneous mixture (d) pure substance **2.3.** (a) pure substance (1 phase) (d) heterogeneous mixture (2 phases) (g) heterogeneous mixture (3 phases) **2.4.** (a) chemically homogeneous, physically homogeneous (d) chemically homogeneous, physically heterogeneous (g) chemically heterogeneous, physically heterogeneous **2.9.** (a) compound (d) no classification possible (g) no classification possible **2.10.** (a) true (d) true (g) true **2.14.** (a) compound (d) mixture **2.15.** (a) false (d) false (g) true **2.16.** (a) 13 (d) 16 **2.17.** (a) hydrogen and helium (d) nitrogen and oxygen (g) hydrogen and oxygen **2.20.** (a) neon (d) argon (g) tin **2.21.** (a) Li (d) B (g) Ni **2.25.** (a) compound (d) compound (g) element **2.27.** (a) 2 (d) 4 (g) 3 **2.28.** (a) one C, two O (d) one Mg, one C, three O (g) three N, twelve H, one B, three O

CHAPTER THREE

3.5. (a) true (d) false **3.10.** 2.2 g **3.15.** (a) electron (d) proton (g) proton and neutron **3.16.** (a) false (d) true **3.18.** (a) number of protons or number of electrons (d) total number of subatomic particles present **3.19.** (a) 24 protons, 29 neutrons, 24 electrons (d) 16 protons, 18 neutrons, 16 electrons (g) 20 protons, 20 neutrons, 20 electrons **3.20.** (a) $^{58}_{28}Ni$ (d) $^{18}_{8}O$ (g) $^{197}_{79}Au$ **3.24.** (a) same number of protons, same number of electrons, differing number of neutrons (d) same number of protons, same number of electrons, differing number of neutrons **3.27.** (a) $^{8}_{5}B$ (d) $^{10}_{5}B$ **3.30.** 6.95 amu **3.32.** 21.9 times **3.36.** (a) Ga (d) Cl (g) Sc **3.37.** (a) $_{11}Na$ and $_{55}Cs$ (d) $_2He$ and $_{10}Ne$ **3.39.** (a) group (d) period (g) period

CHAPTER FOUR

4.1. (a) 84.01 amu (d) 183.20 amu (g) 98.09 amu **4.2.** (a) 354.48 amu (d) 98.96 amu (g) 220.09 amu **4.4.** (a) 1.20×10^{24} molecules H_2O (d) 6.02×10^{23} molecules CO **4.5.** (a) 3.26×10^{23} atoms Cu (d) 3.26×10^{23} molecules $C_9H_8O_4$ **4.7.** (a) 28.01 g CO (d) 34.02 g H_2O_2 (g) 49.01 g NaCN **4.8.** (a) 394 g Au (d) 0.37 g Ag (g) 38.3 g NH_3 **4.9.** (a) 0.179 mole CO (d) 0.0210 mole U (g) 0.156 mole O_2 **4.13.** (a) 2.00 moles S (d) 6.00 moles N (g) 18.0 moles O **4.14.** (a) 6.00 moles (d) 26.0 moles (g) 8.00 moles **4.15.** (a) 6.02×10^{23} atoms B (d) 3.0×10^{23} atoms N **4.16.** (a) 63.55 g Cu (d) 2.9×10^{-21} g Cu **4.17.** (a) 2.50 moles He (d) 6.6×10^{-14} mole P **4.18.** (a) 6.14×10^{22} atoms S (d) 2.11×10^{23} atoms S **4.19.** (a) 16.0 g O (d) 16.0 g O **4.20.** (a) 5.3×10^{-16} g $C_{21}H_{36}O_2$ (d) 2.7×10^{-16} g $C_{21}H_{36}O_2$ **4.21.** (a) 5.42×10^{-3} mole $C_3H_2ClF_5O$ (d) 9.79×10^{-21} C atoms **4.26.** (a) $2H_2 + O_2 \rightarrow 2H_2O$ (d) $2Fe_2O_3 \rightarrow 4Fe + 3O_2$ **4.27.** (a) $CH_4 + 2O_2 \rightarrow CO_2 + 2H_2O$ (d) $C_6H_{12} + 9O_2 \rightarrow 6CO_2 + 6H_2O$ **4.28.** (a) $3PbO + 2NH_3 \rightarrow 3Pb + N_2 + 3H_2O$ (d) $Na_2CO_3 + Mg(NO_3)_2 \rightarrow MgCO_3 + 2NaNO_3$ **4.30.** (a) 0.750 mole Fe (d) 1.12 moles Fe_2O_3 **4.31.** (a) 14.0 moles CO_2 (d) 2.00 moles CO_2 **4.32.** (a) 24.3 g NH_3 (d) 24.3 g NH_3 **4.34.** (a) 57.1 g O_2 **4.36.** (a) 245.2 g $KClO_3$ (d) 766 g $KClO_3$

CHAPTER FIVE

5.2. (a) orbital (d) subshell (g) subshell **5.3.** (a) 5 (d) 50 **5.4.** (a) 8 (d) 2 **5.5.** (a) spherical (d) cloverleaf **5.7.** (a) 10 (d) 32 **5.10.** (a) 2s

(lowest), 4s (highest) (d) 3d (lowest), 5f (highest)
5.12. (a) $1s^22s^22p^5$ (d) $1s^22s^22p^63s^23p^64s^2$ (g) $1s^22s^22p^63s^23p^64s^23d^{10}4p^65s^24d^{10}5p^66s^2$ $4f^{14}5d^{10}6p^6$ **5.15.** (a) $_{13}$Al (d) $_{39}$Y **5.16.** (a) $_{36}$Kr (d) $_{19}$K **5.18.** (a) $_{37}$Rb (d) $_{54}$Xe (g) $_{79}$Au
5.19. (a) $1s^22s^22p^63s^23p^64s^23d^{10}$ (d) $1s^22s^22p^63s^2$ $3p^64s^23d^{10}4p^65s^24d^{10}5p^66s^24f^{14}5d^{10}6p^67s^25f^{14}6d^3$
5.20. (a) noble gas (d) transition element (g) representative element **5.21.** (a) nonmetal (d) nonmetal (g) nonmetal **5.22.** (a) Ca, K, Na, Mg, Fe, Zn, Cu, Mn, Co, Mo, Cr, Sn, V, Ni (d) Ca, K, Na, Mg, Sn (g) Fe, Zn, Cu, Mn, Co, Mo, Cr, V, Ni

CHAPTER SIX

6.1. (a) 2 (d) 2 **6.2.** (a) 1 (d) 2 **6.5.** (a) Be (d) N **6.9.** (a) -3 (d) $+1$ **6.11.** (a) 20 protons and 18 electrons (d) 11 protons and 10 electrons
6.12. (a) $+2$ (d) $+1$ **6.13.** (a) loss of two (d) loss of one **6.17.** (a) $BaCl_2$ (d) BaS **6.18.** (a) MgF_2 (d) LiF **6.19.** (a) Na_2S (d) AlP **6.23.** (a) potassium iodide (d) calcium chloride **6.24.** (a) yes (d) no **6.25.** (a) $+1$ (d) $+4$ **6.26.** (a) iron(II) oxide (d) lead(IV) oxide **6.27.** (a) gold(I) chloride (d) nickel(II) oxide **6.28.** (a) KBr (d) Ba_3P_2
6.29. (a) CoS (d) Au_2S_3 **6.38.** (a) F (d) Mg
6.41. (a) polar covalent (d) nonpolar covalent
6.43. (a) polar bonds, nonpolar molecule (d) polar bonds, polar molecule **6.44.** (a) polar (d) polar
6.45. (a) 8 (d) 7 (g) 6 **6.46.** (a) sulfur tetrafluoride (d) carbon monoxide (g) tetrasulfur dinitride
6.47. (a) ICl (d) SiF_4 (g) H_2O **6.48.** (a) nitrate (d) hydroxide (g) phosphate **6.49.** (a) NH_4^+ (d) ClO_2^- (g) $Cr_2O_7^{2-}$ **6.50.** (a) none (d) nitrate (g) cyanide **6.51.** (a) $Ba(NO_3)_2$ (d) $NaClO_4$ (g) $Co(H_2PO_4)_2$ **6.52.** (a) $Ca(C_2H_3O_2)_2$ (d) Au_2SO_4 (g) $MgCO_3$ **6.53.** (a) sodium sulfate (d) ammonium nitrate (g) copper(I) sulfate **6.54.** (a) $KHCO_3$ (d) $Be(CN)_2$ (g) $AlPO_4$

CHAPTER SEVEN

7.2. (a) gaseous (d) gaseous **7.13.** (a) 0.967 atm (d) 0.816 atm **7.18.** (a) 3.89 L (d) 14.8 L
7.19. (a) 1520 torr (d) 304 torr **7.23.** (a) 5.05 L (d) 14.9 L **7.24.** (a) 328 K (d) 596 K
7.29. (a) 42.4 L (d) 97 L **7.32.** 47 mmHg
7.34. (a) endothermic (d) endothermic
7.35. (a) boiling point (d) decomposition (g) boiling point

CHAPTER EIGHT

8.8. (a) saturated (d) saturated **8.9.** (a) dilute (d) concentrated **8.12.** (a) very soluble (d) very soluble **8.13.** (a) soluble (d) soluble (g) soluble
8.14. (a) 7.10%(m/m) (d) 0.267%(m/m)
8.16. (a) 10.0 g (d) 7.89 g **8.17.** (a) 4.00%(v/v)
8.19. 1.74 mL **8.20.** (a) 25.0 g (d) 13.0 g
8.21. (a) 2.0%(m/v) **8.23.** 0.145%(m/v)
8.25. (a) 6.0 M (d) 0.500 M **8.26.** (a) 85.6 mL (d) 0.01 mL **8.27.** (a) 1.85 g (d) 2.50 g
8.29. 0.00615 M **8.30.** (a) 1 (d) 3
8.31. 0.243 mg **8.33.** (a) 2.9 mEq/L (d) 670 mEq/L **8.34.** (a) 1.2 L **8.35.** (a) 1450 mL (d) 2735 mL **8.36.** (a) 3.0%(m/v) (d) 3.0 mEq/L
8.37. (a) 1.2 M (d) 0.125 M **8.45.** (a) same as (d) greater than **8.46.** (a) 0.2 (d) 0.4
8.50. (a) enlarge in size (d) shrink in size
8.51. (a) hypotonic (d) hypertonic

CHAPTER NINE

9.4. (a) exothermic (d) exothermic **9.13.** (a) 1 (d) 3 (g) 6 **9.24.** (a) right (d) left
9.25. (a) right (d) right **9.26.** (a) right (d) no effect (g) no effect

CHAPTER TEN

10.4. (a) acid $\rightarrow$ HF, base $\rightarrow H_2O$ (d) acid $\rightarrow HCO_3^-$, base $\rightarrow H_2O$ **10.5.** (a) $HClO + H_2O \rightarrow H_3O^+ + ClO^-$ (d) $H_3O^+ + OH^- \rightarrow H_2O + H_2O$
10.7. (a) strong (d) weak **10.8.** (a) strong (d) strong **10.11.** (a) mono (d) mono
10.15. (a) sodium ion (Na^+) and fluoride ion (F^-) (d) potassium ion (K^+) and nitride ion (N^{3-}) (g) lithium ion (Li^+) and carbonate ion (CO_3^{2-}) **10.16.** (a) soluble (d) soluble (g) soluble **10.17.** (a) $Ba(NO_3)_2 \rightarrow Ba^{2+} + 2NO_3^-$ (d) $K_2CO_3 \rightarrow 2K^+ + CO_3^{2-}$ (g) $Mg(C_2H_3O_2)_2 \rightarrow Mg^{2+} + 2C_2H_3O_2^-$
10.18. (a) acid (d) salt (g) acid **10.21.** (a) strong (d) weak **10.22.** (a) strong (d) non (g) non
10.23. (a) yes (d) yes **10.24.** (a) $HCl + NaOH \rightarrow NaCl + H_2O$ (d) $2H_3PO_4 + 3Ba(OH)_2 \rightarrow Ba_3(PO_4)_2 + 3H_2O$ **10.25.** (a) $H_2SO_4 + 2LiOH \rightarrow Li_2SO_4 + 2H_2O$ (d) $2H_3PO_4 + 3Ba(OH)_2 \rightarrow Ba_3(PO_4)_2 + 3H_2O$ **10.27.** (a) 3.3×10^{-12} M (d) 8.3×10^{-4} M **10.29.** (a) acidic (d) basic **10.30.** (a) 4 (d) 11 **10.32.** (a) 1.0×10^{-3} M (d) 1.0×10^{-13} M **10.36.** (a) 1 equiv (d) 1.0 equiv (g) 0.061 equiv **10.37.** (a) 0.500 N (d) 1.85 N
10.38. (a) 0.150 N (d) 10.4 N **10.39.** (a) 0.250 M (d) 0.05 M **10.41.** (a) 12.5 mL (d) 15.00 mL
10.42. (a) 0.0500 equiv (d) 0.0500 equiv
10.44. 36.6 mL **10.49.** (a) F^- (d) $C_2H_3O_2^-$ (g) CO_3^{2-} **10.50.** (a) neutral (d) neutral (g) basic **10.53.** (a) no (d) no
10.54. (a) $HPO_4^{2-} + H^+ \rightarrow H_2PO_4^-$ (d) $HCN + OH^- \rightarrow CN^- + H_2O$

CHAPTER ELEVEN

11.3. (a) oxidized (d) loses   **11.4.** (a) +4 (d) +2 (g) +5   **11.5.** (a) +3 (d) +7 (g) +1
11.6. (a) +3 (d) +6 (g) +4   **11.7.** (a) P (+3), F (−1) (d) Na (+1), S (+6), O (−2) (g) O (−1)
11.8. (a) −2 (d) −2   **11.9.** (a) +1 (d) −1
11.10. (a) Cl_2 (oxidizing agent), H_2 (reducing agent) (d) $FeCl_3$ (oxidizing agent), H_2 (reducing agent) (g) I_2 (oxidizing agent), K_2S (reducing agent)   **11.11.** (a) oxidized (d) neither oxidized nor reduced (g) neither oxidized nor reduced   **11.12.** (a) Al (oxidized, reducing agent), Cl_2 (reduced, oxidizing agent) (d) NH_3 (oxidized, reducing agent), O_2 (reduced, oxidizing agent) (g) Cu (oxidized, reducing agent), NO_3^- (reduced, oxidizing agent)   **11.13.** (a) decomposition (d) double displacement   **11.14.** (a) synthesis (d) single displacement (g) decomposition   **11.15.** (a) metathetical (d) redox (g) redox

CHAPTER TWELVE

12.1. (a) $^{14}_{7}N$ (d) $^{197}_{79}Au$ **12.2.** (a) 12 protons and 12 neutrons (d) 53 protons and 74 neutrons
12.4. (a) $^{200}_{84}Po \rightarrow {}^{4}_{2}\alpha + {}^{196}_{82}Pb$ (d) $^{238}_{92}U \rightarrow {}^{4}_{2}\alpha + {}^{234}_{90}Th$
12.5. (a) $^{10}_{5}Be \rightarrow {}^{0}_{-1}\beta + {}^{10}_{5}B$ (d) $^{25}_{11}Na \rightarrow {}^{0}_{-1}\beta + {}^{25}_{12}Mg$
12.7. (a) $^{0}_{-1}\beta$ (d) $^{234}_{94}Pu$ **12.8.** (a) alpha (d) beta
12.9. (a) $^{4}_{2}\alpha$ (d) $^{17}_{8}O$ **12.10.** (a) $^{9}_{4}Be + {}^{4}_{2}\alpha \rightarrow {}^{1}_{0}n + {}^{12}_{6}C$ (d) $^{246}_{96}Cm + {}^{12}_{6}C \rightarrow {}^{254}_{102}No + 4({}^{1}_{0}n)$
12.14. (a) 1/4 (d) 1/8   **12.16.** 280 years
12.29. (a) fusion (d) both processes (g) fission
12.30. (a) fusion (d) neither

CHAPTER THIRTEEN

13.4. (a) SiO_2 (d) Fe_2O_3 **13.9.** (a) $2C + O_2 \rightarrow 2CO$ (d) $N_2 + O_2 \rightarrow 2NO$

CHAPTER FOURTEEN

14.5. (a) $CH_3-CH_2-CH_2-CH_3$
(d) $CH_3-CH_2-CH-CH_2-CH_3$
              |
             CH_2
              |
             CH_3

14.6. (a) $CH_3-CH-CH_2-CH_3$
                |
               CH_3

(d) CH_3
            |
   $CH_3-C-CH_2-CH_3$
            |
           CH_3

14.7. (a)
H   H   H   H   H
| | | | |
H—C—C—C—C—C—H   (d) C_6H_{14}
| | | | |
H   H   H   H   H

14.8. (a) 2,4− (d) 4−   **14.9.** (a) Hexane (d) 4-ethyl-2,3-dimethylpentane (g) 2,2,3,3-tetramethylbutane

14.10. (a) $CH_3-CH-CH_2-CH_3$
                   |
                  CH_3

(d) $CH_3-CH-CH-CH-CH-CH-CH_2-CH_3$
         | | | |
        CH_3 CH_3 CH_3 CH_3

14.11. (a)
        C
        |
  C—C—C—C—C   2-methylpentane
  5 4 3 2 1

14.13. (a) structural isomers (d) structural isomers (g) neither   **14.14.** (a) structural isomers (d) structural isomers (g) neither

14.15. (a) C—C—C—C—C—C   C—C—C—C
                                        | |
              C—C—C—C—C   C   C
                  |
                  C

C—C—C—C—C   C
| |
C   C—C—C—C
                     |
                     C

(d) 3 (g) 4   **14.17.** (a) C_7-chain (d) C_3-chain

14.16. (a) 0

14.19. (a) —CH—CH_3 (d) —CH—CH_2—CH_3
                | |
               CH_3 CH_3

14.20. (a) 3-isopropylhexane (d) 5-ethyl-2-methyl-5-isopropylheptane   **14.21.** C_nH_{2n} **14.23.** (a) 12 (d) 14   **14.24.** (a) cyclohexane (d) 1,2-dimethylcyclopentane

14.25. (a) [cyclohexane with CH_3, CH_3, CH_3 substituents]   (d) [cyclobutane with CH_3, CH—CH_3 substituent]

14.32. (a) $C_3H_8 + 5 O_2 \rightarrow 3 CO_2 + 4 H_2O$
(d) $2 C_8H_{18} + 25 O_2 \rightarrow 16 CO_2 + 18 H_2O$

14.34. (a)
        Cl
        |
   Cl—C—F
        |
        Cl

14.35. (a) 1,1-dibromoethane

(d) 1,2-difluoroethane

A−4

CHAPTER FIFTEEN

15.2. (a) Saturated; alkane (d) unsaturated; cycloalkene (g) unsaturated; alkene (also contains a cycloalkyl group) **15.3.** (a) C_3H_8 (C_nH_{2n+2}) (d) C_6H_{10} (C_nH_{2n-2}) (g) C_5H_8 (C_nH_{2n-2})
15.4. (a) C_4H_{10} (d) C_5H_{10} (g) C_6H_6 **15.5.** (a) 2-butene (d) 1,3-pentadiene (g) 3-isopropylcyclobutene
15.6. (a) C=C—C—C—C (d) C=C—C—C=C
            | |
            C   C—C

(g) [cyclohexadiene structure]

15.7. (a) C—C=C—C—C   3-methyl-3-hexene
             |
             C—C

(d) [cyclohexene with two methyl groups] 4,5-dimethylcyclohexene

15.10. a, b, c, e

15.12.
	Number of Carbon–carbon Sigma Bonds	Pi Bonds
(a)	4	0
(d)	1	2
(g)	4	2

15.15. d, e, f

15.16. (a) cis-2-pentene (d) cis-1,2-dimethylcyclobutane

15.17. (a) $CH_3—CH_2$ H
                      \\C=C/
                      /   \\
                   CH_3 $CH_2—CH_3$

(d) CH_3 H
        \\C=C/ CH_3
        /   \\ |
       H $CH_2—CH—CH_2—CH_3$

15.18. (2) **15.19.** (a) symmetrical (d) unsymmetrical

15.20. (a) $CH_2=CH_2 + Cl_2 \rightarrow CH_2—CH_2$
                                              | |
                                              Cl   Cl

(d) $CH_2=CH_2 + HBr \rightarrow CH_3—CH_2$
                                            |
                                            Br

15.21. (a) $CH_2=CH—CH_3 + Cl_2 \rightarrow CH_2—CH—CH_3$
                                                          | |
                                                          Cl   Cl

(d) $CH_2=CH—CH_3 + HBr \rightarrow CH_3—CH—CH_3$
                                                       |
                                                       Br

(e) $CH_2=CH—CH_3 + H_2O \xrightarrow{H_2SO_4} CH_3—CH—CH_3$
                                                                 |
                                                                 OH

(f) $CH_2=CH—CH_3 + Br_2 \rightarrow CH_2—CH—CH_3$
                                                      | |
                                                      Br   Br

15.22. (a) $CH_3—CH—CH—CH_3$
                            | |
                            Cl   Cl

(d) [cyclopentane] (g) OH
                          |
                    $CH_3—CH—CH_3$

15.24. (a) $CH_3—CH=CH_2 + H_2O \xrightarrow{H_2SO_4}$
$CH_3—CH—CH_3$
         |
        OH

(d) $CH_2=CH_2 + KMnO_4 + H_2O \rightarrow CH_2—CH_2$
                                                             | |
                                                            OH   OH

15.25. (a) 2 (d) 3   **15.26.** (a) oxidation (d) oxidation

15.28. (a)   F   F
                   \\C=C/
                   /   \\
                  F   F

(d) $CH_2=CH$—[benzene ring]

15.29. (a) $—CH_2—CH_2{+}CH_2—CH_2{+}CH_2—CH_2{+}CH_2—CH_2{+}CH_2—CH_2—$

(d) $—CH_2—CH{+}CH_2—CH{+}CH_2—CH{+}CH_2—CH{+}CH_2—CH—$
              | | | | |
           CH_3 CH_3 CH_3 CH_3 CH_3

15.30. (a) C_nH_{2n-2} **15.32.** (a) hexyne (d) 2-butyne
(g) 1,3-butadiyne   **15.33.** (a) C—C—C≡C—C—C
                                              |
                                              C

(d) C—C≡C—C—C—C   **15.35.** (a) 2 (d) 2
               |
              C—C
               |
               C

(g) 4   **15.36.** (a) $CH_3—CH_3$ (d) $CH_2=CH$
                                                            |
                                                            Cl

CHAPTER SIXTEEN

16.5. (a) saturated (d) aromatic   **16.6.** (a) 12
(d) 6   **16.7.** (a) methylbenzene
(d) 1,2,4-trimethylbenzene

A–5

16.8.

Prefix	Positions on Ring
ortho-	1,2-
meta-	1,3-
para-	1,4-

16.9. (a) *o*-dichlorobenzene (d) *m*-bromoiodobenzene
16.10. (a) 2-phenylbutane (d) 2,4-diphenylpentane
16.11. (a) phenol (d) benzoic acid
16.12. (a) 2-chlorophenol (d) 2-nitroaniline
16.13. (a) phenylcyclohexane (d) 1,3-diphenylbenzene
16.14. (a) $CH_3-CH_2-\text{C}_6\text{H}_4-CH_2-CH_3$

(d) 1,4-diiodo-2,5-diethylbenzene (CH_3-CH_2- ring with I at top and bottom $-CH_2-CH_3$)

16.15. (a) triphenylmethane (Ph)$_3$CH

(d) phenylcyclopentane   **16.17.** (a) 1 (d) 3

16.20. (a) substitution (d) substitution
16.21. (a) bromobenzene (d) chlorobenzene
16.22. (a) benzoic acid (COOH on ring) (d) *p*-chlorotoluene (CH$_3$ and Cl on ring)
16.23. (a) CH$_3$ and NO$_2$ on ring (two isomers: ortho and para) (d) CH$_3$ ring and CH$_2$Cl ring

16.26. $H_3C-\text{C}_6\text{H}_4-CH_3$ (*p*-xylene)

16.27. (a) naphthalene

16.28. (a) 8   **16.29.** (a) 2

CHAPTER SEVENTEEN

17.1. (a) alcohol (d) alcohol (g) phenol
17.4. (a) 2-pentanol (d) 2-ethyl-1-pentanol (2-ethylpentanol) (g) 3-pentanol
17.5. (a) $CH_3-CH_2-\underset{\underset{OH}{|}}{CH}-CH_2-CH_3$

(d) $CH_3-\underset{\underset{OH}{|}}{CH}-CH_2-\underset{\underset{CH_3}{|}}{CH}-CH_3$

(g) cyclobutanol with CH$_3$ substituent   **17.6.** (a) CH_3-CH_2-OH ethanol

(d) $CH_3-\underset{\underset{OH}{|}}{CH}-CH_3$ 2-propanol

(g) $CH_3-\underset{\underset{CH_3}{|}}{\overset{\overset{OH}{|}}{C}}-CH_3$ 2-methyl-2-propanol

17.7. (a) 1,2-propanediol (d) 1,2,4-butanediol
17.8. (a) $\underset{\underset{OH}{|}}{CH_2}-CH_2-CH_2-CH_2-\underset{\underset{OH}{|}}{CH_2}$

(d) $CH_3-\underset{\underset{OH}{|}}{\overset{\overset{OH}{|}}{C}}-CH_2-CH_2-CH_3$

17.9. (a) cyclohexanol (d) *cis*-2-methylcyclohexanol
17.10. (a) $HO-CH_2-\underset{\underset{CH_2-CH_3}{|}}{CH}-CH_3$ 2-methylbutanol

(d) HO—cyclopentane—OH 1,3-cyclopentanediol

17.12. (a) 3-ethylphenol (*m*-ethylphenol) (d) 4-hydroxyphenol (hydroquinone)
17.13. (a) 4-chlorophenol (OH ring Cl) (d) 3-methylphenol (OH ring CH$_3$)

17.16. (a) 4-methylphenol (*p*-cresol)
17.17. (a) 2° (d) 1° (g) 2°
17.18. (a) 1°, 2° (b) 1°, 2°, 1°
17.19. (a) serine $H_2N-\underset{\underset{\underset{\underset{OH}{|}}{CH_2}}{|}}{CH}-CO_2H$

A–6

17.20. (a) CH$_3$—CH$_2$—O---H
                              H---O—H

17.22. (a) 1-pentanol   **17.23.** (a) CH$_3$—CH$_2$—OH
(d)
$$\text{CH}_3\text{—CH}_2\text{—}\underset{\text{OH}}{\text{CH}}\text{—CH}_2\text{—CH}_3$$

17.24. (a) CH$_3$—CH=CH$_2$ (d) CH$_3$—CH=CH$_2$

17.25. (a) CH$_3$—CH—O—CH—CH$_3$
                        | |
                        CH$_3$ CH$_3$

(d) CH$_3$—CH—O—CH—CH$_3$
          | |
          CH$_3$ CH$_3$

17.26. (a)
$$\text{CH}_3\text{—}\underset{\|}{\overset{\text{O}}{\text{C}}}\text{—CH}_3$$

17.28. (a) CH$_3$—CH$_2$—CH$_2$—Cl

(d) CH$_2$—CH—CH$_2$
    | | |
  CH$_3$ Cl   CH$_3$

(g)
$$\text{CH}_3\text{—CH}_2\text{—}\underset{\text{CH}_3}{\overset{\text{CH}_3}{\text{C}}}\text{=CH—CH}_3$$

17.29. (a)
$$\text{CH}_3\text{—}\underset{\text{Cl}}{\text{CH}}\text{—CH}_3$$

(d) CH$_3$—CH$_2$—O—CH$_2$—CH$_3$

17.30.
$$\text{CH}_3\text{—CH}_2\text{—}\overset{\text{O}}{\underset{\|}{\text{C}}}\text{—}\underset{\text{CH}_3}{\text{CH}}\text{—CH}_2\text{—CH}_2\text{—CH}_3$$

17.33.
  CO$_2$H   CO$_2$H
  | |
  CH—OH   C=O
  | |
  CH$_2$ CH$_2$
  | |
  CO$_2$H   CO$_2$H
  malic acid   oxaloacetic acid

17.34. (a) pure alcohol   (d) ethanol containing poisons or harmful substances which render it unfit for consumption   (g) 35% ethanol in water solution

17.35. (a) methanol   (d) methanol   (g) ethylene glycol   **17.39.** (a) hexylresorcinol   (d) catechol derivatives   (g) phenolphthalein   **17.43.** (a) methoxypropane   (d) methoxybenzene   **17.44.** (a) methyl *n*-propyl ether   (d) methyl phenyl ether

17.45. (a) mixed ether   (d) mixed ether

17.47. (a) HO—CH$_2$—CH$_2$—CH—CH$_3$
                                          |
                                        CH$_3$

CH$_3$—CH—CH—CH$_3$ and others
        | |
      OH CH$_3$

17.48. (a) CH$_3$—CH$_2$—CH$_2$—O—CH—CH$_3$
                                            |
                                           CH$_3$

(d) CH$_3$—CH—CH$_2$—CH$_2$—CH$_3$
           |
         O—CH$_2$—CH$_3$

17.53. (a) glycerol   (d) ethylene oxide
17.57. (a) CH$_3$—SH
(d)
                       CH$_3$
                       |
HS—CH$_2$—CH$_2$—CH—CH$_2$—CH$_3$

17.58. (a) CH$_3$—CH$_2$—S—S—CH$_2$—CH$_3$

CHAPTER EIGHTEEN

18.1. a, b, e, f, g

18.6. aldehyde:
$$\text{H—}\overset{\text{O}}{\underset{\|}{\text{C}}}\text{—H}$$
ketone:
$$\text{CH}_3\text{—}\overset{\text{O}}{\underset{\|}{\text{C}}}\text{—CH}_3$$

18.7. (a)
$$\text{CH}_3\text{—CH}_2\text{—CH—CH}_2\text{—C}\overset{\text{O}}{\underset{\text{H}}{\diagdown}}$$

(d)
$$\text{CH}_3\text{—CH}_2\text{—CH}_2\text{—}\underset{\text{CH}_3}{\text{CH}}\text{—}\underset{\text{CH}_3}{\text{CH}}\text{—CH}_2\text{—C}\overset{\text{O}}{\underset{\text{H}}{\diagdown}}$$

(g)
$$\text{CH}_3\text{—CH}_2\text{—}\underset{\text{CH}_3}{\text{CH}}\text{—}\underset{\text{CH}_3}{\text{CH}}\text{—CH}_2\text{—}\underset{\text{CH}_3}{\text{CH}}\text{—C}\overset{\text{O}}{\underset{\text{H}}{\diagdown}}$$

18.8. (a)
$$\text{CH}_3\text{—}\overset{\text{O}}{\underset{\|}{\text{C}}}\text{—}\underset{\text{CH}_3}{\text{CH}}\text{—CH}_2\text{—CH}_3$$

(d)
$$\text{CH}_3\text{—}\underset{\text{CH}_3}{\text{CH}}\text{—}\overset{\text{O}}{\underset{\|}{\text{C}}}\text{—}\underset{\text{CH}_3}{\text{CH}}\text{—CH}_3$$

(g)
$$\text{CH}_3\text{—}\overset{\text{O}}{\underset{\|}{\text{C}}}\text{—CH}_3$$

18.9.
$$\text{CH}_3\text{—}\overset{\text{O}}{\underset{\|}{\text{C}}}\text{—CH}_2\text{—CH}_2\text{—CH}_2\text{—CH}_2\text{—CH}_3$$

$$\text{CH}_3\text{—CH}_2\text{—}\overset{\text{O}}{\underset{\|}{\text{C}}}\text{—}\underset{\text{CH}_3}{\text{CH}}\text{—CH}_2\text{—CH}_2\text{—CH}_3$$

18.10. (a) pentanal
(d) 1-phenyl-2-propanone (phenylacetone)
(g) 3,3-dimethylbutanal   **18.11.** (a) cyclohexanone
(d) cyclohexylethanone (cyclohexyl methyl ketone)
(g) 3-methylcyclohexanone

18.12. (a) Aldehydes do not occur as part of a ring; they must be the last carbon of a chain.

18.13.
$$CH_3-\underset{CH_3}{\overset{}{C}}=O-H-O\overset{}{\underset{H}{-}}$$

18.17. Butanone (methyl ethyl ketone) is an excellent solvent.

18.20. (a) CH_3-CH_2-OH (d) CH_3-CH_2-OH

18.21. (a) $CH_3-\underset{\overset{|}{OH}}{CH}-CH_2-CH_3$

(d) HO—⬡—OH (1,4-cyclohexanediol)

18.22. (a) neither (d) neither (g) hemiacetal

18.23. (a) $CH_3-CH_2-O-\underset{\overset{|}{OH}}{CH}-CH_3$

$CH_3-CH_2-O-\underset{\overset{|}{O-CH_2-CH_3}}{CH}-CH_3$

(d) $CH_3-\underset{\overset{|}{CH_3}}{CH}-O-CH_2$ $-$ $\underset{\overset{|}{OH}}{}$

$CH_3-\underset{\overset{|}{CH_3}}{CH}-O-CH_2-O-\underset{\overset{|}{CH_3}}{CH}-CH_3$

18.24. (a) $CH_3-CH_2-CH_2-\underset{\overset{|}{OH}}{CH}-O-CH_2-CH_3$

(d) glucopyranose ring structure with CH₂OH, OH groups

18.26. (a) aldehyde (d) Ag⁺ is reduced to Ag

18.27. (a) blue changes to a red precipitate. (d) urine

18.28. A: $CH_3-CH_2-\underset{\overset{|}{OH}}{CH}-CH_3$ 2-butanol

B: $CH_3-CH_2-\overset{\overset{O}{\|}}{C}-CH_3$ 2-butanone (negative Benedict's test)

18.30. (a) I: $CH_3-\underset{\overset{|}{OH}}{CH}-CH_3$ II: $CH_3-\overset{\overset{O}{\|}}{C}-CH_3$

(d) I: cyclopentene with CH₃   II: cyclopentane with OH and CH₃

III: N.R. (3° alcohol)   IV: 2-methylcyclopentanone

CHAPTER NINETEEN

19.2. a, c, d, f, h

19.3. (a) hexanoic acid (d) 2-methylbutanoic acid
(g) 4-ethylbenzoic acid

19.4. (a) CH_3-CH_2-COOH (d) $CH_3-\underset{\overset{|}{Cl}}{CH}-COOH$

(g) benzoic acid with CH₃ (o-methylbenzoic acid)

19.5. (a) CH_3-COOH (d) $HOOC-COOH$

19.6. (a) ethanoic acid (d) ethanedioic acid

19.7. (a) $CH_3-\underset{\overset{|}{Cl}}{CH}-COOH$

(d) $CH_3-CH=CH-CH_2-COOH$

(g) $CH_3-CH_2-\overset{\overset{O}{\|}}{C}-COOH$

19.9. 3,4,5-trihydroxybenzoic acid (gallic acid)

19.10. (a) CH_3-COOH

19.11. (a) $CH_3-\underset{O-H----O}{\overset{O----H-O}{C}}C-CH_3$ (dimer)

19.12. (a) $H-COOH$ (d) $CH_3-\underset{\overset{|}{CH_3}}{CH}-\underset{\overset{|}{CH_3}}{CH}-COOH$

19.15. (d) $CH_3-(CH_2)_4-\overset{\overset{O}{\|}}{C}-OH + NH_3 \xrightarrow{heat}$

$CH_3-(CH_2)_4-\overset{\overset{O}{\|}}{C}-NH_2 + H_2O$

19.16. (a) $H_2C=CH-CH_2-CH_2-OH \xrightarrow[H^+]{H_2O} \xrightarrow[H_2O, H^+]{K_2Cr_2O_7}$

$CH_3-\overset{\overset{O}{\|}}{C}-CH_2-COOH$

19.17. (a) CH_3-COO^- acetate
(d) $CH_3-(CH_2)_{14}-COO^-$ hexadecanoate (palmitate)

19.25. (a) 25 mL   **19.26.** (a) 10 meq (d) 10 meq

19.28. a, b, d, e, f, h

19.30. (a) ethyl propanoate (d) *n*-propyl ethanoate
(g) methyl benzoate

19.31. (a) ethyl propanoate (d) *n*-propyl acetate
(g) methyl benzoate

19.32. (a) H—C(=O)—O—CH₃
(d) C₆H₅—CH₂—C(=O)—O—CH₂—CH₃
(g) CH₃—CH₂—C(=O)—O—CH₂—CH₂—Br

19.33. (a) ethyl ethanoate
(d) *n*-propyl #2-hydroxypropanoate
(g) isopropyl methanoate

19.35. CH₃—C(=O)—O—CH₂—CH₂—CH(CH₃)—CH₃

19.39. (a) CH₃—C(=O)Cl + HO—CH₂—CH₂—CH₂—CH₃ ⟶ CH₃—C(=O)—O—CH₂—CH₂—CH₂—CH₃ + HCl

(d) C₆H₅—C(=O)OH + HO—CH₃ ⟶ C₆H₅—C(=O)—O—CH₃ + H₂O

19.43. (a) CH₃—COOH + HO—CH₂—CH₃

19.45. CH₃—C(=O)—O—CH₂—CH₃   ethyl acetate

19.46. (a) CH₃—C(=O)—S—CH₂—CH₃
(d) CH₃—CH₂—CH₂—S—C(=O)—CH₂—CH₂—C(=O)—S—CH₂—CH₂—CH₃

19.49. (—C(=O)—C(=O)—O—CH₂—CH₂—CH₂—O—)ₙ

19.51. (a) CH₃—CH₂—CH₂—CH₂—CH₂—O—S(=O)₂—OH + H₂O

19.54. (a) CH₃—CH₂—CH₂—C(=O)—O—C(=O)—CH₂—CH₂—CH₃
(d) CH₃—(CH₂)₄—C(=O)—O—C(=O)—(CH₂)₄—CH₃

19.56. (a) CH₃—C(=O)—O—CH₃ + CH₃—C(=O)—OH
(d) CH₃—CH₂—C(=O)—O—CH₂—CH₃ + CH₃—CH₂—C(=O)—OH

19.57. (a) CH₃—O—P(=O)(OH)—OH

CHAPTER TWENTY

20.1. a, b, d, e, f, h
20.3. (a) 2° (d) 3° (g) (amide)   **20.4.** (a) 2°
(d) 1° (g) 2°   **20.5.** (a) 3°, 2° (d) 2°, 1°
20.6. (a) 2
20.7. (a) ethylmethylamine (d) methylphenylamine
(g) dimethyl-*n*-propylamine
20.8. (a) 3-Amino-2,4-dimethylpentane
(d) 4-amino-2-pentene (g) 1,2-diaminocyclopentane
20.9. (a) *p*-bromoaniline (d) *N*-ethyl-*N*-methylaniline
(g) *p*-chloroaniline
20.10. (a) CH₃—CH₂—NH₂ (d) CH₃—NH—CH₃
(g) H₂N—C₆H₄—NH₂
20.11. (a) H₂N—CH₂—CH₂—CH₂—NH₂
(d) CH₃—CH(NH₂)—CH₂—CH(Cl)—COOH
(g) C₆H₄(NH—CH₃)(Cl)

20.12. (a) liquid (d) liquid   **20.16.** a, b, c, e, f, g

A–9

20.19. (a) 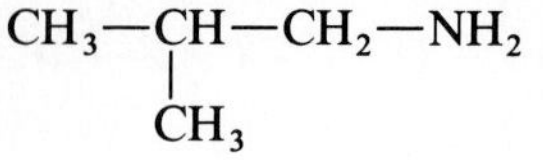 CH₃—CH(CH₃)—CH₂—NH₂ (d) 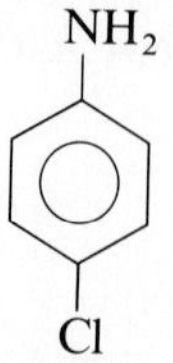 4-chloroaniline (NH₂ on benzene ring with Cl para)

(g) 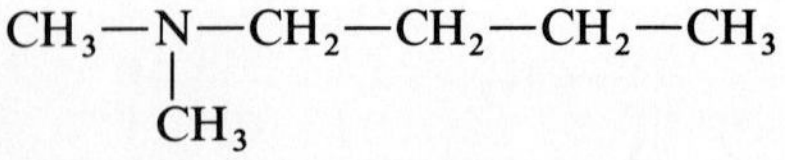 CH₃—N(CH₃)—CH₂—CH₂—CH₂—CH₃

20.22. (a) CH₃—CH₂—$\overset{+}{N}$H₃ + Cl⁻
(d) (CH₃)₃—C—NH—CH₃

20.26. (a) 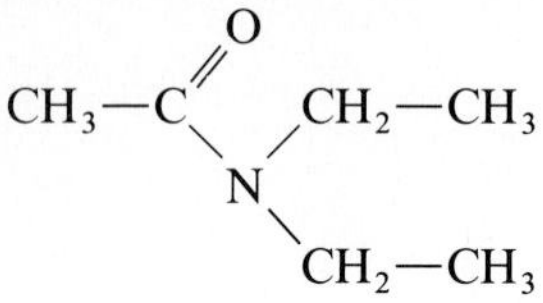 CH₃—C(=O)—N(CH₂—CH₃)(CH₂—CH₃)

(d) 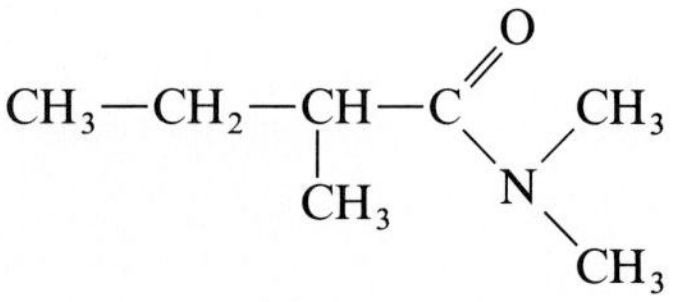 CH₃—CH₂—CH(CH₃)—C(=O)—N(CH₃)(CH₃)

20.27. (a) 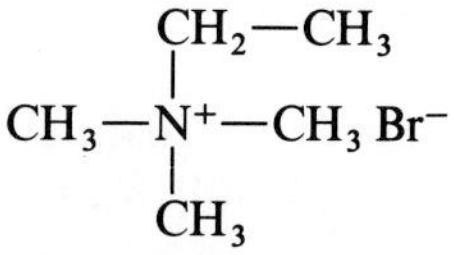 CH₃—N⁺(CH₃)(CH₃)(CH₂—CH₃) Br⁻

(d) 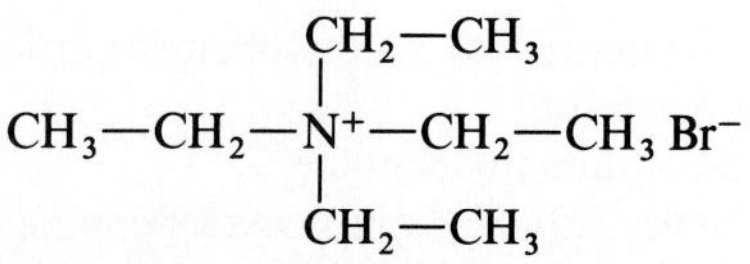 CH₃—CH₂—N⁺(CH₂—CH₃)(CH₂—CH₃)(CH₂—CH₃) Br⁻

20.28. (a) ethyltrimethylammonium bromide
(d) tetraethylammonium bromide

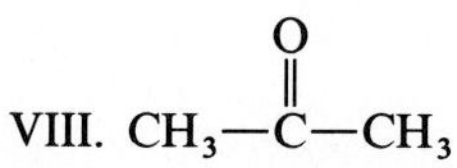

20.29. (a)
I. 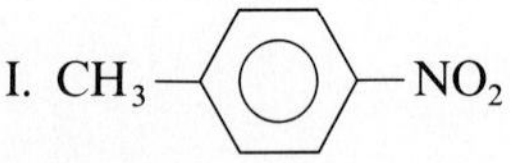 CH₃—C₆H₄—NO₂
II. 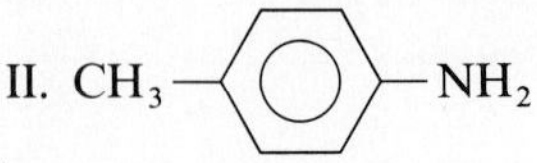 CH₃—C₆H₄—NH₂

(d) VII. 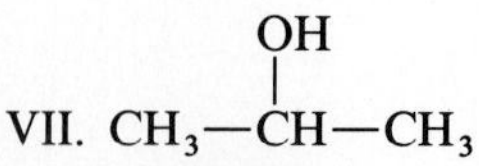 CH₃—CH(OH)—CH₃   VIII. CH₃—C(=O)—CH₃
IX. 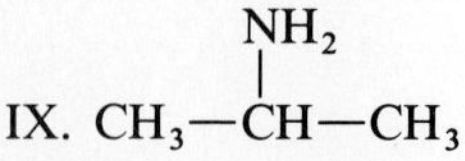 CH₃—CH(NH₂)—CH₃

20.30.

	Size (# of atoms in ring(s))	N-atoms	Double Bonds
(a)	6	1	3
(d)	5	1	0
(g)	9 (2 rings)	1	1

20.31. (a) pyrolidine, pyridine (d) purine (g) none
20.36. (a) coffee (d) tablet   **20.37.** a, b, c, f, g, h
20.38. (a) 1° (d) (amine) (g) 3°
20.39. (a) amide (d) amine (g) amide
20.40. (a) *N*-ethyl-*N*-methylethanamide
(d) *n*-propylformamide (g) *N,N*-dimethylacetamide

20.41. (a) 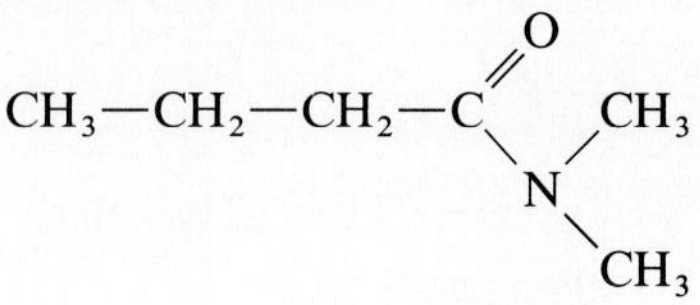 CH₃—CH₂—CH₂—C(=O)—N(CH₃)(CH₃)

(d) 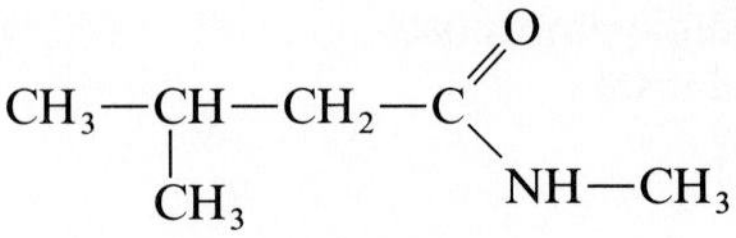 CH₃—CH(CH₃)—CH₂—C(=O)—NH—CH₃

(g) 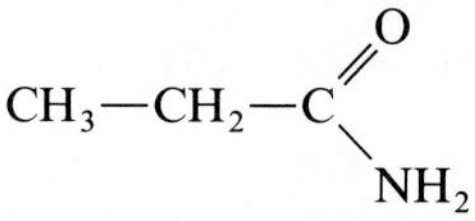 CH₃—CH₂—C(=O)—NH₂

20.46. Quite similar: CH₃—C(=O)—CH₃, HO—C(=O)—OH, H₂N—C(=O)—NH₂

20.50. [ring structure with NH—C(=O), O=C, NH—C(=O), CR₂; urea portion bracketed]

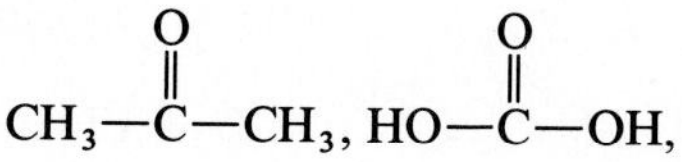

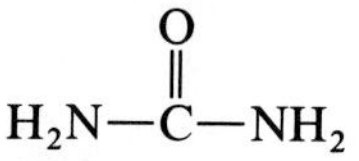

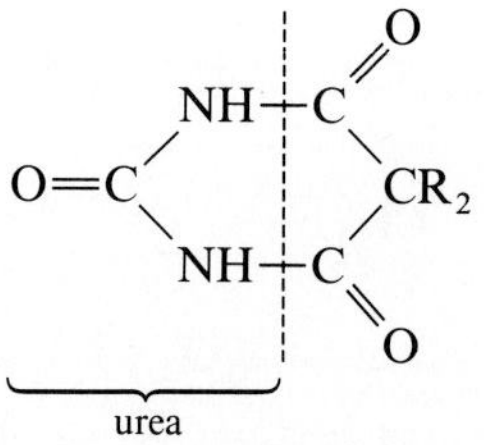

20.53. Amphetamines are "uppers" and barbiturates are "downers". **20.55.** The oriental opium poppy plant

20.61. amide: —C(=O)—NH—   urethane: —O—C(=O)—NH—

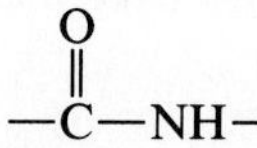

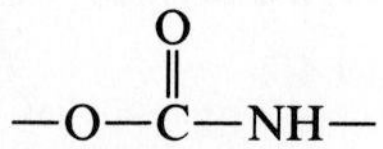

20.63.
(a) 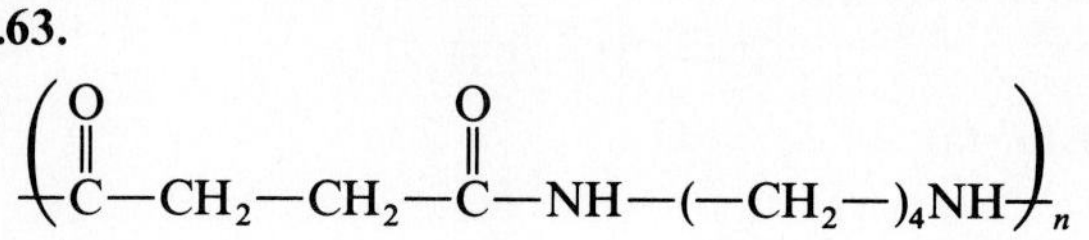 (—C(=O)—CH₂—CH₂—C(=O)—NH—(—CH₂—)₄NH—)ₙ
(d) 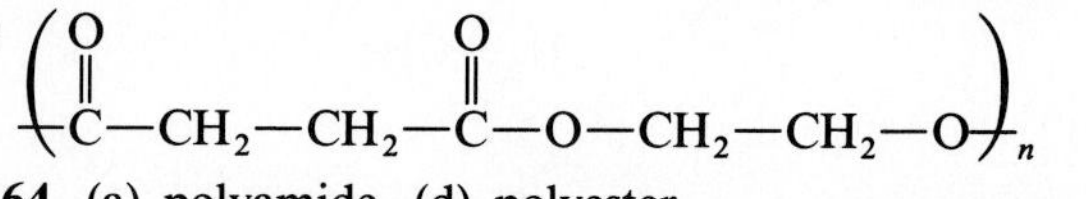 (—C(=O)—CH₂—CH₂—C(=O)—O—CH₂—CH₂—O—)ₙ

20.64. (a) polyamide (d) polyester

CHAPTER TWENTY-ONE

21.4. (a) 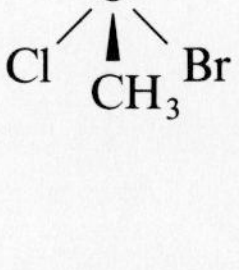 tetrahedral carbon with H, Cl, Br, CH₃   **21.5.** only d

21.7. (a) $CH_3-CH_2-CH_2-*CH-CH_3$
                                    $|$
                                    OH

(d) $H_2N-*CH-COOH$
         $|$
         $*CH-CH_2-CH_3$
         $|$
         CH_3

21.8. (a) $CH_2=CH-\underset{CH_3}{\overset{|}{CH}}=CH_2$ (none)

(d) $CH_3-\overset{*}{CH}-\overset{*}{C}-CH_2-CH_3$
         $\;\;\;\;\;|\;\;\;\;\;|$
         $\;\;\;\;Br\;\;\;Br$
         with CH_3 on the second starred C

(g) $CH_3-\overset{*}{CH}-\overset{*}{CH}-CH_2-CH_3$
              $|\;\;\;\;\;\;|$
              $NH_2\;NH_2$

21.10. (a) pliers (d) out

21.11. (a)   COOH
              $|$
         Cl—————Br
              $|$
              CH_3

21.13. (a) different

21.15. CO_2H
              $|$
         H—————NH_2
              $|$
              CH_3

21.16. a, e   **21.25.** $-53°$

21.26. $+21°$ **21.27.** $-39.6°$
21.29. $26.1°$ **21.35.** b, d   **21.37.** $2^2=4$

21.39. (a)

CH_2OH CH_2OH
H—OH   HO—H
CH_2CH_3 CH_2CH_3

enantiomers

(d) CH_2CH_3
         $|$
    H————Cl
         $|$
    H————H
         $|$
    H————Cl
         $|$
         CH_2CH_3

meso

CHAPTER TWENTY-TWO

22.4. alcohol functional groups (—OH)
22.5. (a) "mono-" means one, single unit; "oligo-" means few (2–10) units   **22.6.** (a) aldose (d) aldose (g) aldose   **22.7.** (a) aldohexose (d) aldohexose (g) aldotriose   **22.8.** (a) 4 (d) 4 (g) 1
22.9. (a) D (d) D (g) L   **22.10.** (a) D-galactose (d) D-talose (g) L-glyceraldehyde
22.12. (a)

$\;\;\;\;\;\;\;\;$CHO
$\;\;\;\;\;\;\;\;|$
H————OH
$\;\;\;\;\;\;\;\;|$
$\;\;\;\;\;\;\;\;CH_2-OH$

D-glyceraldehyde

22.15. (a) diastereomers (d) diastereomers
22.16. (a) $2^4=16$ (d) $2^1=2$ **22.17.** (a) III, IV (d) none of them   **22.19.** (a) both are aldohexoses, but differ in their configurations about carbon number 4 (d) these differ only in their configurations about their penultimate (#5) carbon   **22.20.** (a) D-glucose (d) D-glucose (g) D-fructose

22.24.

CH_2OH with HO, OH, OH on ring carbons and C=O, H group

22.25. (a) carbon #1 (b) carbon #2 (c) carbon #5 (d) carbon #5

22.26. (a) [pyranose ring with CH_2OH, HO, HO, OH, OH] (b) [pyranose ring with CH_2OH, O, OH, HO, HO, OH]

22.28. (a) hemiacetal (d) hemiacetal
22.29. (a) alpha (d) alpha   **22.30.** (a) α-D-glucose (d) α-D-allose

22.34. [α-pyranose structure] α   [β-pyranose structure] β

22.36. the result of mutarotation   **22.37.** all of them   **22.39.** (a) acetal (d) acetal
22.40. (a) alpha (d) beta

22.43.

[α-D-glucose structure] + OH—CH_2—CH_3

α-D-glucose   ethanol

22.50. all of them   **22.52.** D-erythrose

A–11

22.53. D-threose **22.55.** glucose **22.57.** (a) sucrose (d) maltose **22.58.** (a) glucose and fructose (d) two glucose molecules **22.59.** (a) β-cellobiose (d) β-lactose **22.60.** (a) α-1, 6 (d) α-1, 4
22.62. (a) reducing (d) reducing
22.63. (a)

(d)

22.66.

22.69. much less aspartame is needed to sweeten foods **22.71.** (a) amylopectin, glycogen (d) amylopectin, glycogen **22.72.** (a) amylose (d) cellulose
22.77. glucose $\underset{\text{glycogenolysis}}{\overset{\text{glycogenesis}}{\rightleftarrows}}$ glycogen
22.78. 1540 glucose monosaccharides

CHAPTER TWENTY-THREE

23.2. (a) insoluble (d) soluble **23.3.** d, e
23.5. C_{16}: palmitic acid C_{18}: stearic acid C_{20}: arachidic acid **23.9.** b, c, a
23.14. (a)

$$CH_2-O-\overset{\overset{O}{\|}}{C}-(CH_2)_{14}-CH_3$$
$$CH-O-\overset{\overset{O}{\|}}{C}-(CH_2)_{14}-CH_3$$
$$CH_2-O-\overset{\overset{O}{\|}}{C}-(CH_2)_{14}-CH_3$$

23.15. palmitic acid, lauric acid, oleic acid
23.16. (a) Oils have lower melting points than fats.
23.18. (a) lard **23.19.** the number two carbon atom in glycerol **23.21.** (a) glycerol and three fatty acids
23.23. 6 **23.28.** oils; they are more unsaturated
23.30. (a) olive oil
23.31.

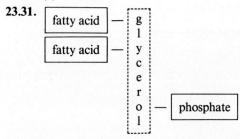

23.32. Phospholipids have only two fatty acids, while triglycerides contain three. **23.36.** Lecithin is a phosphatidyl choline; cephalins are phosphatidyl serines and phosphatidyl ethanolamines. **23.39.** They contain polar "head" groups and nonpolar "tails".
23.40. (a) +1 **23.42.** fatty acid + alcohol
23.44.

$$CH_3-(CH_2)_{14}-\overset{\overset{O}{\|}}{C}-O-CH_2-(CH_2)_{14}-CH_3$$

23.46. (a) wax helps minimize water loss
23.51. sphingosine **23.54.** glycosidic linkage
23.55. (a) sphingomyelins are an important component of myelin, the protective coating that surrounds nerves.
23.56.

23.58. They are soluble in nonplar solvents.
23.59. the liver **23.60.** The body also synthesizes cholesterol. **23.62.** emulsification agents; they aid in digestion of fats and oils **23.66.** arachidonic acid
23.68. Aspirin inhibits prostaglandin synthesis.
23.70. lipids and proteins **23.72.** peripheral and integral proteins **23.74.** more fluid-like or flexible
23.80. both have polar surfaces and nonpolar interiors

CHAPTER TWENTY-FOUR

24.2. C, H, O, N **24.3.** (a) 6 (d) 1 (g) 4
24.4. (a) contractile (d) catalytic (g) structural
24.5. 2.0×10^{-8} moles **24.8.** b, c **24.12.** isoleucine
24.14. A protein is a polymer of at least 51 amino acids.
24.15. (a) threonine (d) cysteine (g) aspartic acid
24.16. (a) asp, glu (see Table 24-3) **24.21.** glycine
24.22.

D-valine   L-valine

24.24.

$$\begin{array}{c} COOH \\ | \\ H-C-NH_2 \\ | \\ CH_2 \\ | \\ COOH \end{array}$$

24.26. (a) $H_3\overset{+}{N}-CH-COO^-$ with side chain $-CH_2-CH(CH_3)-CH_3$ (isobutyl, leucine-like)   (d) $H_3\overset{+}{N}-CH-COO^-$ with side chain $-CH_2-OH$

24.27. It will remain stationary.

24.29. (a) $H_3\overset{+}{N}-CH-COO^-$ with side chain $-CH(CH_3)_2$; pH = 7   **24.30.** (a) high pH

24.31. (a) toward "+" electrode   (d) isoelectric
24.32. yes; pH = 7   **24.36.** 6 amino acids, 5 peptide bonds

24.38.

N-terminus → $H_2N-CH-\underset{\underset{}{\|}}{C}(=O)-NH-CH(CH_3)-\underset{\underset{}{\|}}{C}(=O)-NH-CH(CH_2SH)-\underset{\underset{}{\|}}{C}(=O)-OH$ ← C-terminus

24.41. ser-val-gly   val-gly-ser   gly-ser-val
ser-gly-val   val-ser-gly   gly-val-ser
24.42. 6   **24.43.** opposite direction
24.44. (a) $H_2N-CH(CH_2OH)-C(=O)-NH-CH(CH_2\text{-imidazole})-C(=O)-OH$

(d) $H_2N-CH(CH_2CH_2SCH_3)-C(=O)-NH-CH(CH_2COOH)-C(=O)-OH$

24.45. (a) ser-ala-cys   **24.46.** (a) no net movement
24.47. Polypeptides have fewer than 51 amino acids; proteins contain 51 or more.   **24.28.** the linear sequence of amino acids   **24.50.** no
24.52. pig (ovine)   **24.53.** pig (ovine)

24.57. hydrogen bonds   **24.58.** In both a and b, hydrogen bonds form between carbonyl oxygens and amine hydrogens:

$$\text{C}=\text{O}\text{---}\text{H}-\text{N}$$

24.61. yes   **26.67.** (a) nonpolar   **24.68.** (a) hydrophobic   (d) hydrogen bonds
24.69. (a) $-CH_2-C_6H_5$ and $CH(CH_3)-CH_2-$ with CH_3

(d) $-CH_2-O\text{---}H$ and $H\text{---}O-C_6H_4-CH_2-$

24.72. a protein containing more than one subunit
24.75. no   **24.78.** yes   **24.79.** some polypeptides still remain

24.82. (a) thr-asn-val-lys-ala-ala-trp-gly-lys
24.83. 2°, 3°, 4°   **24.87.** Heat denatures bacterial and viral proteins.   **24.89.** one to reduce the disulfide bonds, one to oxidize them and reform cross links
24.92. AIDS and some drugs   **24.93.** Antigens are foreign proteins which stimulate the production of antibodies.

CHAPTER TWENTY-FIVE

25.5. "-ase"   **25.6.** (a) sucrase   (d) lactose dehydrogenase   **25.8.** (a) fructose   (d) galactose   **25.9.** lactose, cellobiose   **25.11.** The active site is the catalytic site.   **25.15.** Coenzymes are organic cofactors.
25.20. synonymous terms   **25.22.** (a) isomerase   (d) transferase   **25.23.** (a) lipase   (d) transferase
25.23. (a) lipase   (d) dehydrogenase   **25.24.** (a) enzyme acts on only one molecule   (d) enzyme acts only on a D- or on an L-configured substrate
25.25. absolute: urease   linkage: esterase   **25.28.** enzyme specificity   **25.33.** asp or glu (acid groups: —COO—)   **25.37.** Optimum means maximum in this case.   **25.42.** very little change in the rate
25.44. no   **25.47.** The two do not compete for the

A–13

same site. **25.52.** phosphorylation of serine residues **25.54.** to protect the organs where they are synthesized **25.60.** serine **25.63.** Elevated LDH levels are indicative of tissue damage; electrophoretic analysis can identify the organ by its unique LDH pattern.
25.66. Urokinase helps dissolve massive blood clots; treat a rash from poison ivy; dissolve slipped disks in the spine.

CHAPTER TWENTY-SIX

26.2. base, sugar and phosphate **26.3.** DNA contains *deoxy*ribose
26.8. nucleoside: base + sugar
nucleotide: base + sugar + phosphate(s)
26.10. (a) deoxythymidine **26.15.** the sequence of bases in the polymer **26.17.** $5' \to 3'$ **26.19.** hydrogen bonds **26.20.** The bases extend into the interior of the double helix. **26.23.** (a) TGCATA
26.26. (a) 36% **26.28.** The helix must unwind and the two strands separate. **26.31.** New nucleotides are added to the 3' end of the growing strand; hence we say replication occurs in a $5' \to 3'$ direction (see Figure 26–12) **26.32.** One parent strand ends up in each of two daughter DNA molecules. **26.36.** (a) tRNA delivers specific amino acids for protein synthesis
26.37. rRNA
26.40. -A-U-G-C-C-C-A-U-U-G-A-G-C-
26.44. mRNA **26.46.** (a) leu (d) gly **26.47.** T (thymine) is not present in RNA. **26.49.** One codon sequence does not overlap another. **26.51.** met-lys-glu-asp-leu-tyr-pro **26.52.** at least 375 (3 × 125)
26.57. (a) GAA (d) CCC **26.58.** the A-site
26.60. A new tRNA binds to the A-site. **26.61.** AUG, N-formyl methionine **26.63.** no **26.66.** substitution, deletion, insertion **26.68.** (a) gly-tyr-ser-ser-pro (d) gly-ile-(termination) **26.72.** Viruses lack reproductive components present in cells. **26.76.** Viral infections stimulate cells to release interferons, which help protect other cells from further infection. **26.79.** The double strand of DNA is cut at highly selective locations. **26.82.** Plasmid DNA is one source; an isolated gene from another organism or a synthetically-produced segment of DNA is the other.

CHAPTER TWENTY-SEVEN

27.1. vitamin comes from "vital amine"
27.5. RDA = highest amount needed by 95% of the population. Minimum means the smallest amount required for good health. **27.8.** no **27.9.** TPP, an active cofactor for enzymes in carbohydrate metabolism **27.11.** redox reactions **27.12.** hot chocolate **27.15.** biotin **27.16.** cobalamin **27.17.** folic acid **27.18.** pantothenic acid

27.20. "anti-scurvy" or "a-scorbic" vitamin
27.22. excessive vitamins in the body; fat soluble vitamins **27.24.** retinol **27.27.** no **27.29.** tocopherol **27.30.** K_1: green vegetables K_2: intestinal bacteria **27.31.** Vitamin K indirectly assists in blood clotting. **27.33.** (a) water soluble **27.36.** calcium **27.37.** $Al(OH)_3$ or $CaCO_3$; insoluble precipitates form **27.39.** Mg **27.40.** Cu **27.43.** Zn
27.44. yes **27.45.** different organ metabolisms can be independently controlled
27.49.

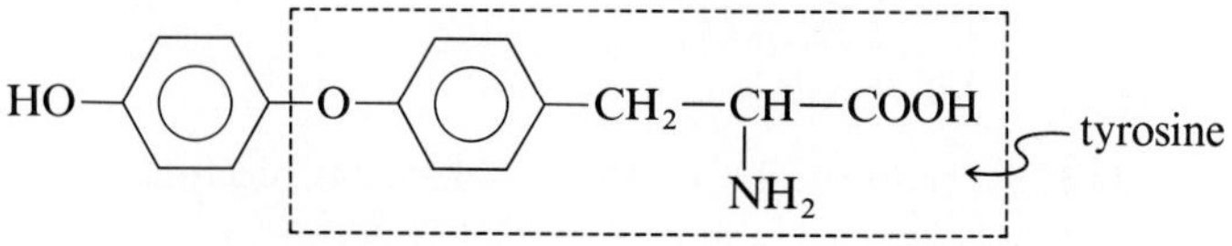

$$HO-\bigcirc-O-\bigcirc-CH_2-\underset{NH_2}{CH}-COOH \quad \leftarrow \text{tyrosine}$$

27.51. Proinsulin → insulin
pro-opiocortin → corticotropin + α-lipotropin
T_4-thyroxine → T_3-thyroxine
27.53. $102 \times 3 = 306$ minimum (plus signaling codons) **27.54.** pro-orticotropin **27.59.** cAMP cannot stimulate both (opposite) processes. **27.62.** development of male characteristics (e.g., facial hair); possible disruption of normal menstrual cycle
27.65. progesterone (see Figure 27-22) **27.67.** DES is a demonstrated carcinogen. **27.68.** current high cost

CHAPTER TWENTY-EIGHT

28.2. (a) "−" **28.4.** Carbon is completely oxidized in CO_2. **28.5.** (a) S^{2-}, S^{4+}, S^{6+} (d) alkenes, aldehydes, carbon dioxide **28.6.** reaction "a"; the three carbons in propane are more reduced than the three carbons in malonic acid **28.9.** $PP_i + H_2O \to 2P_i + $ energy
28.10. yes. Formation of ATP from ADP requires +7.3 kcal; −8.6 kcal is more than enough free energy to couple to ATP formation. **28.13.** phosphoenolpyruvate (−14.8 kcal/mol) **28.15.** The extra hydrogen atom is lost to the surrounding solvent as H^+.
28.20. (a) NAD^+ (oxidation) (d) NADPH (reduction) **28.21.** $52.6/7.3 = 7$ **28.22.** (a) lipids (d) protein **28.23.** Fiber provides bulk in the stool and stimulates bowel action. **28.27.** oxidative phosphorylation **28.30.** This regulates their availability for metabolism. **28.32.** stimulation of lipid synthesis, slowing of lipid oxidation **28.35.** 2,4-DNP stops ATP synthesis **28.36.** Directly or indirectly (through plants) sunlight is our source of energy for all life on earth

28.37. $6CO_2 + 6H_2O \xrightarrow{\text{light}} 6CO_2 + C_6H_{12}O_6$

28.38. Photosynthesis takes place in thylakoid membranes (see Figure 28-13). **28.42.** (a) yes. Light-energized transmembrane potentials can form ATP. (d) yes. until its supply of NADPH is exhausted

CHAPTER TWENTY-NINE

29.2. L-glucose is not allowed into the cell, since its transport system is specific for D-isomers   **29.4.** (a) increased capacity   **29.5.** the brain; primarily the liver   **29.8.** (a) 3 (d) 2   **29.10.** oxidation of glyceraldehyde-3-phosphate to 1,3-diphosphoglycerate
29.12. one glucose molecule yields 2 molecules of PEP via glycolysis
29.13.

	ATP	NADH
(a)	-1	0
(d)	$+2$ (net)	0

29.15. carbon #3 in lactate; carbon #2 in ethanol
29.17. $NADH/NAD^+$; oxidation of glyceraldehyde-3-phosphate   **29.21.** only anaerobic conditions produce alcohol   **29.22.** excessive CO_2 production
29.23. 1715 g   **29.24.** liver and kidneys
29.27. ATP and GTP   **29.28.** -4 **29.29.** reductive biosyntheses   **29.30.** 2 NADPH, 1 CO_2 **29.33.** little ATP is needed, while NADPH needs are high; PFK enzyme   **29.36.** Multiple nonreducing ends provide many sites for the glycogen phosphorylase enzyme.
29.41. (a) 2 (d) 1   **29.42.** hexokinase, PFK, pyruvate kinase   **29.46.** Premature infants are particularly susceptible to hypoglycemia.   **29.48.** Lactate is not the ultimate possible oxidation product of glucose.
29.49. 31%   **29.50.** (a) -2 (d) -1

CHAPTER THIRTY

30.2. inner membrane
30.6.

$$HOOC-CH_2-CH_2-\overset{O}{\underset{\|}{C}}-COOH \rightarrow$$

$$HOOC-CH_2-CH_2-\overset{O}{\underset{\|}{C}}-CoA + CO_2$$

similar enzyme complex, same cofactors, similar keto structure, decarboxylation
30.8. (a) $+2$ (d) 2 (g) 1   **30.9.** carboxylation of pyruvate, utilizing ATP and a biotin cofactor
30.10. The energy of the thioester of succinyl-CoA is utilized to synthesize GTP.

$$\underset{\underset{COOH}{\overset{\overset{O}{\|}}{\underset{CH_2}{\overset{C}{\underset{CH_2}{|}}}}}}{}\overset{S-CoA}{}$$

30.11. (a) 3 (d) 4   **30.12.** (a) immediately lost as CO_2 **30.13.** NADH and $FADH_2$ **30.14.** inner membrane of the mitochondrion   **30.15.** redox potentials   **30.16.** 42%   **30.17.** close to O_2
30.20. The heme iron atom is oxidized (Fe^{3+}) or reduced (Fe^{2+}).   **30.23.** NO_2^- oxidizes Fe^{2+} to Fe^{3+} in heme. This draws CN^- back into the bloodstream, where $S_2O_3^{2-}$ converts it to SCN^-.   **30.24.** obtained from the matrix; deposited into the intermembrane space
30.28. Energy can be obtained from the "push," or tendency for protons to diffuse back into a less concentrated environment.   **30.30.** to power bacterial flagella and photosynthetic ATP synthesis in chloroplasts
30.33. Le Châtelier's principle; everything backs up: ATP, ΔpH, electron flow   **30.35.** Light-driven reactions in photosynthesis remove electrons from H_2O.   **30.37.** Electrons from $FADH_2$ do not pass through site 1.   **30.39.** (a) 38 (d) 15 (e) 39
30.41. Le Châtelier's principle; tightly coupled mitochondria mean high concentrations of H^+ resist more proton pumping and slows e^- transport.

30.43.

	e^- transport	ATP synthesis
(a)	stops	stops
(d)	continues	slows (by 1/3)
(g)	accelerates	accelerates

CHAPTER THIRTY-ONE

31.2. energy storage and insulation   **31.3.** 100 times more   **31.4.** no. On the average, 10% of fat molecules are replaced daily.   **31.8.** HDL, LDL or VLDL; they are classified according to their densities in D_2O
31.11. the mitochondria matrix   **31.13.** both high-energy bonds in ATP are utilized   **31.14.** the β-carbon is oxidized   **31.16.** CoA is attached to the shorter chain each time acetyl-CoA is removed.
31.17. (a) 4 rounds (d) 4 $FADH_2$
31.18. A   **31.21.** ACP (aryl carrier protein)
31.23.

$$CH_3-\overset{O}{\underset{\|}{C}}-CoA + HCO_3^- \xrightarrow[ATP \quad ADP]{carboxylase}$$

acetyl-CoA

$$^-OOC-CH_2-\overset{O}{\underset{\|}{C}}-CoA$$

malonyl-CoA

31.25. malonyl-CoA; every cycle   **31.26.** palmitoyl CoA is a C_{16} fatty acid: (a) 7 (d) 14   **31.27.** pentose phosphate "shunt" or pathway   **31.29.** ester linkage of acetyl-CoA   **31.30.** 78   **31.31.** 387 for three palmitic acid molecules   410 ATP per molecule of tripalmitin

31.33. $(CH_3)_3-\overset{+}{\underset{\underset{CH_2-COO^-}{|}}{N}}-CH_2-O-\boxed{\overset{O}{\underset{\|}{C}}-R}$ ⟵ fatty acid

31.34. Malonyl-CoA inhibits transport of fatty acids into the mitochondria. **31.37.** She is not getting enough carbohydrates in her diet. Extended, frequent periods of ketosis can be injurious to one's health. **31.39.** no

CHAPTER THIRTY-TWO

32.1. periodic replacement of existing proteins. Turnover for insulin is much greater than for connective tissue. **32.5.** protein biosynthesis **32.7.** yes (to a limited extent). Many amino acids can be converted into glycolytic or TCA intermediates whence they can be converted to acetyl CoA (and to fatty acids) or glucose (and glycogen). **32.10.** Tyrosine is formed from hydroxylation of phenylalanine.

32.14.
$$\text{deamination: } R-\overset{\overset{O}{\|}}{C}-COO^- + NH_4^+$$
$$\text{transamination: } R-\overset{\overset{O}{\|}}{C}-COO^- + R-\overset{\overset{NH_3^+}{|}}{C}-COO^-$$

32.16. succinate **32.17.** Transamination yields isoleucine:
$$CH_3-CH-CH_2-\overset{\overset{NH_2}{|}}{C}-COOH$$
$$\underset{CH_3}{|}$$

32.18. (a)
$$HOOC-CH_2-\overset{\overset{NH_2}{|}}{CH}-COOH$$

32.20. Elevated AST can indicate renal infarction; simultaneous elevated BUN values are indicative of kidney malfunction. **32.23.** Transport of citrulline is required for further reactions of the urea cycle.
32.24. NH_3: 82% urea: 45% uric acid: 33%
32.25. amino acid → aspartic acid → argino succinate → arginine → urea **32.26.** (a) 30 (b) 8 (c) 22

32.30. pentose phosphate pathway **32.32.** testosterone (for others see Section 27.8) **32.33.** Low concentrations of thyroid hormone promote protein anabolism; high levels promote catabolism.

CHAPTER THIRTY-THREE

33.1. On the average, females have slightly more adipose tissue than men. **33.3.** Na^+: extracellular fluids K^+: intracellular fluids Cl^-: extracellular fluids
33.4. Blood plasma > interstitial fluid **33.6.** K^+
33.8. fraternity brother: 6.9% sorority sister: 13%
33.9. 5.11 liters **33.13.** centrifugation **33.15.** As much as 50% of total blood proteins escape from the circulatory system; these are returned by the lymphatic system.

33.20.

	bowman's capsule	tubules	distal tubules
(a) water	cap → tubule	tubule → cap	tubule → cap
(d) Na^+	cap → tubule	tubule → cap	tubule → cap

33.21. 1,700 liters passes through kidneys (a) 190 liters = 11% **33.24.** excess glucose draws more water into the urine **33.27.** increased Na^+ and Cl^- ions draw water into the blood and the increased blood volume elevates blood pressure **33.28.** ions
33.30. 7.0–7.8; 7.4 average **33.32.** $\Delta pH = 1$: 10 times different $\Delta pH = 2$: 100 times different
33.33. buffer systems, respiration, and kidney excretion **33.36.** no **33.37.** acidosis; CO_2 may be accumulating in the blood **33.39.** food contents in stomach, pH optimum for absorption, general metabolic state **33.44.** caffeine could increase the rate of excretion for this drug, lowering its concentration in the blood and thereby decreasing its effectiveness

Index

A site, in ribosome, 700
Absolute alcohol, 407
Absorption, lipid, 832–833
Absorption of light by retina of the eye, 352
Accuracy, 5
Acetal formation, in monosaccharides, 543–544
Acetaldehyde, 426–427, 788
Acetals, 433–434
Acetaminophen, 493
Acetanilide, 493
Acetate ion, 449
Acetic acid, 442
   glacial, 442
Acetic anhydride, 460
Acetoacetate, 845
Acetone, 426, 427, 845
Acetophenone, 426
Acetylcholine, 456, 481
Acetylcholinesterase, inhibition of, 663
Acetyl-CoA, from oxidation of fatty acids, 835
Acetyl-CoA, in TCA cycle, 808
Acetylene, 362 (see also Ethyne)
   synthesis of, 363
Achiral, 511
Acid chlorides, 448
Acids
   acetic, 442
   acidic and nonacidic hydrogens in, 231–232
   adipic, 443, 445
   amino, 605–614
   anhydrides of, 460
   Arrhenius, 227–228
   benzoic, 374, 442
   Brønsted–Lowry, 228–229
   caproic, 445
   carboxylic, 440–462
   chlorides of carboxylic, 448
   citric, 443, 445
   dicarboxylic, 442–443
   diprotic, 230
   essential fatty, 573
   esters of inorganic, 459–460
   fatty, 569–573
   formic, 441–442
   glutaric, 443, 445
   lactic, 445
   malic, 443
   malonic, 443, 445
   monoprotic, 230
   neutralization of, 235
   oxalic, 443, 445
   palmitic, 570, 571
   pH scale and, 237–239
   polyprotic, 230–231
   strengths of, 229–230
   strong, 229
   succinic, 443, 445
   titration of, 239–241
   tricarboxylic, 442–443
   triprotic, 231
   weak, 229
Acidosis, 251, 874
Acne, 749
Aconitase, 809
ACP (Acyl carrier protein), 838
ACTH (see Corticotropin, 739–780)
Activation energy, 214
   catalysts and, 218–219
Activators, enzyme, 648
Active site, 647
Active transport, across membranes, 596–597
Acyl carrier protein, 838
   size of, 839
Acyl groups, carried by CoA, 728
Acyl-CoA synthetase, 834
Addictive drugs, 495
Addition reactions
   of alkenes, 354–359
   of alkynes, 364
   symmetrical, 354–356
   unsymmetrical, 355, 356–359
Adenine, 487
   structure of, 677
Adenylate cyclase, activation of, 743
Adenylic acid, 721
Adipic acid, 443, 445
   in synthesis of nylon, 496–497
Adipose tissue, 576
   triglycerides in, 831
ADP, phosphate esters, 462
Adrenal glands, target tissue for corticotropin, 740
Adrenalin, 411, 473
Advil, 442
Agglutinins, 868
AIDS, 638, 709
Alanine, 605
   transport of ammonia, 856
Alanine aminotransferase (ALT), 855
Albinism, 854
Albumin, amino acids in, 608
Alcohol dehydrogenase, 646, 788
Alcohols, 390–420
   absolute, 407
   classification of, 396–397
   common names of, 404
   conversion to amines, 477
   dehydration of, 400–401
   denatured, 407
   hydrogen bonding between, 397–399
   important commonly used, 404–409
   IUPAC nomenclature, 392–396
   oxidation of, 401–403, 430
   physical properties of, 397–399
   preparation of, 399–400
   primary, 396–397
   reactions of, 400–404
   rubbing, 407
   secondary, 396–397
   solubilities in water, 397–399
   substitution reactions of, 403–404
   tertiary, 396–397
   toxicity of ethyl, 788
Alcoholic beverages, alcohol content of, 407
Aldehydes, 422–436
   hydrogen bonding in, 426
   nomenclature of, 425–427
   oxidation of, 434–436
   physical properties of, 426–429
   preparation of, 430
   reduction of, 431–432
   uses of, 428–429
Aldo form of monosaccharides, 545
Aldolase, 783
Aldoses, 531
   D- and L-configurations of, 533
Aldosterone, 592
   control of urine volume, 872
Aldrin, 354
Alkaloids, 495–496
Alkalosis, 251, 874
Alkanes
   combustion of, 331
   general formula of, 312
   halogenation of, 331
   in living organisms, 334
   natural sources of, 327–328
   physical properties, 330
   reactions of, 331
   substitution reactions of, 331
Alkenes, 341–362
   addition reactions of, 354–359
   chemical reactions of, 354–363
   in life processes, 353–354
   IUPAC nomenclature of, 342–346
   physical properties of, 353
   polymerization of, 359–362
   reactivity of, 354–355
Alkyl groups, 319–324
   isobutyl, 323
   isopropyl, 323
   sec-butyl, 323
   tert-butyl, 323
Alkyl halides, substitution in by amines, 476
Alkylammonium ions, notation of, 478
Alkylation, of benzene, 378
Alkylbenzenes, oxidation of, 380
Alkynes, 362–364
   nomenclature of, 363–364
   physical properties of, 364
   reactions of, 364
Allergies, histamine cause of, 484
Allose, 534
Almond flavoring, 429
Alpha radiation
   characteristics of, 271

emission of, 272
symbol for, 271
Alpha-helix, in proteins, 621–622
α_{obs}, 518
ALT (*see* Alanine aminotransferase)
Altrose, 534
Aluminum, characteristics of, 293
Amides, 468–499
   classification of, 490
   formation of, 448, 482
   hydrogen bonding in, 491–492
   hydrolysis of, 492
   IUPAC nomenclature of, 490–491
   peptide bond, 490
   physiological activity of, 493–494
   preparation of, 492
   primary, 492
   properties of, 491
   secondary, 492
   structure of, 490
   tertiary, 492
   uses of, 493–494
Amines, 468–499
   alkyl substitution, 476
   basicity of, 477–480
   heterocyclic, 483–490
   IUPAC nomenclature, 470–472
   occurrences of, 471–475
   preparation from alcohols, 477
   preparation of, 476–477
   primary, 469
   properties of, 471–475
   reduction of nitrile to form, 476
   reduction of nitro group to form, 477
   reductive amination of, 477
   salts of, 478
   secondary, 469
   tertiary, 469
   toxicity of, 473
   uses of, 471–475
Amino acid pool, 851
Amino-acid racemase, 519
Amino acids, 605
   absorption from intestine, 767
   acid–base properties of, 611
   attachment to tRNA, 694
   biosynthesis of, 859–860
   charged forms of, 608, 610–614
   deaminiation, 855
   entry to TCA cycle, 852
   essential, 609, 767
   glucogenic, 853
   ketogenic, 853
   metabolism, 850–861
      carbon skeletons in, 852
      defects in, 854
      regulation of, 860
   side chain, 605
   stereoisomerism in, 610
   structures of, 606–607
   transamination, 855
Aminoacyl site, in ribosome, 700
Aminobenzene (*see* Aniline)
para-Aminobenzoic acid, 664–665 (*see also* PABA)
Ammonia, 469
   basicity of, 477–478
   in urea cycle, 856
   reaction with water, 477
Ammonium ion, 477
Ammonium salt, quaternary, 480–481
Amobarbital (amytal), 494
AMP, 744
   cyclic, 743
Amphetamines, 475
   Benzedrine, 475
   effect on heart rate, 475
   enantiomers of, 510

   in weight control, 475
   methamphetamine, 475
   methoxyamphetamine, 475
Amplification, biological, enzymatic, 744
Amylase, 560
Amylopectin, 559
Amylose, 558, 559
   bonds in, 559–560
Amytal (amobarbital), 494
Anabolic pathways, 770
Anabolic steroids, 746, 860
Anaerobic condition, 786
Analgesics, 495–496
Anderson's disease, 799
Androgens, 738
Anemia, 724, 727, 734
Anesthetics, 412–413
   chloroform, 332
   halothane, 333
Angiotensin II, 873
Anhydride, 460
Aniline, 374, 471, 474
   ionic form at different pH values, 479–480
Anilinium ion, 481
Anomers, 541
Antacids, phosphorus in, 733
Anthracene, 384
Anti-A, B, Rh, 867–868
Antibiotics, 664
   inhibitors of protein synthesis, 705
   penicillin and derivatives of, 666–667
   sulfanilamide, 664–665
Antibodies, 636 (*see* Immunoglobulins)
Anticoagulants, 732
Anticodon, in tRNA, 693
Antifreeze, 408–409
Antigens, 636
   A, B, and Rh, 867–868
Antigen-antibody complex, 636
Antihistamines, 484–485
Antioxidants, 580
Antiparallel, double helix in DNA, 684
Antiport, 597
Antipyretics, 493
Antiseptics, 409
Ants
   formic acid in, 442
   pheromones, 334
APC tablets, 493
Aphrodisiac, secreted by queen bee, 447
Apoenzyme, 648
Apple flavor, 455
Apricot flavor, 455
Aprobarbital, 494
Aqueous ammonia, 477, (*see also* Ammonium hydroxide)
Arabinose, 534
Arachidic acid, 571
Arachidonic acid, 571, 591
Argininosuccinate, 857–858
Aromatic compounds
   heterocyclic, 385–386
   with fused rings, 383–386
Aromatic hydrocarbons, 370–387
   sources of, 385
Aromaticity, 373
Artery grafts, 458
Artificial sweeteners, 555
Aryl group, 391
Ascorbic acid, 728–729
   antioxidant properties of, 580
Aspartame (Nutra-Sweet), 555
   structure of, 617
Aspartate, in urea cycle, 857–858
Aspartate aminotransferase (AST), 855–856
Aspartic acid, 856
Aspirin, 442
   history of, 452–453

   synthesis of, 452
AST (*see* Aspartate aminotransferase)
Atherosclerosis, HDL and LDL levels in, 833
Ativan (lorazepam), 494
Atmosphere
   chemical composition of, 295–296
   regions of, 295–296
Atomic numbers, 58
   number of electrons and, 58–59
   number of protons and, 58–59
   of elements, 59
Atomic theory of matter, 53
   statements of, 53
Atomic weights, 61
   calculation of, 61–62
   use in calculating formula weights, 69–70
Atoms, 40
   limit of chemical subdivision, 40–41
   nucleus of, 57
   size of, 53–54
   subatomic makeup of, 57–58
ATP, 487
   expended to activate fatty acids, 836
   high energy anhydride linkages, 760
   hydrolysis of, 462
   phosphate esters, 462
   universal energy currency in cell, 760
ATP accounting
   glucose, pyruvate through CO_2, 822–825
   glycolysis, 801
   lipid metabolism, 842–843
ATP synthesis
   rates of, 821–822
   via oxidative phosphorylation, 820–822
ATP synthetase complex, 806, 820–822
   in chloroplasts, 822
Atropine, 495
Attractive interactions, in tertiary protein structure, 627
Aufbau diagram, 98
   electron configurations and, 98–100
Aufbau principle, 97
   electron configuration and, 98–100
Autoactivation of enzymes, 664
Auxiliary pigments, in photosynthesis, 773
Avogadro's number, 71
   relationship to the mole, 71
Azide, inhibitor of electron transport, 817

Banana flavor, 455
Barbiturates, 493–494
   structures of common, 494
   synthesis of, 493–494
Bases
   Arrhenius, 227–228
   Brønsted–Lowry, 228–229
   neutralization of, 235
   strengths of, 229–230
   strong, 230
   titration of, 239–241
   weak, 230
Beer, maltose in, 550
Beeswax, 455, 585
Belladonna, 495
Benadryl, 485
Benedict's reagent, 435–436, 546–547
Benzaldehyde, 426, 429
Benzalkonium chloride, 481
Benzedrine, 475
Benzene, 372–373
   alkylation of, 378
   chemical reactions of, 376–381
   derivatives of, 373–377
   halogenation of, 378
   nitration of, 379
   reactivity of, 378–381
   uses of, 382

Benzoate ion, 449
Benzodiazephines, tranquilizer drugs, 494
Benzodiazepines, 493
Benzoic acid, 374, 442
Benzyl group, 376
Bephenium, 481
Beriberi, 721
Beta-pleated sheet
   in immunoglobulins, 637–639
   in proteins, 621–624
   parallel and antiparallel, 621, 623
Beta radiation
   characteristics of, 271
   emission of, 272–273
   symbol for, 271
BHA (*see* Butylated hydroxy anisole)
BHT (*see* Butylated hydroxy toluene)
Bicarbonate, effect on pH in blood, 874
Bile, 589
Bile juice, 832
Bile pigments, oxidation and color of, 590
Bile salts, 589–591
Biochemicals, classification of, 529
Biochemistry, 529
Biomer, 498
Biotin, 725–726
   in gluconeogenesis, 790
Birth control, 747–749
Blood
   AST and ALT enzyme levels in, 856
   cells in, 867
   fatty acid content in, 834
   flow through kidneys, 870
   lipids in, 833
   total volume of, 867
   transfusions, 868
   types, 867–868
Blood clotting, vitamin K in, 732
Blood pressure, 473
   in capillaries, 869
   prostaglandins and, 592
   sodium in relation to, 873
Blood sugar (*see* Glucose)
Blood urea nitrogen, 859
Body builders, muscle production stimulation, 860
Body temperature, prostaglandins and, 592
Boiling, 166
   explained using kinetic molecular theory, 165
Boiling point, 167
   colligative property of solutions, 200–201
   intermolecular force strength and, 168–169
   normal, 167
Bond, peptide, 615
Bonding, delocalized in benzene, 372
Bowel action, stimulation by cellulose, 558, 765
Bowman's capsule, 870
Boyle's Law
   applications of, 155–156
   explained using kinetic molecular theory, 154
   mathematical form, 152–153
Brain, serotonin in human, 488
Bran, 558
Breath analyzers for alcohol, 403
Bromination (*see also* Halogenation)
   as a test for multiple bonds, 355
   of alkanes, 332–333
   of alkenes, 355
bromo- prefix, 333
Bromobenzene, 374
Brompheniramine, 485
Bronchodilator, isoprenaline, 522
Buchner, Eduard, 781
Buffers, 249
   and the human body, 250–251
Bulk, dietary, 558
BUN (*see* Blood urea nitrogen)
Burn victims, artificial skin for, 498

1,3-Butadiene, 361
Butanal, 425–426
Butane, 312
Butanoic acid, 445
1-Butanol, 394
n-Butanol, 394
Butter, 576
t-Butyl alcohol, 394
t-Butyl amine, 470
tert-Butyl group, 323
Butylated hydroxy anisole (BHA), antioxidant, 580
Butylated hydroxy toluene (BHT), antioxidant, 580
Butyraldehyde, 426
Butyric acid, 445

C-14 labels, in TCA cycle, 812
Cadaverine, 473
Caffeine, 487–488, 495
   addiction to, 487–488
Calcium
   as a micronutrient, 733
   characteristics of, 294
Calcium metabolism, interference of DDT in birds, 382–383
Calcium oxalate, solubility of, 450
Calcium propionate, preservative, 450
Caloric content of foods, 765–767
Calorie, 28
Calvin, Melvin, 774
cAMP, 799, (*see* AMP, cyclic)
Cancer cells, in lymphatic system, 870
Candies, 585
Candles, 585
Cannabis sativa, 411
Capillary walls, movement of substances through, 869
Caproic acid, 445
Car polish, 456
Caraway, stereochemistry of, 522
Carbamoyl phosphate, 857
Carbohydrate loading, 561
Carbohydrate metabolism
   regulation of, 799–801
   relationship to carbohydrate, 845
Carbohydrates, 528–562
   dietary sources of, 530
   major source of energy, 765–766
   occurrence of, 530
Carbolic acid (*see* Phenol)
Carbon chains, branched, 318
   normal, 318
Carbon dioxide, 331
   carried by biotin as cofactor, 726
Carbon monoxide, 331
   characteristics of, 297
   effects on human beings, 297–298
   formation of, during combustion, 297
   inhibitor of electron transport, 817
Carbon–oxygen single bonds, in organic compounds, 391–392
Carbonic acid, effect on pH in blood, 874
Carbonic anhydrase, 645
   zinc cofactor, 735
Carbonyl compounds, reduction of, 431
Carbonyl group, 423
   conversion to amine, 477
Carboxylate ion, 448–449
Carboxylbenzene (*see* Benzoic acid)
Carboxylic acids, 440–462
   acidity of, 448–449
   biological effects of certain, 447
   hydrogen bonding in, 446
   neutralization of, 449
   nomenclature of, 443–446
   physical properties of, 446

preparation of, 447–448
   reactions of, 447–448
   salts of, 449–450
Carboxypeptidases, zinc as cofactor, 649
Cardiac output, increase in, 473
Carnauba wax, 585
Carnitine, 834
β-Carotene, 353, 729
Carotic acid, 571
Carrots, β-carotene in, 729
Carvone, 429
   enantiomer of, 522
Catabolic pathways, 770
Catalase, iron in, 734
Catalysts, 218
   activation energy and, 218–219
   effect on reaction rate, 218–219
   enzymes, 645–670
Catechol, 396
Catnip plant, active ingredient in, 455
Cauterization of wounds, 632
Cell, urea cycle locations in, 858
Cell membrane, diagram of, 676
Cell membranes, 593
   fluidity of, 595
   non-polar interior, 595
   polar surfaces of, 594–595
   proteins in, 595–596
   regions of, 594–596
Cellobiose, 550
Cells, fat, 576
Cellulase, 558, 647
Cellulose
   dietary considerations of, 765
   source of cellubiose, 551
Central nervous system
   nicotine stimulant of, 483
   stimulation of, 473
Cephalins, 582
Cereals, carbohydrates in, 765
Cerebrosides, 538, 587
Changes of state, 163
   endothermic, 163
   exothermic, 163
Charles' Law
   applications of, 158
   explained using kinetic molecular theory, 158
   mathematical form, 156–158
Chemical bonds, 113
   types of, 113
Chemical change, 3
   characteristics of, 3–5
Chemical energy, 27, 755–758
Chemical equations, 80
   balanced, 81
      requirements for, 80–81
   calculations involving, 84–86
   coefficients in, 81–83
   conventions for writing, 81
   macroscopic interpretation of, 83–84
   microscopic interpretation of, 83–84
Chemical equilibrium, 219
   catalysts and, 224
   concentration changes and, 222
   conditions necessary for, 219–220
   Le Châtelier's principle and, 221–224
   position of, 221
   pressure changes and, 223–224
   reversible reactions and, 219–221
   temperature changes and, 222–223
Chemical formulas, 45
   calculations involving, 77–79
   macroscopic interpretation of, 75–76
   microscopic interpretation of, 75–76
   parentheses in, 45–46
   subscripts in, 45–46
Chemical properties, 2
   characteristics of, 2–3

Chemical reactions
  classes of, 262–265
  decomposition, 263
  double-displacement, 263
  endothermic, 215
  exothermic, 214–215
  metathetical, 262
  neutralization, 235
  of alkenes, 354–363
  of benzene, 376–381
  oxidation–reduction, 258
  salt hydrolysis, 245–249
  single-displacement, 263
  synthesis, 262
Chemiosmotic hypothesis, 818
Chemistry, 1
  scope of, 1
Chiral carbon atoms, 510–513
Chirality, biological implications of, 522–523
Chloramphenicol, 705–706
Chlordane, 354
Chlordiazepoxide (Librium), 494
Chlorination, of methane, 332
chloro- prefix, 333
Chlorobenzene, 374
Chloroethane, topical freezing agent, 333
Chloroform, 332 (*see also* Trichloromethane)
Chloromethane, 332
Chlorophyll
  in photosynthesis, 772
  magnesium in, 733
  role in photosynthesis, 530
Chloroplast, 772
  diagram of, 772
Cholate (*see* Cholic acid)
Cholecalciferol, 730
Cholesterol, 588–589
  deposition at arterial walls, 833
  heart disease, 833
Cholic acid, 590–591
Choline, 481, 582
Chromatin, 690
Chromosomes, 690
Chylomicrons, 833
Chymotrypsinogen, 664
Cigarettes, nicotine in, 484
Cinnamaldehyde, 429
Cinnamon flavoring, 429
cis-trans isomerization, role in vision, 352
cis-trans isomers, alkenes and cycloalkanes, 349–353
Citral, 429
Citrate, in TCA cycle, 809
Citrate synthetase, control point in TCA cycle, 808
Citric acid, 443, 445
Citrus fruits, acid in, 445
Cleaning agents, oxalic acid to remove rust, 445
Clones, 714
Clove oil, as an antiseptic, 410
Cloverleaf shape, of transfer RNA, 693
Cloves, flavor of, 410
CoA, 786
Coagulation, of proteins, 632
Coagulation factor, 732
Coal tar, 385
Cobalamin, 727
Coca plant, 495
Cocaine, 495
Codeine, 495, 496
Codon, 698
Coefficients, in balanced chemical equations, 81–83
Coenzyme A (CoA), structure of, 728
Coenzymes, 648
Cofactor, 648
Coffee beans, caffeine in, 487–488, 495
Collagen, 621, 624
  hydrolysis of, 631
Collagen protein, strengthening of, 734
Colligative properties
  applications of, 200–201
  boiling point, 200–201
  freezing point, 201
  osmotic pressure, 201–205
  vapor pressure, 200
Collision theory, 213
  reaction rates and, 213–214
  statements of, 213
Colloidal dispersions, 207
  types of, 207–208
Colon cancer, incidence of, 766
Combined-gas law, mathematical form, 158–159
Combustion, of alkanes, 331
Common cold, 709
Competitive inhibition, definition of, 659
Composition, 1
Compounds, 41
  constant composition of, 55–56
  contrasted with mixtures, 41–42
  number of, 41
Concentration
  equivalents, for acids and bases, 241–242
  milliequivalents, 195–198
  molarity, 192–194
  normality, for acids and bases, 241–242
  percentage of solute, 188–192
  units, for solutions, 188–189
Concentration gradient, 780
Condensation polymers, 457
Condensation reaction, in esters and polymers, 457
Condensation reactions, formation of amides, 482
Coniine, 495
Conjugated-double bonds, 346
Contraceptives
  female, 747–749
  male, 749
Conversion factors, 17
  metric–metric, 18
  metric-English, 18–19
Coordinate-covalent bond, 130
Copper
  as a trace element, 734
  excess of, in tissues, 734
Cori cycle, diagram of, 789
Cori's disease, 799
Corn syrup, 550
Corticosterone, 592
Corticotropin, daughter of pro-opiocortin, 739
Cortisone, 429
Cosmetics, 585
  moisturizing agents in, 547
  wax in, 456
Cottonseed oil, 577
Coumarin, 455
Coupled reactions, 758
Coupling factors, in ATP synthesis, 820–821
Covalent bonds
  coordinate, 130
  double, 127
  electron-dot structures and, 125–130
  multiple, 127–129
  nature of, 124–130
  nonpolar, 131
  octet rule and, 125–130
  polar, 131
  polarity and electronegativity, 131–133
  single, 127
  types of, 127–130
Covalent compounds, nomenclature for, 136–138
Covalent modification, 662
Cr-51, in determining blood volume, 867

Creatinine, 873
Cresol, 396
Cristae, 806
Cross-linking, in triple helix of proteins, 621
Cyanide, inhibitor of electron transport, 817
Cyanide poisoning, treatment of, 818
Cyclamate, sodium, 555
Cycle, fatty acid, 836
Cyclic monosaccharides, 539–543
Cycloalkanes, 324–327
  nomenclature of, 325–326
Cycloalkenes, 342
Cyclobutane, 324
Cyclohexadiene, 1,3- and 1,4-, 371
Cyclohexane, 325, 373
Cyclohexanone, 426
1,3,5-Cyclohexatriene (*see* Benzene)
Cyclohexene, 371
Cyclopentane, 324
Cyclopropane, 324
Cysteine, oxidation of, 415
Cytochrome c, 815–816
Cytochrome c oxidase complex, 817
Cytochrome oxidase, iron and copper in, 734
Cytoplasm, diagram of, 676
Cytosine, 487
  structure of, 677

α_D^{20}, 518
D-, (*dextro-*) prefix, 515
Dacron, 409
  458–459
Dalton's law of partial pressures
  applications of, 161–162
  mathematical form, 159–160
Dam, Henrik, 732
Danazol, 749
DAP, 783 (*see also* Dihydroxyacetone phosphate)
Dark reactions, in photosynthesis, 774–775
Daughter molecules of DNA, 687–688
DDT, 382
Dead-end complex, 660
Deamination, oxidative, 855
Debranching of glycogen, 797
Decaying matter, odor of, 473
Decomposition, 171
Decongestants, 474–475
  isoproterenol, 475
  metaproterenol, 475
  phenylephrine, 475
Degeneracy, in genetic code, 699
Dehydration
  BUN values in, 859
  of alcohols, 400–401
Deletion, in DNA mutations, 707
Delocalized bonding, 372
$\Delta G°$ (*see* Standard free energy, 757)
Demerol (meperidine), 496
Denaturation of proteins, 631
  acids and bases, 633
  heat, 632
  heavy metal ions, 633–644
  hydrogen-bonding solvents, 633
  reducing and oxidizing agents, 634–636
Denatured alcohol, 407
Density, 23
  units for, 23
  use as a conversion factor, 23–24
  values, table of, 23
Deoxycholate, sodium, 591
Deoxyribose, 539
  in DNA, 677
Depression, 725
Dermatitis, 724
DES (*see* Diethylstilbestrol)
Detergents, enzymes in, 669

Dextrins, 560
dextro-, (D-) prefix, 515
Dextrorotatory compounds, 517
Dextrose (*see* Glucose)
Diabetes mellitus, 537, 800
    ketoacidosis in, 846–847
Diagnosis, AST and ALT enzymes in, 856
Dialysis, process of, 208–209
1,6-Diaminohexane, in synthesis of nylon, 496–497
Diastereomers, 520–521
    anomers, 541
Diazepam (Valium), 494
Dicarboxylic acids, 442–443
1,1-Dichloroethene, 362
Dichloromethane, 332
Dichlorvos, 460
Dicoumerol, 733
Dietary fats, metabolism of, 833
Diethylamine, 474
Diethylstilbestrol (DES), 747–748
Diffusion
    across membranes, 596–597
    of lipids in cell membranes, 595
Digesting enzymes, protein, 663–664
Digestion, of proteins, 631
Dihydroxyacetone, 535, 536
Dihydroxyacetone phosphate (DAP), 783
Diisopropylphosphofluoridate, modification of serine in enzymes, 663
Dilution
    process of, 198–199
    sample problems involving, 199–200
Dimensional analysis, 19
    use in problem solving, 19–23
Dimethyl ketone (*see* Acetone)
Dimethylamine, 471
DIPF (*see* Diisopropylphosphofluoridate)
Diphenhydramine (Benadryl), 485
1,3-Diphosphoglyceric acid, 784
Dipole–dipole interaction, 168
Disaccharides, 531, 548–555
Disease, lymphatic system in fighting, 870
Diseases, caused by viruses, 709
Disinfectant
    Benzalkonium chloride, 481
    Isopropyl alcohol, 408
    Zephiran chloride, 481
Disinfectants, 409–410
Distillation, fractional of petroleum, 329
Disulfide bonds
    in proteins, 627
    redox reactions during permanent, 635
Disulfides, 414–415
Diuresis, 872
Diuretics, 873
Djerassi, Carl, 747
DNA
    double helix structure of, 683–690
    5' and 3' ends, 682
    lagging strand in replication, 689
    mutations in, 706–708
    recombinant, 711–715
    replication, 687–690
    single stranded, 709
    synthesizer, 715
    tertiary structure of, 686
    viral, 708–711
DNA ligase, 689
DNA polymerase, zinc cofactor, 735
2,4-DNP, uncoupler of proton potentials, 819
L-Dopa, 522
Double bonds, conjugated, 346
Double helix
    antiparallel in DNA, 684
    in DNA structure, 683–690
Downers, 494
1,3-DPG (*see* 1,3-Diphosphoglyceric acid)

Drugs
    absorption and excretion of, 876–878
    addictive, 495
    amine salts of, 478–479
    manufacture of enantiomeric, 669
    methods of administration of, 878

E. coli (*see* Escherichia coli)
Ear wax, 456, 584
EC nomenclature (*see* International Enzyme Commission)
Edema, 873
Efficiency, of energy conversion in glycolysis, 801
Egg, fertilized, implantation of, 747–748
Egg whites, denaturation of, 632
Electrolytes, 233, 865
    characteristics of, 233–234
    in major body fluids, 866
    in the human body, 234
    non, 233
    strong, 234
    weak, 234
Electromagnetic radiation, 516–518
Electron carriers, 762–764
Electron configurations
    Aufbau diagram and, 98
    notation used in, 97–100
    periodic law and, 100–101
    periodic table and, 101–106
    writing, using periodic table, 101–106
Electron orbitals, 95
    notation for, 95
    number of, within a subshell, 95
    order of "filling", 97–100
    shapes of, 95–96
Electron shells, 92
    notation for, 92–93
    number of electrons in, 92–93
    relative energy of, 92–93
Electron subshells, 93
    Aufbau diagrams and, 93–94
    notation for, 93–94
    number of, within a shell, 93
    number of orbitals within, 95
    order of filling, 97–100
Electron transport chain, 813
Electron transport
    inhibitors of, 817–818
    rates of, 821–822
    regulation of rate of, 826
Electron-dot structures, 115
    covalent bonds and, 124–130
    ionic bonds and, 116–121
Electronegativity, 130
    values, table of, 131
Electrons, 57
    characteristics of, 57–58
    location within atom, 57–58
    quantized energy of, 91
    terminology for arrangements of, 92–96
Electrophoresis, 614
Electrostatic attractions, in proteins, 627
Electrostatic interaction, 148
Elements, 41
    abundances of, 43–44
    selected, 291–292
    classification
        as metals and nonmetals, 107–108
        by electronic structure, 106–107
    discovery of, 41–42
    naming of, 43–44
    number of, 41
    symbols for, 44–45
    synthetic
        half-lives for, 278
        preparation of, 277–278

    trace, in biological systems, 734
Embden, Gustav, 781
Embden-Meyerhof pathway, 781
Enamel, in teeth, 735
Enantiomeric drugs, manufacture of, 669
Enantiomers, 509–510
    nomenclature for, 515–516
    projection formulas for, 513–515
Endergonic, 755
Endocrine glands, 735
Endorphins, 741
Energy, 27
    heat, units of, 28–29
    kinetic, 28
    law of conservation of, 27
    levels, for electrons, 92–93
    potential, 27
Energy extraction from food, stages in the, 769
Energy storage
    in fats, 831–832
    in transmembrane potentials, 772
Enol, 545
Enoyl-CoA, 835
Enoyl-CoA hydratase, 835
Environmental Protection Agency (EPA), 354
Enzymatic activity
    control of, 658–663
    covalent modification in, 662–663
    effect of enzyme concentration, 657–658
    effect of pH, 655
    effect of substrate concentration, 655–656
    effect of temperature, 654–655
    factors affecting, 649, 654–668
    modulator subunits in, 661–662
Enzyme-substrate complex, 651
Enzymes, 645–670
    absolute specificity of, 651
    autoactivation of, 664
    categories of, 650
    concentration effect on rate, 657
    conjugated, 647
    degradation, 657
    diseases caused by defects in, 669, 854
    general properties of, 649–651
    induced-fit model, 652–653
    induction, 657
    inhibitors, 658
    lock-and-key model, 651–652
    medical uses of, 668–670
    models of action, 651–654
    protein-digesting, 663–664
    restriction, 712
    simple, 647
    specificity of, 651
    used in diagnosis, 669
Epimers, 538, 541
Epinephrine, 411, 473, 738
    fight or flight hormone, 743
    in regulation of fat metabolism, 834
    stereochemistry of, 522
    target tissue of, 799
Epoxides, 413–414
Equilibrium
    chemical, 219–224
    liquid–vapor, 165–166
    state of, 166
Erythrocytes, 805, 867
Erythromycin, 705
Erythrose, 534
Erythrose-4-phosphate, 792
Erythrulose, 536
Escherichia coli, 712
Essential amino acids, 609, 767, 859
Essential fatty acids, 573, 842
Ester linkage, 451
Esterification reactions, 452
Esters, 440–462
    flavoring agents, 455

hydrolysis of, 456–457
inorganic, of monosaccharides, 548
nomenclature of, 453–454
of glycerol, 570
of inorganic acids, 459–460
preparation of, 451–452
properties of, 454–456
reactions of, 456–460
saponification of, 457
Estradiol, 747
Estrogens, 738, 746–747
Estrone, 592, 747
Ethane, 311
Ethanol
  effects on the human body, 405–406
  important uses of, 405–407
  in alcoholic beverages, 405
  production by fermentation, 788
  production of, 406
  toxicity of, 788
Ethanolamine, 582
Ethene, 342
  structural formula of, 342
Ethers, 390–420
  classification of, 411–412
  hydrogen bonding, 412–413
  mixed, 411
  nomenclature of, 411–412
  properties and uses of, 412–413
  simple, 411
  solubilities of, 412–413
Ethyl acetate, 452
Ethyl alcohol, 392, 394 (*see also* Ethanol)
Ethyl methyl ketone, 426
Ethylamine, 471
Ethylbenzene, 374
Ethylene, 342, 360 (*see also* Ethene)
Ethylene chloride, dry cleaning agent, 333
Ethylene glycol, 408–409, 458, 499
Ethylene-based polymers, 360–361
Ethylene oxide, simplest epoxide, 414
Ethyne, 362
Eugenol, 410
Evaporation, 163
  cooling effect of, 164–165
  explained using kinetic molecular theory, 163
  rate of, 164
Ex-Lax, active ingredient in, 411
Exercise, 560–562
Exergonic, 755
Exons, in RNA, 696
Experiment, 51
  scientific method and, 51–52
Exponent, 8
Exponential notation, 8
  converting to decimal notation, 10–11
  division in, 12–13
  multiplication in, 11–12
  writing numbers in, 9–10
Extracellular fluids, 864–878
Extraction, 412

Facilitated diffusion, 780
  across membranes, 596–597
Facts, 51
  scientific method and, 51–52
FAD, 764 (*see* Flavin adenine dinucleotide)
FADH, formed by fatty acid oxidation, 836–837
$FADH_2$, 764 (*see* Flavin adenine dinucleotide)
Fasting, 560
  ketoacidosis in, 846
Fat cells, 576
Fats, 457, 574
  dietary considerations of, 766
  role of choline in fat transport, 481
Fats and oils, 573

fatty acid composition of various, 575
hydrogenation of, 578–579
hydrolysis of, 577
rancidity of, 579–580
saponification of, 577–578
Fatty acid biosynthesis, 837
Fatty acid cycle, 836
Fatty acids, 450–451, 569–573
  β-oxidation of, 835–837
  biosynthesis of, 837–841
    balanced equation, 841
    pathway, 840
    vs. degradation, 838
  essential, 573, 842
  metabolism, regulation of, 843–845
  mobilization and oxidation of, 833–837
  polyunsaturated, 572
  saturated, 570
  sources of, 571
  transport into mitochondria, 835
  unsaturated, 570–572
  oxidation of, 836
Fatty acyl-ACP, 839
Fatty acyl-CoA, 835
Fatty acyl-CoA dehydrogenase, 835
FDA, 711, 722, 749
Feathers, wax coating on, 456, 584
Feedback inhibition, 661
  negative, by thyroxine, 737
Feedstock, petrochemical, 329
Fehling's reagent, 435–436, 546–547
Female reproductive system, hormones produced by, 747–748
Fermentation, 406, 431, 788
Fertilizer, urea in, 493
Fevers, prostaglandins and, 592
Fiber, dietary 558, 765
  sources of, 558
Fibrin, 621
Fibrinogen, 868
Fibrous proteins, 625
Fight or flight hormone, 743
Filtration, non-selective by kidneys, 872–873
Fish, unpleasant odor of, 479
Flavin adenine dinucleotide (FAD), 723–724
  as an electron carrier, 764
Flavin mononucleotide (FMN), 723–724
Flavoring agents, 455
Floor wax, 585
Flowers, scents of, 354
Fluids
  electrolyte balance in, 873–874
  extra-, intracellular, 865
  pH of various body, 875
Fluoride
  as trace element, 735
  in prevention of tooth decay, 735
fluoro- prefix, 333
FMN (*see* Flavin mononucleotide)
F_0 and F_1 coupling factors, 820–821
Folic acid, 664–665, 727
Follicle-stimulating hormone, 746–748
Food consumption, starches in, 560
Formaldehyde, 426–427
Formalin, 426
Formic acid, 441
Formula, general
  of alkanes, 312
  of alkenes, 342
  of cycloalkanes, 327
Formula weights, 69
  calculation of, from atomic weights, 69–70
Formulas
  condensed structural, 315
  expanded structural, 315
  structural, 314
Fractional distillation, of petroleum, 329
Frame-shift mutations, 707

Free energy, 755
  from oxidation reactions, 759
  independent of path, 757
  positive/negative values of, 755–756
Free energy change, in carbohydrate metabolism, 801
Freon, 333
Fructose, 428, 531, 536, 538
Fructose-1,6-diphosphate, 783
Fructose-6-phosphate, 783
Fruit fresheners, 411
FSH (*see* Follicle-stimulating hormone)
Fumarase, 810
Fumarate, 810
Functional group, 341
Functional groups
  aryl, 391
  benzyl, 376
  carbonyl group, 423
  carboxyl, 441
  ester, 451
  nitrile, 476
  phenyl, 376
  sulfhydryl group, 414
Fused-ring aromatic compounds, 383–386

Galactose, 534, 536, 538, 587
Galactosemia, 552
Galactosyl transferase, 661–662
Gallstones
  cholesterol in, 589
  dissolving of with MTBE, 413
Gamma radiation
  characteristics of, 271
  emission of, 273
  symbol for, 271
Gamma-globulins, 636, (*see* Immunoglobulins)
Gangliosides, 538, 587, 588
Gaseous state
  characteristics of, 147
  explained using kinetic molecular theory, 150
Gelatin, 609
  from hydrolysis of collagen, 631
Gene, 695
  designer, 715
General anesthetic, 413
Genetic code, 697
  degeneracy in, 699
  listing of, 698
Geometric isomers, 348
Geometrical figures, in representation of cycloalkanes, 326
Geraniol, 354
Gibbs, J. Willard, 755
Gibbs free energy (*see* Free energy)
Globular proteins, 625
Glomerulus, nephron, 870
Glucagon, target tissue of, 799
Glucitol, 547
Glucogenic amino acids, 852
Gluconeogenesis, 788
  balanced equation for, 790
Glucose, 428, 531, 534, 536
  absorption, 779
  concentration in blood, 537
  D- and L-configuration of, 542–543
  in diabetes mellitus, 537
  levels in blood and urine, 870–871
  metabolic fates of, 780
  specific rotation of, 541
  testing urine for, 547
  transport, hormone regulation of, 780
Glucose oxidase, 646
Glucose phosphate isomerase, 783
Glucose-1-phosphate, 795–796
Glucose-6-phosphate, 781–782
Glucosides (*see* Glycoside)

Gluose, 534
Glutamate oxaloacetate transaminase, 855
Glutamic acid, 856
   in folic acid, 727
Glutamine, transport of ammonia, 856
Glutaric acid, 443, 445
Glyceraldehyde, 533, 534
   enantiomers of, 513
Glyceraldehyde-3-phosphate, 783
Glycerol, 409, 456
Glycerol phosphate shuttle, 824–825
Glycine, 605
Glycocholate, sodium, 591
Glycogen, 560–562
   debranching of, 797
   granules, 562
   multiple nonreducing ends, 795
   storage form of glucose, 561–562
   synthesis and degradation control, 770–771
Glycogen phosphorylase, 794–795
   control point in metabolism, 799
Glycogenesis, 560, 794
Glycogenolysis, 560–561, 794
Glycolysis, 780
   balanced equation of, 788
   pathway summary, 782
Glycoprotein hormones, 738
Glycosides, 544
Glycosidic bond, 548
α- and β-Glycosidic bonds, 548–550
Glycosphingolipids, 587
Goiter, 735
GOT (see Glutamate oxaloacetate transaminase)
Grain alcohol, 406, (see also Ethanol)
Gram, 15
   conversion factors involving, 19
Granum, 772–773
Grape flavor, 455
Grazing animals
   digestion of cellulose, 558
   selenium toxic to, 735
Grease, dissolving of, 450–451
Greek letters, in acid nomenclature, 446
Group, alkyl, 319
Group specificity, of enzymes, 651
Groups, 63
   periodic table and, 63–64
Growth hormone, pituitary, 711
Guanine, 487
   structure of, 677

H chain, in immunoglobulins, 637
Hair
   denaturation of proteins in, 634
   permanent wave in, 634–636
   proteins in, 608, 621
Hairpin loop, in RNA, 692
Half-life, 275
   calculations involving, 276–277
   synthetic elements and, 278
   values for selected isotopes, 275
Hallucinogens, LSD, 489
Halogenation
   of alkanes, 331
   of alkenes, 355
   of alkynes, 364
   of benzene, 378
Halothane, 333
Hangover symptoms, result of acetaldehyde, 788
HDL, 833, (see Lipoproteins, high density)
Heart attacks
   enzymatic treatment of, 670
   unsaturated fatty acids and, 576
Heart disease
   correlation to fat in diet, 766

HDL and LDL levels in, 833
Heart valve, 458
Heat of fusion, 178
Heat of vaporization, 178
Heavy water, lipoprotein centrifugation in, 833
Helical structure, of amylose, 558
Helix, DNA (see Double helix)
Heme, 484
   in cytochrome c, 815
Hemiacetals, 432–434, 539
Hemiketals, 434, 539
Hemlock, 495
Hemoglobin
   amino acids in, 608
   DNA mutation in, 708
   molecular weight of, 604
   normal vs. sickle-cell, 620
   oligomeric nature of, 629
Hemoglobin S, 620
Hemolytic anemia, 732
Hemorrhage, ascorbic acid in relation to, 729
Hemorrhagic pancreatitis, 664
Henseleit, K., 856
Hepatitis, 709
Heptachlor, 354
2-methyl-Heptadecane, 334
Her's disease, 799
Heroin, 496
Heterocyclic amines, 483–490
Heterocyclic aromatic compounds, 385–386
Heterogeneous mixture, characteristics of, 36–37
Hexachlorobenzene, 381
Hexachlorophene, 410
Hexamethylenediamine, 474
Hexokinase, 781–782
   induced-fit of substrate, 652–653
Hexylresorcinol, 410
High-protein diets, dangers of, 853
Histamine, 484
Histidine, 484
Homogeneous mixture, characteristics of, 37–39
Hookworm, 481
Hormones, 590, 735
   classification and structure of, 738–742
   glycoprotein, 738
   molecular mechanisms of, 742
   regulation of cellular processes, 735
   regulation of fat metabolism, 844
   serotonin, 488
   thyroid, in amino acid metabolism, 861
Housefly, pheromone sex-attractant of, 353
Human body, composition of, 530
Hydration, of alkenes, 355–356
Hydride ions, loss of during oxidation, 758
Hydrocarbons, 308–387
   aromatic, 370–387
   cycloalkanes, 311
   saturated, 308–334
   a summary, 386–387
   unsaturated, 340–364
Hydrogen, characteristics of, 294
Hydrogen atoms, loss of H· during oxidation, 758
Hydrogen bonding
   aldehydes and ketones, 426
   amides, 491–492
   between alcohol molecules, 397–399
   effects of, on properties of water, 177–181
   in carboxylic acids, 446
   in proteins, 627–628
Hydrogen bonds, 168
Hydrogen cyanide, toxicity of, 818
Hydrogen sulfide, 414
Hydrogenation
   of alkenes, 355–356
   of alkynes, 364

of fats and oils, 578–579
of vegetable oils, 355
Hydrohalogenation
   of alkenes, 355–356
   of alkynes, 364
Hydrolysis
   of amides, 492
   of cellulose, 558
   of esters, 456–457
   of fats and oils, 577
   of proteins, 631
   of salts, 245–249
   pH change and, 245–249
Hydrolytic rancidity, 580
Hydrophobic attractions, in proteins, 628
Hydroquinone, 396, 411
Hydrosphere
   characteristics of, 303
   natural hydrologic cycle, 303
Hydroxy-acyl-CoA dehydrogenase, 835
Hydroxyapatite, in teeth enamel, 735
β-Hydroxybutyrate, 845
Hydroxylbenzene (see phenol)
Hypercalcemia, 731
Hyperglycemia, 800
Hyperglycemic, 537
Hypertension, 749
   sodium in relation to, 873
Hypertonic solution, 206
Hypervitaminosis, 730
Hypoglycemia, 537, 801
Hypomagnesemia, 733
Hypopotassemia, 873
Hypothalamus, 736–738
Hypothesis, 52
   scientific method and, 51–52
Hypotonic solution, 206

I.U. (international units), 721
Ibuprofen, 442
Idose, 534
IgG, 636 (see also Immunoglobulins)
   primary structure of, 637
Illegal drugs, amphetamines, 475
Imidazole, 483, 484, 487
Immunoglobulins, 636
   prevent attachment of viruses, 710
Indole, 386, 487, 488
Induced-fit model of enzymes, 652–653
Infectious hepatitis, AST and ALT in diagnosis of, 856
Infertility, in rats, and tocopherol, 732
Inflammation, medicine to relieve, 429
Influenza, 709
Inhibition
   feedback, in enzymes, 661
   of enzyme activity, 658–663
Inhibitors, of electron transport, 817–818
Inner-transition elements, 107
   location within periodic table, 107
Inorganic acids, esters of, 459–460
Inorganic chemistry, 310
Inorganic esters, of monosaccharides, 548
Insect communication (see Pheromones)
Insecticides, 354
   DDT, 382
   Dichlorvos, 460
   Parathion, 460
Insertion, in DNA mutations, 707
Insomnia, 733
Insulin, 618–619, 711, 745
   comparison of, in animals, 619
   regulation of glucose absorption, 780
   target tissues of, 780
Integral proteins, 596
Interferon, 711
Intermembrane space, 806

Intermolecular forces, 168
　hydrogen bonds, 168
　strength of, and boiling point, 168–169
　types of, 168
International Enzyme Commission, 646
Interstitial fluid, 865
Intracellular fluids, 865
Intravenous nourishment, glucose, 537
Introns, in RNA, 696
Invert sugar, 554
Iodination, of fatty acids, 580
Iodine
　as a trace element, 735
　starch indicator, 560
Iodine number, 581
iodo- prefix, 333
Ion product for water, 235
Ionic bonds, 118
　electron-dot structures and, 118–121
　formulas for, 118–121
　octet rule and, 118–121
Ionic compounds, bonding in, 116–118
　containing polyatomic ions, 138–141
　nomenclature for, 122–124
　structure of, 121–122
Ionic solids, characteristics of, 121–122
Ionization
　of amines at different pH values, 479
　of carboxylic acids, 449
Ions, 116
　carboxylate, 448–449
　charges on monoatomic, 116–118
　notation for, 117
Iron
　as a trace element, 734
　characteristics of, 293–294
　ferrous and ferric ions, 758
　in heme, 625
Iron-sulfur enzymes, 734
Isobutane, 312
Isobutyl alcohol, 394
Isobutyl group, 323
Isocitrate, 809
Isocitrate dehydrogenase, 809
Isoelectric point of amino acid, 614
Isoenzymes, 668
　examples of, 668
Isomerism
　optical, 507–510
　stereo, 506–523
　structural, 312
Isomers
　binding to receptor surfaces, 523
　cis-, and *trans-*, 349–353
　geometric, 348
　possible numbers of structural, 314
Isopentane, 314
Isoprenaline, 522
Isopropyl alcohol, 392, 394
Isoproterenol, 475
Isotonic solution, 206
Isotopes, 60
　mass of, table of, 60
　percentage abundances of, table, 60
　symbols for, 60–61
IUPAC nomenclature
　alcohols and phenols, 392–396
　aldehydes, 425–427
　alkanes, 318–324
　alkenes, 342–346
　alkynes, 363–364
　amines, 470–472
　benzene derivatives, 373–377
　carboxylic acids, 443–446
　esters, 453–454
　ethers, 412
　glycosidic bonds, 548–550
　ketones, 425–427

　numbering chains, 320

Jarvik-7 artificial heart, 498
Jojoba wax, 585

Kekule, August, 372
Keratin, 621
Ketal formation, in monosaccharides, 543–544
Ketals, 434
Ketoacidosis, 846
3-Ketoacyl-CoA, 835
Ketogenic amino acids, 853
α-Ketoglutarate, 809
α-Ketoglutarate dehydrogenase, 810
α-Ketoglutaric acid, 856
Ketone bodies, 845
　in the blood, 847
Ketones, 422–436
　hydrogen bonding in, 426
　nomenclature of, 425–427
　physical properties of, 426–429
　preparation of, 430
　reduction of, 431–432
　uses of, 428–429
Ketonuria, 846
Ketose, 531
Ketoses, classification of, 535–536
Ketosis, 846
Kidney malfunction, BUN values in, 859
Kinases, 781
Kinetic energy, 28
Kinetic molecular theory, 147–148
　gaseous state properties and, 150
　liquid state properties and, 149–150
　solid state properties and, 149
　statements of, 147–148
Krebs cycle (*see* Tricarboxylic acid cycle)
Krebs, Hans, 808, 856

L chain, in immunoglobulins, 637
L- *(levo-)* prefix, 515
L-amino oxidase, 520
β-Lactam ring, in penicillin, 666–667
Lactase, 552
Lactate dehydrogenase (LDH), 787
　isoenzyme nature of, 668
　used in diagnosis, 669
Lactic acid, 445, 786
　enantiomers of, 510
Lactose, 551–552
　in human milk, 552
Lactose intolerance, 552
Lactose synthetase
　catalytic subunit of, 661–662
　specificity of, 661–662
Lagging strand, DNA replication, 689
Lard, 576
Lateral diffusion, in cell membranes, 595
Lauric acid, 571
Lauryl alcohol, ester of, 459
Laws, 52
　Boyle's, 152–156
　Charles', 156–158
　combined-gas, 158–159
　conservation of energy, 27
　conservation of matter, 54–55
　constant composition, 55–56
　Dalton's, of partial pressures, 159–162
　periodic, 62–63
　scientific method and, 51–52
LDH (*see* Lactate dehydrogenase)
LDL, 833, (*see* Lipoproteins, low density)
Le Châtelier's principle
　concentration changes and, 222
　in regulation of TCA cycle, 825–826

　keto-enol equilibrium, 546
　prediction of equilibrium position, 221–224
　pressure changes and, 223–224
　statement of, 222
　temperature changes and, 222–223
Lecithin, 582
Lemon oil, 429
Leukocytes, 867
levo-, (L-) prefix, 515
Levorotatory compounds, 517
Levulose (*see* Fructose)
LH (*see* Luteinizing hormone)
Librium (chlordiazepoxide), 494
Lifetime, of blood cells, 867
Ligase, 689
Light, 516–518
　plane-polarized, 516–518
Light reactions, in photosynthesis, 774–775
Light reception by retina, 352
Limburger cheese, odor in, 445
Limonene, 354
Linkage specificity, of enzymes, 651
Linoleic acid, 571, 842
Linolenic acid, 571, 842
Lipase, 832
　hormone regulation of, 834, 844
Lipid bilayer, 594–595
Lipid metabolism, relationship to carbohydrate, 845
Lipids, 568–597
　absorption and transport, 832–833
　classification of, 570
　metabolism of, 830–847
　nonsaponifiable, 569
　saponifiable, 569
Lipoproteins, characteristics of, 833
　low and high density, 833
β-lipotropin, formation of, 739–780
Liquid state, characteristics of, 147
　explained using kinetic molecular theory, 149
Lister, Joseph, 409
Liter, 16
　conversion factors involving, 19
Lithium aluminum hydride, reducing agent, 431–432
Liver
　primary site of ketogenesis, 845–846
　primary site of urea cycle, 856
Livestock, rapid fat growth in, 747–748
Lock-and-key model of enzymes, 651–652
Lorazepam (Ativan), 494
Lotion, glycerol added to, 409
Lotions, 584
LSD (lysergic acid diethylamide), 489
Luminal (phenobarbital), 494
Luteinizing hormone, 746–748
Lymph, 869
　lipids in, 833
Lymph nodes, 870
Lymphatic system, 869–870
Lymphocytes, site of immunoglobulin synthesis, 636
Lysergic acid, 489
Lysozyme, diagram of enzymatic action, 653
Lysyl oxidase, copper in, 734
Lyxose, 534

m- prefix (*see meta-*)
Magnesium, as a micronutrient, 733
　characteristics of, 294
　in chlorophyll, 733
　in photosynthesis, 773
Malate, 810
Malate shuttle, 823–824
Malic acid, 443
Malonate, as a competitive inhibitor, 659
Malonic acid, 443, 445

Malonyl-ACP, 838
Malonyl-CoA, 838, 845
Malt sugar (see Maltose)
Maltase, 550
Maltose, 550
Mannose, 534
Maple syrup urine disease, 854
Margarine, 578–579
Marijuana, 410–411
Markovnikov's rule, 357
Mass, 15
 compared to weight, 15–16
Mass number, 58
 number of neutrons and, 58–59
Matrix, mitochondria, 806
Matter, 2
 changes in, 3–5
 classifications of, 35–39
 composition of, 1
 conservation of, law of, 54–55
 physical states of, 147–150
 properties of, 2–3
 structure of, 1–2
McArdle's disease, 799
Measles, 709
Measurement, 5
 accuracy of, 5–7
 errors in, 5–7
 precision of, 5–7
 systems of, units for, 13–17
Medications
 caffeine in, 487–488
 therapeutic concentrations of, 877
Melanocyte-stimulating hormone, 741
Melting, 170
Membrane proteins, side chains in, 608
Membranes (see also Cell membranes)
Menadione, 732
Menstrual cycle, 747–748
Menthol, 432
Meperidine (Demerol), 496
Mercaptains, 414
Meso compounds, 521
Messenger RNA, 692–693
meta- prefix, 375
Metabolic pathways, 767
Metabolism
 lipid, 830–847
 of carbohydrates, 778–801
 overview of, 754–775
 regulation of, 769–771
 regulation of carbohydrate, 799–801
Metals, 107
 and living organisms, 108–110
 periodic table location of, 108
 properties of, 107–108
Metaproterenol, 475
Meter, 14
 conversion factors involving, 19
Methadone, 474
Methamphetamine, 475
Methane, 311
 chlorination of, 332
Methanol, 392–394
 important uses of, 404–405
 toxicity of, 404–405
Methoxyamphetamine, 475
Methyl alcohol, 392–394, (see Methanol)
Methyl phenyl ketone, 426
Methyl tert-butyl ether (MTBE), 413
Methylamine, 474
Methylbenzene (see Toluene)
Methylene chloride, 332, (see also Dichloromethane)
Metric system
 advantages of, 13
 length units, 14–15
 mass units, 15–16

prefixes, 14
 volume units, 16–17
Meyerhof, Otto, 781
Micelles, 451
Microelectrodes, to measure membrane potentials, 771
Micronutrients, minerals as, 733
Miescher, Friedrich, 675
Milk
 fortified with vitamin D, 730
 human, lactose synthesis in, 662
 lactic acid in sour, 445
Milk sugar, 551, (see lactose)
Milliequivalents, 195
Milliequivalents, sample problems involving, 195–198
Mineral oil, 331
Minerals, 733
 in biological systems, 733
Mirror images, 508–510
Mitchell, Peter, 818
Mitochondria, sites of urea cycle in, 857–858
Mitochondrion, diagram and composition of, 805–806
Mixtures
 characteristics of, 35–39
 heterogeneous, 36–37
 homogeneous, 37–38
 types of, 35–39
Models, molecular
 ball-and-stick, 312–313
 space-filling, 312–313
Models, of enzyme action, 651–654
Modulator subunits, in enzymes, 661
Molar mass, 73
 use of, in calculations, 74–75
Molarity, 192
 sample problems involving, 192–194
Mold inhibitor, 445
Mole, 71
 chemical formulas and, 75–76
 mass of, 74–76
 mole-particle conversion, 72–73
 relationship to Avogadro's number, 71
 use of, as a counting unit, 70–73
Molecules, 40–41
 geometry of, 133–136
 heteroatomic, 41
 limit of physical subdivision, 39–40
 types of, 40–41
Mononucleosis, infectious, 709
Monosaccharides, 531
 biologically important, 535–539
 classification of, 532–535
 cyclic forms of, 539–543
 ester formation in, 548
 oxidation of, 544–547
 reactions of, 543–548
 reduction of, 547–548
Monosodium glutamate (MSG), D- and L-forms of, 522
Morning-after pill, 748
Morpheus, 495
Morphine, 495, 741
 administered as a salt, 495
mRNA, hormone effect on transcription, 746, (see Messenger RNA)
MSG, 522, (see Monosodium glutamate)
MSH (see Melanocyte-stimulating hormone)
Mumps, 709, 711
Muscalure, 353
Muscle, protein in, 621
Muscle cramps, weakness, role of magnesium in, 733
Muscles, injury to, AST in diagnosis of, 856
Muscone, 429
Muscular dystrophy, AST in diagnosis of, 856
Musk, 429

Mutarotation, 541
Mutations, 706
 deletion, insertion, substitution, 707
 in DNA, 706–708
 point, 708
Myelin sheath, 587
Mylar, 459
Mylar film, 409
Myocardial infarction, AST in diagnosis of, 856
Myoglobin, tertiary structure of, 625–626
Myristic acid, 571

NAD^+, structure of, 763–764
NADH, 431
 entry into mitochondria, 823–824
 formed by fatty acid oxidation, 836–837
 in oxidative phosphorylation, 813–814
 structure of, 763–764
NADH–ubiquinone reductase, 817
$NADP^+$, structure of, 763
NADPH
 formed during photosynthesis, 774
 reducing agent in oxidoreductases, 764
 reducing agent for fatty acids, 839
 reduction of glucose, 547
 structure of, 763
 supplied by pentose pathway, 791
Naphthalene, 384
β-Napthylamine, 473, 474
Narcolepsy, treatment of, 475
Narcotic analgesics, 495
Natural gas, thiols responsible for odor of, 414
Nembutal (pentobarbital), 494
Neopentane, 314
Neoprene, 362
Nepetalactone, 455
Nephrons, 870
Nerve cells, transmissions in, 772
Nerve impulse, acetylcholine involved in, 481
Neutral fats, 831
Neutralization, 235
 of carboxylic acids, 449
Neutrons, characteristics of, 57
 location with atom, 57
Niacin, 486, 725
Niacinamide, 725
Nicotinamide adenine dinucleotide (see NADH)
Nicotinamide ring, in NADH, 763
Nicotine, 483–484, 495
Nicotinic acid, 725
Nitration, of benzene, 379
Nitric acid, esters of, 459
Nitriles, reduction of to form amines, 476
Nitrite, in treatment of cyanide poisoning, 818
Nitrobenzene, 374
Nitrogen balance, 852
Nitrogen oxides
 and smog formation, 301–302
 characteristics of, 300–301
 formation of, 301
Nitroglycerin, 459
No-pest strip, pesticide on, 460
Noble-gas elements, location within periodic table, 106–107
Nomenclature
 binary ionic compounds, 122–124
 covalent compounds, 136–138
 ionic compounds with polyatomic ions, 140
Non-competitive inhibitors, 660
Non-polar side chains, in amino acids, 608
Nonelectrolytes, 234
Nonmetals, 108
 and living organisms, 108–110
 periodic table location of, 108
 properties of, 107–108
Nonspontaneous reactions, 756

Nonsuperimposability, 508–510
Norandrolone phenyl propionate, 746
Norepinephrine, 473, 725, 738
Norlutin, 747
Normality, 241
   calculations involving, 242–245
(−) Notation, 517
(+) Notation, 517
Nuclear fission, 285
   atomic bomb and, 286
   characteristics of, 285
   electricity from, 286
Nuclear fusion, 286
   development of, 287
   energy production on the sun, 287
Nuclear medicine
   as a diagnostic tool, 282–284
   as a therapeutic tool, 284–285
   criteria for selecting radionuclides, 282–283
Nuclear reactions, 269
   bombardment, 274–275
   contrasted with chemical reactions, 287
   equations for, 271–274
   transmutation, 274–275
Nucleic acids, 675–715
   in selected viruses, 709
   locations in cell, 675
   primary structure of, 680–683
Nucleosides, 678
   nomenclature of, 679
Nucleotides, 676
   nomenclature of, 679
Nucleus
   characteristics of, 57–58
   diagram of, 676
   location within atom, 57
Nuclide, 269
   daughter, 272
   parent, 272
   radioactive, 270
Nuprin, 442
Nuptial flight of queen bee, 447
Nutra-Sweet (aspartame), 555
Nutrition
   classes of biochemicals in, 765–767
   overview of, 754–775
Nylon
   Nylon 66, 496
   synthesis of, 496–497
   uses and properties of, 496–498

o- prefix *(see ortho-)*
Observations, types of, 5
Observed rotation, 518
Octet rule
   covalent bonds and, 124–130
   ionic bonds and, 116–121
   statement of, 116
Oil
   caraway, 522
   cottonseed, 577
   dissolving of, 450–451
   lemon, 429
   lubricating, 331
   mineral, 331
   of celery, 354
   of ginger, 354
   of wintergreen, 452
   peanut, 578
   spearmint, 429, 522
   vegetable, 457, 576
Oil spill, effect on birds, 584
Oils, 574
Oleic acid, 571
Oligomycin, 828
Oligosaccharides, 531
Olympic Games, urine tests in, 872

Opiate hormones, lipotropin, 739–740
Opium poppy plant, 495
Opsin, 352
Optical activity, 516–518
Optical isomerism, 507–510
Orange flavor, 455
Organic, 309
Organic chemistry, historical perspective, 309
Orinithine, 858–859
Orotic acid, 721
ortho- prefix, 375
Osmolarity, 204
   problems involving, 204–205
Osmosis, 201
   explained using kinetic molecular theory, 201
   process of, 201–205
Osmotic pressure, 202
   importance of in biological systems, 202–204
   in capillaries, 869
Osteoblasts, 729
Osteoporosis, 733
Ovaries, 747
Ovulation, 746–748
Oxalic acid, 442–443
   toxicicity of, 409
Oxaloacetate, 789, 856
   in TCA cycle, 809
Oxidation
   electron loss, 258
   of fatty acids, 833–837
   original meaning of, 257
   oxidation number change, 261
β-Oxidation of fatty acids, 835–837
Oxidation numbers
   determination of, 258–262
   rules for determining, 259
Oxidation reactions
   in living systems, 758–759
   organic compounds, 359
Oxidation–reduction
   important types of processes, 265–266
   recognition of, 264–265
Oxidative phosphorylation, 820
   regulation of rate of, 826
Oxidative rancidity, 580
Oxidizing agents, electron gain, 258
   in oxidation of alcohols, 401–403
   oxidation number change, 261
Oxygen, characteristics of, 292
Oxygen debt, 786
Oxygen transport, red blood cells, 868
Oxytocin, 736
Ozone, formation from nitrogen oxides, 301

P site, in ribosome, 700
p- prefix *(see para-)*
P-32, in determining blood volume, 867
p-orbitals, in benzene, 373
p-orbitals, overlap in pi-bonds, 347
PABA, 664, 727, *(see also* Aminobenzoic acid, *para-)*
Pain relievers
   endorphins, 741
   morphine, 741
Palmitic acid, 570, 571, 831
Pancreatic juice, 832
Pancreatitis, hemorrhagic, 664
Pantothenic acid, 728
para- prefix, 375
Paraben, methyl, 456
Parabens, as mold and yeast inhibitors, 456
Parachutes, nylon used in, 496
Paraffins, 330
Paraffin wax, 585
Parasites
   hookworm, 481
   roundworm, 481

Parathion, 460
Parigamic acid, 721
Parkinson's disease, 522
Pasteur, Louis, 509, 781
PCBs, 382
Peanut butter, 578
Pellagra, 725
Penicillamine, 734
Penicillin, 666–667
   as an enzyme inhibitor, 667
   derivatives of, 666
Penicillinase, 667
Pentane, 314
Pentanone, 425
Pentobarbital (nembutal), 494
Pentose phosphate pathway, 791
   relationship to glycolysis, 793
Penultimate carbon atom, 533
PEP (*see* Phosphoenolpyruvate)
Pepper, black, 495
Pepsin, 646
Peptidases, 663
Peptide bonds, 490, 615
   formation in protein synthesis, 703–706
Peptides, 615
   linear sequence of, 616
Peptidoglycan, 667
Peptidyl site, in ribosome, 700
Percentage of solute, 188
   sample problems involving, 188–192
Periodic law, 62
   electron configurations and, 100–101
   historical development of, 62–63
Periodic table, 63
   diagram of, 64
   electron configurations and, 101–106
   groups, 63–64
   ionic charges and, 118–119
   periods, 64–67
Periods, 63
   periodic table and, 63
Peripheral proteins, 596
Peristalsis, 832
Permanent wave, reactions in, 635–636
Permeability of cell membranes, regulated by hormones, 735
Pernicious anemia, 727
Peroxidase, iron in, 734
Perspective drawings, 513–515
Perspiration, acid in, 445
Petroleum
   composition of crude, 328–329
   products obtained from refining, 329
   refining of, 329
Petroleum jelly, 331
PFK (*see* Phosphofructokinase)
PGE$_{1,2}$ (*see* Prostoglandins)
pH
   effect on amine protonation, 479
   regulation of, in body fluids, 874–875
   values of in various body fluids, 875
pH scale, 237
   hydrogen ion concentration and, 237–238
   problems involving, 238
Pharmacokinetics, 876
Phenacetin, 493
Phenanthrene, 384
Phenobarbital, 494
Phenol, 374, 395
   as a disinfectant, 409
Phenolphthalein, 411
Phenols, 390–420
   IUPAC nomenclature, 392–396
   properties and uses of, 409–410
Phenyl group, 376
Phenylalanine hydroxylase, 669, 854
p-Phenylene diisocyanate, 499
Phenylephrine, 475

β-phenylethylamines, general structure of, 473–474
Phenylketonurea (PKU), 556, 669, 854
Phenylpyruvate, as competitive inhibitor, 854
Pheromones, 334
   housefly, 353
   queen bee secretions, 447
Phosphatases, role in enzyme modification, 663
Phosphate esters, 461–462
Phosphate shunt (*see* Pentose phosphate pathway)
Phosphate transfer potential, compounds with, 760–761
Phosphatidyl choline, 582
Phosphatidyl ethanolamine, 582
Phosphatidyl serine, 582
Phosphoenolpyruvate (PEP), 784
Phosphofructokinase (PFK), 629, 783
   control point in metabolism, 799
Phosphoglucomutase, 795
2-Phosphoglycerate, 784
3-Phosphoglycerate, 784
Phosphoglycerate kinase, 784
Phosphoglycerides, 581–584
   nonpolar tails, 583
   polar heads, 583
Phosphoglyceromutase, 784
Phosphoric acid, esters of, 459, 461
Phosphorus, as a micronutrient, 733
Phosphorus trichloride (PCl_3), substitution in alcohols, 403–404
Photophobia, 724
Photosynthesis, 530, 772
   light and dark reactions in, 774–775
Physical change, 3
Physical properties, 2–3
Pi bond, 347
Pineapple flavor, 455
Piperidine, 485–486
Piperine, 495
Pituitary gland, 736–738
PKU (*see* Phenylketonurea)
Placenta, production of hormones by, 747
Plane of symmetry, 521
Plane-polarized light, 516–518
Plasma, 865
Plasmids, 712
Platelets, 867
Poison ivy and oak, catechols in, 410
Poisoning, nicotine, 484
Polar side chains, in amino acids, 608
Polarimeter, 517
Polarity of bonds, delta notation for, 132–133
   electronegativity differences, 131–133
Polarity of molecule, determining, 133–136
Polarizers, 517
Polio, 711
Pollen, allergies to, 484
Polyamides, 496–499
   nylon, 496
Polyatomic ions, 138
   formulas for compounds containing, 139–140
   names for common, table of, 139
   nomenclature for compounds containing, 140
Polychlorinated biphenyls (PCBs), 382
Polyesters, 457–459
Polyethylene, 360
Polyethylene glycol (PEG), 399
Polymer, 359
Polymerization, of alkenes, 359–362
Polymers, condensation, 457
Polypeptide, 615
Polypropylene, 360
Polysaccharides, 531, 556
   branched, 556, 559–562
   linear, 556–557
   rapid hydrolysis in digestion, 779
Polystyrene, 360
Polyunsaturated oils, 575
Polyurethanes, 498–499
Polyvinylchloride (PVC), 360
Porphyrin, 484
Potassium, characteristics of, 294
   ion exchange in kidneys, 870–872
Potassium dichromate, oxidation of alcohols, 401–403
Potassium permanganate, 358
   oxidation of alcohols, 401–403
Potential energy, 27, 760
Precision, 5
Pregnancy
   BUN values during, 859
   lactose synthetase during, 662
Premature infants
   hemolytic anemia, 732
   susceptible to hypoglycemia, 801
Preservative, calcium propionate, 450
Preservatives, parabens, 456
Pressure, 151
   barometric, 151–152
   units, common, 151–152
Primary alcohols, 396–397
Primary amides, 492
Primary amines, 469
Primary structure of nucleic acids, 680–683
Primary structure of proteins, 618
Primary transcript, in RNA synthesis, 696
Pro-opiocortin, activation of, 739
Proenzymes (*see* Zymogens)
Progesterone, 429, 592
   essential in maintaining pregnancy, 747
Progestins, 747
Prohormones, 739
Proinsulin, 739
   activation of, 739
Projection formulas
   for chiral compounds, 513–515
   rules for drawing, 513–515
Proline, 483
Proof, measure of alcohol concentration, 406
Propanal, 430
Propane, 311
Propanoic acid, 445
Properties, 2
   quantized, 91
Propionaldehyde, 426
Propionic acid, 445
n-Propyl alcohol, 392, 394
Propylene, 360
Prostaglandins, 573, 590–593
Prosthetic group in proteins, 625
Proteins, 602–639
   biological value of selected, 610
   catalytic, 603
   coagulation of, 632
   contractile, 603
   defense, 604
   denaturation of, 631
   digestion of, 631, 766–767
   fibrous, 625
   functions of, 604
   globular, 625
      examples of, 625
   hydrolysis of, 631
   in cell membranes, 595–596
   integral, 596
   levels of structure in, 627
   molecular weights of, 604–605
   nutrient, 603
   oligomeric, 629
   peripheral, 596
   primary structure of, 618
   prosthetic groups in, 625
   quaternary structure in, 629–631
   regulatory, 603
   ribosome site of synthesis, 700–706
   secondary structure of, 621
   sensory, 604
   structural, 603
   synthesis of, 691–708
   tertiary structure of, 625–629
   toxic, 604
   transport, 596–597, 603
Proteolytic enzymes, 663
   in digestion, 766–767
Prothrombin, 868
Proton pumping
   in mitochondria and chloroplasts, 819–820
   in electron transport, 818–822
Protons, 57
   gain or loss by amines, 479
   location within atom, 57
Psicose, 536
Pteridine, in folic acid, 727
Pure substances, 35
   chemical subdivision of, 40–41
   classifications of, 41–42
   physical subdivision of, 39–40
Purine, 386, 487
Purines in DNA and RNA, 677
Puromycin, 705
Putrescine, 473
PVC (*see* Polyvinylchloride)
Pyridine, 386, 485–487
Pyridoxal, 486, 725
Pyridoxal phosphate (PLP), 725
Pyridoxamine, 486, 725
Pyridoxine, 725
Pyrimidine, 386, 485–487
Pyrimidines in DNA and RNA, 677
Pyrrole, 483–484
Pyrrolidine, 483
Pyruvate, 785
   metabolic fates of, 785–786
Pyruvate decarboxylase, 646
Pyruvate dehydrogenase complex, 785–786, 808
   control point in TCA cycle, 825
Pyruvate kinase, 785
   control point in metabolism, 799

QH_2-cytochrome c reductase, 817
Qualitative observation, 5
Quantitative observation, 5
Quaternary ammonium salt, 480–481
Quaternary structure in proteins, 447
Quinone, oxidation of, 758–759

Rabies, 709
Racemic mixture, 519
   resolution of, 520
Racemization, 519
Racker, Ephriam, 820
Radiation
   biological effects of, 279–281
   ion-pair formation and, 278–279
   ionizing effects of, 278–279
   penetrating ability, 279–280
   sources of exposure, 282
Radioactive decay
   alpha particle emission, 272
   beta particle emission, 272–273
   equations for, 271–274
   gamma ray emission, 273
   rate of, 275–277
Radioactive emissions, nature of, 270–271
Radioactive nuclides, 270
   modes of decay, 271–274
   naturally occurring, number of, 275
   synthetic, number of, 275

Radioactivity, artificial, 274–275
Rancidity, 579–580
Raspberry flavor, 455
Rat poison, 733
Reaction rates, 216
   activation energy and, 214
   catalysts and, 218–219
   collision orientation and, 214
   factors that influence, 216–219
   molecular collisions and, 213
   physical nature of reactants and, 216–217
   reactant concentrations and, 217
   temperature and, 217–218
Receptors, epinephrine, 523
Recombinant DNA, 711
   transformation with, 712–713
Red blood cells (*see* Erythrocytes)
Redox reactions, 758, (*see* Oxidation–reduction)
Reducing agent, defined, electron loss, 258
Reducing agent, defined, oxidation number change, 261
Reducing sugars, 547
Reduction, defined, electron gain, 258
Reduction, defined, oxidation number change, 261
Reduction, original meaning of, 258
Reduction potentials, in electron transport, 813–818
Reduction reactions, organic compounds, 359
Refining, of petroleum, 329
Regulation
   fatty acid metabolism, 843–845
   of amino acid metabolism, 860
   of metabolism, 769–771
   of oxidative phosphorylation, 826
   of pH in body fluids, 874–875
   of TCA cycle, 825
Rehydration, BUN values in, 859
Rejection, of transplanted organs, 638–639
Renal infarction, AST in diagnosis of, 856
Replication of DNA, 687–688
Representative elements, 107
   location within periodic table, 107
Resorcinol, 396
Respiration, and Dalton's law, 161–162
Restriction enzymes, 712
Retinal, 352
Retinol, 729
Reverse transcriptase, 710, 713
Rh-factor, 867–868
Rhubarb, poisonous material in leaves of, 445
Riboflavin, 723–724
Ribose, 428, 534, 536, 538–539
   component of ATP, 539
   component of RNA, 539
   deoxy-, 539
   in RNA, 677
Ribose-5-phosphate, 791
Ribosomal RNA, 692
Ribosomes, 700
   role in protein synthesis, 700–706
Ribulose, 536
Rice bran wax, 585
Rickets, 731
RNA
   double stranded, 709
   exons, 696
   introns, 696
   messenger, 692–693, 700–706
   primary transcript in synthesis, 696
   processing of, 696–697
   ribosomal, 692
   synthesis, 694–697
   transfer, 693
   viral, 708–711
RNA polymerase, zinc cofactor, 735
Rotation of polarized light, 517–518

Rotten-egg gas, 414
Roundworm, 481
rRNA (*see* Ribosomal RNA)
Rubber, 361
Rubbing alcohol, 407, (*see also* Isopropyl alcohol)
Running, chemical addiction in, 742

Saccharin, 555
Safrole, 410
Salicylic acid, 452
Salts, 232
   amine, 478
   bile, 589–591
   hydrolysis of, 245
      examples, 245–249
   hydrolytic characteristics of, 247–248
   of carboxylic acids, 449–450
   quaternary ammonium, 480–481
   uses of, common, 233
Saponifiable lipids, 569
Saponification
   fats and oils, 577–578
   of esters, 457
Saran Wrap, 362
Sassafras oil, 410
Saturated fats, linked to heart disease, 833
Sauerkraut, lactic acid in, 445
Scents, of flowers and spices, 354
Scientific method, 51
   procedural steps in, 51–52
   terminology associated with, 51–52
Scurvy, 729
sec-butyl alcohol, 394
sec-butyl group, 323
Secobarbital (seconal), 494
Seconal (secobarbital), 494
Secondary alcohols, 396–397
Secondary amides, 492
Secondary amine, 469
Secondary messenger hypothesis, 743
Secondary structure of proteins, 621
Sedatives, 493
Sedoheptulose-7-phosphate, 792
β-Selinene, 354
Selenium
   as a trace element, 734
   toxicity of, 734–735
Serine, 582
Serotonin, 488, 725
Shampoo, PEG in, 399
Shoe polish, 456
Shortening, 578
Shuttles
   glycerol phosphate, 824–825
   malate, 823–824
Side chain, amino acid, 605
Sigma bond, 346
Significant figures, 7
   determination of, 7–8
Silicon, characteristics of, 292–293
Silk, 498
   protein in, 621
Silver mirror, Tollens' test for aldehydes, 435
Silver proteinate, 633
Simple diffusion across membranes, 596–597
Sites 1,2,3, in electron transport, 814–818
Skin substitute, 498
Smallpox, 709, 711
Soap, 450–451
   micelles, 450–451
Soaps, 578
   produced by saponification, 457
Sodium
   characteristics of, 294
   ion exchange in kidneys, 870–872
Sodium borohydride, reducing agent, 431–432

Sodium lamp, 518
Solid state
   characteristics of, 147
   explained using kinetic molecular theory, 149
Solubility, 183
   rules, based on polarity, 186
   rules, for ionic compounds, 186–187
Solute, 182
Solutions, 182
   acidic, 237
   aqueous, 184
   basic, 237
   buffer, 249–251
   colligative properties of, 200–205
   components of, 181–182
   concentrated, 184
   concentration units, 188–189
   dilute, 184
   dilution of, 198–200
   formation, interparticle interactions, 184
   hypertonic, 205–206
   hypotonic, 205–206
   isotonic, 205–206
   miscibility of, 184
   neutral, 237
   pH scale for, 237–239
   saturated, 183
   types of, 182–183
   unsaturated, 184
Solvent, 182
Soot, 331
Sorbitol, 547
Sorbose, 536
Spearmint, stereochemistry of, 522
Specific dynamic action, 767
Specific gravity, 24
   calculation of, from densities, 24–25
Specific heat, 29
   calculations using, 29–30
Specific rotation, 518
Specificity, of enzymes in chemical reactions, 651
Sphingolipids, 585–586
Sphingomyelins, 586–587
Spices, 354
Spine, treatment with enzymes, 670
spirea plants, 452
Spontaneous reactions, 755
Standard free energy, 757
Starch, 531, 559–560, (*see also* Amylose, Amylopectin)
   amylose, 558
   as a nutrient, 765
   hydrolysis products of, 560
   in world food consumption, 560
Stearate, sodium, 450
Stearic acid, 571
Stereochemical specificity of enzymes, 651
Stereoisomerism, 506–523
   in amino acids, 610
Sterilization
   of medical equipment, 414
   of medical instruments, 481
Steroids, 588–590
   anabolic, 746
   cholesterol, 588–589
   effect on protein concentrations, 746
Stoeckenius, Walther, 828
Streptomycin, 705–706
Stroma, 772–773
Structure, 1
   levels of in proteins, 627
Styrene, 360
Subatomic particles, 56
   arrangement within atom, 57–58
   kinds of, 57
Sublimation, 170
Substitution in DNA mutations, 707

Substitution reactions in alkanes, 331
Substrate, 647
   concentration effect on rate, 654–656
Succinate dehydrogenase
   coenzyme, 723–724
   inhibition of, 659
Succinic acid, 443, 445
Succinyl-CoA, 809
Succinyl-CoA synthetase, 810
Sucrose, 552
   as a nutrient, 765
   per capita consumption of, 552
   specific rotation of, 554
Sugar, amounts added to selected foods, 553
Sugar beets, source of sucrose, 552
Sugar cane, source of sucrose, 552
Sugars
   C_3-, C_4-, C_5-, C_6-, C_7- metabolism, 791–792
   oxidation of, 759
   reducing, 547
Sulfanilamide
   as an antibiotic, 664–665
   derivatives of, 665
Sulfhydryl group, 414
Sulfur
   as a micronutrient, 733
   in rubber, 361
Sulfur dioxide
   and acid rain, 299–300
   characteristics of, 298–299
   effects on human beings, 299
Sulfuric acid, esters of, 459
Sunshine vitamin, 730, (see Cholecalciferol)
Surgery, vitamin K given before, 733
Surgical suture, made of nylon, 496
Sutherland, Earl, 742
Sweet taste, of fructose, 538
Sweeteners, artificial, 555
Sweetness values, comparative of sugars, sweeteners, 556
Symmetrical addition, 355–356
Symmetry, plane of, 521
Symport, 597
Synergistic effect, 494
Synthesis of proteins, 691–708

Table sugar (see Sucrose)
Tagatose, 536
Talose, 534
Target tissues of hormones, 735–737
Taste buds, zinc required for formation of, 735
Taurocholate sodium, 591
Tautomerization, 546
Tautomers, 544–545
TCA cycle, 806–813, (see also Tricarboxylic acid cycle)
   linked to urea cycle, 859
   primary function of, 807
   radioactive labels in, 812
   regulation of, 825
   types of reactions in, 813
Tea leaves, caffeine in, 487–8, 495
Teflon, 360
Temperature scales
   Celsius, 25–26
   Fahrenheit, 25–26
   Kelvin, 25–26
Tendons, proteins in, 608
Terephthalic acid, 458
Termites, 558
Tertiary alcohols, 396–397
Tertiary amides, 492
Tertiary amine, 469
Tertiary structure of proteins, 625
Tertiary structure of DNA, 684–685
Testosterone, 429, 592
   structure of, 746

Tests for alkenes
   bromination, 355
   with potassium permanganate, 358
Tetracycline, 705–706
Tetrafluoroethylene, 360
Tetrahedron, shape of methane, 311
Tetrahydrocannabinol (THC), 410
Tetrahydrofolic acid, 727
Theory, 52
   purposes of, 52
   scientific method and, 51–52
Thermodynamics, 755
Thiamine, 487, 722–733
Thioesters, 453
Thiolase, 835
Thiols, 414–415, 453
Thionyl chloride, 448
Thiosulfate, in treatment of cyanide poisoning, 818
Threonine, chiral carbons in, 512–513
Threose, 534
Throat lozenges, hexylresorcinol in, 410
Thrombin, 646
Thromboses, 749
Thylakoid membranes, ATP synthetase complexes in, 822
Thylakoids, 772–773
Thyme, flavoring agent in, 410
Thymine, 487
   structure of, 677
Thymol, 410
Thyroid hormone, in amino acid metabolism, 861
Thyroid-stimulating hormone, 736
Thyrotropin-releasing factor, 736–737
Thyroxine, 738
   activation of, 739
   controls metabolic rate, 736
   requires iodine for synthesis, 735
Tiger moth, pheromone, 334
Tin, reduction of aromatic nitro group, 477
Tissue plasminogen activator, use in treating heart attacks, 670
Titanium, characteristics of, 295
Titrations, acid-base, 240
   concentrations and, 241–245
Titrations, indicators and, 241–242
TNT (see Trinitrotoluene)
Tobacco, nicotine in, 483–484
Tocopherol, 731
α-Tocopherol, antioxidant properties of, 580
Tollens' reagent, 435–436, 546–547
Toluene, 374
   production from n-heptane, 385
Toluidine, 471
p-Toluidine, 477
Tooth decay
   monosaccharides in, 547
   role of fluoride in preventing, 735
TPA (see Tissue plasminogen activator)
Trace elements, 734–735
   as activators of enzymes, 648
Tranquilizers, 493
   Ativan, 494
   Librium, 494
   Valium, 494
Transaldolase, 792–793
Transaminase, in transport of ammonia, 857
Transamination, 855
Transcript, primary in RNA synthesis, 696
Transcription, 692 (see RNA synthesis)
   hormone effect on rate of, 746
Transfer RNA, 693
   shape of, 695
Transformation, 712–713
Transition elements, 107
   location within periodic table, 107
Transketolase, 792–793

Translation, 692
   in protein synthesis, 700
Transmembrane potentials, 771
   proton, 819
Transmutation reaction, 274
Transpeptidase, inhibition of by penicillin, 667
Transplants, organ, rejection of, 638–639
Transport
   across membranes, 772
   active, 596–597
   fatty acid, into mitochondria, 834
   lipid, 832–833
   of ammonia, 856
Transverse diffusion, in cell membranes, 595
TRF (see Thyrotropin-releasing factor)
Triacylglycerol, 831
Tricarboxylic acid cycle, 806–813, (see also TCA cycle)
Tricarboxylic acids, 442–443
Trichloromethane, 332
Tridecane, 334
Triglycerides, 456, 573–576, 831
   biosynthesis of, 841
   dietary considerations of, 766
   energy, storage of, 576
   melting points of, 575
   mixed, 574
   polyunsaturated, 575
   simple, 574
Triglycerols, 831
Trimethylamine, 474
Trinitrotoluene (TNT), 381
Triose phosphate isomerase, 783
Tripalmitin, 831
Triphosphoric acid, 461
Triple helix, in proteins, 621, 624
tRNA (see Transfer RNA)
Trypsin, 646, 664
TSH, 736, (see Thyroid-stimulating hormone)
Tubule, nephron, 870
Tylenol, 493

Ubiquinone, in electron transport, 814
UDP, role in activation of glucose, 796–797
UDP-glucose, 796
Uncouplers, electron transport, ATP synthesis, 821–822
Undecane, 334
Unsaturated fatty acids, metabolism of, 837
Unsaturated hydrocarbon, 341
Unsymmetrical addition, 355–359
Uppers, 494
Uracil, 487
Urea, 493
   cyclic metabolism of, 856
Urea cycle, 856
   balanced equation of, 856
   linked to TCA cycle, 859
   primary site of, 856
Urease, 647
Urethane linkage, 498–499
Uridine, structure of, 678
Urinary system, 870
Urinary tract, stones, BUN values with, 859
Urinary tract infections, treatment of, 666
Urine
   abnormal constituents in, 872
   formation of, 870–872
   glucose test in, 436
   large volumes of, 872
   testing for glucose in, 547
   tests of, 872
   volume, control of, 872–873
Urokinase, 670
US Recommended Daily Allowances, 722
   listing of, 722
USRDA (see US Recommended Daily

Allowances)
Uterine contractions, prostaglandins and, 592

Vaccinations, 711
Valence electrons, 114
    electron configurations and, 114–115
    electron-dot structures and, 115
Valium (diazepam), 494
Vanilla flavoring, 429
Vanillin, 429
Vapor, 165
Vapor pressure, 166
    and boiling point, 167
    colligative property of solutions, 200
    explained using kinetic molecular theory, 165
    factors affecting magnitude, 166
Vasodilation, prostaglandins and, 592
Vasopressin, 736
    control of urine volume, 872
Vinegar, 442
Vinyl chloride, 346, 360, 362
Viruses, 708
    types of, 708–709
Vital force theory, 309
Vitamin A, 353, (*see* Retinol, 729)
Vitamin B_1 (*see* Thiamine)
Vitamin B_{12} (*see* Cobalamin)
Vitamin B_2 (*see* Riboflavin)
Vitamin B_6 (*see* Pyridoxine)
Vitamin C (*see* Ascorbic acid, 728)
Vitamin D (*see* Cholecalciferol, 730)
Vitamin E (*see* Tocopherol)
Vitamin H (*see* Biotin, 725)
Vitamins, 721–733
    classes of, 722
    water soluble, 722–729
VLDL (*see* Lipoproteins, low density, 833)
Volt meter, to measure membrane potentials, 771
von Gierke's disease, 798
Vulcanization, 362

Water
    abnormal boiling and freezing point of, 177–178
    abundance and distribution of, 175–176
    deionized, 306
    density pattern of, 179–181
    distilled, 306
    fluoridated, 306
    hard, 304
        drinkability of, 305
    ion formation in, 235–237
    soft, 305
    methods of production, 305–306
    thermal properties of, 178–179
Waterproffing, of aquatic birds, 584
Waxes, 455–456, 584–585
    paraffin, 585
    uses and structures of selected, 585
Weight, 15
    compared to mass, 15–16
White blood cells (*see* Leukocytes)
Wilson's disease, 734
Winter, chlorophyll production ceases in, 773
Wintergreen, oil of, 452
Wool, 498
    proteins in, 608

Xylene, 375
Xylitol, 547
Xylose, 534
Xylulose, 536
Xylulose-5-phosphate, 791

Yeast, used in baking to raise bread, 781
Yellow fever, 711

Zephiran chloride, 481
Zinc, as trace element, 735
Zingiberene, 354
Zwitterion, 611
Zymogen granules, 664
Zymogens, activation of, 649

CREDITS

page 2: Larry Mulvehill/Science Source/Photo Researchers; page 4: Nina Howell Starr/Photo Researchers; page 37: Grant Heilman; page 53: courtesy of J.M. Huber Corporation; page 73: Tom Bochsler; page 170: NIH/Science Source/Photo Researchers; page 179: Tom Tajima/The Patriot Ledger; page 206: Russ Kinne/Science Source/Photo Researchers; page 216: UPI/Bettmann Newsphotos; page 281: Harry J. Przekap Jr./Medichrome; page 283: Martin M. Rotker/Taurus Photos; page 284: Stan Levy/Science Source/Photo Researchers; page 293: (a) Grant Heilman; (b) AT&T Technologies, Inc./Public Relations Photo Service; page 299: Westfälisches Amt für Denkmalpflege; page 302: Ellis Herwig/Taurus Photos; page 304: Alfred Pasieka/Taurus Photos; page 328: Joe Munroe/Photo Researchers; page 361: Joseph Nettis/Science Source/Photo Researchers; page 361: Ted Cordingley; page 383: Verna R. Johnson/Photo Researchers; page 383: Wide World Photos; page 385: Tom Bochsler; page 403: Linda K. Moore/Rainbow; page 407: Josephus Daniels/Rapho Guillumente/Photo Researchers; page 408: Jeff Reed/Medichrome; page 411: USDA Soil Conservation Service; page 412: Russ Kinne/Science Source/Photo Researchers; page 435: Tom Bochsler; page 450: Taurus Photos; page 458: Roslin/Medichrome; pages 482 and 485: Robert Harbison; page 497: Stan Levy/Science Source/Photo Researchers; page 498: Dan McCoy/Rainbow; page 537: Photo Researchers; page 547: Mottlee Weissman/Photo Researchers; page 552: Jeff Reed/Medichrome; page 553: Max and Kit Hunn/Photo Researchers; page 557: Biophoto Associates/Photo Researchers; page 561: Jayer Phillips/The Picture Cube; page 561: Don W. Fawcett/Photo Researchers; page 576: David Dempster; page 576: David Phillips/Taurus Photos; page 579: David Dempster; page 584: Leonard Lee Rue III/Photo Researchers; page 593: Don W. Fawcett/Photo Researchers; page 620: NIH photo; page 624: Keith Porter/Photo Researchers; page 634: Jerry Howard/Stock, Boston; page 715: Courtesy of Beckman Instruments, Inc.; page 794: Don W. Fawcett/Photo Researchers.

Atomic Numbers and Atomic Weights

Based on $^{12}_{6}$C. Numbers in parentheses are the mass numbers of the most stable isotopes of radioactive elements.

Element	Symbol	Atomic Number	Atomic Weight	Element	Symbol	Atomic Number	Atomic Weight
Actinium	Ac	89	(227)	Europium	Eu	63	151.96
Aluminum	Al	13	26.98	Fermium	Fm	100	(257)
Americium	Am	95	(243)	Fluorine	F	9	19.00
Antimony	Sb	51	121.75	Francium	Fr	87	(223)
Argon	Ar	18	39.95	Gadolinium	Gd	64	157.25
Arsenic	As	33	74.92	Gallium	Ga	31	69.72
Astatine	At	85	(210)	Germanium	Ge	32	72.59
Barium	Ba	56	137.33	Gold	Au	79	196.97
Berkelium	Bk	97	(247)	Hafnium	Hf	72	178.49
Beryllium	Be	4	9.01	Helium	He	2	4.00
Bismuth	Bi	83	208.98	Holmium	Ho	67	164.93
Boron	B	5	10.81	Hydrogen	H	1	1.01
Bromine	Br	35	79.90	Indium	In	49	114.82
Cadmium	Cd	48	112.41	Iodine	I	53	126.90
Calcium	Ca	20	40.08	Iridium	Ir	77	192.22
Californium	Cf	98	(251)	Iron	Fe	26	55.85
Carbon	C	6	12.01	Krypton	Kr	36	83.80
Cerium	Ce	58	140.12	Lanthanum	La	57	138.91
Cesium	Cs	55	132.91	Lawrencium	Lr	103	(260)
Chlorine	Cl	17	35.45	Lead	Pb	82	207.2
Chromium	Cr	24	52.00	Lithium	Li	3	6.94
Cobalt	Co	27	58.93	Lutetium	Lu	71	174.97
Copper	Cu	29	63.55	Magnesium	Mg	12	24.30
Curium	Cm	96	(247)	Manganese	Mn	25	54.94
Dysprosium	Dy	66	162.50	Mendelevium	Md	101	(258)
Einsteinium	Es	99	(252)	Mercury	Hg	80	200.59
Erbium	Er	68	167.26	Molybdenum	Mo	42	95.94